# ELECTRONIC PROPERTIES OF CARBON NANOTUBES

Edited by **Jose Mauricio Marulanda**

**Electronic Properties of Carbon Nanotubes**
Edited by Jose Mauricio Marulanda

**Published by InTech**
Janeza Trdine 9, 51000 Rijeka, Croatia



**Publishing Process Manager** Viktorija Zgela

**Technical Editor** Teodora Smiljanic

**Cover Designer** InTech Design Team

**Image Copyright** Yellowj, 2010. Used under license from Shutterstock.com

Additional hard copies can be obtained from orders@intechopen.com

Electronic Properties of Carbon Nanotubes, Edited by Jose Mauricio Marulanda
p. cm.
ISBN 978-953-307-499-3

For EU product safety concerns:
IN TECH d.o.o., Prolaz Marije Krucifikse Kozulić 3, 51000 Rijeka, Croatia,
info@intechopen.com or visit our website at intechopen.com.

# Meet the editor

Dr. Jose Mauricio Marulanda has a Doctor of Philosophy in Electrical Engineering from Louisiana State University. He has a Masters and a Bachelor of Science in Electrical Engineering as well. He is currently working as a Senior Hardware Engineer for the Research and Development Center at ABB, Inc. At ABB, he currently conducts research in the area of Electromagnetic Compatibility (EMC). His areas of expertise are Computer Aided Design, VLSI, integrated circuit modeling and design. He has experience with system level validation by qualifying products to be released into the market with zero defects. He has an extensive academic and research experience in VLSI, MOS and bipolar devices, semiconductor device physics, device modeling and carbon nanotubes. His work in carbon nanotubes includes generating models for the electron transport in carbon nanotubes and their implementation as field effect transistors, logic gates, biosensors and interconnects. He has published several papers on carbon nanotubes for peer-reviewed journals and international conferences.

# Contents

# Preface

Carbon nanotubes (CNT) discovered in 1991 have been a subject of intensive research for a wide range of applications. These one-dimensional (1D) graphene sheets rolled into a tubular form have been the target of many researchers around the world. This book concentrates on the semiconductor physics of carbon nanotubes, it brings unique insights into the phenomena encountered in the electronic structure when operating with carbon nanotubes. This book also presents the reader with useful information on the fabrication and applications of these outstanding materials. In the past decades, carbon nanotubes have undergone massive research from countless institutions around the world. The main objective of this book is to give in-depth understanding of the physics and electronic structure of carbon nanotubes.

Readers of this book should have a strong background on physical electronics and semiconductor device physics. Philanthropists and readers with strong background in quantum transport physics and semiconductors materials could definitely benefit from the results presented in the chapters focusing on the different applications of carbon nanotubes.

This book has been outlined as follows: fabrication techniques followed by an analysis on the physical properties of carbon nanotubes, including density of states and electronic structures. Ultimately, the book pursues a significant amount of work in the industry applications of carbon nanotubes. A list of the chapters is given below. It is highly recommended for the reader to go over the following descriptions, as they provide excellent insights into the contents and results of the book's chapters.

*Chapter 1. Assembly of Carbon Nanotube Sheets*
This book starts with presenting the fabrication processes to making CNT sheets. The purity of the nanotubes, the height of the forest, the morphology of the forest, especially the 3D structure by self-assembly during CVD process, and the area density of the nanotubes are the main factors of the draw-ability of the forest.

*Chapter 2. A Close-Packed-Carbon-Nanotube Film on SiC for Thermal Interface Material Applications*
This chapter shows how carbon-nanotube (CNT)/silicon-carbide layered composite materials made by thermal decomposition of SiC are been examined for use as thermal

interface materials. Scanning electron microscope and transmission electron micro-microscope observations revealed that the highly dense, well aligned, and catalyst-and amorphous-free CNTs were strongly adhered to the SiC substrate.

*Chapter 3. Magnetic Carbon Nanotubes: Synthesis, Characterization, and Anisotropic Electrical Properties*
In this chapter, a CNT-inorganic hybrid system is demonstrated, especially, CNT/$\gamma$-$Fe_2O_3$ hybrid materials. It develops the synthesis method of MWCNT/$\gamma$-$Fe_2O_3$ nanostructures via an easy and novel modified sol-gel process. The study shows that NaDDBS molecules are intimately involved in inhibiting the formation of an iron oxide gel.

*Chapter 4. Characterizing Functionalized Carbon Nanotubes for Improved Fabrication in Aqueous Solution Environments*
The utility of XPS for delineating MWNT oxidation kinetics, EXAFS (coupled with XPS and ATR-IR) is demonstrated. This characteristic is useful for elucidating nanoparticle-MWNT interfacial structure, and the dependence of PZC on electron withdrawing/donating character of moieties attached to SWNTs.

*Chapter 5. Fabrication of Carbon Nanotubes for High-Performance Scanning Probe Microscopy*
This chapter attempts to provide a one-stop guide to the fabrication of scanning probes, in hope of assisting future developers in their efforts to exploit the unique properties of CNTs in still more varieties of SPM. It is expected that the coming years will see the variety of CNT SPM probes and the range of applications of CNT SPM probes expand even further.

*Chapter 6. Selective Separation of Single-Walled Carbon Nanotubes in Solution*
In this chapter, it is shown that the separated SWCNTs with uniform structure, e.g. single electronic (m- or s-) type and chirality, have presented better performance than non-separated SWCNTs in nanoelectronics and thin-film devices.

*Chapter 7. Carbon Nanotube AFM Probe Technology*
The application of carbon nanotube probes are shown in the new AFM imaging world. Carbon nanotube probes increase the probe's resolution and longevity, decreasing probe–sample forces, and extending the AFM application fields. This should also have a significant impact in key research areas, such as nanometrology, surface engineering and biotechnology.

*Chapter 8. One-dimensional Crystals inside Single-walled Carbon Nanotubes: Growth, Structure and Electronic Properties*
This chapter shows that by intercalating inorganic compounds into single-walled carbon nanotubes, the electronic properties of SWNTs can be directly modified. Namely, the filling of nanotubes with electron donors (such as metals or metalorganic compounds) can lead to the electronic conductivity of the "1D-crystal@SWNT" composite, which is caused by an increase in the electron density on the nanotube walls within the rigid band structure approximation.

*Chapter 9. Geometric and Spectroscopic Properties of Carbon Nanotubes and Boron Nitride Nanotubes*
In this chapter, the calculated Raman and IR spectra of functionalized (7,0)-SWCNT are provided using the B3LYP functional with the basis sets 6-31G on carbon and hydrogen atoms and 6-311G(d,p) basis set used for oxygen and nitrogen atoms.

*Chapter 10. Detection of Carbon Nanotubes Using Tip-Enhanced Raman Spectroscopy*
As a tip-enhanced spectroscopy, it is shown how TERS system combines SPM (STM, AFM, SFM, SNOM) scanning head, Raman spectroscope, optical microscope, control units and data acquirement into a complicated topography and spectral measuring system.

*Chapter 11. Spectro-Microscopic Study of Laser-Modified Carbon Nanotubes*
In this chapter, a technique that facilitates exclusive chemical modification of the CNTs with controlled locality using a focused laser beam is discussed. With focused laser beam, a morphology modification and structural rearrangement of the CNTs are achieved. With the CNTs housed in a transparent chamber with controlled gaseous environment, we can select the appropriate gas species for the chemical modification

*Chapter 12. Enhanced Control of Carbon Nanotube Properties Using MPCVD with DC Electrical Bias*
This chapter shows an examination of the role of dc electrical bias during microwave plasma-enhanced chemical vapor deposition (MPCVD) synthesis of SWNTs using alignment, spatial density, chirality, and purity as metrics of interest. Furthermore, it is demonstrated that enhanced thermal and electrical transport properties of MWNTs are realized with application of substrate bias during MPCVD synthesis.

*Chapter 13. Spin Dependent Transport Through a Carbon Nanotube Quantum Dot in the Kondo Regime*
An impact of symmetry-breaking perturbations in CNTs in the Kondo regime on transport is discussed. In specific this chapter focuses on the influence of magnetic field and polarizations of electrodes. The conclusions drawn in this chapter can be easily applied to the case of manipulating of orbital degrees of freedom (orbitronics) as well.

*Chapter 14. A Density Functional Theory Study of Chemical Functionalization of Carbon Nanotubes; Toward Site Selective Functionalization*
This chapter briefly presents how computational methods influence a description of its CC bonding network. Results of mono or double functionalization of nanotubes are given. In addition, the use of geometrical constraints of a bismalonate for site-selective addition into a nanotube is shown.

*Chapter 15. Low-Energy Irradiation Damage in Single-Wall Carbon Nanotubes*
The low-energy irradiation damage and its defect characteristics are studied. The damage and recovery are reversible, indicating that the number of carbon atoms is

preserved. The damage has been observed in SWCNTs, but not in multi-wall carbon nanotubes (MWCNTs), suggesting that it is specific to low-dimensional structures or nanostructures. The irradiation often converts the electric properties of a metallic SWCNT into semiconducting, because the defect opens a local band gap in the metallic SWCNT.

*Chapter 16. Exciton Dephasing in a Single Carbon Nanotube Studied by Photoluminescence Spectroscopy*
In this chapter, the temperature and chirality dependence of the photoluminescence (PL) line width of single carbon nanotubes using single SWNT spectroscopy is studied. This is used to clarify the mechanism of exciton dephasing. The PL line width of a single carbon nanotube broadened linearly with increasing temperature, indicating that the line width and exciton dephasing are determined through exciton-phonon interactions, is discribed.

*Chapter 17. A Numerical Study of the Vibration Spectrum for a Double-Walled Carbon Nanotube Model*
This chapter presents numerical results for a double-walled carbon nanotube model with non-conservative boundary conditions, given in the form of two Timoshenko beams coupled through a distributed Van der Waals force.

*Chapter 18. Electronic Band Structure of Carbon Nanotubes in Equilibrium and Non-Equilibrium Regimes*
In this chapter, the concept of chiral vector, chiral angle and the radius of SWCNTs are first discussed and formulated. The authors then proceed to explaining how different symmetries of single walled carbon nanotubes including translational, helical and rotational symmetries affect the electronic band structure.

*Chapter 19. An Alternative Approach to the Problem of CNT Electron Energy Band Structure*
It is intended to discuss in this chapter, TBM (tight binding method), APW (augmented-plane-wave), OPW (orthogonalized-plane-wave) methods and corresponding theoretical concepts. In particularly, conventional band structure models such as the nearly electron approximation (NFA), TBM, APW and OPW models have been used for determining the electron energy band structure.

*Chapter 20. Electronic Structure and Magnetic Properties of N@$C_{60}$-SWCNT*
The design of spin labels for the NMR quantum computer that contained 1D spin chains filling SWCNT with N@$C_{60}$ is proposed. The electronic structure, chemical shift, g-tensor, A-tensor of hfc is influenced by geometrical structure, varied with diameter and chiral index.

*Chapter 21. Carbon Nanotubes Addition Effects on $MgB_2$ Superconducting Properties*
This chapter presents a review of recent developments in the study of the effect of carbon nanotubes (CNT) on the superconducting properties of $MgB_2$ bulk and wire samples, based on the known literature data and our own results.

*Chapter 22. Carbon Nanotube as VLSI Interconnect*
An overview of the exploratory research on CNT as possible VLSI interconnect is presented. The problem of continuing with copper interconnects in highly scaled future technologies are briefly discussed. The work carried out in finding an alternative solution indicates that the CNT based interconnects have the potential to replace copper in future.

*Chapter 23. Carbon Nanotube Based Magnetic Tunnel Junctions (MTJs) for Spintronics Application*
Spintronics devices are the next generation electronic devices with larger capacity and minimum power consumption. After reviewing the basics of MTJs (from fabrication to phenomena), the attention is directed to experimental results from the growth of vertically aligned CNTs and the special methods of fabrication for CNT based spintronics devices.

*Chapter 24. Mechanisms of Single-Walled Carbon Nanotube Nucleation, Growth and Chirality-Control: Insights from QM/MD Simulations*
This chapter reviews recent investigations into the phenomena of SWNT nucleation and growth using state-of-the-art QM/MD methods. The significance of the QM/MD method in this context has therefore been demonstrated. QM/MD simulations of such non-equilibrium, high-temperature processes can provide fundamental knowledge that complements experimental understanding.

*Chapter 25. Multiple Andreev Reflections and Enhanced Quantum Interferences in NbN/Network-Like Carbon Nanotubes/NbN SNS Junctions*
In this chapter, the superconducting proximity effect in S/network-like CNTs/S junctions is observed in the correction of conductance and magneto conductance fluctuations. Enhanced magneto conductance fluctuations similar to UCFs are observed.

*Chapter 26. Quantum Calculation in the Prediction of the Properties of Single-Walled Carbon Nanotubes*
This chapter discusses the calculation of the normal modes using the U Matrix. Ab initio calculations are carried out with GAUSSIAN 98 program at the HF/3-21G level of theory to investigate the effects of polar solvents and different temperatures on the stability of SWCNT in various solvents. Quantum Mechanics (QM) are used to investigate the nature of metals transport and the interaction with single-walled carbon nanotubes (SWCNTs) inters membranes.

*Chapter 27. Single Wall Carbon Nanotubes in the Presence of Vacancies and Related Energy Gaps*
This chapter studies the essential properties of carbon nanotubes, electronic density of states, the coherent potential approximation and the existence of vacancies as they generate the energy gap in carbon nanotubes.

*Chapter 28. STM Observation of Interference Patterns Near the Endcap and its Application to the Chiral Vector Determination of Carbon Nanotubes*
In this chapter, there is a discussion on the STM measurements focusing on the super lattice structures detected near the endcap of the CNT. This is achieved by combining high-resolution STM imaging and several types of simulation methods both for metallic and semiconducting CNT.

*Chapter 29. Liquid Crystal-Anisotropic Nanoparticles Mixtures*
This chapter studies the impact of LC orientation ordering on alignment of carbon nanotubes (CNTs). It comparatively presents the theoretical results obtained in the two anchoring limits: i) the weak anchoring limit of the nematic liquid crystals molecules at the nanotubes' surface, where the nanotubes alignments are caused by the anisotropic interfacial tension of the nanotubes and ii) the strong anchoring limit for which the nematic ordering around nanotubes is apparently distorted (consequently, relatively strong long-range and anisotropic interactions can emerge within the system).

*Chapter 30. Strategies to Successfully Cross-Link Carbon Nanotubes*
In this chapter, there is a review on some of the successful approaches used to cross-link CNTs with a focus on the importance of the chemistry and techniques involved. This chapter shows how, during cross-linking of carbon nanotubes, the formation of highly cross-linked CNT composites, the de-fluorination process is clearly an advantage.

**Acknowledgments**

I would like to thank the authors of the chapters for their excellent contributions in their areas of expertise and for the efforts invested in the publication of their work. I am certainly sure that the material published in this book will be of a great help and genuinely appreciated by students, professors, and researchers all around the world.

**Jose Mauricio Marulanda**

# Part 1

## Synthesis of Carbon Nanotubes

# 1

# Assembly of Carbon Nanotube Sheets

Mei Zhang[1] and Ray Baughman[2]
[1]*Florida State University*
[2]*University of Texas at Dallas*
*The United State of America*

## 1. Introduction

Over the past decades, carbon nanotubes (CNTs) have been actively explored as building blocks for next-generation electronics (Tans et al., 1998; Bachtold et al., 2001; Misewich et al., 2003), optoelectronics, and sensors (Kong et al., 2000; Xia et al., 2003; T. Zhang, 2008), including flexible and transparent devices ( Ju et al., 2007) as well as stretchable devices (Xu et al., 2011). A critical step in constructing CNT-based devices is assembly of CNTs on a substrate or free-standing for device fabrication, which include alignment, density control, and transfer. Scalable and controlled assembly of CNTs synthesized with diverse methods (*e.g.*, solution fabrication or solid-state fabrication methods) on diverse substrates (*e.g.*, silicon, plastics, rubbers, etc) or free-standing presents a major fabrication challenge that must be overcome if CNTs are to be utilized in practical applications. For assembling CNTs into thin films (or called sheet, buckypaper), there are several different methods or processes in different conditions (*e.g.*, solution or solid-state processes) (Hu et al., 2010). Solution processes start with the CNTs in powder form; the powder is dispersed in an appropriate solvent with or without functionalization. The CNT films (buckypapers) are usually made using versions of the ancient art of paper making, by typically long-time filtration of nanotubes dispersed in solvent and peeling the dried nanotubes as a layer from the filter (Rinzler et al., 1998; Endo et al., 2005). Buckypaper normally has a laminar structure with a random orientation of the bundles of the nanotubes in the plane of the film (Berhan et al., 2004). Interesting variations of the filtration route provide ultra-thin nanotube sheets that are highly transparent and highly conductive (Wu et al., 2004; Hu et al., 2004). While filtration-produced sheets are normally isotropic within the sheet plane, sheets having partial nanotube alignment result from applying high magnetic fields during filtration (Fischer et al., 2003). In other important advances, nanotube films have been fabricated by Langmuir-Blodgett deposition (Y. Kim et al., 2003), casting from oleum (Sreekumar et al., 2003), coating (Ago et al., 1999; Dan et al., 2009), and printing (Zhou et al., 2006; Unidym Inc, 2007). The solid-state processes generally have two approaches. One is to synthesize CNTs by floating catalyst chemical vapor deposition (CVD), either to deposit CNTs on a substrate inside the CVD chamber or to collect the CNT aerogel outside of the chamber on a special substrate and then densify it into a film (Y. Li et al., 2004; Martin, 2010). The catalysts are with the CNTs and the CNTs in the film are usually disoriented. The optimized process control lowers the impurities to less than 5 wt% in the film and a 1.2 meter wide and 10 meter long CNT film has been made (Nanocomp Tech. 2010). The other approach is to synthesize CNTs from the catalysts fixed on substrate to form a array either parallel to the substrate (Kong et al., 1998) or perpendicular to the substrate (also called vertical aligned CNT forest) which are fabricated into the CNT films after the synthesis process by the "domino pushing effect" motion (Ding et al., 2008; Pevzner et al., 2010) or drawing CNTs out of the forest (M. Zhang et al., 2004 & 2005). The domino pushing of the CNT forest can efficiently ensure that most of the CNTs are aligned tightly in the film. Well aligned CNT sheets are obtained by drawing CNTs from the forest.

For fabricating sheets that have close to single nanotube properties, long nanotubes are needed. Solution fabrication methods work only for short nanotubes since the ability to disperse nanotubes into a liquid and to fabricate oriented nanotube sheets from liquid dispersions decreases with increasing nanofiber length. Solid-state fabrication methods do not disturb the length of the nanotubes and are the methods that benefit from long nanotubes. In this chapter, two processes for assembling well aligned and super-thin CNT sheets are presented. One is to produce a drawable CNT forest, which has special topology, by CVD and then draw CNTs out of forest to form a free-standing CNT sheet. Another process involves synthesis of a patterned CNT array on substrate by CVD and then knocking them down to form an aligned CNT sheet on substrate with the help of the solvent. The CNT growth and the conditions for making drawable CNT forests are discussed.

## 2. Fabrication of CNT sheet

Since the CNT sheets are fabricated directly from the forests, the synthesis of the CNT forest is an important step. In this chapter, the CNTs are multi-walled CNTs. The CNT forests were synthesized by catalytic CVD using hydrocarbon gas, acetylene or ethylene, as the carbon source (M. Zhang et al., 2004). The following is the basic process and conditions. The catalyst was a ~3 nm thick iron film, which was deposited on a silicon substrate by electron beam evaporation. The substrates were set in the center of a quartz tube furnace. After heating up to 680°C in helium at one atmospheric pressure, 5 molar percent acetylene in helium was introduced at the total flow rate of 580 sccm. Within a few minutes, the dense and vertically aligned CNT forests were grown on substrates. After removing the forest, the substrate is still catalytically active and can be used to grow new forest, indicating a root-growth mode and the presence of the catalysts on the substrate. Based on scanning electron microscopy (SEM), transmission electron microscopy (TEM), and thermo-gravimetric analysis (TGA), the purity of the nanotube forests was very high ( more than 99% carbon in the form of CNTs), with less than 1 wt% iron and amorphous carbon, but more importantly, no carbon particles within the CNTs were observed. The CNT sheets are made from the CNT forests by two approaches, knocking down and drawing.

### 2.1 Knocking down approach

The schematic of the knocking down approach is shown in Fig. 1. The line arrays of thin Fe film was made by patterning the resist on Si substrate following the standard lithography process, depositing the catalyst thin film by electron beam evaporation and then the lifting off the resist mask. The Fe film cracked into nanoparticles as catalysts for CNT growth during the temperature ramp up in CVD process. The CNTs grew away from the catalyst particles and formed a thin wall array during the CVD process (Figs. 2a to 2c). The CNT wall is very thin, so it is transparent (Fig. 2a). The substrate with the array of CNT walls was

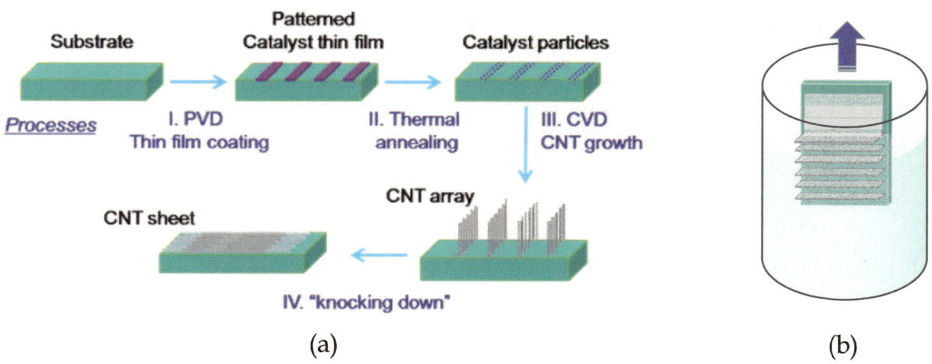


Fig. 1. Schematic experimental processes for making an aligned thin CNT film on substrate (a) and knocking down method (b).

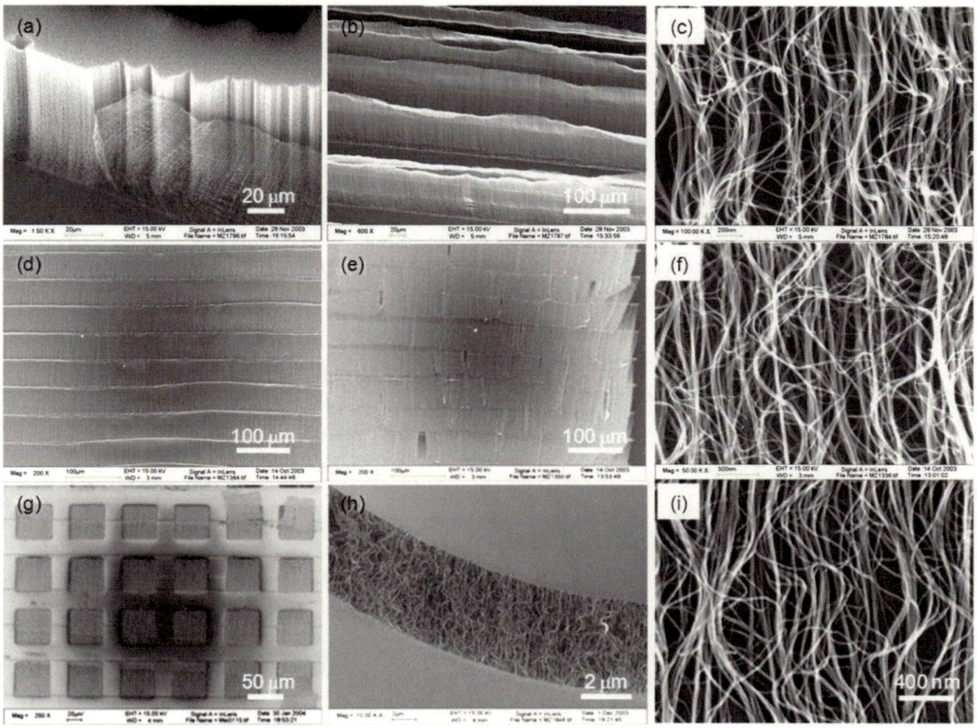


Fig. 2. SEM images show CNTs grown perpendicular to a substrate as a forest and a thin wall (a) and a thin wall array (b). (c) SEM image in high magnification shows the structure of the thin wall. The wall is so thin, it is transparent. (d), (e), and (f) show the thin wall array is "knocked" down and CNTs forms an aligned thin film on substrate. (g) and (h) show that the CNT thin films are etched into desired patterns. After removal of the photo resist, the CNTs have clean surface and keep the original structures. (f) and (i) show the detail of (e) and (h), respectively. The dark circle in (d), (e), and (g) is the shadow of the in-lens detector.

drawn through an acetone solution to horizontally redirect the vertical alignment and dried using nitrogen gas (Fig. 1b and Figs. 2d to 2f). The width of the patterned catalyst lines and the growth time of CVD determined the thickness and the height of the CNT walls. The walls do not shrink in height during flattening and drying. Figure 2e shows that a CNT sheet on substrate is formed from the CNT wall array in which the height of the wall is controlled to just 2 times the gap between two catalyst lines so that the CNT sheet on substrate has the thickness of two overlapping walls. The thickness of the CNT sheet could range from a few hundred nanometers to micrometers through the same process by using forest of different thicknesses, namely by controlling the width of the catalyst lines. Substrate-forest interaction and lateral CNT orientation guided densification only in wall thickness direction to essentially flatten the walls to the substrate with strong adhesion. It is because of the surface tension of the liquids and the strong van der Waals interactions that effectively close the CNTs together when liquids were introduced into the thin forest and dried. This self-assembly transformed CNT walls into highly densely packed CNT thin sheets. The CNT sheet was patterned into arbitrarily shaped CNT islands in desired positions by using standard lithographic processes. Figures 2g and 2h are the SEM images showing the patterned CNT sheet by lithography process. The CNT sheets were etched vertically by oxygen/argon reactive ion etching by using a resist as a mask. Figure 2i is the SEM image of Fig. 2h in high magnification. The surface of the CNTs is clean and the structures of the sheet remain the same, showing that the adhesion with the substrate is sufficient to withstand lithographic processes including heat treatment, immersion into liquids, and drying. The bright and parallel horizontal lines visible in the images are catalyst lines. These lines cannot be totally removed (Fig. 2g). The sheet has to be transferred to another substrate in order for the effects of these lines to be negligible for some applications.

### 2.2 Drawing approach

In drawing approach, the catalyst thin film covers all surface of the Si substrate (Fig. 3). The CNTs are synthesized on top of silicon substrate as a vertically aligned forest by CVD and the transparent nanotube sheets were drawn directly from CNT forests. Draw was initiated using an adhesive strip, like that on a 3M Post-it® Note, or using a blade by cutting into the forest to contact CNTs teased from the forest sidewall. Five-centimeter wide, meter long transparent sheets were made at a meter per minute by hand drawing (M. Zhang et al., 2005).

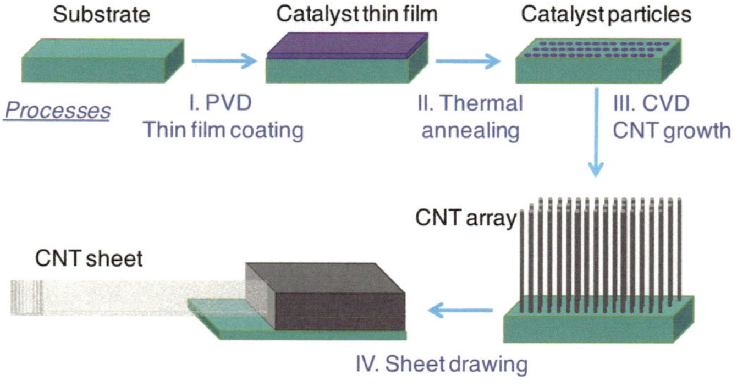


Fig. 3. Schematic experimental processes for fabrication of the free-standing CNT sheet.

Despite a measured areal density of only ~3 μg/cm², these 500 cm² sheets were self-supporting during drawing. Figure 4 shows side-view and top view SEM micrographs of forest and sheet. It indicates that the nanotubes in the forest transition from the highly ordered forest state to a rather disordered intermediate state immediately in front of the forest sidewall, and finally to the highly oriented aerogel state. By taking sequential micrographs through a SEM to form a movie, the process in which the forest nanotubes rotate by about 90° in going from nanotube orientation in the forest to that of the highly oriented state is captured (Kuznetsov et al., 2011).

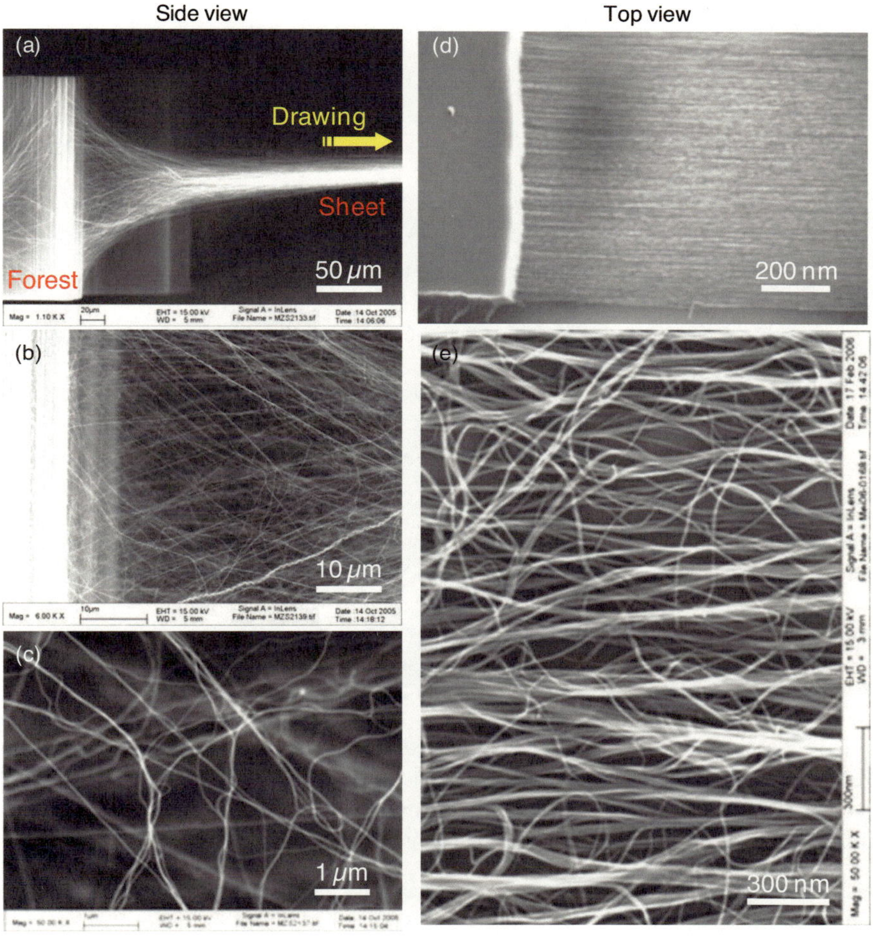


Fig. 4. SEM images show that a CNT sheet is drawing from a forest. (a)-(c) Side view of the forest and the sheet close to the forest. (b) and (c) show the details in (a) and (b), respectively. (d) and (e) Top view of the forest and the CNT sheet.

The forest-drawn CNT sheets can easily be stacked or conveniently assembled into biaxially reinforced sheet arrays. They can be used as conducting layers on non-planar surfaces (Figs. 5a to 5d). These highly anisotropic aerogel sheets can be applied and easily densified into highly oriented sheets having a thickness of 50 nm and a density of ~0.5 g/cm³. We obtain

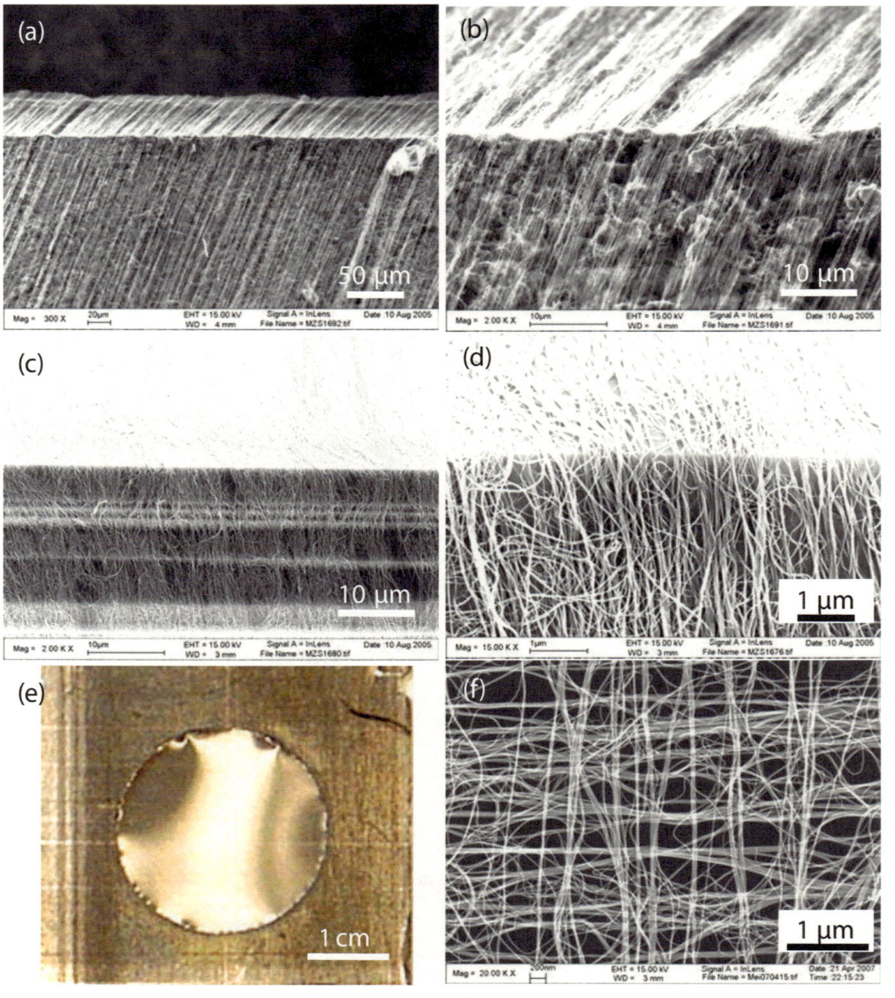


Fig. 5. (a)-(d) SEM images show a CNT sheet covers a substrate with sharp turns. (b) and (d) are the SEM images in higher magnification of the images in (a) and (c), respectively. (e) A photo image shows cross stacked sheets cover a metal plane with a 2.5 cm-diameter hole after liquid densification. The detail structure is shown in (f).

this 360-fold density increase by simply adhering by contact the as-produced sheet to a planar substrate (e.g. glass, many plastics, silicon, gold, copper, aluminum, and steel), vertically immersing the substrate with attached CNT sheet into a liquid (e.g. ethanol) along the nanotube alignment direction, and retracting the substrate from the liquid. Surface tension effects during ethanol evaporation shrink the aerogel sheet thickness to ~50 nm. SEM micrographs taken normal to the sheet plane suggest a decrease in nanotube orientation as a result of densification (Fig. 4e). This observation is deceptive - the collapse of ~20 µm sheets to ~50 nm sheets without changes in lateral sheet dimensions means that out-of-plane deviations in nanotube orientation become in-plane deviations that are noticeable in the SEM micrographs. The aerogel sheets can be effectively glued to a substrate by contacting selected regions with ethanol, and allowing evaporation to densify the aerogel sheet. Adhesion increases because the collapse of aerogel thickness increases contact area between the nanotubes and the substrate.

The aerogel sheets can also be densified into super-thin and free-standing sheet. Figure 5e is a photo image showing that a densified CNT sheet covers a 2.5 cm diameter hole in a metal plane. The super-thin sheet is made by densifing two cross-stacked CNT sheets (Fig. 5f). The nanotube sheets, which combine high transparency with high electronic conductivity, are highly flexible and provide giant gravimetric surface areas. The measured gravimetric strength of orthogonally oriented sheet arrays exceeds that of a high-strength steel sheet (Alive et al., 2009; M. Zhang et al., 2005). These sheets have been used in laboratory demonstrations for microwave bonding of plastics and for making transparent, highly elastomeric electrodes; planar sources of polarized broad-band radiation; conducting appliqués; flexible organic light-emitting diodes; and solar cells (Alive et al., 2009; Ulbricht et al., 2006 & 2007; Williams et al., 2008; M. Zhang et al., 2005).

Many real applications, such as field and thermionic emission electron sources (Kuznetzov et al., 2010; P. Liu et al., 2010; Y. Wei et al., 2008; Xiao et al., 2008; Y. C. Yang et al., 2010), loudspeakers (Alive et al., 2010; Kozlov et al., 2009; Xiao et al., 2008), CNT touch screens (Feng et al., 2010), high strength CNT yarns (Lima et al., 2011; K. Liu et al., 2010; M. Zhang et al., 2004; X. Zhang et al., 2006; Zhong et al., 2010), electrodes for batteries and supercapacitors (H. X. Zhang et al., 2009; R. F. Zhou et al., 2010), CNT/polymer composites (Q. F. Cheng et al., 2010; L. Chen et al., 2009; M. Zhang et al., 2005), and wrappers (Lima et al., 2011) were demonstrated. It is also demonstrated that the CNT sheets can be used as scaffolds for tissue engineering (Galvan-Garcia et al., 2007). It is no doubt that more applications will be developed and practiced.

## 3. Making drawable CNT forest

The CNT draw process does not work for all CNT forests. The experimental results show that the drawability depends strongly on the structural interconnections between CNTs and the network of interconnections between CNT bundles within the forest. The nanotubes in the forest should be intermittently bundled in order to be drawable (Fig. 6). In the forest height direction, this means that a nanotube switches many times from being bundled with a few neighboring nanotubes, to being unbundled, and then to being bundled with a few different neighboring nanotubes. Bundled nanotubes are simultaneously pulled from different elevations in the forest sidewall so that they join with bundled nanotubes that have reached the top and bottom of the forest, thereby minimizing breaks in the resulting fibrils (containing many bundled CNTs) (Figs. 4b and 4c). If there is too little lateral connectivity in the forest, the forest is undrawable because pulling on the forest sidewall just removes a few nanotubes rather than a continuous sheet. If there is too much inter-tube connectivity, only a chunk of forest is extracted before draw terminates.

The interconnections between CNTs and CNT bundles are formed during CNT growth, which are determined by the synthesis process. The CNT synthesis process is a complex process, which is related to the substrate and supporting materials, catalyst materials and their amount, carbon sources and partial pressure (feedstock), carrier gas and gas as an etching agent, total flow rate (gas residual time), process temperature, temperature ramp-up rate and cool-down rate, process pressure, process steps, process time, and many other details, such as history of reaction chamber, contamination, size of chamber and substrate, etc. CNT forests can be produced over a broad range of conditions. However, not every forest is suitable for solid-state fabrication of CNT sheets. This is because the forest needs to meet certain conditions to be a drawable CNT forest as described above. The drawable forest can be fabricated by just using $C_2H_2$/Ar (without $H_2$, water or other agents) and Fe thin film on Si substrate (without buffer layers such as $SiO_2$ and $Al_2O_3$). The buffer layers and other gases as well as their combinations are necessary to control the size and properties of CNTs. The drawable forests can be fabricated under different synthesis conditions. The CNT area density, height of forest, and purity of the CNTs are considered being the parameters for monitoring and controling the interconnections of CNTs in the forest.

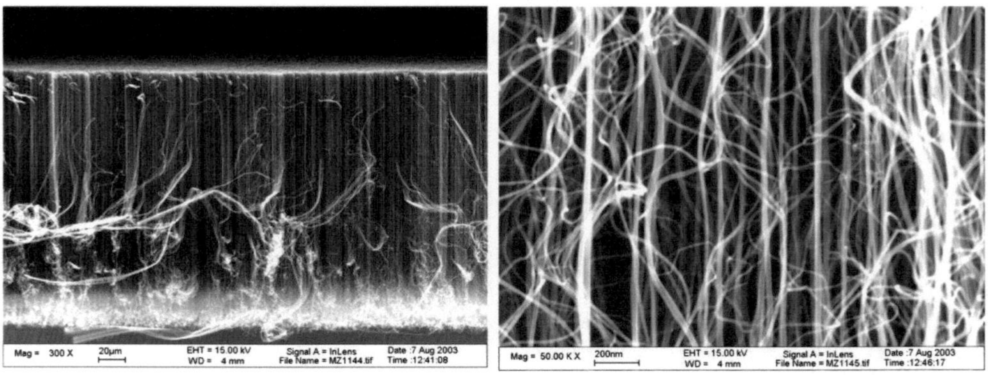

(a) (b)

Fig. 6. SEM images show the side views of a drawable CNT forest (a) and its structure in high magnification (b).

### 3.1 Structures of CNT forest

A single nanotube naturally curves (in bending status) during growth if no external forces exist. The bending stress can come from the nanotube's own weight, interaction with neighbor nanotubes, or limited growing space. As shown in Fig. 7a, a single tube keeps growing straight for a limited length: it falls down to the substrate and turns its growth direction many times during CVD process. A group of CNTs can form a randomly oriented CNT mat or well-aligned CNT arrays, depending on the density of catalyst and their activities under the same synthesis conditions. Figures 7b to 7d show the effects of the number of catalyst particles with similar area density on the formation of a CNT thin sheets array. Figure 7b shows CNT walls that were grown from 0.1 μm wide and 40 μm long catalyst lines patterned by e-beam lithography. There were no external forces during CNT growth. The walls bend when their height is over a certain level. The bending directions and angles depend on each wall's morphologies. The nanotubes within each wall confine the nearest neighbors and attract the outermost nanotubes to their neighbors via van der Waals forces, thereby producing oriented growth. However, the CNTs in such thin walls present random curvatures and are tangled (Fig. 7b) because of the weak confinement in thickness direction. As the thickness of the wall increases, the alignment of the CNTs is improved due to the crowding effect. Figure 7c shows ~100 μm high CNT wall array grown from a 0.5 μm wide and 40 μm long catalyst lines in which nanotubes were better aligned (Fig. 7d). In a forest, the CNTs can have different growth rates, which lead to the structure of the forest. Figure 8 shows three typical structures. In Fig. 8f, more than 80% of the CNTs are not straight: they periodically bend within fixed intervals throughout their entire length. As a result of this regular bending, a wavy structure resulted. It is believed that the wavy structure is formed because there are roughly two groups of catalysts uniformly distributed on substrate: one is more active and results in higher CNT growth rate than the other. Due to van der Waals forces, which stick nanotubes together whenever they touch, the growth rate of the array is limited by the nanotubes with relatively slow growth rate when catalysts stay on the surface of substrate. The nanotubes with higher growth rate are forced to bend periodically. The period of the wave is related to the ratio of growth rates of these two groups (Fig. 8c). When the distribution of the catalyst activity is relatively narrow and the density is high (Fig. 8a), the forest will have the morphology as shown in Fig. 8d. When the distribution of the catalyst activity results in the distribution of growth rate as shown in Fig. 8b and the density is high, the forest will be formed by the straight CNTs which form bundles while the waved CNTs switch between different bundles as the morphologies show in Figs. 6b and 8e. Such structure is believed to be important for assembling CNT sheets by drawing CNTs directly from the forest.

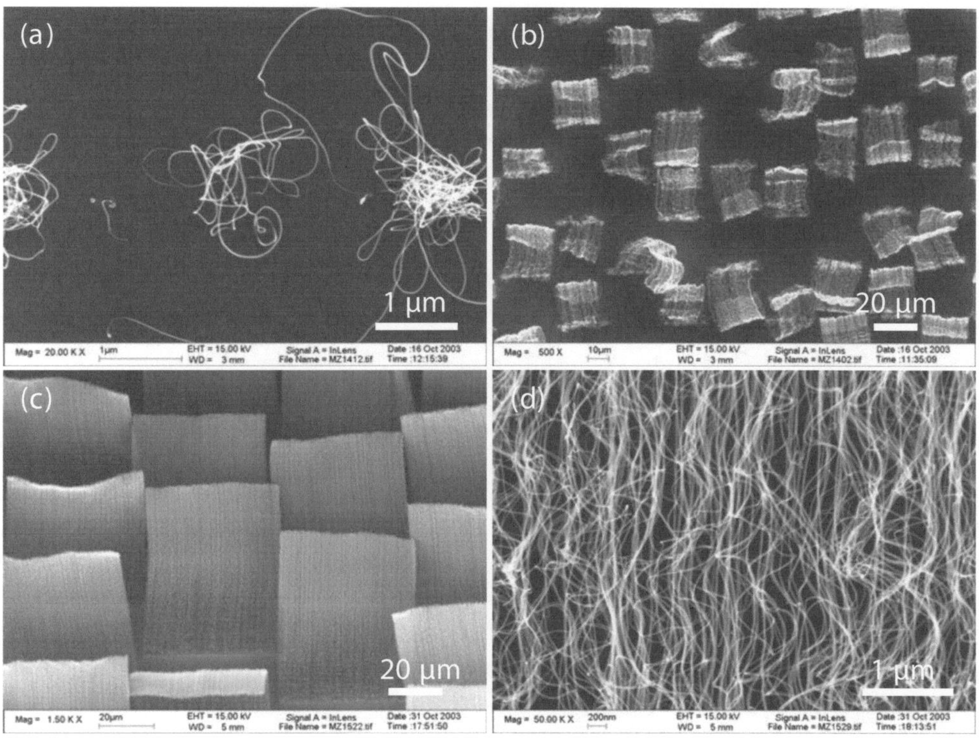


Fig. 7. SEM image of CVT walls grow on patterned (a) very thin, (b) 0.1 μm wide and 40 μm long, and (c) 0.5 μm wide and 40 μm long catalyst lines. (d) SEM image of the CNTs in a wall in (c).

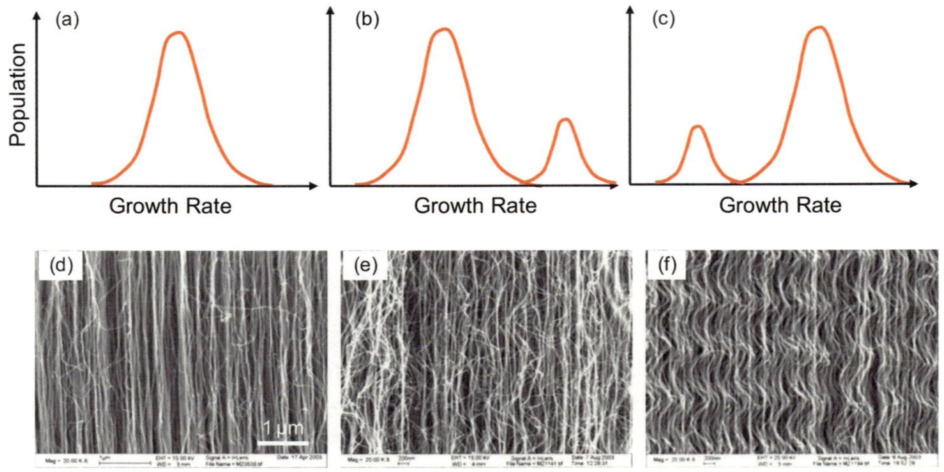


Fig. 8. (a) to (c) Schematic of the population of the CNTs related with the growth rate and (d) to (f) typical SEM images of the CNT forests. (d), (e), and (f) are in the same scale and they are corresponding to (a), (b), and (c), respectively.

### 3.2 CNT area density

As shown in Fig. 7, CNT area density (number of nanotubes per square centimeter) in the forest is a key factor to establishing the interconnections between CNTs in the forest and the drawability of the forest. The schematic illustration of drawability to the area density is shown in Fig. 9. If the area density is very low, CNTs will lay on the substrate randomly. The CNTs can grow in the out-of-plane direction and form a vertical aligned forest when its area density exceeds a threshold value. If the forest has very high density, the CNTs in the forest will form big bundles and the forest will not be drawable.

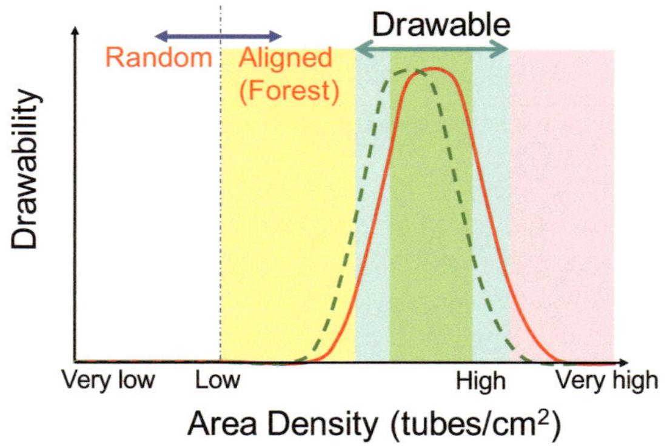


Fig. 9. Schematic illustration showing the relationship between the forest drawability and the area density of CNTs. The dash line corresponds to the forest with higher height.

Experimentally, the area density of the forest is calculated from counting the root of the nanotubes on substrate after removal the forest from the substrate. Figure 10 shows the surfaces of the substrates after removing the forest. Each circle dot in the images is the root of a nanotube.

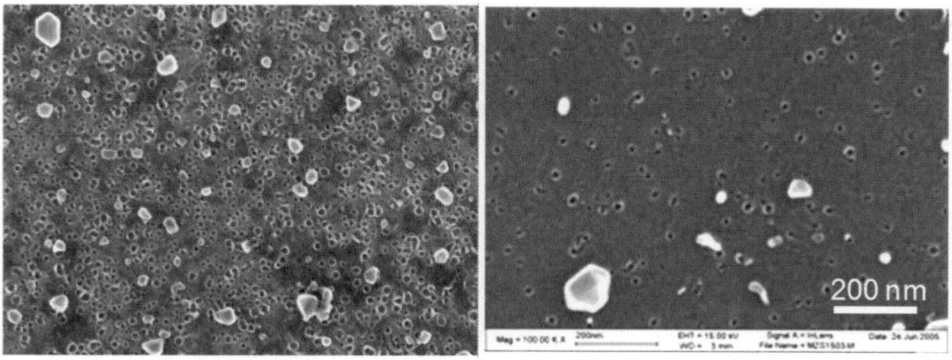


(a) (b)

Fig. 10. SEM images show the surfaces of the substrate after removal CNT forests. (a) and (b) show the typical surfaces of the drawable forest and the un-drawable forest, respectively. Each circle in the image is a root of a CNT. The bright dots are the by-products of the CVD process. Two images are in the same scale.

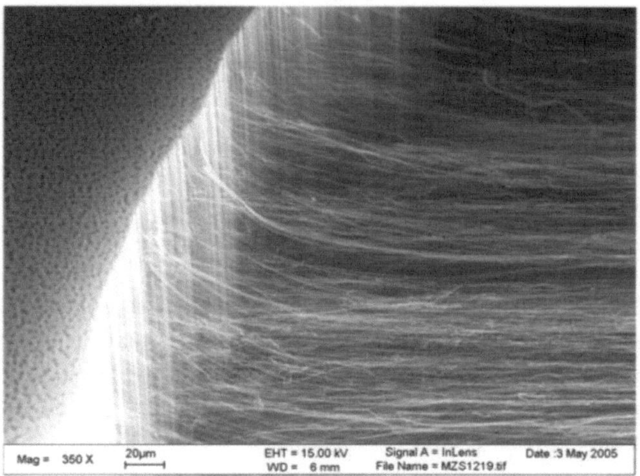


Fig. 11. SEM image of a drawable forest. There are holes with ~ 1 µm diameter distributed in the forest.

Drawable forests must have a high enough area density for CNTs in the forest to form interconnections. The required area density of the forest is related to the diameter of the nanotubes. When the forest is formed by CNTs with ~10 nm diameters, the well-drawn forest has the area density $\sim 10^{11}/cm^2$, and the undrawable forest has an area density of less than $10^{10}/cm^2$. The area density needs to be higher for the forest with thinner CNTs.
The effect of area density can be compensated by adjusting the height of the forest (see section 3.4). The area density can also be lowered by controlling the distribution of catalysts through patterning. Figure 11 shows the SEM image of a drawable forest. There are uniformly distributed holes due to the missing catalysts on the substrate. The less CNT interconnections in the holes results in the easy draw of the too densely packed CNT forest.

### 3.3 Purity of the CNTs

Generally, the drawable forests need to be clean. Amorphous carbon (a-C) deposited on CNTs during CVD process might not be avoidable. A proper amount a-C might be helpful in CNT interconnections. However, too much a-C will increase the locking between CNTs and CNT bundles, which will cause the breaking of fibrils during sheet draw. The results of TGA measurement of the drawable forest and TEM observation of the CNT sheet are shown in Fig. 12.

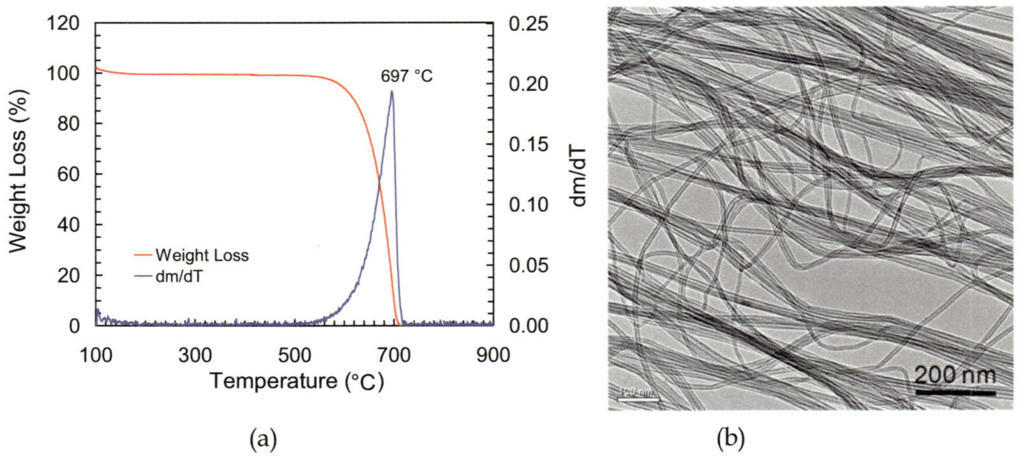


Fig. 12. (a) TGA data shows the thermal stability and the purity of the CNTs and (b) TEM image of a CNT sheet.

### 3.4 Height of the forest

As described above, the interconnections of CNTs in the forest play an important role in the drawability of the forest. The CNTs form small bundles, each consisting of a few nanotubes, in the forest with individual nanotubes moving in and out of different bundles. The three-dimensional connectivity caused by intermittently switched bundling is believed to be important for the drawing process. The too-long and too-short CNT forests are not suitable for the solid-state process because the interconnections there either too much or too little for continuously pulling CNTs out of the forest. Figure 13a is the schematic relationship between drawability and length of the forest. CNTs can be drawn from a ~20 μm high forest as shown in Fig. 13b. Experiments demonstrated that the good drawablity has been obtained in up to 500 μm high forests formed by nanotubes ~10 nm in diameter. Two-millimeter high forests formed by tubes 30~50 nm in diameter are demonstrated to be drawable Inoue, 2011.

Since the drawability is determined by the interconnections of CNTs in the forest, the very-short and very-high forests could be drawable by adjusting other parameters. For example, very-short forest could be drawable if the area density is high enough and relatively more interconnections occur along their length. For high forests, lowering the area density and the interconnections between CNTs and their bundles along their length will create the same effect (Fig. 9).

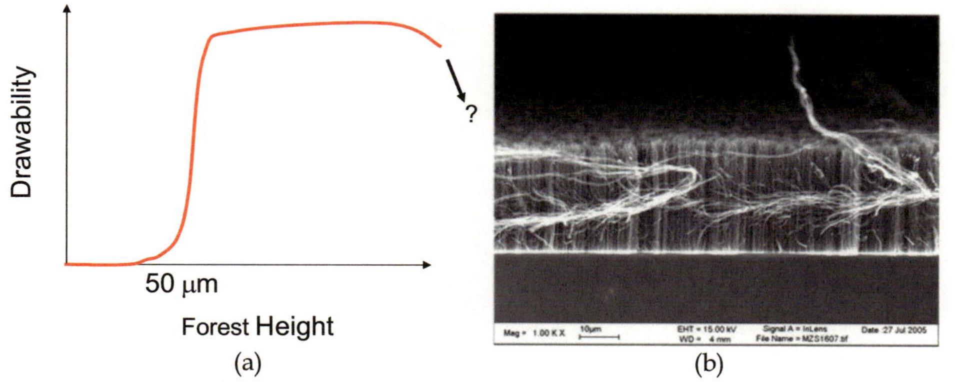


Fig. 13. (a) Schematic of the possible relationship between drawability and forest height. (b) SEM image shows that the CNTs can be drawn continuously from a 20 μm high CNT forest.

The forest-sheet conversion rate and the thickness of as-produced sheet depend on the height and the density of the forest, as well as the bundling level of tubes in the forest. For forests having similar topology, the highest forests were easiest to draw into sheets - most probably because increasing nanotube length increases inter-fibril mechanical coupling within the sheets, and produce a higher forest-sheet conversion rate. A one centimeter length of 245 μm high forest converts to about a three-meter-long free-standing CNT sheet. By adjusting the height and density of the forest; converting a one centimeter length forest to over a ten meter long sheet has been achieved. The thickness of the as-produced CNT sheet increased with increasing forest height and was ~18 μm in SEM images of a sheet drawn from a 245 μm high forest.

### 3.5 Process temperature

Temperature is another important parameter, which determines other synthesis parameters for CNT growth and the quality of the CNTs. The intensity ratio of the G band ($I_G$ at ~1580 cm$^{-1}$) and D band ($I_D$ at ~1350 cm$^{-1}$) in Raman spectra and the initial burning temperature of the CNTs obtained from the TGA are used to evaluate the quality of the CNTs. G band and D band were originated from the Raman active in-plane atomic displacement $E_{2g}$ mode and disorder-included features due to the finite particle size effect or lattice distortion, respectively (Tuinstra & Koenig, 1970). The increase of $I_G/I_D$ indicates that the degree of long-range ordered crystalline perfection of the CNTs increases. The better the crystallization of the graphene layers of the CNTs, the higher the initial burning temperature of the CNTs. Maintaining a spatial homogeneous temperature during the growth process was demonstrated a critical factor for fabricating long CNTs with consistent electrical characteristics (X. Wang et al., 2009). Many researches show that the quality of the CNTs becomes better as the growth temperature increases (Y. T. Lee et al., 2002; C. J. Lee et al., 2001; K. Kim et al., 2005; X. Feng et al., 2009). However, other parameters for CNT growth must be optimized if increasing temperature since the temperature increase has direct influences on the formation and the activity of the catalysts and the feedstock of the carbon atoms. The partial pressure of the hydrocarbon gas usually needs to be lowered at a higher process temperature.

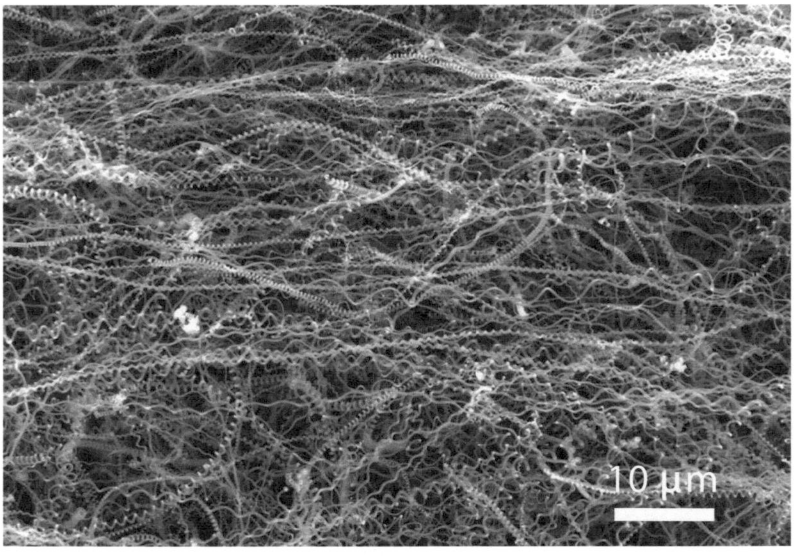


Fig. 14. SEM image of large amount of helically coiled carbon nanostructures. Each coil grows with its own diameter and pitch.

## 4. Conclusion

Individual carbon nanotubes are like minute bits of string, and many trillions of these invisible strings must be assembled together to make useful macroscopic articles. This chapter presents the fabrication processes to making CNT sheets. The purity of the nanotubes, the height of the forest, the morphology of the forest, especially the 3D structure by self-assembly during CVD process, and the area density of the nanotubes are the main factors of the drawability of the forest. There are no inherent limitations on either sheet width or length, and no special difficulties arise in maintaining sheet quality during the draw. Currently, the sheets are drawn at up to 2 m/s from special CNT forests (Alive et al., 2009; Lima, et al., 2011) and the width at up to 20 cm (C. Feng et al., 2010). This solid-state process is scalable for continuous, high-rate production. Extension of the technologies of solid-state sheet fabrication to longer CNTs, as well as to a few walled or single walled nanotubes, are important because longer nanotube lengths will enable properties improvements for active devices by means of enabling closer approach of sheet properties to those of individual nanotubes and the conductivity, transparency, and strength of the sheet could be improved by using thinner nanotubes.

CNTs can also be in the coiled structure (Amelinckx et al., 1994 ; Dunlap, 1992; M. Zhang & J. Li, 2009). Figure 14 shows the helically coiled CNTs grown on iron-coated indium tin oxide substrate by catalytic CVD (M. Zhang et al., 2000). More than 95% of the wires are in helical structures. The coils have various diameters and pitches. They grow out of the substrate and maintain their self-organization well during growth. If the coiled CNT forest is drawable, the sheet will have interesting properties that the straight CNTs could not provide. There is plenty of room to further improve the processes and properties of the CNT sheets.

## 5. Acknowledgment

This work was supported or supported partly by National Science Foundation, the Air Force Research Laboratory, Texas Advanced Technology Program, Robert A. Welch Foundation, and the Strategic Partnership for Research in Nanotechnology Consortium in Texas.

## 6. References

Ago, H.; Petritsch, K.; Shaffer, M. S. P.; Windle, A. H.; & Friend, R. H. (1999). Composites of carbon nanotubes and conjugated polymers for photovoltaic devices. *Adv. Matls.* Vol. 11, pp. 1281-1285.

Amelinckx, S.; Zhang, X. B. ; Bernaerts, D.; Zhang, X. F. ; Ivanov, V. ; & Nagy, J. B. (1994). A Formation Mechanism for Catalytically Grown Helix-Shaped Graphite Nanotubes. *Science,* Vol. 29, pp. 635-639.

Alive, A.; Lima, M. D.; Fang, S. & Baughman R. H. (2010). Underwater sound generation using carbon nanotube projectors. *Nano Lett.*, Vol. 10, No. 7, pp. 2374-2380.

Aliev, A. E.; Oh, J.; Kozlov, M. E.; Kuznetsov, A. A.; Fang, S. L.; Fonseca, A. F.; Ovalle, R.; Lima, M. D.; Haque, M. H.; Gartstein, Y. N.; Zhang, M.; Zakhidov, & Baughman, R. H. (2009). Giant-Stroke, Superelastic Carbon Nanotube Aerogel Muscles. *Science,* Vol. 323, pp. 1575-1579.

Bachtold, A.; Hadley, P.; Nakanishi, T.; & Dekker,C. (2001). Logic Circuits with Carbon Nanotube Transistors. *Science,* Vol. 294, pp. 1317-1320.

Berhan, L.; Yi, Y. B.; Sastry, A. M.; Munoz, E.; Selvidge, M.; & Baughman, R. (2004). Mechanical properties of nanotube sheets: Alterations in joint morphology and achievable moduli in manufacturable materials. *J. Appl. Phys.* Vol. 95, pp. 4335-4346.

Chen, L.; Liu C.; Wang, J.; Zhang, W.; Hu, C.; & Fan, S. (2009). Auxetic Materials with Large Negative Poisson's Ratios Based on Highly Orientied Carbn Nanotube Structures. Appl. Phys. Lett. Vol. 94, pp. 253111.

Cheng, Q.F.; Wang, J.P.; Wen, J.J.; Liu, C.H.; Jiang, K.L.; Li, Q.Q.; & Fan, S.S. (2010). Carbon nanotube/epoxy composites fabricated by resin transfer molding. *Carbon,* Vol. 48, pp. 260-266.

Dan, B. ; Irvin, G. C. ; & Pasquali, M. (2009). Continuous and Scalable Fabrication of Transparent Conducting Carbon Nanotube Films. *ACS nano,* Vol. 3, No. 4, pp. 835-843.

Dunlap, B. I. (1992). Connecting carbon tubules. *Phys. Rev. B,* Vol. 46, pp. 1933-1936.

Endo, M.; Muramatsu, H.; Hayashi, T.; Kim, Y. A.; Terrones, M.; & Dresselhaus, M. S. (2005). Nanotechnology: 'Buckypaper' from coaxial nanotubes. *Nature,* Vol. 433, p. 476.

Feng, X.; Liu, K.; Xie, X.; Zhou, R.; Zhang, L.; Li, Q.; Fan, S.; & Jiang, K. (2009). Thermal Analysis Study of the Growth Kinetics of Carbon Nanotubes and Epitaxial Graphene layers on Them. *J. Phys. Chem.* Vol. 113, PP. 9623-9631.

Feng, C.; Liu, K.; Wu, J.; Liu, L.; Cheng, J.; Zhang, Y.; Sun, Y.; Li, Q.; Fan, S.; Jiang, K. (2010). Flexible, stretchable, transparent conducting films made from superaligned carbon nanotubes. *Adv. Funct. Mater.* Vol. 20, pp. 885-891.

Fischer, J. E.; Zhou, W.; Vavro, J.; Llaguno, M. C.; Guthy, C.; Haggenmueller, R.; Casavant, M. J.; Walters, D. E.; & Smalley, R. E. (2003). Magnetically aligned single wall carbon nanotube films: Preferred orientation and anisotropic transport properties. *Journal of Applied Physics,* Vol. 93, pp. 2157-2163.

Galvan-Garcia, P.; Keefer, E. W.; Yang, F.; Zhang, M.; Fang, S. L.; Zakhidov, A. A.; Baughman, R. H. & Romero, M. I. (2007). Robust Cell Migration and Neuronal Growth on Pristine Carbon Nanotube Sheets and Yarns. *Journal of Biomaterials Science,* Vol. 18, pp. 1245-1261.

Hu, L.; Hecht, D. S.; & Gruener, G. (2010). Carbon Nanotube Thin Films: Fabrication, Properties, and Applications. *Chem. Rev.* Vol. 110, pp. 5790–5844.

Hu, L.; Hecht, D. S.; & Gruener, G. (2004). Percolation in transparent and conducting carbon nanotube networks. *Nano Letters,* Vol. 4, pp. 2513-2517.

Inoue, Y.; Suzuki, Y.; Minami, Y.; Muramatsu, J.; Shimamura, Y.; Suzuki, K.; Ghemes, A.; Okada, M.; Sakakibara, S.; Mimura, H.; & Naito K. (2011). Anisotropic Carbon Nanotube Papers Fabricated from Multiwalled Carbon Nanotubes Webs. *Carbon,* Vol. 49, pp. 2437-2443.

Ju, S. Y.; Facchetti, A.; Xuan, Y.; Liu, J.; Ishikawa, F.; Ye, P. D.; Zhou, C. W.; Marks, T. J.; Janes, D. B. (2007). Fabrication of Fully Transparent Nanowire Transistors for Transparent and Flexible Electronics. *Nat. Nanotechnol.,* Vol. 2, pp. 378– 384.

Kim, K.; Kim, K. J.; Jung, W. S.; Bae, S. Y.; Park, J.; Choi, J.; & Choo, J. (2005). Investigation on the temperature-dependent growth rate of carbon nanotubes using chemical vapor deposition of ferrocene and acetylene. *Chemical Physics Letters,* Vol. 401, pp. 459-464.

Kim, Y.; Minami, N.; Zhu, W. H.; Kazaoui, S.; Azumi, R.; & Matsumoto, M. (2003). Langmuir-Blodgett films of single-wall carbon nanotubes: layer-by-layer deposition and in-plane orientation of tubes. *Jpn. J. Appl. Phys., Part 1,* Vol. 42, pp. 7629-.

Kong, J. ; Soh, H. T. ; Cassell, A. M.; Quate, C. F.; & Dai, H. (1998). Synthesis of individual single-walled carbon nanotubes on patterned silicon wafers. *Nature,* Vol. 395, pp. 878-881.

Kong, J.; Franklin, N. R.; Zhou, C.; Chapline, M. G.; Peng, S.; Cho, K.; & Dai, H. (2000). Nanotube Molecular Wires as Chemical Sensors. *Science,* Vol. 287, pp. 622-625.

Kozlov, M. E.; Haines, C. S.; Oh, J.; Lima, M. D.; & Fang, S. (2009). Sound of carbon nanotube assemblies. *J. Appl. Phys.,* Vol. 106, pp. 124311.

Kuznetzov, A. A.; Lee, S. B.; Zhang, M.; Baughman, R. H.; & Zakhidov, A. A. (2010). Electron field emission from transparent multiwalled carbon nanotube sheets for inverted field emission displays. *Carbon,* Vol. 48, pp. 41-46.

Kuznetsov, A. A.; Fonseca, A. F.; Baughman, R. H.; & Zakhidov, A. A. (2011). Structural Model for Dry-Drawing of Sheets and Yarns from Carbon Nanotube Forests. *ACS Nano,* Vol. 5 (2), pp. 985–993.

Lee, C. J.; Park, J.; Huh, Y.; & Lee, J. Y. (2001). Temperature effect on the growth of carbon nanotubes using thermal chemical vapor deposition. *Chemical Physics Letters,* Vol. 343, pp. 33-38.

Lee, Y. T.; Park, J.; Choi, Y. S.; Ryu, H. & Lee, H. J. (2002). Temperature-Dependent Growth of Vertically Aligned Carbon Nanotubes in the Range 800−1100 °C. *J. Phys. Chem. B,* Vol. 106, No. 31, pp 7614–7618.

Li, Y.; Kinloch, I. A.; & Windle, A. H. (2004). Direct spinning of carbon nanotube fibers from chemical vapor deposition synthesis. *Science,* Vol. 304, pp. 276-278.

Lima, D.; Fang, S.; Lepró, X.; Lewis, C.; Ovalle-Robles, R.; Carretero-González, J.; Castillo-Martínez, E.; Kozlov, M. E.; Oh, J.; Rawat, N.; Haines, C. S.; Haque, M. H.; Aare, V.; Stoughton, S.; Zakhidov, A. A.; & Baughman, R. H. (2011). Biscrolling Nanotube Sheets and Functional Guests into Yarns. *Science* Vol. 331, pp. 51-55.

Liu, K.; Sun, Y. H.; Zhou, R. F.; Zhu, H. Y.; Wang, J. P.; Liu, L.; Fan, S. S.; & Jiang, K. L. (2010). Carbon Nanotube Yarns with High Tensile Strength Made by a Twisting and Shrinking Method. *Nanotechnology*, Vol. 21, pp. 045708.

Martin, C. (2010). A Carbon Nano-Wired World. *R & D magazine, Vol. 52, No. 3, June, 2010, p. 40.*

Misewich, J. A.; Martel, R.; Avouris, Ph.; Tsang, J. C.; Heinze, S. & Tersoff, J. (2003). Electrically Induced Optical Emission from a Carbon Nanotube FET. *Science,* Vol. 300, pp. 783-786.

Nanocomp Technologies Inc. (2011). http://www.nanocomptech.com/html/nanocomp-technology.html

Pevzner, A.; Engel, Y.; Elnathan, R.; Ducobni, T.; Ben-Ishai, M.; Reddy, K.; Shpaisman, N.; Tsukernik, A.; Oksman, M.; & Patolsky, F. (2010). Knocking Down Highly-Ordered Large-Scale Nanowire Arrays. *Nano Lett.*, Vol. 10, pp. 1202– 1208.

Rinzler, A. G.; Liu, J.; Dai, H.; Nikolaev, P.; Huffman, C.B.; Rodr´ıguez-Mac´ıas, F.J.; Boul, P.J.; Lu, A.H.; Heymann, D.; Colbert, D.T.; Lee, R.S.; Fischer, J.E.; Rao, A.M.; Eklund, P.C.; & Smalley, R.E. (1998). Large-scale purification of single-wall carbon nanotubes: process, product, and characterization. *Applied Physics* A, Vol. 67, pp. 29-37.

Sreekumar, T. V.; Liu, T.; Kumar, S.; Ericson, L. M.; Hauge, R. H.; & Smalley, R. E. (2003). Single-wall carbon nanotube films. *Chemistry of Materials,* Vol. 15, pp. 175-178.

Tans, S. J.; Verschueren, A. R. M.; & Dekker, C. (1998). Room-temperature transistor based on a single carbon nanotube. *Nature,* Vol. 393, pp. 49-52.

Ulbricht, R.; Lee, S. B.; Inoue, K.; Zhang, M.; Fang, S.; Baughman, R. H.; & Zakhidov, A. A. (2007). Transparent Carbon Nanotube Sheets as 3-D Charge Collectors in Organic Solar Cells. *Solar Energy Materials and Solar Cells,* Vol. 91, pp. 416-419.

Ulbricht, R.; Jiang, X.; Lee, S. B.; Inoue, K.; Zhang, M.; Fang, S.; Baughman, R. H.; & Zakhidov, A. A. (2006). Polymetric Solar Cells with Oriented and Strong Transparent Carbon Nanotube Anode. *Phys. Stat. Sol. B,* Vol. 243, pp. 3528-3532.

Unidym Inc. (2007). http://www.unidym.com.

Wang, D.; Song, P.; Liu, C.; Wu, W.; & Fan, S. (2008). Highly oriented carbon nanotube papers made of aligned carbon nanotubes. *Nanotechnology,* Vol. 7, pp. 075609.

Wang, X.; Li, Q.; Xie, J.; Jin, Zhong.; Wang, J.; Li, Y.; Jiang, K.; & Fan, S. (2009). Fabrication of ultralong and electrically uniform single-walled carbon nanotube on clean substrates. *Nano Lett.* Vol. 9, No. 9, pp. 3137-3141.

Wei, Y.; Liu, L.; Liu, P.; Xiao, L.; Jiang, K.; & Fan, S. (2008). Scaled fabrication of single-nanotube-tipped ends from carbon nanotube micro-yarns and their field emission applications. *Nanotechnology,* Vol. 19, pp. 475707.

Williams, C. D.; Robles, R. O.; Zhang, M.; Li, S.; Baughman, R. H.; & Zakhidov, A. A. (2008). Multiwalled carbon nanotube sheets as transparent electrodes in high brightness organic light-emitting diodes. *Applied Physics Letters,* Vol. 93, pp. 183506.

Wu, Z. C.; Chen, Z. H.; Du, X.; Logan, J. M.; Sippel, J.; Nikolou, M.; Kamaras, K.; Reynolds, J. R.; Tanner, D. B.; Hebard, A. F.; & Rinzler, A. G. (2004). Transparent, conductive carbon nanotube films. *Science*, Vol. 305, pp. 1273-1276.

Xia, Y. N.; Yang, P. D.; Sun, Y. G.; Wu, Y. Y.; Mayers, B.; Gates, B.; Yin, Y. D.; Kim, F.; Yan, Y. Q. (2003). One-Dimensional Nanostructures: Synthesis, Characterization, and Applications. Adv. Mater. Vol. 15, pp. 353- 389.

Xiao, L.; Liu, P.; Liu, L.; Jiang, K.; Feng, X.; Wei, Y.; Qian, L.; Fan, S.; Zhang, T. (2008). Barium-Functionalized Multiwalled Carbon Nanotube Yarns as Low-Work-Function Thermionic Cathodes. *Appl. Phys. Lett.*, Vol. 92, pp. 153108.

Xiao, L.; Chen, Z.; Feng, C.; Liu, L.; Bai, Z. Q.; Wang, Y.; Qian, L.; Zhang, Y. Y.; Li, Q. Q.; Jiang, K. L.; Fan, S. S. (2008). Flexible, Stretchable, Transparent Carbon Nanotube Thin Film Loudspeakers. *Nano Lett.*, Vol. 8, pp. 4539-4545.

Xu, F.; Lu, W.; & Zhu, Y. (2011). Controlled 3D Buckling of Silicon Nanowires for Stretchable Electronics. *ACS Nano*, Vol. 5, pp. 672- 678.

Zhang, M.; Nakayama, Y.; & Pan, L. (2000). Synthesis of Carbon Tubule Nanocoils in High Yield Using Iron-Coated Indium Tin Oxide as Catalyst. *Jpn. J. Appl. Phys.*, Vol. 39, pp. L1242-L1244.

Zhang, M.; Atkinson, K. R.; & Baughman, R. H. (2004). Multifunctional carbon nanotube yarns by downsizing an ancient technology. *Science*, Vol. 306, pp.1358-1361.

Zhang, M.; Fang, S.; Zakhidov, A. A.; Lee, S. B.; Aliev, A. E.; Williams, C. D.; Atkinson, K. R.; & Baughman, R. H. (2005). Strong, Transparent, Multifunctional Carbon Nanotube Sheets. *Science*, Vol. 309, pp. 1215-1219.

Zhang, M. & Li, J. (2009). Carbon Nanotube in Different Shapes. *Materials Today*, Vol. 12, pp. 12-18.

Zhang, T.; Mubeen, S.; Myung, N.; & Deshusses, M. (2008) Recent progress in carbon nanotube-based gas sensors. *Nanotechnology*, Vol. 19, pp. 332001–14.

Zhang, H.; Feng, C.; Zhai, Y.; Jiang, K.; Li, Q.; & Fan, S. (2009). Cross-Stacked Carbon Nanotube Sheets Uniformly Loaded with $SnO_2$ Nanoparticles: A Novel Binder-Free and High-Capacity Anode Material for Lithium-Ion Batteries. *Adv. Mater.* Vol. 21, pp. 2299–2304.

Zhang, X.; Jiang, K.; Feng, C.; Liu, P.; Zhang, L.; Kong, J.; Zhang, T.; Li, Q.; & Fan, S. (2006). Spinning and Processing Continuous Yarns from 4-Inch Wafer Scale Super-Aligned Carbon Nanotube Arrays. *Adv. Mater.*, Vol. 18, pp. 1505- 1510.

Zhong, G.; Hofmann, S.; Yan, F.; Telg, H.; Warner, J. H.; Eder, D.; Thomsen, C.; Milne, W. I.; & Robertson, J. (2009). Acetylene: A Key Growth Precursor for Single-Walled Carbon Nanotube Forests. *J. Phys. Chem. C*, Vol. 113, pp. 17321-17325.

Zhong, X. H.; Li, Y. L.; Liu, Y. K.; Qiao, X. H.; Feng, Y.; Liang, J.; Jin, J.; Zhu, L.; Hou, F.; Li, J. Y. (2010). Continuous Multilayered Carbon Nanotube Yarns. *Adv. Mater.*, Vol. 22, pp. 692- 696.

Zhou, Y.; Hu, L.; & Gruner, G. (2006). A method of printing carbon nanotube thin films. *Appl. Phys. Lett.* Vol. 88, pp. 123109.

Zhou, R.; Meng, C.; Zhu, F.; Li, Q.; Liu, C.; Fan, S.; & Jiang, K. (2010). High-performance supercapacitors using a nanoporous current collector made from super-aligned carbon nanotubes. *Nanotechnology*, Vol. 21, pp. 345701.

2

# A Close-Packed-Carbon-Nanotube Film on SiC for Thermal Interface Material Applications

Wataru Norimatsu[1], Chihiro Kawai[2] and Michiko Kusunoki[3]
[1]*Graduate School of Engineering, Nagoya University*
[2]*Electronics & Materials R & D Laboratories, Sumitomo Electric Industries, Ltd.*
[3]*EcoTopia Science Institute, Nagoya University*
*Japan*

## 1. Introduction

Carbon nanotubes (CNT) are predicted to have an extremely high thermal conductivity along the tube axis of 6600 W/mK (Berber, 2000) while experimental results (Fujii et al., 2005; Kim et al., 2001) have produced values of 2000~3000 W/mK. However, only a few studies have investigated CNTs for use as thermal interface materials (TIMs) (Biercuk et al., 2002; Liu et al., 2004; Huang et al., 2005). In this study, we focus on the heat-release structure of a central processing unit (CPU). Figure 1 shows the schematic diagram of the typical heat-release structure.

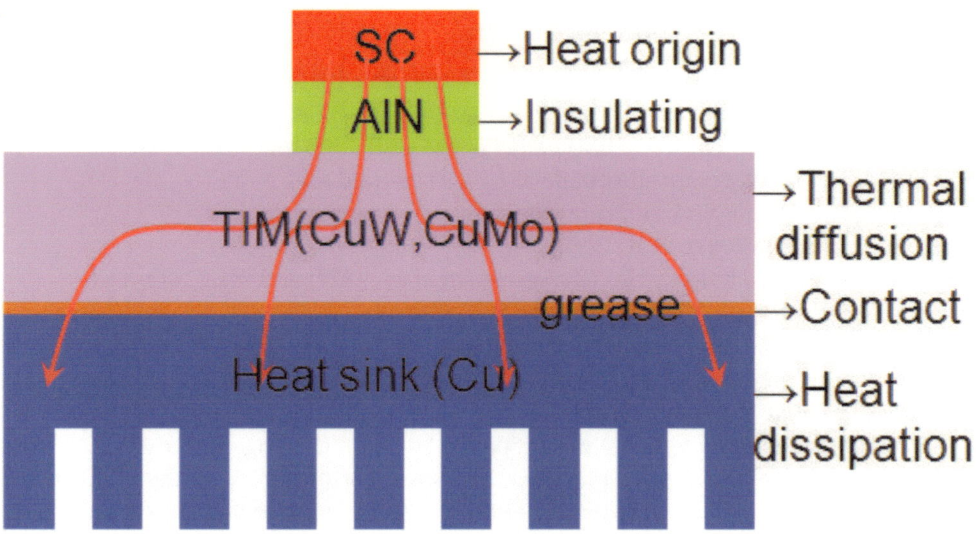


Fig. 1. Schematic diagram of the typical heat-release structure.

In the typical structure, TIMs such as Cu-W alloys, Cu-Mo alloys, Al-SiC solid solution and so on, are used to dissipate heat efficiently from the heat source, such as a semiconductor (SC), to heat sinks. Recently, a rapid increase in the heat generated by various electronic devices requires an urgent development of high-performance TIMs. The performance of the actual TIMs depends not only intrinsic thermal conductivity, but also on the thermal resistance. Here, the thermal resistance of TIMs is given by the sum of their intrinsic thermal resistance and the contact thermal resistance between the TIM and the contacting material. It is thus important for thermal management to reduce the contact resistance on the contact face as well as to employ highly thermal conductive materials. Actually, in order to reduce the contact resistance, a layer of grease ranging from 50 to 100 μm thick is applied to the contact face between the TIM and the contacted material. The grease intrudes into the microscopic undulations and surface roughnesses of the contacted material, and eliminates any air gap, leading to an increase of the contacted area and a consequent decrease of the contact thermal resistance. However, the total thermal resistance does not decrease efficiently with the use of such a thick grease layer, since the intrinsic thermal conductivity of grease is low.

To reduce the total thermal resistance, thermally conducting AlN or Ag particles are dispersed into the grease, but they lead to higher costs and the thermal conducting paths are interrupted in the particle-dispersed structure. On the basis of the high thermal conductivity of CNTs, some investigators have investigated various uses of to reduce total resistance; dispersing CNTs into plastics or embedding an aligned CNT array in a silicone elastomer, resulting in thermal conductivity enhancement (Biercuk et al., 2002; Liu et al, 2004; Huang; 2005). However, it is difficult to ensure a continuous thermal conducting path in the dispersed materials, and in the aligned materials, the low CNT density limits the overall heat conductance, the total thermal conductivity in these structures is reduced to the order of 1 W/mK. That is, workers have not succeeded in utilizing the extremely high thermal conductivity of CNTs.

## 2. Features of CNT/SiC composite material

We have reported that a vertically-aligned CNT film can be formed on SiC by a surface decomposition method (Kusunoki at al., 1997; 1999; 2000; 2002; 2005). Figure 2 shows the transmission electron microscope (TEM) image of the CNT/SiC composite material.

The features of CNT/SiC composite obtained by this method are,

1. high-density, well-aligned, and catalyst-free,
2. flexible CNT tips (Miyake et al., 2007) ,
3. high thermal conductivity of SiC (Burgemeister et al., 1979) and the CNTs, and
4. high adhesive strength of CNTs with the SiC substrate.

In other words, a CNT/SiC composite material meets the requirements for TIMs. In this study, then, we investigate the heat transfer characteristics of CNT/SiC composite materials made by the surface decomposition of SiC.

The CNT/SiC composite materials were prepared by the following procedure. A 6H-SiC single-crystal substrate and a polycrystalline SiC substrate obtained by the CVD method were cut into 10 × 10 × 0.25 $mm^3$ pieces, and their both sides of surface were polished. Carbon nanotube films with thicknesses of 1 and 4 μm were formed on the two sides of the SiC substrate by heating the substrate in a vacuum of about $10^{-4}$ Torr at temperatures of 1700 and 1900 °C, respectively. Observations of the microstructure of the CNT/SiC composites were carried out using a scanning electron microscope (SEM) and a JEM-2010-type transmission electron microscope (TEM).

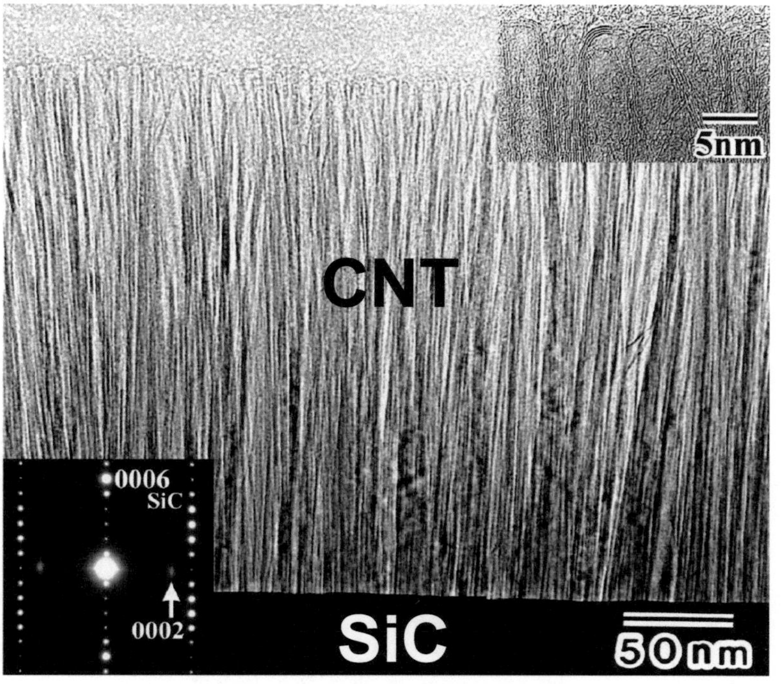


Fig. 2. TEM image and the corresponding electron diffraction pattern of CNT/SiC composite. High-magnification image around the CNT cap is inserted.

## 3. Thermal transport properties of CNT/SiC composite materials

Thermal resistance measurements were performed using an apparatus based on the American Society of Testing Materials (ASTM) Method D5470. A schematic diagram of the measurement system is shown in Fig. 3.

The prepared CNT/SiC composite material was sandwiched between three types of copper holders and a load ranging from 750 to 3750 g/cm$^2$ was applied from above. The surfaces of the copper holders were (1) $R_z$ = 0.03 μm, and the flatness is ± 0.3 μm in 10 × 10 mm$^2$, (2) $R_z$ = 1.0 μm, and the flatness of ± 0.3 μm, and (3) $R_z$ = 1.0 μm, and the flatness of ± 15 μm. The upper holder was heated by a ceramic heater. Thermocouples were inserted in copper holders, and the temperatures at different positions were measured as shown in the right side of the Figure. The temperatures $T_h$, at the top, and $T_c$ at the bottom of the sample, were obtained by extrapolating the temperature gradient. The thermal resistance $R_t$ is given by the following equation, where $Q$ is the supplied heat value.

$$R_t = (T_h - T_c) / Q \quad [\mathrm{K/W}] \tag{1}$$

For comparison, the thermal resistances of 100 μm thick samples of commercial silicone grease (G747, Shin-Etsu Chemical Co., Ltd., thermal conductivity of 1.09 W/mK) and grease with Ag particles (GR-SG014, TIMELY Co. Ltd., thermal conductivity of 9.0 W/mK) applied to the holders were also measured.

Figure 4(a) and (b) show, respectively, the SEM image of the CNT tips and the TEM image of the interface between the CNT and the SiC.

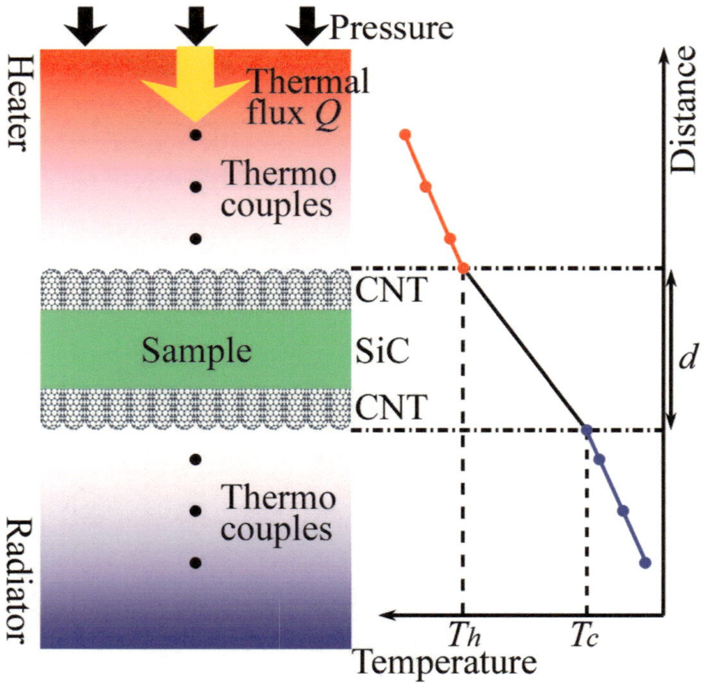


Fig. 3. Schematic diagram of the thermal resistance measurement system.

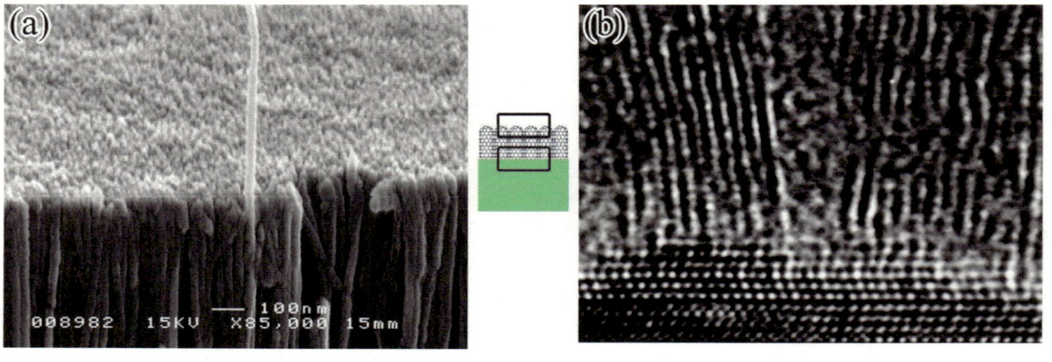


Fig. 4. (a) Scanning electron microscope image of the CNT tips. (b) High-resolution transmission electron microscope image showing the interface between the CNT and SiC.

As is seen in Fig. 4(a), aligned CNT bundles with a diameter of about 30 nm are closely packed and well aligned. We have reported that the planar density of CNTs obtained by surface decomposition of SiC is estimated to be about $3 \times 10^4$ μm$^{-2}$ (Kusunoki et al., 1997). This density is more than one hundred times as high as that of CNTs grown by conventional CVD method (Lee et al., 1999; Pan et al., 1999). The high-resolution TEM image shows the presence of $(0002)_{graphite}$ lattice fringes almost perpendicular to the SiC surface. These fringes correspond to the walls of multiwalled nanotubes. This image demonstrates that CNTs are strongly adhered to the SiC substrate at the atomic scale. Another feature of our

CNTs is the absence of catalyst and amorphous layers. We therefore expected that the thermal interface resistance between CNTs and SiC will be sufficiently low to be useful.
In Fig. 5, the thermal resistances of different samples are shown for an applied pressure in this case fixed at 3750 g/cm², with the flattest sample holder (1).

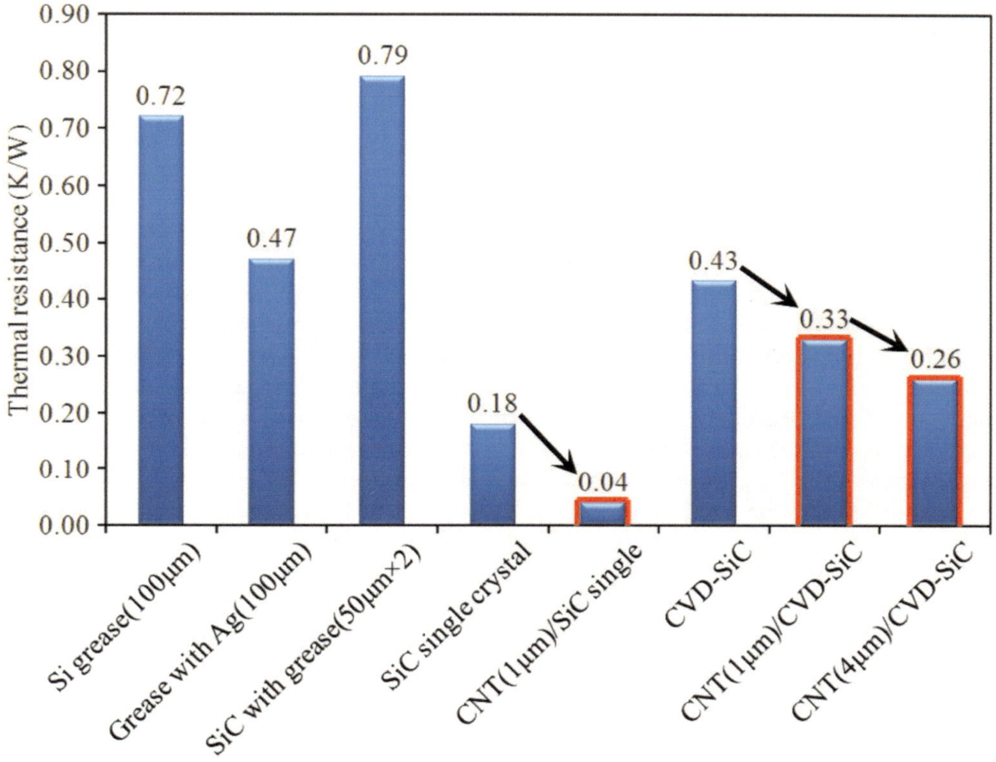


Fig. 5. Thermal resistances of various samples. Results obtained from the samples with CNTs are emphasized by the red lines.

The thermal resistances of silicon grease and grease with Ag particles were 0.72 and 0.47 K/W, respectively. The resistance of the pristine SiC single crystal was 0.18 K/W, reflecting the high thermal conductivity of the SiC crystal. The low resistance of pristine SiC indicates the high potential of SiC crystal itself as a TIM. It should be mentioned here that the intrinsic thermal conductivity of SiC crystal is about 360 W/mK, which is comparable to that of copper (403 W/mK) or silver (428 W/mK) (Burgemeister et al., 1979). In order to make a comparison practical TIMs, then, we measured the thermal resistance of the pristine SiC single crystal coated with silicon grease to a thickness of 50 μm. The result obtained was 0.79 K/W. This increase is attributed to the thick coating of grease. On the other hand, the thermal resistance of the 1μm-CNT/SiC single crystal sample was 0.04 K/W, which is a remarkably low value. This observation suggests that the contact resistance is drastically reduced by the use of CNTs instead of grease.
The thermal resistances of a polycrystalline SiC substrate produced by the CVD method (CVD-SiC), 1μm-CNT/CVD-SiC, and 4μm-CNT/CVD-SiC were 0.43, 0.33, and 0.26 K/W,

respectively. These results indicate that the resistance of CVD-SiC is higher than that of single crystal SiC because of the thermal scattering due to the presence of grain boundaries, but the resistance can be reduced by forming CNTs. We then concluded that CNT formation on both single-crystal and polycrystalline SiC can result in a high performance TIM.

As was mentioned above, it is important for thermal management to reduce the contact thermal resistance as well as adopting the use of thermal conductive materials. The thermal resistance values of CNT/SiC composites containing the contact resistance were quite low, which arises from the large contact area between CNTs and the contacted materials. We then altered the contact conditions by applying a load to the contact face, and remeasured their thermal resistance values. Figure 6(a) shows the pressure dependence of thermal resistance using the 1μm-CNT/SiC single crystal sample.

The resistance value rapidly decreases with increasing pressure, reaching saturation above 3750 g/cm$^2$. This result indicates that an increase of the pressure leads to an increase of the contact area. Schematic diagrams of the contacting conditions in the cases of low and high pressures are shown in Fig. 6(b). When the pressure is low, point contacts between the CNT tips and the contacted material raise the contact thermal resistance. At a high pressure, the flexibility of the CNT tips can lead to an increase of the contact area, which lowers the contact resistance drastically. Similar flexible buckling phenomena of these CNTs were previously reported (Miyake et al., 2007). When further pressure is applied, the contact area does not increase, giving rise to the saturation behavior above 3750 g/cm$^2$. In addition, it is quite important for the application that CNTs have never been peeled off from the SiC substrate during iterations of loading and unloading under such condition.

We here attempt the conversion of the measured thermal resistance into the thermal conductivity $k$ in order to compare with the previous reports on TIMs. The measured thermal resistance $R_t$ is given by the sum of the intrinsic thermal conductivity component $d/k$ and the contact thermal resistance $R_{contact}$.

$$R_t \cdot A\ [m^2W/K] = d/k + R_{contact} \tag{2}$$

Here, $d$ is the thickness of the sample, and $A$ is the sectional-area of the sample. The intrinsic thermal conductivity is then given by the inverse of the gradient of the thickness-resistance graph, while the contact thermal resistance is given by the intercept of the same graph. Therefore, we should obtain both the intrinsic thermal conductivity and contact thermal resistance components by this method. However, it is difficult to distinguish the intrinsic thermal conductivity from the contact thermal resistance in CNT/SiC composite materials because there are three components (CNT, SiC, and CNT) and four interfaces. We therefore define the practical thermal conductivity $k'$ as follows.

$$k'\ [W/mK] = d/(R_t \cdot A) = (Q \cdot d)/\{(T_h - T_c) \cdot A\} \tag{3}$$

Here, the sample thickness and area are fixed to $d$ = 250 μm and $A$ = 100 mm$^2$. It should be noted that the practical thermal conductivity $k'$ can be used only in comparing materials having the same thickness and area. It is of particular importance that the practical conductivity $k'$ includes the contact resistance component, in order for the practical conductivity $k'$ to reflect the performance of the actual TIMs directly. The obtained practical thermal conductivities of Si grease, grease with Ag particles, SiC single crystal with grease, SiC single crystal, and 1μm-CNT/SiC single crystal were, respectively, 1.39, 2.13, 4.43, 13.9,

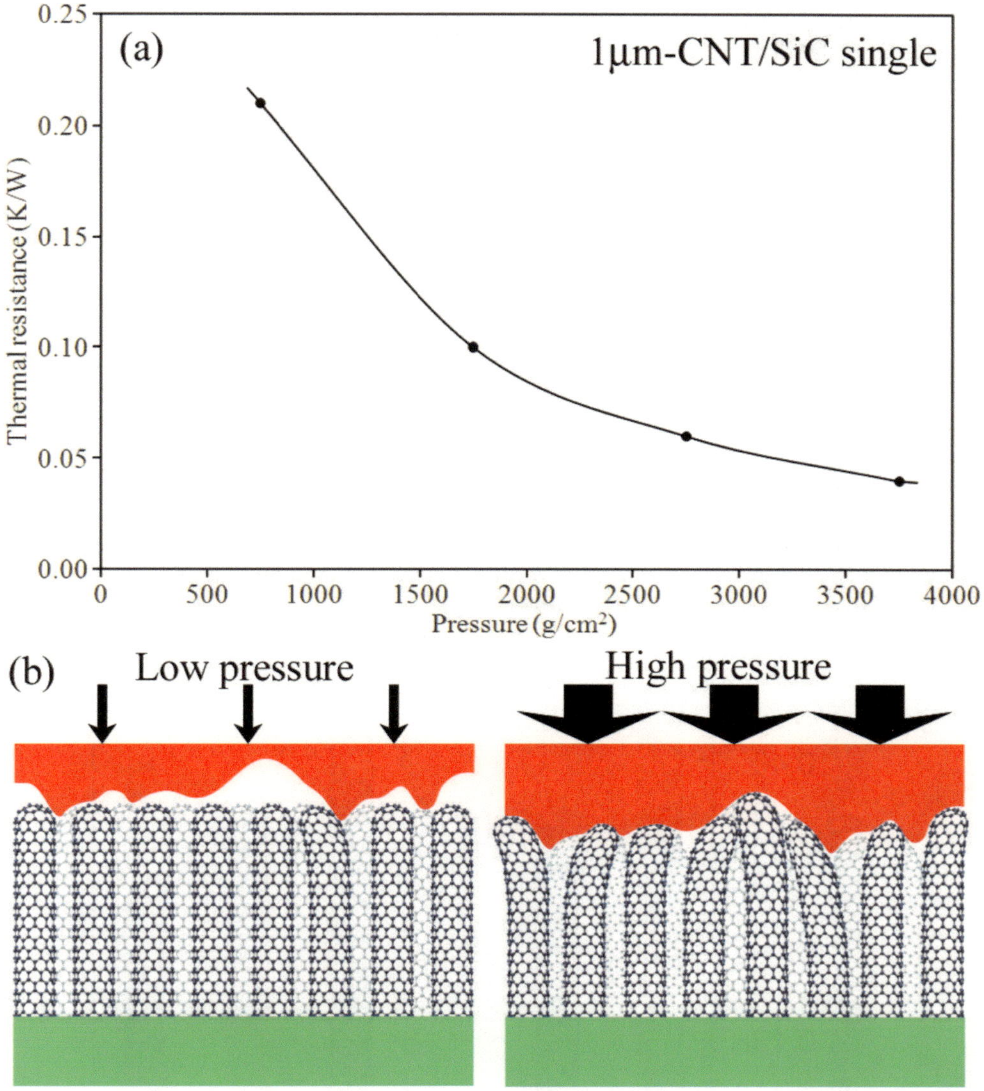


Fig. 6. (a) Pressure dependence of the thermal resistance of the 1μm-CNT/SiC single crystal sample. (b) Schematic diagrams of the contact interface in low pressure and high pressure scenarios.

and 62.5 W/mK. The important item to note here is that the *k*' value of the SiC single crystal with grease is almost equivalent to that of typically TIMs already in use. The *k*' value of 1μm-CNT/SiC was then more than 14 times as high as that of a typical TIM. According to published papers, *k*' values of dispersed-CNT and aligned-CNT were 0.50 and 1.00 W/mK, respectively (Biercuk et al., 2002; Liu et al., 2004: Huang et al., 2005). Our *k*' value is more than 60 times as high as these. This excellent value is attributed to the extremely dense, and well-aligned CNTs, the flexibility of CNT tips, the high thermal conductivities of CNT and SiC, and the strong adhesion between the CNTs and SiC. The flexible CNT tips and strong adhesion in the CNT/SiC interface drastically reduce the contact thermal resistance, while the dense CNTs and thermally conductive SiC transport heat quite efficiently. Although a SiC single crystal costs too much for use in a real device, CNT/SiC composites using low-price polycrystalline SiC substrates produced by a CVD method also have sufficiently high thermal conductivities to be effective TIMs.

The above mentioned experimental values were obtained using sample holder (1), of which the surface roughness $R_z$ was 1.0μm and was without the macroscopic surface undulation. Actually, it is not realistic to use the heat sinks with the mirror finished surface due to the high cost performance. In Figures 7 and 8, we show the results using the sample holders of (2) the roughness of $R_z$ = 1.0 μm, and the flatness of ± 0.3 μm, and (3) $R_z$ = 1.0 μm, and the flatness of ± 15 μm, respectively.

In Fig. 7, it is understood that the longer CNTs on the surface of SiC can reduce the contact thermal resistance also for the holders with rough surface. Figure 8 also tells us that CNTs effectively lowers the contact resistance to 0.6 k/W with the rough and undulant contacting surface. However, thermal resistance of about 0.6 K/W is the close value to that of the grease shown in Fig. 5. This is due to the large surface undulation which induces the little contact area and cannot be overcome by CNTs with 4 μm length. In order to increase the contact area, we formed porous graphite region with the thickness of about 60 μm between CNTs and SiC by progressing thermal decomposition of SiC additionally (Kawai et al. 2009). A TEM image of the graphite region is shown in Figure 9.

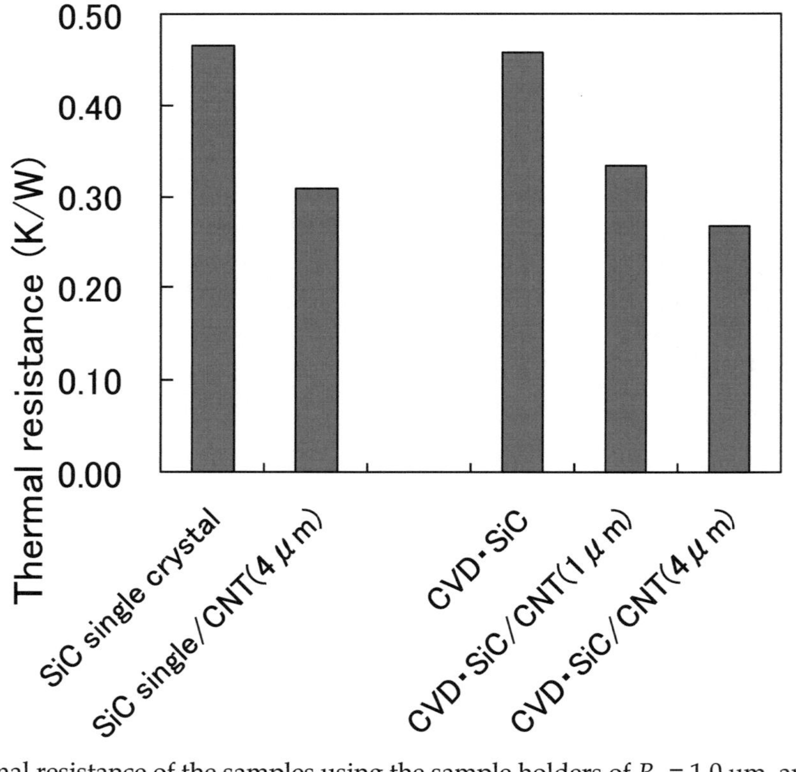


Fig. 7. Thermal resistance of the samples using the sample holders of $R_z$ = 1.0 μm, and the flatness ± 0.3 μm.

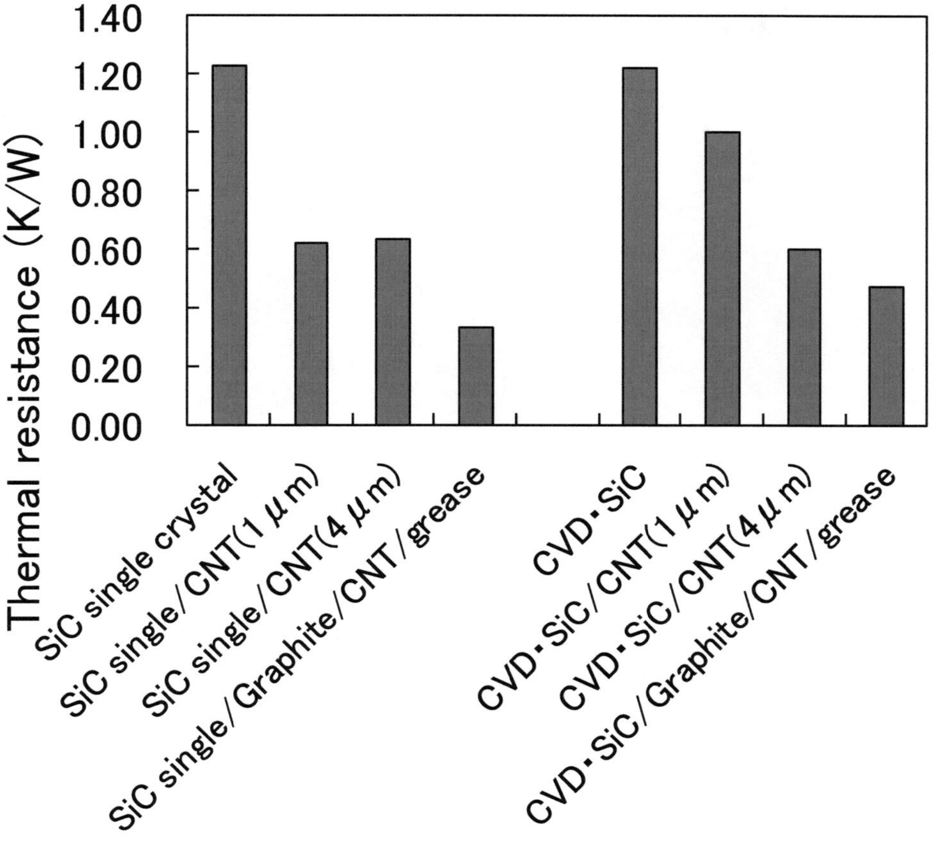


Fig. 8. Thermal resistance of the samples using the sample holders of $R_z$ = 1.0 μm, and the flatness ± 15 μm.

As shown in the image, there are bright area and the plate-like area with the dark contrast. The dark plate-like region consists of the several dozen of graphene layers with the area of several hundred square nanometers, which arrange in the horizontal and the perpendicular directions. These graphite layers surround the hollow area with several hundred square nanometers, indicating the high deformability under pressure. In addition, we used grease together with the sample (Kawai et al., 2009). These results are also shown in Figure 8. The value for SiC single crystal / Graphite / CNT / grease is 0.33 K/W, and 0.47 K/W for CVD-SiC / Graphite / CNT / grease. These favorable values were obtained by the following mechanism, as shown in Figure 10.

The hollow space in graphite can include grease. By applying pressure, highly deformable graphite changes its shape and follows the contacting undulation. In addition to that, grease included in graphite exudes into the interface between CNTs and the contacting material, and helps to increase the contacting area, leading to the decrease in thermal resistance. It is then concluded that the use of CNT / graphite / SiC together with grease can decrease the thermal resistance in the actually used heat-release structure with rough and undulant surface.

Fig. 9. TEM image of porous graphite formed between CNTs and SiC.

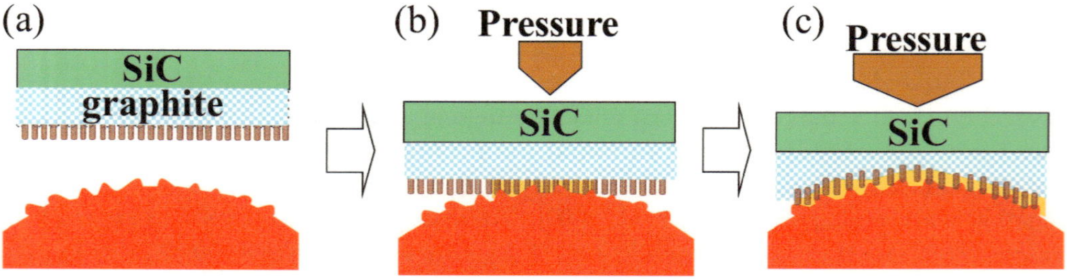


Fig. 10. Schematic diagram of contacting phenomena of CNT / Graphite / SiC with the rough and undulant surface using grease (orange color).

## 4. Summary

We have demonstrated the excellent performance of CNT/SiC composites as thermal interface materials. There was a more than 14 times enhancement in the practical thermal conductivity with 1μm-CNT/SiC single crystal compared with conventional TIMs. This high-quality performance results primarily from a combination of the high thermal conductivities of CNT and SiC and the reduction of contact thermal resistance due to the flexibility of the CNT tips. Based on the obtained results, one of the ideal heat-release structures is shown in Figure 11.

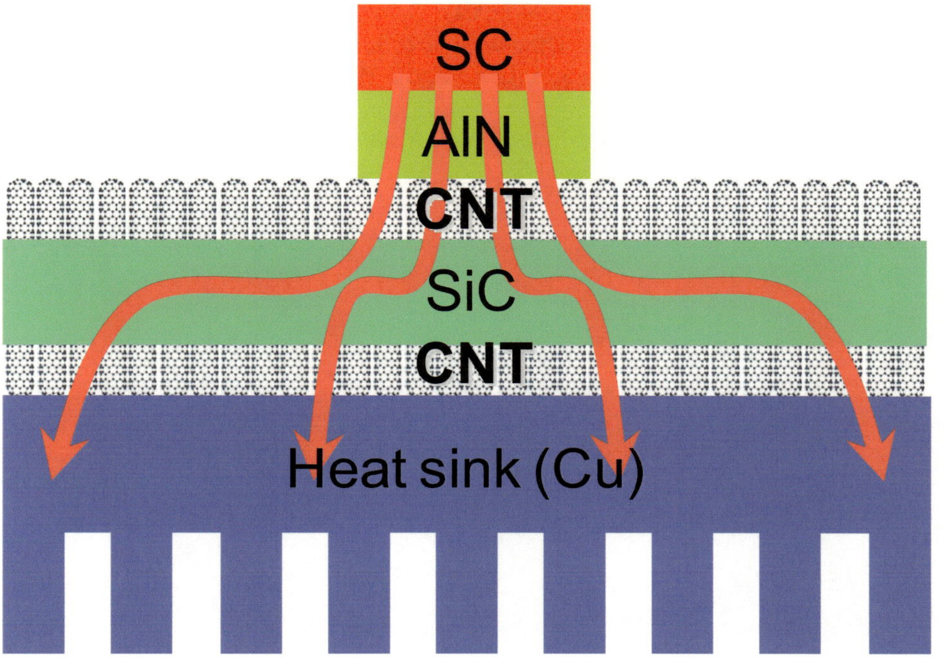


Fig. 11. Possible heat-release structure using the CNT/SiC composite material.

CNT/SiC layered composite materials may be one of the most efficient thermal interface materials which can solve efficiently current serious thermal problems not only in the Si-semiconductor industry, but also in the automobile industry.

## 5. References

Berber, S., Kwon, Y-K. & Tomanek, D. (2000). *Phys. Rev. Lett.* 84, 4613.

Biercuk, M. J., Llaguno, M. C., Radosavljevic, M., Hyun, J. K., Johnson, A. T. & Fischer, J. E. (2002). *Appl. Phys. Lett.* 80: 2767.

Burgemeister, E. A., von Muench, W. & Pettenpaul, E. (1979). *J. Appl. Phys.* 50: 5790.

Fujii, M., Zhang, X., Xie, H., Ago, H., Takahashi, K., Ikuta, T., Abe, H. & Shimizu, T. (2005). *Phys. Rev. Lett.* 95: 065502.

Huang, H., Liu, C. H., Wu, Y., & Fan, S. S. (2005). *Adv. Mater.* 17: 1652.

Kim, P., Shi, L., Majumdar, A. & McEuen, P. L. (2001). *Phys. Rev. Lett.* 87: 215502.

Kusunoki, M., Rokkaku, M. & Suzuki, T. (1997). *Appl. Phys. Lett.* 71: 2620.

Kusunoki, M., Suzuki, T., Kaneko, K. & Ito, M. (1999). *Phil. Mag. Lett.* 79: 153.

Kusunoki, M., Suzuki, T., Hirayama, T., Shibata, N. & Kaneko, K. (2000). *Appl. Phys. Lett.* 77: 531.

Kusunoki, M., Suzuki, T., Honjo, C., Hirayama, T. & Shibata, N. (2002). *Chem. Phys. Lett.* 366: 458.

Kusunoki, M., Honjo, C. Suzuki, T. & Hirayama, T. (2005). *Appl. Phys. Lett.* 87: 103105.

Lee, C. J., Kim, D. W., Lee, T. J. Choi, Y. C., Park, Y. S., Lee, Y. H., Choi, N. S. Lee, W. B., Park, G-S. & Kim, J. M. (1999). *Chem. Phys. Lett.* 312: 461.

Liu, C. H. , Huang, H., Wu, Y. & Fan, S. S. (2004). *Appl. Phys. Lett.* 84: 4248.

Miyake, K., Kusunoki, M. Usami, H., Umehara, N. & Sasaki, S. (2007). *Nano Lett.* 7: 3285.

Pan, Z. W., Xie, S. S., Chang, B. H., Sun, L. F., Zhou, W. Y. & Wang, G. (1999). *Chem. Phys. Lett.* 299: 97.

PCT/JP2009/004200, Kawai, C., Norimatsu, W. & Kusunoki, M.

3

# Magnetic Carbon Nanotubes: Synthesis, Characterization and Anisotropic Electrical Properties

Il Tae Kim and Rina Tannenbaum*
*Georgia Institute of Technology*
*United States*

## 1. Introduction

Carbon nanotubes (CNTs) have been the focus of extensive research in recent years due to their exceptional mechanical, thermal, and electrical properties (Treacy et al., 1996; Lourie et al., 1998; Yu et al., 2000; Lukic et al., 2005). As a result of their nanoscale dimensions and high surface area, CNTs could also be considered as efficient templates for the assembly and tethering of nanoparticles on their surface (Grzelczak et al., 2006). The decoration of CNTs with various compounds and various structures could increase their surface functionality and the tunability of their properties, such as their electrical and magnetic characteristics (Korneva et al., 2005; Kuang et al., 2006). Recent reports described the attachment of various inorganic nanoparticles to either the external surface of the CNTs, or to the internal surface of the CNT cavity, through several experimental methods (Han et al., 2004; Qu et al., 2006). In this context, it is important to note that the control of the size of these tethered nanoparticles is of primary importance for the purpose of tailoring the physical and chemical properties of these hierarchical materials.

Iron oxide nanoparticles, such as magnetite and maghemite, have been of technological and scientific interest due to their unique electrical and magnetic properties. These nanoparticles can be used in such diverse fields as high-density information storage and electronic devices (Sun et al., 2000; Pu et al., 2005; Yi et al., 2006; Jia et al., 2007; Wan et al., 2007). Maghemite, $\gamma$-$Fe_2O_3$, is the allotropic form of magnetite, $Fe_3O_4$ (Rockenberger et al., 1999; Pileni et al., 2003; Sun et al., 2004). These two iron oxides are crystallographically isomorphous. The main difference is the presence of ferric ions only in $\gamma$-$Fe_2O_3$, and both ferrous and ferric ions in $Fe_3O_4$. As a result, while the magnetic properties of $Fe_3O_4$ are superior, $\gamma$-$Fe_2O_3$ is more stable, since the iron cannot be further oxidized under ambient conditions. This renders $\gamma$-$Fe_2O_3$ nanoparticles easier to work with, especially in the presence of organic solvents and organic ligands, and consequently, they have been widely used for magnetic storage in a variety of fields such as floppy disks and cassette tapes. However, maghemite-CNT nanohybrid materials have not been studied as extensively as magnetite-CNT nanohybrid materials, with the exception of several few examples (Sun et al., 2005; Youn et al., 2009).

*All correspondences should be addressed to rinatan@mse.gatech.edu or 404-385-1235.

The alignment of CNTs in a variety of matrices can be used to reinforce, intensify, and enhance some of the properties of the resulting systems, as well as introduce various degrees of anisotropy into the properties of the desired nanomaterials (Kimura et al., 2002; Garmestani et al., 2003). The alignment of CNTs in a suspension under a magnetic field requires that the energy produced by the torque acting on a magnetically-anisotropic segment exceeds the thermal energy of that particular segment, such that: $\delta U \sim B^2 n \delta\chi > kT$ , where $B$ is the field strength, $n$ is the number of carbon atoms in the segment, and $\delta\chi$ is the magnetic anisotropy (Fisher et al., 2003). However, due to the low magnetic susceptibility of CNTs, their alignment by the application of an external magnetic field requires a relatively high magnetic field (Camponeschi et al., 2007). This drawback could be eliminated by enhancing the magnetic susceptibility of carbon nanotubes via the tethering of magnetic nanoparticles onto their surface. In zero field, the magnetic moments of the maghemite nanoparticles randomly point in different directions, resulting in a vanishing net magnetization. However, if a sufficient homogeneous magnetic field is applied, the magnetic moments of the nanoparticles align in parallel, and the resulting dipolar interactions are sufficiently large to overcome thermal motion and to reorient the magnetic CNTs.
In this chapter, we describe and report a convenient approach for the decoration of CNTs with near-monodisperse maghemite nanoparticles by employing a novel and simple modified sol-gel process (in-situ process) with an iron salt as precursor, followed by calcination. The resulting hybrid nanomaterials are superparamagnetic at room temperature and are conducive to facile alignment under relatively low magnetic fields. Subsequently, the nanohybrid materials, i.e. the magnetized carbon nanotubes, were incorporated into a polymer matrix and aligned by the application of a magnetic field, forming polymer composites with an aligned filler phase. It is therefore expected that the composites formed in this manner would exhibit anisotropic mechanical and electrical properties that would depend on and correlate with the parallel and perpendicular direction to the magnetic field that has been applied and under which the alignment has taken place.

## 2. Experimental details

### 2.1 Synthesis of maghemite-MWCNT nanohybrid materials

Pure-MWCNTs were first dispersed in a solution mixture of concentrated $H_2SO_4$ and $HNO_3$ with the volume ratio of 3:1. The suspension was ultra-sonicated for 3 hrs at room temperature. After that, the concentration of the suspension was diluted up to 50% and filtered with a PTFE membrane (0.45 µm pore size) with the aid of a vacuum pump. Carboxylated MWCNT (MWCNT-COOH) was washed with de-ionized water several times to reach neutral pH and dried under vacuum at 50 °C overnight. The synthesis of maghemite-MWCNT was performed by first adding 0.65 g $Fe(NO_3)_3 \cdot 9H_2O$ to 20 ml of absolute ethanol (100% purity) and stirring until the $Fe(NO_3)_3 \cdot 9H_2O$ was dissolved completely. Subsequently, this iron salt solution was added to a suspension of oxidized MWCNTs with a mass ratio of 4:1 ($Fe(NO_3)_3 \cdot 9H_2O$ : MWCNTs mass ratio of 4:1), stirred, and sonicated for 3 hrs. Twenty ml of 1.2 mM of NaDDBS were added to the solution and stirred for 30 min. Then, 1.2 ml of propylene oxide was added as a gelation agent and stirred for 30 min. The mixture was then placed in a Fisher Scientific iso-temperature oven for drying for 3 days at 100 °C. The resulting powder products were washed with ethanol several times and dried at 50 °C. The calcination of these powders was performed in a furnace under argon atmosphere at both 500 °C and 600 °C for 2 hrs. The overall strategy for the preparation of MWCNT/γ-$Fe_2O_3$ is shown in Figure 1 (Kim et al., 2010).

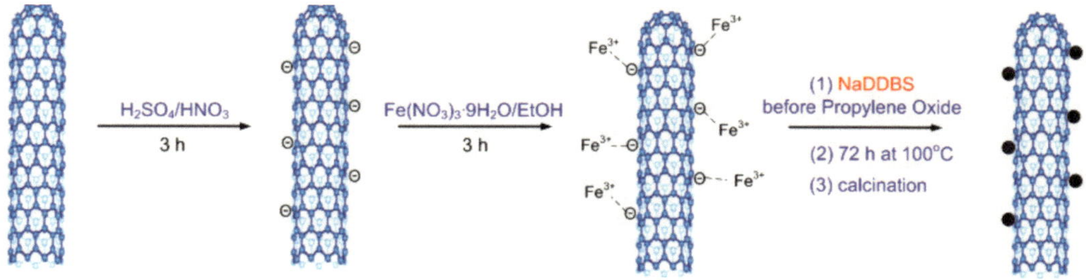


Fig. 1. Schematic representation for the preparation of nanohybrid materials, MWCNT/γ-$Fe_2O_3$ via a modified sol-gel technique (Reprinted with permission from Kim et al., *J. Phys. Chem. C* 2010, 114, 15, 6944-6951, Copyright 2010 ACS).

### 2.2 Fabrication of polymer nanocomposites with aligned feature

Various weight percents of magnetic multi-walled carbon nanotubes (m-MWCNTs) were dispersed in a small amount of ethanol with sonication for 1 hr. Epoxy resin (PR2032) was added to the suspension and mixed with a mechanical stirrer for 30 min in order to obtain optimal dispersion. After that, the nanocomposite solution was sonicated to evaporate entire solvent at 50 °C. The curing agent (PH3660) was added into the solution, mixed, and degassed under vacuum. The solution was immediately poured into a mold, and a 0.3 T magnetic field was applied for 1 hr at room temperature, for 1 hr at 60 °C, and for another 1 hr at 60 °C without a magnetic field. The nanocomposite was post-cured at 60 °C for 6 hrs in the iso-temperature oven (Kim et al., 2011).

### 2.3 Characterization

The dried samples were ground into a fine powder using a ceramic mortar and pestle. Tiny amounts of samples were rarified with KBr powder, ground, and pressed in a KBr pellet with a punch and die. A Nicolet Nexus 870 spectrometer scanned the range from 4000 to 400 $cm^{-1}$ with a resolution of 2 $cm^{-1}$ and data spacing of 0.964 $cm^{-1}$. XRD measurements were performed using an X'pert Pro Alpha-1 (wavelength of 1.54 Å). XRD peaks were collected from $2\theta = 0°$ to 90° with a step size of 0.02°. XPS scans of powder samples were taken using a Surface Science Laboratories SSX-100 ESCA spectrometer using monochromatic Al $K\alpha$ radiation (1486.6 eV). Raman spectra were recorded in the range of 200-2000 $cm^{-1}$ at ambient temperature using a WITEC Spectra Pro 2300I spectrometer equipped with an Ar-ion laser, which provided a laser beam of 514 nm wavelength. The magnetic properties of MWCNTs were measured using a 5.5 T Quantum Design Superconducting Quantum Interface Device (SQUID) magnetometer. The alignment of the sample was conducted by a magnet (GMW-5403) at 0.3 T. The morphology and aligned feature of as-prepared samples were also characterized using SEM (LEO 1530). TEM samples were prepared by placing a droplet of solution onto a TEM grid, and for the observation of aligned features, samples were micro-tomed into 100 nm thick slices using a diamond knife and placed on a TEM grid. These samples were analyzed using a Hitachi HF2000, 200 kV transmission electron microscopy.

The electrical conductivity data of as-prepared composites were collected using impedance analyzer (Solartron Instruments SI 1260 with dielectric interface 1296) for the frequency range 0.1 Hz ~ 1 MHz. All the data were collected under an AC voltage of 0.1 V. Contact was achieved by silver painting the two ends of the samples, and then using coaxial probers on a probe station attached to the impedance analyzer (Peng et al., 2008).

## 3. Decoration of carbon nanotubes with magnetic nanoparticles and the characteristics of the resulting hybrid nanostructures

A variety of methods to form nanohybrid materials on the surface of CNTs have been reported. Correa-Duarte group (Correa-Duarte et al., 2005) coated CNTs with iron oxide nanoparticles (magnetite/maghemite) via a layer by layer (LBL) assembly technique and aligned CNT chains in relatively small external magnetic fields. Subsequently, the resulting magnetic CNT structures could be used as building blocks for the fabrication of nanocomposite materials. Cai group (Wan et al., 2007) decorated CNTs with magnetite nanoparticles in liquid polyols. As a result, these nanoparticles could have significant potential for application in the fields of sensors. In addition, Gao group (Jia et al., 2007) initiated the self-assembly of magnetite particles along MWCNTs via a hydrothermal process. The resulting materials feature nanoparticle beads along the CNT surface, rendering this as an appropriate material to be used as a functional device.

The maghemite-CNT nanocomposite systems also have been reported even though research has not been studied as extensively as magnetite-CNT system. Liu group (Sun et al., 2005) decorated MWCNTs with maghemite via the pyrolysis of ferrocene at different temperatures. This product is expected to provide an efficient way for the large-scale fabrication of magnetic CNT composites. Jung group (Youn et al., 2009) decorated single-wall CNTs (SWCNTs) with iron oxide nanoparticles along the nanotube via a magneto-evaporation method. The nanotubes were aligned vertically on ITO surfaces, suggesting the possibility of rendering this process adequate and cost-effective for mass production. The method described in this work consisted of the use of an iron-oleate complex, oleic acid, and truncated SWCNTs to create iron oxide nanoparticles. The research also demonstrated the anisotropic properties of vertically aligned SWCNTs in a nanocmoposite by comparing current densities of the aligned and non-aligned CNTs.

Keeping pace with these researches' streaming, we have developed the MWCNT/$\gamma$-$Fe_2O_3$ nanohybrid materials. As a first step, the MWCNTs were carboxylated in order to introduce negative charges on their surface, which in turn will interact with Fe (III) ions present in a strong acid solution. This process was also coupled with sonication to ensure dispersion of the MWCNTs in the suspension. The x-ray photoelectron spectroscopy (XPS) wide-survey (Fig. 2a) and high resolution spectra (Fig. 2b) reveal not only the presence of carbon-carbon bonding of MWCNTs at 285 eV binding energy but also the formation of a carbonyl moiety consistent with carboxylated groups at 288 eV binding energy. Nucleation sites for the iron oxide were generated at the CNT surface due to the electrostatic interaction between Fe (III) ions and the carboxylate surface groups of acid-treated CNTs. In this system, the occurrence of gelation was inhibited by the addition of a surface active molecule, sodium dodecylbenzenesulfonate (NaDDBS), before the addition of propylene oxide, which is a gel promoter. The surfactant interfered in the growth stage of the iron oxide nanoparticles (gel phase) and prevented the formation of a gel. This occurred because the NaDDBS molecules had already coordinated to the iron (III) centers due to the attraction between the negatively-charged hydrophilic head of the surfactant and the positively-charged iron (Matarredona et al., 2003; Camponeschi et al., 2008). Therefore, due to the presence of the NaDDBS molecules, no aggregates of $\gamma$-$Fe_2O_3$ were formed but rather the nanoparticles remained individually isolated and dispersed along the length of the CNTs.

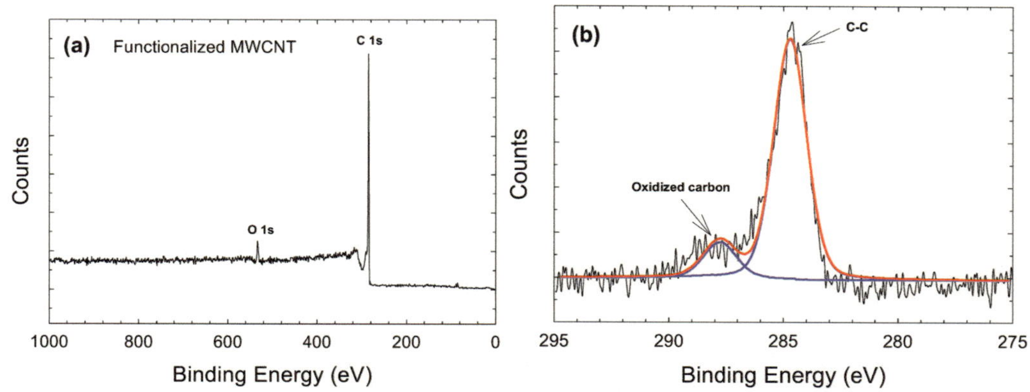


Fig. 2. (a) The XPS survey spectrum of functionalized MWCNTs. (b) The high-resolution XPS spectrum of C1s. (Adapted with permission from Kim et al., *J. Phys. Chem. C* 2010, 114, 15, 6944-6951, Copyright 2010 ACS).

X-ray diffraction patterns of MWCNT containing iron oxide nanoparticles calcinated at different temperatures with the initial $Fe(NO_3)_3 \cdot 9H_2O$ : MWCNTs mass ratio of 4:1 and 2:1 demonstrate the high crystalline nature of the nanoparticles as shown in Figure 3. The diffraction peak at $2\theta$ = 26° can be confidently indexed as the (002) reflection of the MWCNTs, similar to that of pure MWCNTs. The other peaks in the range of 20° < $2\theta$ < 80° correspond to the (220), (311), (400), (422), (511), (440), and (533) reflections of maghemite ($\gamma$-$Fe_2O_3$) and/or magnetite ($Fe_3O_4$). When the mass ratio of $Fe(NO_3)_3 \cdot 9H_2O$ and MWCNTs increases from 2:1 to 4:1, the intensity of the carbon (002) reflection decreases. Also, when calcination temperature increases from 500 °C to 600 °C, the crystal structure of the product becomes better-defined. Because XRD patterns of maghemite and magnetite are practically identical (Sun et al., 2005), x-ray diffraction alone cannot be used to distinguish between the two phases. Therefore, we employed additional experimental techniques to discern between these two phases.

The FTIR spectrum of the product of this modified sol-gel process shows the presence of well-crystallized iron oxide nanoparticles after calcination at 600 °C as shown in Figure 4. Maghemite ($\gamma$-$Fe_2O_3$) has an inverse spinel structure and therefore, it can be seen as an iron-deficient form of magnetite. If the powder is not heat-treated, a weak peak from 800 to 400 $cm^{-1}$ is shown. This is evidence of an amorphous iron oxide phase with minimal long-range order typical of maghemite or magnetite. However, after calcination, IR bands show strong peaks at 576 and 460 $cm^{-1}$, which correspond to a partial vacancy ordering in the octahedral positions in the maghemite crystal structure (White et al., 1967; deFaria et al., 1997; Millan et al., 2007).

X-ray photoelectron spectroscopy (XPS) as well as Raman spectroscopy confirmed that the iron oxide nanoparticles formed were indeed maghemite and not magnetite. After the formation of oxidized MWCNTs decorated with iron oxide nanoparticles followed by

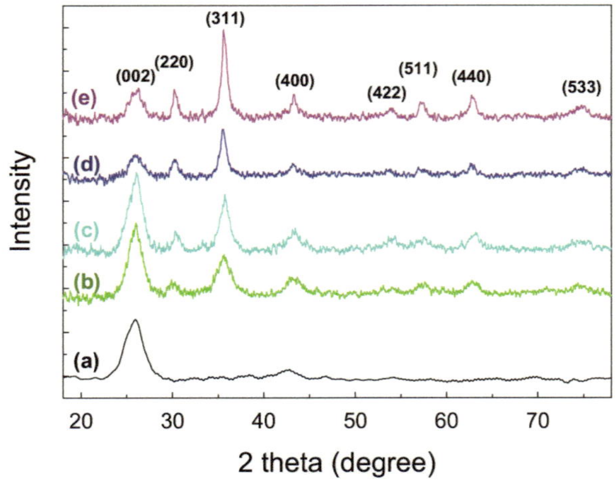


Fig. 3. XRD patterns of MWCNT/γ-$Fe_2O_3$ nanostructures fabricated with two different mass ratios of $Fe(NO_3)_3 \cdot 9H_2O$ and MWCNTs: (a) MWCNT; (b) 2:1 at 500 °C; (c) 2:1 at 600 °C; (d) 4:1 at 500 °C; (e) 4:1 at 600 °C (Reprinted with permission from Kim et al., *J. Phys. Chem. C* 2010, 114, 15, 6944-6951, Copyright 2010 ACS).

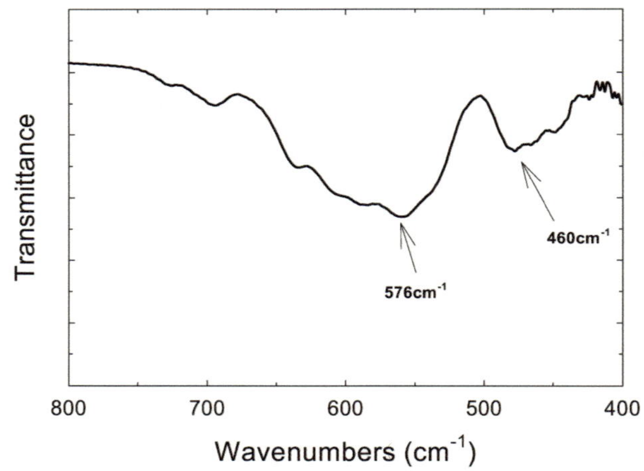


Fig. 4. FTIR spectrum of MWCNT/γ-$Fe_2O_3$ after calcination at 600 °C (Reprinted with permission from Kim et al., *J. Phys. Chem. C* 2010, 114, 15, 6944-6951, Copyright 2010 ACS).

calcination at 600 °C, Figure 5 shows XPS characteristic iron peaks in addition to carbon and oxygen. The position of the Fe (2p3/2) and Fe (2p1/2) peaks were marked at 711.3 and 724.4 eV, respectively, which are in good agreement with the values reported for γ-$Fe_2O_3$ in the literature (Hyeon et al., 2001; Sun et al., 2005). Therefore, this suggests the formation of γ-$Fe_2O_3$ in our samples. Raman spectroscopy can also effectively distinguish between maghemite and magnetite nanoparticles. The strong peak at ~1350 $cm^{-1}$ can be assigned to the D band of MWCNTs, while another dominant peak at ~1576 $cm^{-1}$ can be ascribed the G band of MWCNTs as shown in Figure 6 (Jorio et al., 2003). In contrast to magnetite, the maghemite bands are not well-defined, but rather consist of several broad peaks around

350, 500, and 700 $cm^{-1}$, which are unique to these species and are absent in other types of iron oxide nanoparticles (deFaria et al., 1997). This supports the conclusion that the nanoparticles bound at the walls of the MWCNTs are maghemite and not magnetite.

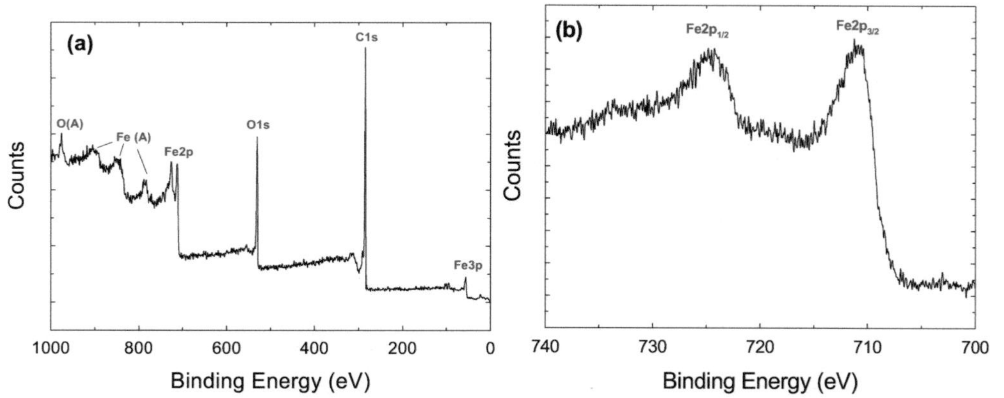


Fig. 5. (a) The XPS survey spectrum of MWCNT/$\gamma$-$Fe_2O_3$. (b) The high-resolution XPS spectrum of Fe 2p bands (Adapted with permission from Kim et al., *J. Phys. Chem. C* 2010, 114, 15, 6944-6951, Copyright 2010 ACS).

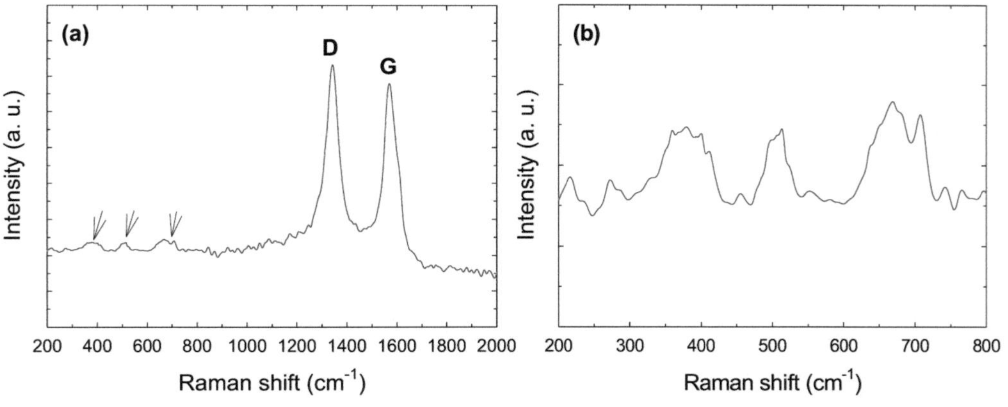


Fig. 6. (a) The Raman spectrum of MWCNT/$\gamma$-$Fe_2O_3$ nanostructure prepared at 600 °C with the mass ratio of 4:1. (b) The detailed Raman spectrum of the same sample in the 200-800 $cm^{-1}$ spectral range (Reprinted with permission from Kim et al., *J. Phys. Chem. C* 2010, 114, 15, 6944-6951, Copyright 2010 ACS).

Scanning electron microscopy (SEM) and transmission electron microscopy (TEM) images of MWCNTs/$\gamma$-$Fe_2O_3$ confirmed that $\gamma$-$Fe_2O_3$ was attached to the walls of the MWCNTs as shown in Figure 7. The high-resolution transmission electron microscopy (HRTEM) image of a nanoparticle (Figure 7(b)) illustrates the maghemite interlayer spacing of the (311) lattice plane of approximately 0.25 nm (Hyeon et al., 2001). Furthermore, the inset image of Figure 7(b) shows the electron diffraction patterns of maghemite, indicating the high crystallinity of the maghemite nanoparticles. At a mass ratio of 4:1 between the $Fe(NO_3)_3 \cdot 9H_2O$ precursor and the MWCNTs, the particle size increased with increasing temperature from 500 °C to

600 °C, and the average sizes were 10.1 nm and 10.8 nm, respectively as shown in Figure 7(c) and (d). Similarly, when the mass ratio of $Fe(NO_3)_3 \cdot 9H_2O$ precursor and MWCNT was 2:1, the average particle sizes as a result of the increased temperature were 7.9 nm and 8.4 nm, respectively (Figure 7(e) and 7(f)), which also slightly increased with increasing temperature. This result indicated that both a higher mass ratio between the $Fe(NO_3)_3 \cdot 9H_2O$

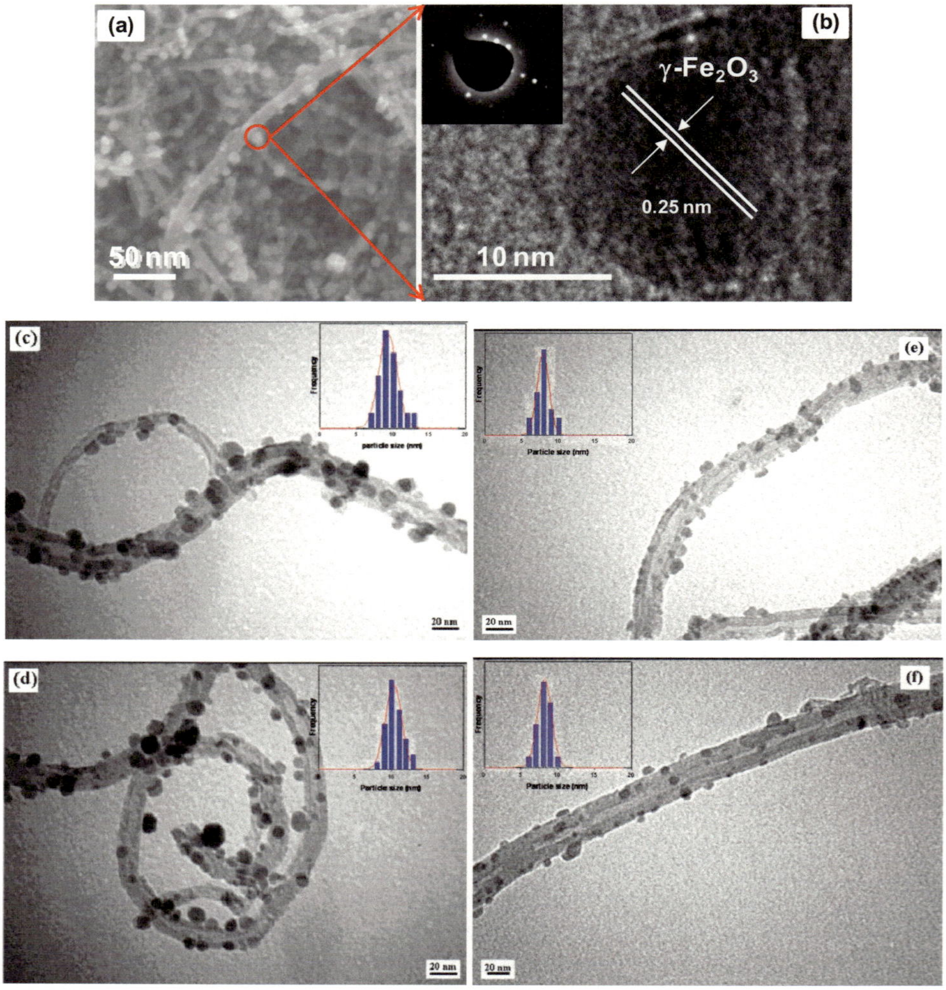


Fig. 7. (a) SEM image of MWCNT/$\gamma$-$Fe_2O_3$ hybrid structures prepared with 4:1 mass ratio of iron salt and MWCNT; (b) High resolution TEM image of maghemite. Inset shows diffractions of a single maghemite nanoparticle. TEM images of MWCNT/$\gamma$-$Fe_2O_3$ prepared with 4:1 mass ratio of iron salt and MWCNT: (c) High-resolution image prepared at 500 °C; (d) High magnification image prepared at 600 °C. TEM images of MWCNT/$\gamma$-$Fe_2O_3$ prepared with 2:1 mass ratio; (e) High magnification image prepared at 500 °C; (f) High magnification image prepared at 600 °C (Adapted with permission from Kim et al., *J. Phys. Chem. C* 2010, 114, 15, 6944-6951, Copyright 2010 ACS, and Kim et al., *Carbon* 2011, 49, 1, 54-61, Copyright 2011 Elsevier ).

precursor and the MWCNT and increasing temperature led to larger nanoparticles, and therefore, we can conclude that particle size could be controlled by the precursor to MWCNT mass ratio and temperature.
Chemical analysis using EDS during the TEM analysis showed the presence of Fe, O, and C in the maghemite-MWCNT system as shown in Figure 8, and the calculated atomic ratio of Fe and O was close to 2:3, which suggested the formation of $\gamma$-$Fe_2O_3$.

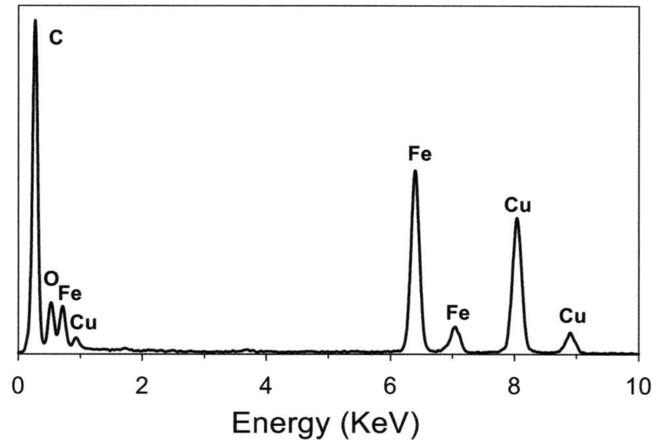


Fig. 8. Energy dispersion spectrum (EDS) of the MWCNT/$\gamma$-$Fe_2O_3$ hybrid material (Adapted with permission from Kim et al., *J. Phys. Chem. C* 2010, 114, 15, 6944-6951, Copyright 2010 ACS).

The magnetic properties of the as-prepared MWCNTs/$\gamma$-$Fe_2O_3$ nanocomposites were measured using Superconducting Quantum Interference Device (SQUID) magnetometer. The magnetization hysteresis loops were measured in fields between ±50 kOe at room temperature as shown in Figure 9(a). The saturation magnetization ($M_s$) of the samples obtained is below 2 emu/g, which is considerably smaller than that of bulk iron ($M_s$ = 222 emu/g) as shown in Table 1. Coercivity is below 10 Oe, which is larger than that of bulk iron ($H_c$ = 1 Oe). The conclusion drawn from the measurement of magnetic properties is that both samples, having different ratios between $Fe(NO_3)_3 \cdot 9H_2O$ precursor and MWCNT, exhibit superparamagnetic behavior at room temperature. This should be mainly attributed to the small size of $\gamma$-$Fe_2O_3$ nanoparticles that were formed in the presence of MWCNTs (Pascal et al., 1999). This result is in good accordance with the TEM observation of the small sizes of the maghemite nanoparticles mentioned above.
The magnetic attraction of our sample was also tested by placing a magnet near a vial containing the maghemite-MWCNT nanostructures as shown in Figure 9(c) and 9(d). Our samples can be easily dispersed in solution and form a stable suspension. When a magnet approaches the vial, magnetic carbon nanotubes are attracted toward the magnet. This phenomenon illustrates that the maghemite nanoparticles that are anchored on the surface of the MWCNTs impart to the composite material a magnetic response similar to that observed with magnetite.
This novel method for the magnetization of carbon nanotubes through the tethering of magnetic iron oxide nanoparticles with controlled size and site distribution would open up

a slew of new opportunities for applications in which the alignment of CNTs is not only desired, but is actually required. While many groups have studied strategies to align MWCNT/$Fe_3O_4$ nanostructures under external magnetic fields due to their strong magnetic properties, very little attention has been devoted to MWCNT/$\gamma$-$Fe_2O_3$ conjugate nanomaterials. Therefore, we would like to show that this latter system also exhibits similar interesting properties and can constitute a facile gateway to MWCNT alignment processes under tight morphological control and relatively low magnetic fields, resulting in enhanced anisotropic electrical conductivity behavior, in the following sections.

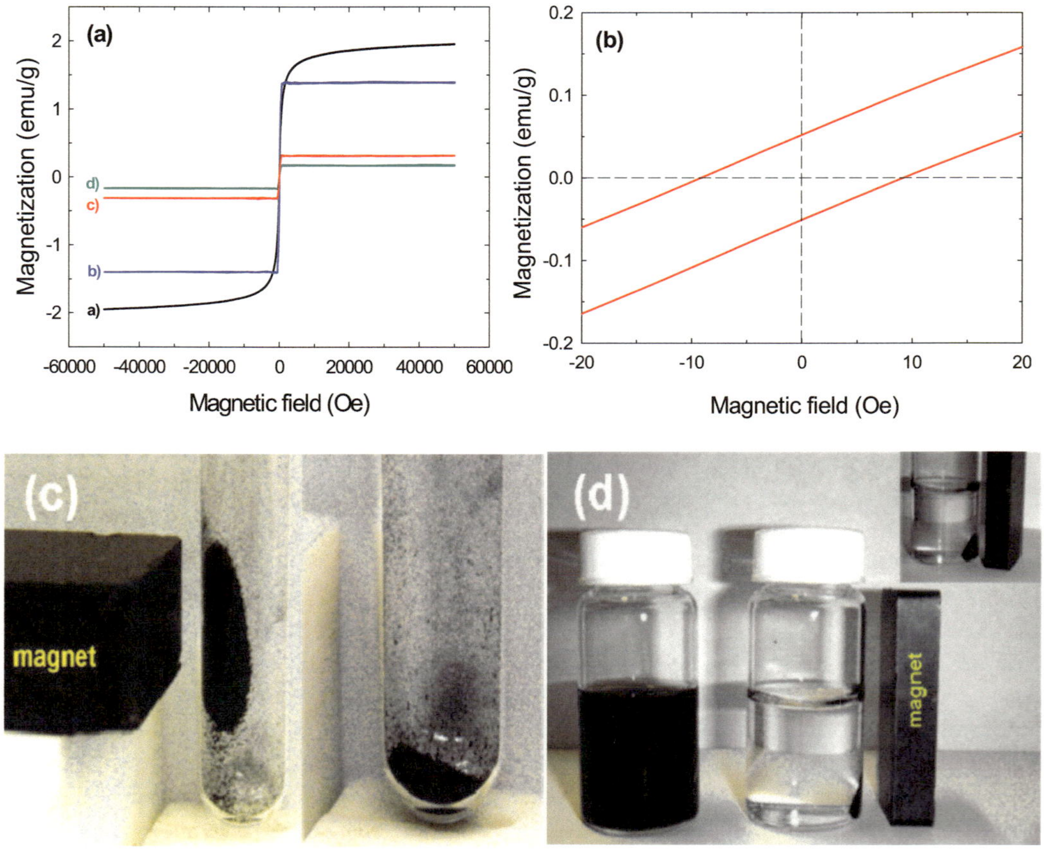


Fig. 9. (a) Magnetization vs. applied magnetic field for the magnetic carbon nanotubes prepared at different mass ratios and temperatures: 4:1 mass ratio of $Fe(NO_3)_3 \cdot 9H_2O$ and MWCNT at a) 500 °C, b) 600 °C, and 2:1 mass ratio of $Fe(NO_3)_3 \cdot 9H_2O$ and MWCNT at c) 500 °C, d) 600 °C. (b) The enlarged hysteresis loop of the MWCNT/$\gamma$-$Fe_2O_3$ structures formed from a 4:1 mass ratio of $Fe(NO_3)_3 \cdot 9H_2O$ and MWCNT calcinated at 600 °C. The photographs of magnetic carbon nanotubes (c) in the presence (left image) and in the absence (right image) of a magnet and (d) suspended in ethanol in the absence (left image) and in the presence (right image) of an externally-placed magnet (Adapted with permission from Kim et al., *J. Phys. Chem. C* 2010, 114, 15, 6944-6951, Copyright 2010 ACS, and Kim et al., *Carbon* 2011, 49, 1, 54-61, Copyright 2011 Elsevier).

| Magnetic properties | Calcination temperature (°C) | 2:1 | 4:1 |
|---|---|---|---|
| $M_s$ (emu/g) | 500 | 0.3 | 2.0 |
| | 600 | 0.2 | 1.4 |
| $H_c$ (Oe) | 500 | 4.8 | 2.8 |
| | 600 | 6.3 | 9.6 |

Table 1. Magnetic properties as a function of both different mass ratio of $Fe(NO_3)_3 \cdot 9H_2O$ and MWCNT and different calcination temperatures.

## 4. Alignment strategies of carbon nanotubes in polymer matrices

Alignments of CNTs by electric, shear induced field, and magnetic field were reported previously by several groups (Chen et al., 2001; Nagahara et al., 2002). Bauhofer group (Martin et al., 2005) successfully demonstrated the application of AC electric fields allowing both the alignment of carbon nanofibers in epoxy resin and their connection into a network. Zhu group (Zhu et al., 2009) studied electric field aligned MWCNT/epoxy nanocomposites with a sample size of up to several centimetres using fast UV polymerization, showing significant anisotropic properties for storage modulus and electrical conductivity.

For the characterization of aligned composite systems using shear induced field, we probed the effects of shear flow on the alignment of dispersed SWCNTs in polymer solutions as a previous study (Camponeschi et al., 2006). The sample solutions were placed in the 8.5 mm gap between the outer cylinder and the spindle, as shown Figure 10. In turn, the spindle was allowed to rotate for one week at several different angular velocities ranging from 12 to 100 rpm. TEM samples were taken in situ from the solutions flowing in circular motion in the gap between the outer cylinder and inner cylinder as shown in Figure 10(b).

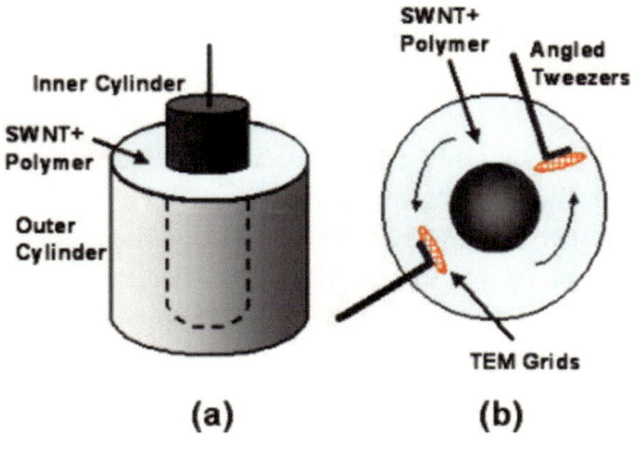


Fig. 10. (a) Concentric cylinder arrangement in the Brookfield viscometer. (b) TEM sample retrieval and preparation (Reprinted with permission from Camponeschi et al., *Langmuir* 2006, 22, 4, 1858-1862. Copyright 2006 ACS).

In this experimental set up, for systems in which effective dispersion of the carbon nanotubes was achieved by the combined action of both NaDDBS and Carboxymethylcellulose (CMC). The only system in which tube alignment was observed was for the NaDDBS/CMC/SWCNT solution that was subjected to shear stresses at the highest angular velocity used in the experiments as shown in Figure 11.

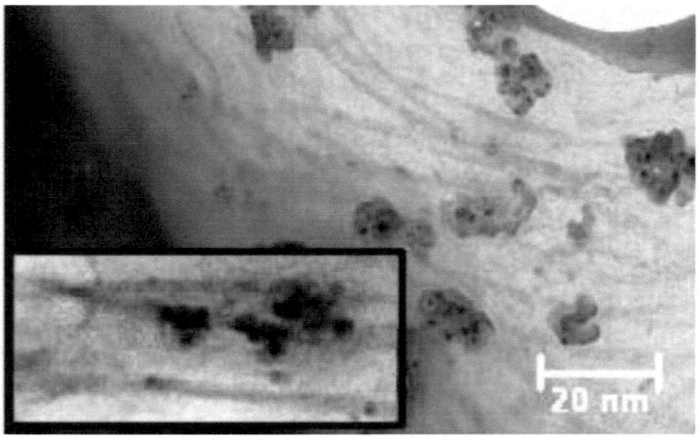


Fig. 11. Oriented carbon nanotubes dispersed with NaDDBS and CMC and subjected to shear flow at 100 rpm. The inset image is a 4-fold magnification of the larger image showing the local orientation of the surface modified SWCNT (Reprinted with permission from Camponeschi et al., *Langmuir* 2006, 22, 4, 1858-1862. Copyright 2006 ACS).

A high magnetic field is an efficient and direct ways to align carbon nanotubes. Tanimoto group have found that a high magnetic field of 7 T aligns arc-grown MWCNTs (Fujiwara et al., 2001). They dried a MWCNT dispersion in methanol under a constant magnetic field and observed the MWCNTs alignment parallel to the field. This result was explained by the difference between the diamagnetic susceptibilities parallel ( $\chi_{//}$ ) and perpendicular ( $\chi_{\perp}$ ) to the tube axis; if $|\chi_{\perp}|$ is larger than $|\chi_{//}|$, a MWCNT tends to align parallel to the magnetic field by overcoming thermal energy (Ajiki et al., 1993; Fujiwara et al., 2001). More recently, Steinert and Dean (Steinert & Dean, 2009) obtained solution cast PET-carbon nanotube composite films by applying a magnetic field, resulting in increased conductivity with the increase of the applied magnetic field. Furthermore, in our previous study (Camponeschi et al., 2007), we prepared magnetically aligned carbon nanotube composite systems; thus, carbon nanotubes were aligned parallel to the direction of magnetic field, resulting in enhanced mechanical properties. However, due to the low magnetic susceptibility of carbon nanotubes, their alignment by the application of an external magnetic field requires a relatively high magnetic field. This draw-back could be solved by enhancing the magnetic susceptibility of carbon nanotubes by tethering magnetic nanoparticles on their surface, as we developed MWCNT/$\gamma$-$Fe_2O_3$ hybrid materials.
The samples for SEM were prepared by dispersing as-prepared nanostructures in water solution with surfactant, sonicating for 30 min, and then depositing the samples onto silicon wafer under an external field. Figure 12 shows the SEM images of magnetic carbon nanotubes. When a droplet of dispersed hybrid materials in a water solution was dried under the magnetic field, the surface-modified MWCNT were aligned easily as shown in

Figure 12(a). However, when the nanocomposite solution was dried without applying magnetic field, the surface-modified MWCNT did not exhibit alignment features.

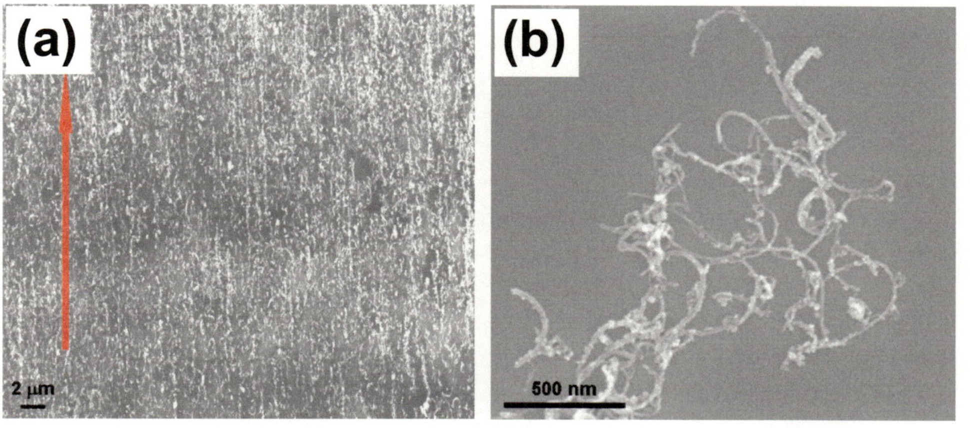


Fig. 12. (a) SEM image of aligned magnetic carbon nanotube hybrid materials parallel to the direction of magnetic field. (b) SEM image of magnetic carbon nanotube hybrid materials that were not subjected to a magnetic field (Reprinted with permission from Kim et al., *Carbon* 2011, 49, 1, 54-61.2011. Copyright 2011 Elsevier).

The TEM images of composites in which surface-modified MWCNT (m-MWCNT) and unmodified MWCNT were embedded in epoxy matrices are shown in Figure 13(a) through 13(d). We first compared the alignment features of the MWCNT/epoxy nanocomposite and the m-MWCNT/epoxy nanocomposite systems, under the same experimental conditions, i.e. the same strength of the externally-applied magnetic field (0.3 T). Figure 13(a) and 13(b), representing MWCNT/epoxy composites with 0.5 wt% MWCNT and 1.0 wt% MWCNT, respectively, did not reveal any alignment features of filler phase in the polymer matrix under the externally-applied magnetic field. However, in the case of the m-MWCNT/epoxy nanocomposite systems also having 0.5 wt% m-MWCNT and 1.0 wt% m-MWCNT and shown in Figure 13(c) and 13(d), respectively, it is obvious that the m-MWCNTs embedded in the epoxy matrix have indeed aligned parallel to the direction of magnetic field (0.3 T). Comparing the alignment features of aligned m-MWCNT hybrid materials and aligned m-MWCNT/epoxy composites (Figure 12(a), 13(c), and 13(d)), it becomes evident that the m-MWCNT hybrid materials in the absence of a polymer matrix show better alignment, fact which could be attributed to the viscosity of the polymer matrix during processing. Therefore, we can conclude that the m-MWCNT hybrids can be aligned under a relatively weak magnetic field even when embedded in a polymer matrix. This alignment is expected to directly affect the anisotropic conductivity of the resulting epoxy composites, as will be shown in the subsequent section (Figure 14 and 15). The bundling of the m-MWCNTs in the polymer matrix, as observed in the inset in Figure 13(c), may be attributed to the anisotropic nature of the dipolar interactions of the iron oxide nanoparticles near the ends of the carbon nanotubes, i.e. the near-linear stacking of the north and south poles of the m-MWCNT in the polymer matrix, resulting in their observed end-to-top connectivity (Butter et al., 2003; Correa-Duarte et al., 2005).

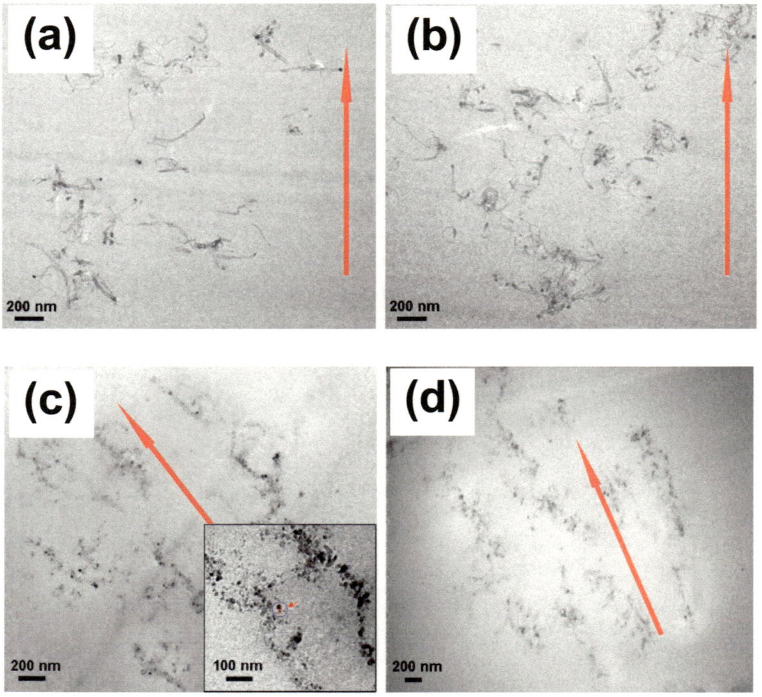


Fig. 13. (a) TEM image of MWCNT/epoxy composites with 0.5 wt% filler loading. (b) TEM image of MWCNT/epoxy composites with 1.0 wt% filler loading. (c) TEM images of m-MWCNT/epoxy composites with 0.5 wt% filler loading. Inset shows the end-to-top connectivity between two m-MWCNTs under an external magnetic field. (d) TEM image of m-MWCNT/epoxy composites with 1.0 wt% filler loading (Adapted with permission from Kim et al., *Carbon* 2011, 49, 1, 54-61. Copyright 2011 Elsevier).

## 5. Anisotropic electrical conductivity of composite system

The electric conductivities of the m-MWCNT/epoxy composites were measured at a series of different frequencies, from 0.1 Hz to 1 MHz. The real and imaginary parts of the impedance (Z′ and Z″) were collected, and the magnitude of the AC conductivity (σ) was calculated using equations:

$$Z = Z' + iZ''$$

$$\sigma = \frac{1}{\sqrt{Z'^2 + Z''^2}} \frac{L}{A}$$

where, $i$ is the imaginary unit, $L$ is the path length along the measurement direction, and $A$ is the electrode cross-sectional area. Figure 14 shows various conductivities of a series of m-MWCNT/epoxy nanocomposite samples containing various degrees of content of m-MWCNT in the polymer matrix. At the same magnetic field (0.3 T), the conductivity increased with increasing m-MWCNT content in composites. In the case of 0.1 wt% filler content, the nanocomposite exhibited dielectric behavior because the low mass-fraction of m-MWCNT made it difficult to form interconnected m-MWCNT networks that would have facilitated electron flow. However, for m-MWCNT contents of 0.5 wt% and higher, the samples exhibited increased conductivity as a function of increased mass-fraction of the m-MWCNT.

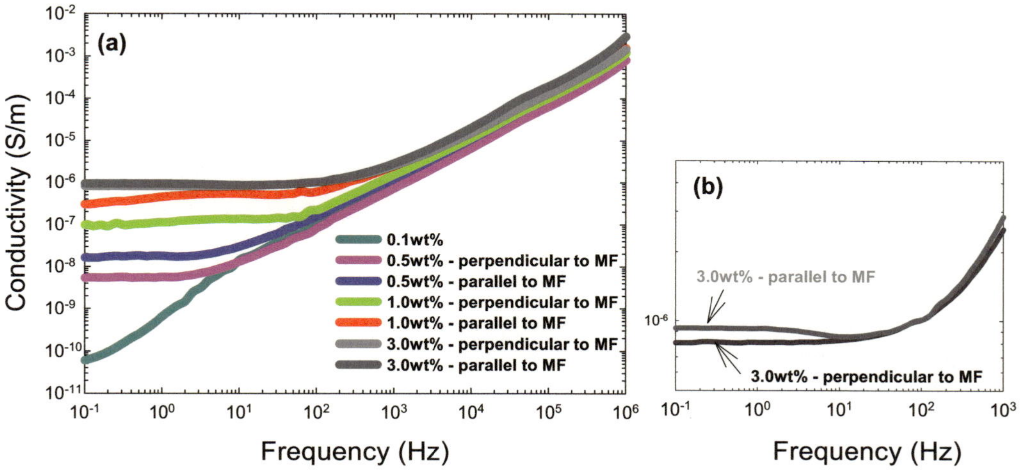


Fig. 14. (a) The conductivity of m-MWCNT/epoxy composites as a function of frequency for different mass loading of m-MWCNT as measured in the direction parallel to the magnetic field and perpendicular to the magnetic field. (b) The magnified region of a nanocomposite with a 3.0 wt% filler loading (Adapted with permission from Kim et al., *Carbon* 2011, 49, 1, 54-61. Copyright 2011 Elsevier).

Percolation theory predicts a critical concentration or percolation threshold where the material converts from a capacitor to a conductor (Weber et al., 1997; Ounaies et al., 2003). In order to determine the percolation threshold of the aligned system, the volume conductivity data could be fitted to a power law in terms of volume fraction of m-MWCNT.

$$\sigma_c \propto (v - v_c)^t$$

where $\sigma_c$ is the composite conductivity, $v$ is the m-MWCNT volume fraction in the composite, $v_c$ is the critical volume fraction, and $t$ is the critical exponent. We assumed that the density of m-MWCNT is the same as that of unmodified-MWCNT (2.1 g/cm$^3$), since both mass and volume of m-MWCNT increase similarly. The inset in Figure 15 shows the plot of $\sigma_c$ as a function of $v - v_c$ for the parallel measurements. The linear fit to the data generated a straight line with $v_c$ = 0.2 vol% (corresponding to 0.4 wt%), which gives a good fit.

When we compared the results of samples in which conductivity was measured in the direction of the m-MWCNT alignment (parallel to the magnetic field) and perpendicular to the m-MWCNT alignment (perpendicular direction to the magnetic field) for the same mass fraction of m-MWCNT, we observed that the conductivity measured parallel to the magnetic field was higher than that measured perpendicular to the magnetic field, indicating a cooperative effect due to the alignment of the m-MWCNTs in the polymer

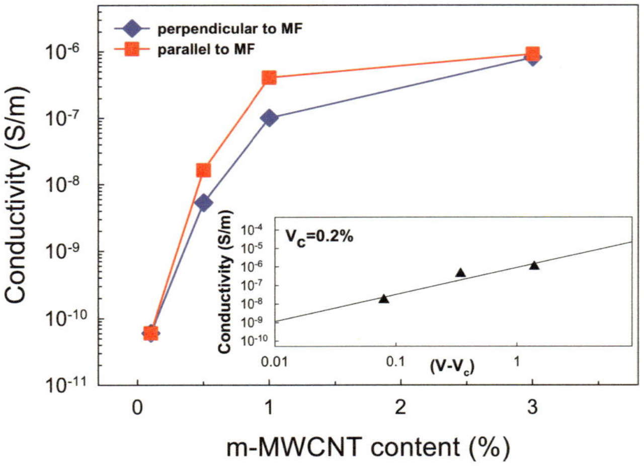


Fig. 15. The conductivity of m-MWCNT/epoxy composites as a function of different mass loading of m-MWCNT measured in the direction parallel to the magnetic field and perpendicular to it. Inset shows percolation equation fit to the experimental conductivity data obtained parallel to the direction of the magnetic field (Adapted with permission from Kim et al., *Carbon* 2011, 49, 1, 54-61. Copyright 2011 Elsevier).

matrix, as was previously shown in Figure 13(c) and 13(d). Figure 15 shows the variation of the conductivities extracted from the plateau region at low frequency as a function of m-MWCNT mass fractions in the epoxy nanocomposite for both the parallel and perpendicular directions with respect to the magnetic field. The measured conductivities are summarized in Table 2.

We would like to note that for 3.0 wt% m-MWCNT sample, even though the conductivity in the parallel direction was somewhat larger than that in the perpendicular direction (see Figure 14(b)), the values obtained were, nevertheless, quite similar. This is most likely due to the following factors: (a) We assumes that the viscosity of the composite solution containing 3.0 wt% m-MWCNT is higher than for other compositions as evidenced by the superior alignment of m-MWCNT without polymer matrix to that of m-MWCNT/epoxy composites,

| m-MWCNT content (wt%) | Conductivity (S/m) | | Conductivity ratio of parallel and perpendicular |
|---|---|---|---|
| | Parallel to MF | Perpendicular to MF | |
| 3.0 | $1.0 \times 10^{-6}$ | $8.5 \times 10^{-7}$ | 1.2 |
| 1.0 | $4.1 \times 10^{-7}$ | $1.0 \times 10^{-7}$ | 4.1 |
| 0.5 | $1.6 \times 10^{-8}$ | $5.3 \times 10^{-9}$ | 3.0 |
| 0.1 | $6.0 \times 10^{-11}$ | | 1.0 |

Table 2. The conductivity of m-MWCNT/epoxy composites in the directions that were parallel and perpendicular to the externally-applied magnetic field as a function of m-MWCNT content.

as discussed in a previous section. By introducing higher mass fractions of the carbon nanotubes into the polymer solution, the viscosity of the system could be further increased, fact which could then handicap with the alignment process. Therefore, we can conclude that when the magnetic field was applied to the 3.0 wt% m-MWCNT sample, the alignment of the decorated carbon nanotubes was not as effective as in the less concentrated samples, and hence, the differences between the conductivities in the parallel and the perpendicular directions were not as pronounced, mainly due to the higher viscosity of the solution. (b) In addition, the conductivity of 3.0 wt% m-MWCNT sample (measured in either direction) was not much higher than the conductivity of the 1.0 wt% m-MWCNT sample (see Figure 15). Tethered iron oxide (maghemite) nanoparticle has high resistivity (Mei et al., 1987). Hence, the higher viscosity of the 3.0 wt% m-MWCNT sample may lead to the formation of iron oxide rich regions, resulting in a decrease of the conductivity.

## 6. Anisotropic response m-CNT-epoxy composite system to compression

The initial goal of using a magnetic field on carbon nanotube composites was to promote the alignment of the carbon nanotubes, which would improve the mechanical properties of the composite, particularly in the direction of the alignment. The effects on the glass transition temperature should be relatively simple to predict since they are well documented (Akima et al., 2006; Bliznyuk et al., 2006; Dou et al., 2006; Lanticse et al., 2006; Park et al., 2006). It is expected that increasing the extent of alignment and orientation of the carbon nanotubes and epoxy matrix chains (Al-Haik et al., 2004) will result in an increase in the glass transition temperature of the composite (Ajayan et al., 1994; Akima et al., 2006; Bliznyuk et al., 2006; Dou et al., 2006; Lanticse et al., 2006; Park et al., 2006). Table 3 summarizes the $T_g$ values for the various samples tested.

| Fe:CNT | Magnetic field (Tesla) | $T_g \pm 5.2$ (°C) |
|---|---|---|
| 0:0 | 0 | 54.9 |
| 0:0 | 0.4 | 63.8 |
| 0:0 | 0.8 | 68.7 |
| 2:1 | 0 | 41.6 |
| 2:1 | 0.4 | 65.6 |
| 2:1 | 0.8 | 75.1 |
| 4:1 | 0 | 45.5 |
| 4:1 | 0.4 | 68.9 |
| 4:1 | 0.8 | 86.0 |

Table 3. The glass transition temperature of the m-CNT/epoxy nanocomposites measured in samples subjected to various external magnetic fields.

The glass transition temperature of the epoxy matrix increased with increasing magnetic field and implies that there is some molecular orientation/alignment occurring in the epoxy (Garmestani et al., 2003). When the magnetic CNTs are introduced into the matrix, the glass transition temperature of the nanocomposite decreased, probably due to the plasticizing effect of the NaDDBS molecules associated with the $Fe_2O_3$ nanoparticles tethered to the surface of the CNTs. However, in the samples in which the magnetic CNTs were aligned when the nanocomposite was subjected to an external magnetic field, the general trend showed that an increase in the extent of alignment caused an increase in $T_g$, as expected.

Preliminary compression data that illustrate the effect of a magnetic field on the modulus of the epoxy matrix naocomposites are shown in Figure 16.

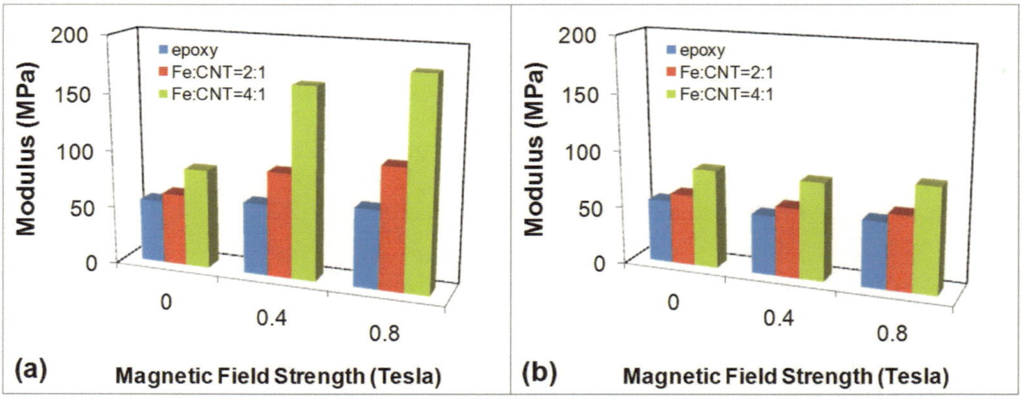


Fig. 16. The moduli of the m-CNT/epoxy nanocomposites that were subjected to an external magnetic field measured at room temperature. (a) Measured in the direction that is parallel to the applied magnetic field; (b) Measured in the direction that is perpendicular to the applied magnetic field .

The modulus was measured at room temperature both in the direction that is parallel to the applied magnetic field (Figure 16a) and in the direction that is perpendicular to the applied magnetic field (Figures 16b). The presence of the externally-applied magnetic field had little effect on the pure epoxy. It has been shown in previous work that the orientation and possible alignment of epoxy chains is indeed possible, but only at very high magnetid fields (Garmestani et al., 2003). Hence, the moduli of pure epoxy are constant, irrespective of the magnitude of the magnetic field applied on the samples and of the direction of the measurement. Conversely, the moduli observed for the epoxy filled with the m-CNTs indeed increase with the increase in the magnetic field, mainly for measurements conducted parallel to the direction of the magnetic field. Moreover, higher concentrations of the iron oxide nanoparticles tethered to the surface of the CNTs result in considerably higher moduli, particularliy in the direction parallel to the magnetic field, probably due to an increase in the susceptibility of the m-CNTs to the applied magnetic field.

## 7. Summary and outlook

In this chapter, we have demonstrated on CNT-inorganic hybrid system, especially, CNT/$\gamma$-$Fe_2O_3$ hybrid materials. We developed the synthesis method of MWCNT/$\gamma$-$Fe_2O_3$ nanostructures via an easy and novel modified sol-gel process. Our study shows that NaDDBS molecules are intimately involved in inhibiting the formation of an iron oxide gel. As a result, well-defined and well-dispersed maghemite nanoparticles can be obtained. In addition, the particle size of these nanoparticles could be precisely modulated by changing the temperature and the mass ratio of the $Fe(NO_3)_3 \cdot 9H_2O$ precursor and MWCNTs. Finally, tethered $\gamma$-$Fe_2O_3$ magnetic nanoparticles on the surface of MWCNTs imparted superparamagnetic properties to the composite material.
Due to the acquired magnetic property of the m-MWCNTs, they could be aligned either alone or embedded in a polymer matrix by the application of only a relatively weak

magnetic field. Conductivity measurements performed on m-MWCNT/epoxy composites showed that the conductivity of the m-MWCNT/epoxy composites increased with increasing m-MWCNT contents with low percolation threshold (~0.4–0.5 wt% m-MWCNT loading). Moreover, the conductivity measured in the direction parallel to the magnetic field was higher than that measured in the direction perpendicular to it. However, the alignment of a nanocomposite sample having a loading of 3.0 wt% m-MWCNT was not as effective as samples with lower nanofiller content because of the higher solution viscosity in the more concentrated samples. This hurdle could, in principle, be overcome by either applying a stronger magnetic field or selecting other polymer matrices with low solution viscosity.

In summary, our facile magnetic functionalization method could be effectively applied for the development of conductive films, composites with conductive polymers, and bio-based composites with aligned features. Furthermore, we suggest that this maghemite-CNT hybrid material may be used for biomedical applications such as drug delivery or special medical applications such as cancer diagnosis in the not-so-distant future (Sincai et al., 2001; Sousa et al., 2001).

## 8. Acknowledgement

This work was supported in part by grants from NSF, Division of Engineering, award No. ECCS-0535382 and by the Air Force/Bolling AFB/DC MURI, award No. F49620-02-1-0382. Il Tae Kim was supported by a Paper Science and Engineering (PSE) Graduate Fellowships from the Institute of Paper Science and Technology (IPST) at the Georgia Institute of Technology. The authors are indebted to Drs. Erin Camponeschi, Hamid Garmestani, Karl Jacob and Allen Tannenbaum for their invaluable contributions and stimulating input.

## 9. References

Ajayan, P. M.; Stephan, O.; Colliex, C. & Trauth, D. (1994). Aligned carbon nanotube by cutting a polymer resin-nanotube composite. *Science*, 265, pp.1212.

Akima, N.; Iwasa, Y.; Brown, S.; Barbour, A. M.; Cao, J.; Musfeldt, J. L.; Matsui, H.; Toyota, N.; Shiraishi, M.; Shimoda, H. & Zhou, O (2006). Strong anisotropy in the far-infrared absorption spectra of stretch-aligned single-walled carbon nanotubes. *Adv. Mater.*, 18, 9, pp. 1166-1169.

Ajiki, H. & Ando, T. (1993). Magenetic-properties of carbon nanotubes, *J. Phys. Soc. Jpn.*, 62, 7, pp. 2470-2480.

Al-Haik, M. S.; Garmestani, H.; Li, D. S.; Hussaini, M. Y.; Sablin, S. S.; Tannenbaum, R. & Dahmen, K (2004). Mechanical properties of magnetically oriented epoxy. *J. Polym. Sci. Polym. Phys.*, 42, pp. 1586-1600.

Bliznyuk, V. N.; Singamaneni, S.; Sanford, R. L. Chiappetta, D.; Crooker, B. & Shibaev, P. V. (2006). Matrix mediated alignment of single wall carbon nanotubes in polymer composite films. *Polymer*, 47, 11, pp. 3915-3921.

Butter, K.; Bomans, P. H.; Frederik, P. M.; Vroege, G. J. & Philipse, A. P. (2003). Direct observatioin of dipolar chains in ferrofluids in zero field using cryogenic electron microscopy, *J. Phys.: Conds. Matter.*, 15, 15, pp. S1451-S1470.

Camponeschi, E.; Florkowski, B.; Vance, R.; Garrett, G.; Garmestani, H. & Tannenbaum, R. (2006). Uniform directional alignment of single-walled carbon nanotubes in viscous polymer flow, *Langmuir*, 22, 4, pp. 1858-1862.

Camponeschi, E.; Vance, R.; Al-Haik, M.; Garmestani, H. & Tannenbaum, R. (2007). Properties of carbon nanotube-polymer composites aligned in a magnetic field, *Carbon*, 45, 10, pp. 2037-2046.

Camponeschi, E.; Walker, J.; Garmestani, H. & Tannenbaum R. (2008). Surfactant effects on the particle size of iron (III) oxides formed by sol-gel synthesis, *J. Non-Cryst. Solids*, 351, 34, pp. 4063-4069.

Chen, X. Q.; Saito, T.; Yamada, H. & Matsushige, K. (2001). Aligning single-wall carbon nanotubes with an alternating-current electric field, *Appl. Phys. Lett.*, 78, 23, pp. 3714-3716.

Correa-Duarte, M.A.; Grzelczak, M.; Salgueirino-Maceira, V. ; Giersig, M. ; Liz-Marzan, L. M.; Farle, M.; Sierazdki, K. & Diaz, R. (2005). Alignment of carbon nanotubes under low magnetic fields through attachment of magnetic nanoparticles, *J. Phys. Chem. B*, 109, 41, pp. 19060-19063.

deFaria, D. L. A.; Silva, S. V. & deOliveira, M. T. (1997). Raman microspectroscopy of some iron oxides and oxyhydroxides, *J. Raman Spectrosc.*, 28, 11, pp. 873-878.

Dou, S. X.; Yeoh, W. K.; Shcherbakova, O.; Wexler, D.; Li, Y.; Ren, S. M.; Munroe, P.; Chen, S . K.; Tan, S. K.; Glowacki, B. A. & MacManus-Driscoll, J. P. (2006). Alignment of carbon nanotube additives for improved performance of magnesium diboride superconductors. *Adv. Mater.*, 18, 6, pp. 785-788.

Fischer, J. E.; Zhou, W.; Vavro, J.; Llaguno, M. C.; Guthy, C. & Haggenmueller, R. (2003). Magnetically aligned single wall carbon nanotube films: Preferred orientation and anisotropic transport properties, *J. Appl. Phys.*, 93, 4, pp. 2157-2163.

Fujiwara, M.; Oki, E.; Hamada, M.; Tanimoto, Y.; Mukouda, I. & Shimomura, Y. (2001). Magnetic orientation and magnetic properties of a single carbon nanotube, *J. Phys. Chem. A*, 105, 18, pp. 4383-4386.

Garmestani, H.; Al-Haik, M. S.; Dahmen, K.; Tannenbaum, R.; Li, D.; Sablin, S. S. & Hussaini, M. Y. (2003). Polymer-mediated alignment of carbon nanotubes under high magnetic fields, *Adv. Mater.*, 15, 22, pp. 1918-1921.

Grzelczak, M.; Correa-Duarte, M. A. & Liz-Marzan, L. M. (2006). Carbon nanotubes encapsulated in wormlike hollow silica shells, *Small*, 2, 10, pp. 1174-1177.

Han, L.; Wu, W.; Kirk, F. L.; Luo, J.; Maye, M. M.; Kariuki, N. N.; Li, Y. H.; Wang, C. & Zhong, C. J. (2004). A direct route toward assembly of nanoparticle-carbon nanotube composite materials, *Langmuir*, 20, 14, pp. 6019-6025.

Hyeon, T.; Lee, S. S.; Park, J.; Chung, Y. & Bin, N. H. (2001). Synthesis of highly crystalline and monodisperse maghemite nanocrystallites without a size-selection process, *J. Am. Chem. Soc.*, 123, 51, pp. 12798-12801.

Jia, B.; Gao, L. & Sun, J. (2007). Self-assembly of magnetite beads along multiwalled carbon nanotubes via a simple hydrothermal process, *Carbon*, 45, 7, pp. 1476-1481.

Jorio, A.; Pimenta, M. A.; Souza, A. G.; Saito, R.; Dresselhaus, G.; & Dresselhaus, M. S. *New J. Phys.*, 5, 139.

Kim., I. T.; Nunnery, G.; Jacob, K.; Schwartz, J.; Liu, X. & Tannenbaum, R. (2010). Synthesis, characterization, and alignment of magnetic carbon nanotubes tethered with maghemite nanoparticles, *J. Phys. Chem. C*, 114, 15, pp. 6944-6951.

Kim, I. T.; Tannenbaum, A. & Tannenbaum, R. (2011). Anisotropic conductivity of magnetic carbon nanotubes embedded in epoxy matrices, *Carbon*, 49, 1, pp. 54-61.

Kimura, T.; Ago, H.; Tobita, M.; Ohshima, S.; Kyotani, M. & Yumura, M. (2002). Polymer composites of carbon nanotubes aligned by a magnetic field, *Adv. Mater.*, 14, 19, pp. 1380-1383.

Korneva, G.; Ye, H.; Gogotsi, Y.; Halverson, D.; Friedman, G.; Bradley, J. C. & Kornev, K. (2005). Carbon nanotubes loaded with magnetic particles, *Nano Lett.*, 5, 5, pp. 879-884.

Kuang, Q.; Li, S. F.; Xie, Z. X.; Lin, S. C.; Zhang, X. H.; Xie, S. Y.; Huang, R. B. & Zheng, L. S. (2006). Controllable fabrication of $SnO_2$-coated nanotubes by chemical vapor multiwalled carbon deposition, *Carbon*, 44, 7, pp. 1166-1172.

Lanticse, L. J.; Tanabe, Y.; Matsui, K.; Kaburagi, Y.; Suda, K.; Hoteida, M.; Endo, M. & Yasuda, E. (2006). Shear-induced preferential alignment of carbon nanotubes resulted in anisotropic electrical conductivity of polymer composites. *Carbon*, 44, 14, pp. 3078-3086.

Lourie, O.; Cox D. M. & Wagner H. D. (1998). Buckling and collapse of embedded carbon nanotubes, *Phys. Rev. Lett.*, 81, 8, pp. 1638-1641.

Lukic, B.; Seo J. W.; Bacsa R. R.; Delpeux, S.; Beguin, F.; Bister, G.; Fonseca, A.; Nagy, J. B.; Kis, A.; Jeney, S.; Kulik A. J. & Forro L. (2005). Catalytically grown carbon nanotubes of small diameter have a high Young's modulus, *Nano Lett.*, 5, 10, pp. 2074-2077.

Martin, C. A.; Sandler, J. K. W.; Windle, A. H.; Schwarz, M. K.; Bauhofer, W.; Schulte, K. & Shaffer, M. S. P. (2005). Electric field-induced aligned multi-wall carbon nanotube networks in epoxy composites, *Polymer*, 46, 3, pp. 877-886.

Matarredona, O.; Rhoads, H.; Li, Z. R.; Harwell, J. H.; Balzano, L. & Resasco, D. E. (2003). Dispersion of single-walled carbon nanotubes in aqueous solutions of the anionic surfactant NaDDBS, *J. Phys. Chem. B*, 107, 48, pp. 13357-13367.

Mei, Y.; Zhou, Z. J. & Luo, H. L. (1987). Electrical-resistivity of rf-sputtered iron-oxide thin-films, *J. Appl. Phys.*, 61, 8, pp. 4388-4389.

Millan, A.; Palacio, F.; Falqui, A.; Snoeck, E. ; Serin, V. ; Bhattacharjee, A. ; Ksenofontov, V. ; Gutlich, P. & Gilbert, I. (2007). Maghemite polymer nanocomposites with modulated magnetic properties, *Acta Mater.* 55, 6, pp. 2201-2209.

Nagahara, L. A.; Amlani, I.; Lewenstein, J. & Tsui, R. K. (2002). Directed placement of suspended carbon nanotubes for nanometer-scale assembly, *Appl. Phys. Lett.*, 80, 20, pp. 3826-3828.

Ounaies, Z.; Park, C.; Wise, K. E.; Siochi, E. J. & Harrison, J. S. (2003). Electrical properties of single wall carbon nanotube reinforced polyimide composites, *Compos. Sci. Technol.*, 63, 11, pp. 1637-1646.

Park, C.; Wilkinson, J.; Banda, S.; Ounaies, Z.; Wise, K. E.; Sauti, G.; Lillehei, P. T. & Harrison, J. S. (2006). Aligned single-wall carbon nanotube polymer composites using an electric field. *J. Poly. Sci. B: Poly. Phys.*, 44, 12, pp. 1751-1762.

Pascal, C.; Pascal, J. L.; Favier, F.; Moubtassim, M. L. E. & Payen, C. (1999). Electrochemical synthesis for the control of gamma-$Fe_2O_3$ nanoparticle size. Morphology, microstructure, and magnetic behavior, *Chem. Mater.*, 11, 1, pp. 141-147.

Peng, C. Q.; Thio, Y. S. & Gerhardt, R. A. (2008). Conductive paper fabricated by layer-by-layer assembly of polyelectrolytes and ITO nanoparticles, *Nanotechnology*, 19, 50, 505603.

Pileni, M. P. (2003). The role of soft colloidal templates in controlling the size and shape of inorganic nanocrystals, *Nature Mater.*, 2, 3, pp. 145-150.

Pu, H. T. & Jiang, F. J. (2005). Towards high sedimentation stability: magnetorheological fluids based on CNT/$Fe_3O_4$ nanocomposites, *Nanotechnology*, 16, 9, pp. 1486-1489.

Qu, L. T.; Dai, L.& Osawa, E. (2006). Shape/size-controlled syntheses of metal nanoparticles for site-selective modification of carbon nanotubes, *J. Am. Chem. Soc.*, 128, 16, pp. 5523-5532.

Rockenberger, J.; Scher, E. C. & Alivisatos, A. P. (1999). A new nonhydrolytic single-precursor approach to surfactant-capped nanocrystals of transition metal oxides, *J. Am. Chem. Soc.*, 121, 49, pp. 11595-11596.

Sincai, M.; Ganga, D.; Bica, D. & Vekas, L. (2001). The antitumor effect of locoregional magnetic cobalt ferrite in dog mammary adenocarcinoma, *J. Magn. Magn. Mater.*, 225, 1-2, pp. 235-240.

Sousa, M. H.; Rubim, J. C. ; Sobrinho, P. G. & Tourinho, F. A. (2001). Biocompatible magnetic fluid precursors based on aspartic and glutamic acid modified maghemite nanostructures, *J. Magn. Magn. Mater.*, 225, 1-2, pp. 67-72.

Steinert, B. W. & Dean, D. R. (2009). Magenetic field alignment and electrical properties of solution cast PET-carbon nanotube composite films, *Polymer*, 50, 3, pp. 898-904.

Sun, S. H.; Murray, C. B.; Weller, D.; Folks, L. & Moser, A. (2000). Monodisperse FePt nanoparticles and ferromagnetic FePt nanocrystal superlattices, *Science*, 287, 5460, pp. 1989-1992.

Sun, S. H.; Zeng, H.; Robinson, D. B.; Raoux, S.; Rice, P. M.; Wang, S. X. & Li, G. X. (2004). Monodisperse MFeO4 (M = Fe, Co, Mn) nanoparticles, *J. Am. Chem. Soc.*, 126, 1, pp. 273-279.

Sun, Z.; Liu, Z.; Wang, Y.; Han, B.; Du, J. & Zhang, J. (2005). Fabrication and characterization of magnetic carbon nanotube composites, *J. Mater. Chem.*, 15, 42, pp. 4497-4501.

Sun, Z. Y.; Yuan, H. Q.; Liu, Z. M.; Han, B. X. & Zhang, X. R. (2005). A highly efficient chemical sensor material for $H_2S$: alpha-$Fe_2O_3$ nanotubes fabricated using carbon nanotube templates, *Adv. Mater.*, 17, 24, pp. 2993-2997.

Treacy, M. M. J.; Ebbesen, T. W. & Gibsoj, J. M. (1996). Exceptionally high Young's modulus observed for individual carbon nanotubes, *Nature*, 381, 6584, pp. 678-680.

Yi, D. K.; Lee, S. S. & Ying, J. Y. (2006). Synthesis and application of magnetic nanocomposite catalysts, *Chem. Mater.*, 18, 10, pp. 2459-2461.

Youn, S.C.; Jung, D.; Ko, Y. K.; Jin, Y. W.; Kim, J. M. & Jung, H. (2009). Vertical alignment of carbon nanotubes using the magneto-evaporation method, *J. Am. Chem. Soc.*, 131, 2, pp. 742-748.

Yu, M. F.; Files, B. S.; Arepalli, S. & Ruoff, R. (2000). Tensile loading of ropes of single wall carbon nanotubes and their mechanical properties, *Phys. Rev. Lett.*, 84, 24, pp. 5552-5555.

Wan, J.; Cai, W.; Feng, J.; Meng, X. & Liu, E. (2007). In situ decoration of carbon nanotubes with nearly monodisperse magnetite nanoparticles in liquid polyols, *J. Mater. Chem.*, 17, 12, pp. 1188-1192.

Weber, M. & Kamal, M. R. (1997). Estimation of the volume resistivity of electrically conductive composites, *Polym. Compos.*, 18, 6, pp. 711-725.

White, W. B. & Deangeli, B. A. (1967). Interpretation of vibrational spectra of spinels, *Spectrochim. Acta, Part A*, A 23, 4, pp. 985-995.

Zhu, Y. F.; Ma, C.; Zhang, W.; Zhang, R. P.; Koratkar, N. & Liang, J. (2009). Alignment of multiwalled carbon nanotubes in bulk epoxy composites via electric field, *J. Appl. Phys.*, 105, 5, 054319.

# 4

# Characterizing Functionalized Carbon Nanotubes for Improved Fabrication in Aqueous Solution Environments

Charles C. Chusuei and Mulugeta Wayu
*Chemistry Department, Middle Tennessee State University, Murfreesboro*
*USA*

## 1. Introduction

The rediscovery of carbon nanotubes (Iijima, 1991) has inspired extensive research activity. These materials have extremely high surface areas, large aspect ratios, remarkably high mechanical strength, and can have electrical and thermal conductivities that are similar to that of copper (Ebbesen et al., 1996). They come in two forms: single-walled carbon nanotubes (SWNTs) and multiwalled carbon nanotubes (MWNTs). SWNTs have diameters ranging from 1.2 to 1.4 nm. MWNTs have larger overall diameters, with sizes depending on the number of concentric walls within the structure. Like graphite, carbon nanotubes are relatively non-reactive, except at the nanotube caps which are more reactive due to the presence of dangling bonds. The reactivity of the carbon nanotube side walls' $\pi$-system can also be influenced by tube curvature or chirality (Okpalugo et al., 2005). In particular, their remarkable structure-dependent properties have attracted great attention due to their potential applications in heterogeneous catalysis (Planeix et al., 1994), use as substrates for destruction of cancer cells (Kam et al., 2005) and applications for biological and chemical sensing (Poh et al., 2004). Carbon nanotubes require chemical modification in aqueous solution environments to make them more amenable for attachment of reactive surface species. In the case of attaching metal nanoparticles to the carbon surface, functionalization is necessary to avoid agglomeration of the metal. Sensor applications involve the tethering of chemical moieties with specific recognition sites for the detecting ultra-trace analytes (Dai, 2002). Surface functionalization is also necessary for depositing high-loading, catalytically active metal nanoparticles on them (Xing et al, 2005).
Great attention has been paid to attaching functional groups onto carbon nanotube surfaces (Holzinger et al., 2001; Kim et al., 2004; Chen et al., 2005; Park et al., 2006) and probing the electronic structure resulting from post-nanotube-synthesis preparations. To understand the changes that result from surface functionalization strategies, well-defined characterization of the carbon nanotube's surface chemistry and structure is needed. The ability to get an accurate detailed picture of the tethered functional groups that attach to the solid surface using aqueous solution preparation methods is important for controlling carbon nanotube surface composition composition.

We have developed an array of analytical methods to probe the surface composition of carbon nanotubes during various stages of nanomaterial synthesis in our laboratory. Summarized herein are three case studies. In the first study, sonochemically functionalized MWNTs were probed by X-ray photoelectron spectroscopy (XPS) revealing a consecutive, first-order attachment mechanism. In the second study, extended X-ray absorption fine structure (EXAFS) and attenuated total reflection infrared (ATR-IR) spectroscopy were used to examine tethered Pt nanoparticles on functionalized MWNTs. In the third study, we functionalized high pressure carbon monoxide (HiPco) SWNTs to produce carboxylic acid (COOH-SWNT), maleic anhydride (MA-SWNT), and nitroso (NO-SWNT) attached SWNTs in order to examine the effects of the tethered groups on the solid surface point-of-zero charge (PZC). The PZC is defined as the aqueous solution pH value at which the degree of surface protonation and hydroxylation are equal, which results in an electrostatically neutral charge at the electrical double layer interface (Brown et al., 1999). SWNTs were used in the PZC studies due to their relative ease for surface functionalization with specific moieties.

## 2. Experimental

In the first case study, MWNTs produced from chemical vapor deposition were obtained from Nanolab, Inc. (Waltham, MA). The as-purchased MWNTs (95% purity, ~30 nm in diameter) were put into a mixture solution of $HNO_3$ and $H_2SO_4$ in an Erlenmeyer flask. The concentrations of both acids were 8.0 M. The flask was placed in an ultrasonic bath (Fisher Scientific, 130 W and 40 kHz) maintained at 60 °C. Sonication was performed for 1, 2, 4 and 8 hrs. The sonochemically treated MWNTs were then separated from the acids in a centrifuge (Thermal IEC Centra CL2), and thoroughly washed using doubly distilled, deionized water prior to analysis (Xing et al., 2005).

The chemical oxidation states and surface compositions of the resulting sonochemically treated MWNTs and Pt electrocatalysts were analyzed by XPS using an ion-pumped Perkin-Elmer PHI ESCA 560 system using a PHI 25-270AR double pass cylindrical mirror analyzer. An Mg Kα anode operated at 15 kV and 250 W with photon energy of $h\nu$= 1253.6 eV was used. The base pressure of the chamber after a bake out was ~$1x10^{-10}$ Torr. The operating pressure during XPS scans did not exceed $5x10^{-8}$ Torr. The C 1s core level at 284.4 eV, corresponding with the carbon nanotube oxidation state (Suzuki et al., 2002), was used to charge reference the XP spectra. XPS data were curve fitted using CasaXPS VAMAS processing software version 2.2 (Devon, United Kingdom) with a Shirley background subtraction and 70%-to-30% Gaussian-Lorentzian line shapes.

In the second case study, MWNT-Pt nanoparticle structural analysis was performed using EXAFS. Finely dispersed Pt nanoparticles (3.5 nm in diameter) tethered onto MWNTs were prepared via sonicating MWNTs in $HNO_3/H_2SO_4$ for 2 hrs followed by reducing the Pt salt precursor, $K_2PtCl_4$ (Xing, 2004). Spectra were obtained from the 12-BM BESSRC Advanced Photon Source (APS) beamline at the Argonne National Laboratory and the X18B beamline at the National Synchrotron Light Source (NSLS) at the Brookhaven National Laboratory to analyze the Pt $L_{III}$ edge (11.564 keV) of the Pt nanoparticles tethered to the carbon nanotube surface. A spectrum of a 5 $\mu$m thick Pt foil was taken in the transmission mode for absolute energy calibration. The absorption edge of the data obtained for each sample, plotted as $\chi$

(*E*), was checked to ensure alignment prior to plotting in *k* space. Spectra of the dried MWNT-Pt nanoparticles samples and $PtO_2$ standard were obtained in the fluorescence mode. Anhydrous $PtO_2$ (99.95% purity, metals basis) obtained from Alfa Aesar, 8 mg of which were diluted in 170 mg of boron nitride, was used as received for the reference spectrum. In the beamline, a double crystal Si (111) monochromator was used for energy selection. Spectra taken in the fluorescence mode used a 13-element Ge detector. The ion chambers employed had a 8:2 gas mixture of $N_2$-to-Ar. Data were processed using the IFEFFIT library of numerical XAS algorithms written in Perl programming that utilizes the *ab initio* EXAFS code, FEFF 6.01 (Newville, 2001; Ravel et al., 2005). Further analysis by ATR-IR and Raman spectroscopies were performed as described by Hull et al. (2006).

The experimental procedure for the third case study (McPhail et al., 2009) was as follows: (1) COOH-SWNTs were prepared by refluxing in $H_2SO_4/HNO_3$ according to Lu et al. (2007). A 30.0 mL solution of 3:1 concentrated nitric-to-sulfuric acid ratio was added to a 100-mL round-bottom glass flask along with 60.1 mg of p-SWNTs. The mixture was refluxed at 338 K for 12 hrs, with constant magnetic stirring, under $N_2$ atmosphere. (2) NO-SWNTs were prepared using an electrochemical functionalization procedure based on the description made by Wang et al. (2005). SWNT sheets were prepared by sonicating them in 1% Triton X-100 (The Chemistry Store.com Inc.; St. Cayce, SC) solution followed by vacuum filtration with Millipore Teflon filter paper (0.2 $\mu$m pore size). The solid was dried on the filter paper and lifted off as a single sheet to be used as a working electrode in the electrochemical functionalization. Remaining surfactant was removed by annealing at 120 °C for 2 hrs, 300 °C for 1 hr, and 800 °C for 30 minutes under inert Ar atmosphere. Complete removal of the Triton X-100 required additional electrochemical oxidation of the nanotubes, which occurred during the nitrosylation. Electrochemical nitrosylation took place in 6 M $KNO_2$ with a SWNT sheet as the working electrode, AgCl as the reference electrode, and Pt wire as the counter electrode. The system was sparged with Ar gas and run for 6 hrs at 2.0 V using a 273 EG&G Princeton Applied Research potentiostat. (3) The MA-SWNTs adduct was prepared via Diels-Alder addition reaction with maleic anhydride as the dienophile (Dewar et al., 1970), in good agreement with *ab initio* predictions for Diels-Alder additions to SWNT sidewalls (Mercuri et al., 2009). The synthesis involved adding 50.0 mg of the HiPco p-SWNTs to an excess of maleic anhydride dissolved in chloroform. This mixture was placed in a 100-mL round-bottom flask and refluxed at 323 K for 12 hrs, with constant stirring, under $N_2$ atmosphere. All of the functionalization procedures were carried out for prolonged periods (12 hrs) to ensure saturation of the SWNTs with their respective moieties. The SWNTs were then recovered by vacuum filtration using a Buchner funnel and Fluoropore polytetrafluoroethylene (PTFE) membrane filters with a 0.2 $\mu$m pore size. Samples were rinsed with copious amounts of Millipore $H_2O$. Samples were then dried in a vacuum desiccator and saved in glass vials for further analysis.

Isoelectric point measurements at the solid-liquid interface were made on MA-SWNT, NO-SWNT, COOH-SWNT and p-SWNT surfaces using a method described by Park and Regalbuto (1995). Twelve solutions in the range of pH = 1.0-12.0 were made using dilute aqueous solutions of NaOH and HCl. A 1.8 mL aliquot of each solution was pipetted into polyethylene vials and allowed to equilibrate for 1 hr. The initial pH of each solution was then recorded. A 2.0 mg amount of the SWNTs to be examined were added to each vial, which were then capped and shaken with a vortex mixer to settle the SWNTs. After an additional 12-hr equilibration period, the final pH at the SWNT solid surface was measured for each vial using a spear-tip semisolid electrode. Finally, initial pH values versus final pH values were plotted.

## 3. Results and discussion

The sonochemically functionalized MWNTs were characterized and quantified by XPS. XPS is an effective surface sensitive method for quantifying the extent (or level) of surface oxidation (Huefner, 2003). The distribution of oxygen containing functional groups (-C-O-, -C=O, and O-C=O) is also often characterized by deconvoluting the C 1s spectral envelope to obtain quantitative information, based on differences in XPS binding energy (BE) (Datsyuk et al., 2008).

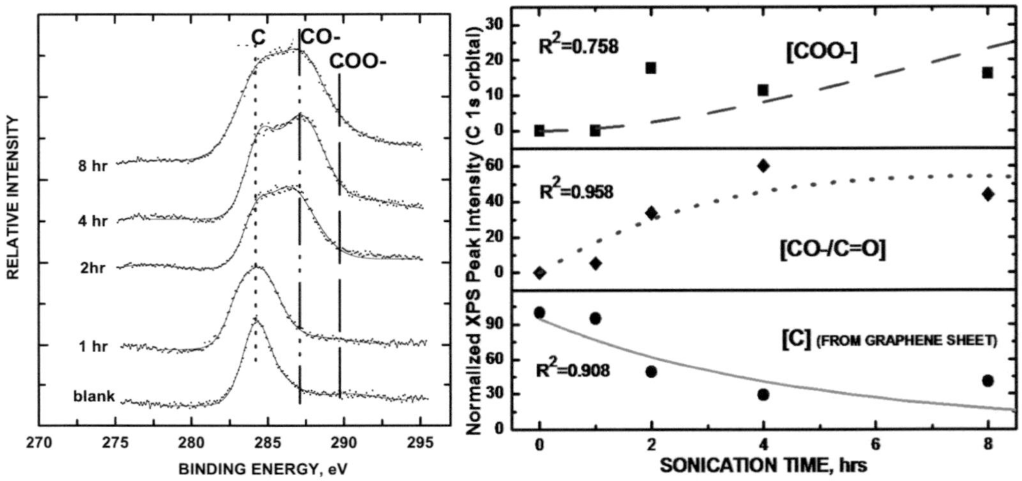


Fig. 1. (Left-hand panel) XPS stackplot of C 1s core level of the carbon nanotubes at varying sonochemical treatment times; (Right-hand panel) Kinetic model of sonochemically treated carbon nanotube oxidation process based on deconvoluted XPS C 1s integrated peak areas for C (on the MWNT graphene sheet), CO- and COO- oxidation states.

Fig. 1 (left-hand panel) shows the narrow scan spectra of the C 1s region of sonochemically treated and untreated MWNTs. The XPS spectrum shows distinct carbon peaks, representing the major constituents of the oxidized MWNT surface. The dominant peak structure for the C 1s core level at a BE of 284.4 eV corresponds to the bare, untreated MWNT surface (Ago et al., 1999; Suzuki et al., 2002). C 1s core level shifts at 287.6 and 288.3 eV indicate that the moieties consist of CO-/C=O and COO- respectively, in agreement with literature values reported for these groups tethered onto the MWNTs (Langley et al., 2005). Intensities of the high BE states increased due to oxidation as sonication ensued. The CO-/C=O and COO- concentrations were quantified relative to the graphitic carbon peak. The C 1s line broadening with extra feature developments were attributed to the surface oxidation of MWNTs where C atoms bond to more O atoms as a result of the sonochemical treatment. The population of the oxidized groups (CO-, C=O, and COO-) relative to the MWNT carbon were quantified via plotting the sum of their C 1s peak areas relative to that of the graphitic MWNT carbon as a function of sonochemical treatment time. The increase in surface oxidation measured from the integrated C 1s peak areas of the ([CO-] + [COO-])/[C] tracks well with the overall increase in XPS atomic percent oxygen. A greater uptake of oxygen by the surface carbon atoms corresponds to a higher population density of $CO_x$ functional groups detected by the XPS.

The kinetic model for the oxidation process is shown in Fig. 1 (right-hand panel). A stochastic addition mechanism obeying a consecutive 1st-order mechanism was revealed. Here, we report the first detailed mechanistic delineation of the carbon nanotube oxidation process. Evolution of the high binding energy peak intensities during sonication shows a consecutive, single-step first order O-attachment mechanism, leading to the carboxylate. This scheme is consistent with a report made by (Chiang et al., 2011) showing CO to be an intermediate species, which could be oxidized quickly to other forms, usually COO under acidic environment. Sonication creates defect sites on the sidewalls that allow for O atom attachment (Li et al., 2006). Differential equations describing the mechanism are as follows:

$$\frac{d[C]}{dt} = -k_1[C]; \frac{d[CO-]}{dt} = k_1[C] - k_2[COO]; \frac{d[COO]}{dt} = k_2[CO].$$

Least squares fittings show rate constants of $k_1$ (C→CO) = 2.11(±0.15) x$10^{-2}$ $s^{-1}$ and $k_2$ (CO→COO) = 7.3(±1.2) x$10^{-2}$ $s^{-1}$. The overall correlation coefficient ($R^2$) value = 0.923 indicated the goodness of fit for our kinetic model. $R^2$ values for individual concentration profiles of C, CO and COO are shown in Fig. 1 (right-hand panel). We thus demonstrate that stochastic functionalization on MWNTs is possible under aqueous solution conditions. In contrast, gas phase kinetic studies have typically been limited to a single-step reaction (Brukh et al., 2007).

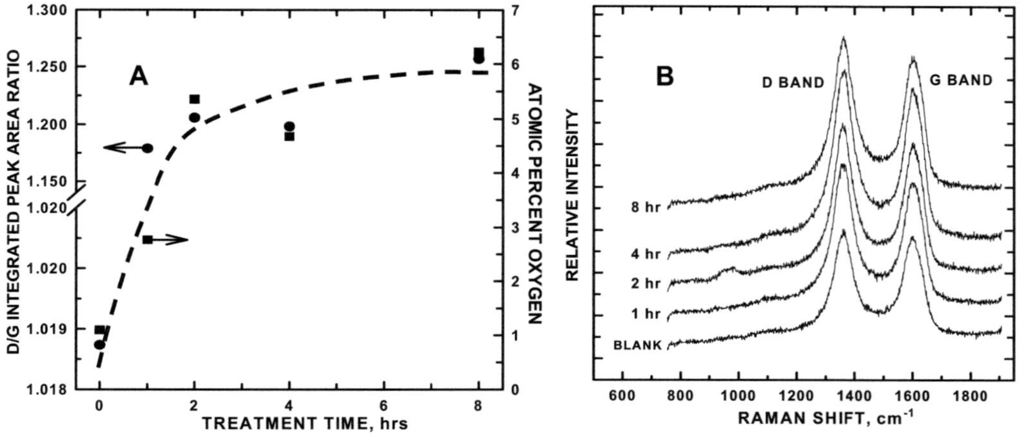


Fig. 2. (A) A plot of the uptake of D-to-G integrated peak area ratios (•, left-hand axis) and atomic percent oxygen (■, right-hand axis) versus sonochemical treatment time; (B) Raman shifts showing the emergence of the D and G bands of sonochemically treated MWNTs before deposition of Pt nanoparticles. The dashed line serves as a guide to the eye, denoting functional group saturation at 2 hrs.

In examining the Raman D-to-G integrated peak area ratios (Fig. 2A; left-hand axis), the disordered $sp^3$ state increased with longer sonochemical treatment. The largest increase occurred between 0 and 1 hr of sonication with a plateau reached at 2 hrs. Noteworthy is the fact that the plateau of the relative Raman D-to-G band intensities (left-hand axis) coincided with a plateau of the atomic percent mole fractions of oxygen (right-hand axis), obtained from normalizing XPS high-resolution energy scans of the O 1s core level (Fig. 2A; right-hand axis), at 2 hrs. The population of $sp^3$-hybridized carbon increased relative to the $sp^2$-hybridized carbon during sonication, accompanying the creation of sidewall defects to which the functional groups attached. Thus, the groups covalently bonded to the surface with moieties directly forming from C atoms within the graphene sheets. The growth rate of $sp^3$-to-$sp^2$ Raman intensities with sonication time (Fig. 2B) was also consistent with that of our consecutive 1st-order kinetic model. The density of the surface functional groups was directly (albeit not linearly) related to sonication time; 2 hrs of sonication resulted in optimal Pt nanoparticle dispersion. Upon deposition of the Pt nanoparticles, the Raman line shapes and relative D-to-G band intensities remained unchanged. The presence of these peaks verified that the carbon nanotubes remained largely intact during the oxidation procedure and after deposition of Pt nanoparticles.

To examine the local structure of the nanoparticles, EXAFS was performed on the Pt deposited on the 2-hr sonochemically treated carbon nanotubes. The Pt-CNT samples were examined as a dry powder-like form instead of aqueous solution phase to get a stronger signal. EXAFS is an oscillatory feature in the X-ray absorption above the absorption edge of the target atoms and is defined as the fraction deviation in the absorption coefficient:

$$\chi(k) = \sum \frac{N_j f_j(k) \exp[-2k^2 \sigma_j^{\,2}]}{kR_j^{\,2}} \sin[2kR_j + \delta_j(k)]$$

with $R$ being the distance from the target to neighboring atom. $N$ is the coordination number of the neighboring atom, and $\sigma^2$ is the disorder of the neighbor distance (i.e., the Debye-Waller factor). The photoelectron wave number $k = [2m\,(E-E_o)/\hbar^2]^{1/2}$, $f\,(k)$ is the scattering amplitude, and $\delta\,(k)$ is the phase shift. Oscillations arise from the photoelectron wave backscattering from the nearest neighbor atoms. Assuming a cuboctahedron structure, predicted for the Pt nanoparticles of the size (3.5 nm in diameter) deposited on a carbon surface, there would be ~1500 atoms per cluster on the 20% Pt loaded, 2-hr sonochemically treated MWNT surface. Though the majority of the bonding emanated from bulk Pt-Pt interactions according to XPS (*vide infra*), EXAFS oscillations indicate Pt coordination to lower molecular weight atoms. The only two low-molecular-weight atoms present that can interact with Pt are C and O. The formation of Pt-C in the cluster was unlikely because temperatures in excess of 560°C are required to form the carbide (Kojima et al., 1982; Lamber et al., 1993). Since EXAFS scattering is sensitive only to the first few atomic shells and given the size of the Pt NPs, $PtO_x$ appears to be present only at the top most surface layers of the cluster. A majority of the Pt atoms were in the metallic (zero) oxidation state, consistent with the observed XPS Pt 4f core level shift. Upon deposition of the Pt nanoparticles, the Raman line shapes and relative D-to-G band intensities remained unchanged. The presence of these peaks verified that the MWNTs remained largely intact during sonochemical treatment and after deposition of the Pt nanoparticles.

The XPS Pt $4f_{7/2}$ core level of the Pt-CNT (not shown; prepared using a 2-hr sonochemical treatment), referencing the graphitic C 1s orbital at a BE equal to 284.4 eV (Ago et al., 1999;

Suzuki et al., 2002), had a BE= 71.4 eV, indicating that Pt was predominantly in the metallic (zero) oxidation state (Fleisch et al., 1986). XPS signals from the C 1s and O 1s and Pt 4f levels and from no other elements were observed. The asymmetry observed in the $4f_{5/2}$ level at ~78 eV indicated a small population of PtO or $PtO_2$, which was masked by much larger signal from metallic Pt. The lack of insufficient signal from the Pt oxide ($PtO_x$) hampered precise determination of the stoichiometric proportions of PtO and $PtO_2$. Hence, EXAFS was needed for clearer structural elucidation.

In comparing the FTs of the EXAFS Pt $L_{III}$ edge of Pt nanoparticles deposited on the -COO- and -C=O functionalized MWNTs, the first nearest neighbor atom was observed at ~1.78 Å in the Pt-MWNT sample instead of the expected distance of ~2.78 Å for Pt-Pt interactions in its zero oxidation state (Fig. 3A). The latter distance was observed for a standard $PtO_2$ powder used for comparison. This result was consistent with the XPS core level shift for the Pt $4f_{7/2}$ orbital observed at 71.4 eV, denoting metallic Pt (Fleisch et al., 1986). The low $R$ value feature at $R$ = 1.01 Å was an artifact of imperfect background subtraction due to the intense Pt $L_{III}$ white line at the absorption edge. The dotted line spectrum was that of the reference foil, obtained in the transmission mode for comparison. The strong peaks between 2 and 3 Å correspond to the interaction occurring between the first nearest neighbor (1NN) Pt atoms. Peak positions of the FT at $R$=2.12 and 2.70 Å were in good agreement with positions observed by others for Pt foil (Frenkel et al., 2001; Zhang et al., 2004). In comparing $R$ values of Pt-MWNT samples, the predominant 1NN interaction in these samples was

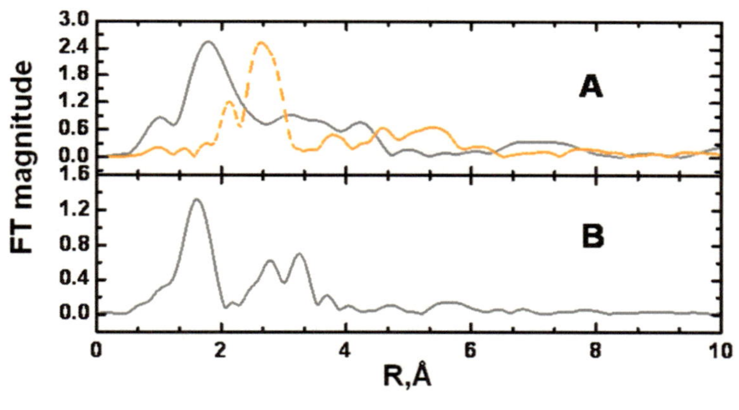


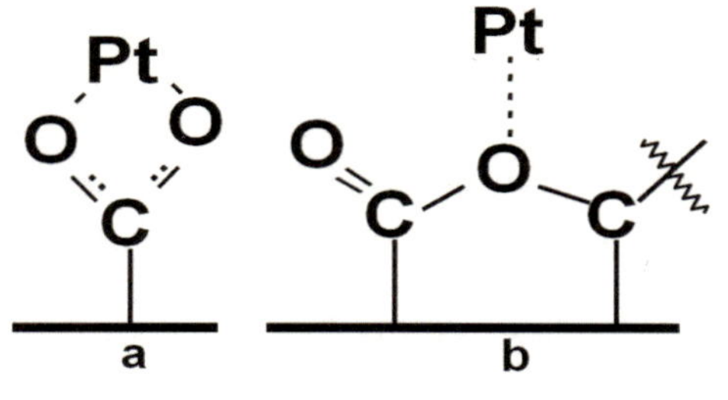


Fig. 3. FTs of EXAFS oscillations plotted in R space of (A) Pt-CNTs (solid gray line; a Pt foil scan is shown as a dotted, orange spectral line); (B) $PtO_2$ standard; and proposed (a) and (b) structures for Pt-MWNT coordination.

clearly not with Pt-Pt, denoted by the ~1.78 Å position. A FT of a reference $PtO_2$ powder is shown in Fig. 3B. The location of its 1NN, signifying Pt-O, is seen at $R$ =1.59 Å. EXAFS and XPS data indicated the presence of $PtO_x$ on the Pt-MWNT surface, but not in the bulk of the nanoparticles.

Fig. 4 shows ATR-IR difference spectra of 2 hr sonochemically treated carbon nanotubes before and after Pt nanoparticles were tethered to these surfaces. The C-O ester features, denoted by peaks (2) and (3) in the 2-hr sonicated MWNTs with no Pt deposited, were replaced by a broad single band with a center at 1092 $cm^{-1}$ after Pt nanoparticle attachment. This change in IR envelope shape indicated a strong interaction of ester O with the Pt nanoparticles. The carbonyl O band at 1700 $cm^{-1}$ (before Pt nanoparticle deposition) were replaced by two peaks absorbing at 1712 and 1629 $cm^{-1}$, indicative of Pt nanoparticles interactions with carbonyl O. Pt binding with the carbonyl O was evident from the absorbance shift from a single feature at 1700 $cm^{-1}$ to two peaks at 1712 and 1629 $cm^{-1}$. Bands from the ester C-O stretches were still present with vibrational stretches at 1160 $cm^{-1}$. From *Fig. 4*, it was clear that the Pt loading of the oxidized MWNTs dramatically altered the absorbance signal from the C-O stretches in the 1300-to-900 $cm^{-1}$ region. The carbonyl C=O signal at 1700 $cm^{-1}$ was less affected although there was a shift to higher frequency at 1712 $cm^{-1}$ along with the emergence of another stretch at 1629 $cm^{-1}$, indicative of multiple binding sites for the Pt nanoparticles. Hence, ATR-IR data showed that the ester O peaks present before tethering Pt nanoparticles were radically altered after the Pt deposition, denoting their involvement in the coordination of the Pt nanoparticles to create the nanostructure. Based on this IR result and the EXAFS analysis, we propose two Pt-MWNT surface structures. Attachment can occur via carboxylate ions in which the O atoms effectively have equal bond order and participation in the Pt binding in the form of COO(Pt) (Fig. 3a). Pt nanoparticles can also coordinate to ester O atoms bound to the carbon nanotube surface, bridging between two carbons and serving as a binding site for the Pt nanoclusters in the form of C(=O)CO(Pt) (Fig. 3b). According to Petroski and El-Sayed (2003), because the d band of Pt is close to the Fermi level, electron density to form new bonds would come from the C=O group rather than the Pt. Hence, shifts in the C=O stretch would be sensitive to coordination with Pt (peak 1 in Fig. 4) as observed.

In our final case study, variations in the measured PZC were seen between differently functionalized SWNT structures (McPhail et al., 2009). Fig. 5 shows a plot of final versus initial pH values of solutions to which various SWNT samples were added. A plateau (horizontal dashed lines) in the plot indicates the PZC for each specifically-functionalized carbon nanotube.

The PZC values in this series of functionalized carbon nanotubes indicated a relatively acidic surface, amenable for adsorption of anionic (metal nanoparticle) precursors. The PZC values for the SWNTs were in ascending order: COOH-SWNTs (1.2) < MA-SWNTs (2.0) < p-SWNTs (3.5) < NO-SWNTs (7.5). Lowering of the p-SWNTs PZC compared to other studies (Matarredona et al., 2003) was attributed to our use of smaller radius (~0.7 nm) SWNTs. The COOH groups, due to its acidity, lowered the PZC to a greater extent than the MA groups (by 0.8 pH units). The PZCs were found to be tunable within 6.3 pH units by functionalizing them with various moieties of different electron withdrawing/donating character. The moieties markedly affected the PZCs. There is an obvious correlation of PZC with electron distribution, emanating from attached moieties along the SWNTs sidewalls.

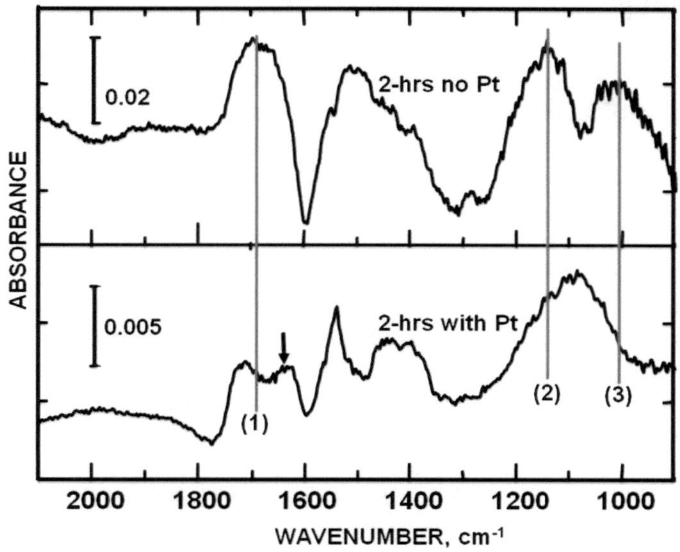


Fig. 4. ATR-IR difference spectra of 2100-900 cm$^{-1}$ region of 2-hr sonochemically treated carbon nanotubes before and after Pt nanoparticle deposition. Untreated carbon nanotubes were used for background subtraction. Vibrations from (1) carbonyl and ester (2) asymmetric and (3) symmetric stretches are noted for comparison. The arrow denotes a new frequency signifying coordination with Pt nanoparticles.

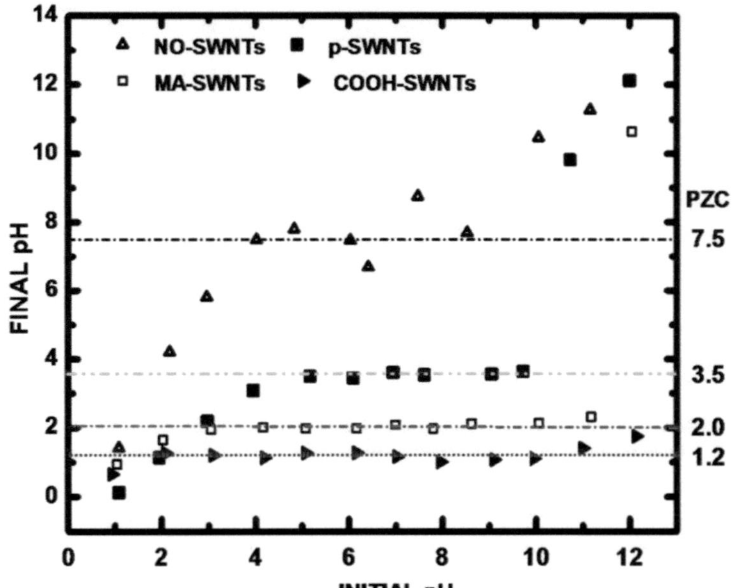


Fig. 5. The point of zero charge (PZC) values of NO-SWNTs, p-SWNTs, MA-SWNTs and COOH-SWNTs are denoted by horizontal lines.

In the context of electrophilic aromatic substitution (EAS) reactions, nitroso groups are known to be electron withdrawing, maleic anhydride groups are lightly electron releasing, and carboxylic acid groups are strongly electron releasing, which can be quantitatively described by Hammett sigma constants ($\sigma$). Since carbon nanotubes are essentially aromatic, peri-condensed benzenoids (composed of $sp^2$ carbons, arranged in a graphite-like hexagonal pattern) that have aromatic character (Linert et al., 2007; Lukovits et al., 2007) and are used to fabricate hierarchical structures (Zorbas et al., 2005), it is appropriate to explore how $\sigma$ relates to our observed PZC measurements. Also known as the "substituent constant," $\sigma$ determines the effect that a given substituent will have on the equilibrium and rate constants for the disassociation of benzoic acids. The $\sigma$ parameter takes into account resonance, field, and inductive effects of the substituent. The result is a value whose magnitude gives the relative strength of a substituent's effect on the electronic distribution of a benzoic acid. Standard tables show $\sigma$ values (for the meta- position) of 0.71, 0.39, and 0.35 for nitro, acetoxy, and carboxylic acid groups, respectively (Hansch et al., 1991; Carey, 2002). Larger $\sigma$ values denote greater electron-withdrawing character. The MA and COOH groups, which are the least electron withdrawing (i.e., more electron releasing) lowered the PZC, relative to p-SWNTs, while NO, the most electron withdrawing moiety, raised the PZC. Variations in the electron releasing/withdrawing character of the substituents correlate well with the observed PZC trend. In our previous work (McPhail et al., 2009), we postulated that the PZC was dependent on SWNT electronic structure.

Here, we note a new observation: greater $\sigma$ values coincide with a greater propensity to be hydroxylated, thereby increasing the PZC. The greater electron donating character of the moiety led to an increased degree of surface hydroxylation. Quantitatively, the $\sigma$ values of the substituents show the same increasing trend as that of the experimentally measured PZCs for each corresponding, functionalized SWNT (Fig. 5).

## 4. Conclusions

In summary, we have demonstrated the utility of XPS for delineating MWNT oxidation kinetics, EXAFS (coupled with XPS and ATR-IR) for elucidating nanoparticle-MWNT interfacial structure, and the dependence of PZC on the electron withdrawing/donating character of moieties attached to SWNTs. Sonication of MWNTs is a facile functionalization technique as it lowers the surface activation energy barrier resulting in low temperature functionalization and reduction in surface physical damage. The process greatly reduces the functionalization time to as low as 2 hrs. Sonochemical treatments tend to create dangling bonds on the surfaces of carbon nanotubes, which progressively oxidize to hydroxyl (OH), carbonyl (CO), and carboxyl (COOH) functional groups (Al-Aqtash and Vasiliev, 2009). Kinetic studies uncovered a stochastic functionalization mechanism involved in the preparation of MWNTs for nanoparticle attachment. EXAFS, coupled with XPS and ATR-IR data, was pivotal in the elucidation of ester-like O atoms found to play an important role in synthesizing Pt nanoparticle-MWNT structures. Controlled surface functionalization on SWNTs can influence its PZC, an important variable for Coulombic attachment of structures onto the surface. The above described surface analytical methods, performed on MWNTs and SWNTs as benchmarks, may well be applicable for examining aqueous solution functionalization processes on newly emerging carbon nanomaterials, i.e., graphene and graphene oxides (Liu et al., 2008; Geim, 2009; Yan and Chou, 2010), for advanced technological applications.

## 5. Acknowledgements

We gratefully acknowledge support from the Faculty Research Creative Activity Committee (FRCAC) of Middle Tennessee State University awarded in 2011.

## 6. References

Ago, H.; Kugler, T.; Cacialli, F.; Salaneck, W. R.; Shaffer, M. S. P.; Windle, A. H.; Friend, R. H. (1999). Work Functions and Surface Functional Groups of Multiwall Carbon Nanotubes. *J. Phys. Chem. B*, 103, 8116-8121.

Al-Aqtash, N.; Vasiliev, I. Y. (2009). Ab initio Study of Carboxylated Graphene. *J. Phys. Chem. C*, 113, 12970-12975.

Brown, G.E., Jr.; Henrich, V.E.; Casey, W.H.; Clark, D.L.; Eggleston, C.; Felmy, A.; Goodman, D.W.; Grätzel, M.; G. E. Maciel; McCarthy, M.I.; Nealson, K.; Sverjensky, D.A.;Toney, M.F.; Zachara, J.M. (1999) Chemical Interactions of Metal Oxide-Aqueous Solution Interfaces," *Chem. Rev.*, 99, 77-174.

Brukh, R.; Mitra, S. (2007). Kinetics of Carbon Nanotube Oxidation. *J. Mater. Chem.*, 17, 619-623.

Carey, F. A. (2002). *Advanced Organic Chemistry, Part A: Structure and Mechanisms*, 4th ed.; Kluwer Academics/Plenum Publishers: New York.

Chen, G.-X.; Kim, H.-S.; Park, B. H.; Yoon, J.-S. (2005). Controlled Functionalization of Multiwalled Carbon Nanotubes with Various Molecular-Weight Poly (l-lactic acid). *J. Phys. Chem. B*, 109, 22237-22243.

Chiang, Y.-C.; Lin, W.-H.; Chang, Y.-C. (2011). The Influence of Treatment Duration on Multi-walled Carbon Nanotubes Functionalized by $H_2SO_4/HNO_3$ Oxidation. *Appl. Surf. Sci.*, 257, 2401-2410.

Dai, H. (2002). Carbon Nanotubes: Synthesis, Integration, and Properties. *Acc. Chem. Res.*, 35, 1035-1044.

Datsyuk, V.; Kalyva, M.; Papagelis, K.; Parthenios, J.; Tasis, D.; Siokou, A.; Kallitsis, I.; Galiotis, C. (2008). Chemical Oxidation of Multiwalled Carbon Nanotubes. *Carbon*, 46, 833-840.

Dewar, M. J. S.; Pyron, R. S. (1970). Nature of the Transition State in Some Diels-Alder Reactions. *J. Am. Chem. Soc.*, 92, 3098-3103.

Ebbesen, T.W.; Lezec, H.J.; Hiura, H.; Bennett, J.W.; Ghaemi, H.F.; Thio, T. (1996). Electrical Conductivity of Individual Carbon Nanotubes. *Nature*, 382, 54-56.

Fleisch, T. H.; Mains, G. J. (1986). Photoreduction and Reoxidation of Platinum Oxide Surfaces. *J. Phys. Chem.*, 90, 5317-5320.

Frenkel, A. I.; Hills, C. W.; Nuzzo, R. G. (2001). A View from the Inside: Complexity in the Atomic Scale Ordering of Supported Metal Nanoparticles. *J. Phys. Chem. B*, 105, 12689-12703.

Geim, A.K. (2009). Graphene: Status and Prospects. *Science*, 324, 1530-1534.

Hansch, C.; Leo, A.; Taft, R. W. (1991). A Survey of Hammett Substituent Constants and Resonance and Field Parameters. *Chem. Rev.*, 91, 165–195.

Holzinger, M., Vostrowsky, O., Hirsch, A., Hennrich, F., Kappes, M., Weiss, R.; Jellen, F. (2001). Sidewall Functionalization of Carbon Nanotubes. *Ang. Chem. Inter. Ed.*, 40, 4002–4005.

Huefner, S. (2003). *Photoelectron Spectroscopy*, Springer: Berlin.

Hull, R. V.; Li, L.; Xing, Y.; Chusuei, C. C. (2006). Pt Nanoparticle Binding on Functionalized Multiwalled Carbon Nanotubes. *Chem. Mater.*, 18, 1780-1788.

Iijima, S. (1991). Helical Microtubules of Graphitic Carbon. *Nature*, 354, 56-58.

Kam, N.W.S.; O'Connell, M.; Wisdom, J.; Dai, H. (2005). Carbon Nanotubes as Multifunctional Biological Transporters and Near-infrared Agents for Selective Cancer Cell Destruction." *Proc. Natl. Acad. Sci. U.S.A.*, 102, 11600-11605.

Kim, B.; Sigmund, W. M. (2004). Functionalized Multiwall Carbon Nanotube/Gold Nanoparticle Composites. *Langmuir*, 20, 8239-8242.

Kojima, I.; Miyazaki, E.; Iwao, Y. (1982). Field Emission Study of VIII Transition Metals. III. Adsorption of Ethylene and Acetylene on Platinum. *Appl. Surf. Sci.*, 10, 27-41.

Lamber, R.; Jaeger, N.I. (1993). Electron Microsopy Study of the Interaction of Nickel, Palladium and Platinum with Carbon. III: Formation of a Substitutional Platinum-carbide in Ultrafine Platinum Particles. *Surf. Sci.*, 289, 247-254.

Langley, L. A.; Villanueva, D. E.; Fairbrother, D. H. (2005). Quantification of Surface Oxides on Carbonaceous Materials. *Chem. Mater.*, 18, 169-178.

Li, J.-L.; Kudin, K. N.; McAllister, M. J.; Prud'homme, R. K.; Aksay, I. A.; Car, R. (2006). Oxygen-Driven Unzipping of Graphitic Materials. *Phys. Rev. Lett.*, 96, 176101.

Linert, W.; Lukovits, I. (2007). Aromaticity of Carbon Nanotubes. *J. Chem. Info. Model.*, 47, 887-890.

Liu, L.; Ryu, S.; Tomasik, M. R. ;Stolyarova, E.; Jung, N.; Hybertsen, M. S.; Steigerwald, M. L.; Brus, L. E. and Flynn, G.W. (2008). Graphene Oxidation: Thickness-Dependent Etching and Strong Chemical Doping. *Nano Lett.*, 8, 1965-1970.

Lu, X.; Imae, T. (2007). Size-Controlled In Situ Synthesis of Metal Nanoparticles on Dendrimer-Modified Carbon Nanotubes. *J. Phys. Chem. C*, 111, 2416-2420.

Lukovits, I.; Kármán, F.; Nagy, P.M.; Kálmán, E. (2007). Aromaticity of Carbon Nanotubes. *Croat. Chem. Acta*, 80, 233–237.

Matarredona, O.; Rhoads, H.; Li, Z.; Harwell, J. H.; Balzano, L.; Resasco, D. E. (2003). Dispersion of Single-Walled Carbon Nanotubes in Aqueous Solutions of the Anionic Surfactant NaDDBS. *J. Phys. Chem. B*, 107, 13357-13367.

Mawhinney, D.B.; Naumenko, V.; Kuznetsova, A.; Yates, J., John T. (2000). Infrared Spectral Evidence for the Etching of Carbon Nanotubes: Ozone Oxidation at 298 K. *J. Am. Chem. Soc.*, 122, 2383-2384.

McPhail, M.R.; Sells, J.A.; He, Z.; Chusuei. C. C. (2009). Charging Nanowalls: Adjusting the Carbon Nanotube Isoelectric Point via Surface Chemical Functionalization. *J. Phys. Chem. C*, 113, 14102-14109.

Mercuri, F.; Sgamellotti, A. (2009). First-principles Investigations on the Functionalization of Chiral and Non-chiral Carbon Nanotubes by Diels-Alder Cycloaddition Reactions. *Phys. Chem. Chem. Phys.*, 11, 563-567.

Newville, M. (2001). IFEFFIT: Interactive XAFS Analysis and FEFF Fitting. *J. Synch. Rad.*, 8, 322-324.

Okpalugo, T. I. T.; Papakonstantinou, P.; Murphy, H.; McLaughlin, J.; Brown, N. M. D. (2005). High Resolution XPS Characterization of Chemical Functionalised MWCNTs and SWCNTs. *Carbon*, 43, 153-161.

Park, J.; Regalbuto, J. R. (1995). A Simple, Accurate Determination of Oxide PZC and the Strong Buffering Effect of Oxide Surfaces at Incipient Wetness. *J. Colloid Inter. Sci.*, 175, 239-252.

Park, M. J.; Lee, J. K.; Lee, B. S.; Lee, Y.-W.; Choi, I. S.; Lee, S.-g. (2006). Covalent Modification of Multiwalled Carbon Nanotubes with Imidazolium-Based Ionic Liquids: Effect of Anions on Solubility. *Chem. Mater.*, 18, 1546-1551.

Petroski, J.; El-Sayed, M.A. (2003). FTIR Study of the Adsorption of the Capping Material to Different Platinum Nanoparticle Shapes. *J. Phys. Chem. A*, 107, 8371-8375.

Planeix, J.M.; Coustel, N.; Coq, B.; Brotons, V.; Kamblar, P.S.; Dutartre, R.; Geneste, P.; Bernier, P.; Ajayan, P.M. (1994). Application of Carbon Nanotubes as Supports in Heterogeneous Catalysis. *J. Am. Chem. Soc.*, 116, 7935-7936.

Poh, W. C.; Loh, K. P.; Zhang, W. D.; Sudhiranjan; Ye, J.-S.; Sheu, F.-S. (2004). Biosensing Properties of Diamond and Carbon Nanotubes. *Langmuir*, 20, 5484-5492.

Ravel, B.; Newville, M. (2005). ATHENA, ARTEMIS, HEPHAESTUS: Data Analysis for X-ray Absorption Spectroscopy Using IFEFFIT. *J. Synchrotron Radiat.*, 12, 537-541.

S. Lee, et al. (2010). Characterization of Multi-walled Carbon Nanotubes Catalyst Supports by Point of Zero Charge, *Catal. Today*, doi:10.1016/j.cattod.2010.10.031

Suzuki, S.; Watanabe, Y.; Ogino, T.; Heun, S.; Gregoratti, L.; Barinov, A.; Kaulich, B.; Kiskinova, M.; Zhu, W.; Bower, C.; Zhou, O. (2002). Electronic Structure of Carbon Nanotubes Studied by Photoelectron Spectromicroscopy. *Phys. Rev. B*, 66, 035414-1-035414-4.

Wang, Y.; Malhotra, S. V.; Owens, F. J.; Iqbal, Z. (2005). Electrochemical Nitration of Single-wall Carbon Nanotubes. *Chem. Phys. Lett.*, 407, 68-72.

Xing, Y. (2004). Synthesis and Electrochemical Characterization of Uniformly-Dispersed High Loading Pt Nanoparticles on Sonochemically-Treated Carbon Nanotubes. *J. Phys. Chem. B*, 108, 19255-19259.

Xing, Y.; Li, L.; Chusuei, C. C.; Hull, R. V. (2005). Sonochemical Oxidation of Multiwalled Carbon Nanotubes. *Langmuir*, 21, 4185-4190.

Zhang, J.; Hongling, Z.; Qing, Q.; Yang, Y.; Li, Q.; Liu, Z.; Guo, X.; Du, Z. (2003). Effect of Chemical Oxidation on the Structure of Single-Walled Carbon Nanotubes. *J. Phys. Chem. B*, 107, 3712-3718.

Zhang, Y.; Toebes, M. L.; van der Eerden, A.; O'Grady, W. E.; de Jong, K. P.; Koningsberger, D. C. (2004). Formation, Characterization, and Magnetic Properties of $Fe_3O_4$ Nanowires Encapsulated in Carbon Microtubes. *J. Phys. Chem. B*, 108, 18509-18519.

Zorbas, V.; Smith, A. L.; Xie, H.; Ortiz-Acevedo, A.; Dalton, A. B.; Dieckmann, G.; Draper, R. K.; Baughman, R. H.; Musselman, I. H. (2005). Importance of Aromatic Content for Peptide/Single-walled Carbon Nanotube Interactions. *J. Am. Chem. Soc.*, 127, 12323-12328.

5

# Selective Separation of Single-Walled Carbon Nanotubes in Solution

Hongbo Li and Qingwen Li
*Suzhou Institute of Nano-Tech and Nano-Bionics, CAS*
*P. R. China*

## 1. Introduction

SWCNT can be conceptualized by wrapping a one-atomic-layer thick graphene into a hollow cylinder. As shown in Figure 1, the wrapping way is represented by the chiral vector (n, m), which denote the number of unit vectors along two directions in the crystal lattice of graphene sheet.[1] Because of the symmetry and unique electronic structure of graphene, the structures of SWCNTs strongly affect their electrical properties. In particular, their band gaps can vary from zero to about 2 eV and electrical conductivity can be in a range of a metal or semiconductor. For a given (n,m) nanotube, when n = m, the nanotube is metallic; when n − m is a multiple of 3, the SWCNTs are semiconducting with geometry-dependent band gaps.

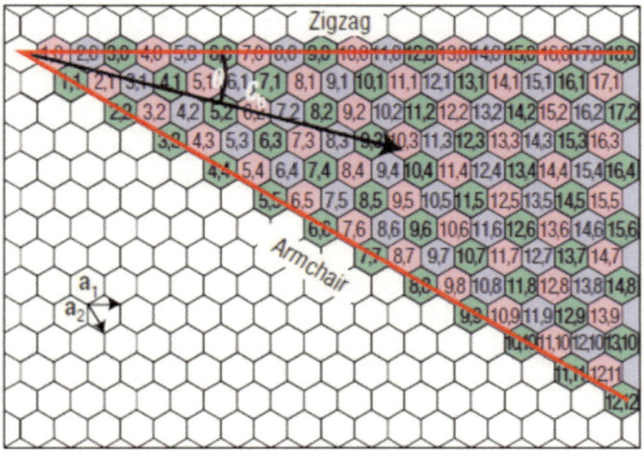


Fig. 1. A SWCNT is formed by rolling up a graphene along a chiral vector. The circumference of the SWCNT is determined by a chiral vector $Ch = na_1 + ma_2$, where (n,m) are integers known as the chiral indices and $a_1$ and $a_2$ are the unit vectors of the graphene lattice. Reproduced from ref.1.

Metallic SWCNTs (m-SWCNTs) may show ballistic behaviors and are ideal conducting connectors and electrodes for electronic devices due to their excellent electron transport behavior, optical transparency and flexibility.[2-5] Semiconducting SWCNTs (s-SWCNTs), with mobilities as high as 79 000 $cm^2$ $V^{-1}$ $s^{-1}$,[6] are considered to be among the most promising candidates for next-generation high-performance field-effect transistors (FETs). The unique electronic properties of SWCNTs make them the ideal building blocks for nanoelectronics and thin film devices.[7, 8] Unfortunately, the established CNT synthethetic methods often lead to the growth of the mixture of nanotubes, typically with 1/3 m-tubes and 2/3 s-tubes, hard to selectively harvest the CNTs with a specific electronic type (m- / s-) or with a desired chirality, which has greatly hindered their widespread applications. Therefore, the mass m/s separation of SWCNTs is of great significance for their advanced applications in high-performance nanoelectronic devices. Chiral separation is also required for the uniform property of electronic devices. Since none of current synthetic techniques produce identical populations of SWCNTs, over the past two decades, many post-synthetic approaches have therefore been developed in an attempt to separate monodisperse SWCNTs into different fractions by their electronic types, diameters and chiralities. This chapter provides a brief introduction of the recent progress in the separation of SWCNT, in particular with an emphasis on the chromatographic approach.

## 2. Several important post-synthetic separation approaches

One of the earliest developed enrichment techniques of SWCNTs is based on the selective removal strategy with bulk SWCNT powders and films. The enrichment of m- or s- type SWCNTs can be obtained by removing the other type (usually m-SWCNTs) through electrical breakdown[9], plasma[10, 11], irradiation[12, 13] effects or chemical reagents[11, 14-17]. These "dry" physical or chemical removal techniques are scalable and compatible with existing semiconductor processing for SWCNT-based nano-devices. However, as the selective removal is an irreversible process, the removed fraction cannot be used any more. Hence in recent years, separation of monodisperse SWCNTs, such as selective dispersion, dielectrophoresis, centrifugation, and chromatography has been greatly developed. SWCNTs can be thus sorted by electronic types and chiralities to some extent. Great efforts are still required to further understand the separation mechanism and therefore able to optimize the SWCNT sorting at lower cost and higher purity.

### 2.1 Selective functionalization

Both theoretical calculations and experimental studies have shown that the electronic properties, diameters and chiralities of SWCNTs may affect their interaction with some compounds, containing small organic molecules, surfactants, DNA and conjugated polymers. Such tan interaction in return leads to the differentiation and sorting of SWCNTs of different types.

Metallic SWCNTs are usually believed to have a higher reactivity than semiconducting SWCNTs due to the higher electron density of states (DOS) at their Fermi levels[14] and smaller ionization potential[18] . By covalent reaction on the sidewalls of SWCNTs, one type of SWCNTs (generally m-SWCNTs) tends to be functionalized first, leading to its selective dispersion in a solution and leaving the other one insoluble in the pristine form. As illustrated in Figure 2, Strano et al. first reported that diazonium reagents preferred to react with m-fractions rather than s-fraction, making s-SWCNTs well suspended in aqueous solution, and achieving SWCNT manipulation under controlled conditions.[14] Selectivity is dictated by the availability of electrons near the Fermi level to stabilize a charge-transfer transition state preceding bond formation. Later Doyle et al. found that several 4-substituted benzenediazonium salts, Ar-R (Ar = $N_2$+-$C_6H_4$ and R = Cl, $NO_2$, OMe) at pH 10 showed the highest reactivities for s-SWCNTs with the largest band gaps.[19] Additionally, nitronium ions[20, 21] and osmium tetroxide ($OsO_4$)[21] were also found to react preferentially with m-SWCNTs due to their larger electron density near the Fermi level. However, azomethine ylides derived from trialkylamine-N-oxides exhibited selective reactions with s-SWCNTs by cycloaddition, which was achieved by preorganizing the starting N-oxides on the nanotube surface prior to generating the reactive ylide.[22] Separation of m-SWCNTs from functionalized s-SWCNTs was also successfully accomplished by inducing solubilization of s-SWCNTs in the presence of lignoceric acid. Although covalent functionalization seems an easy approach for the separation of m- or s-SWCNTs, it remains a challenge as a result of uncontrollable reaction selectivity and possible structural damages on CNTs themselves.

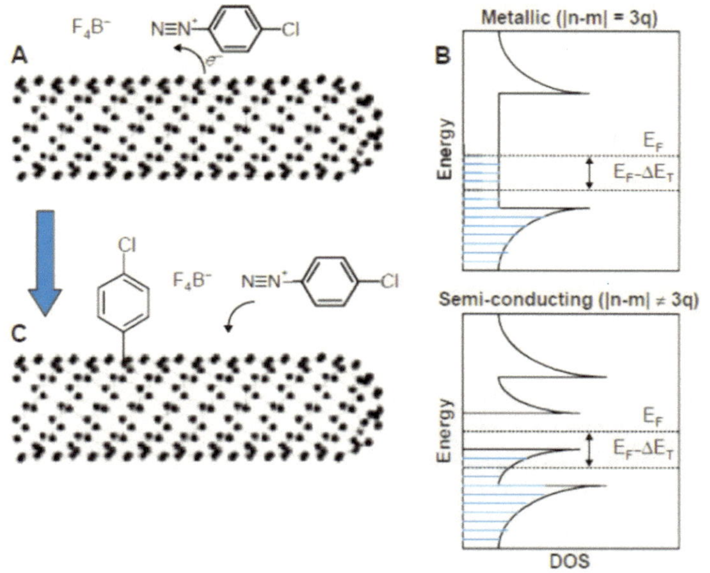


Fig. 2. (A) Diazonium reagents extract electrons, forming $N_2$ gas and leaving a stable C−C covalent bond at the nanotube surface. (B) The extent of electron transfer is dependent on the density of states in that electron density near $E_F$ leads to higher initial activity for metallic and semimetallic nanotubes. (C) A functionalized nanotube may exist as a delocalized radical cation, which tends to receive electrons from neighboring nanotubes or react with fluoride or diazonium salts. Reproduced from ref.14.

In comparison with covalent functionality, non-covalent functionality is a more popular separation method, as it is a non-destructive process. By interacting with the hydrophobic surface of SWCNTs or matching their chiral structures, some organic molecules may selectively adsorb onto or wrap along SWCNT species. In 2003, Chattopadhyay and coworkers reported that octadecylamine (ODA) had a higher affinity to carboxyl functionlized s-SWCNTs than m-fractions. S-SWCNTs were thus retained in the supernatant due to the increased solubility by stronger ODA adsorption, whereas m-SWCNTs were selectively precipitated.[23] As such, through the selective interaction of amine with s-SWCNTs, a self-sorted s-SWCNT network could be fabricated by spin-coating SWCNT solution on amine-functionalized surfaces,[24] where the chirality separation of nanotubes and simultaneous control of density and alignment may be achieved in one step. The field-effect transistors fabricated based on this method showed an on/off ratio as high as 900,000.

Interestingly, in the case of neutral SWCNTs without carboxyl group, both theoretical calculations and experiments have shown that the amine groups prefer to selectively interact with m-SWCNTs. Maeda et al. showed that m-SWCNTs could be highly concentrated to 87% by applying a dispersion-centrifugation process in a tetrahydrofuran solution of propylamine.[25] Similar results were also obtained in the m/s separation of SWCNTs by using bromine[26] and porphyrins[27]. The porphyrin and its derivates tend to attach onto the sidewalls of s-SWCNTs, making them enriched in the supernatant. While due to the formation of charge-transfer complex between bromine and m-SWCNTs, m-SWCNTs can be thus effectively sorted out by centrifugation.

In addition to m/s recognition, control on the structures of dispersants may also lead to the selective separation of SWCNTs by diameter and chirality. Ortiz-Acevedo et al. proposed a novel approach to coat SWCNTs with reversible cyclic peptides (RCPs) that covalently wrap around CNTs through the oxidation of thiols incorporated into the peptide backbone.[28] By controlling the length of the RCPs, they demonstrated limited diameter-selective solubilizations of SWCNTs. Some aromatic polymers were also found to selectively solubilize certain nanotube species. As illustrated in Figure 3, Nish et al. reported that the polymer poly(9,9-dioctylfluorenyl-2,7-diyl) was very selective to nanotubes.[29] They suggested that the SWCNT–polymer bonding was strongly influenced by both the relative orientation of the polymer chain to the nanotube structure, and the possible charge transfer that occur from metallic tubes and lead to changes in the conformation of the polymer.

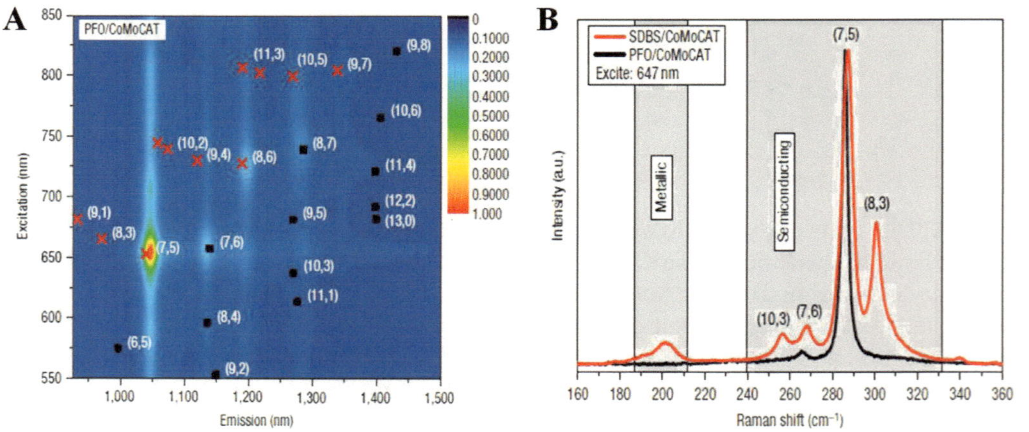


Fig. 3. (A) The PLE map shows the strength of the PL emission as a false color plot, with the different species labeled by their (n,m) indices. The emission is dominated by a very prominent (7,5) peak. (B) A comparison between the Raman spectra, taken using a 647 nm light source, of surfactant-wrapped SWCNTs (red line) and PFO wrapped SWCNTs (black line), normalized with respect to the semiconducting (7,5) nanotube. Reproduced from ref.29.

Small organic moluecules are also designed for the diameter- selective and chiral separation. Tromp et al. used oligo-acene adducts as a diameter-selective molecular anchor for separation and functionalization of SWCNTs.[30] SWCNT field effect transistors fabricated from diameter-sorted SWCNTs showed remarkably improved electrical properties compared to non-sorted SWCNTs. More recently, Wang et al. reported that the enantiomers of SWCNTs can be separated based on molecular recognition with chiral diporphyrin nanotweezers.[31] The chiral nanotweezers consisting of phenanthrene spacer and two chiral porphyrins may discriminate the diameter and handedness of CNTs simultaneously by taking into account the relationship between the (n, m) selectivity and the structures of previously reported chiral nanotweezers. Owing to the relatively narrow cleft made by two porphyrins, the nanotweezers showed high selectivity toward (6,5)-SWCNTs possessing the smallest diameter among the major components of SWCNTs grown from CoMoCAT.. In addition, the single enantiomer of (6,5)-SWCNTs could be enriched through the molecular recognition with 1. These results imply that it is crucial for the selective functionalization of SWCNTs to generate a broad variety of molecular anchors and functional backbones with excellent diameter selectivity.

### 2.2 Dielectrophoresis

Due to different dielectric constants of m- and s-SWCNTs under an alternating current (a.c.) electric field, SWCNTs can be sorted by their electronic types via dielectrophoresis. Early in 2003, Krupke et al. developed an alternating current dielectrophoretic method to separate m-SWCNTs from s-SWCNTs in suspension.[32] As shown in Figure 4A, when SWCNT suspension was placed on an array of microelectrodes and applied with an alternating current, an opposite movement of m- and s-tubes occurred along the electric field gradient. M-SWCNTs were attracted toward the microelectrode array, leaving s-SWCNTs in the solvent. Principlly, a complete separation between metallic and semiconducting SWCNTs may be possible if all tubes in the suspension are dispersed as individual tubes (no bundles). However, this method was only available for the sorting of m- and s-SWCNTs on a small scale.

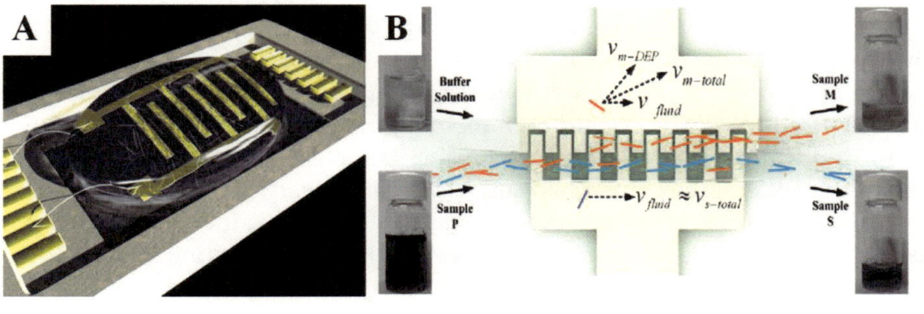


Fig. 4. (A) Schematic of dielectrophoresis of SWCNT suspension by using a microelectrode array. Reproduced from ref.32. (B) The continuous separation mechanism based on dielectrophoresis by using an H-shaped microfluidic channel. Reproduced from ref.33.

Later, Shin et al. developed this technique and established a nondestructive, scalable method for SWCNT separation by using an H-shaped microfluidic channel, where highly pure m-SWCNTs were continuously extracted from a suspension, as shown in Figure 4B.[33]

Two laminar streams were generated in an H-shaped microfluidic channel with two inlets and two outlets. The flow conditions were carefully controlled to minimize diffusive and convective transport across the boundary between the two flows. Dielectrophoretic force from the embedded electrode at the junction extracted m-SWCNTs from a stream of nanotube suspension toward the other stream of buffer solution without nanotubes. Thus, the highly pure m-SWCNTs were obtained simultaneously at separate outlets. However, the separation efficiency and purity of s-SWCNTs is still difficult to be enhanced. Moreover, this method is only compatible with the m/s separation, ineffective for the chiral separation.

### 2.3 Centrifugation

Centrifugation is a process widely used in industry and in laboratory settings for the separation of mixtures driven by the centrifugal force. More-dense components in the mixture migrate away from the axis of the centrifuge, while less-dense components in the mixture migrate towards the axis. Increasing the effective gravitational force on a test tube, more dense components will be precipitated on the bottom of the tube. The less dense components are collected in the supernatant solution. SWCNTs are usually synthesized in a highly mixed status with a broad distribution of lengths, diameters and chiralities, leading to the sorting a tough issue. However, centrifugation is found effective for the separation of SWCNTs. To date with this simple technique, the SWCNTs can be sorted by the m/s, diameter and chirality.

#### 2.3.1 Density gradient ultracentrifugation

Density gradient ultracentrifugation (DGU) is a technique commonly utilized to separate and isolate different sub-cellular components, DNA from RNA, and even different sequences of DNA by their compositions. Arnold et al. firstly described the sorting of DNA-wrapped SWCNTs by DGU,[34] this technique has been well developed for SWCNT sorting with multiple purposes, such as separation by m/s, diameter and chirality, even the separation of the mirror-image isomers of seven (n,m) species.[35-41] In this process, SWCNT suspension is added to centrifuge tubes containing liquid mixtures arranged to form a spatially varying density profile. Under strong centrifugation, SWCNT species are separated by migrating to regions matching their individual densities. Hence, the sorting efficiency is greatly dependent on the dispersants and DGU processing.

Salts like sodium cholate (SC) or cosurfactants are often added to optimize the fine sorting of monodispersed SWCNTs. Arnold et al reported in 2006 that the sorting of SWCNTs by diameter, bandgap and electronic type using structure-discriminating surfactants to engineer subtle differences in their buoyant densities.[35] Density gradients were formed from aqueous solutions of a non-ionic density gradient medium, iodixanol. Gradients were created directly in centrifuge tubes by one of two methods: by layering of discrete steps and subsequent diffusion into linear gradients or by using a linear gradient maker. As shown in Figure 5, they have isolated the SWCNTs of narrow diameter distributions, more than 97% of which are within a 0.02-nm-diameter range. Furthermore, by using cosurfactants, they obtained bulk quantities of SWCNTs of predominantly a single electronic type. By a hydrodynamic model, Nair et al. described the motion of surfactant-suspended single-walled carbon nanotubes in a density gradient.[42] The theoretical results predicted that the number of surfactant molecules adsorbed on each nanotube determined its effective density and, hence, its position in the gradient after centrifugation has been completed.

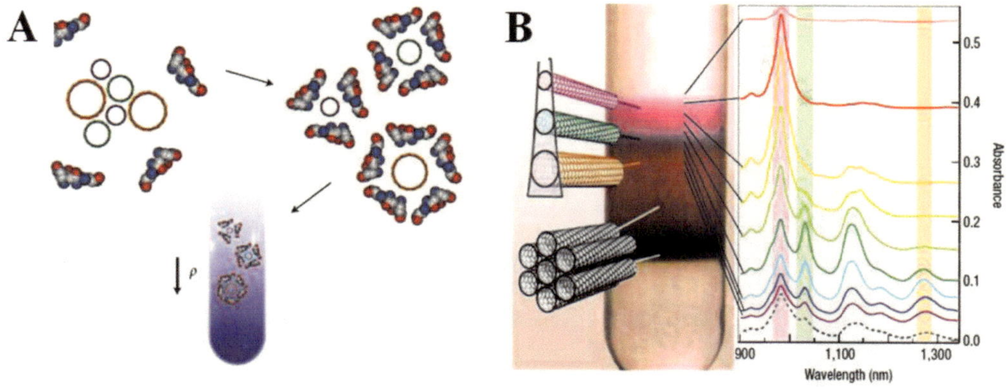


Fig. 5. Sorting of SWCNTs by electronic type, diameter and bandgap using DGU. (A) Schematic of surfactant encapsulation and sorting; (B) Photographs and optical absorbance (1 cm path length) spectra after separation using DGU. Visually, the separation is made evident by the formation of coloured bands of SC encapsulated, CoMoCAT-grown SWCNTs (7–11 A°) sorted by diameter and bandgap. Bundles, aggregates and insoluble material sediment to lower in the gradient. The spectra indicate SWCNTs of increasing diameter are more concentrated at larger densities. Three diameter ranges of semiconducting SWCNTs are maximized in the third, sixth and seventh fractions (highlighted by the pink, green and light brown bands). These have chiralities of (6,5), (7,5) and (9,5)/(8,7), and diameters of 7.6, 8.3 and 9.8/10.3 A° respectively. Reproduced from ref.35.

In 2008, Yanagi et al. achieved the separation of m- and s-SWCNTs using sucrose as a gradient medium in sucrose-DGU.[36] By lowering the temperature during sucrose-DGU and tuning the concentrations of the surfactants, m- and s-SWCNT samples were obtained with high purity, estimated to be 69% and 95%, respectively, from their optical absorption spectra. They pointed out that the temperature during centrifugation was also an important parameter that improved the m/s separation capability. Recently Antaris et al. found that nonionic, biocompatible block copolymers were useful to isolate m- and s-SWCNTs using DGU.[40] Separations conducted with different Pluronic block copolymers revealed that Pluronics with shorter hydrophobic chain lengths led to the purity levels for s-fraction sorting higher than 99% when Pluronic F68 was used. In contrast, X-shaped Tetronic block copolymers showed a specific affinity to m-SWCNTs, yielding metallic purity levels of 74% for Tetronic 1107.

In addition to m/s separation, by further tuning the surfactant component, enhanced diameter-dependent and chiral sorting of SWCNTs can be achieved by DGU. Zhao et al. used sodium deoxycholate (DOC) and sodium dodecyl sulfate (SDS) as cosurfactant encapsulating agents to form a DOC-restricted SDS wrapping morphology around the SWCNTs and thus 97% pure isolation of (6,5) SWCNTs was achieved.[37] Interestingly, via optimizing surfactant structure or DGU processing, enantiomer separation of SWCNTs is obtained. Green et al. reported that by using chiral surfactant, such as sodium cholate, left- and right-handed SWCNTs can be discriminated.[38] This sorting strategy can be employed for simultaneous enrichment of SWCNTs by handedness and roll-up vector having diameters ranging from 0.7 to 1.5 nm.

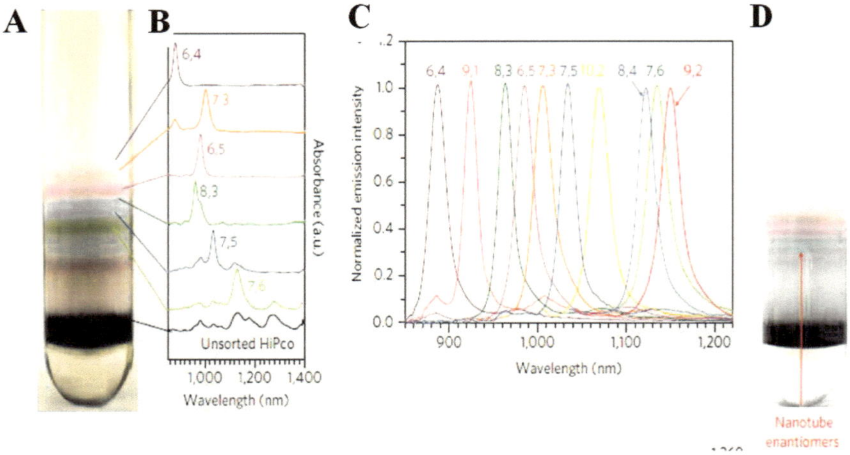


Fig. 6. Sorting of HiPco SWCNTs by (n,m) structure using single-step nonlinear DGU. (A) Image of a centrifuge tube containing HiPco SWCNTs sorted by one 18-h nonlinear DGU run at 268,000g (max). (B) Near-infrared absorption spectra of the marked colored layers. (C) Photoluminescence spectra of 10 separated fractions excited at the $E_{22}$ peak of each main (n,m) component. (D) Image of a centrifuge tube showing resolved pairs of enantiomer bands sorted from HiPco SWCNTs by nonlinear DGU with sodium cholate. Reproduced from ref.41.

However, simultaneously sorting many SWCNTs species in a single step still is a bottleneck. Recently, Ghosh et al. made a breakthrough for SWCNT sorting.[41] They showed that highly polydisperse HiPco SWCNTs were readily sorted in a single step to enable fractions enriched in any of ten different (n,m) species by introducing nonlinear density gradients. Furthermore, minor variants of the method allowed separation of the mirror-image isomers of seven (n,m) species. They prepared centrifuge tubes with nonlinear, S-shaped gradients designed to have very small variations of density with depth at densities typical of suspended nanotubes. A photograph of such a tube after centrifugation shows extensive color banding (Figure 6A). Absorption spectra of the separated fractions (Figure 6B) clearly indicate that different (n,m) species have been sorted into distinct layers. From top to bottom, the bands are identified as (6,4), (7,3), (6,5), (8,3), (7,5) and (7,6). The (6,5) band is purple in color; (8,3) and (7,6) are green; and (7,5) is blue. Further evidence of effective sorting is displayed in Figure 6C, which shows normalized emission spectra of ten fractions separated from DGU-processed samples containing single or co-surfactants, each excited at the $E_{22}$ peak of its dominant species. The enriched species are (6,4), (7,3), (6,5), (9,1), (8,3), (9,2), (7,5), (8,4), (10,2) and (7,6). By adding a co-surfactant to sodium cholate in their nonlinear DGU method, they achieved effective and highly reproducible single-step enantiomeric sorting of several (n,m) species in HiPco samples. Further they found optimal enantiomer separation with a surfactant mixture of 0.7% sodium cholate plus 0.175% sodiumdodecyl sulphate in a slight variant of the nonlinear DGU gradient described earlier. The enantiomers of (6,5), (8,3), (8,4), (6,3) and (6,4) SWCNTs were able to be separated by this technique. This novel approach promised a scalable, relatively simple, and refined separation method of SWCNTs.

### 2.3.2 Normal speed centrifugation

In comparison with DGU usually set the speed at about 2x10$^5$ g, normal centrifugation is found also available for the sorting of SWCNTs at low speed of about 2x10$^4$ g. More recently, Tanaka et al. reported a rapid and scalable method for the separation of metallic and semiconducting SWCNTs by normal speed centrifugation.[43] When SDS dispersed SWCNT suspension was mixed with liquid agarose gel and then centrifugated, it was found that m-fraction was favorably enriched in supernatant, and s-fractions were retained in the precipitated gel. Such separation effect resulted from the selective interaction of s-SWCNTs with agarose gel. Upon centrifugation, s-SWCNTs were selectively trapped in the gel, whereas metallic nanotubes remained in the free state with SDS micelles in the solution as illustrated in Figure 7A. The effective separation is greatly dependent on the dispersant and gel. It is also found that the purity of the m- and s-SWCNTs obtained by centrifugation can be improved by optimizing the gel concentration and composition of agarose.

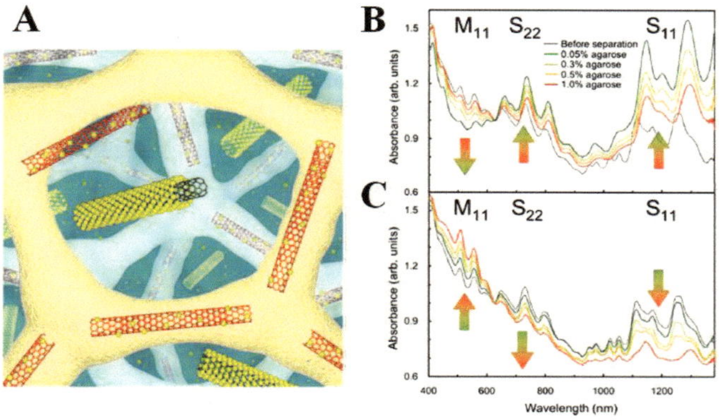


Fig. 7. (A) Model of MS separation using agarose gel. Red, s-SWCNTs; beige, agarose gel matrix; green, m-SWCNTs; yellow, SDS. Spectral results of (B) Gel fraction and (C) Solution fraction separated by centrifugation method at various concentrations of agarose (0.05-1.0%). Reproduced from ref.43.

The purity of s-SWCNTs in the compressed gel could be increased by decreasing agarose concentration in the starting gel (Figure 7B), while the purity of the metallic tubes collected in the supernatant tended to increase with agarose concentration up to approximately 1.0% (Figure 7C), beyond which the metallic nanotube purity increased only slightly. When the separation was repeated for the solution fraction concentrated m-SWCNTs, m-SWCNTs could be further enriched. Due to the selective entrapment of s-SWCNTs in agarose gel, the effective m/s separation could be also realized by simple frozen, thawed, and squeezed procedures. Although this method was reported to be readily scalable and not restricted by equipment limitations, the melting and mixing of excessive agarose with SWCNT dispersion made the sorted SWCNTs containing agraose impurities, which were hard to be removed.

### 2.4 Chromatography

Chromatography is a popular method used for purification of individual chemical or biological compounds from their mixtures. Different compounds exhibit different physicochemical properties, leading them to behave diversely between mobile and stationary phases. Based on the similar properties of SWCNTs to biological macromolecules in sizes and surface properties, earlier from 1998, researchers have made efforts to separated SWCNTs by the chromatography. Due to the diversity of the stationary and mobile phases, SWCNTs dispersed in solution can be sorted following different separation mechanisms, such as size-exclusion chromatography (SEC)[44-50], anion exchange chromatography (IEC)[8, 51-54] and electrokinetic chromatography[46, 50, 55-57]. By tuning the stationary phase and eluents, the dispersed SWCNTs have been successively sorted by length, m/s, diameter and chirality.

### 2.4.1 Size-exclusion chromatography

Among numerous chromatographic methods, gel filtration chromatography, or gel permeation chromatography is widely applied in the efficient and low-cost separation of biological macromolecules.[58] Its separation is based on differences in the sizes or weights of the analytes, which govern their access to the pore beads packed in a column. In general, the smaller analytes can enter the pores more easily and therefore spend more time in these pores, increasing their retention time. Conversely, larger analytes spend little if any time in the pores and are eluted quickly. It can be thus inferred that the pore sizes of gel beads for the column packing play a critical role in the separation of an analyte with a desired range of molecular weights. Since the lengths of SWCNTs in suspensions prepared by ultrasonication are in a wide range from 50 nm to 1000 nm, SWCNTs are separated according to their lengths by the size-exclusion chromatography. Many porous packing media as the stationary phases have been used.

In 1998, Duesberg et al. reported that carbon nanospheres, metal particles, and amorphous carbon could be efficiently removed by size exclusion chromatography when applied to surfactant stabilized dispersions of SWCNT raw material.[44] In addition, length separation of the tubes was achieved. 1wt% sodium dodecylsulfate (SDS) solution was used to disperse and stabilize SWCNTs. Controlled-pore glass (CPG) with an average pore size of 300 nm (CPG 3000Å, Fluka) was packed in the column. Different fractions of SWCNTs were eluted out sequentially with 0.25wt% SDS aqueous solution. Later, Farkas et al. accordingly undertook to length sort of cut SWCNTs by size exclusion chromatography (SEC) using a HPLC system,[45] promising that efficient length separation with good resolution is feasible on a preparative scale. Further by using three silica-based column resins in series with pore sizes of 2000, 1000, and 300 Å, Huang et al. demonstrated that DNA dispersed SWCNTs with very narrow length distribution could be sorted out.[47] The atomic force microscopy revealed that the average length decreased monotonically from >500 nm in the early fractions to <100 nm in the late fractions, with length variation ≤ 10% in each of the measured fractions.

Polysaccharide-based porous beads were also applied for the separation of SWCNTs. Heller and Arnold et al. reported that by the gels of sephacryl S-500, the concomitant length and diameter separation of SWCNTs were achieved.[46] As shown in Figure 8, separation by diameter was concomitant with length fractionation, and nanotubes that were cut shortest also possessed the greatest relative enrichments of large-diameter species. They demonstrated that the longer sonication time led to an increase in the electrophoretic mobility of CNTs in the gels and thus determined the degree of both length and diameter separation of the nanotubes.

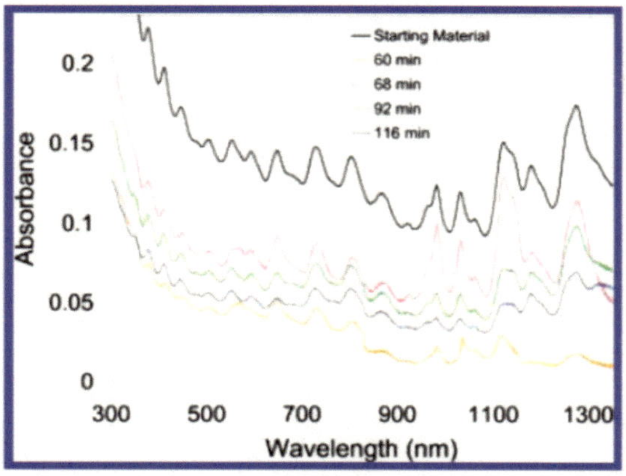


Fig. 8. Absorption spectra of selected fractions of nanotubes probe-tip sonicated for 3 h and separated by a size exclusion column. The spectra show changes in the concentrations of the nanotube species with respect to elution time. Reproduced from ref.46.

The stationary phase played an important role in the chromatographic separation of carbon nanotubes, Moshammer et al first demonstrated that by size exclusion chromatography of Sephacryl S-200 gel, SDS-dispersed SWCNTs could be fractionated according to electronic structure type.[50] More recently, Liu et al achieved the m/s separation of SWCNTs by using agarose derived filling gel.[59] Sepharose 2B gel (a bead-formed cross-linked agarose gel matrix, GE Healthcare, bead size range 60-200 $\mu$m) was used. When SWCNTs dispersed in SDS solution were applied to the top of the gel column, from Figure 9A, m- and s-SWCNTs could be sorted with a two-step elution using SDS and sodium deoxycholate (DOC) solution, respectively. Importantly, by the successive addition of DOC solutions with concentrations ranging from 0.05 to 2 wt % and fractional collection at each concentration,

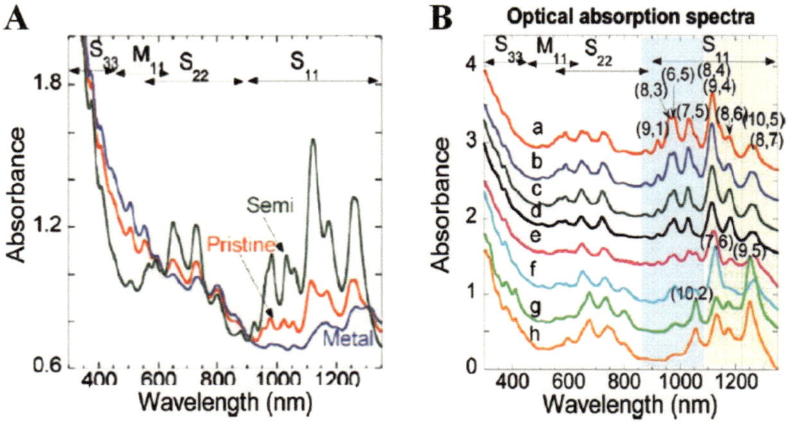


Fig. 9. (A) Optical absorption spectra (normalized at 620 nm). (B) Optical absorption spectra of the selectively enriched s-SWCNTs fractions by fractional collection with DOC eluants of different concentrations: (a) DOC 0.05 wt %, fraction 1, (b) DOC 0.05 wt %, fraction 3, (c) DOC 0.05 wt %, fraction 6, (d) DOC 0.05 wt %, fraction 9, (e) DOC 0.1 wt %, (f) DOC 0.25 wt %, (g) DOC 0.5 wt %, and (h) DOC 2 wt %. Reproduced from ref.59.

they found that smaller-diameter enriched s-SWCNTs were eluted first with the DOC solution at lower concentration and the larger-diameter enriched s-SWCNTs preferred to be eluted at higher DOC concentrations, as displayed in Figure 9B. Thus, diameter-selective enrichment of semiconducting fraction was achieved. These results indicate that agarose gel is effective for simultaneous sorting of CNTs by their electronic types and diameters, which predicts the SWCNT separation in a simple, low-cost and scalable way.

### 2.4.2 Ion-exchange chromatography

Ion exchange chromatography is a process that allows the separation of ions and polar molecules based on their charge. It can be used for almost any kind of charged molecule including large proteins, small nucleotides and amino acids. The surface of stationary phase displays ionic functional groups that interact with analyte ions of opposite charge. In comparison with size-exclusion chromatography, by anion exchange chromatography, more refined separation of SWCNTs can be achieved, not only by m/s, and diameter, but also by chirality. Earlier in 2003, Ming Zheng's group reported that bundled SWCNTs were effectively dispersed in water by their sonication in the presence of single-stranded DNA (ssDNA) and demonstrated that DNA-coated carbon nanotubes could be fractionized with different electronic structures by ion-exchange chromatography, as shown in Figure 10A and B.[51] A strong anion-exchange column HQ20 (Applied Biosystems) functionalized by quarterized polyethyleneimine was chosen, which is expected to bind to the negatively charged phosphate groups of DNA. A linear salt gradient (0 to 0.9 M NaSCN in 20 mM MES buffer at pH 7) at a flow rate of 2 ml/min was used to fractionalize SWCNTs. Further by a systematic search of the ssDNA sequence with d(GT)n, n = 10 to 45, they found that wrapping of carbon nanotubes (CNTs) by ssDNA was sequence-dependent.[52] The electrostatics of the DNA-CNT hybrid depends on tube diameter and electronic properties, enabling nanotube separation by anion exchange chromatography. Optical absorption and Raman spectroscopy showed that the early fractions were enriched in the smaller diameter and metallic tubes, whereas the fractions collected later were enriched in the larger diameter and semiconducting tubes.

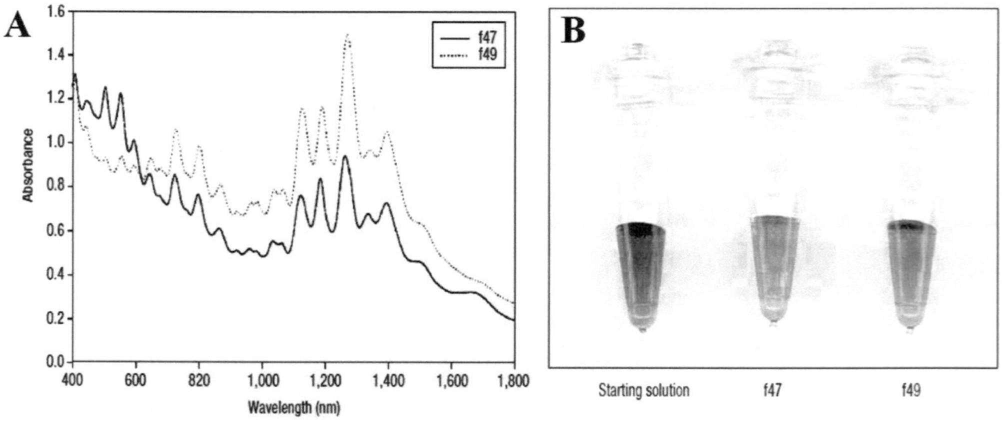


Fig. 10. Separation of DNA-CNT by anion exchange chromatography. (A) Electronic absorption spectra of two fractions f47 and f49. (B) Visual comparison of DNA-CNT solutions of the starting material, f47 and f49. Reproduced from ref.51.

However the separation resolution by this approach was not perfect. They proposed that a major issue was probably related to the broad distribution of tube length, since the dispersed CNTs were randomly cut during the long-term ultrasonic process, resulting in the tubes ranging from 50 to 1000 nm in length. Hence, in 2007, they combined the SEC with IEC to improve the separation resolution of small diameter SWCNTs.[53] They first narrowed the length distribution of SWCNTs by conducting SEC separation and then performed a chiral separation by the IEC. (9,1) tubes were separated from the same diameter but different chirality and much more abundant (6,5) species. Such exquisite separation was suggested as a result of chirality dependent interactions between DNA-wrapped SWCNTs and the IEX resin. These interactions could be electrostatic in nature, arising from chirality-dependent DNA-wrapping, and/or electrodynamic in nature, originating from chirality-dependent van der Waals forces.

Similar separation results were also reported by the group of Hongjie Dai.[8] With separated SWCNT fraction, they fabricated FET devices with s-tubes of small diameters, achieving the high on-/off-current ($I_{on}/I_{off}$) ratios up to 105 owing to s-SWCNTs with only a few (n,m) chiralities in the fraction. This was the first time that chemically separated SWCNTs were used for short channel, all-semiconducting SWCNT electronics dominant by just a few (n,m)s.

After single-chiral tube specie was separated by SEC-IEC technique, by designing an effective search of a DNA library of $10^{60}$ in size, as shown in Figure 11A, all 12 major single-chirality semiconducting species were separated from a synthetic mixture by Ming Zheng group.[54] They identified more than 20 short DNA sequences, each of which recognizes and enables chromatographic purification of a particular nanotube species from the mixture. Recognition sequences exhibit a periodic purine–pyrimidines pattern, which can undergo hydrogen-bonding to form a two-dimensional sheet, and fold selectively on nanotubes into a well-ordered three-dimensional barrel, as illustrated in Figure 11B-D. They proposed that the ordered two-dimensional sheet and three dimensional barrel provided the structural basis for the observed DNA recognition of SWCNTs.

It is clearly indicated that for the surface properties of dispersed SWCNTs, which may be tailored by dispersants, and the interaction of dispersed CNTs with stationary phase are both critical issues to determine the selectivity and efficiency of chromatographic methods for SWCNT sorting. Ss-DNA appears more selective than SDS for the recognition of SWCNTs of different structures; however, it's very costly and currently difficult to be applied for a large-scale separation.

### 2.4.3 Electrokinetic chromatography

Electrokinetic chromatography is a well-established chromatography technique performed under electric field, taking the advantages of electrically driven force and the tunable selectivity of stationary phase. Earlier in 2003, Doorn et al. found that capillary electrophoresis (CE) could be performed on polymer-stabilized bundles and SDS suspensions of HiPco SWCNTs.[55] They showed, for poly(vinylpyrrolidone) (PVP)-stabilized tube bundles, that separations resulted in the fractions containing bundles of different electronic properties, which were dependent on bundle sizes.. CE on SDS dispersed CNT suspensions separated their large aggregates from smaller bundles and produced a relatively pure fraction of individual isolated nanotubes. Isolation of the aggregates from individual nanotubes more likely is attributed to the differences in molecular weight or diameter, leading to different migration behaviors under electric field.

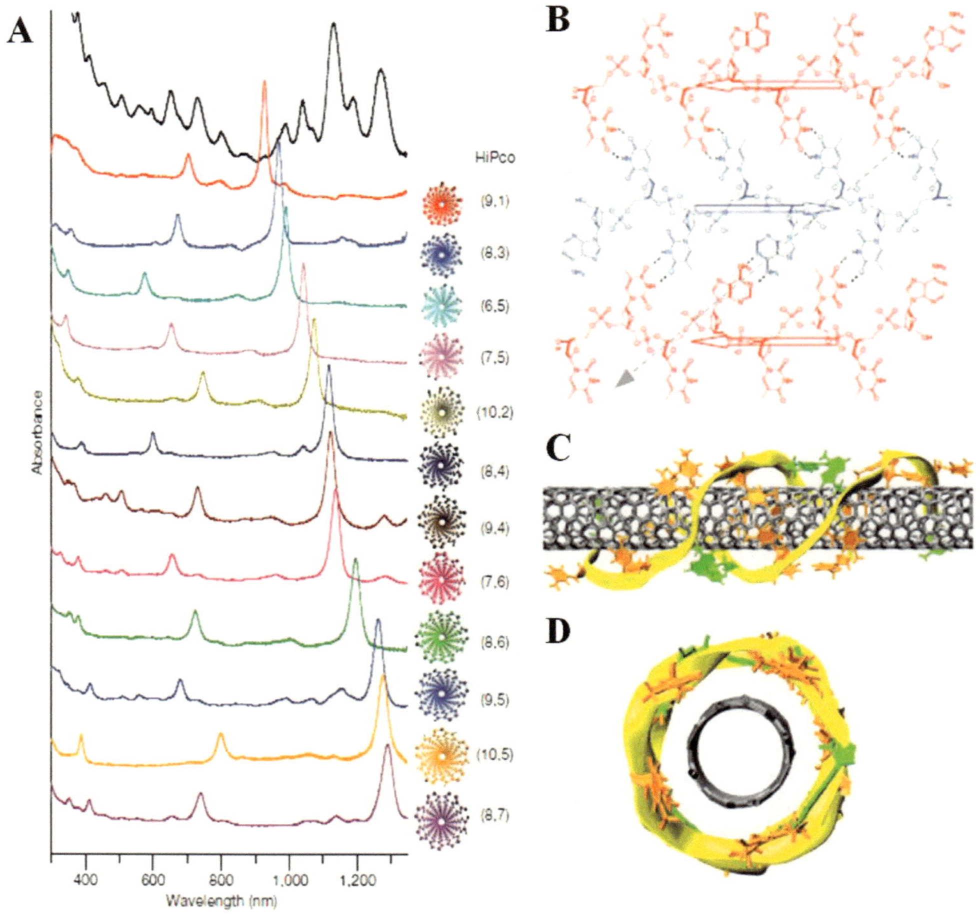


Fig. 11. (A) Optical absorption spectra and atomic structures. Ultraviolet–visible–near-infrared absorption spectra of 12 purified semiconducting SWCNTs (ranked according to the measured $E_{11}$ absorption wavelength) and the starting HiPco mixture. The structure of each purified SWCNT species (viewed along the tube axis) and its (n,m) notation are given at the right side of the corresponding spectrum. (B) A 2D DNA sheet structure formed by three anti-parallel ATTTATTT strands. The dotted lines between bases indicate hydrogen bonds. The open arrow in each strand denotes 5′ to 3′ direction. The dashed grey arrow (top right to bottom left) represents the roll-up vector along which the DNA barrel in c is formed. (D) A DNA barrel on a (8,4) nanotube formed by rolling up a 2D DNA sheet composed of two hydrogen-bonded anti-parallel ATTTATTTATTT strands. (D) The structure in c viewed along the tube axis. Color coding: orange, thymine; green, adenine; yellow ribbons, backbones. Reproduced from ref.54.

Agarose gel electrophoresis, where agarose gel is the stationary phase, was successfully applied to sort SWCNTs by length, diameter and electronic type. In 2004, Heller et al. performed the concomitant length and diameter separation of SWCNTs by the gel electrophoresis.[46] HiPco SWCNTs were suspended in sodium cholate hydrate.

Electrophoresis was performed in a 7 ×10 cm, 1% agarose gel in TAE buffer (trisacetate-EDTA) with 50 mM sodium cholate at 100 V. Nanotube fractions were removed from the gel via electroelution by creating a second set of eight wells in the gel 4.5 cm from the original 40 μL sample wells. Material was pipetted out of the second set of wells after 30 min of electrophoresis and repeatedly after an additional 5 min of applied potential to obtain six fractions. Highly resolved fractions of nanotubes with average lengths between 92 and 435 nm were sorted out. They demonstrated that nanotubes that have been cut shortest also possess the greatest relative enrichments of large-diameter species.

In 2008, by modulating the dispersant and electrophoresis condition, Tanaka et al. performed a better m/s separation of SWCNTs with agarose gel electrophoresis.[56] As illustrated in Figure 12, when the SWCNTs were isolated with SDS and embedded in agarose gel, only the m-SWCNTs were separated from the starting gel under an electric field. Such method is available for the sorting of other kinds of SWCNTs produced by arc and laser ablation methods. It affords a solution containing 70% pure m-SWCNTs and leaves a gel containing 95% pure s-SWCNTs. SDS and agarose played a synergic role for such separation effect.

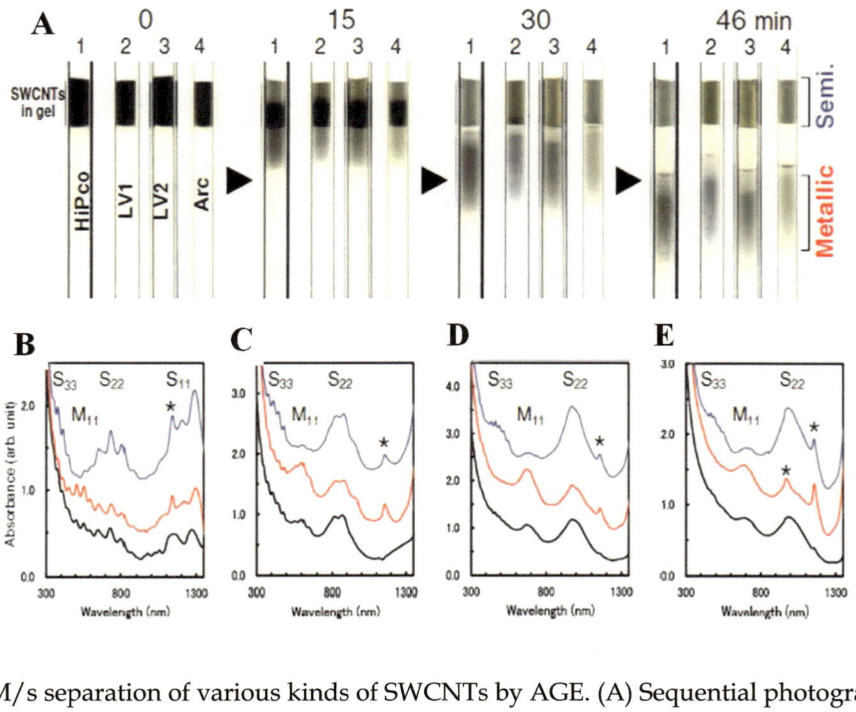


Fig. 12. M/s separation of various kinds of SWCNTs by AGE. (A) Sequential photographs showing the progress of separation. Tube 1, HiPco SWCNTs (1.0 ± 0.3 nm); tube 2, LV1 SWCNTs (1.2 ± 0.1 nm); tube 3, LV2 SWCNTs (1.4 ± 0.1 nm); and tube 4, Arc SWCNTs (1.4 ± 0.1 nm). Agarose gel concentrations of gel samples are 0.2% (HiPco, LV1, and LV2 SWCNTs) and 0.5% (Arc SWCNTs). (b–e) Absorption spectra of separated SWCNTs: (B) HiPco-, (C) LV1-, (D) LV2-, and (E) Arc-nanotubes. Blue and red spectra indicate semiconducting and metallic fractions, respectively. Results for the SWCNT dispersion before separation (black line). The peaks (970 and 1160 nm) derived from water are indicated by asterisks. Reproduced from ref.46.

Similar m/s separation results were also reported by Moshammer et al.[50] They proposed that in the SDS-dispersed "starting" suspensions, s-SWCNTs were primarily in the form of small bundles whereas m-SWCNTs were predominantly suspended as individual tubes. They thought that the selective dispersion ability of SDS led to the different mobilities of m- and s-SWCNTs during the gel electrophoresis and finally sorted SWCNTs into two fractions. However such an explanation seems kind of intuitive and the detailed mechanism is still not clear. Considering that SDS dispersed SWCNTs are charged and dominant in micelles, by employing thionine (TN) as a probing molecule and monitoring color changes and absorption bands of TN molecules, we roughly estimate the migration and adsorption properties of SDS micelles in the gel as well as on SWCNT surfaces during the electrophoresis.[57] Figure 13A illustrated the electrophoresis results of SDS gel and SDS-SWCNT gels with and without the TN probe, respectively. It was observed that although TN is a positively charged dye, when dissolved in 1% SDS solution, TNs interact with SDS micelles to form negatively charged SDS-TN micelles and migrate toward the anode under the electric field. The electrophoresis of SDS-SWCNT gel with TNs also resulted in the separation of SWCNTs into two fractions. The staying fraction in the initial gel showed a Spanish green color, pretty similar to that observed in the staying fraction of SDS-SWCNT gel without TNs. The moving fraction, however, showed two color regions, a light blue region (arrow direction), which migrated faster and

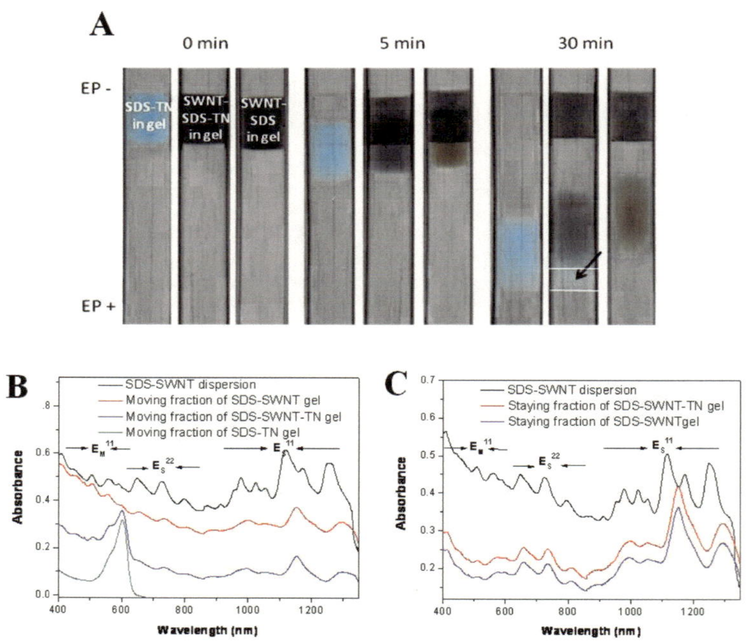


Fig. 13. (A) Sequential photographs of SDS-TN gel, SWCNT-SDS gel, and SWCNT-SDS-TN gel performed by agarose gel electrophoresis at 0, 5, and 30 min, respectively. (B) Optical absorption spectra of initial SDS-SWCNT dispersion and the moving fractions collected from the three kinds of gels after agarose gel electrophoresis shown a; (C) Optical absorption spectra of initial SDS-SWCNT dispersion as well as the staying fractions collected from SDS-SWCNT gels with and without TNs after agarose gel electrophoresis. Reproduced from ref.57

displayed color similar to the case of SDS-TN, and a gray-blue region (the region above arrow), different from the color obtained with SDS-SWCNTs without TN probe. By analyzing their corresponding absorption spectra (Figure 13B and C), we proposed that the unique interaction of s-SWCNTs with agarose and the exfoliation of some SDS molecules from SDS-s-SWCNT entities lead to the m/s separation of SWCNTs. Understanding the role of SDS in the separation may help us to further optimize the purification of each fraction and develop a more effective and low-cost separation strategy.

## 3. Applications of separated SWCNTs

Separated m- and s- SWCNTs offer many unique opportunities for a variety of technological applications.[60] Regarding metallic nanotubes, in principle, they may carry an electrical current density of $4\times10^9$ A/cm$^2$, which is more than 1000 times greater than the copper. Since in 1997 first electrical devices based on metallic SWCNTs were fabricated,[2, 5] their potential applications in nanocircuitry, conductive nanocomposites and transparent conductive films have been widely investigated. The electrical conductive performance in the films from the separated m-SWCNTs was consistently much better than that in the films from as-purified SWCNTs.[61-63] For examples, employing the DGU method, Hersam and co-workers enriched m-SWCNTs of different diameter ranges for transparent conductive films.[61] For enriched metallic HiPco SWCNTs, the resulting film exhibited a sheet resistance of ~231 Ω/square for 75% optical transmittance at 550 nm, in comparison with ~1340 Ω/square in the reference film of the same optical transmittance from nonseparated HiPco SWCNTs. The films of enriched m-SWCNTs from laser-ablation- and arc-discharge-produced nanotube samples generally exhibited better performance, with less than 140 Ω/square sheet resistance at optical transmittances of over 70% in the visible and near-IR spectral regions.

The separated m-SWCNTs can also enhance the transparent conductive performance in composite films with conductive polymers.[63, 64] Wang et al. prepared the composite films of enriched m-SWCNTs with poly(3,4-ethylenedioxythiophene)/poly-(styrenesulfonate) (PEDOT/PSS) in various compositions, by the spraying method.[63] The sheet resistance indicated that the composite films prepared from enriched m-SWCNTs were consistently and substantially better in device performance than those with nonseparated SWCNTs, which could be applied in electronic devices such as organic light-emitting diodes (OLEDs).

In comparison to m-SWCNTs at a similar energy level, s-SWCNTs have a large density of electronic states and present diameter-dependent transition bands. S-SWCNTs are capable of carrying a high current with an electron/hole mobility. Hence, s-SWCNTs are widely investigated due to the pursued applications in field-emission transistors (FETs).[9, 11, 15] To maximize the performance of FETs, it is desirable to obtain SWCNTs with similar diameter, chirality and thus band-gaps, and connect them in parallel to build each FET device for sufficient on-currents and reproducible device characteristics. The lengths of SWCNTs should also be controllable to meet the requirement of desired channel length. The well-refined separation of s-SWCNTs according to their length, diameter, and chirality are very significant for the application of SWCNTs in nanoelectronics.

S-SWCNTs separated from post-synthetic methods have already been used in FETs. For example, with a small-diameter s-SWCNT fraction separated by the SEC-IEC chromatography, Zheng et al. fabricated FET devices composed of separated s-SWCNTs in parallel, with high on-/off-current ($I_{on}/I_{off}$) ratios up to $10^5$ owing to s-SWCNTs with only a few (n,m) chiralities in the fraction.[8] In addition, nanoelectronic devices composed of single chirality-enriched (10,5) tubes were fabricated with $I_{on}/I_{off}$ ratios as high as $10^6$.[65] The performance of the FETs made with the slightly bigger (10,5) tubes was better than the previously reported results using (7,6) and (8,4) tubes, even larger single-chirality s-SWCNTs were preferred in order to further improve the device performance. Similarly, the s-SWCNTs from agarose gel-based separation were also used in thin-film FETs, with performance better than that in devices fabricated with non-separated SWCNTs. [43] Recently Lee et al. demonstrated that the CoMoCat SWCNTs treated with diazonium salts and purified by DGU could be used to fabricate solution-processable FET devices with a full semiconductor device yield.[66] By increasing the network thickness, the effective mobility of the devices could be raised to ~10 $cm^2 V^{-1}s^{-1}$ while keeping the on–off ratio higher than 5 000. The removal of impurities was found to be essential for achieving high-on–off-ratio devices. It was easier to achieve a full semiconductor device yield using the CoMoCat SWCNTs, which were very small in diameter and where the difference between chiralities were significantly large in terms of the reactivity with diazonium salts.

## 4. Conclusions and outlook

The separated SWCNTs with uniform structure, eg single electronic (m- or s-) type and chirality, have presented better performance than non-separated SWCNTs in nanoelectronics and thin-film devices. For real industry applications of SWCNTs, a simple, low-cost, high-purity and scale-up separation technique is highly demanded. So far by post-synthetic techniques, especially by density gradient ultracentrifugation and chromatography technique, SWCNTs have been separated according to their lengths, electronic types, diameters and chirality in a certain scale. For the density gradient ultracentrifugation method, the relative expensive equipment and the careful process make it less competitive than a well-established chromatography method. Considering the effective roles of stationary phase and mobile phase, we propose that the chromatography will offer more opportunities for the refined separation in a simple, lower-cost and scalable way for the industrial applications of SWCNTs.

## 5. Acknowledgements

Funding from the 973 Project (2011CB932600-G), Hundred Talent Program for Q. Li and Knowledge Innovation Program (No. KJCX2.YW.M12) by Chinese Academy of Science, International Collaboration Project (No. 2009DFB50150), National Basic Research Program (No. 2010CB934700) by Ministry of Science and Technology and National Natural Science Foundation of China (No. 20903069 and 21073223) are gratefully acknowledged.

## 6. References

[1] Mintmire, J. W.; White, C. T., Universal density of states for carbon nanotubes. *Phys Rev Lett* 1998, *81* (12), 2506-2509.

[2] Tans, S. J.; Devoret, M. H.; Dai, H. J.; Thess, A.; Smalley, R. E.; Geerligs, L. J.; Dekker, C., Individual single-wall carbon nanotubes as quantum wires. *Nature* 1997, *386* (6624), 474-477.

[3] Frank, S.; Poncharal, P.; Wang, Z. L.; de Heer, W. A., Carbon nanotube quantum resistors. *Science* 1998, *280* (5370), 1744-1746.

[4] Wu, Z. C.; Chen, Z. H.; Du, X.; Logan, J. M.; Sippel, J.; Nikolou, M.; Kamaras, K.; Reynolds, J. R.; Tanner, D. B.; Hebard, A. F.; Rinzler, A. G., Transparent, conductive carbon nanotube films. *Science* 2004, *305* (5688), 1273-1276.

[5] Bockrath, M.; Cobden, D. H.; McEuen, P. L.; Chopra, N. G.; Zettl, A.; Thess, A.; Smalley, R. E., Single-electron transport in ropes of carbon nanotubes. *Science* 1997, *275* (5308), 1922-1925.

[6] Zhou, X. J.; Park, J. Y.; Huang, S. M.; Liu, J.; McEuen, P. L., Band structure, phonon scattering, and the performance limit of single-walled carbon nanotube transistors. *Phys Rev Lett* 2005, *95* (14),146805-146808.

[7] Snow, E. S.; Novak, J. P.; Campbell, P. M.; Park, D., Random networks of carbon nanotubes as an electronic material. *Appl Phys Lett* 2003, *82* (13), 2145-2147.

[8] Zhang, L.; Zaric, S.; Tu, X. M.; Wang, X. R.; Zhao, W.; Dai, H. J., Assessment of chemically separated carbon nanotubes for nanoelectronics. *J Am Chem Soc* 2008, *130* (8), 2686-2691.

[9] Collins, P. C.; Arnold, M. S.; Avouris, P., Engineering carbon nanotubes and nanotube circuits using electrical breakdown. *Science* 2001, *292* (5517), 706-709.

[10] Hassanien, A.; Tokumoto, M.; Umek, P.; Vrbanic, D.; Mozetic, M.; Mihailovic, D.; Venturini, P.; Pejovnik, S., Selective etching of metallic single-wall carbon nanotubes with hydrogen plasma. *Nanotechnology* 2005, *16* (2), 278-281.

[11] Zhang, G. Y.; Qi, P. F.; Wang, X. R.; Lu, Y. R.; Li, X. L.; Tu, R.; Bangsaruntip, S.; Mann, D.; Zhang, L.; Dai, H. J., Selective etching of metallic carbon nanotubes by gas-phase reaction. *Science* 2006, *314* (5801), 974-977.

[12] Huang, H. J.; Maruyama, R.; Noda, K.; Kajiura, H.; Kadono, K., Preferential destruction of metallic single-walled carbon nanotubes by laser irradiation. *J Phys Chem B* 2006, *110* (14), 7316-7320.

[13] Zhang, Y. Y.; Zhang, Y.; Xian, X. J.; Zhang, J.; Liu, Z. F., Sorting out semiconducting single-walled carbon nanotube Arrays by preferential destruction of metallic tubes using xenon-lamp irradiation. *J Phys Chem C* 2008, *112* (10), 3849-3856.

[14] Strano, M. S.; Dyke, C. A.; Usrey, M. L.; Barone, P. W.; Allen, M. J.; Shan, H. W.; Kittrell, C.; Hauge, R. H.; Tour, J. M.; Smalley, R. E., Electronic structure control of single-walled carbon nanotube functionalization. *Science* 2003, *301* (5639), 1519-1522.

[15] An, L.; Fu, Q. A.; Lu, C. G.; Liu, J., A simple chemical route to selectively eliminate metallic carbon nanotubes in nanotube network devices. *J Am Chem Soc* 2004, *126* (34), 10520-10521.

[16] Wang, Y.; Liu, Y. Q.; Li, X. L.; Cao, L. C.; Wei, D. C.; Zhang, H. L.; Shi, D. C.; Yu, G.; Kajiura, H.; Li, Y. M., Direct enrichment of metallic single-waited carbon nanotubes induced by the different molecular composition of monohydroxy alcohol homologues. *Small* 2007, *3* (9), 1486-1490.

[17] Zhang, H. L.; Liu, Y. Q.; Cao, L. C.; Wei, D. C.; Wang, Y.; Kajiura, H.; Li, Y. M.; Noda, K.; Luo, G. F.; Wang, L.; Zhou, J.; Lu, J.; Gao, Z. X., A Facile, Low-Cost, and Scalable Method of Selective Etching of Semiconducting Single-Walled Carbon Nanotubes by a Gas Reaction. *Adv Mater* 2009, *21* (7), 813-816.

[18] Lu, J.; Nagase, S.; Zhang, X. W.; Wang, D.; Ni, M.; Maeda, Y.; Wakahara, T.; Nakahodo, T.; Tsuchiya, T.; Akasaka, T.; Gao, Z. X.; Yu, D. P.; Ye, H. Q.; Mei, W. N.; Zhou, Y. S., Selective interaction of large or charge-transfer aromatic molecules with metallic single-wall carbon nanotubes: Critical role of the molecular size and orientation. *J Am Chem Soc* 2006, *128* (15), 5114-5118.

[19] Doyle, C. D.; Rocha, J. D. R.; Weisman, R. B.; Tour, J. M., Structure-dependent reactivity of semiconducting single-walled carbon nanotubes with benzenediazonium salts. *J Am Chem Soc* 2008, *130* (21), 6795-6800.

[20] An, K. H.; Park, J. S.; Yang, C. M.; Jeong, S. Y.; Lim, S. C.; Kang, C.; Son, J. H.; Jeong, M. S.; Lee, Y. H., A diameter-selective attack of metallic carbon nanotubes by nitronium ions. *J Am Chem Soc* 2005, *127* (14), 5196-5203.

[21] Banerjee, S.; Wong, S. S., Selective metallic tube reactivity in the solution-phase osmylation of single-walled carbon nanotubes. *J Am Chem Soc* 2004, *126* (7), 2073-2081.

[22] Menard-Moyon, C.; Izard, N.; Doris, E.; Mioskowski, C., Separation of semiconducting from metallic carbon nanotubes by selective functionalization with azomethine ylides. *J Am Chem Soc* 2006, *128* (20), 6552-6553.

[23] Chattopadhyay, D.; Galeska, L.; Papadimitrakopoulos, F., A route for bulk separation of semiconducting from metallic single-wall carbon nanotubes. *J Am Chem Soc* 2003, *125* (11), 3370-3375.

[24] LeMieux, M. C.; Roberts, M.; Barman, S.; Jin, Y. W.; Kim, J. M.; Bao, Z. N., Self-sorted, aligned nanotube networks for thin-film transistors. *Science* 2008, *321* (5885), 101- 104.

[25] Maeda, Y.; Kimura, S.; Kanda, M.; Hirashima, Y.; Hasegawa, T.; Wakahara, T.; Lian, Y. F.; Nakahodo, T.; Tsuchiya, T.; Akasaka, T.; Lu, J.; Zhang, X. W.; Gao, Z. X.; Yu, Y. P.; Nagase, S.; Kazaoui, S.; Minami, N.; Shimizu, T.; Tokumoto, H.; Saito, R., Large-scale separation of metallic and semiconducting single-walled carbon nanotubes. *J Am Chem Soc* 2005, *127* (29), 10287-10290.

[26] Chen, Z. H.; Du, X.; Du, M. H.; Rancken, C. D.; Cheng, H. P.; Rinzler, A. G., Bulk separative enrichment in metallic or semiconducting single-walled carbon nanotubes. *Nano Lett* 2003, *3* (9), 1245-1249.

[27] Li, H. P.; Zhou, B.; Lin, Y.; Gu, L. R.; Wang, W.; Fernando, K. A. S.; Kumar, S.; Allard, L. F.; Sun, Y. P., Selective interactions of porphyrins with semiconducting single-walled carbon nanotubes. *J Am Chem Soc* 2004, *126* (4), 1014-1015.

[28] Ortiz-Acevedo, A.; Xie, H.; Zorbas, V.; Sampson, W. M.; Dalton, A. B.; Baughman, R. H.; Draper, R. K.; Musselman, I. H.; Dieckmann, G. R., Diameter-selective solubilization of single-walled carbon nanotubes by reversible cyclic peptides. *J Am Chem Soc* 2005, *127* (26), 9512-9517.

[29] Nish, A.; Hwang, J. Y.; Doig, J.; Nicholas, R. J., Highly selective dispersion of singlewalled carbon nanotubes using aromatic polymers. *Nat Nanotechnol* 2007, *2* (10), 640-646.

[30] Tromp, R. M.; Afzali, A.; Freitag, M.; Mitzi, D. B.; Chen, Z., Novel strategy for diameter-selective separation and functionalization of single-wall carbon nanotubes. *Nano Lett* 2008, *8* (2), 469-472.

[31] Wang, F.; Matsuda, K.; Rahman, A. F. M. M.; Peng, X. B.; Kimura, T.; Komatsu, N., Simultaneous Discrimination of Handedness and Diameter of Single-Walled Carbon Nanotubes (SWNTs) with Chiral Diporphyrin Nanotweezers Leading to Enrichment of a Single Enantiomer of (6,5)-SWNTs. *J Am Chem Soc* 2010, *132* (31), 10876-10881.

[32] Krupke, R.; Hennrich, F.; von Lohneysen, H.; Kappes, M. M., Separation of metallic from semiconducting single-walled carbon nanotubes. *Science* 2003, *301* (5631), 344- 347.

[33] Shin, D. H.; Kim, J. E.; Shim, H. C.; Song, J. W.; Yoon, J. H.; Kim, J.; Jeong, S.; Kang, J.; Baik, S.; Han, C. S., Continuous Extraction of Highly Pure Metallic Single-Walled Carbon Nanotubes in a Microfluidic Channel. *Nano Lett* 2008, *8* (12), 4380-4385.

[34] Arnold, M. S.; Stupp, S. I.; Hersam, M. C., Enrichment of single-walled carbonnanotubes by diameter in density gradients. *Nano Lett* 2005, *5* (4), 713-718.

[35] Arnold, M. S.; Green, A. A.; Hulvat, J. F.; Stupp, S. I.; Hersam, M. C., Sorting carbon nanotubes by electronic structure using density differentiation. *Nat Nanotechnol* 2006, *1* (1), 60-65.

[36] Yanagi, K.; Iitsuka, T.; Fujii, S.; Kataura, H., Separations of Metallic and Semiconducting Carbon Nanotubes by Using Sucrose as a Gradient Medium. *J Phys Chem C* 2008, *112* (48), 18889-18894.

[37] Zhao, P.; Einarsson, E.; Xiang, R.; Murakami, Y.; Maruyama, S., Controllable Expansion of Single-Walled Carbon Nanotube Dispersions Using Density Gradient Ultracentrifugation. *J Phys Chem C* 2010, *114* (11), 4831-4834.

[38] Green, A. A.; Duch, M. C.; Hersam, M. C., Isolation of Single-Walled Carbon Nanotube Enantiomers by Density Differentiation. *Nano Res* 2009, *2* (1), 69-77.

[39] Bonaccorso, F.; Hasan, T.; Tan, P. H.; Sciascia, C.; Privitera, G.; Di Marco, G.; Gucciardi, P. G.; Ferrari, A. C., Density Gradient Ultracentrifugation of Nanotubes: Interplay of Bundling and Surfactants Encapsulation. *J Phys Chem C* 2010, *114* (41), 17267- 17285.

[40] Antaris, A. L.; Seo, J. W. T.; Green, A. A.; Hersam, M. C., Sorting Single-Walled Carbon Nanotubes by Electronic Type Using Nonionic, Biocompatible Block Copolymers. *Acs Nano* 2010, *4* (8), 4725-4732.

[41] Ghosh, S.; Bachilo, S. M.; Weisman, R. B., Advanced sorting of single-walled carbon nanotubes by nonlinear density-gradient ultracentrifugation. *Nat Nanotechnol* 2010, *5* (6), 443-450.

[42] Nair, N.; Kim, W. J.; Braatz, R. D.; Strano, M. S., Dynamics of surfactant-suspended single-walled carbon nanotubes in a centrifugal field. *Langmuir* 2008, *24* (5), 1790- 1795.

[43] Tanaka, T.; Jin, H.; Miyata, Y.; Fujii, S.; Suga, H.; Naitoh, Y.; Minari, T.; Miyadera, T.; Tsukagoshi, K.; Kataura, H., Simple and Scalable Gel-Based Separation of Metallic and Semiconducting Carbon Nanotubes. *Nano Lett* 2009, *9* (4), 1497-1500.

[44] Duesberg, G. S.; Muster, J.; Krstic, V.; Burghard, M.; Roth, S., Chromatographic size separation of single-wall carbon nanotubes. *Applied Physics a-Materials Science & Processing* 1998, *67* (1), 117-119.

[45] Farkas, E.; Anderson, M. E.; Chen, Z. H.; Rinzler, A. G., Length sorting cut single wall carbon nanotubes by high performance liquid chromatography. *Chem Phys Lett* 2002, *363* (1-2), 111-116.

[46] Heller, D. A.; Mayrhofer, R. M.; Baik, S.; Grinkova, Y. V.; Usrey, M. L.; Strano, M. S., Concomitant length and diameter separation of single-walled carbon nanotubes. *J Am Chem Soc* 2004, *126* (44), 14567-14573.

[47] Huang, X. Y.; McLean, R. S.; Zheng, M., High-resolution length sortin and purification of DNA-wrapped carbon nanotubes by size-exclusion chromatography. *Anal Chem* 2005, *77* (19), 6225-6228.

[48] Yang, Y. L.; Xie, L. M.; Chen, Z.; Liu, M. H.; Zhu, T.; Liu, Z. F., Purification and length separation of single-walled carbon nanotubes using chromatographic method. *Synthetic Met* 2005, *155* (3), 455-460.

[49] Arnold, K.; Hennrich, F.; Krupke, R.; Lebedkin, S.; Kappes, M. M., Length separation studies of single walled carbon nanotube dispersions. *Phys Status Solidi B* 2006, *243* (13), 3073-3076.

[50] Moshammer, K.; Hennrich, F.; Kappes, M. M., Selective Suspension in Aqueous Sodium Dodecyl Sulfate According to Electronic Structure Type Allows Simple Separation of Metallic from Semiconducting Single-Walled Carbon Nanotubes. *Nano Res* 2009, *2* (8), 599-606.

[51] Zheng, M.; Jagota, A.; Strano, M. S.; Santos, A. P.; Barone, P.; Chou, S. G.; Diner, B. A.; Dresselhaus, M. S.; McLean, R. S.; Onoa, G. B.; Samsonidze, G. G.; Semke, E. D.; Usrey, M.; Walls, D. J., Structure-based carbon nanotube sorting by sequence-dependent DNA assembly. *Science* 2003, *302* (5650), 1545-1548.

[52] Zheng, M.; Jagota, A.; Semke, E. D.; Diner, B. A.; Mclean, R. S.; Lustig, S. R.; Richardson, R. E.; Tassi, N. G., DNA-assisted dispersion and separation of carbon nanotubes. *Nat Mater* 2003, *2* (5), 338-342.

[53] Zheng, M.; Semke, E. D., Enrichment of single chirality carbon nanotubes. *J Am Chem Soc* 2007, *129* (19), 6084-6085.

[54] Tu, X. M.; Manohar, S.; Jagota, A.; Zheng, M., DNA sequence motifs for structure-specific recognition and separation of carbon nanotubes. *Nature* 2009, *460* (7252), 250-253.

[55] Doorn, S. K.; Strano, M. S.; O'Connell, M. J.; Haroz, E. H.; Rialon, K. L.; Hauge, R. H.; Smalley, R. E., Capillary electrophoresis separations of bundled and individual carbon nanotubes. *J Phys Chem B* 2003, *107* (25), 6063-6069.

[56] Tanaka, T.; Jin, H. H.; Miyata, Y.; Kataura, H., High-Yield Separation of Metallic and Semiconducting Single-Wall Carbon Nanotubes by Agarose Gel Electrophoresis. *Appl Phys Express* 2008, *1* (11),114001.

[57] Li, H. B.; Jin, H. H.; Zhang, J.; Wen, X. N.; Song, Q. J.; Li, Q. W., Understanding the Electrophoretic Separation of Single-Walled Carbon Nanotubes Assisted by Thionine as a Probe. *J Phys Chem C* 2010, *114* (45), 19234-19238.

[58] Sviridov, D.; Meilinger, B.; Drake, S. K.; Hoehn, G. T.; Hortin, G. L., Coelution of other proteins with albumin during size-exclusion HPLC: Implications for analysis of urinary albumin. *Clin Chem* 2006, *52* (3), 389-97.

[59] Liu, H.; Feng, Y.; Tanaka, T.; Urabe, Y.; Kataura, H., Diameter-Selective Metal/Semiconductor Separation of Single-wall Carbon Nanotubes by Agarose Gel. *J Phys Chem C* 2010, *114* (20), 9270-9276.

[60] Baughman, R. H.; Zakhidov, A. A.; de Heer, W. A., Carbon nanotubes - the route toward applications. *Science* 2002, *297* (5582), 787-792

[61] Green, A. A.; Hersam, M. C., Colored semitransparent conductive coatings consisting of monodisperse metallic single-walled carbon nanotubes. *Nano Lett* 2008, *8* (5), 1417-1422.

[62] Miyata, Y.; Yanagi, K.; Maniwa, Y.; Kataura, H., Highly stabilized conductivity of metallic single wall carbon nanotube thin films. *J Phys Chem C* 2008, *112* (10), 3591- 3596.

[63] Wang, W.; Fernando, K. A. S.; Lin, Y.; Meziani, M. J.; Veca, L. M.; Cao, L.; Zhang, P.; Kimani, M. M.; Sun, Y. P., Metallic single-walled carbon nanotubes for conductive nanocomposites. *J Am Chem Soc* 2008, *130* (4), 1415-1419.

[64] De, S.; Lyons, P. E.; Sorel, S.; Doherty, E. M.; King, P. J.; Blau, W. J.; Nirmalraj, P. N.; Boland, J. J.; Scardaci, V.; Joimel, J.; Coleman, J. N., Transparent, Flexible, and Highly Conductive Thin Films Based on Polymer - Nanotube Composites. *Acs Nano* 2009, *3* (3), 714-720.

[65] Zhang, L.; Tu, X. M.; Welsher, K.; Wang, X. R.; Zheng, M.; Dai, H. J., Optical Characterizations and Electronic Devices of Nearly Pure (10,5) Single-Walled Carbon Nanotubes. *J Am Chem Soc* 2009, *131* (7), 2454-2455.

[66] Lee, C. W.; Han, X. D.; Chen, F. M.; Wei, J.; Chen, Y.; Chan-Park, M. B.; Li, L. J., Solution- Processable Carbon Nanotubes for Semiconducting Thin-Film Transistor Devices. *Adv Mater* 2010, *22* (11), 1278-1282.

6

# Fabrication of Carbon Nanotubes for High-Performance Scanning Probe Microscopy

Ian Thomas Clark and Masamichi Yoshimura
*Surface Science Laboratory, Toyota Technological Institute, 2-12-1 Hisakata, Tempaku, Nagoya*
*Japan*

## 1. Introduction

The benefits of using carbon nanotubes (CNTs) as probes for scanning probe microscopy (SPM) have been recognized for many years.[1] Since the initial report on the fabrication and use of CNT SPM probes, many accounts of the fabrication of CNT SPM probes, and demonstrations of the superior performance of these probes have been published. Significant progress has been made since the initial studies by Dai et al.,[1, 2] although a method for fabricating CNT SPM probes that fully exploits the desirable properties of the CNT and is truly reproducible and cost-effective remains elusive. In the present work, we detail the specific properties of CNTs that make them appropriate for various SPM methods, and review methods for the fabrication of CNT SPM probes. Our review of fabrications methods includes a review of methods in the literature, as well as recent, previously unpublished methods developed in our laboratory. The goal of this work is to provide a concise and up-to-date guide to aid researchers in the further development of CNT SPM probes.

SPM now encompasses several tens of distinct techniques, and is arguably the most widely used method to obtain real-space information on surfaces and solid-fluid interfaces; in the past several years SPM has been applied even to fluid-fluid interfaces, such as imaging living cells in buffer solution.[3] The basic block diagram of a general SPM instrument is presented in Fig. 1. In all variants of SPM, a probe is brought close to the surface or interface of interest, and rastered across the surface/interface while one or more interactions between the probe and the surface/interface is transduced and recorded. In many, but not all cases one of the transduced interactions is used in a feedback loop to maintain a constant distance between the probe and the surface/interface. Scanning tunneling microscopy (STM), the first SPM method reported,[4] transduces the quantum mechanical tunnel current from a sharp metal probe biased relative to a sample. Since the introduction of STM the list of SPM methods has grown dramatically. A non-exhaustive list of commonly used SPM methods includes several varieties of atomic force microscopy (AFM), which map surface/interface morphology by transducing the short-range van der Waals and capillary forces between the surface/interface and probe,[5-8] magnetic force microscopy (MFM), which maps the stray magnetic field above a surface/interface by transducing the force on a magnetic probe due to a magnetic or current-carrying sample,[9-11] electrostatic force microscopy (EFM), which maps the electrostatic field above a surface/interface by transducing the force on a conductive probe biased with a DC voltage relative to the sample,[12, 13] and Kelvin force microscopy (KFM), which maps the contact potential by transducing the AC component of the capacitive force on a conductive probe biased with both AC and DC components relative to the sample.[14, 15] The demands on the probe vary widely from method to method; however, as we will describe here, for many of these methods, carbon nanotubes make an excellent probe material.

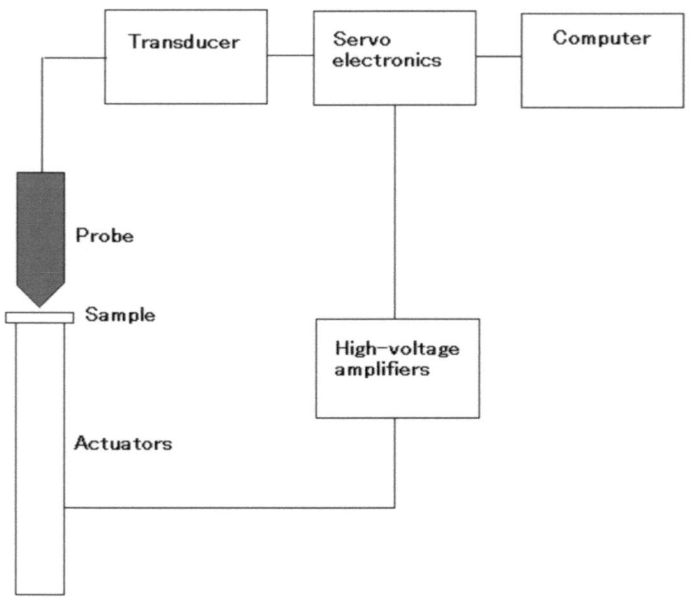


Fig. 1. Block diagram of a generic SPM instrument. Servo electronics are frequently, but not always used to maintain a constant sample-probe distance.

Just as "SPM" describes a family of related microcopy methods rather than a single method, "CNT" describes a family of related nanostructures rather than a specific nanostructure; CNTs vary widely in number of walls and diameter. Single-wall CNTs (SWNTs) and individual walls of multi-wall CNTs (MWCNTs) also differ in helicity - the manner in which the graphene lattice is wrapped to form the cylindrical CNT wall. The electronic and mechanical properties of a CNT depend, sometimes strongly, on these geometric attributes.[16] CNTs with a certain diameter and number of walls can be grown by choosing an appropriate growth method from the myriad of published CNT growth methods.[17] With modern purification methods, even SWNTs of a specific helicity can be obtained.[18] Further, catalytic chemical vapor deposition (CVD) growth methods can be used to make CNTs encapsulating metal particles and nanowires,[19-21] and methods of plating the exterior of CNTs with metals and electrically insulating layers have also been published;[22, 23] these metal/CNT and insulator/CNT hybrid structures naturally have different electronic, mechanical and magnetic properties. Developers of CNT SPM probes therefore have a great deal of control over the properties of the CNTs used in probe manufacture. In an effort to guide CNT SPM probe developers to the optimal CNT variety for their specific SPM application, the following three sections relate CNT attributes to electronic/magnetic/mechanical properties, and further relate those properties to the design requirements of probes for various varieties of SPM.

## 2. The benefits of CNTs for SPM applications

### 2.1 Geometric considerations and probe convolution

It was recognized many years ago that the *shape* of CNTs is almost ideal for use as a SPM probe. Generally, the goal of SPM is to map some property of a surface/interface in real space; however, the image acquired by SPM is not a map of that property of the sample, but rather a convolution of the properties of the sample and the SPM probe. This is illustrated with an idealized contact-mode AFM experiment in Fig. 2: as the pyramidal probe of Fig. 2(a) is rastered over the rectangular protrusion, it traces a path that is sloped at the angle of the pyramidal probe, rather than vertical like the actual profile of the protrusion. The SPM "image" reflects both the shape of the surface and the shape of the probe. As is illustrated in Fig. 2(b), the pseudo-one-dimensional nature of CNTs can minimize erroneous features in the SPM image due to probe convolution.

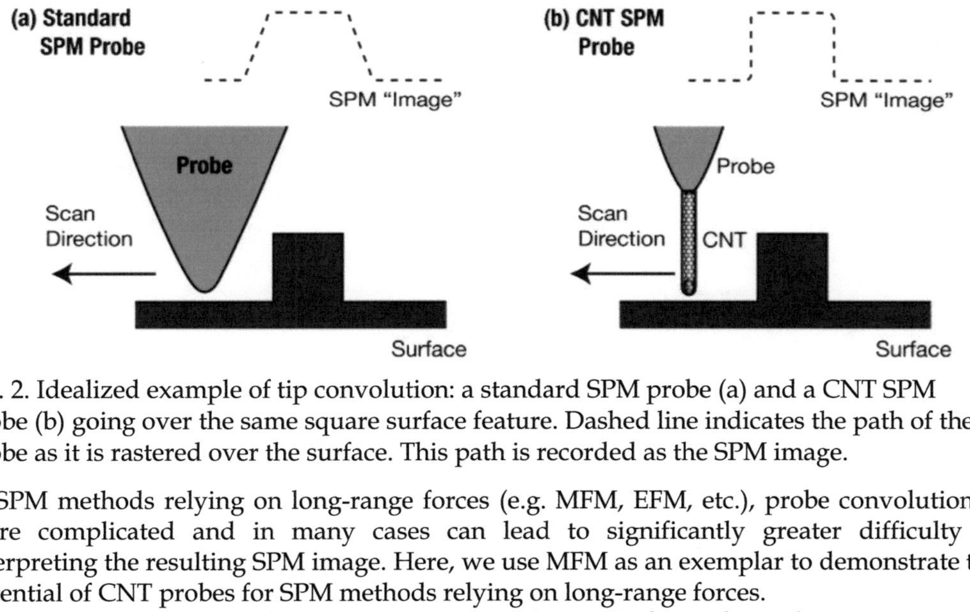


Fig. 2. Idealized example of tip convolution: a standard SPM probe (a) and a CNT SPM probe (b) going over the same square surface feature. Dashed line indicates the path of the probe as it is rastered over the surface. This path is recorded as the SPM image.

In SPM methods relying on long-range forces (e.g. MFM, EFM, etc.), probe convolution is more complicated and in many cases can lead to significantly greater difficulty in interpreting the resulting SPM image. Here, we use MFM as an exemplar to demonstrate the potential of CNT probes for SPM methods relying on long-range forces.

To a first approximation, the magnetic interaction energy of a probe with a magnetization $\mathbf{M}_{probe}(\mathbf{r})$ at the point $\mathbf{r}$ with the magnetic field induced by the sample, $\mathbf{H}_{sample}(\mathbf{r})$ can be expressed as

$$E = \int_{probe\ volume} \mathbf{H}_{sample}(\mathbf{r}) \cdot \mathbf{M}_{probe}(\mathbf{r})\, dV , \qquad (1)$$

where the integral is taken over the volume of the probe. As in AFM, the probe is mounted on a cantilever driven at a frequency near its resonant frequency. The force gradient due to the magnetic interaction (i.e. the second $z$ derivative of interaction energy) weakens or enhances the restoring force of the cantilever. The resulting shift in resonance frequency is detected as a shift in the relative phases of the driving force and the motion of the cantilever.

In practice the cantilever is typically kept a few tens of nanometers above the surface during MFM phase mapping to prevent interference of the stronger but shorter-range van der Waals and capillary forces.

Several factors complicate analysis of these MFM phase maps: first, magnetic field decay above the sample depends on the lateral wavelength of the magnetization field of the sample. More rigorously, a sample with a magnetization field

$$\mathbf{M} = \mathbf{M}_0 \sin(2\pi x / l) , \tag{2}$$

where $x$ is in the plane of the sample surface, will induce a magnetic field that decays exponentially away from the surface with the characteristic decay length $l$. Viz,

$$\mathbf{H}_{\text{sample}} \propto \exp(-z^2 / l) . \tag{3}$$

If Eq. 2 is considered to be a Fourier component of an arbitrary magnetization field, it is clear from Eq. 3 that lateral variations in the magnetic field above the sample are effectively a low-pass-filtered version of lateral variations in the magnetization field since smaller-wavelength components of the magnetic field decay more rapidly away from the surface.

A second complicating factor can be seen from Eq. 1: the phase shift mapped in MFM does not arise from a de facto surface-surface interaction as it does in AFM, but rather the interaction between the stray magnetic field and the entire volume of the magnetic material of the probe; the shape of the entire probe, rather than just the probe apex, contributes to the convolution. To illustrate, we compare a typical commercial MFM probe, which is made by depositing a few-nanometer layer of magnetic material over an approximately pyramidal silicon probe, as shown in Fig. 3(a), with a CNT-encapsulated magnetic nanoparticle MFM

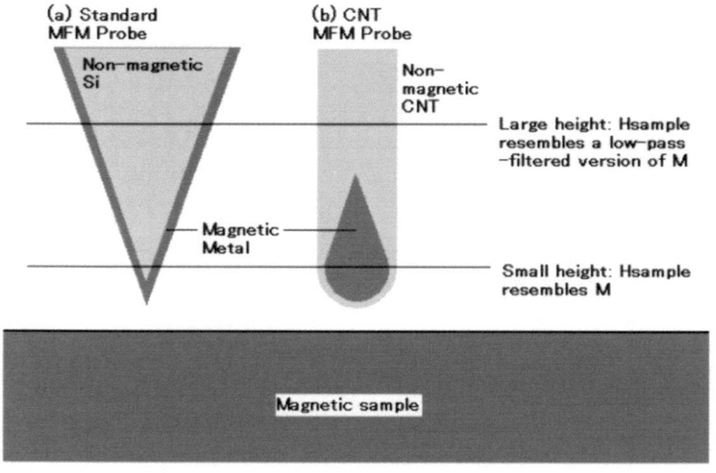


Fig. 3. Cartoon of a (a) standard MFM probe composed of a layer magnetic material over a pyramidal, non-magnetic probe, and (b) a CNT MFM probe composed of a magnetic nanoparticle encased in the tip of a CNT interacting with the stray magnetic field induced by a magnetic sample. Note that the CNT SPM probe has a larger fraction of its magnetic material closer to the sample surface, where the stray magnetic field is a more faithful representation of the samples magnetization field.

probe, like those recently reported by Tanaka et al. (fabrication method discussed below).[24] The magnetic catalyst particle used to grow these CNT probes is in the tip of the CNT and has a characteristic "carrot"-shape as illustrated in Fig. 3(b). The amount of magnetic material in a probe between in a differentially thin slice between $z$ and $z+dz$ increases monotonically with increasing $z$ (i.e. increasing distance from the sample surface) from the tip of the probe for the standard probe. For the CNT probe, this value increases rapidly in the first few nanometers from the tip, but then decreases with increasing $z$, going to zero at tens to a few hundreds of nanometers from the CNT tip. Since this differentially thin slice of the probe between $z$ and $z+dz$ interacts with the sample-induced magnetic field at $z$, and the low pass filtering effect of the sample-induced magnetic field increases with increasing $z$, it is clear that less material at larger $z$ is preferential for high-resolution MFM imaging. Unlike AFM where a minimal lateral dimension of the probe is desirable, in MFM both a minimum lateral and vertical dimension are desirable. Since a finite volume of magnetic material is needed in the probe, one must in general balance the criteria of low lateral and low vertical dimension. In light of this analysis, due largely to its small effective vertical dimension, CNT MFM probes are preferable to standard MFM probes. Indeed, we have found that even fairly wide CNT MFM probes (tube radius as large as 60 nm) provide significantly higher resolution MFM images than significantly sharper (tip radius of curvature of approximately 15 nm) standard MFM probes.[24, 25] While MFM is the only example of its kind provided here CNT SPM probes have been proposed for EFM and other SPM methods relying on long-range forces based on similar geometric arguments.

### 2.2 Mechanical considerations

The mechanical properties of CNTs have been celebrated for many years and have been the subject of hundreds of experimental and theoretical investigations, as reviewed most recently by Wang et al,[26] and Shokrieh et al.[27] As is often cited, the Young's modulus and tensile strength of CNTs are five and 120 times greater, respectively, than those of 304 stainless steel. Although these simple numbers indicate that CNTs are in general mechanically robust, the mechanical properties of CNTs depend on their diameter, number of walls, and aspect ratio. To understand how the mechanical properties of CNTs are beneficial in SPM applications as well as to select the optimal CNT length, diameter, and number of walls for a given SPM method, it is important to consider the forces applied to the SPM probe during operation.

Tip "crashes" are a common problem in almost all forms of SPM; mistakes made in approaching the probe to the sample, environmental vibrations large enough to transmit through the instrument's vibration isolation systems, or rastering the probe across a surface feature too steep for the servo circuit to avoid, or too large for the servo actuators to overcome can all result in a probe-sample collision that applies extremely large transient stresses to the probe apex breaking brittle probe materials or plastically deforming metal probe materials. While fracture and plastic deformations of CNTs have been observed in experimental[28-30] and theoretical[31, 32] investigations, the strains leading to fracture or plastic deformation are unlikely to be encountered in even the most violent tip crash. Unlike more common probe materials, both SWNTs and MWCNTs can be deformed to a remarkable extent, and regain their original shape when the stress is relieved.[33-38] In our experience, CNT probe destruction during tip crash is generally the result of failure of the CNT/probe junction, rather than failure of the CNT itself. To minimize stress on the CNT/probe junction, the thinnest and longest and therefore most compliant CNT compatible with the particular SPM application should be used. As described above, long, thin CNTs also minimize tip convolution for most SPM methods.

In noncontact (NC)- and TappingMode™ (TM)-AFM, and most methods based on NC- and TM-AFM (e.g. EFM, MFM, KFM, etc.), the forces applied to the probe under normal operation are extremely small. Thus, thin, high-aspect-ratio CNTs can be used for probes. In practice, above a certain aspect ratio the resolution is found to decrease with increasing CNT length,[39] an effect that most authors attribute to thermally excited vibrations in the CNT;[39, 40] however, the extremely high elastic modulus of CNTs results in a very low susceptibility to thermal vibrations, and excellent resolution has been reported for CNT SPM probes with aspect ratios of more than 50. Interestingly, Carbon Nanotube SPM Probes commercially available from NanoSensors™ use 1.2- to 2.4-nanometer-diamter SWNTs or DWNTs extending 50 to 750 nm beyond the end of the probe handle, the shortest and widest of which still have an aspect ratio comparable to the slenderest CNT AFM probes reported in the literature.

CM-AFM places substantially stricter requirements on the mechanical properties of the probe. Since the probe stays in contact with the surface as it is rastered, lateral as well as normal forces are applied to the probe. Continuous contact also results in much faster wear of the probe tip, especially when imaging high elastic modulus materials. Wear is expected to be a problem for any probe material; however, since CNTs have the highest elastic modulus of any known material they should be least susceptible to wear. Further, the pseudo-two-dimensional nature of the graphitic sheets that comprise the CNT surfaces generally present no asperities even on the atomic scale, making them even less susceptible to wear.

While the slenderness of CNTs is expected to provide reduced tip convolution in CM-AFM as it does in other methods, it is also expected to result in a more flexible probe - a source of concern because of the non-vanishing lateral forces applied to the probe tip during CM-AFM. However, based on scanning anodic oxidation experiments, Dai et al. reported that flexure of the CNT is problematic only at scanning rates significantly faster than those generally used in CM-AFM, even for very slender CNT AFM probes,[2] again due to the CNT's extremely high elastic modulus. While there is significant controversy regarding the bending rigidity of MWCNTs,[34] recent theoretical studies suggest that for the same outer diameter, a greater number of walls does indeed increase bending stiffness. Further, unlike single-, and few-wall CNTs, which soften at relatively low strains due to large-scale buckling, more dense CNTs (i.e. CNTs with a greater number of walls), do not soften significantly at modest strains because the inner tubes resist buckling. We expect that by selecting denser MWCNTs, or CNT growth methods that favor denser MWCNTs, even slenderer CNT CM-AFM probes (or alternatively, faster scan rates) than those reported by Dai et al. can be realized.

CNT probes also have great potential for SPM-based nanoindentation. As in other scanning probe methods, the large aspect ratio gives CNT nanoindentation probes access to deep narrow recesses. Reproducible indentation measurements from the bottom of pits in the surface of a commercial polycarbonate DVD were reported by Akita and co-workers.[41, 42] Further, the extremely high elastic modulus of CNTs makes them ideal for indentation applications. As in the case of CM-AFM, dense MWCNTs are expected to provide the best performance, since the presence of inner tubes reduces the compressive compliance, and increases flexural stiffness, thereby increasing the load necessary for Euler or shell buckling of the CNT.

Perhaps the most compelling benefit of using CNTs as nanoindentation probes, however, stems from the atomic-scale near-perfection of their tips. Traditional nanoindentation probes are made by mechanically grinding synthetic diamond. The grinding process yields a tip that is rough and irregular on the few-angstrom to few-nanometer scale. For extremely shallow indents, morphological irregularities result in a significant uncertainty in the projected area of the probe-sample contact area, complicating or even precluding quantitative analysis of the nanoindentation data. The smooth, hemispherical morphology of CNT tips, in conjunction with their high elastic modulus, is expected to reduce these uncertainties, possibly allowing quantitative determination of mechanical properties from few nanometer, or even sub-nanometer indents. The ability to probe mechanical properties to the sub-nanometer length scale would be of great benefit to the study of the tribological properties of industrially important thin films, such as gate insulators in VLSI circuits, and ultrathin diamond-like carbon films used in hard drives.

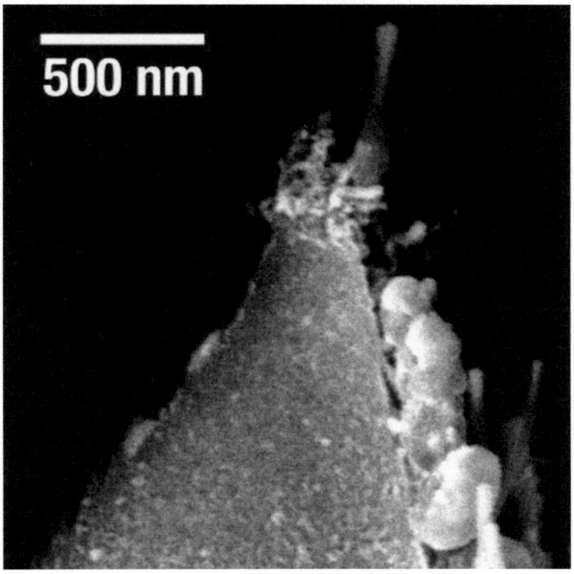


Fig. 4. SEM image of a single-CNT MFM probe fabricated using an ion-shadowed MPECVD fabrication method.

### 2.3 Electronic considerations

The electronic properties of CNTs generally make them a good probe material for SPM methods that require a conductive probe. CNTs are capable of carrying tremendous current densities.[16] Several examples of CNT probes used successfully in EFM, scanning anodic oxidation lithography[2] and nanoscale surface conductivity measurements.[22, 43] While it is well known that most synthetic methods produce ~50% semiconducting, and further that even metallic CNTs generally have a very sparse density of states (DOS) near the (vacuum) Fermi level, empirically it is found that CNT SPM probes function very well for STM.[1, 22, 44,

[45] The excellent performance of CNT SPM probes is somewhat confounding since materials with a sparse DOS near the Fermi level generally make poor SPM probes due to the small number of states available for electrons to tunnel out of and into. While further study is warranted, the absence of this problem in the case of CNT STM probes maybe attributed to a larger local DOS (LDOS) at the CNT tip due to broken bonds or the presence of the defects that are necessary to form the closed CNT tip.[46] In some cases, residual metal catalyst particles at the probe tip may also be responsible for broadening the LDOS near the probe tip.[43]

## 3. CNT SPM probe fabrication methods

Mounting a CNT - a definitively nanometer-scale object - on a millimeter-scale SPM probe (hereafter called the probe handle), is a non-trivial challenge. Approaches to this problem generally fall into one of two categories: direct attachment (DA), in which a previously grown CNT is attached to the tip of a SPM probe, and catalyst deposition / CVD methods (CDCVD), in which catalyst is deposited on a SPM probe and the CNTs are subsequently grown by CVD. Both categories have benefits and drawbacks. Modern DA methods are highly reproducible and offer good control of CNT position, angle, and length, but are tremendously time consuming. CDCVD methods are generally less reproducible, but much faster and less expensive. In the following, we describe the existing variations of DA methods, followed by a description of the existing methods of CDCVD methods. Each of these sections is organized roughly in order of increasing technical complexity, rather than chronologically. For completeness, we note that CNT SPM probe fabrication by magnetopheretic,[47] and dielectropheretic[48] attraction of solvated CNTs to the apex of the probe handle immersed in the CNT solution have also been reported; however, these methods have not been explored to the extent that CDCVD and DA methods have, being the subject of a single publication each. We thus, opt to direct interested readers to the pertinent references rather than describing them at length here.

### 3.1 Probe fabrication by DA methods

The first, and simplest DA CNT SPM probe fabrication method is that reported by Dai et al. in their seminal 1998 Science paper.[1] In this method, bundles of arc-discharge-grown CNTs are secured to the probe handle with an acrylic adhesive. While the adhesion between the bundles and the probe handle has not, to our knowledge, been studied in detail, some authors suggest that sputtering Cr onto the CNT SPM probe strengthens the CNT bundle / probe handle junction.[49] Optical microscopy is used to align the CNT bundles with probe handle during fabrication. Although the authors attest that in most cases one or two CNTs extends down below the rest of the bundle, this method does not produce single CNT probes.[1, 2, 39, 49] In a similar approach, which does purportedly frequently yield single-CNT SPM probes, the probe handle is coated with a metal film to increase its conductivity, biased at -5 V ~-25 V relative to the CNT boule, and translated toward the CNT boule under the view of an optical microscope until a flash is observed.[50] While the physics of this fabrication method was left largely unexplored, the authors suggest that electrostatic attraction is responsible for attracting the CNTs to the apex of the probe handle. The arc discharge evidenced by the flash is believed to follow the attached CNT burning through due to Joule heating at a defect within the CNT or at a CNT-CNT junction. This burning serves to sever the probe CNT from the boule, leaving, in some cases, an isolated CNT on the probe apex.

Alignment of CNTs with probe handle using microtranslators mounted in an scanning electron microscope (SEM) chamber, as first reported by Akita et al.,[51, 52] is a greater technical challenge relative to the method of Dai et al.,[1] and is indeed more time consuming, but the results are significantly more consistent and reproducible. In the original report by Akita et al., CNTs were first dispersed in a solvent, and aligned perpendicular to the edge of a blade by dielectrophoresis. This provided a sparse array of highly aligned CNTs. The CNT-decorated blade and the probe handle were then mounted on separate translators inside the SEM chamber. The probe handle was aligned with, and then brought into contact with a CNT. Although the van der Waals force between the CNT and the probe handle is typically strong enough to hold the CNT in place, amorphous carbon is often deposited over the CNT/probe handle contact via electron-stimulated deposition by scanning the SEM's electron beam over a very small area around the junction. This method for amorphous carbon deposition requires a hydrocarbon background pressure of a few microtorr. Hydrocarbon contamination from a diffusion pump provides sufficient carbon feedstock, but carbon-containing molecules may have to be intentionally leaked into the chamber for the case of low-pressure turbo-molecular pump (TMP)-pumped chambers. The final step is generally to cut the CNT of the newly fabricated CNT SPM probe free of the blade by stopping the electron beam over the CNT.[53] In a variant of this method used to make MFM probes, however, the probe is brought within one micrometer of the base of a CNT, and the CNT dislodged from the substrate and attached to the probe handle by applying a DC potential to the probe handle.[54] This obviates the need for cutting of the CNT - an important consideration in the fabrication of MFM probes since the magnetic particle resides in the tip of the CNT. In spite of their inherent cost and complexity, SEM-based DA methods are currently the most reliable, reproducible and versatile method of CNT fabrication. In addition to AFM probes,[51-53, 55] and MFM probes,[54] nanoindenter probes,[35, 41, 42] four-point surface conductivity probes,[22] and STM probes[45] have all been fabricated by various SEM-based DA methods.

### 3.2 Probe fabrication by CDCVD methods

Surprisingly, the simplest CDCVD CNT SPM probe fabrication method - simply depositing a layer of catalytically active metal on the probe handle and exposing to CvD conditions - was among the last reported CDCVD methods.[24, 56, 57] In the first report of simple CDCVD fabrication by Wongwiriyapan et al., the authors deposited 0.5 nm~2 nm Fe films atop 5 nm~10 nm Al buffer layers on W STM tips, and grew CNTs using thermal CVD with a $CH_4$ carbon source. A significant drawback to thermal CVD is that the resulting CNTs are randomly oriented. Our group opted instead for microwave plasma enhanced CVD (MPECVD).[24, 43, 57, 58] In our MPECVD apparatus, a 200 V bias is applied between the sample holder and a second electrode held ~10 cm above the sample holder (sample holder negative). The narrow velocity distribution of the ions that induce the CNT growth ensure that the CNTs grow toward the upper electrode with an angular distribution of approximately three degrees.[59] Using MPECVD with a 15-nm Co catalyst films deposited on commercial AFM cantilevers, Tanaka et al. fabricated high-resolution MFM probes.[24] Sakamoto et al. fabricated SPM probes decorated with Pd nanowire/MWCNT heterostructures using MPECVD with a thin Pd catalyst film deposited on AFM cantilevers. In both of the previously mentioned reports, the probe handle and the lower face of the cantilever were covered with an array of CNTs. Methods for tip-selective employing MPECVD will be described below.

Efforts to place the catalyst, and thus the subsequent CNTs, selectively at the tip of the probe handle by exploiting the electric field enhancement at the sharply curved tip were first reported by Hafner et al,[40] and Cheung et al.[60] This method used dielectropherisis to attract Fe or FeMo impregnated alumina particles to the near-tip area of the probe handle, and subsequent thermal CVD with an ethylene feedstock for CNT growth. The resulting few-nanometere-diameter CNTs grew along the probe handle surface due to the strong van der Waals attraction, until they reached the probe apex. Since the energy to bend the CNT around the sharp tip of the probe handle exceeded the energy of the van der Waals attraction, the CNTs tended to extend out from the tip of the probe handle. Due to the non-oriented growth of thermal CVD, however, the length distribution of the CNT SPM probes was quite large, necessitating trimming. Tung, et al. exploited the enhanced electric field around the apex of a W STM tip to selectively electroplate Co catalyst near the apex.[58] MPECVD growth yielded CNTs highly aligned along the long axis of the probe handle. Due to the controllable growth direction of the CNTs, CNT length can be controlled by adjusting growth time, and no trimming is necessary. The drawback of Tung's method (and all MPECVD-based methods) for some SPM applications is that MPECVD-grown CNTs tend to exhibit large diameters (15 nm~100 nm depending on catalyst size and chemical composition), compared to CNTs grown by thermal CVD.

A number of other methods for selectively depositing catalyst near the probe handle apex, and forcing the CNTs to grow in the desired direction during CVD have been reported. Charles Lieber's group at Harvard University developed and refined a method commonly known as pore growth wherein electrochemical etching was used to make pores in the tip of a Si AFM probe parallel to the long axis of the probe handle. Catalyst deposited in the pores yielded CNTs that were forced to grow along the pore direction during CVD growth.[39, 61, 62] For SSRM applications, we recently presented a method for the fabrication of nano-brushes of coaxial Pd nanowire/MWCNT heterostructures using a custom-fabricated area-selective electroplating cell to deposit Pd catalyst on the top ~1 μm of a commercial Si AFM probe, and the ion-induced alignment properties of MPECVD to force the Pd/MWCNTs to grow parallel with the probe handle long axis.[43] More recently, we have developed a tip-selective variation on the method of Tanaka et al. described above. In this method, a FeNi alloy catalyst is deposited only on the forward face of a commercial Si AFM probe. The following MPECVD grows highly aligned, FeNi-nanoparticle-containing CNTs. While catalyst is deposited over a few square microns of the probe handle surface, the catalyst film at the probe handle apex blocks ions coming out of the plasma, preventing them from striking portions of the catalyst film away from the apex. This ion shadowing effect can be used to grow a single CNT at the probe handle apex with ~70% reproducibility. The alignment and length of the CNT at the probe handle apex can be controlled purely by growth conditions. A representative probe of this type is shown in Fig. 3.

Because of the scalability of CDCVD methods a number of authors have attempted to develop mass-production CDCVD CNT SPM probe fabrication methods. Yenilmez et al. reported a method in which a wafer of silicon AFM probes was spin coated with a few-micron layer of photoresist leaving only the pyramidal probe tips above the resist layer.[63]

Thereafter, a suspension of FeMo-impregnated alumina particles was spin coated on the wafer, the photoresist removed, and CNTs grown using thermal CVD with a methane carbon source. Unfortunately, this wafer-scale method produces CNTs that extend 1μm~10μm beyond the end of the probe handle, necessitating that every CNT SPM probe be individually shortened by electrical discharge.[1]

## 4. Summary and conclusions

Due to their simple shape, high aspect ratio, exceptional mechanical and electronic properties CNTs have for many years been an extremely tempting material for scanning probe microscopists. Although the inherent complexities and uncertainties of CNT SPM probe fabrication have thus far precluded the sort of uniform wafer-scale production that would lead to affordable and readily available commercial CNT SPM probes, the 15 years since Dai et al.'s seminal paper, has seen the development of a wide array of serial, but extremely effective CNT SPM probe fabrication methods, and the application of these probes to wide array of SPM methods and a wide array of samples. In the preceding text, we have attempted to provide a one-stop guide to this body of work, in the hope of assisting future developers in their efforts to exploit the unique properties of CNTs in still more varieties of SPM. We expect that the coming years will see the variety of CNT SPM probes and the range of applications of CNT SPM probes expand even further.

## 5. References


[1] H. Dai, J. H. Hafner, A. G. Rinzler, D. T. Colbert and R. E. Smalley, Nature 384 (6605), 147-150 (1996).

[2] H. Dai, N. Franklin and J. Han, Applied Physics Letters 73 (11), 1508 (1998).

[3] C. A. Putman, K. O. v. d. Werf, B. G. d. Grooth, N. F. v. Hulst and J. Greve, Biophysical Journal 67 (4), 1749–1753 (1994).

[4] G. Binnig, H. Rohrer, C. Gerber and E. Weibel, Physical Review Letters 49 (1), 57–61 (1982).

[5] G. Binnig, H. Rohrer and C. Gerber, Physical Review Letters 56 (9), 930–933 (1986).

[6] G. Meyer and N. M. Amer, Applied Physics Letters 53 (12), 1045 (1988).

[7] F. J. Giessibl, Reviews of Modern Physics 75 (3), 949–983 (2003).

[8] Q. Zhong, D. Inniss, K. Kjoller and V. B. Elings, Surface Science Letters 290 (1-2), L688-L692 (1993).

[9] Y. Martin and H. K. Wickramasinghe, Applied Physics Letters 50 (20), 1455-1457 (1987).

[10] J. J. Sáenz, N. García, P. Grütter, E. Meyer, H. Heinzelmann, R. Wiesendanger, L. Rosenthaler, H. R. Hidber and H. J. Güntherodt, Journal of Applied Physics 62 (10), 4293 (1987).

[11] U. Hartmann, Annual Review of Materials Science 29, 53-87 (1999).

[12] Y. Martin, D. W. Abraham and H. K. Wickramasinghe, Applied Physics Letters 52 (13), 1103 (1988).

[13] L. N. Kantorovich, A. I. Livshits and M. Stoneham, Journal of Physics: Condensed Matter 12 (6) (2000).

[14] M. Nonnenmacher, M. P. O'Boyle and H. K. Wickramasinghe, Applied Physics Letters 58 (25), 2921- (1991).
[15] W. Melitza, J. Shena, A. C. Kummela and S. Lee, Surface Science Reports 66 (1), 1-27 (2011).
[16] M. S. Dresselhaus, G. Dresselhaus and P. C. Eklund, *Science of fullerenes and carbon nanotubes*. (Academic Press, 1996).
[17] M. Meyyappan, Journal of Physics D: Applied Physics 42, 213001 (2009).
[18] M. S. Arnold, A. A. Green, J. F. Hulvat, S. I. Stupp and M. C. Hersam, Nature Nanotechnology 1, 60 - 65 (2006).
[19] L. Jankovi, D. Gournis, P. N. Trikalitis, I. Arfaoui, T. Cren, P. Rudolf, M.-H. Sage, T. T. M. Palstra, B. Kooi, J. D. Hosson, M. A. Karakassides, K. Dimos, A. Moukarika and T. Bakas, Nano Letters 6 (6), 1131–1135 (2006).
[20] R. Li, X. Sun, X. Zhou, M. Cai and X. Sun, Journal of Physical Chemistry C 111 (26), 9130–9135 (2007).
[21] C. Müller, A. Leonhardt, M. C. Kutz and B. Büchner, Journal of Physical Chemistry C 113 (7), 2736–2740 (2009).
[22] S. Yoshimoto, Y. Murata, K. Kubo, K. Tomita, K. Motoyoshi, T. Kimura, H. Okino, R. Hobara, I. Matsuda, S.-i. Honda, M. Katayama and S. Hasegawa, Nano Letters 7 (4), 956-959 (2007).
[23] Z. Deng, E. Yenilmez, J. Leu, J. E. Hoffman, E. W. J. Straver, H. Dai and K. A. Moler, Applied Physics Letters 85 (25), 6263-6265 (2004).
[24] K. Tanaka, M. Yoshimura and K. Ueda, Journal of Nanomaterials 2009, 147204 (2009).
[25] I. T. Clark, Y. Matsuoka and M. Yoshimura, in preperation.
[26] C. M. Wang, Y. Y. Zhang, Y. Xiang and J. N. Reddy, Applied Mechanics Reviews 63 (3), 030804 (2010).
[27] M. M. Shokrieh and R. Rafiee, Mechanics of Composite Materials (2010).
[28] O. Lourie, D. M. Cox and H. D. Wagner, Physical Review Letters 81 (8), 1638–1641 (1998).
[29] J. F. Waters, L. Riester, M. Jouzi, P. R. Guduru and J. M. Xu, Applied Physics Letters 85 (10), 1787-1789 (2004).
[30] J. F. Waters, P. R. Guduru, M. Jouzi, J. M. Xu, T. Hanlon and S. Suresh, Applied Physics Letters 87 (10), 103109 (2005).
[31] P. Zhang, P. E. Lammert and V. H. Crespi, Physical Review Letters 81 (24), 5346–5349 (1998).
[32] D. Srivastava, M. Menon and K. Cho, Physical Review Letters 83 (15), 2973–2976 (1999).
[33] M. R. Falvo, G. J. Clary, R. M. Taylor, V. Chi, J. F.P. Brooks, S. Washburn and R. Superfine, Nature 389, 582-584 (1997).
[34] P. Poncharal, Z. L. Wang, D. Ugarte and W. A. d. Heer, Science 283, 1513-1516 (1999).
[35] S. AKITA, H. NISHIJIMA, T. KISHIDA and Y. NAKAYAMA, Japanese Journal of Applied Physics 39, 3724-3727 (2000).
[36] S. D. Solares, Y. Matsuda and W. A. G. III, Journal of Physical Chemistry B 109, 16658-16664 (2005).
[37] Y.-R. Jeng, P.-C. Tsai and T.-H. Fang, Applied Physics Letters 90 (16), 161913 (2007).
[38] X. Duan, C. Tang, J. Zhang, W. Guo and Z. Liu, Nano Letters 7 (1), 143–148 (2007).

[39] S. S. Wong, J. D. Harper, P. T. Lansbury and C. M. Lieber, Journal of the American Chemical Society 120, 603 (1998).
[40] J. H. Hafner, C. L. Cheung and C. M. Lieber, Journal of the American Chemical Society 121 (41), 9750-9751 (1999).
[41] S. Akita, H. Nishijima, T. Kishida and Y. Nakayama, Japanese Journal of Applied Physics 39, 7086-7089 (2000).
[42] S. Akita and Y. Nakayama, Japanese Journal of Applied Physics 40, 4289-4291 (2001).
[43] I. T. Clark, G. Rius, Y. Matsuoka and M. Yoshimura, Journal of Vacuum Science and Technology B 28 (6), 1148-1152 (2010).
[44] A. Pasquini, G. B. Picotto and M. Pisani, Sensors and Actuators A 123-124, 655–659 (2005).
[45] N. Isomura, X. Wu and Y. Watanabe, Journal of Chemical Physics 131, 164707 (2009).
[46] A. D. Vita, J.-C. Charlier, X. Blase and R. Car, Applied Physics A 68, 283-286 (1999).
[47] A. Hall, W. G. Matthews, R. Superfine, M. R. Falvo and S. Washburn, Applied Physics Letters 82 (15), 2506-2508 (2003).
[48] H. W. Lee, S. H. Kim and Y. K. Kwak, Reviews of Scientific Instruments 76, 046108 (2005).
[49] J. Li, A. M. Cassell and H. Dai, Surface and Interface Analysis 28, 8-11 (1999).
[50] R. M. D. Stevens, N. A. Frederick, B. L. Smith, D. E. Morse, G. D. Stucky and P. K. Hansma, Nanotechnology 11, 1-5 (2000).
[51] S. Akita, H. Nishijima, Y. Nakayama, F. Tokumasu and K.Takeyasu, Journal of Physics D: Applied Physics 32, 1044-1048 (1999).
[52] H. Nishijima, S. Kamo, S. Akita, Y. Nakayama, K. I. Hohmura, S. H. Yoshimura and K.Takeyasu, Applied Physics Letters 74, 4061–4063 (1999).
[53] J. Martinez, T. D. Yuzvinsky, A. M. Fennimore, A. Zettl, R. Garcıa and C. Bustamante, Nanotechnology 16, 2493–2496 (2005).
[54] T. Arie, H. Nishijima, S. Akita and Y. Nakayama, Journal of Vacuum Science and Technology B 18 (1), 104-106 (2000).
[55] K. I. Hohmura, Y. Itokazu, S. H. Yoshimura, G. Mizuguchi, Y.-s. Masamura, K. Takeyasu, Y. Shiomi, T. Tsurimoto, H. Nishijima, S. Akita and Y. Nakayama, Journal of Electron Microscopy 49 (3), 415-421 (2000).
[56] W. Wongwiriyapan, S.-i. Honda, T. Mizuta, T. Ohmori, T. Murakami, K. Kisoda, H. Harima, J.-G. Lee, H. Mori, K. Oura and M. Katayama, Japanese Journal of Applied Physics 45 (3A), 1880-1882 (2006).
[57] T. Sakamoto, C.-C. Chiu, K. Tanaka, M. Yoshimura and K. Ueda, Journal of Nanomaterials 2009, 851290 (2009).
[58] F.-K. Tung, M. Yoshimura and K. Ueda, Journal of Nanomaterials 2009, 612549 (2009).
[59] I. T. Clark, Y. Matsuoka and M. Yoshimura, in preperation.
[60] C. L. Cheung, J. H. Hafner, T. W. Odom, K. Kim and C. M. Lieber, Applied Physics Letters 76 (21), 3136-3138 (2000).
[61] C. L. Cheung, J. H. Hafner and C. M. Lieber, PNAS 97 (8), 3809–3813 (2000).

[62] A. T. Woolley, C. Y. Cheung, J. H. Hafner and C. M. Lieber, Chemistry and Biology 7, R193-R204 (2000).
[63] E. Yenilmez, Q. Wang, R. J. Chen, D. Wang and H. Dai, Applied Physics Letters 80 (12) (2002).

7

# Carbon Nanotube AFM Probe Technology

Z. W. Xu[1], F. Z. Fang[1*] and S. Dong[2]
[1]*State Key Laboratory of Precision Measuring Technology & Instruments, Centre of MicroNano Manufacturing Technology, Tianjin University*
[2]*Center for Precision Engineering, Harbin Institute of Technology*
*China*

## 1. Introduction

The invention of atomic force microscopy (AFM) is having a great impact on various areas, such as nano metrology, materials science, surface science and biology (Binnig et al., 1986). The lateral resolution of AFM is mainly determined by the probe's shape and physical property, especially the geometry and dimension of the probe end. Conventional AFM probe is pyramidal shape by micro-fabrication. The pyramidal probe would result in image resolution degradation by severe probe broaden effect, especially for the structures with higher aspect ratio, such as gratings and structures in MEMS.

To broaden the AFM applications, researchers pursue new kind probes that have longer lifetime, higher resolution, and better mechanical property. Carbon nanotubes (CNT) show many excellent properties, such as, high aspect ratio, high Young's modulus, excellent elastic buckling property, and electrical and thermal conductivity (Iijima, 1991). The above characteristics make carbon nanotube be ideal as probes in AFM. Carbon nanotubes have demonstrated considerable potential as AFM probes after the first CNT AFM probe was invented in 1996 (Dai et al., 1996).

This chapter would introduce the history of carbon nanotube AFM probes, including the CNT probes' farbication and configuration optimization, the CNT probes' image artefact and its elimination study, the applications of these new kind probes and researches to improve their performance.

## 2. CNT probes' fabrication

There are two key processes in the fabrication of carbon nanotube AFM probe: attachment and modification. Firstly, the CNT must be fixed to the end of ordinary AFM probe to be the new imaging point; and then configuration or function modification to optimize the CNT probe's properties.

### 2.1 CNT attachment

#### 2.1.1 Direct manipulation method

Since carbon nanotubes were first applied as AFM probes in 1996 by Dai (Dai et al., 1996), various methods have been developed for their fabrications. The earliest method was to

---

*Corresponding email: fzfang@gmail.com

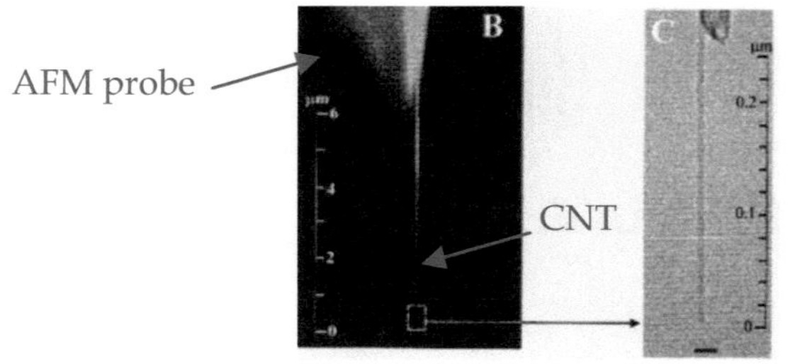


Fig. 1. SEM and TEM images of the first CNT AFM probe (Dai et al., 1996)

employ precise manipulation by picking and sticking a multi-walled carbon nanotube (MWNT) bundle to the silicon probe with an acrylic adhesive under direct view of optical microscope. This method can not be employed to well control the CNT probe's orientation. Further research extended this method using nanomanipulator in scanning electron microscope (SEM), and nanotube with smaller diameter can be selected and fixed to Si probe in SEM by electron beam induced carbon deposition (Nishijima et al., 1999) or Pt depostion (Fang et al., 2009).

Welding method was then developed for fixing CNT to Si probe end (Fang et al., 2007 ; Stevens et al., 2000). First, silicon probe and carbon nanotube were brought into a close distance by manipulating two microtranslators under direct view of inverted optical microscope. When carbon nanotube and silicon probe were in close proximity, an electric field of less than 20V was applied between them. The nanotube was attracted to the silicon probe and favorably aligned with the apex of silicon tip. Then the microtranslators are manipulated to make the protruding nanotube contact with the apex of silicon probe. The applied voltage is further increased between them to 30-60V until the nanotube was energetically disassociated by applying voltages and weld to the end of silicon probe. The carbon nanotubes grown in CVD method have small defects, at which points the resistance locally heats the nanotube until it is oxidized and divided. Fig. 2 is the schematic illustration of fabrication method.

The main drawback of the direct manipulation method is time consuming, and it is not an appropriate method for wafer-scale fabrication.

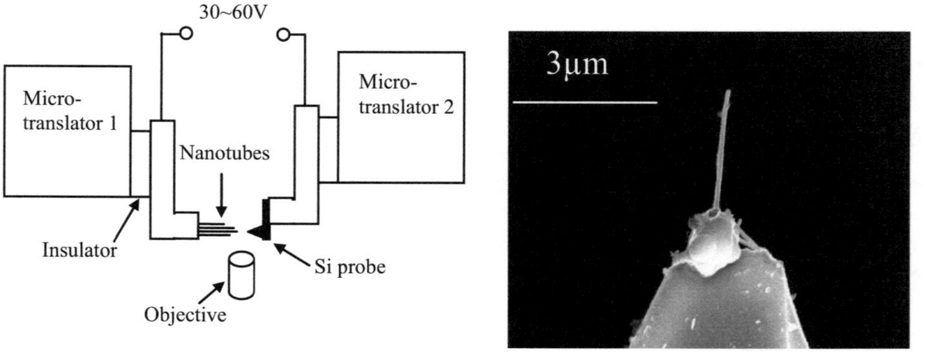


Fig. 2. Schematic illustration of welding process and CNT probe fabricated (Fang et al., 2007)

### 2.1.2 Chemical vapor deposition (CVD) method

Chemical vapor deposition (CVD) method was used to grow carbon nanotubes on a catalyst deposited silicon probe surface in 1999 (Hafner et al., 1999). The CNT's growth length can be

controlled by adjusting the growth time (Edgeworth et al., 2010). Ye reported an innovative approach that combined chemical vapor deposition with nanopatterning and traditional silicon micromachining technologies to large-scale grow CNT probe through effective nanocatalyst protection and release before and after the microfabrication (Ye et al., 2004).
The key advantage of the CVD direct growth method is the possible route to batch fabrication. It was predicted that nanotube bundle AFM probes could achieve reproducibly be wafer-scale fabrication by this technique in the near future [Wilson & Macpherson, 2009]. While, the reproducible production of CNT probes with individual single-walled carbon naontube (SWNT) is still the great challenge. How to accurately control the CNT probe's orientation becomes a key problem for the CVD method.

### 2.1.3 Pick-up method

In order to fabricate SWNT AFM probe, pick-up method was proposed in 2001 (Hafner et al., 2001). The SWNT could be picked up by the ordinary AFM probe with van der Waals forces during AFM probe scanning the SWNT substrate with nanotubes grow upwards. The pick-up method is flexible, while this method needed isolated and vertically aligned carbon nanotube samples.

### 2.1.4 Other approaches

Approaches have also been proposed to fabricate CNT probe in CNT solution under external fields, such as, using dielectrophoresis (Tang et al., 2005), alternating magnetic field (Hall et al., 2003). Tamara directly grew CNTs on the apex of AFM probe by selective heating of the catalyst under the microwave irradiation in the presence of ethanol vapor (Tamara et al., 2010). CNT probes were assembled by transplanting a CNT bearing polymeric carrier to a AFM cantilever (Kim et al., 2009).

### 2.2 Configuration modification

There are three key factors that determine the resolution and properties of the CNT probe, i.e., radius r, length L, and orientation θ (CNT orietation angle with respect to the sample surface). It is vital that a precise control of these parameters allows the optimization of CNT probe to meet different applications' demand.

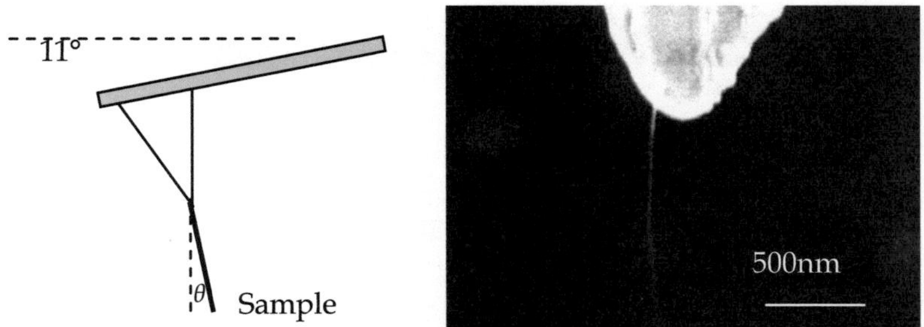


Fig. 3. Illustration of the CNT probe's thermally induced oscillations

A nanotube probe would be very gentle in lateral direction if the nanotube's aspect ratio ($L/r$) is too large. Fig. 3 shows a scanning electron microscope (SEM) image of a long CNT

AFM probe. The blurring of the lower portion of the CNT is due to the thermally induced oscillations.

The lateral force constant ($k_l$) of the CNT probe can be derived as the following

$$k_l = \frac{3YI}{L^3} = \frac{3\pi Y}{4} \cdot \frac{r^4}{L^3} \quad (1)$$

where $I$ is the stress moment over the cross-section of the nanotube, $Y$ is the Young's modulus of the carbon nanotube. The CNT probe bending response will dominate for a probe angle of even a few degrees with respect to the sample surface.

### 2.2.1 Orientation modification

The CNT probe's orientation should be vertical to sample surface theoretically for high resolution imaging. However, it is hard to accurately control the CNT probe's orientation during the CNT attachment process for most methods.

External electric field method was effective for CNT probe's orientation control (Fang et al., 2007). In the welding method, nanotube was attracted to the opposite Si probe and favorably aligned with the apex of probe under external applying voltage, as shown in Fig. 4. The attraction is due to the induced dipole moment in the nanotubes under external voltage. The largest strength of the electric field appears at the probe's apex, which will cause the CNT alignment with the apex. In the plasma enhanced chemical vapor deposition (PECVD) method, the CNT probe's orientation was controlled by an electric field present in the plasma discharge to align the nanotubes parallel to the electric field (Ye et al., 2004).

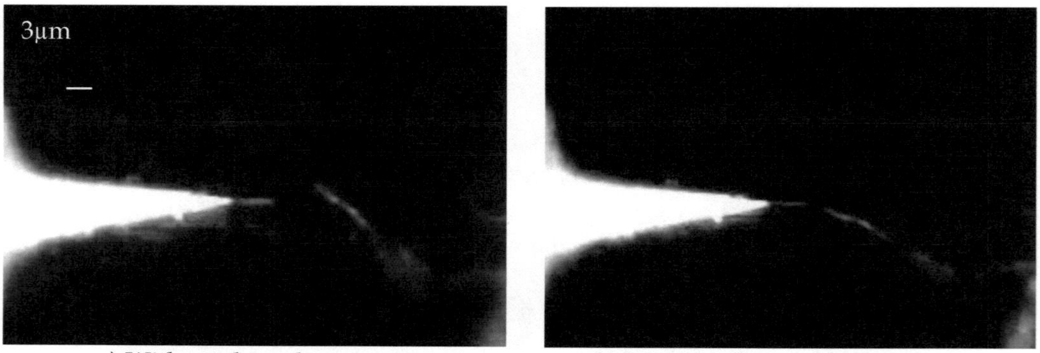


a) Without the voltage b) With a voltage of 30V

Fig. 4. The protruding nanotube favorably aligned with the probe apex (Fang et al., 2007).

Focused ion beam (FIB) irradiation method was recently proposed to modify the CNT probe orientation with high accuracy and reproducibility (Deng et al., 2006; Park et al., 2006). For the CNT probe's orientation alignment process, repeated single scans of the ion beam imaging were applied. The CNT probe was gradually aligned towards the FIB irradiation direction under every FIB scan and finally aligned parallel to the FIB irradiation direction, as shown in Fig. 5. The variables that affect the FIB-CNT interaction process include ion accelerating voltage, ion beam current, exposure time, and ion scan size (or the scan pixel density). Optimized combinations of these variable values can be made to achieve acceptable results (Fang et al., 2009).

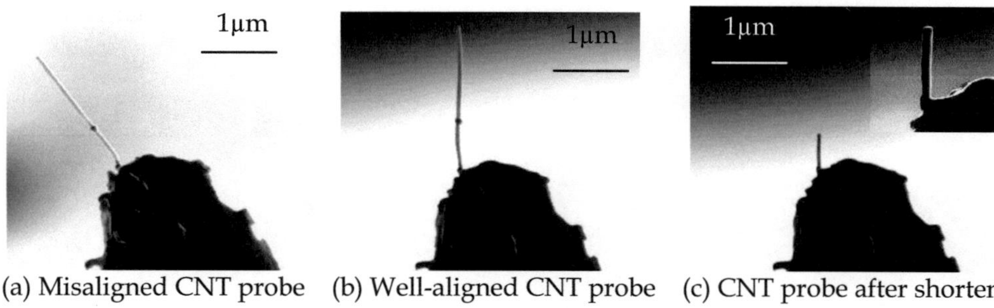


(a) Misaligned CNT probe (b) Well-aligned CNT probe (c) CNT probe after shortening

Fig. 5. Nanotube probe's orientation and length are optimized by FIB irradiation and milling processes. The FIB alignment parameters in Figure 5(b) are 30kV and 50pA in the ion acceleration voltage and the ion current, respectively (Fang et al., 2009).

The CNT orientation changes are mainly due to the strain induced by the FIB irradiation and the CNT's excellent plastic ability. Experimental and simulation results have demonstrated that the carbon nanotubes can be considered as self-healing materials under electron or ion irradiation (Krasheninnikov & Nordlund, 2010). Two mechanisms govern the CNT defect annealing (Krasheninnikov et al., 2002). The first mechanism is vacancy healing through dangling bond saturation and by forming nonhexagonal rings. The second one is the migration of carbon interstitials and vacancies, followed by Frenkel pair recombination.

### 2.2.2 Length modification

The CNT probe's length should be chosen and modified for different applications. Shortening the CNT probe can decrease the thermal oscillation amplitude. Therefore, short CNT probe is required for the CNT probe's high resolution imaging or applicatons with high lateral force constant requirement, such as friction study (Lai et al., 2010). On the contrary, long CNT probe is useful for the high aspect ratio imaging, e.g., biological samples.

The first method to shorten the CNT probe was realized on AFM by electrical etching (Dai et al., 1996). Then the method of using electron bombardment under the electric field was applied in the direct manipulation methods under optical microscope or SEM, which can observe and control the shortening process in real-time (Fang et al., 2007).

A new 'nanoknife' method was proposed to precisely cut and sharpen CNT probe by local vaporization of carbon resulting from Joule heating (Wei et al., 2007). The 'nanoknife' was a short carbon nanotube adhered to a metal tip. The 'nanoknife' cutting process is controllable and repeatedly, as shown in Fig. 6. In the cutting process, a DC voltage of 5-10 V was applied between the CNT probe and the 'nanoknife', the 'nanoknife' was then manipulated to contact with the CNT probe at the selected position. When the contact was made, the CNT probe would be precisely cut at the contact position (Wei et al., 2009). The cutting position can be precisely controlled using the nanomanipulators. It is found that the cutting process happened within 0.01 second by measuring the current going through the contact.

In recent years, Focused Ion Beam milling method was used to precisely shorten the CNT probes (Fang et al., 2009). The C-C chemical bonds of CNT would be broken during the high energy ion milling. The end of the CNT probe after FIB shortening is found to be a round end with fullerene-like cap, which is independent with the FIB's parameters, as shown in Fig. 5 (c). Since the FIB's beam energy is a Gaussian distribution, the Gaussian tail would produce many Carbon dangling bonds at the irradiation sites, where the dangling bonds would spontaneously close into a graphitic dome after shortening (Charlier et al., 1997).

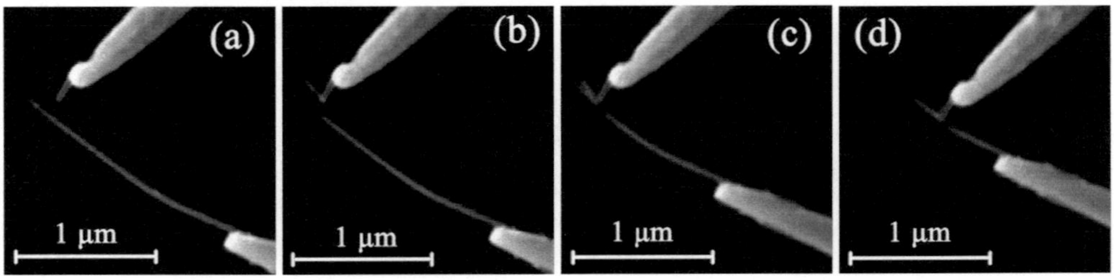


Fig. 6. SEM images showing a CNT (the lower one) is cut by a CNT 'nanoknife' repeatedly and becomes shorter and shorter (Wei et al., 2007).

### 2.2.3 High aspect ratio

Shortening the CNT probe can be useful to increase its resolution and stiffness. While a short CNT probe is not always the best choice for any applications. For high aspect ratio imaging, such as grating and biological sample imaging, the CNT probe should possess sufficient aspect ratio to avoid the imaging degradation by probe broadening effect. However, the force required to bend or buckle the CNT probe would greatly decrease as its aspect-ratio increase, which would be prone to result in image artefact at the sample steep positions. A strategy to improve the mechanical integrity of CNT probe is to coat it with a thin layer film, such as polymers, Au or carbon (Yum et al., 2010).

For high aspect ratio imaging, the best CNT probe configuration should have sufficient probe lateral stiffness with portion CNT protruding at the probe end, which can get high image resolution and good probe rigidity. Method of using electron beam locally induced Pt deposition in-site to strengthen the nanotube probe with the nanotube end free of deposition was proposed (Xu et al., 2009).

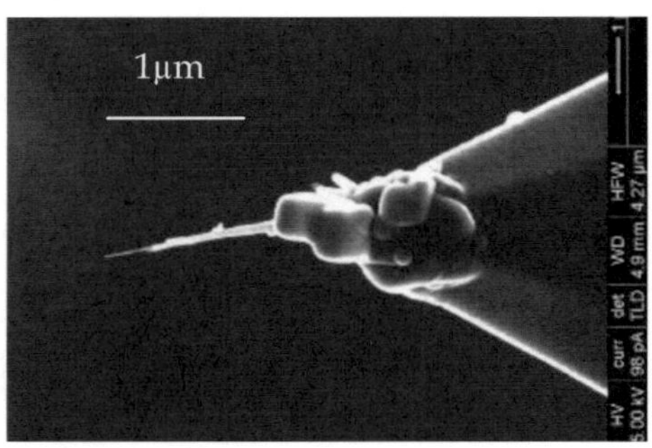


Fig. 7. CNT probe for high aspect ratio imaging (Xu et al., 2009).

Besides the configurations' demands, the attachment force of CNT probe should be large enough for versatile applications under different conditions. The probe-sample interaction forces for ordinary AFM probe are typically 10–100 nN in air. The bounding force of CNT probe can be roughly calculated by measuring the Si probe's deflection before and after the CNT is cut by nanomanipulation (Xu et al., 2009). The bonding force for CNT and Si probe with Pt depostion fix method is found larger than 500nN. The direct manipulation method and CVD growth method would have sufficient bounding strength than other methods.

Although the CNT probes have been developed, but they are still not widely adopted by the AFM users. The key issue that restricts the CNT probe's wide applications is that the CNT probe is much more expensive in comparison with ordinary tapping-mode AFM probe, which is mainly due to its complex fabrication technics and less productivity. It is vital to optimize the CNT probe's fabrication process with an accessible price and meeting versatile application demands. Moreover, the CNT probe's complex mechnical response during its working would also be an obstacle for its wide applications. For example, the CNT probe would be apt to produce image artefact during the high aspect ratio image. The CNT probe-sample interaction and its mechanical response are discussed in the following section.

## 3. CNT probe-sample interaction

The CNT probe is generally used in tapping-mode AFM (TM-AFM), where CNT probe intermittent contacts with sample surface and oscillating with high frequency during its scanning. Comparing with the ordinary AFM tappingmode probe, many new phenomena would appear when using a CNT probe. It is necessary to study the CNT probe interaction with sample and explain the new phenomena for CNT probe in advance.

### 3.1 Introductions on tip-sample interaction force and phase angle

Fig. 8 depicts schematically the forces acting between the AFM probe and sample as a function of the probe-sample separation D (Clemens et al., 2006). As D decreases, the AFM probe firstly detects the attractive force. If the magnitude of the attractive force gradient *dF/dD* exceeds the probe's force constant *k* (point B), then the AFM probe cantilever will be unstable and jump into contact with the sample (point B'). The position B is called "snap into contact point". As the separation further decreases, the CNT probe-sample interactions would become repulsive force.

Thus in the experiment, the ability to precisely track the force versus distance curve for all tip-to-surface separations, mainly depends on the AFM probe's force constant. For example, the unstable phenomenon of "snap to contact" would appear if the CNT probe's lateral force constant is too low. On the contrary, if the probe force constant *k* is always larger than the sample force gradient *dF/dD*, the cantilever-dependent instability can be practically eliminated, thus enable a faithful measurement of the probe-to-sample interactions (Landman et al., 1990).

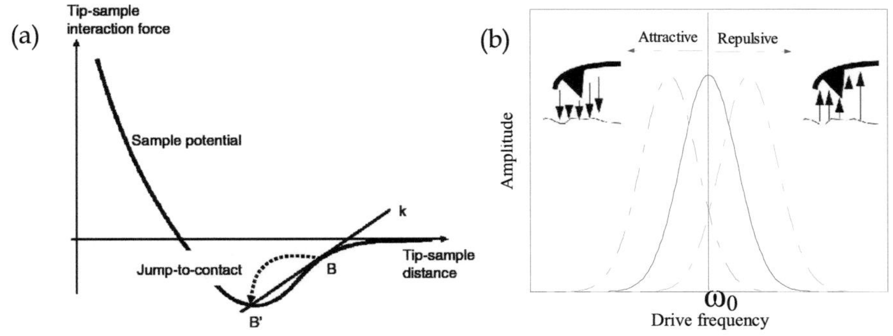


Fig. 8. Interaction force curve as the tip approaches the sample (a) (Clemens et al., 2006) and the AFM probe's resonance frequency shift under different forces (b).

The AFM probe-sample interaction forces variations can be sensitively detected by the cantilever's resonance frequency shift and phase angle. Any additional tip-sample force gradient would produce a shift of the cantilever's resonant frequency for the probe in tappingmode AFM. Attractive forces make the cantilever "softer" effectively, reducing the cantilever resonant frequency. In contrast, repulsive forces make the cantilever "stiffer" effectively, increasing the resonant frequency (Fang et al., 2008).
As a function of the driving frequency $\omega$, the phase angle $\varphi$ of the cantilever oscillation relative to the signal sent to the piezoelectric driver can be derived as the following,

$$\tan\varphi = \frac{\omega\omega_0 / Q}{\omega_0^2 - \omega^2} \quad (2)$$

where the $\omega_0$ is the probe's resonance frequency and Q is the quality constant of vibrating cantilever. From equation (2), the probe phase angle increases from 0° to 180° and the phase angle is 90° under the resonance frequency.
When the cantilever is far enough from the sample, there is no interaction between the AFM tip and the sample. Thus, the probe phase angle is 90° when the driving frequency is chosen as the cantilever resonance frequency according to equation (2). As the probe approaching the sample surface, the attractive forces first acting on the probe would decrease the probe's resonance frequency, where attractive forces make the cantilever "softer" effectively. Subsequently, the probe driving frequency becomes larger than the probe resonance frequency. Therefore, the probe phase angle is larger than 90° when the probe suffers attractive forces. Conversely, the phase angle is less than 90° when the probe works in repulsive forces region.
Therefore, the phase angle can be used to sensitively detect the probe-sample interactions. By mapping the phase of the cantilever oscillation during the AFM scan, phase imaging goes beyond simple topographical mapping to detect variations in composition, adhesion, friction, viscoelasticity, and other properties, as shown in Fig. 9. The phase images have been used to present the hydrophobic nature of single CNT by using the CNT probe with different AFM image parameters (Fang et al., 2008).

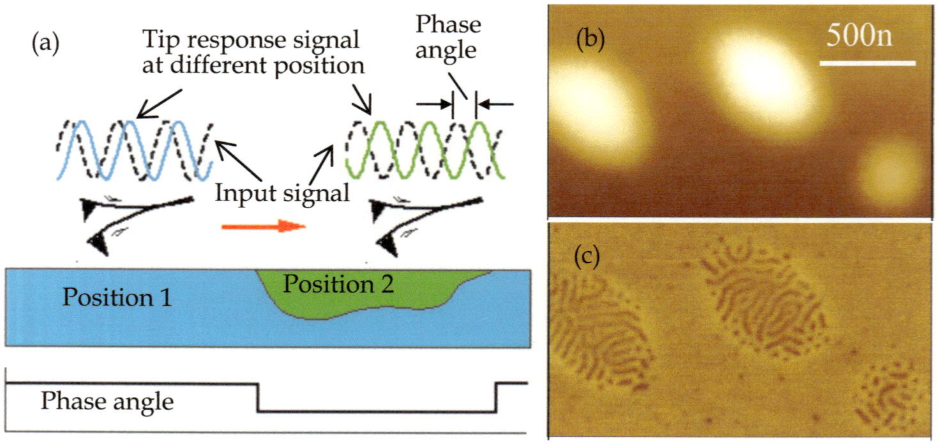


Fig. 9. (a) Phase angle in TM-AFM (b) topography and (c) phase images of copolymer. The height scale is 10nm and the phase angle scale is 20° (Fang et al., 2008).

### 3.2 Image artefact

For ordinary TM-AFM Silicon probe, its force constant is large enough to avoid the instability of "snap to contact". However, in the case of CNT probe, it is very gentle in lateral directions and the CNT would elastically buckle if its lateral force exceeds the threshold value. In this section, we will discuss what the CNT probe's mechanical responses and feedback would be if the CNT probe's instability occurred. And the research works of how to overcome the instability would also present.

#### 3.2.1 Artefact at the steep positions

When the CNT probe scans sample positions with a great surface slope, such as a grating step, the nanotube would be nearly parallel with the grating edge, where the lateral force exterted to the CNT probe would increase dramatically comparing with scanning the flat positions (Akita et al., 2000b). Akita has done the CNT probe force curve (force-versus-distance curve) measurement performed near the pit edge and on a plane surface, as denoted in Fig. 10. The amplitude at a plane surface monotonically decreases with the probe-sample distance, which is similar to the conventional Si probes. The CNT probe's oscillation is recovered at $z < -40$ nm due to the reproducible buckling of the nanotube. While for the case near the pit wall, the probe's oscillation is stopped abruptly, which indicates that the CNT probe snaps to contact with sample. These results from the trapping of the nanotube probe by attraction between the sidewalls of the nanotube and the pit. The adhesion force of the nanotube suffered near the pit edge is tested about 10nN (Akita et al., 2000a). At this point, it is impossible to control the position of the probe height for the AFM observation because that there is no distance dependence of the amplitude on the force curve. Therefore, the AFM image of the CNT probe at the pit edge is unstable with many ripple and wavelike distortion.

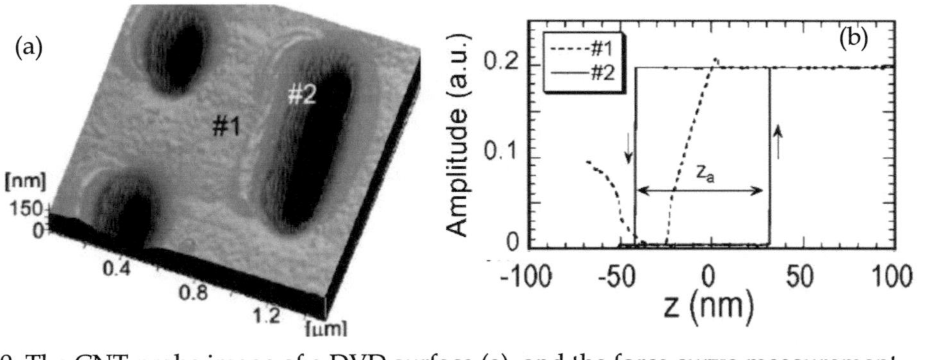


Fig. 10. The CNT probe image of a DVD surface (a), and the force curve measurement performed near the pit edge and on a plane surface (b) (Akita et al., 2000a).

It was found that the ripple and wavelike distortion is more easily occurred at the pit's left edge than the right edge in general, as shown in Fig. 10. This are mainly attributed to the probe having a tilt angle of 11° when it scans a sample in AFM as shown in Fig. 3, where the vibrating probe suffers different repulsive forces at different pit edges (Fang et al., 2008). What is the formation mechanism for the ripple and wavelike distortion at the grating edge? The phase angle data combining with height and amplitude data have been proposed to explain the image artifact, which is resulting from the CNT bent-adhesion-separation repeated process (Fang et al., 2008; Strus et al., 2005), as analyzed in Fig. 11.

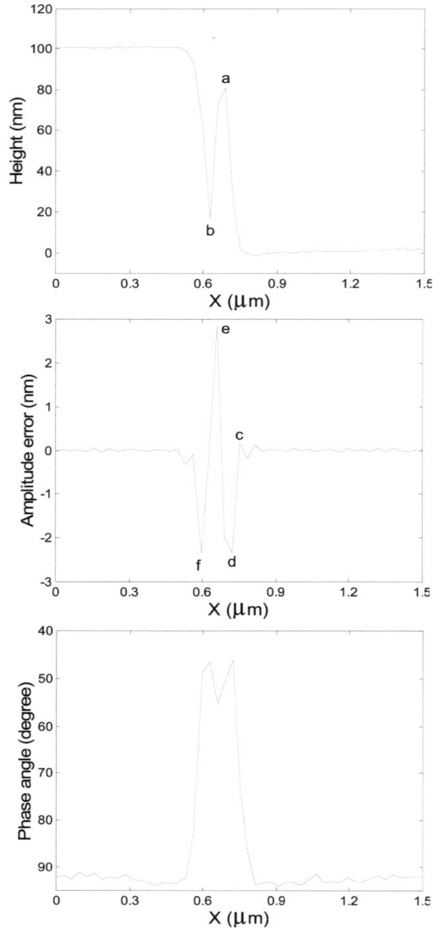


Fig. 11. Sectional view of height, amplitude and phase angle of the CNT probe image artifacts (Fang et al., 2008).

When the CNT probe contacts with the grating edge, the CNT–trench wall adhesion would bend the CNT probe and then reduce the vibration amplitude of the cantilever, as position d indicates in Fig. 11. The AFM controller perceives the sudden probe amplitude drop as a surface height increase and responds by quickly moving the probe away from the sample. At position *a*, the CNT has finally been broken free from the trench wall, and the cantilever vibrates at its free amplitude. There is about a $10^{\circ}$ phase angle decrease around position *a* in the phase angle curve at the bottom of Fig. 11, which confirms that the CNT probe really moves away from the trench wall. In order to restore the original scan amplitude, the AFM controller again lowers the probe. At position *b*, the CNT probe contacts with the trench wall again and the amplitude also drops due to the bending of the CNT. Therefore, the image artifact at the trench wall results from the CNT probe bending–adhesion–separation repeated process (Fang et al., 2008). When the ripple and wavelike AFM results appear at the steep positions using CNT probe, we should carefully judge whether it is the real sample morphology or the image artefact resulting from the CNT probe's instability.

The CNT probe can be treated as a series system of the ordinary AFM probe combining with a end-fixed carbon nanotube. The effective force constant for this series system can be calculated as followed:

$$K_{eff} = \frac{F}{X} = \frac{F}{\frac{F}{K_s} + \frac{F}{K_c}} = \frac{K_s K_c}{K_s + K_c} \quad (3)$$

Therefore, the force constant of the Si probe would also affect the CNT probe's effective stiffness. Under many CNT probe fabrication experiments, the lateral force constant requirement of stable working for the CNT probe attaching to the triangular-shape AFM probe is found relative larger than the one to the rectangular shape probe. This is mainly due to the triangular shape AFM probe's less resistance ability to the lateral force (J. E. Sader & R. C. Sader, 2003).

### 3.2.2 CNT-sample boundary conditions

In general, CNT probes are harmonically excited near resonance and dynamically operated in either an attractive or repulsive regime using tapping-mode AFM. When CNT probes are operated in the net repulsive regime, CNTs have the potential to buckle, bend, adhere, and slide (Strus & Raman, 2009). Many studies highlighted that the CNT probe-sample boundary conditions and its mechanical responses would play a vital role in the CNT probe's stability.

A thermal noise forcing method was used to investigate the mechanical response of CNT AFM probes under the free sliding and pinned CNT contacts cases, as shown in Fig. 12. Generally, thermal noise forcing produces probe oscillation amplitudes only about angstrom. Therefore, the thermal noise spectrum is very sensitive to tiny variations of the contact properties between the CNT probe and the surface, thus providing an accurate picture of the CNT probe's mechanical response (Buchoux et al., 2009). Molecular simulations were proposed to explain the effects of CNT's slip and snap-to-contact on the resolutions (Solares & Chawla, 2010).

Strus suggested that the identification of CNT pinning and slipping during intermittent contact can be based on phase contrast imges and energy-dissipation spectroscopy, as shown in Fig. 12. The pinning or slipping of the CNT depends on the scan surface and the CNT probe's stiffness. It is found that CNT probe is easier slipping on the graphite surface than on the graphene oxide and silicon oxide surfaces (Strus & Raman, 2009).

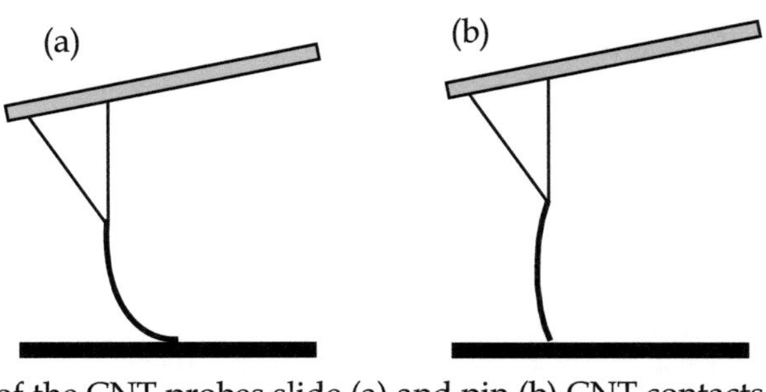


Fig. 12. Illustration of the CNT probes slide (a) and pin (b) CNT contacts cases.

The CNT probe's slipping effect was experimental studied (Fang et al., 2009). It is found that the CNT probe is apt to slide if it scans with a large vibration amplitude. There are many dark lines resulting from CNT's slide in the AFM image as arrows indicated in Fig. 13(b). Due to CNT's hydrophobic property and its high aspect ratio, the capillary force between the CNT probe and the sample surface is much smaller. Therefore, the CNT probe does not need to work at large amplitudes to break the capillary force as the Si probe does and it can work well under low scanning amplitude of less than 30 nm.

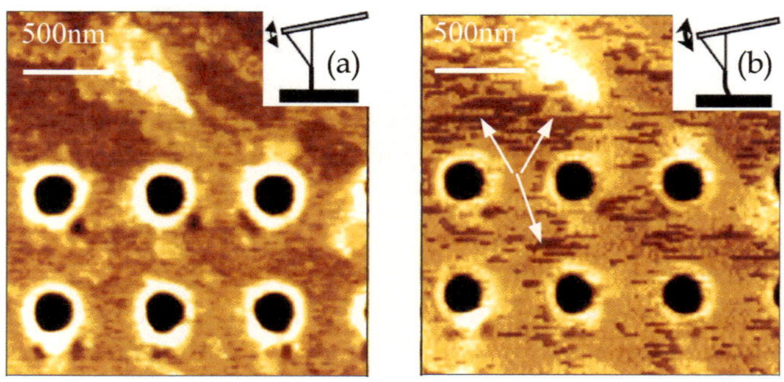


Fig. 13. Nanoholes images captured by CNT probe with 18nm (a) and 40nm (b) scan amplitudes (Fang et al., 2009).

The CNT slipping is an obstacle for accurate AFM topography measurements, while it can be treated as a potential method of analyzing surface composition by enhancing the probe-sample frictions contrast. AFM peeling force spectroscopy and theoretical studies have been used to quantitatively investigate the physics of adhesion and stiction of carbon nanotubes on different material substrates, such as, HOPG, polymer (Strus et al., 2008).

### 3.3 Artefact elimination

Studies have been conducted to lessen and avoid the CNT probe's image artefact. Larger scan amplitude ratio was employed to depress the ringing artefact, by increased from 66% to 96% of the 34 nm unconstrained vibration amplitude (Strus et al., 2005). Akita found that the CNT probe's image instability near the wall of the DVD surface pits can be avoided by slightly increasing the probe's free tapping amplitude to avoid the pit's attractions (Akita et al., 2000a). Solares has theoretically explored that spectral inversion method and dual-frequency-modulation method can mitigate the imaging artifacts when using short nanotube probes (<100 nm in length) in air. The two methods are capable of performing simultaneous imaging and force spectroscopy (Solares & Chawla, 2010).

The most effective method to overcome the CNT probe's image artefact is to strengthen the CNT probe, as discussed in the Section 2.2. For example, several micrometers long multiwalled CNT probes were prepared and strengthened with carbon molecular layers to overcome mechanical instabilities (Vakarelski et al., 2007).

## 4. CNT probe's applications

### 4.1 High resolution imaging

CNT's excellent properties make CNT probe be a favorable choice for AFM probe in topographic imaging. SWNT probe with several nanometers' radius shows great image resolution over conventional Si probe (Cheung et al., 2000). CNT probes have also been showed higher image resolution for the high aspect ratio samples, such as, grating, cellular samples, etc (Dai et al., 1996; Fang et al., 2007; Hafner, 1999, 2001; Stevens et al., 2000). It is found that not only the CNT probe's radius would influence its resolution, its end configuration also showed impact on CNT probe's image ability (Fang et al., 2009).

CNT has a much higher Young's modulus and exceptional elastic buckling property, which results in the much better wear resistance ability of the CNT probes relative to the Si probes. The excellent wear resistance property makes CNT probes have consistent image resolution, which is very important for applications, such as, online inspection.

Electrostatic force microscopy (EFM) is one technique sensitive to the long-range electrostatic forces, which are useful in the study of semiconductor devices. Because of the long-range nature of electrostatic force, interpretation of EFM images is often complicated as the probe beyond the probe apex would also affect the EFM image results. This effect can be greatly reduced using high aspect ratio CNT probes (Wilson & Macpherson, 2009). For similar reasons CNT probes are also suitable for high resolution magnetic force microscopy. Iron-filled carbon nanotubes (Fe-CNTs) have been proposed for the application for magnetic force microscopy (Wolny et al., 2008).

### 4.2 Nanofabrication

Carbon nanotube probe is not only a tool for imaging, but also for nanofabrication, as shown in Fig. 14. AFM nanolithography can be widely applied in fields such as data storage and device fabrication. Due to nanotube's excellent properties, carbon nanotube probes have been explored in fabricating oxide nanostructures on silicon surfaces (Dai & Franklin, 1998), a p-GaAs(100) surface (Huang et al., 2006), a metallic tantalum film (Choi et al., 2007).

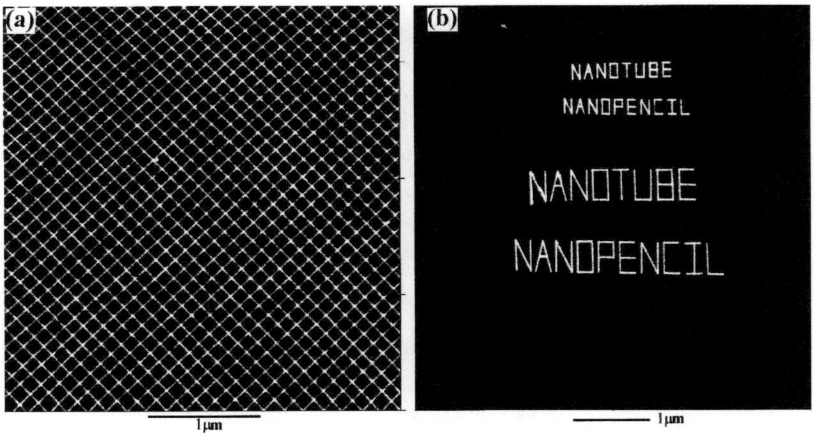


Fig. 14. AFM image of 2 nm tall, 10 mm wide, and 100 nm spaced silicon-oxide (light) lines fabricated by a nanotube tip using anodic oxidation method (Dai & Franklin, 1998).

It was found that hydrophobicity is a key factor in the improved reliability of CNT probes on the nanoscale oxide fabrication over conventional Si probes (Kuramochi et al., 2007). Okazaki studied the nano-gratings direct nanolithography on organic polysilane PMPS films using CNT probe (Okazaki et al., 2000).

### 4.3 Biology application

Due to its excellent properties, CNT probe has been widely used in the biological application. It produces lower damage to the proteins than the Si probe during AFM scanning, as shown in Fig. 15 (Fang et al., 2007). CNT probe has been proven more suitable for imaging soft biological samples, such as, DNA, protein (Cheung et al., 2000).

Technology for introducing biomolecules and nanoparticles into living cells with minimal invasiveness is a key task to study the physical properties and biochemical interactions that govern the cell's behavior (Chen et al., 2007). Micro glass pipettes have been used for intracellular analysis research. However, due to their relatively large sizes and rigid structure, these probes cause severe damage to the cells (Niu et al., 2011).

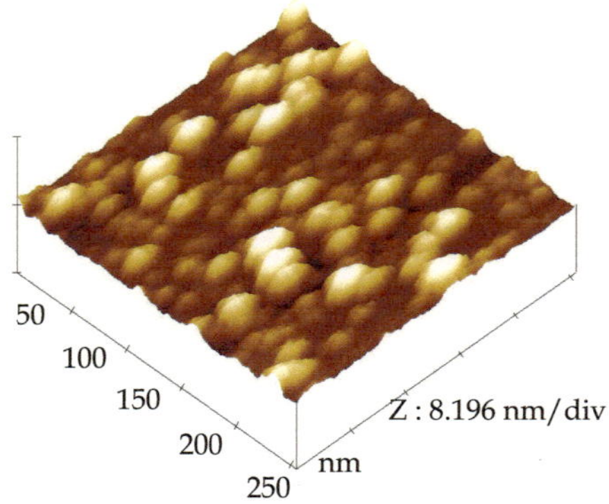


(a) Measured by CNT probe

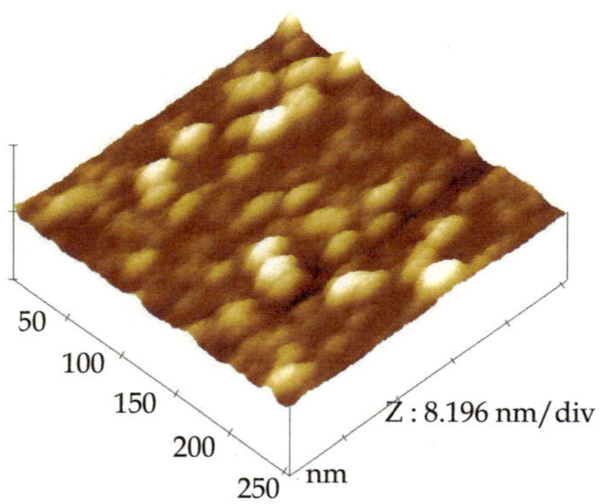


(b) Measured by Si probe

Fig. 15. AFM analyses of IgG protein. The Si probe has created a larger probe indentation to the soft proteins than the CNT probe (Fang et al., 2007).

Due to its nanoscopic dimensions, high mechanical strength, and functionalizable surfaces, CNT probe has been used as a "nanoneedle" in a nanoscale cell injection system to deliver cargo into cells. A multiwalled CNT probe was functionalized with cargo via a disulfide-based linker. Penetration of cell membranes with this "nanoneedle" was controlled by the

AFM. The following reductive cleavage of the disulfide bonds within the cell's interior resulted in the release of cargo inside the cells, after which the nanoneedle was retracted by the AFM control, as shown in Fig. 16 (Chen et al., 2007). The CNT probe "nanoneedle" technique causes little membrane or cell damage. Other biomolecules such as DNA and RNA, or synthetic structures such as polymers and nanoparticles can be delivered into cells in a similar fashion (Chen et al., 2007; Vakarelski et al., 2007).

Gold nanoparticle-decorated CNT probes are used to study intracellular environments in situ using surface-enhanced Raman spectroscopy (SERS), as shown in Fig. 17 (Niu et al., 2011). The high aspect ratio of CNTs displaces less intracellular volume than ordinary glass pipettes or optical fibers cell probes, thus perturbing the cell much less, especially when measurements performed over long periods of time. This might be a useful method for tip-enhanced Raman spectroscopy (TERS) application using CNT probe.

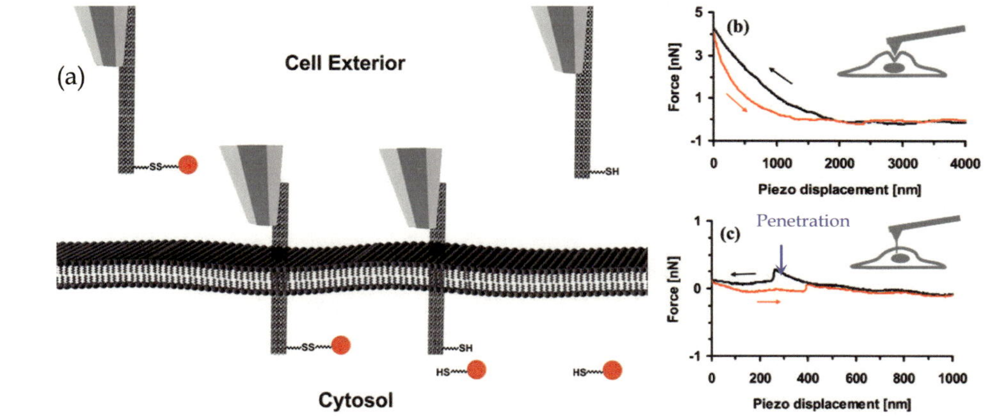


Fig. 16. Schematic of the CNT probe nanoinjection procedure. A CNT probe with cargo attached to the CNT surface via a disulfide linker penetrates a cell membrane (a) (Chen et al., 2007). Force vs piezo displacement curves for the indentation of a living cell using a pyramidal Si probe (b) and a C/Au-coated nanotube probe (c) (Vakarelski et al., 2007).

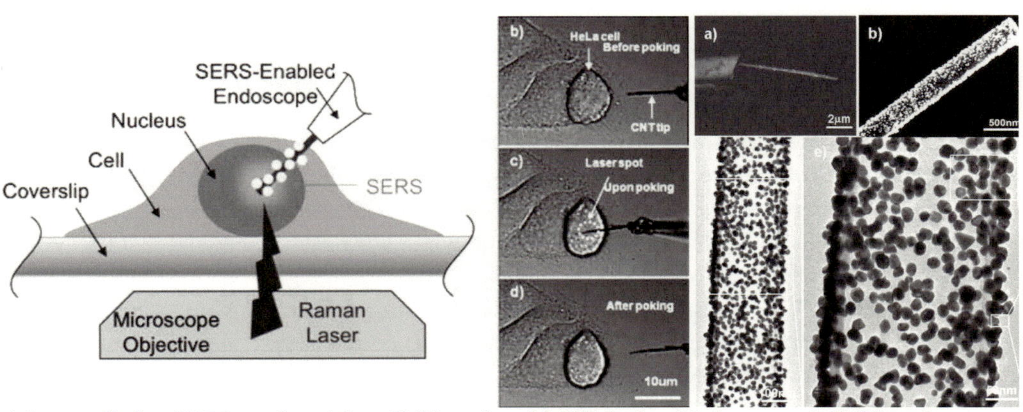


(a) Intracellular SERS study with a CNT probe (b) SEM of gold-decorated CNT probe

Fig. 17. Intracellular SERS study with a nanoparticle-decorated CNT probe (Niu et al., 2011).

## 5. Summary and outlook

Carbon nanotube probes open up a new AFM imaging world, increasing the probe's resolution and longevity, decreasing probe-sample forces, and extending the AFM application fields. It would also have a significant impact in key research areas, such as, nanometrology, surface engineering and biotechnology.

Future main attentions in the carbon nanotube AFM probes can be summarized as follows:

- CNT probe fabrication technics' optimization: the technics of high efficient fabrication with low cost; the fabrication's reproducibility and controllability.
- CNT probe imaging accuracy calibration: the CNT-sample boundary conditions and CNT's mechanical response effect studies on the CNT probe imaging resolutions in nanometer or angstrom levels; CNT probe's accuracy comparison with the others precision measurement techniques.
- CNT probe's unique applications: TERS; the biological application, such as, functionalized CNT probe as nanoneedle for intracellular studies and drug delivery.

## 6. Acknowledgement

The authors would thank the persons who have given kind helps in our CNT probe's researches, Dr. L. Q. Guo from Harbin Institute of Technology, Prof. Jason H. Hafner from Rice University, Prof. Hongjie Dai from Stanford University, Dr. Jason Cleveland from Asylum Research and Dr. M. C. Strus from Purdue University. Thanks go for Mr. Haifeng Gao and Miss Lihua Xu from Tianjin University for helping the typesetting.

The research work is supported by '111' project by the State Administration of Foreign Experts Affairs and the Ministry of Education of China (B07014), National High Technology Research and Development Program of China (863 No. 2009AA044305& 2009AA044204), National Natural Science Foundation of China (No. 50905126, 90923038&50935001), Ministry of Industry and Information Technology (No. 2011ZX04014-071), Natural Science Foundation of Tianjin(No. 10JCYBJC06400), and National Basic Research Program of China (973 Program, Grant No.2011CB706700).

## 7. References

Akita, S., Nishijima, H. & Nakayama, Y. (2000a). Influence of stiffness of carbon-nanotube probes in atomic force microscopy. *J. Phys. D: Appl. Phys.*, Vol.33, No.21, (November 2000), pp. 2673–2677, ISSN 0022-3727

Akita, S., Nishijima, H., Kishida, T. & Nakayama, Y. (2000b). Influence of Force Acting on Side Face of Carbon Nanotube in Atomic Force Microscopy. *Jpn. J. Appl. Phys.*, Vol.39, (March 2000), pp. 3724-3727, ISSN 0021-4922

Binnig, G., Quate, C. & Gerber. C. (1986). Atomic Force Microscope. *Phys. Rev. Lett.*, Vol.56, No.9, (March 1986), pp. 930~933, ISSN 0031-9007

Buchoux, J., Aim´e, J. P., Boisgard, R., Nguyen, C. V., Buchaillot, L. & Marsaudon, S. (2009). Investigation of the carbon nanotube AFM tip contacts: free sliding versus pinned contact. *Nanotechnology*, Vol.20, No.47, (November 2009), 475701 (8pp), ISSN 0957-4484

Charlier, J.C., Vita, A.D., Blasé, X. & Car, R. (1997). Microscopic Growth Mechanisms for Carbon Nanotubes. *Science*, Vol.275, (January 1997), pp. 647-649, ISSN 0036-8075

Chen, X., Kis, A., Zettl, A. & Bertozzi, C. R. (2007). A cell nanoinjector based on carbon nanotubes. *PNAS*, Vol.104, No.20, (May 2007), pp. 8218–8222

Cheung, C. L., Hafner, J. H. & Lieber, C. M. (2000). Carbon nanotube atomic force microscopy tips: Direct growth by chemical vapor deposition and application to high-resolution imaging. *Proc. Natl Acad. Sci. USA*, Vol.97, No.8, (April 2000), pp. 3809–3813, ISSN 0027-8424

Choi, J. S., Bae, S., Ahn, S. J., Kim, D. H., Jung, K. Y., Han, C., Chung, C. C. & Lee, H. (2007). Feasibility of multi-walled carbon nanotube probes in AFM anodization lithography. *Ultramicroscopy*, Vol.107, No.10-11, (October 2007), pp. 1091–1094, ISSN 0304-3991

Clemens, M. F., Sunil, D. G., Kim, S. & Kim S. G. (2006). A Nanoscanning Platform for Bio-Engineering: an in-Plane Probe with Switchable Stiffness. *Nanotechnology*, Vol.17, No.4, (February 2006), pp. S69~S76, ISSN 0957-4484

Dai, H. J., Hafner, J. H., Rinzler, A.G., Colbert, D. T., & Smalley, R. E. (1996). Nanotubes as Nanoprobes in Scanning Probe Microscopy. *Nature*, Vol.384, (November 1996), pp. 147~151, ISSN 1476-1122

Dai, H. J. & Franklin, N. (1998). Exploiting the Properties of Carbon Nanotubes for Nanolithography. *Applied Physics Letters*, Vol.73, No. 11, (September 1998), pp. 1508~1510, ISSN 0003-6951

Deng, Z. F., Yenilmez E., Reilein A., Leu, J., Dai, H. J. & Moler, K. A. (2006). Nanotube Manipulation with Focused Ion Beam. *Applied Physics Letters*, Vol.88, No.2, (January 2006), 023119 (3 pp), ISSN 0003-6951

Edgeworth, J. P., Burt, D. P., Dobson, P. S., Weaver, J. M. R. & Macpherson, J. V. (2010). Growth and morphology control of carbon nanotubes at the apexes of pyramidal silicon tips. *Nanotechnology*, Vol.21, No.10, (March 2010), 105605 (10pp), ISSN 0957-4484

Fang, F. Z., Xu, Z. W., Dong, S. & Zhang, G. X. (2007). High aspect ratio nanometrology using carbon nanotube probes in atomic force microscopy. *Annals of the CIRP*, Vol.56, No.1, (2007), pp. 533~536, ISSN 0007-8506

Fang, F. Z., Xu, Z. W. & Dong, S. (2008). Study on Phase Images of Carbon Nanotube Probe in Atomic Force Microscopy. *Measurement Science and Technology*, Vol.19, No.5, (May 2008), pp. 055501 (7 pp), ISSN 0957-0233

Fang, F. Z., Xu, Z. W., Zhang, G. X. & Hu, X. T. (2009). Fabrication and configuration of carbon nanotube probes in atomic force microscopy. *Annals of the CIRP*, Vol.58, No.1, (April 2009), pp. 455~458, ISSN 0007-8506

Hafner, J. H., Cheung, C. L. & Lieber, C. M. (1999). Growth of Nanotubes for Probe Microscopy Tips. *Nature*, Vol.398, No.6730, (April 1999), pp. 761~762, ISSN 1476-1122

Hafner, J. H., Cheung, C. L., Oosterkamp, T. H. & Lieber, C. M. (2001). High-Yield Assembly of Individual Single-Walled Carbon Nanotube Tips for Scanning Probe Microscopies. *Journal of Physical Chemistry B*, Vol.105, No.4, (February 2001), pp. 743~746, ISSN 1520-6106

Hall, A., Matthews, W. G., Superfine, R., Falvo, M. R. & Washburn, S. (2003). Simple and Efficient Method for Carbon Nanotube Attachment to Scanning Probes and Other Substrates. *Appl. Phys. Lett.*, Vol.82, No.15, (April 2003), pp. 2506~2508, ISSN 0003-6951

Huang, W. P., Cheng, H. H., Jian, S. R., Chuu, D. S., Hsieh, J. Y., Lin, C. M. & Chiang, M. S. (2006). Localized electrochemical oxidation of p-GaAs(100) using atomic force microscopy with a carbon nanotube probe. *Nanotechnology*, Vol.17, No.15, (August 2006), pp. 3838–3843, ISSN 0957-4484

Iijima, S. (1991). Helical Microtubules of Graphitic Carbon. *Nature*, Vol.354, (November 1991), pp. 56~58, ISSN 1476-1122

Kim, S., Lee, H. W. & Kim, S. G. (2009),. Transplanting assembly of carbon-nanotube-tipped atomic force microscope probes. *Applied Physics Letters*, Vol.94, No.19, (May 2009), 193102, ISSN 0003-6951

Krasheninnikov, A. V., Nordlund, K. & Keinonen, J. (2002). Production of defects in supported carbon nanotubes under ion irradiation. *Phys. Rev. B*, Vol.65, No.16, (April 2002), 165423, ISSN 1098-0121

Krasheninnikov, A. V. & Nordlund, K. (2010). Ion and electron irradiation-induced effects in nanostructured materials. *J. Appl. Phys.*, Vol.107, No.7, (April 2010), 071301, ISSN 0021-8979

Kuramochi, H., Tokizaki, T.,Yokoyama, H. & Dagata, J. A. (2007). Why nano-oxidation with carbon nanotube probes is so stable: I. Linkage between hydrophobicity and stability. *Nanotechnology*, Vol.18, No.13, (April 2007), 135703 (6pp), ISSN 0957-4484

Lai, W. C., Chin,S. C., Chang, Y. C., Chen, L. Y. & Chang, C. S. (2010). Lattice-resolved frictional pattern probed by tailored carbon nanotubes. *Nanotechnology*, Vol.21, No.5, (February 2010), 055702 (9pp), ISSN 0957-4484

Landman, U., Luedtke, W. D., Bumham, N. A. & Colton, R. J. (1990). Atomistic mechanisms and dynamics of adhesion, nanoindentation, and fracture. *Science*, Vol.248, No.4594, (April 1990), pp. 454-611, ISSN 0036-8075

Nishijima, H., Kamo, S., Akita, S. & Nakayama, Y. (1999). Carbon-Nanotube Tips for Scanning Probe Microscopy: Preparation by a Controlled Process and Observation of Deoxyribonucleic Acid. *Appl. Phys. Lett*, Vol.74, No.26, (June 1999), pp. 4061~4063, ISSN 0003-6951

Niu, J. J., Schrlau, M. G., Friedman, G. & Gogotsi, Y. (2011). Carbon Nanotube-Tipped Endoscope for In Situ Intracellular Surface-Enhanced Raman Spectroscopy. *Small*, Vol.7, No.4, (February 2011), pp. 540-545, ISSN 1613-6810

Okazaki, A., Kishida, T., Akita, S., Nishijima, H. & Nakayama, Y. (2000). Direct Nanolithography of Organic Polysilane Films Using Carbon Nanotube Tips. *Jpn. J. Appl. Phys.*, Vol.39, No.12B, (December 2000), pp. 7067~7069, ISSN 0021-4922

Park B. C., Jung K. Y., Song W. Y., Beom-Hoan O., Ahn S. J. (2006). Bending of a carbon nanotube in vacuum using a focused ion beam, *Advanced Materials*, Vol.18, No.1, (January 2006), pp. 95-98, ISSN 0935-9648

Sader, J. E. & Sader, R. C. (2000). Susceptibility of Atomic Force Microscope Cantilevers to Lateral Forces: Experimental Verification. *Appl. Phys. Lett.*, Vol.83, No.15, (October 2000), pp. 3195~3197, ISSN 0003-6951

Solares, S. D. & Chawla, G. (2010). Exploration of AFM Imaging Artifacts Occurring at Sharp Surface Features When Using Short Carbon Nanotube Probes and Possible Mitigation With Real-Time Force Spectroscopy. *Journal of Manufacturing Science and Engineering,* Vol.132, No.3, (June 2010), 030904(14pp), ISSN 1087-1357

Stevens, R. M. D., Frederick, N. A., Smith, B. L., Morse, D. E., Stucky, G. D. & Hansma, P. K. (2000). Carbon Nanotubes as Probes for Atomic Force Microscopy. *Nanotechnology.* Vol.11, No.1, (March 2000), pp.1~5, ISSN 0957-4484

Strus, M. C., Raman, A., Han, C. S & Nguyen, C. V. (2005). Imaging artefacts in atomic force microscopy with carbon nanotube tips. *Nanotechnology,* Vol.16, No.11, (November 2005), pp. 2482–2492, ISSN 0957-4484

Strus, M. C., Zalamea, L., Raman, A., Pipes, R. B., Nguyen, C. V. & Stach, E. A. (2008). Peeling Force Spectroscopy: Exposing the Adhesive Nanomechanics of One-Dimensional Nanostructures. *Nano Letters,* Vol. 8, No.2, (February 2008), pp. 544-550, ISSN 1530-6984

Strus, M. C. & Raman, A. (2009). Identification of multiple oscillation states of carbon nanotube tipped cantilevers interacting with surfaces in dynamic atomic force microscopy. *Physical Review B,* Vol.80, No.22, (December 2009), 224105, ISSN 1098-0121

Tamara, S. D., Stephanie, H. & Ulrich, S. S. (2010). Microwave-Assisted Fabrication of Carbon Nanotube AFM Tips. *Nano Lett.,* Vol.10, No.10, (October 2010), pp. 4009–4012, ISSN 1530-6984

Tang, J., Yang, G., Zhang, Q., Parhat, A., Maynor, B., Liu, J., Qin, L. C. & Zhou, O. (2005). Rapid and Reproducible Fabrication of Carbon Nanotube AFM Probes by Dielectrophoresis. *Nano Letters,* Vol.5, No.1, (January 2005), pp. 11~14, ISSN 1530-6984

Vakarelski, I. U., Brown, S. C., Higashitani, K. & Moudgil & B. M. (2007). Penetration of Living Cell Membranes with Fortified Carbon Nanotube Tips. *Langmuir,* Vol.23, No.22, (October 2007), pp. 10893-10896, ISSN 0743-7463

Wei, X. L., Chen, Q., Liu, Y. & Peng, L. M. (2007). Cutting and Sharpening Carbon Nanotubes Using a Carbon Nanotube 'Nanoknife'. *Nanotechnology,* Vol.18, No.18, (May 2007), 185503, ISSN 0957-4484

Wei, X. L., Jiang, A. N., Gao, S. & Chen, Q. (2009). Length Control and Sharpening of Carbon Nanotube Scanning Probe Microscope Tips Using Carbon Nanotube "Nanoknife". *J. Nanosci. and Nanotechnol.,* Vol.9, No.2, (February 2009), pp. 1258–1262, ISSN 1533-4880

Wilson, N. R., & Macpherson, J. V. (2009). Carbon Nanotube Tips for Atomic Force Microscopy. *Nature nanotechnology,* Vol.4, No.8, (August 2009), pp. 483~491, ISSN 0957-4484

Wolny, F., Weissker, U., Mühl, T., Leonhardt, A., Menzel, S., Winkler, A. & Büchner, B. (2008). Iron-filled carbon nanotubes as probes for magnetic force microscopy. *Journal of Applied Physics,* Vol.104, No.6, (September 2008), 064908, ISSN 0021-8979

Xu, Z.W., Fang, F.Z. & Hu, X.T. (2009). Fabrication of Carbon Nanotube Probes in Atomic Force Microscopy. *Advanced Materials Research,* Vol.76-78, (June 2009), pp. 497-501, ISSN 1022-6680

Ye, Q., Cassell, A. M., Liu, H. B., Chao, K. J., Han, J. & Meyyappan, M. (2004). Large-Scale Fabrication of Carbon Nanotube Probe Tips for Atomic Force Microscopy Critical Dimension Imaging Applications. *Nano Letters*, Vol.4, No.7, (July 2004), pp. 1301~1308, ISSN 1530-6984

Yum, K., Wang, N. & Yu, M. F. (2010). Nanoneedle: A multifunctional tool for biological studies in living cells. *Nanoscale*, Vol.2, No.3, (March 2010), pp. 363–372, ISSN 2040-3372

# Part 2

# Electronic Properties and Structure

# 8

# One-Dimensional Crystals inside Single-Walled Carbon Nanotubes: Growth, Structure and Electronic Properties

Andrei Eliseev[1], Lada Yashina[2,3],
Marianna Kharlamova[1] and Nikolay Kiselev[4]
[1]*Moscow State University, Department of Materials Science*
[2]*Moscow State University, Department of Chemistry*
[3]*Rare Metals Institute "GIREDMET"*
[4]*Institute of Crystallography RAS*
*Russia*

## 1. Introduction

Single-walled carbon nanotubes (SWNTs) discovered in 1993 are currently among the most exciting and promising nanostructures (Bethune et al., 1993; Iijima & Ichihashi, 1993). They arouse huge interest due to their unique atomic structure, outstanding chemical and electronic properties (thermal and electric conductivity), as well as mechanical characteristics (high values of Young's modulus, tensile and compressive strengths, high cracking resistance etc.). SWNTs possess the maximum geometric anisotropy factors among the nanostructures known so far. The unique properties of carbon nanotubes (CNTs) are governed not only by their unusual tubular structure, but also by the fact that they are virtually devoid of any structural defects (Dresselhaus et al., 1995; Iijima, 1991; Saito et al., 1992). As a result, CNTs are of a great importance for development of nanoelectronics elements (logical gates, memory devices, emitters, and nanowires), nanoelectromechanical systems, nanocomposite fillers (aimed at increasing strength and functionality of bulk materials), probe tips for scanning probe microscopy etc. One of the major areas of SWNTs technological application has been the development of a new generation of field-effect transistors (Tans et al., 1998).

The electronic properties of defect-free SWNTs are extremely sensitive to the nanotube's geometric structure (Avouris et al., 2007; Saito et al., 1992), which depends to a great extent on the chiral vector; this may be regarded both as an advantage and a serious drawback of this material. So far, no efficient methods have been developed for the preparation and isolation of SWNTs with a desired chirality (Hou et al., 2008; Odom et al., 2000). For this reason, many attempts have been undertaken to develop methods that would allow separating the array of SWNTs into semiconducting and metallic nanotubes and/or modify the electronic properties of SWNTs without their separation by chirality (Chaturvedi et al., 2008; Eliseev et al., 2009a; Monthioux et al., 2006).

Modification of nanotubes allows direct adjustment of their electronic properties. One of the simplest ways to controlled modification of the SWNTs is filling of the nanotube channels with appropriate substances (Brown et al., 2001). Encapsulation of a substance into a nanotube can either lead to a complete amendment of the nanotube's band structure (in case the encapsulated substance interacts intensively with the nanotube walls, e.g. in fluorinated SWNTs), or only to a shift of the electron density within the rigid band structure approximation (Sceats et al., 2006; Sloan et al., 2002a). In the simplest case, if an electron donor with the Fermi level located higher than that of the SWNT is encapsulated into metallic nanotubes, the electron density at the nanotube walls, as well as the nanomcomposite conductivity increase, whereas an electron acceptor with the Fermi level located lower than that of the SWNT would cause the nanocomposite transition into the semiconducting state (Chaturvedi et al., 2008; Rahman et al., 2005; Weissmann et al., 2006). Therefore, this approach based on electron transfer upon the introduction of electron-donor or electron-acceptor compounds (metals, semiconductors, dielectrics) into the channels of single-walled nanotubes allows controlling the electronic structure of the SWNTs, as well as creating the p – n-junctions inside a single nanotube if the channels are partially filled (e.g. if a nanotube is half-filled).

The synthesis of filled nanotubes was first reported by Ajayan and Iijima in 1993; they used multi-walled nanotubes as "molecular containers" for lead (Ajayan & Iijima, 1993). These experimental results confirmed the theory-based conclusions on the existence of sufficiently strong capillary forces inside carbon nanotubes, which may retain gases and liquids inside the channels (Pederson & Broughton, 1992). Later on, other researchers developed and employed this approach for filling carbon nanotubes with a variety of metal halides [$M^{I}I$ ($M^{I}$ = Li, Na, K, Cs, Rb, Ag), $M^{II}I_2$ ($M^{II}$ = Ca, Cd, Co, Sr, Ba, Fe, Pb, Hg), $M^{III}I_3$ ($M^{III}$ = La, Ce, Pr, Nd, Gd), $(Te/Sn)I_4$, $Al_2I_6$, $AgCl_xBr_yI_z$, $M^{I}Cl$ ($M^{I}$ = Na, Cs, Ti), $M^{II}Cl_2$ ($M^{II}$ = Cd, Fe, Co, Pd), $M^{III}Cl_3$ ($M^{III}$ = La, Nd, Sm, Eu, Gd, Tb), $M^{IV}C1_4$ ($M^{IV}$ = Hf, Th, Zr, Pt), $Al_2Cl_6$, $(Th/V)Cl_6$], elemental forms (S, Se, Te, $I_2$, Cs, Re, Bi, Pt, Au, Ru, Fe, Ag), fullerenes ($C_{60}$, $C_{70}$, $C_{80}$), endofullerenes ($Gd@C_{82}$), a $(KCl)_x(UCl_4)_y$, oxides ($Re_xO_y$, $V_2O_5$, $Sb_2O_3$, $CrO_3$, PbO, $UO_2$, $ZrO_2$, $MoO_2$, NiO, CdO, $La_2O_3$), metals (Pd, Pt, Cu, Ag, Au), hydroxides (KOH, CsOH), and chalcogenides (SnSe, HgTe and $CdBr_{2-x}Te_x$) (Chaturvedi et al., 2008; Cohen, 2001; Corio et al., 2004; Eliseev et al., 2009a; Fagan et al., 2005; Govindaraj et al., 2000; Kataura et al., 2002; Monthioux, 2002; Monthioux et al., 2006; Sceats et al., 2006; Sloan et al., 2000a).

At present, several methods are used for filling carbon nanotubes with various substances, which fall into two large groups: filling of nanotubes during their growth (i.e. the *in situ* methods) and encapsulation from the gas or liquid phases into cavities of pre-formed carbon nanotubes (i.e. the *ex situ* methods) (Monthioux et al., 2006).

## 2. Filling of single-walled carbon nanotubes during their growth (*in situ* methods)

The simplest of all the approaches that have been proposed to date for the nanotubes encapsulation is filling of SWNTs in the course of their catalytic growth *(in situ)*. Currently two methods are applied that employ the *in situ* strategy for the encapsulation of inorganic compounds into the nanotubes: catalytic chemical vapour deposition (CCVD) of hydrocarbons and arc-discharge synthesis (Monthioux et al., 2006).

Arc-discharge synthesis of carbon nanotubes filled with various compounds is performed using graphite rods electrodes, a compound-containing anode (usually metals are encapsulated using this approach), and a catalyst. This approach was used to prepare single-walled carbon nanotubes for the first time (Bethune et al., 1993; Iijima & Ichihashi,

1993). To the present day, there has been a number of works on the application of the arc-discharge synthesis for filled multi-walled nanotubes preparation. In most of the cases, a number of substance is incorporated into the NTs in the carbide form (Cr, Mn, Fe, Ni, Pd, Y, Gd, Dy, Yb, La, Ce). The use of elements that do not form carbides or an accurate control of specific synthesis conditions allows encapsulating elemental compounds (Se, Ge, Sb, Cr, Mn, Co, Cu, Re, Au, Sm, Gd, Dy, Yb) (Ajayan & Ebbesen, 1997; Beguin et al., 2006). It was also demonstrated that the presence of sulfur in the graphite anodes in catalytic amounts is of key importance for the formation of filled nanotubes (Demoncy et al., 1998). Most likely it provides liquid phase at the surface of the nanotube channels (i.e. due to the metal-sulfur eutectic), which in its turn ensures encapsulation of the selected substance into the nanotubes' cavities. However, large temperature gradients in the cathode area, which lead to non-uniform nanotubes filling, make control of the filling process impossible. Another disadvantage of this approach is that it does not allow filling of nanotubes with transition metals, since in this case metal—carbon solid solutions and various carbides are formed.

In order to avoid formation of carbide species the catalytic CVD method may be employed. In this case pyrolysis of the carbon source should be accompanied by simultaneous sublimation or decomposition of metal-containing compounds (usually carbonyls or metallocenes) (Monthioux et al., 2006). Most often nanotubes filled with transition metals (Fe, Co, Ni, Cu) used as catalysts for SWNT growth are produced by this technique (Leonhardt et al., 2003). The use of the CCVD method for the preparation of the "1D-crystal@SWNT" nanocomposites is limited due to the need of strict temperature control and restricted number of carbon source - guest precursor combinations. Thus, the CCVD and arc-discharge synthesis are complementary in terms of the initial compounds choice.

The *in situ* approaches do not allow filling of nanotubes with any unstable species and complex chemical compounds (i.e. metal oxides, metal salts), since these methods require maintaining relatively high temperatures and reducing conditions throughout the synthesis. The major disadvantage of the *in situ* strategy for filling of single-walled nanotubes is its low efficiency: the yield of filled SWNTs does not exceed several percent. These drawbacks have facilitated the development of the *ex situ* approaches to filling of SWNTs, which are described below.

## 3. Filling of pre-synthesized carbon nanotubes (*ex situ* methods)

The filling of pre-synthesized nanotubes (i.e. the *ex situ* method) is considered to be the most universal approach to encapsulated nanotubes preparation. This technique enables filling single-walled nanotubes with virtually any chemical compounds from either gas or liquid phases (depending on the aggregate state of the encapsulated compound at the moment of its contact with the nanotube) (Eliseev et al., 2009a; Monthioux et al., 2006). This approach consists of several steps, the first of them being the opening of the SWNT ends.

### 3.1 Opening of nanotube ends

In order to fill carbon nanotubes using the *ex situ* methods, first their ends should be opened, which is performed using two main approaches, i.e. thermal treatment of the NTs in an oxidative gaseous medium (either dry air or oxygen) or treatment with liquid oxidation agents, such as concentrated acids ($HNO_3$, $H_2SO_4$, $HNO_3$-$H_2SO_4$), hydrogen peroxide, potassium permanganate, osmium tetraoxide or HF-$BF_3$ mixture (Ajayan & Iijima, 1993; Monthioux et al., 2001; Seraphin et al., 1993). Concentrated acids also allow removing catalytic particles and various contaminants (amorphous carbon, polyaromatic compounds, and graphite particles).

In fact, the oxidation involves both the ends and the walls of carbon nanotubes. For example, treatment of SWNTs with an acid was shown to result in the lateral defects formation (one defect per each 5 nm of the nanotube) (Zhang et al., 2003). Unlike that of multi-walled nanotubes, partial oxidation of single-walled nanotubes leads to formation of "hole" defects, through which substances may penetrate into the tubes both through their ends and walls.

A comparative study of various methods for the SWNT opening demonstrated that thermal oxidation tends to be a more efficient approach than acidic treatment (Brown et al., 2001). Since products formed during the acidic treatment can react with the carbon atoms of the SWNTs, the use of gaseous oxidants is preferrable to avoid contaminations (Monthioux et al., 2001). Oxidation in air for about 30 min at 300-500°C seems to be the optimum choice (Fig.1). For such treatment, the opening of nanotubes is practically complete. In this process the sample loses approximately 40% of its mass (Zhang et al., 2003).

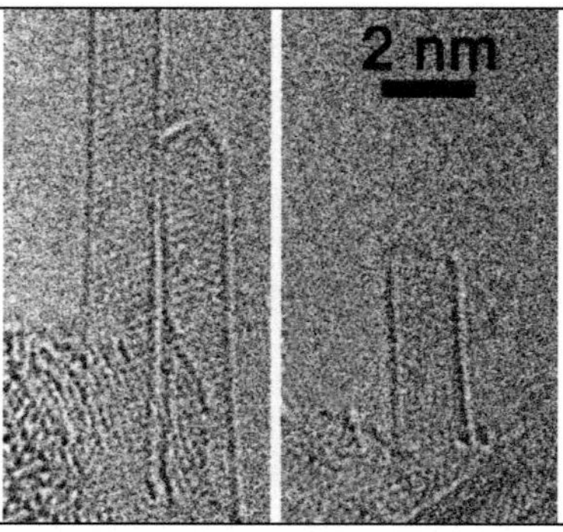


Fig. 1. HRTEM images of closed nanotubes and SWNTs opened by oxidation in air at 500°C

### 3.2 Filling of single-walled nanotubes from the gas phase

Notwithstanding multiple studies of multi-walled nanotubes filling from the gas phase, the filling of SWNTs remains much less investigated (Chancolon et al., 2006; Eliseev et al., 2009a). As a rule, filling of carbon nanotubes from the gas phase is carried out in vacuum at high temperatures. A sealed tube is heated up to or above the vaporization (or sublimation) temperature of an encapsulated material. In order to synthesize the "1D-crystal@SWNT" nanocomposites, the lowest possible temperatures should be used to avoid (or minimize) deencapsulation. During the NTs annealing, the vapor of the encapsulated compound undergoes capillary condensation and thus penetrates into the nanotube, where it crystalizes during subsequent cooling.

This two-step technique is widely used in order to fill carbon nanotubes with various fullerenes (for instance, $C_{60}$, Fig. 2), which have high affinities to nanotube surfaces and high vapor pressures (approx. $3x10^{-4}$ Torr at 500°C) (Pan et al., 2002; Smith & Luzzi, 2000). The fullerenes encapsulation occurs through the ends and wall defects of the SWNTs (Jeong et al., 2003). The encapsulation process depends strongly on the temperature and time of the NT treatment; partial vapor pressure of the introduced compound can also play a certain role (Smith & Luzzi, 2000). This process is rather time consuming (takes about 2 days), however, it enables homogeneous and complete (virtually 100%) filling of SWNTs.

SWNTs can be successfully encapsulated from the gas phase not only with fullerenes, but also with endofullerenes ($M_x@C_n$) or doped fullerenes (Hirahara et al., 2000; Okazaki et al., 2003). As a rule, endofullerenes are synthesized in advance by the arc-discharge method with some metal added to the graphite anode (Okazaki et al., 2003). Subsequently, a mixture of opened SWNTs and endofullerenes is annealed at 400-500°C in an evacuated tube for several days (Okazaki et al., 2003). This method was used to fill the SWNTs with a variety of fullerenes ($C_{60}$, $C_{70}$, $C_{80}$, $C_{84}$, $C_{78}$, $C_{90}$), doped fullerenes (Cs, K, $FeCl_3$), and endofullerenes ($N@C_{60}$, $La_2@C_{80}$, $Sc_3N@C_{80}$, $Er_xSc_3$-$xN@C_{80}$, $Dy_3N@C_{80}$, $Gd@C_{82}$, $La@C_{82}$, $La_2@C_{82}$, $Dy@C_{82}$, $Sm@C_{82}$, $Sc_2@C_{84}$ @ $Gd_2@C_{92}$) (Monthioux et al., 2006).

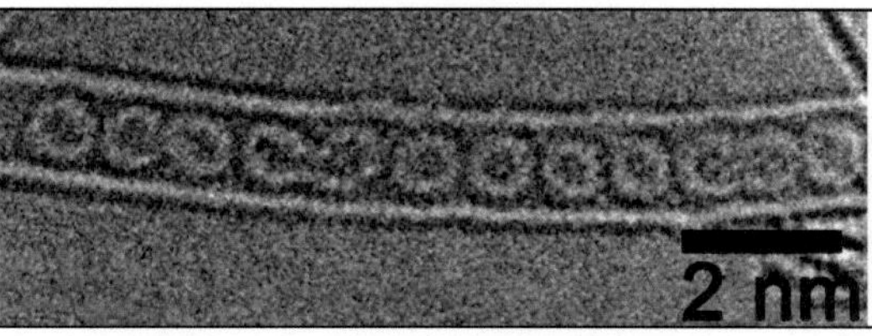


Fig. 2. HRTEM image of an SWNT filled with $C_{60}$ fullerenes from the gas phase

The principle disadvantage of this approach is a limited choice of compounds that can be encapsulated. First, the compound's vaporization (or sublimation) temperature should be below 1000°C to ensure it does not interact with carbon and cause closure of the nanotube ends. Second, the compound (as a rule, a volatile oxide or a salt) should undergo sublimation in molecular form, which substantially limits the number of suitable compounds. Another serious limitation of this method that the cluster thus formed are usually discrete, and they block the internal tube volume, whereas for practical reasons composites with continuous filling are required.

### 3.3 Filling of single-walled nanotubes from the liquid phase

Filling of SWNTs from the liquid phase is performed using the so-called capillary method, which involves impregnation of opened nanotubes with solutions or melts of selected compounds (Eliseev et al., 2009a; Monthioux et al., 2006). The use of melts is preferable, since it excludes contamination of composites with the solvent and also eliminates the necessity of filtration, which makes the formed nanocomposite denser. The excess of the encapsulated compound that remains at the SWNT's outer surface can be potentially removed by washing the sample or heating it under dynamic vacuum conditions.

The retraction of liquids into single-walled nanotubes takes place only if a number of conditions is met (Zhang et al., 2003). First, carbon nanotubes must be opened at least from one end. Second, the liquid phase must efficiently wet the SWNT surface, which limits its surface tension to 130-170 mN m$^{-1}$. This excludes the possibility to fill SWNTs with any melts that have high surface tensions, but, on the other hand, allows employing the majority of organic and inorganic solvents including water ($\gamma$= 72 mN m$^{-1}$ at 25°C) and benzene ($\gamma$ = 28.9 mN m$^{-1}$). Third, the melting (or decomposition) point of the encapsulated material should be below 1100°C in order to prevent the SWNT closure and destruction during the composite synthesis.

### 3.3.1 Encapsulation from suspensions or solutions

Filling of nanotubes from solutions was first implemented in 1994 in order to encapsulate NiO and $UO_{2+x}$ nanoparticles into multi-walled carbon nanotubes (Tsang et al., 1994). In the subsequent years, it was applied to fill MWNTs with Ag, Au, Pt, Pd metal particles, etc. (Cohen, 2001; Govindaraj et al., 2000; Satishkumar et al., 1996). In 1998, this approach was employed by a research group from the University of Oxford headed by Sloan to encapsulate single-walled nanotubes with metal Ru nanoparticles (Sloan et al., 1998).

At the moment, aqueous solutions of metal chlorides or nitrates (e.g. $RuCl_3$, $AgNO_3$, and $Fe(NO_3)_2$) are most often employed to encapsulate the SWNTs by the *ex situ* approach from the liquid phase (Chen et al., 1997; Monthioux, 2002; Monthioux et al., 2006) (Fig. 3) Another popular solvent is nitric acid, which is used due to its low surface tension (43 $mN \cdot m^{-1}$), allows avoiding a separate opening procedure (Zhang et al., 2003). As a rule, the second step upon the SWNTs' treatment with a solution is thermal treatment or hydrogenation in an $H_2$ flow at 150–450°C for several hours; this leads to the formation of metal or oxide nanoparticles inside the SWNTs.

It is worth noting that filling of SWNTs with inorganic compounds using suspensions or solutions may be employed for a wide variety of substances (e.g. metals, oxides, chlorides, fullerenes and endofullerenes). However, this method has a number of limitations and disadvantages (Zhang et al., 2003). First, due to the procedure's nature the SWNT channels may be contaminated with the solvent, the products of its interaction with the nanotube walls and/or the encapsulated compound. Second, the encapsulated substance is distributed non-uniformly within the CNT channel, and the filling is far from being complete due to the solvent molecules encapsulation into the SWNTs. Indeed, when the solvent is removed and/or gaseous products are formed during the thermal treatment, individual cluster particles with 2 to 100 nm in diameter may be formed with the maximum yield of approximately 25% – 30% (Zhang et al., 2003). It should also be mentioned that nanoparticles formed in such a manner are most often polycrystalline, while from the practical perspective the single-crystalline nanoparticles inside the SWNTs are of major importance.

Fig. 3. HRTEM image of an SWNT filled with Fe nanoparticles introduced from $Fe(NO_3)_2$ solution at room temperature with subsequent annealing at 300°C

### 3.3.2 Encapsulation from melts

Continuous and uniform filling of SWNTs was successfully accomplished by the *ex situ* approach involving filling of nanotubes from melts. This technique provides a 2-3 times larger encapsulation yield as compared to the filling from suspensions and solutions (Eliseev et al., 2009a; Monthioux et al., 2006; Sloan et al., 2002b). The method is based on the melts penetration into the single-walled nanotube channels due to capillary forces.

The encapsulation procedure is usually performed under vacuum conditions at temperatures 10-100°C higher than the melting point of the quest material, which is followed by the slow cooling of the system in order to allow crystallization of the encapsulated particles. As a rule, metal halides as well as substances with low melting points are encapsulated using this approach, since they meet all the requirements to the introduced materials, i.e. low surface tension (<170 $mN \cdot m^{-1}$) and melting point (<1100 °C) (see Table 1) (Brown et al., 2003).

For the first time, Ajayan and Iijima (1993) employed the *ex situ* introduction of melts into the NT channels to fill multi-walled carbon nanotubes with the PbO particles (Ajayan & Iijima, 1993). The encapsulation yield was 90%, which is approximately twice the filling yield achieved by other methods (Ajayan & Iijima, 1993). Later Ajayan demonstrated that this approach can be successfully used to fill the SWNTs without performing preliminary opening of the nanotube ends (Xu et al., 2000).

The *ex situ* introduction of inorganic substances from melts was used to fill the SWNTs with nanoparticles of various metal halides [$M^{I}I$ ($M^{I}$ = Li, Na, K, Cs, Rb, Ag), $M^{II}I_2$ ($M^{II}$ = Ca, Cd, Co, Sr, Ba, Fe, Pb, Hg), $M^{III}I_3$ ($M^{III}$ = La, Ce, Pr, Nd, Gd), $(Te/Sn)I_4$, $Al_2I_6$, $AgCl_xBr_yI_z$, $M^{I}Cl$ ($M^{I}$ = Na, Cs, Ti), $M^{II}Cl_2$ ($M^{II}$ = Cd, Fe, Co, Pd), $M^{III}Cl_3$ ($M^{III}$ = La, Nd, Sm, Eu, Gd, Tb), $M^{IV}C1_4$ ($M^{IV}$ = Hf, Th, Zr, Pt), $Al_2Cl_6$, $(Th/V)Cl_6$], elemental forms (S, Se, Te, $I_2$, Cs, Re, Bi, Pt, Au, Ru, Fe, Ag), fullerenes ($C_{60}$, $C_{70}$, $C_{80}$), endofullerenes ($Gd@C_{82}$), a $(KCl)_x(UCl_4)_y$ mixture, oxides ($Re_xO_y$, $V_2O_5$, $Sb_2O_3$, $CrO_3$, PbO, $UO_2$), hydroxides (KOH, CsOH), and chalcogenides (SnSe, HgTe and $CdBr_{2-x}Te_x$) (Brown et al., 2003; Carter et al., 2006; Dujardin et al., 1994; Eliseev et al., 2009a; Flahaut et al., 2006a; Monthioux et al., 2006; Sloan et al., 1999).

Table 1 lists the surface tension and the melting points of a number of substances encapsulated into the SWNTs, as well as the filling temperatures and yields (Brown et al., 2003; Eliseev et al., 2009a; Monthioux et al., 2006; Xu et al., 2000). According to the TEM data, the SWNT channels filling yield for encapsulation with inorganic compounds was 50% to 90%.

In most cases, the encapsulated nanoparticles were found in the form of one-dimensional nanocrystals within the SWNT. Based on the analysis of the TEM micrographs (Fig. 4 a,c), structural models of the one-dimensional nanocrystals may be proposed (Fig. 4 b,d) (Sloan et al., 2000b; Sloan et al., 2002a).

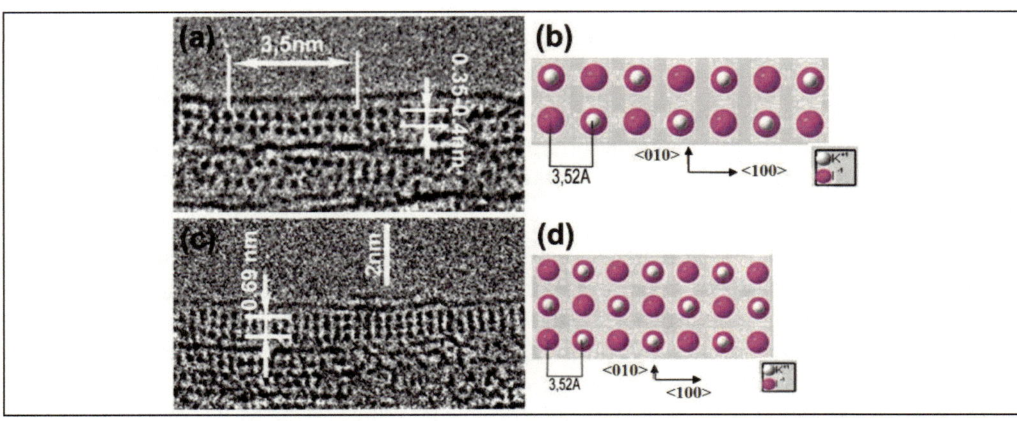


Fig. 4. HRTEM images of the KI nanocrystals inside the SWNT channel (a,c) and atomic models of the 1D crystals (b,d)

| Material | $\gamma$ (mN · m$^{-1}$) | $T_{melt}$, °C | $T_{int}$, °C | Loading factor (%) | Material | $\gamma$ (mN · m$^{-1}$) | $T_{melt}$, °C | $T_{int}$, °C | Filling yield (%) |
|---|---|---|---|---|---|---|---|---|---|
| AgCl | 113–173 | 560 | 560–660 | 40-50 | K | 117 | 336 | | |
| AgBr | 151 | 432 | 532–590 | 40-50 | KCl | 93 | 771 | 870 | |
| $AgBr_{0.2}Cl_{0.8}$ | 173 | 410 | 510 | 40-50 | $(KCl)_x(UCl_4)_y$ | 44-65 | 335, 562 | 435, 662 | < 10 |
| AgI | 171 | 455 | 555 | 80-90 | KI | 70 | 681 | 781 | 60-80 |
| Al | 860 | 660 | – | | $LaCl_3$ | 109 | 860 | 910 | 20-40 |
| $BaI_2$ | 130 | 740 | 840 | < 10 | LiI | 94 | 449 | 549 | 20-30 |
| $Bi_2O_3$ | 200 | 825 | | | NaI | 81 | 661 | 761 | 10-20 |
| $CaI_2$ | 83 | 784 | 884 | < 10 | $NdCl_3$ | 102 | 784 | 834 | 20-40 |
| Cs | 67 | 29 | | | Pb | 470 | 327 | – | |
| CsI | 69 | 627 | 727 | 30-40 | PbO | 132 | 886 | | 80-90 |
| CuCl | – | 430 | 530 | 30-50 | $Re_2O_3$ | 32 | 220 | 250 | 50-60 |
| CuBr | – | 492 | 590 | 60-80 | Rb | 77 | 39 | | |
| CuI | – | 606 | 705 | >90 | RbI | 70 | 647 | 747 | 60-70 |
| $EuCl_3$ | – | 850 | 860 | 20-40 | S | 61 | 115 | 165 | 20-30 |
| $FeCl_2$ | – | 674 | 774 | | Se | 97 | 221 | 320 | 20-40 |
| $FeBr_2$ | – | 684 | 784 | | Te | 190 | 450 | 520 | 20-40 |
| $FeI_2$ | – | 587 | 687 | 50-60 | SnTe | – | 807 | 907 | 60-70 |
| $CoBr_2$ | – | 678 | 778 | | $TbCl_3$ | – | 588 | 638 | 20-40 |
| Ga | 710 | 30 | – | | $SmCl_3$ | – | 686 | 706 | 20-40 |
| $GdCl_3$ | 92 | 609 | 659 | 20-40 | $UCl_4$ | 27 | 590 | 690 | < 10 |
| Hg | 490 | −38 | – | | $V_2O_5$ | 80 | 690 | | |
| HF | 117 | | | | $YbCl_3$ | – | 854 | 904 | 20-40 |
| $HNO_3$ | 43 | | | | $ZrCl_4$ | 1.3 | 437 | 487 | 50-70 |

Table 1. Surface tension values and melting points of encapsulated materials, synthesis temperatures of the "1D-crystal@SWNT" nanocomposites (encapsulation from melt) and encapsulation yields

The *ex situ* encapsulation from melt has a number of advantages in comparison with other filling techniques. Among them are the possibility to use a wide range of substances to fill the SWNTs, the simplicity of approach's, the composites uniformity, high loading factor (up to 90%), and high crystallinity of the synthesized nanoparticles. Another benefit is the absence of any solvent and/or by-products (oxides, carbides) contamination in the "1D-crystal@SWNT" systems. This makes the *ex situ* filling of single-walled nanotubes from melt the most efficient approach for the "1D-crystal@SWNT" systems synthesis that has been developed so far.

### 3.3.3 Direct synthesis within the SWNTs

Notwitstanding its efficiency, the encapsulation from melts has a number of limitations. This approach cannot be employed for compounds with surface tension values exceeding 170 mN·m$^{-1}$ and melting points above 1100°C. This essentially limits the possibilities to directly introduce many metals (due to high $\gamma$ values of their melts) and covalent compounds (e.g. metal oxides or chalcogenides) into the SWNT channels. On the other hand, encapsulation of metallic nanoparticles into semiconducting SWNTs or semiconductor crystals into metallic tubes is expected to result in the most significant changes in the electronic structures of the single-walled nanotubes (Eliseev et al., 2009b; Kramberger et al., 2009a; Shiozawa et al., 2009a). Moreover, low-dimensional broadband $A^{II}B^{VI}$ and $A^{IV}B^{VI}$ semiconductors with an exciton diameter of several nanometers (e.g. exciton diameter of CdS is 4.8 nm) are of interest both for the quantum-size effects studies and possible application in light emitters, sensors, solar cells etc. (Cao et al., 2004).

Although direct filling of nanotubes with semiconducting materials having high melting points failed, such materials can be synthesized directly in the internal channels of the SWNTs (Eliseev et al., 2009b). First, the nanotube channels are filled with molten metal iodides ($ZnI_2$, $CdI_2$, $PbI_2$) using the capillary approach at temperatures exceeding the substances' melting points by 100°C ($T_{melt}$ = 446, 388 and 412°C, respectively). The obtained $MI_2$@SWNT composites undergo sulfidation, selenation or telluration when treated with molten chalcogens:

$$MI_2@OCHT + 2\,X_{liquid} \rightarrow MX@OCHT + \tfrac{1}{2}\,I_{2\,gas} + \tfrac{1}{4}\,X_nI_{m\,gas} \quad (X = S, Te, X_nI_m = 2\,S_2I_2 \text{ or } Te_4I_4), \tag{1}$$

$$MI_2@OCHT + X_{liquid} \rightarrow MX@OCHT + I_{2\,gas} \quad (X = Se). \tag{2}$$

Slow cooling of the chalcogenides synthesized in the SWNT channels results in their crystallization in the form of one-dimensional crystals. It is assumed that the chalcogen atoms are transported into the single-walled nanotubes filled with the iodides ($MI_2$), and the $I_2$ and $X_nI_m$ gaseous molecules formed in the reactions (1) and (2) are removed through defects in the SWNT walls (Eliseev et al., 2009b). This suggestion is in line with the observed destruction of one-dimensional crystals in the SWNT channels upon the system's irradiation with high-energy electrons, that leads to escape of the encapsulated compounds through the defects in the nanotube walls and their subsequent decomposition on the NT's external surfaces (Hutchison et al., 2008).

The approach based on chemical reactions within the SWNT channels was successfully employed to synthesize the 1D crystals of semiconducting $A^{II}B^{VI}$ and $A^{IV}B^{VI}$ compounds (A = Zn, Cd, Pb; B = S, Se, Te) (Fig. 5) with high melting points (up to 1750°C) and the band gaps of 0.3 to 3.7 eV (Eliseev et al., 2009b).

## 4. Structures and properties of the "1D-crystal@SWNT" nanocomposites

The "1D-crystal@SWNT" nanostructures arouse huge interest for theoretical studies and possible applications. Since this research area has been under development only for the past 10 years, the information on the properties of the "1D-crystal@SWNT" composites is rather scarce (Eliseev et al., 2009a; Monthioux et al., 2006). The same is true for the simulation of structures and electronic properties of encapsulated carbon nanotubes (Kramberger et al., 2009a; Shiozawa et al., 2009a). The experimental studies focus mainly on the possibility to alter the SWNTs band gaps by filling them with electron-donor or electron-acceptor

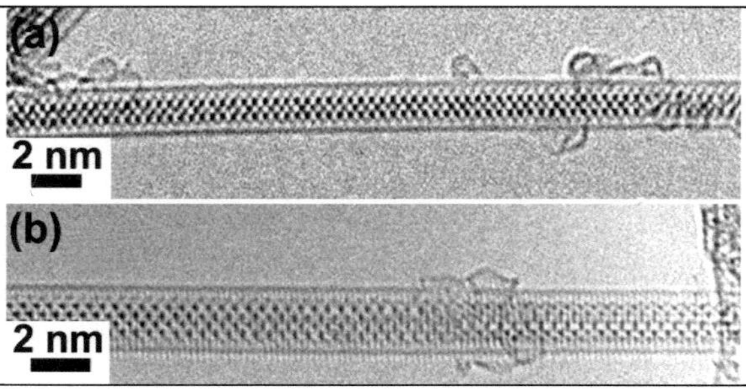


Fig. 5. HRTEM image of the $PbTe_xI_{2-2x}$@SWNT composite (two types of the observed projections)

compounds or changes in the atomic and electronic structure of the encapsulated compounds due to forced structural deformations of the materials inside the single-walled nanotubes (Shiozawa et al., 2008).

The "1D-crystal@SWNT" nanocomposite structures are most efficiently studied by high resolution transmission electron microscopy. Since a "1D-crystal@SWNT" nanocomposite is a single-walled carbon nanotube with a 1D-crystal inside, the HRTEM images display either crystallographically ordered atomic arrangements or columns of several atoms (see fig. 3). Historically, HRTEM was first implemented to study the "1D-KI@SWNT" nanocomposites (Meyer et al., 2000; Sloan et al., 2000b). In these systems one-dimensional KI crystals with 2x2 or 3x3 atoms in diameter were observed depending on the tube diameter. Peapods can also be considered as the "1D-crystal@SWNT" nanocomposites (Smith et al., 1998).

It stands to mention that in the SWNTs of various diameters the encapsulated compounds (for instance, KI, $PbI_2$, CuI, HgTe) can form structures that differ by the number of atoms in the nanotube cross section or by symmetry (Carter et al., 2006; Flahaut et al., 2006a; Kataura et al., 2002; Kiselev et al., 2008). The structure of such nanocrystals differs substantially from the bulk substances both with regards to their symmetry, bond lengths, and bond angles. The distortion of geometric parameters is obviously caused by the steric limitations of the one-dimensional nanocrystals within the nanotube walls, as well as by the adjustment of nanoparticle structures to the internal diameter of the SWNTs channel (Eliseev et al., 2009a; Kataura et al., 2002). For some systems the spatial confinement of an intercalated substance by the channel walls leads to the formation of a 1D-crystal with a structure that is not typical for the 3D-crystals of the same compound. For instance, 3D-AgBr is cubic, while the 1D-crystals formed within the SWNTs have a hexagonal structure that is absent from the Equilibrium phase diagram related to bulk state. In certain situations entirely new structures with unusual properties (including a five-fold symmetry) can be formed ((Ebbesen, 1996; Kirkland et al., 2005; Mittal et al., 2001; Sloan et al., 2002c).

In fact, such deviations pertain to the choice in the nanocrystal's crystallographic orientation with respect to the nanotube axis (different crystallographic directions of the bulk substance can coincide with the long axis of a one-dimensional nanocrystal), reduction of the coordination number for atoms within the encapsulated inorganic substance, the difference between the crystal lattice parameters for the one-dimensional crystal and those for the bulk material, as well as the formation of nanocrystals with a crystal lattice that is not typical for the bulk materials (Eliseev et al., 2009a; Kataura et al., 2002; Monthioux et al., 2006).

The crystallographic orientation of the nanocrystals relative to the nanotube axis changes due to the demand to preserve the stoichiometry of the encapsulated compound to the highest possible extend, whereas the reduction of the coordination number and the change in the nanocrystal's lattice parameters along the radial direction are apparently explained by the spatial limitations of the nanocrystal within the SWNT walls (Mittal et al., 2001). The nanocrystal's lattice parameters along the nanotube axis are less distorted, since the nanocrystal experiences virtually no steric limitations along this direction. Nevertheless, a number of studies indicate the extention/compression of the 1D crystal lattice along the SWNT axis. In particular, this effect was observed for the KI, CuI, Ag, $Sb_2O_3$, KI nanocrystals and the $(C_{60})_n$ fullerene chains (Friedrichs et al., 2002; Meyer et al., 2000; Sloan et al., 2002b). The observed lattice distortion can be as high as 14%. The compression (or stretching) of the lattice parameter along the nanotube axis is likely to be caused by the stretching (or contraction) of the unit cell within the SWNT channel in the radial direction; the cell volume remains unchanged.

Analysis of possible one-dimensional crystal configurations formed within single walled carbon nanotubes indicates that the structure of a crystal is mostly governed by the SWNT diameter (Table 2). To the moment a number of crystal geometries unusual for bulk case have been reported in the SWNT channels. As most of these crystals are not easily described in terms of traditional space groups, here we would utilize the following notation:

1. One-dimensional unit cell is further denoted as $(A_xB_y)_n/L$, where *n* corresponds to a number of molecular formula units, and *L* to the number of layers in the unit cell.
2. To describe the lattice of a one-dimensional crystal, additional symmetry notation is involved based on Bravais lattices with a rotation axis $C_n$ (or an inverse rotation axis $S_n$) aligned along the tube using a P letter for primitive, an C for base-centered and an F for face- centered structures. In several cases the structures could be described as a dense packing.
3. When the anion/cation diameter ratio for the 1D crystal is substantial, there is a number of vacant cationic positions in the structure, and cations can easily migrate inside of the channel. In this case the cationic positions cannot be easily determined and symmetry notation is applicable only to anion sublattice. Nevertheless, few detailed studies performed with microscopy image simulation allow distinguishing between different cationic forms of one-dimensional crystals (fig. 6).
4. For a number of crystals the formation of helix structures can be observed. The most common examples include $I_2$@SWNT, RbI@SWNT and $H_2O$@SWNT (Chen et al., 2009; Fan et al., 2000; Kirkland et al., 2005; Liu & Wang, 2005). For these crystals we use the $1/N(A_xB_y)_n/L$ notation, where N is a number of unit cells per period of rotation, i.e. $N = \phi/360$ with $\phi$ – rotation angle (distortion) for a single unit cell.

The most interesting structural changes are observed for low-diameter SWNTs, where the lattice constant is comparable to the nanotube diameter. Moreover, the structural deviations are more frequently observed for crystals with complex (non-primitive) structures in the bulk. Obviously, the packing rules for non-interacting atoms within SWNT channel should comply with simple geometric considerations (Table 3). These kinds of structures are typical for inert metals crystallized within the SWNT. However, due to a lack of informative HRTEM studies only few examples can be found in the literature (Govindaraj et al., 2000; Jeong et al., 2003).

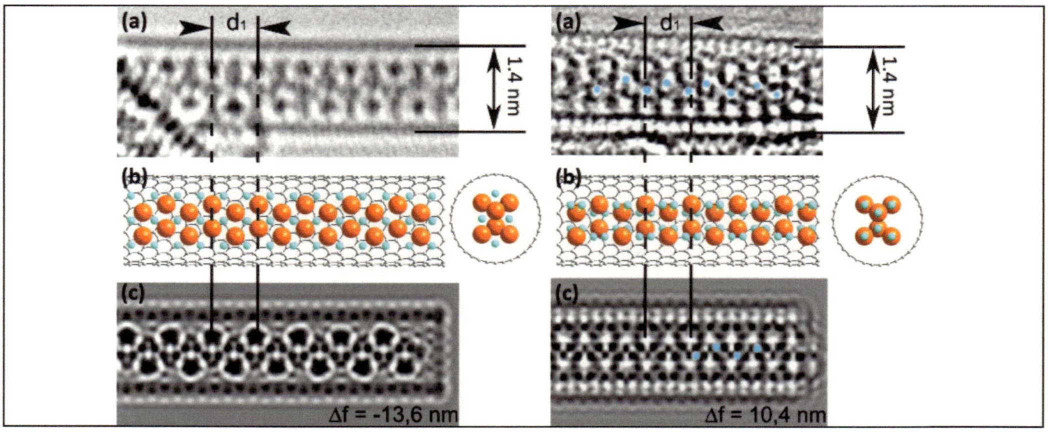


Fig. 6. HRTEM images (a, d) structural models (b, e) and TEM image simulation results (c, f) for two types of the CuI one-dimensional nanocrystal within a 1.4 nm SWNT channel. Copper atoms are located in the tetrahedral (a-c) and octahedral (d-f) sites of a two-layer hcp $(CuI)_6/2$ structure

Another simple case is ionic crystals, such as alkali metal halides, for example, KI. All the projections for KI crystal resolve in a simple cubic lattice with a (001) axis aligned along the tube channel. This structural motif is least subject to confinement by a nanotube: for low-diameter (below 1.5 nm) tubes $(MX)_4/2$ crystals with a face centered cubic structure (a 2×2 crystal) are formed within the channel, $(MX)_9/2$ *fcc* (a 3×3 crystal) is typical for tube diameters of 1,5 – 1,7 nm, while for greater diameters lattices of bulk NaCl type are formed. In this case the major structural changes are the variation of the interatomic distances and the cell volume expansion. It is worth noting that CsI intercalated into SWNTs also has a NaCl-type structure, despite the fact that the bulk substance has a *bcc* structure.
Most of less-ionic crystals having dense packing of anions in their bulk structures (2- or 3-layer hexagonal lattices) tend to form a two-layer hexagonal packing within the tube. This case is characteristic for copper and silver halides, which have been recently studied in detail (Eliseev et al., 2010; Eliseev et al., 2011). Crystals with covalent bonding tend to preserve some structural fragments in one-dimensional crystals. Atomic coordination numbers can play a certain role in defining the crystallographic orientation of the nanocrystals relative to the nanotube axis (Kiselev et al., 2008).
The degree of crystallinity and the structure of the one-dimensional nanocrystals within the nanotube channel depend strongly on the ratio between the structural parameters of the encapsulated compound and the internal diameter of the SWNT channel. For example, for the CuHal@SWNT and AgHal@SWNT (Hal = Cl, Br, I) systems the degree of crystallinity for nanoparticles incorporated into the SWNT channels increases when passing from the metal chlorides to iodides, which is presumably caused by the increase in the halogen ion radius when passing from Cl to I [$r(Cl^-)$ = 1-67, $r(Br^-)$ = 1-82, $r(I^-)$ =2.06 Å], which causes the increase in the M-Hal bond lengths in this series (Eliseev et al., 2010; Eliseev et al., 2011). As a result, in the CuCl@SWNT composite the CuCl single crystal does not completely fill the nanotube's channel, which results in substantial mobility of the Cu and Cl atoms; this, in its turn, leads to a decrease in the degree of crystallinity. At the same time, in the CuI@SWNT composite the size of the CuI crystal fits well the diameter of the tube, which results in the enhanced degree of crystallinity.

| Compound | $d_1$ nanotube, nm | Type of packing | Cell formulae | Crystal symmetry | $d_2$ crystal | Expansion | | $d_2/(d_1 - d_{001}$ graphite) |
|---|---|---|---|---|---|---|---|---|
| | | | | | | across tube axis | along tube axis | |
| KI | 1,4 | 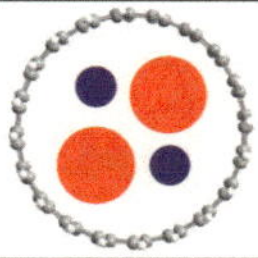 | $(MI)_4/2$ Helix 1/20 | P$\bar{4}$ | 1,06 | +14% | +14% | 0,995 |
| RbI | 1,4 | | | | 1,00 | +6,4% | -1,8% | 0,957 |
| CsI | 1,4 | | | | 1,11 | +7,1% | +2,9% | 1,042 |
| LiI | 1,6 |  | $(MI)_9/2$ | F4 (*fcc*) | 1,13 | +25% | +25% | 0,893 |
| KI | 1,6 | | | | 1,31 | +7,4% | +4,9% | 1,036 |
| CsI | 1,6 | | | | 1,37 | +9,3% | +2,3% | 0,988 |
| NaI | 2,5 | 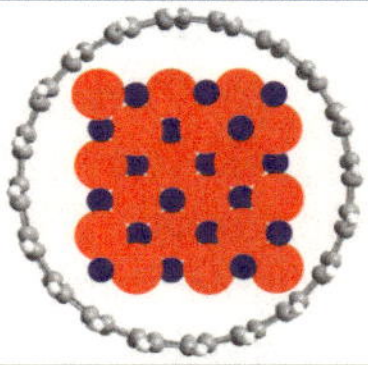 | $(NaI)_{36}/2$ | P$\bar{4}$ | 2,00 | +4% | +4,7% | 0,924 |
| AgBr | 1,4 |  | $(MHal)_6/2$ | P$\bar{2}mm$ | 0,98 | -2% | +4,9% | 0,920 |
| AgI | 1,4 | | | | 1,11 | -4,5% | -0,7% | 1,042 |
| CuBr | 1,4 | | | | 1,00 | 0,0% | +2,1% | 0,957 |
| CuI | 1,4 | | | | 1,06 | -2,8% | +2,3% | 0,995 |
| $BaI_2$ | 1,4 |  | | P2 | 1,21 | - | - | 1,136 |
| $PbI_2$ | 1,6 | | | | 1,43 | - | - | 1,130 |
| HgTe | 1,36 | 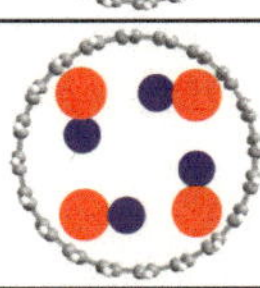 | $(HgTe)_4/2$ | P$\bar{4}$ | 0,99 | - | - | 0,966 |
| SnSe | 1,4 | 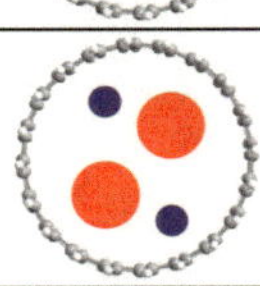 | $(MChal)_4/2$ | P$\bar{4}$ | 0,91 | - | - | 0,854 |
| SnTe | 1,35 | | | | 0,84 | +10,3 % | -16,7% | 0,828 |
| SnTe | 1,35 | 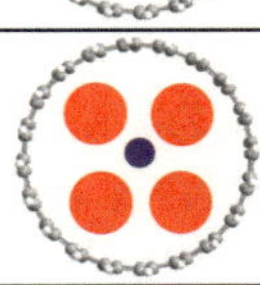 | $(SnTe)_5/2$ | C4 | 1,08 | +4,8% | -7,0% | 1,064 |
| CdS | 1,4 | 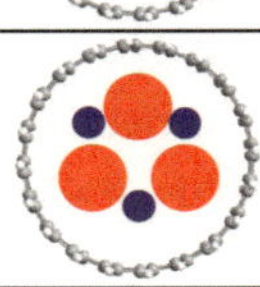 | $(CdS)_6/2$ | P$\bar{6}$ | 0,86 | +6,4% | +9,5% | 0,832 |

Table 2. Structural parameters of the "1D-crystal@SWNT" nanocomposites resolved by HRTEM

| D/d | | D/d | | D/d | |
|---|---|---|---|---|---|
| 1,00 - 2,00 | 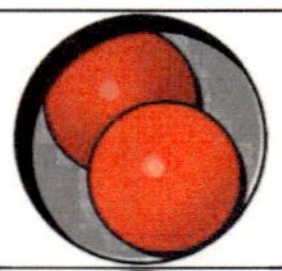 | 2,15 - 2,41 | 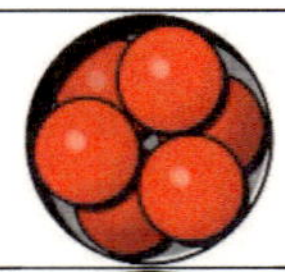 | 2,70 - 3,00 | 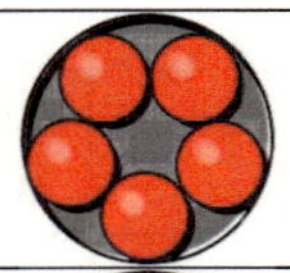 |
| 2,00 - 2,15 | 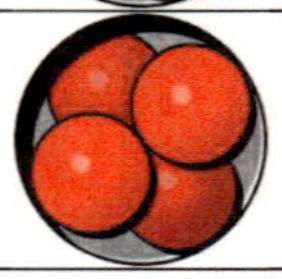 | 2,41 - 2,70 | 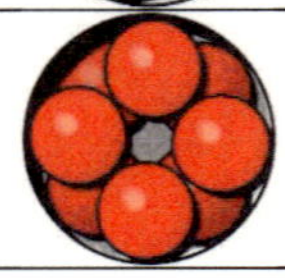 | 3,00 - 3,30 | 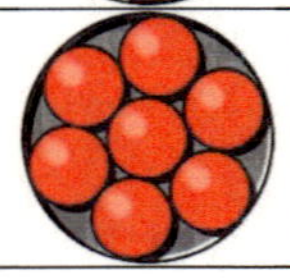 |

Table 3. Geometry of non-interacting atomic packing within one-dimensional cavity

Apparently, the nanocrystal structure can change upon variation of the nanotube diameter (e.g. for CuI@SWNT and $PbI_2$@SWNT) (Flahaut et al., 2006a; Kiselev et al., 2008). Recently it has been experimentally determined that copper and silver bromides $(MHal)_6/2$ can change their symmetry from $P\bar{2}mm$ in the (19,0) and (18,0) SWNTs to $P\bar{2}m$ in the (17,0) SWNTs due to contraction of the <111> and extension of the <112> axis (Kiselev et al., 2011).

In summary, the total confinement effect should be considered with relation to the nanocrystal size/SWNT diameter ratio. According to the available data, the following cases of one-dimensional nanocrystal/SWNT size misfit should be considered:

1. $d_{cryst}/(d_{SWNT} - d_{graphite}(00l)) < 0,8$
   The crystals are highly unstable and tend to undergo phase transformations that yield more stable configuration. These structures could be observed as intermediates during deintercalation caused by electron beam irradiation (Hutchison et al., 2008).
2. $0,9 < d_{cryst}/(d_{SWNT} - d_{graphite}(00l)) < 1$
   This is the most common case. The crystal lattice is slightly expanded to fit the SWNT channel diameter well. Usually it is associated with a reduction of the lattice parameter along the tube channel at constant unit cell volume. Twisting of the lattice can appear in case of significant host-guest interaction. In a number of cases the lattice can both expand along and across the tube, thus providing an evidence for a strong crystal-SWNT interaction.
3. $d_{cryst}/(d_{SWNT} - d_{graphite}(00l)) \sim 1$
   If the nanotube diameter fits the nanocrystal size well, the crystal is stable even under electron beam irradiation (Hutchison et al., 2008). Slight variation of the crystal cell parameters can occur while the cell volume remains constant.
4. $1 < d_{cryst}/(d_{SWNT} - d_{graphite}(00l)) < 1,1$
   A slight confinement for cubic and hexagonal phases is observed in this case. One-dimensional crystal parameter can be reduced up to 10% in the cross-section accompanied by distortion of the crystal lattice and increase of the lattice parameter along the SWNT. In this case some deformation of the SWNT can occur. For layered structures the crystal can be stable even for higher ratios due to possible SWNT deformation, as it was observed for $PbI_2$@SWNT (Flahaut et al., 2006b). Slight deformation of the tube can also be observed for more symmetric crystals (Sloan et al., 2002a). For chiral tubes some twisting of the crystal within the channel is expected.
5. $1,2 < d_{cryst}/(d_{SWNT} - d_{graphite}(00l))$
   The crystals are highly unstable and tend to undergo phase transformation yielding a more stable configuration. These structures can be observed as intermediates during deintercalation caused by electron beam irradiation (Hutchison et al., 2008).

Most structures of one-dimensional nanocrystals within the SWNTs were shown to be metastable (Eliseev et al., 2009a). This was demonstrated for metal halides nanoparticles (i.e. AgBr, AgI, $CoCl_2$) formed within the SWNTs. After removal of the carbon nanotube shell, these particles lose their nanowire morphology (Bendall et al., 2006). Dynamic behavior was reported for the $ZrCl_4$, $Re_xO_y$, and CuI nanoparticles within the SWNT channels upon electron beam irradiation (Brown et al., 2001; Costa et al., 2005; Kiselev et al., 2008). The $ZrCl_4$ nanoparticles form clusters, the $Re_xO_y$ particles evenly rotate inside the single-walled nanotubes, while one-dimensional CuI crystals deencapsulate onto the nanotube surface and decompose yielding metallic copper. Live HRTEM study of the 1D-CuI@SWNT nanocomposite brought two phenomena to light: (1) the CuI crystal rotation and oscillation within the nanotube channel and (2) its emergence from the nanotube through macro-defects (Hutchison et al., 2008). In a series of images (some shown in Fig. 7) a small fragment of the CuI 1D-crystal (about 5.1 nm long) within the nanotube moved and rotated. By the end of the scanning process, the crystal was destroyed. The reason for the crystal movement and escape was charge generation at the SWNT walls in concert with the and $OH^-$ generation; these particles interact with the SWNT walls, thus producing defects. The observed effects enabled to propose a technique for controllable nanocrystal deencapsulation. In case of its application for partial crystal removal the method provides a p-n-junction within a single SWNT.

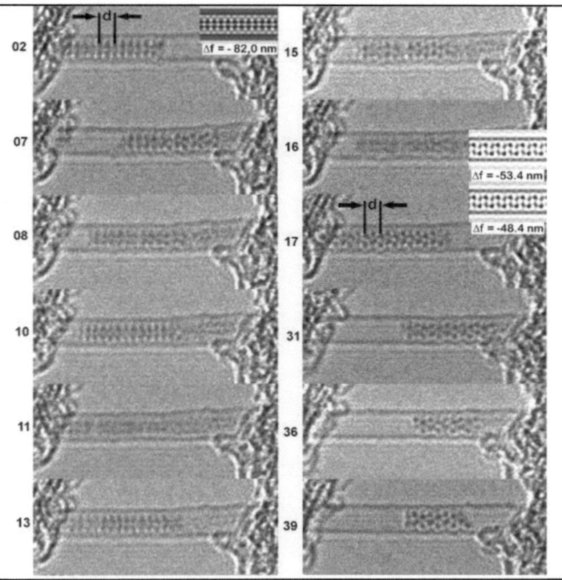


Fig. 7. Oscillation and rotation of the 1D-CuI nanocrystal within the SWNTs

## 5. Electronic structure of the "1D-crystal@SWNT" composites

For the 1D crystal encapsulated into a nanotube, the interaction between the 1D crystal and the template may play a crucial role and therefore can govern the structure and electronic properties of such a system. The 1D crystal – SWNT interaction may include the formation of local chemical bonds, a template-induced distortion of the crystal structure and the bond geometry, as well as non-local effects (i.e. charge transfer etc). If there are no local chemical bonds formed between the 1D-crystal and the nanotube, the charge transfer can be described within the rigid bands model, which considers the interaction in terms of a "doping" effect with the corresponding increase (*n*-doping) or decrease (*p*-doping) of the SWNT's Fermi level. Tuning of the doping level modifies a number of crucial fundamental properties, e.g. allowed optical transitions, the phonon spectrum, the bulk conductance, the internal charge neutrality level, and the collective free charge carrier response. To a first approximation, the charge transfer may be predicted using a difference in the work function between the SWNTs and the crystal and the equilibrium distance from the outer atoms of 1D crystal to the SWNT, as it has been recently demonstrated for graphene (Khomyakov et al., 2009). There are two weak points for such a prediction. First, the 1D crystal WF apparently differs from that for the bulk material, however, it may be obtained from the quantum chemical modeling relatively easily. Second, the 1D crystal geometry can undergo changes once embedded into the SWNT. It should be noted that the exact *ab initio* modeling of the 1D crystal@SWNT nanocomposites is complicated by the period uncertainly, thus a fragment consisting of a very large number of atoms must be retranslated. The doping level can be determined experimentally by direct WF measurements, optical absorption spectroscopy, XPS/UPS, Raman (including electrochemical charging behavior) or XANES spectroscopy, EELS, as well as theoretically by *ab initio* quantum chemical calculations.

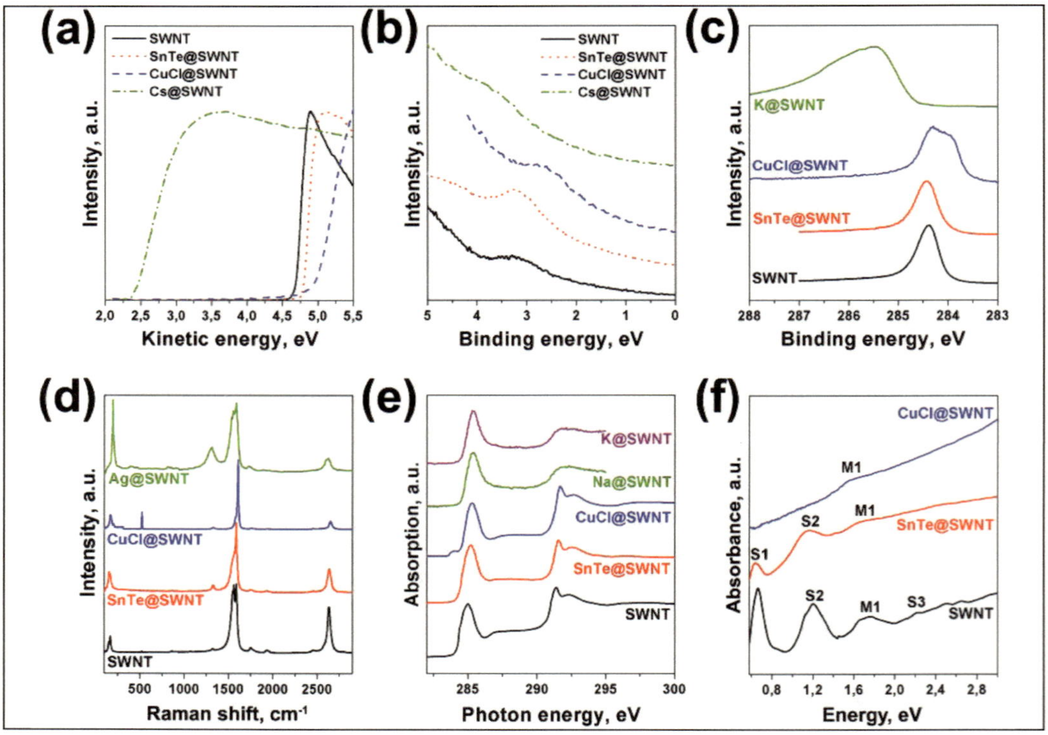


Fig. 8. Secondary electrons cutoff spectra (a), valence band spectra (b), C 1*s* photoemission spectra (c), Raman spectroscopy data (d), C 1s X-ray absorption spectra (e), optical absorption data (f) for pure SWNTs and SWNTs intercalated by different compounds (Corio et al., 2004; Eliseev et al., 2011; Kramberger et al., 2009b; Liu et al., 2003; Suzuki et al., 2000; Yashina et al., 2011)

*Quasi-free-standing 1D crystals within the SWNTs*

Free-standing one-dimensional crystals with 3-5 atoms in diameter currently attract attention due to their unique properties, in particular, size-dependent quantum effects like van Hove singularities etc. Better understanding of the 1D crystal physics can be achieved through development of inert substrates or templates that allow minimizing the crystal-template/substrate interaction. Single-walled carbon nanotubes have been considered as a promising template for growing 1D-crystals due to their chemical inertness towards most inorganic substances, as well as their well-studied electronic properties.

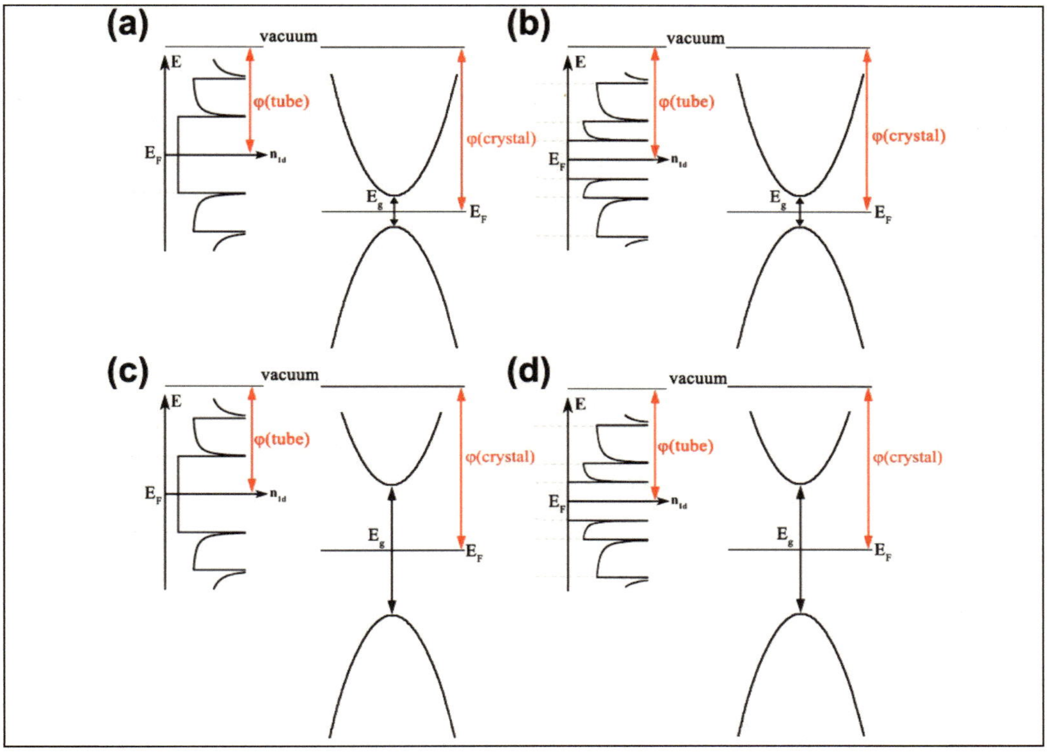


Fig. 9. Semiconductor/SWNT DOS scheme, illustrating possible charge transfer between the guest compounds and the SWNTs: (a) narrow gap semiconductor/metallic SWNT, (b) narrow gap semiconductor/semiconducting SWNT, (c) wide gap semiconductor/metallic SWNT, (d) wide gap semiconductor/semiconducting SWNT

For the quasi-free-standing 1D crystals obtained in such a manner, the Fermi-level shift is absent if the work function of the 1D crystal is nearly the same as that of the SWNT (within the range of ±0.3 eV). This condition may be met for narrow gap semiconductors. Among them, tin telluride (SnTe), an $A^4B^6$ semiconductor, was used as a model case for studying the 1D-crystal formation effect (Yashina et al., 2011) (Fig. 8). The C 1*s* photoemission spectra and Raman spectroscopy data indicate that there are minor differences between the carbon binding energies and the C-C bond vibrations for pristine nanotubes and the nanocomposite (Fig. 8c, d). Both optical absorption and the C 1s X-ray absorption spectra obtained for SnTe@SWNT (shown in Fig. 8e and f) proved that the valence band structure of the composite does not essentially change due to the SnTe intercalation. Only minor influence of the intercalated crystal on the metallic SWNTs was detected by Raman spectroscopy resulted in a slight increase of WF of the metallic nanotubes (Fig. 8a and b). DFT modeling of this system also confirmed a very small (but non-zero) interaction between the $(SnTe)_5/2$ one-dimensional crystal and the nanotube with a diameter of 1.34 nm (Yashina et al., 2011). This suggests that the SnTe@SWNT nanocomposite is a well-suited model system to study the physics of quasi-free-standing 1D-crystals. One may expect similar behavior for other cubic narrow-gap semiconductors, i.e. PbTe, PbSe, and PbS.

The possibility of charge transfer between the encapsulated compound and the SWNT walls generally depends on the filler's band structure (fig. 9). If the guest compound is not chemically bonded to the nanotube, it is possible to adjust the electron density at the SWNT walls by encapsulating narrow gap semiconductors with a desired energy position of the conductance band. To estimate the electron transfer efficiency, one can use the work function and the band gap values for the guest material taking into account the density of states at the bottom of the semiconductor conduction band and the nanotube walls (fig. 10). Contrary, encapsulation of noninteracting wide gap semiconductors with work functions similar to those of SWNT will not result in any charge transfer, thus giving quasi-free-standing 1D crystals within the SWNT channel.

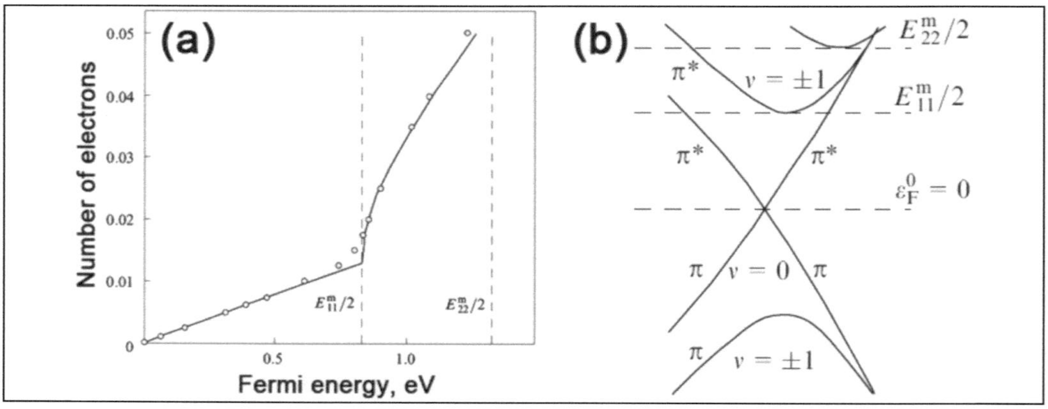


Fig. 10. (a) Charge per carbon atom versus the Fermi energy shift plot for (9,9) SWNT, (b) (9,9) SWNT electronic structure near K-point

*Acceptor doping of SWNT*

Acceptor doping of the SWNTs can be performed by filling nanotubes with a number of inorganic substances, i.e. Se, Te (Chernysheva et al., 2008), $FeCl_3$ (De Blauwe et al., 2009; Liu et al., 2004a), $FeCl_2$, $FeBr_2$, $FeI_2$ (Kharlamova et al., 2009), AgCl, AgBr, AgI (Eliseev et al., 2010), CuCl, CuBr, CuI (Eliseev et al., 2011), $CoBr_2$ (Kharlamova et al., 2010) etc. Acceptor doping causes a corresponding shift of the C 1s line towards lower BEs in XPS, the upshift of the G-mode in the Raman spectra and the appearance of a new empty state in the NEXAFS C 1s spectra (Fig. 8). Similar to the donor doping (see below), the S1 peak in the optical absorption spectrum disappears, so that the Fermi level shifts into the valence band, as evidenced by the disappearance of optical transitions and the emergence of an additional pre-peak in the C 1s core-level excitations.

A new approach to detail studies of the doping effect is Raman spectroscopy applied at electrochemical charging. A good illustration of this technique is the CuI@SWNT nanocomposite studies (Eliseev et al., 2011). Raman spectroscopy maps obtained for the CuX@SWNT composites under electrochemical charging illustrate a shift of the Kohn anomaly positions to more negative potentials (fig. 11). For CuCl@SWNT, it moves to $V_{bias}$ = –1.1 V as compared to $V_{bias}$ = –0,4 V for raw SWNTs. Since the position of the Kohn anomaly corresponds to the equivalent electron density and the Fermi level position in both the pristine SWNTs and the nanocomposite, the relative bias shifts of the Kohn anomaly can be used as an evaluation tool for the Fermi level position changes. This assumption allows direct determination of the Fermi level downshift in the composites, which equals –0.7 eV for CuCl@SWNT. This means that the first Van Hove singularities of SWNTs are emptied due to the acceptor effect of copper halide, which is in good agreement with the optical absorbance spectroscopy results. The mechanism proposed for the acceptor behavior of copper halides involves a formation of a new level, which takes place due to the Cu-SWNT interaction (as observed by XAS), and binding of SWNT conduction π-electrons with copper *3d* electrons at these localized states. This leads to the corresponding Fermi level downshift for the SWNTs. Besides, the dependence charge transfer efficiency on the nanotubes chirality was also reported (Eliseev et al., 2011).The behavior and the mechanism are applicable to most of acceptor dopants.

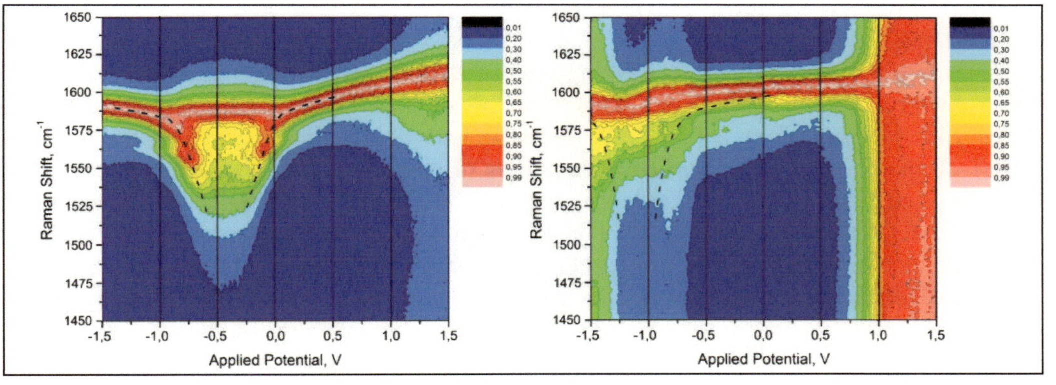


Fig. 11. G-region Raman scattering maps for charged pristine SWNTs (a) and CuCl@SWNT (b) nanocomposites illustrating the Kohn anomaly shift upon intercalation of the CuX compounds. The maps were normalised to the maximum scattering intensity in the 1450-1650 $cm^{-1}$ region to improve the data presentation. Dashed lines indicate the softening phonon branches just for illustration purposes

*Donor doping and acceptor-to-donor doping transformation*

Donor doping of SWNTs is performed by their filling with certain metals or metallorganic compounds (Kramberger et al., 2009a; Kramberger et al., 2009b). Most opportunities to tune the electronic properties of SWNTs are provided by alkali metals doping. In a series of publications (Kramberger et al., 2009a; Kramberger et al., 2009b), the donor effect of the metal intercalated from the vapor phase was reported. Gradual potassium intercalation was achieved by vacuum metal deposition on the SNWT bundles; this opens a way to tickle the delicate balance of the internal charge transfer and the local WF (Kramberger et al., 2009a).

However, it should be noted that in this case the filling of the SWNT cavity probably takes place concurrently with the intertube space filling within the bundle. To overcome this problem, an elegant approach was proposed, which includes filling of the SWNTs from the vapor phase with large metalloorganic compounds like FeCp2 (Shiozawa et al., 2008) and CeCp2 (Shiozawa et al., 2009b) with their subsequent transformation upon heating. For example, annealing in vacuum converts the encapsulated cerocene into the nano-structured cerium within the nanotubes; at the same time, the cyclopentadiene rings form an internal wall of the nanotube, thus it becomes double-walled. As a consequence of the increased electron doping of the outer tubes by the encapsulated cerium chains, the Fermi level upshift is observed (Shiozawa et al., 2009b).

Another approach was implemented for rare-earth nanowires within the SWNTs (Ayala et al., 2011). Quantum $ErCl_3$ nanowires templated inside carbon nanotubes tailored under high temperature and vacuum, were studied within a combined XAS and resonant photoemission approach. It was shown that the wire formation occurs spontaneously inside the tubes at thermal heating. The salt filling undergoes a chemical transformation upon high temperature heating in vacuum, which leads to the formation of elemental Er nanowires inside the SWNTs. The XAS and the XPS core level spectroscopy data for the filler and the SWNT, as well as from RES-PES across the *4d* and *3d* edges indicate that both the bonding environment and the hybridization change upon thermal annealing. During the annealing of the SWNTs doped with $ErCl_3$, which initially illustrates an acceptor behavior, it transforms into an n-doped SWNT due to conversion to metallic Er.

*P-n junctions*

Local variation of the WF allows development of nanoelectronics elements, i.e. transistors, due to the possibility to create a p-n junction within an individual nanotube by acceptor and donor filling of its different parts. The example was demonstrated in (Kato et al., 2009) where ultimate one-dimensional heterojunctions of electron donor and acceptor materials were realized within the inner hollow space of a SWNT. The heterojunction structures of Cs/I and Cs/C60 inside the SWNTs yield the air-stable rectifying performance. Clear tunneling currents through the p-n junction barrier were detected only for Cs/I@SWNTs, which is explained by the difference in the depletion layer structures. Based on potential calculations, symmetrical and asymmetrical depletion layers were found to be formed in Cs/I@SWNTs and Cs/C60@SWNTs, respectively. Low temperature measurements also supply evidence of asymmetric depletion layer formation in Cs/C60@SWNTs.

*Local interactions between the 1D crystal and the SWNT*

Due to the SWNTs inertness, the formation of strong chemical bonds with the 1D crystal is hardly possible. Nevertheless, relatively weak local interactions may be observed rather frequently. This was already demonstrated for a number of metal crystals like Ba (Liu et al., 2004b) and Ag (Borowiak-Palen et al., 2006) intercalated into the SWNT bundles. In these cases, new energy levels were observed in the C 1s NEXAFS spectra at photon energies corresponding to the lower π-resonance for Ba or between the π - and σ-resonances for Ag. In case of Ba, at lower doping pure ionic charge transfer from Ba to the nanotubes occurs similar to the alkali-metal-doped SWNTs. However, at high doping levels a simple ridgid band filling of the SWNT's conduction band is not applicable anymore, and one has to take into account hybridization between the Ba and the C states within the intercalation compound. This is in contrast to alkali-metal-intercalated nanotubes where a purely ionic charge transfer is observed. At high Ba content the covalent interaction results in a greater

lattice expansion than expected for $Ba^{2+}$. The suppression of the π-plasmon and a splitting of the π*-states lead to the appearance of an additional peak in the C 1s spectra. Regarding the charge carrier plasmon, for fully doped samples the plasmon energy is approximately two times higher than for the alkali-metal-doped SWNTs, although the dc limit of the optical conductivity is of the same order.
The local interactions issue can be clarified through detailed structural study of the local atomic geometry using EXAFS (Eliseev et al., 2011). CuX nanocrystals were found to bond to the nanotube walls with a length of 2 Å; and copper to carbon coordination increased in the sequence CuI@SWNT → CuBr@SWNT → CuCl@SWNT. In CuCl@SWNT, copper is "bonded" to four carbon atoms and only one Cl atom. These interactions are clearly indicated by XPS, NEXAFS, HERFD XAS, XES, EXAFS, optical absorption and Raman spectroscopy data (Eliseev et al., 2011; Generalov et al., 2010).
Thus encapsulation of electron donor or acceptor compounds into SWNT channels results in the changes in the electronic structure due to both charge transfer between SWNT wall and the guest compound and the chemical interaction between the filler and the carbon nanotubes. This interaction is realized through hybridisation of the $2p_z$ π–orbitals of carbon with *p*- or *d*-orbitals of filler, forming new localised states. The absolute energy positions of these states remain virtually the same for different fillers, but the total charge transfer increases with an increase of the electron affinity of the intercalated compound.

## 6. Conclusions

By intercalating inorganic compounds into single-walled carbon nanotubes, the electronic properties of SWNTs can be directly modified. Namely, the filling of nanotubes with electron donors (such as metals or metalorganic compounds) can lead to the electronic conductivity of the "1D-crystal@SWNT" composite, which is caused by an increase in the electron density on the nanotube walls within the rigid band structure approximation. Alternatively, intercalation of SWNTs with electron acceptors leads to an electron density transfer from the SWNT walls to the one-dimensional crystal, which results in the system transition into a semiconducting state. To control the charge transfer efficiency upon encapsulation, one should consider the work functions and the band gap values for the guest material taking into account the density of state at the bottom of the semiconductor conduction band and the nanotube walls. Moreover, one should always take into account possible chemical interaction between the filler and the carbon nanotubes. This interaction is realized through hybridisation of the $2p_z$ п–orbitals of carbon with *p*- or *d*-orbitals of filler, forming new localised states. Depletion of electrons from the SWNT walls to these localised states will result in decrease of electron density on the conjugated п–orbitals of graphene layer and the corresponding Fermi level downshift. Encapsulation of noninteracting wide gap semiconductors with work functions similar to those of the SWNT will not result in any charge transfer, thus yielding quasi-free-standing 1D crystals within SWNT channel.
This approach enables control of the SWNTs' electronic structures and development of ground-breaking nanosystems such as quantum threads, single nanotubes with a p — n-transition, and nanocables (for conductor intercalation into semiconducting nanotubes) (Zhao & Xie, 2003; Zhou et al., 2006). In this context, nanocomposites based on single-walled nanotubes are considered to be extremely promising for nanoelectronics applications, especially for creating the p — n-junctions inside a single nanotube.

In one of the recent studies a possibility to develop efficient electrodes for symmetric supercondensers based on the $CrO_3$@SWNT composites was demonstrated; high charging rates can be achieved for such electrodes due to reactions between the $CrO_x$ nanocrystals and the electrolyte (Lota et al., 2007). Nanotubes filled with the electron donor compound nanocrystals display low photoelectric work functions, which may be used to manufacture field effect emitters for modern electroluminescent tubes and X-ray minitubes. The above-described experiments are just the very first examples of how the unique properties of the intercalated SWNTs can be used, and the actual applicability of such nanocomposites should be assessed in more detail. The fundamental and practical advancement in this sphere requires further development of the intercalation techniques for various compounds, as well as of the analytical methods and theoretical approches for studying such nanocomposites.

## 7. Acknowledgment

The authors express their thanks to the research group of the Laboratory of Electron Microscopy, A V Shubnikov Institute of Crystallography of the RAS, for fruitful discussions and for providing research results. The work was supported by RFBR Grant No. 09-03-00817, FASI No. 02.513.12.3012 and P2307.

## 8. References

Ajayan, P.M.Iijima, S. (1993). Capillarity-induced filling of carbon nanotubes. *Nature*, Vol.361, No.6410, (January 1993), pp. 333-334, ISSN 0028-0836

Ajayan, P.M.Ebbesen, T.W. (1997). Nanometre-size tubes of carbon. *Reports on Progress in Physics*, Vol.60, No.10, (October 1997), pp. 1025-1062, ISSN 0034-4885

Avouris, P.; Chen, Z.H. & Perebeinos, V. (2007). Carbon-based electronics. *Nature Nanotechnology*, Vol.2, No.10, (October 2007), pp. 605-615, ISSN 1748-3387

Ayala, P.; Kitaura, R.; Nakanishi, R.; Shiozawa, H.; Ogawa, D.; Hoffmann, P.; Shinohara, H. & Pichler, T. (2011). Templating rare-earth hybridization via ultrahigh vacuum annealing of ErCl3 nanowires inside carbon nanotubes. *Physical Review B*, Vol.83, No.8, (February 2011) ISSN 1098-0121

Beguin, F.; Flahaut.E.; Linares-Solano, A. & Pinson, J. (2006). Surface properties, porosity, chemical and electrochemical applications. *Lect.Notes Phys.*, Vol.667,(February 2006), pp. 495-549, ISSN 0075-8450

Bendall, J.S.; Ilie, A.; Welland, M.E.; Sloan, J. & Green, M.L.H. (2006). Thermal stability and reactivity of metal halide filled single-walled carbon nanotubes. *Journal of Physical Chemistry B*, Vol.110, No.13, (April 2006), pp. 6569-6573, ISSN 1520-6106

Bethune, D.S.; Klang, C.H.; de Vries, M.S.; Gorman, G.; Savoy, R.; Vazquez, J. & Beyers, R. (1993). Cobalt-catalysed growth of carbon nanotubes with single-atomic-layer walls. *Nature*, Vol.363, No.6430, (June 1993), pp. 605-607, ISSN 0028-0836

Borowiak-Palen, E.; Ruemmeli, M.H.; Gemming, T.; Pichler, T.; Kalenczuk, R.J. & Silva, S.R.P. (2006). Silver filled single-wall carbon nanotubes - synthesis, structural and electronic properties. *Nanotechnology*, Vol.17, No.9, (May 2006), pp. 2415-2419, ISSN 0957-4484

Brown, G.; Bailey, S.R.; Sloan, J.; Xu, C.G.; Friedrichs, S.; Flahaut, E.; Coleman, K.S.; Hutchison, J.L.; Dunin-Borkowski, R.E. & Green, M.L.H. (2001). Electron beam induced in situ clusterisation of 1D ZrCl4 chains within single-walled carbon nanotubes. *Chemical Communications*, No.9, (September 2001), pp. 845-846, ISSN 1359-7345

Brown, G.; Bailey, S.R.; Novotny, M.; Carter, R.; Flahaut, E.; Coleman, K.S.; Hutchison, J.L.; Green, M.L.H. & Sloan, J. (2003). High yield incorporation and washing properties of halides incorporated into single walled carbon nanotubes. *Applied Physics A-Materials Science & Processing*, Vol.76, No.4, (March 2003), pp. 457-462, ISSN 0947-8396

Cao, J.; Sun, J.L.; Hong, J.; Li, H.Y.; Chen, H.Z. & Wang, M. (2004). Carbon nanotube/CdS core-shel nanowires prepared by a simple room-temperature chemical reduction method. *Advanced Materials*, Vol.16,(January 2004), pp. 84-87, ISSN 1521-4095

Carter, R.; Sloan, J.; Kirkland, A.I.; Meyer, R.R.; Lindan, P.J.D.; Lin, G.; Green, M.L.H.; Vlandas, A.; Hutchison, J.L. & Harding, J. (2006). Correlation of structural and electronic properties in a new low-dimensional form of mercury telluride. *Physical Review Letters*, Vol.96, No.21, (June 2006) ISSN 0031-9007

Chancolon, J.; Archaimbault, F.; Pineau, A. & Bonnamy, S. (2006). Filling of carbon nanotubes with selenium by vapor phase process. *Journal of Nanoscience and Nanotechnology*, Vol.6, No.1, (January 2006), pp. 82-86, ISSN 1533-4880

Chaturvedi, P.; Verma, P.; Singh, A.; Chaudhary, P.K.; Harsh & Basu, P.K. (2008). Carbon nanotube - Purification and sorting protocols. *Defence Science Journal*, Vol.58, No.5, (September 2008), pp. 591-599, ISSN 0011-748X

Chen, S.M.; Kobayashi, K.; Miyata, Y.; Imazu, N.; Saito, T.; Kitaura, R. & Shinohara, H. (2009). Morphology and Melting Behavior of Ionic Liquids inside Single-Walled Carbon Nanotubes. *Journal of the American Chemical Society*, Vol.131, No.41, (October 2009), pp. 14850-14856, ISSN 0002-7863

Chen, Y.K.; Chu, A.; Cook, J.; Green, M.L.H.; Harris, P.J.F.; Heesom, R.; Humphries, M.; Sloan, J.; Tsang, S.C. & Turner, J.F.C. (1997). Synthesis of carbon nanotubes containing metal oxides and metals of the d-block and f-block transition metals and related studies. *Journal of Materials Chemistry*, Vol.7, No.3, (March 1997), pp. 545-549, ISSN 0959-9428

Chernysheva, M.V.; Kiseleva, E.A.; Verbitskii, N.I.; Eliseev, A.A.; Lukashin, A.V.; Tretyakov, Y.D.; Savilov, S.V.; Kiselev, N.A.; Zhigalina, O.M.; Kumskov, A.S.; Krestinin, A.V. & Hutchison, J.L. (2008). The electronic properties of SWNTs intercalated by electron acceptors. *Physica E-Low-Dimensional Systems & Nanostructures*, Vol.40, No.7, (May 2008), pp. 2283-2288, ISSN 1386-9477

Cohen, M.L. (2001). Superconductivity in fullerene systems. *Nanonetwork Materials: Fullerenes, Nanotubes and Related Systems*, Vol.590,(March 2001), pp. 297-304, ISSN 0094-243X

Corio, P.; Santos, A.P.; Santos, P.S.; Temperini, M.L.A.; Brar, V.W.; Pimenta, M.A. & Dresselhaus, M.S. (2004). Characterization of single wall carbon nanotubes filled with silver and with chromium compounds. *Chemical Physics Letters*, Vol.383, No.5-6, (January 2004), pp. 475-480, ISSN 0009-2614

Costa, P.M.F.J.; Sloan, J.; Rutherford, T. & Green, M.L.H. (2005). Encapsulation of RexOy clusters within single-walled carbon nanotubes and their in tubulo reduction and sintering to Re metal. *Chemistry of Materials*, Vol.17, No.26, (December 2005), pp. 6579-6582, ISSN 0897-4756

De Blauwe, K.; Kramberger, C.; Plank, W.; Kataura, H. & Pichler, T. (2009). Raman response of FeCl(3) intercalated single-wall carbon nanotubes at high doping. *Physica Status Solidi B-Basic Solid State Physics*, Vol.246, No.11-12, (December 2009), pp. 2732-2736, ISSN 0370-1972

Demoncy, N.; Stephan, O.; Brun, N.; Colliex, C.; Loiseau, A. & Pascard, H. (1998). Filling carbon nanotubes with metals by the arc-discharge method: the key role of sulfur. *European Physical Journal B*, Vol.4, No.2, (July 1998), pp. 147-157, ISSN 1434-6028

Dresselhaus, M.S.; Dresselhaus, G. & Saito, R. (1995). Physics of carbon nanotubes. *Carbon*, Vol.33, No.7, (July 1995), pp. 883-891, ISSN 0008-6223

Dujardin, E.; Ebbesen, T.W.; Hiura, H. & Tanigaki, K. (1994). Capillarity and Wetting of Carbon Nanotubes. *Science*, Vol.265, No.5180, (September 1994), pp. 1850-1852, ISSN 0036-8075

Ebbesen, T.W. (1996). Wetting, filling and decorating carbon nanotubes. *Journal of Physics and Chemistry of Solids*, Vol.57, No.6-8, (June 1996), pp. 951-955, ISSN 0022-3697

Eliseev, A.A.; Kharlamova, M.V.; Chernysheva, M.V.; Lukashin, A.V.; Tretyakov, Y.D.; Kumskov, A.S. & Kiselev, N.A. (2009a). Preparation and properties of single-walled nanotubes filled with inorganic compounds. *Russian Chemical Reviews*, Vol.78, No.9, (May 2009a), pp. 833-854, ISSN 0036-021X

Eliseev, A.A.; Chernysheva, M.V.; Verbitskii, N.I.; Kiseleva, E.A.; Lukashin, A.V.; Tretyakov, Y.D.; Kiselev, N.A.; Zhigalina, O.M.; Zakalyukin, R.M.; Vasiliev, A.L.; Krestinin, A.V.; Hutchison, J.L. & Freitag, B. (November 2009b). Chemical Reactions within Single-Walled Carbon Nanotube Channels. *Chemistry of Materials*, Vol.21, No.21, (November 2009b), pp. 5001-5003, ISSN 0897-4756

Eliseev, A.A.; Yashina, L.V.; Brzhezinskaya, M.M.; Chernysheva, M.V.; Kharlamova, M.V.; Verbitsky, N.I.; Lukashin, A.V.; Kiselev, N.A.; Kumskov, A.S.; Zakalyuhin, R.M.; Hutchison, J.L.; Freitag, B. & Vinogradov, A.S. (2010). Structure and electronic properties of AgX (X = Cl, Br, I)-intercalated single-walled carbon nanotubes. *Carbon*, Vol.48, No.10, (August 2010), pp. 2708-2721, ISSN 0008-6223

Eliseev, A.A.; Yashina, L.V.; Verbitskii, N.I.; Kharlamova, M.V.; Chernysheva, M.V.; Lukashin, A.V.; Kiselev, N.A.; Kumskov, A.S.; Freitag, B.; Brzhezinskaya, M.M.; Vinogradov, A.S.; Zubavichus, Y.V.; Kleimenov, E. & Nachtegaal, M. (2011). Structure and electronic properties of CuX(X=Cl, Br, I)-intercalated single-wall carbon nanotubes: «1D crystal» - SWNT bonding and interactions. *Carbon* (2011), p. in print, ISSN 0008-6223

Fagan, S.B.; Filho, A.G.S.; Filho, J.M.; Corio, P. & Dresselhaus, M.S. (April 2005). Electronic properties of Ag- and CrO3-filled single-wall carbon nanotubes. *Chemical Physics Letters*, Vol.406, No.1-3, (April 2005), pp. 54-59, ISSN 0009-2614

Fan, X.; Dickey, E.C.; Eklund, P.C.; Williams, K.A.; Grigorian, L.; Buczko, R.; Pantelides, S.T. & Pennycook, S.J. (2000). Atomic arrangement of iodine atoms inside single-walled carbon nanotubes. *Physical Review Letters*, Vol.84, No.20, (May 2000), pp. 4621-4624, ISSN 0031-9007

Flahaut, E.; Sloan, J.; Friedrichs, S.; Kirkland, A.I.; Coleman, K.S.; Williams, V.C.; Hanson, N.; Hutchison, J.L. & Green, M.L.H. (2006a). Crystallization of 2H and 4H PbI2 in carbon nanotubes of varying diameters and morphologies. *Chemistry of Materials*, Vol.18, No.8, (April 2006a), pp. 2059-2069, ISSN 0897-4756

Flahaut, E.; Sloan, J.; Friedrichs, S.; Kirkland, A.I.; Coleman, K.S.; Williams, V.C.; Hanson, N.; Hutchison, J.L. & Green, M.L.H. (2006b). Crystallization of 2H and 4H PbI2 in carbon nanotubes of varying diameters and morphologies. *Chemistry of Materials*, Vol.18, No.8, (April 2006b), pp. 2059-2069, ISSN 0897-4756

Friedrichs, S.; Meyer, R.R.; Sloan, J.; Kirkland, A.I.; Hutchison, J.L. & Green, M.L.H. (2002). Complete characterization of an (Sb2O3)(n)/SWNT inclusion composite. *Physics of the Solid State*, Vol.44, No.3, (July 2002), pp. 463-466, ISSN 1063-7834

Generalov, A.V.; Brzhezinskaya, M.M.; Puttner, R.; Vinogradov, A.S.; Chernysheva, M.V.; Eliseev, A.A.; Kiselev, N.A.; Lukashin, A.V. & Tretyakov, Y.D. (2010). Electronic Structure of CuI@SWCNT Nanocomposite Studied by X-Ray Absorption Spectroscopy. *Fullerenes Nanotubes and Carbon Nanostructures*, Vol.18, No.4-6, (March 2010), pp. 574-578, ISSN 1536-383X

Govindaraj, A.; Satishkumar, B.C.; Nath, M. & Rao, C.N.R. (2000). Metal nanowires and intercalated metal layers in single-walled carbon nanotube bundles. *Chemistry of Materials*, Vol.12, No.1, (January 2000), pp. 202-205, ISSN 0897-4756

Hirahara, K.; Suenaga, K.; Bandow, S.; Kato, H.; Okazaki, T.; Shinohara, H. & Iijima, S. (2000). One-dimensional metallofullerene crystal generated inside single-walled carbon nanotubes. *Physical Review Letters*, Vol.85, No.25, (December 2000), pp. 5384-5387, ISSN 0031-9007

Hou, P.X.; Liu, C. & Cheng, H.M. (2008). Purification of carbon nanotubes. *Carbon*, Vol.46, No.15, (December 2008), pp. 2003-2025, ISSN 0008-6223

Hutchison, J.L.; Grobert, N.; Zakalyukin, R.M.; Eliseev, A.A.; Chernisheva, M.V.; Kumnskov, A.S.; Grigoriev, Y.V.; Krestinin, A.V.; Freitag, B. & Kiselev, N.A. (2008). The behaviour of 1D CuI crystal@SWNT nanocomposite under electron irradiation. *Electron Microscopy and Multiscale Modeling, Proceedings*, Vol.999,(July 2008), pp. 79-92, ISSN 0094-243X

Iijima, S. (1991). Helical microtubules of graphitic carbon. *Nature*, Vol.354, No.6348, (November 1991), pp. 56-58, ISSN 0028-0836

Iijima, S.Ichihashi, T. (1993). Single-shell carbon nanotubes of 1-nm diameter. *Nature*, Vol.363, No.6430, (June 1993), pp. 603-605, ISSN 0028-0836

Jeong, G.H.; Farajian, A.A.; Hirata, T.; Hatakeyama, R.; Tohji, K.; Briere, T.M.; Mizuseki, H. & Kawazoe, Y. (2003). Encapsulation of cesium inside single-walled carbon nanotubes by plasma-ion irradiation method. *Thin Solid Films*, Vol.435, No.1-2, (July 2003), pp. 307-311, ISSN 0040-6090

Kataura, H.; Maniwa, Y.; Abe, M.; Fujiwara, A.; Kodama, T.; Kikuchi, K.; Imahori, H.; Misaki, Y.; Suzuki, S. & Achiba, Y. (2002). Optical properties of fullerene and non-fullerene peapods. *Applied Physics A-Materials Science & Processing*, Vol.74, No.3, (March 2002), pp. 349-354, ISSN 0947-8396

Kato, T.; Hatakeyama, R.; Shishido, J.; Oohara, W. & Tohji, K. (2009). P-N junction with donor and acceptor encapsulated single-walled carbon nanotubes. *Applied Physics Letters*, Vol.95, No.8, (August 2009) ISSN 0003-6951

Kharlamova, M.V.; Brzhezinskaya, M.M.; Vinogradov, A.S.; Suzdalev, I.P.; Maksimov, Yu.V.; Ishmennik, V.K.; Novichikhin, S.V.; Krestinin, A.V.; Yashina, L.V.; Lukashin, A.V.; Tretyakov, Yu.D. & Eliseev, A.A. (2009). The Formation and Properties of One-dimensional FeHal2 (Hal=Cl, Br, I) Nanocrystals in Channels of Single-walled Carbon Tubes. *Nanotechnol.Russ.*, Vol.4, No.9-10, (October 2009), pp. 634-646, ISSN 1993-4068

Kharlamova, M.V.; Eliseev, A.A.; Yashina, L.V.; Petukhov, D.I.; Liu, C.P.; Wang, C.Y.; Semenenko, D.A. & Belogorokhov, A.I. (2010). Study of the electronic structure of single-walled carbon nanotubes filled with cobalt bromide. *JETP Letters*, Vol.91, No.4, (February 2010), pp. 196-200, ISSN 0021-3640

Khomyakov, P.A.; Giovannetti, G.; Rusu, P.C.; Brocks, G.; van den Brink, J. & Kelly, P.J. (2009). First-principles study of the interaction and charge transfer between graphene and metals. *Physical Review B*, Vol.79, No.19, (May 2009) ISSN 1098-0121

Kirkland, A.I.; Meyer, M.R.; Sloan, J. & Hutchison, J.L. (2005). Structure determination of atomically controlled crystal architectures grown within single wall carbon nanotubes. *Microscopy and Microanalysis*, Vol.11, No.5, (October 2005), pp. 401-409, ISSN 1431-9276

Kiselev, N.A.; Zakalyukin, R.M.; Zhigalina, O.M.; Grobert, N.; Kumskov, A.S.; Grigoriev, Y.V.; Chernysheva, M.V.; Eliseev, A.A.; Krestinin, A.V.; Tretyakov, Y.D.; Freitag, B. & Hutchison, J.L. (2008). The structure of 1D CuI crystals inside SWNTs. *Journal of Microscopy-Oxford*, Vol.232, No.2, (November 2008), pp. 335-342, ISSN 0022-2720

Kiselev, N.A.; Kumnskov, A.S.; Zakalyukin, R.M.; Vasiliev, A.L.; Chernisheva, M.V.; Eliseev, A.A.; Krestinin, A.V.; Freitag, B. & Hutchison, J.L. (2011). The structure of nanocomposites 1D cationic conductor crystal@SWNT. *Journal of Microscopy-Oxford*, Vol.in print,(January 2011) ISSN 0022-2720

Kramberger, C.; Rauf, H.; Knupfer, M.; Shiozawa, H.; Batchelor, D.; Rubio, A.; Kataura, H. & Pichler, T. (2009a). Potassium-intercalated single-wall carbon nanotube bundles: Archetypes for semiconductor/metal hybrid systems. *Physical Review B*, Vol.79, No.19, (May 2009a) ISSN 1098-0121

Kramberger, C.; Rauf, H.; Knupfer, M.; Shiozawa, H.; Batchelor, D.; Kataura, H. & Pichler, T. (2009b). Electronic and optical properties of alkali metal doped carbon nanotubes. *Physica Status Solidi B-Basic Solid State Physics*, Vol.246, No.11-12, (December 2009b), pp. 2693-2698, ISSN 0370-1972

Leonhardt, A.; Ritschel, A.; Kozhuharova, R.; Graff, A.; Muhl, T.; Huhle, R.; Monch, I.; Elefant, D. & Schneider, C.M. (2003). Synthesis and properties of filled carbon nanotubes. *Diamond and Related Materials*, Vol.12, No.3-7, (March 2003), pp. 790-793, ISSN 0925-9635

Liu, X.; Pichler T.; Knupfer M. & Fink J. (2003). Electronic and optical properties of alkali-metal-intercalated single-wall carbon nanotubes. *Physical Review B*, Vol.67, No.12, (March 2003) ISSN 1098-0121

Liu, X.; Pichler, T.; Knupfer, M.; Fink, J. & Kataura, H. (2004a). Electronic properties of FeCl3-intercalated single-wall carbon nanotubes. *Physical Review B*, Vol.70, No.20, (November 2004a) ISSN 1098-0121

Liu, X.; Pichler, T.; Knupfer, M. & Fink, J. (2004b). Electronic properties of barium-intercalated single-wall carbon nanotubes. *Physical Review B*, Vol.70, No.24, (December 2004b) ISSN 1098-0121

Liu, Y.C.Wang, Q. (2005). Transport behavior of water confined in carbon nanotubes. *Physical Review B*, Vol.72, No.8, (August 2005) ISSN 1098-0121

Lota, G.; Frackowiak, E.; Mittal, J. & Monthioux, M. (2007). High performance supercapacitor from chromium oxide-nanotubes based electrodes. *Chemical Physics Letters*, Vol.434, No.1-3, (January 2007), pp. 73-77, ISSN 0009-2614

Meyer, R.R.; Sloan, J.; Dunin-Borkowski, R.E.; Kirkland, A.I.; Novotny, M.C.; Bailey, S.R.; Hutchison, J.L. & Green, M.L.H. (2000). Discrete atom imaging of one-dimensional crystals formed within single-walled carbon nanotubes. *Science*, Vol.289, No.5483, (August 2000), pp. 1324-1326, ISSN 0036-8075

Mittal, J.; Monthioux, M.; Allouche, H. & Stephan, O. (2001). Room temperature filling of single-wall carbon nanotubes with chromium oxide in open air. *Chemical Physics Letters*, Vol.339, No.5-6, (May 2001), pp. 311-318, ISSN 0009-2614

Monthioux, M.; Smith, B.W.; Burteaux, B.; Claye, A.; Fischer, J.E. & Luzzi, D.E. (2001). Sensitivity of single-wall carbon nanotubes to chemical processing: an electron microscopy investigation. *Carbon*, Vol.39, No.8, (May 2001), pp. 1251-1272, ISSN 0008-6223

Monthioux, M. (February 2002). Filling single-wall carbon nanotubes. *Carbon*, Vol.40, No.10, (February 2002), pp. 1809-1823, ISSN 0008-6223

Monthioux, M.; Flahaut, E. & Cleuziou, J.P. (2006). Hybrid carbon nanotubes: Strategy, progress, and perspectives. *Journal of Materials Research*, Vol.21, No.11, (November 2006), pp. 2774-2793, ISSN 0884-2914

Odom, T.W.; Huang, J.L.; Kim, P. & Lieber, C.M. (2000). Structure and electronic properties of carbon nanotubes. *Journal of Physical Chemistry B*, Vol.104, No.13, (April 2000), pp. 2794-2809, ISSN 1089-5647

Okazaki, T.; Shimada, T.; Suenaga, K.; Ohno, Y.; Mizutani, T.; Lee, J.; Kuk, Y. & Shinohara, H. (2003). Electronic properties of Gd@C-82 metallofullerene peapods: (Gd @ C-82)(n)@SWNTs. *Applied Physics A-Materials Science & Processing*, Vol.76, No.4, (March 2003), pp. 475-478, ISSN 0947-8396

Pan, C.; Chandrasekharaiah, M.S.; Agan, D.; Hauge, R.H. & Margrave, J.L. (2002). Determination of sublimation pressures of a fullerene (C60/C70) solid solution. *The Journal of Physical Chemistry*, Vol.96, No.16, (May 2002), pp. 6752-6755, ISSN 0022-3654

Pederson, M.R.Broughton, J.Q. (1992). Nanocapillarity in Fullerene Tubules. *Physical Review Letters*, Vol.69, No.18, (November 1992), pp. 2689-2692, ISSN 0031-9007

Rahman, M.M.; Kisaku, M.; Kishi, T.; Roman, T.A.; Dino, W.A.; Nakanishi, H. & Kasai, H. (2005). Electric and magnetic properties of Co-filled carbon nanotube. *Journal of the Physical Society of Japan*, Vol.74, No.2, (February 2005), pp. 742-745, ISSN 0031-9015

Saito, R.; Saito, R.; Fujita, M.; Fujita, M.; Dresselhaus, G.; Dresselhaus, G.; Dresselhaus, M.S. & Dresselhaus, M.S. (1992). Electronic structure of chiral graphene tubules. *Applied Physics Letters*, Vol.60, No.18, (May 1992), pp. 2204-2206, ISSN 1077-3118

Satishkumar, B.C.; Govindaraj, A.; Mofokeng, J.; Subbanna, G.N. & Rao, C.N.R. (1996). Novel experiments with carbon nanotubes: Opening, filling, closing and functionalizing nanotubes. *Journal of Physics B-Atomic Molecular and Optical Physics*, Vol.29, No.21, (November 1996), pp. 4925-4934, ISSN 0953-4075

Sceats, E.L.; Green, J.C. & Reich, S. (2006). Theoretical study of the molecular and electronic structure of one-dimensional crystals of potassium iodide and composites formed upon intercalation in single-walled carbon nanotubes. *Physical Review B*, Vol.73, No.12, (March 2006) ISSN 1098-0121

Seraphin, S.; Zhou, D.; Jiao, J.; Withers, J.C. & Loutfy, R. (1993). Yttrium carbide in nanotubes. *Nature*, Vol.362, No.6420, (April 1993), p. 503, ISSN 0028-0836

Shiozawa, H.; Pichler, T.; Kramberger, C.; Gruneis, A.; Knupfer, M.; Buchner, B.; Zolyomi, V.; Koltai, J.; Kurti, J.; Batchelor, D. & Kataura, H. (2008). Fine tuning the charge transfer in carbon nanotubes via the interconversion of encapsulated molecules. *Physical Review B*, Vol.77, No.15, (April 2008) ISSN 1098-0121

Shiozawa, H.; Pichler, T.; Kramberger, C.; Rummeli, M.; Batchelor, D.; Liu, Z.; Suenaga, K.; Kataura, H. & Silva, S.R.P. (2009a). Screening the Missing Electron: Nanochemistry in Action. *Physical Review Letters*, Vol.102, No.4, (January 2009a) ISSN 0031-9007

Shiozawa, H.; Kramberger, C.; Rummeli, M.; Batchelor, D.; Kataura, H.; Pichler, T. & Silva, S.R.P. (2009b). Electronic properties of single-walled carbon nanotubes encapsulating a cerium organometallic compound. *Physica Status Solidi B-Basic Solid State Physics*, Vol.246, No.11-12, (December 2009b), pp. 2626-2630, ISSN 0370-1972

Sloan, J.; Hammer, J.; Zwiefka-Sibley, M. & Green, M.L.H. (1998). The opening and filling of single walled carbon nanotubes (SWTs). *Chemical Communications*, No.3, (February 1998), pp. 347-348, ISSN 1359-7345

Sloan, J.; Wright, D.M.; Woo, H.G.; Bailey, S.; Brown, G.; York, A.P.E.; Coleman, K.S.; Hutchison, J.L. & Green, M.L.H. (1999). Capillarity and silver nanowire formation observed in single walled carbon nanotubes. *Chemical Communications*, No.8, (April 1999), pp. 699-700, ISSN 1359-7345

Sloan, J.; Dunin-Borkowski, R.E.; Hutchison, J.L.; Coleman, K.S.; Clifford Williams, V.; Claridge, J.B.; York, A.P.E.; Xu, C.; Bailey, S.R.; Brown, G.; Friedrichs, S. & Green, M.L.H. (2000a). The size distribution, imaging and obstructing properties of C60 and higher fullerenes formed within arc-grown single walled carbon nanotubes. *Chemical Physics Letters*, Vol.316, No.3-4, (January 2000a), pp. 191-198, ISSN 0009-2614

Sloan, J.; Novotny, M.C.; Bailey, S.R.; Brown, G.; Xu, C.; Williams, V.C.; Friedrichs, S.; Flahaut, E.; Callender, R.L.; York, A.P.E.; Coleman, K.S.; Green, M.L.H.; Dunin-Borkowski, R.E. & Hutchison, J.L. (2000b). Two layer 4 : 4 co-ordinated KI crystals grown within single walled carbon nanotubes. *Chemical Physics Letters*, Vol.329, No.1-2, (October 2000b), pp. 61-65, ISSN 0009-2614

Sloan, J.; Kirkland, A.I.; Hutchison, J.L. & Green, M.L.H. (2002a). Integral atomic layer architectures of 1D crystals inserted into single walled carbon nanotubes. *Chemical Communications*, No.13, (April 2002a), pp. 1319-1332, ISSN 1359-7345

Sloan, J.; Friedrichs, S.; Meyer, R.R.; Kirkland, A.I.; Hutchison, J.L. & Green, M.L.H. (2002b). Structural changes induced in nanocrystals of binary compounds confined within single walled carbon nanotubes: a brief review. *Inorganica Chimica Acta*, Vol.330,(March 2002b), pp. 1-12, ISSN 0020-1693

Sloan, J.; Grosvenor, S.J.; Friedrichs, S.; Kirkland, A.I.; Hutchison, J.L. & Green, M.L.H. (2002c). A one-dimensional BaI2 chain with five- and six-coordination, formed within a single-walled carbon nanotube. *Angewandte Chemie-International Edition*, Vol.41, No.7, (March 2002c), p. 1156-+, ISSN 1433-7851

Smith, B.W.; Monthioux, M. & Luzzi, D.E. (1998). Encapsulated C60 in carbon nanotubes. *Nature*, Vol.396, No.6709, (November 1998), pp. 323-324, ISSN 0028-0836

Smith, B.W.; Luzzi, D.E. (2000). Formation mechanism of fullerene peapods and coaxial tubes: a path to large scale synthesis. *Chemical Physics Letters*, Vol.321, No.1-2, (April 2000), pp. 169-174, ISSN 0009-2614

Suzuki, S.; Bower C.; Watanabe Y.; Zhou O. (2000). Work functions and valence band states of pristine and Cs-intercalated single-walled carbon nanotube bundles. *Applied Physics Letters*, Vol.76, No.26, (June 2000), pp. 4007-4009, ISSN 0003-6951

Tans, S.J.; Verschueren, A.R.M. & Dekker, C. (1998). Room-temperature transistor based on a single carbon nanotube. *Nature*, Vol.393, No.6680, (May 1998), pp. 49-52, ISSN 0028-0836

Tsang, S.C.; Chen, Y.K.; Harris, P.J.F. & Green, M.L.H. (1994). A simple chemical method of opening and filling carbon nanotubes. *Nature*, Vol.372, No.6502, (November 1994), pp. 159-162, ISSN 0028-0836

Weissmann, M.; Garcia, G.; Kiwi, M.; Ramirez, R. & Fu, C.C. (2006). Theoretical study of iron-filled carbon nanotubes. *Physical Review B*, Vol.73, No.12, (March 2006) ISSN 1098-0121

Xu, C.G.; Sloan, J.; Brown, G.; Bailey, S.; Williams, V.C.; Friedrichs, S.; Coleman, K.S.; Flahaut, E.; Hutchison, J.L.; Dunin-Borkowski, R.E. & Green, M.L.H. (2000). 1D lanthanide halide crystals inserted into single-walled carbon nanotubes. *Chemical Communications*, No.24, (November 2000), pp. 2427-2428, ISSN 1359-7345

Yashina, L.V.; Eliseev, A.A.; Kharlamova, M.V.; Volykhov, A.A.; Egorov, A.V.; Savilov, S.V.; Lukashin, A.V.; Puttner, R. & Belogorokhov, A.I. (2011). Growth and Characterization of One-Dimensional SnTe Crystals within the Single-Walled Carbon Nanotube Channels. *The Journal of Physical Chemistry C*, Vol.115, No.9, (February 2011), pp. 3578-3586, ISSN 1932-7447

Zhang, Z.X.; Pan, Z.Y.; Wei, Q.; Li, Z.J.; Zang, L.K. & Wang, Y.X. (2003). Mechanics of nanotubes filled witn C60, C36 and C20. *International Journal of Modern Physics B*, Vol.17,(October 2003), pp. 4667-4674, ISSN 1793-6578

Zhao, J.J.Xie, R.H. (2003). Electronic and photonic properties of doped carbon nanotubes. *Journal of Nanoscience and Nanotechnology*, Vol.3, No.6, (December 2003), pp. 459-478, ISSN 1533-4880

Zhou, Z.; Zhao, J.J.; Chen, Z.F.; Gao, X.P.; Lu, J.P.; Schleyer, P.V. & Yang, C.K. (2006). True nanocable assemblies with insulating BN nanotube sheaths and conducting Cu nanowire cores. *Journal of Physical Chemistry B*, Vol.110, No.6, (February 2006), pp. 2529-2532, ISSN 1520-6106

# 9

# Geometric and Spectroscopic Properties of Carbon Nanotubes and Boron Nitride Nanotubes


Metin Aydin[1] and Daniel L. Akins[2]

[1]*Department of Chemistry, Faculty of Art and Sciences, Ondokuz Mayıs University, Samsun*

[2]*Center for Analysis of Structures and Interfaces (CASI), Department of Chemistry, The City College of The City University of New York, New York*

[1]*Turkey*

[2]*USA*


## 1. Introduction

Carbon, the first element in Group 4A, is a nonmetal and has $1s^2$ $2s^22p^2$ electronic configuration, in which four valence electrons allow it to form a number of hybridized atomic orbitals. Therefore, carbon atoms in the elemental substances bonds to each other covalently by the sharing of electron pairs, in which the covalent bonds have directional properties. This in turn provides carbon capability to adapt into various molecular and crystalline structures. The natures of these bonds underlie the varied chemical properties and physical properties of the carbon allotropes. Pure carbon-based materials are not only diamonds (as shown in Figure1.1a), and graphite (Figure 1.1b) but also fullerenes (Figure 1.1c), carbon nanotubes (CNT) (see Figure 1.2), and amorphous carbon. These allotropes have been considered some of the most important materials in nanotechnology.

The unique properties of single-walled carbon nanotubes originate from their distinctive structure, which is composed of C-C bonds more closely related to that in graphite rather than in diamond. specifically, despite the fact that diamond has a coordination number of four with $sp^3$ hybridization, the $sp^2$ hybridization in graphite links carbon atoms in a two-dimensional (2D) layer of hexagons that lead to each layer in graphite being a planar structure in the ideal cases. In the latter case, each carbon atom contributes three electrons to the sigma bonds within the plane and has one electron left in the $p_z$ orbitals. These $p_z$ obitals cooperatively allow the electron to delocalize over the entire plane, giving rise to a molecular orbital that is perpendicular to the plane of graphene, which allows the fourth valence electron in carbon atoms to move freely on the plane. Within the layers, the carbon-carbon bond distance is similar to the bond length in benzene (the carbon atoms are strongly bound to each other and carbon-carbon distance is about 0.14 nm), leading to a very large inplane value for Young's modulus. The distance between layers (about 0.34 nm) is too large to permit significant orbital overlap; layers are bounded to each other mainly by weak long-range Van der Waals type interactions. The weak interlayer coupling gives graphite the property of a seemingly very soft material, the property that allows using graphite in a pen for writing. In diamond, all valence electrons are localized around the carbon atoms. These structural differences produce profound affects on the electrical properties of graphite and diamond: graphite is a semimetal, and diamond is an insulator with a band gap around 6 eV.

Since carbon nanotubes' (CNTs) discovery in 1991 [1], they have received extensive attention due to many unique physical and chemical properties. Considering that CNTs are the subject of one of the most important areas of research in nanotechnology due, to their unique properties and potential for precious profitable applications – varying from electronics to chemical process control — a massive amount of effort has been invested on investigation of nanotubes' active components. Indeed, their discovery has resulted in a variety of technological uses, namely, nanotechnology, functional nanodevices, [2,3,4] materials science, heat conduction, [5,6] electronics, [7,8] molecular memories,[9] optics, [10,11,12] unique electrical properties, transistors, electrically excited single-molecule light sources, [13,14,15,16] DNA functionalization, [17,18] high-performance adsorbent electrode material for energy-storage device, [19] and proteins.[20,21] Most of the research has been invested to understand their optical and structural properties as well as the development and advancement of carbon nanotubes.

Carbon nanotubes (CNTs) are in the structural family of fullerene. They are molecules composed entirely of carbon atoms. The carbon atoms are usually arranged in a hexagonal pattern, bonded together with extremely strong covalent bonds. A carbon nanotube bundle can be visualized as a finite number of carbon graphite shells arranged around a hollow center axis with a constant spacing of around 0.34 nm, with a tubular diameter normally ranging up to several nanometers from 1.4 nm, and with lengths up to several microns. This number of graphite shells is the basis for the key division in carbon nanotubes, such as single-walled nanotubes (SWNTs) and multi-walled nanotubes (MWNTs). The concept of either single-walled or multi-walled carbon nanotubes is important since the properties change significantly with respect to each other. One of the goals of the research focused on CNTs is to understand how well electrons flow through carbon nanotubes. As it turns out, many of the results were inconsistent.

In general, multi-walled nanotubes have individually different properties, even if they are made under the same experimental conditions. Because of uncontrollable inconsistencies between each individual nanotube, it is difficult construct a consistent theory of electric flow because of structural defects. In general, the amounts of structural defects in any tube significantly effects the flow of electrons in any other CNT. These defects are not yet fully understood, therefore, many investigators have investigated how to eradicate these defects, so as to obtain the highest performance for nanotube conductance. Furthermore, in optics, strong Coulomb effects [14,15,16,22,23,24] in carbon nanotubes bring about formation of the exciton states that are allegedly a "bright" (allowed one-photon electronic transition) and "dark" (forbidden one-photon transition), and dramatically decrease the efficiency of one-photon light emission via trapping of the carriers by "dark" excitons [13,22, 25,26,27,28,33]. However, a proper use of these "bright" and "dark" exciton states, which have distinctively different recombination times may benefit the use of which quantum coherence [29,30,31,32] and multiphoton schemes of excitation potentially not only allowing one to efficiently manipulate the dark states, but may also create conditions for efficient light generation in different frequency regions; i.e., producing "slow" or "fast" light, thus implementing quantum light storage media with a negative refractive index,[33,34] and other quantum-optical regimes [33,35,36]. Possible quantum-optical carbon nanotube devices have a potential for suitable performance at elevated temperatures, because the binding energies of excitons in single-walled nanotubes (SWCNTs) are up to hundreds of meV [23, 24, 16, 37, 38, 39, 40].

CNTs are not only used for electron transport, but also used as an ideal carbon fiber [41] and as a storage units for hydrogen for use in hydrogen fuel cells,[41,42]. From a mechanical perspective, the strength of CNTs illustrates their material properties, where the majority defects can be neglected, that make it possible to view each nanotube as an ideal carbon fiber. In nature, the strong C-C bonds in CNTs are some of the strongest bonds, thus making a carbon nanotube one of the strongest materials in nature. For instance, for an ideal carbon nanotube, Young modulus of 1000 GPa and tensile strength of 60 GPa have been measured from individual structure, which is an order of magnitude higher than ordinary engineering plastics. High electrical and thermal conductivity have also been determined experimentally, [43,44,45,46], with value close to or better than metals. With such a combination of properties and a product form compatible with modern polymer processing technologies, the possibilities of creating new engineered materials are enourmous [48][47]. The use of carbon nanotubes for antistatic and conductive applications in polymers is already a commercial reality and is growing in sectors such as electronics and the automotive industry. The loading for achieving electrical percolation with multi-wall carbon nanotubes (MWCNTs) may be lower than with conductive carbon black grades.

Carbon nanotubes can be synthesized by various methods, including arc-discharge, laser ablation, and chemical vapor deposition [24]. In the arc-discharge method, carbon atoms are evaporated by a plasma of helium gas ignited by high currents passing through a carbon anode and cathode assembly. The carbon atoms nucleate on a metal catalyst and grow to several micrometers in length. A similar principle is adopted in the laser ablation method where intense laser pulses are used to ablate a carbon target containing metal catalysts. During laser ablation, a flow of inert gas is passed through the growth chamber to carry the nanotube downstream to a cold collection finger. The produced SWCNTs mostly crystallize in the form of ropes having tens of individual nanotubes close-packed into hexagons via Van der Waals interactions. In chemical vapor deposition, a flowing hydrocarbon gas is decomposed by metal catalysts. The precipitation of carbon from the saturated phase in metal particles leads to the formation of a tubular carbon solid.

Boron nitride nanotubes (BNNTs) are among the nanocomposites, which have been synthesized successfully [48,49,50] following the syntheses of carbon nanotubes (CNTs). The electronic properties of boron nitride nanotubes differ from carbon nanotubes. While carbon nanotubes can be either metallic or semiconducting, depending on their chirality and the diameter [51], all boron nitride nanotubes (BNNTs) are found to be semiconducting materials having the large band gaps[53], with band gaps only weakly depending on the diameter, chirality, and the number of the walls of the tube. Because of their semiconducting behavior, BNNTs are very interesting materials for application in nanoscale devices and are considered remarkable alternatives to CNTs [52,53]. Modification of the electronic properties of nanotubes by doping and functionalization is also an important subject for nano-devices. The doped nanotubes may exhibit a dramatic change with regard to the isolated nanotube. Furthermore, because of the strong interactions between electrons and holes in BNNTs [54,55], the excitonic effects in are more important than in CNTs. Bright and dark excitons in BNNTs qualitatively alter the optical response [57].

The optical properties of carbon nanotubes are uniquely connected with the absorption, photoluminescence, and Raman scattering of carbon nanotubes. Spectroscopic methods are used for the characterization of large quantities of nanotubes without damaging them. Optical absorption, photoluminescence and Raman spectroscopies permit a reliable characterization for the quality of nanotube such as chirality, size, and structural defect of the produced nanotubes. These characterization shed light on other properties, such as optical, mechanical, and electrical properties. Raman spectroscopy, or light scattering, is a powerful technique used to investigate their spectroscopic and structural properties and provides significant insight into the fundamental physical processes that occur. Even though a large number of phonon modes in the Raman spectra of carbon nanotubes would be expected, most are Raman inactive due to the selection rules, as a consequence of the high symmetry properties of the nanotubes. The Raman spectrum of a carbon nanotube exhibits a few characteristic modes that can be used to determine the size of nanotubes and to classify the type of the nanotubes, such as semiconducting and metallic. For example, in the low frequency region, one type is called the radial breathing mode (RBM), which is in the radial direction with the same phase as a result of the vibration of the entire tube. This mode is strongly diameter dependent. The RBM is used to determine the size of the nanotube. There are two characteristic Raman bands that lie in the range of 1300-1650 $cm^{-1}$, which are called tangential modes. The line shape of these Raman modes may be used to classify whether the nanotube is metallic or semiconducting. These Raman modes in the high energy region are also slightly diameter dependent.

Theoretical calculations may be very useful to understand the nature of optical properties as well as their storage mechanisms. For a better understanding of the physical and optical properties of nanotubes, theoretical calculation may be needed to specify the material properties due to the dependence of their spectroscopic properties on the size of the nanotubes. The electronic structures of the single-walled boron nitride nanotubes (SWBNNTs) were theoretically investigated by Rubio *et al.* [56] using the tight binding approximation. All SWBNNTs were found to be semiconducting materials with the band gaps greater than 2 eV. BNNTs with larger diameters have a larger band gap, with a saturation value corresponding to the band gap of a hexagonal boron nitride [56,57]. The local-density-functional (LDA) calculations [58] indicated that the folding of a hexagonal boron nitride sheet into a BNNT is energetically more favorable than to form a CNT from a graphite sheet [57]. Molecular dynamics simulations have been applied to explore the interactions between molecules and BNNTs in addition to structural and thermal stability of the BNNTs [59]. It has been reported that the (5, 5) boron nitride nanotube transport water, although the (5, 5) CNT is not able to do so because of much larger energy barrier. The Van der Waals interactions between water molecules and nitrogen atoms decrease the energy barrier of the BNNTs. For reason, BNNTs with the small diameter were suggested as an aspirant for a synthetic aquaporin-1 water channel [57]. A variety of subjects of BNNTs have been investigated by theoretical methods, including hydrogen storage [60,61,62,63,64,65, 66], magnetism [67,68,69,70,71,72,73], phonon characteristics [74,75,76], stability [77,78,79], molecular dynamics [80,81,82,83], field-electron emission [84], scanning tunneling microscopy simulation [85], electron transport [86], symmetry breaking [87], work function [88], spin-splitting [89] and quantum computing [90,91].

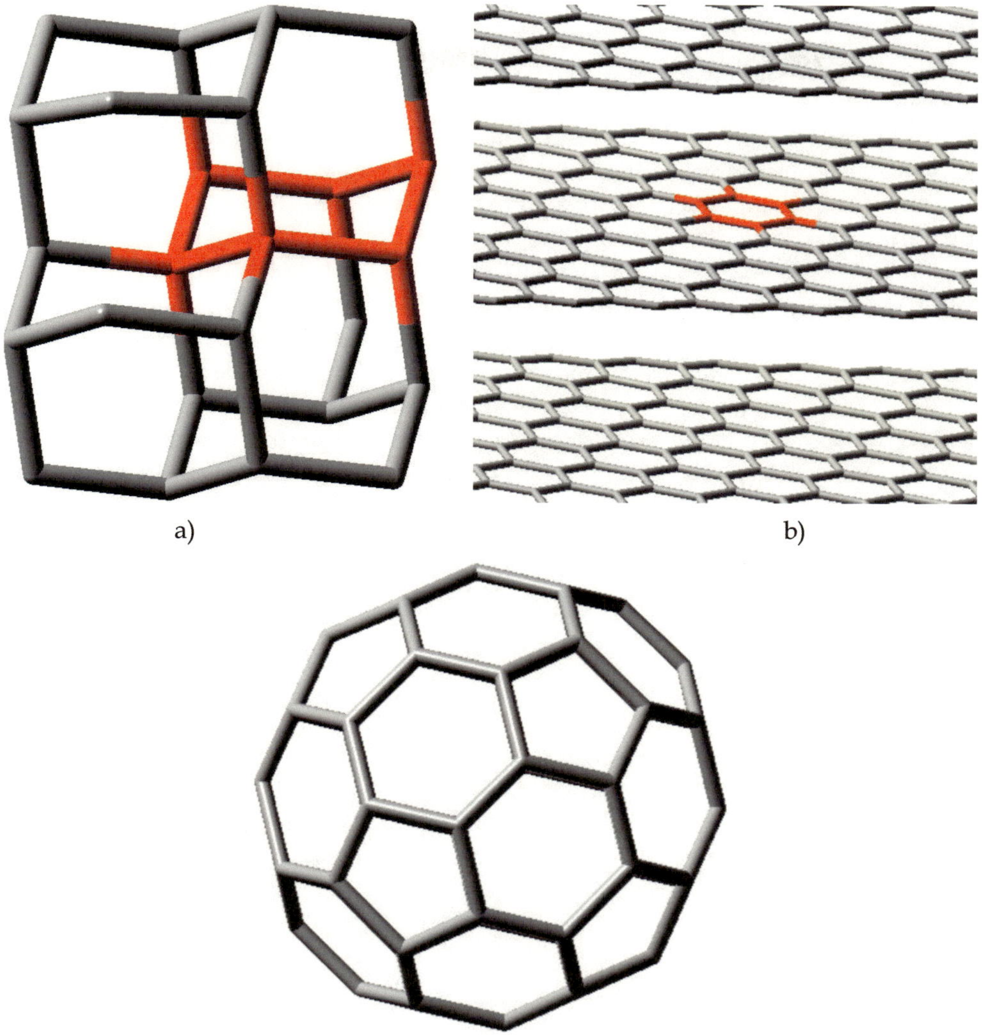

a) b) c)

Fig. 1.1. a. Diamond, b. Graphite, c. $C_{60}$ fullerenes

Fig. 1.2. (5,5)-single-wall carbon nanotube (SWCNT)

## 2. Geometrical structure

As is well known, a carbon nanotube can be visualized as being formed by rolling up a well-defined projected area within the hexagonal lattice of a graphene sheet in a seamless fashion such that all carbon-carbon (C-C) valences are satisfied, and the direction in which the roll up is performed transforms into the circumference of the tube. The projected area is in fact a homomorphic representation of a particular carbon nanotube.
There are many possible ways to roll up a graphene sheet into a cylinder with the hexagons to be completed. Figure 2.3A illustrates such an example of tube construction. The roll-up vector is also termed the chiral vector ($\vec{C}$) and is defined as $n\vec{a}_1 + m\vec{a}_2$, where $\vec{a}_1$ and $\vec{a}_2$ are the basis vectors the graphene lattice; n and m are the so-called chiral indices. By using unit vectors $\vec{a}_1$ and $\vec{a}_2$ with the chiral indices n and m, geometric parameters of a carbon nanotube can be defined. In general, an infinite number of nanotube geometries are possible, with specific nanotubes characterized by chiral indices (n,m), which, in turn, define the chiral angle (θ) and tube diameter ($d_t$); the latter is also dependent on the C-C bond length of the hexagonal lattice. For n = m, the nanotube is said to have the "armchair" conformation; for n ≠ 0 and m = 0, the conformation is called "zigzag"; while for n ≠ 0 and m ≠ 0 the conformation is termed "chiral." A seen in Figure 2.3E-F, double-walled carbon nanotubes (DWCNTs) are in the border between single-walled carbon nanotubes (SWCNT) and multi-walled carbon nanotubes (MWCNT).

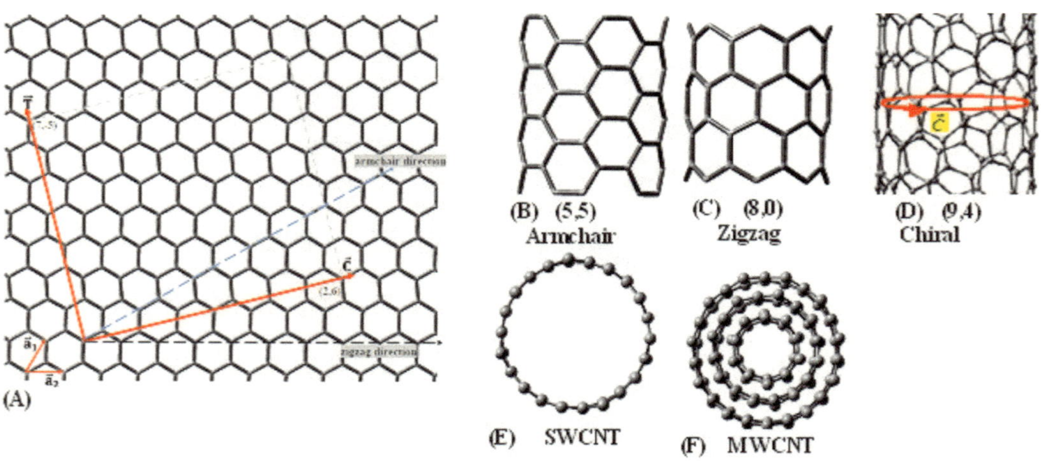


Fig. 2.3. The graphene honeycomb sheet and the schematically represented nanotube that can be obtained by rolling up the graphene layer along the chiral vector $\vec{C}$ turns into the circumference of the cylinder, and the translation vector $\vec{T}$ is aligned along the cylinder axis and represents the unit cell of a carbon nanotube.

As mentioned above, the chiral vector $\vec{C}$ is defined by the use of the real space unit vectors $\vec{a}_1$ and $\vec{a}_2$ of the hexagonal lattice as follows:

$$\vec{C} = n\vec{a}_1 + m\vec{a}_2 \tag{2.1}$$

Here the angle between the unit vectors $\vec{a}_1$ and $\vec{a}_2$ is 60 degree and $|\vec{a}_1| = |\vec{a}_2| = a = \sqrt{3}a_{CC}$ where $a_{CC}$($\cong$ 0.142 nm) is the average C-C bond distance in two dimensional network (2D) of the graphene lattice and n and m are integer numbers. The magnitude of chiral vector $\vec{C}$ can be given by equation 2.2

$$|\vec{C}| = a\sqrt{n^2 + m^2 + nm} \tag{2.2}$$

Since the single wall carbon nanotube is formed by rolling up the sheet along the direction of the chiral vector $\vec{C}$, the circumference of cylindrical carbon nanotube equal to the length of the chiral vector $\vec{C}$. For a given pair of integers (*n, m*), the diameter $d_t$ of a carbon nanotube may be derived from the length of the chiral vector $\vec{C}$ as follows:

$$d_t = \frac{|\vec{C}|}{\pi} = \frac{a\sqrt{n^2+m^2+nm}}{\pi} \tag{2.3}$$

The angle (θ) between the chiral vectors $\vec{C}$ and the primitive lattice translation vector $\vec{a}_1$is called as chiral angle, which is in the range of 0 to 30 degree and can be expressed by the pair of integers (*n, m*), and defined by using the scalar product of the vectors $\vec{a}_1$ and $\vec{C}$:

$$\cos(\theta) = \frac{\vec{a}_1.\vec{C}}{|\vec{a}_1||\vec{C}|} = \frac{2n+m}{2\sqrt{n^2+m^2+nm}} \tag{2.4}$$

The chiral angle θ can be given in tangential form as well by the vector and scalar product of the vectors $\vec{a}_1$ and $\vec{C}$ as follow:

$$\tan(\theta) = \frac{|\vec{a}_1 \otimes \vec{C}|}{|\vec{a}_1.\vec{C}|} = \frac{\sqrt{3}m}{2n+m} \tag{2.5}$$

Furthermore, it may be useful to specify the unit cell of a carbon nanotube (CNT), which is a part of a nanotube containing non-equivalent atoms. CNT can be naturally defined as a cylinder built on two orthogonal vectors $\vec{C}$ and $\vec{T}$, where $\vec{T}$ is the translation vector parallel to the nanotube axis and perpendicular to $\vec{C}$ as seen in Figure 2.3A. The translation vector $\vec{T}$ can be expressed in term of the basis vectors $\vec{a}_1$ and $\vec{a}_2$ as follows:

$$\vec{T} = t_1\vec{a}_1 + t_2\vec{a}_2 \equiv (t_1, t_2) \tag{2.6}$$

Since the $\vec{T}$ and $\vec{C}$ vectors are orthogonal to each other and their scalar product should be zero,

$$\vec{T}.\vec{C} = (t_1\vec{a}_1 + t_2\vec{a}_2).(n\vec{a}_1 + m\vec{a}_2) = t_1(n + m/2) + t_2(n/2 + m) = 0 \tag{2.7}$$

$$t_1 = -\left(\frac{n+2m}{2n+m}\right)t_2 \tag{2.8}$$

In Figure 2.3A the unit cell of the graphite is the rhombus specified by the vectors $\vec{a}_1$ and $\vec{a}_2$ and containing two carbon atoms. The number of hexagons (N) per unit cell of a carbon nanotube can be obtained dividing the area of the nanotube unit cell ($|\vec{T} \otimes \vec{C}|$) by the area of the graphite unit cell ($|\vec{a}_1 \otimes \vec{a}_2|$, the area of the rhombus is equal to the area of the hexagon):

$$N = \frac{|\vec{T}\otimes\vec{C}|}{|\vec{a}_1\otimes\vec{a}_2|} = \frac{|(t_1\vec{a}_1+t_2\vec{a}_2)\otimes(n\vec{a}_1+m\vec{a}_2)|}{|\vec{a}_1\otimes\vec{a}_2|} = |(mt_1 - nt_2)|\frac{|\vec{a}_1\otimes\vec{a}_2|}{|\vec{a}_1\otimes\vec{a}_2|} = |(mt_1 - nt_2)| \quad (2.9)$$

$$mt_1 - nt_2 = \pm N \quad (2.10)$$

Substituting Eq. 2.8 into 2.10 and by setting $d_R = \frac{2(n^2+m^2+nm)}{N}$, we found that

$$t_1 = \pm\frac{n+2m}{d_R} \text{ and } t_2 = \mp\frac{2n+m}{d_R} \quad (2.11)$$

Here $d_R$ is called the greatest common divisor (GCD) and may be defined in terms of the greatest common divisor of n and m as given by Eq. 2.12,

$$d_R = \begin{cases} GCD(2n+m, n+2m) \\ \text{or} \\ 3M, \text{if } n-m=3l \\ M, \text{if } n-m \neq 3l \\ \text{here } l = 0,1,2,\ldots \end{cases} \quad (2.12)$$

## 2.2 Electronic structure of SWNTs

Before examining the electronic properties of carbon nanotubes, we briefly discuss the electronic structure of graphene investigated which also underline the band structure of the nanotubes. For more detail see Ref.92. Graphene is characterized by two types of chemical bonds which are linked to the $sp^2$ hybridization of the carbon atomic orbitals. The energy band of the three strong $\sigma$- bonds within the honeycomb lattice are far from the Fermi level. Therefore, these bonds do not have any contribution to the electronic transport properties of graphene and nanotubes.

The remaining $p_z$ orbitals, pointing out of the $\sigma$-bond plane, cannot couple with the $\sigma$- states for symmetry arguments. The lateral interaction with neighboring $p_z$ orbitals creates delocalized $\pi$- (bonding) and $\pi^*$ (antibonding) states, which determine the energy bands around the Fermi energy $E_F$.

The two-dimensional graphene lattice in real space can be formed by translating one unit cell of graphene defined by two basis vectors: $\vec{a}_1 = \frac{a}{2}(\sqrt{3}\hat{x} + \hat{y})$ and $\vec{a}_2 = \frac{a}{2}(\sqrt{3}\hat{x} - \hat{y})$, where $|\vec{a}_1| = |\vec{a}_2| = a = \sqrt{3}a_{CC}$ is the length of the basis vector and $a_{CC}(\approx 0.142 \text{ nm})$ is the nearest neighbor C-C bonding distance. The unit cell of graphene consists of two atoms, A and B, as seen in Fig. 2.3a. A and B atoms form two complementary, hexagonal sublattices. The basis vectors $\vec{b}_1$ and $\vec{b}_2$ in the reciprocal lattice as shown in Fig. 2.3b, satisfies the condition $\vec{a}_i.\vec{b}_j = 2\pi\delta_{ij}$ allows one to find the reciprocal-lattice vectors $\vec{b}_1$ and $\vec{b}_2$: $\vec{b}_1 = \frac{2\pi}{a}\left(\frac{\hat{x}}{\sqrt{3}} + \hat{y}\right)$ and $\vec{b}_2 = \frac{2\pi}{a}\left(\frac{\hat{x}}{\sqrt{3}} - \hat{y}\right)$.

A tight binding approximation (TBA) takes into consideration the interactions between only the nearest neighbor atoms for the band structure of graphene (Figure 2.3c), the remaining interactions are ignored, and only one $p_z$ orbital per carbon atom is allowed for in calculations. It should be pointed out that tight-binding approximation gives reliable results at the energy range near the Fermi level of the graphene sheet, which is the region of interest for electronic transport [93]. By considering two independent wave functions (adhering to the Bloch theorem) for the electrons corresponding to the two complementary sublattices and neglecting the overlap matrix elements $S = \langle\psi(p_z, A)|\psi(p_z, B)\rangle$ one obtains the following relation for the energy dispersion of the bands in graphene:

$$E^{\pm}(\vec{k}) = \pm\gamma\sqrt{3 + 2[\cos(\vec{k}.\vec{a}_1) + \cos(\vec{k}.\vec{a}_2) + \cos(\vec{k}.\vec{a}_1 - \vec{k}.\vec{a}_2)]} \quad (2.13)$$

Where $\gamma = 2.9 \pm 0.2$ eV is the interaction energy between the nearest neighbor atoms A and B. The $\vec{k} = k_x\hat{x} + k_y\hat{y}$ vectors, which belong to the first Brillouin zone (BZ) of the hexagon, as seen in Fig. 2.3b, represent the ensemble of available electronic momenta.

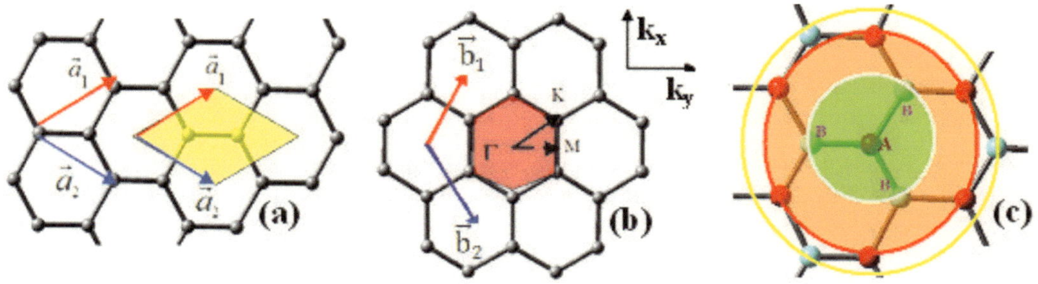


Fig. 2.3. **a)** Unit cell of graphene containing two atoms A and B defining two complementary sublattices; **b)** the hexagonal Brillouin zone (BZ) of graphene with the reciprocal lattice vectors $\vec{b}_1$ and $\vec{b}_2$: $\vec{b}_1 = \frac{2\pi}{a}\left(\frac{\hat{x}}{\sqrt{3}} + \hat{y}\right)$ and $\vec{b}_2 = \frac{2\pi}{a}\left(\frac{\hat{x}}{\sqrt{3}} - \hat{y}\right)$. The $\Gamma, K$ *and* $M$ represent the high symmetry points; **c)** Atom A (in red color) has three the nearest atoms B (in light blue colour), six second nearest atoms A and three third nearest atoms B as shown in the shaded area in the figure. Here $\vec{a}_1$ and $\vec{a}_2$ are the unit cell vectors: $\vec{a}_1 = \frac{a}{2}(\sqrt{3}\hat{x} + \hat{y})$ and $\vec{a}_2 = \frac{a}{2}(\sqrt{3}\hat{x} - \hat{y})$.

Figure 2.4a-b shows a three-dimensional plot of the energy dispersion *E*(*k*) along the high symmetry directions of the BZ, defined by the Γ, M and K points. The conduction and valence bands touch and are degenerate at the six K points at the corners of the first BZ, thus allowing the classification of graphene as a semimetal. Three out of the six K points are equivalent due to the spatial symmetry of the hexagonal lattice, thus two distinguishable points remain called K and K'. The bonding bands are completely filled and the antibonding bands are empty at zero temperature. The undoped state, where the Fermi surface contains only the six K points, is called the "charge neutrality point". In the region of the Fermi, Taylor expansion of Eq.2. 13 provides a simplified linear energy dispersion relation:

$$E(\vec{k}) = \frac{\sqrt{3}a\gamma}{2}|\vec{k} - \vec{k}_F| \quad (2.14)$$

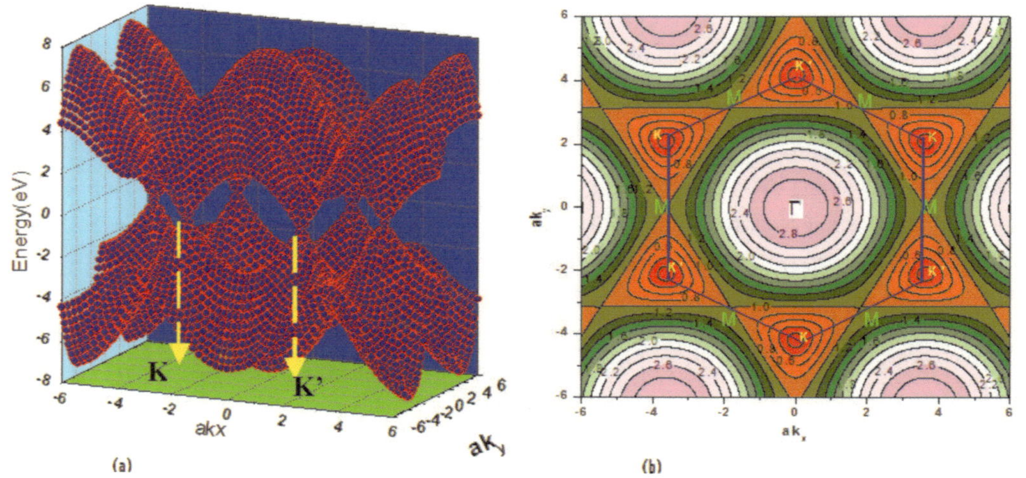


Fig. 2.4. The energy dispersion of the E(k) for the graphene: a) indicates three-dimensional plot and b) the counter plot.

By using Zone-folding approximation, the energy dispersion relation of CNTs can be obtained from the simplified linear energy dispersion relation of the graphene near the Fermi level since a carbon nanotube consisting of one graphene sheet rolled into a cylinder. By applying periodic boundary conditions along the circumference of a SWNT, it is easy see that, whereas the allowed wave vectors in direction perpendicular to the tube axis are quantized, the wave vectors are parallel to the nanotube axis remain continuous since the nanotube has an infinite length. The application of periodic boundary conditions around the tube circumference brings about the following restrictions on the allowed wavefunctions:

$$\psi_k(\vec{r}+\vec{C}) = \exp(i\vec{k}.\vec{C})\,\psi_k(\vec{r}) = \psi_k(\vec{r}) \tag{2.15}$$

where vectors $\vec{k}$ and $\vec{C}$ and are respectively allowed wave vector undertaken the tube surface and circumference of the nanotube. This is the first equality arising from the Bloch theorem. Thus, the periodic boundary condition imposed along the circumference direction restricted the wave vectors to:

$$\vec{k}.\vec{C} = 2\pi q;\; q\text{=}0, 1, 2, \tag{2.16}$$

The energy dispersion relation near the Fermi-points is the most fascinating: If the distance between the high symmetry points, Γ and K, is defined by the vector $\vec{k}_F = \frac{1}{3}(\vec{b}_1 - \vec{b}_2)$, then the component of the $\vec{k}_F$ on the circumferential direction ($\vec{C} = n\vec{a}_1 + m\vec{a}_2$) can be obtained by a scalar product of these two vectors $\vec{k}_F$ and $\vec{C}$, with using $\vec{b}_i.\vec{a}_j = 2\pi\delta_{ij}$, as follow:

$$\vec{k}_F.\vec{C} = \frac{1}{3}(\vec{b}_1 - \vec{b}_2).(n\vec{a}_1 + m\vec{a}_2) = 2\pi\left(\frac{n-m}{3}\right) \tag{2.17}$$

If the origin of the reciprocal lattice is placed to the Fermi point, the distance between $\vec{k}_F$ and one of the allowed states at $\vec{k}$ can be given in the new coordinate system:

$$\Delta\vec{k} = \vec{k} - \vec{k}_F = \vec{k}_\perp + \vec{k}_{||} \tag{2.18}$$

where $\vec{k}_{\perp}$ is along the circumference direction $\vec{C}$, and $\vec{k}_{||}$ is along the nanotube axis and $\vec{k}_{\perp}$ is perpendicular to $\vec{k}_{||}$. The scalar product of the Eq 2. 18 with the chiral vector $\vec{C}$, with using Eqs. 2. 16 and 2. 17, we could easily reach one of the important equation, which indicate that component of the vector $\Delta\vec{k}$ along the circumference direction, $|\vec{k}_{\perp}| = |\Delta\vec{k}_{C}^{q}|$, is quantized and given by:

$$|\vec{k}_{\perp}| = |\Delta\vec{k}_{C}^{q}| = \frac{\Delta\vec{k}.\vec{C}}{|\vec{C}|} = \frac{(\vec{k}.-\vec{k}_{F}).\vec{C}}{|\vec{C}|} = \frac{2}{3d_t}[3q-(n-m)] \tag{2.19}$$

where $d_t$ is the diameter of the nanotube.

The general form of the energy dispersion relation at vicinity of Fermi level for the 1D nanotube can be obtained by substituting Eq. 2. 19 into Eq. 2. 14 as follow:

$$E(\Delta\vec{k}) = \frac{\sqrt{3}a|\gamma|}{2}|\Delta\vec{k}| = \frac{\sqrt{3}a|\gamma|}{2}\sqrt{|\Delta\vec{k}_{C}^{q}|^2 + |\vec{k}_{||}|^2} \tag{2.20}$$

From equation 2. 20, it is obvious that the nanotube can be either metallic or semiconducting, depending on whether (n-m) is the multiple of 3. For instance, at the Fermi level (at the corner of hexagon, K point), if an allowed $\vec{k}$ line cross the graphene $\vec{k}_F$ point, $\vec{k}-\vec{k}_F = 0$ and we find a selection rule on the metallic and semiconducting nanotube as 3q = n-m, which indicates that the SWNT is metallic and $E(\Delta\vec{k}) = \frac{\sqrt{3}|\gamma|}{2}|\vec{k}_{||}|$. If $\vec{k}-\vec{k}_F \neq 0$, $3q \neq$ n-m, which denotes that the SWNT is semiconducting and $E(\Delta\vec{k}) \cong \frac{\sqrt{3}a|\gamma|}{2}|\Delta\vec{k}_{C}^{q}| = \frac{2\sqrt{3}a|\gamma|}{d_t}q$.

One-dimensional (1D) density-of-states (DOS) of the nanotubes may be defined as the number of accessible electronic states for a given energy interval. The DOS has a great importance for many physical phenomena such as optical absorption-emission, conductivity, etc. Furthermore, the one-dimensional DOS generate the resonant Raman scattering in addition to intensive interband transitions in the spectra of the optical absorption and emission. The DOS is known to depend strongly on the dimension of the system[92]. One-dimensional density-of-states may be calculated by using Eq. 2. 21 [92]:

$$DOS(E) = \frac{2}{N}\sum_{q}^{N}\frac{\delta(E_q(\vec{k})-E)dE}{\left|\frac{dE_q(\vec{k})}{d\vec{k}}\right|} \tag{2.21}$$

Since the energy dispersion vicinity of Fermi level is linear, then, the density of states (DOS) of metallic nanotube is constant at Fermi level and inversely proportional to the diameter of the nanotube: $DOS(E_F) = \frac{a}{2\pi^2\gamma d_t}$

The average energy position of the peaks depends on the nanotube diameter, which is defined by the linear dispersion approximation (as discussed above). Figure 2.5 provides calculated energy dispersion and DOS for (10, 10) carbon nanotube.

## 3. Results and discussion

**Computational methods**: The ground state geometry of the single-walled carbon nanotubes (SWCNTs), double walled carbon nanotubes (DWCNTs), the single-walled boron nitride nanotubes (SWBNNTs) and functionalized-SWCNTs were optimized without symmetry restriction on the initial structures. Both structure optimization and vibrational analysis calculations were implemented by using DFT with functionals, specifically, B3LYP, in which the exchange functional is of Becke's three parameter type, including gradient correction, and the correlation correction involves the gradient-corrected functional of Lee, Yang and Parr. The basis set of split valence type 6-31G, as contained in the Gaussian 03 software package,[94] was used. The results of the calculations did not produce any imaginary frequencies. The vibrational mode descriptions were made on the basis of calculated nuclear displacements using visual inspection of the animated normal modes (using GaussView03) [94], to assess which bond and angle motions dominate the mode dynamics for the nanotube. The DFT method was chosen because it is computationally less demanding than other approaches as regards inclusion of electron correlation. Moreover, in addition to its excellent accuracy and favorable computation expense ratio, the B3LYP calculation of Raman frequencies has shown its efficacy in numerous earlier studies performed in this laboratory and by other researchers, often proving itself the most reliable and preferable method for many molecular species of intermediate size, including anions and cations[95].

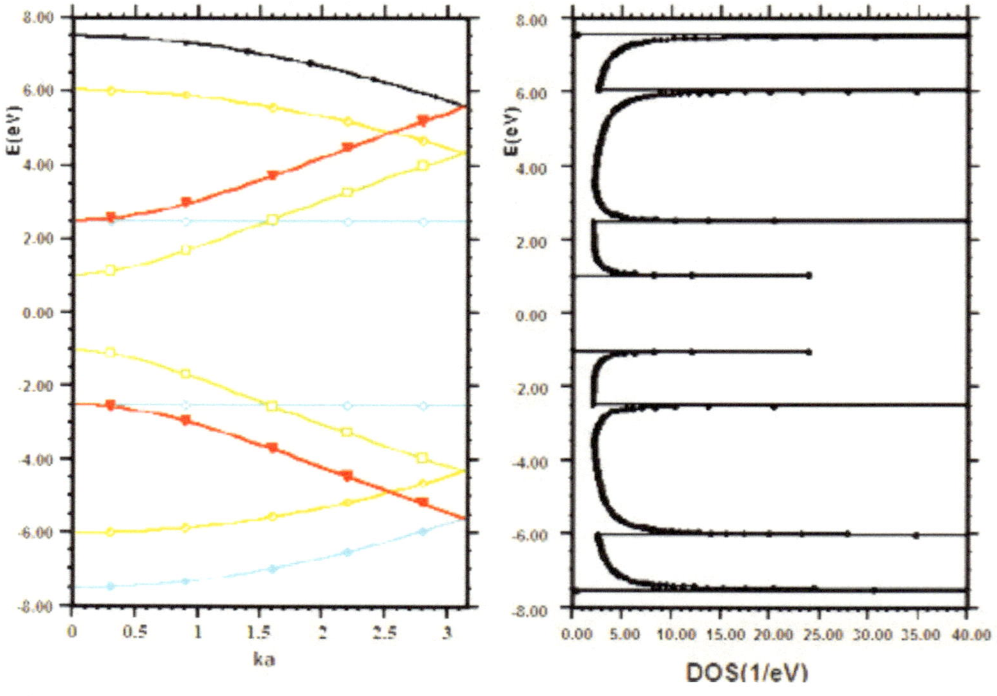


Fig. 2.5. Calculated energy dispersion relation and the density of states (DOS) for a (10,10) nanotube.

In our calculations hydrogen atoms have been placed at the end points of the unit cells. Furthermore, the time-dependent density functional theory at TD-B3LYP level were applied to calculate the vertical electronic transitions for the SWCNTs, SWBNNTs and functionalized (7,0)- and (10,0)-SWCNTs.

### 3.1 Structure results

Calculated averaged C-C bond distances within the single wall carbon nanotube (SWCNT) for the ($n$,0)-SWCNTs with $n$ = 6 to 19 for the ($n$,n)-SWCNTs with $n$ = 3 to 10 were not only found to be diameter dependent, but also dependent on length of the tube as seen in Table 3.1.1 and 3.1.2 and Figure 3.1.1. The calculated averaged B-N bond distances within the armchair and zigzag boron nitride nanotubes (BNTs) also indicated similar diameter- and size-dependence (see Table 3.1.1 and 3.1.2). The plot of the calculated averaged bond distances for the (*n*,0)- and (n,n)-type nanotubes indicated that the bond distances within the nanotube slightly decreases and stabilizes with increasing diameter of the isolated nanotube; the solid curves in Figure 3.1.1 is a fit to the calculated energies using a functional form that depends inversely on nanotube diameter: fitting equations are given in Eq. 1a-1e. As seen in Figure 3.1.1, for the (n,0)-SWCNTs, the plot of the averaged CC bond distance versus the tube diameter indicated that the CC bond distance does not only decrease with increasing tube diameter as expected, but also the bond distances of (0,2n) and (0,2n+1)-type SWCNTs are well separated. This circumstance is also observed in the folding (curvature) energies (Eq2a-2b) and in the Raman spectra of (0,2n)- and (0,2n+1)-type SWCNTs. However, the calculations for the rest of the nanotube studied here did not clearly indicate this observation.

$$R_{CC}((2n,0);\text{in nm}) = 0.1422 - \frac{0.0004}{d_t} + \frac{0.0004}{d_t^2} \tag{3.1.1a}$$

$$R_{CC}((2n+1,0);\text{in nm}) = 0.1422 - \frac{0.0003}{d_t} + \frac{0.0003}{d_t^2} \tag{3.1.1b}$$

$$R_{CC}((n,n);\text{in nm}) = 0.1421 - \frac{0.0003}{d_t} + \frac{0.0003}{d_t^2} \tag{3.1.1c}$$

$$R_{BN}((n,n);\text{in nm}) = 0.1449 - \frac{0.0002}{d_t} + \frac{0.0003}{d_t^2} \tag{3.1.1d}$$

$$R_{BN}((n,0);\text{in nm}) = 0.1433 - \frac{0.0003}{d_t} + \frac{0.0003}{d_t^2} \tag{3.1.1e}$$

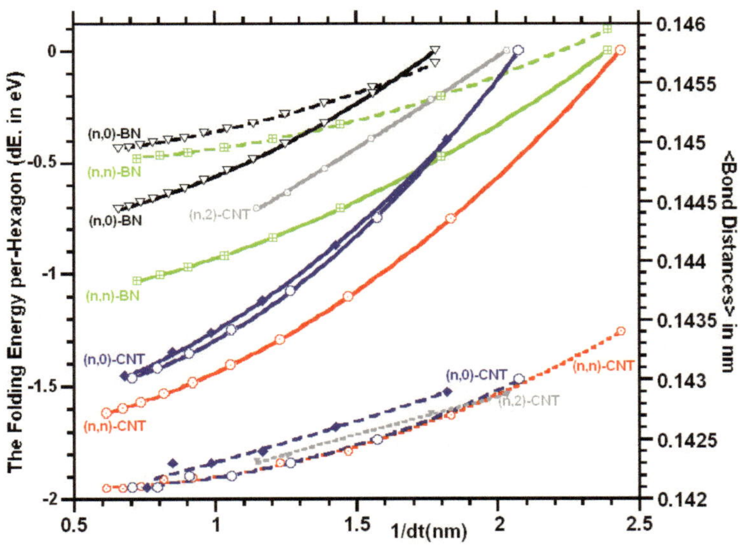


Fig. 3.1.1. The calculated folding (curvature) energies of (n,0)-, (n,n)-SWCNTs/BN-nanotubes and (n,2)-SWCNTs referenced to their the nanotube with the smallest diameter and boron nitride nanotubes with their corresponding bond distances.

Figure 3.1.1 provides the plot of the calculated global energy per-hexagon for the (n,0)- and (n,n)-SWCNTs/SWBNNTs referenced to their corresponding (6,0)- and (3,3)-SWCNT/SWBNNTs energies, respectively. The result of the calculations suggested that the curvature energy (or folding energy) of the nanotubes rapidly decrease and stabilizes with increasing the diameter of the isolated nanotube; the solid curves in Figure 3.1.1 is a fit to the calculated energies using a functional form that depends inversely on nanotube diameter. The fitting parameters are given in equations. 3.1.2a-2e. The plot of the calculated folding energy for the (n,2)-SWCNTs, with n=5 to 10, exemplified similar fitting equations as given in Equation 3.1.2f.

$$\Delta E((n,0)-CNTs) := E_{(n,0)} - E_{(6,0)} = -1.64 - \frac{0.04(eV.nm)}{d_t(nm)} + \frac{0.36(eV.(nm)^2)}{d_t^2(nm)^2} \quad (3.1.2a)$$

$$\Delta E((n,0)-CNTs) = E_{(n,0)} - E_{(6,0)} = -1.52 - \frac{0.25(eV.nm)}{d_t(nm)} + \frac{0.74(eV.(nm)^2)}{d_t^2(nm)^2} \quad (3.1.2b)$$

$$\Delta E((n,n)-CNTs) = E_{(n,n)} - E_{(3,3)} = -1.71 - \frac{0.02(eV.nm)}{d_t(nm)} + \frac{0.30(eV.(nm)^2)}{d_t^2(nm)^2} \quad (3.1.2c)$$

$$\Delta E((n,0)-BNNT) = E_{(n,0)} - E_{(7,0)} = -0.82 + \frac{0.26(eV.(nm)^2)}{d_t^2(nm)^2} \quad (3.1.2d)$$

$$\Delta E((n,n)-BNNTs) = E_{(n,n)} - E_{(3,3)} = -1.17 + \frac{0.06(eV.nm)}{d_t(nm)} + \frac{0.18(eV.(nm)^2)}{d_t^2(nm)^2} \quad (3.1.2e)$$

$$\Delta E((n,2)-CNTs) = E_{(n,0)} - E_{(6,0)} = -1.54 - \frac{0.67(eV.nm)}{d_t(nm)} + \frac{0.04(eV.(nm)^2)}{d_t^2(nm)^2} \quad (3.1.2f)$$

In Eq. 3.1.2a-3.1.2f, $E_{n,m}$ stand for the folding energy (in eV) of the (*n*,m)-type isolated single-wall carbon and boron nitride nanotubes and $d_t$ corresponds to their tube diameter in nm unit. Clearly, larger diameter SWCNTs/SWBNNTs can be more easily formed than the smaller diameter ones in gas phase. For example, while energy differences for the (7,0) through (12,0)-SWCNT, relative to the (6,0)-SWCNT, rapidly approaches a limiting value, one notes that from the (11,0)- up to the (16,0)-SWCNT a limiting value is nearly reached and changes in energy approach zero. Of course, this is to be expected since the smaller diameter, the more strained are the $sp^2$ hybridized sigma bonds; or stated another way, the smaller the diameter the more altered from planarity must be the $sp^2$ hybridized orbitals. It should be noted that the formation of the nanotube in gas phase and in any environment might be different. The size of the nanotube in an environment also depends on cavity size of the environment owing to the electrostatic interactions between the tube and its neighboring. When comparing the folding energy of the (n,0)-type nanotubes with that of the (n,n)-type nanotube, the calculations indicated that folding of the zigzag nanotubes are more easily than that of the armchair-nanotubes for both carbon nanotubes and boron nitride nanotubes. This observed circumstance is expected because of the geometrical reason, for instance, while three of the sigma bonds as a result of the $sp^2$-hybridized orbitals in the armchair nanotubes are folded, only two of three $sp^2$-hybridized orbitals in the zigzag-form of the nanotubes are folded and one of them along the nanotube axis without folding. This assessment also is found in the bond distances such as the CC bond distances in the (n,n)-type nanotubes are longer than in the (n,0)-type nanotubes as seen in Figure 3.1.1 and Table 3.1.1 or equations 3.1.1a-3.1.1e. Based on part of the calculations, when comparing the carbon nanotubes with the boron nitride nanotubes, as seen in Figure 3.1.1 and equations 3.1.2a-3.1.2e, the formation of the carbon nanotubes is more easily than the boron nitride nanotubes.

| | SWCNTs | | | | | | SWBNNTs | | | |
|---|---|---|---|---|---|---|---|---|---|---|
| n | (0, n) | | (n, n) | | (2, n) | | (0, n) | | (n, n) | |
| | <CC> | dt | <CC> | dt | <CC> | dt | <BN> | dt | <BN> | dt |
| 3 | | | 0.143 | 0.411 | | | | | 0.146 | 0.418 |
| 4 | | | 0.143 | 0.545 | | | | | 0.145 | 0.555 |
| 5 | | | 0.142 | 0.680 | 0.143 | 0.493 | | | 0.145 | 0.693 |
| 6 | 0.143 | 0.473 | 0.142 | 0.815 | 0.143 | 0.567 | | | 0.145 | 0.831 |
| 7 | 0.143 | 0.552 | 0.142 | 0.951 | 0.143 | 0.644 | 0.146 | 0.562 | 0.145 | 0.969 |
| 8 | 0.142 | 0.628 | 0.142 | 1.087 | 0.143 | 0.721 | 0.146 | 0.642 | 0.145 | 1.107 |
| 9 | 0.143 | 0.708 | 0.142 | 1.222 | 0.142 | 0.797 | 0.145 | 0.721 | 0.145 | 1.245 |
| 10 | 0.142 | 0.785 | 0.142 | 1.357 | 0.142 | 0.875 | 0.145 | 0.801 | 0.145 | 1.383 |
| 11 | 0.142 | 0.864 | 0.142 | 1.492 | | | 0.145 | 0.880 | | |
| 12 | 0.142 | 0.941 | 0.142 | 1.628 | | | 0.145 | 0.960 | | |
| 13 | 0.142 | 1.020 | | | | | 0.145 | 1.040 | | |
| 14 | 0.142 | 1.098 | | | | | 0.145 | 1.120 | | |
| 15 | 0.142 | 1.177 | | | | | 0.145 | 1.199 | | |
| 16 | 0.142 | 1.253 | | | | | 0.145 | 1.279 | | |
| 17 | 0.142 | 1.332 | | | | | 0.145 | 1.359 | | |
| 18 | 0.142 | 1.410 | | | | | 0.145 | 1.439 | | |
| 19 | 0.141 | 1.476 | | | | | 0.145 | 1.518 | | |

Table 3.1.1. The calculated averaged CC and BN bond distances and diameters of the (n, m)-type nanotube in nm.

We also examined the dependence of the curvature energy on the length of the zigzag (n,0)-SWCNTs referenced to the (6,0)-SWCNT, defined as $\Delta E = \mathrm{E}_{(n,0)}(\mathrm{m}) - \mathrm{E}_{(6,0)}(\mathrm{m})$. Where m indicates the number of hexagon along the tube axis, m = 1 to 4, and the chiral index n is ranging from 6 to 11. As seen in Figure 3.1.2, the calculated folding energy per hexagon for each m is compatible with each other. The estimated fitting parameters, given in equations 3.1.3a-3.1.3d, are almost the same. When we examined the folding energy per hexagon as function of the number of hexagon along the tube axis (m) for a given (n,0)-SWCNT, for example, (6,0)-SWCNT with m= 1 to 4, we found that the folding energy slightly increases (less than 0.1 eV) and stabilizes with increasing the number of hexagon along tube axis (m) of the isolated zigzag nanotube. This consequence also can be easily seen in equations 3.1.3a to 3.1.3d. Table 3.1.2 provides the calculated averaged CC bond distances and diameters of the zigzag and armchair isolated single-wall carbon nanotubes. The calculated averaged CC bond distances also slightly increases and stabilizes with increasing length of the tube. It can be seen that dependence of the geometric parameters and folding energies on the diameter and length of the nanotube is as a result of the deformation in the $sp^2$-hybridized orbitals of the CC bonds.

$$\mathrm{E}_{(n,0)}(\mathrm{m}=1) - \mathrm{E}_{(6,0)}(\mathrm{m}=1) = -1.51 - \frac{0.16(\mathrm{eV.nm})}{\mathrm{d_t(nm)}} + \frac{0.42(\mathrm{eV.(nm)^2})}{\mathrm{d_t^2(nm)^2}} \tag{3.1.3a}$$

$$E_{(n,0)}(m=2) - E_{(6,0)}(m=2) = -1.51 - \frac{0.28(eV.nm)}{d_t(nm)} + \frac{0.48(eV.(nm)^2)}{d_t^2(nm)^2} \quad (3.1.3b)$$

$$E_{(n,0)}(m=3) - E_{(6,0)}(m=3) = -1.51 - \frac{0.17(eV.nm)}{d_t(nm)} + \frac{0.43(eV.(nm)^2)}{d_t^2(nm)^2} \quad (3.1.3c)$$

$$E_{(n,0)}(m=4) - E_{(6,0)}(m=4) = -1.51 - \frac{0.15(eV.nm)}{d_t(nm)} + \frac{0.41(eV.(nm)^2)}{d_t^2(nm)^2} \quad (3.1.3d)$$

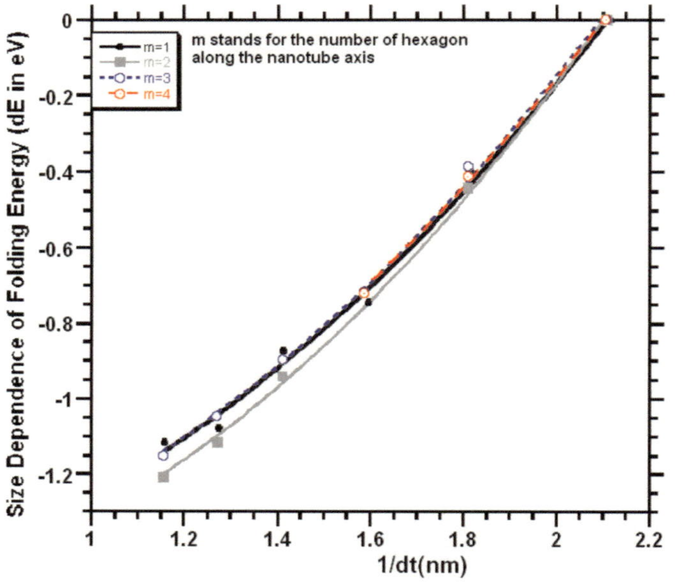


Fig. 3.1.2. The calculated size (the number of hexagon along the nanotube axis (m)) dependence of the folding energies for the (n,0)-SWCNTs, n=6 to 11, referenced to the (6,0)-SWCNT with the same number of m.

| | SWCNTs | | | | | | | | | | | |
|---|---|---|---|---|---|---|---|---|---|---|---|---|
| | (4,4) | | (0,6) | | (0,7) | | (0,8) | | (0,9) | | (0,10) | |
| m | <CC> | dt | <CC> | dt | <CC> | dt | <CC> | dt | <CC> | dt | <CC> | dt |
| 1 | 0.141 | 0.540 | 0.143 | 0.473 | 0.143 | 0.552 | 0.142 | 0.627 | 0.143 | 0.707 | 0.142 | 0.785 |
| 2 | 0.142 | 0.544 | 0.143 | 0.474 | 0.143 | 0.553 | | | 0.143 | 0.709 | 0.143 | 0.787 |
| 3 | 0.143 | 0.545 | 0.144 | 0.475 | 0.143 | 0.553 | 0.143 | 0.631 | 0.143 | 0.709 | 0.143 | 0.787 |
| 4 | 0.143 | 0.546 | 0.144 | 0.475 | 0.143 | 0.553 | 0.143 | 0.631 | | | | |
| 5 | 0.143 | 0.546 | 0.144 | 0.475 | 0.143 | 0.553 | | | | | | |
| 6 | 0.143 | 0.546 | | | | | | | | | | |
| 7 | 0.143 | 0.547 | | | | | | | | | | |
| 8 | 0.143 | 0.547 | | | | | | | | | | |
| 9 | 0.143 | 0.547 | | | | | | | | | | |
| 10 | 0.143 | 0.547 | | | | | | | | | | |
| 11 | 0.143 | 0.547 | | | | | | | | | | |

| | SWCNTs | | SWBNNT | |
|---|---|---|---|---|
| | (0,11) | | (0,8) | |
| m | <CC> | dt | <CC> | dt |
| 1 | 0.142 | 0.864 | 0.145 | 0.642 |
| 2 | 0.143 | 0.865 | 0.146 | 0.642 |
| 3 | 0.143 | 0.865 | 0.146 | 0.642 |
| 4 | | | 0.146 | 0.642 |
| 5 | | | 0.146 | 0.643 |

Table 3.1.2. The dependence of calculated averaged CC and BN bond distances on the length and diameters (dt) of the nanotubes in nm. Where m indicates the number of hexagon along the nanotube axis.

The calculated full natural bond orbital analysis (NBO) indicates that three of the four valence electrons of the carbon atoms in SWCNTs are $sp^2$-hybridized in the one-dimensional (1D) network, with ~34% s and ~66% $p_{xy}$ character, and the forth electron is ~ 100% $p_z$ in character. As expected, each carbon atom contributes three electrons to the sigma bonds within the surface of the CNT and has one electron left in the $p_z$ orbitals that is delocalized over the entire surface. Figure 3.1.3A provides the calculated electron density in some of the upper occupied molecular orbitals and lower unoccupied molecular orbitals. It is to be noted that the HOMO and LUMO are pure nonbonding π-orbitals (resulting from $p_z$ atomic orbitals).

### 3.2 DWCNT

The DFT technique, at same level of the theory, was performed to calculate the Raman and IR spectra for (n,0)&(2n,0)-DWCNTs, (6,0)&(12,0), (7,0)&(14,0) and (8,0)&(16,0), as well as their inner- and outer-shell diameters and electron densities in gas phase. The key conclusions of these calculations on DWCNTs are summarized below. The diameter dependence of the curvature energies of the DWCNTs reference to the global energies of their corresponding inner- and outer-SWCNTs is well fitted by a Lannard-Jones potential expression as given in equation 3.2.1,

$$\Delta E(eV) = E[(2n,0)\&(n,0)] - E(2n,0) - E(n,0) = 3.676\left\{\left(\frac{0.702(nm)}{D_t(nm)}\right)^6 - \left(\frac{0.702(nm)}{D_t(nm)}\right)^{12}\right\} \quad (3.2.1)$$

which may be interpreted such as a van der Waals type intertube interactions for DWCNTs. Where $\frac{1}{D_t} = \frac{1}{d_t(Inner\ \ Shell)} - \frac{1}{d_t(Outer\ Shell)}$. A comparison the diameters of the inner- and outer-shells of the DWCNTs with their corresponding SWCNTs diameters showed that the averaged inner-shells diameters decrease (~ -0.08Å) and averaged outer-shells diameters increased as much as 0.25Å. These changes also found in the averaged C-C bond distances; about -0.014, 0.004 and 0.009 Å in inner-shells and 0.044, 0.028 and 0.023 Å in outer-shells for (6,0)&(12,0), (7,0)&(14,0) and (8,0)&(16,0)-DWCNTs, respectively, reference to their corresponding averaged C-C bond distances for the SWCNTs. These predictions explicitly indicate the existence of intertube interactions in DWCNT systems, which may be expressed by a van der Waals type interaction, not like chemical bonding interactions in the ground state. Furthermore, Figure 3.2.1B provides the calculated electron density of (0,6)&(0,12)-DWCNT showed that first four highest occupied molecular orbitals (from HOMO to HOMO-3 with the $A_{1u}$, $A_{2g}$ and $2E_{1g}$ symmetries, respectively) belong to the outer-shell and the next highest occupied molecular orbitals from HOMO-4 to HOMO-24 include both inner- and outer-shells of (0,6)&(0,12)-DWCNT. The lowest unoccupied molecular orbital, LUMO ($E_{1u}$) lies about 0.780 eV above the HOMO ($A_{1u}$) belongs to the outer-shell and the next one ($B_{2u}$) belongs to the inner-shell and lies 0.849 eV above the HOMO ($A_{1u}$). The calculated electron density also indicated that an intratube (inner and outer tube) interaction may possibly take place in the excited state: the LUMO+7 with $A_{2u}$ symmetry and 2.494 eV above the HOMO ($A_{1u}$), LUMO+8 ($E_{1u}$ and 2.557 eV), LUMO+10 ($E_{1g}$ and 2.563 eV) and LUMO+15 ($E_{1g}$ and 3.637 eV). The intratube CC σ-bonding interaction in the excited state might lead to a probable intertube charge transfer, which can be observed by a significant change in the tangential modes (TM) range of Raman spectra when the tube excited to its intratube charge transfer state. The TM may not only provide information about the metallic or semiconducting character of nanotubes, but also on the inner-outer tube (intratube) changes transfer. In addition, very recently, Resonant Raman measurements,[96] photoemission measurements and theoretical calculations provide a evidence of charge transfer between the inner- and outer-shells of DWCNTs.

Based on this scenario, the small sized-DWCNTs can be used as energy conversion systems as a consequence of charge transfer between intershells. This illustration also can reflect on the intensity of the Raman bands at the excitation energy where the charge transfer takes place between inner- and outer-shells. For (0,7)&(0,14)- and (0,8)&(0,16)-DWCNTs, the plotted electron densities did not signify any intratube CC antibonding in the excited state up to 4 eV above the ground state.

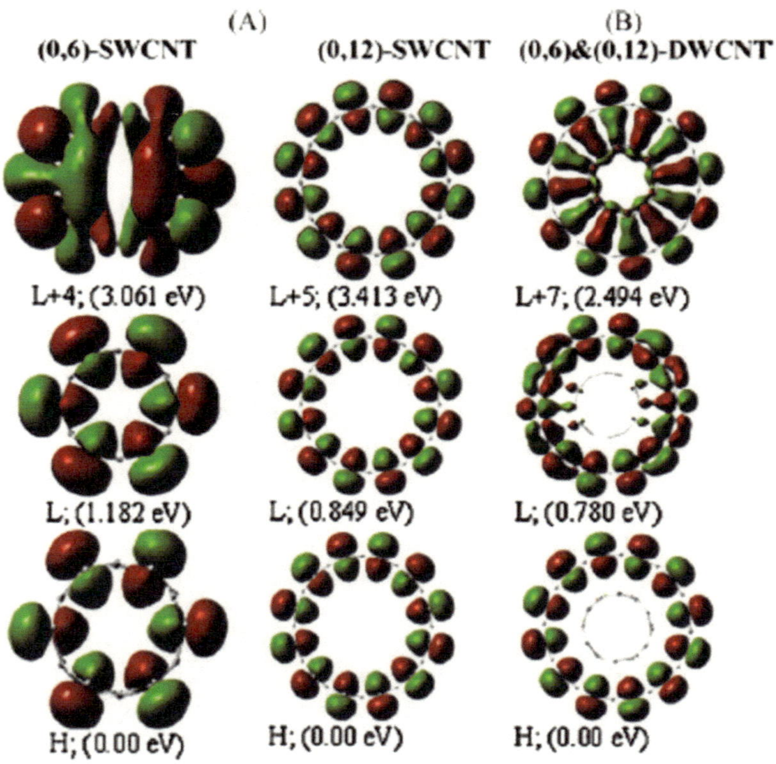


Fig. 3.2.1. Calculated electron densities in the lowest HOMO and LUMO states: (A) for the (0,6)- and (0,12)-SWNTs and (B) for the (0,6)&(0,12)-DWNT.

### 3.3 Raman spectra

We calculated Raman spectra of the ($n$, m)-nanotubes with $n$ = 6 to 19 for the zigzag nanotube, n = m = 3 to 12 for the armchair nanotube, and n = 6 to 10 and m= 2 for the chiral nanotube; for the zigzag and armchair single-wall boron nitride nanotubes (SWBNNTs) chiral index (n, m) ranging from n = 7 to 19 and n = m = 3 to 10, respectively. For the (n,0)-

SWCNTs with n = 6 to 11 and the (8, 0)-boron nitride nanotubes, the length dependence of the Raman spectra was also calculated at the same level of the DFT. The calculated Raman spectra of the carbon and boron nitride nanotubes illustrate similar structure. Therefore, in this chapter, we will comprehensively discuss the (n,0)-SWCNTs only.

### 3.3.1 Raman spectra of (n,0)-SWCNTs

Figure 3.3.1 provides the calculated Raman spectra of the zigzag (*n*,0)-SWCNTs with *n* ranging from 6 to 19. Table 3.3.1 provides vibrational mode assignments and frequencies, while Figure 3.3.2A provides diagrams of the atomic motions associated with the vibrational frequencies for the (11,0)-SWCNT, used as a representative case. All assignments to motions of atoms or groups of atoms in Table 3.3.1 have been accomplished through use of vibration visualization software (specifically, GaussView03). The calculated Raman spectra indicate that there are two additional Raman bands besides the RBM in the low frequency region. The frequencies of these latter bands are also found to depend on SWCNT diameter, as shown in Figure 3.3.1A, and Figure 3.3.1B indicate the calculated vibrational frequencies as a function diameters of SWCNTs. As seen in Figure 3.3.1B, the RBM (with $A_{1g}$ symmetry) and two other Raman bands (with $E_{1g}$ and $E_{2g}$ symmetries) have frequencies that inversely depend on a nanotube's diameter. A linear fit to the calculate RBM frequency dependence on nanotube diameter is provided; a linear equation, $\omega_{RBM}(A_{1g}) = 12.04(cm^{-1}) + \frac{221.4(nm.cm^{-1})}{d_t(nm)}$, is in excellent agreement with the empirically determined expression[97]; indeed, a popular one has the value 12.5 for the offset constant and 223.5 for the constant shift parameter that appear on the r-h-s of Eq. 2. Even though a linear equation reproduces the RBMs within about ± 3 $cm^{-1}$ error ranges for the large size SWCNTs, it overestimates RBMs for $(0,n<7)$-SWCNTs which have diameters smaller than 0.55 nm such as about 14 $cm^{-1}$ for (0,6)-SWCNT. In actual fact, this is not so surprising, because the C-C-C bond strain rapidly increases with decreasing CNTs diameters (as seen in Figure 3.1.1), the plot of curvature energy per hexagon of the isolated-SWCNTs. Therefore, the best fitting may be obtained using a quadratic fitting equation as given in Equation 3.3.1, which reproduces the RBMs within ± 2 $cm^{-1}$ error ranges when comparing with the calculated Raman spectra of the SWCNTs from (0,6) to (0,19) by the DFT technique.

$$\omega_{RBM}(A_{1g}) = -14.12(cm^{-1}) + \frac{269.67(nm.cm^{-1})}{d_t(nm)} - \frac{20.24(nm^2.cm^{-1})}{[d_t(nm)]^2} \tag{3.3.1}$$

Even though the corresponding linear fit to the two other accompanying calculated low frequency bands ($\omega_{BD}$ of $E_{1g}$ symmetry and $\omega_{ED}$ of $E_{2g}$ symmetry) as functions of the inverse of the SWCNT diameter may be given linear equations such as: $\omega_{BD}(E_{1g}) = 51.9(cm^{-1}) + \frac{136.8(nm.cm^{-1})}{d_t(nm)}$ and $\omega_{ED}(E_{2g}) = -43.8(cm^{-1}) + \frac{80.9(nm.cm^{-1})}{d_t(nm)}$, the best fittings are provided in Equations 3.3.3a-b and 3.3.4, respectively. It is to be noted that both the $E_{1g}$ and $E_{2g}$ bands have lower frequencies than the RBM, with the $E_{2g}$ band being of lowest frequency (see Figure 3.3.1B). These two latter bands are labeled as BD for bonding deformation and ED for elliptical deformation, which derives from the predominate motions that define vibrational mode motions, as ascertained with the vibration visualization software mentioned earlier (see Table 3.3.1). Furthermore, the calculated $E_{2g}$ band for (0,2n)-SWCNTs is not only diameter dependent, but is also dependent on whether the number of hexagons formed in the circumference direction of the SWCNTs (0,2n) type or (0,2n+1) type, respectively are even or odd. As shown in Figure 3.3.1B, the $E_{2g}$ band for (0,2n) and (0,2n+1) type SWCNTs is well separated with decreasing tube diameter, but they again join together at the large tube diameters. Therefore, in order to obtain a more precise fitting equation for this Raman band (of symmetry $E_{2g}$), we obtained two fitting equations as given in Eq. 3.3.3a for (0, 2n) type SWCNTs and Eq. 3b for (0,2n+1) type SWCNTs.

$$\omega_{BD}\left(E_{1g};(2n,0)\right)=4.15(cm^{-1})+\frac{227.82(nm.cm^{-1})}{d_t(nm)}-\frac{33.60(nm^2.cm^{-1})}{[d_t(nm)]^2} \quad (3.3.3a)$$

$$\omega_{BD}\left(E_{1g};(2n+1,0)\right)=-0.43(cm^{-1})+\frac{224.23(nm.cm^{-1})}{d_t(nm)}-\frac{37.34(nm^2.cm^{-1})}{[d_t(nm)]^2} \quad (3.3.3b)$$

$$\omega_{ED}\left(E_{2g};(n,0)\right)=-0.31(cm^{-1})+\frac{1.13(nm.cm^{-1})}{d_t(nm)}+\frac{33.23(nm^2.cm^{-1})}{[d_t(nm)]^2} \quad (3.3.4)$$

Moreover, as seen in Figure 3.3.1B, for large sized SWCNTs, the $\omega_{RBM}(A_{1g})$ and $\omega_{BD}(E_{1g})$ mode frequencies converge. The calculated frequency separation between the RBM and BD to found to be 0, 5, 9 and 14 $cm^{-1}$, respectively, when $n$ has the values 27, 24, 22 and 20. Thus, one can anticipate the (27,0)-SWCNT would have unresolvable RBM and BD bands. We can anticipate, in general, that the acquisition of Raman spectra for experimental samples consisting of large diameter SWCNT with the purpose of characterizing the sample in terms of electronic properties and purity may be complicated by the existence of this BD band, which can lead to apparent broadening of bands as well as the presence of additional bands that may lead to the erroneous conclusion that more than one type of SWCNT is present in the sample.

As regards other general conclusions that can be drawn from our calculations for the SWCNTs, we note that the lowest frequency $\omega_{ED}(E_{2g})$ mode may not be observable for large diameter nanotube due to Rayleigh scattering; however, our calculation suggests that (6,0) and (7,0) zigzag SWCNTs, with computed $\omega_{ED}$'s of 145.5 and 110.8 $cm^{-1}$, should be resolvable from Rayleigh scattering. Also, we have found that calculated Raman bands in the mid-frequency region exit nearly size-independent peak positions. While, as indicated in Table 3.3.1 or Figure 3.3.1C and 3.3.1E, in the high-frequency region there are three Raman bands of symmetries $E_{1g}/E_{2g}/A_{1g}$ that lie close to one another. Raman bands $A_{1g}$ symmetry essentially remain constant for (2n,0) type CNTs (with band position 1526 ± 0.5 $cm^{-1}$, see Fig. 3.3.1E), but that for (2n+1,0) type CNTs are diameter dependent (Fig. 3.3.1C), $A_{1g}$(1486-1525 $cm^{-1}$ or $1526(cm^{-1})+\frac{3.55(nm.cm^{-1})}{d_t(nm)}-\frac{14.86(nm^2.cm^{-1})}{[d_t(nm)]^2}$). We further observe that the $E_{1g}$ (~1547 ± 25 $cm^{-1}$) and $E_{2g}$ (~1532 ± 12 $cm^{-1}$) Raman modes first approach one another in frequency then separates as one calculates these frequencies for increase diameter of the SWCNT. These shifts in the peak positions may result from the nanotube curvature effect. The curvature energy (as given by Eq. 1) of the nanotube brings about dissimilar force constants along the nanotube axis and the circumference direction. Therefore, the nanotube geometry causes a force constant reduction along the tube axis compared to that in the circumferential direction. As a result, the curvature effect might play crucial role in the shift of the peak positions of the G-band as well as the RBM band as mentioned earlier. Furthermore, we might note that these Raman bands, since they are overlapping, might pose a problem in assessing, based on the shape of bands in this transverse vibrations region, whether SWCNT samples are metallic or semiconducting. In addition, the calculated Raman band positions for bands around 1330 cm-1, the disordered graphite region are found to be slightly size dependent, exhibiting a small blue shift with increasing diameter of the SWCNTs.

| | (19,0) | (18,0) | (17,0) | (16,0) | (15,0) | (14,0) | (13,0) | (12,0) | (11,0) | (10,0) | (9,0) | (8,0) | (7,0) | (6,0) | Assignment |
|---|---|---|---|---|---|---|---|---|---|---|---|---|---|---|---|
| $E_{1g}$ | 1586.9<br>1507.7 | 1508.7 | 1510.8 | 1525.6 | 1536.0 | 1542.1 | 1550.8 | 1558.3 | 1564.6 | 1571.2 | 1575.6 | 1576.5 | 1574.0 | 1550.2 | Skeletal deformation due to the asymmetric stretching of the C-C-C bonds: $\nu_{as}$(C-C-C) |
| $E_{2g}$ | 1525.5 | 1537.3 | 1537.6 | 1539.3 | 1533.1 | 1541.8 | 1533.6 | 1544.2 | 1536.8 | 1543.7 | 1534.3 | 1533.8 | 1515.6 | 1498.3 | Skeletal deformation due to the stretching of the C-C bonds along the tube axis: $\nu$(C-C); asymmetric stretching of the C-C-C bonds: $\nu_{as}$(C-C-C). Note that wagging of CH bonds also occurs |
| $A_{1g}$ | 1524.3 | 1527.9 | 1542.7 | 1527.5 | 1521.2 | 1527.6 | 1518.2 | 1527.7 | 1512.9 | 1528.2 | 1503.4 | 1527.8 | 1486.2 | 1526.6 | Skeletal deformation due to the stretching of the C-C bonds, $\nu$(C-C-C) and $\nu$(C-C) along the tube axis |
| $E_{2g}$ | 1399.1 | 1402.7 | 1408.3<br>weak | 1410.5 | 1411.7 | 1417.5 | 1417.9 | 1422.9 | 1421.6 | 1423.8 | 1421.3 | 1415.9 | 1411.7 | 1375.8 | Skeletal deformation due to asymmetric and symmetric stretching of the C-C-C bonds: $\nu$(C-C-C) and $\nu_s$(C-C-C), and C-C bond stretching along the nanotube axis, $\nu$(C-C) |
| $E_{2g}$ | 472.2 | 463.9 | 450.9 | 464.9<br>457.8 | 477.1 | 471.0<br>462.6<br>442.2 | 479.7 | 464.2 | 483.3 | 467.1<br>461.2 | 489.1 | 478<br>454.4 | 500<br>463<br>461.2 | 402.0 | Bending deformation of the C-C-C bonds in the radial direction. Also contribution from the wagging of the CH bonds along tube circumference |
| $A_{1g}$ | 158.9 | 167.6 | 177.0 | 188.2 | 200.3 | 214.3 | 230.3 | 248.7 | 270.3 | 295.8 | 326.8 | 362.9 | 410.2 | 457.1 | Radial breathing mode of CNT |
| $E_{1g}$ | 132.9 | 147.4 | 148.5 | 163.9 | 163.9 | 183.6 | 184.9 | 208.0 | 209.9 | 238.5 | 242.2 | 278.3 | 284.4 | 332.4 | Bending deformation of the CNT due to expansion of the CNT along diagonal axis, with the motion of two end groups being in opposite directions |
| $E_{2g}$ | 15.4 | 17.1 | 18.9 | 21.9 | 24.7 | 28.3 | 32.9 | 38.4 | 46.1 | 54.7 | 68.4 | 83.7 | 110.8 | 145.5 | Stretching of the CNT along its diameter, resulting in an elliptical shape |

Table 3.3.1 DFT-calculated Raman vibrational frequencies (in cm_1) and assignments for (n,0)-CNT at the B3LYP/6-31G level.

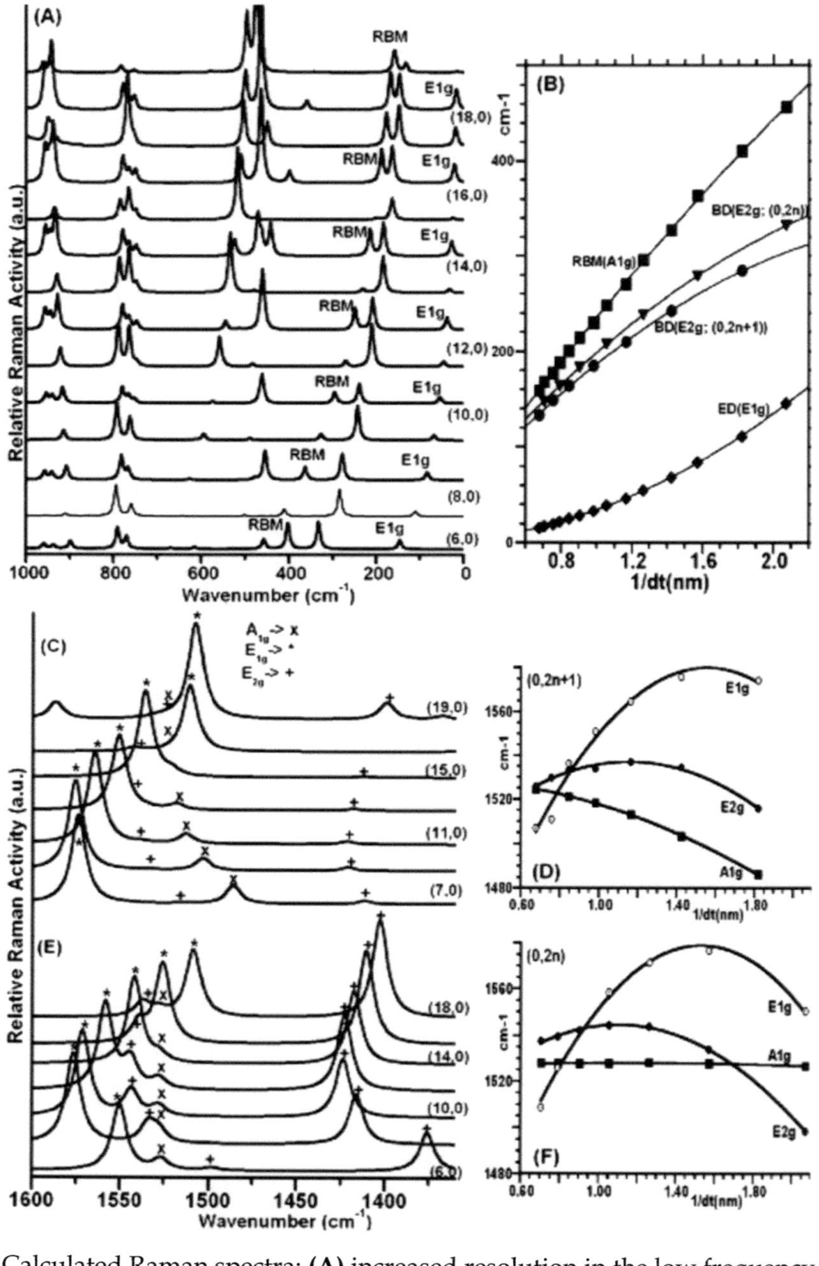


Fig. 3.3.1. Calculated Raman spectra: **(A)** increased resolution in the low frequency region, showing diameter dependence of the calculated Raman band frequencies for the (*n*, 0)-SWCNTs, *n* = 0 to 19; **(B)** the plots of the frequencies of vibrational modes of symmetries $A_{1g}$, $E_{1g}$ and $E_{2g}$ versus 1/dt; in the high energy region: **(C)** for (2n+1,0)-SWCNTs, **(D)** the plots of the frequencies of vibrational modes of symmetries $A_{1g}$, $E_{1g}$ and $E_{2g}$ versus 1/dt;; **(E)** for (2n,0)-SWCNTs, **(F)** the plots of the frequencies of vibrational modes of symmetries $A_{1g}$, $E_{1g}$ and $E_{2g}$ versus $1/d_t$;. Where the signs; x, * and + stand for $A_{1g}$, $E_{1g}$ and $E_{2g}$, respectively.

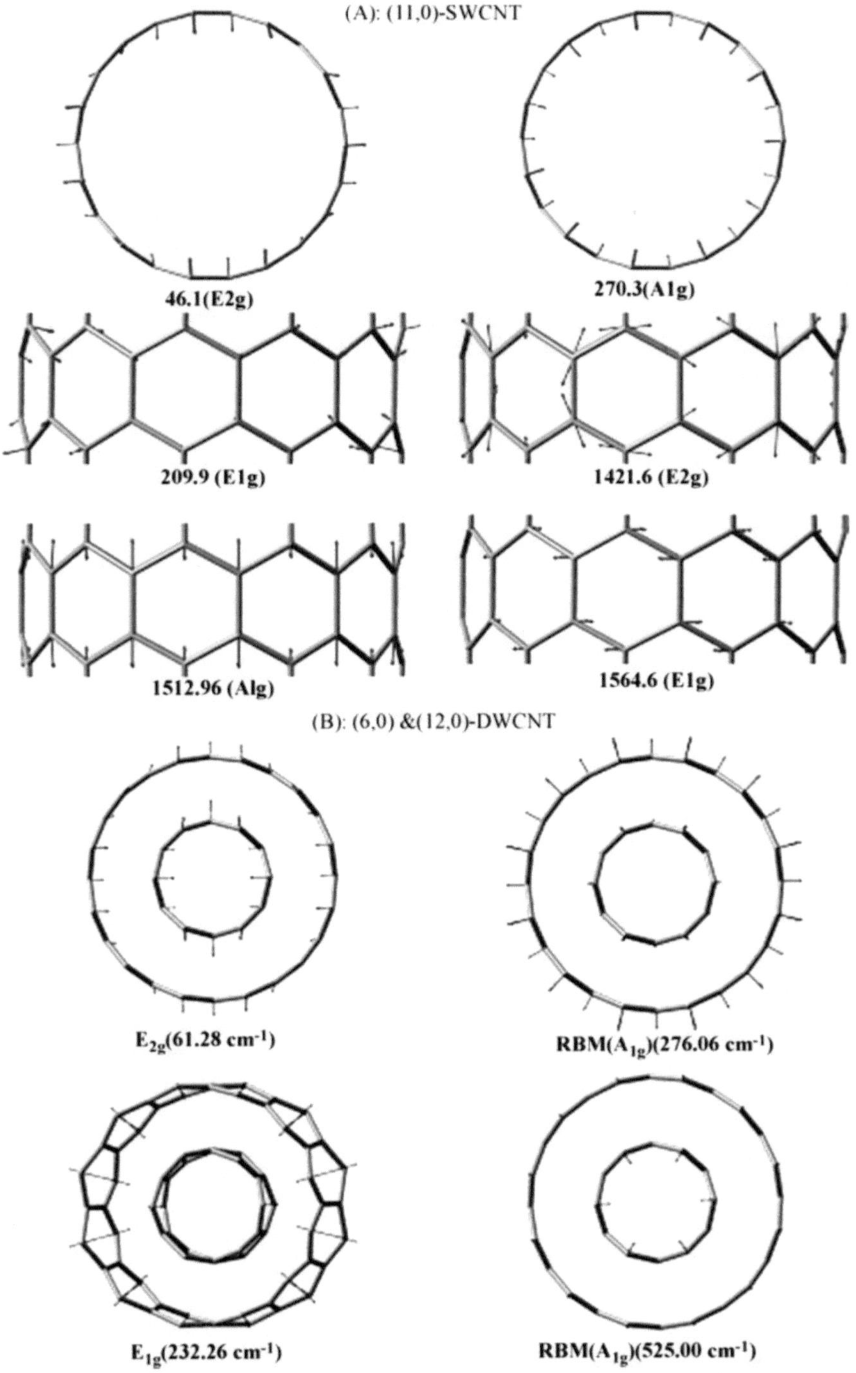


Fig. 3.3.2. Calculated molecular motions for some vibrational bands of the (11,0)-SWNT and (6,0)&(12,0)-DWNT. The nuclear motions of the other SWNTs studied are provided in Table 3.3.1.

### 3.3.2 Raman spectra of the (n,n) and (n,2)-SWCNTs

The calculated Raman spectra of armchair (n, n)- and (n,2) SWCNTs exhibited similar structure to these of zigzag-SWCNTs. Figure 3.3.3A provide the calculated Raman spectra of the (n,n)-SWCNTs. As seen in the Figure 3.3.3A-B, RBM($A_{1g}$), ED($E_{1g}$) and BD($E_{2g}$) bands of vibrational frequencies in the Raman spectra are strongly diameter dependent, but there is no significant change in the calculated Raman bands with the symmetry $E_{2g}$ for even and number of chiral index n as observed in the Raman spectra of the (n,0)-SWCNTs (see Figure 3.3.1A-B). Figure 3.3.3B shows the calculated vibrational frequencies as a function of the tube's diameter. As given in Figure 3.3.3A, the RBM (with $A_{1g}$ symmetry) and two other Raman bands (with $E_{1g}$ and $E_{2g}$ symmetries) have frequencies that inversely depend on a nanotube's diameter. A linear fit to the calculate RBM($A_{1g}$), ED($E_{1g}$) and BD($E_{2g}$) frequency dependence on nanotube diameter is given in equations 3.3.5a-c;

$$\omega_{RBM}(A_{1g}) = -12.13(\text{cm}^{-1}) + \frac{270.52(\text{nm.cm}^{-1})}{d_t(\text{nm})} - \frac{54.34(\text{nm}^2.\text{cm}^{-1})}{[d_t(\text{nm})]^2} \quad (3.3.5a)$$

$$\omega_{BD}(E_{2g}) = 24.72(\text{cm}^{-1}) + \frac{136.81(\text{nm.cm}^{-1})}{d_t(\text{nm})} - \frac{5.33(\text{nm}^2.\text{cm}^{-1})}{[d_t(\text{nm})]^2} \quad (3.3.5b)$$

$$\omega_{ED}(E_{1g}) = -0.79(\text{cm}^{-1}) + \frac{6.56(\text{nm.cm}^{-1})}{d_t(\text{nm})} + \frac{22.10(\text{nm}^2.\text{cm}^{-1})}{[d_t(\text{nm})]^2} \quad (3.3.5b)$$

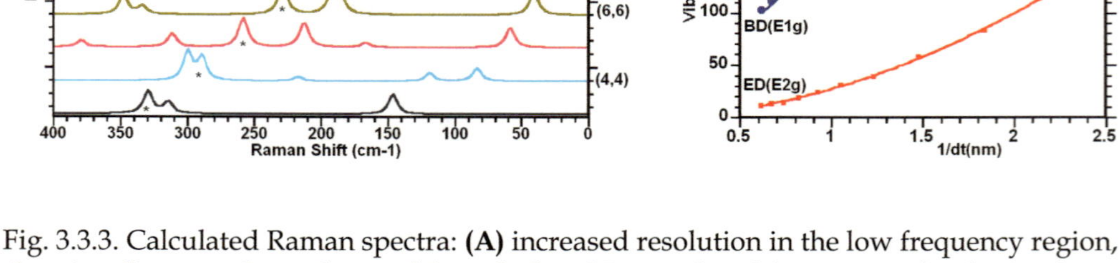


Fig. 3.3.3. Calculated Raman spectra: **(A)** increased resolution in the low frequency region, showing diameter dependence of the calculated Raman band frequencies for the (*n*, n)-SWCNTs, *n* = 3 to 12; **(B)** the plots of the frequencies of vibrational modes of symmetries $A_{1g}$, $E_{1g}$ and $E_{2g}$ versus 1/dt.

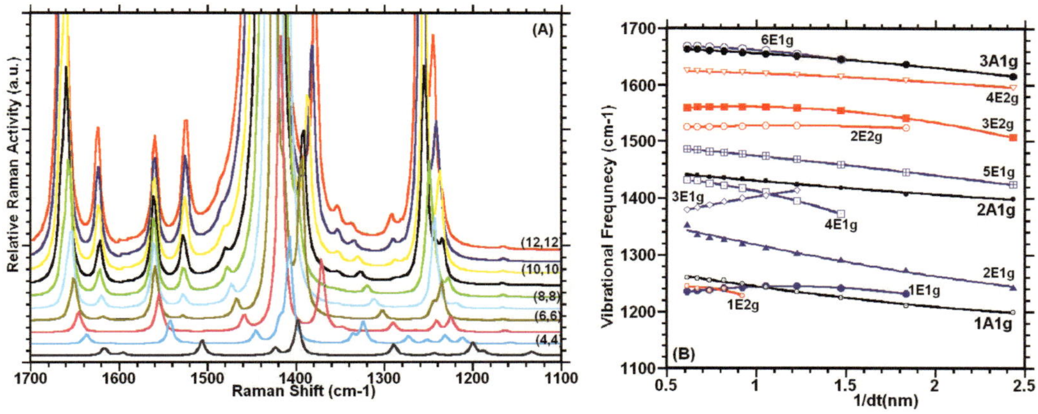


Fig. 3.3.4. Calculated Raman spectra: **(A)** increased resolution in the high frequency region, showing to some extent diameter dependence of the calculated Raman band frequencies for the (n, n)-SWCNTs, n = 3 to 12; **(B)** the plots of the frequencies of vibrational modes of symmetries $A_{1g}$, $E_{1g}$ and $E_{2g}$ versus 1/dt.

In the high energy region, the calculated Raman spectra exhibit many Raman active spectra in range of 1200-1700 cm-1 as seen in Figure 3.3.4A, which are also slightly diameter dependent. As seen in Figure 3.3.4B, a linear fit to the calculate $A_{1g}$, $E_{1g}$ and $E_{2g}$ frequency dependence on the diameter of the nanotube is given in equations 3.3.6a-l;

$$\omega(3A_{1g}) = 1670.16(\text{cm}^{-1}) - \frac{8.06(\text{nm.cm}^{-1})}{d_t(\text{nm})} - \frac{5.76(\text{nm}^2.\text{cm}^{-1})}{[d_t(\text{nm})]^2} \quad (3.3.6a)$$

$$\omega(6E_{1g}) = 1662.50(\text{cm}^{-1}) + \frac{24.47(\text{nm.cm}^{-1})}{d_t(\text{nm})} - \frac{24.77(\text{nm}^2.\text{cm}^{-1})}{[d_t(\text{nm})]^2} \quad (3.3.6b)$$

$$\omega(4E_{2g}) = 1627.18(\text{cm}^{-1}) - \frac{1.88(\text{nm.cm}^{-1})}{d_t(\text{nm})} - \frac{4.47(\text{nm}^2.\text{cm}^{-1})}{[d_t(\text{nm})]^2} \quad (3.3.6c)$$

$$\omega(3E_{2g}) = 1541.74(\text{cm}^{-1}) + \frac{43.73(\text{nm.cm}^{-1})}{d_t(\text{nm})} - \frac{23.28(\text{nm}^2.\text{cm}^{-1})}{[d_t(\text{nm})]^2} \quad (3.3.6d)$$

$$\omega(2E_{2g}) = 1515.45(\text{cm}^{-1}) + \frac{21.33(\text{nm.cm}^{-1})}{d_t(\text{nm})} - \frac{9.08(\text{nm}^2.\text{cm}^{-1})}{[d_t(\text{nm})]^2} \quad (3.3.6e)$$

$$\omega(5E_{1g}) = 1502.66(\text{cm}^{-1}) - \frac{24.66(\text{nm.cm}^{-1})}{d_t(\text{nm})} - \frac{3.20(\text{nm}^2.\text{cm}^{-1})}{[d_t(\text{nm})]^2} \quad (3.3.6f)$$

$$\omega(2A_{1g}) = 1460.21(\text{cm}^{-1}) - \frac{32.55(\text{nm.cm}^{-1})}{d_t(\text{nm})} + \frac{2.90(\text{nm}^2.\text{cm}^{-1})}{[d_t(\text{nm})]^2} \quad (3.3.6g)$$

$$\omega(4E_{1g}) = 1437.00(\text{cm}^{-1}) + \frac{16.41(\text{nm.cm}^{-1})}{d_t(\text{nm})} - \frac{41.16(\text{nm}^2.\text{cm}^{-1})}{[d_t(\text{nm})]^2} \quad (3.3.6h)$$

$$\omega(3E_{1g}) = 1389.42(\text{cm}^{-1}) - \frac{81.32(\text{nm.cm}^{-1})}{d_t(\text{nm})} + \frac{8.94(\text{nm}^2.\text{cm}^{-1})}{[d_t(\text{nm})]^2} \quad (3.3.6i)$$

$$\omega(2E_{1g}) = 1329.12(\text{cm}^{-1}) + \frac{94.23(\text{nm.cm}^{-1})}{d_t(\text{nm})} - \frac{20.70(\text{nm}^2.\text{cm}^{-1})}{[d_t(\text{nm})]^2} \quad (3.3.6j)$$

$$\omega(1A_{1g}) = 1295.48(cm^{-1}) - \frac{58.27(nm.cm^{-1})}{d_t(nm)} + \frac{7.64(nm^2.cm^{-1})}{[d_t(nm)]^2} \quad (3.3.6k)$$

$$\omega(1E_{1g}) = 1201.23(cm^{-1}) + \frac{73.99(nm.cm^{-1})}{d_t(nm)} - \frac{31.24(nm^2.cm^{-1})}{[d_t(nm)]^2} \quad (3.3.6l)$$

The calculated Raman bands with symmetry labeled as $3A_{1g}$ and $6E_{1g}$ are almost overlapping and other bands also close to one other. As observed in low frequency region, the diameter dependence of the Raman band in high energy region is somewhat different than these found in the Raman spectra of the (n,0)-SWCNTs.

The calculated Raman spectra of the (n,2)-SWCNTs (see Figure 3.3.5), with n = 5 to 10, did not only exhibit the dependence on the diameter of the nanotube, but also relatively complicated than the calculated Raman spectra of the armchair- and zigzag-type carbon nanotubes. In order to provide a reliable fitting equations for the diameter dependence of the RBM($A_{1g}$), ED($E_{1g}$) and BD($E_{2g}$) bands of vibrational frequencies in low energy region, the number of calculated Raman spectra for the chiral nanotube is not large enough; therefore, based on the present data, tentative fitting equations for the diameter dependence of these three Raman bands in the low energy region may be given by equations 3.3.7a-c.

$$\omega_{RBM}(A_{1g}) = -12.11 + \frac{233.62(nm.cm^{-1})}{d_t(nm)} + \frac{0.64(nm^2.cm^{-1})}{[d_t(nm)]^2} \quad (3.3.7a)$$

$$\omega_{BD}(E_{2g}) = 20.12 + \frac{94.37(nm.cm^{-1})}{d_t(nm)} + \frac{10.29nm^2.cm^{-1})}{[d_t(nm)]^2} \quad (3.3.7b)$$

$$\omega_{ED}(E_{1g}) = 2.45 + \frac{11.63(nm.cm^{-1})}{d_t(nm)} + \frac{21.29(nm^2.cm^{-1})}{[d_t(nm)]^2} \quad (3.3.7c)$$

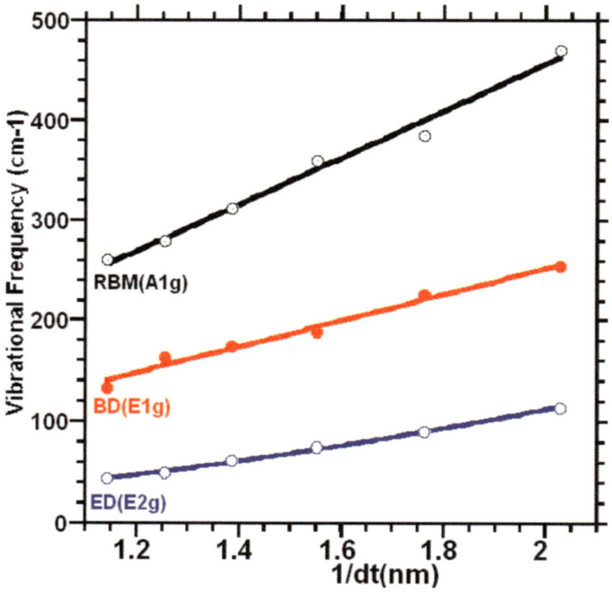


Fig. 3.3.5. The plots of the frequencies of vibrational modes of symmetries $A_{1g}$, $E_{1g}$ and $E_{2g}$ versus $1/d_t$ for the chiral (n, 2)-SWCNTs with n = 5 to 10.

### 3.3.3 Dependence of the RBM Band on the length of nanotube

We calculated Raman spectra of the zigzag (n,0)-SWCNTs as a function of m (number of hexagon along the nanotube axis or length of the tube), where n = 6 to 11 and m = 1 to 5, at same level of the DFT. The results of the calculations showed that the RBMs($A_{1g}$) is also slightly dependent on the m (length of the nanotube) even if its dependence is not as strongly dependent on the diameter. When we plotted the RBMs as function of the m with a desired nanotube, for instance, for the (6,0)-SWCNT, the dependence of the calculated RBMs on the m are not compatible with one other, except for (10,0) and (11,0)-SWCNT. For these two latter SWCNTs, the RBMs slightly decreased with increasing value of m. On the contrary, as seen in Figure 3.3.6A, the plot of the calculated RBMs as function of diameter for a desired length (m), each curve for the RBMs data is in excellent agreement with one other for each of the number of hexagon along the tube axis (m) as shown in Figure 3.3.6C. The fitting equations for each set of the m are given by equations 3.3.8a-d. The fitting parameters for each set of the m are well consistent with each other. It should be noticed that for the fitting equations, we set the constant parameter to -14.1 cm$^{-1}$ and excluded two RBMs of frequencies at 498.9 cm$^{-1}$ and 445.9 cm$^{-1}$ for the (6,0)- and (7,0)-SWCNT with the m=3, respectively, in order to obtain a best fitting to calculated data.

$$\omega_{RBM}(A_{1g}; m = 1) = -14.1 + \frac{274.93(nm.cm^{-1})}{d_t(nm)} - \frac{24.57(nm^2.cm^{-1})}{[d_t(nm)]^2} \quad (3.3.8a)$$

$$\omega_{RBM}(A_{1g}; m = 2) = -14.1 + \frac{257.22(nm.cm^{-1})}{d_t(nm)} - \frac{14.03(nm^2.cm^{-1})}{[d_t(nm)]^2} \quad (3.3.8b)$$

$$\omega_{RBM}(A_{1g}; m = 3) = -14.1 + \frac{248.95(nm.cm^{-1})}{d_t(nm)} - \frac{9.69(nm^2.cm^{-1})}{[d_t(nm)]^2} \quad (3.3.8c)$$

$$\omega_{RBM}(A_{1g}; m = 4) = -14.1 + \frac{304.09(nm.cm^{-1})}{d_t(nm)} - \frac{34.53(nm^2.cm^{-1})}{[d_t(nm)]^2} \quad (3.3.8d)$$

Figure 3.3.6B-C and Table 3.3.2 provides the calculated ED($E_{1g}$) mode of frequencies and the calculations indicated that the ED($E_{1g}$) mode of frequencies are almost length independent. Furthermore, we calculated length dependence of the Raman spectra for the (4,4)-SWCNTs. The calculations showed that the length dependence of the RBM bands slightly change with the number of hexagon along the nanotube axis.

| | RBM($A_{1g}$) in cm$^{-1}$ | | | | | | ED($E_{1g}$) in cm$^{-1}$ | | | | | |
|---|---|---|---|---|---|---|---|---|---|---|---|---|
| m | (6,0) | (7,0) | (8,0) | (9,0) | (10,0) | (11,0) | (6,0) | (7,0) | (8,0) | (9,0) | (10,0) | (11,0) |
| 1 | 457,1 | 410,2 | 362,9 | 326,8 | 295,8 | 270,3 | 145.5 | 110.8 | 83.7 | 68.4 | 54.7 | 46.1 |
| 2 | 467,3 | 414,0 | | 319,0 | 291,8 | 266,0 | 139.38 | 109.82 | | 67.6 | 54.4 | 44.94 |
| 3 | 498,9 | 445,9 | 356,6 | 325,2 | 284,5 | 262,1 | 145.12 | 106.97 | 88.6 | 66.99 | 56.26 | 45.76 |
| 4 | 476,7 | 426, | 383,3 | | | | 146.24 | 105.27 | 88.24 | | | |
| 5 | 484,4 | 421,6 | | | | | 142.67 | 105.78 | | | | |

Table 3.3.2. The calculated length (the number of hexagon along the nanotube axis, m = 1 to 5) dependence of the RBM($A_{1g}$) and ED($E_{1g}$) modes at B3LYP/6-31G level of the DFT for the zigzag (n,0)-SWCNTs, with n = 6 to 11.

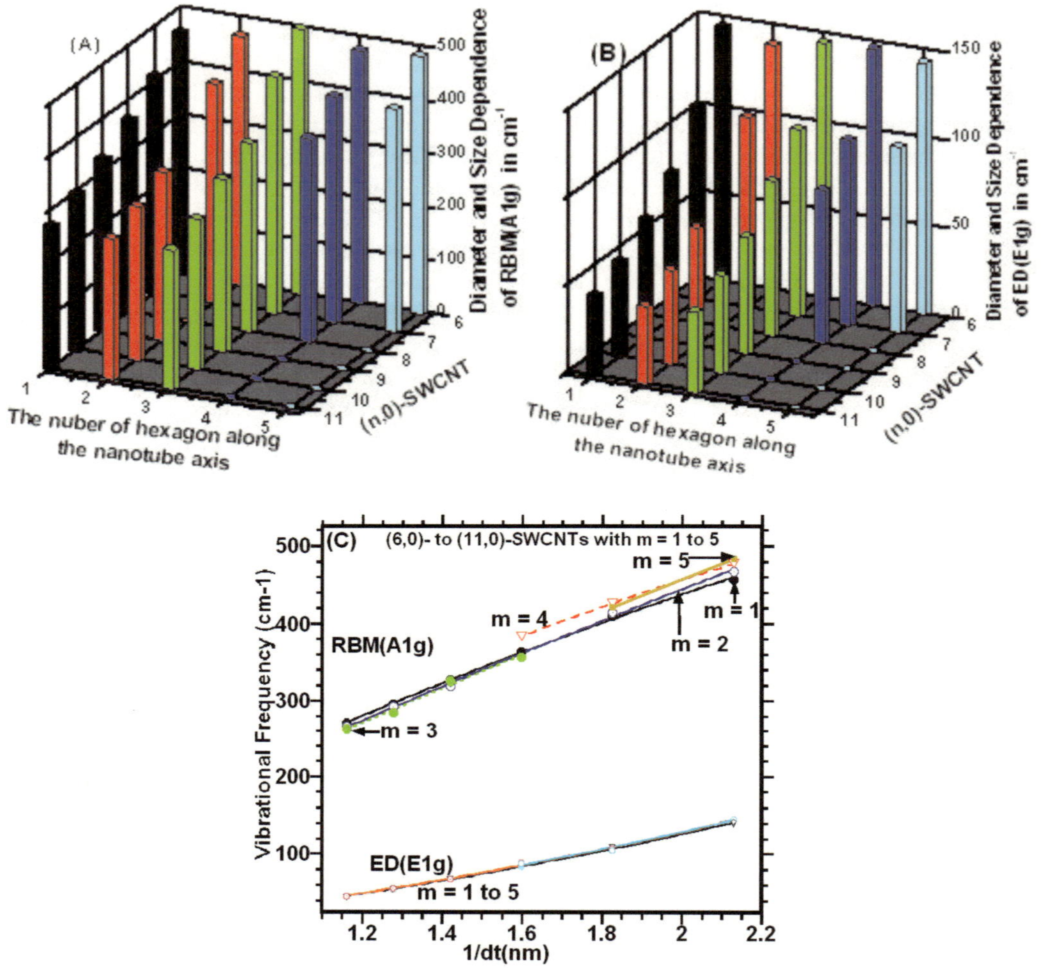


Fig. 3.3.6. The calculated dependence of the Raman modes on the diameter and length (the number of hexagon (m =1 to 5) along the nanotube axis ) of the nanotubes: (A) for the RBM($A_{1g}$); (B) for the ED($E_{1g}$) for the zigzag (n, 0)-SWCNTs with n = 6 to 11; and (C) illustrates the plot of the RBM and ED modes as function of the m.

### 3.3.4 Raman spectra for DWCNTs

Figures 5 provides the calculated Raman spectra of (0,6)&(0,12) and (0,7)&(0,14)-DWCNTs in the low energy region. The calculations showed that the frequencies of the RBMs and tangential modes (TMs) of DWCNT significantly differ from those calculated for SWCNT. The calculated Raman spectra of these DWCNTs exhibited two RBM modes resulting from the radial motion of the inner- and outer-shells, in-phase and out-of-phase, as seen in Figure 3B, and both of these RBM modes are strongly diameter dependent. A large gap between RBMs of DWCNT decreases with increasing diameter of the inner- and outer-shells. Comparing these calculated RBMs with their corresponding ones in the SWCNTs spectra, we noticed that the RBMs at 457 $cm^{-1}$ in the Raman spectrum of (6,0)-SWCN and at 249 $cm^{-1}$

in the (12,0)-SWCNT spectrum are respectively blue-shifted to 525 $cm^{-1}$ and 276 $cm^{-1}$ in the Raman spectrum of (6,0)&(12,0)-DWCNT. Additionally, the RBMs for (7,0)-SWCNT (410 $cm^{-1}$) and (14,0)-SWCNT (214 $cm^{-1}$) are respectively upward shifted to 450 $cm^{-1}$ and 237 $cm^{-1}$ in the spectrum of (7,0)&(14,0)-DWCNT. The relative distances between RBMs in the spectrum of (n,0)&(2n,0)-DWCNTs are larger than the distances between corresponding RBMs in Raman spectra of (n,0)- and (2n,0)-SWCNTs. Even though we do not have enough data for the calculated RBMs for different sized DWCNTs, a tentative equation for the RBMs may be given by equations below;

$$\omega_{inner}(RBM) = 62cm^{-1} + \frac{138(nm.cm^{-1})}{d_t(nm)} + \frac{39(nm^2.cm^{-1})}{(d_t(nm))^2} \tag{3.3.9a}$$

$$\omega_{outer}(RBM) = 45cm^{-1} + \frac{240(nm.cm^{-1})}{d_t(nm)} - \frac{30(nm^2.cm^{-1})}{(d_t(nm))^2} \tag{3.3.9b}$$

Where $d_t$ stand for the shell diameter. In the high frequency region, Figure 5 provides the calculated Raman modes (with $E_{1g}$ / $A_{1g}$ / $E_{2g}$ symmetries). When we compare these tangential bands with their band position in the corresponding SWCNT spectra, it can be seen that they are downward shifted (relative to SWCNTs). The animations of the normal modes showed that the strong Raman peaks are mostly resulting from the nuclear motions of the outer-shell.

### 3.4.1 (n,0)- and (n,n)-boron nitride nanotube

Blase *et al* [98] have theoretically studied the possibility of obtaining boron nitride nanotubes (BNNTs) and initiated an exploration on their characteristic properties [99,100,101], by investigating the connection between hexagonal boron nitride sheets [102] and graphite. Even though a hexagonal boron nitride sheet is made up of alternating atoms of boron and nitrogen, boron nitride surprisingly yields atomic structures like those of graphite. Moreover, the crystallographic parameters of hexagonal boron nitride sheets and graphite are nearly identical. Like CNTs, BNNTs with different chiralities can be formed by rolling up a hexagonal sheet of boron nitride in different chiral directions.

BNNTs have many properties similar to those of CNTs [103,104]. For example, chemical inertness [105], high heat conduction, piezoelectricity [106], exceptional elastic properties [107,108,109,110], high mechanical strength [111,112,113,114],and diameter dependence of the Raman bands [99]. In contrast, whereas CNTs are highly toxic for the human body, and can be semiconductors and conductors depending on their chirality, BNNTs show good biocompatibility and are always semiconducting with a large band gap [99], regardless of their chirality.

The diameter dependence of the vibrational frequency of SWBNNTs has been investigated by many researchers [115]. Sanchez-Portal and Hernandez [116] calculated the dependence of the vibrational properties of the SWBNNTs using TB approximation. By using first principles, Wirtz and Rubio [117] calculated the phonon-dispersion relation for Raman and infrared active modes of (n,0)- and (n,m)-type of the BNNTs. A comparative study of the radial breathing mode was also conducted [118] showed that the frequencies of BNNTs are lower than those of CNTs whose diameter are close to that of BNNTs. Furthermore, the surface buckling that is sole to BNNTs was observed in consequence of the different hybridizations of boron and nitrogen in the bent nanotube surfaces [108,117,119]. Even though numerous theoretical studies have been done on the diameter dependence of the

RBM modes of frequency, a systematic study has not yet been undertaken, although BNNTs has applications in a broad range of fields, such as nanoresonators, nanosensors, actuators and transducers. [120,121,122]. Up to now, SWBNNTs have been especially considered as circular nano-cylinders like SWCNTs. However, Chowdhury *et al*[99] pointed out that this may not be always true as the iconicity of the BN bonds responsible for the surface buckling in hoop direction, could also lead to distortion of the optimized configuration in the axial direction when BNNTs are used with practical end constraints.
The present work aims to achieve a realistic configuration of the optimized SWBNNTs structure, and conduct a detailed study on geometric and spectroscopic properties. These properties include diameter dependence of the Raman and IR spectra of SWBNNTs at same level of the density functional theory (DFT) as used for SWCNTs.

### 3.4.2 Raman spectra of the (n,0)- and (n,n)-SWBNNTs

We calculated the spectroscopic and geometric parameters of the (n,0)-SWBNNTs with n = 7 to 19 and (n,n)-SWBNNTs (n=3 to 10) at the same level of the DFT used for the SWCNTs. The assignments to motions of atoms or groups of atoms have been carried out through use of vibration visualization software (GaussView03) that is similar to the assignment made for the (n,0)-SWCNTs as seen in Table 3.3.1, and therefore, are given here. As found for the CNTs, the calculated Raman spectra of the (n,0)- and (n,n)-SWBNNTs in the low frequency region indicated that three of the Raman bands of frequencies are strongly diameter dependent. If the motion of the boron and nitrogen atoms within the nanotube is due to the stretching of the BNNT along its diameter, this result in an elliptical shape abbreviated as ED mode. If it is due to the bending deformation of the BNNT, which results from expansion of the nanotube along diagonal axis, with the motion of two end groups being in opposite directions, this is abbreviated as ED mode. Finally, if the motion of the atoms is in the radial direction in the same phase, this is referred to radial breathing mode, RBM. The calculations exhibited by these Raman bands are also strongly diameter dependent as shown in Figure 3.4.1A-B and Figure 3.4.2A-B for the (n,0)- and (n,n)-SWBNNTs, respectively. The best fit were obtained using a quadratic fitting equation, given by Eq. 3.4.1a-c and Eq. 3.4.2a-c for the (n,0)- and (n,n)-SWBNNTs, respectively. These reproduced the calculated RBMs within $\sim \pm 0.6$ cm$^{-1}$ error ranges, when compared to their corresponding calculated Raman spectra by the DFT technique.

$$\mathrm{RBM}_{(n,0)} = -14.34 + \frac{242.57(\mathrm{nm.cm^{-1}})}{d_t(\mathrm{nm})} - \frac{24.94(\mathrm{nm^2.cm^{-1}})}{[d_t(\mathrm{nm})]^2} \quad (3.4.1a)$$

$$\mathrm{BD}_{(n,0)} = -0.51 + \frac{214.21(\mathrm{nm.cm^{-1}})}{d_t(\mathrm{nm})} - \frac{38.23(\mathrm{nm^2.cm^{-1}})}{[d_t(\mathrm{nm})]^2} \quad (3.4.1b)$$

$$\mathrm{ED}_{(n,0)} = -9.81 + \frac{21.41(\mathrm{nm.cm^{-1}})}{d_t(\mathrm{nm})} + \frac{15.55(\mathrm{nm^2.cm^{-1}})}{[d_t(\mathrm{nm})]^2} \quad (3.4.1c)$$

$$\mathrm{RBM}_{(n,n)} = -27.51 + \frac{270.53(\mathrm{nm.cm^{-1}})}{d_t(\mathrm{nm})} - \frac{75.46(\mathrm{nm^2.cm^{-1}})}{[d_t(\mathrm{nm})]^2} \quad (3.4.2a)$$

$$\mathrm{BD}_{(n,n)} = 1.09 + \frac{153.87(\mathrm{nm.cm^{-1}})}{d_t(\mathrm{nm})} - \frac{14.25(\mathrm{nm^2.cm^{-1}})}{[d_t(\mathrm{nm})]^2} \quad (3.4.2b)$$

$$\mathrm{ED}_{(n,n)} = -17.13 + \frac{34.04(\mathrm{nm.cm^{-1}})}{d_t(\mathrm{nm})} + \frac{7.13(\mathrm{nm^2.cm^{-1}})}{[d_t(\mathrm{nm})]^2} \quad (3.4.2c)$$

Furthermore, as mentioned in Section 3.3.1, the calculated BD band for (n,0)-SWCNTs is not only diameter dependent, but is also dependent on whether the number of hexagons formed in the circumference direction of the SWCNTs (0,2n) type or (0,2n+1) type, respectively are even or odd. However, the calculated ED band for the SWBNNTs did not exhibit dependence on even and odd number of the hexagon in the circumference direction of the (n,0)-SWBNNTs. Moreover, as seen in Figure 3.4.1A-B and 3.4.2A-B, for large sized SWBNNTs, the $\omega_{RBM}$ and $\omega_{BD}$ mode frequencies converge. The calculated frequency separation between the RBM and BD to found to be 11 $cm^{-1}$ for the (19,0)-SWBNNT and 20 $cm^{-1}$ for the (10,10)-SWBNNTs. It should be noticed that the RBM and BD bands would have unresolvable for the (n>20,0)-SWBNNT, such that found for the (n,0)-SWCNTs.

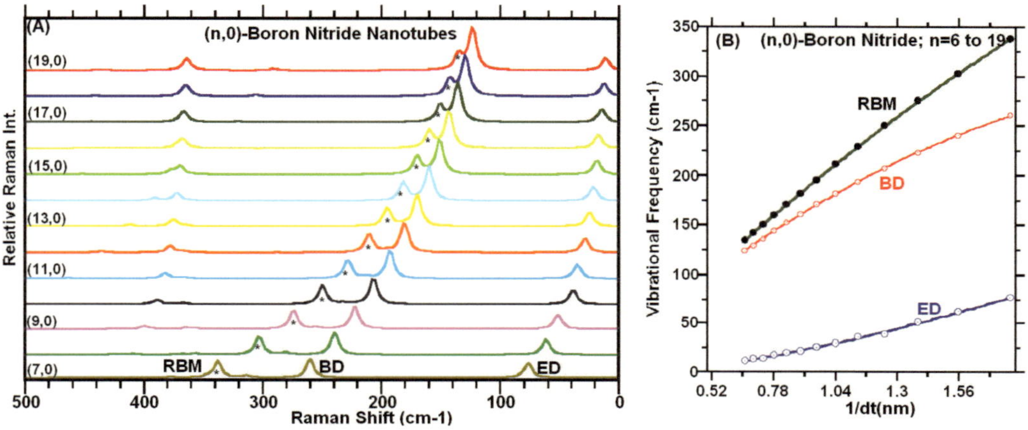


Fig. 3.4.1. Calculated Raman spectra: (A) increased resolution in the low frequency region, showing diameter dependence of the calculated Raman band frequencies for the (*n*, 0)-SWBNNTs, *n* = 7 to 19; (B) the plots of the frequencies of vibrational modes (ED, BD and RBM) versus 1/dt.

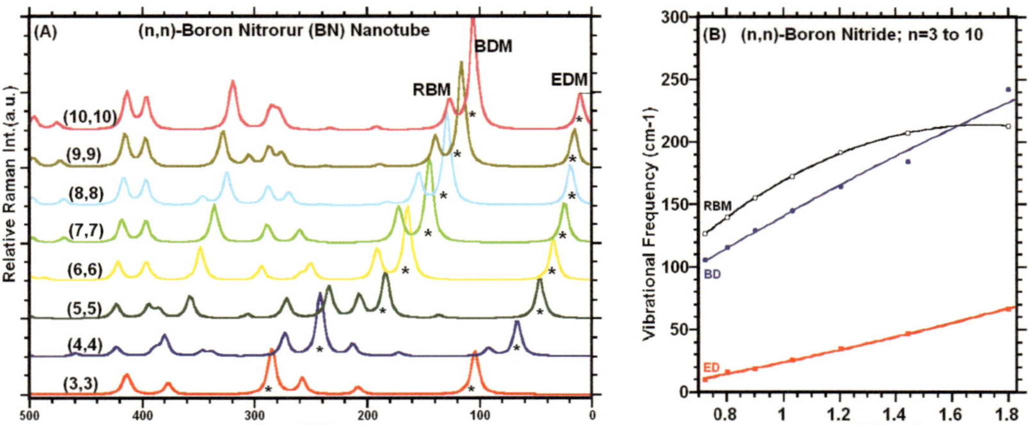


Fig. 3.4.2. Calculated Raman spectra: **(A)** increased resolution in the low frequency region, showing diameter dependence of the calculated Raman band frequencies for the (*n*, n)-SWBNNTs, *n* = 3 to 10; **(B)** the plots of the frequencies of vibrational modes (ED, BD and RBM) versus 1/dt.

We can anticipate, as mentioned earlier for the (n,0)-SWCNTs, in general, that the acquisition of Raman spectra for experimental samples consisting of large diameter SWBNNT with the purpose of characterizing the sample in terms of electronic properties and purity may be complicated by the existence of this BD band, which can lead to apparent broadening of bands as well as the presence of additional bands that may lead to the erroneous conclusion that more than one type of the nanotube is present in the sample.
Also, we have found that calculated Raman bands of the (n,0)- and (n,n)-SWBNNTs in the mid-frequency region (500-1000 $cm^{-1}$) exit almost diameter-independent peak positions. While, as seen in Figure 3.4.3A, in the high-frequency region there are three major Raman bands that are diameter-dependent; two of them lie between 1400-1550 cm-1 region and another one lie between 1200-1300 cm-1 region. However, the Raman peaks at about 1200 cm-1 remain almost constant. The latter two Raman approach one another with decreasing the tube diameter. For the (n,n)-SWBNNTs, the pattern of the Raman peak in the region of 1000-1550 $cm^{-1}$ are similar to these of the (n,0)-SWBNNTs. While the Raman peak at about 1170 cm-1 remain almost constant, the other major peaks indicated the dependence on the diameter of the (n,n)-SWBNNTs as shown in Figure 3.4.3B. As discussed earlier, the curvature energy (as given by Eq. 3.1.1) of the nanotube brings about dissimilar force constants along the nanotube axis and the circumference direction. Therefore, the nanotube geometry causes a force constant reduction along the tube axis compared to that in the circumferential direction. As a result, based on the calculations, the curvature effect might play a crucial role in the shift of the peak positions of the G-band as well as the RBM band. Furthermore, It should be noticed that these three major peaks (in the region of ~1200-1500 $cm^{-1}$) consist of more than one Raman peak with the different symmetry like $A_g$ and $E_g$.

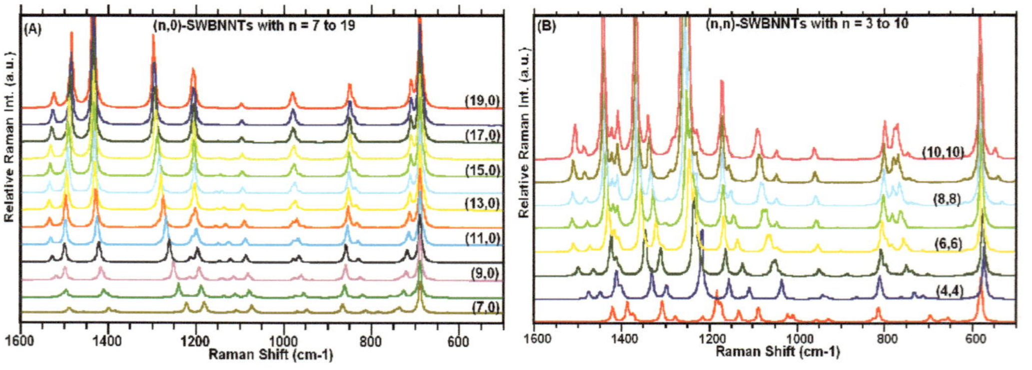


Fig. 3.4.3. Calculated Raman spectra: increased resolution in the high frequency region, showing diameter dependence of the calculated Raman band frequencies: **(A)** for the (*n*, 0)-SWBNNTs, $n = 7$ to 19 and **(B)** for the (*n*, n)-SWBNNTs, $n = 3$ to 10.

Since the atomic motion for each symmetry is in different directions, one (for instance, with symmetry of $A_g$ or $E_g$) might be enhanced or slightly shifted by a functional group connected to the nanotube by means of covalently bonding or electrostatic interactions. This enhancement or shift in the Raman active peak(s) position might provide some useful information about the structure of functionalized-nanotube, for instance, which of the side of a given functional group covalently bonded or interacts with the nanotube surface or functional group attached to the nanotube along the tube axis or perpendicular to it. This is important because the absorption of the nanotube in the perpendicular direction is much greater than absorption in the parallel direction; therefore, the availability of photons with polarization parallel to the nanotube axis is greater than that of photons with polarization perpendicular to it. It means that nanotube creates a local electric field aligned along the axis of the nanotube and facilitating the interaction of functional group with the nanotube. Even it may provide charge transfer mechanism for the functionalized-nanotube as the charge transfer from nanotube to molecule as observed between the molecule and nanoparticles by surface enhanced Raman spectroscopy (SERS).[123]

### 3.5.1 IR spectra of (n,0)- and (n,n)-SWCNTs/SWBNNTs

Figures 3.5.1A-B provide calculated IR spectra for the (*n*,0)- and (n,n)-SWCNTs, respectively, where *n* ranges from 6 to 19 for the zigzag and from 3 to 12 for the armchair nanotube. As evidenced in Figure 3.5.1A-B, there are two IR active vibrational modes, of $A_{2u}$ and $E_{1u}$ symmetries, whose frequencies are strongly diameter dependent. Least squares fits to the computed frequencies as functions of diameter for the zigzag- and armchair-SWCNTs are shown in Figure 3.5.2A. The fit to the data, for the two bands (of $A_{2u}$ and $E_{1u}$ symmetries) are respectively given by Eqs. 3.5.1a-b for the zigzag- and by the Eqs. 3.5.1c-d for the armchair-SWCNTs. The principal motions these bands correspond to, have been determined by the visualization software mentioned earlier. For the $A_{2u}$ band, the wagging of the SWCNT is along its radial direction, where the motion of the two end groups move in opposite directions. For the $E_{1u}$ band, the principal motion involves wagging of the SWCNT along its circumference direction. Our short hand notation for these two principal vibrations are the subscripts shown on the frequencies, where RW and CW, respectively, specify radial and circumference wagging.

$$\omega_{RW}(A_{2u}; (n,0)) = -1.85 + \frac{209.38(nm.cm^{-1})}{d_t(nm)} - \frac{3.71(nm^2.cm^{-1})}{[d_t(nm)]^2} \quad (3.5.1a)$$

$$\omega_{CW}(E_{(1u)}; (n,0)) = 7.33 + \frac{322.62(nm.cm^{-1})}{d_t(nm)} - \frac{8.46(nm^2.cm^{-1})}{[d_t(nm)]^2} \quad (3.5.1b)$$

$$\omega_{RW}(A_{2u}; (n,n)) = 265.42 + \frac{9.48(nm.cm^{-1})}{d_t(nm)} + \frac{35.46(nm^2.cm^{-1})}{[d_t(nm)]^2} \quad (3.5.1c)$$

$$\omega_{CW}(E_{(1u)}; (n,n)) = 367.66 + \frac{39.55(nm.cm^{-1})}{d_t(nm)} + \frac{5.51(nm^2.cm^{-1})}{[d_t(nm)]^2} \quad (3.5.1d)$$

It is to be noted (for completeness sake), that in the high frequency vibrational region of Figure 3.5.1A-B, as revealed upon close inspection, there are three IR vibrational mode of frequencies ~1520 $cm^{-1}$ ($E_{1u}$ symmetry), ~1365 $cm^{-1}$ ($E_{1u}$) and ~1332 $cm^{-1}$ ($A_{2u}$) whose positions show weak dependence on diameter; additionally, in the mid frequency region, there is one IR-mode of frequency ~780 $cm^{-1}$($A_{2u}$) that is also weakly diameter dependent. Hence, the strong diameter dependence of the two low frequency bands may be useful for determining structural indices of SWCNT samples. A more detailed description of the assignments for bands in the IR spectra for the various (n,0)-SWCNTs is contained in Table 3.5.1.

Figure 3.5.3A-B provide the calculated IR spectra of the zigzag-(n,0) and armchair-(n,n) single-wall boron nitride nanotube (SWBNNTs) exhibits similar band structure to their corresponding SWCNTs with the same chiral index (n,0) and (n,n), when comparing the calculated IR spectra of the (n,0)- and (n,n)-SWBNNTs with the (n,0)- and (n,n)-SWCNTs (see Figure 3.5.1A-B). Least squares fits to the computed frequencies as functions of diameter for the zigzag- and armchair-SWBNNTs are shown in Figure 3.5.2B. The fits to the data, for the two bands ($\omega_{RW}$ and $\omega_{CW}$) are respectively given in Eqs. 3.5.2a-b for the (n,0)- and given in the Eqs. 3.5.2c-d for the (n,n)-SWBNNTs.

$$\omega_{RW}(A_{2u};(n,0)) = -12.40 + \frac{239.27(nm.cm^{-1})}{d_t(nm)} - \frac{23.64(nm^2.cm^{-1})}{[d_t(nm)]^2} \quad (3.4.2a)$$

$$\omega_{CW}(E_{(1u)};(n,0)) = -38.67 + \frac{397.76(nm.cm^{-1})}{d_t(nm)} - \frac{81.00(nm^2.cm^{-1})}{[d_t(nm)]^2} \quad (3.4.2b)$$

$$\omega_{RW}(A_{2u};(n,n)) = 104.50 + \frac{167.86(nm.cm^{-1})}{d_t(nm)} - \frac{12.14(nm^2.cm^{-1})}{[d_t(nm)]^2} \quad (3.4.2c)$$

$$\omega_{CW}(E_{(1u)};(n,n)) = 298.17 - \frac{48.00(nm.cm^{-1})}{d_t(nm)} + \frac{62.56(nm^2.cm^{-1})}{[d_t(nm)]^2} \quad (3.4.2d)$$

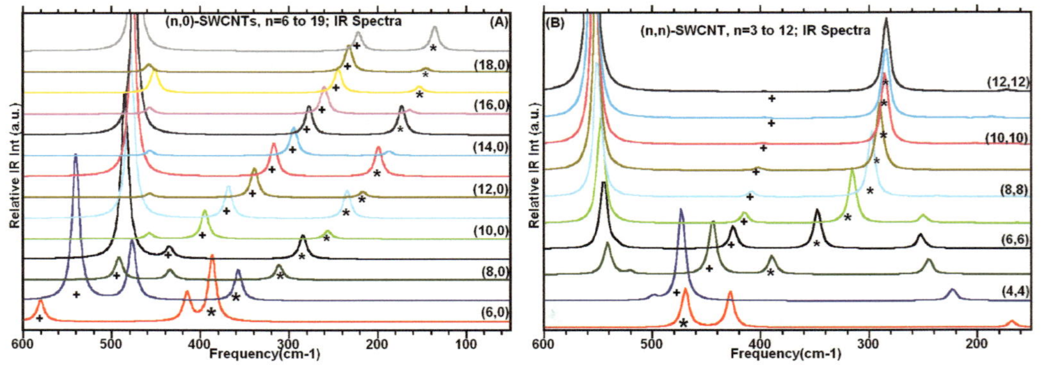


Fig. 3.5.1 Calculated IR spectra (A) for the (n,0)-SWCNTs, with n varying from 0 to 19; (B) for the (n,n)-SWCNTs with n = 3 to 12. Where signs * and + stand for the IR active modes of frequencies; $\omega_{RW}(A_{2u})$ and $\omega_{RW}(A_{2u}$, respectively.

Furthermore, for DWCNTs (not shown here), the calculated IR bands with $A_{2u}$ symmetries at 387 and 216 cm$^{-1}$ for (0,6)- and (0,12)-SWCNTs are blue-shifted to 395 and 253 cm$^{-1}$ in IR spectrum of (0,6)&(0,12)-DWCNTs, respectively. The IR bands with $A_{2u}$ symmetries at 358 and 187 cm$^{-1}$ in spectra of (0,7)- and (0,14)-SWCNTs correspond to the bands at 411 and 247 cm$^{-1}$ in spectrum of (0,7)&(0,14)-DWCNT. Furthermore, the IR bands with $E_{1u}$ symmetries at 580 and 339 cm$^{-1}$ for (0,6)- and (0,12)-SWCNTs are also blue-shifted to 617 and 350 cm$^{-1}$ in the spectrum of (0,6)&(0,12)-DWCNT, respectively. The bands (with $E_{1u}$ symmetries) at 541 and 295 cm$^{-1}$ in spectra of (0,7)- and (0,14)-SWCNTs correspond to the bands at 567 and 300 cm$^{-1}$ in the spectrum of (0,7)&(0,14)-DWCNT, respectively.

### 3.6.1 Calculated vertical transitions of the SWCNTs and SWBNNTs

As mentioned earlier in Section 2, the chiral index (n,m) of the nanotube indicates whether the tube is metallic or semiconducting. When the (n-m) = 3q (q = 0, 1,2 ,3 ...), all the tubes are metallic, if not , semiconducting. Therefore, all armchair tubes are metallic. The calculated density of states (DOS) of the nanotube using tight binding approximation (TBA) does not comprise one smooth band, but shows a number of side-bands with spikes around the Fermi level separated from each other [92,124], which are called the van Hove singularities. The side-bands represent 1D channel for conduction along the nanotube. The side-bands and their sharp onsets have been observed in their tunneling DOS obtained from scanning tunneling spectroscopic measurements [125].

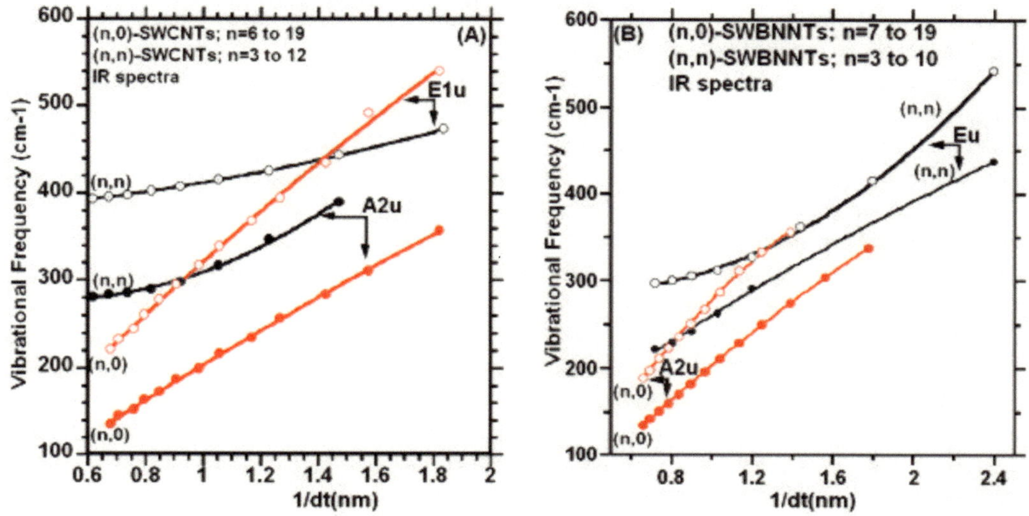


Fig. 3.5.2 The plots of the dependence of calculated IR frequencies, in the low frequency region, as a function of nanotube diameter for vibrational modes of symmetries $A_{2u}$ and $E_{1u}$; (A) for the SWCNTs and (B) for the SWBNNTs.

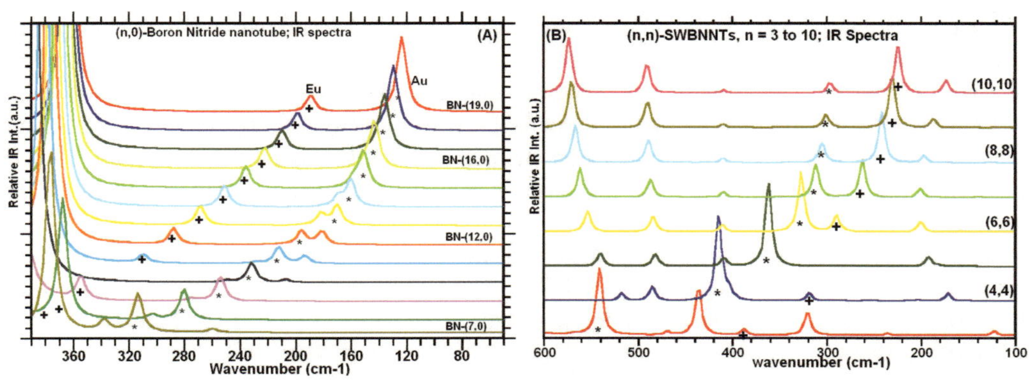


Fig. 3.5.3 (A) Calculated IR spectra of the (n,0)-SWNTs, with n varying from 7 to 19; (B) illustrate for the (n,n)-SWBNNTs with n = 3 to 10.

For larger-diameter tubes, the singularities move close together and combine, thus making the DOS similar to that of a graphene layer. In contrast, for smaller-diameter nanotubes, the singularities are well separated specially near the Fermi level and can give initial and final states for a resonance Raman scattering (RRS) process. In a simplified representation, neglecting all the curvature effects and considering the linear dispersion of $\pi$ and $\pi^*$ bands of graphene only, the energy gap between q[th] van Hove singularity in valence and conduction bands for metallic tubes[126] is $E_{qq}^{M} = \frac{6q\gamma a_{CC}}{d_t} = 3E_{qq}^{S}\ (q = 1,2.3, \ldots\ )$ and $E_{qq}^{S} = \frac{2q\gamma a_{CC}}{d_t}$ for semiconducting tubes as discussed in Section 2. In the case of metallic tubes, the linear crossing of π and π* bands near the Fermi level yields a small constant DOS leading to sharp van Hove singularities at energies away from the Fermi energy. In contrast, for semiconducting tubes, the DOS at Fermi energy is zero with sharp van Hove singularities away from the Fermi energy [92,127].

The electronic structures of the SWBNNTs were theoretically investigated by Rubio *et al.* [56] using the tight binding approximation (TBA). All SWBNNTs were found to be semiconducting materials with band gaps larger than 2 eV. BNNTs with larger diameters have a larger band gap, with a saturation value corresponding to the band gap of a hexagonal boron nitride (BN) as calculated by Rubio *et al.*[56] or see the Figure 4 in Ref.[57]. The subsequent studies based on local-density-functional (LDA) calculations proved that it is energetically more favorable to fold a hexagonal boron nitride sheet into a BNNT than to forma CNT from a graphite sheet [58]. This observation is also supported by our calculations at B3LYP/6-31G level of DFT as discussed in Section 3. Based on the band-folding analysis, SWBNNTs are semiconductors with large-gap for the zigzag- and armchair-SWBNNTs. This is due to the strong hybridization effects may take place because tube curving reduces the BNNT's band gap remarkably. For the (n, 0) BNNTs, with n > 12 the hybrid state was found not to play any role in determining tube gaps. The gaps are steady at around 4 eV as indicated by LDA. In another calculation using density functional theory, the band gaps of BNNTs were eventually saturated at 5.03 eV with an increase in diameter [128,129,130,131,132]. Based our calculated dipole allowed vertical electronic transitions, for the (n, 0)-BNNTs with n > 11, the gabs are steady around 5.9 eV, see Figure 3.6.1B and Table 3.6.4. For the (n,n)-SWBNNTs, with n>10, the band gaps slightly increases with increasing tube diameter and seem to be steady at around 5.84 eV according to our DFT calculations as seen Figure 3.6.1B and Table 3.6.4. It should be noted that there are many forbidden electronic transition lie below the allowed one. Based on our experience on the calculated electronic transitions for many organic compounds at B3LYP level and at the BLYP level of DFT for the transition metals [95], the calculated value of the electronic transitions are somewhat higher than their experimental values, as much as 0.4 eV. Therefore, the calculated values of the electronic band gabs for the nanotubes might be reasonable.

Various methods have been developed to tune the electronic structure of BNNTs is nicely discussed by Chunyi Zhi *et al.*[57], for instance, applying electric field [133,134,135,136,137,138] or strain [139], or chemical methods, like doping [140,141,142,67], introducing defects [143,144,145] or surface modification [60,146,147]. The methods and species chosen are summarized in Table 2 in Ref. 57. Physical methods could directly reduce the band gap of BNNTs, while chemical methods were found to tune the band gap by introducing localized energy levels inside the gap [57]. For example, by applying an 2 V/nm electric field, the band gap of a (12, 12) BNNT was directly reduced from 4.5 eV to about 2.5 eV [135], while F-doping induced an unoccupied localized state in the gap of BNNTs [148]. In addition, chemical modification methods of BNNTs were sometimes found to enrich the properties of BNNTs, for instance, polarization field was induced by chemical adsorption [149] or ferromagnetism appeared in doped BNNTs.

For electronic structure of multi-walled BNNTs (MWBNNTs), it has been reported that there is hybridization between p and s states of inner and outer tubes [150,151]. This effect leads to the top valence and bottom conduction bands localization on the outer and inner tubes, respectively [57]. The band gap of DWBNNTs is slightly narrower than that of the inner tubes. This is because the downward shifts of the p states of the inner tube are larger than those of the outer tube. Furthermore, in the interwall region the peculiar charge redistribution is induced by the near-free-electron states of BNNTs [57,152]. The fluorine doping can significantly modify the DWBNNTs' inter-wall interactions and thus both tube walls turn into effective conducting channels. This leads to a remarkably improved electrical transport in them [151].

There are strong interactions between electrons and holes in BNNTs [54,55]. The excitonic effects were indeed shown to be more important in BNNTs than in CNTs. Bright and dark excitons in BNNTs alter the optical response qualitatively. For example, the absorption spectrum of the (8, 0) BNNT is dominated by a peak at 5.72 eV, due to an exciton with a binding energy of 2.3 eV. The binding energy for the first excitonic peak is more than 3 eV for the (2, 2) nanotube, which will fast converge to around 2.1 eV of the hexagonal single BN sheet as a result of strongly localized nature of this exciton [57], which is consistent with the results by the Louie's group [54]: a 2.3 eV binding energy for a (8, 0) BNNT.

In this section, we provided the calculated vertical electronic transitions of the (n,0)- and (n,n)-type nanotubes using time-dependent DFT methods at the TD-B3LYP/6-31G level. Table 3.6.1 and Table 3.6.2 provide the calculated electronic transitions from (7,0) to (11,0) and (14,0) and from (3,3) to (10,10) for the SWCNTs, respectively. The transitions from the (7,0) to (14,0) and from (3,3) to (10,10) for the zigzag- and armchair-SWBNNTs are given in Table 3.6.3 and Table 3.6.4, respectively.

The diameter dependence of the calculated dipole allowed electronic transitions are shown in Figure 3.6.1A-B. As seen in Figure 3.61A, the allowed electronic transitions lie in the range from 0.5 eV to 3.0 eV for the (n,0)-SWCNTs, with n = 7 to 14, and lie in the range from 2.8 eV to 4.4 eV for the (n,n)-SWCNTs, with n=3 to 10. Furthermore, the plot of the calculated dipole allowed electronic transitions showed that while the allowed transitions rapidly decrease and stabilize (around 0.5 eV) with increasing the diameter of the (n,0)-SWCNTs, for the armchair SWCNTs, the electronic transitions are diameter dependent, but also these transitions converge and become constant (around 3 eV) with increasing tube diameters of the (n,n)-SWCNTs. For the SWBNNTs, the result of the calculations indicated that the dipole allowed electronic transitions are not only diameter dependent, but also the electronic transitions lie in the range from 5.7 eV to 6.6 eV as seen in Figure 6.3.1B, which is higher than the calculated electronic transitions for the SWCNTs. This large band gap in SWBNNTs relative to SWCNTs is not so surprising since the B-N bond contains a significant ionic component. This polarity can extremely alter both molecular and solid-state electronics as well as optical properties of the system by modifying the character of the frontier molecular orbitals [57]. Furthermore, the calculated band gap of BNNTs between 5.7 and 6.6 eV is independent to tube chirality. This provides good electrical insulation, in spite of the fact that CNTs can be a metal or a narrow band-gap semiconductor. This discrepancy in electronic structure results in different luminescence emission: SWBNNTs have violet or ultraviolet luminescence under excitation by electrons or photons, while SWCNTs can emit infrared light and the wavelengths depend on their chiralities.

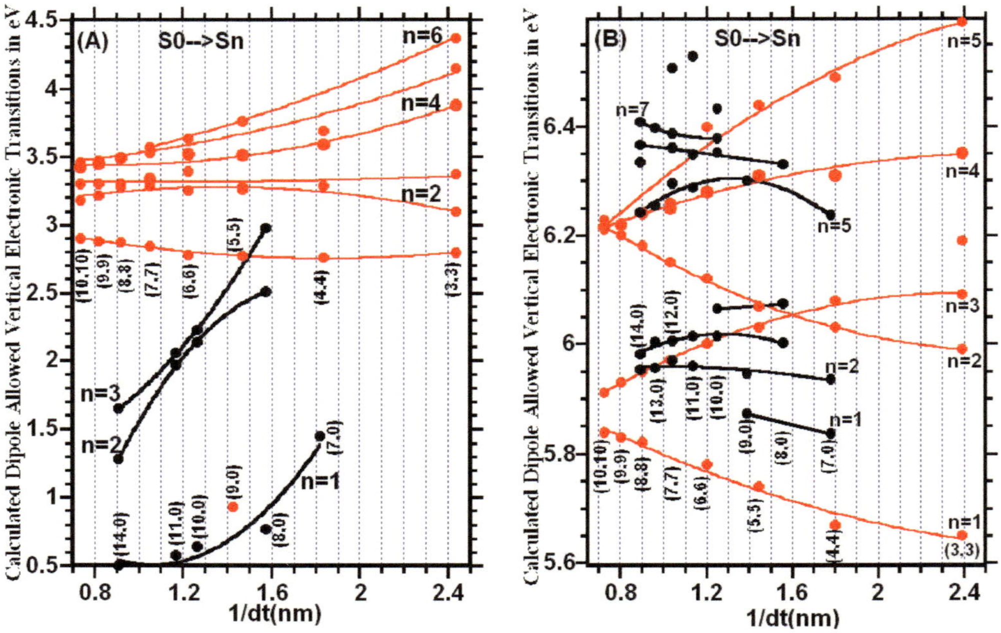


Fig. 3.6.1. Diameter dependence of calculated dipole allowed vertical electronic transitions of: a) SWCNTs and b) SWBNNTs at B3LYP/6-31G level of DFT.

| | (7,0) | | | (8,0) | | | (9,0) | | | (10,0) | | | (11,0) | | | (14,0) | | |
|---|---|---|---|---|---|---|---|---|---|---|---|---|---|---|---|---|---|---|
| **m** | **TB** | **DFT** | **f** | **TB** | **DFT** | **f** | **TB** | **DFT** | **f** | **TB** | **DFT** | **f** | **TB** | **DFT** | **f** | **TB** | **DFT** | **f** |
| **1** | | 0.91 | | 1.04 | 0.77 | 0.02 | | 0.56 | | 1.10 | 0.64 | 0.01 | | 0.36 | | 0.758 | 0.50 | 0.0042 |
| **2** | | 1.25 | | | 1.46 | | | 0.80 | | | 1.09 | | | 0.51 | | | 0.63 | |
| **3** | | 1.25 | | | 1.46 | | | 0.80 | | | 1.09 | | | 0.52 | | | 0.63 | |
| **4** | 1.42 | 1.45 | 0.06 | | 2.44 | | | 0.93 | 0.04 | | 1.77 | | 0.98 | 0.58 | 0.02 | | 0.89 | |
| **5** | | 2.52 | | 2.40 | 2.51 | 0.36 | | 2.32 | | | 2.04 | | | 1.80 | | | 1.10 | |
| **6** | | 2.53 | | | 2.51 | 0.36 | | 2.32 | | | 2.04 | | | 1.80 | | | 1.10 | |
| **7** | | 2.93 | | | 2.53 | | | 2.50 | | 2.20 | 2.14 | 0.50 | 1.78 | 1.97 | 0.50 | | 1.29 | 0.0872 |
| **8** | | 2.93 | | | 2.00 | | | 2.50 | | | 2.14 | 0.50 | | 1.97 | 0.50 | 1.436 | 1.65 | 0.7917 |
| **9** | | 3.02 | | | 2.78 | | | 2.64 | | | 2.23 | 0.12 | | 2.06 | 0.12 | | 1.65 | 0.7925 |
| **10** | | 3.02 | | | 2.78 | | | 2.64 | | | 2.53 | | | 2.06 | | | 1.95 | |
| **11** | | 3.08 | | | 2.77 | | | 2.72 | | | 2.75 | | | 2.14 | | | 1.96 | |
| **12** | | 3.08 | | | 2.98 | 0.13 | | 2.72 | | | 2.75 | | | 2.14 | | | 2.20 | |
| **dt** | | 0.56 | | | 0.64 | | | 0.71 | | | 0.79 | | | 0.87 | | | 1.10 | |

Table 3.6.1. DFT-calculated vertical singlet–singlet transitions ($\mathbf{S_0} \rightarrow \mathbf{S_m}$ in eV) of the (n,0)-SWCNTs, n = 7 to 11, at the B3LYP/6-31G level. Oscillator strengths (f) and results of the calculated energy from the tight-binding (TB) approximation for each nanotube are provided for comparison. The value of $\gamma$ was taken to be 2.7 eV in the TB calculations. The $d_t$ stands for diameter in nm.

| | (3,3) | | (4,4) | | (5,5) | | (6,6) | | (7,7) | | (8,8) | | (9,9) | | (10,10) | |
|---|---|---|---|---|---|---|---|---|---|---|---|---|---|---|---|---|
| **m** | **DFT** | **f** | **DFT** | **f** | **DFT** | **f** | **DFT** | **f** | **DFT** | **f** | **DFT** | **f** | **DFT** | **f** | **DFT** | **f** |
| **1** | 1.35 | | 1.84 | | 2.15 | | 2.36 | | 2.58 | | 2.69 | | 2.77 | | 2.83 | |
| **2** | 2.40 | | 2.76 | 0.0033 | 2.77 | 0.0072 | 2.78 | 0.0106 | 2.84 | 0.0108 | 2.87 | 0.0132 | 2.88 | 0.0153 | 2.90 | 0.0173 |
| **3** | 2.40 | | 2.77 | | 2.89 | | 2.93 | | 2.99 | | 2.99 | | 2.98 | | 2.98 | |
| **4** | 2.79 | 0.0003 | 2.77 | | 2.90 | | 2.93 | | 2.99 | | 2.99 | | 2.98 | | 2.98 | |
| **5** | 3.08 | | 3.22 | | 3.26 | 0.1197 | 3.25 | 0.1118 | 3.25 | | 3.18 | | 3.14 | | 3.11 | |
| **6** | 3.08 | | 3.22 | | 3.26 | | 3.29 | | 3.29 | 0.1042 | 3.28 | 0.0036 | 3.22 | 0.0081 | 3.18 | 0.0197 |
| **7** | 3.10 | 0.0082 | 3.25 | | 3.27 | | 3.29 | | 3.34 | 0.0012 | 3.28 | 0.0036 | 3.22 | 0.0081 | 3.18 | 0.0197 |
| **8** | 3.10 | 0.0082 | 3.25 | | 3.28 | 0.0007 | 3.29 | | 3.34 | 0.0012 | 3.30 | 0.0963 | 3.25 | | 3.20 | |
| **9** | 3.37 | 0.1209 | 3.29 | 0.1281 | 3.29 | | 3.29 | | 3.34 | | 3.30 | | 3.25 | | 3.20 | |
| **10** | 3.82 | | 3.59 | 0.0040 | 3.50 | | 3.34 | | 3.34 | | 3.30 | | 3.30 | 0.0878 | 3.30 | 0.0779 |
| **11** | 3.88 | 0.0001 | 3.59 | 0.0040 | 3.51 | 0.0001 | 3.39 | 0.0002 | 3.35 | | 3.35 | | 3.35 | | 3.34 | |
| **12** | 3.88 | 0.0001 | 3.69 | 0.0003 | 3.51 | 0.0005 | 3.39 | 0.0002 | 3.35 | | 3.35 | | 3.35 | | 3.34 | |
| **13** | 3.94 | | 3.69 | 0.0003 | 3.58 | | 3.52 | 0.0017 | 3.53 | 0.0099 | 3.49 | 0.0294 | 3.45 | 0.0328 | 3.39 | |
| **14** | 3.95 | | 3.76 | | 3.58 | | 3.52 | 0.0017 | 3.53 | 0.0099 | 3.49 | 0.0294 | 3.45 | 0.0328 | 3.39 | |
| **15** | 3.95 | | 3.85 | | 3.74 | | 3.63 | | 3.57 | 0.0132 | 3.51 | | 3.45 | | 3.42 | 0.0359 |
| **16** | 3.99 | | 3.96 | 0.0954 | 3.76 | 0.0606 | 3.63 | 0.0450 | 3.57 | 0.0132 | 3.51 | | 3.45 | | 3.42 | 0.0359 |
| **17** | 3.99 | | 3.96 | 0.0954 | 3.77 | 0.0658 | 3.63 | 0.0450 | 3.58 | | 3.51 | | 3.45 | | 3.42 | |
| **18** | 4.15 | 0.1277 | 3.99 | | 3.82 | | 3.68 | | 3.64 | | 3.54 | | 3.47 | 0.0164 | 3.46 | 0.1043 |
| **19** | 4.15 | 0.1275 | 3.99 | | 3.82 | | 3.68 | | 3.64 | | 3.54 | | 3.47 | 0.0165 | 3.46 | 0.1043 |
| **20** | 4.37 | 0.0009 | 3.99 | | 3.85 | | 3.72 | | 3.66 | | 3.58 | | 3.53 | | 3.48 | |
| **dt** | 0.41 | | 0.55 | | 0.68 | | 0.81 | | 0.95 | | 1.09 | | 1.22 | | 1.36 | |

Table 3.6.2. DFT-calculated vertical singlet–singlet transitions ($\mathbf{S_0} \rightarrow \mathbf{S_m}$ in eV) and oscillator strengths (f) of the (n,n)-SWCNTs, n = 3–10, at the B3LYP/6-31G level. The $d_t$ stands for diameter in nm.

Moreover, the properties of triplet states in the nanotubes may play a circular role in the number of fundamental physics phenomena such as singlet-triplet splitting in low-dimensional materials is a degree of electronic correlation strength and exchange effects [153], like that of exciton binding energy. Also, how relaxation of photo-excitations takes place in CNTs is not well known as a result of being weakly-emissive materials [154,155], this properties are called as the 'dark' singlet excitons below the optically allowed states [156].

The calculated singlet-triplet electronic energy levels of the (7,0)-SWCNT is given in Figure 3.6.2A. As seen in the figure, for the (7,0)-SWCNT, while the calculations exhibited only one dipole allowed electronic transition ($S_0 \rightarrow S_4$) below about 3 eV; however, there are many forbidden or very weak electronic transitions (oscillator strength less than 0.0001) in this energy region. The distance between these electronic transitions are ~0.47 eV between $S_1$ and $S_4$ electronic energy levels and ~0.15 eV between $S_2$($\sim S_3$) and $S_4$ states. The calculations indicated similar situations for the (8,0)-SWCNT. The distance between the energy levels of the first excited singlet state ($S_1$) and second triplet state ($T_2$) is about 0.05 eV. As consequence, these results may imply that there would be an intersystem crossing (IC) from $S_4$ to $S_2$ and

$S_1$ electronic energy levels, followed by an intersystem crossing (ISC) process to the second the lowest triplet energy level ($T_2$). This process may lead to quenching the florescence by nonradiative decay.

| | (3,3) | | (4,4) | | (5,5) | | (6,6) | | (7,7) | | (8,8) | | (9,9) | | (10,10) | |
|---|---|---|---|---|---|---|---|---|---|---|---|---|---|---|---|---|
| **m** | **DFT** | **f** | **DFT** | **f** | **DFT** | **f** | **DFT** | **f** | **DFT** | **f** | **DFT** | **f** | **DFT** | **f** | **DFT** | **f** |
| **1** | 5.28 | | 5.45 | | 5.61 | | 5.71 | | 5.75 | | 5.76 | | 5.77 | | 5.76 | |
| **2** | 5.50 | | 5.67 | 0.0107 | 5.74 | 0.0029 | 5.78 | 0.0005 | 5.80 | | 5.82 | 0.0001 | 5.83 | 0.0003 | 5.84 | 0.0006 |
| **3** | 5.50 | | 5.68 | | 5.79 | | 5.84 | | 5.86 | | 5.87 | | 5.87 | | 5.87 | |
| **4** | 5.65 | 0.0174 | 5.68 | | 5.79 | | 5.84 | | 5.86 | | 5.87 | | 5.87 | | 5.87 | |
| **5** | 5.99 | 0.0127 | 5.94 | | 5.94 | | 5.93 | | 5.94 | | 5.95 | 0.5283 | 5.93 | 0.7425 | 5.91 | 0.9893 |
| **6** | 5.99 | 0.0127 | 5.94 | | 5.99 | | 6.00 | 0.2125 | 5.97 | 0.3482 | 5.95 | 0.5284 | 5.93 | 0.7424 | 5.91 | 0.9893 |
| **7** | 6.09 | 0.0802 | 6.00 | | 5.99 | | 6.00 | 0.2125 | 5.97 | 0.3512 | 5.97 | | 5.99 | | 5.97 | |
| **8** | 6.12 | | 6.03 | 0.1050 | 6.03 | 0.1173 | 6.01 | | 6.01 | | 6.00 | | 5.99 | | 5.97 | |
| **9** | 6.12 | | 6.08 | 0.0532 | 6.03 | 0.1171 | 6.01 | | 6.01 | | 6.00 | | 6.00 | | 6.00 | |
| **10** | 6.13 | | 6.08 | 0.0532 | 6.07 | 0.1338 | 6.12 | 0.1620 | 6.08 | | 6.05 | | 6.02 | | 6.00 | |
| **11** | 6.19 | 0.0418 | 6.13 | | 6.14 | | 6.12 | | 6.08 | | 6.05 | | 6.02 | | 6.02 | |
| **12** | 6.19 | 0.0418 | 6.13 | | 6.14 | | 6.12 | | 6.15 | 0.1907 | 6.18 | 0.2230 | 6.14 | | 6.10 | |
| **13** | 6.35 | 0.0421 | 6.30 | | 6.17 | | 6.16 | | 6.18 | | 6.18 | | 6.14 | | 6.10 | |
| **14** | 6.35 | 0.0421 | 6.30 | | 6.17 | | 6.16 | | 6.18 | | 6.20 | | 6.15 | | 6.12 | |
| **15** | 6.37 | | 6.31 | 0.1307 | 6.31 | 0.2425 | 6.28 | 0.4262 | 6.24 | | 6.20 | | 6.15 | | 6.12 | |
| **16** | 6.37 | | 6.31 | 0.1307 | 6.31 | 0.2423 | 6.28 | 0.4262 | 6.24 | | 6.20 | | 6.20 | 0.2572 | 6.21 | 1.3747 |
| **17** | 6.39 | | 6.39 | | 6.35 | | 6.33 | | 6.24 | | 6.20 | | 6.22 | | 6.21 | 1.3750 |
| **18** | 6.53 | | 6.45 | | 6.42 | | 6.40 | 0.2929 | 6.25 | 0.6546 | 6.24 | 0.8986 | 6.22 | | 6.21 | 0.2906 |
| **19** | 6.59 | 0.0100 | 6.49 | 0.1349 | 6.44 | 0.2154 | 6.40 | 0.2928 | 6.26 | 0.6538 | 6.24 | 0.8979 | 6.22 | 1.1396 | 6.23 | |
| **20** | 6.59 | 0.0100 | 6.49 | 0.1349 | 6.44 | 0.2173 | 6.41 | | 6.32 | | 6.31 | | 6.22 | 1.1396 | 6.23 | 0.0005 |
| **$d_t$** | **0.42** | | **0.56** | | **0.69** | | **0.83** | | **0.97** | | **1.11** | | **1.25** | | **1.38** | |

Table 3.6.3. DFT-calculated vertical singlet–singlet transitions ($\mathbf{S_0} \rightarrow \mathbf{S_m}$ in eV) and oscillator strengths (f) of the (n,n)-SWBNNTs, n = 3–10, at the B3LYP/6-31G level. The $d_t$ stands for the diameter in nm.

For the (8,0)-SWCNT, the calculations indicated that there are two dipole allowed electronic transitions below 3 eV. Dipole allowed second electronic state ($S_2$) lies about 0.03 eV above the second triplet excited state ($T_2$), see Figure 3.6.2A. If an ISC process between the $S_2$ and $T_2$ electronic states take place, then, there would be a $T_2 \rightarrow T_1$ transition followed by a triplet-singlet electronic transition $T_1 \longrightarrow S_0$, see Appendix in Section 4 for detailed discussions for the singlet-triplet transitions. As a consequence, as mentioned in Ref. 153 that the triplet states play any crucial role for the nonradiative decay of photo-excitations and this property is called the "dark" singlet excitons below the optically allowed states [153,156].
Similarly, the calculated vertical transitions for the (8,0)- and (4,4)-SWBNNTs, up to ~6 eV, indicated that there are many forbidden or very weak electronic transitions (oscillator strength (f) less than 0.0001) that lie below the first dipole allowed singlet-singlet electronic transition. The first dipole allowed singlet-singlet electronic transition has almost the same energy with the spin forbidden singlet-triplet electronic transitions in addition to many dipole forbidden singlet-singlet transitions below the first dipole allowed electronic transitions as seen in Figure 3.6.2B. Thus, these calculated findings may suggest that there

would be a IC and/or ISC processes for the (4,4)- and (8,0)-SWBNNTs, which might lead to a nonradiative decay.

| | (7,0) | | (8,0) | | (9,0) | | (10,0) | | (11,0) | | (12,0) | | (13,0) | | (14,0) | |
|---|---|---|---|---|---|---|---|---|---|---|---|---|---|---|---|---|
| **m** | **DFT** | **f** | **DFT** | **f** | **DFT** | **f** | **DFT** | **f** | **DFT** | **f** | **DFT** | **f** | **DFT** | **f** | **DFT** | **f** |
| **1** | 5.05 | | 5.61 | | 5.77 | | 5.87 | | 5.82 | | 5.83 | | 5.84 | | 5.82 | |
| **2** | 5.05 | | 5.61 | | 5.86 | | 6.01 | | 5.92 | | 5.95 | | 5.95 | | 5.92 | |
| **3** | 5.67 | | 5.69 | | 5.86 | | 6.01 | | 5.92 | | 5.95 | | 5.95 | | 5.92 | |
| **4** | 5.67 | | 5.80 | | 5.87 | 0.0215 | 6.01 | 0.0481 | 5.95 | 0.1456 | 5.97 | 0.0871 | 5.96 | 0.1029 | 5.95 | 0.1456 |
| **5** | 5.68 | | 5.87 | | 5.87 | 0.0215 | 6.01 | 0.0481 | 5.95 | 0.1456 | 5.97 | 0.0871 | 5.96 | 0.1029 | 5.95 | 0.1456 |
| **6** | 5.68 | | 6.00 | 0.0255 | 5.95 | 0.0135 | 6.07 | 0.0251 | 5.98 | 0.0233 | 6.01 | 0.0239 | 6.00 | 0.0186 | 5.98 | 0.0233 |
| **7** | 5.73 | | 6.00 | 0.0255 | 6.06 | | 6.24 | | 6.07 | | 6.19 | | 6.11 | | 6.07 | |
| **8** | 5.84 | 0.0095 | 6.02 | | 6.06 | | 6.24 | | 6.07 | | 6.19 | | 6.11 | | 6.07 | |
| **9** | 5.84 | 0.0094 | 6.02 | | 6.06 | | 6.34 | | 6.24 | 1.4646 | 6.30 | 0.8777 | 6.26 | 0.0094 | 6.24 | 1.4646 |
| **10** | 5.87 | | 6.04 | | 6.06 | | 6.34 | | 6.24 | 1.4646 | 6.30 | 0.8777 | 6.26 | 0.0094 | 6.24 | 1.4646 |
| **11** | 5.87 | | 6.07 | 0.0164 | 6.09 | | 6.35 | | 6.33 | 0.0700 | 6.36 | 0.0256 | 6.26 | 1.3216 | 6.33 | 0.0700 |
| **12** | 5.90 | | 6.11 | | 6.09 | | 6.35 | | 6.33 | 0.0700 | 6.39 | 0.0168 | 6.26 | 1.3217 | 6.33 | 0.0700 |
| **13** | 5.90 | | 6.11 | | 6.16 | | 6.35 | 0.4783 | 6.37 | 0.0226 | 6.39 | 0.0168 | 6.36 | | 6.37 | 0.0226 |
| **14** | 5.93 | 0.0180 | 6.14 | | 6.16 | | 6.35 | 0.4784 | 6.40 | | 6.43 | | 6.37 | | 6.40 | |
| **15** | 6.04 | | 6.14 | | 6.28 | | 6.38 | 0.0317 | 6.40 | | 6.51 | 0.5031 | 6.37 | | 6.40 | |
| **16** | 6.04 | | 6.16 | | 6.28 | | 6.43 | | 6.41 | 0.2283 | 6.51 | 0.5031 | 6.40 | 0.0557 | 6.41 | 0.2283 |
| **17** | 6.09 | | 6.16 | | 6.30 | 0.0918 | 6.43 | 0.1102 | 6.41 | 0.2283 | 6.52 | | 6.45 | | 6.41 | 0.2283 |
| **18** | 6.09 | | 6.17 | | 6.30 | 0.0918 | 6.43 | 0.1102 | 6.43 | | 6.52 | | 6.45 | | 6.43 | |
| **19** | 6.24 | 0.1460 | 6.18 | | 6.35 | | 6.50 | | 6.48 | | 6.59 | 0.0012 | 6.50 | | 6.48 | |
| **20** | 6.24 | 0.1468 | 6.33 | 0.2616 | 6.35 | | 6.62 | | 6.48 | | 6.59 | 0.0012 | 6.60 | | 6.48 | |
| $d_t$ | **0.56** | | **0.64** | | **0.72** | | **0.80** | | **0.88** | | **0.96** | | **1.04** | | **1.12** | |

Table 3.6.4. DFT-calculated vertical singlet–singlet transitions ($\mathbf{S_0} \rightarrow \mathbf{S_m}$ in eV) and oscillator strengths (f) of the (n,0)-SWBNNTs, n = 7–14, at the B3LYP/6-31G level. The $d_t$ stands for the diameter in nm.

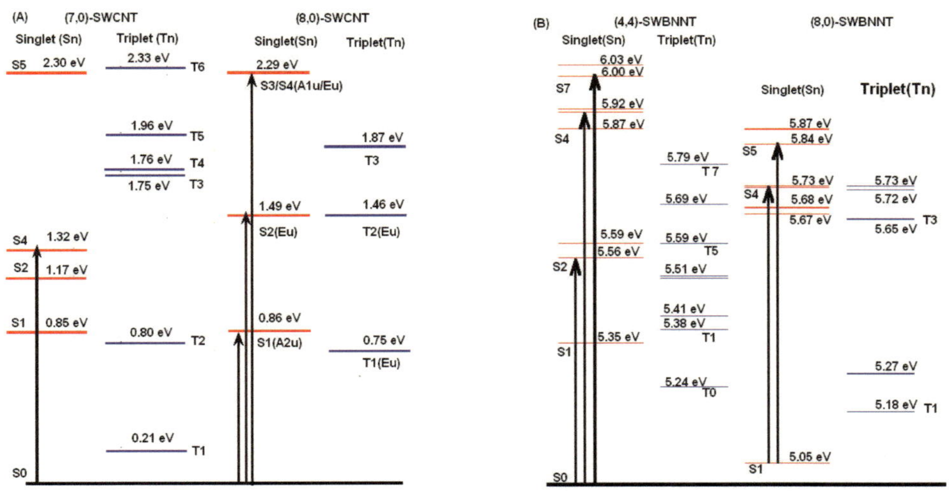


Fig. 3.6.2. Calculated singlet and triplet vertical electronic transitions: (A) for the (7,0)- and (8,0)-SWCNT and (B) for the (4,4)- and (8,0)-SWBNNTs at TD-B3LYP/6-31G(d,p) level of DFT. The vertical solid lines indicate dipole allowed vertical electronic transitions.

### 3.7 Functionalization of single-wall carbon nanotubes

Single-walled carbon nanotubes (SWNTs) demonstrate useful properties for different prospective applications counting miniature biological devices, such as used as electrodes for detecting biomolecules in solutions. Furthermore, the electrical properties of SWNTs are sensitive to surface charge transfer and changes in the surrounding electrostatic environment, undergo severe changes by adsorptions of desired molecules or polymers [157,158]. SWNTs are subsequently promising for chemical sensors for detecting molecules in the gas phase and biosensors for probing biological processes in solutions. Nevertheless, significant effort is necessary to realize interactions between nanotubes and organic molecules or biomolecules and how to impart explicitness and selectivity to nanotube-based bioelectronic devices [159].

Functionalizated carbon nanotube with inorganic and biological macromolecules like deoxyribonucleic acid (DNA) makes possible the formation of hybrid materials with interesting properties. Biological functionalization, particularly deoxyribonucleic acid (DNA) functionalization has attracted much scientific attention because of the possible development of sensitive and ultrafast detection systems for molecular electronics. As a result of the existence of a large number of delocalized $\pi$-electrons on its bases, DNA can be used as molecular wire [160]. Furthermore, the functionalization of CNTs with DNA molecules magnifies the CNT solubility in organic media and promotes application and development in DNA based nanobiotechnology. Also, the functionalization character of CNTs with DNA molecules may be used to distinguish metallic CNT from semiconducting CNTs. DNA chains have various functional structural groups available for covalent interaction with CNTs for construction of DNA-based devices through the sequence-specific pairing interactions. Functionalized carbon nanotubes (CNT) are proficient for biomedical applications [161,162]. They can be used for biosensing [163] or act as nano-heaters [164], temperature sensors [165] and drug-carrier systems for therapy and diagnosis at the cellular level [166]. Functionalization of the outer surface of CNT with biomolecules such as nucleic acids, proteins, peptides and polymers makes possible their definite internalization into the cell [167,168,169]. On the other hand, the acceptable uptake mechanism remains a contentious problem since it may depend on cell type, bio-functionalizated scheme, size of the nanotube and other factors [170,171,172,173].

Müller et al.[174] have reported that the Raman signal of functionalized carbon nanotube, specially the intensity of the radial breathing mode, suffer from its chemical functionalization. They concluded that chemical reaction appears to be diameter selective under certain reaction conditions, possibly accompanied by an effect related to the tube species. Sayes *at el.*[175] used Raman spectroscopy and thermogravimetric analysis to analysis the phenylated-SWNTs (SWNT-phenyl-$SO_3H$ and SWNT-phenyl-$SO_3Na$). They have reported that Raman spectroscopy provide a direct evidence of covalent sidewall functionalization. They mentioned that Raman spectrum of the starting purified SWNTs shows a small disorder mode (D-band) at 1290 cm−1 (in Fig. 2A of Ref.[175]). The spectra of the least, medium and most functionalized samples exhibit progressively increasing disorder modes relative to the large tangential modes (G-band) at ~1590 $cm^{-1}$.

In this section, we provided the calculated the calculated Raman and IR spectra of functionalized (7,0)-SWCNT, using the B3LYP functional with the basis sets 6-31G on carbon and hydrogen atoms and 6-311G(d,p) basis set used for oxygen and nitrogen atoms. All functional groups used here (-phenyl-$SO_3H$, carboxy (-$CO_2H$), and 3-methoxy-6-methyl-4-[(2-nitro-4-methylphenyl) azo] benzene) are covalently bonded to the (7,0)-SWCNT. The calculated Raman spectrum of the functionalized carbon nanotube is given in Figure 3.7.1,

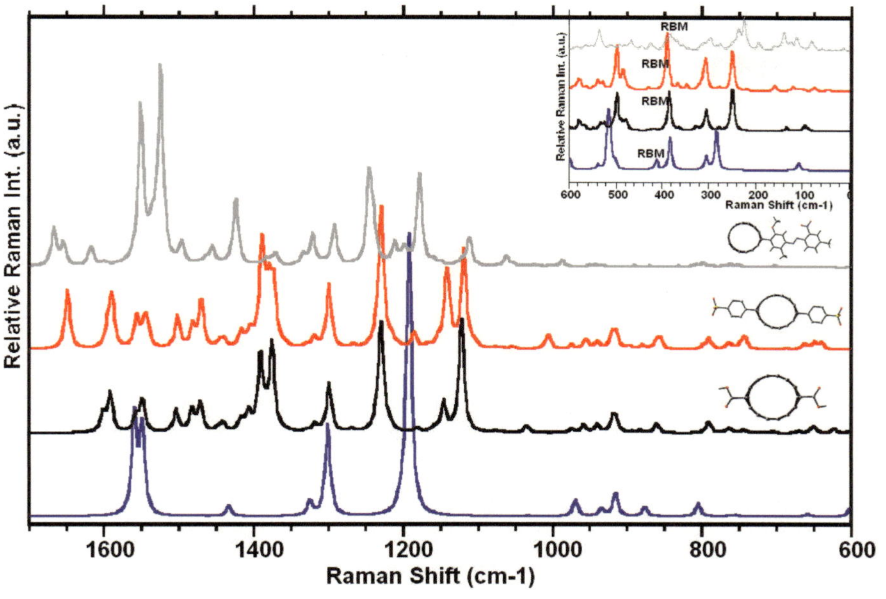


Fig. 3.7.1. Calculated Raman spectra of functionalized and isolated (7,0)-SWCNT.

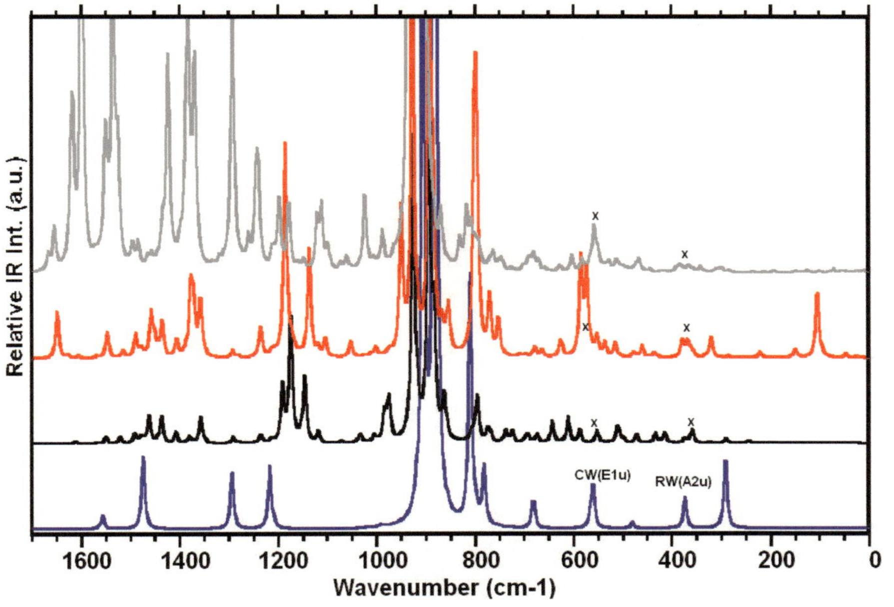


Fig. 3.7.2. Calculated IR spectra of functionalized and isolated (7,0)-SWCNT.

with the spectrum of the (7,0)-SWCNTs for comparison. As shown in the figure, there is slight shift in the peak positions, however, the relative intensities of specially G-, D- and RBM modes changed with the functional group, which is consistent with the experimental observations as discussed above. It should be pointed out that the calculated Raman spectrum was for the nonresonance case; however, the experimental measurements usually

| | (7,0)-SWCNT | carboxy (-$CO_2H$)-(7,0)-SWCNT | phenyl-$SO_3H$-(7,0)-SWCNT | 3-methoxy-6-methyl-4-[(2-nitro-4-methylphenyl) azo] benzene-(7,0)-SWCNT |
|---|---|---|---|---|
| RBM(A1g) | 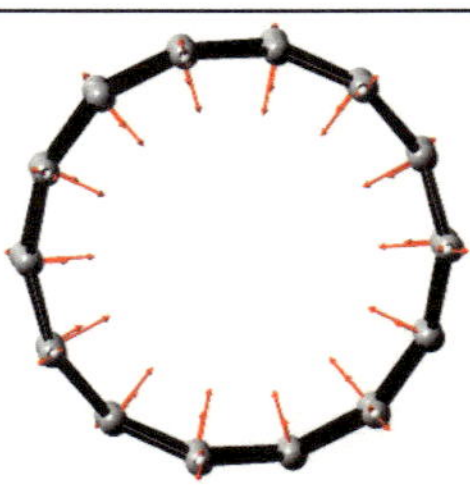 414.04 | 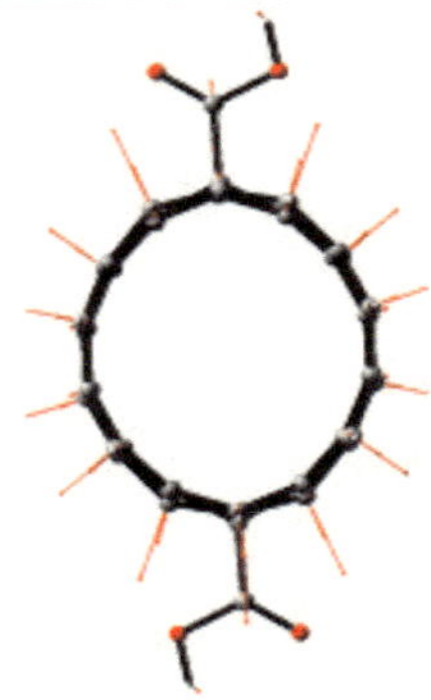 385.78 | 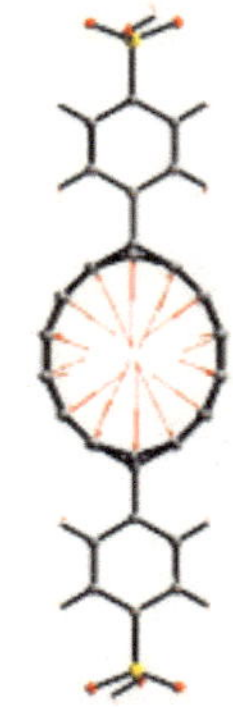 390.48 | 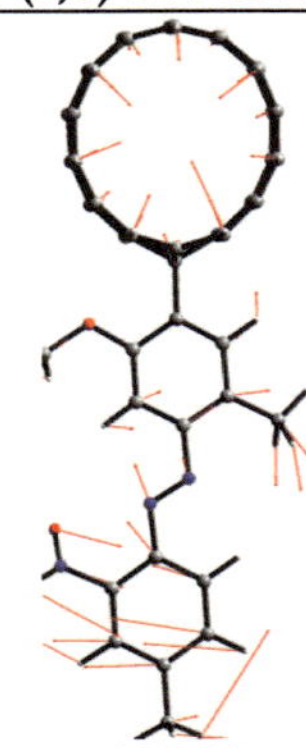 385.82 |
| BD($E_{1g}$) |  308.03 | 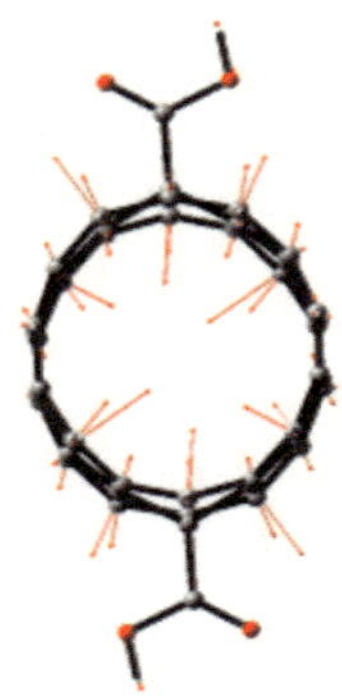 306.27 |  307.27 | 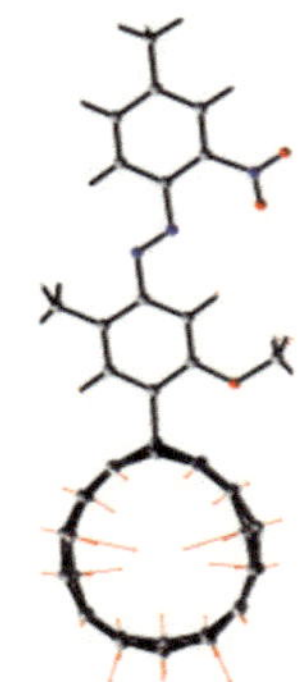 311.05 |
| ED($E_{2g}$) |  109.82 | 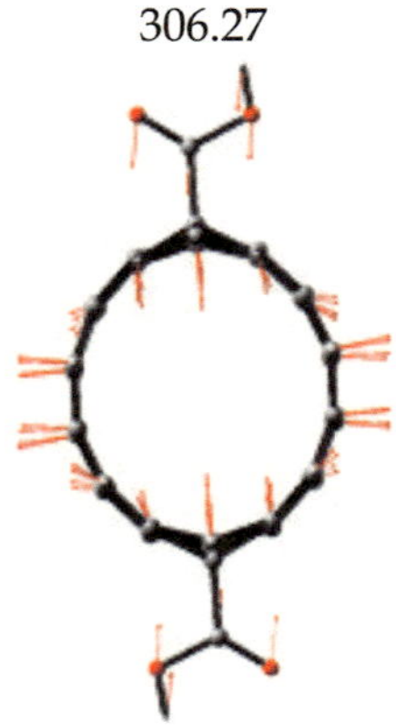 95.26 |  120.83 | 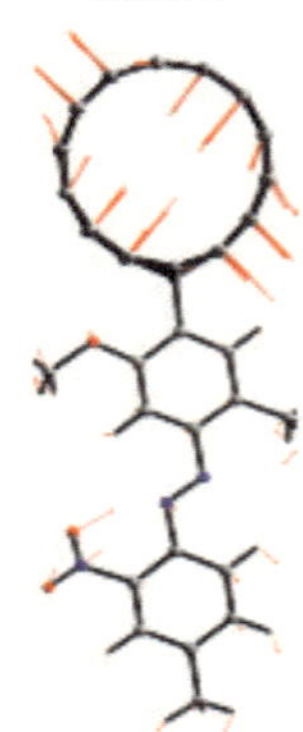 111.98 |

Fig. 3.7.3. Calculated molecular motions for selected vibrational bands of the functionalized (7,0)-SWCNT and the calculated values of vibrational frequencies in $cm^{-1}$.

at the resonance case. Therefore, the relative intensities of corresponding peaks in the observed spectrum at the resonance may be expected to somewhat differ in intensity reference to their nonresonance spectrum. Figure 3.7.2 provides the calculated IR spectra of the functionalized (7,0)-SWCNT with these functional group. The IR spectra of these functionalized-CNT showed that the IR spectrum of isolated (7,0)-SWCNT differ than its functionalized structure. Figure 3.7.3 shows the vibrational motion for the selected vibrational modes of the frequency in low energy region.

The calculated electronic transitions of the functionalized (7,0)-SWCNT with various functional groups showed that the functionalized (7,0)-SWCNT produce many dipole allowed electronic transition in low energy region, while the isolated (7,0)-SWCNT exhibited only one allowed transition in the same region (see Table 3.7.1). The calculated dipole allowed vertical electronic transition for the 3-methoxy-6-methyl-4- [(2-nitro-4-methylphenyl) azo] benzene functionalized (7,0)-SWCNT (with the calculated electron densities in molecular orbitals as seen in Figure 3.7.4) suggested that there might be a charge transfer (CT) mechanism as result of the transitions from the HOMO of the SWCNT to the LUMO of the molecule.

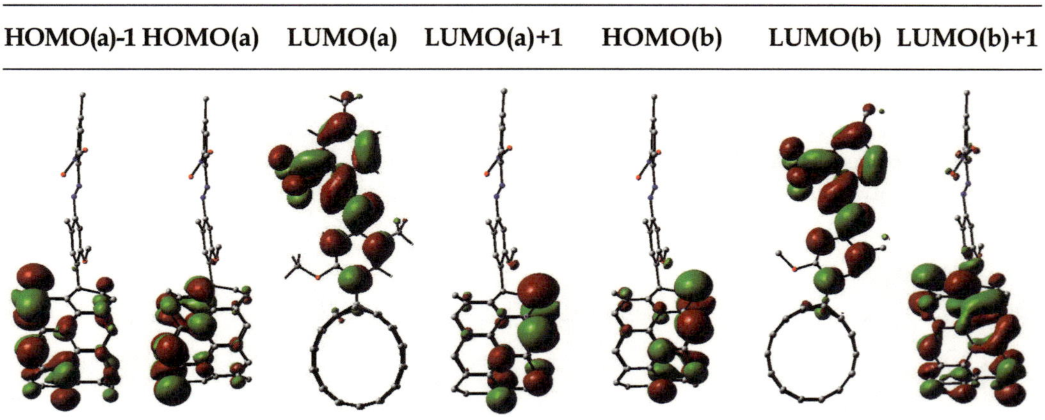


Fig. 3.7.4. Plot of the electron densities in the HOMOs and LUMOs for the 3-methoxy-6-methyl-4- [(2-nitro-4-methylphenyl) azo] benzene-(7,0)-SWCNT system.

Furthermore, we also calculated dipole allowed vertical electronic transitions for two of the 3-methoxy-6-methyl-4- [(2-nitro-4-methylphenyl) azo] benzene molecules covalently bonded to (7,0)-SWCNT, see Table 3.7.2. Figure 3.7.5 provides the calculated electron densities in molecular orbitals. The results of calculated dipole allowed vertical electronic transitions suggested that the $\mathbf{S_0} \rightarrow \mathbf{S_3}$ transition is favorable candidate for the charge transfer from the (7,0)-SWCNT to the 3-methoxy-6-methyl-4- [(2-nitro-4-methylphenyl) azo] benzene molecule in consequence of the transitions from the highest occupied molecular orbital of the CNT to the lowest unoccupied molecular orbital of the 3-methoxy-6-methyl-4- [(2-nitro-4-methylphenyl) azo] benzene molecules as shown in Figure 3.7.5. It should be noted that the spin forbidden electronic transition ($\mathbf{S_0} \rightarrow \mathbf{T_2}$; 0.728 eV) lies between the $\mathbf{S_0} \rightarrow \mathbf{S_2}$(0.766 eV) and $\mathbf{S_0} \rightarrow \mathbf{S_3}$(0.776) dipole allowed electronic transitions as seen in Table 3.7.2. This result indicates a possibility of the ISC process for this doubly functionalized (10,0)-SWCNT system.

| 0→n | H→L | CI | eV | f | 0→n | H→L | CI | eV | f |
|---|---|---|---|---|---|---|---|---|---|
| 0→1 | H(b) →L(b) | 0.29 | 1.30 | 0.0012 | 0→6 | H(a) →L(a) | 0.32 | 1.54 | 0.0011 |
| | H(b) →L(b)+1 | 0.93 | | | | H(a) →L(a)+2 | 0.67 | | |
| 0→2 | H(b) →L(b) | 0.94 | 1.44 | 0.0003 | 0→7 | H(a)-1→L(a) | 0.86 | 1.55 | 0.0021 |
| | H(b) →L(b)+1 | -0.28 | | | | H(a) →L(a)+2 | 0.28 | | |
| 0→3 | H(a)-1→L(a)+2 | 0.69 | 1.47 | 0.0030 | 0→8 | H(a)-1→L(a)+1 | 0.80 | 1.56 | 0.0025 |
| | H(b)-1→L(b)+3 | 0.16 | | | | H(a) →L(a)+2 | 0.50 | | |
| 0→4 | H(a) →L(a) | 0.80 | 1.51 | 0.0008 | 0→9 | H(a)-7→L(a) | 0.64 | 1.57 | 0.0003 |
| | H(a) →L(a)+1 | 0.54 | | | | H(a)-1→L(a) | 0.38 | | |
| 0→5 | H(a) →L(a) | -0.47 | 1.52 | 0.0042 | 0→10 | H(b) →L(b)+1 | 0.12 | 1.65 | 0.0055 |
| | H(a) →L(a)+1 | 0.79 | | | | H(b) →L(b)+2 | 0.95 | | |

Table 3.7.1. Calculated vertical doublet-doublet electronic transitions ($D_0 \rightarrow D_n$) energies and oscillator strengths (f) of the functionalized (7,0)-SWCNT at TD-B3LYP/6-31G(d,p) level of the theory. Where the functional group, 3-methoxy-6-methyl-4- [(2-nitro-4-methylphenyl) azo] benzene, is covalently bonded to the (7,0)-SWCNT (see Fig. 3.7.4). While the upper case letters H and L indicate the highest occupied molecular orbitals (HOMO) and the lowest unoccupied molecular orbitals (LUMO), respectively, the lower case letters (a) and (b) stand for the alpha and beta spin states, respectively. CI represents configurationally interaction coefficient.

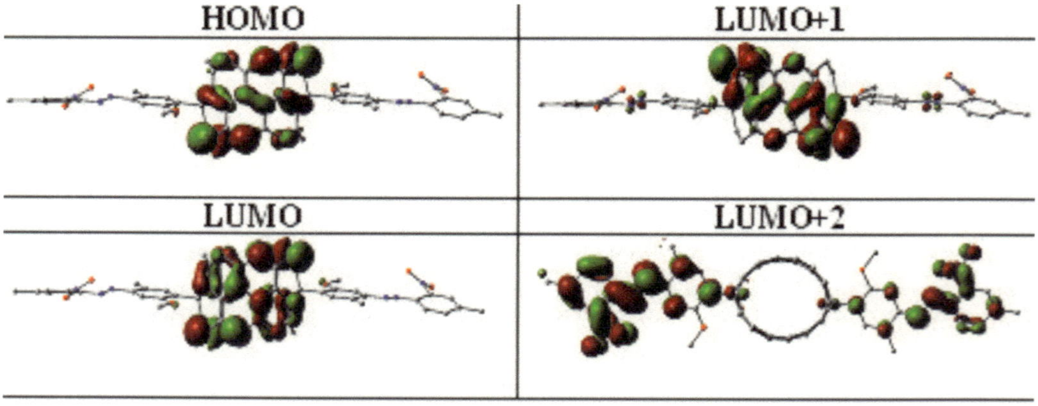


Fig. 3.7.5. Plot of the electron densities in the HOMOs and LUMOs for the 3-methoxy-6-methyl-4- [(2-nitro-4-methylphenyl) azo] benzene-(7,0)-SWCNT system.

We optimized the functionalized (10,0)-SWCNT, with four of the 3-methoxy-6-methyl-4- [(2-nitro-4-methylphenyl) azo] benzene molecules which covalently bonded to (10,0)-SWCNT. Because of the technique reason, we could not calculate the electronic transitions. However, when we plot the electron density in the molecular orbitals as seen in Figure 3.7.6, the HOMO, LUMO and LUMO+1 belongs to (10,0)-SWCNT, the molecular orbitals from

| $S_0 \rightarrow S_n$ | HOMO→LUMO | CI | eV | f | $S_0 \rightarrow T_n$ | HOMO-->LUMO | CI | eV |
|---|---|---|---|---|---|---|---|---|
| $S_0 \rightarrow S_1$ | HOMO→LUMO+1 | 0.64 | 0.375 | | $S_0 \rightarrow T_1$ | HOMO-1→LUMO | -0.15 | 0.174 |
| $S_0 \rightarrow S_2$ | HOMO→LUMO | 0.36 | 0.766 | 0.0232 | | HOMO→LUMO+1 | 0.94 | |
| | HOMO→LUMO+2 | -0.23 | | | $S_0 \rightarrow T_2$ | HOMO-2→LUMO | 0.13 | 0.728 |
| $S_0 \rightarrow S_3$ | HOMO→LUMO+2 | 0.65 | 0.776 | 0.0009 | | HOMO→LUMO+4 | 0.78 | |
| | HOMO→LUMO+3 | 0.23 | | | $S_0 \rightarrow T_3$ | HOMO→LUMO+1 | 0.11 | 0.758 |
| $S_0 \rightarrow S_4$ | HOMO→LUMO | -0.29 | 0.810 | 0.0407 | | HOMO→LUMO+2 | 0.69 | |
| | HOMO→LUMO+2 | -0.10 | | | $S_0 \rightarrow T_4$ | HOMO→LUMO+3 | 0.69 | 0.775 |
| | HOMO→LUMO+3 | 0.50 | | | | HOMO→LUMO+4 | 0.14 | |
| $S_0 \rightarrow S_5$ | HOMO-2→LUMO | -0.14 | 0.988 | 0.0070 | | | | |
| | HOMO→LUMO+4 | 0.63 | | | | | | |
| $S_0 \rightarrow S_6$ | HOMO-1→LUMO | 0.67 | 1.062 | | | | | |

Table 3.7.2. Calculated vertical singlet-singlet ($S_0 \rightarrow S_n$) and singlet-triplet ($S_0 \rightarrow T_n$) electronic transitions for two of the 3-methoxy-6-methyl-4- [(2-nitro-4-methylphenyl) azo] benzene molecule covalently bonded to (7,0)-SWCNT. Where The calculations were performed at TD-B3LYP/6-31G(d,p) level of the DFT. CI stands for configurationally interaction coefficient.

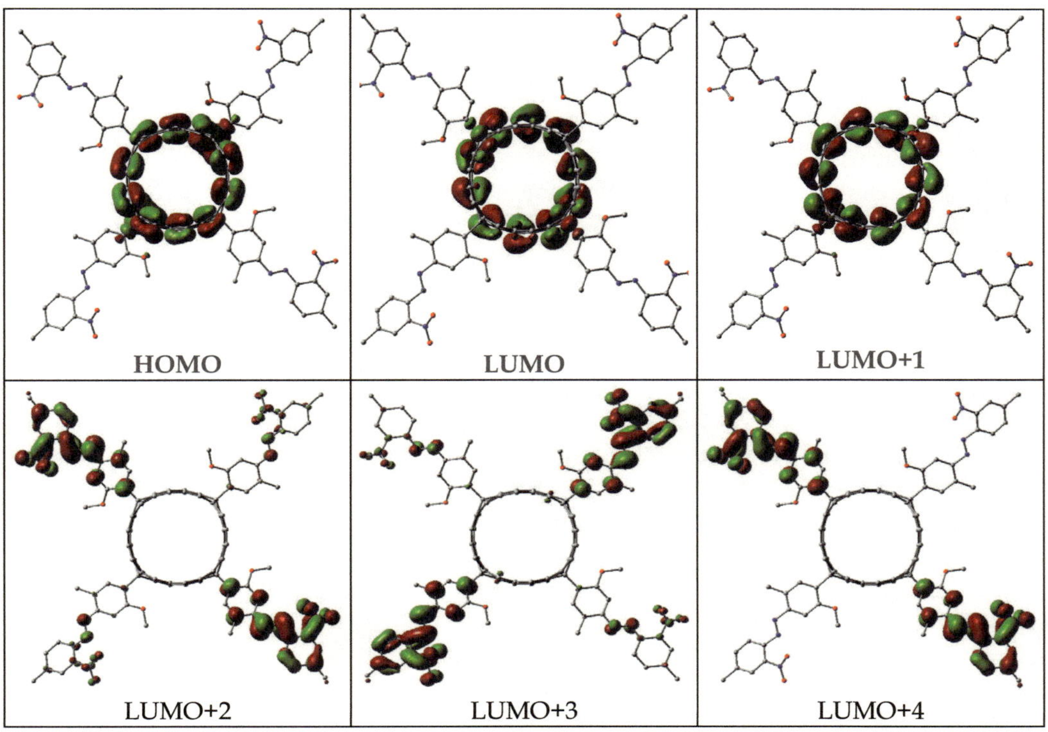


Fig. 3.7.6. Plot of the electron densities in the HOMOs and LUMOs for the 3-methoxy-6-methyl-4- [(2-nitro-4-methylphenyl) azo] benzene-(7,0)-SWCNT system.

LUMO+2 to LUMO+5 belongs to the 3-methoxy-6-methyl-4- [(2-nitro-4-methylphenyl) azo] benzene molecules. This results again suggested the charge transfer from the (10,0)-SWCNTs to the molecule in low energy region.

## 4. Acknowledgment

We wish to thank Dr. Abdullah Cavus, Dr. Nathan Stevens, and Miss Oya Onar for their assistance and suggestions in this work.

## 5. References

[1] S. Iijima, Nature. 354 (1991) 56.
[2] Nepal D, Sohn JI, Aicher WK, et al. Biomacromolecules 6(6) (2005) 2919.
[3] Karajanagi SS, Yang HC, Asuri P, et al. Langmuir, 22(4) (2006) 1392.
[4] Hod Finkelstein, Peter M. Asbeck, Sadik Esener, 3rd IEEE Conference on Nanotechnology (IEEE-NANO), vol.1, (2003) 441.
[5] D.S. Wen, Y.L. Ding, International Journal of Heat and Mass Transfer 47 (2004) 5181.
[6] H. Masuda, A. Ebata, K. Teramae, N. Hishiunma, Netsu Bussei (Japan) 4 (1993) 227.
[7] Carissa S. Jones, Xuejun Lu, Mike Renn, Mike Stroder, and Wu-Sheng Shih. Microelectronic Engineering; DOI: 10.1016/j.mee.2009.05.034.
[8] S.M. Jung, H.Y. Jung, J.S. Suh. Sensors and Actuators B. 139 (2009) 425.
[9] T. Rueckes, K. Kim, E. Joselevich, G.Y. Tseng, C.L. Cheung and C.M. Lieber, *Science* 289 (2000) 94.
[10] M. J. O'Connell *et al.*, Science 297 (2002) 593.
[11] S. M. Bachilo, M. S. Strano, C. Kittrell, R. H. Hauge, R. E. Smalley, and R. B. Weisman, Science 298 (2002) 2361.
[12] A. Hartschuh, H. N. Pedrosa, L. Novotny, and T. D. Krauss, Science 301 (2003) 1354.
[13] J. Maultzsch, R. Pomraenke, S. Reich, E. Chang, D. Prezzi, A. Ruini, E. Molinari, M. S. Strano, C. Thomsen1, and C. Lienau, Phys. Stat. Sol. (b) 243, No. 13 (2006) 3204.
[14] E. Chang, G. Bussi, A. Ruini, and E. Molinari, Phys. Rev. Lett. 92 (2004) 196401.
[15] C. D. Spataru, S. Ismail-Beigi, L. X. Benedict, and S. G. Louie, Phys. Rev. Lett. 92 (2004) 077402.
[16] V. Perebeinos, J. Tersoff, and P. Avouris, Phys. Rev. Lett. 92 (2004) 257402.
[17] Won-Il Park, Hun-Sik Kim, Soon-Min Kwon, Young-Ho Hong, Hyoung-Joon Jin, Carbohydrate Polymers. 77 (2009) 457.
[18] Lingjie Meng, Chuanlong Fu, and Qinghua Lu, Natural Science, 19 (2009) 801.
[19] C.T. Hsieh, Y.T. Lin, Microporous Mesoporous Mater. 93 (2006) 232.
[20] Davis JJ, Coleman KS, Azamian BR, et al. Chem Eur J., 9(16) (2003) 3732.
[21] Poenitzsch VZ, Winters DC, Xie H, et al. J Am Chem Soc, 129(47) (2007) 14724.
[22] H. Zhao and S. Mazumdar, Phys. Rev. Lett. 93 (2004) 157402.
[23] C. L. Kane and E. J. Mele, Phys. Rev. Lett. 93 (2004) 197402.
[24] T. Ando, J. Phys. Soc. Jpn. 66 (1997) 1066.
[25] F. Wang, G. Dukovic, L. E. Brus, and T. F. Heinz, Science 308 (2005)838.

[26] E. B. Barros, R.B. Capaz, A.Jorio, G.G. Samsonidze, A.G. Souza Filho, S. Ismail-Beigi, C. D. Spataru, S. G. Louie, G.Dresselhaus, and M.S. Dresselhaus," Phys. Rev. B 73 (2006) 241406.
[27] H. Zhao , S. Mazumdar C.-X. Sheng, M. Tong, and Z. V. Vardeny, Phys. Rev. B 73 (2006) 075403.
[28] H.Kishida, Y.Nagasawa, S.Imamura, and A.Nakamura, Phys Rev Lett 100 (2008) 097401.
[29] N.G.Kalugin, Y.Rostovtsev, Opt.Lett.31, (2006) 969.
[30] M. Jain, H.Xia, G. Y. Yin, A. J. Merriam, and S. E. Harris, Phys.Rev.Lett.77 (1996) 4326.
[31] O.Kocharovskaya, Ya.I.Khanin, ZhETPh 90 (1986) 1610.
[32] V.A.Sautenkov, Y.V.Rostovtsev, C.Y.Ye, G.R.Welch, O.Kocharovskaya, and M.O.Scully, Phys. Rev. A 71 (2005) 063804.
[33] Elefterios Lidorikis and Andrea C. Ferrari, ACSNANO, 3(5) (2009) 1238.
[34] Y. Wang, X. Wang, J. Rybczynski, D. Z. Wang, K. Kempa, and Z. F. Ren, Appl. Phys. Lett. 86 (2005) 153120.
[35] Oliver Kiowski, Katharina Arnold, Sergei Lebedkin, Frank Hennrich, and Manfred M. Kappes, Phys. Rev. Lett. 99 (2007) 237402.
[36] R. M. Russo, E. J. Mele, C. L. Kane, I. V. Rubtsov, M. J. Therien, and D. E. Luzzi, Phys. Rev. B 74 (2006) 041405.
[37] J. Maultzsch, R. Pomraenke, S. Reich, E. Chang, D. Prezzi, A. Ruini, E. Molinari, M. S. Strano, C. Thomsen, and C. Lienau, Phys. Rev. B 72 (2005) 241402(R).
[38] T. G. Pedersen, Phys. Rev. B 67 (2003) 073401.
[39] T. Ogawa and T. Takagahara, Phys. Rev. B 44 (1991) 8138.
[40] H. B. Zhao and S. Mazumdar, Phys. Rev. Lett. 93 (2004) 157402.
[41] J Moreno, K Kasai, M David, H Nakanishi and H Kasai. *J. Phys. Condens. Matter*, 21 (2009) 064219.
[42] B.K. Pradhan, A.R. Harutyunyan, D. Stojkovic, J.C. Grossman, P. Zhang, M.W. Cole, V. Crespi, H. Goto and J. Fujiwara., J. Mater. Res.,17(9) (2002).
[43] P. Kim, L. Shi, A. Majumdar, and P.L. McEuen, *Phys. Rev. Lett.*, 87 (2001) 215502-1.
[44] J. Hone, M. Whitney, C. Piskoti, and A. Zettl, Phys. Rev. Lett. B., 59(4) (1999) R2514.
[45] Motoo Fujii, Xing Zhang, Huaqing Xie, Hiroki Ago, Koji Takahashi, Tatsuya Ikuta,Hidekazu Abe, and Tetsuo Shimizu3, Phys. Rev. Lett. B. 95 (2005) 065502.
[46] Quoc Ngo, Brett A. Cruden, Alan M. Cassell, Megan D. Walker, Qi Ye, Jessica E. Koehne, M. Meyyappan, Jun Li, and Cary Y. Yang, Mat. Res. Soc. Symp. Proc. 812 (2004) F3.18.1
[47] G. S. Choi, K. H. Son, D. J. Kim, Microelectronic Engineering, 66(1/4) (2003) 206.
[48] N.G. Chopra, et al., Science. 269 (1995) 966.
[49] D. Golberg, Y. Bando, W. Han, K. Kurashima, T. Sato, Chem. Phys. Lett., 308 (1999) 337.
[50] C. C. Tang, Y. Bando, T. Sato, K. Kurashima, Chem. Commun. 12 (2002) 1290.
[51] M. S. Dresselhaus, G. Dresselhaus, P. C. Eklund, Science of Fullerenes and Carbon Nanotubes, Academic Press, San Diego, CA (1996).
[52] D. Zhang, R. Q. Zhang, Chem. Phys. Lett., 371 (2003) 426.
[53] M. Mirzaei, N. L. Hadipour, Physica E, 40 (2008) 800.

[54] C.H. Park, C.D. Spataru, S.G. Louie, Physical Review Letters 96 (2006) 126105.
[55] L. Wirtz, A. Marini, A. Rubio, Physical Review Letters 96 (2006) 126104.
[56] A. Rubio, J.L. Corkill, M.L. Cohen, Physical Review B 49 (1994) 5081.
[57] Chunyi Zhi, Yoshio Bando, Chengchun Tang, Dmitri Golberg Materials Science and Engineering R 70 (2010) 92.
[58] X. Blase, A. Rubio, S.G. Louie, M.L. Cohen, Europhysics Letters 28 (1994) 335.
[59] C.Y. Won, N.R. Aluru, Journal of the American Chemical Society 129 (2007) 2748.
[60] G. Mpourmpakis, G.E. Froudakis, Catalysis Today 120 (2007) 341.
[61] Z. Zhou, J.J. Zhao, Z.F. Chen, X.P. Gao, T.Y. Yan, B. Wen, P.V. Schleyer, Journal of Physical Chemistry B 110 (2006) 13363.
[62] X.J. Wu, J.L. Yang, J.G. Hou, Q.S. Zhu, Journal of Chemical Physics 124 (2006) 54706.
[63] J.R. Cheng, R. Ding, Y. Liu, Z.F. Ding, L.B. Zhang, Computational Materials Science 40 (2007) 341.
[64] I. Cabria, M.J. Lopez, J.A. Alonso, Nanotechnology 17 (2006) 778.
[65] E. Durgun, Y.R. Jang, S. Ciraci, Physical Review B 76 (2007) 073413.
[66] S.H. Jhi, D.J. Roundy, S.G. Louie, M.L. Cohen, Solid State Communications 134 (2005) 397.
[67] G.Y. Gou, B.C. Pan, L. Shi, Journal of the American Chemical Society 131 (2009) 4839.
[68] R.Q. Wu, L. Liu, G.W. Peng, Y.P. Feng, Applied Physics Letters 86 (2005) 122510.
[69] F. Lii, Z.H. Zhu, M.W. Zhao, Y.Y. Xia, Journal of Physical Chemistry C 112 (2008) 16231.
[70] F. Li, Z.G. Zhu, X.D. Yao, G.Q. Lu, M.W. Zhao, Y.Y. Xia, Y. Chen, Applied Physics Letters 92 (2008) 102515.
[71] J.B. Wu, W.Y. Zhang, Chemical Physics Letters 457 (2008) 169.
[72] J.B. Wu, W.Y. Zhang, Solid State Communications 149 (2009) 486.
[73] G. Rahman, S.C. Hong, Journal of Nanoscience and Nanotechnology 8 (2008) 4711.
[74] L. Wang, J. Lu, L. Lai, W. Song, M. Ni, Z.X. Gao, W.N. Mei, Journal of Physical Chemistry C 111 (2007) 3285.
[75] P. Saxena, S.P. Sanyal, Physica E 24 (2004) 244.
[76] I. Savic, N. Mingo, D.A. Stewart, Physical Review Letters 101 (2008) 165502.
[77] G.Y. Gou, B.C. Pan, L. Shi, Journal of Physical Chemistry C 112 (2008) 19353.
[78] J. Song, J. Wu, Y. Huang, K.C. Hwang, Nanotechnology 19 (2008) 445705.
[79] J.T. Tanskanen, M. Linnolahti, A.J. Karttunen, T.A. Pakkanen, Chemphyschem 9 (2008) 2390.
[80] J.W. Kang, H.J. Hwang, Journal of Physics 16 (2004) 3901.
[81] C.Y. Won, N.R. Aluru, Journal of Physical Chemistry C 112 (2008) 1812.
[82] X.Y. Li, W. Yang, B. Liu, Nano Letters 7 (2007) 3709.
[83] W.H. Moon, H.J. Hwang, Materials Letters 58 (2004) 2331.
[84] V. Meunier, C. Roland, J. Bernholc, M.B. Nardelli, Applied Physics Letters 81 (2002) 46.
[85] M.D. Ganji, A. Mohammadi-Nejad, Physics Letters A 372 (2008) 4839.
[86] F.W. Zheng, G. Zhou, Z.R. Liu, J. Wu, W.H. Duan, B.L. Gu, S.B. Zhang, Physical Review B 78 (2008) 205415.
[87] M. Ishigami, J.D. Sau, S. Aloni, M.L. Cohen, A. Zettl, Physical Review Letters 97 (2006) 176804.

[88] J. Zhang, K.P. Loh, M. Deng, M.B. Sullivian, J.W. Zheng, P. Wu, Journal of Applied Physics 99 (2006) 104309.
[89] K.S. Ryu, Y.I. Kim, J.K. Jung, Y. Chen, C.H. Lee, Journal of Nanoscience and Nanotechnology 8 (2008) 5193.
[90] Z. Zhou, J. Zhao, Z. Chen, X. Gao, T. Yan, B. Wen, P.V.R. Schleyer, Journal of Physical Chemistry B 110 (2006) 13363.
[91] M.B. Belonenko, N.G. Lebedev, Technical Physics 54 (2009) 338.
[92] Saito R, et al., *Physical Properties of Carbon Nanotubes. Imperial College Press*, London; 1998
[93] L. Yang, M. P. Anantram, J. Han, and J. P. Lu, *Physical Review B*, vol. 60, (1999) 13874.
[94] Gaussian, Inc., Carnegie Office Park-Bulding 6, Pittsburgh, PA106, USA.
[95] (a) C. Guo, M. Aydin, H. R. Zhu, D. L. Akins, J. Phys. Chem. B 106 (2002) 5447; (b) H, Guo, X. Zhang, M. Aydin, W. Xu, H. R. Zhu, D. L Akins, Journal of Molecular Structure 689 (2004) 153; (c) M. Aydin, F. Jean-Mary, N. Stevens, D. L.Akins, J. Phys. Chem. B 108(2004) 9695; (d) W. Xu, M.Aydin, S. Zakia, D. L. Akins, J. Phys. Chem. B 108 (2004) 5588; (e) M. Aydin, J. R.Lombardi, J. Phys. Chem. A. 113 (2009) 2809; (f) Metin Aydin, Photofragmentation Spectroscopy. VDM Publishing House Ltd., 2009.
[96] R. Pfeiffer, F. Simon, H. Kuzmany, and V. N. Popov, Phys. Rev. B 72 (2005) 161404(R).
[97] A. Lan, Y. Zhang, X. Zhang, Z. Iqbal, H. Grebel, Chem. Physics Lett. 379 (2003) 395.
[98] Blase X, Rubio A, Louie S and Cohen M L Europhys. Lett. 28 (1994) 335
[99] R Chowdhury, C Y Wang, S Adhikari and F Scarpa Nanotechnology 21 (2010) 365702
[100] Chopra N, Luyken R, Cherrey K, Crespi V, Cohen M, Louie S and Zettl A *Science* 269 (1995) 966.
[101] Golberg D, Bando Y, Tang C and Zhi C *Adv. Mater.* 19 (2007) 2413–32
[102] Won C Y and Aluru N R *J. Phys. Chem.* C 112 (2008) 1812.
[103] Santosh M, Maiti P K and Sood A K *J. Nanosci. Nanotechnol.* 9 (2009) 5425.
[104] Yuan J and Liew K M *Nanotechnology* 19 (2008) 445703
[105] Zhi C Y, Bando Y, Tang C C, Huang Q and Golberg D *J. Mater. Chem.* 18 (2008) 3900.
[106] Oh E S *Mater. Lett.* 64 (2010) 859.
[107] Xu F, Bando Y, Golberg D, Ma R, Li Y and Tang C *J. Chem. Phys.* 119 (2003) 3436.
[108] Moon W H and Hwang H J *Physica* E 23 (2004) 26.
[109] Verma V, Jindal V K and Dharamvir K *Nanotechnology* 18 (2007) 435711
[110] Song J, Wu J, Huang Y and Hwang K C *Nanotechnology* 19 (2008) 445705
[111] Suryavanshi A, Yu M, Wen J, Tang C and Bando Y *Appl. Phys. Lett.* 84 (2004) 2527.
[112] Chopra N and Zettl A *Solid State Commun.* 105 (1998) 297.
[113] Zhi C, Bando Y, Tang C, Honda S, Kuwahara H and Golberg D *J. Mater Res.* 21 (2006) 2794.
[114] Huang Q, Bando Y, Xu X, Nishimura T, Zhi C, Tang C, Xu F, Gao L and Golberg D *Nanotechnology* 18 (2007) 485706.
[115] M. Aydin and.L. Akins Vib. Spectrosc. 53 (2010) 163.

[116] Sanchez-Portal D and Hernandez E *Phys. Rev.* B 66 (2002) 235415

[117] Wirtz L and Rubio A *IEEE Trans. Nanotechnol.* 2 (2003) 341.

[118] Akdim B, Pachter R, Duan X F and Adams W W *Phys. Rev.* B 67 (2003) 245404

[119] Zhang Z H, Guo W L and Dai Y T *J. Appl. Phys.* 105 (2009) 084312

[120] Jensen K, Kim K and Zettl A *Nat. Nanotechnol.* 3 (2008) 533

[121] Chowdhury R, Adhikari S and Mitchell J *Physica* E 42 (2009) 104

[122] Li C Y and Chou T W *Appl. Phys. Lett.* 84 (2004) 5246–8

[123] Chang, J.; Cañamares, M. V.; Aydin, M.; Vetter, W.; Schreiner, M.; Xu, W. ve Lombardi. J. R., Journal of Raman Spect., 40 (2009) 1557.

[124] Shankar Ghosh, Pallavi V. Teredesai, and A. K. Sood *Pure Appl. Chem.*, 74 (9) (2002) 1719.

[125] P. Delaney, H. J. Choi, I. Ihm, S. G. Louie, M. L. Cohen. *Nature (London)* 391 (1998) 466.

[126] C. S. Sunder, P. Ch. Sahu, V. S. Sastry, G. V. N. Rao, V. Sridharan, M. Premila, A. Bharathi, Y. Hariharan, T. S. Radhakrishnan, D. V. S. Muthu, A. K. Sood. *Phys. Rev. B* 53 (1996) 8180.

[127] M. J. Peters, L. E. McNeil, J. P. Lu, D. Kahn. *Phys. Rev. B* 61 (2000) 5939.

[128] A.S. Barnard, I.K. Snook, S.P. Russo, Journal of Materials Chemistry 17 (2007) 2892.

[129] Y.M. Chou, H.W. Wang, Y.J. Lin, W.H. Chen, B.C. Wang, Diamond and Related Materials 18 (2009) 351.

[130] A.V. Osadchy, E.D. Obraztsova, S.V. Terekhov, V.Y. Yurov, Journal of Theoretical Physics Letters 77 (2003) 405.

[131] G.Y. Guo, S. Ishibashi, T. Tamura, K. Terakura, Physical Review B 75 (2007) 245403.

[132] X.M. Li, W.Q. Tian, X.R. Huang, C.C. Sun, L. Jiang, Journal of Nanoparticle Research 11 (2009) 395.

[133] K.H. Khoo, M.S.C. Mazzoni, S.G. Louie, Physical Review B 69 (2004) 201401.

[134] C. Attaccalite, L. Wirtz, A. Marini, A. Rubio, Physica Status Solidi B 244 (2007) 4288.

[135] C.W. Chen, M.H. Lee, S.J. Clark, Nanotechnology 15 (2004) 1837.

[136] S.L. Hu, Z.Y. Li, X.C. Zeng, J.L. Yang, Journal of Physical Chemistry C 112 (2008) 8424.

[137] C.H. Park, S.G. Louie, Nano Letters 8 (2008) 2200.

[138] H.P. Lan, L.H. Ye, S.A. Zhang, L.M. Peng, Applied Physics Letters 94 (2009) 183110.

[139] Z.G. Wang, Z. Li, D.M. Cheng, European Physical Journal 46 (2009) 20601.

[140] R.J. Baierle, T.M. Schmidt, A. Fazzio, Solid State Communications 142 (2007) 49.

[141] R.X. Wang, R.X. Zhu, D.J. Zhang, Chemical Physics Letters 467 (2008) 131.

[142] L. Lai, W. Song, J. Lu, Z.X. Gao, S. Nagase, M. Ni, W.N. Mei, J.J. Liu, D.P. Yu, H.Q. Ye, Journal of Physical Chemistry B 110 (2006) 14092.

[143] Y.F. Li, Z. Zhou, D. Golberg, Y. Bando, P.V. Schleyer, Z.F. Chen, Journal of Physical Chemistry C 112 (2008) 1365.

[144] P.N. Dpsilayachkov, D.V. Makaev, Journal of the Physics and Chemistry of Solids 70 (2009) 180.

[145] T.M. Schmidt, R.J. Baierle, P. Piquini, A. Fazzio, Physical Review B 67 (2003) 113407.

[146] Y.F. Li, Z. Zhou, J. Zhao, Journal of Chemical Physics 127 (2007) 184705.
[147] Z.H. Zhang, W.L. Guo, Journal of the American Chemical Society 131 (2009) 6874.
[148] Z. Zhou, J.J. Zhao, Z.F. Chen, P.V. Schleyer, Journal of Physical Chemistry B 110 (2006) 25678.
[149] J. Zhang, K.P. Loh, P. Wu, M.B. Sullivan, J.W. Zheng, Journal of Physical Chemistry C 112 (2008) 10279.
[150] V.A. Margulis, E.E. Muryumin, E.A. Gaiduk, Physical Review B 78 (2008) 035415.
[151] H.T. Liu, G. Zhou, Q.M. Yan, J. Wu, B.L. Gu, W.H. Duan, D.L. Zhao, Physical Review B 75 (2007) 125410.
[152] S. Okada, S. Saito, A. Oshiyama, Physical Review B 65 (2002) 165410.
[153] Tretiak, S. Nano Lett., 7(8) (2007) 2201.
[154] Gambetta, A.; Manzoni, C.; Menna, E.; Meneghetti, M.; Cerullo, G.; Lanzani, G.; Tretiak, S.; Piryatinski, A.; Saxena, A.; Martin, R. L.; and Bishop, A. R. Nature Phys., 2 (2006) 515.
[155] Korovyanko, O. J.; Sheng, C. X.; Vardeny, Z. V.; Dalton, A. B.; and Baughman, R. H. Phys. Rev. Lett., 92 (2004) 017403.
[156] Capaz, R. B.; Spataru, C. D.; Ismail-Beigi, S.; and Louie, S. G. Phys. Rev. B 74 (2006) 121401.
[157] Shim, M.; Javey, A.; Kam, N. W. S.; Dai, H. *J. Am. Chem. Soc. 123 (*2001) 11512.
[158] Kong, J.; Franklin, N.; Zhou, C.; Chapline, M.; Peng, S.; Cho, K.; Dai, H. *Science 287 (*2000) 622.
[159] Moonsub Shim, Nadine Wong Shi Kam, Robert J. Chen, Yiming Li, and Hongjie Dai *Nano Lett.*, Vol. 2, No. 4, (2002)
[160] Ambarish Paul and Baidurya Bhattacharya, Materials and Manufacturing Processes, 25 (2010) 891
[161] Lacerda L, Bianco A, Prato M and Kostarelos K *Adv. Drug Deliv. Rev.* 58 (2006) 1460
[162] Bianco A, Kostarelos K and Prato M *Expert Opin. Drug Deliv.* 5 (2008) 331
[163] Wang J and Lin Y H *TRAC Trends Anal. Chem.* 27 (2008) 619.
[164] Gannon C J *et al Cancer* 110 (2007) 2654
[165] Vyalikh A *et al Nanomedicine* 3 (2008) 321
[166] Liu Z *et al Cancer Res.* 68 (2008) 6652
[167] Lu Q *et al Nano Lett.* 4 (2004) 2473
[168] Kam N W and Dai H *J. Am. Chem. Soc.* 127 (2005) 6021
[169] Kam N W S, Liu Z A and Dai H J *Angew. Chem. Int. Edn* 45 (2006) 577
[170] Lacerda L, Raffa S, Prato M, Bianco A and Kostarelos K *Nano Today* 2 (2007) 38
[171] Kostarelos K *et al Nat. Nanotechnol.* 2 (2007) 108
[172] Becker M L *et al Adv. Mater.* 19 (2007) 939
[173] C Lamprecht, I Liashkovich, V Neves, J Danzberge, E Heiste, M Rangl, H M Coley, J McFadden, E Flahaut, H J Gruber, P Hinterdorfe, F Kienberge and A Ebne *Nanotechnology* 20 (2009) 434001
[174] M. Müller, J. Maultzsch, D. Wunderlich, A. Hirsch, and C. Thomsen phys. stat. sol. (b) 244, No. 11, (2007), 4056.

[175] Christie M. Sayes, Feng Liang, Jared L. Hudsona, Joe Mendez, Wenhua Guob, Jonathan M. Beach, Valerie C. Moore, Condell D. Doyle, Jennifer L. West, W. Edward Billups, Kevin D. Ausman, Vicki L. Colvin. Toxicology Letters 161 (2006) 135.

# 10

# Detection of Carbon Nanotubes Using Tip-Enhanced Raman Spectroscopy

Jia Wang, Xiaobin Wu, Rui Wang and Mingqian Zhang
*State Key Laboratory of Precision Measurement Technology and Instruments, Department of Precision Instruments, Tsinghua University, Beijing China*

## 1. Introduction

As a kind of new nano-materials, carbon nanotubes (CNTs) are of the particular properties, in the aspects of physics, mechanics, electronics, optics and so on. They indicate the remarkable potential applications in the nano-electronic devices, composite materials, scanning probe microscopy (SPM), field emission, hydrogen storage and environment protection. The research on CNTs has become a hot spot in the fields of nano-science and nano-technology, especially in nano-materials (Iijima, 1991; Yu et al., 2000; Frank et al., 1998; Chiang et al., 2001).

The commonly used methods for the measurement and characterization of CNTs include the scanning electron microscopy (SEM), transmission electron microscopy (TEM), scanning probe microscopy (SPM), Raman spectroscopy, infrared spectroscopy and so on. Electron microscopy and SPM are commonly used in nowadays research on nano-science. And both of them provide direct characterization of the topography and nano-structure of CNTs with comparatively high spatial resolution. However, they fail to provide the corresponding specimens' chemical information, which is also required in the research and identifications of nano-materials and nano-structures.

Raman spectroscopy is one of the spectral analysis methods used to detect vibrational, rotational, and other low-frequency modes information of the specimen structures or molecular structures, and is usually employed to obtain the structural information of CNTs (Rao et al., 1997; Dresselhaus et al., 2002; Dresselhaus et al., 2005). However, this spectral method is subject to two obstacles. One is the diffraction-limited spatial resolution, and the other is its inherent small Raman cross section and weak signal.

As facing the challenges of nano-scale measurement, a new characterization method, which can simultaneously obtain topography characteristics with nanometer resolution and spectral information from nanometer localized specimen surface is desiderated.

Tip-Enhanced Raman Spectroscopy (TERS) is an emerging technology developed to realize the aim mentioned above in the recent years (Stöckle et al, 2000; Hayazawa et al., 2000; Hartschuh et al., 2003; Pettinger et al., 2004). It spans the main obstacles of conventional Raman spectroscopy by tactfully combining the near-field optics advantage of nanometric spatial resolution and spectroscopy of obtaining chemical information. Also the Raman signal from nanometer localized specimen surface is considerably enhanced with the tip-enhanced optical technology (TEOT) and the signal-to-noise ratio (SNR) is consequently

improved. Especially, in TERS, the topography and spectral information are obtained simultaneously. By correspondingly analyzing the two images, the specific recognition of the feature distribution can be realized.
With the development in recent years, TERS has been applied to the characterization of nano-materials(Hayazawa et al, 2003; Huihong et al., 2006; Hartschuh et al., 2003), biological specimens(Anderson et al., 2003; Watanabe et al., 2004; Bailo & Deckert, 2008), dye molecules(Hayazawa et al., 2000; Pettinger et al., 2004; Watanabe et al., 2005), and semiconductor materials (Sun et al., 2003; Lee et al., 2007; Saito et al., 2008; Sun & Shen, 2001). In January 2008, the topic meeting, entitled as "Tip Enhanced Raman and Fluorescence Spectroscopy (TERFS): Challenges and Opportunities", was held in the National Physical Laboratory (NPL), London, Britain. Over a hundred of scientists and scholars from eleven countries and a number of equipment manufacturers attended this meeting. Research results and experiences were shared and the challenges and prospects of TERS were discussed. TERS has been demonstrated as a promising tool in the characterization of nano-materials and nano-structures, and provides powerful approach for the research of nano-science. It surely has the potential to expand the knowledge and comprehension of the phenomena and laws in the nano-scale.
However, the technique challenges involved in TERS system is more complicated than the SPM setup or the Raman spectroscope individually. At present, several companies can provide the commercial TERS system. Yet, as an emerging spectral analysis approach TERS is still in the developing process, and the integrated commercial systems are often difficult to adapt to the rapidly boosting research and detection needs in different fields. The thorough understanding of the TERS principle and key technologies is very beneficial whether to master the operation of commercial systems, or to buildup their own TERS systems upon specific detection needs.
In this chapter, TERS Measuring principle and system are firstly introduced and some key techniques are discussed in section two. CNT specimen preparation is introduced in section three. Carbon nanotube, single-walled carbon nanotube (SWNT) and multi-walled carbon nanotube (MWNT) specimens are detected by the TERS and results are discussed in section four. Some conclusions are presented and future works are discussed in the CNT detection using TERS in the section five.

## 2. TERS measuring system

Combining a SPM and Raman spectroscopy, TERS is of the ability to simultaneously obtain topography and corresponding spectral information of the specimen with high spatial resolution and high sensitivity. It provides powerful and promising tool for the characterization and research of nano-scale materials. The TERS systems are generally composed of the SPMs, the illumination/collection configurations and the Raman spectroscope. In this section, TERS's working principle, measuring system and some key techniques will be introduced and discussed.

### 2.1 Principle of TERS

Raman spectroscopy is one of the common spectral analysis methods used to detect molecular vibration, rotation, other low-frequency modes and chemical information of specimen and is usually employed to obtain the structure and vibration information of CNTs (Rao et al., 1997; Dresselhaus et al., 2002; Dresselhaus et al., 2005).

Two obstacles have to be faced when try to use conventional Raman spectroscopy to characterize CNTs. Firstly, the resolution of conventional Raman spectroscopy is limited by the optical diffraction limit to about half-wavelength. It corresponds to an zone of about 200×200 nm. Thus, the conventional Raman spectroscopy can only obtain the average spectral information of the chemical composition from the whole illuminated area of the nanometer specimen. Meanwhile the measurement and characterization of basic nano-element and nano-composition is vital in nano-material research. Secondly, another difficulty is the low detection sensitivity as nanometer localized spectroscopy is detected. Since the cross section of Raman scattering is quite small, the spectral signal is too weak to be directly detected. Additionally, the quantity of sample in nano-materials research is rather limited. It consequently results in the extremely weak response signal even using a strong exciting illumination beam. Thus, according to the conventional approach, the spectral detection requires long acquisition time and only obtains poor signal-to-noise ratio result. It is crucial to efficiently enhance the optical signal to obtain desirable sensitivity and SNR.

In sum, characterization of CNTs faces the major challenges with the Raman spectral approach are to obtain a high-resolution beyond the diffraction limit localized spectral signal with sufficient sensitivity and corresponding topography information in nanometer scale of the specimen.

As shown in Fig.1, the incident beam with appropriate wavelength and polarization state is focused at the apex of the nano-scale metallic tip. Excited by the incident laser, an enhanced electric field is generated in the vicinity of a metallic tip. The mechanism of the enhancement can be attributed to the localized surface plasmon resonance (LSPR) effect (Bohren & Huffman, 1998) and the lightning-rod effect (Novotny et al., 1997). The metallic tip can be regarded as a nano-scale light source with quite high power density. Consequently, Raman signal from the localized position of the specimen excited by the enhanced field very close to the tip is markedly enhanced. Since the enhancement of the Raman intensity is proportional to the 4th power of the electric field enhancement (Kerker et al., 1980), the Raman signal enhancement obtained in TERS experiment is usually up to $10^3$~$10^6$. The enhanced spectral signal is scattered and converted into the far-field by the tip. Through the collection optics, the Raman signal is guided into the Raman spectroscope to be further analyzed.

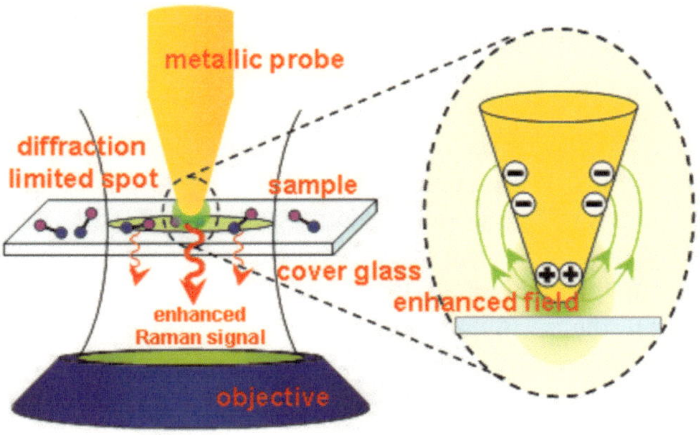


Fig. 1. Concept of tip-enhanced Raman spectroscopy (TERS) shows a strongly enhanced optical field generated at the apex of sharp metal tip of SPM by the external illumination (R. Wang, 2010c).

By means of scanning the tip over the specimen at a certain distance, tens nanometers above the surface, the corresponding topography and Raman information from specimen can be obtained simultaneously.

TERS is a skillful combination of the SPM (or SNOM) and Raman spectroscopy, and an improved variant of the surface enhanced Raman spectroscopy (SERS). It may be imagined that the rough metallic surface excited and enhanced in SERS is scaled down to a nanometer "hot-spot" scattering Raman spectral signal enhanced from the tip apex (Otto, 2002). In this situation, the distance and relative position between the "hot-spot" and the specimen is precisely controllable. Thus, the irksome measurement uncertainty and non-repeatability in the former SERS detection caused by the random distributed enhancing "hot-spots" on the rough metal substrate is avoided. Overall, the single effectively enhancing "hot-spot" of TERS provides an access to the repeatable characterization and quantitative analysis of the low density CNTs content specimen. Additionally, it provides a better understanding of the enhancement mechanism. TERS not only has the nanometer spatial resolution, but also has Raman spectral analysis performance.

The efficient excitation of electromagnetic-field enhancement requires the incident light to match with surface plasmon wave vector. To satisfy this condition, some factors of the enhancing system should be taken into consideration, including metal materials, size, tip shape, excitation wavelength, beam polarization state, and incident angle and surrounding refractive index (Martin & Girard, 1997; Krug et al., 2002; Neacsu et al., 2005). As theoretically predicted and experimentally proved, gold and silver are most suitable materials for generating enhancement under visible light illumination. Furthermore, according to the lightning-rod effect, the most contribution to the local field enhancement owes to the longitudinal electric components which polarization is along the axis of the tip rather than the horizontal E-components perpendicular to the axis of the tip. Therefore, in order to obtain effective enhancement, it is ideal to use a longitudinal E-component to excite at the tip apex.

### 2.2 System of TERS

As shown in Fig.2, a typical TERS system consists of a SPM, illumination/collection optics and a Raman spectroscope and other mechanical, optical and electronic devices in an integrated detection system. The separation between the tip and the specimen is regulated and as well as kept in few to tens nanometers during the scanning detection. The illumination optics focuses the incident light on the tip apex to excite the enhanced local field and Raman signal. The localized spectral signal is scattered and converted by the tip apex and then collected with the collecting optics in the far-field. In fact, the obtained signal is mixture that includes the tip-enhanced near-field Raman signal and the far-field background signal. Then, all of collected Raman signal are guided to the spectroscope to be analyzed. In TERS, the far-field Raman signal is regarded as the background noise, because it contains the spectral information of the whole illuminated area rather than only the nanometer zone beneath the tip. A common method to deduct the far-field background is to withdraw the tip from the focus to detect the far-field Raman signal and subtract it from the mixture signal. The fundamental of TERS system design is to raise the efficiency of signal excitation, collection and improve the SNR, that is, to increase the near-field signals while suppress the background noise. By means of point-by-point scanning the tip over the specimen, the corresponding topography and Raman signal of the specimen can be obtained simultaneously.

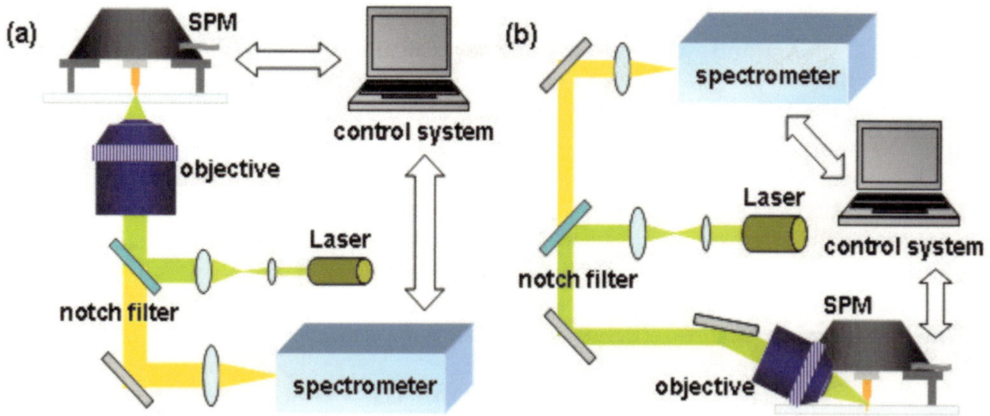


Fig. 2. Schematic diagrams of TERS systems in transmission mode (a) and in reflection mode (b).

#### 2.2.1 Illumination/collection configurations

According to the illumination/collection configurations, TERS systems can be divided into transmission-mode ones and reflection-mode ones.

Considering the structural and spatial constrains, the illumination and collection configurations often shares the same one-objective-lens-based optical path. The configuration of the transmission-mode TERS system usually employs an inverted microscope. The incident laser is tightly focused with a high numerical aperture (N.A.) objective on the tip apex and the backward-scattered tip-enhanced Raman scattering signal is collected with the same objective. The transmission-mode TERS is of ability to reduce the far-field background signal with a high N.A. objective and the SNR is considerably improved and the system construction is relatively simple. Nevertheless, the applicability of transmission-mode TERS is restricted. It can only be suitable to detect transparent or sparsely distributed specimens.

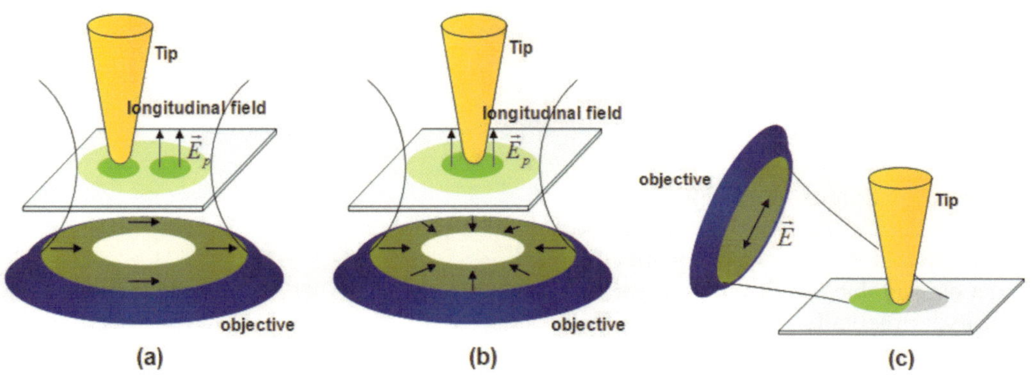


Fig. 3. Schematic diagrams of tip-enhanced Raman spectroscopy using (a) linearly polarized light (or beam) annular illumination; (b) radially polarized annular illumination; (c) linearly polarized side illumination.

The reflection-mode TERS systems are suitable for detecting any specimens regardless of the transparency. This kind of systems often utilizes a side-illumination/collection configuration (Fig. 3(c)). The incident light pass through a long working distance objective located at the side of the SPM scanning head and is focused on the tip apex. In this configuration, only the long working distance objective with lower N.A. (NA<0.6) can be used. The not-so-tightly focused laser oblique incident on the specimen surface and illuminate a relatively large elliptical area. The undesirable increased illuminated area leads to stronger far-field background noise. Therefore, the side-illumination reflection-mode TERS suffers from the lower SNR and signal collection efficiency.
One of illumination problems is how to generate the strongest longitudinal electric components to excite the most efficiently the field enhancement close to the tip. One of the solutions is to choose appropriate polarization light as the exciting beam (Hayazawa et al., 2004; Saito et al., 2005; Anderson et al., 2006). The possible choices include linear polarization, circular polarization, radial polarization and azimuthal polarization light. It will be discussed in detail in the following section.

### 2.2.2 Distance control and topography measurement

Intend to enhance the localized Raman signal and obtain high resolution near-field optical information, a sharp metallic tip is precisely located in the near-field region of the sample under test and controlled at a certain few nanometers' distance from the sample surface. Generally, the distance is maintained within 10 nm. In TERS system, the SPM technology is employed to control the tip-specimen separation. During the scanning detection, the constant separation is maintained by the feedback control from SPM. Furthermore, the height of the specimen surface is measured and consequently the topography data can be obtained.
Generally, there are three types of SPMs often used as components to build up TERS systems: scanning tunneling microscope (STM), atomic force microscope (AFM), and shear-force microscope (SFM).

## 2.3 Key technology in TERS system

Typical TERS system combines the SPM, illumination/collection optics, Raman spectroscope with other mechanical, optical and electronic devices. There are certain key technologies involved in this integrated detection system, including the fabrication of TERS tips, selection of SPM, dual scanner/dual closed-loop controller, illumination with radially polarized beam, illumination/collection configuration based on parabolic mirror, characterizing optical properties of the tips with white light, and substrates.

### 2.3.1 Fabrication of TERS tips

It is indubitable that the tip is the most crucial element in TERS deciding both spatial resolution and Raman enhancement efficiency.
The most suitable metal materials for tip are silver and gold, which are proven theoretically and experimentally for TERS systems. They both can provide most effective tip-enhancement under visible light illumination. Silver is capable of providing stronger enhancement, because its imaginary part of permittivity is smaller (Johnson & Christy, 1972). However, since the Ag tips are rapidly oxidized in air, they need to be protected properly and used timely. On the other hand, gold tips are chemically inert against oxygen and stable against radicals.

The fabrication of metal-coated AFM tips used in TERS system (AFM-TERS tips) is often based on the commercial Si or Si3N4 AFM tips. And the physical coating method (evaporation, deposition, or sputtering) is utilized to metallize the AFM tip. It is need to note that the stress of the metal coating might bring in the warping problem of the AFM cantilever and affect the performance of the tip. Typically, the TERS tips acquired via this method are unlikely to provide significant enhancement and are frequently being scraped during scanning (Kharintsev et al., 2007).

The coating of the AFM tip can also be realized via chemical approach such as silver mirror reaction (Saito et al., 2002; J. J. Wang et al., 2005). However, the reactants may contaminate the cantilever and worse, yet still attach to the tip apex. Another issue concerning AFM-TERS tips is that the Raman signal of the inner material of the TERS tips (Si or Si3N4) is also being enhanced and exists as the noise. Especially in the research on semi-conductor materials, this issue should be pay closer attention to (Lee et al., 2007).

The preparation of STM-TERS tips is comparatively mature. The tips are fabricated from high purity single-crystal gold (Ren et al., 2004) or silver wires. The metal wires are shaped with electrochemistry etching. This method is capable of providing gold STM-TERS tips with effective enhancement and small tip apex size. And the success rate (tolerant) is high. However, it is not so good to fabricate silver wire tip (Pettinger et al., 2002).

The preparation method of SFM-TERS tips is more flexible. Addition to the previously mentioned etched gold tips (Hartschuh & Novotny, 2002), the etched tungsten tips with gold or silver coating (Sun & Shen, 2001), are also suitable to the SFM in TERS system. In recent years, the method demonstrates obvious advantage of attaching gold or silver nano-particles to the apex of the fiber probe (Kalkbrenner et al., 2001). By optionally selecting the size of the nano-particles, the optical property of the TERS tip can be designed to obtain strong Raman enhancement. In addition, since the refractive index of fiber is lower, far-field Raman scattering enhanced causes by the body of the tip can be effectively suppressed.

In addition to these mentioned methods above, in order to achieve precise control of the size and structure of the tip for improving the enhancement efficiency, focused ion beam etching (FIB) has been used to produce TERS tip (Frey et al., 2002). But since the FIB processing is complex and costly, it has not being widely used.

### 2.3.2 Selection of SPM

As shown in Fig.4, STM, AFM, and SFM are the three types of SPMs commonly used as components to build up TERS systems. They have different feedback mechanisms and respective technique features. Although, they are all able to regulate effectively the tip-specimen separation and obtain the topography. The TERS systems based on the different types are of the different technical properties. Therefore, the TERS systems based on appropriate SPM components can be selected in dependence to the sample and experimental condition. The performance properties of the STM, AFM, or SFM in TERS systems are list in Table 1.

### 2.3.3 Dual scanner/ dual closed-loop controller

Different from common SPMs, our homemade TERS system has a configuration of dual-scanner/dual closed-loop controller in which the two scanners and two closed-loop controllers are able to work independently.

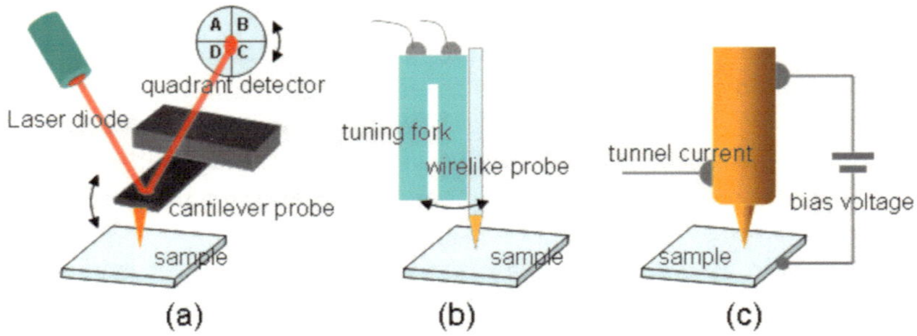


Fig. 4. The SPM equipment for TERS: (a) Atom force microscope; (b) Shear-force microscope; and (c) Scanning tunnel microscope. (R. Wang, 2010c)

| SPM | Feedback types | Advantages | Disadvantages |
|---|---|---|---|
| AFM | Optical lever feedback | 1) without limit to sample types and environment;<br>2) easy to operate, convenient to be combined with other AFM imaging methods;<br>3) able to study pressure-induced spectral changes by tip (Watanabe et al.,2004; Saito et al.,2008). | 1) easy to damage metallic coating of tip;<br>2) low-enhancement factor;<br>3) stray light from optical lever interferes spectral measurement. |
| SFM | Tuning fork/Shear force feedback | 1) without limit to sample types and environment, little damage on samples, especially for biological sample in liquid;<br>2) easy to fabricate tip;<br>3) convenient to be combined with inverted and upright microscope respectively. | 1) complicated operation;<br>2) poor lateral resolution. |
| STM | Tunneling current feedback | 1) high spatial resolution and high control precision;<br>2) working in gap mode with high detection sensitivity and high-resolution spectral imaging;<br>3) mature fabricating technique to produce tips. | 1) only conducting samples or ultra-thin samples distributed on conducting substrate;<br>2) mostly working in reflection mode due to opacity of sample or substrate.<br>3) surface undulation of sample no more than a few hundred nanometers usually |

Table 1. Comparisons of TERS systems based on AFM, SFM and STM

#### 2.3.3.1 Dual scanner

The scanning types of SPM can be divided into two types, sample stage scanning and tip head scanning. For the former type SPM, a scanning stage on which the sample is attached to is driven by the scanner and moves in a raster pattern below the tip. The tip only moves along its axis to adjust the tip-specimen separation. For latter type SPM, the tip is always amounted on a 3-D movable scanning head, and the sample is kept still. The progressive scanning and the tip-specimen separation are both controlled by the tip scanning controller. Conventional SPM usually works with one controller in one particular mode. However, SNOM and TERS systems work more complexly than the conventional ones. For SNOM or TERS detection, it needs to appropriately couple the incident light beam to illuminate and excite the sample in a certain condition\manner, which may relate to wavelength, incident angle, focused beam, specific polarization. Additionally, appropriate optical configuration for collecting the scattering, transmission, or reflection light in 3-D space or a specific direction is also required.

However, it is difficult to meet various kinds of detection demands to the diversity of specimens with single scanning type systems. For example, if the CNTs specimen is put on the sample stage, and the optical axis of the illumination configuration is coaxial with the tip axis, stage scanning mode is then appropriate for the optical alignment principle. But in other conditions the tip scanning is requested, such as detecting the SPP waveguide which needs to be precisely coupled with external fixed optical path for excitation and collection. Although the tip might be possible to slightly deviate from the optical axis during the scanning, the impact of the nanometer deviation can be negligible.

Therefore, it is preferable to employ the dual-scanner configuration which can provide the sample stage scanning and the tip head scanning in TERS system respectively. Its advantages include more precise control, arbitrarily choosing sample, and scanning mode to any different requirements of sample detection, illumination alignment, or optical coupling mode. It is convenient to tip-illumination alignment and ensures the optimal detection results.

The common requirements to the dual-scanner configuration:

a. Accurate, real-time, and closed-loop measuring/ to control the tip-sample separation;
b. Precise alignment of the tip apex with the excitation focus;
c. Nanometer accuracy, high repeatability, point-by-point scanning in raster mode over the whole detected region to obtain the sample information.

#### 2.3.3.2 Dual closed-loop controller

In order to ensure two independent scanners to work with high-precision positioning and high repeatability two independent closed-loop controllers should be equipped in a TERS system. Nanometer accurate positioning and scanning are usually realized with a piezoelectric actuator. The measuring time for TERS is usually much longer than SNOM imaging and in result the higher stability and repeatability are required in the TERS system.

Firstly, the piezoelectric scanner might drift a certain distance with ambient temperature fluctuation. Thus, the real-time closed-loop control is required to correct and eliminate the impact of drifting, and improve the accuracy and repeatability. It is of significant importance to collecting, processing, and analyzing the Raman information and the topography data correspondingly.

Another important issue to be considered is that the certain integration time is needed to each detected point to acquire Raman spectrum data in TERS detection. Since the Raman signal is extremely weak, even with the highest sensitivity spectral detector, it still needs to spend certain integration time. Meanwhile, the location and the topography data at the point are recorded. It is distinct from the regular SPMs scanning point-by-point and line-by-line with relative high frequency, the scanning speed of TERS detection is rather slow or even stops at each detected point. It is evident that to measure the topography and point-by-point spectra with TERS requires much longer detection time than the regular SPMs. Therefore, TERS system demands the higher stability and repeatability which facilitate the tip and sample maintaining the relative position during the detection process.

Popular commercial SPM systems are usually only equipped with a 3-D controller. At the mean time, it is capable of closed-loop to control only one of the two scanners, either the scanner of the tip or the scanner of the sample stage. But the other not-under-controlled scanners may drift or disturbed by the external environmental perturbations such as the mechanical vibration. These all lead to the spatial mismatch of the detected spectral data and morphology data or even drive the tip out of the detected region. It seriously affects the results of experiments and the corresponding identification upon the Raman mapping and the relative topography data.

Therefore, two independent controllers should be used to realize the closed-loop control of the tip scanning head and the sample stage scanning precisely and respectively in TERS systems. Under closed-loop control, the 3-D scanning and positioning with high accuracy and repeatability can be realized by the two scanners respectively, and the internal drift of the scanner and the external perturbations of the environment can be effectively reduced. It ensures the spectral detection, topography detection, and the spatial corresponding data collection of the Raman spectra and the topography of the CNTs and other specimens. It is quite benefit for the corresponding analysis of the TERS result to recognize and identify the molecular structure, chemical information and characteristics of the sample.

### 2.3.4 Illumination with radially polarized beam

As mentioned above, in the lighting-rod effect, as the incident optical beam with the electric component parallel to the tip axis illuminates on the tip, the charge is driven to the foremost of the tip and forms a large surface charge accumulation at the tip apex. It means that the efficient enhancement of the local electric field at the tip apex will arise with illumination by the longitudinal E-component or incident polarization parallel to the tip axis (Novotny et al., 1997). While the polarization of the incident beam is perpendicular to the tip axis, the tip apex remains uncharged. So it does not actively contribute to the local field enhancement , also it may bring in the background noise.

Therefore, in order to obtain effective enhancement, it is crucial to form a longitudinal (parallel to the tip axis) E-component to efficiently arise the lighting-rod effect at the tip apex. The E-field distribution at the focus is determined by both the incident beam polarization and the optical configuration. It is crucial to make them to match to each other in the experimental setup.

How to generate experimentally a longitudinal E-component parallel to the tip axis?

In side-illumination situation, a linear polarization beam in TM mode illuminates the sample in a large incident angle. The most of the E-component of the incident optical field can be oriented parallel to the tip axis. And the strong longitudinal field excites effectively the strong field enhancement based on the lighting-rod effect.

In transmission-mode TERS systems, the incident beam is focused with high N.A. objective lens from the bottom to the tip apex as shown in Fig.3 (a) and (b). As the beam is linearly polarized (e.g. along x-axis), the distribution of the longitudinal field presents as two symmetric side lobes while the center is almost zero at the focus is shown in Fig.5(b). The reason is that the phase of the longitudinal E-component is out in the focus. If the tip is positioned at the center, the near-field enhancement is inefficient and the far-field noise is introduced by the transverse field illumination. To improve the enhancement and decrease the background noise, the tip has to be adjusted to either of the side lobes in the side of the focal center.

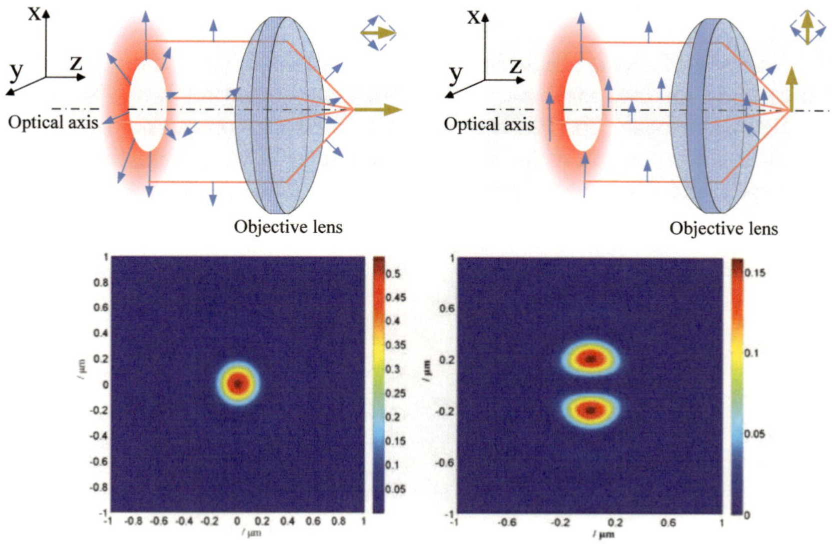


Fig. 5. Longitudinal field distributions at the focuses of radial polarization (a) and linear polarization (b) beam focused by high N.A. objective.

Fortunately, a doughnut radially polarized incident beam focused by high N.A. objective is found to be more suitable in this kind of configurations. The first reason is that the longitudinal E-component is in-phase and causes enhancement at the focus as shown in Fig.5 (a). The second reason is that the symmetry of the doughnut radial polarization beam is more suitable to the axial or cylindrical symmetric optical system. No matter where one looks at, in any cross-section in the optical axis the field distribution is quite symmetric. The E-vector vibration directions of the radially symmetric rays are pointed against to each other. As focused by an objective at the focus with symmetrical incident angle, their E-vector vibration direction is in-phase in the longitudinal orientation and constructive interference. Especially when the radially polarized beam is tightly focused by a high N.A. objective, the stronger longitudinal field expected theoretically is effectively formed at the focus, Fig. 5(b). Therefore, the local enhancement can be actively excited by the longitudinal field as the tip entering the focus. Additionally, with radially polarized illumination, the longitudinal field area just locates at the center of the focus. It facilitates more convenient realization of optimal tip-longitudinal field alignment , coupling, efficient signal excitation and collection.

### 2.3.5 Illumination/collection configuration based on parabolic mirror

One of the significant advantages of reflection-mode TERS systems is that the detectable sample is not restricted no matter to transparency or opaque sample. This not only expands measuring range of specimens, but also allows more flexibility to the choice of substrates. The appropriate substrate can further promote the enhancement or reduce the background noise in TERS. But with side-illumination/collection configuration, only a long working distance objective with comparative low N.A. (typically smaller than 0.6) can be utilized. This results a strong far-field background noise and low collection efficiency.

Of late years, an illumination/collection configuration with a parabolic mirror (PM) has been introduced into the reflection-mode TERS system (Steidtner & Pettinger, 2007; Sackrow et al., 2008). This novel solution is considerable and promising in the research and industrial field of TERS.

The reflective surface of PM is shaped to an axial symmetric paraboloid. According to the geometric properties of the parabolic surface and the reflection law, the incident beam which is parallel to the axis of the PM is reflected by the reflector and converged to the focus of the parabolic surface. Instead of an objective lens, the PM converges the parallel incident light at the tip apex, and the enhanced Raman signal is collected with the same PM (Fig. 6).

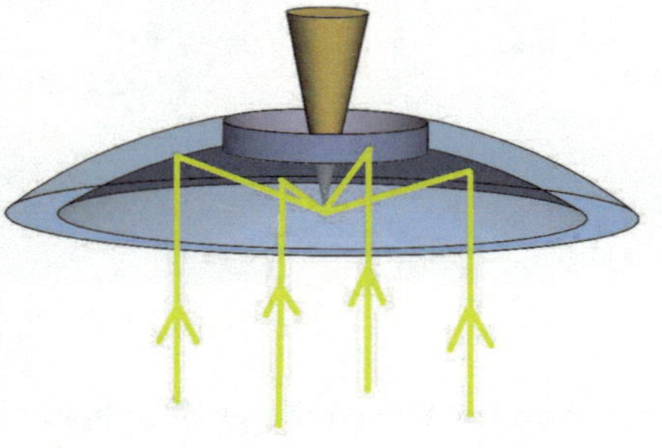

Fig. 6. Schematic diagrams of tip-enhanced Raman spectroscopy based on a parabolic mirror.

PM based reflection-mode TERS system has the following advantages:

a. In the PM based configuration, the focusing of the incident light is realized by reflection, instead of refraction in other objective lens based configurations. Because focusing with refection fundamentally avoid the chromatic aberration, regardless of the wavelength the incident parallel beam is focused at the fixed point. This facilitates the PM base TERS system, which can optionally employ different wavelength laser as excitation light (D. Zhang et al., 2009).
b. The N.A. of the PM used in this configuration can be up to approximately 1. With the high N.A. parabolic mirror, the incident light can be tightly focused at the tip apex to enhance the near-field Raman scattering. Still, the collection of the signal is also effective with the high N.A. parabolic mirror.
c. Unlike the side-collection one, the PM based configuration efficiently collects the signal from the whole 3-D space around the detected position.

d. The PM based TERS system works in reflection-mode, and neither the incident light nor the enhanced signal needs to transmit through the specimen. Therefore, it is capable of detecting any specimen regardless of the transparency or opaque.

It is also noticed that the optical system is cylindrically symmetric about the optical axis as well as the tip axis, like in the transmission-mode TERS. In order to generate a stronger longitudinal field at the focal region to excite the enhancement, a radially polarized donut-shaped incident beam should be utilized. Besides the more effective tip-enhancement result from the stronger longitudinal field, the far-field background noise is reduced simultaneously due to the smaller focal region.

But there are still several difficulties concerning the PM based TERS system. Firstly, this configuration requires the parabolic mirror with a higher precision surface shape. The PM which critically determines the illumination and collection efficiency is one of the most crucial components of the optics. In addition, the amount of far-field noise is related to the size of the focus, which depends on the shape of the reflector. Secondly, it is difficult to precisely align and couple the optical circuit. So it needs high skill and patience to adjust carefully.

### 2.3.6 Substrates

The TERS system rarely requires for the special preparation of the specimen or the specific experimental environment (air, liquid or vacuum). But it must be considered to reduce the far-field background noise and improve SNR. Yet, the highly dispersed thin film of specimen tends to demonstrate better TERS detection result. The choice of substrates on which the specimens dispersed is fairly tolerated.

Generally, the glass, mica, silicon or other commonly-used flat materials can all be utilized as the substrates in the TERS system based on AFM or SFM. In the STM based TERS system, single crystal metal or a metal coated surface is used. Numerous studies show that the material and microstructures of the substrate significantly affect the excitation of LSPs in the tip-specimen-substrate system. Considering that, only depending on the enhancement of the tip apex is hard to achieve the single molecule detection sensitivity and nano-meter spatial resolution, it would be ideal to further increase the enhancement factors and improving the confinement by appropriately bringing in the suitable substrate with nanometer structures.

In AFM or SFM based TERS system, using a SERS substrate can further enhance the Raman signal and quench the fluorescence. This technique is known as the SERS-TERS mode (Hayazawa et al., 2001). This kind of substrates can be prepared by depositing randomly distributed gold or silver nano-particles on the original substrate. As for STM based TERS system, the atomic-level smooth single-crystal gold surface is commonly used as the substrate. Under this composition namely gap-mode (Ren et al., 2005), an electromagnetic field enhancement can be greatly excited at the gap between the tip and the specimen.

Additionally, the smooth gold substrate eliminates the SERS enhancement induced with local surface fluctuations and provides an access to better understanding of the mechanism of the tip-enhancement effect.

## 2.4 Evaluation method of performances of TERS systems

Generally, there are 3 indicators used to evaluate the performance of a TERS system, including contrast, enhancement factor (EF), and the lateral resolution of the Raman mapping.

Contrast is defined as the ratio of near field tip-enhanced Raman intensity to the far-field Raman signal intensity, and is given by Eq. (1) (Schmid et al., 2007; Tarun et al., 2009):

$$C=\frac{I_{near}}{I_{far}}=\frac{I_{total}-I_{far}}{I_{far}}=\frac{I_{total}}{I_{far}}-1 \tag{1}$$

in which, the $I_{near}$ indicates the intensity of the near-field Raman signal from the region just below the tip apex, $I_{far}$ denotes the far-field Raman signal intensity as the tip is withdrawn, and $I_{total}$ is the TERS experiment detected Raman intensity, which is the mixture of the near-field and far-field Raman signal intensity. The contrast can be regarded as the SNR of the TERS detection. It reflects the excitation/collection efficiency and the detection ability of the system.

In certain studies, the ability of the tip induced Raman signal amplification tends to be more concerned, and it is known as the enhancement factor. Taking into account of the difference between the sizes of the regions that give rise to the near-field and far-field Raman signal, the enhancement factor can be expressed as (Schmid et al., 2007):

$$EF=contrast\times\frac{V_{focus}}{V_{tip}}=contrast\times\frac{A_{focus}\cdot h_{focus}}{A_{tip}\cdot h_{tip}}=contrast\times\frac{d^2{}_{focus}}{d^2{}_{tip}}\times\frac{h_{focus}}{h_{tip}} \tag{2}$$

where $V_{tip}$, $A_{tip}$, $h_{tip}$, $d_{tip}$ and $V_{focus}$, $A_{focus}$, $h_{focus}$, $d_{focus}$ represents the near-field on the tip apex and far-field illuminated at the focus sample volume, the cross-sectional area of the illumination, radius and penetration depth of the illumination respectively. These parameters mentioned above are difficult to be accurately measured in experiment and are estimated based on the experimental conditions generally.

EF can be regarded as the normalized contrast according to the near-field/far-field signal excitation regions. It is more suitable for comparing with the electromagnetic simulation to analytically investigate the performance of the tip and the enhanced mechanisms.

Since only the Raman signal from the region just below the tip apex can be enhanced, the effect in the near-field region is estimated in accord with the size of the tip apex whose diameter is only tens of nanometers generally. But the diameter of the laser focusing spot is usually in submicron region. As a result of the great difference between the size of the near-field and the far field illumination, the experimentally obtained EF is usually about $10^3$~$10^6$. There are still some gaps in the comparison of the electromagnetic simulation expected and experiments.

Presently, the reported EF obtained in experiment can be as high as $10^9$, and some research has achieved the single-molecule lever detection (Hayazawaet al., 2006; W. H. Zhang et al., 2007; Steidtner & Pettinger, 2008). Different forms of the formula should be applied to calculate the EF of samples with different sizes and transparency. The schematic of samples with different thickness and the corresponding EF estimation formula is shown in Fig.7. For thin samples (Fig. 7(a)), since the penetration depth of the near-field and the far-field illumination are the same, the excitation region is determined just by the far-field focal spot radius or the tip radius respectively. For thicker Raman active material (Fig. 7 (b)), the impact of the different penetration depth has to be taken into consideration. Due to the far-field illuminated sample volume is much larger than the tip excitation region, when the EF of the tip-enhancement is small the near-field signals may be submerged in the far-field background noise, and results in low contrast. The phenomenon has been confirmed in experiments (Mehtani et al., 2005). Strictly speaking, for the nano-dots and nano-wires, the TERS enhancement factors should also be estimated by using different formulas. Due to the differences between samples, the calculation of the EF is influenced by subjective factors. And the contrast, which is calculated just based on the experimentally measured values is of more practical significance. Nevertheless, the EF is still very necessary for the research on the electromagnetic enhancement mechanism of TERS. In 2005, Pettinger et al (Pettinger et al., 2005) ingeniously designed an experiment to avoid the subjective factors mentioned above. By comparing the photobleaching time of the single-layer Malachite green dye with and without tip enhancement, the more accurate TERS enhancement factor is obtained as about $6\times10^{6}$.

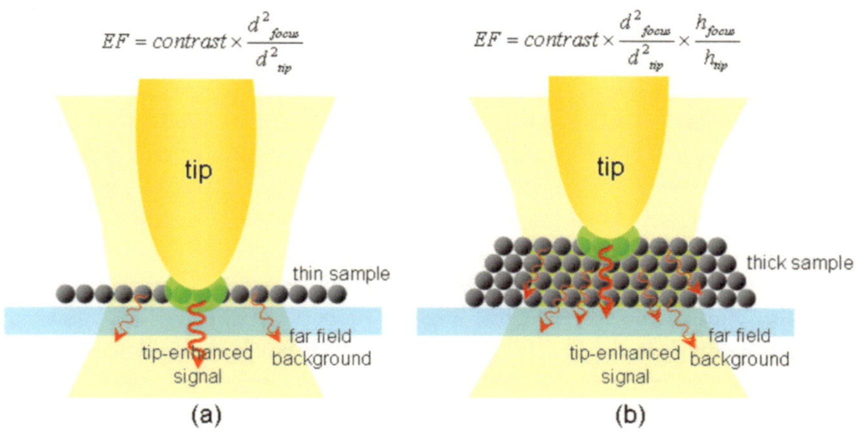


Fig. 7. Comparison of enhancement factor calculation (a) thin sample and (b) thick sample.

The lateral spatial resolution of TERS spectral imaging is mainly determined by the radius of the tip apex. Since the localized electromagnetic field enhancement is highly confined to the tip apex, the lateral resolution, spectral resolution may even be better than the resolution of topography. Nowadays, the highest lateral resolution of TERS image reported is less than 15 nm (Anderson et al., 2005). For improving the lateral resolution the topography and spectral imaging, tips with smaller radius should be used. Moreover, these tips are capable of contributing more to the tip-enhancement and field enhancement (Downes et al., 2008).

## 3. Preparation of CNTs specimen

Specimen preparation is an important link of applying the TERS method to the detection of practical specimens. It will finally determine whether the optimal measurement results can be obtained as well.

### 3.1 Requirement of specimen preparation

The in-situ non-destructive spectral detection with high resolution and high sensitivity may be accomplished with TERS system. Since the Raman spectrum directly reflects the chemical information of samples, the specimen in TERS detection is dispensed with extra label. In general, the specimen preparation, especially for the preparation of CNTs samples, needs to satisfy following requirements:

a. The smooth substrate surface. The fluctuations of the substrate surface should be less than the sample-size scale.
b. Firmly attaching and immobilizing the specimen to the substrate surface.
c. Transparency. When using transmission-mode TERS system to detect the sample, the specimen on the substrate should be transmission. The specimen can be highly dispersed or diluted to obtain good transparency. However, in reflection-mode TERS system the detected specimen is regardless to its transparency.
d. conductivity (STM-TERS)

According to specific samples and measuring conditions, there might be certain additional requirements for the specimen preparation.

### 3.2 Common preparation methods for CNTs specimen

In our transmission-mode TERS system, the laser illuminates the sample from the bottom and goes through the specimen to focus on the tip apex and locally excited Raman scattering at the specimen surface. Additionally, the enhanced Raman signal also has to transmit through the specimen to be collected. Therefore, the detected specimen is optimal to be of good transparency. The CNTs specimen can be highly dispersed into a thin film. Meanwhile, the sample is required not to chemically react with the substrate or the tip and maintaining stabile properties in the whole measuring process. The specimen should also be firmly attached on the substrate surface to avoid relative movement during the scanning. In order to obtain high-resolution images, highly dispersed samples are generally used. Therefore, the transmission-mode TERS system is more suitable for detecting sparse thin films of CNTs samples with good transparency. For example, the specimen preparation methods used in our transmission-mode TERS detection are presented here.

#### 3.2.1 Preparation method for single wall carbon nanotube sample (SWNTs)

The SWNTs specimens used in our experiments are bundles of SWNTs. The SWNTs were synthesized by arc-discharge method. After purification, the high purity (> 95%) SWNTs powder can be obtained. Then 0.01mg powder was dissolved in 10mL ethanol. The solution was evenly dispersed by ultrasonic dispersing method. One droplet of the solution was taken and dripped on a 120μm thickness cover slip. After air-curing, the transparent CNTs specimen has been prepared as well as for detection (Guan, 2006; Li et al., 2004).

#### 3.2.2 Preparation method for double wall carbon nanotube sample (DWNTs)

The transparent thin DWNTs membrane was prepared following the series of processing (Wei et al., 2006): firstly, the synthesized DWNTs were soaked in the $H_2O_2$ solution with 30% Concentration for 72 hours. Then, HCl solution (37%) was added to remove the amorphous carbon, catalyst particles and other impurities. The up to 90% (by mass fraction) purity DWNTs were obtained. By dripping several drops of ethanol or acetone, the purified macro-membrane of DWNTs were rapidly spreading to ultra-thin film and floating on the surface of the solution. The film of DWNTs can be easily collected with a cover slip. The flat and transparent thin film of DWNTs was obtained, and after air-curing the highly dispersed DWNTs were immobilized on the substrate surface.

## 4. Detection of CNTs

The TERS technology has been widely developed and applied in the fields of physics, chemistry, material science, and biology in nanometer scale. Especially, the detection of CNTs will be presented and discussed as an emphasis of this section.

### 4.1 Applications of TERS

The significant advantage of TERS is that it can provide corresponding topography and Raman mapping of the nano-material specimen with high spatial resolution and detection sensitivity. Since 2000, TERS has been experimentally proved practicable (Stöckle et al., 2000; Hayazawa et al., 2000; Anderson, 2000; Pettingeret al., 2000) and shows the potential of nanometric spectral detection. Then, the research on TERS's applications has been being advanced continually.

Up to now, it has been applied to dye molecule detection (Steidtner & Pettinger, 2008; W. H. Zhang et al., 2007), semi-conductor material determination(Sun et al., 2003; Lee et al., 2007; Saito et al., 2008; Sun & Shen, 2001) , biological specimen identification(Anderson et al., 2003; Watanabe et al., 2004; Bailo & Deckert, 2008), and nanao-material characterization (Hayazawa et al., 2000; Hartschuh et al., 2003; J. J. Wang et al., 2005; Qian et al., 2006).

It is distinct from the single molecule level or single-molecule state distributed specimen detection, TERS approach aims on the real single molecule, individual molecule Raman spectral detection. In 2008, Pettinger et al. directly measured the spatial location and the corresponding Raman spectra of the single BCB molecule absorbed on Au (111) surface using STM-TERS system in ultra-high vacuum (Steidtner & Pettinger, 2008). The experiment shows that under resonance Raman excitation, the enhancement factor is to $10^6$. It is enough to satisfy the requirement of the single molecule detection sensitivity of the dye molecules in the TERS characterization.

The in situ spectrum detection of biological samples is another research hotspot in TERS applications in recent years. Since the Raman spectra directly reflect the molecular structure information of the sample, the specimen in TERS detection is dispensed with extra label. Meanwhile, TERS is capable of obtaining high resolution in situ non-destructive detection with high sensitivity. In 2003, Anderson et al. (Anderson et al., 2003) applied TERS technique to the detection of drosophila compound eyes, and measured the fine structures of the eye surface and the near-field Raman spectra at different positions. In 2006, Hayazawa et al. (Hayazawa et al., 2006) measured the near-field Raman spectroscopy of the adenine nano-crystal sample. Compared with the standard far-field spectrum, slight frequency shift of the TERS spectra is reviewed. In 2008, Bailo et al. (Bailo & Deckert, 2008) detected the topography of single-stranded cytosine RNA and its tip-enhanced Raman spectra at several different positions. The single-base detection sensitivity of the TERS system is indirectly proved. The author further noted that although the TERS system is not yet sufficient to the spectral imaging with the single-base spatial resolution, TERS technology is still expected to directly sequence the DNA or RNA samples by using certain detection and data processing method. TERS research on live Biological macromolecules and virus has been commenced. Recently, Deckert et al. (Cialla et al., 2009) measured the TERS signals of the single tobacco mosaic virus at different positions, and the characteristic Raman shifts were identified. Budich et al. (Budich et al., 2008) also reported the TERS characterization of the Staphylococcus epidermidis cell wall in liquid environment. These applications all indicate the great potentials of the TERS technique in biological and life science research.

The demand of nano-material characterization is a major motivation to promote the development of TERS. CNT is a kind of the typical 1-D nano-materials. Much attention has been paid to the research and detection of CNT, Due to its outstanding physical, mechanical, thermal and electrical properties along with application potentials. In 2003, Hayazawa et al. (Hayazawa et al., 2003) applied TERS method to detect the SWNTs specimen and obtained the diameter and chirality of the CTNs by analizing the TERS reaults. Harschuh et al. (Hartschuh et al., 2003) also utilized TERS to characterize SWNTs and simultaneously acquired the topography and the TERS mapping. Comparing the features in the two images, the spatial distribution of the CNTs can be accurately recognized. In this detection, the spatial resolution of TERS mapping reached up to 23 nm, even better than that of topography, which is 29 nm. With the development of the tip preparation and the TERS realization, better than 15 nm TERS resolution can be achieved nowadays (Hartschuh et al., 2005; Anderson et al., 2005; Huihong et al., 2008: 2243-6). In 2005, Hartschuh et al. measured the TERS and tip-enhanced photoluminescence (PL) mapping of SWNTs with better than 15 nm resolution (Hartschuh et al., 2005). The structural defects and optical properties of CNTs may be further revealed with the combination of TERS detection and tip-enhanced photoluminescence (PL). In 2006, Kawata et al. imaged SWNTs bundle within 700 nm×700 nm region. It validated that the radial breathing mode (RBM) feature inducing the Raman shift corresponding to the vibrational mode of the graphitic layer in the radial direction is sensitively and directly depended on the diameters of the tube. By imaging a serial of TERS mappings according to different RBM Raman shifts, the distribution of SWNTs with different diameters can be obtained separately. Lately, Kawata's group reported another way to further improving the resolution of TERS. In this method, the CNTs below the tip are locally pressed by the tip apex, and the slight Raman frequency shift induced by the pressure can be accurately detected. The technology is employed in nanometer metrology and spatial resolution is to 4 nm (Yano et al., 2009).

Carbon nanotubes are categorized as single-walled nanotubes (SWNTs), double-walled nanotubes (DWNTs) and multi-walled nanotubes (MWNTs). According to the difference in sizes, structures and features of them, SWNTs and MWNTs will be measured. Also experimental examples will be provided and discussed respectively in the following passages.

### 4.2 Detection of SWNTs

Generally speaking, the Raman peaks of SWNTs spectra represent the major vibration modes in CNTs, the radial breathing mode (RBM), graphite-like mode (G-mode corresponding to G-band), and defect-related mode (D-mode, D-band)(Dresselhaus et al., 2005; Rao et al., 1997).

The Raman shifts related to RBM is at the low frequency region of the spectrum. They are associated with the radial direction vibration modes of the nanotubes and directly reflect the diameters of the SWNTs. The G-mode vibrations of SWNTs are induced by the planar vibrations of the carbon atoms. The G-mode vibrations bring in Raman shifts in G-band, which is generally around 1600 $cm^{-1}$. Also most graphite-like materials all generate the G-band in Raman spectra. The D-mode of CNTs is related to the out-of-order defects of the CNTs or the amorphous carbon in the sample. The D-mode vibrations of SWNTs result in the Raman shifts around 1300 $cm^{-1}$. The ratio of G-band to D-band is an important term for the evaluation of the purity of the SWNTs specimen (Eklund et al., 1995).

#### 4.2.1 Conventional Raman spectrum measurement of SWNTs

For comparison and calibration, the conventional far-field Raman signal of the SWNTs specimen was measured before the TERS detection. The photon-counter integrated time is set 100ms in the Raman spectroscope, Renishaw RM2000. The result is shown in Fig. 8(a). From the far-field Raman spectrum, several Raman shift feature peaks of the SWNTs specimen can be distinguished. One is the G-band corresponding to the planer vibration. It breaks up to several characteristic peaks as the fold in Brillouin Zone and mainly represents as $G^+$ band in the high frequency region (1592 $cm^{-1}$) and G- band in the low frequency region (1569$cm^{-1}$). Second is the D-band at 1347 $cm^{-1}$, which is resulted from amorphous carbon or out-of-order defects in SWNTs sample. The purity of SWNTs sample can be evaluated based on the ratio of G band and D band (Dresselhaus et al., 2005). As the experimental result indicated, the ratio of G and D band of the specimen is 26.

Another characteristic Raman shift corresponding to RBM was detected in low frequency region and accumulated 3 times with the integrated time, 10 seconds. Considering that the sample is bundle of SWNTs and the tube-tube interaction in the bundle, the relationship between the diameter $d_t$ (nm) of the nanotube and the RBM band $\omega_{RBM}$ ($cm^{-1}$) can be expressed as following (Rao et al.,2001),

$$\omega_{RBM} = 224/d_t + 10 \tag{3}$$

This is a modified formula for a bundle of SWNTs slightly different from isolated SWNTs. As the detected RBM frequency is 170$cm^{-1}$ (Fig. 8 (b)), the diameter of the SWNTs is estimated to be 1.4 nm.

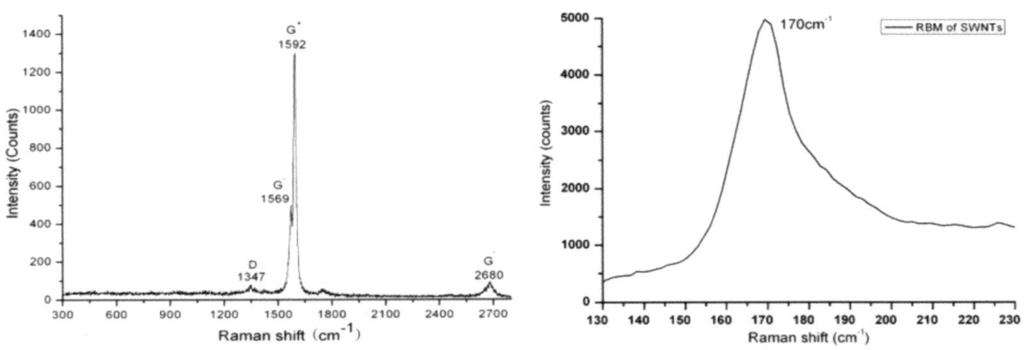


Fig. 8. (a) Raman spectrum of SWNTs (b) RBM of SWNTs

#### 4.2.2 Topographic and TERS measurement of SWNTs bundle

The SWNTs specimen is detected with a transmission-mode TERS system based on an AFM. Firstly, the topography of the SWNTs specimen is obtained by AFM (Fig.9). It can be indicated that in the sample most SWNTs exist in the form of bundles due to the action of Van der Waals force. On the SWNTs bundle one position marked with cross is selected as the interest point of TERS detection. Then, the position of the specimen is relatively moved to the near-field region and just below the metallized tip apex. The TERS spectrum of the SWNTs specimen is shown in Fig.10 (a). Then, the far-field Raman spectrum at the same position is detected ( Fig.10 (b) ), as the tip is withdrawn from the near-field region of the specimen (Wu et al, 2009).

Fig. 9. Topographic image of a Single SWNT bundle with diameter 100nm, by TERS.

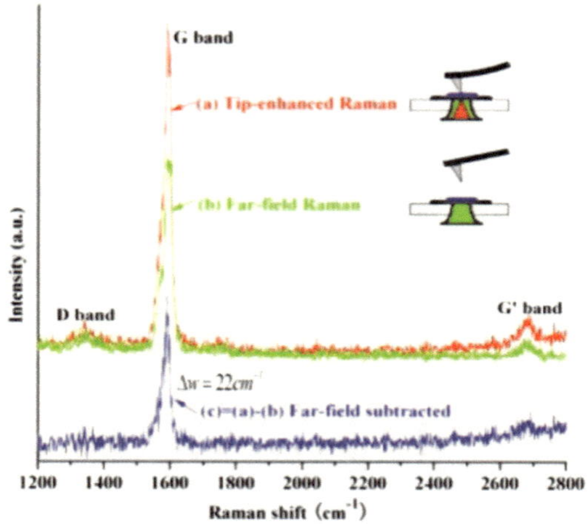


Fig. 10. TERS and far-field Raman spectra of a single SWNT bundle.

Comparing the TERS spectrum with the far-field Raman spectrum, it can be noticed that the frequency of G-band Raman peak is not obviously shifted and the shape is not evidently changed. In addition, the characteristic G-band is considerably enhanced as a result of the tip-enhanced electromagnetic field, while the background noise is not obviously increased. This can be explained as that the noise is mainly comes from the far-field stray light in the illuminated area and not Raman scattering localized and enhanced by the tip. While Raman signal from the localized zone is greatly enhanced by the tip-enhanced electromagnetic field highly confined to the tip's apex. It results in the higher SNR detection results. The pure near-field Raman signal of SWNTs can be obtained as shown in Fig.10(c), by subtracting the far-field signal from the TERS spectrum. The enhancement factor is calculated to be 230 according to the formula given in Fig. 7 (a) for thin film specimens. The tip enhancement factor will be much larger than that value, because the practical diameter of far-field laser spot is always larger than the size of the diffraction-limit used in the estimation. It indicates adequately that TERS technology is of novel and excellent performances in detection of weak Raman signal in nanometer localized region.

### 4.3 Detection of MWNTs

Multi-walled nanotubes (MWNTs) are formed with multiple coaxial rolled layers of graphite. Due to the structures and compositions, they have similar Raman shifts with SWNTs, but still with several distinct features. Additionally, being the simplest manifestation of MWNTs, the double-walled carbon nanotubes (DWNTs) are considered to be an ideal model for researching the interaction between graphite layers, as well as the connections between the SWNTs and MWNTs.

#### 4.3.1 Detection of DWNTs

A DWNT has the structure of two coaxial hollow cylinders formed by two layers of rolled-up graphite. The typical interlayer space between the inner and the outer walls, ranges from 0.33 nm to 0.42 nm. DWNTs maintain several outstanding properties of SWNTs, yet have some distinct ones, such as higher stability and particular electronic and optical properties (Gennaro et al., 2009). In Raman spectra of DWNTs, the graphite-like band (G-band) region arose by the graphitic layers is regarded as one of the feature Raman shifts of CNTs. And the RBM frequency corresponding to the radial direction vibration of the nanotubes is sensitively and directly depending on the diameters of the inner and the outer walls.

In this practical example, DWNTs specimen was investigated using our home-made transmission-mode TERS system (Wu et al., 2010). The Raman enhancement factor was calculated. The tip-enhanced Raman spectra especially in the G-band and the RBM regions are obtained. A tip-pressure induced RBM band shift of the DWNT is observed, and the experimental results are discussed.

In this research, a backward-scattering TERS system with a golden AFM tip is used. The radius of the tip apex is about 30 nm. The DWNTs specimen is prepared using the method mentioned in section 3.2.2. And the DWNTs used here were synthesized by chemical vapor deposition (CVD) process. In TERS detection, the excitation laser (532 nm) after expanding and collimating is focused on the tip by a high-NA objective (NA = 0.95). Also the tip-enhanced Raman scattering from a confined region of the specimen is efficiently excited. Then the Raman scattering is collected by the same objective lens and channeled into the Raman spectroscope to be analyzed.

In advance, the topography of the DWNTs specimen was detected (Fig. 11(a)). The height cross section along the dashed line is shown in Fig. 11(b). It shows that the heights of the 3 figures are all about 10 nm. Judging from the height, the two arc-like bundles each contains

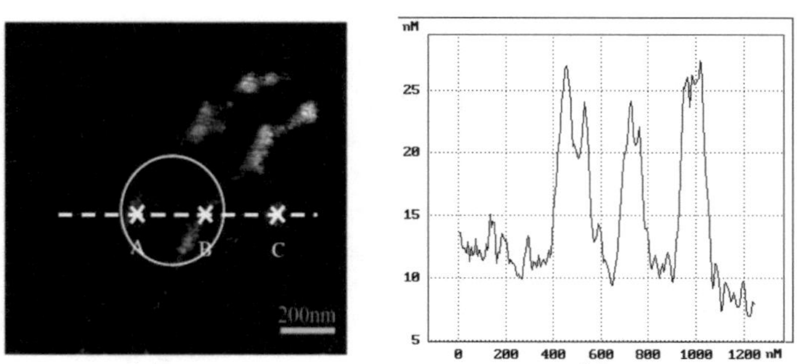


(a) (b)

Fig. 11. (a) Topographic image of the DWNTs specimen (scanning area: 1.2 μm × 1.2 μm). (b) Height cross section along the dashed line in (a).

estimated about 10 to 15 DWNTs. After acquiring the topography of the specimen, the region-of-interest could be narrowed down and defined to the three figures. Then, the single-point TERS detection was carried out at each interested position. The detected TERS spectrum of position A of the specimen is shown in Fig. 12. For comparison, the far-field Raman spectrum was detected as the tip was withdrawn from the near-field region. The separation between the tip and the sample surface was about 1.5 μm then.

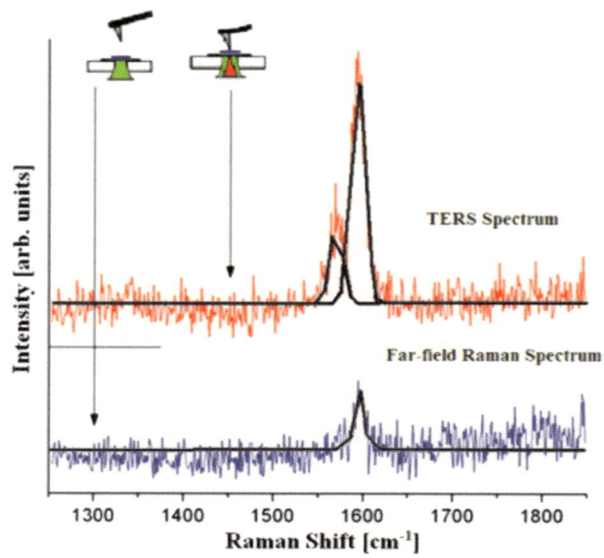


Fig. 12. Far-field Raman spectrum and TERS spectrum of DWNTs at position A in Fig. 11(a).

Compared with the far-field spectrum, the TERS spectrum indicates that a strong enhancement at the G-band is obtained. With Lorentz fitting, it can be observed that the G band in the TERS spectrum splits mainly into two peaks: 1590 cm$^{-1}$ ($G^+$) and 1564 cm$^{-1}$ ($G^-$). However, in the far-field Raman spectrum the $G^-$ band may be too weak to be distinguished from the noise. According to the Eq. (2) given in section 2.4, the EF is calculated to be $3.6 \times 10^2$.

With the tip fixed at the laser focus, the specimen is moved along the dashed line shown in Fig. 11(a). The TERS spectra of positions A, B, and C are detected in succession (shown in Fig. 13). TERS spectra of positions A and B both show similar Raman peaks in the G-band and almost no signals in the D-band. However, at position C the Raman peak in the D-band is strong and the G band signal is weak. By comparative analysis of the topography and the TERS spectra it can be ratiocinated that the two arc-like lines are bundles of DWNTs, while the granule at position C is amorphous carbon respectively. The Raman signal from C includes the D-band from amorphous carbon and the weak G-band signal might from DWNTs in the vicinity of C point illuminated area in the far-field.

For further characterization of DWNT specimen, the Raman spectra in the RBM region were detected and analyzed to estimate the diameters of the DWNTs' inner and the outer walls, respectively. The Raman shift in RBM region can be directly determined from the diameter of the nanotube. Thus, it's quite useful in estimating the sizes of the inner and the outer walls of DWNTs and the corresponding deformation of DWNTs under radial pressure.

Since the interaction between the inner and the outer walls and the force exerted between two DWNTs will both induce upshift to the frequencies of the Raman shifts in RBM band (Bandow et al., 2002), the formula $\omega_{RBM} = 224/\ d_t$ with about 10% higher RBM frequency modification is adopted here to calculate the corresponding diameter of the nanotubes, as Eq. (3) above similar to SWNTs.

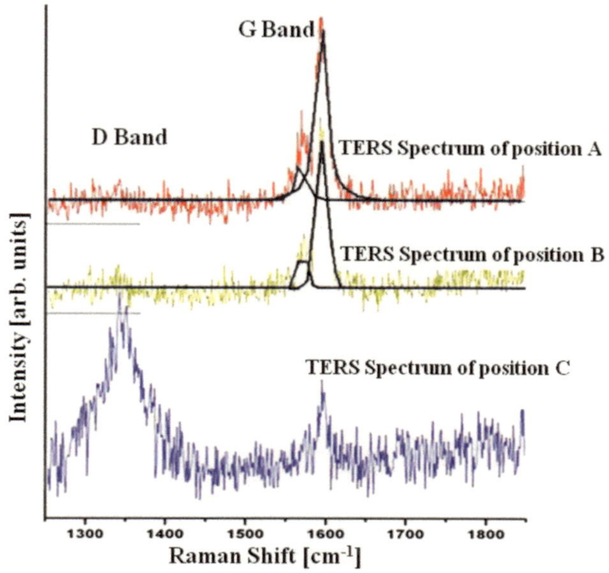


Fig. 13. TERS spectra at the 3 positions (A, B, and C) marked correspondingly in Fig. 11(a)

The experimental obtained RBM band consists of two groups of Raman shift peaks, the range from 140 $cm^{-1}$ to 160 $cm^{-1}$ and 250 $cm^{-1}$ to 270 $cm^{-1}$, which can be observed in Fig.15 (far-field) and Fig.16 (tip-enhanced). Since the RBM frequency depends inversely on the nanotube diameter, the two groups of peaks correspond to the outer and the inner walls of DWNTs respectively. The calculated diameters of the nanotubes are shown in Table 2. The outer walls' diameters vary between 1.51 nm and 1.69 nm while the inner walls' diameters vary between 0.89 nm to 0.93 nm.

In this experiment, the topography of another selected area of the DWNTs sample was detected by using AFM (shown in Fig. 14) before the Raman spectral investigation. Two arc-like bundles of DWNTs, whose heights were 3 nm and 6 nm respectively, were observed in the detected area. With the tip withdrawn from the near-field of the sample, the far-field Raman spectrum of the DWNTs was obtained (as shown in Fig. 15). Then, for further detection of the tip-enhanced RBM of the DWNTs, the AFM golden tip was approached to the near-field of the sample, and the scanning stage was precisely moved to align the interested DWNT bundle with the tip's apex. After eliminating the far-field background, the tip-enhanced RBM of position A and B are shown in Fig. 16(a) and (b) respectively.

Comparing the tip-enhanced RBM with the far-field one, one can draw two conclusions. First, the excitation area is highly confined. Seven peaks in the RBM region are detected from the far-field Raman spectrum while 5 and 7 peaks are detected at the position A and B in TERS, respectively.

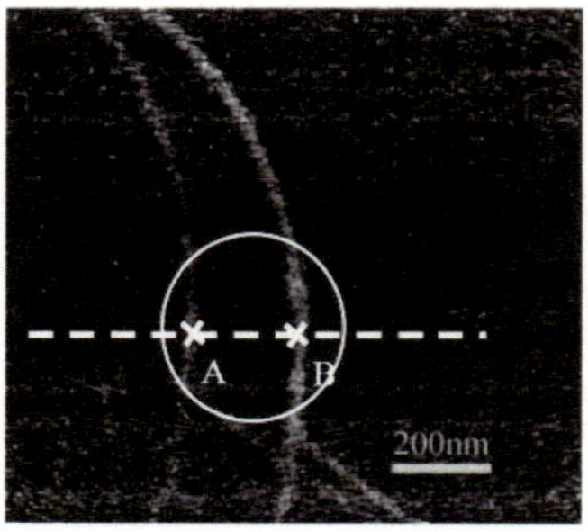


Fig. 14. Topographic image of DWNTs.

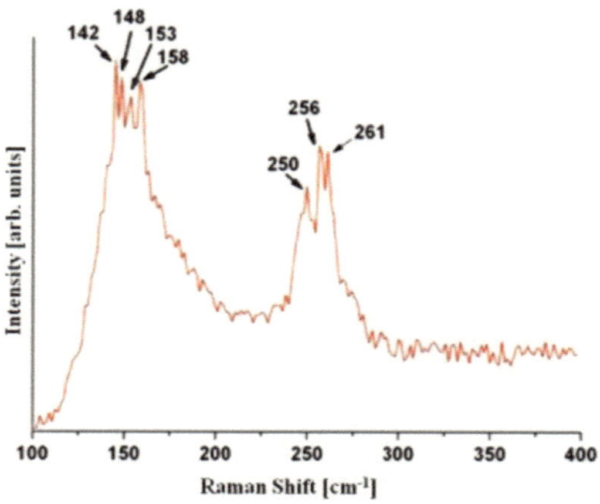


Fig. 15. Far-field Raman spectrum of a DWNT in the RBM region.

| outer wall | | inner wall | | interlayer space (nm) |
|---|---|---|---|---|
| Raman shift ($cm^{-1}$) | Diameter (nm) | Raman shift ($cm^{-1}$) | Diameter (nm) | |
| 142 | 1.69 | 250 | 0.93 | 0.38 |
| | | 256 | 0.91 | 0.39 |
| | | 261 | 0.89 | 0.40 |
| 148 | 1.62 | 250 | 0.93 | 0.345 |
| | | 256 | 0.91 | 0.355 |
| | | 261 | 0.89 | 0.365 |
| 153 | 1.57 | 261 | 0.89 | 0.34 |
| 158 | 1.51 | - | - | - |

Table 2. Correspondence of the far-field RBM frequencies and the diameters of the DWNTs.

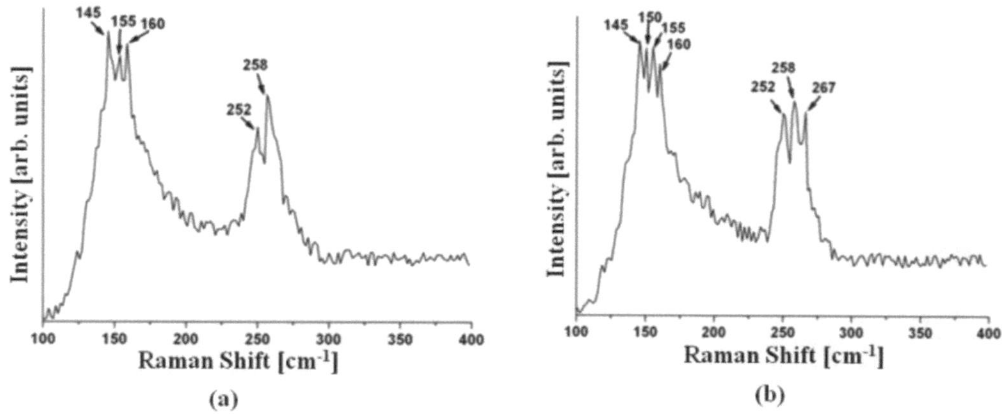


Fig. 16. The tip-enhanced RBM from (a) A point and (b) B respectively after subtracting the far-field background.

| Far-field ($cm^{-1}$) | Position A ($cm^{-1}$) | Position B ($cm^{-1}$) | Peak shift ($cm^{-1}$) |
|---|---|---|---|
| 142 | 145 | 145 | 3 |
| 148 | - | 150 | 2 |
| 151 | 155 | 155 | 2 |
| 158 | 160 | 160 | 2 |
| 250 | 252 | 252 | 2 |
| 256 | 258 | 258 | 2 |
| 261 | - | 267 | 6 |

Table 3. Contrast of the far-field and the tip-enhanced Raman shift of RBM.

As the Raman shifts in RBM bands directly reflect the diameters of the nanotubes, this indicates that the left bundle includes more assembled forms of inner and outer walls than the right bundle at A. This also demonstrates that, in the TERS method, the Raman spectrum can be efficiently enhanced in a highly-confined volume and that the RBM of each bundle can be detected separately. Second, the slight shift in RBM is observed and is related to the deformation of DWNTs caused by the tip pressure. The tip-enhanced RBM bands at A and B reflect 2 to 3 $cm^{-1}$ upshifts compared with the initial ones shown in the far-field Raman spectrum. Considering the systematic residuals and random error in experiments, the measurement was taken several times with the scanning stage aligning the position of interest in the mode, and the tip scans from left to right and then from right to left. The results obtained for the tip-enhanced RBM bands are all in good agreement, and it indicates that the Raman shift phenomenon is repeatedly measured. The peak shift between the tip-enhanced RBM bands and the far-field RBM bands is related to the deformation of DWNTs and may be attributed to a combination of the tip mechanical pressure and the interaction between the two walls of the DWNTs.

For further experimental investigation on the dependences of the band shift and the intensity of TERS on the pressure should be carried out. That would provide further qualitative and quantitative analyses of the external pressure effects and the internal force interactions of the DWNTs. Also, the stress distribution in the DWNTs should be obtained by TERS mapping measurement. It can be assumed that the interacting force between the two walls of the DWNT under tip pressure can be achieved by measuring the tip-enhanced RBM bands shift.

### 4.3.2 Detection of MWNTs

In this detection, a single MWNT was investigated with our home-made side-illumination reflection-mode TERS system (R. Wang et al., 2010a). An Ag coated optical fiber probe fabricated in the chemical method was used. The experimental results indicate that the measuring system has a spatial resolution better than 70 nm, and the enhancement factor is about $5\times10^3$.

The laser beam form a Nd:YAG laser, wavelength 532 nm is focused on the probe tip by a long work distance objective (50×, NA=0.42, WD=20.5mm, Mitutoyo) at the incident angle of 65° relative to the axis of the probe. The polarization of the incident light is along the axis of the probe to get stronger enhanced field on the tip apex. The separation between the tip and the sample is kept constant and less than 5 nm by the shear-force feedback control. The Raman scattering is collected by the same objective. Then it is steered into the Raman spectroscope, Renishaw RM2000, and detected by a cooled CCD detector. The measured wavenumber range from 4000 to 100 $cm^{-1}$ is determined by the band width of the edge filter.

In this experiment, the samples under test were water dispersible MWNTs (Shenzhen Nanotech Port Co. Ltd) with the tube diameters of 20 ~ 40 nm. After further diluted with deionized water and dispersed with ultrasonic, the sample solution (~ 0.2 wt. %) was dropped on the surface of the silicon wafer and dried.

A topographic image of a single MWNT was obtained, as shown in Fig. 17(a). A cross section along the dash-dot line in (a) is shown in Fig. 17(b). It is clearly indicated that the height of the MWNTs is about 40 nm. The diameter of the single MWNT is markedly broadened to 100 nm due to the geometry of the tip. Three points A1, A2, and A3 with the intervals of ~70 nm are chosen on the three crosses and the TERS information on the three points is detected respectively.

The TERS spectra with tip and far-field Raman spectrum without tip are displayed in Fig. 17(c). The two peaks (about 1300 $cm^{-1}$ and 1600$cm^{-1}$) are corresponding to the defect-related mode (D mode) and graphite-like mode (G mode) respectively. Among the TERS spectra at A1, A2 and A3 the distinct differences can be observed. Point A1 and A3 locate brims of the tube respectively and it leads Raman peaks unobvious. Point A2 just locates on the carbon tube and there is a stronger Raman peaks in the TERS spectrum. It demonstrates experimentally that the spatial resolution is better than 70 nm with the TERS system.

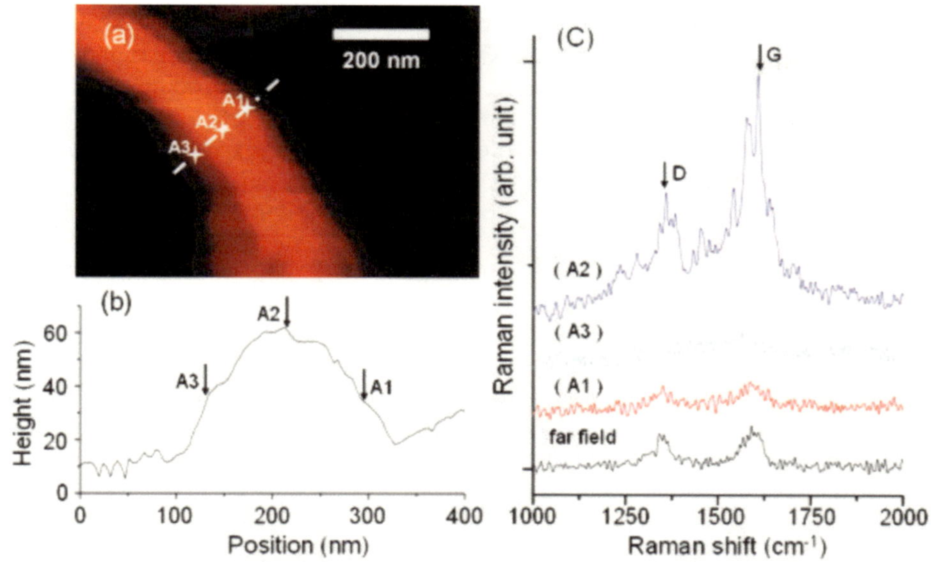


Fig. 17. TERS measurement results of a single MWNT. (a) topographic image of a single MWNT; (b) corresponding height-cross-sections along three measuring points marked in (a); (c) tip-enhanced Raman spectra of the MWNT at different points and the far field Raman spectrum of the whole area.

The comparison between the TERS Raman intensity at point A2 and the far-field Raman intensity shows that the former has 5-6 times intensity than the latter. Considering the radius of the tip and the size of laser focusing spot, the tip-enhancement factor is about $5\times10^3$. Although both tip-enhanced Raman spectra and far-field Raman spectra show the inherent Raman peaks of MWNTs, spectrum at A2 seem to have more fine features. It is probably from the structure defect or the interaction between the tip and the MWNTs.

### 4.4 TERS mapping

One of the significant superiority of TERS is that it provides the topography and the Raman spectral mapping simultaneously and the two images can be correspondingly recognized. To obtain the spectral signal from the interested localized region in nanometer scale on the sample and further investigate the chemical information from the nano-structure in the area, the TERS mapping is useful and necessary. In TERS mapping, the metallized tip scans point by point over the interested area of the specimen and the full spectra in each detected point are acquired. Then the specific Raman peaks or bands, which directly reflect the feature of the interested material structure are picked out, and subsequently set as reference to identify the composites and the molecule structure in the Raman spectral mapping. The spectrum of each position is analyzed and intensity of the Raman peaks is marked in the corresponding position of the mapping. T. Yano et al utilized TERS to measure the distribution of SWNTs with a spatial resolution far beyond the diffraction limit. The RBM of SWNTs in the near-field Raman spectra is corresponding to the diameters of various SWNTs in the immediate vicinity of the tip. The near-field Raman imaging of the RBM provided a super-resolved color mapping corresponding to the diameter distribution of SWNTs within a bundle (Yano et al., 2006).

In experiment, the detection of Raman spectrum needs a few seconds to a few minutes accumulation time to obtain the spectrum signal with the higher ratio of signal and noise. With the increase of the TERS mapping area, the number of points to be detected grows up rapidly. Thus, intending to obtain a high resolution TERS mapping in a comparatively large area, may require up to an hour or even more. Therefore, the long time stability of the TERS system is quite critical (as discussed in section 2.3.3).
The TERS mapping of a Cr grating on a Si wafer and its corresponding topography is presented as shown in Fig.18. The measuring area is 300 nm × 1200 nm and the sampling point of TERS mapping is 4 points × 12 points with the interval about 100nm. The accumulation time at each point is ten seconds, and the total detecting time of TERS is about 25 minutes. The picked reference Raman peak is 520 $cm^{-1}$, which corresponds with Si. Thus, the TERS mapping shows the distribution of Si in the detected area. As shown in Fig.18, the spatial relationship between the topography and the Raman mapping demonstrates that the TERS mapping can provide the correspondent recognition of physical and chemical information (R. Wang, 2010b).

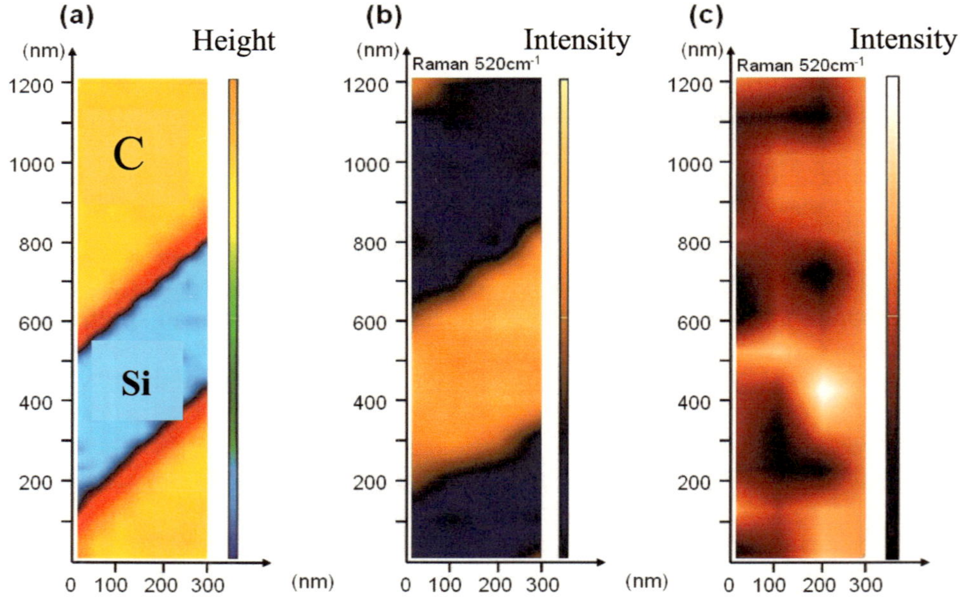


Fig. 18. (a) The topography (b) the TERS mapping and (c) far-field Raman mapping of the Si/Cr grating.

## 5. Conclusions

As a tip-enhanced spectroscopy, TERS system combines SPM (STM, AFM, SFM, SNOM) scanning head, Raman spectroscope, optical microscope, control units and data acquirement into a complicated topography and spectral measuring system. TERS is employed not only to detect Raman spectrum from nanometer localized volume, but also topographic data corresponding to the former in nanometer accuracy simultaneously. TERS mapping and spectral imaging become the important analysis method in nano-material science research.

TERS has become the promising tool in the nanometric Raman spectrum recognition and identification.

There is no doubt that TERS is a strong tool in the measurement and characterization of CNTs. Some SWNT bundle samples are measured and characterized with the TERS system. The topography of SWNTs is obtained, and SWNT bundles, diameter 100nm, were synchronously measured with the TERS. The experimental results reveal that Raman spectroscopic detection beyond the optical diffraction limits has been realized. The most TERS spectrum data obtained is from three pieces of SWNTs, based on the analysis to the TERS data. Comparing TERS and far-field Raman spectrum data, the results indicate that the TERS enhancement factor is over to 230.

The DWNTs and MWNTs samples are also measured with the TERS system. The experimental results indicate that the Raman signal can be efficiently enhanced in a highly-confined volume. The RBM of each bundle may be detected separately. The Raman peaks shift between the tip-enhanced RBM bands and the far-field RBM bands is related to the deformation of DWNTs. It may be attributed to a combining process of the tip mechanical pressure and the interaction between the two walls of the DWNTs. It demonstrates experimentally that the spatial resolution is better than 70 nm and the enhancement factor is about $5\times10^3$.

The TERS results demonstrate experimentally that TERS technology is not only of the spatial and optical resolution beyond the diffraction limits, but also has ability to realize nanometer localization spectral detection. They have the accurate corresponding relationship and have many benefits to the spectral recognition in the nanometer localize space in the potential applications.

TERS technology has been paid much attention to and widely used in nano-science and nano-technology. Even though many achievements have been obtained with TERS, it must be noted that the TERS technology is being developed and not yet perfect. Some issues need to be further investigated in the coming research.

It has been found that, there are some slight differences between TERS spectrum and far-field micro-Raman spectrum. Up to now, the clear conclusions and common views on the mechanism have not yet been reached, although some people tried to explain it based on the gradient field effect (Ayars et al., 2000), tip polarization/depolarization effect (Gucciardi et al., 2008), optical antenna effect (Novotny, 2008), tip thermal effect (Downes et al., 2006), and pressure effect (Watanabe et al., 2004) and so on. Theoretically, it is a quite complicated issue to build an exquisite physical model and explain thoroughly the mechanism of TERS. The selection rule of TERS needs to be built under the system of nanometer metallized tip and molecule localized electromagnetic field coupling.

Some following issues on TERS might be studied in the future.

a. TERS will work under the vacuum and liquid environment (Steidtner & Pettinger, 2008). It is helpful to explore the TERS sensitivity limit and spatial resolution. It will make one to understand deeply the TERS theoretical mechanism. TERS will play an important role in life science as it gets the tip-enhanced spectrum under liquid environment.
b. TERS will be used in semiconductor inspection. It may be used in inspections for stress distribution and fission process in nanometer scale (Zhu et al., 2007a, 2007b; Georgi et al., 2007). The polarization control techniques may be used to reduce the far-field background to improve the contrast (Lee et al., 2007).

c. New and optimizing tip preparation methods will massively produce high quality tips with uniform geometry, firm metal coating, controllable optical and resonance properties, time stability, and durability to satisfy the excitation conditions (Cui et al., 2007; Yeo et al., 2006).
d. With appropriate polarization light, for example radial polarized light, as an exciting beam as mentioned above, it will further improve the spatial resolution and tip enhancement factors. It can realize the atom resolution and real single piece of SWNTs measurement.
e. Instrumentation and commercialization of TERS will be accomplished in the combination with more type probe microscopes and spectroscopes and special optical devices, such as parabolic mirror.

In summary, the TERS technology with excellent performances will become an important measurement and characterization tool in the field of nano-materials and nano-structures.

## 6. Acknowledgement

The authors gratefully thank all people who gave us helps in the discussions, fabrications and inspections for specimen preparations. This work was financed by the National Natural Science Foundation of China for the support under the grant 60427003, and the National Basic Research Program of China (973 Program) Grant, project research on optical detection in nanometric scale, No. 2007CB936801.

## 7. References

Anderson, M. S. (2000). Locally enhanced Raman spectroscopy with an atomic force microscope. *Applied Physics Letters*, Vol. 76, No. 21, pp. (3130-3132), ISSN 0003-6951

Anderson, M. S. & Gaimari, S. D. (2003). Raman-atomic force microscopy of the ommatidial surfaces of Dipteran compound eyes. *Journal of Structural Biology*, Vol. 142, No. 3, pp. (364-368), ISSN 1047-8477

Anderson, N.; Hartschuh, A.; Cronin, S. & Novotny, L. (2005). Nanoscale vibrational analysis of single-walled carbon nanotubes. *Journal of the American Chemical Society*, Vol. 127, No. 8, pp. (2533-2537), ISSN 0002-7863

Anderson, N.; Bouhelier, A. & Novotny, L. (2006). Near-field photonics: tip-enhanced microscopy and spectroscopy on the nanoscale. *Journal of Optics A: Pure and Applied Optics*, Vol. 8, No. 4, pp. (227–233), ISSN 0950-0340

Ayars, E. J.; Hallen, H. D. & Jahncke, C L. (2000). Electric field gradient effects in Raman spectroscopy. *Physical Review Letters*, Vol. 85, No. 19, pp.(4180-4183), ISSN 0031-9007

Bailo, E. & Deckert, V. (2008). Tip-enhanced Raman spectroscopy of single RNA strands: Towards a novel direct-sequencing method. *Angewandte Chemie-International Edition*, Vol. 47, No. 9, pp. (1658-1661), ISSN 1521-3773

Bandow, S.; Chen, G.; Sumanasekera, G. U.; Gupta, R.; Yudasaka, M.; Iijima, S. & Eklund, P. C. (2002). Diameter-selective resonant Raman scattering in double-wall carbon nanotubes. *Physical Review B*, Vol. 66, No. 7, pp.( 75416-1-8), ISSN 1098-0121

Bohren, C. F. & Huffman, D. R. (1998). Absorption and Scattering of Light by Small Particles. Wiley, ISBN: 978-0-471-29340-8, New York.

Budich, C.; Neugebauer, U.; Popp, J. & Deckert, V. (2008). Cell wall investigations utilizing tip-enhanced Raman scattering. *Journal of Microscopy-Oxford,* Vol. 229, No. 3, pp. (533-539), ISSN 0022-2720

Chiang, I. W.; Brinson, B. E.; Smalley, R. E.; Margrave, J. L., & Hauge, R. H. (2001). Purification and Characterization of Single-Wall Carbon Nanotubes. *The Journal of Physical Chemistry B,* Vol. 105, No. 35, pp. (1157-1161), ISSN 1520-6106

Cialla, D.; Deckert-Gaudig, T.; Budich, C.; Laue, M.; Moller, R.; Naumann, D.; Deckert, V. & Popp, J. (2009). Raman to the limit: tip-enhanced Raman spectroscopic investigations of a single tobacco mosaic virus. *Journal of Raman Spectroscopy,* Vol. 40, No. 3, pp. (240-243), ISSN 0377-0486

Downes, A.; Salter, D. & Elfick A. (2006). Heating effects in tip-enhanced optical microscopy. *Optics Express,* Vol. 14, No. 12, pp.(5216-5222), ISSN 1094-4087

Downes, A.; Salter, D. & Elfick, A. (2008). Simulations of tip-enhanced optical microscopy reveal atomic resolution. *Journal of Microscopy,* Vol. 229, No. 2, pp. (184-188), ISSN 0022–2720

Dresselhaus, M. S.; Dresselhaus, G.; Jorio, A.; Souza, F. A. G. & Saito R. (2002). Raman spectroscopy on isolated single wall carbon nanotubes. *Carbon,* Vol. 40, No. 12, pp. (2043-2061), ISSN 0008-6223

Dresselhaus, M. S.; Dresselhaus, G.; R. Saito and A. Jorio (2005). Raman Spectroscopy of Carbon Nanotubes. *Physics Reports,* Vol. 409, No. 2pp.( 47-99), ISSN 0370-1573

Eklund, P. C.; Holden, J. M. & Jishi, R. A. (1995). Vibrational Modes of Carbon Nanotubes; Spectroscopy and Theory. *Carbon,* Vol. 33, No. 7, pp.( 959–972), ISSN 0008-6223

Frank, S.; Poncharal, P.; Wang, Z. L. & Heer, W. A. (1998). Carbon Nanotube Quantum Resistors. *Science,* Vol. 280, No. 5370, pp. (1744-1746), ISSN 0036-8075

Frey, H. G.; Keilmann, F.; Kriele, A. & Guckenberger, R. (2002). Enhancing the resolution of scanning near-field optical microscopy by a metal tip grown on an aperture probe. *Applied Physics Letters,* Vol. 81, No. 26, pp. (5030-5032), ISSN 0003-6951

Gennaro, P.; Marc, C. & Razvigor, O. (2009). High resolution probing of multi wall carbon nanotubes by tip enhanced Raman spectroscopy in gap-mode. *Chemical Physics Letters,* Vol. 469, No. 1-3, pp.( 161-165), ISSN 0009-2614

Georgi, C.; Hecker, M. & Zschech, E. (2007). Raman intensity enhancement in silicon-on-insulator substrates by laser deflection at atomic force microscopy tips and particles. *Applied Physics Letters,* Vol. 90, No. 17, pp.( 171102-1 - 171102-3), ISSN 0003-6951

Guan, L. (2006) Filling of Carbon Nanotube: Studies on Structure and Properties of Materials Confined in Nanospace. Peking University, Beijing

Gucciardi, P. G.; Lopes, M.; Deturche, R.; Julien, C.; Barchiesi, D. & de la Chapelle, M. L. (2008). Light depolarization induced by metallic tips in apertureless near-field optical microscopy and tip-enhanced Raman spectroscopy. *Nanotechnology,* Vol. 19, No. 21, pp.(215702), ISSN 0957-4484

Hartschuh, A. & Novotny, L. (2002). Near-field Raman spectroscopy using a sharp metal tip. *Technical Digest. Summaries of papers presented at the Quantum Electronics and Laser Science Conference,* Conference Edition (IEEE Cat. No.02CH37338),: vol.1 | (271+40 suppl.) pp. (39-40), ISBN 1-55752-708-3

Hartschuh, A.; Sánchez, E. J.; Xie, X. S. & Novotny, L. (2003). High-Resolution Near-Field Raman Microscopy of Single-Walled Carbon Nanotubes. *Physical Review Letters,* Vol. 90, No. 9, pp. (095503-1-095503-4), ISSN 0031-9007

Hartschuh A., Qian H. H., Meixner A. J., Anderson N., & Novotny L. (2005). Nanoscale optical imaging of excitons in single-walled carbon nanotubes. *Nano Letters,* Vol. 5, No. 11, pp. (2310-2313), ISSN 1530-6984

Hayazawa, N.; Inouye, Y.; Sekkat, Z. & Kawata, S. (2000). Metallized tip amplification of near-field Raman scattering. *Optics Communications,* Vol. 183, No. 1-4, (September 2000), pp. (333-336), ISSN 0030-4018

Hayazawa, N.; Inouye, Y.; Sekkat, Z. & Kawata, S. (2001). Near-field Raman scattering enhanced by a metallized tip. *Chemical Physics Letters,* Vol. 335, No. 5-6, pp. (369-374), ISSN 0009-2614

Hayazawa, N.; Yano, T.; Watanabe, H.; Inouye, Y. & Kawata, S. (2003). Detection of an individual single-wall carbon nanotube by tip-enhanced near-field Raman spectroscopy. *Chemical Physics Letters,* Vol. 376, No. 1-2, pp. (174-180), ISSN 0009-2614

Hayazawa, N.; Saito, Y. & Kawata, S. (2004). Detection and characterization of longitudinal field for tip-enhanced Raman spectroscopy. *Applied Physics Letters,* Vol. 85, No., 25, pp. (6239-6241), ISSN 0003-6951

Hayazawa, N.; Watanabe, H.; Saito, Y. & Kawata, S. (2006). Towards atomic site-selective sensitivity in tip-enhanced Raman spectroscopy. *Journal of Chemical Physics,* Vol. 125, No. 24, pp. (244706), ISSN 0021-960

Iijima, S. (1991). Helical microtubules of graphitic carbon. *Nature,* Vol. 354, No. 6348, pp. (56-58), ISSN 0028-0836

Johnson, P.B. & Christy, R.W. (1972). Optical Constants of the Noble Metals. *Phys. Rev. B,* Vol. 6, No. 12, pp. (4370-4379), ISSN 1098-0121

Kalkbrenner, T.; Ramstein, M.; Mlynek, J. & Sandoghdar, V. (2001). A single gold particle as a probe for apertureless scanning near-field optical microscopy. *Journal of Microscopy-Oxford,* Vol. 202, pp. (72-76), ISSN 0022-2720

Kerker, M.; Wang, D. S. & Chew, H. (1980). Surface enhanced raman-scattering (SERS) by molecules adsorbed at spherical-particles. *Applied Optics,* Vol. 19, No. 24, pp. (4159-4174), ISSN 0003-6935

Kharintsev, S. S.; Hoffmann, G. G.; Dorozhkin, P. S.; With, G. de & Loos, J. (2007). Atomic force and shear force based tip-enhanced Raman spectroscopy and imaging. *Nanotechnology,* Vol. 18, No. 31, pp. (315502-9pp), ISSN 0957-4484

Krug, J. T. ; Sanchez, E. J. & Xie, X. S. (2002). Design of near-field optical probes with optimal field enhancement by finite difference time domain electromagnetic simulation. *Journal of Chemical Physics,* Vol. 116, No. 24, pp. (10895-10901), ISSN 0021-9606

Lee, N.; Hartschuh, R. D.; Mehtani, D.; Kisliuk, A.; Maguire, J. F.; Green, M.; Foster, M. D. & Sokolov, A. P. (2007). High contrast scanning nano-Raman spectroscopy of silicon. *Journal of Raman Spectroscopy,* Vol. 38, No. 6, pp. (789-796), ISSN 0377-0486

Li, H.; Feng, L.; Guan, L.; Shi, Z. & Gu, Z. (2004). Synthesis and purification of single-walled carbon nanotubes in the cottonlike soot. *Solid State Communications,* Vol. 132, No. 3-4, pp. (219-224), ISSN 0038-1098

Martin, O. J. F. & Girard, C. (1997). Controlling and tuning strong optical field gradients at a local probe microscope tip apex. *Applied Physics Letters,* Vol. 70, No. 6, pp. (705-707), ISSN 0003-6951

Mehtani, D.; Lee, N.; Hartschuh, R. D.; Kisliuk, A.; Foster, M. D.; Sokolov, A. P. & Maguire, J. F. (2005). Nano-Raman spectroscopy with side-illumination optics. *Journal of Raman Spectroscopy,* Vol. 36, No. 11, pp. (1068-1075), ISSN 0377-0486

Neacsu, C. C.; Steudle, G. A. & Raschko, M. B. (2005). Plasmonic light scattering from nanoscopic metal tips. *Applied Physics B-Lasers and Optics,* Vol. 80, No. 3, pp. (295-300), ISSN 0946- 2171

Novotny, L.; Bian, R. X. & Xie, X. S. (1997). Theory of nanometric optical tweezers. *Physical Review Letters,* Vol. 79, No. 4, pp. (645-648), ISSN 0031-9007

Novotny, L. (2008). Nano-optics - Optical antennas tuned to pitch. *Nature,* Vol. 455, No. n.d., pp. (887-887), ISSN 0028-0836

Otto, A. (2002). What is observed in single molecule SERS, and why?. *Journal of Raman Spectroscopy,* Vol. 33, No. 8, pp. (593-598), ISSN 0377-0486

Pettinger, B.; Picardi, G.; Schuster, R. & Ertl, G. (2000). Surface Enhanced Raman Spectroscopy: Towards single molecule spectroscopy. *Electrochemistry,* Vol. 68, pp. (942-949), ISSN 0021-891X

Pettinger, B.; Picardi, G.; Schuster, R. & Ertl G. (2002). Surface-enhanced and STM-tip-enhanced Raman spectroscopy at metal surfaces. *Single Molecules,* Vol. 3, No. 5-6, pp. (285-294), ISSN 1438-5163

Pettinger, B.; Ren, B.; Picardi, G.; Schuster, R. & Ertl, G. (2004). Nanoscale Probing of Adsorbed Species by Tip-Enhanced Raman Spectroscopy. *Physical Review Letters,* Vol. 92, No. 9, pp. (0961011- 0961014), ISSN 0031-9007

Pettinger, B.; Ren, B.; Picardi, G.; Schuster, R. & Ertl, G. (2005). Tip-enhanced Raman spectroscopy (TERS) of malachite green isothiocyanate at Au(111): bleaching behavior under the influence of high electromagnetic fields. *Journal of Raman Spectroscopy,* Vol. 36, No. 6-7, pp. (541-550), ISSN 0377-0486

Qian, H.; Gokus, T.; Anderson, N.; Novotny, L.; Meixner, A. J. & Hartschuh, A. (2006). Near-field imaging and spectroscopy of electronic states in single-walled carbon nanotubes. *Physica Status Solidi (B),* Vol. 243, No. 13, pp. (3146-3150), ISSN 0370-1972

Qian, H.; Georgi, C.; Anderson, N.; Green, A. A.; Hersam, M. C.; Novotny, L. & Hartschuh, A. (2008). Exciton transfer and propagation in carbon nanotubes studied by near-field optical microscopy. *Physica Status Solidi B,* Vol. 245, No. 10, pp. (2243-2246), ISSN 0377-0486

Rao, A. M.; Richter, E.; Bandow, S.; Chase, B.; Eklund P. C.; Williams, K. A.; Fang, S.; Subbaswamy, K. R.; Menon, M.; Thess, A.; Smalley, R. E.; Dresselhaus, G. & Dresselhaus, M. S. (1997). Diameter-Selective Raman Scattering from Vibrational Modes in Carbon Nanotubes. Science, Vol. 275, No. 5297, pp. (187-191), ISSN 0036-8075

Rao, A. M.; Chen, J.; Richter, E.; Schlecht, U.; Eklund, P. C.; Haddon, R. C.; Venkateswaran, U. D.; Kwon, Y. K. & Tománek, D. (2001). Effect of van der Waals Interactions on the Raman Modes in SingleWalled Carbon Nanotubes. *Physical Review Letters*, Vol. 86, No. 17, pp.(3895-3898), ISSN 0031-9007

Ren, B.; Picardi, G. & Pettinger, B. (2004). Preparation of gold tips suitable for tip-enhanced Raman spectroscopy and light emission by electrochemical etching. *Review of Scientific Instruments*, Vol. 75, No. 4, pp. (837-841), ISSN 0034-6748

Ren, B.; Picardi, G.; Pettinger, B.; Schuster, R. & Ertl, G. (2005). Tip-enhanced Raman spectroscopy of benzenethiol adsorbed on Au and Pt single-crystal surfaces. *Angewandte Chemie-International Edition*, Vol. 44, pp. (139-142), ISSN 1433-7851

Sackrow, M.; Stanciu, C.; Andreas, M. Lieb & Meixner A. J. (2008). Imaging Nanometre-Sized Hot Spots on Smooth Au Films with High-Resolution Tip-Enhanced Luminescence and Raman Near-Field Optical Microscopy. *Chem. Phys. Chem.*, Vol. 9, No. 2, pp. (316-320), ISSN 1439-4235

Saito, Y.; Wang, J. J.; Smith, D. A. & Batchelder, D. N. (2002). A simple chemical method for the preparation of silver surfaces for efficient SERS. *Langmuir*, Vol. 18, No. 8, pp. (2959-2961), ISSN 0743-7463

Saito, Y.; Hayazawa, N.; Kataura, H.; Murakami, T.; Tsukagoshi, K.; Inouye, Y. & Kawata, S. (2005). Polarization measurements in tip-enhanced Raman spectroscopy applied to single-walled carbon nanotubes. *Chemical Physics Letters*, Vol. 410, No. 1-3, pp. (136-141), ISSN 0009-2614

Saito, Y.; Motohashi, M.; Hayazawa, N. & Kawata, S. (2008). Stress imagining of semiconductor surface by tip-enhanced Raman spectroscopy. *Journal of Microscopy*, Vol. 229, No. n.d., pp. (217-222), ISSN 0022-2720

Schmid, T.; Yeo, B. S.; Zhang, W. & Zenobi, R. (2007). Name of paper Use of tip-enhanced vibrational spectroscopy for analytical applications in chemistry, biology, and materials science. In: Kawata S, V. M. Shalaev, ed. Tip enhancement. Oxford: Elsevier, pp. (115-155), ISSN 1871-0018

Steidtner, J. & Pettinger, B. (2007). High-resolution microscope for tip-enhanced optical processes in ultrahigh vacuum. *Review of Scientific Instruments*, Vol. 78, No. 103104, pp. (1-8), ISSN 0034-6748

Steidtner, J. & Pettinger, B. (2008). Tip-enhanced Raman spectroscopy and microscopy on single dye molecules with 15 nm resolution. *Physical Review Letters*, Vol. 100, No. 23, pp. (236101-1-236101-4), ISSN 0031-9007

Stöckle, M.; Doug, S. Y.; Deckert, V. & Zenobi, R. (2000). Nanoscale chemical analysis by tip-enhanced Raman spectroscopy. *Chemical Physics Letters*, Vol. 318, pp. (131-136), ISSN 0009-2614

Sun, W. X & Shen, Z X. (2001). A practical nanoscopic Raman imaging technique realized by near-field enhancement. *Materials Physics and Mechanics*, Vol. 4, No. 1, pp. (17-21), ISSN 1605-8119

Sun, W. X. & Shen, Z. X. (2003). Near-field scanning Raman microscopy using apertureless probes. *Journal of Raman Spectroscopy*, Vol. 34, No. 9, pp. (668-676), ISSN 0377-0486

Tarun, A.; Hayazawa, N. & Kawata, S. (2009). Tip-enhanced Raman spectroscopy for nanoscale strain characterization. *Analytical and Bioanalytical Chemistry,* Vol. 394, pp. (1775–1785), ISSN 1618-2642

Wang, J. J.; Saito, Y.; Batchelder, D. N.; Kirkham, J.; Robinson, C. & Smith, D. A. (2005). Controllable method for the preparation of metalized probes for efficient scanning near-field optical Raman microscopy. *Applied Physics Letters,* Vol. 86, No. 26, pp. (2631111 - 2631113), ISSN 0003-6951

Wang, R.; Wang, J.; Hao, F.; Zhang, M. & Tian, Q. (2010a). Tip-enhanced Raman spectroscopy with silver-coated optical fiber probe in reflection mode for investigating multiwall carbon nanotubes. *Applied Optics,* Vol. 49, No. 10, pp. (1845-1848), ISSN 1559-128X

Wang, R. (2010b). Research on the Tip-Enhanced Raman Spectroscopy in Reflection Mode Based on Silver-Coated Optical Fiber Probe, Tsinghua University, Beijing, China

Wang, R.; Hao, F.; Zhang, M.; Wang, J.; Yu, J. & Tian, Q. (2010c). Tip-Enhanced Raman Spectroscopy and System Design. *Laser & Optoelectronics Progress,* Vol. 47, No. n.d., pp. (031601-1-031601-10), ISSN 1006-4125

Watanabe, H.; Hayazawa, N.; Inouye, Y. & Kawata, S. (2005). DFT vibrational calculations of Rhodamine 6G adsorbed on silver: Analysis of tip-enhanced Raman spectroscopy. *Journal of Physical Chemistry B,* Vol. 109, No. 11, pp. (5012-5020), ISSN 1520-6106

Watanabe, H.; Ishida, Y.; Hayazawa, N.; Inouye, Y. & Kawata, S. (2004). Tip-enhanced near-field Raman analysis of tip-pressurized adenine molecule. *Physical Review B,* Vol. 69, No. 15, pp. (15541801-15541811), ISSN 1098-0121

Watanabe, H.; Ishida, Y.; Hayazawa, N.; Inouye, Y. & Kawata, S. (2004).Tip-enhanced near-field Raman analysis of tip-pressurized adenine molecule. *Physical Review B*, Vol. 69, No. 15, pp.( 155418-1- 155418-11), ISSN 1098-0121

Wei, J. Q.; Zhu, H. W.; Li, Y. H.; Chen, B.; Jia, Y.; Wang, K.; Wang, Z.; Liu, W.; Luo, J.; Zheng, M.; Wu, D.; Zhu, Y. & Wei, B. (2006). Ultrathin Single-Layered Membranes from Double-Walled Carbon Nanotubes. *Advanced Materials,* Vol. 18, No. 13, pp. (1695-1700), ISSN 0935-9648

Wu, X.; Wang, J. , Wang R. , Xu J. , Tian Q. & Yu J.(2009). Detection of Single-Walled Carbon Nanotube Bundles by Tip-Enhanced Raman Spectroscopy. Spectroscopy and Spectral Analysis, Vol. 29, No. 10 ,pp2681-2685, ISSN 1000-0593

Wu, X.; Zhang, M.; Wang, J. & Tian, Q. (2010). Experimental Research on Double-walled Carbon Nanotubes using Tip-enhanced Raman Spectroscopy. *Journal of the Korean Physical Society,* Vol. 56, No. 4, pp.(1103-1108), ISSN 0374-4884

Yano, T.; Verma, P. & Kawata S. (2006). Diameter-selective near-field Raman analysis and imaging of isolated carbon nanotube bundles. *Applied Physics Letters,* Vol. 88, No. 9, pp.(093125), ISSN 0003-6951

Yano, T.; Verma, P.; Saito, Y.; Ichimura, T. & Kawata, S. (2009). Pressure-assisted tip-enhanced Raman imaging at a resolution of a few nanometers, *Nature Photonics,* Vol. 3, No. 8, pp.(473-477), ISSN 1749-4885

Yeo, B. S.; Zhang, W. H.; Vannier, C. & Zenobi, R. (2006). Enhancement of Raman signals with silver-coated tips. *Applied Spectroscopy*, Vol. 60, No. 10, pp.(1142-1147), ISSN 0003-7028

Yu, M. F.; Files, B. S.; Arepalli, S. & Ruoff, R. S. (2000). Tensile Loading of Ropes of Single Wall Carbon Nanotubes and their Mechanical Properties. *Physical Review Letters*, Vol. 84, No. 24 , pp. (5552-5555), ISSN 0031-9007

Zhang, D.; Wang, X.; Braun, K.; Egelhaaf, H. J.; Fleischer, M.; Hennemann, L.; Hintz, H.; Stanciu, C.; Brabec, C. J.; Kern, D. P. & Meixner, A. J. (2009). Parabolic mirror-assisted tip-enhanced spectroscopic imaging for non-transparent materials. *Journal of Raman Spectroscopy*, Vol. 40, No. 10, pp. (1371–1376), ISSN 0377-0486

Zhang, W. H.; Yeo, B. S.; Schmid, T. & Zenobi, R. (2007). Single molecule tip-enhanced Raman spectroscopy with silver tips. *Journal of Physical Chemistry*, Vol. 111, No. 4, pp. (1733-1738), ISSN 1520-6106

Zhu, L.; Atesang, J.; Dudek, P.; Hecker, M.; Rinderknecht, J.; Ritz, Y.; Geisler, H.; Herr, U.; Greer, R. & Zschech, E. (2007a). Experimental challenges for approaching local strain determination in silicon by nano-Raman spectroscopy. *Materials Science-Poland*, Vol. 25, No. 1, pp.(19-31), ISSN 0137–1339

Zhu, L.; Georgi, C.; Hecker, M.; Rinderknecht, J.; Mai, A.; Ritz, Y. & Zschech, E. (2007b). Nano-Raman spectroscopy with metallized atomic force microscopy tips on strained silicon structures. *Journal of Applied Physics*, Vol. 101, No. 10, pp.( 104305-1-104305-6), ISSN 0021-8979

# 11

# Spectro-Microscopic Study of Laser-Modified Carbon Nanotubes

Cheng-Hao Chuang[1], Chorng-Haur Sow[2] and Minn-Tsong Lin[3*]

[1,3]*Department of Physics, National Taiwan University, Taipei*
[2]*Department of Physics, National University of Singapore*
[3]*Institute of Atomic and Molecular Sciences, Academia Sinica, Taipei*
[1,3]*Taiwan*
[2]*Singapore*

## 1. Introduction

In 1991, Iijima discovered the needle-like tubes insides the fullerene production through the arc-discharge evaporation method.(1) Using transmission electron microscopy (TEM), a new type of the coaxial tube, namely carbon nanotube (CNT), was observed for the first time. One-dimensional (1D) CNTs are rolled graphitic sheet with a few nanometer in diameter. The other groups demonstrated that it is possible to generate different graphitic layers axially, such as single-wall CNTs (SWCNTs), double-wall CNTs (DWCNTs), and multi-wall CNTs (MWCNTs).(2) The electronic property of these CNTs was observed to be either metallic or semiconducting.(3) As a result, scientists have devoted much attention to study the intrinsic property of different rolled models.(4)

In the case of chemical environment, the $sp^2$-hybridized carbon sheet establishes the strongly covalent C-C bond, resulting in a high mechanical strength 100 times than that of steel.(5) The outstanding mechanical property is suitable for the tip of atomic force microscopy (AFM).(6) Besides, the chemical inertness built by the stable C-C bond leads to a perfect honeycomb structure with very few defect and gas adsorption. Only a few percentage of gas adsorption gives rise to the huge change of electronic transport property on CNTs surface. Different sensitivities of transportation with various gas species are very useful to identify for gas sensor applications.(7) The large surface area and its hollow interior lead to the potential application in the hydrogen-stored battery utilization. For the advantage of recyclable ability and the enhancement of metallic particles, CNTs is an extremely valuable for energy storage.(8) Besides, it is important for nano-electric devices due to its superior thermal conductivity (6000 $W\ m^{-1}K^{-1}$ > diamond) and current capacity ($10^9\ A/cm^2$ 100 times than that of copper wire).(9) Despite of these exceptional characteristics, it is hard to produce a specific type of CNTs and manipulate its electronic structure to fit the diverse requirements. Therefore, the chemical functionalization route that links the specific chemical group to the tubular surface for its chemical and mechanical tuning offers an attractive post-synthesis technique to tailor the properties of the CNTs to suit the applications. In other words, it offers the functions of CNTs base multiplied by a new potential performance as an integrated device.(10) The types

*mtlin@phys.ntu.edu.tw

of chemical modification are divided into the physisorption and chemisorption on the surface or at the defective site.(11) Owing to the weak interaction between the molecule species and the $\pi$-conjugate surface of CNTs, the physisorption method to control electronic structure is not suitable for the exceedingly environmental demand. On the other hand, chemisorbed molecules attach on the surface randomly and thus not assign to the desirable active position for effective hybrid utilization.

In terms of the chemical modification in field emission experiments, plasma effect,(12) heat process,(13) and laser effect(14) are found to be effective methods to improve its emission performance. This is attributed mostly to chemical doping and morphological change. However, detailed knowledge of its local electronic structure remains unknown. Recently, a focused laser system has been developed to trim the CNTs arrays to form three-dimensional structure.(15; 16) After trimming the CNTs array by focused laser, the turn-on electric field decreases and its emission current density increases in a field emission experiment.(14) However, the correlation between laser irradiation and electronic structure of CNTs has not been discussed. Researchers found the larger electron density of state (DOS) at tips of CNTs, which is attributed to the influences of lattice defect, deformed carbon cage, or dangling bond.(17; 18) The sufficient valence electrons at tips plus a strong elasticity, hence, could play an important role in the field emission performance.(19) The electron beam from the tips is more stable and efficient than the other emission systems.

In this work, we present a technique that facilitates exclusive chemical modification of the CNTs with controlled locality using a focused laser beam. With focused laser beam, we can achieve morphological modification and structural rearrangment of the CNTs. With the CNTs housed in a transparent chamber with controlled gaseous environment, we can select the appropriate gas species for the chemical modification. In addition of gas species selection, it can broaden the CNT-based application by choosing chemical bonding species. The electronic structure and its chemical modification of CNTs are suitable to be resolved by X-ray photoelectron spectroscopy (XPS). Due to the localized surface functionalization, Scanning PhotoElectron Microscopy (SPEM) equipped with XPS provides the chemical and elemental information with the spatial resolution of 100 nm. The precise information of electronic structure is investigated by micro-XPS and converted to a mapping image for the spatially resolved chemical analysis by SPEM.(20; 21)The enhanced chemical shift of 0.9 eV in C 1s state is found in the air-treated CNTs. The chemical modifications of nitrogen-treated and oxygen-treated CNTs reveal the less up-shift of C 1s state by 0.6 and 0.2 eV, respectively. In nitrogen-treated and oxygen-treated CNTs, the appearance of C-N (C-O) bond and surface-adsorbed nitrogen (oxygen) gas are both observed on the top area of CNTs. The presence of chemisorbed gas atoms accounts for the origin of up-shift of C 1s state in $N_2$ and $O_2$-treated CNTs. Furthermore the chemical shifts dependent on different gaseous environments shall involve the combination of structural defect and the mixing gas molecules. VB spectrum in air-treated CNTs exhibits the DOS transition from the mixing of C 2s and C 2s/2p-$\sigma$ band to C 2p-$\pi$ band, which is driven by laser irradiation. In the case of nitrogen-treated (oxygen-treated) CNTs, the spectral features of N 2s (O 2s) band in VB spectra are attributed to the appearance of C-N (C-O) chemical bond of N 1s (O 1s) state. In terms of physisorbed gas contribution, the surface-adsorbed nitrogen molecules in as-grown CNTs bring about a larger rearrangement of DOS than the consequence of nitrogen-treated CNTs. However, in oxygen-treated CNTs, laser irradiation induces the same tendency of DOS transition, which is consistent with the behavior of the air-treated CNTs. By controlling gas molecule species and laser beam, we can modify the chemical composition of CNTs at the specific region effectively. Our findings will open up a new field of the tuneable electronic structure for the advanced nano-electric applications.

## 2. CNTs fabrication and its chemical modifications: A new way to add mulitfunctional properties of CNTs

### 2.1 CNTs fabrication

MWCNTs and SWCNTs can be synthesized by high pressure arc discharge method, laser ablation, or chemical vapor deposition (CVD).(1; 22–24) Our MWCNTs sample is grown on Si substrate by plasma-enhanced CVD (PECVD).(25) After thin $SiO_2$ layer on Si sample is cleaned, Fe layer is deposited on flesh Si sample by the means of radio frequency magnetron sputtering. In the PECVD chamber, the resistance heating plate is used to raise the temperature of Fe/Si specimen up to 750 °C. The mixture of $C_2H_2$ and $H_2$ gas with the flow rate of 15 sccm (standard cubic centimeters per minute) and 60 sccm respectively are introduced into the PECVD chamber. These species are the reactive carbon sources for the construction of CNTs synthesis. The high annealing temperature causes the surface tension of Fe layer to decrease, the Fe layer is then divided into Fe particles with different diameters.(26) The annealed Fe surface works as the active catalyst to absorb the $C_2H_2/H_2$ gas into the particle. While the carbon atoms diffuse across the Fe particle to the opposite side, the self-organized nanotube with the hexagonal lattice is built vertically on the Si sample.(27) According to different fabrication methods, the electronic character of CNTs shows either metallic or semiconducting relying on the chirality with tube diameter or the grown condition.(3) Therefore, it reflects the importance of manufacture process in determining the electronic structure of CNTs and hence their applications.

Advanced applications of CNTs include chemical sensors,(7) nanoelectronic devices, (28) and field-emission devices.(29) In a field-emission experiment performed by Zhao *et al.*,(30) after UV or infrared light irradiation, as-grown CNTs sample showed an enhancement of emitted current and the decrease of its turn-on electric field. The possible reasons for the improved emission performance were attributed to the C-C bond breaking, structural damage, surface cleaning, or thermal effect. While the photon energy (PE) of incident light is larger than the energy of chemical bond (e.g. C-C bond = 6.29 eV, O-O bond = 5.16 eV, C-O bond = 3.2 eV, C-H bond = 3.5 eV),(31; 32) the bond breaking can occur by the photoelectron excitation effect. Even the covalent C-C bond insides CNTs has a chance to be broken if it allows the multi-photon excitation to happen. For example, Nd:YAG laser with PE of 4.66 eV is able to damage C-C bond, which results in the dangling bond.(30) Although the infrared light with PE by 1.18 eV can not break any chemical bond, it is observed the promotion in the performance of field-emission experiment. On the other hand, it can work well due to photothermal effect on the surface contamination.(33) These findings call for the detailed investigation of the mechanism that can account for the observed improvement in field emission results. Hence, it is worthwhile to probe the spatially-resolved electronic properties of CNTs after the treatment of laser trimming.

### 2.2 Chemical modification assisted by the focused laser technology

We introduce the He-Ne laser beam into the optical microscope for advanced application. Figure 1(a) and (b) show pictures of the optical microscope-focused laser beam system used in this work. Figure 1(b) emphasizes the scanning X-Y stage and objective lens in the optical microscope, which are essential for the creation of three-dimensional pattern using the focused laser beam. A schematic diagram of the set-up is shown in Figure 1(c). The He-Ne laser beam

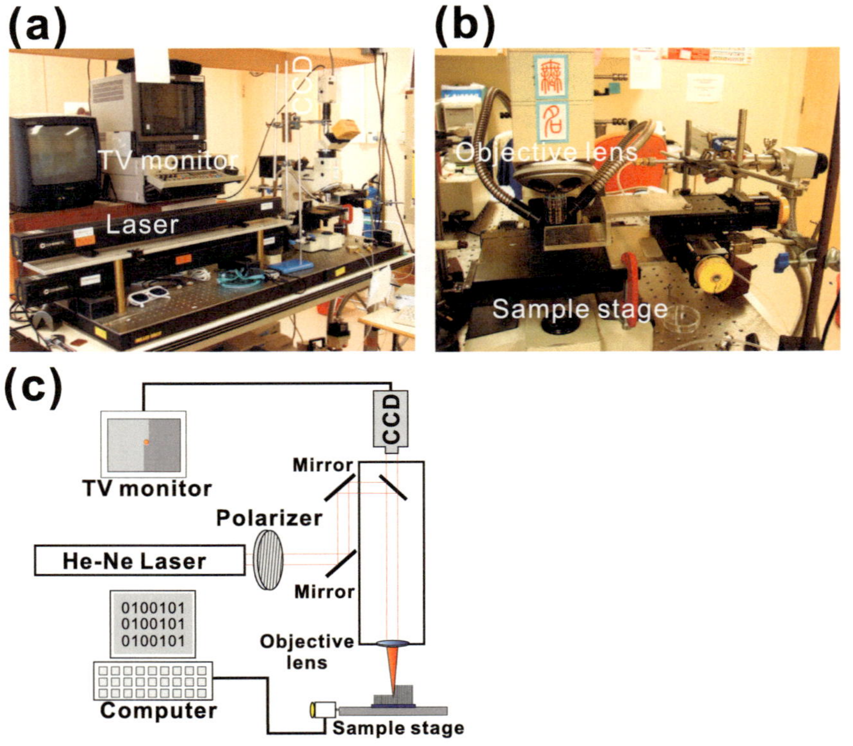


Fig. 1. (a)(b) Pictures of the optical microscope-focused laser beam system. (c) Outline of optical path inside the microscope and the computer-controlled pattern which is monitored by CCD simultaneously. The objective lens focusses the laser beam from a He-Ne laser (26 mW max) to a beam spot of $\sim 3\ \mu$m. The focused laser beam effectively trims away the CNTs and create a wide variety of micropatterns on the CNTs sample.

(632.8 nm) is directed into the optical microscope by controlling two reflecting mirrors. The beam splitter reflects the beam to the objective lens (magnification 50 $\times$) for focusing, and also allow the scattering light from the surface to be recorded. The optical image of laser-cutting pattern during laser incidence is simultaneously monitored by TV screen. Through the objective lens with a numberical aperture of 0.95, the laser beam is concentrated on its focal point with the micrometer size. The sample stage could adjust the z-axis position of the sample to bring the sample into focus. The computer-controlled X-Y stage offers an extensive capacity to form two-dimensional pattern or even three-dimensional structure of CNTs.(15) The laser beam through the tunable polarizer lens and mirrors is achieved to the maximum power of 26 mW.

In order to understand how gas molecule and morphological change affect the electronic structure of the CNTs, focused laser beam trimming of the CNTs can be conducted with the CNTs in vacuum or controlled gaseous environment. This was achieved with a transparent mini-chamber in the experiment that provides the vacuum and selective gaseous environments. Focused laser beam locally creates chemically active CNTs array, and allows the gaseous molecule to bond. By controlling the pressure and gas flow rate, the laser system could be operated with the sample housed in different gaseous environments.(15; 16)

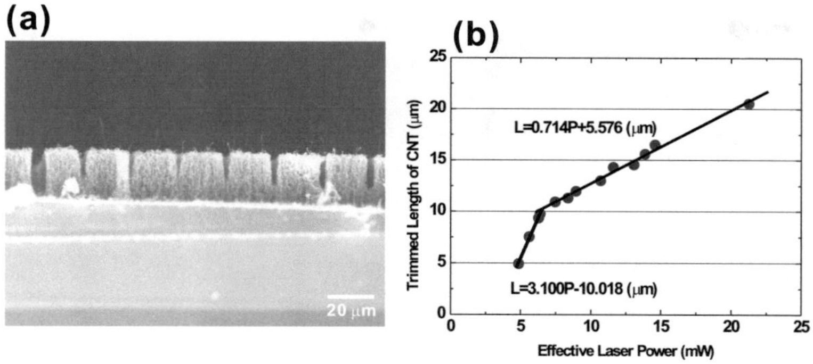


Fig. 2. (a) A cross-sectional SEM image of laser-trimmed CNTs. (b) Trimmed length of CNTs array as a function of laser power.

Therefore, the unique combination of gas cell environment with the focused laser allows us to probe the role of chemical bond inside CNTs.

## 3. Using laser beam to create activation sites in CNTs array

### 3.1 Different laser powers

Figure 2(a) shows the SEM image of CNTs array trimmed by different laser powers, which demonstrates the correlation between the laser power and the trimmed length of CNTs. The original CNTs array is about 20 $\mu$m in length. With the laser beam maintains the same focal point, the trimmed length of CNTs depends on the incident laser power. The laser power is precisely varied by a polarizer which is insected in the optical train of the laser beam as shown in Figure 1(c). Figure 2(b) exhibits the result of trimmed length as a function of laser power. The fitting result shows two linear equations separated at the transition point of 6.5 mW, although the detailed mechanism for such a transition is not clear at the moment. We believe that the untrimmed CNTs represent highly activated CNTs ready to form chemical functionalizaed species with the gaseous species inside the sample chamber. This is the origin of the excellent modification of electronic structure of CNTs array.

Figure 3(a) shows the cross-sectional SEM image of as-grown CNTs array before laser irradiation. Subsequently, the CNTs sample is irradiated by laser beam with different laser powers (7.8 mW, 10.8 mW, 15.0 mW, and 21.0 mW) in vacuum environment. Figure 3(b)~(e) reveal the laser impact on the residual part of CNTs array. The laser power of 7.8 mW ($>$ 3.1 mW minimum power) is capable of trimmed length at least 10 $\mu$m long. The morphology of the remaining CNTs after laser trimming exhibits an obvious increase in the diameter of nanotube on the top region of CNTs array. In Figure 3(c), while incident laser beam raises to the power by 10.8 mW, the tube size tends to get larger at the top region of CNTs array. It is believed that the injected laser beam causes the some parts of CNTs array to vanish, and its residual energy is available to be absorbd by CNTs array.

As the He-Ne laser offers the excited PE of only 1.96 eV, it is almost impossible to break C-C bond or C-O bond.(31) Hence, we believe the photothermal effect caused by laser beam shall dominate the trimming mechanism and its diameter-increased tendency. It causes the local annealing effect on its surface morphology and opens its structure for further modification. Meanwhile, the amount of laser-generated energy transport through the nanotubes is more than that dispersed inside the vacuum environment. It is observed that the larger power the

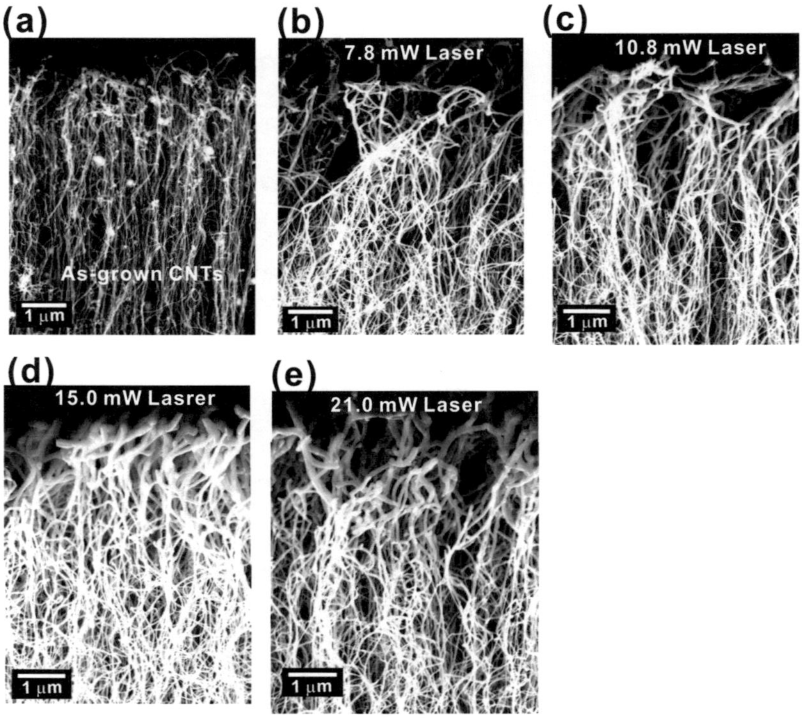


Fig. 3. (a) SEM image of as-grown CNTs. (b)∼(e) SEM images of modified CNTs with laser power of 7.8, 10.8, 15.0, and 21.0 mW, respectively.

laser beam imposes, the more enlarged diameter the segment of nantube shows. Although the major part of absorbed energy is able to make CNTs disappear by gas vaporization, the residual energy can make the tube size increase. The beginning of enlarged diameter should come from the structural reconstruction by the thermal melting effect. Previous study indicates that individual nanotube at annealing temperature of 1500 °C can coalesce a lot of nanotubes as a diameter-doubled tube.(34) It is apparent that the tube diameter at the bottom region is smaller than that at the top region. Therefore, the localized modification of lattice structure is exclusively found at the top region of CNTs array. It would be an advanced method to trim the morpholgy and tune its chemical performance reliably.

## 4. Spatially resolved chemical mapping of CNTs

### 4.1 Scanning PhotoElectron Microscopy (SPEM)

Owing to the powerful function of SPEM, it has been built at several synchrotron centers, e.g. NSRRC in Taiwan, PAL in South Korea, ALS in USA, and ELETTRA in Italy. In Taiwan, the SPEM is set at U5-SGM undulator beamline at NSRRC in Hsinchu. The photon energy of monochromatic X-ray could be operated from 60 to 1500 eV using a spherical grating monochromator (SGM).(20) The installed gratings (ruling density of 285, 400, 800, and 1600 l/mm) are used to provide the monochromator X-ray source. Due to the top-up mode of storage ring operated in NSRRC, the incident photon flux could maintain the same brightness during measurement. In Figure 4, the schematic describes the layout of U5 beamine and SPEM. The X-ray beam, which is collimated by the refocusing mirror and pinhole, overfill the

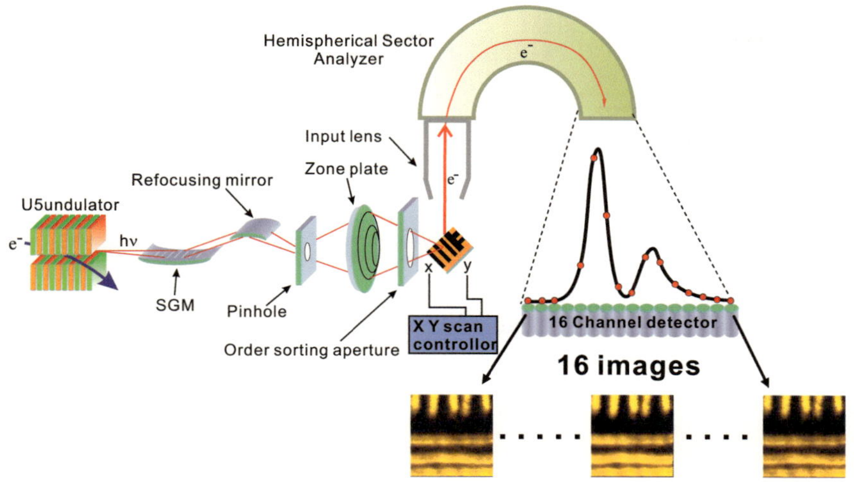


Fig. 4. Outline of SPEM equipment. The incident X-ray beam through the grating, refocusing mirror, and focusing optics is able to display the focused X-ray. The emitted photoelectron is acquired and collected by the 16-channel detector of HSA analyzer.

Fresnel zone plate (ZP) to produce the diffractive light. Order sorting aperture (OSA) behind the ZP could select the first-order diffractive beam for the purpose of given focusing X-ray. The spatial resolution and the focus length are ultimately dependent on the performance of ZP and OSA.(21) The optimal lateral resolution of X-ray beam is estimated about 100 nm. Photoelectron excited by X-ray irradiation is collected by a 279.4-mm diameter hemispherical sector electron analyzer (HSA) with Omni V small-area lens and 16-channel detector (Physical Electronics). In order to acquire the maximum amount of photoelectrons, HSA with the acceptance angle of 7° is mounted at the angle of 54.7° with respect to the incident X-ray beam. This magic incident angle could receive a maximum intensity of photoelectrons due to the angular dependence of orbital angular symmetry. The 16-channel module using the same pass energy of HSA is capable of recording the energy distribution curve (EDC) of photoelectrons simultaneously. Adding the X-Y scanning function of flexure stage, SPEM could acquire a two-dimensional chemical image with 16 different BEs.(35; 36) The sample stage could be precisely probed at the certain location in the mapping image with the help of the readout device. The calibration of photon energy is measured from the 4f state of bulk Au sample or C 1s state of highly ordered pyrolytic graphite (HOPG) sample during the experiment.

### 4.2 Design of controllable gaseous environment insides laser-irradiatied CNTs system

As-grown CNTs are trimmed off by the focused laser in various gaseous environments (air, vacuum, $N_2$, and $O_2$). The modified morphology resembles the edge of sawtooth because the laser beam passes the focusing lens obliquely, as illustrated in Figure 5(a). As-grown and laser-modified area are manufactured together in one sample in order to study their respective electronic structure under the same experimental condition. Figure 5(b) shows the cross-sectional SEM image of modified CNTs. The inset of Figure 5(b) represents the magnified image of the sawtooth-shaped morphology. Figure 5(c) shows the C 1s SPEM image obtained by collecting photoelectrons of C 1s state. The top area of CNTs array, which genetrates the bright contrast, is dependent on the shadowing effect related to the acceptance angle of analyzer and highly intensive signal from the selected chemical state. Also, the

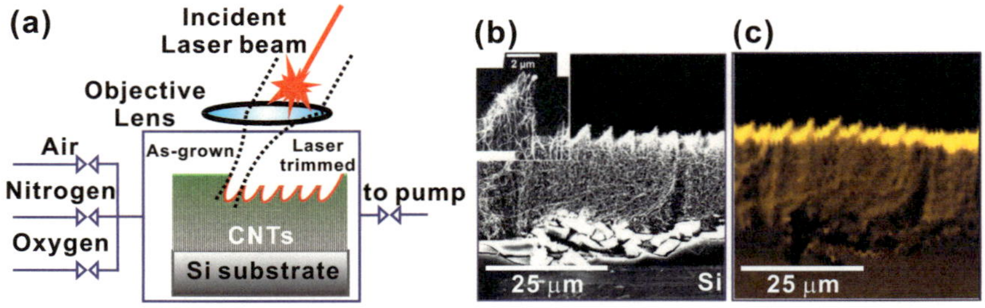


Fig. 5. (a) Schematic diagram of oblique focused laser beam to trim CNTs in the selected gaseous environment. (b)(c) Cross-sectional SEM image and corresponding SPEM image, respectively. The inset of (b) shows the magnified SEM image.

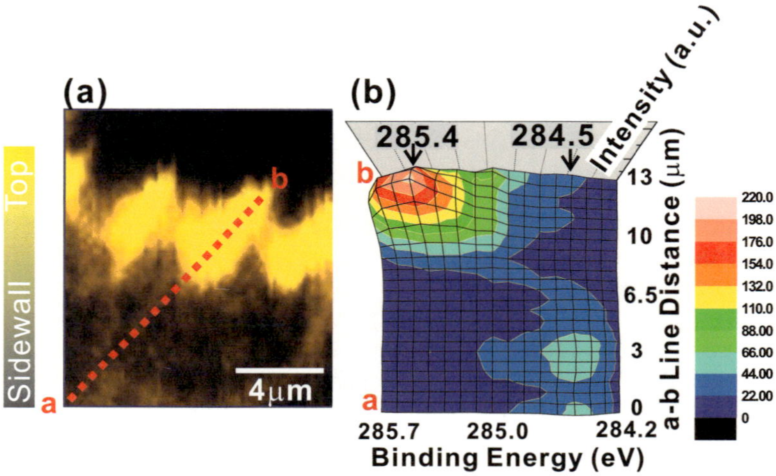


Fig. 6. (a) Cross-sectional SPEM image of laser-modified CNTs in air environment. The image corresponds to the intensity summation over 16 channels around C 1s state. (b) C 1s spectra taken along a-b line denoted in (a). The distance along a-b line is the ordinate and the binding energy of the 16-channel represents the abscissa. The colorimeter on the right side of (b) reflects the emission intensity of C 1s state collected from the individual channels. The chemical shift of 0.9 eV is observed between the laser-modified and sidewalls region.

intensity contrast on the sidewalls area is ascribed from the shadowing effect between the neighboring bundles. The cracks of the Si substrate in Figure 5(b)(c) clearly indicate that they are obtained from the same region of the sample.

## 5. Electronic structure evolution of laser-modified CNTs

### 5.1 Chemical shift of C 1s state of air-treated CNTs

Figure 6(a) shows a cross-sectional SPEM image of the laser-modified CNTs. The sample is treated in ambiance with laser power 19.3 mW. This mapping image stems from the intensity summation of the 16 SPEM images corresponding to the C 1s BE range between 284.2 and 285.7 eV. The C 1s spectra extracted along a-b line, as denoted in Figure 6(a), are analyzed to reveal its position-dependent electronic structures. These spectra are the intensity summation

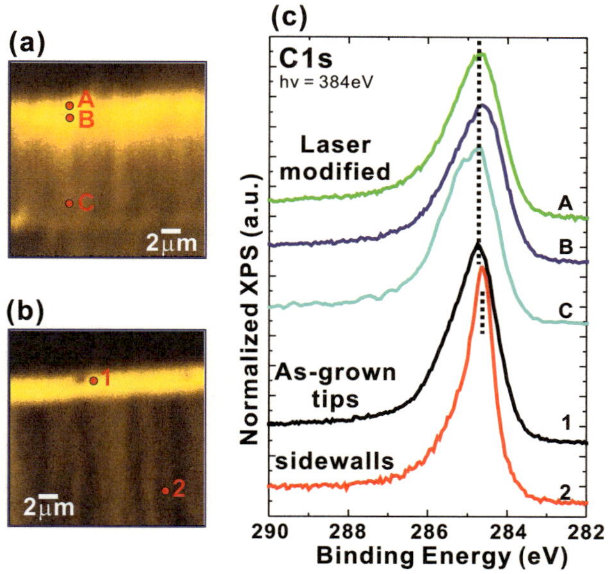


Fig. 7. (a) Cross-sectional SPEM C 1s image of CNTs for laser modification in vacuum. (b) Cross-sectional C 1s mapping image of as-grown CNTs. (c) Photoelectron spectra of C 1s state from as-grown area (position 1 and 2) and the area modified in vacuum (position A, B, and C).

over 400 nm $\times$ 400 nm square selected along a-b line. The shadowing effect can result in the higher signal intensity, instead of changing its EDC. Figure 6(b) reflects the 16-channel curves relative to C 1s state with 1.5 eV range. The ordinate is the position along the a-b line; abscissa is the BE of the 16 channels, the photoelectron intensity is represented by different colors. With the aid of this representative method, we can easily visualize the quantities of the spatially distributed chemical shifts with their relative intensity. Near the point "a" (around sidewalls region), the C 1s state shows the lowest BE with 284.5 eV, which is identical to the sidewalls of as-grown CNTs.(17) Around the point "b" (modified top region) the C 1s peak shifts to the highest BE with 285.4 eV. It is surprising that the chemical shift of C 1s state between sidewalls (a) and top (b) is as large as 0.9 eV. But in the pristine CNTs experiment,(18) the C 1s peak in the top region is higher than that in the sidewalls region only by 0.2 eV.

### 5.2 Role of coexisted gas molecule and laser assistance in chemical configuration of CNTs

As both adsorbed gas molecules and morphological change occur during laser irradiation, we design the vacuum experiment to clarify its correlation. Figure 7(a)(b) show the cross-sectional SPEM images, acquired around C 1s state with BE 282.5 $\sim$ 288.5 eV, with and without laser treatment in vacuum ($10^{-2} \sim 10^{-3}$ torr), respectively. Figure 7(c) shows the C 1s spectra of various locations denoted in Figure 7(a)(b). C 1s spectra obtained from the modified (A $\sim$ C ) and as-grown top (1) areas show the same BE position. Compared to the spectrum measured at as-grown sidewalls area (2), it shows the slight up-shift of C 1s peak ($<$ 0.1 eV). The loss of gaseous molecule could reduce its possiblity of chemical reaction at the carbon site, while the laser beam breaks the covalent bond of hexagonal C-C structure. It means the morphological change alone is not sufficient to induce the up-shift of C 1s state, although it has ever been exposed to the air environment after laser irradiation.

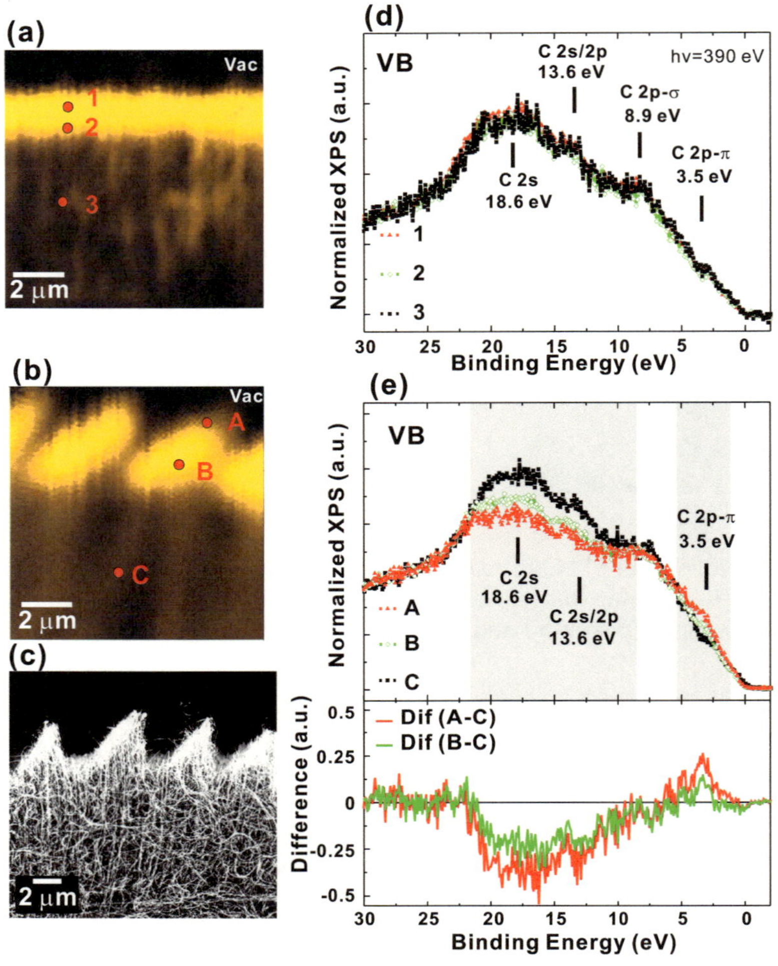


Fig. 8. (a)(d) Cross-sectional SPEM image of as-grown CNTs and air-treated CNTs, respectively. (b)(e) Spatially-resolved VB spectra from position 1 $\sim$ 3 of as-grown CNTs and position A $\sim$ C of air-treated CNTs. (c) SEM image relative to SPEM image of (b). The bottom figure of (e) provides the proof of DOS arrangement between 2p-$\pi$ and 2p-$\sigma$ electrons.

### 5.3 Rearrangment of DOS in air-treated CNTs

Figure 8(a) shows the cross-sectional SPEM image of as-grown CNTs, which is summarized by 16-channel images (C 1s BE 290.0 $\sim$ 278.0 eV). Figure 8(d) exhibits the position-relative VB spectra for the top region (1) and sidewalls region (2 and 3). The normalization process of VB spectra is critical for the density of state (DOS) comparison, thus the integrated intensity around BE 32.0 $\sim$ 31.0 eV is set to the same unity for the assumption of uniform scattering. The broad spectral feature around BE 18.6 eV is attributed to the formation of C 2s band, and the other feature around BE 13.6 eV is related to the mixing state of C 2s and C 2p band.(37) The C 2p-$\sigma$ band of sp$^2$-hybridized carbon is identified around BE 8.9 eV, and the other C 2p-$\pi$ band is assigned to BE 3.5 eV near to the Fermi edge.(38; 39)

As denoted "1" in Figure 8(a), the VB spectrum performs the DOS behavior where is near to the top of CNTs. While probing at the position 2 and 3, it shows the electronic behavior of sidewalls. The bright yellow area reflects more intensive signal of C 1s state from the top region (1) than the sidewalls region (2 and 3). The reason for the appearance of some stripes on the sidewalls area is the shadowing effect. According to the meaning of individual bands in VB spectra, we observe the similar DOS composition in the position 1 $\sim$ 3. In view of lattice structure, the close carbon cage at the end of nanotube involves the pentagon-related curvature and the dangling bond of unpaired $\pi$ bond, which should make a change in the configuration of electronic structure. Thus, some groups have indicated that the uniquely structural characteristic at the tip stands for the DOS enhancement near the Fermi level.(17; 18) However, in our VB result we do not find any significant difference around BE 10.0 $\sim$ 0 eV between the top and sidewalls. That maybe result from different CNTs species or the sensitivity of probing positions around top region. The electronic performance between top and sidewalls rarely show any difference in C 2s, C 2s/2p, and C 2p-$\sigma$ band. That is understood that the chemical binding mechanism happens to the outside electrons near Fermi level.

Figure 8(b)(c) show the cross-sectional SPEM and SEM image of trimmed CNTs. This sample is trimmed in ambient environment, as the same fabrication process in Figure 6(a). The sawtooth-shaped CNTs array results from the shift of incident axis of laser beam. The similar characteristic between chemical and morphological image manifests the ability to identify the spatially-resolved electronic structure. VB spectra, e.g. the position A and B for the top area and the position C for the sidewalls area, are exhibited in Figure 8(e). The bottom inset of Figure 8(e) is used to compare the spectral difference of top (A and B) and sidewalls (C). The adopted parameters for the experiments, i.e. incident X-ray angle, photon energy, and pass energy of HSA, are set to the same for comparsion. The VB performance measured at the position C in Figure 8(e) presents the same DOS distribution as that probed at top and sidewalls region of as-grown CNTs in Figure 8(d). VB spectra at position A and B appear more C 2p-$\pi$ electrons between BE 5.2 $\sim$ 2.1 eV than that at the position C. Meanwhile, the spectral features around BE 21.0 $\sim$ 9.0 eV reflect the negative difference of A - C and B - C curve. Compared with the reference of sidewalls region (C), the transition from C 2s and C 2s/2p band to C 2p-$\pi$ is observed at the irradiated top regions (A and B). This decreasing BE range including C 2s band and mixing C 2s/2p band accounts for the removal of carbon atoms or/and re-hybridization of DOS. The increasing DOS around C 2p-$\pi$ state is ascribed to the production of structural defect with dangling bond or/and the gas hybridization effect. Because the unpaired $\pi$ electron at the defect site behaves as the reactive site, the surrounding gas molecules could participate in its chemical hybridization. The transition of DOS in VB result is in agreement with the up-shift of C 1s state by 0.9 eV, which is given between the top and sidewalls region.

## 6. Variable chemical binding reactions

Laser-irradiated CNTs array in vacuum does not account for the direct correlation of the morphological change. Thus, high chemical shift in air-treated CNTs is attributed to the gas contribution involving the laser-induced reconstruction. In other words, the light irradiation is able to induce the chemical activity of $sp^2$-bonded carbon with the gas atom. In the sections below, we will discuss the $N_2$ and $O_2$-treated CNTs for the electronic modification and chemical functionalization.

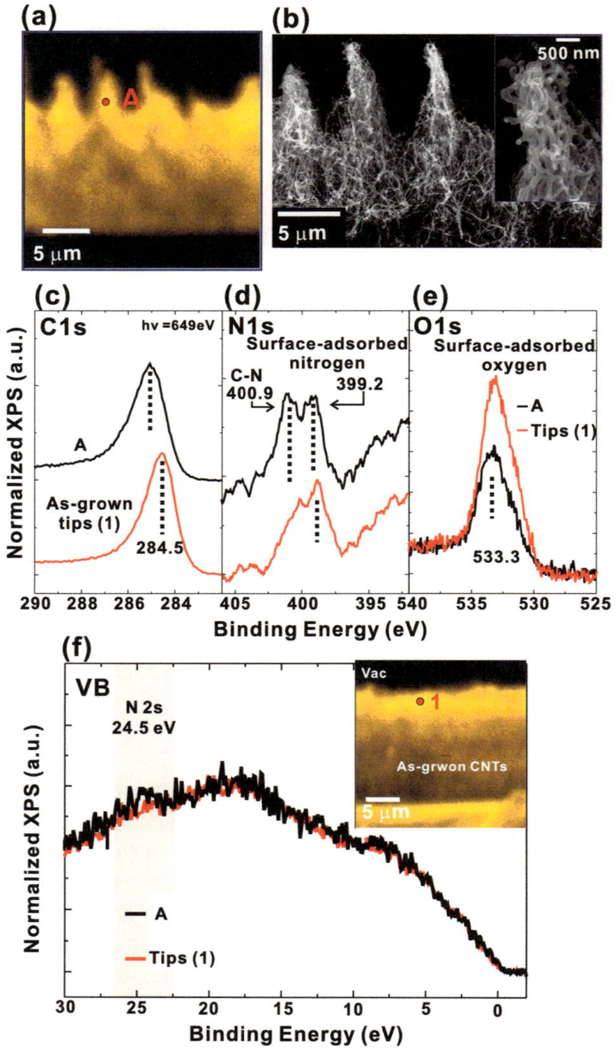


Fig. 9. (a)(b) Cross-sectional SPEM and SEM image of nitrogen-treated CNTs, respectively. The inset of (b) shows the morphology of laser-irradiated CNTs array. (c)-(f) Position-related C 1s, N 1s, O 1s and VB spectra. The inset of (f) shows the SPEM image of as-grown CNTs immersed in nitrogen environment. The sort of individual spectra, denoted A and as-grown tips, are used to compare its electronic change by laser exposure.

### 6.1 Nitrogen-treated CNTs

Figure 9(a)(b) present the SPEM and SEM image of the nitrogen-treated CNTs sample, respectively. The gas cell is filled with nitrogen gas (300 $\sim$ 500 torr) during the laser trimming process. The mapping image reveals the triangle shape of laser-trimmed CNTs similar to the morphology image taken by SEM. The high magnifcation SEM image as shown in the right corner of Figure 9(b) reveals the enlarged diameter of nanotubes, compared with the tube diameter in the sidewalls region. The position-resolved photoemission spectra (C 1s, N 1s,

O 1s, and VB), measured with photon energy by 649 eV, are exhibited in Figure 9(c) $\sim$ (f). The inset picture of Figure 9(f) shows the SPEM image of as-grown CNTs area, which is also transfered to the nitrogen-filled gas cell for comparsion. The character "A" and "1" on the top region of CNTs array are used to mark the position investigated by SPEM.

In Figure 9(c), C 1s spectrum obtained from the top region (poistion marked "A" in Fig. 9(a)) reveals the up-shift of C 1s state by 0.6 eV, as compared with as-grown CNTs (poistion marked "1" in inset of Fig. 9(f)). Figure 9(d) shows N 1s spectra with and without laser irradiation. The spectra have been smoothed without loss of energy resolution. The peak at 400.9 eV is related to C-N bond, the other one at 399.2 eV is ascribed to surface-adsorbed nitrogen(40) because both laser modified and as-grown areas have been immersed in nitrogen and ambiance environment. Two nitrogen peaks located at BE of 400.9 and 399.2 eV are found in the spectrum obtained after the CNTs has been irradiated by laser. Actually, the substitutional N atoms in the hexigonal carbon sheet could account for the up-shift of C 1s state and enhanced C-N bond (400.9 eV).(41)

It is also important to see the chemical reaction with oxygen species in N-treated CNTs, as shown in Figure 9(e). The oxygen feature at BE 533.3 eV is ascribed to the physically adsorbed oxygen molecule.(42) It makes sense that lower signal of physical oxygen absorption in the top region (A) is corresponding to the replacement action by C-N bond during laser irradiation.

In Figure 9(f), VB spectrum acquired from the top region of nitrogen-treated CNTs (A) shows the similar performance as that of the top region of as-grown CNTs (1), except for the spectral feature at BE 24.1 eV. It is assigned to N 2s band under the concession of C-N configuration.(43) However, the VB spectrum of as-grown CNTs (1) is quite different from the previous data of as-grown CNTs in Figure 8(b). It is found that VB spectra of the as-grown CNTs (1) after the nitrogen and ambient environment is quite sensitive to the surface contamination even without laser incidence. Surface-adsorbed features in N 1s state for the peak at BE 399.2 eV and in O 1s state for the peak of BE 533.2 eV are observed in both CNTs samples, which may determine the similar performance of VB spectra. In fact, due to enhanced C-N bond at BE 400.9 eV, the nitrogen-treated CNTs (A) reveals more intense feature in N 2s band than that of as-grown CNTs (1).

In short, the chemical shift of C 1s state by 0.6 eV is assigned to strong attachment of nitrogen gas, which is ascribed to the enhanced C-N bond in N 1s state. The surface-adsorbed nitrogen does not make the reorganization of C 1s state happen due to physical absorption. On the contrary, the DOS composition in VB spectra is very sensitive to the surface nitrogen contamination, which causes almost the same behavior. The DOS distribution with or without laser irradiation does not exhibit the difference. In terms of the existence of C-N bond, it is obtained that N 2s band in VB specteum is slight enhanced by laser irradiation.

### 6.2 Oxygen-treated CNTs

Since nitrogen-treated CNTs have shown the possibility to modify chemical property, the introduction of oxygen gas is the alternative route for engineering the electronic structure effectively. CNTs sample is irradiated by laser beam with laser power of 14.8 mW within oxygen environment (300 $\sim$ 600 torr). As-grown CNTs, existed on the rest of laser-irradiated sample, are exposed to the same oxygen environment and ambient transport process, too. Figure 10(a)(b) reveal the SPEM and SEM image of oxygen-treated CNTs, respectively. The inset of Figure 10(b) highlights some melting balls and fusion of nanotube bundles after laser irradiation. Figure 10(c)(d)(e) exhibit C 1s, O 1s, and VB spectra for the individual positions of oxygen-treated (position "B" and "C" marked in Fig. 10(a)) and as-grown (position marked "2"

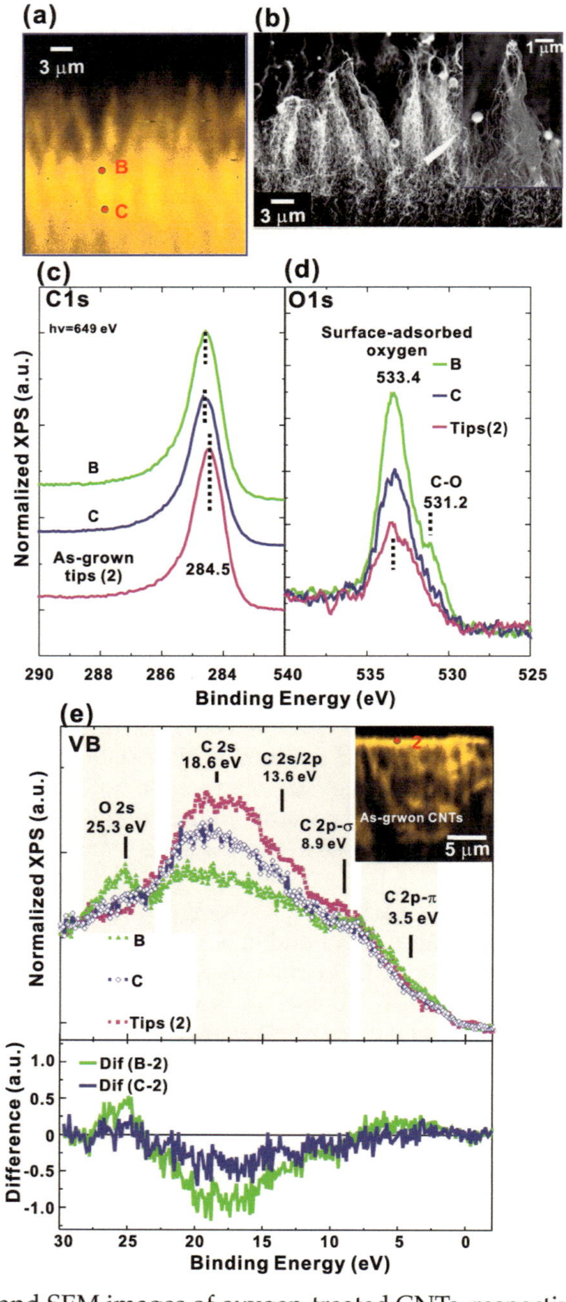


Fig. 10. (a)(b) SPEM and SEM images of oxygen-treated CNTs, respectively. The inset of (b) magnifies the laser-irradiated morphology, which suffered from local heating dose and oxygen gas. (c)~(e) Position-related C 1s, N 1s, and VB spectra. The inset of (e) shows the SPEM image of as-grown CNTs without the laser irradiation. Difference between as-grown and laser-irradiated spectra demonstrates the modification of DOS in VB range, as shown in the bottom figure of (e).

in inset of Fig. 10(e)) CNTs. The measured positions are also marked in the mapping image of Figure 10(a) and the inset of Figure 10(e).

In Figure 10(c), the C 1s peak (B and C) after laser irradiation shifts toward higher BE, however only by 0.2 eV relative to the top region of as-grown CNTs (position "2"). In Figure 10(d) O 1s spectrum (B) exhibits two oxidation states located at BE 533.4 and 531.2 eV. The weak feature at BE 531.2 eV is related to C-O bond, and the other feature at BE 533.4 eV is ascribed to the adsorbed oxygen linked to carbon.(42) Physically adsorbed oxidation is also seen in N-treated CNTs (Figure 9(e)), because of the ambient transport process. The intensity of C-O bond at the top region (B) is larger than that near sidewalls (C), due to the energy localization. Meanwhile, the amounts of adsorbed oxygen at the modified regions (B and C) are more than that top region of as-grown CNTs (2), individually. It is considered that opening the tubular network of CNTs is possbile to add the buliding site of C-O bond and has the side effect to enhance surface-adsorbed oxygen.

VB spectra of as-grown area (2) and oxygen-treated area (B and C) are illustrated in Figure 10(e). The bottom part of Figure 10(e) is the spectral difference between oxygen-treated (B and C) and as-grown area (2). It is found that the broad range between BE 22.1 $\sim$ 8.0 eV in the oxygen-treated CNTs (B and C) shows the decreasing tendency compared with the reference spectrum of as-grown area (2). This spectral range includes the C 2s band at BE 18.6 eV, mixing C 2s/2p band at BE 13.6 eV, and C 2p-$\sigma$ band at BE 3.5 eV. Apparently, the change between "B" and "2" is larger than that between "C" and "2", becuase of the rearrangement of DOS and morphological change. The C 2p-$\pi$ band around BE 3.5 eV close to Fermi level presents the slight enhancement at the top region of oxygen-treated CNTs (B), but at the sidewalls region (C) it does not. Dangling bond and defect site may result in the increasing intensity of 2p-$\pi$ band after laser irradiation. Indeed, the O 2s-derived band at BE 25.3 eV is obviously raised in the oxygen-treated CNTs (B). By contrast, the resultant spectrum of the sidewalls region (C) appears less enhanced around O 2s-derived band. The appearance of C-O state in O 1s state and O 2s band in VB spectrum could count on the formation of C-O chemical bond.

The surface-adsorbed oxygen feature at BE 533.4 eV demonstrates the intensity enhancement from the position of sidewalls region (C) to top region (B). It seems that laser impact generates the structural defect and the activity site to bond oxygen atom. Therefore, the physisorbed oxygen amount on the surface should be induced by the influence of laser irradiation. The removal of carbon atoms and oxygen contamination may result in the decreasing intensity and re-hybridization around C 2s and C 2s/2p band (BE 22.1 eV $\sim$ 8.0 eV). Meanwhile, C 2p-$\pi$ band around BE 8.0 $\sim$ 2.0 eV is increased while the C-O chemical bond occurs, in particular the case of top region (B). In fact, the formation of chemical C-O state increases the extra factor inside the DOS re-distribution; therefore, without this, like the case of the sidewalls region (C), it shows the middle modification due to the less impact of dangling bond and chemical C-O bond.(44; 45) The structural defect and dangling bond of CNTs after laser trimming should be the important factors in the DOS reconstruction.(46) It means the eliminated amount of C 2s and C 2s/2p band and increased C 2p-$\pi$ intensity is in agreement to the result of air-treated CNTs. In fact, the behavior of C 2p-$\pi$ electrons of oxygen-treated CNTs is observed for a little different, because the mixing contribution between air environment and laser-induced defect play the important role in C 2p-$\pi$ electrons.

## 7. Discussion

The dangling bonds and topological defects are supposed to be increased upon laser trimming.(29) Naturally, laser trimming causes the breakage of the C-C bond in CNTs and

the gas molecules nearby can bond to the C atoms possessing dangling bonds. The existence of stable C-N and C-O bonds on the surface of nanotubes was expected to become metallic, independent of the tubular diameter and chirality.(7; 47) Both $N_2$ and $O_2$ molecules can be bonded with CNTs assisted by laser and result in the different chemical shifts (0.6 and 0.2 eV). The highest chemical shift (0.9 eV) in the air-treated CNTs may be attributed to the mixing gas contribution, which needs to be further studied. On the other hand, the result of vacuum-treated CNTs shows the slight up-shift of C 1s state ($< 0.1$ eV). However, even after the gas treatment, a lot of defects are still expected to exist at the laser trimmed surface of the gas-treated CNTs. These defects may also contribute to the observed C 1s chemical shift besides the gas interaction. A shift of C 1s state is observed at higher BE shoulder as a result of the reconstruction of chemical environment between the structural defect and gas atom.
In terms of VB spectra, the air-treated CNTs yields the local modification of DOS at the top region of CNTs array, where results in the raising C 2p-$\pi$ electron and decreasing C 2s and C 2p-$\sigma$ electron. After studying the influence of gaseous environments on laser trimming of CNTs, the laser irradiation plus oxygen contribution dominate the reorganization of DOS. The DOS change between $\pi$ and $\sigma$ band is corresponding to the existence of C-O bond and generation of structural defect. Furthermore, the surface-adsorbed contamination of $N_2$ molecules can largely influence the DOS distribution. While we consider the laser-induced defect and dangling bond after laser irradiation, the bonded gas molecules at the defective sites are both available at the trimmed surface. This kind of combination causes the obvious modification of DOS distribution in VB.
Hence, the dangling bond of defective site and gas contamination shall take the essential contribution of the reconstruction of DOS. It is believed that laser irradiation is a viable technique for functionalizing CNTs incorporating with N and O atoms. By controlling the type of gas molecules used, we can change and tune the electronic character of CNTs for the development of electronic device. Laser technology provide the stage of localized chemical modification for the novel applications.

## 8. Acknowledgement

Authors greatly acknowledge the support and fruitful discussion of Dr. Chia-Hao Chen, Dr. Yanwu Zhu, Prof. Shangjr Gwo, Prof. Shien-Der Tzeng, Dr. Yin-Ming Chang, Mr. Chih-Wei Peng, Mr. Sheng-Syun Wong, Mr. Po-Tsong Wu, and Ms. Soi-Chan Lei. This work was supported by the National Science Council of Taiwan under grant no. NSC 98-2120-M-002-010.

## 9. References

[1] Iijima, S. (1991). Helical Microtubules of Graphitic Carbon. *Nature*, Vol. 354, pp. 56-58, Doi:10.1038/354056a0

[2] Thess, A.; Lee, R.; Nikolaev, P.; Dai, H.; Petit, P.; Robert, J.; Xu, C.; Lee, Y. H.; Kim, S. G.; Rinzler, A. G.; Colbert, D. T.; Scuseria, G. E.; Tomanek, D.; Fischer, J. E. & Smalley, R. E. (1996). Crystalline Ropes of Metallic Carbon Nanotubes. *Science*, Vol. 273, pp. 483-487. Doi:10.1126/science.273.5274.483

[3] Wildöer, J. W. G.; Venema, L. C.; Rinzler, A. G.; Smalley, R. E. & Dekker C. (1998). Electronic Structure of Atomically Resolved Carbon Nanotubes. *Nature*, Vol. 391, pp. 59-62. Doi:10.1038/34139

[4] Hamada, N.; Sawada, S. & Oshiyama, A. (1992). New One-dimensional Conductors: Graphitic Microtubules. *Phys. Rev. Lett.*, Vol. 68, pp. 1579-1581. Doi:10.1103/PhysRevLett.68.1579

[5] Demczyk, B. G.; Wang Y. M.; Cumings J.; Hetman, M.; Han, W.; Zettl, A. & Ritchie, R. O. (2002). Direct Mechanical Measurement of the Tensile Strength and Elastic Modulus of Multiwalled Carbon Nanotubes. *Mater. Sci. Eng. A*, Vol. 334, pp. 173-178. Doi: 10.1016/S0921-5093(01)01807-X

[6] Dai, H. J.; Hafner, J. H.; Rinzler, A. G.; Colbert, D. T. & Smalley, R. E. (1996). Direct Mechanical Measurement of the Tensile Strength and Elastic Modulus of Multiwalled Carbon Nanotubes. *Nature*, Vol. 384, pp. 147-150. Doi:10.1038/384147a0

[7] Collins, P. G.; Bradley, K.; Ishigami, M.; & Zettl, A (2000). Extreme Oxygen Sensitivity of Electronic Properties of Carbon Nanotubes. *Science* Vol. 287, pp. 1801-1804. Doi: 10.1126/science. 287.5459.1801

[8] Yoo, E.; Gao, L.; Komatsu, T.; Yagai, N.; Arai, K.; Yamazaki, T.; Matsuishi, K.; Matsumoto, T.; & Nakamura J. (2004). Atomic Hydrogen Storage in Carbon Nanotubes Promoted by Metal Catalysts. *J. Phys. Chem. B*, Vol. 108, No. 49, pp. 18903-18907. Doi: 10.1021/jp047056q

[9] Berber, S.; Kwon, Y. K.; & Tomanek D. (2000). Unusually High Thermal Conductivity of Carbon Nanotubes. *Phys. Rev. Lett.*, Vol. 84, pp. 4613-4616. Doi: 10.1103/PhysRevLett.84.4613

[10] Klein, K. L., Melechko, A. V.; McKnight, T. E.; Retterer, S. T.; Rack, P. D.; Fowlkes, J. D.; Joy, D. C.& Simpson, M. L. (2008). Surface Characterization and Functionalization of Carbon Nanofibers. *J. Appl. Phys.*, Vol. 103, pp. 061301-061327. Doi:10.1063/1.2840049

[11] Peng, S. & Cho, K. (2000). Chemical Control of Nanotube Electronics. *Nanotechnology*, Vol. 11, pp. 57-60. Doi: 10.1088/0957-4484/11/2/303

[12] Gohel, A.; Chin, K. C.; Zhu, Y. M.; Sow, C. H.; & Wee, A. T. S. (2005). Field Emission Properties of $N_2$ and Ar Plasma-treated Multi-wall Carbon Nanotubes. *Carbon*, Vol. 43, pp. 2530-2535.

[13] Lee, H. J.; Lee, Y.D.; Moon, S. I.; Cho, W. S.; Lee, Y. H.; Kim, J. K.; Hwang, S. W.; & Ju, B. K. (2000). Enhanced Surface morphologies of Screen-printed Carbon Nanotube Films by Heat Treatment and Their Field-emission Properties. *Carbon*, Vol. 44, pp. 2625-2630.

[14] Saurakhiya, N.; Zhu, Y. W.; Cheong, F. C.; Ong, C. K.; Wee, A. T. S.; Lin, J. Y.; & Sow, C. H. (2005). Pulased Laser Deposition-assisted Patterning of Aligned Carbon Nanotubes Modified by Focused Laser Beam for Efficient Field Emission. *Carbon*, Vol. 43, pp. 2128-2133.

[15] Lim, K. Y.; Sow, C. H.; Lin, J.; Cheong, F. C.; Shen, Z. X.; Thong, J. T. L.; Chin, K. C. & Wee, A. T. S., (2003). Laser Pruning of Carbon Nanotubes as a Route to Static and Movable Structures. *Adv. Mater.*, Vol. 15, pp. 300-303. Doi: 10.1002/adma.200390072

[16] Cheong, F. C.; Lim, K. Y.; Sow, C. H.; Lin, J.; & Ong, C. K. (2003). Large Area Patterned Arrays of Aligned Carbon Nanotubes via Laser Trimming. *Nanotechnology*, Vol. 14, No. 4, pp. 433-437. Doi: 10.1088/0957-4484/14/4/305

[17] Suzuki, S.; Watanabe, Y.; Ogino, T.; Heun, S.; Gregoratti, L.; Barinov, A.; Kaulich, B.; Kiskinova, M.; Zhu, W.; Bower C. & Zhou, O. (2002) Electronic Structure of Carbon Nanotubes Studied by Photoelectron Spectromicroscopy. *Phys. Rev. B*, Vol. 66, pp. 035414. Doi: 10.1103/PhysRevB.66.035414

[18] Chiou, J. W.; Yueh, C. L.; Jan, J. C.; Tsai, H. M.; Pong, W. F.; Hong, I. H.; Klauser, R.; Tsai, M. H.; Chang, Y. K.; Chen, Y. Y.; Wu, C. T.; Chen, K. H.; Wei, S. L.; Wen, C. Y.; Chen,

l. C.; & Chuang, T. J. (2002). Electronic Structure of the Carbon Nanotube Tips Studied by X-ray-absorption Spectroscopy and Scanning Photoelectron Microscopy. *Appl. Phys. Lett.* Vol. 81, pp. 4189-4192. Doi:10.1063/1.1523152

[19] de Heer, W. A.; Châtelain, A.& Ugarte D., (1995). A Carbon Nanotube Field-Emission Electron Source. *Science*, Vol. 270, No. 5239, pp. 1179-1180. Doi:10.1126/science.270.5239.1179

[20] Hong, I. H.; Lee, T. H.; Yin, G. C.; Wei, D. H.; Juang, J. M.; Dann, T. E.; Klauser, R.; Chuang, T. J.; Chen, C. T.; & Tsang, K. L., (2001). Performance of the SRRC Scanning Photoelectron Microscope. *Methods Phys. Res. A*, Vol. 467-468, pp. 905-908.

[21] Klauser, R.; Hong, I. H.; Lee, T. H.; Yin, G. C.; Wei, D. H.; Tsang, K. L.; Chuang, T. J.; Wang, S. C.; Gwo, S. & Zharnikov, M. (2001). Zone-plate-based Scanning Photoelectron Microscopy at SRRC: Performance and Applications. *Surface Review and Letters*, Vol. 9, No. 1, pp. 213-222.

[22] Journet, C.; Maser, W. K.; Bernier, P.; Loiseau, A.; de la Chappelle, M. L.; Lefrant, S.; Deniard, P.; Lee, R. & Fischer, J. E. (1997). Large-scale Production of Single-walled Carbon Nanotubes by the Electric-arc Technique. *Nature*, Vol. 388, pp. 756-758.

[23] Thess, A.; Lee, R.; Nikolaev, P.; Dai, H. J.; Petit, P.; Robert, J,; Xu, C. H.; Lee, Y. H.; Kim, S. G.; Rinzler, A. G.; Colbert, D. T.; Scuseria, G. E.; Tomanek, D.; Fischer, J. E. & Smalley R. E. (1996). Crystalline Ropes of Metallic Carbon Nanotubes. *Science*, Vol. 273, pp. 483. Doi:10.1126/science.273.5274.483

[24] Iijima, S. & Ichihashi, T. (1993). Single-shell Carbon Nanotubes of 1-nm Diameter. *Nature*, Vol. 363, pp. 603-605. Doi:10.1038/363603a0

[25] Wang, Y. H.; Lin, J.; Huan, C. H. A. & Chen, G. S. (2001). Synthesis of Large Area Aligned Carbon Nanotube Arrays from $C_2H_2$ - $H_2$ Mixture by rf Plasma-enhanced Chemical Vapor Deposition. *Appl. Phys. Lett.*, Vol. 79, pp.680-682. Doi:10.1063/1.1390314

[26] Chhowalla, M.; Teo, K. B. K.; Ducati, C.; Rupesinghe, N. L.; Amaratunga, G. A. J.; Ferrari, A. C.; Roy, D.; Robertson, J. & Milne, W. I. (2001). Growth Process Conditions of Vertically Aligned Carbon Nanotubes Using Plasma Enhanced Chemical Vapor Deposition. *J. Appl. Phys.*, Vol. 90, pp. 5308-5317. Doi:10.1063/1.1410322

[27] Fan, S.; Chapline, M. G.; Franklin, N. R.; Tombler, T. W.; Cassell, A. M. & Dai, H. (1999). Self-oriented Regular Arrays of Carbon Nanotubes and Their Field Emission Properties. *Science*, Vol. 283, pp. 512-514. Doi:10.1126/science.283.5401.512

[28] Collins, P. G.; Arnold, M. S. & Avouris, P. (2001). Engineering Carbon Nanotubes and Nanotube Circuits Using Electrical Breakdown. *Science*, Vol. 292, pp. 706-709. DOI:10.1126/science.1058782

[29] Rinzler, A. G.; Hafner, J. H.; Nikolaev, P.; Lou, L.; Kim, S. G.; Tománek, D.;Nordlander, P.; Colbert, D. T. & Smalley, R. E. (1995). Unraveling Nanotubes: Field Emission from an Atomic Wire. *Science*, Vol. 269, pp. 1550-1553. Doi:10.1126/science.269.5230.1550

[30] Zhao, W. J.; Kawakami, N.; Sawada, A. & Takai, M. (2003). Field Emission from Screen-printed Carbon Nanotubes Irradiated by Tunable Ultraviolet Laser in Different Atmospheres. *J. Vac. Sci. Technol. B*, Vol. 21, pp. 1734-1737. Doi:10.1116/1.1587136

[31] Yavas, O.; Suzuki, N.; Takai, M.; Hosono, A. & Kawabuchi, S. (1998). Laser Cleaning of Field Emitter Arrays for Enhanced Electron Emission. *Appl. Phys. Lett.*, Vol. 72, pp. 2797-2780. Doi:10.1063/1.121461

[32] Rochanachirapar, W.; Murakami, K.; Yamasaki, N.; Abo, A.; Wakaya, F.& Takai, M. (2005). Influence of Gas Atmosphere during Laser Surface Treatment of CNT Cathode. *J. Vac. Sci. Technol. B*, Vol.23, pp.762-765. Doi:10.1116/1.1868698

[33] Hwang, J. D.; Chen, k. F.; Chan, L. H. & Chang, Y. Y. (2006). Using Infrared Laser to Enhance Field Emission of Carbon Nanotube. *Appl. Phys. Lett.*, Vol. 89, pp. 33103-33106. Doi:10.1063/1.2222337

[34] Nikolaev, P.; Thess, A.; Rinzler, A. G.; Colbert, D. T. & Smalley, R. E. (1997). Diameter Doubling of Single-wall Nanotubes. *Chem. Phys. Lett.*, Vol. 266, pp. 422-426.

[35] Chuang, C. H.; Chen, C. H.; Chang, Y. M.; Peng, C. W.; Wong, S. S.; Tzeng, S. D.; Gwo, S.; Zhu, Y.; Sow, C. H. & Lin, M. T. (2007). Enhanced Chemical Shift of Carbon Nanotube from Laser Assisted Gas Incorporation. *Appl. Phys. Lett.*, Vol. 91, pp. 183101. Doi:10.1063/1.2801698

[36] Klauser, R., Hong, I. H.; Su, H. J.; Chen, T. T.; Gwo, S.; Wang, S. C.; Chuang, T. J. & Gritsenko, V. A.(2001). Oxidation States in Scanning-probe-induced $Si_3N_4$ to $SiO_x$ Conversion Studied by Scanning Photoemission Microscopy. *Appl. Phys. Lett*, Vol. 79, pp. 3143-3145. Doi:10.1063/1.1415415

[37] Larciprete, R.; Goldoni, A.; Lizzit, S.& Petaccia, L. (2005). The Electronic Properties of Carbon Nanotubes Studied by High Resolution Photoemission Spectroscopy. *Appl. Sur. Sci.*, Vol. 248, pp. 8-13. doi:10.1016/j.apsusc.2005.03.023

[38] Schiessling, J.; Kjeldgaard, L.; Rohmund, F.; Falk, L. K. L.; Campbell, E. E. B.; Nordgren, J. & Brühwiler, P. A. (2003). Synchrotron Radiation Study of the Electronic Structure of Multiwalled Carbon Nanotubes. *J. Phys.: Condens. Matter*, Vol. 15, pp.6563-6579. Doi: 10.1088/0953-8984/15/38/022

[39] Bianconi, A.; Hagström, S. B. M. & Bachrach, R. Z. (1977). Photoemission Studies of Graphite High-energy Conduction-band and Valence-band States using Soft-x-ray Synchrotron Radiation Excitation. *Phys. Rev. B*, Vol. 16, pp. 5543-5548. Doi:10.1103/PhysRevB.16.5543

[40] Kim, T. Y.; Lee, K. R.; Eun, K. Y.; Oh, K. H.; Carbon Nanotube Growth Enhanced by Nitrogen Incorporation. *Chem. Phys. Let.*, Vol. 372, pp. 603-607.

[41] Lim, S. H.; Elim, H. I.; Gao, X. Y.; Wee, A. T. S.; Ji, W.; Lee j. Y. & Lin, J. (2006). Electronic and Optical Properties of Nitrogen-doped Multiwalled Carbon Nanotubes. *Phys. Rev. B*, Vol. 73, pp. 045402. Doi:10.1103/PhysRevB.73.045402

[42] Valentini, L.; Lozzi, l.; Picozzi, S.; Cantalini, C.; Santucci, S. & Kenny, J.M. (2004). Adsorption of Oxidizing Gases on Multiwalled Carbon Nanotubes. *J. Vac. Sci. Technol. A*, Vol. 22, pp. 1450. Doi:10.1116/1.1705588

[43] Ray, S. C.; Pao, C. W.; Tsai, H. M.; Chiou, J. W.; Pong, W. F.; Chen, C. W.; Tsai, M. H.; Papakonstantinou, P.; Chen, L. C. & Chen, K. H. (2007). A Comparative Study of the Electronic Structures of Oxygen- and Chlorine-treated Nitrogenated Carbon Nanotubes by X-ray Absorption and Scanning Photoelectron Microscopy. *Appl. Phys. Lett.*, Vol.91, pp. 202102. Doi:10.1063/1.2807275

[44] Bittencourt, C.; Felten, A.; Douhard, B.; Ghijsen, J.; Johnson, R. L; Drube, W. & Pireaux, J. J. (2006). Photoemission Studies of Gold Clusters Thermally Evaporated on Multiwall Carbon Nanotubes. *Chem. Phys.*, Vol. 328, pp. 385-391. Doi:10.1016/j.chemphys.2006.07.041

[45] Lim, S. C.; Jo, C. S.; Jeong, H. J.; Shin, Y. M.; Lee, Y. H.; Samayoa, I. A.; & Choi, J. (2002). Effect of Oxidation on Electronic and Geometric Properties of Carbon Nanotubes. *Jpn. J. Appl. Phys.*, Vol.41, No. 9, pp. 5635-5639. Doi: 10.1143/JJAP.41.5635

[46] Belavin, V. V.; Bulusheva, L. G. & Okotrub, A. V. (2004). Modifications to the Electronic Structure of Carbon Nanotubes with Symmetric and Random Vacancies. *Int. J. Quantum Chem.*, Vol. 96, pp. 239-246. Doi: 10.1002/qua.10629

[47] Miyamoto, Y.; Cohen, M. L. & Louie, S. G. (1997). Theoretical Investigation of Graphitic Carbon Nitride and Possible Tubule Forms. *Solid State Commun.* Vol.102, pp. 605-608. Doi:10.1016/S0038-1098(97)00025-2

# 12

# Enhanced Control of Carbon Nanotube Properties Using MPCVD with DC Electrical Bias

Placidus Amama[1,2*], Matthew Maschmann[1,3*] and Timothy Fisher[1,4]
[1]*Air Force Research Laboratory, Materials and Manufacturing Directorate, AFRL/RXB*
[2]*University of Dayton Research Institute*
[3]*Universal Technology Corporation*
[4]*Purdue University*
*USA*

## 1. Introduction

The engineering properties of carbon nanotubes (CNTs) allow for an extraordinarily large potential application space including thermal management, integrated circuits, mechanical reinforcement, and medical devices, among others. CNTs are generally characterized by the quantity of concentric graphene shells comprising their cylindrical wall structure. CNTs consisting of a single graphene cylinder are characterized as single-walled CNTs (SWNTs), while multiple concentric graphene cylinders are called multi-walled CNTs (MWNTs). Typical diameters for SWNTs are approximately 1−3 nm, while MWNTs diameters range from approximately 2 nm to greater than 100 nm. The unique atomic arrangement of a SWNT dictates that each atom resides on both the interior and exterior of the structure, with the atomic orientation defined by a chiral vector relating the fully traversed perimeter of the SWNT to the unit vectors of a graphene sheet. Two thirds of SWNT chiralities are electrically semi-conducting, exhibiting an electronic band gap inversely proportional to their diameter, while the remaining third are metallic. Though the transport properties of MWNTs are degraded relative to SWNTs by the wall-to-wall interactions, they may still exceed the properties of traditional macroscale materials such as copper or aluminum. Despite the advantageous properties offered by CNTs, their integration into functional materials and devices in a manner that maximizes their benefit remains a significant technical and engineering challenge. Specific applications may demand a unique blend of characteristics such as diameter, alignment, purity, density, and chirality to maintain proper operation. *In situ* morphology and orientation control of CNTs during synthesis represents a promising path towards selectivity of these device-specific requirements, especially for applications requiring CNTs with engineered properties to be synthesized directly on a fucntionalized substrate. We examine the role of dc electrical bias during mircrowave plasma-enhanced chemical vapor deposition (MPCVD) synthesis of SWNTs using alignment, spatial density, chirality, and purity as metrics of interest. Further, we demonstrate enhanced thermal and electrial transport properties of MWNTs realized with application of substrate bias during MPCVD synthesis.

---

* contributed equally to this work.

## 2. Plasma enhanced Chemical Vapor Deposition

There is a wide range of methods for producing CNTs such as laser ablation, arc discharge, pyrolysis, and chemical vapor deposition (Huczko, 2002; Rakov, 2000). Chemical vapor deposition (CVD) has emerged as the method of choice for producing CNTs because of its simplicity, flexibility and affordability as well as the potential for scalability and the precise control of CNT properties. In a typical CVD growth process, a carbon precursor such as a hydrocarbon gas is heated to 750-900°C in the presence of a suitable catalyst (e.g., Fe, Co, and Ni), and if all the other reaction conditions such as catalyst particle size, nature of the catalyst support, and gas composition are optimized, nucleation and growth of CNTs proceeds. The CVD process is highly unique because CNTs grow from catalyst 'seeds,' and several studies have shown that there is an intimate relationship between the catalyst properties and the nanotube properties (Amama, et al., 2005a; Hofmann, et al., 2003). In other words, with proper control of the catalyst properties, it is possible to grow CNTs of controlled properties via CVD (Amama, et al., 2007; Amama, et al., 2010; Crouse, et al., 2008; Maschmann, et al., 2005; Zhu, et al., 2010). As such, an appreciable amount of research is underway to fully understand catalyst evolution during synthesis (Amama, et al., 2010; Amama, et al., 2009; Kim, et al., 2010). The CVD process typically involves either thermally driven gas phase decomposition of a hydrocarbon gas (thermal CVD) or both thermal and plasma decomposition (plasma-enhanced CVD).

MPCVD growth has received significant attention mainly because of the potential for low temperature CNT synthesis required for compatiblity with standard nanofabrication and CMOS processes and the ability to produce highly graphitized, vertically aligned CNTs (Amama, et al., 2006a; Meyyappan, et al., 2003). A distinguishing feature of the MPCVD process is the presence of a highly reactive plasma environment, which enhances the decomposition of the hydrocarbon feedstock during CNT growth. The generation of highly energetic ions by the plasma and their subsequent transport to the growth surface are two critical factors that influence the growth properties (Yen, et al., 2005). Using the wide parameter space of the MPCVD, a key advantage over other CVD processes, low temperature growth (Amama, et al., 2006a; Boskovic, et al., 2002), CNT alignment (Maschmann, et al., 2006d), chiral (Li, et al., 2004) and diameter control (Amama, et al., 2006b) have been demonstrated. In many plasma-enhanced CVD studies, the plasma source used is microwave energy which is characterized by high plasma density with a resonant field that is able to concentrate the plasma, ensuring that significant electron loss to the surrounding does not occur (Yen, et al., 2005). The plasma intensity is controlled by the microwave power while the ion flux directed at the substrate may be controlled by a dc bias voltage applied to the growth substrate. These parameters operate indepdently in MPCVD and are capable of substantially altering the properties of CNTs.

## 3. Effect of DC electrical bias during SWNT synthesis

### 3.1 Negative polarity electrical bias

Strict vertical alignment of CNTs may be of significant advantage for many applications, including electron emitters, mechanical enhancement, and high-density electronics. Although direct synthesis of vertically aligned or vertically oriented SWNT arrays has been commonly reported in the literature (Iwasaki, et al., 2005; Maruyama, et al., 2005; Murakami, et al., 2004; Zhong, et al., 2005), most refer to a general packing of SWNTs of subsequent density that the growth front advances and remains in plane with the originating growth substrate. Closer examination of these arrays clearly reveals that individual CNTs within the array exhibit significant waviness and inconsistent orientation with respect to the substrate. Gravity assisted thermal CVD synthesis has reportedly resulted in freestanding vertical SWNTs by orienting the growth substrate upside down such that the SWNT growth direction corresponds to the direction of the gravitation field during synthesis (Yeh, et al., 2006). The vertical orientation using the gravity assisted technique seems to diminish for growth times greater than 1.5 minutes, as the free tips of sufficiently long SWNTs contact and are retained by growth substrate due to thermal vibrations (Yeh, et al., 2007). Others have designed catalyst systems embedded within a modified porous anodic alumina (PAA) template. SWNTs originating from within isolated vertical pore will follow the pore axis toward its opening, resulting in a vertical orientation (Maschmann, et al., 2006a; Maschmann, et al., 2006b). Though this technique successfully aligns individual SWNTs within the confined vertical pores, the free ends of SWNTs emerging from the pores adhere strongly to the top horizontal PAA surface rather than maintaining vertical alignment. SWNT functionalization within the vertical pore structure has been achieved (Franklin, et al., 2009a; Franklin, et al., 2009b), though the misaligned SWNTs on the top surface of the template may be unattractive for some applications. Assembly of freestanding vertical SWNTs from solution post-synthesis is also achievable through electrophoresis into predefined vertical vias etched in a silicon nitride mask (Goyal, et al., 2008). This technique yields variable SWNT deposition with respect to overall occupancy of pores and the number of SWNTs deposited per occupied pore, and SWNTs requires magnesium nitrate hexahydrate to encourage improved substrate adhesion. Though these techniques successfully generate vertically aligned SWNTs, each requires either substrate manipulation or significant catalyst processing, which may be undesirable or impractical for practical application.

Chiral selectivity, with respect to metallic or semiconducting behaviour, is important for optimal operation of many types of devices. Metallic CNTs are obviously well suited for applications requiring high current carrying capacity, such as electrical interconnects (Close, et al., 2008; Kreupl, et al., 2002); however, they may also be advantageous in devices requiring high sensitivity to small electrical potential changes, such as electro-chemical biological sensors (Claussen, et al., 2009). Field effect transistors utilizing semiconducting SWNT channels have been extensively studied and have been found to exhibit ballistic electronic transport even at room temperature operation (Franklin and Chen, 2010). Application of SWNT transistors in electronics offer obvious dimensional and efficiency advantages, and significant research continues in this area with respect to device processing and characterization. The strong preferential growth of semiconducting (Li, et al., 2004) or metallic (Harutyunyan, et al., 2009) SWNTs to population densities greater than 90% chirality selectivity have been reported in the literature by utilizing remote RF plasma and control of gas composition during annealing, respectively. We demonstrate the preferential selectivity of both vertical alignment and semiconducting chirality through the use of negative polarity substrate bias applied during SWNT synthesis using MPCVD.

To investigate the influence of DC electrical bias on SWNT synthesis, a SEKI AX5200S MPCVD reactor with electrically grounded chamber walls, shown schematically in Fig. 1. A hollow stainless steel rod contacts the bottom surface of an otherwise electrically isolated graphite heater stage and delivers a dc potential via a voltage-controlled current source (Sorensen DCS600-1.7E). A K-type thermocouple embedded in the rod monitored the stage temperature, while the growth substrate surface temperature was measured using a dual wavelength pyrometer (Williamson model 90). The silicon growth substrate rested on a 5.08-cm diameter, 3.30-mm thick molybdenum puck used to concentrate the plasma directly above the sample.

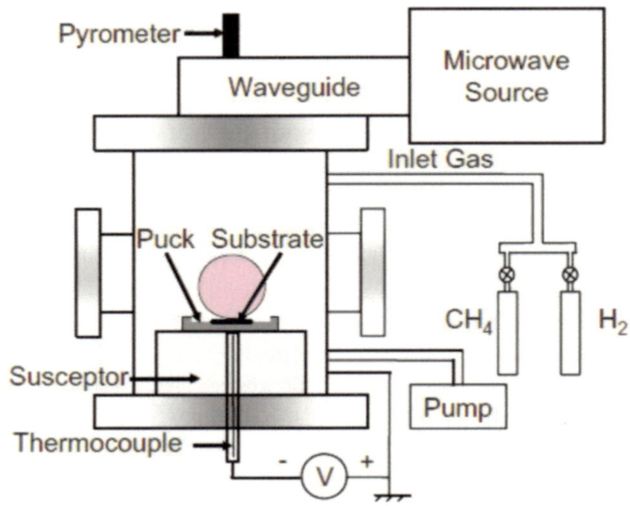


Fig. 1. Schematic of microwave plasma-enhanced chemical vapor deposition chamber.

An MgO supported Co catalyst was utilized for each SWNT synthesis. The catalyst particles were prepared by a wet mechanical mixing and combustion synthesis procedure using a solution of molybdenum, cobalt nitrate hexahydrate, and magnesium nitrate to produce bimetallic Mo/Co catalyst particles embedded in a nanoporous MgO support (Maschmann, et al., 2006d; Maschmann, et al., 2006c). The susceptor was first heated to 900°C in 50 sccm of flowing hydrogen at a pressure of 10 Torr. A dc substrate bias between 0 and -250 V was applied gradually to the substrate at a rate of approximately - 25 V/second after ignition of a 200 W microwave plasma. Methane was then introduced at a flow rate of 5 sccm to initiate CNT growth. Each synthesis was 20 minutes in duration. The surface temperature of the substrate recorded by the pyrometer was approximately 770°C and relatively insensitive to the applied bias.
Characterization of the SWNT product was performed using a Hitachi S-4800 field emission scanning electron microscope (SEM) and Senterra micro-Raman spectrometer. Laser excitation wavelengths of 533 and 785 nm were selected for recording Raman spectra, with at least ten locations examined for each sample. SEM characterization was utilized to assess SWNT relative alignment with respect to the growth substrate, SWNT length, density, and diameter estimates of individual SWNTs and SWNT bundles. Multi-excitation wavelength Raman spectra analysis allowed for quantification of SWNT quality, diameter distributions, and relative trends with respect to SWNT chirality.
The application of negative bias strengthens the electric field inherently present in the plasma sheath region immediately above the substrate, thereby accelerating the impingement of positively charged ions, such as $H^+$, towards the substrate. A plasma sheath is established as a result of the large mobility mismatch between ions and free electrons

generated within the plasma. The relatively low mass of electrons allows them to acquire a translational speed many times greater than that of the relatively heavy ions and accelerate away from the central concentrated plasma sphere located above the substrate.

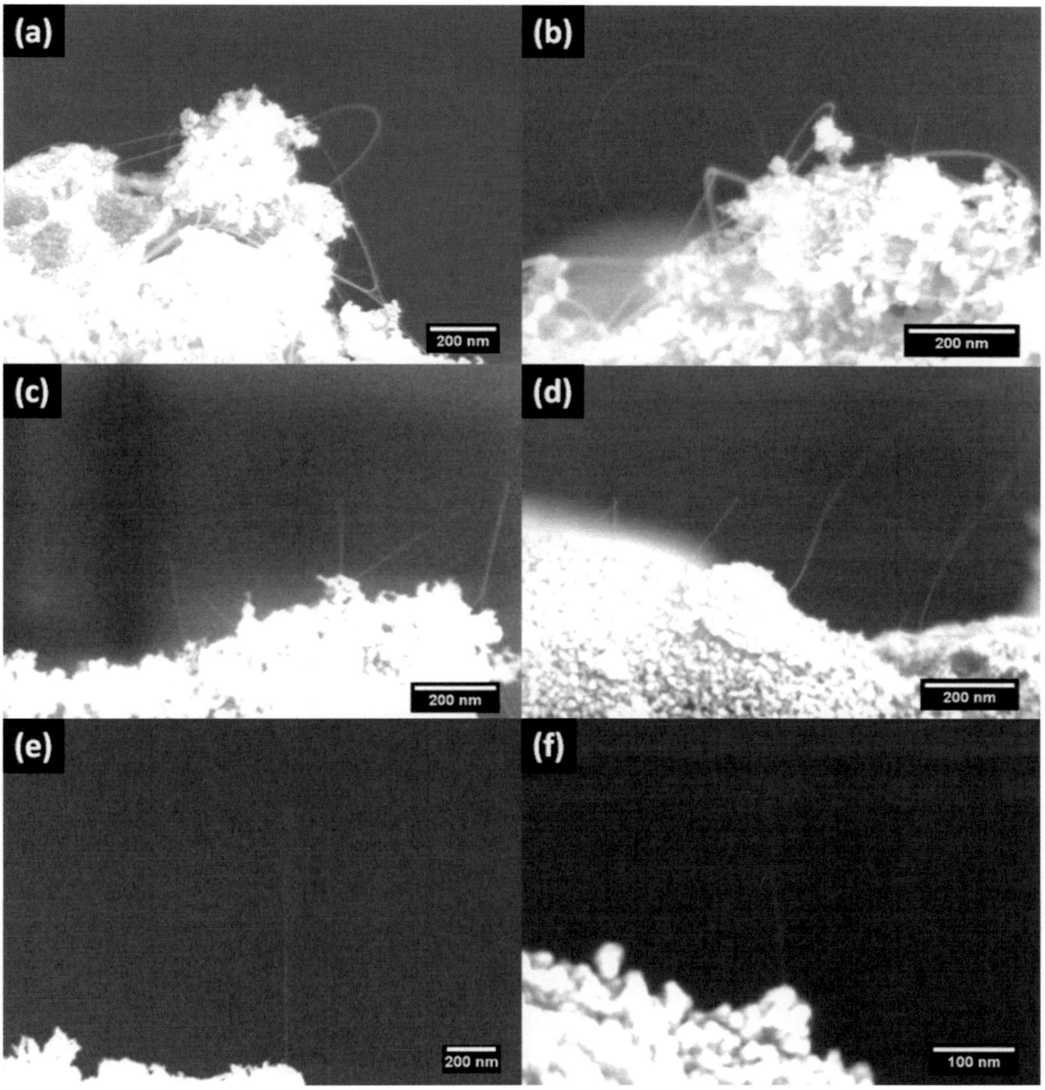


Fig. 2. Cross sectional SEM micrographs of SWNTs synthesized under negative polarity substrate bias in MPCVD at (a) 0V, (b) -50V, (c) -100V, (d) -150V, (e) -200V, and (f) -250V. From (Maschmann, et al., 2006d)

The bulk plasma is therefore electron deficient, setting up a net positive charge with respect to chamber walls, and an electric field is generated between the plasma and the surrounding surfaces. The highly anisotropic polarization of CNTs (Benedict, et al., 1995) establishes an interaction force between the CNT and the enhanced electric field near the growth substrate.

The magnitude of interaction is of sufficient magnitude to orient SWNTs (Peng, et al., 2003; Ural, et al., 2002; Zhang, et al., 2001) and multi-walled CNTs (Jang, et al., 2003; Merkulov, et al., 2001; Meyyappan, et al., 2003) along electric field lines *in situ* during CVD synthesis as well as during post-synthesis processing procedures (Kamat, et al., 2004; Yamamoto, et al., 1998).

Negative polarity substrate bias was systematically varied between 0 and -250V in 50V increments (Maschmann, et al., 2006d). Cross-sectional SEM analysis revealed distinct trends with respect to both SWNT spatial density and orientation relative to the growth substrate, as seen in Fig. 2a-f. SWNTs grown in the absence of applied bias or at -50V had a tendency to form large diameter bundles that generally followed the profile of the MgO support particles. No preferential growth perpendicular to the growth substrate was observed. The SWNTs synthesized at -100V and -150V, however, demonstrated a strong tendency to break free of the support particle in favor of a vertical orientation, normal to that of the support particles. SWNTs grown at these bias levels also tended to form bundles, with many longer SWNT bundles formed vertically oriented loops. A decrease in overall spatial density relative to the synthesis preformed without bias may also be discerned. At the greater bias magnitudes of -200V and -250V, a strong preference to vertical alignment is observed, in addition to a marked decrease in SWNT spatial density. Very few SWNTs were observed along the perimeter of the catalyst support particles, as is typically observed when bias is omitted from synthesis. Freestanding SWNTs with lengths of several microns were frequently observed, though the free tips of these SWNTs were often obscured by thermal vibrations.

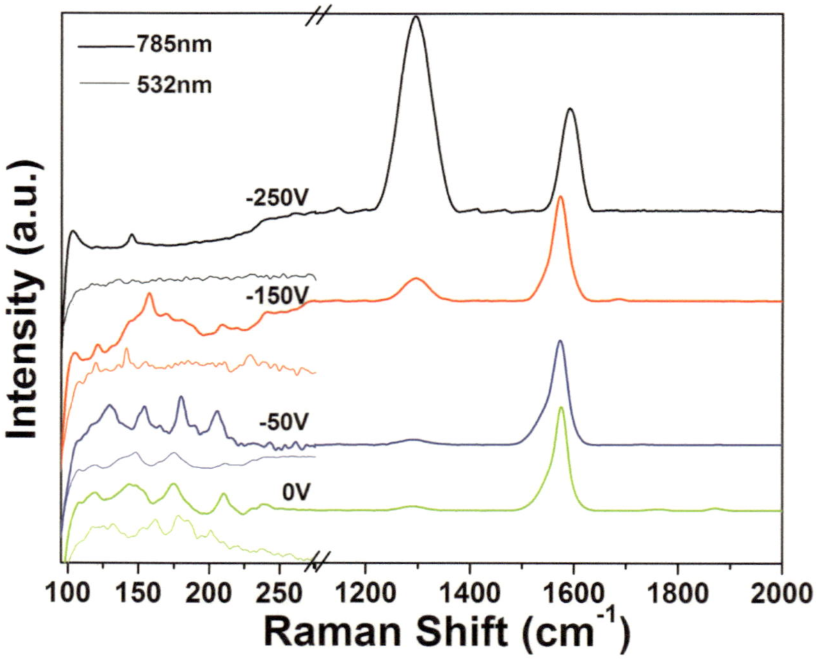


Fig. 3. Raman spectra of SWNTs synthesized using negative substrate bias in MPCVD. From (Maschmann, et al., 2006d).

Raman spectra of the SWNTs synthesized with negative polarity bias revealed another trend not readily observable from SEM observation, likely due to SWNT bundling and the inherent resolution limitations of the SEM. The radial breathing mode (RBM) distributions gradually skewed to lower frequency Raman shifts with increased levels of negative bias. Because SWNT diameter is inversely proportional to RBM frequency (Bachilo, et al., 2002; Rao, et al., 1997), the RBM distributions shift suggests a trend towards larger diameter SWNTs (up to 2.5 nm) as negative bias is increased. Locating the RBM peaks relative to excitation wavelengths on a Kataura plot (not shown) indicates that SWNTs synthesized without bias are a mix of metallic and semiconducting chiralities (Maschmann, et al., 2006d). Magnitudes of negative bias at and above -150V shift the measured RBM frequencies into bands of exclusively semiconducting chiralities. Additionally, a decreasing trend in the G- to D-band ratio is a further indication of a decreased spatial density observed by SEM and is perhaps an indication of an increased occurrence of SWNT wall defects. The Lorentzian lineshape of the G-band obtained from SWNTs under high levels of negative bias further support the RBM trend indicating a high concentration of semi-conducting SWNTs (Brown, et al., 2001; Pimenta, et al., 1998). The predominance of larger diameter SWNTs and corresponding decrease in SWNT density is thought to be a result of enhanced $H^+$ ion bombardment, which is known to preferentially etch small diameter SWNTs (Zhang, et al., 2005). Metallic SWNTs may have also been burned up as a result of transmitting a high current density.

### 3.2 Positive polarity electrical bias

Application of dc bias that is positive with respect to chamber walls is believed to decrease the magnitude of the electric field within the plasma sheath region near the growth substrate. $H^+$ ions, generated in abundance within the plasma, therefore attain a lower translational velocity before encountering the growth substrate. In fact, because the substrate in this configuration is the surface of greatest potential relative to the grounded chamber, the ions are instead more readily attracted toward the chamber walls. The mitigation of potentially harmful $H^+$ ion bombardment on the growth substrate is examined by varying the magnitude of positive polarity dc electrical bias during MPCVD SWNT synthesis, similarly to the methodology described in the previous section.

Substrate bias was varied between 0 and +200V in 50V increments while maintaining otherwise standard synthesis conditions. Bias levels of +250V or greater were attempted, but consistently led to plasma instabilities and were not further examined. Within the bias range of 0 and +100V, only incremental increases in SWNT spatial density were observed. SEM micrographs obtained from samples synthesized within this range of biases, shown in Fig. 4 (a) and (b), reveal SWNT bundles spanning tens of microns in length and tens of nanometers in diameter. No preferential vertical alignment of SWNTs is observed for these samples using cross-sectional SEM imaging (not shown). Larger biases of +150V and +200V resulted in dramatic increases in SWNT density, with a significant population of large-diameter SWNT ropes observed uniformly coating the support particle surfaces. Figures 4 (c-e) show typical SEM micrographs of SWNT products synthesized at +150 and +200V. The diameters of SWNT ropes often exceed 50 nm, with smaller feeder bundles ranging between 10-25 nm. Cross-sectional SEM analysis of these samples (Figure 4e) reveals that a small fraction of isolated SWNTs are freestanding and oriented in the direction normal to the support particle. Within the resolution limitations of the SEM, the vertical SWNTs synthesized at

+200V bias appear to be smaller in diameter than the vertical SWNTs synthesized using negative bias (Fig. 3c-f). The hypothesized weakened ion bombardment, even relative to the neutral 0V bias case, may encourage the synthesis of CNTs of all orientations that may otherwise be etched by $H^+$ ions, allowing these SWNTs to escape the bundling effect encountered by SWNTs that follow the profile of the support particles.

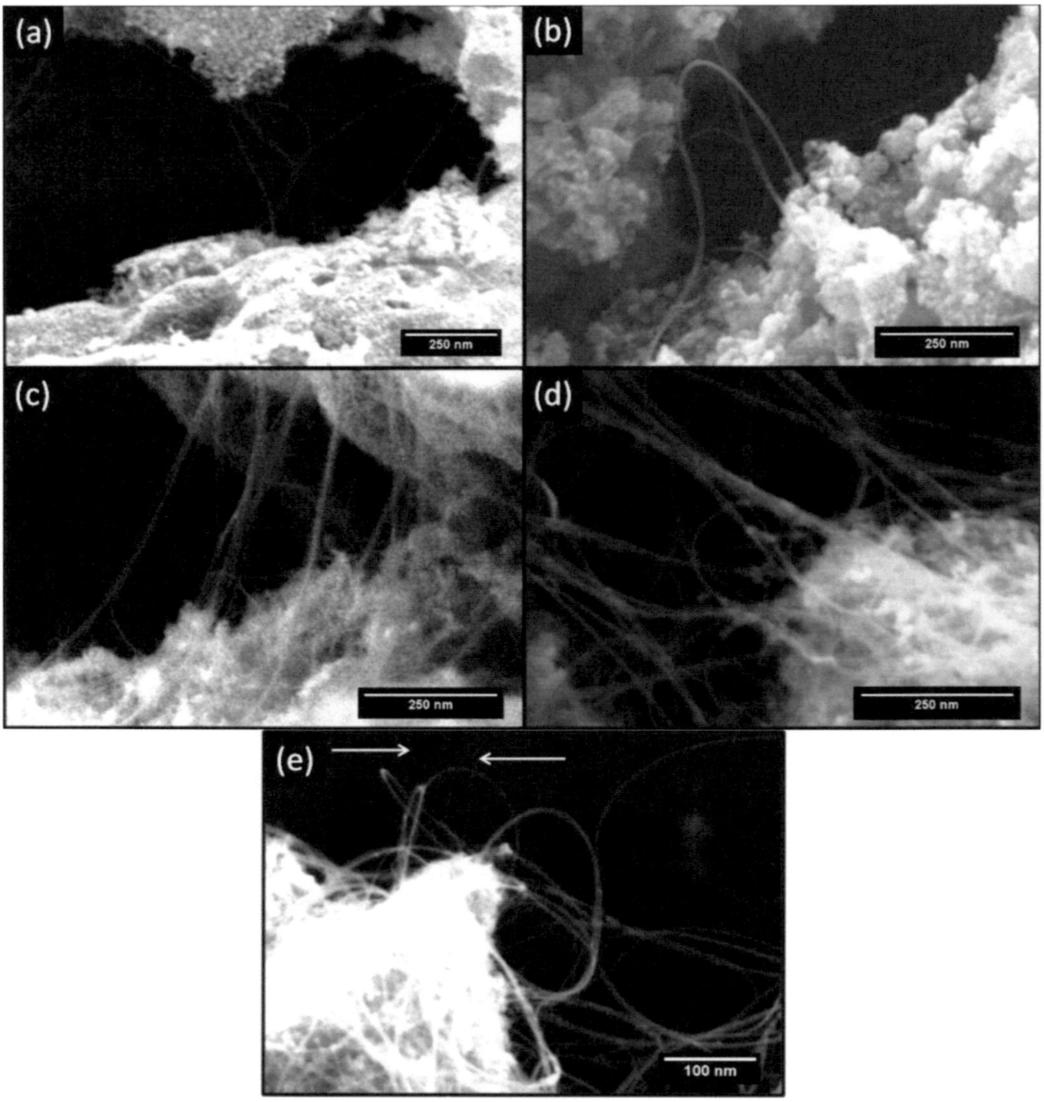


Fig. 4. SEM micrographs of SWNTs synthesized at (a) 0V, (b) +50, (c) +150, and (d-e) +200V substrate bias in MPCVD. Arrows indicate the presence of freestanding vertical SWNTs.

Raman spectroscopy yields further insights into the SWNTs produced using positive bias. While negative bias resulted in a shift in RBM peaks towards lower frequencies, the application of positive bias resulted in a shift in RBM peaks towards higher frequencies, as shown in Fig. 5. RBMs in the range of 100 - 200 $cm^{-1}$ are present for all levels of positive bias for both 785 and 532 nm excitation wavelengths, but RBM frequencies greater than 250 $cm^{-1}$ emerge at bias levels above +150V. Employing a 785 nm excitation wavelength, a RBM peak at 259 $cm^{-1}$ emerges at +150V, while a peak at 261 $cm^{-1}$ is present at +200V. Using a 532 nm excitation wavelength, a RBM peak at 251 $cm^{-1}$ emerges at +200V. In terms of SWNT diameter distribution, the presence of these RBMs indicates the emergence of SWNTs with diameters less than 1 nm (Bachilo, et al., 2002; Rao, et al., 1997). As mentioned previously, this effect may be attributed to decreased $H^+$ ion bombardment which tends to preferentially etch smaller diameter SWNTs. A mixture of metallic and semiconducting chiralities exist, based on the location of RBM peaks on a Kataura plot (not shown), indicating that no chiral selectivity is attained using positive polarity bias.

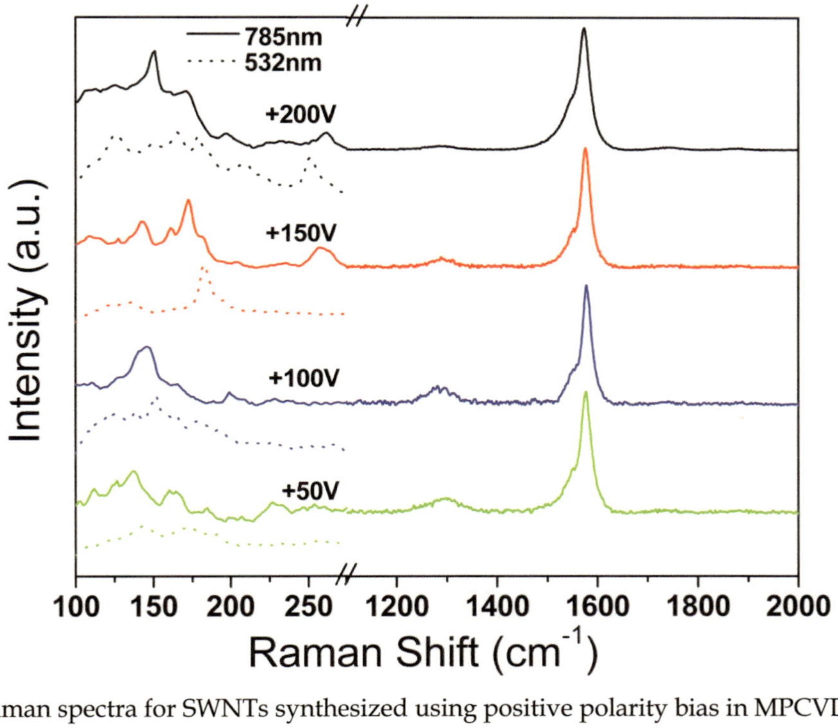


Fig. 5. Raman spectra for SWNTs synthesized using positive polarity bias in MPCVD.

Examination of the G-band further indicates a significant difference in composition of SWNTs grown under negative and positive bias. A Breit-Wanger-Fano line shape, appearing as a shoulder on the G-band at approximately 1550 $cm^{-1}$ in Fig. 5, is indicative of metallic SWNTs (Brown, et al., 2001; Pimenta, et al., 1998) and is absent in G-bands obtained for SWNTs grown using negative bias (Fig. 3). Additionally, the G- to D-band ratios are substantially greater when utilizing positive bias. While application of negative bias attracts and accelerates $H^+$ ions to the growth substrate, thus damaging SWNT walls, the application of positive bias appears to adequately decrease the incoming velocity of $H^+$ ions to the substrate and may protect SWNTs from excessive ion bombardment. Consequently, the ratio of G- to D-band ratio for SWNTs grown using positive applied bias increased from approximately 10 for samples grown without bias to approximately 40 for those grown at

+200 V. Such a high ratio indicates a large quantity of high-quality SWNTs with little amorphous carbon.

## 4. Influence of DC electrical bias during MWNT synthesis

The influence of dc bias voltage during MPCVD synthesis of MWNTs from dendrimer-templated $Fe_2O_3$ nanoparticles will be discussed with respect to the the resulting thermal and electrial transport properties of MWNT arrays. The DC bias values exmained range from -200 to +200V, in 100V increments using similar experimental techniques discussed in the previous sections. The electrical resistance of the MWNTs were measured by obtaining the slope of I-V characterization of randomly selected individual MWNTs across lithographically defined Au/Ti electrodes. Five individual MWNTs were studied for each level of dc bias. The thermal performance was assessed by utilizing the MWNT arrays as a thermal interface. Thermal resistance of the CNT interface material was determined using a photoacoustic technique (Cola, et al., 2007). The thermal resistance measurement was performed at a single interface pressure of 10 psi. Three MWNT array interfaces from each synthesis bias level were produced and characterized.

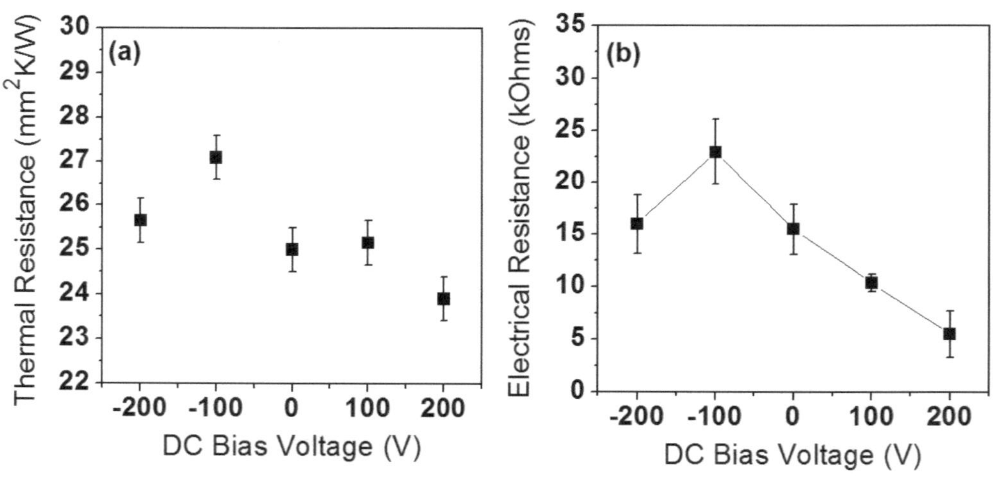


Fig. 6. Measured thermal interface resistance of MWNT arrays determined using a photoacoustic technique (a) and electrical resistance of individual MWNTs (b) as a function of dc bias voltage used during growth in the MPCVD. From (Amama, et al., 2008)

Figure 6 exhibits the electrical and thermal resistance values as a function of applied substrate bias during MPCVD synthesis. Similar trends with respect to substrate bias exist among the data sets, suggesting that similar phenomena during synthesis may be affecting both thermal and electrical transport. MWNTs grown under positive dc bias (+200V) demonstrate the lowest resistances, while the highest resistances were observed for MWNTs grown under negative dc bias voltage (-100V). The lowest thermal interface resistance (23.9 $mm^2/K/W$) was observed for MWNT arrays grown under a dc bias voltage of +200 V while MWNT arrays grown at -100 V showed the highest thermal interface resistance (27.1 $mm^2/K/W$). Similarly, the lowest electrial resistance (5.5 kOhms) was attained at +200V, while the greatest electrical resistance (23 kOhms). The electrical resistance data exhibits a nearly linear decrease with respect to applied positive polarity bias, the thermal resistance observed at +100V was statistically equivalent to that observed at 0V bias. It is possible that the defect density present in MWNTs may contribute to the observed variation in electrical resistances as shown previously (Lan, et al., 2007).

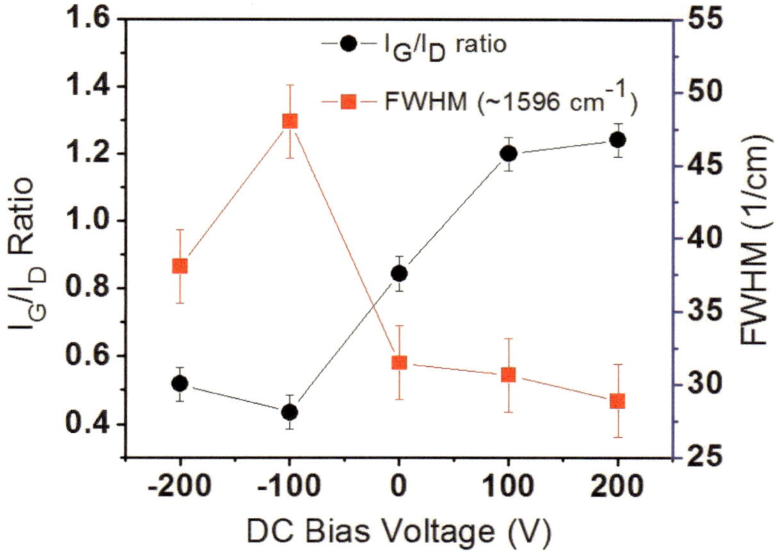


Fig. 7. Raman spectroscopy data obtained from MWNT arrays synthesized in MPCVD using dc substrate bias. $I_G/I_D$ ratio represents the relative peak intensity ratio of the G-band to D-band, while the FWHM data is measured relative to the G-bank peak. From (Amama, et al., 2008).

The thermal and relectrical resistance trends are consistent with those exhibited by the G- to D-band ratio measured via Raman spectroscopy for the MWNT samples. As seen in Fig. 7, the relative ratio of the well graphitized carbon (G-band) to disordered carbon (D-band) steadily increases as a function of positive polarity dc bias. The ratio maxima occurs at +200V, consistent with the minimal thermal and electrical resistance measurmements. The minima at -100V corresponds to the maximum observed thermal and electrical resistance. The observed behavior of the $I_G/I_D$ ratio is consistent with the full width at half maximum (FWHM) of the G-band at ~1596 $cm^{-1}$. We hypothesize that negative dc bias voltage accelerated $H^+$ ions, introducing defects on the CNTs. This effect is most pronounced for MWNTs grown under -100 V. The relatively consistent trend between the measured resistance data and the Raman spectra data gives further evidence of this hypothesis. Biasing the substrate positively, on the other hand, reduces electric field near the substrate, reducing the bombardment of $H^+$ and other positively charged hydrocarbon ions generated in the plasma from the CNTs.

## 5. Conclusion

The parameter space for MPCVD synthesis of CNTs is vast, allowing a user a high level of fidelity with respect to control of CNT structure and morphology. The application of substrate bias independently from plasma power and other growth parameters is a unique and robust feature of MPCVD that enables *in situ* control of CNT alignment, quality, density, and chirality and extends the potential application space for plasma-grown CNTs. We have demonstrated that both the polarity and magnitude of the applied bias dictate the resulting CNT yield. Negative polarity bias lends itself to vertical orientation and is a means to preferentially synthesize larger diameter semiconducting SWNTs. Conversely, positive polarity bias dramatically increases the SWNT quality and yield while resulting in a mix of metallic and semiconducting chiralities. To the detriment of the technique, however, the quality metrics seem exclusive to a given bias polarity. For example, the synthesis of high density, vertical freestanding SWNTs has, to date, been a challenge through variation of bias alone, and more research is required to fully optimize the capabilities of applied bias during SWNT synthesis. For MWNT synthesis, the alignment capability of negative polarity bias is well established, though the application of positive polarity bias remains relatively unexplored. We observe that positive polarity bias at levels greater than +100V during MPCVD synthesis appears to demonstrate a protective role, partially shielding CNTs from harmful ion bombardment. As a result, MWNTs exhibit enhanced thermal and electrical conductivity. The degree of freedom offered by substrate bias during MPCVD synthesis offers a tremendous extension to traditional CNT synthesis capabilities and potential inroads to myriad applications requiring strict control of SWNT or MWNT properties.

## 6. References


Amama, P. B.; Lim, S.; Ciuparu, D.; Yang, Y.; Pfefferle, L. & Haller, G. L. (2005a). Synthesis, Characterization, and Stability of Fe-MCM-41 for Production of Carbon Nanotubes by Acetylene Pyrolysis. *J. Phys. Chem. B*, 109, 7, pp. 2645-2656

Amama, P. B.; Ogebule, O.; Maschmann, M. R.; Sands, T. D. & Fisher, T. S. (2006a). Dendrimer-assisted low-temperature growth of carbon nanotubes by plasma-enhanced chemical vapor deposition. *Chem. Commun. (Cambridge, U. K.)*, 27, pp. 2899-2901

Amama, P. B.; Maschmann, M. R.; Fisher, T. S. & Sands, T. D. (2006b). Dendrimer-Templated Fe Nanoparticles for the Growth of Single-Wall Carbon Nanotubes by Plasma-Enhanced CVD. *J. Phys. Chem. B*, 110, 22, pp. 10636-10644

Amama, P. B.; Cola, B. A.; Sands, T. D.; Xu, X. F. & Fisher, T. S. (2007). Dendrimer-assisted controlled growth of carbon nanotubes for enhanced thermal interface conductance. *Nanotechnology*, 18, 38, pp. 385303-385306

Amama, P. B.; Lan, C.; Cola, B. A.; Xu, X.; Reifenberger, R. G. & Fisher, T. S. (2008). Electrical and Thermal Interface Conductance of Carbon Nanotubes Grown under Direct Current Bias Voltage. *J. Phys. Chem. C*, 112, 49, pp. 19727-19733

Amama, P. B.; Pint, C. L.; Mcjilton, L.; Kim, S. M.; Stach, E. A.; Murray, P. T.; Hauge, R. H. & Maruyama, B. (2009). Role of water in super growth of single-walled carbon nanotube carpets. *Nano Lett.*, 9, 1, pp. 44-49

Amama, P. B.; Pint, C. L.; Kim, S. M.; Mcjilton, L.; Eyink, K. G.; Stach, E. A.; Hauge, R. H. & Maruyama, B. (2010). Influence of Alumina Type on the Evolution and Activity of Alumina-Supported Fe Catalysts in Single-Walled Carbon Nanotube Carpet Growth. *ACS Nano*, 4, 2, pp. 895-904

Bachilo, S. M.; Strano, M. S.; Kittrell, C.; Hauge, R. H.; Smalley, R. E. & Weisman, R. B. (2002). Structure-Assigned Optical Spectra of Single-Walled Carbon Nanotubes. *Science*, 298, 5602, pp. 2361-2366

Benedict, L. X.; Louie, S. G. & Cohen, M. L. (1995). Static polarizabilities of single-wall carbon nanotubes. *Phys. Rev. B*, 52, 11, pp. 8541-8549

Boskovic, B. O.; Stolojan, V.; Khan, R. U. A.; Haq, S. & Silva, S. R. P. (2002). Large-area synthesis of carbon nanofibres at room temperature. *Nat. Mater.*, 1, 3, pp. 165-168

Brown, S. D. M.; Jorio, A.; Corio, P.; Dresselhaus, M. S.; Dresselhaus, G.; Saito, R. & Kneipp, K. (2001). Origin of the Breit-Wigner-Fano lineshape of the tangential G-band feature of metallic carbon nanotubes. *Phys. Rev. B*, 63, 15, pp. 155411-155418

Claussen, J. C.; Franklin, A. D.; Ul Haque, A.; Porterfield, D. M. & Fisher, T. S. (2009). Electrochemical Biosensor of Nanocube-Augmented Carbon Nanotube Networks. *ACS Nano*, 3, 1, pp. 37-44

Close, G. F.; Yasuda, S.; Paul, B.; Fujita, S. & Wong, H. S. P. (2008). A 1 GHz integrated circuit with carbon nanotube interconnects and silicon transistors. *Nano Lett.*, 8, 2, pp. 706-709

Cola, B. A.; Xu, J.; Cheng, C.; Xu, X.; Fisher, T. S. & Hu, H. (2007). Photoacoustic characterization of carbon nanotube array thermal interfaces. *J. Appl. Phys.*, 101, 5, pp. 054313-054319

Crouse, C. A.; Maruyama, B.; Colorado Jr, R.; Back, T. & Barron, A. R. (2008). Growth, New Growth, and Amplification of Carbon Nanotubes as a Function of Catalyst Composition. *J. Am. Chem. Soc.*, 130, 25, pp. 7946-7954

Franklin, A. D.; Sayer, R. A.; Sands, T. D.; Fisher, T. S. & Janes, D. B. (2009a). Toward surround gates on vertical single-walled carbon nanotube devices. *J. Vac. Sci. Technol., B*, 27, 2, pp. 821-826

Franklin, A. D.; Sayer, R. A.; Sands, T. D.; Janes, D. B. & Fisher, T. S. (2009b). Vertical Carbon Nanotube Devices With Nanoscale Lengths Controlled Without Lithography. *Nanotechnology, IEEE Transactions on*, 8, 4, pp. 469-476

Franklin, A. D. & Chen, Z. (2010). Length scaling of carbon nanotube transistors. *Nat Nano*, 5, 12, pp. 858-862

Goyal, A.; Liu, S.; Iqbal, Z.; Fetter, L. A. & Farrow, R. C. (2008). Directed self-assembly of individual vertically aligned carbon nanotubes. *J. Vac. Sci. Technol., B*, 26, pp. 2524-2528

Harutyunyan, A. R.; Chen, G.; Paronyan, T. M.; Pigos, E. M.; Kuznetsov, O. A.; Hewaparakrama, K.; Kim, S. M.; Zakharov, D.; Stach, E. A. & Sumanasekera, G. U. (2009). Preferential Growth of Single-Walled Carbon Nanotubes with Metallic Conductivity. *Science*, 326, 5949, pp. 116-120

Hofmann, S.; Ducati, C.; Robertson, J. & Kleinsorge, B. (2003). Low-temperature growth of carbon nanotubes by plasma-enhanced chemical vapor deposition. *Appl. Phys. Lett.*, 83, 1, pp. 135-137

Huczko, A. (2002). Synthesis of aligned carbon nanotubes. *Appl. Phys. A*, 74, 5, pp. 617-638

Iwasaki, T.; Zhong, G.; Aikawa, T.; Yoshida, T. & Kawarada, H. (2005). Direct Evidence for Root Growth of Vertically Aligned Single-Walled Carbon Nanotubes by Microwave Plasma Chemical Vapor Deposition. *J. Phys. Chem. B*, 109, 42, pp. 19556-19559

Jang, Y.-T.; Ahn, J.-H.; Ju, B.-K. & Lee, Y.-H. (2003). Lateral growth of aligned mutilwalled carbon nanotubes under electric field. *Solid State Commun.*, 126, 6, pp. 305-308

Kamat, P. V.; Thomas, K. G.; Barazzouk, S.; Girishkumar, G.; Vinodgopal, K. & Meisel, D. (2004). Self-Assembled Linear Bundles of Single Wall Carbon Nanotubes and Their Alignment and Deposition as a Film in a dc Field. *J. Am. Chem. Soc.*, 126, 34, pp. 10757-10762

Kim, S. M.; Pint, C. L.; Amama, P. B.; Hauge, R. H.; Maruyama, B. & Stach, E. A. (2010). Catalyst and catalyst support morphology evolution in single-walled carbon nanotube supergrowth: Growth deceleration and termination. *J. Mater. Res.*, 25, 10, pp. 1875-1885

Kreupl, F.; Graham, A. P.; Duesberg, G. S.; Steinhogl, W.; Liebau, M.; Unger, E. & Honlein, W. (2002). Carbon nanotubes in interconnect applications. *Microelectron. Eng.*, 64, 1-4, pp. 399-408

Lan, C.; Amama, P. B.; Fisher, T. S. & Reifenberger, R. G. (2007). Correlating electrical resistance to growth conditions for multiwalled carbon nanotubes. *Appl. Phys. Lett.*, 91, 9, pp. 093105-093107

Li, Y.; Mann, D.; Rolandi, M.; Kim, W.; Ural, A.; Hung, S.; Javey, A.; Cao, J.; Wang, D.; Yenilmez, E.; Wang, Q.; Gibbons, J. F.; Nishi, Y. & Dai, H. (2004). Preferential Growth of Semiconducting Single-Walled Carbon Nanotubes by a Plasma Enhanced CVD Method. *Nano Lett.*, 4, 2, pp. 317-321

Maruyama, S.; Einarsson, E.; Murakami, Y. & Edamura, T. (2005). Growth process of vertically aligned single-walled carbon nanotubes. *Chem. Phys. Lett.*, 403, 4-6, pp. 320-323

Maschmann, M. R.; Amama, P. B. & Fisher, T. S. (2005). Effect of DC Bias on Microwave Plasma Enhanced Chemical Vapor Deposition Synthesis of Single-Walled Carbon Nanotubes. *ASME Conference Proceedings*, 0-7918-4223-1 Orlando, Florida, (November, 2005)

Maschmann, M. R.; Franklin, A. D.; Amama, P. B.; Zakharov, D. N.; Stach, E. A.; Sands, T. D. & Fisher, T. S. (2006a). Vertical single- and double-walled carbon nanotubes grown from modified porous anodic alumina templates. *Nanotechnology*, 17, 15, pp. 3925-3929

Maschmann, M. R.; Franklin, A. D.; Scott, A.; Janes, D. B.; Sands, T. D. & Fisher, T. S. (2006b). Lithography-Free in Situ Pd Contacts to Templated Single-Walled Carbon Nanotubes. *Nano Lett.*, 6, 12, pp. 2712-2717

Maschmann, M. R.; Amama, P. B.; Goyal, A.; Iqbal, Z.; Gat, R. & Fisher, T. S. (2006c). Parametric study of synthesis conditions in plasma-enhanced CVD of high-quality single-walled carbon nanotubes. *Carbon*, 44, 1, pp. 10-18

Maschmann, M. R.; Amama, P. B.; Goyal, A.; Iqbal, Z. & Fisher, T. S. (2006d). Freestanding vertically oriented single-walled carbon nanotubes synthesized using microwave plasma-enhanced CVD. *Carbon*, 44, 13, pp. 2758-2763

Merkulov, V. I.; Melechko, A. V.; Guillorn, M. A.; Lowndes, D. H. & Simpson, M. L. (2001). Alignment mechanism of carbon nanofibers produced by plasma-enhanced chemical-vapor deposition. *Appl. Phys. Lett.*, 79, 18, pp. 2970-2972

Meyyappan, M.; Delzeit, L.; Cassell, A. & Hash, D. (2003). Carbon nanotube growth by PECVD: a review. *Plasma Sources Sci. Technol.*, 12, 2, pp. 205-216

Murakami, Y.; Chiashi, S.; Miyauchi, Y.; Hu, M.; Ogura, M.; Okubo, T. & Maruyama, S. (2004). Growth of vertically aligned single-walled carbon nanotube films on quartz substrates and their optical anisotropy. *Chem. Phys. Lett.*, 385, 3-4, pp. 298-303

Peng, H. B.; Ristroph, T. G.; Schurmann, G. M.; King, G. M.; Yoon, J.; Narayanamurti, V. & Golovchenko, J. A. (2003). Patterned growth of single-walled carbon nanotube arrays from a vapor-deposited Fe catalyst. *Appl. Phys. Lett.*, 83, 20, pp. 4238-4240

Pimenta, M. A.; Marucci, A.; Empedocles, S. A.; Bawendi, M. G.; Hanlon, E. B.; Rao, A. M.; Eklund, P. C.; Smalley, R. E.; Dresselhaus, G. & Dresselhaus, M. S. (1998). Raman modes of metallic carbon nanotubes. *Phys. Rev. B*, 58, 24, pp. R16016

Rakov, E. G. (2000). Methods for preparation of carbon nanotubes. *Russ. Chem. Rev.*, 69, 1, pp. 35-52

Rao, A. M.; Richter, E.; Bandow, S.; Chase, B.; Eklund, P. C.; Williams, K. A.; Fang, S.; Subbaswamy, K. R.; Menon, M.; Thess, A.; Smalley, R. E.; Dresselhaus, G. & Dresselhaus, M. S. (1997). Diameter-Selective Raman Scattering from Vibrational Modes in Carbon Nanotubes. *Science*, 275, 5297, pp. 187-191

Ural, A.; Li, Y. & Dai, H. (2002). Electric-field-aligned growth of single-walled carbon nanotubes on surfaces. *Appl. Phys. Lett.*, 81, 18, pp. 3464-3466

Yamamoto, K.; Akita, S. & Nakayama, Y. (1998). Orientation and purification of carbon nanotubes using ac electrophoresis. *J. Phys. D: Appl. Phys.*, 31, 8, pp. L34

Yeh, C. M.; Chen, M. Y.; Syu, J. S.; Gan, J. Y. & Hwang, J. (2006). Effect of gravity on the growth of vertical single-walled carbon nanotubes in a chemical vapor deposition process. *Appl. Phys. Lett.*, 89, 3, pp. 033117-033119

Yeh, C. M.; Chen, M. Y.; Gan, J.-Y.; Hwang, J.; Lin, C. D.; Chao, T. Y. & Cheng, Y. T. (2007). Effects of time on the quality of vertically oriented single-walled carbon nanotubes by gravity-assisted chemical vapour deposition. *Nanotechnology*, 18, 14, pp. 145613

Yen, J. H.; Leu, I. C.; Lin, C. C. & Hon, M. H. (2005). Synthesis of well-aligned carbon nanotubes by inductively coupled plasma chemical vapor deposition. *Applied Appl. Phys. A: Mater. Sci. Process.*, 80, pp. 415-421

Yen, J. H.; Leu, I. C.; Lin, C. C. & Hon, M. H. (2005). Synthesis of well-aligned carbon nanotubes by inductively coupled plasma chemical vapor deposition. *Appl. Phys. A*, 80, pp. 415-421

Zhang, G.; Mann, D.; Zhang, L.; Javey, A.; Li, Y.; Yeilmez, E.; Wang, Q.; Mcvittie, J. P.; Nishi, Y.; Gibbons, J. & Dai, H. (2005). Ultra-high-yield growth of vertical single-walled carbon nanotubes: Hidden roles of hydrogen and oxygen. *Proc. Natl. Acad. Sci.*, 102, 45, pp. 16141-16145

Zhang, Y. G.; Chang, A. L.; Cao, J.; Wang, Q.; Kim, W.; Li, Y. M.; Morris, N.; Yenilmez, E.; Kong, J. & Dai, H. (2001). Electric-field-directed growth of aligned single-walled carbon nanotubes. *J. Appl. Phys. Lett.*, 79, 19, pp. 3155-3157

Zhong, G. F.; Iwasaki, T.; Honda, K.; Furukawa, Y.; Ohdomari, I. & Kawarada, H. (2005). Very High Yield Growth of Vertically Aligned Single-Walled Carbon Nanotubes by Point-Arc Microwave Plasma CVD. *Chem. Vap. Deposition*, 11, 3, pp. 127-130

Zhu, Z.; Jiang, H.; Susi, T.; Nasibulin, A. G. & Kauppinen, E. I. (2010). The Use of NH3 to Promote the Production of Large-Diameter Single-Walled Carbon Nanotubes with a Narrow (n,m) Distribution. *J. Am. Chem. Soc.*, 133, 5, pp. 1224-1227

# 13

# Spin Dependent Transport Through a Carbon Nanotube Quantum Dot in the Kondo Regime

Stanisław Lipiński and Damian Krychowski
*Institute of Molecular Physics, Polish Academy of Sciences*
*Poznań, Poland*

## 1. Introduction

Carbon nanotubes (CNTs) are thin, hollow cylinders, which can be envisioned as being rolled up from graphene - a two-dimensional honeycomb lattice with a carbon atom in each site. Multiwall carbon nanotubes were discovered by Japanese scientist Sumio Iijima in 1991 (Iijima, 1991) and, two years later, individual single wall carbon nanotubes (SWCNTs) were reported (Iijima & Ichihashi, 1993). Their diameter is as little as 1 nanometer. Carbon nanotubes provide the ultimate limit for microelectronic miniaturization. Soon after discovery carbon nanotubes attracted tremendous interest from fundamental science and technological perspectives and thousands of papers have been published on this subject (for a review see e.g. (Dresselhaus et al., 2001)). Amazing are mechanical properties of CNTs e.g. their high strength and high flexibility. Since the C-C bonds within CNTs are one of the strongest bonds in the nature these tubes are predicted to be by far the strongest fibres that can be made. SWCNT was tested to have tensile strength of order of 60 GPa (Wei, 2003). Nanotubes exhibit also exceptional electrical properties, some of them will be discussed in this chapter. The unique electrical properties of CNTs stem from unusual electronic properties of graphene (Novoselov, 2005; Wallace, 1947). The $sp^2$ hybridization between one $s$-orbital and two $p$-orbitals leads to a trigonal planar structure with a formation of a $\sigma$-bond between carbon atoms what is responsible for robustness of the lattice in all carbon allotropes. The unaffected $p$-orbital, which is perpendicular to the planar structure binds covalently with neighboring carbon atoms leading to the formation of half filled $\pi$-bands. The low energy band structure of graphene has two bands that join at the corners of the Brillouin zone (two distinct Fermi points at $K$ and $K'$, called Dirac points (Fig. 1a)). The name of these points reflects the fact that the energy bands disperse linearly away from the touching points, so the dispersion relation for electrons (holes) is described by an isotropic cone that opens upward (downward) near $K$, $K'$ points (massless fermions). Graphene is a gapless semiconductor. This is changed when rolling up a piece of graphene to form a carbon nanotube. From the bandstructure of graphene one can obtain the bandstructure of nanotube by imposing appropriate boundary conditions along the circumference. Quantization of momentum in the circumferential direction is given by condition $\pi d \cdot k_{\perp} = 2\pi i$ ($i = 1, 2...$). Spacing in $k_{\perp}$ is thus $\Delta k_{\perp} = 2/d$, where $d$ is the diameter of the tube. The quantization cuts discrete slices of two-dimensional Dirac dispersion of graphene. Each line corresponds to a 1D subband for conduction along the nanotube (Fig. 1b, c). Depending on the way graphene is rolled up, carbon nanotube can be either metallic or semiconducting. The former occurs if a slice happens to pass through

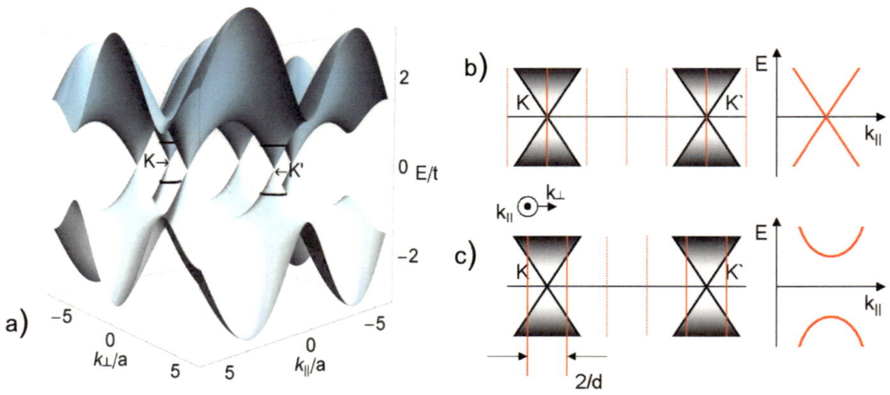


Fig. 1. a) Band structure of graphene. The valence and conduction states meet in $K$ and $K'$ points. b) Quantization of energy in metallic carbon nanotube. The vertical lines represent $k_\perp$ values intercepting the dispersion cones at $K$ and $K'$. c) Quantization of energy in semiconducting CNT.

the Dirac point (Fig. 1b). Metallic SWNT have a Fermi velocity $v_F \approx 8 \cdot 10^5$ m/s that is comparable to typical metals. Metallic nanotubes have conductivities and current densities that exceed the best metals. Semiconducting SWNTs (Fig. 1c) have a bandgap $E_g \approx 0.9$ eV/d. The inverse dependence of the gap on diameter reflects the similar dependence on diameter of the separation of 1D subbands. Semiconducting tubes have mobilities and transconductances that meet or exceed the best semiconductors.

The research presented in this chapter is addressed to spintronics, a new rapidly developing field of electronics, which exposes the role of spin in controlling current flowing through nanoscopic systems. Spintronic systems have promising potential applications e.g. in reprogrammable logic devices and quantum computing because of long coherent lifetime of spin degree of freedom, fast data processing speed and low dissipation (Das Sarma, 2001). The most spectacular commercial impact of this field to date has been in the area of spin valves used in magnetic hard disk drivers. The principle of operation of such spin valve is based on magnetoresistive effect ($\mathcal{GMR}$), for which the Nobel Prize was awarded in 2007 to Albert Fert and Peter Grünberg. In their experiments it was found (Baibich et al., 1988; Binasch et al., 1989) that the resistivity of non-magnetic spacer layers sandwiched between ferromagnetic films changes unexpectedly largely with the relative alignment of the magnetizations in the films. Later similar effects have been observed in semiconductor tunnel junctions (Tanaka et al., 2001) and molecular systems (Xiong et al., 2004) and the term tunnel magnetoresistance ($\mathcal{TMR}$) has been coined. In this case transport from one ferromagnetic electrode to another occurs not by simple extended state conduction, but by quantum tunneling across a non-magnetic region. In this context, carbon nanotubes are particularly interesting, because they exhibit long spin lifetime and can be contacted with ferromagnetic materials (Cottet et al., 2006). Electronic transport through a nanotube depends on the contacts with electrodes. At low temperatures the properties of short tubes weakly coupled with electrodes are dominated by strong correlations. Of special interest in this respect is Kondo effect - a formation of many-body dynamical singlet between a localized spin and delocalized conduction electrons of electrodes (Hewson, 1993). The tunability of nanostructures have allowed studies of Kondo effect in nonequilibrium. Since carbon nanotubes posses in addition to spin also orbital degeneracy, it becomes possible to realize in these systems highly symmetric Kondo effect, where both spin and orbital pseudospin are quenched. The increase of degeneracy corresponds to enhancement of Kondo temperature, what is important for potential applications. In the following, we will discuss an impact of symmetry-breaking perturbations in CNTs in the Kondo regime on transport, in specific we will focus on the influence of magnetic field and polarizations of electrodes. The conclusions drawn in this chapter can be easily adopted also to the case of manipulating of orbital degrees of freedom (orbitronics).

## 2. Carbon nanotube quantum dot

When two metallic electrodes are deposited on top of a CNT, tunnel barriers develop at the nanotube-metal interfaces (Fig. 2). When the resistance of the two barriers becomes comparable to or is larger than the quantum resistance ($R_Q = h/(2e^2)$), the island becomes strongly separated. A finite length $L$ between the electrodes results in quantized of energy levels $\Delta E = (hv_F)/2L$. Discrete level structure is a consequence of quantum confinement effect. Carbon nanotube quantum dot (CNT-QD) is a zero-dimensional island tunnel-coupled to metal leads. Schematic side view of CNT-QD is presented on Fig. 2. With the nanotube

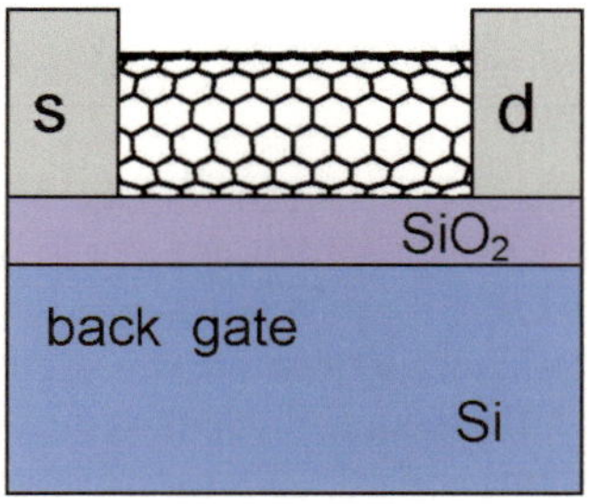


Fig. 2. Schematic side view of carbon nanotube quantum dot.

lying on an oxidized Si substrate, a natural way of gating quantum dot is to apply a voltage to the doped Si. The Si then acts as a back gate affecting the whole dot. For small dots the charging energy $E_C = e^2/(2C)$ becomes important, because the dots have very small capacitances (capacitance of a tube $C/L \sim ln(1/d)$). A typical diameter of SWCNT is $d \sim 1$ nm and $L$ is of order of few hundreds of nm, what gives $E_C$ of order of a few meV. Of the same order, but usually larger is single particle level spacing $\Delta E$ ($\Delta E \sim 1.7$ meV/L[$\mu$m]). The transmission of the contacts determines the relevant regime for charge transport. Depending on the ratio between tunnel induced broadening of energy levels $\Gamma$ and charging energy $E_C$, three regimes can be distinguished:

1. $\Gamma \ll E_C$ - closed QD, charging effects dominate transport (Coulomb blockade regime)
2. $\Gamma \leq E_C$ - intermediate coupling , increasing role of higher order tunneling processes (Kondo regime)
3. $\Gamma \gg E_C$ - open QD, interaction effects do not play the role and transport is dominated by interference (Fabry-Perot regime)

In the first regime for temperatures lower than charging energy the electrons will enter the dot one by one yielding the well known Coulomb blockade (CB) oscillations of the transport as a function of gate voltage $V_g$ (Fig. 3b). The addition of each extra electron to the dot requires charging energy. This leads to a ladder of discrete addition levels $\mu(N)$ indicating the energy required to add the Nth electron. The Coulomb blockade can be lifted by changing

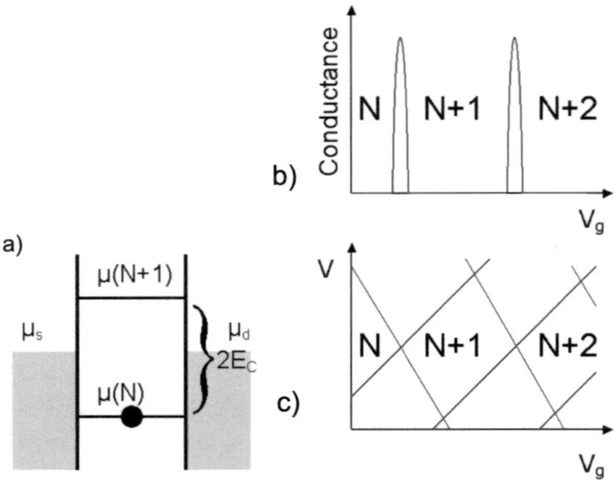


Fig. 3. a) Energy diagram for a quantum dot b) Coulomb oscillations of conductance versus gate voltage. On the peak the number of electrons can fluctuate between $N$ and $N+1$, in between transport is blocked. c) Coulomb diamonds in the differential conductance versus transport voltage and gate voltage.

the gate voltage or by changing the source-drain voltage $V$ (Fig. 3c). In the $V_g$ -$V$-plane the diamond-shaped regions are observed, in which current is blocked. In a wider range of gate voltage than the one presented on Fig. 3b, apart from charging energy also single particle level spacing separation can be extracted from a nonuniform distribution of conductance peaks. In general the distance between the CB peaks is determined by a sum of charging and quantum confinement energy. In an ideal semiconducting nanotube, sets of four electronic states can be grouped together into a shell. All four states are degenerate, with two choices for spin and two for orbital. The orbital degeneracy can be intuitively viewed to originate from two equivalent ways electrons can circle the graphene cylinder, that is clockwise and anticlockwise. For semiconducting tubes the fourfold periodicity of addition energy is observed (Buitelaar et al., 2002; Liang et al., 2002) and for metallic tubes, where only spin degeneracy is present, twofold shell filling was reported (Cobden & Nygård, 2002). Transport in regime 1 is governed by sequential tunneling (first order tunneling processes), which gives rise to current only at the Coulomb peaks. In the opposite limit of high transparency (regime 3) nanotube acts as an electron wave guide creating resonances at certain energies. Such system can be regarded as an open quantum dot with resonances corresponding to the broad energy levels of the dot (Liang et al., 2001). In the intermediate regime (2) electron number on the dot is still fixed, but significant cotuneling is allowed leading to finite conductance in the valleys between Coulomb peaks (Kondo effect).

### 3. Exotic spin-orbital Kondo effect

Kondo effect was first discovered in noble metals containing a small concentration of magnetic impurities, where a logarithmic increase of resistivity was observed at low temperatures (de Hass et al., 1933). It was explained by Japanese theorist Jun Kondo in 1964 (Kondo, 1964). Magnetic impurities embedded in metallic hosts cause anomalous resonant scattering of conduction electrons. The many-body dynamical singlet between a localized spin and delocalized conduction electrons is formed. This effect has become one of the most extensively

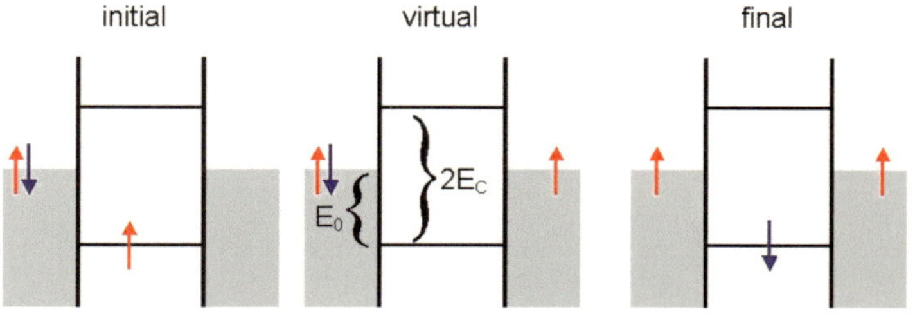


Fig. 4. Spin-flip cotunneling. By going through the different states, the dot spin can effectively be flipped.

studied many body problems in the field of theoretical solid state physics in the last decades. Kondo effect has been also observed in a wide range of nanoscopic systems including semiconductor-based QDs (Cronewett et al., 1998; Goldhaber-Gordon et al., 1998) and molecular systems (Liang et al., 2002; Park et al., 2002). Thanks to versatility of nanoobjects this effect can be studied in a variety of situations, offering quite novel physics. A path of studying this many-body phenomenon in a controlled way and in out-of equilibrium situations has been opened (De Franceschi et al., 2002; Grobis et al., 2008; Paaske et al., 2006). Due to different geometry of QDs coupled to the leads compared to geometry of impurity in an alloy, instead of logarithmic increase of resistance at low temperatures, for QDs a similar increase of conductance is observed. In a quantum dot all electrons have to travel through the device as there is no path around it. Kondo resonance formed at the Fermi energy mixes the states from both leads increasing the conductance. The essence of the Kondo spin screening process in QD is illustrated on Fig. 4, where the spin flip (exchange) cotuneling process is presented. The virtual spin flips at the dot are caused by tunneling processes off the dot with a given spin followed by tunneling of electron of opposite spin on the dot. Classically these processes are forbidden by energy conservation. The intermediate virtual state (Fig. 4) is allowed to exist for a very short time $t \sim h/E_0$ by Heisenberg uncertainty principle. The final state has the same energy as the initial and the sequence of processes shown on Fig. 4 is known as elastic cotunneling (Averin et al., 1992). Adding many spin-flips processes of higher order coherently, the spin-flip rate diverges. The spin-flip processes resonantly enhance around a characteristic temperature called Kondo temperature $T_K$ and change the energy spectrum of the system generating at the Fermi energy many-body resonance known as Kondo resonance. A consequence of its occurrence is an earlier mentioned logarithmic increase of the conductance and its saturation, in the case of symmetric coupling to the leads conductance reaches value $(2e^2)/h$. Electrons are transmitted perfectly through the dot due to location of Kondo resonance at the Fermi energy (Fig. 6). The Kondo temperature sets the temperature, respectively the voltage or magnetic field scale above which the Kondo resonance is suppressed. $T_K$ can be estimated from the width of the Kondo resonance and deduced from the temperature dependence of linear conductance. The Kondo effect can also occur replacing the spin by orbital (Sasaki et al., 2004) or charge (Holleitner et al., 2004; Wilhelm et al., 2002) degrees of freedom. The necessary condition for the occurrence of this effect is the same degeneracy of the states in the electrodes and in the QD and conservation of spin or pseudospin in tunneling processes. For the two-fold degenerate states the allowed symmetry operations are rotations in spin space (SU(2)). Spin and orbital degeneracies can also occur simultaneously leading to highly symmetric Kondo state (SU(4)). SU(4) group

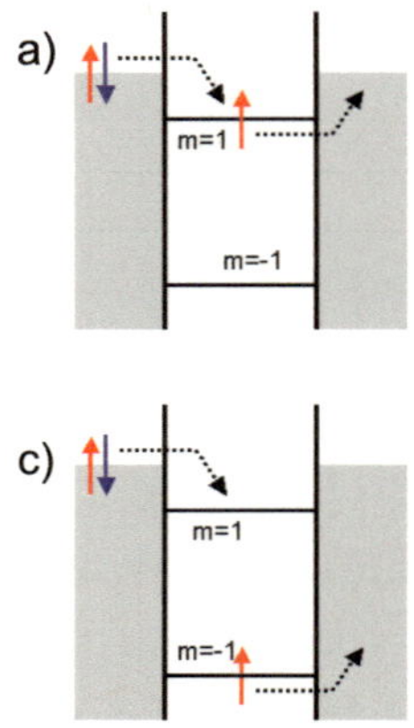


Fig. 5. Cotunneling processes leading to quenching of spin and orbital pseudospin in SU(4) quantum dot a) spin-flip fluctuation b) orbital fluctuation c) spin-orbital fluctuation.

characterizes the rotational invariance in spin and orbital space. The simultaneous screening of orbital and spin degrees is caused by tunneling processes causing spin, orbital pseudospin and spin-orbital fluctuations (Fig. 5). In this case orbital psudospins play exactly the same role as spins. The spectral density of SU(4) Kondo system shows a peak slightly shifted from the Fermi energy, it is pinned at $\omega \sim T_K^{SU(4)}$ (Fig. 6). The many-body peak is also

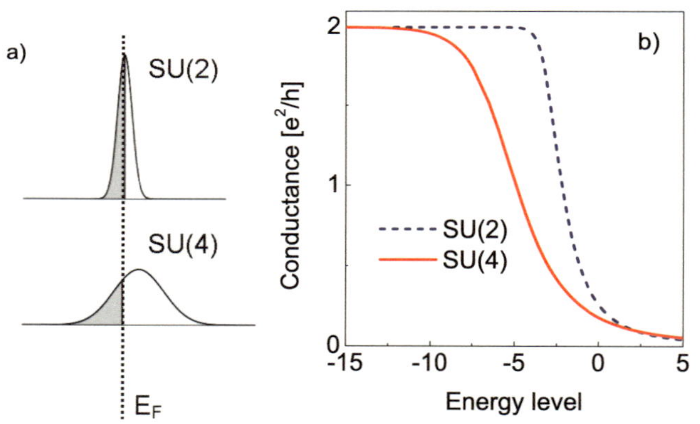


Fig. 6. Comparison of SU(4) and SU(2) Kondo effects a) Schematic view of Kondo resonances. SU(4) resonance is wider and shifted from the Fermi level. b) Conductance versus site energy.

much broader than that for the SU(2) Kondo effect, what means exponential enhancement of Kondo temperature. This makes these systems interesting for practical applications. Fig.6 shows a schematic picture of Kondo resonances for both symmetries and the corresponding dependencies of conductance on dot energy. The scattering phases at $E_F$ are $\delta_{SU(2)} = \pi/2$ and $\delta_{SU(4)} = \pi/4$ respectively, and the zero temperature linear conductances $\mathcal{G}^{SU(2)} = 2(e^2/h)\sin^2(\delta_{SU(2)}) = \mathcal{G}^{SU(4)} = 4(e^2/h)\sin^2(\delta_{SU(4)}) = 2(e^2/h)$. Since in both cases total conductance reaches the same value one cannot reliably distinguish between SU(2) and SU(4) Kondo effect in the unitary limit based on the conductance analysis alone. It is worth to mention that for higher degeneracies total conductances diminish, what is easy to check putting the values of phase shifts $\delta_{SU(2N)} = \pi/2N$. SU(4) Kondo effect in nanoscopic systems has been first observed for vertical QDs (Sasaki et al., 2004), but most spectacular evidence of this phenomena has been reported by Jarillo-Herrero et al. for carbon nanotubes (Jarillo-Herrero et al., 2005). Fig. 7 presents how formation of Kondo resonance manifests in the differential conductance. The conductance exhibits a pronounced enhancements in regions for 1 and 3 electrons. The estimated Kondo temperature for the discussed system was $T_K = 7.7$ K what can be ascribed to the enhanced degeneracy. A confirmation that the observed phenomena is really spin-orbital Kondo effect is the influence of finite magnetic field on the low-energy Kondo behavior of conductance. For perpendicular magnetic field, where only spin degeneracy is removed the splitting of high conductance line into three lines has been observed (Makarowski et al., 2007), for SU(2) symmetry only two lines should be visible. For the field applied parallel to the nanotube axis, when spin-orbital degeneracy is removed a splitting of the Kondo resonance into four peaks results (Fig. 7). A few

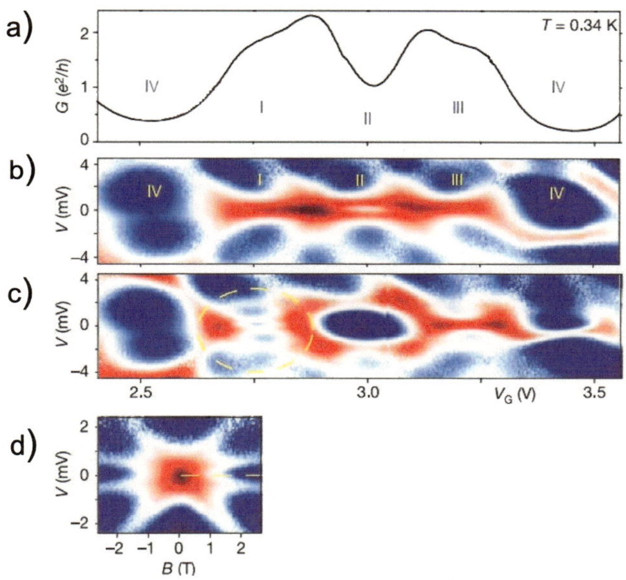


Fig. 7. Experimental evidence of SU(4) Kondo effect in semiconducting carbon nanotube quantum dot. a) Linear conductance versus gate voltage at 0.34 K. Latin numbers classify the valleys according to the occupations at the dot. b) Color-scale plot of differential conductance versus gate and transport voltages at zero magnetic field. Conductance increases from blue to red. The enhanced linear conductance signals the occurrence of Kondo effect. c) Same as b) but for CNT-QD in axial magnetic field $h_{||} = 1.5$ T. d) Fourfold splitting of the Kondo peak as a function of field. Adapted from ((Jarillo-Herrero et al., 2005)).

observations of spin-orbital Kondo effect in CNTs (Grove-Rasmussen et al., 2007; Wu et al., 2009) and several interesting theoretical papers on SU(4) Kondo problem have been published very recently (Büsser & Martins, 2007; Choi et al., 2005; Galpin et al., 2006; Lim et al., 2006; Lipiński & Krychowski, 2005; Mizumo et al., 2009). The unusual strongly correlated Fermi liquid state, where spin and orbital degrees of freedom are totally entangled is interesting for quantum computing and storage technology. Doubling of storage density is expected, because each 4-state bit is exactly equivalent to two 2-state bits. The exotic spin-orbital Kondo effect occurs if the orbital quantum number is conserved during tunneling. If the leads to the dot are formed within the same nanotube this requirement is fulfilled, otherwise some mixing in the orbital channels may occur. It is surprising therefore that SU(4) Kondo effect is observed with metallic electrodes attached. Some authors suggest (Choi et al., 2005) that, in these systems, the orbital quantum number is still conserved during higher order tunneling events, probably because the CNT-QD is coupled to the nanotube section underneath the contacts, where the carriers dwell for some time before moving into the metal. Even for small mixing of orbital channels (mixing smaller than the Kondo energy) the SU(4) Kondo description can still serve as a reasonable first insight into physics of these systems.

## 4. Model and formalism

CNT-QD exhibits four-fold shell structure in the low energy spectrum. In the present considerations we restrict to the single shell, what is justified for short nanotubes at low temperatures, because the level spacing in this case is larger than thermal energy. We take into account only the top most occupied shell and treat other electrons as inert core. The dot is modeled by two-orbital Anderson impurity model:

$$\mathcal{H} = \sum_{k\alpha m\sigma} \epsilon_{k\alpha m\sigma} c^{+}_{k\alpha m\sigma} c_{k\alpha m\sigma} + \sum_{k\alpha m\sigma} t_{\alpha}(c^{+}_{k\alpha m\sigma} d_{m\sigma} + h.c) + \sum_{m\sigma} E_{m\sigma} d^{+}_{m\sigma} d_{m\sigma} + \sum_{m} \mathcal{U} n_{m+} n_{m-} + \sum_{\sigma\sigma'} \mathcal{U}' n_{1\sigma} n_{-1\sigma'} \quad (1)$$

where $m = \pm 1$ numbers the orbitals, the leads channels are labeled by $(m, \alpha)$, $\alpha = L, R$. $E_{m\sigma} = E_0 + eV_g + g\mu_B h + g\mu_{orb} h \cos(\theta)$, we set $|e| = g = \mu_B = k_B = \hbar = 1$. $\theta$ specifies orientation of magnetic field $h$ relative to the nanotube axis, $\mu_{orb}$ is the orbital moment. The first term of (1) describes electrons in the electrodes, the second describes tunneling to the leads, the third represents the dot site energy and the last two terms account for intra ($\mathcal{U}$) and interobrbital ($\mathcal{U}'$) Coulomb interactions. We will consider carbon nanotubes coupled to electrodes which can be either nonmagnetic or ferromagnetic. The spin polarization of the leads $P_\alpha$ is defined by spin-dependent densities of states $\varrho_{\alpha m\sigma}$ as $P_\alpha = (\varrho_{\alpha+} - \varrho_{\alpha-})/(\varrho_{\alpha+} + \varrho_{\alpha-})$. The spin-dependent coupling strength to the lead $\Gamma$ is described by $\Gamma = \sum_\alpha \Gamma_\alpha = \sum_{km\alpha\sigma} \pi t^2_\alpha \varrho_{\alpha m\sigma}$. In the following, the wide conduction-band approximation with the rectangular density of states is used $\rho_{\alpha m\sigma}(\epsilon) = \rho_{\alpha\sigma} = 1/(2\mathcal{D}_{\alpha\sigma})$ for $|\epsilon| < \mathcal{D}_{\alpha\sigma}$, $\mathcal{D}_{\alpha\sigma}$ is the half bandwidth.

We are interested in nonequilibrium properties e.g. in current flowing through the dot. A common tool used in the description of transport characteristics are nonequilibrium Green's functions of Keldysh type, defined as a path-ordered product of annihilation and creation operators on a closed time contour which begins and ends at the same point. For a review of the techniques of the nonequilibrium Green's functions and its applications in electronic transport we refer the reader to (Haug & Jauho, 1998). In our discussion, instead of performing contour integration we adopt approximation known as Ng ansatz (Ng, 1996), which allows construct approximate nonequilibrium function for interacting system from the knowledge of equilibrium functions and nonequilibrium characteristics of the corresponding noninteracting system. The exact values of the latter can be easily found. Commonly used nonequilibrium Green's functions are lesser $G^{<}_{m\sigma,m\sigma}(t-t') = i\langle d^{+}_{m\sigma}(t') d_{m\sigma}(t)\rangle$ and greater $G^{>}_{m\sigma,m\sigma}(t-t') = -i\langle d_{m\sigma}(t) d^{+}_{m\sigma}(t')\rangle$ functions. They are linked with ordinary retarded or advanced Green's functions $G^{R}_{m\sigma,m\sigma}(t-t') = -i\theta(t-t')\langle[d_{m\sigma}(t), d^{+}_{m\sigma}(t')]\rangle$, $G^{A}_{m\sigma,m\sigma}(t-t') =$

$i\theta(t'-t)\langle[d_{m\sigma}(t), d^{+}_{m\sigma}(t')]\rangle$ through the relation $\mathbf{G}^{>} - \mathbf{G}^{<} = \mathbf{G}^{\mathbf{R}} - \mathbf{G}^{\mathbf{A}}$. Accordingly linked are also corresponding self energies $\mathbf{\Sigma}^{>} - \mathbf{\Sigma}^{<} = \mathbf{\Sigma}^{\mathbf{R}} - \mathbf{\Sigma}^{\mathbf{A}}$ (Keldysh requirement). Ng ansatz assumes linearity of lesser self-energy $\mathbf{\Sigma}^{<}$ and lesser self energy of the corresponding noninteracting system $\mathbf{\Sigma}^{<} = \mathbf{\Lambda}\mathbf{\Sigma}^{<(0)}$, where $\mathbf{\Sigma}^{<(0)} = \sum_{\alpha} 2if_{\alpha}(\omega)\mathbf{\Gamma}_{\alpha}$. Coefficient, or in general case matrix $\mathbf{\Lambda}$, can be found from Keldysh requirement. Ng approximation is exact for noninteracting particles, and it preserves continuity of current condition in the steady-state limit (Ng, 1996).

Now a few words about approximations used in treating the many-body problem. Kondo effect is a consequence of strong electron correlations present in the system. For a correct description of physics in this range crucial is a preservation of dot electron-conduction electron correlations. The retarded Green's functions used in our analysis are found from the equation of motion method (EOM). EOM consists of differentiating the Green's functions with respect to time which generates the hierarchy of equations with higher order GFs. In order to truncate the series of equations, we use at the third step of the chain of equations the self-consistent procedure proposed by Lacroix (Lacroix, 1998), which approximates the GFs involving two conduction-electron operators by:

$$
\begin{aligned}
\langle\langle c^{+}_{k\alpha m'\sigma'} d_{m'\sigma'} c_{k\alpha m\sigma} | d^{+}_{m\sigma}\rangle\rangle &\simeq \langle c^{+}_{k\alpha m'\sigma'} d_{m'\sigma'}\rangle \langle\langle c_{k\alpha m\sigma} | d^{+}_{m\sigma}\rangle\rangle \\
\langle\langle c^{+}_{k\alpha m'\sigma'} c_{k\alpha m'\sigma'} d_{m\sigma} | d^{+}_{m\sigma}\rangle\rangle &\simeq \langle c^{+}_{k\alpha m'\sigma'} c_{k\alpha m'\sigma'}\rangle \langle\langle d_{m\sigma} | d^{+}_{m\sigma}\rangle\rangle
\end{aligned} \tag{2}
$$

Knowing $G^{R}$ and $G^{A}$ one calculates self energies $\Sigma^{R}$ , $\Sigma^{A}$. The advantage of EOM method in comparison to other many-body techniques e.g. slave boson formalism often used in the analysis of Kondo limit (Coleman, 1987), is that EOM works in the whole parameter space except only the close vicinity of Kondo fixed point and it accounts not only for spin or pseudospin fluctuation, but also for charge fluctuations. This is of importance in analysis of systems with finite charging energy.

Let us now give few formulas determining quantities we study. Current flowing through CNT-QD in the $(m\sigma)$ channel $\mathcal{I}_{m\sigma} = (\mathcal{I}_{Lm\sigma} - \mathcal{I}_{Rm\sigma})/2$ is calculated from the time evolution of the occupation numbers $\hat{N}_{\alpha} = \sum_{km\sigma} c^{+}_{k\alpha m\sigma} c_{k\alpha m\sigma}$:

$$
\begin{aligned}
\mathcal{I}_{\alpha}(t) = &\; -e\langle d\hat{N}_{\alpha}(t)/dt\rangle = i(e/\hbar)\sum_{km\sigma}[t_{\alpha}\langle c^{+}_{k\alpha m\sigma}(t) d_{m\sigma}(t)\rangle - h.c.] = \\
&= \sum_{km} t_{\alpha}[G^{<}_{m\sigma,k\alpha m\sigma}(t) - G^{<}_{k\alpha m\sigma,m\sigma}(t)]
\end{aligned} \tag{3}
$$

The thermal averages are expressed by the lesser Green's function as:

$$
\langle c^{+}_{k\alpha m\sigma} d_{m\sigma}\rangle = \int \frac{d\omega}{2\pi i} G^{<}_{m\sigma,k\alpha m\sigma}(\omega) \tag{4}
$$

Conductances are defined as $\mathcal{G}_{\sigma} = d\mathcal{I}_{\sigma}/dV = \sum_{m} d\mathcal{I}_{m\sigma}/dV$. The useful quantities characterizing the spin-dependent transport are polarization of conductance $\mathcal{PC} = (\mathcal{G}_{+} - \mathcal{G}_{-})/(\mathcal{G}_{+} + \mathcal{G}_{-})$ and tunnel magnetoresitance ($\mathcal{TMR}$). Tunnel magnetoresistance is defined as the relative difference of differential conductances for parallel (P) and antiparallel (AP) configurations of polarizations of the leads $\mathcal{TMR} = (\mathcal{G}^{\mathcal{P}} - \mathcal{G}^{\mathcal{AP}})/\mathcal{G}^{\mathcal{AP}}$. The spin transport is characterized by spin current. In general case apart from longitudinal component $\mathcal{I}^{z}$, which is easily expressible by the difference of charge currents for opposite spin channels $\mathcal{I}^{z} = \mathcal{I}_{+} - \mathcal{I}_{-}$, also transverse (spin flip currents) are required. $\mathcal{I}^{x} = Re[\mathcal{I}^{+}]$ and $\mathcal{I}^{y} = Im[\mathcal{I}^{+}]$. $\mathcal{I}^{+} = (\mathcal{I}^{+}_{L} - \mathcal{I}^{+}_{R})/2$ can be expressed similarly as Eq.(3) by $\mathcal{I}^{+}_{\alpha}(t) =$

$2\sum_{km} t_\alpha [G^<_{m-,k\alpha m+}(t) - G^<_{k\alpha m-,m+}(t)]$. To supplement transport characteristics we will also present discussion of the shot noise. The shot noise reveals information of transport which are not accessible by knowledge of conductance alone, for example about the correlations. The temporal fluctuations of the current are defined as:

$$\mathcal{S}_{\alpha m\sigma\nu m'\sigma'}(t-t') = \langle[\Delta\hat{\mathcal{I}}_{\alpha m\sigma}(t), \Delta\hat{\mathcal{I}}_{\nu m'\sigma'}(t')]_+\rangle = \langle[\hat{\mathcal{I}}_{\alpha m\sigma}(t), \hat{\mathcal{I}}_{\nu m'\sigma'}(t')]_+\rangle - 2\cdot\mathcal{I}_{\alpha m\sigma}(t)\mathcal{I}_{\nu m'\sigma'}(t') \quad (5)$$

where $\Delta\hat{\mathcal{I}}_{\alpha m\sigma}(t)$ is the fluctuation of the current operator around its average value. At very small bias ($eV < k_B T$) noise is dominated by thermal noise. The thermal noise is related to fluctuations in the occupations of the leads due to thermal excitation, and vanishes at zero temperature. Contribution to the noise, we are interested in - shot noise is an unavoidable temporal fluctuation of current caused by the discreteness of the electronic charge (Blanter & Büttiker, 2000). Current is not a continuous flow, but a sum of discrete pulses in time, each corresponding to the transfer of an electron through the system. If the electrons are transmitted randomly, independently of each other the transfer of them can be described by Poissonian statistics. Deviations from the Poissonian noise appear to be due to correlations between electrons. A convenient means to assess how correlations affect shot noise is the Fano factor $\mathcal{F}$ defined as the ratio between the actual shot noise $\mathcal{S}$ and the Poissonian noise $\mathcal{F} = \mathcal{S}/(2e\mathcal{I})$.

Let us close this section by a few words on the energy scale of the effects examined. In the full symmetric case SU(4) (equal coupling to the leads and equal inter and intraorbital interactions) the behavior of the model is governed by four parameters, two of them specify the dot: orbital energy $E_0$ and single electron charging energy $\mathcal{U}$. Another two parameters characterize the coupling to the leads - $\Gamma$, and the leads themselves - the half bandwidth $\mathcal{D}$. Tunnel barrier widths and source and drain capacitances change with the number of electrons at the dot and consequently both lead - dot coupling and charging energy change with the gate voltage. By gate voltage one can directly control the site energy. The value of $\mathcal{U}$ can be inferred from the size of Coulomb diamonds, for semiconducting CNT-QDs it takes values of order of tens meV (Babić et al., 2004; Jarillo-Herrero et al., 2004). Intermediate coupling strength required in the Kondo range corresponds to $\Gamma$ of order of several meV (Jarillo-Herrero et al., 2005; Makarowski et al., 2007). $\mathcal{D}$ is the largest energy scale in our problem and it is of order of tens of meV. Choosing parameters of CNT-QD within the above intervals of parameters gives estimation of Kondo temperature in the range of several Kelvin, what agrees with characteristic temperatures observed in these systems. Typical diameter $d$ of SWNCT is of order of several nm. Orbital magnetic moment $\mu_{orb}$ scales with CNT diameter and can be estimated from the slopes between two Coulomb peaks that correspond to the addition energy of the electrons to the same orbital. We assume $\mu_{orb} = 10\mu_B$, which corresponds to the diameter $d = 2.9$ nm. In the following pictures, all the energies are given in units of $\Gamma$ and similarly other quantities in accordance with the earlier chosen sets of the units ($|e| = q = \mu_B = k_B = \hbar = 1$). The bandwidth $\mathcal{D}$ is assumed $\mathcal{D} = 50$.

## 5. SU(4) Kondo effect in carbon nanotube quantum dot

Until recently, the prospect of using the Kondo effect in spintronic applications have been very poor because the required temperatures for semiconducting QDs lie in mK range. The use of single wall carbon nanotubes as quantum dots has pushed the Kondo temperatures to the range of several K. In experiment of Jarillo-Herrero et al. (JH) the reported Kondo temperature was $T_K \sim 7.7$ K (Jarillo-Herrero et al., 2005) and in (Makarowski et al., 2007) Kondo temperature as high as $T_K \sim 15$ K has been reported. Other properties which make CNT-QDs ideal candidates for electronic applications are long spin lifetimes and the fact that Kondo effect can be seen over a very wide range of gate voltages encompassing hundreds of Coulomb oscillations (Nygård et al., 2000). SU(4) Kondo effect in carbon nanotube is an example of many-body effect occurring for entangled degrees of freedom. In case of CNTs spin degrees of freedom are entangled with both chiralities of the nanotube. The SU(4) group is the minimal group allowing such spin-orbital entanglement and which guarantees rotational invariance both in spin and orbital spaces. In this section we investigate the SU(4) Kondo

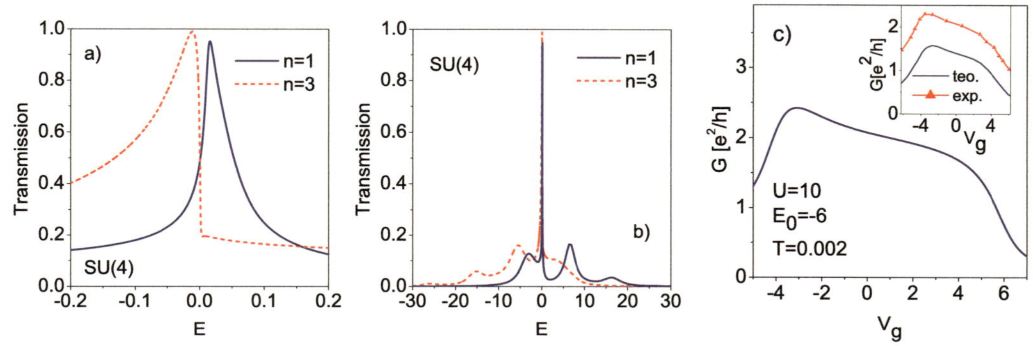


Fig. 8. a) Comparison of Kondo transmissions of CNT-QD in the range of single $n = 1$ and triple $n = 3$ electron occupations of the dot ($E_0 = -6, \mathcal{U} = 10$). b) Same as a) but shown in extended scale. The high energy peaks correspond to charge fluctuations. c) Gate dependence of linear conductance in the $n = 1$ range. Inset shows a similar dependence for CNT-QD parametrized by $E_0 = -6.25, \mathcal{U} = 12.5$ compared to the experimental curve ((Jarillo-Herrero et al., 2005)).

effect in the one and three electron valleys. For $n = 1$ both the total spin $S^z = (n_+ - n_-)/2$ and orbital pseudospin $T^z = (n_1 - n_{-1})/2$ are quenched due to spin-orbital fluctuations. For $n = 3$ the concepts of total spin or pseudospin is easier to understand replacing the electron occupations in definition of $S^z$ or $T^z$ by hole occupations. At very low temperatures the simple tunneling picture breaks down, scattering processes of any order contribute to the transport. The result is many-body state, which couples to the electrodes with a very high transmission probability, which as it is seen from Fig. 8 approaches one. The spin orbital fluctuations are also influenced by charge fluctuations. For infinite $\mathcal{U}$ the only charge fluctuations are ($n = 0 \leftrightarrow n = 1$), but for finite $\mathcal{U}$, which is the case considered, additional fluctuations ($n = 1 \leftrightarrow n = 2$) come into play. The role of these fluctuations for may-body processes is the larger the closer the corresponding charge fluctuation peaks are to the Fermi energy. Apart form the Coulomb peak ($\omega = E_0 + \mathcal{U}$) corresponding to fluctuation into the doubly occupied state, also a track of fluctuations into higher occupancy is visible. For $n = 1$ the Kondo peak occurs slightly above the Fermi level and dot occupation for the single spin-orbital channel is $n_{ms} \sim \frac{1}{4}$. The shift of the Kondo peak away from the Fermi level can be understood from Friedel sum rule (Langreth, 1966), which neglecting charge fluctuation perturbation, gives in this case scattering phase shift at $E_F$ $\delta \sim \pi/4$. Accordingly, the linear conductance at zero temperature $\mathcal{G}(0) = 4(e^2/h)\sin^2(\delta) = 2(e^2/h)$. Charge fluctuations slightly modify this picture, but as it is seen conductance (transmission) is still close to the unitary limit. For $n = 3$ much broader Kondo resonance is formed below the Fermi level and electron occupation per spin-orbital channel accounts $n_{ms} \sim 3/4$ (phase shift $\delta \sim (\frac{3}{4})\pi$). Charge fluctuations in both cases play different roles, what reflects in considerably different widths of Kondo resonances. Figure 8c shows the calculated gate dependence of linear conductance in the single dot occupancy. Values close to the unitary limit are observed for deep dot levels and a drop of the conductance is visible when gate voltage moves the system closer to the mixed valence range. We also show in the inset a similar gate dependence of conductance calculated for another choice of parameters, which nicely reproduces the shape of experimental gate dependences reported in (Jarillo-Herrero et al., 2005) (compare conductance in region I on Fig. 7a). This fitting to experiment has been reported by us earlier in (Krychowski & Lipiński, 2009). As it is seen the calculated conductance is underestimated in comparison to experimental value by a constant value of $0.5(e^2/h)$ in the whole gate range. This background contribution can be ascribed to possibly additional non-Kondo conductance channel present in the system and neglected in our analysis.

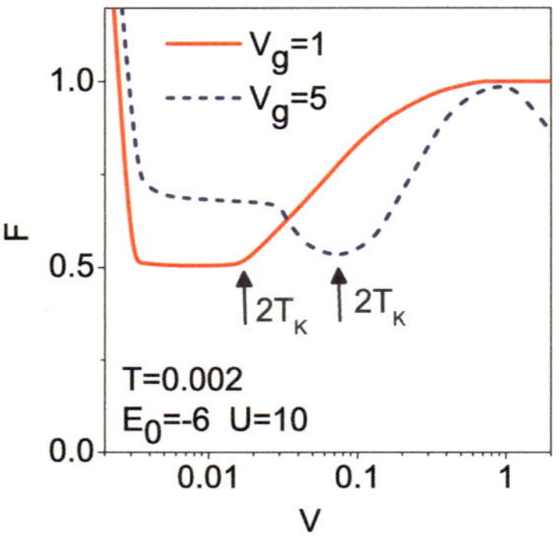


Fig. 9. Bias dependence of shot noise Fano factor. Low bias noise suppression induced by Kondo correlations is not complete ($\mathcal{F} = 1/2$).

Summarizing, in fully symmetric SU(4) state the four states at the dot are degenerate. Quantum fluctuations between these states induced by coupling to the leads (coupled fluctuations in spin and orbital sectors) result in formation of highly correlated Fermi liquid state, where spin and orbital degrees of freedom are totally entangled. Kondo resonance is no longer peaked at $E_F$ and Kondo temperature is largely enhanced in comparison to SU(2) systems. The linear conductance cannot reliably distinguish between SU(2) and SU(4) Kondo effects in the unitary limits (see Figure 6). As we will discuss in the next section, the field evolution of conductance is different for two cases, and this allows identify the type of Kondo effect. Distinction between two cases can be done also analyzing the shot noise. For SU(2) symmetry, the shot noise vanishes ($\mathcal{F} = 0$). Results presented on Fig. 9 display that due to the entanglement the SU(4) system remains noisy in the Kondo range, the Fano factor does not vanish. This fact has been recently observed in CNT-QDs (Delattre et al., 2009). For $V \sim 2T_K$ Fano factor takes the value $\mathcal{F} = \frac{1}{2}$ for the deep dot level position ($V_g = 1$) or slightly higher in the range closer to mixed valence ($V_g = 5$). In the latter case also the limit of constant value of $\mathcal{F}$ is not preserved in the whole low bias range due to a shallow dip in the Kondo peak introduced by charge fluctuations. At extremely small bias a rapid increase of $\mathcal{F}$ is observed and it is due to the fact that noise is dominated by thermal noise in this case. Fano factor is $(2T)/V$ in this range due to fluctuation-dissipation theorem and divergent at $V = 0$. For high voltages $V > 2T_K$ an increase of $\mathcal{F}$ is visible due to the weakening of Kondo correlations.

Maximum of $\mathcal{F}(V)$ for curve corresponding to $V_g = 5$ and the following decrease of $\mathcal{F}$ is due to Coulomb charge fluctuations. For the deep dot level the influence of these fluctuations is less significant and a drop of bias dependence of Fano factor is observed for still higher voltages beyond the presented range.

## 6. Effect of magnetic field

In the following sections we will discuss impact of symmetry-breaking perturbations on transport through CNT-QD in the Kondo regime. Since our study is addressed to spintronics we will analyze the effect of magnetic field and polarizations of electrodes. Field perpendicular to the nanotube axis breaks only the spin degeneracy and parallel field breaks both spin and orbital degeneracy. When an axial magnetic field is applied to CNTs, the electronic states are modified by an Aharonov-Bohm phase (A-B). The A-B phase affects electron states differently depending on the orbital quantum number. Quantization condition for momentum in the circumferential direction is now generalized to:

$$\pi d \cdot k_{\perp} + (2\pi)\frac{\varphi}{\varphi_0} = (2\pi)i \quad (i = 1, 2..., \ h_{||} \neq 0), \tag{6}$$

where $(2\pi)\frac{\varphi}{\varphi_0}$ is A-B phase acquired by the electrons while traveling the nantube circumference ($\varphi = h_{||}\pi d^2/4$ and $\varphi_0$ is the flux quantum). The splitting of the states numbered by opposite orbital numbers can be expressed similar to Zeeman splitting in terms of orbital magnetic moment by the term $m\mu_{orb}$ included in definition of Hamiltonian (1), $\mu_{orb} = (e|v_F|d)/4$. Orbital moment scales with CNT diameter and is typically one order of magnitude larger than Bohr magneton. This fact is the reason for strong magnetic field anisotropy of conductance of CNT-QD, the orbital pseudospin is more susceptible to magnetic field than the real spin. Already at small axial fields a considerable change of conductance is observed. For the shallow site energy, when unperturbed Kondo peak is noticeably shifted from Fermi level the field induced emergence of orbital satellites on both sides of the main peak and a shift of one of them towards the Fermi level (Fig. 10c) results in the observed initial increase of the linear conductance. Further increase of the field reflects in the decrease of conductance, and it happens when the lower satellite moves below $E_F$. For high magnetic field also spin splitting is visible (Fig. 10d). For perpendicular orientations the orbital motion is unaffected by the field, and only spin splitting results in high magnetic fields, for small fields only a slight shift of the peak towards Fermi level and its broadening leading to the increase of DOS at $E_F$, what leads to the corresponding increase of conductance. For deeper site energies the similar reconstructions of the Kondo peak in the field results in the decrease of conductance in both cases (Fig. 10b), because the unperturbed resonance is closely located to $E_F$. Figs. 10e, f present finite bias and magnetic field differential conductance maps. For axial field four lines and for perpendicular three lines of high conductance are visible. A pair of inner lines on Fig. 10e corresponds to orbital conserving fluctuations for both spin channels and the outer lines reflect orbital and simultaneous spin and orbital fluctuations. The latter two processes are not resolved for the assumed value of coupling to the leads and temperature. This picture qualitatively reflects the field dependence observed in (Jarillo-Herrero et al., 2005) (compare Fig. 7c, d). The difference between the $n = 1$ and $n = 3$ behaviors in the parallel field has been reported in (Jarillo-Herrero et al., 2005; Makarowski et al., 2007). For the first region the four line structure has been observed whereas for the second region only three lines were visible. This fact has been interpreted in (Galpin et al., 2010) as a consequence of spin-orbit interaction,

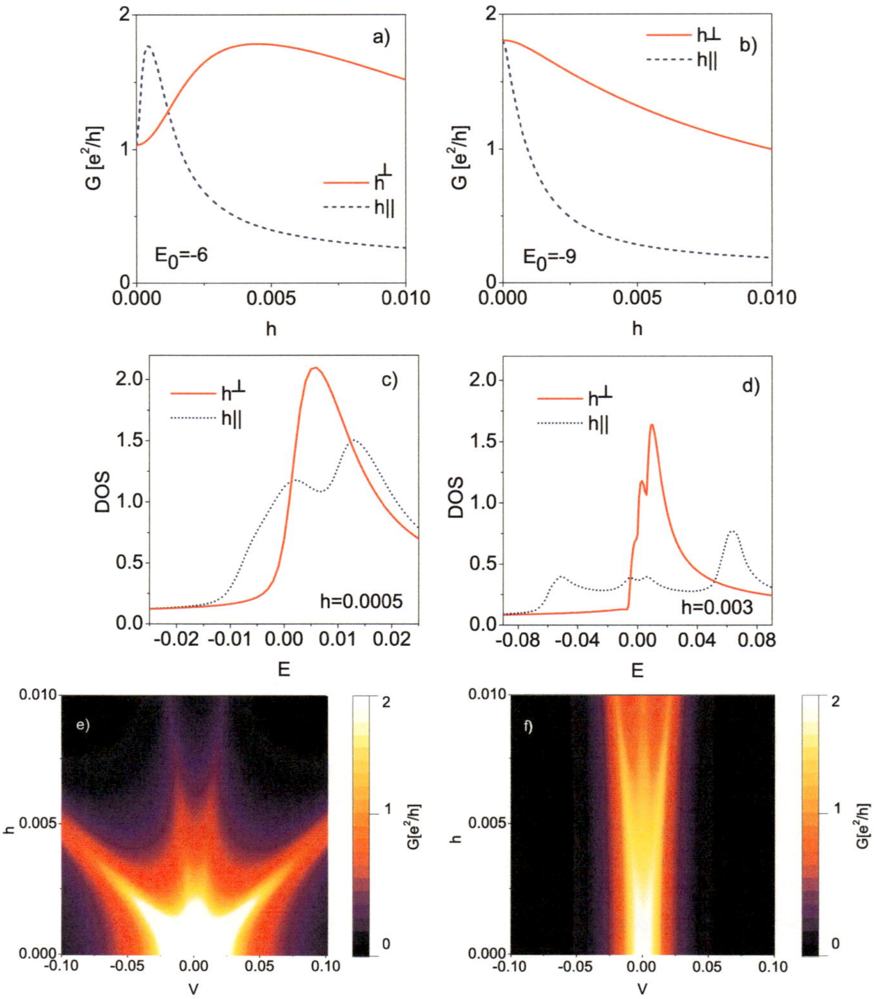


Fig. 10. Magnetic field anisotropy of conductance of CNT-QD in the Kondo regime. a) Magnetic field dependence of conductance for parallel and perpendicular orientations for the dot energy $E_0 = -6$ ($\mathcal{U} = 15$). b) Same as a) but for the deeper dot energy $E_0 = -9$. c) DOS of CNT-QD ($E_0 = -6$) for small magnetic field. d) Same as c) for high magnetic field e) Color-scale plots of differential conductance versus axial field and bias voltage ($E_0 = -9$). f) Same as e) but for perpendicular field.

but alternatively one can think that it is only a consequence of difference of the widths of unperturbed Kondo peaks resulting from different roles played by charge fluctuations in both cases (compare e.g. Fig. 8a). For $n = 3$ at moderate fields the lines corresponding to different spin orientations are not resolved from the main peak. A careful look at the JH conductance maps also supports this point of view, since also in their pictures only three lines are visible for low values of axial field. Typically three lines structure occurs for the perpendicular field orientation. The central line corresponds to the orbital fluctuations and the outer lines are due to spin mixing fluctuations. The experimental confirmation of such behavior can be

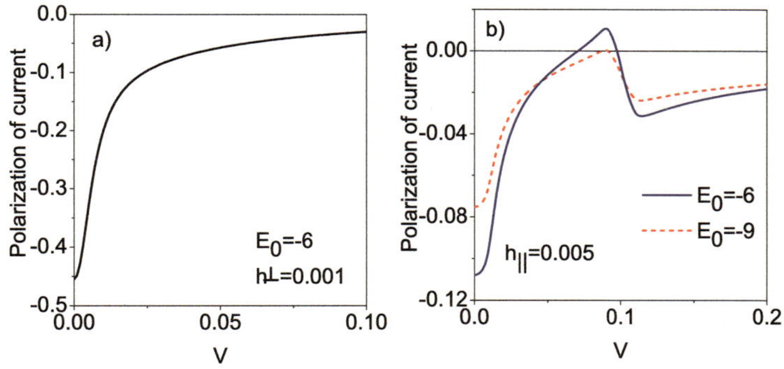


Fig. 11. Bias dependences of polarization of currents in a) perpendicular b) parallel magnetic fields.

found in (Makarowski et al., 2007). More detailed comparison to experiment including the relevant parameters and considering the asymmetry of the leads has been published by us in (Krychowski & Lipiński, 2009).

Current flowing in magnetic field is spin polarized. Fig. 11 presents an example of polarization of current $(\mathcal{I}_+ - \mathcal{I}_-)/(\mathcal{I}_+ + \mathcal{I}_-)$ corresponding to colorscale plot of differential conductance presented at Fig. 10. Substantial polarization of current is observed for small bias. Polarization can change its sign when spin-orbital Kondo satellites enter transport window. For high voltages ($eV \gg g\mu_B h$) current becomes unpolarized.

## 7. Kondo spin filter

Spin filter is a device that filters electrons by their spin orientation. Controlling the spin degree of freedom is currently an important challenge in spintronics. In particular in quantum information technology spin filters can be used for initialization and readout of spin quantum bits (Loss & DiVincenzo, 1998). Recently we have shown (Krychowski et al., 2007), that

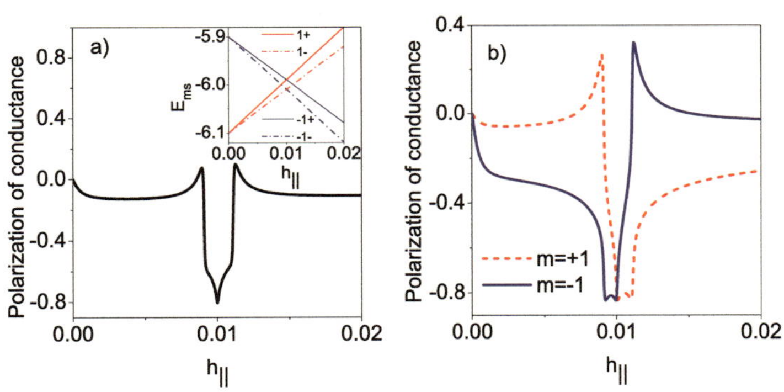


Fig. 12. Spin polarization of conductance of CNT-QD characterized by orbital level mismatch $\Delta_{orb} = 0.1$ in axial magnetic field. Inset illustrates recovery of orbital degeneracy. b) Orbital resolved spin polarizations corresponding to the picture a).

CNT-QD characterized by orbital level mismatch $\Delta_{orb} = E_{1\sigma} - E_{-1\sigma} \neq 0$ can serve as efficient spin filter operating in the low field range. Orbital mismatch occurs e.g. in nanotubes with torsional deformation. Fig. 12 presents polarization of total conductance and orbital resolved polarizations versus axial magnetic field. The idea of the proposed spin filtering mechanism is explained in the inset of Fig. 12a. Magnetic field is exploited to tune spin-polarized states into orbital dengeracy. Axial field might recover the orbital degeneracy either within the same spin sector or with mixing of spin channels. In the former case ($h_{||} = \Delta_{orb}/(2\mu_{orb})$) almost the same polarizations of conductance are observed for both orbital channels what results in large total polarization, whereas for the latter case ($h_{||} = \Delta_{orb}/(2\mu_{orb} + 1)$) the spin polarizations of different orbital sectors have opposite signs. Taking $\Delta_{orb} = 0.1$ meV gives estimation for the required fields for filtering 86 mT.

## 8. Kondo spin valve

The simplest two-terminal spintronic device is spin valve, on which the read heads of hard drives and MRAMs are based. Fig. 13 presents spin valve in which carbon nanotube is attached to two ferromagnetic electrodes. The technology of coupling CNTs to ferromagnetic electrodes e.g. to Co, Fe, Ni , NiPd leads is well elaborated (Cottet et al., 2006) and interesting experimental results have been published, also in the Kondo range (Hauptmann at al., 2008). These data concern however SU(2) symmetry in metallic carbon nanotubes contacted to ferromagnetic leads. There is also rich theoretical literature on Kondo effect in quantum dot coupled to magnetic leads, but with exception of our recent paper (Lipiński & Krychowski, 2010) all of them also apply only to SU(2) symmetry (e.g. (Bułka & Lipiński, 2003; Choi et al., 2004; Martinek et al., 2003; Sergueev et al., 2002; Świrkowicz et al., 2006)). In the following we discuss SU(4) case. To control the transport the dependence on the relative orientation of magnetic moments of the leads is exploited. Polarization of electrodes breaks the spin

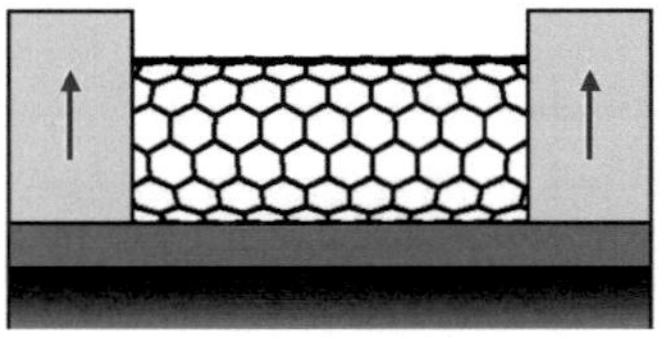

Fig. 13. Schematic view of spin valve. CNT-QD contacted to ferromagnetic electrodes, polarizations of which might be oriented either parallel or antiparallel.

degeneracy. Different tunneling rates for up and down spin electrons result in different widths of Kondo peaks for each spin channel. There is however also more crucial impact of polarization on Kondo resonance, which manifests most strongly close to the charge degeneracy points. Charge fluctuations are spin-dependent and they induce an effective exchange field. To find the spin splitting we use, following (Martinek et al., 2005) the Haldane scaling approach (Haldane, 1978), where charge fluctuations are integrated out, but effectively introduce spin dependent renormalization of the effective dot energies. We do not write here the resulting analytic form of exchange splitting ($\Delta_{exch.} = E_{m+} - E_{m-}$) , which can be found e.g. in (Martinek et al., 2005), but only present on Fig. 14f an example of its gate dependence. We see that not only the magnitude of this splitting, but also the sign changes with gate voltage. Controlling the spin degree of freedom by purely electrical means is currently an important challenge of spintronics. In contrast to an applied magnetic field, it acts rapidly and allows very localized addressing. To control the transport in spin valves the dependence on the relative orientation of magnetic moments of the leads is exploited. Figs. 14a, b present bias dependencies of $\mathcal{TMR}$ for negative and positive exchange fields. Linear $\mathcal{TMR}$ reaches for

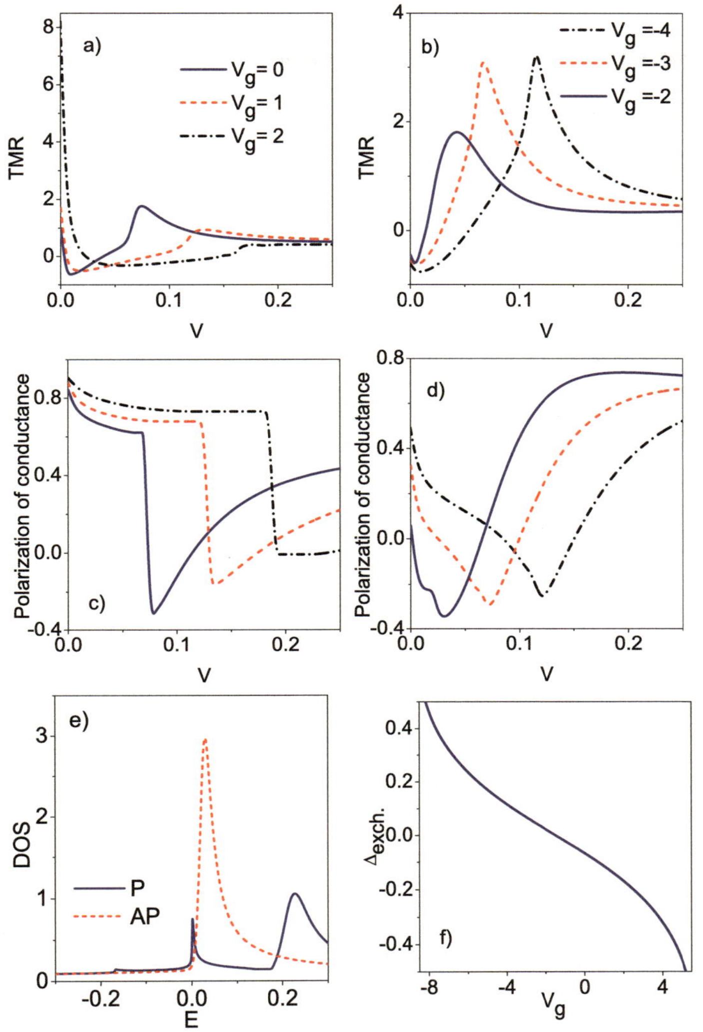


Fig. 14. Bias dependencies of tunnel magnetoresistance for negative (a) and positive (b) exchange splitting. Compare Fig. 14f. c,d) Polarizations of conductance for parallel configuration. Marking of the curves corresponds to the same gate assignment as on Figs a, b) e) Density of states of CNT-QD for parallel and antiparallel configurations f) Gate dependence of exchange splitting for $P = 0.6$, dot parameters are $E_0 = -6, \mathcal{U} = 15$.

$V_g = 2$ ($\Delta_{exch.} = -0.2$) giant value of 800%. Positive value of linear $\mathcal{TMR}$ means dominance of DOS at the Fermi level for parallel orientation of polarizations of electrodes $\varrho_P(E_F)$ relative to DOS for antiparallel configuration $\varrho_{AP}(E_F)$. This is the case presented e.g. on Fig. 14e. In the opposite case ($\Delta_{exch.} > 0$) AP density of states (or transmission) dominates over P transmission what results in negative (inverse $\mathcal{TMR}$). Important message following from the above observation is that one can control $\mathcal{TMR}$ electrically by the change of gate voltage. The sharpness of the peak of DOS for P configuration is the reason of the observed dramatic change of $\mathcal{TMR}$ in the low voltage range leading even to a change of sign (Fig. 14a). For

P configuration the three peak structure is visible for sufficiently high exchange field. The satellites are resolved from the main peak if $|\Delta_{exch.}| > T_K$. The central peak corresponds to orbital fluctuations and the side peaks to spin and spin-orbital fluctuations. For parallel orientation differential conductance sharply increases close to bias voltage equal to exchange splitting $V \sim \Delta_{exch.}$ and it reflects in the occurrence of peaks of $\mathcal{TMR}$. For high voltages magnetoresistance saturates and reaches value close to Juliére limit for uncorrelated electrons (Julliére, 1975). A small difference is a consequence of the influence of charge fluctuations occurring for higher energies (Juliére limit for $P = 0.6$ is $P^2(1 - P^2) = 0.56$). Figures 14c, d present bias dependencies of polarizations of conductance for parallel configuration. Minima of PC coincide with maxima of $\mathcal{TMR}$ and occur for voltages equal to exchange splitting. Due to the asymmetric shape of Kondo satellites (Fig. 14e) and reverse of positions of up and down peaks with the change of the sign of exchange field, modification of character of bias dependence of PC is observed. For positive exchange splitting a gradual change of

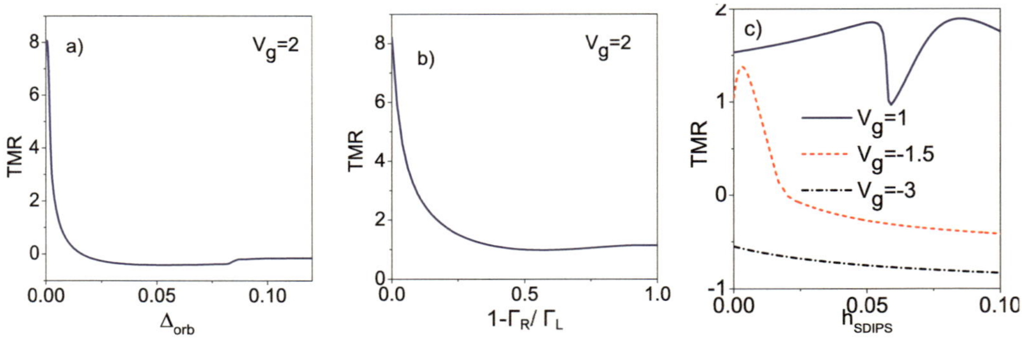


Fig. 15. a) TMR dependence on the orbital level mismatch $\Delta_{orb}$ ($V_g = 2$). b) TMR dependence on asymmetry of coupling $1 - \Gamma_R/\Gamma_R$ ($V_g = 2$). c) Influence of interfacial spin scattering on magnetoresistance of CNT-QD ($P = 0.6, E_0 = -6, \mathcal{U} = 15$).

PC is seen near $V \sim \Delta_{exch.}$ whereas for negative a sharp jump occurs. The corresponding bias dependence of polarization of current, not presented here, would resemble the one discussed earlier for magnetic field (Fig. 11). The question arises how robust is $\mathcal{TMR}$ against geometrical disturbances, since the full symmetric case is not easily accessible in experiment due to residual symmetry breaking perturbations. Two examples of influence of nonmagnetic perturbations on $\mathcal{TMR}$ are shown on Fig. 15 - the effect of level mismatch and impact of asymmetry in the coupling. It is seen that in order to record giant $\mathcal{TMR}$ values ideal nanotubes symmetrically coupled to the leads are desired. Before concluding this section, let us make a remark on another possible contribution to the spin splitting of the conductance peaks. The interface between a ferromagnet and quantum dot can scatter electrons with spin parallel or antiparallel to the magnetization of the lead with different phase shifts. This spin dependence of interfacial phase shift (SDIPS) can significantly modify spin dependent transport in the peculiarity $\mathcal{TMR}$. Phenomenologically one can introduce the effect of SDIPS as Zeeman splitting induced by an effective field $h_{SDIPS}$ (Cottet & Choi, 2006). This can be justified physically on the following ground. For a double barrier system, the ferromagnetic exchange field makes the confinement potential of electrons on the dot spin dependent. This naturally induces a spin dependence of orbital energies (Cottet et al., 2006). Orbital energy $E_{m\sigma}$ Eq. (1) should be supplemented therefore by the term $g\sigma\mu_B h_{SDIPS}$. The introduced effective field depends on the configuration of electrodes and it vanishes for AP configuration with symmetrical coupling. The effect of interfacial spin scattering on $\mathcal{TMR}$ have been discussed by us in (Krychowski & Lipiński, 2008) for the case of vanishing exchange splitting. Here we present the example of calculated magnetoresistance when both exchange splitting and $h_{SDIPS}$ have been taken into account (Fig. 15c). It is seen, that spin activity of the interface can considerably modify $\mathcal{TMR}$ and even change its sign. For gate voltage $V_g = 1$ exchange field is negative and thus acts contrary to the positive spin scattering field what leads to the minimum of $\mathcal{TMR}$ when the two fields compensate. For vanishing exchange splitting ($V_g = -1.5$) in the range of small spin scattering fields $\mathcal{TMR}$ is determined by the difference of transmission rates for opposite spin channels, for higher values of $h_{SDIPS}$, $\mathcal{TMR}$ is determined almost entirely by this field and slightly decreases. Curve corresponding to $V_g = -3$ is an example where the action of one of the fields is amplified by another field and it results in strengthening of inverse $\mathcal{TMR}$. In the simple discussion presented above it was assumed that spin scattering field is gate independent. In general case both exchange and interfacial spin scattering fields are tunable with the gate voltage and the angle between ferromagnetic polarizations and this property could be exploited for manipulating spins and current flowing through the dot.

## 9. Spin currents

We extend our analysis by considering the spin-flip processes in the dot which mix the spin channels. They are represented by a perturbation:

$$\mathcal{H}' = \sum_m \mathcal{R}(d^+_{m+} d_{m-} + h.c) \tag{7}$$

Spin flips may be caused e.g., by transverse component of a local magnetic field. These processes are assumed to be coherent, in the sense that spin-flip strength $\mathcal{R}$ involves reversible transitions. Before discussion of spin currents let us first refer to the influence of spin flip processes on magnetoresistance, which is similar to the earlier mentioned effect of interfacial scattering field. The effect of spin flip transitions is illustrated on Fig. 16. Spin-flip makes

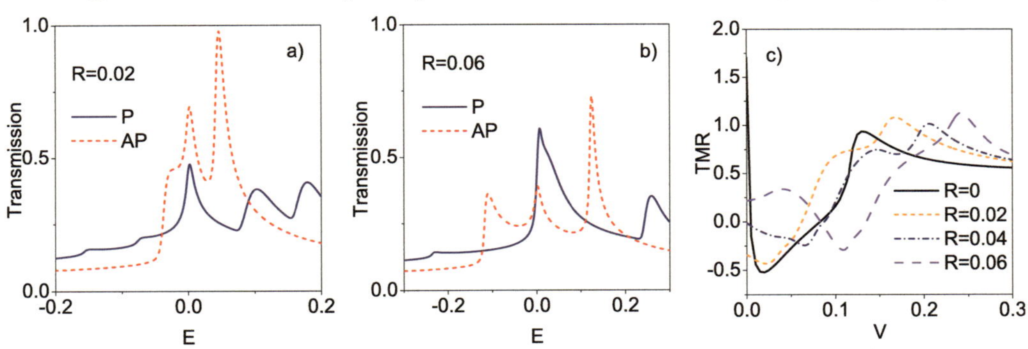


Fig. 16. Impact of spin-flip processes on $\mathcal{TMR}$. a) Transmissions for both spin polarization configurations for the weak spin flip scattering b) Same as a), but for strong spin-flip scattering. c) $\mathcal{TMR}$ for different values of spin-flip amplitude ($P = 0.6$, $E_0 = -5$, $\mathcal{U} = 15$).

the alignment of the lead polarizations less important. The resulting equilibration of the spin population leads to weakened magnetoresistance effect. The detailed behavior of $\mathcal{TMR}$ depends on the relative size of exchange splitting and spin flip transition amplitude. For the specific case shown on Fig. 16 linear $\mathcal{TMR}$ first rapidly decreases and changes sign with the increase of $\mathcal{R}$ and then again becomes positive for larger $\mathcal{R}$. This can be understood looking at the plot of the evolution of the corresponding transmissions for parallel and antiparallel configurations (Figs 16a, b). For AP configuration for strong enough spin-flip amplitude the three peak structure is visible with satellites located roughly at $\omega \approx T_K \pm 2\mathcal{R}$. For P configurations the structure is richer, in the case of both large exchange splitting and strong spin-flip scattering the enhanced transmission is expected for $\omega \approx T_K \pm \Delta_{exch.} \pm 2\mathcal{R}$, but whether the separate peaks are well resolved or not depends on the relative strength of these two perturbations and Kondo energy. For $\mathcal{R} = 0.02$ AP transmission at $E_F$ dominates over the P transmission and negative value of linear $\mathcal{TMR}$ is observed. For $\mathcal{R} = 0.06$ the opposite situation occurs and positive $\mathcal{TMR}$ is seen. The oscillating character of bias dependence of $\mathcal{TMR}$ reflects entering of the succeeding transmission peaks into the transport window. Whether the satellites mark on $\mathcal{TMR}$ curve as a distinct maximum or minimum or only as an inflection point, or are not visible at all, depends on the height of transmission peaks and their mutual separation on the energy scale. So far we have discussed a flow of spin polarized

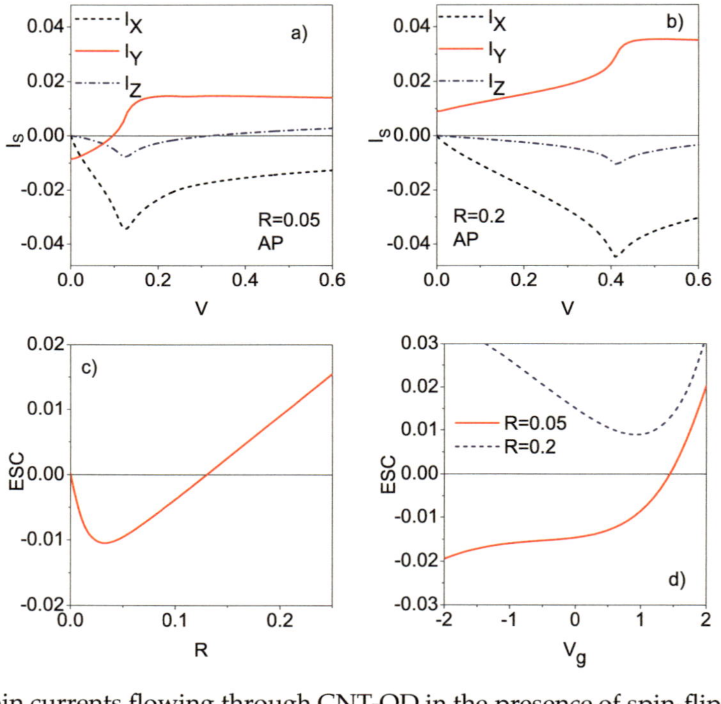


Fig. 17. a,b) Spin currents flowing through CNT-QD in the presence of spin-flip scattering for antiparallel configuration of polarizations of the leads. c) Dependence of equilibrium spin current $\mathcal{I}^y$ on spin-flip scattrering amplitude for AP configuration. d) Gate dependence of equilibrium spin current $\mathcal{I}^y$ for AP configuration ($P = 0.6$, $E_0 = -5$, $\mathcal{U} = 15$).

current through carbon nanotube quantum dot induced by presence of magnetic field or polarization of electrodes. More recently, there has been an increasing interest in generation of pure spin current without an accompanying charge current. Spintronic devices such as transistors (Žutić et al., 2004) require spin currents, just as conventional electronic devices require charge currents. The attractive attribute of spin current is that it is associated with a flow of angular momentum, which is a vector quantity. This feature allows information to be sent across nanoscopic structures. As opposed to charge current a spin current is invariant under time reversal. This property determines the low dissipative or even dissipativeless spin transport (Shen, 2008). Figs. 17a, b show the examples of spin currents calculated for AP configuration. Spin flip perturbation mixes the spin channels and therefore beyond the longitudinal current $\mathcal{I}^z$ also transverse currents appear. The observed minima or rapid change of spin currents occur near $V \sim 2\mathcal{R}$, what corresponds to maxima of AP transmission for this energy (compare Fig. 16a, b). Interesting observation is the occurrence of equilibrium spin current ($\mathcal{ESC}$). For the case discussed it is $\mathcal{I}^y$ component of spin current, which does not vanish for zero bias. The action of spin flip term is equivalent to the operation of magnetic field $h_x$ in $x$-direction. Spin torque $h_x \times \sigma$ acts along $y$-axis, but is oriented in the opposite directions for right and left moving electrons. In consequence, the charge flow in opposite directions is associated with opposite $y$ component of the spin. This happens in equilibrium, where charge current vanishes. Figures 17c, d illustrate that the magnitude and the sign of $\mathcal{ESC}$ is tunable by a gate voltage or magnetic field. It is worth to mention that the discussed system would also generate equilibrium spin current if only one ferromagnetic lead is connected to the dot with spin flips and such a system can be viewed as spin battery (Brataas et al., 2002; Hai et al., 2009). To supplement spin dependent transport characteristics we present on Fig. 18 spin-opposite shot noise. It characterizes the degree of correlations between charge transport events in opposite spin channels. In the absence of spin-flip scattering the currents of spin-up electrons and spin-down electrons are independent, and the cross-correlations between different spin-currents vanish. For simplicity we discuss only the case of parallel orientation of polarizations, when spin cross correlations are characterized by one element $\mathcal{S}_{L+L-}$. The positive values of $\mathcal{S}_{L+L-}$ means mutual amplification of currents in opposite spin

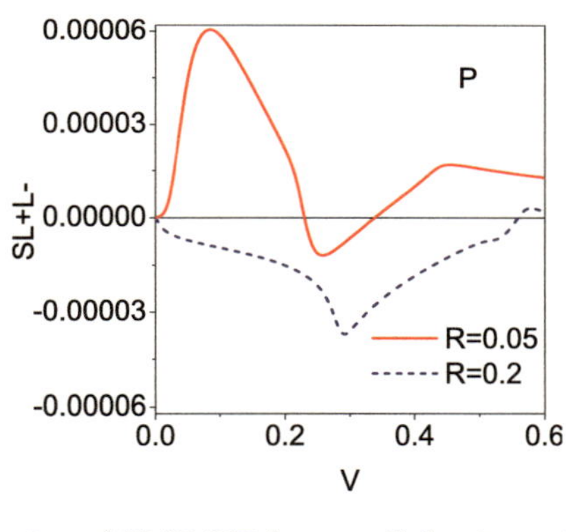


Fig. 18. Spin-opposite shot noise of CNT-QD for parallel orientation of polarizations of the leads $P = 0.6$, $E_0 = -5, \mathcal{U} = 15$.

channels and negative value their mutual weakening. Spin cross-correlations are determined by interference of spin rising and lowering transmissions, which are energy dependent. The decisive factors lowering shot noise is the Pauli exclusive principle and Coulomb interaction. These effects are in general case mixed. Pauli principle however acts only on electrons with the same spin and therefore spin-opposite noise probes interaction induced correlations only.

## 10. Conclusions

Carbon nanotubes present an ideal systems for spintronic applications due to long spin lifetimes and elaborated technology of coupling of them to ferromagnetic electrodes. The

domain sizes are much larger than electrode-nanotube interface and therefore CNTs are supposed to probe a single ferromagnetic domain. This allows for a precise spin control of transport through CNT-QD. Examining transport for different tunneling rates allows understanding of different transport regimes. For weak and intermediate transparency, strong correlations become decisive. Due to higher-order tunneling processes Kondo resonance is formed. This many-body resonance is much narrower than atomic or charge resonances and therefore transport control in this range requires much smaller effective fields. Disadvantage of working in this regime is necessity of use low temperatures. In carbon nanotubes however not only spin but also pseudospin is engaged resulting in highly symmetric SU(4) Kondo effect with considerably increased characteristic temperature. SU(4) Kondo effect is interesting, as the ground state involves a non-trivial entanglement of charge and orbital degrees of freedom. The description of this entanglement is not only a challenge for nanoscopic systems, important from application point of view, but its understanding sheds also a light on similar effects in bulk systems, where additional degeneracy results form crystal symmetry. In ideal semiconducting CNTs the orbital pseudospin and the real spin are indistinguishable, they start to differ in magnetic field. Studying quantum dots allows to analyze a rich aspects of Kondo physics owing to the tunability of relevant parameters of the dots and the ability for driving the system out of equilibrium in different ways.
The aim of this chapter was to give an overview of our current research on the impact of symmetry breaking perturbations on spin polarized transport in CNTs in the Kondo regime. We discussed the effect of magnetic field, polarizations of electrodes and real spin flip processes on the dot. The proposals of low field Kondo spin filter, Kondo spin valve and spin battery have been given. To supplement transport characteristics we have also presented short analysis of spin dependent shot noise. For spintronic devices, such as spin transistors it is necessary and beneficial to investigate the spin-current noise. In these devices, the spin currents rather than the charge currents are used as the carrier information. Investigation of spin-current noise has received very little attention up to now. Noise experiments are difficult to perform, since one needs to detect the shot noise over the background $1/f$ noise caused by fluctuations in the physical environment and measurement equipment. Relatively high Kondo temperature in CNTs is an advantage in possible measuring of shot noise in these systems, because high currents can be applied. To date, only one experiment on the noise in spin-orbital Kondo range of CNT-QD has been carried, and this result indicates that due to entanglement SU(4) Kondo systems remains noisy even in the unitary conductance limit (Delattre et al., 2009). There is still a lack of spin-resolved shot noise measurements, which are interesting because they give unambiguous probe of the electronic interactions. Such experiments seem to be within the reach of present-day measuring techniques, e.g. by spin filtering methods (Frolov et al., 2009), or detecting magnetization fluctuations. in the leads which senses the spin current noise via spin-transfer torque (Foros et al., 2005).

## 11. References

Averin, D. V. & Nazarov, Yu. V. (1992). *Single Charge Tunneling - Coulomb Blokade Phenomena in Nanostructures*, Plenum Press and NATO Scientific Affairs Division, New York

Babić, B., Kontos, T. & Schönenberger, C. (2004). Kondo effect in carbon nanotubes at half filling. *Phys. Rev. B*, 70, pp. 235419-1-9

Baibich, M. N., Broto J. M., Fert A., Nguyen Van Dau, F. & Petroff F., Etienne P., Creuzet, G., Friederich, A. & Chazelas J. (1988). Giant magnetoresistance of (001)Fe/(001)Cr Magnetic Superlattices. *Phys. Rev. Lett*, 61, pp. 2472-2475

Blanter, Ya. M. & Büttiker, M. (2000). Shot noise in mesoscopic conductors. *Phys. Rep.* 336, pp. 1-166

Binasch G., Grünberg, P., Saurenbach, F. & Zinn W. (1989). Enhanced magnetoresistance in layered magnetic structures with antiferromagnetic interlayer exchange. *Phys. Rev. B*, 39, pp. 4828-4830

Brataas, A., Tserkovnyak, Y., Bauer, G. E. W. & Halperin, B. I. (2002). Spin battery operated by ferromagnetic resonance. *Phys. Rev. B*, 66, pp. 060404-1-4

Buitelaar, M. R., Bachtold, A., Nussbaumer T., Iqbal, M. & Schönenberger, C. (2002). Multiwall carbon nanotubes as quantum dots. *Phys. Rev. Lett.*, 88, pp. 156801-1-4

Bułka, B. R. & Lipiński, S. (2003). Coherent electronic transport and Kondo resonance in magnetic nanostructures. *Phys. Rev. B*, 67, pp. 024404-1-8

Büsser, C. A. & Martins, G. B. (2007). Numerical results indicate a half-filling SU(4) Kondo state in carbon nanotubes. *Phys. Rev. B*, 75, pp. 045406-1-7

Choi, M. S., Sanchez, D. & Lopez, R. (2004). Kondo effect in a quantum dot coupled to ferromagnetic leads: A numerical renormalization group analysis. *Phys. Rev. Lett*, 92, pp. 056601-1-4

Choi, M., López, R., Aguado, R. (2005). SU(4) Kondo effect in carbon nanotubes. *Phys. Rev. Lett.* 95, pp. 067204-1-4

Cobden, D. H. & Nygård, J. (2004). Shell filling in closed single-wall carbon nanotube quantum dots. *Phys. Rev. Lett.*, 89, pp.046803-1-4

Coleman, P. (1987). Mixed valence as an almost broken symmetry. *Phys. Rev. B* 35, 5072-5116

Cottet, A. & Choi, M. S. (2006). Magnetoresistance of a quantum dot with spin-active interfaces. *Phys. Rev. B* 74, pp. 235316-1-9

Cottet, A., Kontos, T., Sahoo, S., Man, H. T., Choi, M. S., Belzig, W., Bruder, C., Morpurgo, A. F. & Schönenberger, C. (2006). Nanospintronics with carbon nanotubes. *Semicond. Sci. Technol.*, 21, pp. 78-95

Cronewett, S. M., Oesterkamp, T. H. & Kouwenhoven, L. P. (1998). A tunable Kondo effect in quantum dots. *Science* 291, pp. 540-544

Das Sarma, S. (2001). Spintronics. *American Scientist*, 89, pp. 516-523

De Franceschi, S., Hanson, R., van der Wiel, W. G., Elzerman, J. M., Wijpkema, J. J., Fujisawa, T., Tarucha, S. & Kouwenhoven (2002). Out-of-equilibrium Kondo effect in mesoscopic devices. *Phys. Rev. Lett.* 89, pp. 156801-1-4

de Haas, W. J, de Boer, J. & van den Berg G. J. (1933). The electrical resistance of gold, copper and lead at low temperatures. *Commun. Kamerlingh Onnes Lab.*, No. 233b; *Physica*, 1, pp. 1115-1124

Delattre, T., Feuillet-Palma, C., Herman, L. G., Morfin, P., Berroir, J. M., Fève, G., Plaçais, B., Glattli, D. C., Choi, M. S., Mora, C. & Kontos, T. (2009). Noisy Kondo impurities. *Nature Physics*, 5, pp. 208-212

Dresselhaus, M. S., Dresselhaus, G. & Avouris, Ph. (2001). *Carbon Nanotubes. Synthesis, Structure, Properties, and Applications*, Springer-Verlag, Berlin

Foros, J., Brataas, A., Tsevkovnyak, Y. & Bauer, G. E. W. (2005). Magnetization noise in magnetoelectronic nanostructures. *Phys. Rev. Lett.*, 95, pp. 016601-1-4

Frolov, S. M., Venkatesan, A., Yu, W., Folk, J. A. & Wegscheider, W. (2009). Electrical Generation of Pure Spin Currents in a Two-Dimensional Electron Gas. *Phys. Rev. Lett.* 102, pp. 116802-1-4

Galpin, M. R., Logan, D. E., Krishnamurthy, H. R. (2006). Dynamics of capacitively coupled double quantum dots. *J. Phys.: Condens. Matter*, 18, pp. 6571-6583

Galpin, M. R., Jayatilaka, F. W., Logan, D. E. & Anders F. B. (2010). Interplay between Kondo physics and spin-orbit coupling in carbon nanotube quantum dots. *Phys. Rev. B*, 81, pp. 075437-1-14

Goldhaber-Gordon, D., Shtrikman, H., Mahlul, D., Abuschmagder, D., Meiraev, U. & Kastner, M. A. (1998). Kondo effect in a single-electron transistor. *Nature* 391, pp. 156-159

Grobis, M., Rau, I. G., Potok, R. M., Shtrikman, H. & Goldhaber-Gordon, D. (2008). Universal scaling in nonequilibrium transport through a single channel Kondo dot. *Phys. Rev. Lett.*, 100, pp. 246601-1-4

Grove-Rasmussen, K., Jørgensen, H. J. & Lindelof, P. E. (2007). Fabry-Perot interference, Kondo effect and Coulomb blokade in carbon nanotubes. *Physica E*, 40, pp. 92-98

Hai, P. N., Ohya, S., Tanaka, M., Barnes, S. E. & Maekawa, S. (2009). Electromotive force and huge magnetoresistance in magnetic tunnel junctions. *Nature*, 458, pp. 489-492

Haldane, F. D. M. (1978). Scaling theory of the asymmetric Anderson Model. *Phys. Rev. Lett.*, 40, 416-419

Haug, H. & Jauho, A. P. (1998).*Quantum Kinetics in Transport and Optics of Semiconductors*, Springer, Berlin

Hauptmann, J. R., Paaske, J. & Lindelof, P. E. (2008). Electric-field-controlled spin reversal in a quantum dot with ferromagnetic contacts. *Nature Physics*, 4, pp. 373-376

Hewson, A. C. (1993). *The Kondo problem to heavy fermions*, pp.1-472, Cambridge University Press, ISBN 978-0521599474, Cambridge

Holleitner, A., Chudnovskiy, A., Pfannkuche, D., Eberl, K. & Blick R. H. (2004). Pseudospin Kondo correlations versus hybridized molecular states in double quantum dots. *Phys. Rev. B*, 70, pp. 075204-1-5

Iijima, S. (1991). Helical microtubules of graphitic carbon. *Nature*, 354, pp. 56-58

Iijima, S. & Ichihashi, T. (1993). Single-shell carbon nanotube of 1-nm diameter. *Nature*, 363, pp. 603-605

Jarillo-Herrero, P., Sapmaz, S., Dekker, C., Kouwenhoven, L. P. & van der Zant, H. S. J. (2004). Electron-hole symmetry in a semiconducting carbon nanotube quantum dot. *Nature*, 429, 389-392

Jarillo-Herrero, P., Kong, J., van der Zant, H. S. J., Dekker, C., Kouvenhoven, L. P. & De Franceschi, S. (2005). Orbital Kondo effect in carbon nanotubes. *Nature*, 434, pp. 484-488

Julliére, M. (1975). Tunneling between ferromagnetic films. *Phys. Rev. Lett.*, 54A, pp. 225-226

Kondo, J. (1964). Resistance minimum in dilute magnetic alloys. *Prog. Theor. Phys.*, 32, pp.37-49

Krychowski, D., Lipiński, S. & Krompiewski., S. (2007). Spin dependent transport through a carbon nanotube quantum dot in magnetic field. *J. Alloys and Compd.*, 442, pp. 379-381

Krychowski, D. & Lipiński, S. (2008). Tunnel magnetoresistance in carbon nanotube quantum dots. *Acta Phys. Pol. A*, 113, 545-548

Krychowski, D. & Lipiński, S. (2009). Kondo effect in carbon nanotube quantum dot in a magnetic field. *Acta Phys. Pol. A*, 115, pp. 293-29

Lacroix, C. (1998). Density of states for the Anderson model. *J. Phys. F: Metal Phys.*, 11, pp. 2389-2397

Langreth, D. C. (1966). Friedel sum rule for Anderson's model of localized impurity states. *Phys. Rev.*, 150, pp. 516-518

Liang, W., Bockrath, M., Bozovic, D., Hafner, J. H., Tinkham, M. & Park, H. (2001). Fabry-Perot interference in a nanotube electron waveguide. *Nature*, 411, pp. 665-669

Liang, W., Bockrath, M. & Park, H. (2002). Shell filling and exchange coupling in metallic single-walled carbon nanotubes. *Phys. Rev. Lett.*, 88, pp. 126801-1-4

Liang, W. J., Shores, M. P., Bockrath, M., Long, J. R. & Park (2002). Kondo resonance in a single-molecule transistor. *Nature* 417, pp. 725-729

Lim, J. S., Choi, M., Choi, M. Y., López, R. & Aguado, R. (2006). Kondo effect in carbon nanotubes: From SU(4) to SU(2) symmetry. *Phys. Rev. B* 74, pp. 205119-1-16

Lipiński, S. & Krychowski, D. (2005). Orbital Kondo effect and spin polarized transport through quantum dots. *Phys. Status. Solidi B* 242, pp. 206-209

Lipiński, S. & Krychowski, D. (2010). Spin-polarized current and shot noise in a carbon nanotube quantum dot in the Kondo regime. *Phys. Rev. B*, 81, pp. 115327-1-19

Loss, D. & DiVincenzo, D. P. (1998). Quantum computation with quantum dots. *Phys. Rev. A*, 57, 120-126

Makarowski, A., Zhukov, A., Liu, J. & Finkelstein, G. (2007). SU(2) amd SU(4) Kondo effect in carbon nanotube quantum dots. *Phys. Rev. B*, 75, 241407(R)-1-4

Martinek, J., Utsumi, Y., Imamura, H., Barnaś, J., Maekawa, S., König, J. & Schön, G. (2003). Kondo Effect in Quantum Dots Coupled to Ferromagnetic Leads. *Phys. Rev. Lett.* 91, pp. 127203-1-4

Martinek, J., Sindel, M., Borda, L., Barnaś, Bulla, R., König, J., Schön, G., Maekawa, S. & von Delft, J. (2005). Gate-controlled spin splitting in quantum dots with ferromagnetic leads in the Kondo regime. *Phys. Rev. B.*, 72, pp. 121302-1-4

Mizumo, M., Kim, E. H. & Martins G. B. (2009). Transport and strong-correlation phenomena in carbon nanotube quantum dots in a magnetic field. *J. Phys.: Condens. Matter* 21, pp. 292208-1-5

Ng, T. K. (1996). AC response in the nonequilibrium Anderson impurity model. *Phys. Rev. Lett.*, 76, pp. 487-490

Novoselov, K. S., Geim, A. K., Morozov, S. V., Jiang, D., Katsnelson, M. I., Grigorieva, I. V., Dubonos, S. V. & Frisov, A. A.(2005). Two-dimensional gas of massless Dirac fermions in graphene. *Nature*, 438, pp. 197-200

Nygård, J., Cobden, D. H. & Lindelof, P. E. (2000). Kondo physics in carbon nanotube. *Nature*, 408, pp. 342-346

Paaske, J., Rosch, A., Wölfe, P., Mason, N., Marcus, C. M. & Nygård, J. (2006). Non-equilibrium singlet-triplet KOndo effect in carbon nanotubes. *Nature Physics* 2, pp. 460-464

Park, J., Pasupathy, A. N., Goldsmith, J. I., Chang, C., Yaish, Y., Petta, J. R., Rinkoski, M., Sethna, J. P., Abruna, H. D., McEuen, P. L. & Ralph, D. C. (2002). Coulomb blokade and the Kondo effect in single-atom transistor. *Nature* 417, pp. 722-725

Sasaki, S., Amaha, S., Asakawa, N., Eto, M. & Tarucha S. (2004). Enhanced Kondo effect via tuned orbital degeneracy in a spin 1/2 artificial atom. *Phys. Rev. Lett.* 93, pp. 017205-1-4

Sergueev, N., Sun, Q. F., Guo, H., Wang, B. G. & Wang, J. (2002). Spin-polarized transport through a quantum dot: Anderson model with on-site Coulomb repulsion. *Phys. Rev. B*, 65, 165303-1-9

Shen, S. Q. (2008). Spintronics and spin current. *AAPPS Bulletin*, Vol. 18, No. 5, pp. 29-36

Świrkowicz, R., Wilczyński, M. & Barnaś, J. (2006). Spin-polarized transport through a single-level quantum dot in the Kondo regime. *J. Phys.: Condens. Matter.*, 18, pp. 2291-1-9

Tanaka, M. & Higo Y. (2001). Large tunneling magnetoresistance in GaMnAs/AlAs/GaMnAs ferromagnetic semicontuctor tunnel junction. *Phys. Rev. Lett.*, 87, pp. 026602-1-4

Wallace, P. R. (1947). The band theory of graphite. *Phys. Rev.*, 71, pp. 622-634
Wei, Ch., Cho K. & Srivastava D. (2003). Tensile strength of carbon nanotube under realistic temperature and strain rate. *Phys. Rev. B*, 67, pp. 115407-1-6
Wilhelm, U., Schmid, J., Weis, J. & v. Klitzing, K. (2002). Experimental evidance for spinless Kondo effect in two electrostically coupled quantum dot system. *Phisics E*, 14, pp. 385-390
Wu, F., Danneau, R., Queipo, P., Kauppinen, E., Tsuneta, T. & Hakonen, P. J. (2009). Single-walled carbon nanotube weak links in Kondo regime with zero-field splitting. *Phys. Rev. B*, 79, pp. 073404-1-4
Xiong, Z. H., Wu, D., Valy Vardeny, Z. & Shi, J. (2004). Giant magnetoresistance in organic spin-valves. *Nature*, 427, pp. 821-824
Žutić, I., Fabian, J. & Das Sarma, S. (2004). Spintronics: Fundamentals and applications. *Rev. Mod. Phys.*, 76, pp. 323-410

14

# A Density Functional Theory Study of Chemical Functionalization of Carbon Nanotubes - Toward Site Selective Functionalization

Takashi Yumura
*Department of Chemistry and Materials Technology, Kyoto Institute of Technology, Matsugasaki, Sakyo-ku, Kyoto*
*Japan*

## 1. Introduction

Single walled carbon nanotubes exhibit metallic or semiconducting characters, depending on how to roll-up graphene sheet (Iijima, 1991; Iijima & Ichihashi, 1993; Dresselhaus et al., 1996; Saito et al., 1998). In addition of the unique electronic properties, functionalization of nanotubes can expand the versatility because of enhancing their solubility as well as feature expansion by attached functional groups (Hirsch, 2002; Haddon, 2002; Niyogi et al., 2002; C.A. Dyke & Tour, 2004; Tasis et al., 2006; Prato et al, 2008; Vzquez & Prato, 2009; Strano et al., 2009; Karousis et a;., 2010). The functionalization is roughly categorized into two types; noncovalent- and covalent-functionalization. In covalent functionalization, only highly reactive reagents are available. In a pioneering paper discussing the covalent bond formation, Haddon et al. reported that nanotubes were successfully functionalized by a divalent carbene-derivative $CCl_2$ (Chen et al., 1998; Chen et al., 1998; Kamarás et al., 2003). In the experiments, its divalent atom binds into two carbon atoms of a nanotube surface. After the publication, different types of cyclopropanized nanotube were generated by a single Bingel reaction (Coleman et al., 2003; Worsley et al., 2004; Umeyama et al., 2007). Note that Bingel reaction has been also utilized for selective functionalization of fullerenes (Diederich & Thilgen, 1996). Another type of covalent functionalization of nanotubes is radical additions. In this case, radical species, whose precursors are alkyhalides, diazonium salts, alky-lithium and so on, can make one covalent bond with a nanotube surface (Holzinger et al., 2001; Bahr & Tour , 2001; Bahr et al., 2001; Bahr et al., 2001; Ying et al., 2003; Saini et al., 2003; Peng et al., 2003; Stevens et al., 2003; Peng et al., 2003).
From a viewpoint of the electronic properties of nanotubes, covalent functionalization can modulate their band structures. If their bands in the vicinity of the Fermi level are significantly perturbed upon the functionalization, one can detect a sizable change of physically observable values, such as the conductivity. Recent theoretical studies evaluated the conductance of monovalent or divalent functionalized nanotubes (Park et al., 2006; Lee & Marzari, 2006; Lpez-Bezanilla et al., 2009). The theoretical findings suggested that divalent functionalization can preserve the conductance of its pristine nanotube, whereas monovalent case destroys it (Park et al., 2006; Lpez-Bezanilla et al., 2009). The theoretical

results inspired experimentalists to measure the conductance of functionalized nanotubes. In fact, changes of the conductance of nanotubes upon the functionalization were observed experimentally (Goldsmith et al., 2007; Kanungo et al., 2009).

Although our knowledge on functionalization of nanotubes has been expanded, "site-selective" functionalization in either divalent or monovalent manners has been still a challenging target in nanotube chemistry. The difficulty in achieving site selective functionalization comes from high reactivity of functional groups. Of course, higher reactive species are necessary to attach to a rigid tube-bond. In an opposed action they would attack randomly at almost equivalent tube-bonds. If one can control binding sites for carbene-derivatives on a nanotube surface to pattern the functional groups, such functionalization can open the door toward designing novel tube-based nanomachines. For example, Globus et al. proposed that "paddle wheels" can be constructed conceptually by addition of benzynes onto a nanotube circumferentially, whereas the addition in the axial direction results in "gear teeth" (Han & Globus, 1997; Globus et al., 1998). Accordingly, selective functionalization of nanotubes is now required to design the unique nanomachines.

For the purpose of achieving site-selective functionalization of nanotubes, it is indispensable to elucidate how functional groups interact with an outer or inner surface of a nanotube, as well as how the resultant interactions perturb its $sp^2$ bonding frameworks. In this direction, recent quantum chemistry calculations would help to obtain an insight of changes of nanotube surfaces upon the functionalization. In particular, recent advances in *ab initio* density functional theory (DFT) calculations, where Khon-Sham equations including electron correlation are solved iteratively (Hohenberg & Kohn, 1964; Kohn & Sham, 1965; Parr & Yang, 1996), allow us to accurately calculate large-scale systems like nanotubes. Enhancing the reliability of DFT calculations is due to improvement in the quality of exchange-correlation functionals (Becke, 1988; Lee et al., 1988; Becke, 1992; Becke, 1992; Becke, 1993; Stephens et al., 1994; Vosko et al., 1980). Actually, recent DFT methods can provide reliable data, such as energies and force in materials. Thus one can utilize the DFT calculations to look at CC bondings in nanotubes "on an atomic level", and to devise a plausible strategy toward site-selective functionalization of nanotubes.

Because of retaining the unique electronic properties of nanotubes after the divalent functionalization, we discuss in this chapter the properties of nanotubes functionalized by carbene-derivatives from a viewpoint of computational chemistry based on DFT calculations. The aim in this study is finding out an approach to how site-selective functionalization of nanotubes can be achieved. This chapter is organized as follows. In section 2.1, we briefly present how computational methods influence a description of its CC bonding network by investigating the addition of carbene into naphthalene as a simple test case. After the discussion, results of mono- or double functionalization of nanotubes are given. First, we discuss chemical reactivity of inner or outer tube surface toward carbene with the focus on surface modification by the addition in section 2.2. In sections 2.3, we propose two strategies for site-selective double addition into a nanotube, based on the findings in section 2.2. In section 2.3.1, we investigate whether surface modification induced by the inner carbene addition can direct the next carbene into a specific site. On the other hand, another strategy is introduced in section 2.3.2; the use of geometrical constraints of a bismalonate for site-selective addition into a nanotube. Finally we conclude this chapter in Section 3.

## 2.1 Choice of computational methods

Prior to discussing functionalization of nanotubes by carbene-derivatives, we briefly explain computational methods suitable to accurately describe the interactions. As a simple and relevant test case, we focused on the addition of carbene into naphthalene with ten $\pi$ electrons (Choi & Kertesz, 1998). In the test case, we employed three different calculations; Hartree-Fock (HF), the second-order Møller-Plesset perturbation theory (MP2) (Szabo & Ostlund, 1996), and density functional theory (DFT) methods. In the DFT calculations, GGA-based PW91 functional and hybrid B3LYP functional are considered as a standard functional. When the divalent atom of carbene (labeled by C11) attacks at the bridgehead atoms of naphthalene (labeled by C1 and C6), the valence tautomerization of 1,6-methano[10]annulene occurs, as shown in the upper of Figure 1.

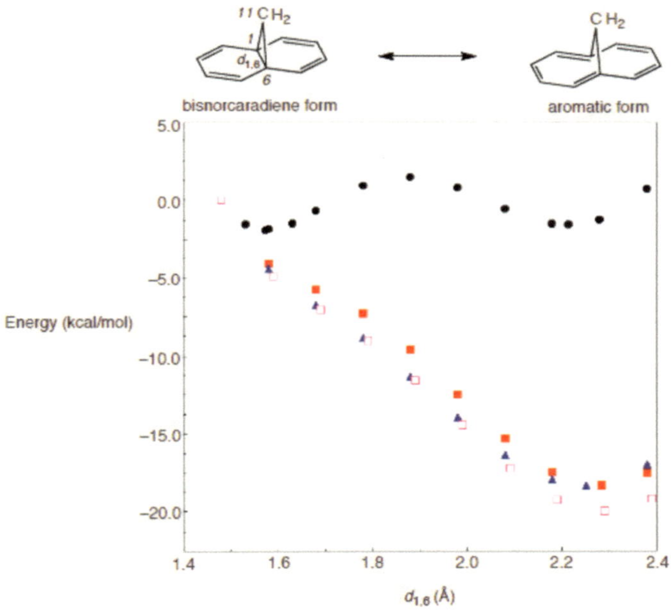


Fig. 1. Total energies of 1,6-methano[10]annulene as a function of the $d_{1,6}$ value, obtained from Hartree-Fock (black circles), second order Møller-Plesset perturbation theory (MP2) (blue triangles), and density functional theory calculations. PW91 and B3LYP functionals (purple open and red closed squares, respectively) are considered in the DFT calculations. The basis set used is the 6-31G* basis set.

In the tautomerization, there are two types of structure; one is bisnorcaradiene form, and the other is aromatic form. As shown in Figure 1, the two isomers can be distinguished by the separation between the C1 and C6 atoms (the $d_{1,6}$ bond); the bisnorcaradiene form has a shorter $d_{1,6}$ bond, whereas the aromatic form has a longer $d_{1,6}$ bond. Reflecting the $d_{1,6}$ values, the two tautomers have different number of $\pi$ electrons in naphthalene-moiety. In the aromatic form, ten $\pi$ electrons remain in its naphthalene-moiety, while two $\pi$ electrons transfer from naphthalene to form two CC bonds in the bisnorcaradiene form.

The $d_{1,6}$ value is not only a geometrical parameter, but it is key in relative stability of the two tautomers of 1,6-methano[10]annulene. Figure 1 displays total energies of 1,6-methano[10]annulene against the $d_{1,6}$ value, depending on computational methods (HF,

MP2, and DFT (B3LYP and PW91 (Perdew & Wang, 1992) methods). As shown in Figure 1, there are two local minima that are energetically equivalent in the HF calculations. The minima correspond to the bisnorcaradiene and aromatic forms. However, we obtained from DFT and MP2 calculations only an aromatic form as a stable conformation. The calculated $d_{1,6}$ values are around 2.3 Å, which can reproduce that obtained experimentally (2.235 Å) (Bianchi et al., 1980). The DFT and MP2 methods include electron correlation, but the HF method does not include. Thus, the differences indicate that electron correlation should be important in relative stability of the two tautomers, and describing the geometrical change of naphthalene induced by the interactions with carbene.

In general, MP2 calculations are more time-consuming than DFT calculations. Since nanomaterials, such as nanotubes, contain few hundred of atoms in the unit cell, MP2 methods are not doable to estimate the properties of nanomaterials. Actually our systems introduced in this chapter contain up to 240 carbon atoms of a nanotube plus a few functional groups in the unit cell in periodic boundary condition (PBC) calculations. Instead, DFT methods, which can describe their properties in a relatively accurate manner, are less time-consuming. Accordingly, we will investigate interactions between carbene-derivatives and a nanotube surface by using DFT methods with B3LYP or PW91 functionals.

### 2.2 Addition of carbene into a nanotube, and concomitant surface modification

Let us first discuss the addition of carbene into the outer (Chen et al., 2004; Bettinger, 2004; Zhao et al., 2005; Bettinger, 2006; Zheng et al., 2006; Yumura & Kertesz, 2007; Yumura et al., 2007; Lee & Marzari, 2008) or inner (Yumura & Kertesz, 2007; Yumura et al., 2007) surface of the (5,5) or (10,10) nanotube, whose optimized geometries are seen in Figures 2 and 3.

As mentioned above, the divalent atom of carbene binds into a CC bond of $sp^2$ systems. The (5,5) and (10,10) nanotubes have two types of CC bond (a CC bond orthogonal or slanted relative to the tube axis). Accordingly, there are two types of optimized structure for an inner (outer) carbene binding into an armchair nanotube. Figures 2 and 3 show the preferences of an orthogonal site over a slanted site as the single carbene addition, irrespective of an inner or outer addend. In terms of the geometrical features, a slanted bond retains after the addition of an inner or outer carbene, as shown in Figures 2 and 3 (Yumura & Kertesz, 2007; Yumura et al., 2007). When carbene binds into an orthogonal site, different behaviors were found between the inner and outer surfaces. The inner addition retains the orthogonal bond, whereas the outer addition breaks it. The different behaviors come from more rigorous restriction of surface modification toward the tube center. This is understandable from a strain analysis based on π-orbital axial vector (POAV) (Haddon, 1998), where the strain energy is proportional to the square of a POAV value. In fact, a POAV value would increase by the surface modification toward the center. Then the strain energy would be more pronounced, if carbene freely binds into the inner surface rather than into the outer surface. As a result, the concave nanotube surface prohibits inner carbene from freely binding into the inner surface to keep the CC bond at the binding site.

Breaking or retaining the CC bond at the binding site of a nanotube is reminiscent of the valence tautomerization of 1,6-methano[10]annulene. From an analogy to 1,6-methano[10]annulene whose tautomers have different number of π electrons, we can understand the modification of a nanotube upon the carbene addition obtained from DFT

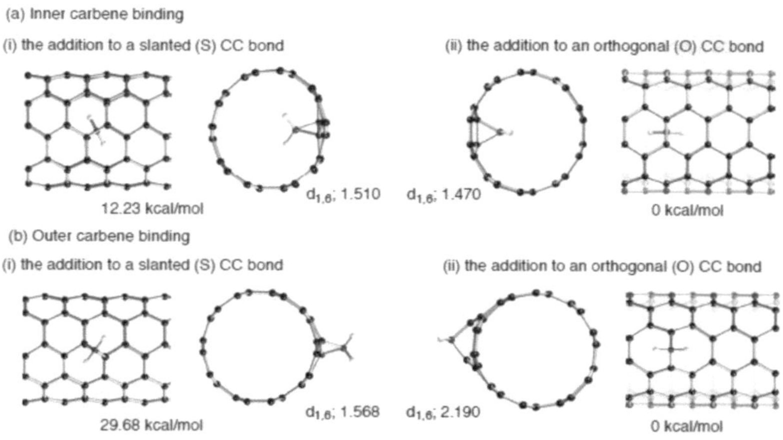


Fig. 2. Optimized structures for the addition of carbene into an inner or outer surface of the (5,5) nanotube. Their relative energies are also given in kcal/mol, lengths of the CC bond at the binding site (the $d_{1,6}$ values) in Å. These values were obtained from PW91 PBC calculations (Yumura & Kertesz, 2007).

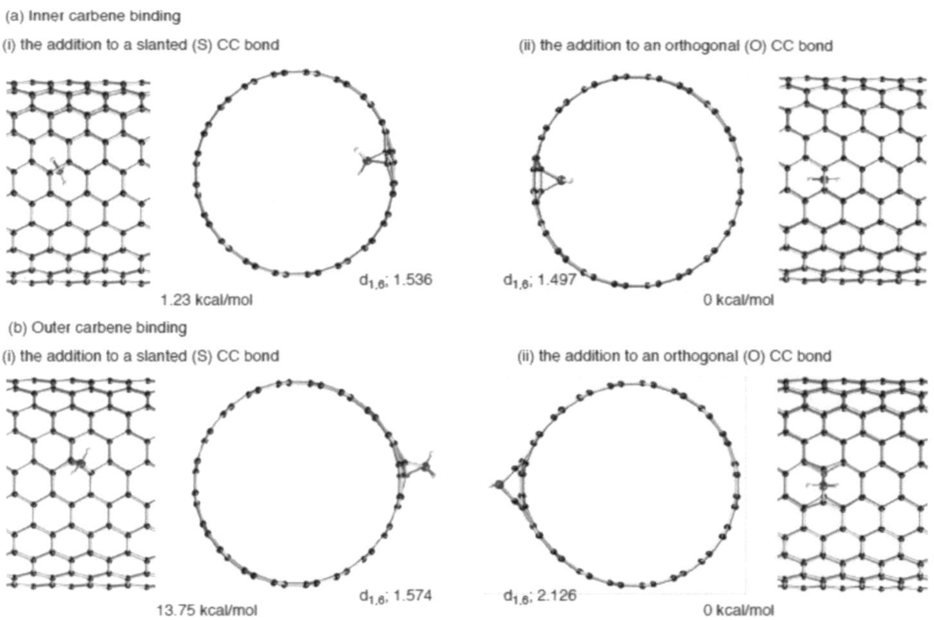


Fig. 3. Optimized structures for the addition of carbene into an inner or outer surface of the (10,10) nanotube. Their relative energies are also given in kcal/mol, lengths of the CC bond at the binding site (the $d_{1,6}$ values) in Å. These values were obtained from PW91 PBC calculations (Yumura et al., 2007).

calculations. To interpret the surface modification, we applied a basic concept in organic chemistry, the Clar concept (Clar, 1972). Based on Clar valence bond (VB) representation, all $\pi$ electrons in the pristine armchair nanotubes can be given by aromatic sextet (Matsuo et al.,

2003; Ormsby & King, 2004; Ormsby & King, 2007). In the inner addition into a nanotube where the $d_{1,6}$ CC bond retains, two π electrons migrate from the nanotube. Due to the π electron migration, the CC bondings of the nanotube attached by inner carbene are significantly perturbed. Based on the Clar representation, the perturbed surface consists of two butadiene (B) patterns near the binding site plus quinonoid (Q) pattern in the circumferential direction (Figure 4(a)).

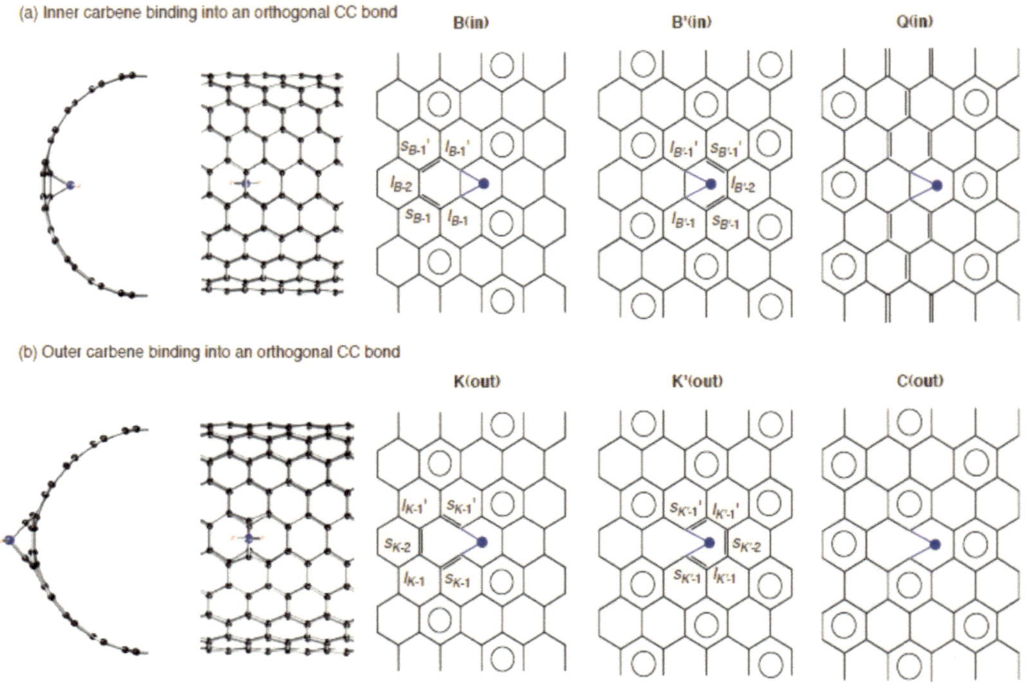


Fig. 4. Clar representation of modified nanotube surface upon the functionalization by inner (a) or outer (b) carbene. Aromatic sextets are given by circles. Note that pristine armchair nanotubes have all π electrons represented by aromatic sextets.

Basically, the DFT-optimized geometries for the nanotubes attached by inner $CH_2$ are explainable from the Clar representation, as shown in Table 1 and Figure 5. Actually, bond-length alternations in the circumferential direction follow the Q pattern in Figure 5, but the alternations are significant only near the binding site. Similar perturbation patterns can be seen in a defected nano-peapod (Figure 6) where a defected $C_{60}$ ($C_1$-$C_{59}$) makes two covalent bonds with the inner surface of the (10,10) nanotube (Yumura et al., 2007). Intrinsic reaction coordination (IRC) analyses based on B3LYP calculations suggest that $C_{60}$ directly converts into a $C_1$-$C_{59}$ defect as the initial intermediate with a barrier of 8.4 eV (Yumura et al., 2007). Experimentally, the formation of defective fullerenes has been actually observed under electron irradiation in transmission electron microscopy measurements (TEM) (Urita et al., 2004; Sato et al., 2006). Thus, one can utilize the activation of guest molecules (e.g. by heating-treatment or electron irradiation) to locally perturb a host nanotube surface from inside out.

| | $l_{B-1}$ | $s_{B-1}$ | $l_{B-2}$ | $s_{B-1}'$ | $l_{B-1}'$ | $l_{B'-1}$ | $s_{B'-1}$ | $l_{B'-2}$ | $s_{B'-1}'$ | $l_{B'-1}'$ |
|---|---|---|---|---|---|---|---|---|---|---|
| (5,5) | 1.471 | 1.402 | 1.470 | 1.402 | 1.471 | 1.470 | 1.403 | 1.471 | 1.401 | 1.470 |
| (10,10) | 1.468 | 1.399 | 1.454 | 1.399 | 1.468 | 1.469 | 1.399 | 1.454 | 1.398 | 1.469 |

Table 1. Optimized bond-lengths (Å) within the two six-membered rings nearest to the site for the binding of inner carbene along the axis of a (5,5) or (10,10) tube according to PW91 PBC calculations. Definition of bond types is given in Figure 4.

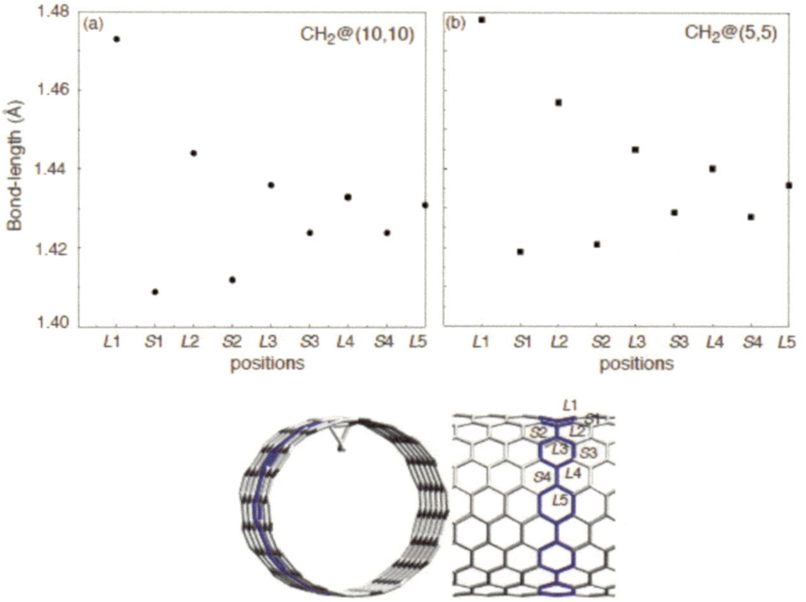


Fig. 5. Bond length alternations of quinonoid patterns in the circumferential direction of the (5,5) or (10,10) nanotube functionalized by inner carbene. The bond lengths were obtained from PW91 PBC calculations. L and S indicate longer and shorter bonds in the quinonoid pattern in Figure 4, respectively (Yumura & Kertesz, 2007; Yumura et al., 2007).

In contrast, the outer carbene addition cleaves the $d_{1,6}$ CC bond at an orthogonal site. In this situation, $\pi$ electrons do not migrate, instead $\sigma$ electrons participate in the formation of the new covalent bonds. As a result, Clar patterns of the nanotube remain almost unchanged after the outer addition, except for the two six-membered rings nearest to the binding site in the axial direction, where Kekulé patterns appear (Table 2). The surface modification patterns in the inner or outer carbene additions are completely different from those in the alkyl radical. additions (Yumura, 2011). In the outer radical additions, $D_{3h}$-like deformation patterns appear due to the formation of one covalent bond. In contrast, an inner radical prefers energetically to separate from a tube rather than to form one covalent bond together with modifying the surface structure. This result indicates the inertness of the inner surface of a nanotube, because the destabilization due to the surface modification dominates the stabilization due to the weak covalent bond formation. Lower reactivity of inner surface toward H or F atom than outer surface was also reported Chen et al. (Chen et al., 2003)

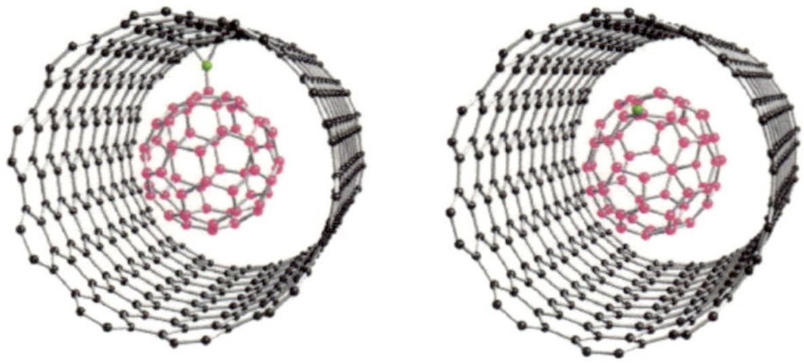

Fig. 6. Two types of optimized geometry for defected nanopeapod ($C_1$-$C_{59}$@(10,10) tube). Two covalent bonds are formed between a defected fullerene ($C_1$-$C_{59}$) and the nanotube in the left-hand side compound, whereas the right-hand side compound does not have such inner covalent bonds. The reactive $C_1$ atom in the defected fullerene is given by green, and the $C_{59}$ moiety by purple. The optimized structures were obtained from PBC PW91 calculations (Yumura et al., 2007) as well as a B3LYP cluster approach (Yumura et al., 2007).

| | $s_{K\text{-}1}$ | $l_{K\text{-}1}$ | $s_{K\text{-}2}$ | $l_{K\text{-}1}'$ | $s_{K\text{-}1}'$ | $s_{K'\text{-}1}$ | $l_{K'\text{-}1}$ | $s_{K'\text{-}2}$ | $l_{K'\text{-}1}'$ | $s_{K'\text{-}1}'$ |
|---|---|---|---|---|---|---|---|---|---|---|
| (5,5) | 1.414 | 1.451 | 1.444 | 1.452 | 1.410 | 1.410 | 1.451 | 1.444 | 1.452 | 1.411 |
| (10,10) | 1.414 | 1.444 | 1.448 | 1.445 | 1.415 | 1.414 | 1.445 | 1.448 | 1.445 | 1.413 |

Table 2. Optimized bond-lengths (Å) within the two six-membered rings nearest to the site for the binding of outer carbene along the axis of a (5,5) or (10,10) tube according to PW91 PBC calculations. Definition of bond types is given in Figure 4.

### 2.3 Site-selective double additions

Considering the surface modification created by the single carbene addition, let us devise plausible strategies for enhancing site-selectivity of double divalent additions. The CC bonds in pristine nanotubes are nearly equivalent in terms of the length and electron density. Because of the featureless CC bonds in nanotubes, carbene derivatives cannot select a specific CC bond as a binding site. On the other hand, the inner carbene addition into a nanotube makes a great difference in the seamless $sp^2$ bondings; double- or single-bond characters appear only near the binding site. Thanks to increasing or decreasing electron population on CC bonds at the limited region, carbene can bind selectively into a CC bond with richer electron density. Therefore, "local" modification of a nanotube by the inner carbene addition can play an important role in site preferences for the second addition, as shown in Scheme 1 (Yumura & Kertesz, 2007). In this situation, we do not need to consider repulsion between the two carbenes.

Scheme 1

In contrast, such site-preferences for the addition of two outer carbenes into a nanotube are not expected. The most important reason is that a nanotube functionalized by outer carbene has Clar patterns plus Kekulé patterns only near the binding site. The double bonds in Kekulé patterns may be a good candidate for the binding site of the second carbene. However, the second carbene cannot add to a double CC bond of the six-membered rings with Kekulé patterns, because of repulsion between the two carbenes. Thus, conformational restrictions in functional groups with divalent atoms are necessary for site-selective functionalization of nanotubes. Here, we consider a bismalonate with a 2,3-butanediol tether as a functional group possible to bind selectively into a nanotube, drawing a direct line with fullerene chemistry reported by the Diderlich groups (Kessinger et al., 2000). Note that the bismalonate can selectively bind into fullerene to form only cis-3 adduction, as shown in Scheme 2.

Scheme 2

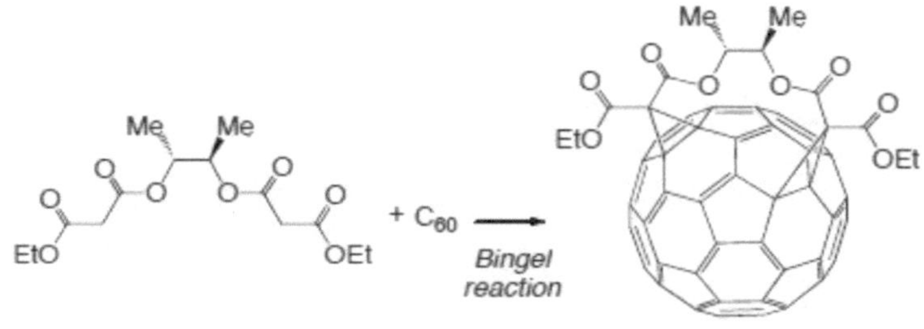


Thus, we investigate the addition of the bismalonate into a nanotube as another plausible approach (Yumura & Kertesz, 2009). In the following sections, we separately discuss the two strategies for the site-selective functionalization of nanotubes.

### 2.3.1 Cooperativity of double carbene additions

In this section, we discuss whether the surface modification induced by the first inner addition influences site-preferences for the second addition. To quantitatively look at the double addition (Scheme 1), two key values are the binding energy for the double additions ($E_{double\text{-}bind}$), and the interaction energy ($E_{interact}$), defined as follows,

$$E_{double\text{-}bind}(X) = E_{total}(C11(X)\text{–}NT\text{–}C11(0)) - E_{total}(NT) - E_{total}(substituent(s)) \quad (1)$$

and

$$E_{interact}(X) = E_{double\text{-}bind}(X) - E_{single\text{-}bind}(X) - E_{single\text{-}bind}(0) \quad (2)$$

where X and 0 represent sites for the second and first carbene bindings, respectively, and C11(X)–NT–C11(0) represents a nanotube attached by two divalent C11 atoms of carbenes or carbene-derivatives, and $E_{single\text{-}bind}(X)$ and $E_{single\text{-}bind}(0)$ are the binding energies for the single additions of the second and first carbenes, respectively. The interaction energy ($E_{interact}$) indicates how much the first addition can have the power to have an impact on site preferences for the second attachment through the concomitant surface modification. A negative (positive) $E_{interact}$ value indicates that the second addition into a site X is stabilized (destabilized) by the first inner addition. When two attached carbenes are far, the two carbenes do not exert influence each other. Accordingly, an $E_{interact}$ value would be negligible.

Here we estimated the $E_{double\text{-}bind}$ and $E_{interact}$ values at various sites for the second attachment after the first inner carbene addition into an orthogonal bond (Yumura & Kertesz, 2007). Considering the local modification by the first attachment, we choose possible sites (X) for the second binding site (Figure 7). We define an orthogonal binding site for the second $CH_2$ molecule (outer surface) as $O_p(q)$ relative to the binding site $O_0(0)$ for the first $CH_2$ molecule (inner surface). Here $p$ represents the order of the carbon belt of $O_p(q)$ site, lying in a plane perpendicular to the axis, from the $O_0(0)$ site with respect to the armchair framework in the axial direction, and $q$ represents the order of the orthogonal bond $O_p(q)$ with respect to the $O_0(0)$ site in the circumferential direction. In slanted binding sites, the $S_{p1-p2}(q)$ sites are between the $p_1$-st and $p_2$-nd carbon belts.

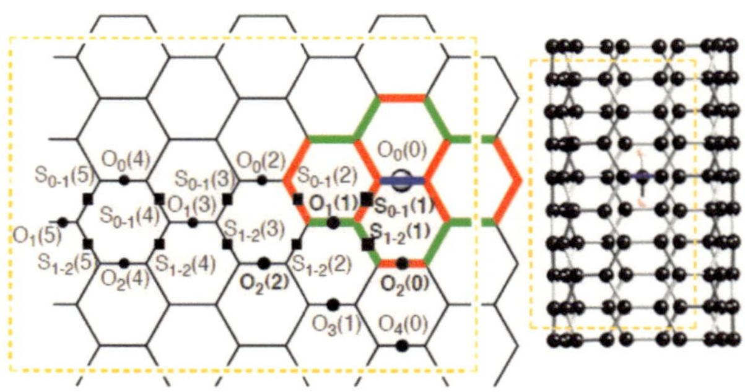


Fig. 7. Definition of sites for the second divalent C11 atom after the first divalent addition. The first addition site is given by $O_0(0)$. Based on Figure 4, double bond characters created by the first attachment are given by green, and single bond characters by red.

As a result of DFT calculations, we found that some configurations of the double addition have a negative $E_{interact}$ value (Yumura & Kertesz, 2007). The most noticeable $E_{interact}$ value was calculated at the $S_{1\text{-}2}(1)$ site to be around -25 kcal/mol. More importantly the effects of the first attachment on stabilizing the second attachment are limited near the first binding site. For example, energetically stable configurations in Figure 8 have significant $E_{interact}$ values ranging from -8.0 to -5.2 kcal/mol. Then the stabilized sites for second attachment have double-bond characters or are between double bonds. In other words, the second carbene attacks at a specific CC bond on which electron population enhances by the first attachment, as shown in Figures 4 and 7. The results strongly indicate cooperative behaviors between the first inner and second outer carbenes through the local surface modification. Thus DFT calculations demonstrate that the locally perturbed surface by the inner attachment, whose CC bond lengths change by up to 0.048 Å, has a strong impact of site-preferences for the second attachment.

Contrary, similar cooperative effects cannot be found between two carbenes binding into the outer surface of the (10,10) nanotube, based on DFT calculations (Yumura & Kertesz, 2009).

In the double outer additions, possible 20 sites for the binding of second carbene were considered. The $E_{doube\text{-}bind}$ and $E_{interact}$ values are plotted in Figure 9 as a function of the separation between the two divalent C11 atoms. As shown in Figure 9(a), we found that second orthogonal bindings are energetically favorable relative to slanted bindings, judging from the $E_{double\text{-}bind}$ values. Furthermore, DFT calculations found that four orthogonal sites have substantial $E_{interact}$ values with a negative sign in Figure 9(b) (-5.7 ~ -2.5 kcal/mol)

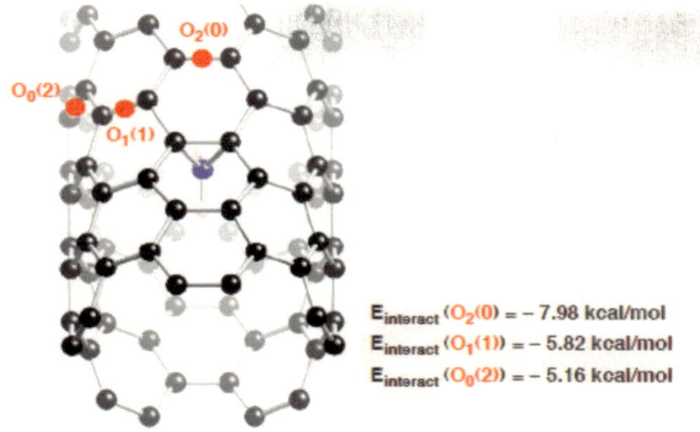


Fig. 8. Stable conformations of the binding of the second carbene into the outer surface of the (5,5) nanotube after the first inner addition at an orthogonal bond. The divalent atom of the first attachment is given by blue, and relatively stable sites for the second attachment by blue. The $E_{interact}$ values were obtained from PW91 PBC calculations (Yumura & Kertesz, 2007).

(Yumura & Kertesz, 2007). However, the $E_{interact}$ values are less pronounced than those in the second addition after the first inner addition. Despite negative $E_{interact}$ values at some sites, we found significant positive $E_{interact}$ values at a certain site, meaning that the first carbene prohibits the second carbene from approaching into the sites. Of course, the destabilized effects are due to steric repulsion between the two carbenes. The results suggest that the second carbene cannot bind into a specific CC bond from the others, being in sharp contrast to the inner addition. This is reasonable because Clar patterns of the pristine nanotube remain almost unchanged after the first outer addition.

### 2.3.2 Site-selective addition of a bismalonate into nanotube

Toward enhancing site-selectivity for the addition of carbene-derivatives into the outer surface of a nanotube, some conformational restrictions of carbene-derivatives would be necessary. To find out a suitable carbene-derivative in this direction, fullerene chemistry can provide helpful hints. Here we pay attention to a successful case of regioselective functionalization of fullerene, reported by Kessinger et al. (Kessinger et al., 2000) They used diethylbutane-2,3-diyl bismalonate as a functional group of fullerene, as shown in Scheme 2. According to their report, a flexible spacer of 2,3-butanediol should be responsible for the selective functionalization. With respect to Bingel reaction, Gao et al. investigated the possible mechanism for the reaction between $CCl_3^-$ and $C_{60}$, and then compared it with the corresponding carbene reaction mechanism (Gao et al., 2009). According to the B3LYP calculations, both mechanisms are competitive in energy. In addition, Bettinger also analyzed carbene reaction between $CCl_2$ and a finite-length nanotube, and found that a transition state for the reaction lies only a few kcal/mol above the dissociation limit toward the nanotube and $CCl_2$ (Bettinger, 2006).

Following the previous experimental and theoretical studies, we recently performed extensive DFT calculations to elucidate the stability of 12 configurations of cycloaddition of diethylbutane-2,3-diyl to the outer surface of a nanotube (Yumura & Kertesz, 2009). Then,

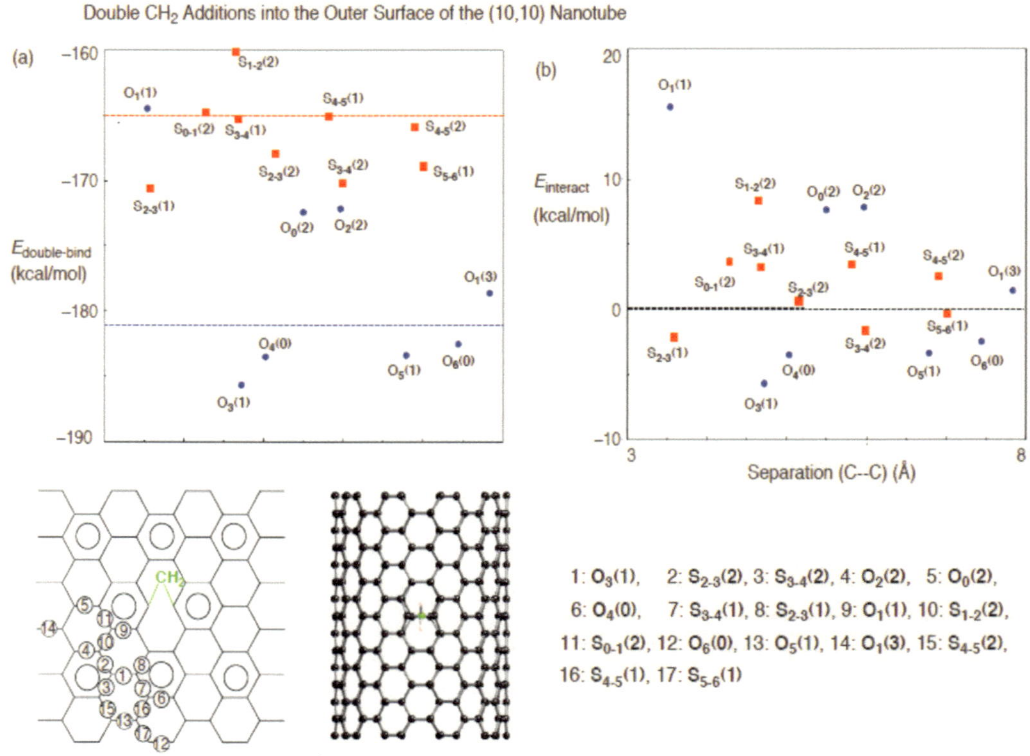


Fig. 9. The $E_{double\text{-}bind}$ (a) and $E_{interact}$ (b) values in (10,10) nanotube attached by two outer carbenes as a function of the separation between the divalent C11 atoms. Orthogonal additions are given by blue, and slanted additions by red.

we cannot unfortunately obtain a transition state for the carbene mechanism due to computational limitation. In the analysis, we used for simplicity a model of the bismalonate in Chart 1 where terminal Et groups are replaced with the H atoms. DFT calculations found two energetically stable configurations of bismalonate functionalized nanotubes in Figure 10.

The energy difference between the two configurations is 6.5 and 4.7 kcal/mol for the finite-length and infinite-length model calculations, respectively. The energy difference is more pronounced by double addition of outer carbenes. The results suggest that replacement of $CH_2$ by bismalonate enhances site-selectivity of bisfunctionalization of the nanotube.

There are several geometrical factors affecting the binding of a bismalonate with the 2,3-butanediol tether. One of the most important factors is, of course, conformational restriction due to the existence of the 2,3-butanediol tether. The restriction is understandable, because the rotation around the dihedral angle O-C-C-O ($\theta$) of the 2,3-butanediol tether costs energy, as shown in Figure 11 displaying its total energy changes against $\theta$. In fact, Figure 11 shows that there is a barrier of ~5.0 kcal/mol between two local minima in the range of $-140 < \theta < 140$. The rotation of the 2,3-butanediol tether also links to the separation between the two C11 atoms, and thus the tether constraint prohibits the two C11 atoms from freely binding into the nanotube.

Chart 1 Bismalonate with 2,3-butanediol tether and its model.

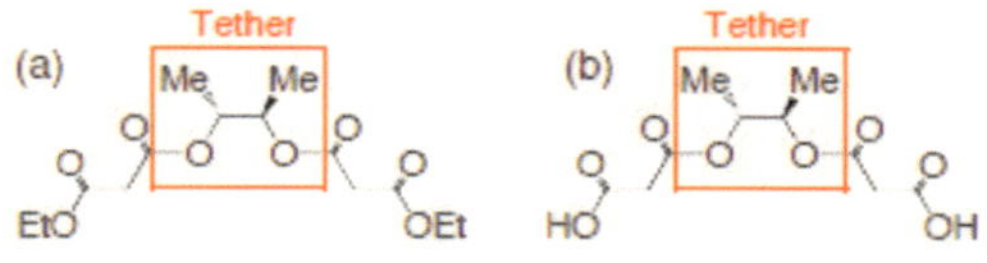


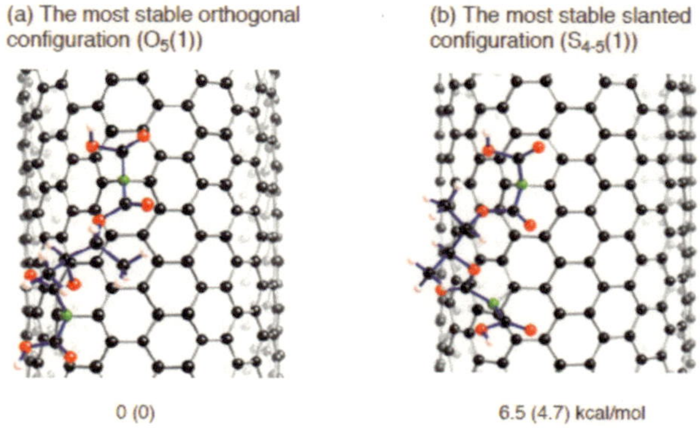


Fig. 10. Two stable conformations of the binding of the bismalonate with 2,3-butanediol tether into the outer surface of the (10,10) nanotube. The divalent atoms are given by green, oxygen atoms by red, and hydrogen atoms by white. Bonds of the bismalonate are given by blue. The optimized geometries were obtained by PW91 calculation based on PBC and cluster-model approaches. Relative energy obtained from a cluster approach is given, and that from PBC approach in parenthesis (Yumura & Kertesz, 2009).

Another factor differentiating the bismalonate addition from the double carbene addition is that two carboxy substituents affect the aromatic-bismalonate tautomerization in terms of energetics. To look at the importance of binding orientations of carboxy groups, we optimized dicarboxyl-methanonanotube and 11,11-dicarboxy-1,6-methano[10]annulene, where dicarboxycarbene binds into the (10,10) nanotube and naphthalene, respectively. Scanning total energy in 11,11-dicarboxy-1,6-methano[10]annulene along the $d_{1,6}$ bond (Figure 12) indicates that types of the two carboxyl orientations strongly influence the stability of the aromatic and bismalonate forms: double minimum at two carboxyl groups twisted with respect to a mirror plane, whereas single minimum at two carboxyl groups being symmetric.

A different view of Figure 12 suggests that at a longer $d_{1,6}$ bond range, the symmetric form is only available, whereas at a shorter $d_{1,6}$ bond range twisted and symmetric forms are allowed. Thus, the binding orientations (twisted or symmetric forms), determined by the rotation of 2,3-butanediol tether, closely relate with whether the $d_{1,6}$ bond opens or not.

Similar tendencies can be found in the optimized structures for dicarboxy-methanonanotubes. In fact, the optimized structure for the binding of the C11 atom into an orthogonal bond of the (10,10) nanotube has a breaking $d_{1,6}$ bond, and at the same time the two carboxy groups are symmetric, as shown in Figure 13. In contrast, the slanted case with a retaining $d_{1,6}$ bond has a twisted form. Similarly Lee and Marzari found such relationship, and theoretically proposed orientation of two $NO_2$ groups attached to the C11 atom can control the $d_{1,6}$ bond length on the $sp^2$ surface of fullerene and a nanotube (Lee & Marzari 2008).

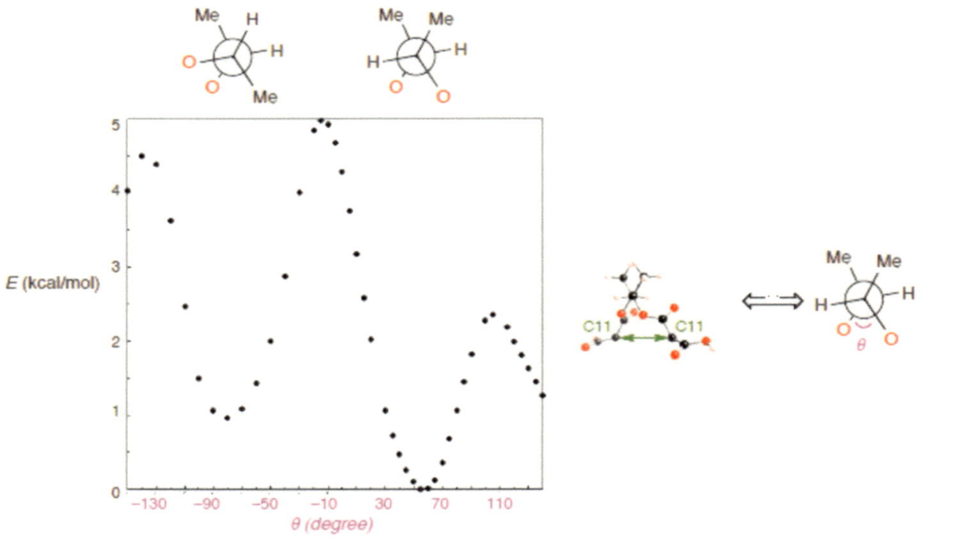


Fig. 11. The energy changes of a free bismalonate with 2,3-butanediol tether upon the rotation around dihedral angle θ of the tether. The energy values were obtained at PW91/6-31G* level of theory (Yumura & Kertesz, 2009).

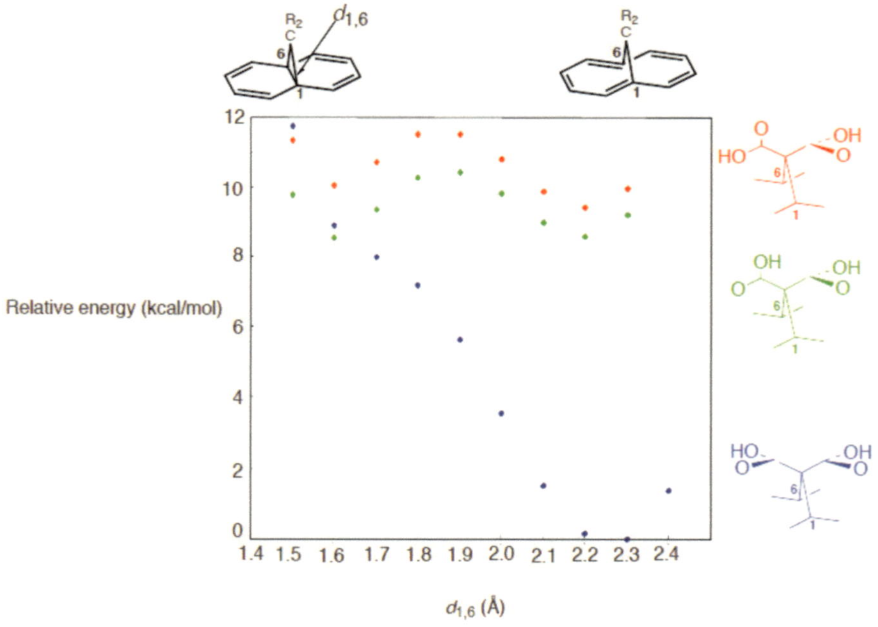


Fig. 12. Total energies of 11,11-dicarboxy-1,6-methano[10]annulene as a function of the $d_{1,6}$ value, obtained at PW91/6-31G* level of theory (Yumura & Kertesz, 2009). In the blue structure, the two carboxy groups are symmetric with respect to a mirror plane, whereas in the green and red structures, the groups are twisted.

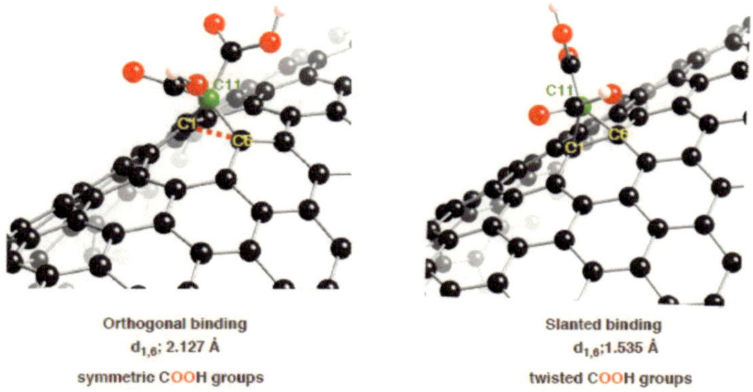


Fig. 13. Two types of optimized structure for dicarboxy-methanonanotube. We can distinguish between the two structures by where the divalent C11 atom is bound; an orthogonal or a slanted bond. The divalent atoms are given by green, and oxygen atoms by red. The optimized geometries were obtained by PW91 calculation based on PBC and cluster-model approaches (Yumura & Kertesz, 2009).

Of course, such relationship cannot exist in the double $CH_2$ addition, and accordingly the binding orientations play a crucial role in the site-selective bismalonate addition. We see quantitatively the effects of the two factors in enhancing the selectivity by estimating the $\Delta E_{bind}$ values, given in followings.

$$\Delta E_{bind} = E_{double\text{-}bind}\,(\text{bismalonate}) - E_{double\text{-}bind}\,(CH_2) \quad (3)$$

The $\Delta E_{bind}$ values are displayed as a function of the separation between the two C11 atoms in Figure 14.

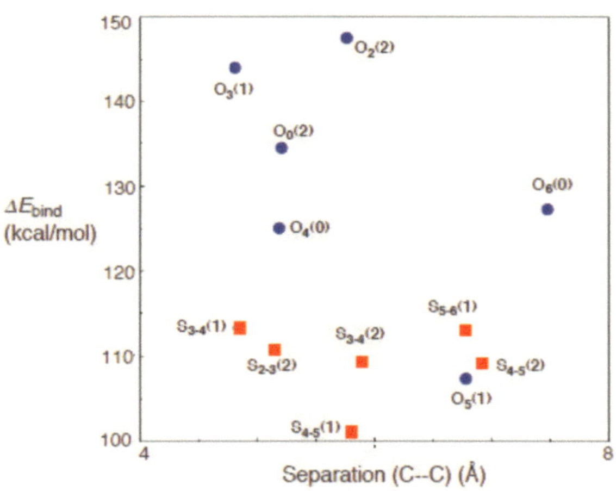


Fig. 14. The differences in the binding energy between the bismalonate and double carbene functionalization ($\Delta E_{bind}$). The $\Delta E_{bind}$ values were obtained from a cluster approach at PW91/6-31G*//PW91/3-21G level of theory (Yumura & Kertesz, 2009).

As shown in Figure 14, more significant $\Delta E_{bind}$ values were obtained in orthogonal orientations for the second attachment rather than slanted orientations. Since a positive $\Delta E_{bind}$ value indicates that a certain orientation is destabilized by the replacement of $CH_2$ with bismalonate, preferences of an orthogonal site over a slanted site as a second attachment are significantly weakened in the bismalonate functionalization, except for the $O_5(0)$ configuration in Figure 10(a). As a result, site-selectivity can be enhanced by the bismalonate functionalization compared with the double carbene addition (Yumura & Kertesz 2009). The DFT calculations clearly demonstrate that geometrical constraints of carbene derivative are key in the addition of its C11 atoms into a specific sites.

## 3. Conclusion

Density functional theory (DFT) calculations were employed to devise a plausible strategy for site-selective functionalization of nanotubes by carbene-derivatives. An accurate description of CC bondings of a nanotube functionalized by their divalent carbons was obtained with the aid of large-scale DFT calculations. Then, Clar valence bond (VB) concept, a basic concept in organic chemistry, can help to interpret the disruption of a nanotube surface by carbene-functionalization obtained from DFT calculations, and thus the concept is a useful tool to find out an approach to site-selective functionalization of nanotubes.

The most important DFT finding is that one inner carbene can have the power to locally perturb a nanotube surface by making two covalent bonds. The locally modified surface consists on two butadiene and one quinonoid patterns. In contrast, an outer carbene does not have such power, and accordingly the outer addition cannot disrupt Clar patterns in a pristine nanotube. The differences in the surface modification between the inner and outer bindings originate from whether a CC bond at the binding site opens or not. Note that retaining the CC bond at the binding site for inner carbene is due to more rigorous restriction of surface modification toward the tube center.

Considering different surface modification by the first attachment, we propose two approaches to how functional groups containing a divalent atom can selectively bind into a nanotube. In one approach, one can utilize "local" modification by the first inner attachment to control a site for the second outer attachment. In this situation, the second carbene selectively binds into a CC bond whose electron population enhances upon the first addition. Similar CC bonds with double-bond characters do not emerge on a nanotube functionalized by one outer carbene. Thus geometrical restrictions of a functional group are indispensable for site-selective functionalization of outer surface of a nanotube. For example, a bismalonate with 2,3-butanediol tether is a possible candidate for a functional group that can site-selectively add to a nanotube. For enhancing site-selectivity of functionalization of nanotubes, there are at least two key factors due to geometrical constraints of the bismalonate; a barrier for the rotation around the dihedral angle O–C–C–O of the 2,3-butanediol tether as well as orientations of its carboxy groups attaching divalent atoms. As a result, two divalent atoms of the bismalonate cannot bind freely into a nanotube to enhance the site-selectivity. In the above findings, subtle geometrical changes of nanotube surfaces as well as functional groups are key in site-specific functionalization of nanotubes. Thus, large-scale DFT treatment at relatively accurate manners is required to investigate the nanotube functionalization and to construct nanotube-based building blocks in nano-devices.

## 4. Acknowledgement

The author thanks Prof. Miklos Kertesz at Georgetown University (USA) for valuable discussion on the surface modification induced by carbene bindings. The project is partially supported by a Grant-in-Aid for Young Scientists (B) from the Japan Society for the Promotion of Science (JSPS) at Kyoto Institute of Technology (No. 22710088).

## 5. References

Bahr, J.L. & Tour, J.M. (2001). Highly functionalized carbon nanotubes using in situ generated diazonium compounds. *Chem. Mater.*, 13, 3823-3824.

Bahr, J.L.; Yang, J.; Kosynkin, D.V.; Bronikowski, M.J.; Smalley, R.E. & Tour, J.M. (2001). Functionalization of carbon nanotubes by electrochemical reduction of aryl diazonium salts: A bucky paper electrode. *J. Am. Chem. Soc.*, 123, 6536-6542.

Bahr, J.L.; Mickelson, E.T.; Bronikowski, M.J.; Smalley, R.E. & Tour, J.M. (2001). Dissolution of small diameter single-wall carbon nanotubes in organic solvents? *Chem. Commun.*, 193-194.

Becke, A.D. (1988). Density-functional exchange-energy approximation with correct asymptotic behavior. *Phys. Rev. A*, 38, 3098–3100.

Becke, A.D. (1992). Density-functional thermochemistry. I. the effect of the exchange-only gradient correction. *J. Chem. Phys.*, 96, 2155-2160.

Becke, A.D. (1992). Density-functional thermochemistry. II. the effect of the Perdew-Wang generalized-gradient correlation Correction. *J. Chem. Phys.*, 97, 9173-9177.

Becke, A.D. (1993). Density-functional thermochemistry. III. the role of exact exchange. *J. Chem. Phys.*, 98, 5648-5652.

Bettinger, H.F. (2004). Effects of finite carbon nanotube length on sidewall addition of fluorine atom and methylene. *Org. Lett.*, 6, 731-734.

Bettinger, H.F. (2006). Addition of carbenes to the sidewalls of single-walled carbon nanotubes. *Chem. Euro. J.*, 12, 4372-4379.

Bianchi, R.; Pilati, T. & Simonetta, M. (1980). Structure of 1,6-Methano[10]annulene. *Acta Crystallogr. B*, 36, 3147-3148.

Chen, J.; Hamon, M.A.; Hu, H.; Chen, Y.; Rao, A.M.; Eklund, P.C. & R.C. Haddon, (1998). Solution properties of single-walled carbon nanotubes. *Science*, 282, 95-98.

Chen. Y.; Haddon, R.C.; Fang, S.; Rao, A.M.; Eklund, P.C.; Lee, W.H.; Dickey, E.C.; Grulke, E.A.; Pendergrass, J.C.; Chavan, A.; Haley. B.E. & Smalley, R.E. (1998). Chemical attachment of organic functional groups to single-walled carbon nanotube material. *J. Mater. Res.*, 13, 2423-2431.

Chen, Z.F.; Nagase, S.; Hirsch, A.; Haddon, R.C.; Thiel, W. & Schleyer, P.v.R. (2004). Side-wall opening of single-walled carbon nanotubes (SWCNTs) by chemical modification: A critical theoretical study. *Angew. Chem. Int. Ed.*, 43, 1552-1554.

Chen, Z.F.; Thiel, W. & Hirsch, A. (2003). Reactivity of the convex and concave surfaces of single-walled carbon nanotubes (SWCNTs) towards addition reactions: Dependence on the carbon-atom pyramidalization. *ChemPhysChem*, 4, 93-97.

Choi, C.H. & Kertesz, M. (1998). A new interpretation of the valence tautomerism of 1,6-methano[10]annulene and its derivatives. *J. Phys. Chem. A*, 102, 3429-3437.

Clar, E. (1972). The aromatic sextet; Wiely: London.

Coleman, K.S.; Bailey, S.R.; Fogden, S. & Green, M.L.H. (2003). Functionalization of single-walled carbon nanotubes via the Bingel reaction. *J. Am. Chem. Soc.*, 125, 8722-8723.

Diederich, F. & Thilgen, C. (1996). Covalent fullerene chemistry. *Science*, 271, 317-324.

Dresselhaud, M.S.; Dresselhaus, G. & Eklund, P.C. (1996). Science of fullerenes and carbon nanotubes; Academic Press: San Diego.

Dyke, C.A. & Tour, J.M. (2004). Covalent functionalization of single-walled carbon nanotubes for materials applications. *J. Phys. Chem. A*, 108, 11151-11159.

Gao, X.; Ishimura, K.; Nagase, S. & Chen, Z.F. (2009). Dichlorocarbene addition to $C_{60}$ from the trichloromethyl anion: carbene mechanism or Bingel mechanism? *J. Phys. Chem. A*, 113, 3673-3676.

Globus, A.; Bauschlicher, C.; Han, J.; Jaffe, R.; Levit, C. & Srivastave, D. (1998). Machine phase fullerene nanotechnology. *Nanotechnology*, 9, 192-199.

Goldsmith, B.R.; Coroneus, J.G.; Khalap, V.R.; Kane, A.A.; Weiss, G.A. & Collins, P.G. (2007). Conductance-controlled point functionalization of single-walled carbon nanotubes. *Science*, 315, 77-81.

Haddon, R.C. (1998). Organometallic chemistry of fullerenes: $\eta^2$- and $\eta^5$-(п)complexes. *J. Comp. Chem.*, 19, 139-143.

Haddon, R.C. (2002). Carbon nanotubes. *Acc. Chem. Res.* 35, 997-997.

Han, J. & Globus, A. (1997). Molecular dynamics simulations of carbon nanotube-based gears. *Nanotechnology*, 8, 95-102.

Hirsch, A. (2002). Functionalization of single-wall carbon nanotubes. *Angew. Chem. Int. Ed.*, 41, 1853-1859.

Hohenberg, P. & Kohn, W. (1964). Inhomogeneous electron gas. *Phys. Rev.*, 136, B864-B871.

Holzinger, M.; Vostrowsky, O.; Hirsch, A.; Hennrich, F.; Kappes, M.; Weiss, R. & Jellen, F. (2001). Sidewall functionalization of carbon nanotubes. *Angew. Chem. Int. Ed.*, 40, 4002-4005.

Iijima, S. (1991). Helical microtubules of graphitic carbon. *Nature*, 354, 56-58.

Iijima, S. & Ichihashi, T. (1993). Single-shell carbon nanotubes of 1-nm diameter *Nature*, 363, 603-605.

Kamarás, K.; Itkis, M.E.; Hu, H.; Zhao, B. & Haddon, R.C. (2003). Covalent bond formation to a carbon nanotube metal. *Science*, 301, 1501-1501.

Kanungo, M.; Lu, H.; Malliaras, G.G. & Blanchet, G.B. (2009). Suppression of metallic conductivity of single-walled carbon nanotubes by cycloaddition reactions. *Science*, 323, 234-237.

Karousis, N.; Tagmatarchis, N. & Tasis, D. (2010). Current progress on the chemical modification of carbon nanotubes. *Chem. Rev.*, 110, 5366–5397.

Kessinger, R.; Thilgen, C.; Mordasini, T. & Diederich, F. (2000). Optically active macrocyclic cis-3 bis-adducts of $C_{60}$: Regio- and stereoselective synthesis, exciton chirality coupling, and determination of the absolute configuration, and first observation of exciton coupling between fullerene chromophores in a chiral environment. *Helv. Chim. Acta*, 83, 3069-3096.

Kohn, W. & Sham, L. J. (1965). Self-consistent equations including exchange and correlation effects. *Phys. Rev.*, 140, A1133-A1138.

Lee, C.; Yang, W. & Parr, R.G. (1988). Development of the colle-salvetti correlation-energy formula into a functional of the electron density. *Phys. Rev. B*, 37, 785–78.

Lee, Y.S. & Marzari, N. (2006). Cycloaddition functionalizations to preserve or control the conductance of carbon nanotubes. *Phys. Rev. Lett.*, 97, 116801.

Lee, Y.S. & Marzari, N. (2008). Cycloadditions to control bond breaking in naphthalenes, fullerenes, and carbon nanotubes: A first-principles study. *J. Phys. Chem. C*, 112, 4480-4485.

Lpez-Bezanilla, A.; Triozon, F.; Latil, S.; Blase, X. & Roche, S. (2009). Effect of the chemical functionalization on charge transport in carbon nanotubes at the mesoscopic scale. *Nano Lett.*, 9, 940-944.

Matsuo, Y.; Tahara, K. & Nakamura, E. (2003). Theoretical studies on structures and aromaticity of finite-length armchair carbon nanotubes. *Org. Lett.*, 5, 3181-3184.

Niyogi, S.; Hamon, M.A.; Hu, H.; Zhao, B.; Bhowmik, P.; Sen, R.; Itkis, M.E. & Haddon, R.C. (2002). Chemistry of single-walled carbon nanotubes. *Acc. Chem. Res.*, 35, 1105-1113.

Ormsby, J.L. & King, B.T. (2004). Clar Valence bond representation of π-bonding in carbon nanotubes. *J. Org. Chem.*, 69, 4287-4291.

Ormsby, J.L. & King, B.T. (2007). The regioselectivity of addition to carbon nanotube segments. *J. Org. Chem.*, 72, 4035-4038.

Park, H.; Zhao, J. & Lu, J.P. (2006). Effects of sidewall functionalization on conducting properties of single wall carbon nanotubes. *Nano Lett.*, 6, 916-919.

Parr, R.G. & Yang, W. (1996). Density-functional theory of atoms and molecules; Oxford University Press: Oxford.

Peng, H.; Alemany, L.B.; Margrave, J.L. & Khabashesku, V.N. (2003). Sidewall carboxylic acid functionalization of single-walled carbon nanotubes. *J. Am. Chem. Soc.*, 125, 15174-15182.

Peng, H.; Reverdy, P.; Khabashesku, V.N. & Margrave, J.L. (2003). Sidewall functionalization of single-walled carbon nanotubes with organic peroxides. Chem. Commn., 362-363.

Perdew, J.P. & Wang, Y. (1992). Accurate and simple analytic representation of the electron-gas correlation energy. *Phys. Rev. B*, 45, 13244.

Prato, M.; Kostarelos, K.; Bianco, A. (2008). Functionalized carbon nanotubes in drug design and discovery. *Acc. Chem. Res.*, 41, 60-68.

Saini, R.K.; Chiang, I.W.; Peng, H.; Smalley, R.E.; Billups, W.E.; Hauge, R.H.; Margrave, J.L. (2003). Covalent sidewall functionalization of single wall carbon nanotubes. *J. Am. Chem. Soc.*, 125, 3617-3621.

Saito, R.; Dresselhaus, G. & Dresselhaus, M.S. (1998). Physical properties of carbon nanotubes.; Imperial College Press: London.

Sato, Y.; Yumura, T.; Suenaga, K.; Urita, K.; Kataura, H.; Kodama, T.; Shinohara, H. & Iijima S. (2006). Correlation between atomic rearrangement in defective fullerenes and migration behavior of encaged metal ions. *Phys. Rev. B*, 73, 233409.

Stevens, J.L.; Huang, A.Y.; Peng, H.; Chiang, I.W.; Khabashesku, V.N. & Margrave, J.L. (2003). Sidewall amino-functionalization of single-walled carbon nanotubes through fluorination and subsequent reactions with terminal diamines. *Nano Lett.*, 3, 331-336.

Stephens, P.J.; Devlin, F.J.; Chabalowski, C.F. & M.J. Frisch, (1994). Ab initio calculation of vibrational absorption and circular dichroism spectra using density functional force fields. *J. Phys. Chem.*, 98, 11623-11627.

Strano, M.S.; Boghossian, A.A.; Kim, W.J.; Barone, P.W.; Jeng, E.S.; Heller, D.A.; Nair, N.; Jin, H.; Sharma, R. & Lee, C.Y. (2009). The chemistry of single-walled nanotubes. *MRS BULLETIN*, 34, 950-961.

Szabo, A. & Ostlund, N,S, (1996). Modern Quantum Chemistry.; Dover Publishing: Mineola, New York.

Tasis, D.; Tagmatarchis, N.; Bianco, A. & Prato, M. (2006). Chemistry of carbon nanotubes. *Chem. Rev.*, 106, 1105-1136.

Umeyama, T.; Tezuka, N.; Fujita, M.; Matano, Y.; Takeda, N.; Murakoshi, K.; Yoshida, K.; Isoda, S. & Imahori, H. (2007). Retention of intrinsic electronic properties of soluble single-walled carbon nanotubes after a significant degree of sidewall functionalization by the Bingel reaction. *J. Phys. Chem. C*, 111, 9734-9741.

Urita, K.; Sato, Y.; Suenaga, K.; Gloter, A.; Hashimoto, A.; Ishida, M.; Shimada, T.; Shinohara, H. & Iijima, S. (2004). Defect-induced atomic migration in carbon nanopeapod: Tracking the single-atom dynamic behavior. *Nano Lett., 4*, 2451-2454.

Vosko, S.H.; Wilk, L. & Nusair, M. Accurate spin-dependent electron liquid correlation energies for local spin density calculations: a critical analysis. (1980). Can. J. Phys. 58, 1200-1211.

Vzquez, E. & Prato, M. (2009). Carbon nanotubes and microwaves: Interactions, responses, and applications. ACS Nano, 3, 3819-3824.

Worsley, K.A.; Moonoosawmy, K.R. & Kruse, P. (2004). Long-range periodicity in carbon nanotube sidewall functionalization. *Nano Lett.*, 4, 1541-1546.

Ying, Y.; Saini, R.K.; Liang, F.; Sadana, A.K. & Billups, W. E. (2003). Functionalization of carbon nanotubes by free radicals. *Org. Lett.*, 5, 1471-1473.

Yumura, T. & Kertesz, M. (2007). Cooperative behaviors in carbene additions through local modifications of nanotube surfaces. *Chem. Mater.*, 19, 1028-1034.

Yumura, T.; Kertesz, M. & Iijima, S. (2007). Confinement effects on site-preferences for cycloadditions into carbon nanotubes. *Chem. Phys. Lett.*, 444, 155-160.

Yumura, T.; Kertesz, M. & Iijima, S. (2007). Local modifications of single-wall carbon nanotubes induced by bond formation with encapsulated fullerenes. *J. Phys. Chem. B*, 111, 1099-1109.

Yumura, T. (2011). Chemically reactive species remain alive inside carbon nanotubes: a density functional theory study. *Phys. Chem. Chem. Phys.*, 13, 337-346.

Yumura, T. & Kertesz, M. (2009). Roles of conformational restrictions of a bismalonate in the interactions with a carbon nanotube. *J. Phys. Chem. C*, 113, 14184-14194.

Zhao, J.J.; Chen, Z.F.; Zhou, Z.; Park, H.; Schleyer, P.v.R. & Lu, J.P. (2005). Engineering the electronic structure of single-walled carbon nanotubes by chemical functionalization. *ChemPhysChem*, 6, 598-601.

Zheng, G.S.; Wang, Z.; Irle, S. & Morokuma, K. (2006). Origin of the linear relationship between $CH_2$/NH/O-SWNT reaction energies and sidewall curvature: armchair nanotubes. *J. Am. Chem. Soc.*, 128, 15117-15126.

# 15

# Low-Energy Irradiation Damage in Single-Wall Carbon Nanotubes

Satoru Suzuki
*NTT Basic Research Laboratories, NTT Corporation, 3-1, Morinosato Wakamiya, Atsugi, Kanagawa Japan*

## 1. Introduction

Single-wall carbon nanotubes (SWCNTs) are one of the most promising materials for future nano-electronics, because of their unique quasi-one-dimensional structures and excellent electric and mechanical properties. They also have very high chemical stability, owing to their robust $sp^2$-bonding carbon network (graphene) with no dangling bonds. Because of the structural robustness, low-energy (typically 10 eV-20 keV) electron and photon irradiation in a vacuum had been generally assumed not to cause damage to SWCNTs when the energy is smaller than the knock-on threshold. In fact, analytical tools that use low-energy electrons or photons, such as scanning electron microscopy (SEM), had been commonly used for characterization of SWCNTs without serious concerns.

In 2004, however, we reported that electron irradiation in a SEM caused severe damage (low-energy irradiation damage) in SWCNTs produced by both thermal chemical vapor deposition and laser ablation methods (Suzuki et al., 2004b). Other techniques using low-energy electrons and vacuum-ultraviolet (VUV) light or soft x-rays (especially high-brilliance synchrotron radiation light), such as low-energy electron microscopy (LEEM) and photoemission spectroscopy, also inevitably damage SWCNTs. Therefore, paying attention to the low-energy irradiation damage is practically important for those who study SWCNTs. For example, when we measure the Raman and photoluminescence (PL) spectra and electric properties and take SEM images of the same SWCNTs, the SEM observations should be done last. Doing the high-resolution SEM observation first would inevitably cause severe damage and tremendously affects the following measurements.

The low-energy irradiation damage and its defect characteristics are also physically interesting. In this chapter, we will review the physical and chemical property changes induced by the damage, and the defect properties, which are significantly different from those of other types of damage. We will examine the defect-induced metal-semiconductor transition of the room-temperature electric properties and discuss its mechanism. We will also summarize other types of damage, which are often confused with the low-energy irradiation damage, focusing on the differences between them.

Before continuing to the main text, I must briefly explain how I compare spectra obtained form the same SWCNT sample. In many studies of the physical or chemical treatment of SWCNTs and graphene, spectra are often normalized to the maximum peak height. In contrast, when I show irradiation- and annealing-induced changes of Raman and PL spectra, the spectra are never normalized. That is, I obtain the spectra under the same condition to the best of my ability and directly compare the raw spectra. This methodology has been applied in all of our related reports, unless otherwise mentioned. With arbitrary spectral normalization, we would no longer be able to discuss the reversibility of the damage and recovery, which is a very important characteristic of low-energy irradiation damage.

## 2. What is low-energy irradiation damage?

We define low-energy irradiation damage as damage solely caused by irradiation of low-energy particles, where low-energy means that the energy is much smaller than the threshold energy of knock-on damage. Thus, the mechanism of the damage is completely different from knock-on damage. Moreover, we discriminate low-energy irradiation damage and secondary damage caused by the irradiation, such as damage by radicals. Irradiation by both electrons and photons irradiation was found to damage SWCNTs. However, other particles, such as atoms and ions, or quasi-particles such as plasmons, may also cause the damage.

## 3. Property changes caused by low-energy irradiation damage

### 3.1 Raman and PL spectra

In a Raman spectrum, a SWCNT shows the so-called G band (tangential mode) and disorder-induced D band, which are characteristic of a graphene sheet. The D band is ideally inactive and its appearance is evidence of symmetry breaking. The intensity ratio of the G and D bands is often utilized as an indicator of the degree of crystallinity. Another very important mode of SWCNTs is the radial breathing mode (RBM), which is often used for diameter evaluation. For a general review of Raman spectroscopy of CNTs, see (Dresselhaus et al., 2005), for example. Like other types of damage, low-energy irradiation damage generally decreases the G band and RBM intensities (There are some exceptions at the edges of the resonance window, as discussed below) and the G/D intensity ratio and increases the D band intensity, as shown in Figs. 1(a) and (b). Generally, the decrease of intensity is more prominent for the RBM than for the G band. Considering that the detectable Raman intensity from individual SWCNTs is owing to the resonance enhancement effect, the disappearance of Raman spectra is probably due to a reduction of the resonance enhancement. The initially divergent joint density of states, which is a characteristic of one-dimensional systems, would be considerably broadened by the formation of defects. Low-energy irradiation damage causes almost no broadening or almost no shift of the Raman peaks including the D band (Suzuki et al., 2010), although significant D band broadening due to gas-phase reaction has been observed (Yang et al., 2006. Zhang et al., 2006).

When SWCNTs are moderately damaged (or considerably recover from severe damage), originally hidden non-resonant RBM peaks sometimes appear. At the excitation wavelength of 785 nm, metallic and semiconducting SWCNTs are usually observed at about 150-160 and 200-240 cm$^{-1}$, respectively. In Fig. 2(a), however, the moderately damaged (partially recovered) SWCNTs show a sharp peak at 182 cm$^{-1}$ in the off-resonance region. The metallic SWCNTs at 156 cm$^{-1}$, which were initially not strongly excited in this sample, also became more prominent in the moderately damaged sample. Further damage extinguishes these peaks again, as also shown in the figure. Similar off-resonant RBM peaks are also often observed in doped SWCNTs grown from boron- and nitrogen-containing feedstocks, as shown in Fig. 2(b) (Suzuki & Hibino, 2011). I think that the defects slightly shift the absorption energy or broaden the absorption edge and this makes the originally off-resonant peak resonant. Similarly, complicated behavior of the RBM intensity with increasing damage is observed at the edge of the resonance window (Suzuki & Kobayashi, 2007a).

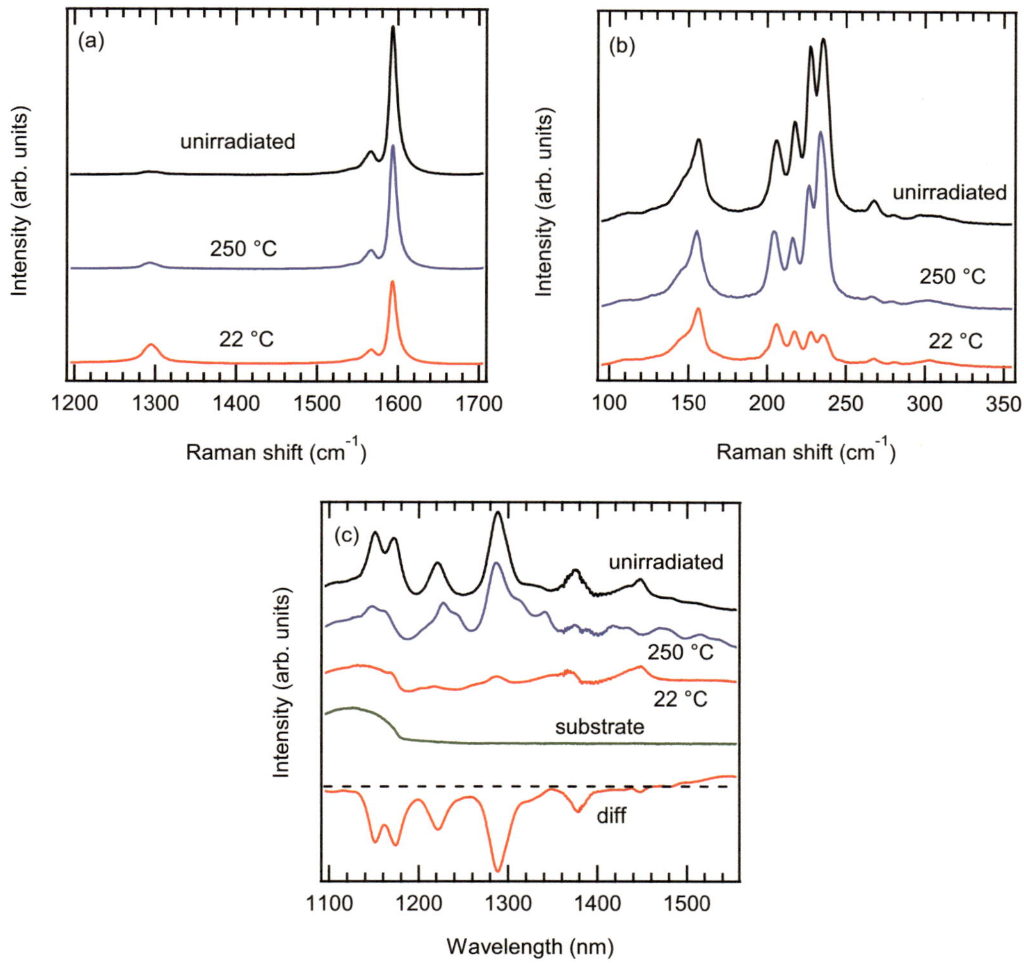


Fig. 1. (a) G and D band, and (b) RBM regions of Raman spectra, and (c) PL spectra of unirradiated SWCNTs and of SWCNTs irradiated at 250 and 22 °C. The irradiated electron energy and dose were 20 keV and $5.7 \times 10^{16}$ cm$^{-2}$. The excitation wavelength was 785 nm.

These results mean that the Kataura plot is modified by the defects.

The PL peak intensity of suspended semiconducting SWCNTs is more sensitively decreased than the Raman peak intensity, as shown in Fig. 1(c). In addition, broad spectral intensity newly appears at the longer wavelength side when the extent of the damage is moderate.

Severe damage finally extinguishes all spectral intensities in Raman (including the D band (Suzuki et al., 2005a)) and PL spectra (Suzuki & Kobayashi, 2007b). However, note that, in marked contrast to the damage caused by knock-on collisions and by radicals, the low-energy irradiation damage itself never eliminates a SWCNT.

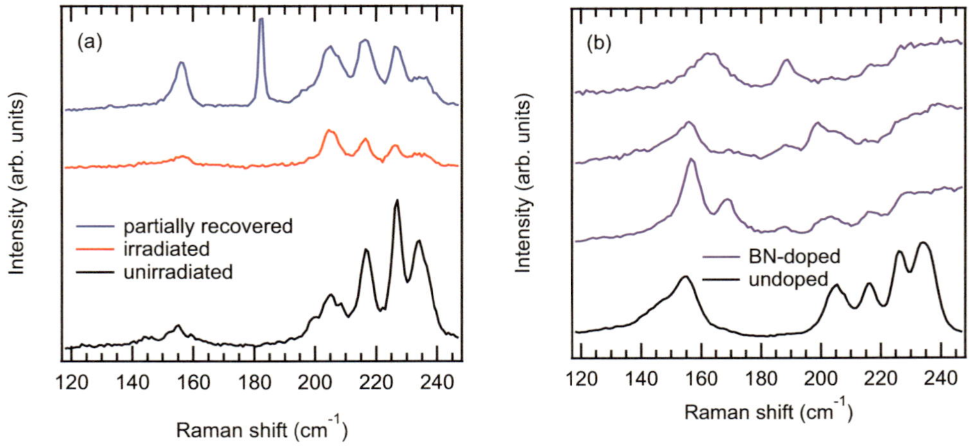


Fig. 2. (a) RBM spectra of unirradiated and electron-irradiated SWCNTs and partially recovered SWCNTs. The electron energy and irradiation dose were 20 keV and $6.3\times10^{16}$ cm$^{-2}$, respectively. The irradiated and considerably damaged SWCNTs were partially recovered by annealing in Ar atmosphere at 350 °C. The wavenuber range of 160-200 cm$^{-1}$ is the off-resonance region and a peak is rarely observed there, initially. (b) RBM spectra of undoped and BN-doped SWCNTs. The doped SWCNTs also often exhibit peaks in the off-resonance region. The excitation wavelength was 785 nm.

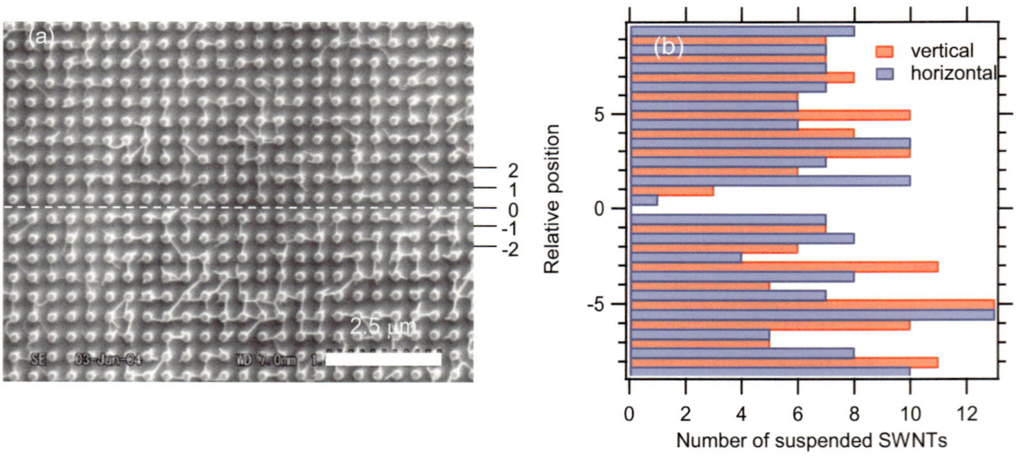


Fig. 3. (a) SEM image of the SWNT sample after eliminating the electron-irradiated SWCNTs. The irradiation was done along the dashed line. The electron energy and the local dose were 1 keV, and $1.5\times10^{19}$ cm$^{-2}$, respectively. Note that the elimination was done by selective combustion in air, not by the irradiation itself. (b) Position dependence of the number of SWCNTs suspended between neighboring pillars in (a). Here, diagonally suspended SWCNTs were neglected.

### 3.2 Chemical stability

The low-energy irradiation itself does not cut a SWCNT. However, the damage significantly decreases the chemical tolerance of SWCNTs, because the irradiation-induced defects make the sidewall chemically active. Therefore, we can selectively eliminate the irradiated SWCNTs by heating in air, as shown in Fig. 3 (Suzuki et al., 2005a). A part of the SWCNT sample was intensively irradiated in a SEM using the line scan mode along the dashed line. Then, the sample was heated in air at 420 °C for 30 m. The irradiated SWCNTs were selectively eliminated by combustion due to the reduced chemical tolerance. Note that a irradiation dose that is too high often has an entirely opposite effect, because the irradiation-induced contaminants on the SWCNT surfaces protect the SWCNTs from oxygen. As shown in (b), the irradiation effects almost completely disappear about 600 nm from the irradiation line. Such a high spatial resolution can be easily obtained using a convergent electron beam.

There have been many attempts to functionalize SWCNTs by using other molecules or metal particles. In many cases, defects are intentionally created to functionalize the sidewall, which is originally inert (Yan et al., 2005). Low-energy irradiation damage could also be applied for spatially selective functionalization with electron beam lithography.

### 3.3 Electric properties

The electric properties are much more sensitively changed by low-energy irradiation damage than Raman and PL spectra. Moderate irradiation can convert a metallic field effect transistor (FET) into semiconducting. I will discuss this remarkable phenomenon in sec. 5. Here, I focus on the intensive irradiation effects I have studied by in-situ electric measurements during electron irradiation in a SEM equipped with piezo-actuated micro-probes for electric measurements (Suzuki, 2011). The device used here consists of two SWCNTs (A branch is seen between the electrodes) suspended between the drain and source electrodes (height: 300 nm), as shown in Fig. 4(a). The high-magnification SEM image was taken after all experiments had been completed. Otherwise, the conductivity of the SWCNTs would almost vanish. Fig. 4(b) shows the results of in-situ electric measurements during irradiation. The whole SWCNTs were first irradiated by an electron beam using the normal SEM observation mode. The SEM observation gradually decreased the conductivity. Then, at ~42.76 s, they were intensively irradiated by using the line scan mode. This irradiation decreased the conductivity by two orders of magnitude in only a few seconds. As shown in Fig. 4(c), a very abrupt current decrease occurred at least within the initial 44 ms, which is the time resolution of the measurements. The gate voltage characteristics of the device before and after the irradiation are shown in Fig. 4(d). The irradiation decreased the two-probe conductivity by four to five orders of magnitude in the whole gate voltage range. Considering that the initial resistivity of the device would be dominated by the contact resistance between the SWNTs and electrodes, the intrinsic conductivity decrease would be much larger. Thus, intensive irradiation finally makes a SWCNT almost insulating. Similar results had been observed in previous works by another group (Marquardt et al., 2008. and Vijayaraghavan et al., 2010) and in our early work (Suzuki and Kobayashi., 2005), in which conventional on-substrate SWCNT devices were used. Here, I would like to get remind the readers again that even intensive irradiation does not cut a SWCNT. In fact, a SWCNT can be completely recovered by annealing, as shown later in sec. 4.2.

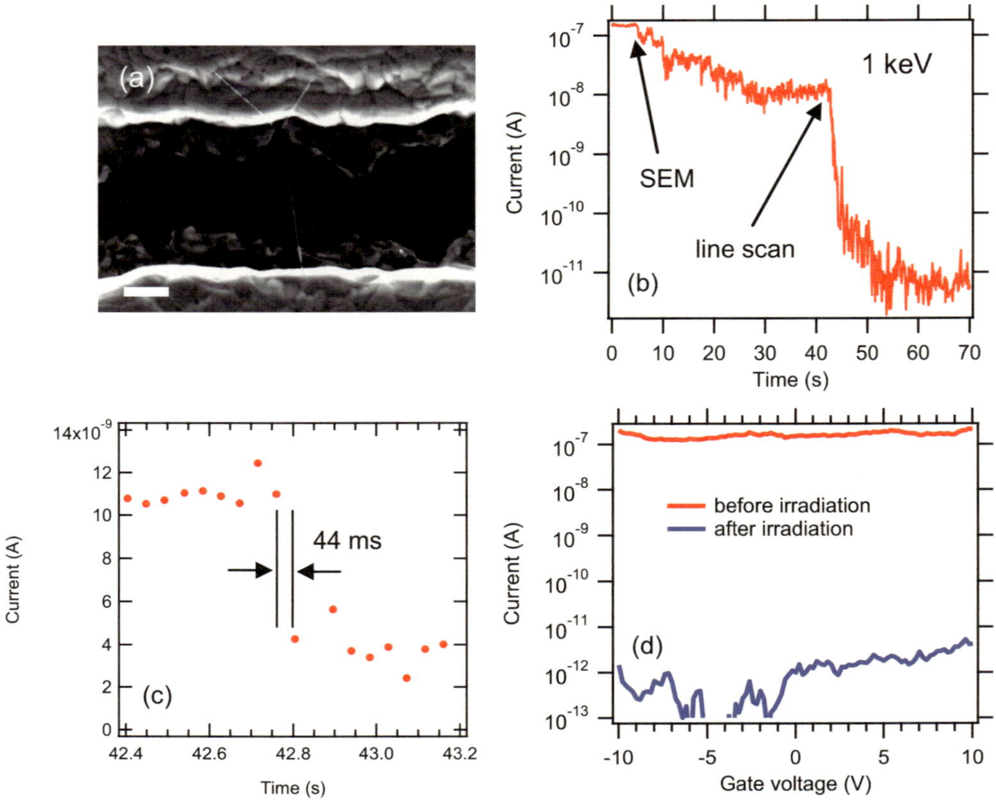


Fig. 4. (a) SEM image of a suspended SWNT device obtained after experiments. The substrate acts as a back-gate electrode. Scale bar: 200 nm. (b) Drain current during SEM observation and line scans. The drain voltage was set to 0.1 V, and the substrate (back-gate) was grounded. The electron energy was 1 keV. During the SEM observation, the irradiation dose rate was $1.2 \times 10^{13}$ $cm^{-2}s^{-1}$ on average in the observation area. The dose rate during the lines scan was ~$6 \times 10^{16}$ $cm^{-2}s^{-1}$. (c) Time evolution of the drain current just before and after the line scan. The data collection was done using the sampling mode of a semiconductor parameter analyzer (Agilent 4156C) and the time interval is not always the same. (d) Gate characteristics of the device before and after the electron irradiation. The drain voltage was 0.1 eV.

## 4. Characteristics of low-energy irradiation damage

### 4.1 Energy dependence

The low-energy irradiation damage has been observed in an energy range of several electron-volts to 25 keV. One important characteristic of the damage is that, in general, a lower energy is more destructive than a higher energy (Suzuki et al., 2004b), as shown in Fig. 5(a) (except for the energies below ~20 eV, where the optical absorption and electron energy loss spectra strongly reflect the specific electronic density of states of graphene). This can be well understood by the fact that the interaction (the cross section of electronic excitation) between a SWCNT and an incident electron (photon) is generally larger at a

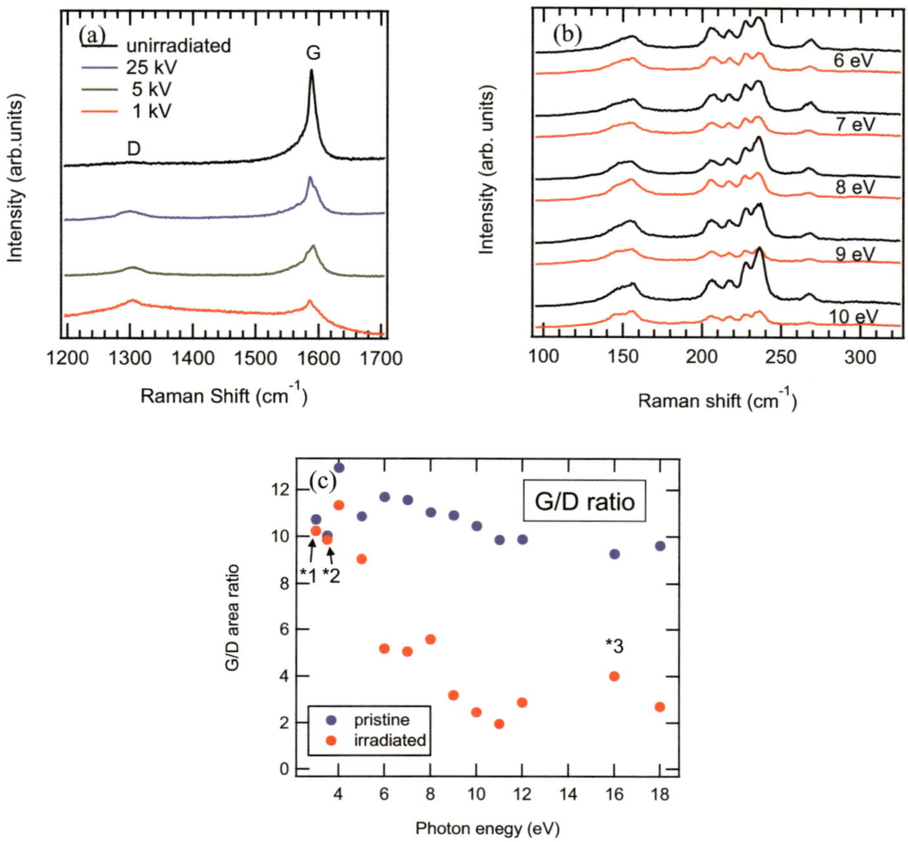


Fig. 5. (a) G and D band spectra of pristine and electron-irradiated SWCNTs, showing the acceleration voltage dependence of electron irradiation damage. The irradiation dose was $2.7\times10^{18}$ cm$^{-2}$. (b) RBM spectra of pristine SWCNTs (black lines) and SWCNTs illuminated by 6-10 eV photons (red). Because the spectra were originally slightly sample-dependent, spectra before and after the irradiation are shown for each photon energy. (c) G/D area ratio of pristine SWCNTs and SWCNTs irradiated at various photon energies. In (b) and (c), the photon dose was $5\times10^{17}$ cm$^{-2}$ except for *2 and *3 in (c). The excitation wavelength was 785 nm. *1) higher order light cut by a pyrex window. *2) very intense unmonochromatized light through a pyrex window (hν≤3.5 eV). *3) The photon dose was $2.5\times10^{17}$ cm$^{-2}$.

smaller energy. A high-energy electron or photon easily penetrates a SWCNT without any electric excitation. An exception is when the photon energy is tuned to near the C 1s absorption edge, for which severer damage was observed than at the energy below the absorption threshold. However, the resonance effect does not seem to be very prominent, as will be shown in Fig. 11(b).

The existence of the threshold energy is also expected for the low-energy irradiation damage. As seen in the RBM spectra shown in Fig. 5(b), distinct spectral intensity decreases are observed at ~6 eV or larger for semiconducting SWCNTs with diameters of about 1.2-1.0 nm at 200-240 cm$^{-1}$ (Suzuki & Kobayashi, 2008). Decreases in the G/D ratio are also clearly observed at 6 eV or higher, as shown in Fig. 5(c). Notably, the damage in metallic SWCNTs with diameters of ~1.6 nm at 140-160 $cm^{-1}$ is not clear up to 8 eV. This is because of the diameter dependence of low-energy irradiation damage, as discussed in sec. 4.3. In electron (hole) tunneling injection studies using a scanning tunneling microscope (STM), slightly lower threshold energy of ~ 4 eV has been observed (Yamada et a., 2009). The reason for the discrepancy may be that the threshold energy depends on the diameter, chirality, or detailed defect structures. Anyway the threshold energy of the low-energy irradiation damage seems to be several electron volts.

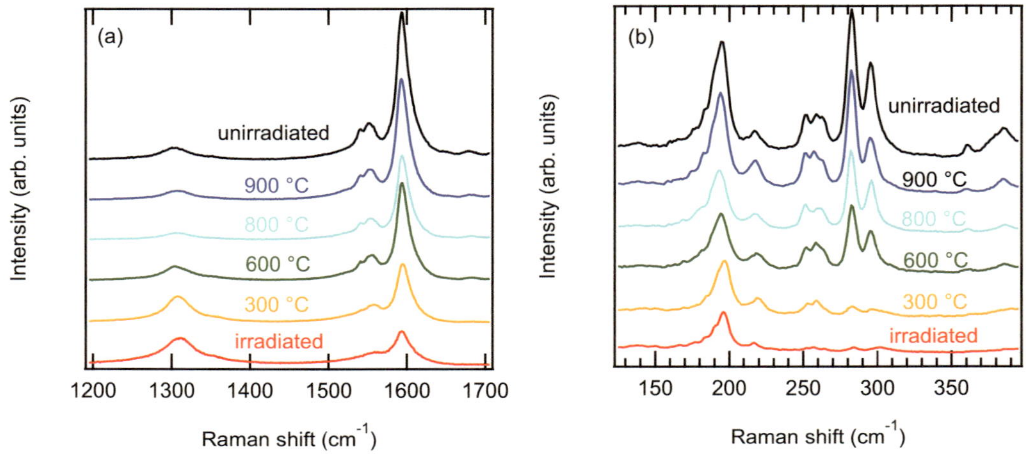


Fig. 6. (a) G and D band, and (b) RBM spectra of unirradiated, photon-irradiated, and annealed SWCNTs. The SWCNTs were irradiated by unmonochromatized synchrotron radiation light (hν≤1 keV) up to a dose of $8\times10^{20}$ $cm^{-2}$. The excitation wavelength for the Raman measurements was 633 nm.

### 4.2 Reversible damage and recovery

Probably the most important characteristic of the low-energy irradiation damage is the reversibility of the damage and recovery. Figure 6 shows Raman spectra of SWCNTs before and after VUV light irradiation and of SWCNTs annealed at 300, 600, 800, 900 °C. The irradiation caused severe damage and drastically decreased the G/D ratio and the RBM intensities. All the RBM peaks above 200 $cm^{-1}$ corresponding to a diameter less than ~1.2 nm almost completely disappeared. However, the annealing at 300, 600, and 800 °C gradually recovered the spectra, and at 900 °C, all the peaks including the once disappeared peaks are almost fully recovered. I would like to point out once again that the spectra were not normalized at all. Thus, the results reveal that not only the spectral shape but also the spectral intensity itself is almost fully recovers by annealing.

The reversible damage and recovery is also observed in the electric properties. As mentioned in sec. 3.3, the low-energy irradiation damage almost extinguishes the electric conductivity. However, the extinguished conductivity is also fully recoverd by annealing, as shown in Fig. 7(a). An originally metallic SWCNT device was intensively irradiated in a SEM using the line scan mode. The irradiation extinguished the conductivity and made the SWCNT almost insulating. However, the conductivity was fully recovered by annealing in a vacuum at 300 °C. Moreover, the reversibility can be observed repeatedly, as shown in Fig. 7(b). The complete reversibility of the electric properties has also been observed by another group, although they attribute the conductivity decrease and recovery to substrate charging and its release (Marquardt et al., 2008. and Vijayaraghavan et al., 2010). I will discuss this issue later in sec. 7.4.

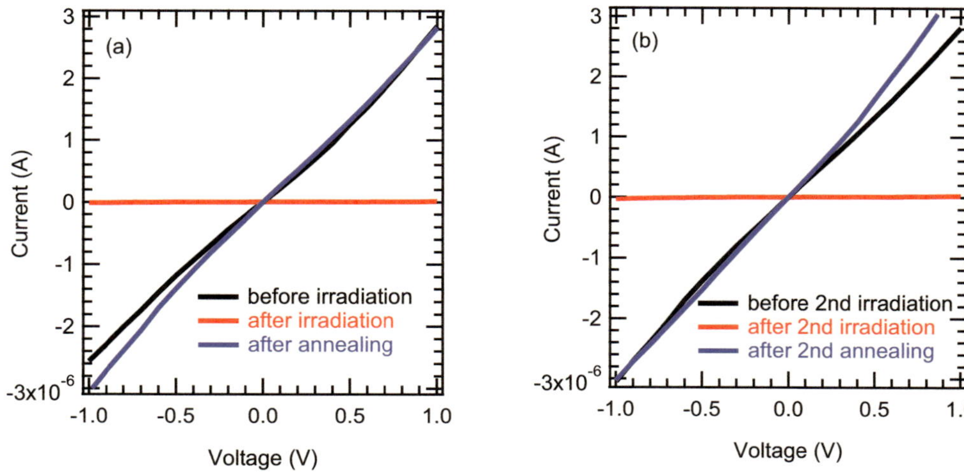


Fig. 7. (a) Electric properties of a SWCNT device before and after the first electron irradiation and after annealing. The SWCNT was intensively irradiated by electrons of 20 keV up to a dose of $1\times10^{20}$ cm$^{-2}$. Then, the irradiated device was annealed at 300 °C in Ar atmosphere for 30 min. (b) The electric properties of the same SWCNT device before and after the second irradiation and after the second annealing. The SWCNT was intensively irradiated again by 20-keV electrons up to a dose of $1.7\times10^{20}$ cm$^{-2}$. After the irradiation, the SWCNT was annealed at 300 °C for 30 min and fully recovered again.

The reversibility of the damage and recovery indicates that the damage is not accompanied by a reduction of carbon atoms and that the number of carbon atoms is preserved. Recently, Mera et al. directly measured ion desorption from SWCNTs under soft X-ray illumination (Mera et al, 2010). They also excluded emission of carbon atoms from the SWCNTs.

### 4.3 Diameter dependence

Another important characteristic of the low-energy irradiation damage is that strong diameter dependence is observed when the irradiation is done at room temperature or above (Suzuki and Kobayashi, 2006). For example, in the RBM and PL spectra shown in Figs. 1, 2, 5, and 6, we can clearly see the diameter dependence of the damage; that is, thinner SWCNTs are more severely damaged. Especially, in Fig. 1(c), we can see a large difference in the extent of the damage due to very small diameter difference. The SWNTs observed at about 1151 and 1172 nm can be assigned to (12,1) and (11,3) tubes having diameters of 0.995 and 1.014 nm, respectively. These two peaks were considerably weakened by the photon irradiation at 250 °C. On the other hand, the occurrence of the damage was not obvious for thicker SWNTs after the same irradiation dose at 250 °C. The peaks at 1224 nm is assigned to (10,5) tubes having a diameter of 1.050 nm. The diameter difference between the considerably damaged (11,3) and hardly damaged (10,5) tubes is only 0.037 nm. It is very interesting that such a small diameter difference results in the distinctly different sensitivity to the irradiation.

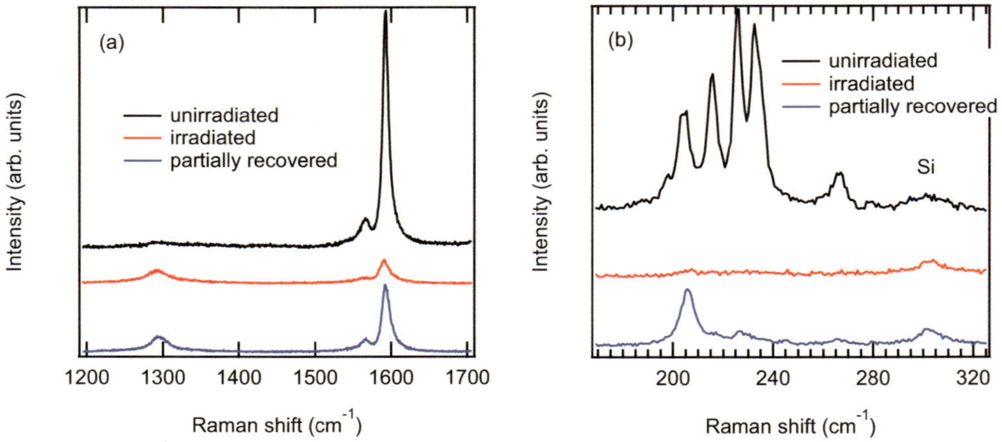


Fig. 8. (a) G and D band and (b) RBM spectra of pristine, irradiated, and partially recovered SWCNTs. The SWCNTs were irradiated by 1-keV electrons up to a dose of $8\times10^{17}$ cm$^{-2}$. Then, the damaged SWCNTs were partially recovered by annealing at 400 °C in Ar atmosphere for 30 min. The excitation wavelength was 785 nm. In (b), the hump at ~300 cm$^{-1}$ is from the Si substrate.

The diameter dependence is observed in the recovery process, as shown in Fig. 8. The electron irradiation largely decreased the G/D ratio (a) and once almost completely extinguished all of the RBM peaks (b). The sample was partially recovered by annealing at 400 °C. Then, only the peak at about 205 cm$^{-1}$, corresponding to the thickest SWNTs among the initially observed ones, significantly recovered.

The diameter dependence of damage is more or less also observed in knock-on damage (Krasheninnikov & Nordlund, 2010) and damage by radicals (Yang et al., 2006, Zhang et al., 2006b). However, the diameter dependence of low-energy irradiation damage is more prominent, as mentioned above. The damage caused by knock-on collision and radicals also occurs in thick MWCNTs and graphite, but the low-energy irradiation damage has not. Also noteworthy is that the diameter dependence is not prominent when SWCNTs are irradiated at low temperature, as shown in the next section.

### 4.4 Temperature dependence of the damage

Severer damage is observed at lower temperatures (Suzuki & Kobayashi, 2007a). As shown in Fig. 1, the irradiation at 250 °C results in much less damage than at 22 °C. The temperature dependence is seen at lower temperatures, and less damage is observed at -27 °C than at -267 °C (6 K), as shown in Fig. 9. These results suggest that low-energy irradiation-induced defects can be healed even at -27 °C. In fact, the electric conductivity of irradiated SWCNTs gradually recovers at room temperature, as shown in Fig. 10 (Suzuki & Kobayashi, 2007). The temperature dependence of the damage is completely opposite to that observed in gas phase reactions (An et al., 2002, Zhang et al., 2006a).

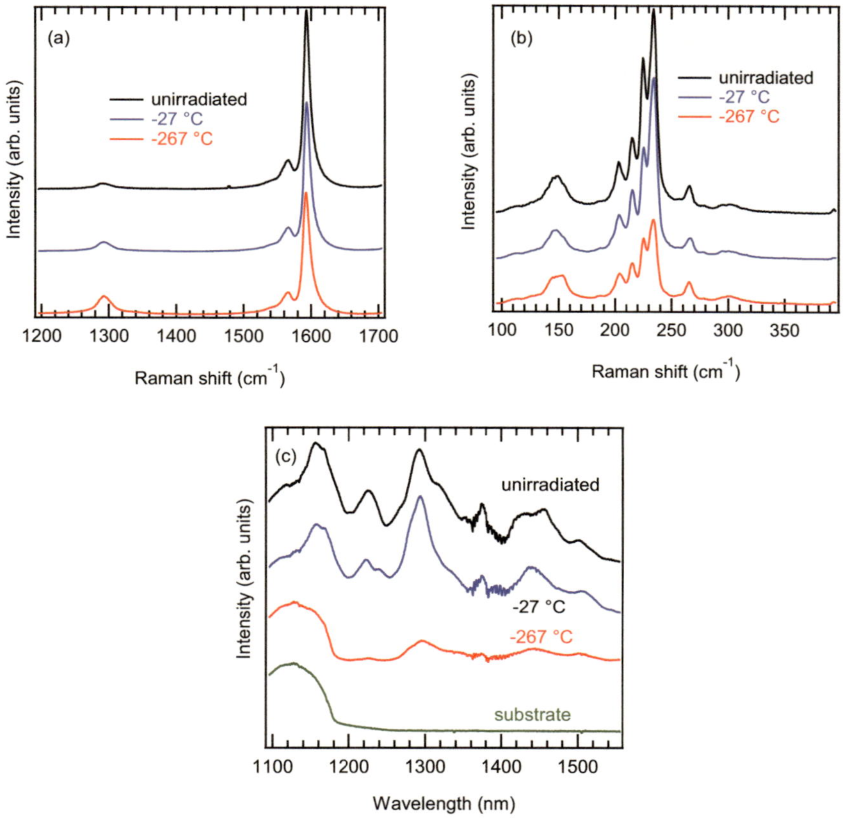


Fig. 9. (a) G and D band, (b) RBM, and (c) PL spectra of pristine SWCNTs and of SWCNTs irradiated at -27, and -267 °C. The SWCNTs were irradiated by 40-eV photons up to a dose of $7.2 \times 10^{17}$ cm$^{-2}$. The excitation wavelength was 785 nm. In (b), the hump at ~300 cm$^{-1}$ is from the Si substrate.

Fig. 9(b) also shows that the diameter dependence of the damage is less prominent at -27 °C, meaning that the recovery of all the observed SWCNTs is almost completely forbidden at this temperature, regardless of the diameter. In other words, the less damage in a thicker SWCNT observed at room temperature and above is a consequence of the fact that a defect created in the thicker SWCNT can be more quickly healed by the thermal energy at the irradiation temperature. The diameter dependence can mainly be ascribed to the diameter dependence of the defect healing rather than to that of defect creation.

#### 4.5 Activation energy of defect healing

The recovery of the damage at room temperature or below suggests that the activation energy of the defect healing is quite small. We have proposed a simple method for determining the activation energy of defect healing in SWCNTs (Suzuki et al., 2010). For example, recovery curves of the G/D ratio can be used for the analysis. In Fig. 11(a), we show Raman spectra of SWCNTs before and after irradiation and after 2, 14, and 64 min annealing at 240 °C. The annealing gradually recovered the irradiated SWCNTs. From these measurements, we obtained recovery curves of the G/D ratio at several temperatures, as shown in Fig. 11(b). The reason for the relatively small values of the G/D ratio is that we adopted the area ratio rather than the peak height ratio, in order to decrease static errors.

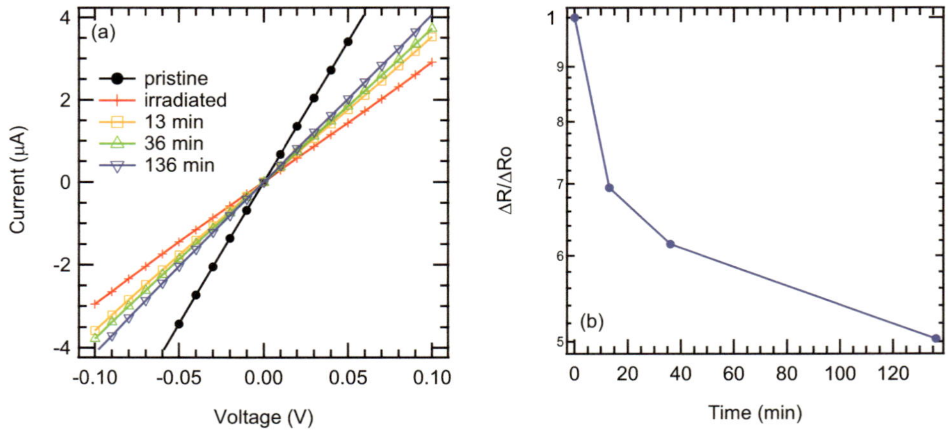


Fig. 10. (a) Current-voltage characteristics of a SWCNT device before and just after electron irradiation and 13, 36, 136 min after the irradiation. The electron energy and irradiation dose were 20 keV and $1.8\times10^{15}$ $cm^{-2}$,respectively. (b) Time evolution of the irradiation-induced resistance (ΔR) at 22 °C. The initial value of the irradiation-induced resistance (ΔRo) was 19.8 kΩ.

The activation energy $E_a$ of defect healing is given by

$$E_a = \frac{k}{\left(T_1^{-1} - T_2^{-1}\right)} \ln \frac{t_1}{t_2} \quad (1)$$

where $t_1$ and $t_2$ are annealing times during which the G/D ratio $R$ increases from $R_1$ to $R_2$. Here, it is sufficient that the G/D ratio is just a monotonic function of the defect density (We do not have to assume a specific relation between the G/D ratio and defect density, such as that the G/D ratio is inversely proportional to the defect density). The activation energy seems to depend on the extent of the damage. In the region of $2.8 \leq R \leq 3.0$ (heavily damaged), we obtained an activation energy value of 1.4 ±0.2 eV, as shown in Fig. 11(c), from the recovery curves at 140 and 120 °C and eq. (1), whereas at $6 \leq R \leq 6.5$ (lightly damaged), a smaller value of 0.7 ±0.2 eV was obtained. I would like to mention that these values may be affected by gas absorption at defect sites, because the SWCNTs were once exposed to air after the irradiation. Anyway, the values are small enough for the defects to be healed at moderate temperatures.

Interestingly, although a partial recovery of low-energy irradiation damage at room temperature is easily observed in the electric properties (Fig. 10), it has not been observed in Raman spectra. Once we obtain the activation energy value, we can estimate the recovery curve at a given temperature $T_3$. Using the recovery curve $R_1$ ($R_2$) at $T_1$ ($T_2$), the recovery curve at $T_3$ is given by

$$R_3(t_3) = R_{1,2}\left(t_3 \exp\left[\frac{E_a}{k}\left(T_{1,2}^{-1} - T_3^{-1}\right)\right]\right). \quad (2)$$

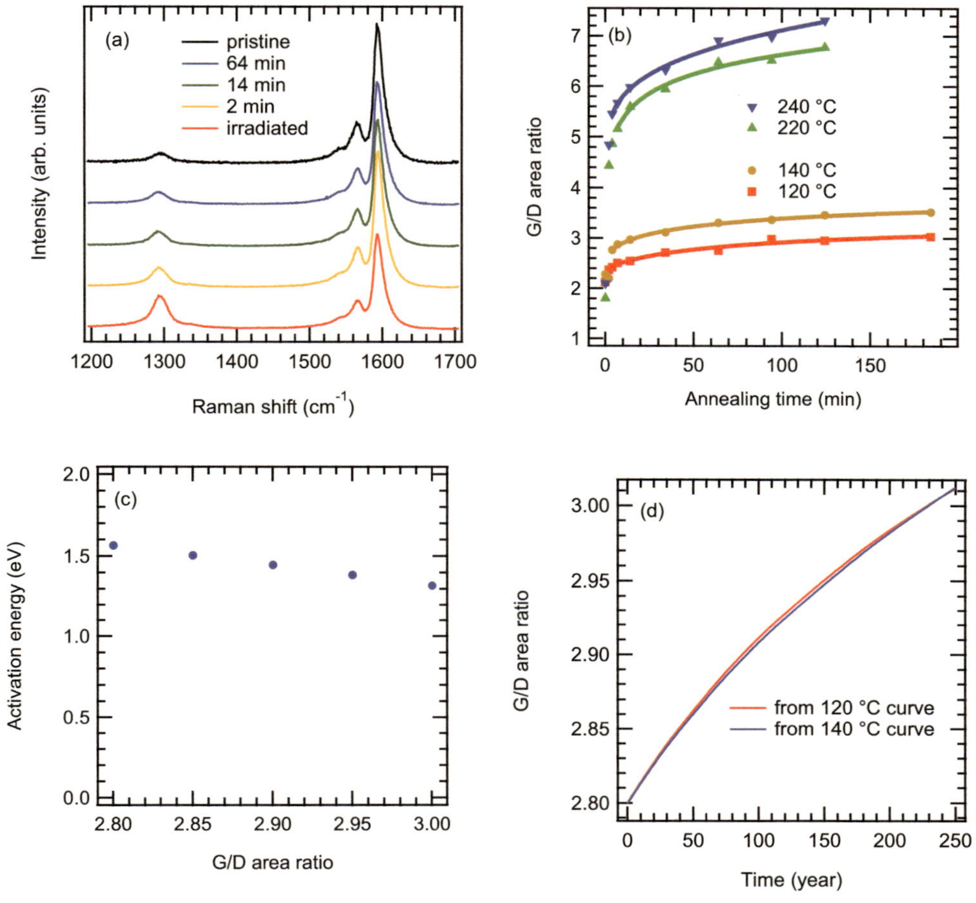


Fig. 11. (a) Raman spectra of SWCNTs before and after electron irradiation and after annealing at 240 °C. The electron energy and dose were 20 keV, and $8\times10^{16}$ $cm^{-2}$, respectively. The excitation wavelength was 785 nm. (b) Recovery curves of the G/D area ratio otbained at several annealing temperatures. (c) Activation energy of the defect healing obtained from the recovery curves at 140 and 120 °C. (d) Simulated recovery curves of the G/D ratio at room temperature (20 °C) obtained from the recovery curves at 120 and 140 °C in (b), respectively.

Fig. 11(d) shows the recovery curves at 20 °C simulated from the experimental recovery curves at 140 and 120 °C and eq. (2). The two independently obtained curves are almost consistent. Note that the unit of the horizontal axes is "year". Recovery of the G/D ratio from 2.8 to 3.0 at 20 °C would take about 230 years. Similarly, the recovery from 6.0 to 6.5 at 20 °C was estimated to take about 7 years (Suzuki et al., 2010). Thus, the recovery would be much too slow to observe at room temperature in usual experiments. The very long recovery time at room temperature is a consequence of the relatively slow recovery at elevated temperatures in Raman spectra. On the other hand, the recovery of the electric properties is much more rapid. Annealing at 300 °C for 30 min often results in recovery of conductivity of several orders of magnitude, as already shown in Fig. 7.

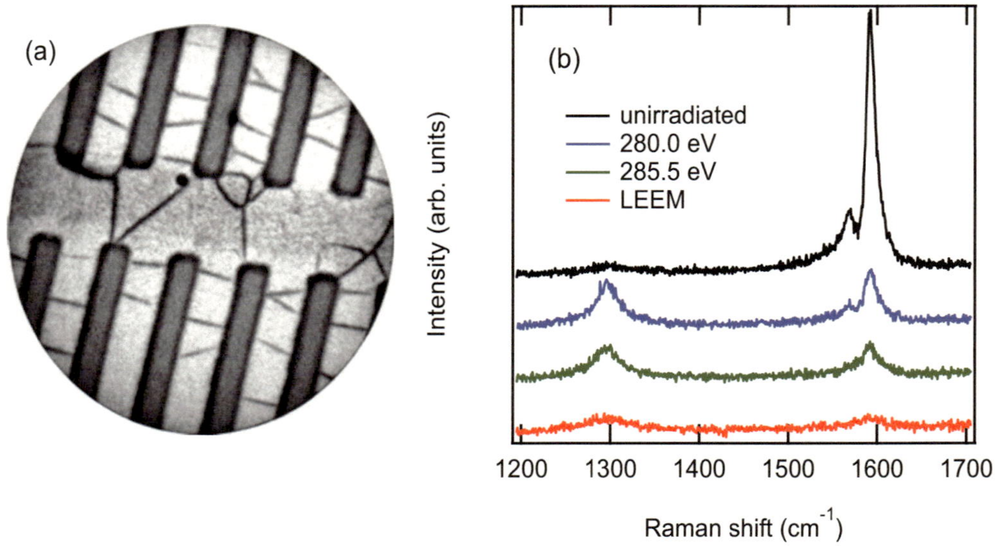


Fig. 12. (a) LEEM image of suspended SWNTs on a patterned Si substrate. (b) G and D band Raman spectra of unirradiated and soft X-ray-illuminated (280. 0 and 285.5 eV) SWCNTs and electron-irradiated (~20 eV) SWCNTs. The excitation wavelength was 785 nm. The soft X-ray illumination was done at BL-27SU at SPring-8, Hyogo, Japan. The electron irradiation was a consequence of LEEM observation. The soft X-ray and electron irradiations were done after thorough degassing.

Though we evaluate the activation energy from the G/D ratio here, any other quantity that is a monotonic function of the defect density can basically be used for the analyses. This method can also be used to analyze other kinds of defects and desorption barriers of chemisorbed atoms or molecules on SWCNTs.

### 4.6 Occurrence of the damage in an ultra-high vacuum

As I mentioned in sec. 2, the low-energy irradiation damage is caused by the irradiation itself. We found that the damage does not depend on the remnant gas pressure at ~$10^{-4}$ Pa or below (Suzuki et al., 2008). The low-energy irradiation damage has been observed in a surface-science-grade ultra-high vacuum (UHV) of ~$1\times10^{-8}$ Pa by VUV light illumination (Suzuki & Kobayashi, 2006a. Suzuki & Kobayashi, 2007a, Mera et al., 2009. Mera et al., 2010.), electron beam irradiation (Arima et al., 2009.), and electron (hole) injection from a STM tip (Berthe et al., 2007. Yamada et al., 2009). The damage occurs as ever when SWCNTs are thoroughly degassed in a UHV before irradiation. Other examples of occurrence of the damages in UHV surface analysis systems are shown in Fig. 12(b). The SWCNTs were irradiated by electrons during LEEM observation [Fig. 12(a)] or by soft X-rays at a photoemission spectroscopy beamline attached to a synchrotron radiation ring. The damage to SWCNTs is especially severe in LEEM observation using very low-energy electrons of several tens electron volts, due to the energy dependence of the damage (sec. 4.1). In an UHV, no irradiation-induced change is observed even in high-energy-resolution photoemission spectroscopy (Suzuki et al., 2004a), indicating that chemical reactions with gas molecules are negligible. Nevertheless, very severe damage is observed in Raman spectra.

### 4.7 Structure dependence

A low-energy electron and photon can easily dissociate a small molecule (for example, photodissociation). On the other hand, such low-energy irradiation damage (or structural change) is not commonly observed inside the bulk of a metal or semiconductor. Actually, it has not been reported for graphite. Very interestingly, even among CNTs, the damage has been reported for SWCNTs but not for MWCNTs. An electron irradiation experiment in an SEM has shown that the irradiation causes no reduction of the conductivity of MWCNTs with a diameter of ~10 nm (Bachtold et al., 1998. Hobara et al., 2004). The irradiation conditions used in those studies (4 $C{\cdot}cm^{-2}$ of 20-keV and 20 $C{\cdot}cm^{-2}$ of 10-keV electrons) roughly correspond to 10 to 1000 fold of a value that can cause a SWCNT conductivity decrease of a few orders of magnitude (Suzuki, 2011). Thus, the damage seems to be specific to SWCNTs or thin CNTs with a diameter of ~1 nm. Even among SWCNTs, the extent of the damage strongly depends on the diameter: Thinner SWCNTs are more severely damaged, as discussed in section 4.3.

The diameter dependence of the damage may explain the occurrence of the damage in SWCNTs and its absence in MWCNTs and graphite. Considering that the damage strongly depends on the diameter among SWCNTs, it would be possible that a MWCNT of 10-nm diameter is no longer damaged by low-energy irradiation at room temperature. If the occurrence and the absence of the damage originate in the diameter difference, we can expect that strain in the sidewall plays an essential role in the defect formation or its stabilization. Alternatively, it is interesting to view the occurrence and absence of the damage in terms of dimensionality. Graphite, in which the damage does not occur, is a three-dimensional material, and a SWCNT, in which the damage occurs, is a one-dimensional material. Notably, it has been well established that structural changes occur in zero-dimensional fullerenes by photon and electron irradiation (Zhao et al., 1994. Onoe et al., 2003). This is generally described as "polymerization" instead of damage, because the irradiation causes chemical bonds to form between neighboring fullerenes. The structural change can be reversibly restored by annealing, exactly like the low-energy irradiation damage of a SWCNT. The electronic states, which spread in the whole crystal in a bulk material, should be localized in low-dimensional materials or nanomaterials, and the degrees of freedom of atomic movement should become larger. Thus, in low-dimensional materials or nanomaterials, local structural change would easily occur with low-energy irradiation and the defect structure would be stabilized (See also sec. 6).

In terms of the relation between the damage and structure, it is very interesting to explore whether the damage occurs in graphene, which is a two-dimensional material and can be considered to be a SWCNT of infinite diameter. Zhou et al. reported that soft x-ray illumination damages graphene, on the basis of their C 1s x-ray absorption and Raman spectroscopy results (Zhou et al., 2009). Very interestingly, the illumination effects increased with a decreasing number of layers of exfoliated graphene and were negligible even for monolayer epitaxial graphene on SiC, which has a relatively strong interaction with the substrate. These results suggest that low dimensionality is strongly related to the low-energy irradiation damage.

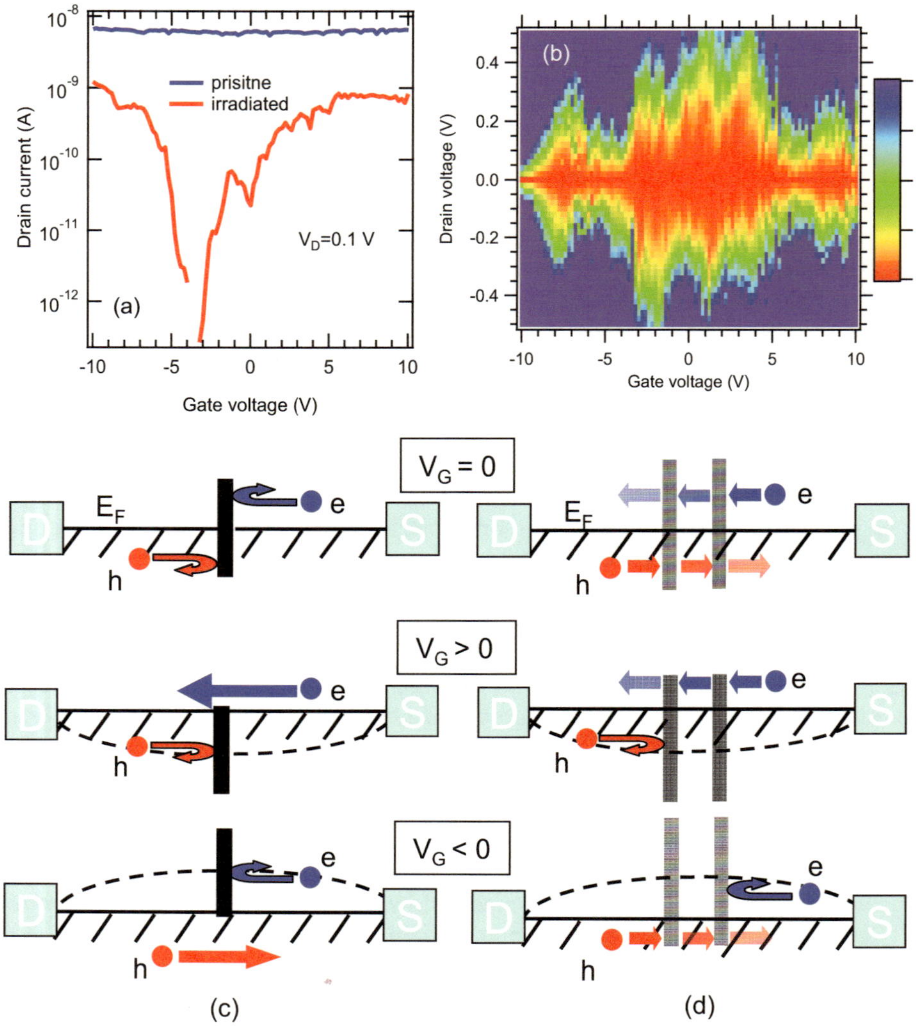


Fig. 13. (a) Room-temperature gate voltage characteristics of a SWCNT device before and after electron irardiation. The SWCNT was once scanned by an electron beam of 100 pA. The electron energy and scan speed were 20 keV and 400 nm/s, respectively. (b) Room-temperature Coulomb diamond characteristics of a SWCNT device before and after electron irradiation. The SWCNT was once scanned by an electron beam of 500 pA. The electron energy and scan speed were 20 keV and 400 nm/s, respectively. (c) Schematic explanations of the defect-induced semiconductng properites. (d) Schematic explanations of the defect-induced Coulomb oscillation properites.

## 5. Metal-semiconductor transition of a SWCNT-FET induced by defects

As mentioned in sec. 3.3, intensive irradiation finally makes a SWCNT almost insulating. However, when the damage is moderate, a metal-semiconductor transition of the electric properties is often observed. In our early study, we irradiated the whole device in a SEM and observed the conversion of the electric properties at 28 K (Vijayaraghavan et al., 2005). Further irradiation caused an increase of the nominal band gap observed in the low-temperature electric properties. More recently, we succeeded in converting the room-temperature device characteristics from metallic to semiconducting by local irradiation using an electron beam lithography system (Suzuki et al., 2008). Before the irradiation, the device characteristics were almost gate-independent, which is a common feature of a metallic SWCNT. A part of a metallic SWCNT was once scanned by an electron beam. Then, the room-temperature gate characteristics of the device were converted to semiconducting, as shown in Fig. 12(a). After the irradiation, ambipolar semiconducting gate characteristics were clearly observed.

Room-temperature Coulomb oscillations have been observed when defects form a small dot in a SWCNT channel (Matsumoto et al., 2003). The low-energy irradiation damage can also be used to fabricate such small dots intentionally. As shown in Fig. 12(b), after irradiation, multi-dot Coulomb oscillation properties are sometimes observed at room temperature.

A schematic model of a possible mechanism for the irradiation-induced semiconducting properties is shown in Fig. 12(c) (Kanzaki et al., 2007. Suzuki et al., 2008). The temperature dependence of device characteristics after irradiation shows that an energy barrier for carriers is formed in the SWCNT channel. The barrier height observed in the electric properties reaches about 0.6 eV, when irradiation-induced semiconducting properties are observed at room temperature. Recently, a STM study more directly showed that a local band gap is actually formed in a metallic SWCNT by a carrier injection-induced defect (Yamada et al., 2009). This suggests that the defect-induced local band gap opening is the origin of the barrier. The carrier transport is inhibited by the barrier at the gate voltage of around 0 V. The device still turns on at large gate voltage. This can be reasonably explained in terms of gate-induced band bending in a metallic SWCNT. The density of states near the Fermi level of a metallic SWCNT is very small. Owing to the small density of states near the Fermi level, we can bend the band by applying gate voltage and reduce the effective barrier height for an electron. At sufficiently large gate voltage, the device will turn on. Thus, the metal-semiconductor transition is explained by the defect-induced barrier formation and gate-induced band bending. When Coulomb oscillation is observed, the defects seem to act as tunneling barriers, as schematically shown in Fig. 12(d). In this case, tiny multi-dots divided by the defects seem to be formed in the vicinity of the irradiated part.

The defect-induced conversion of the electric properties seems to be caused by defects formed by other methods. In fact, conversion of the electric properties from metallic to semiconducting also occurs when defects are induced by plasma treatment of metallic SWCNT-FETs. More interestingly, the defect-induced semiconducting electric properties well explain the fact that the ratio of "semiconducting" SWCNTs that act as FETs has been reported to strongly depend on the growth method (Suzuki et al., 2008, Mizutani et al., 2009). The plasma-enhanced CVD method has been reported to produce preferentially semiconducting SWCNTs, and the ratio of semiconducting SWCNTs has been reported to reach about 90 % (Li et al., 2004. Ohnaka et al., 2006) or even 97 % (Mizutani et al., 2009). On the other hand, for the laser ablation method, which generally produces high-quality SWCNTs, the semiconducting SWCNT ratio was evaluated to be quite small, about 30 % (Li et al., 2004). I think that the growth method dependence of the "semiconducting" SWCNT ratio is mainly due to the growth method dependence of defect density. Distinguishing whether the electronic structure is metallic or semiconducting by electric measurements may be inconclusive, especially when the SWCNT shows "semiconducting" properties.

## 6. Mechanism of the low-energy irradiation damage

The low-energy irradiation damage can be caused by 10-eV photons, which have very small momentum. This indicates that the momentum of an incident particle would have no essential role in the damage, which is in remarkable contrast to knock-on damage. Thus, the defect formation would be due to bond breaking, which follows an electronic excitation by the energy of the incident particle. An energy of ~10 eV is still high enough to cut C-C bonds. Thus, it is reasonable that low-energy irradiation creates a defect with finite probability if the defect structure is stable and the lifetime is long enough. A simple example of this kind of structural change is photodissociation of a molecule. A bond breaking following electronic excitation can easily dissociate a small molecule In a bulk crystal, on the other hand, even the breaking of several bonds would result in immediate re-bonding without any structural change because an atom has very little freedom of displacement due to the existence of surrounding atoms. The situation in a SWCNT is one between a molecule and bulk. More than one bond breaking would be necessary to stabilize the defect. Among related carbon materials, low-energy electron and photon irradiation-induced structural change (polymerization) is known to occur for fullerenes. On the other hand, the damage has not been reported for graphite or MWCNTs, as discussed in sec. 4.7.

Interestingly, Yamada et al. (Yamada et al., 2009) have proposed, on the basis of thier STM results, that carrier injection first creates primary defects whose lifetime is very short (<50 ms). Most of them are quickly annihilated and the structure is restored. However, in rare cases, a primary defect fails to recover and a stable defect is created. The quantum efficiency of the primary defect formation was evaluated to be $2\times10^{-10}$ at a bias voltage of 3.5 V near the defect creation threshold.

The detailed atomic structure of a low-energy irradiation-induced defect is not clear at present. A detectable change has not been observed even with microscopy techniques, such as SEM. This is one of the main reasons that the low-energy irradiation had not been recognized for such a long time, although SEM had been commonly used for characterizing SWCNTs since their discovery. Our previous TEM observation showed that the tube wall is not clearly destroyed regardless of severe damage (Suzuki et al., 2005b). The Stone-Wales defect, which is formed by a C-C bond rotation, is consistent with the conservation of the number of carbon atoms. However, the stable structure seems to contradict the relatively small activation energy and healing at a moderate temperature or even at room temperature or below. Another possible defect is a vacancy in the tube wall with a migratory C adatom on the surface. The observed activation energies (0.7-1.4 eV, sec. 4.5) are very close to the C adatom migration energies, which are theoretically predicted to depend on the SWCNT diameter and to be 0.6 to 1.3 eV (Krasheninnikov et al., 2004). However, a high-resolution TEM observation has shown that annihilation of the vacancy and migratory adatom is governed by the recombination barrier rather than by the adatom migration barrier itself (Hashimoto et al., 2004). The existence of such a vacancy-adatom defect (with the adatom bounded in the vicinity of the vacancy) was also strongly suggested by a scanning tunneling microscopy study (Lee et al., 2005). The vacancy-adatom defect is simply formed by breaking two bonds of a C atom. The recombination barrier of the vacancy-adatom defect in SWCNTs has been calculated to be ~1-2 eV (Okada, 2007), which is rather close to the observed activation energies. Determining the presice defect structure is a future issue.

## 7. Other types of damage and irradiation-induced phenomena

The low-energy irradiation damage is often confused with other types of damage and irradiation-induced phenomena. Here, I would like to summarize differences in the defect characteristics of low-energy irradiation damage and other damages.

### 7.1 Knock-on damage

Knock-on damage is caused by ballistic ejection of an atom from a solid by an incident particle. Thus, the damage is accompanied by a loss of SWCNT mass. The displacement energy (kinetic energy at which an atom can escape from the solid.) in a SWCNT is considered to depend on the diameter and to be 15-20 eV. However, when the incident particle is an electron whose mass is much smaller than a carbon atom, the threshold energy becomes ~80 keV from the energy and momentum conservation laws. The threshold energy for a photon will be much larger. Thus, the energy at which knock-on damage is observed is much larger than that at which the low-energy irradiation damage is normally observed. Knock-on damage occurs in MWCNTs and graphite. For a recent review focusing on carbon and other nanomaterials, see (Krasheninnikov, 2010).

### 7.2 Contaminant effects

Low-energy irradiation often causes hydrocarbon contaminants to adhere to the sample surface. The contaminant adhesion is prominent in a conventional SEM, in which a UHV is usually unavailable. At the electron energy where severe damage is observed (1 keV or smaller), severe contaminant adhesion is also observed. This is because the interaction both between electrons and a SWCNT and between electron and hydrocarbon gases are strong at such a low energy. However, effects caused by the damage and by the contaminant adhesion can be easily distinguished by annealing. The contaminants do not sublimate even at ~900 °C, whereas the low-energy irradiation damage can be recovered by annealing at moderate temperatures, as discussed in sec. 4. In our experiments, it is rather difficult to detect contaminant effects because the spectra or electric properties of pristine SWCNTs and SWCNTs recovered from the damage are almost identical (see Figs. 6 and 7), although I do not deny that the contaminants cause some so-called environmental effects (Ohno, 2010).

### 7.3 Damage by radicals

Irradiation in remnant gases can also cause radical-induced etching, which is the opposite of the contaminant adhesion. For example, the cutting of MWCNTs has been clearly demonstrated under gas atmosphere formed by intentional gas bleeding (Yuzvinsky et al., 2005). In this way, the damage by radicals is generally accompanied by etching, which eventually cuts and eliminates CNTs. Thus, this damage can not be fully recovered. In some literatures, reversible chemisorption and desorption on SWCNT or graphene has been suggested. However, spectra before chemisorption and after desorption by annealing are often compared after arbitrary normalization. This damage also occurs in thick MWCNTs and graphite, although some diameter dependence of the damage is observed (Yang et al., 2006, Zhang et al., 2006b). Metallic SWCNTs are preferentially damaged by radicals (Yang et al., 2006. Zhang et al., 2006b), although such preference has not been observed for low-energy irradiation damage. Moreover, severer damage is observed at higher temperatures due to more activated chemical reactions (An et al., 2002. Zhang, 2006a). This is entirely opposite to the low-energy irradiation damage (sec. 4.4). Of course, the radical effect

becomes severer at higher pressures and negligible in a UHV. Reader should recall that low-energy irradiation damage occurs in a UHV after thorough degassing, although no indication of chemical reaction is observed (sec. 4.6).
In a standard SEM in which the adhesion of contaminants occurs, the radical effects seem to be less important. Otherwise, the contaminants would not adhere due to etching. According to my experience, intensive irradiation in a conventional SEM can not cut or eliminate even a SWCNT (Fig. 7). I suppose that the radical effects are largely suppressed due to the adhesion of contaminants, which would protect the sample surface (sec. 3.2).

### 7.4 Substrate charging effects

There has been an attempt to explain the electric property changes by irradiation-induced charging of the substrate (back-gate dielectric) just under the SWCNT (Marquardt et al., 2008. Vijayaraghavan et al., 2010). They observed a large conductivity decrease of on-substrate SWCNT devices by electron irradiation in a SEM. Furthermore, they reversibly and repeatedly recovered the electric properties by applying a high bias voltage (~10 V) to the SWCNT. The phenomena they observed seem to be essentially the same as ours. However, they ascribed the conductivity decrease to a local band gap opening caused by irradiation-induced charging of the dielectric $SiO_2$ layer just under the SWCNT. Actually, theoretical calculations predict that a uniform (Li et al., 2003) or inhomogeneous (Rotkin & Hess, 2004) electric field can open a gap in a metallic SWCNT of specific chiralities. Marquardt et al. (Marquardt et al., 2008) and Vijayaraghavan et al. (Vijayaraghavan et al., 2010) think that electron irradiation in a SEM causes such local and inhomogeneous charging of the dielectric. In their model, the recovery is explained by a release of trapped charges in the vicinity of the SWCNT caused by the high-bias voltage applied to the drain electrodes. In a conventional on-substrate device, a large electric field may be produced by irradiation-induced charging, considering that the field strength is inversely proportional to the square of the distance.
However, the substrate charging model does not at all explain the fact that the irradiation-induced conductivity decrease is as ever observed for suspended SWCNTs, as shown in Fig. 4. The theoretical calculations have predicted that an extraoridinarily high electric field of ~1 $Vnm^{-1}$ barely opens a band gap of several ten milli-electron volts. It is very unlikely that such a high electric field is formed at SWCNTs suspended 300 nm above the substrate. In fact, a simulation has been performed under a condition where the gate voltage was applied from a metal tip located only 0.5 nm from a SWCNT (Rotkin & Hess, 2004). Similarly, this model cannot explain the degradation of Raman and PL spectra of suspended SWCNTs (Figs. 1 and 9). Moreover, electron (hole) injection-induced band gap opening has been observed for a metallic SWCNT lying on a metal substrate, which does not have charge trap sites (Yamada et al., 2009). This model does not explain the observed band gap value, either. The calculations show that the maximum value of the field-induced band gap is at most ~0.1 eV, which is not sufficient to explain the almost insulating properties observed at room temperature. In fact, an energy barrier of ~0.6 eV was observed for a SWCNT whose room-temperature electric properties were converted from metallic to semiconducting by irradiation (sec. 5). Finally, it should be noted that the irradiation-induced conductivity decrease has been observed in all measured SWCNTs, whereas, in the theoretical calculations, band gaps open only in SWCNTs having certain chiralities. Considering that the irradiation-induced physical property changes can recover at a moderate temperature (~300 °C [Fig. 7]), the high bias-induced recovery observed in refs. 7 and 8 seems to be due

to annealing by Joule heating. I do not deny an electric field-induced band gap opening in a metallic SWNT. However, I do not think that such a high or inhomogeneous electric field is produced by simple SEM observation or line scans.

### 7.5 Irradiation-induced heating effects

Low-energy electron and photon irradiation may increase the temperature of the irradiated SWCNTs. However, the heating effect itself does not explain the low-energy irradiation damage at all because less damage is observed at higher temperatures, as shown in Fig. 1. Originally, SWCNTs are thermally very stable materials. Thus, at least under usual conditions, irradiation-induced heating itself would not damage the SWCNTs in a vacuum, if ever. In practice, damage is often observed during Raman measurements in air when the excitation laser power is too large. However, this is not low-energy irradiation damage, but instead would be combustion, because this damage is not observed in a vacuum or an inert gas atmosphere.

## 8. Conclusion

I have shown that low-energy electron and photon irradiation solely damages SWCNTs. The low-energy irradiation damage extinguishes the characteristic optical and electric properties and reduces chemical tolerance. Thus, we have to pay attention to the damage when we use analytical tools that use low-energy electrons (SEM, LEEM etc.) and VUV light or soft X-rays (photoemission spectroscopy using bright light). The defects have some unique properties. The damage and recovery are reversible, indicating that the number of carbon atoms is preserved. The damage strongly depends on diameter. That is, thinner SWCNTs are more severely damaged. The damage has been observed in SWCNTs but not in MWCNTs, suggesting that it is characteristic of low-dimensional structures or nanostructures. The activation energy of the defect healing depends on the extent of the damage and was evaluated to be about 0.7 to 1.4 eV. Because of the relatively small activation energy, the defects can be healed even at room temperature or below, and less damage occurs at higher temperatures. I also showed that the irradiation-induced defects can convert the room temperature electric properties of a metallic SWCNT to semiconducting. The conversion can be explained by the local band gap opening caused by the defect and gate-voltage-induced band bending in the metallic SWCNT. Energetically, the low-energy is still sufficiently larger than the C-C bond energy and can therefore break the bonds. Future studies should address the detailed defect structure.

## 9. Acknowledgment

This work has been done through cooperation of many coworkers. I thank all of my coworkers for their cooperation and assistance in this work.

## 10. References

An, K. H.; Heo, J. G.; Jeon, K. G.; Bae, D. J.; Jo, C.; Yang, C. W.; Park, C. Y.; Lee, Y. H.; Lee, Y. S. & Chung Y. S. (2002). X-ray photoemission spectroscopy study of fluorinated single-walled carbon nanotubes. Appl. Phys. Lett. Vol. 80, No. 22 (April 2002), pp. 4235-7.

Arima, S.; Lee, S.; Mera, Y.; Ogura, S.; Fukutani, K.; Sato, S.; Tohji, J. & Maeda, K. (2009). Electron-stimulated defect formation in single-walled carbon nanotubes studied by hydrogen thermal desorption spectroscopy. *Appl. Surf. Sci.* Vol. 256, No. 4 (June 2009), pp. 1196-9.

Bachtold, A.; Henny, M.; Terrier, C.; Strunk, C.; Schonenberger, C.; Salvetat, J. P.; Bonard, J. M. & Forro, L. (1998). Contacting carbon nanotubes selectively with low-ohmic contacts for four-probe electric measurements. *Appl. Phys. Lett.* Vol. 73, No. 2 (July 1998), pp. 274-6.

Berthe, M.; Yoshida, S.; Ebine, Y.; Kanazawa, K.; Okada, A.; Taninaka, A.; Takeuchi, O.; Fukui, N.; Shinohara, H.; Suzuki, S.; Sumitomo, K.; Kobayashi, Y.; Grandidier, B.; Stievenard, D. & Shigekawa, H. (2007). Reversible defect engineering of single-walled carbon nanotubes using scanning tunneling microscopy. Nano Lett. Vol. 7, No. 12 (November 2007), pp. 3623-7.

Chen, B. H.; Wei, J. H.; Lo, P. Y.; Pei, Z. W.; Chao, T. S.; Lin, H. C. & Huang, T. Y. (2006). Novel method of converting metallic-type carbon nanotubes to semiconducting-type carbon nanotube field-effect transistors. *Jpn. J. Appl. Phys.* Vol. 45, No. 4B (April 2006), pp. 3680-5.

Dresselhaus, M. S.; Dresselhaus, G.; Saito, R. & Jorio, A. (2005). Raman spectroscopy of carbon nanotubes. *Phys. Rep.*, Vol. 409, (2005), pp. 47-99.

Hashimoto, A.; Suenaga, K.; Gloter, A.; Urlta, K. & Iijima, S. (2004). Direct evidence for atomic defects in graphene layers. Nature Vol. 430, No. 7002 (August 2004), pp. 870-3.

Hobara, R.; Yoshimoto, S.; Ikuno, T.; Katayama, M.; Yamauchi, N.; Wongwiriyapan, W.; Honda, S.; Matsuda, I.; Hasegawa, S. & Oura, K. (2004). Electric transport in multiwalled carbon nanotubes contacted with patterned electrodes. Jpn. J. Appl. Phys. Vol. 43, No. 8B (July 2004), pp. L1081-4.

Kanzaki, K.; Suzuki, S.; Inokawa, H.; Ono, Y.; Vijayaraghavan, A. & Kobayashi, Y. (2007). Mechanism of metal-semiconductor transition in electric properties of single-walled carbon nanotubes induced by low-energy electron irradiation. J. Appl. Phys. Vol. 101, No. 3 (Feburuary 2007), pp. 034317-1-4.

Krasheninnikov, A. V.; Nordlund, K.; Lehtinen, P. O.; Foster, A.; Ayuela, S. & Nieminen, R. M. (2004). Adsorption and migration of carbon adatoms on zigzag carbon nanotubes. Carbon Vol. 42, No. 5-6, (January 2004), pp. 1021-5.

Krasheninnikov, A. V. & Nordlund, K. (2010). Ion and electron irradiation-induced effects in nanostructured materials. J. Appl. Phys. Vol. 107, No. 7 (April 2010), pp. 071301-1-70.

Li, Y.; Rotkin, V. R. & Ravaioli U. (2003). Electronic response and bandstructure modulation of carbon nanotubes in a transverse electric field. Nano Lett. Vol. 3, No. 2 (November 2003), pp. 183-7.

Marquardt, C. W.; Dehm, S.; Vijayaraghavan, A.; Blatt, S.; Hennrich, F. & Krupke, R. (2008). Reversible metal-insulator transitions in metallic single-walled carbon nanotubes. *Nano Lett.* Vol. 8, No. 9 (August 2008), pp. 2767-72.

Lee, S.; Kim, G.; Kim, H.; Choi, B.; Lee, J.; Jeong, B. W.; Ihm, J.; Kuk, Y. & Kahng, S. J. (2005). Paired gap states in a semiconducting carbon nanotube: Deep and shallow levels. Phys. Rev. Lett. Vol. 95, No. 16 (October 2005), pp. 166402-1-4.

Matsumoto, K.; Kinoshita, S.; Gotoh, Y.; Kurachi, K.; Kamimura, T.; Maeda, M.; Sakamoto, K.; Kuwahara, M.; Atoda, N. & Awano, Y. (2003). Jpn. J. Appl. Phys. Vol. 42, No. 4B (Janurary 2003), pp. 2415-8.

Mera, Y.; Harada, Y.; Arima, S.; Hata, K.; Shin, S. & Maeda, K. (2009). Defects generation in single-walled carbon nanotubes induced by soft x-ray illumination. *Chem. Phys.Lett.* Vol. 473, No. 1-3 (March 2009), pp. 138-41.

Mera, Y.; Fujikawa, T.; Ishizaki, K.; Xiang, R.; Shimomi, J.; Maruyama, S.; Kakiuchi, T.; Mase, K. & Maeda, K. (2010). Ion desorption from single-walled carbon nanotubes induced by soft x-ray illumination. *Jpn. J. Appl. Phys.* Vol. 49, No. 10 (October 2010), pp. 105104-1-5.

Mizutani, T.; Ohnaka, H.; Okigawa, Y.; Kishimoto, S. & Ohno, Y. (2009). A study of preferential growth of carbon nanotubes with semiconducting behavior grown by plasma-enhanced chemical vapor deposition. *J. Appl. Phys.* Vol. 106, No. 7 (October 2009), pp. 073705-1-5.

Ohnaka, H.; Kojima, Y.; Kishimoto, S.; Ohno, Y & Mizutani, T. (2006). Fabrication of carbon nanotube field effect transistors using plasma-enhanced chemical vapor deposition grown nanotubes. *Jpn. J. Appl. Phys.* Vol. 45, No. 6B (June 2006), pp. 5485-9.

Ohno, Y. (2010). Environmentaleffects on photoluminescence of single-walled carbon nanotubes. In: Carbon Nanotubes, InTech, ISBN 978-953-307-054-4.

Okada, S. (2007). Energetics and electronic structures of carbon nanotubes with adatom-vacancy defects. Chem. Phys. Lett. Vol. 447, No. 4-6 (September 2007), pp. 263-7.

Onoe, J.; Nakayama, T.; Aono, M. & Hara T. (2003). Structural and electrical properties of an electron-beam-irradiated C60 film. Appl. Phys. Lett. Vol. 82, No. 4 (January 2003), pp. 595-7..

Rotkin, V. R. & Hess, K. (2004). Possibility of a metallic field effect transistor. *Appl. Phys. Lett.* Vol. 84, No. 16 (February 2004), pp. 3139-41.

Suzuki, S.; Watanabe, Y.; Ogino, T.; Homma, Y.; Takagi, D.; Heun, S.; Gregoratti, L.; Barinov, A. & Kiskinova, M. (2004a). Observation of single-walled carbon nanotubes by photoemission microscopy. *Carbon,* Vol. 42, No. 3, (January 2004), pp. 559-L63.

Suzuki, S.; Kanzaki, K.; Homma, Y. & Fukuba, S. (2004b). Low-acceleration-voltage electron irradiation damage in single-walled carbon nanotubes. *Jpn. J. Appl. Phys.,* Vol. 43, No. 8B, (July 2004), pp. L1118-L1120.

Suzuki, S.; Takagi, D.; Homma, Y.; Kobayashi, Y. (2005a). Selective removal of carbon nanotubes utilizing low-acceleration-voltage electron irradiation damage. *Jpn. J. Appl. Phys.* Vol. 44, No. 4 (January 2005), pp. L133-5.

Suzuki, S.; Fukuba, S.; Kanzaki, K.; Homma, Y. & Kobayashi, Y. (2005b). Spatially selective removal of carbon nanotubes for fabricating nanotube circuits. *Proceedings of 5th IEEE Conference on Nanotechnology.* Nagoya, July 2005.

Suzuki, S. & Kobayashi, Y. (2005). Conductivity decrease in carbon nanotubes casued by low-acceleration-voltage electron irradiation. *Jpn. J. Appl. Phys.* Vol. 44, No. 49 (November 2005), pp. L1498-L1501.

Suzuki, S.; Maeda, F. & Kobayashi, Y. (2006). Photon-induced damage creation in carbon nanotubes. *30th Fullerenes Nanotube General Symposium.* Nagoya, January 2006.

Suzuki, S. & Kobayashi, Y. (2006a). Diameter dependence of low-energy electron and photon irradiation damage in single-walled carbon nanotubes. *Chem. Phys. Lett.* Vol. 430, No. 1-3 (September 2006), pp. 370-4.

Suzuki, S. & Kobayashi, Y. (2006b). Processing and electric property control of carbon nanotubes by low-energy electron irradiation. *NTT Technical. Review* Vol. 4, No. 11 (November 2006), pp. 25-30.

Suzuki, S. & Kobayashi, Y. (2007a). Healing of low-energy irradiation-induced defects in single-walled carbon nanotubes at room temperature. *J. Phys. Chem. C* Vol. 111, No. 12 (March 2007), pp. 4524-8.

Suzuki, S. & Kobayashi, Y. (2007b). Low-energy irradiation damage in single-walled carbon nanotubes. *Mater. Res. Soc. Symp. Proc.* Vol. 994, San Francisco, (April 2007), pp. F04-02-1-12.

Suzuki, S. & Kobayashi, Y. (2008). Threshold energy of low-energy irradiation damage in single-walled carbon nanotubes. *Jpn. J. Appl.Phys.* Vol. 47, No. 4 (April 2008), pp. 2040-3.

Suzuki, S.; Hashimoto, J.; Ogino, T. & Kobayashi, Y. (2008). Electric property control of carbon nanotubes by defects. *Jpn. J. Appl.Phys.* Vol. 47, No. 4 (April 2008), pp. 3292-5.

Suzuki, S. Yamaya, K.; Homma, Y.; Kobayashi, Y. (2010). Activation energy of healing of low-energy irradiation-induced defects in single-wall carbon nanotubes. *Carbon* Vol. 48, No. 11 (May 2010), pp. 3211-7.

Suzuki, S. (2011). Origin of the electric property change of a single-wall carbon nanotube caused by low-energy irradiation: defects or substrate charging? *e-J. Surf. Sci. & Nanotechnol.* Vol. 9, (March 2011), pp. 103-6.

Suzuki, S. & Hibino, H. (2011). Characterization of doped single-wall carbon nanotubes by Raman spectroscopy. *Carbon* Vol. 49, No. 7 (January 2011), pp. 2264-72.

Vijayaraghavan, A.; Marquardt, C. W.; Dehm, S.; Hennrich F. & Krupke, R. (2010). Imaging defects and junctions in single-walled carbon nanotubes by voltage-contrast scanning electron microscopy. *Carbon* Vol. 48, No. 2 (September 2009), pp. 494-500.

Yamada, K.; Sato, H.; Komaguchi, T.; Mera, Y. & Maeda, K. (2009). Local opening of a large bandgap in metallic single-walled carbon nanotubes induced by tunnel injection of low-energy electrons. *Appl. Phys. Lett.* Vol. 94, No. 25 (June 2009), pp. 253103-1-3.

Yan, Y. H.; Chan-Park, M. B.; Zhou, Q.; Li, C. M.; Yue, C. Y. (2005). Functionalization of carbon nanotubes by argon plasma-assisted ultraviolet grafting. *Appl. Phys. Lett.* Vol. 87, No. 21 (November 2005), pp. 213101-1-3.

Yang, C. M.; An, K. H.; Park, J. S.; Park, K. A.; Lim, S. C.; Cho, S. H.; Lee, Y. S.; Park, W.; Park, C. Y. & Lee. Y. H. (2006). Preferential etching of metallic single-walled carbon nanotubes with small diameter by fluorine gas. Phys. Rev. B Vol. 73, No. 7 (February 2006), pp. 075419-1-7.

Yuzvinsky, T. D.; Fennimore, A. M.; Nickelson, W.; Esquivias, C. & Zettl, A. (2005). Precision cutting of nanotubes with a low-energy electron beam. Appl. Phys. Lett. Vol. 86, No. 5 (January 2005), pp. 053109-1-3.

Zhang, G.; Qi, P.; Wang, X.; Lu, Y.; Mann, D.; Li, X. & Dai, H. (2006a). Hydrogenation and hydrocarbonation and etching of single-walled carbon nantoubes. *J. Am. Chem. Soc.* Vol. 128, No. 18 (April 2006), pp. 6026-7.

Zhang, G.; Qi, P.; Wang, X.; Lu, Y.; Li, X.; Tu R. Bangsaruntip, S.; Mann, D.; Zhang, Li. & Dai, H. (2006b). Selective etching of metallic carbon nanotubes by gas-phase reaction. Science Vol. 314, No. 5801 (November 2006), pp. 974-7.

Zhao, Y. B.; Poirier, D. M.; Pechman, R. J. & Weaver, J. H. (1994). Electron stimulated polymerization of solid C60. Appl. Phys. Lett. Vol.64, No. 5 (January 1994), pp. 577-9.

Zhou, S. Y.; Girit, C. O.; Scholl, A.; Jozwiak, C. J.; Siegel, D. A.; Yu, P.; Robinson, J. T.; Wang, F. & Zettl, A. (2009). Instability of two-dimensional graphene: Breaking sp2 bonds with soft x rays. Phys. Rev. B Vol. 80, No. 12 (September 2009), pp. 121409-1-4.

16

# Exciton Dephasing in a Single Carbon Nanotube Studied by Photoluminescence Spectroscopy

Kazunari Matsuda
*Institute for Chemical Research, Kyoto University, Uji, Kyoto*
*Institute of Advanced Energy, Kyoto University, Uji, Kyoto*
*Japan*

## 1. Introduction

The lattice vibration strongly affects the dynamics of electrons and excitons in the solid. In bulk semiconductor materials, the loss of coherence (i.e., dephasing) of electrons and excitons occurs by phonon scattering on very fast time scale of sub-picosecond order. The dephasing time and mechanism depend strongly on the dimensionality and size of the electronic states in low-dimensional semiconductors, such as one-dimensional (1D) quantum wires and zero-dimensional quantum dots (Gammon et al., 1996 & Braun et al., 1997). A single-walled carbon nanotube (SWNT) is a prototypical 1D electronic system and the optical properties of SWNTs have attracted a great deal of attention, both from the perspective of their fundamental physics (Iijima 1991, O'Connell et al., 2002, & Bachilo et al., 2002), and for their optoelectronic device applications. Potential applications include photoluminescence (PL) (O'Connell et al., 2002 & Bachilo et al., 2002), electroluminescence (Misewich et al., 2003), nonlinear optical switching media (saturable absorbers) for ultrafast lasers (Rozhin et al., 2006), and fluorescent biolabeling (Leeuw et al., 2007). These optical properties are due to the creation of very stable excitons by enhanced Coulomb interactions of the order of several-hundred meV, even at room temperature (Ando 1997, Kane et al., 2003, Spataru et al., 2004, Wang et al., 2005, & Maultzsch et al., 2005). The exciton dynamics are dominated by both the exciton dephasing and the energy relaxation processes, in which the exciton-phonon interactions play an important role (Bradley et al., 2007). However, the exciton dephasing time and mechanism in SWNTs is still only poorly understood and under many discussions. Thus, we describe about detail mechanism of the exciton dephasing in a single carbon nanotube.

## 2. Optical transition of carbon nanotube

In 2002, O'Connell reported that the micelle encapsulated semiconducting SWNTs show strong PL signals (O'Connell et al., 2002). The first observation of PL signals is due to the isolation of individual SWNT preventing from bundling between semiconducting and metallic carbon nanotubes. Lefebvre also fabricated suspended SWNTs between patterned Si pillars and also observed strong PL signals from unprocessed SWNTs (Lefebvre et al., 2003). The black line of Figure 1 shows the typical macroscopic (ensemble averaged) PL spectrum from micelle encapsulated SWNTs dispersed in $D_2O$ solution. The many PL peaks arising from various types of SWNTs with different chirarities (diameters) are observed in the macroscopic PL spectrum. The optically excited e-h pair in SWNTs forms bound hydrogen like exciton state due to strong Coulomb interaction and recombination of exciton (not free e-h pair) is the origin PL signals in the SWNTs. The shape of each PL (or absorption) peak contains physical information including the exciton dynamics. The shape of ensemble averaged PL spectrum shows the near Gaussian function, which means that each SWNT with same chiral indicies is in the different environment (local strain, dielectric constant, and ....). The exciton transition (absorption and PL) of each SWNT is affected from the Gaussian distributed inhomogeneous environment, causes fluctuation of the energy. The PL broadening arising from the inhomogenity is called as "inhomogeneous broadening", which covers the intrinsic optical and electronic properties of SWNTs. In contrast, the effect of the inhomogeneous broadening is eliminated in the PL spectra from a single SWNT (red line in Fig. 1). The sophisticated optical spectroscopic technique, called as a single carbon nanotube spectroscopy, enables us to observe PL spectrum from a single SWNT (Hartschuh et al., 2003, Htoon et al., 2003, Lefebvre et al., 2004, & Matsuda et al., 2005). The red solid line of Figure 1 shows the PL spectrum from a single SWNT. The PL of a single SWNT shows very clear single peak and very narrow linewidth, which provide us the fruitful information of intrinsic properties of SWNT. The intrinsic linewidth in the PL spectrum of a single SWNT is called as homogeneous linewidth. The homogeneous linewidth contains the dynamical properties of exciton in the SWNT, discussed after in detail.

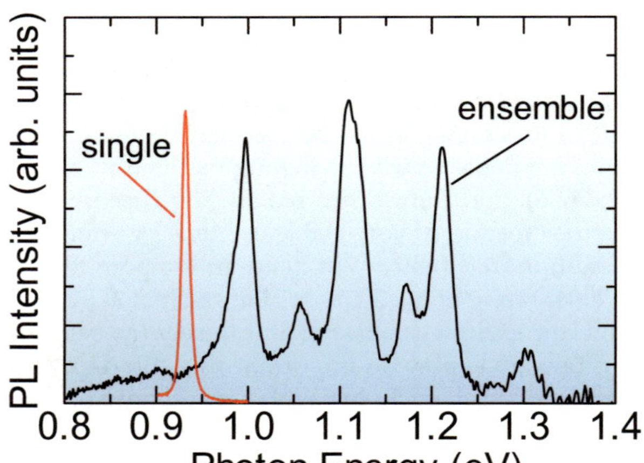


Fig. 1. Ensemble averaged PL spectrum (balck solid line) and PL spectrum of a single SWNT (red line).

### 2.1 Homogeneous linewidth broadening of optical spectrum

In the quantum two-level systems, the optical spectra broaden and have finite linewidths due to finite lifetime of the excited state according to the Heisenberg relationship. There are two kinds of excited state lifetime, energy relaxation time $T_1$ and phase relaxation (dephasing) time $T_2$. In the molecular systems, the phase relaxation of the excited state occurs due to the violation of the excited state such as molecular collisions and so on. In the solid systems, the thermal vibration of lattice, i.e. phonon perturbs the electronic states, leading to the phase relaxation of excited states (dephasing). In contrast, the energy relaxation occurs radiative and nonradiative transition from the excited states to the ground state. The linewidth of the optical spectra, homogeneous broadening Γ, can be described as,

$$\Gamma = \hbar(1/T_1 + 2/T_2), \tag{1}$$

where $\hbar(=h/2\pi)$ is a Plank constant. Using this relationship, the information on the dephasing time can be obtained from homogeneous broadening of the optical spectra. Also, in the time-domain measurement, the dephasing time can be measured by using nonlinear optical spectroscopy such as photon echo (or four-wave mixing) experiments (Graham et al., 2011).

## 3. Experimental technique of single carbon nanotube imaging and spectroscopy

The samples used for single SWNT spectroscopy were isolated SWNTs synthesized on patterned Si substrates by an alcohol catalytic chemical vapor deposition method (Maruyama et al., 2002). The Si substrates were patterned with parallel grooves typically from 300 nm to a few μm in width and 500 nm in depth using an electron-beam lithography technique. The isolated SWNTs grow from one side toward the opposite side of the groove. We prepared several SWNT samples by changing the growth temperature (650-850 °C) and time (30 sec-10 min). The average number density of isolated SWNTs in the sample is 0.1-1/μm$^2$ (Inoue et al., 2006, Matsuda et al., 2008, & Matsunaga et al., 2008).

Single SWNT PL measurements were carried out from 300 (room temperature) to 5 K using a home-built variable temperature confocal microscope setup as shown in Fig. 2. The SWNT samples mounted on a stage were excited with a continuous-wave He-Ne laser (1.959 eV) and femtosecond pulsed Ti:$Al_2O_3$ Laser, and the laser beam was focused on the sample

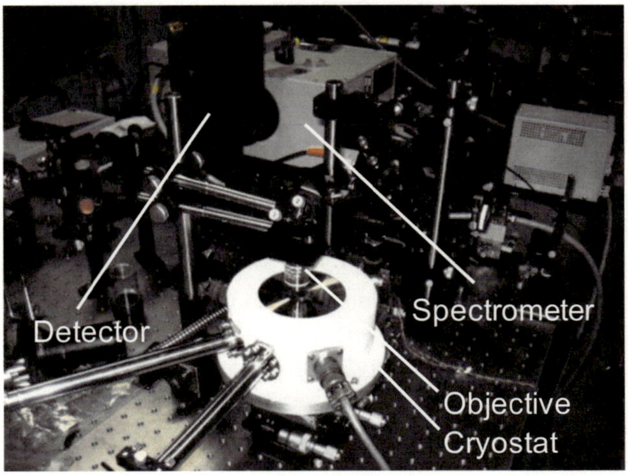


Fig. 2. Picture of experimental setup of low temperature optical microscope for a single SWNT spectroscopy.

surface through a microscope objective (NA 0.8). The PL signal from SWNTs was spectrally dispersed by a 30-cm spectrometer equipped with a liquid-nitrogen ($LN_2$)-cooled InGaAs photodiode array (spectral range: 0.78-1.38 eV) or a $LN_2$-cooled charge coupled device (spectral range: ≥1.20 eV). The spectral resolution of our system is typically less than 0.7 meV. The detector accumulation time was typically between 30 sec and 1 min. We also obtained PL images of the luminescent SWNTs on the sample, using a Si avalanche photodiode or electron multiple (EM) CCD camera.

Figure 3 shows a typical PL image of isolated semiconducting SWNTs at room temperature. The monitored PL energy range is typically between 1.18 and 1.37 eV. Several spatially isolated bright spots can be seen in the 3D image. Each bright spot represents the PL signal from a single luminescent SWNT because the PL spectrum obtained just on each bright spot has only a single PL peak, as shown later. The spot size estimated from the spatial profiles of the PL intensity is about 500 nm.

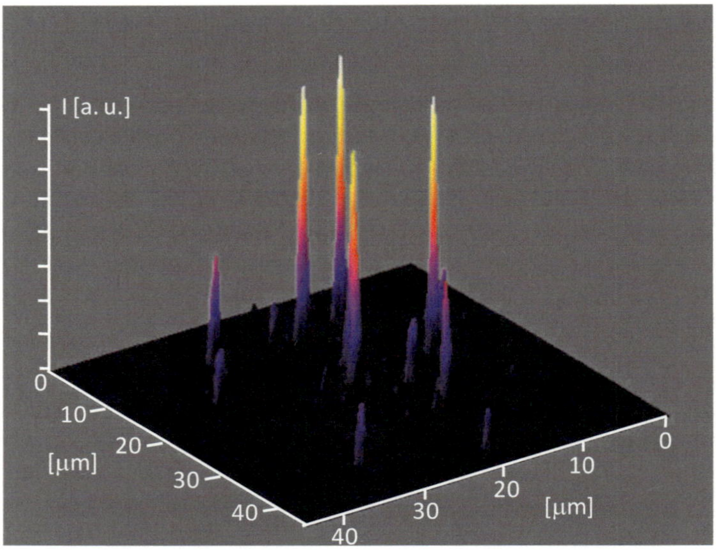


Fig. 3. 3D plots of PL image of single SWNT at room temperature.

### 3.1 Single carbon nanotube spectra

We obtained PL spectra from about 180 different isolated SWNTs with a variety of chiral indices as shown in Fig. 4. The chiral index assignment for each PL spectrum is based on the previous data of the emission energies (Lefebvre, Fraser, et al., 2004). There is a distribution of emission energy (~ten meV) even within the same chiral index. Only a single sharp peak can be seen in each spectrum. This is certain evidence that the PL comes from a single SWNT. In addition, the PL linewidth tends to become broader with an increase of the PL peak energy. Figures 5(a) and (b) show, in more details, the typical PL spectra of single isolated SWNTs with relatively small and large diameter [(a): (10, 6), $d$ = 0.83 nm, (b): (7, 5), $d$ = 1.11 nm,] in the sample.

The red solid lines in Figs. 5(a) and (b) represent single Lorentzian functions and approximately reproduce the experimentally obtained PL spectra. In addition, it is found

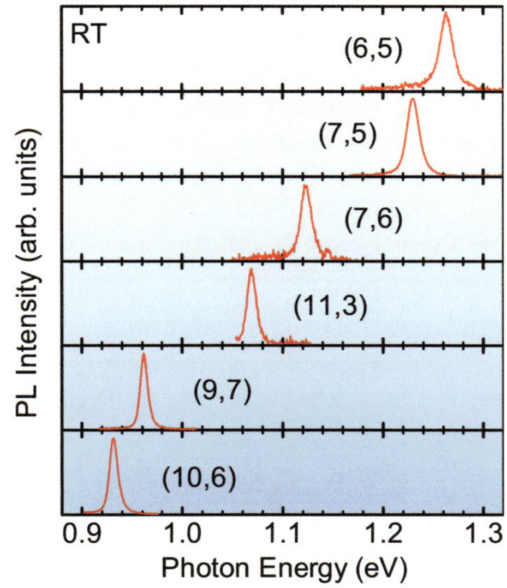


Fig. 4. PL spectra of single SWNT with different chiral index (*n*, *m*) (Reprinted with permission from [Inoue et al, 2006]. Copyright, American Physical Society).

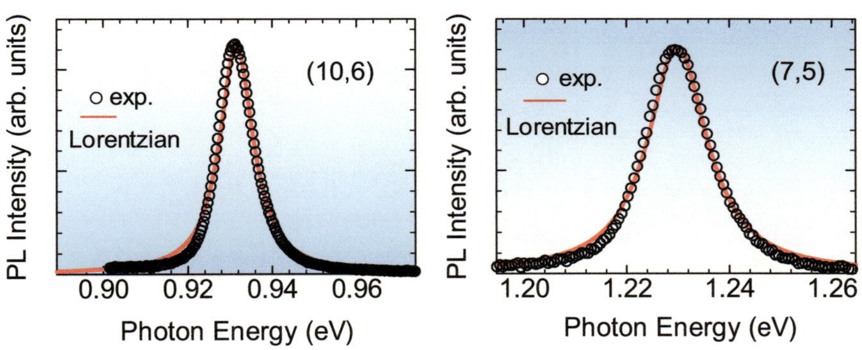


Fig. 5. (a), and (b) PL spectra of single SWNT with chiral index of (10, 6) and (7, 5) at room temperature (Reprinted with permission from [Inoue et al, 2006]. Copyright, American Physical Society).

that the Stokes shift of SWNTs was very small (less than 5 meV) for each nanotube species by comparison between absorption and PL spectra of ensemble SWNTs in gelatin matrices. These facts show that the observed PL peaks correspond to the zero-phonon lines of free excitons and the spectral linewidth of the PL spectra is determined by the homogeneous broadening.

### 3.2 Dephasing mechanism of exciton state

Figure 6. shows temperature dependence of PL spectra of single SWNTs from 40 to 297 K. Solid red lines correspond to fitted Lorentzian functions. The SWNT has a chiral index of (9, 8), based on the emission energies reported (Lefebvre, Fraser, Homma et al., 2004). The PL spectral shape can be approximately fitted by a single Lorentzian function at entire temperature range. The linewidths (full-width at half-maximum, FWHM) of the Lorentzian functions are almost dominated by the exciton dephasing time due to exciton-phonon interactions, because the exciton lifetime is longer than 30 ps, and the contribution to the linewidth from the exciton lifetime (energy relaxation time) is negligibly small (Hirori et al., 2006, & Gokus et al., 2008). From the homogeneous linewidth, we calculated the exciton dephasing time at room temperature as about 120 fs for (9, 8) nanotubes. The order of the experimentally obtained dephasing time is almost consistent with the Ab initio calculation (Bradley et al., 2007).

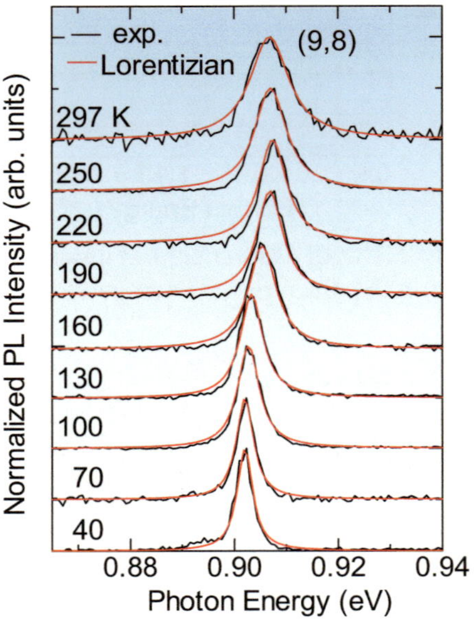


Fig. 6. Temperature dependence of PL spectra of a single SWNT from 40 to 300 K (Reprinted with permission from [Yoshikawa et al, 2008]. Copyright, American Physical Society).

The PL spectra become narrower when the temperature decreases as shown in Fig. 6. The linewidth of a single SWNT corresponding to a homogeneous linewidth is plotted as a function of temperature in Fig. 7. The temperature dependence of the homogeneous linewidth is usually described as (Karaiskaj et al., 2007):

$$\Gamma \approx \Gamma_0 + A \cdot T + \frac{B}{\exp(\hbar\omega / k_B T) - 1}, \tag{2}$$

where $\Gamma_0$ is the residual linewidth at $T$=0 K including the instrumental spectral resolution, and is the energy of the high frequency phonon modes in SWNTs, such as the radial breathing mode (RBM) and averaged phonon mode. The coefficients $A$ and $B$ are exciton-phonon coupling constants for low- and high-frequency modes, respectively. The solid line in Fig. 7 shows the linewidth fitted using Eq. (2). The temperature dependence of the homogeneous linewidth shows almost linear behaviour over a wide temperature range. The value of $A$ obtained for the (9, 8) nanotube is 0.020 meV/K. The contribution of the high-frequency modes to the linewidth broadening is much smaller than that of the low-frequency acoustic phonon mode and negligibly small.

We obtained PL spectra at 300 K from more than 200 different isolated SWNTs with a variety of chiral indices. The homogeneous linewidth as a function of the nanotube diameter $d$ from the PL spectra of single carbon nanotubes is shown in Fig. 8. The smallest values are plotted as the intrinsic linewidth for nanotubes with the same chiral index, because the extrinsic factors (defects, impurities, and so on) cause broadening to a residual linewidth. The PL linewidth at 300 K clearly increases as the diameter decreases.

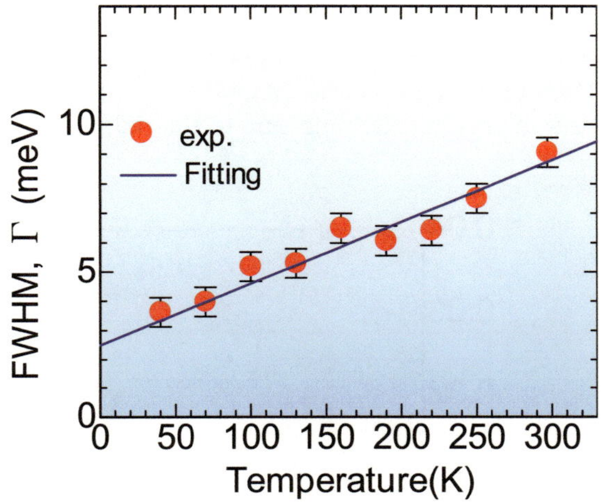


Fig. 7. Temperature dependence of linewidth broadening and theoretically fitted result using Eq. (2) (Reprinted with permission from [Yoshikawa et al, 2008]. Copyright, American Physical Society).

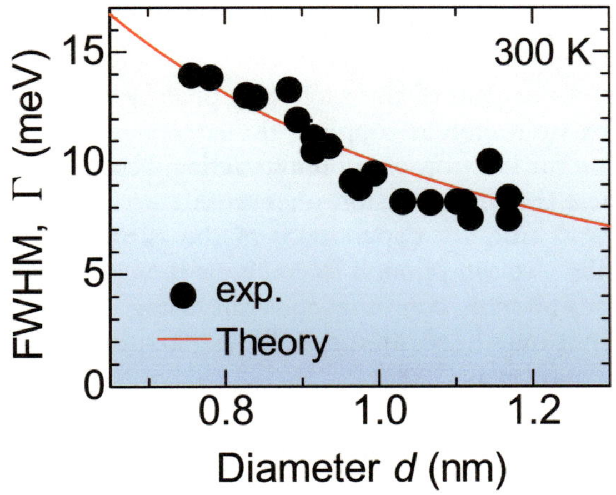


Fig. 8. Diameter dependence of homogeneous linewidth (Reprinted with permission from [Yoshikawa et al, 2008]. Copyright, American Institute of Physics).

The exciton-phonon coupling constant $A$ for low energy phonon modes is plotted in Fig. 9. The diameter dependence of $A$ is well consistent with that of the homogeneous linewidth at 300 K in Fig. 8. These observations indicate that the diameter dependence of the PL linewidth observed at 300 K is not due to the residual linewidth and the nonlinear temperature term in Eq. (2), which is caused by high-frequency modes such as the RBM and averaged phonon mode. From the diameter dependence of the homogeneous linewidth and the exciton-phonon coupling constant $A$, it can be seen that the exciton-phonon interaction is stronger and the exciton dephasing time shorter in smaller tubes. Figure 10 shows the chirality dependence of the homogeneous linewidth at 300 K and exciton-phonon coupling constant $A$. The data are plotted for SWNTs of around 1.0 nm indiameter (within the range from 0.9-1.2 nm). Both the homogeneous linewidth and the exciton-phonon coupling constant are almost constant, and neither show the clear chiral angle θ dependence. However, there is the scattering of the data due to the diameter dependence, as discussed below.

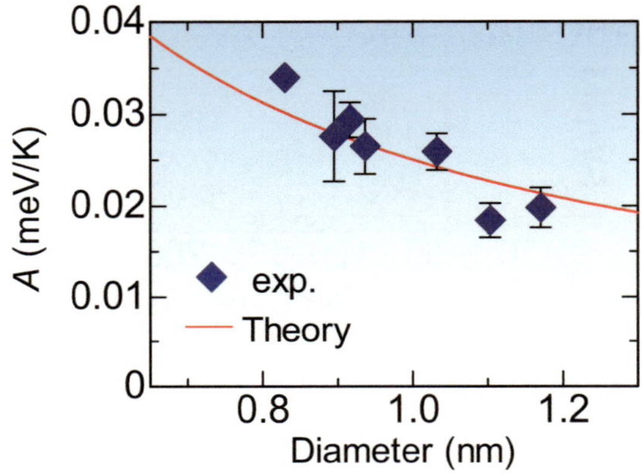


Fig. 9. Diameter dependence of linewidth broadening factor $A$ and theoretically calculated result (Reprinted with permission from [Yoshikawa et al, 2008]. Copyright, American Institute of Physics).

We will discuss the mechanism of the exciton dephasing from the diameter and chirality dependence of the exciton-phonon coupling (homogeneous linewidth) in detail. Here we refer the discussion on the electron-phonon interaction, because there have been no detailed quantitative theoretical studies of exciton-phonon interactions. As described in Jiang et al., 2007, the diameter and chirality dependence of the exciton-phonon interactions show a similar tendency to the electron-phonon interactions (Jiang et al., 2005, & Jiang et al., 2007). The exciton (electron)-phonon coupling constant using the reduced exciton (electron)-phonon matrix element may be written as follows (Suzuura et al., 2002, Jiang et al., 2005, Jiang et al., 2007, & Popov et al., 2006),

$$A \propto \frac{\sqrt{M_0}}{d} \cdot \sum_j \left| g_j \right|^2 E^{-1/2}, \tag{3}$$

where $M_0$ is the exciton effective mass, $E$ is the exciton energy, and $g_j$ is the reduced matrix-element of the exciton-phonon interaction. As both $M_0$ and $E$ are inversely proportional to the diameter *d*, the exciton-phonon coupling in Eq. (2) has $1/d$ dependence. The diameter dependence ($A \propto 1/d$) arises from the specific characteristics of the exciton-phonon interaction in 1D carbon nanotubes.

The linear temperature dependence of the linewidth noted above suggests that the phonon modes that contribute to linewidth broadening are low-energy acoustic modes, such as TW (twisting) and LA (longitudinal acoustic). The chirality dependences of the predicted reduced exciton (electron)-phonon matrix elements for TW and LA phonon modes are described by $g_{TW}=\cos 3\theta$ and $g_{LA}=0.66\sin 3\theta$, respectively (Suzuura et al., 2002). Using the reduced matrix-elements, we calculated the diameter dependence of the exciton-phonon coupling constant from Eq. (2). The results are indicated by the red solid line in Figs. 8 and 9. The calculated diameter dependence of the homogeneous linewidth from Eqs. (2) and (3) is also shown in Figs. 8 and 9. We assumed that the residual linewidths are constant (2.0 meV) in the theoretically calculated curve, because they are almost constant independent of the diameter, and the variation of the values is small (within ~1 meV) for each single nanotube. The theoretical calculated line in Fig. 9 agrees closely with the experimental results.

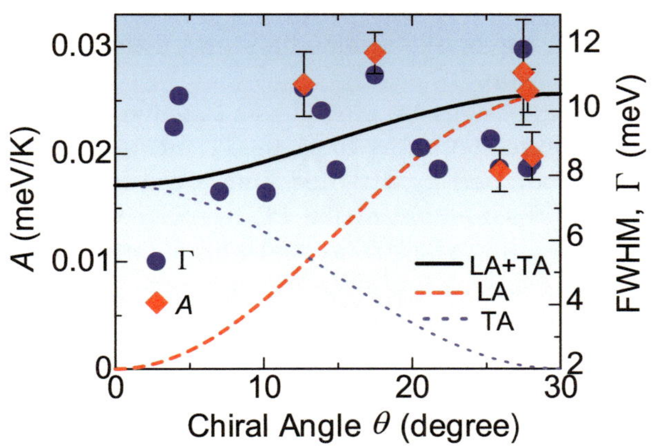


Fig. 10. Chiral angle dependence of linewidth broadening factor *A* and FWHM at room temperature. (solid line) (Reprinted with permission from [Yoshikawa et al, 2008]. Copyright, American Institute of Physics).

Figure 10 also shows the calculated chirality dependence of the exciton-phonon coupling constant (homogeneous linewidth) for SWNTs of around 1 nm in diameter, indicated by the solid line. The contributions of the two phonon modes LA and TW are indicated as dashed and dotted lines, respectively. The calculated result indicated by the solid line is in agreement with the experimental results of the chirality dependence of the exciton-phonon coupling constant (homogeneous linewidth). In the small (large) chiral angle nanotubes, the TW (LA) mainly contributes to the exciton-phonon coupling constant (homogeneous linewidth). These observations indicate that both the LA and TA modes contribute to the exciton dephasing for carbon nanotubes.

### 3.3 Dephasing exciton state via exciton-exciton interaction

We studied temperature dependence of the PL spectra of an isolated SWNT excited with a He-Ne laser. The spectra were measured under lower excitation conditions (less than < 100 μW). The PL spectra clearly narrow when the temperature decreases (Lefebvre et al., 2004). The exciton energy changes from 5 to 90 K (~1 meV) is much smaller than that in other compound semiconductor quantum wire (12 meV from 5 to 90 K in GaAs wires) in the weak exciton-phonon coupling regime. This weak temperature dependence is consistent with the experimental fact of narrow PL spectra with Lorentizian lineshape in Fig. 5. Both the dephasing time and the energy relaxation time of the excitons contribute to homogeneous linewidth broadening. The measured PL lifetimes of excitons are considerably longer (≥20 ps) from 10 to 300 K (Perebeinos et al., 2005, & Hirori et al., 2006) and this contribution to the linewidth is negligibly small (≤ 0.01 meV) under lower excitation conditions. Thus, the temperature-dependent linewidths are determined approximately by the dephasing time of the exciton due to the exciton-phonon interactions. Based on the homogeneous linewidth, the exciton-dephasing time is evaluated from 350 fs at 90 K to more than 940 fs at 5 K.

Figure 11 shows PL spectra obtained from a typical single SWNT [assigned chiral index: (11, 4)] at 30 K and at various excitation intensities of 1.72-eV and 150-fs laser pulses. Each spectrum has a single peak located at 0.941 eV. Even in the high-excitation region above about 15 pJ per pulse, the PL spectrum shows a single peak, without a change in the peak energy. Furthermore, additional spectral structures are not observed in any lower energy regions. Similarly, the PL bands due to biexcitons and inelastic exciton scattering are not observed in this spectral region.

Spectrally integrated PL intensities are plotted as a function of excitation laser intensity in Fig. 12. In a low excitation region below 10 pJ, the PL intensity grows almost linearly with excitation intensity (as indicated by a dotted line). Conversely, in the higher excitation intensity region (> 20 pJ), saturation of the PL intensity is clearly apparent. We show normalized PL spectra for a single SWNT, excited with different intensities, on an expanded energy scale in Fig. 13. The spectral linewidth broadens with increasing excitation intensity.

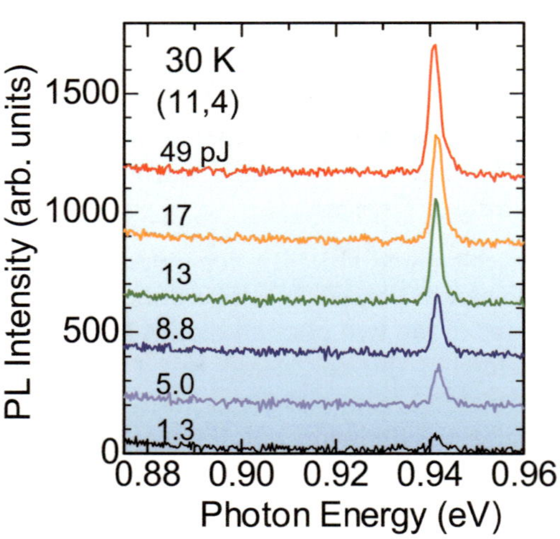


Fig. 11. Excitation power dependence of PL spectra of single SWNT at 30 K (Reprinted with permission from [Matsuda et al, 2008]. Copyright, American Physical Society).

The homogeneous linewidth is plotted in Fig. 14 as a function of excitation intensity, with the FWHM of the linewidth broadening nonlinearly. This broadening cannot be explained by laser heating effects because the nonlinear broadening behavior contradicts the linear temperature dependence of the linewidth broadening [see Fig. 7]. This excitation intensity-dependent linewidth broadening indicates that the exciton dynamics are strongly affected by the multiple excitons present in a SWNT.

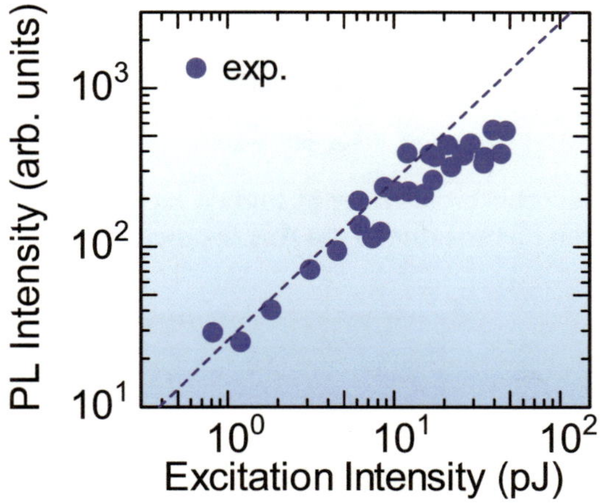


Fig. 12. Integrated PL intensity as a function of excitation intensity. Dotted line corresponds to linear dependence (Reprinted with permission from [Matsuda et al, 2008]. Copyright, American Physical Society).

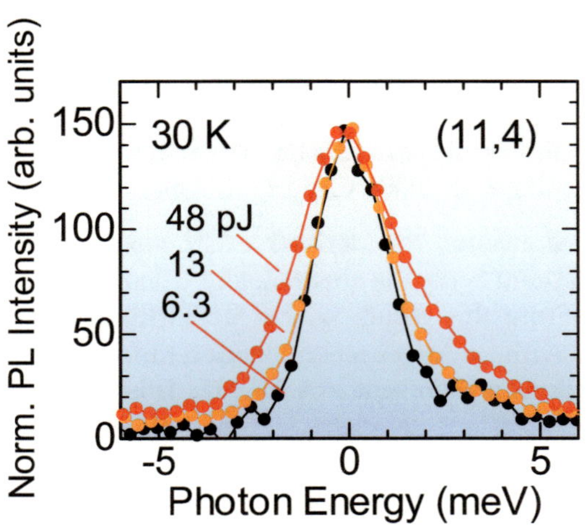


Fig. 13. Normalized PL spectra of a single SWNT on an expanded energy scale excited with different intensities (Reprinted with permission from [Matsuda et al, 2008]. Copyright, American Physical Society).

Under higher excitation conditions, when additional homogeneous linewidth broadening and saturation of the PL intensity occur, more than one exciton is created in a SWNT, leading to a remarkable scattering process between excitons. Exciton-exciton annihilation occurs due to the rapid Auger nonradiative recombination process through strong Coulomb interactions (Wang et al., 2004, Ma et al., 2005, Haung et al., 2006, & Wang et al., 2006).

The PL intensity saturation at higher excitation regions, as shown in Fig. 12, can be explained by the opening of the nonradiative relaxation path due to the exciton-exciton annihilation.

The homogeneous linewidth $\Gamma$ in Fig. 14 is determined by both the excitation power independent exciton-phonon interaction term $\Gamma_{ex\text{-}ph}$ and the excitation power dependent Auger term $\Gamma_{ex\text{-}ex}$,

$$\Gamma = \Gamma_{ex-ph} + \Gamma_{ex-ex}. \tag{6}$$

The homogeneous linewidth broadening at higher excitation regions can be explained by the shortening of exciton lifetime due to the Auger process.

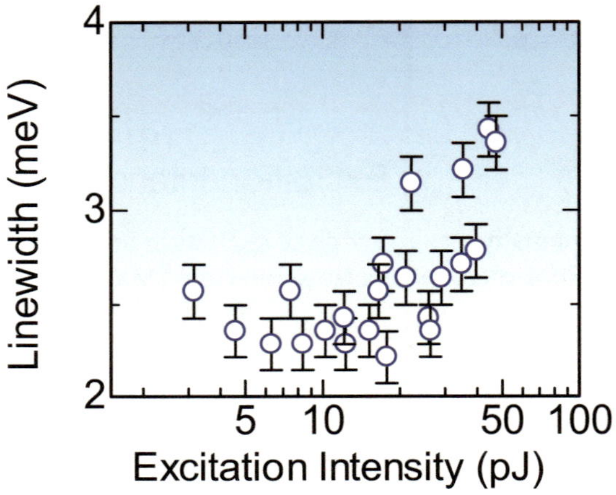


Fig. 14. Homogeneous linewidths as a function of excitation intensity (Reprinted with permission from [Matsuda et al, 2008]. Copyright, American Physical Society).

From the experimental results, the derived exciton-exciton annihilation constant of 1.6 ps/μm is almost consistent with the theoretically calculated value based on perturbation theory (Wang 2006). From this result, it was found that the multiparticle Auger process occurs very efficiently with an Auger recombination time estimated at 800 fs for ~1 μm-long SWNT when two excitons are present in a SWNT. This very short, in comparison to bulk semiconductors, time constant is characteristics of low-dimensional systems, and is comparable to the exciton-dephasing time at low temperature. These results indicate that the exciton dephasing is limited under high excitation conditions, not by the exciton-phonon interactions, but by the exciton-exciton interactions.

## 4. Conclusion

In this chapter, we studied the temperature and chirality dependence of the PL linewidth of single carbon nanotubes using single SWNT spectroscopy to clarify the mechanism of exciton dephasing. The PL linewidth of a single carbon nanotube broadened linearly with increasing temperature, indicating that the linewidth and exciton dephasing are determined through exciton-phonon interactions. From the chirality dependence of the PL linewidth, we concluded that exciton dephasing is caused by both the longitudinal acoustic and twisting phonon modes. Saturation of the PL intensity and broadening of the PL linewidth in a single SWNT occur simultaneously with an increase in excitation laser intensity. Our findings show that the rapid exciton-exciton annihilation through multiparticle Auger recombination broadens the homogenous linewidth.

## 5. Acknowledgment

The author thanks Y. Miyauchi, Y. Murakami, S. Maruyama, T. Inoue, K. Yoshikawa, R. Matsunaga, and Y. Kanemitsu for collaborations and discussions. This study is partially supported by Grant-in-Aid from MEXT, Japan.

## 6. References

Ando, T. (1997). Excitons in Carbon Nanotubes. *Journal of the Physical Society of Japan,* Vol. 66, No. 4, pp. 1066-1073.

Bachilo, S. M.; Strano, M. S.; Kittrell, C.; Hauge, R. H.; Smalley, R. E. & Weisman, R. B. (2002). Structure-assigned Optical Spectra of Single-walled Carbon Nanotubes. *Science,* Vol. 298, pp. 2361-2366.

Braun, W.; Bayer, M.; Forchel, A.; Zull.; H. Reithmaier J. P.; Filin, A. I.; Walck S. N. & Reinecke T. L. (1997). Excitonic Wave Packets in $In_{0.135}$ $Ga_{0.865}$ As/GaAs Quantum Wires. *Physical Review B,* Vol. 55, No. 15, pp. 9290–9293.

Bradley, F. H.; Kamisaka, H.; Yamashita K. & Prezhdo, Oleg V. (2007). Ab Initio Study of Vibrational Dephasing of Electronic Excitations in Semiconducting Carbon Nanotubes. *Nano Letters,* Vol. 7, No. 11, pp. 3260-3265.

Gammon, D.; Snow, E. S.; Shanabrook, B. V.; Katzer, D. S. & Park D. (1996). Fine Structure Splitting in the Optical Spectra of Single GaAs Qantum Dots. *Physical Review Letters,* Vol. 76, No. 16, pp. 3005-3008.

Gokus, T.; Hartschuh, A.; Harutyunyan, H.; Allegrini, M.; Hennrich, F.; Kappes, M.; Green, A. A.; Hersam, M. C.; Araújo, P. T. & Jorio A. (2008). Exciton Decay Dynamics in Individual Carbon Nanotubes at Room Temperature. *Applied Physics Letters,* Vol. 92, No. 15, 153116-1-153116-3.

Graham, M. W.; Ma, Y-Z.; Green, A. A.; Hersam, M. C. & Fleming, G. R. (2011). Pure Optical Dephasing Dynamics in Semiconducting Single-walled Carbon Nanotubes. *Journal of Chemical Physics,* Vol. 134, No. 3, pp. 034504-1-034504-13.

Hartschuh, A.; Pedrosa, H. N.; Novotny L. & Krauss, T. D. (2003). Simultaneous Fluorescence and Raman Scattering from Individual Single-walled Carbon Nanotubes. *Science,* Vol. 301 No. 5638, pp. 1354-1356.

Hirori, H.; Matsuda, K.; Miyauchi Y.; Maruyama S. & Kanemitsu Y. (2006). Exciton Localization of Single-walled Carbon Nanotubes Revealed by Femtosecond Excitation Correlation Spectroscopy. *Physical Review Letters,* Vol. 97, No. 25, pp. 257401-1-257401-4.

Htoon, H.; O'Connell, M. J.; Cox, P. J. Doorn, S. K. & Klimov V. I. (2002). Low Temperature Emission Spectra of Individual Single-walled Carbon Nanotubes: Multiplicity of

Subspecies within Single-species Nanotube Ensembles. *Physical Review Letters,* Vol. 93, No. 2, pp. 027401-1-027401-4.

Huang L. & Krauss, T. D. (2006). Quantized Bimolecular Auger Recombination of Excitons in Single-Walled Carbon Nanotubes. *Physical Review Letters,* Vol. 96, No. 5, 057407-1-057407-4.

Iijima, S. (1991). Helical Microtubules of Graphitic Carbon. *Nature,* Vol. 354, pp. 56-58.

Jiang, J.; Saito, R.; Gruneis, A.; Chou, S. G.; Samsonidze, G. G.; Jorio, A.; Dresselhaus G. & Dresselhaus, M. S. (2005). Photoexcited Electron Relaxation Processes in Single-wall Carbon Nanotubes. *Physical Review B,* Vol. 71, No. 4, 045417-1-045417-9.

Jiang, J.; Saito, R.; Sato, K.; Park, J. S.; Samsonidze, Ge. G.; Jorio, A.; Dresselhaus, G. & Dresselhaus, M. S. (2007). Exciton-photon, Exciton-phonon Matrix Elements and Resonance Raman Intensity of Single Wall Carbon Nanotubes. *Physical Review B,* Vol. 75, No. 3, pp. 035405-1-035405-10.

Inoue, T.; Matsuda, K.; Kanemitsu, Y.; Murakami, Y. & Maruyama S. (2006). Diameter Dependence of Exciton-phonon Interaction in Individual Single-walled Carbon Nanotubes Studied by Microphotoluminescence Spectroscopy. *Physical Review B,* Vol. 73, No. 23, pp. 233401-1-233401-4.

Kane, C. L., & Mele, E. J. (2003). Electron Interactions and Scaling Relations for Optical Excitations in Carbon Nanotubes. *Physical Review Letters,* Vol. 93, No.19, pp. 197402-1-197402-4.

Karaiskaj, D., & Mascarenhas, A. (2007) Role of Electron-phonon Interactions and External Strain on the Electronic Properties of Semiconducting Carbon Nanotubes. *Physical Review B,* Vol. 75, No. 11, pp. 115426-1-115426-4.

Leeuw, T. K; Reith, R. M.; Simonette R. A.; Harden M. E.; Cherukuri, P.; Tsyboulski, D. A.; Beckingham, K. M. & Weisman R. B. (2007). Single-walled Carbon Nanotubes in the Intact Organism: Near-IR Imaging and Biocompatibility Studies in Drosophila. *Nano Letters,* Vol. 7, No. 9, pp. 2650-2654.

Lefebvre, J.; Homma, Y. & Finnie P. (2003). Bright Band Gap Photoluminescence from Unprocessed Single-walled Carbon Nanotubes. *Physical Review Letters,* Vol. 93, No. 21, pp. 217401-1-217401-4.

Lefebvre, J.; Fraser, J. M.; Homma, Y. & Finnie P. (2004). Photoluminescence from Single-walled Carbon Nanotubes: a Comparison Between Suspended and Micelle-encapsulated Nanotubes. *Applied Physics A,* Vol. 78, No. 8, pp. 1107-1110.

Lefebvre, J.; Fraser, J. M.; Finnie, P. & Homma, Y. (2004). Photoluminescence from an Individual Single-walled Carbon Nanotube. *Physical Review B,* Vol. 69, No. 7, pp. 075403-1-075403-5.

Lefebvre, J.; Finnie, P. & Homma, Y. (2004). Temperature-dependent Photoluminescence from Single-walled Carbon Nanotubes. *Physical Review B,* Vol. 70, No. 4, 045419-1-045419-8.

Ma, Y-Z.; Valkunas, L.; Dexheimer, S. L.; Bachilo, S. M. & Fleming, G. R. (2005). Femtosecond Spectroscopy of Optical Excitations in Single-Walled Carbon Nanotubes: Evidence for Exciton-Exciton Annihilation. *Physical Review Letters,* Vol. 94, No. 15, pp. 157402-1-157402-4.

Maruyama, S.; Kojima, R.; Miyauchi Y.; Chiashi S. & Kohno M. (2002). Low-temperature Synthesis of High-purity Single-walled Carbon Nanotubes from Alcohol. *Chemical Physics Letters,* Vol. 360, No. 3-4, pp. 229-234.

Matsuda, K.; Kanemitsu, Y.; Irie, K.; Saiki, T.; Someya, T.; Miyauchi, Y. & Maruyama, S.; (2005). Photoluminescence Intermittency in an Indivitual Single-walled Carbon Nanotube at Room Temperature. *Applied Physics Letters,* Vol. 86, No. 12, pp. 123116-1-123116-3.

Matsuda, K.; Inoue, T.; Murakami, Y.; Maruyama, S. & Kanemitsu, Y. (2008). *Physical Review B,* Vol. 77, No. 19, pp. 193405-1-193405-4.

Matsunaga, R.; Matsuda, K. & Kanemitsu Y. (2008). Evidence for Dark Excitons in a Single Carbon Nanotube due to the Aharonov-Bohm Effect. *Physical Review Letters,* Vol. 101, No. 14, pp. 147404-1-147404-4.

Maultzsch, J.; Pomraenke, R.; Reich, S.; Chang, E.; Prezzi, D.; Ruini, A.; Molinari, E.; Strano, M. S.; Thomsen, C. & Lienau, C. (2005). Two-photon Photoluminescence and Exciton Binding Energies in Single-walled Carbon Nanotubes. *Physical Review B,* Vol. 72, No. 24, pp. 241402(R)-1-241402(R)-4.

Misewich, J. A.; Martel, R.; Avouris, Ph.; Tsang, J. C.; Heinze, S. & Tersoff, J. (2003). Electrically Induced Optical Emission from a Carbon Nanotube FET. *Science,* Vol. 300, No. 5620, pp. 783-786.

O'Connell, M. J.; Bachilo, S. M.; Huffman, C. B.; Moore, V. C.; Strano, M. S.; Haroz, E. H.; Rialon, K. L.; Boul, P. J.; Noon, W. H.; Kittrell, C.; Ma, J.; Hauge, R. H.; Weisman R. B. & Smalley R. E. (2002). Band Gap Fluorescence from Individual Single-walled Carbon Nanotubes. *Science,* Vol. 297, No. 5581, pp. 593-596.

Perebeinos, V.; Tersoff, J. & Avouris, P. (2005). Radiative Lifetime of Excitons in Carbon Nanotubes. *Nano Letters,* Vol. 5, No. 12, 2495-2499.

Popov V. N. & Lambin, P. (2006). Intraband Electron-phonon Scattering in Single-walled Carbon Nanotubes. *Physical Review B,* Vol. 74, No. 7, pp. 075415-1-075415-4.

Rozhin, A. G.; Sakakibara, Y.; Namiki, S.; Tokumoto, M.; Kataura, H. & Achiba, Y. (2006). Sub- 200-fs Pulsed Erbium-doped Fiber Laser using a Carbon Nanotube-polyvinylalcohol Mode Locker. *Applied Physics Letters,* Vol. 88, No.5, pp. 051118-1-05118-3.

Spataru, C. D.; Ismail-Beigi, S.; Benedict, L. X. & Louie, S. G. (2004). Excitonic Effects and Optical Spectra of Single-Walled Carbon Nanotubes. *Physical Review Letters,* Vol. 92, No. 7, pp. 077402-1-077402-4.

Suzuura H. & Ando, T. (2002). Phonons and Electron-phonon Scattering in Carbon Nanotubes. *Physical Review B,* Vol. 65, No. 23, pp. 235412-1-235412-4.

Wang, F.; Dukovic, G.; Knoesel, E.; Brus, L. E. & Heinz, T. F. (2004). Observation of Rapid Auger Recombination in Optically Excited Semiconducting Carbon Nanotubes. *Physical Review B,* Vol. 70, No. 24, pp. 241403(R)-1-241403(R)-4.

Wang, F.; Dukovic, G.; Brus, L. E. & Heinz T. F. (2005). The Optical Resonances in Carbon Nanotubes Arise from Excitons. *Science,* Vol. 308, Vol. 5723, pp. 838-841.

Wang, F.; Wu, Y.; Hybertsen, M. S. & Heinz, T. F. (2006) Auger Recombination of Excitons in One-Dimensional Nanostructures. *Physical Review B,* Vol. 73, No. 24 pp. 245424-1-245424-5.

Yoshikawa K.; Matsunaga, R; Matsuda, K. & Kanemitsu, Y. (2009). Mechanism of Exciton Dephasing in a Single Carbon Nanotube Studied by Photoluminescence Spectroscopy. *Applied Physics Letters,* Vol. 94, No. 9, pp. 093109-1-093109-3.

# 17

# A Numerical Study of the Vibration Spectrum for a Double-Walled Carbon Nanotube Model

Marianna A. Shubov[1]*and Matthew P. Coleman[2]
[1]*Department of Mathematics & Statistics, University of New Hampshire, Durham*
[2]*Department of Mathematics and Computer Science, Fairfield University, Fairfield*
*USA*

## 1. Introduction

In this paper we offer a series of new results devoted to the numerical analysis of a double-walled carbon nanotube model. This model is given in the form of two coupled Timoshenko beams connected through the distributed Van der Waals force (Gibson et al., 2007; Ru, 2000). Typically, nanotubes can be modeled as quantum systems and studied by a molecular simulations approach, or as classical systems (such as flexible beams, shells membranes (Mahan, 2002; Pantano et al., 2003; 2004; Wang et al., 2004; 2005)), or as specific hybrid models (Wang, 2005). The choice of model in any situation involves a tradeoff in that, while molecular models may yield more accurate results, implementing them is extremely time and labor intensive, which is not the case for models from continuum mechanics.

The scientific and engineering communities have acknowledged the very desirable properties of carbon nanotubes (CNTs) and their potential use in wide-ranging applications. The author of (Jamieson, 2000) argues that nanotechnology, mainly due to CNTs, may impact technology more than did the silicon revolution. Depending on the atomic structure, CNTs have electrical properties that can range from those of metals to those of semiconductors. The mechanical properties of CNTs are also unique. They possess exceptionally high specific stiffness and specific strength; they are extremely elastic, being able to bend through a complete 360° without noticeable damage. The application potential for materials with these properties is almost limitless.

Developing mathematical models for CNTs is of critical importance. Such models must be verified and quantified by performing and analyzing experiments. As we have mentioned, two groups of models exist: molecular simulation models and continuum mechanics models. Continuum models are generally based on traditional engineering models such as beams, shells, or membranes. The nanotubes are treated as continuous materials with definite geometries and common material properties such as *Young's modulus*. In contrast, molecular models consider each atom, and mathematically define the interactions among the atoms. Based on their work on atomic simulations of CNTs, the authors of (Jakobson et al., 1996) provide a justification for incorporating continuum mechanics models into CNTs study,

*Corresponding author: Marianna A. Shubov, Department of Mathematics & Statistics, University of New Hampshire, Durham

stating that "The laws of continuum mechanics are amazingly robust and allow one to treat even intrinsically discrete objects only a few atoms in diameter."

The most commonly used models are the following: the Euler–Bernoulli beam model, Timoshenko beam model, and flexible shell and membrane models. Typically, many models for multi-walled nanotubes allow for independent wall movement, and the wall interaction is a function of the local wall separation distance.

Vibration of a double-walled carbon nanotube (DWCNT) generated by a nonlinear interlayer Van der Waals force is studied in (Xu et al., 2006). The results indicate that the nonlinear factors of the Van der Waals force, on the one hand, have little effect on the coaxial free vibrations. On the other hand, these nonlinear factors greatly affect noncoaxial free vibrations. As is indicated in (Qian et al., 2002), although carbon nanotubes can have diameters only several times larger than the length between carbon atoms, continuum models have been found to describe their mechanical behavior very accurately, in many circumstances.

Our analysis of an initial boundary-value problem models small transversal vibrations of a double-walled carbon nanotube. The system of equations is similar to the ones mentioned in a number of papers (see references (Gibson et al., 2007; Jakobson et al., 1996; Pantano et al., 2003; Qian et al., 2002; Ru, 2001; Wang et al., 2006; Xu et al., 2006; Yoon et al., 2003)). The physical system consists of two nested nanotubes interacting through the distributed Van der Waals force; each nanotube is modeled as a Timoshenko beam with specific parameters. As pointed out in (Wang et al., 2006), "Unlike the Euler–Bernoulli beam model, the Timoshenko beam model allows for the effects of transverse shear deformation and rotary inertia. These effects become significant for carbon nanotubes with small length-to-diameter ratios that are normally encountered in applications."

The model is given in the form of two coupled Timoshenko beams (i.e., in the form of four coupled hyperbolic partial differential equations). The system is equipped with a set of nonself-adjoint boundary conditions involving four independent complex parameters. Indeed, all other articles treating the Timoshenko model consider only the traditional energy-conserving boundary conditions, thus our treatment is a generalization of their work (as these latter conditions are just limiting special cases of the nonself-adjoint conditions treated herein). An asymptotic analysis of the eigenspectrum for this problem was performed in (Shubov & Rojas-Arenaza, 2010a;b;c), under certain simplifying assumptions. We must mention that the assumptions are somewhat restrictive—indeed, they cannot be satisfied by a physical double-walled carbon nanotube system. However, even for this simplified case, the necessary computations were extremely complex and cumbersome, and it is unclear if the more general problem even is tractable.

Regardless, this special case is a valid and interesting mathematical problem whose behavior should be quite similar to the more general physical problem. Thus, we feel that a study of the vibration spectrum for this case certainly will shed light on the spectrum of the more general problem, particularly by our choosing values for the physical parameters that are similar to those for physical carbon nanotubes.

The paper is organized as follows. In Section 2, we introduce the general mathematical model, perform separation of variables and rewrite the special case of the model treated in (Shubov & Rojas-Arenaza, 2010a) in dimensionless form. In Section 3, we present the asymptotic results derived in (Shubov & Rojas-Arenaza, 2010a). The Legendre-tau spectral method is described in Section 4, and in Section 5 we present our numerical results and comparison with the asymptotic results predicted by (Shubov & Rojas-Arenaza, 2010a).

## 2. The mathematical model

We consider the system consisting of two Timoshenko beams coupled through the van der Waals force, as given in (Shubov & Rojas-Arenaza, 2010a;b;c):

$$\sigma A_1 W_{1tt}(x,t) + k_1 G A_1[\Phi_{1x}(x,t) - W_{1xx}(x,t)] = -C[W_2(x,t) - W_1(x,t)] \quad (1)$$

$$\sigma I_1 \Phi_{1tt}(x,t) - E I_1 \Phi_{1xx}(x,t) + k_1 G A_1[\Phi_1(x,t) - W_{1x}(x,t)] = 0 \quad (2)$$

$$\sigma A_2 W_{2tt}(x,t) + k_2 G A_2[\Phi_{2x}(x,t) - W_{2xx}(x,t)] = C[W_2(x,t) - W_1(x,t)] \quad (3)$$

$$\sigma I_2 \Phi_{2tt}(x,t) - E I_2 \Phi_{2xx}(x,t) + k_2 G A_2[\Phi_2(x,t) - W_{2x}(x,t)] = 0. \quad (4)$$

For boundary conditions, the left end of each beam is free, while the right end of each is subject to the standard set of two-parameter boundary conditions:

$$W_{1x}(0,t) - \Phi_1(0,t) = \Phi_{1x}(0,t) = 0 \quad (5,6)$$

$$W_{2x}(0,t) - \Phi_2(0,t) = \Phi_{2x}(0,t) = 0 \quad (7,8)$$

$$k_1 G A_1[\Phi_1(L,t) - W_{1x}(L,t)] = \sigma I_1 \alpha_1 W_{1t}(L,t) \quad (9)$$

$$E\Phi_{1x}(L,t) = -\sigma \beta_1 \Phi_{1t}(L,t) \quad (10)$$

$$k_2 G A_2[\Phi_2(L,t) - W_{2x}(L,t) = \sigma I_2 \alpha_2 W_{2t}(L,t) \quad (11)$$

$$E\Phi_{2x}(L,t) = -\sigma \beta_2 \Phi_{2t}(L,t). \quad (12)$$

Here, $0 \leq x \leq L$ where $L$ is the length of each beam, and $t \geq 0$. $W_i(x,t)$ is the transverse displacement of beam $i$, $\Phi_i(x,t)$ is the bending angle of beam $i$, $i = 1,2$. The physical and geometrical constants are as follows: $\sigma$ is the mass per unit volume; $E$, Young's modulus; $G$, the shear modulus; $A_i$, the uniform cross-sectional area of beam $i$; $I_i$ the uniform area moment of inertia of beam $i$; and $k_i$; the shear connection factor for beam $i$. We note that $E = 2(1+\nu)G$, where $\nu$ is the Poisson's ratio.
Further, we note the following:

$$\alpha_i = \beta_i = 0 \Rightarrow \text{ right end of beam } i \text{ is free} \quad (13)$$

$$\alpha_i = \beta_i = \infty \Rightarrow \text{ right end of beam } i \text{ is clamped} \quad (14)$$

$$\alpha_i = \infty, \beta_i = 0 \Rightarrow \text{ right end of beam } i \text{ is simply-supported} \quad (15)$$

$$\alpha_i = 0, \beta_i = \infty \Rightarrow \text{ right end of beam } i \text{ is roller-supported.} \quad (16)$$

We separate variables by letting

$$W_j(x,t) = e^{-i\omega t} w_j(x),$$
$$\Phi_j(x,t) = e^{-i\omega t} \phi_j(x),$$

$j = 1, 2$, and, following the notation in (Shubov & Rojas-Arenaza, 2010a), the system (1)–(12) becomes

$$\omega^2 w_1(x) = \widehat{k}_1[\phi_1'(x) - w_1''(x)] + C_1[w_2(x) - w_1(x)] \tag{17}$$

$$\omega^2 \phi_1(x) = -\frac{E}{\sigma}\phi_1''(x) + \tilde{k}_1[\phi_1(x) - w_1'(x)] \tag{18}$$

$$\omega^2 w_2(x) = \widehat{k}_2[\phi_2'(x) - w_2''(x)] - C_2[w_2(x) - w_1(x)] \tag{19}$$

$$\omega^2 \phi_2(x) = -\frac{E}{\sigma}\phi_2''(x) + \tilde{k}_2[\phi_2(x) - w_2'(x)] \tag{20}$$

$$w_1'(0) - \phi_1(0) = 0 \tag{21}$$

$$\phi_1'(0) = 0 \tag{22}$$

$$w_2'(0) - \phi_2(0) = 0 \tag{23}$$

$$\phi_2(0) = 0 \tag{24}$$

$$\tilde{k}_1[\phi_1(L) - w_1'(L)] = -i\omega\alpha_1 w_1(L) \tag{25}$$

$$\frac{E}{\sigma}\phi_1'(L) = i\omega\beta_1\phi_2(L) \tag{26}$$

$$\tilde{k}_2[\phi_2(L) - w_2'(L)] = -i\omega\alpha_2 w_2(L) \tag{27}$$

$$\frac{E}{\sigma}\phi_2'(L) = i\omega\beta_2\phi_2(L). \tag{28}$$

Here, we have

$$\tilde{k}_i = \frac{k_i G A_i}{\sigma I_i}, \quad \widehat{k}_i = \frac{k_i G}{\sigma}, \quad C_i = \frac{C}{\sigma A_i}, \qquad i = 1, 2.$$

Again, following (Shubov & Rojas-Arenaza, 2010a), we consider the special case

$$\tilde{k}_1 = \tilde{k}_2 = \tilde{k}, \quad \widehat{k}_1 = \widehat{k}_2 = \widehat{k}. \tag{29}$$

We must note that these conditions cannot hold for a physical double-walled carbon nanotube (e.g., the shape factors must be different, $k_1 \neq k_2$). However, without these assumptions, the asymptotic treatment of the problem becomes extremely difficult, and possibly intractable. Thus, at this point in time, this particular special case is the only one for which there are analytical results with which to compare. We now cast the problem in dimensionless form. Following (Traill-Nash & Collar, 1953) and, more appropriately, (Coleman & Schaffer, 2010), we introduce dimensionless quantities as follows:

$$\begin{aligned}
&\widehat{x} = \frac{x}{L}, \quad \widehat{w}_i(\widehat{x}) = \frac{1}{L}w_i(x), \quad \widehat{\phi}_i(\widehat{x}) = \phi_i(x), \qquad i = 1, 2, \\
&\lambda = \sqrt{\frac{\sigma\tilde{k}}{E\widehat{k}}}L^2\omega, \quad \gamma_1 = \frac{\widehat{k}}{\tilde{k}L^2}, \quad \gamma_2 = \frac{E}{\sigma\tilde{k}L^2}, \\
&\alpha_i' = \frac{1}{\sigma A\tilde{k}L}\sqrt{\frac{E\widehat{k}}{\sigma\tilde{k}}}\,\alpha_i, \quad \beta_i' = \frac{1}{\sigma A\tilde{k}L^3}\sqrt{\frac{E\widehat{k}}{\sigma\tilde{k}}}\,\beta_i, \qquad i = 1, 2, \\
&C_i' = \frac{L^2}{\widehat{k}}C_i = \frac{L^2}{k_i G A_i}C, \qquad i = 1, 2.
\end{aligned} \tag{30}$$

We abuse notation, and use $x, w_i, \phi_i$ instead of $\widehat{x}, \widehat{w}_i$ and $\widehat{\phi}_i$, and the resulting dimensionless system is

$$
\begin{aligned}
&-\gamma_2\lambda^2 w_1(x) = -\phi_1'(x) + w_1''(x) - C_1'[w_2(x) - w_1(x)] && (31)\\
&-\gamma_1\gamma_2\lambda^2\phi_1(x) = \gamma_2\phi_1''(x) - \phi_1(x) + w_1'(x) && (32)\\
&-\gamma_2\lambda^2 w_2(x) = -\phi_1'(x) + w_2''(x) + C_2'[w_2(x) - w_1(x)] && (33)\\
&-\gamma_1\gamma_2\lambda^2\phi_2(x) = \gamma_2\phi_2''(x) - \phi_2(x) + w_2'(x), \qquad 0 < x < 1, && (34)\\
&w_1'(0) - \phi_1(0) = 0 && (35)\\
&\phi_1'(0) = 0 && (36)\\
&w_2'(0) - \phi_2(0) = 0 && (37)\\
&\phi_2'(0) = 0 && (38)\\
&\phi_1(1) - w_1'(1) + i\alpha_1'\lambda w_1(1) = 0 && (39)\\
&\gamma_2\phi_1'(1) - i\beta_1'\lambda\phi(1) = 0 && (40)\\
&\phi_2(1) - w_2'(1) + i\alpha_2'\lambda w_2(1) = 0 && (41)\\
&\gamma_2\phi_2'(1) + i\beta_2'\lambda\phi(1) = 0. && (42)
\end{aligned}
$$

## 3. Asymptotic estimation of vibration spectrum

The first-order asymptotic estimation of the vibration frequencies for problem (31)–(42) is given in Theorem 2.5, of (Shubov & Rojas-Arenaza, 2010a); we present the results here, but in dimensionless form.

**Theorem** (Shubov, Rojas-Arenaza). *Assume that the boundary parameters $\alpha_i'$ and $\beta_i'$, $i \geq 1,2$, satisfy the following conditions*

$$
\alpha_1' \neq \alpha_2', \quad \beta_1' \neq \beta_2', \quad \alpha_i' \neq \sqrt{\gamma_2}, \quad \beta_i' \neq \sqrt{\gamma_1}\,\gamma_2, \quad \textit{and}
$$
$$
\left|\frac{\alpha_i' - \sqrt{\gamma_2}}{\alpha_i' + \sqrt{\gamma_2}}\right| \neq \left|\frac{\beta_i' - \sqrt{\gamma_1}\,\gamma_2}{\beta_i' + \sqrt{\gamma_1}\,\gamma_2}\right|.
$$

*Then, the set of frequencies $-i\lambda$ of system* (31)–(42) *splits into the following four separate branches:*

$$
-i\lambda_n^{(1)} = \frac{1}{2\sqrt{\gamma_2}}\left[\log\frac{1-\alpha_1''}{1+\alpha_1''} + 2n\pi i\right] + O\left(\frac{1}{n}\right), \tag{43}
$$
$$
-i\lambda_n^{(2)} = \frac{1}{2\sqrt{\gamma_1}}\left[\log\frac{1-\beta_1''}{1+\beta_1''} + 2n\pi i\right] + O\left(\frac{1}{n}\right), \tag{44}
$$
$$
-i\lambda_n^{(3)} = \frac{1}{2\sqrt{\gamma_2}}\left[\log\frac{1-\alpha_2''}{1+\alpha_2''} + 2n\pi i\right] + O\left(\frac{1}{n}\right), \tag{45}
$$
$$
-i\lambda_n^{(4)} = \frac{1}{2\sqrt{\gamma_1}}\left[\log\frac{1-\beta_2''}{1+\beta_2''} + 2n\pi i\right] + O\left(\frac{1}{n}\right), \qquad n = \pm 1, 2, 3, \ldots \tag{46}
$$

*where*

$$
\alpha_i'' = \frac{1}{\sqrt{\gamma_2}}\alpha_i', \quad \beta_i'' = \frac{1}{\sqrt{\gamma_1}\,\gamma_2}\beta_i', \qquad i = 1, 2. \qquad \square
$$

We note that log represents the complex logarithm, $\log z = \ln |z| + i \arg z$. We note also the important fact that the Van der Waals force between the two tubes does not appear in the first-order approximation.

## 4. The legendre-tau spectral method

We compare the asymptotic results of the Theorem with a numerical approximation of the spectrum using the Legendre-tau spectral method (Gottlieb et al., 1984). This entails transforming problem (1)–(12) to one on the interval $-1 \leq x \leq 1$ by letting $x \to \frac{2}{L}x - 1$. Assuming there will be no confusion, we keep the original variables $x, w_i, \phi_i, i = 1, 2$, and the resulting system is

$$\omega^2 w_1(x) = \frac{2\widehat{k}}{L}\phi_1'(x) - \frac{4\widehat{k}}{L^2} w_1''(x) + C_1[w_2(x) - w_1(x)] \tag{47}$$

$$\omega^2 \phi_1(x) = -\frac{4E}{\sigma L^2}\phi_1''(x) + \widetilde{k}\phi_1(x) - \frac{2\widetilde{k}}{L} w_1'(x) \tag{48}$$

$$\omega^2 w_2(x) = \frac{2\widehat{k}}{L}\phi_2'(x) - \frac{4\widehat{k}}{L^2} w_2''(x) - C_2[w_2(x) - w_1(x)] \tag{49}$$

$$\omega^2 \phi_2(x) = -\frac{4E}{\sigma L^2}\phi_2''(x) + \widehat{k}\phi_2(x) - \frac{2\widehat{k}}{L} w_2'(x), \qquad -1 \leq x \leq 1, \tag{50}$$

$$\frac{2}{L} w_1'(-1) - \phi_1(-1) = 0 \tag{51}$$

$$\phi_1'(-1) = 0 \tag{52}$$

$$\frac{2}{L} w_2'(1) - \phi_2(-1) = 0 \tag{53}$$

$$\phi_2'(-1) = 0 \tag{54}$$

$$\widetilde{k}\phi_1(1) - \frac{2\widehat{k}}{L} w_1'(1) = -i\omega\alpha_1 w_1(1) \tag{55}$$

$$\frac{\partial E}{\sigma L}\phi_1'(1) = i\omega\beta_1\phi_1(1) \tag{56}$$

$$\widetilde{k}\phi_2(1) - \frac{2\widetilde{k}}{L} w_2'(1) = -i\omega\alpha_2 w_2(1) \tag{57}$$

$$\frac{\partial E}{\sigma L}\phi_2'(1) = i\omega\beta_2\phi_2(1). \tag{58}$$

We let

$$w_1(x) = \sum_{n=0}^{N} a_n P_n(x), \qquad \phi_1(x) = \sum_{n=0}^{N} b_n P_n(x),$$

$$w_2(x) = \sum_{n=0}^{N} c_n P_n(x), \qquad \phi_2(x) = \sum_{n=0}^{N} d_n P_n(x),$$

where $P_n$ is the Legendre polynomial of degree $n$.

We then compare coefficients of $x^n$, for $n = 0, 1, \ldots, N-2$, in each of the equations resulting from (47)–(50) and, including the 8 equations resulting from boundary conditions (51)–(58), the result is a system of $4N+4$ equations in the $4N+4$ unknowns $a_n, b_n, c_n, d_n, n = 0, 1, \ldots, N$, and the parameter $\omega$. We may rewrite the system in the form

$$(\omega^2 A + \omega B + C)(a_0, \ldots, a_N, b_0, \ldots, b_N, c_0, \ldots, c_N, d_0, \ldots, d_N)^T = 0, \tag{59}$$

where $A, B$ and $C$ are $(4N+4) \times (4N+4)$ matrices. Then, the vibration spectrum consists of those numbers $-i\omega$, where $\omega$ is a latent value of (59), i.e., where $\omega$ satisfies

$$\det(\omega^2 A + \omega B + C) = 0. \tag{60}$$

It is easy to show that $\omega$ satisfies (60) if and only if $\omega$ is an eigenvalue of the $(8N+8) \times (8N+8)$ matrix

$$\begin{bmatrix} -A^{-1}B & -A^{-1}C \\ I & 0 \end{bmatrix},$$

where $I$ is the $(4N+4) \times (4N+4)$ identity matrix and 0 the $(4N+4) \times (4N+4)$ 0-matrix. In practice, $A$ is often singular—indeed, that is the case here. We remedy the situation by letting

$$\omega = \frac{\zeta - 1}{\zeta + 1},$$

yielding the equation

$$\det(\zeta^2 X + \zeta Y + Z),$$

where $X, Y, Z$, of course, are $(4N+4) \times (4N+4)$ matrices $X$ is nonsingular, so we may proceed by finding the eigenvalues of

$$\begin{bmatrix} -X^{-1}Y & -X^{-1}Z \\ I & 0 \end{bmatrix}$$

and transforming back.

## 5. Comparison of numerical and asymptotic results

Assumptions (29) imply that $k_1 = k_2$ and $A_1/I_1 = A_2/I_2$. While, as mentioned above, this means that we are not looking at a double-walled tube, these assumptions have the advantage of allowing us better to see the effect that the damping parameters and Van der Waals force have on the imaginary parts—i.e., the actual "frequency" parts—of the eigenfrequencies, as we shall see below.

Form our physical and geometrical parameters, we choose the carbon nanotube data given in (Wang et al., 2006). Thus, we have $E$ =1 TPa, $G$ =.4 TPa, $A$ =2.3090706 nm$^2$, $I$ =.459649366 nm$^4$ and $\rho$ =2.3 g/cm$^3$, and with a Van der Waals constant of $C$ =.06943 TPa. Further, from our previous work, we have seen that, as the value of the slenderness ratio $L/d$ increases, one must go further out along the spectrum in order to find agreement with the asymptotic results. Thus we choose $L$ =2.5 nm, resulting in $L/d$ =2.85714286.

The dimensionless parameters then become

$$\gamma_1 = .03185$$
$$\gamma_2 = .0652925$$
$$C' = .5729492131.$$

For the damping constants, there is nothing in the literature to guide our choices. However, we can see that, if each $\alpha_i'' < 1$ and each $\beta_i'' < 1$ in (43)–(46), the asymptotic behavior of the imaginary parts of the eigenfrequencies will behave as though both right ends are free; similarly, if the arguments in the logs all are negative, the behavior will be as if both right ends are clamped. (Of course, there are many more possibilities; however, "clamped" and "free" are the most common types, so, due to space limitations, we restrict ourselves to these two cases. Also, we mention that the critical cases $\alpha'' = 1$ and $\beta'' = 1$ are studied in (Coleman & Schaffer, preprint), for the single Timoshenko beam.) Further, our choices are guided by the wish to see clearly the separation of the spectrum into branches.

To study the case where the right ends are free-like, we choose our dimensionless damping parameters to be

$$\alpha_1' = .2, \quad \beta_1' = .01, \quad \alpha_2' = .1, \quad \beta_2' = .001. \tag{61}$$

For clamped-like, we choose:

$$\alpha_1' = .3, \quad \beta_1' = .013, \quad \alpha_2' = 2, \quad \beta_2' = .02. \tag{62}$$

For all of our numerical examples, we have performed computations at $N = 180$, 200 and 220 Legendre polynomials, and we see that all results have converged to at least 10 decimal places.

1) For our first example, we consider the case with damping parameters given by (61) and with no Van der Waals force. This will give us a baseline for later examples, and will allow us to see how the spectrum separates into four branches. The results can be seen in Tables 1A and 1B, where we actually separate the frequencies into their four branches. First, however, we must note that the branching is an asymptotic phenomenon, thus one needs to go out along the spectrum before it can be seen. As mentioned earlier, for larger values of $L/d$, one must go very far out before one sees the branching starting to occur. Here, we begin to see the branching and agreement with the asymptotic results pretty clearly after about the 4th or 5th eigenfrequency of each branch. For the first few, however, it may not even make sense to assign them to a branch; thus, while we do so by making our best guess, we mark them with $*$ to denote the fact that this assignment is problematic.

Table 1A, then, lists the first 40 eigenfrequencies, and the 50th, 60th, 70th, 80th, 90th and 100th eigenfrequencies, of each $\alpha$-branch. The final column lists the asymptotic approximations for the imaginary parts, and the line at the bottom gives the asymptotic approximations for the real parts. Table 1B does the same, but for the $\beta$-branches.

As mentioned, in both tables the frequencies seem clearly to have split into branches, based on the real parts, well before the 10th frequency. By the 100th frequency in each branch, we have at least a three-decimal place match between the numerical and asymptotic real parts, and a four-decimal place match between the numerical and asymptotic imaginary parts.

One item of note: we see that the first frequency of the $\alpha$-branch predicted by the asymptotic results does not appear. As we shall see, it appears that this frequency may have been "damped out" by the boundary damping.

2) For Example 2, we use the damping parameters given in (62), and Tables 2A and 2B are analogous to Tables 1A and 1B, respectively. Here, it is not clear how to deal with the first few entries in each table. However, they separate into branches very quickly. In Table 2A we see that, by the 100th frequency, we have at least a three-decimal place match between the numerical and asymptotic real parts, and a three-decimal- place match between the numerical and asymptotic imaginary parts. In Table 2B, by the 100th frequency we see a four-decimal place match between the numerical and asymptotic frequencies. Meanwhile, for

the $\beta_2$ branch, the numerical and asymptotic real parts match to three decimal places. For the $\beta_1$ branch, the match is not as good (two decimal places), though they still clearly seem to be converging.
For the remaining examples we introduce the Van der Waals force. Specifically, we wish to see what happens to the spectrum as the Van der Waals constant increases from 0 to about twice the value of the physically realistic value of $C' = .5729492131$. Thus, we consider what happens for the values

$$C' = 0, \quad .25, \quad .5, \quad .75 \quad \text{and} \quad 1.$$

3) For Examples 3 and 4, we look at two cases without boundary damping. Example 3 considers the case where the right ends are free, that is, for which

$$\alpha_1' = \beta_1' = \alpha_2' = \beta_2' = 0;$$

while Example 4 considers the right ends to be clamped, i.e.,

$$\alpha_1' = \beta_1' = \alpha_2' = \beta_2' = \infty.$$

We note that, in Examples 3 and 4, all numerical real parts are of absolute value $< 1.0E - 10$. The results for Example 3 can be found in Tables 3A and 3B. In Table 3A, we list the imaginary parts of the first 40 frequencies. The first column represents the double $\alpha$- and $\beta$-branches, identical for $C' = 0$. Introducing $C' > 0$ leads to the splitting of these pairs. What is striking is that, for each pair of frequencies, one decreases as the value of $C'$ increases, while the other is unaffected. (Indeed, it turns out that each of the even-numbered frequencies is unchanged to 13 decimal places!) Secondly, as we go out along the spectrum, the first member of each pair is less affected by the Van der Waals force, so that, when we get to the 39th–40th pair, they agree to three decimal places. (We look more closely at this phenomenon in Table 3B.)
Further, in comparing these results with those of Example 1, we see that the first predicted frequencies, missing in Table 1A, do appear here. Thus, as mentioned, it appears that the first pair was damped out via the boundary damping in Example 1, and that only one of these seems to be damped out by the inclusion of the Van der Waals force. Further, by comparing the first column of Table 3A with the results of Example 1, it is clear that the damping also affects the imaginary or "frequency" parts of the eigenfrequencies.
In Table 3B, we list the 49th–50th, 99th–100th, 149th–150th, 199th–200th, 249th–250th, 299th–300th, 349th–350th and 399th–400th eigenfrequencies, both numerical and asymptotic, for the case $C' = 1$ (i.e., corresponding to the last column in Table 3A). We see still closer agreement between the entries in each pair, and very close agreement with the asymptotics, as well. (Note that we list the branch for each eigenfrequency.) (Of course, the numbering here is very different from the numbering in Examples 1 and 2; e.g., the 40th entry in Table 3A corresponds to the 12th entry in Table 1A.)
4) The results of Example 4 are given in Tables 4A and 4B, in the same format as Tables 3A and 3B, respectively. In Table 4A, we see that the matching between the members of each pair is quite similar to that occurring in Table 3A. And again here, we see in Table 4B still closer agreement in each pair, and with the asymptotic results.
5) Example 5 is combination of Examples 1 and 3, and Example 6 is a combination of Examples 2 and 4. Example 5 looks at the damped system with the free-like parameters in (61), for the Van der Waals constant with values $C' = 0, .5$ and 1. The results are given in Tables 5A and 5B. In Table 5A, we proceed as in Table 3A, by listing the first 40 eigenfrequencies, although here we consider only the three values of $C'$. We see here that, for each pair, both

imaginary parts are affected by the Van der Waals force. However, we still see the closer matching of each pair as we go out along the spectrum. Meanwhile the real parts (damping rates) also are affected by the Van der Waals force, although there does not seem to be a noticeable pattern in that, in some cases it increases, while for others it decreases; in particular, there seems to be no branch-related pattern. Table 5B, then, is analogous to Table 3B, again using only the Van der Waals constant $C' = 1$. For the imaginary parts, the results are quite similar to those given in Table 3B. Meanwhile, the effect of the van der Waals on the real parts is diminished, as well, with the exception of the $\beta_2$-branch. However, this must be due to the fact that the $\beta_2$ damping rates are an order of magnitude smaller than the other damping rates.

6) In Table 6A, we proceed as in Table 4A, by listing the first 40 eigenfrequencies, but again only considering the three values of $C'$. We see again that, for each pair, both imaginary parts are affected by the Van der Waals force. Again we see the closer matching of each pair as we go out along the spectrum. Indeed, the last few pairs match more closely than the undamped pairs in Table 4A. The real parts behave quite the same as in Table 5A. Table 6B, then, is analogous to Table 4B, once more using only the Van der Waals constant $C' = 1$. Again, the imaginary parts behave quite similarly to those in Table 4B, and the real parts behave quite similarly to those in Table 5B.

In closing, we should mention that, although the results in (Shubov & Rojas-Arenaza, 2010b) show that the system is nonconservative, we have been unable to find any unstable eigenfrequencies in our numerical investigations.

| | Numerical | | | | Asymptotic (Im) |
|---|---|---|---|---|---|
| | $\alpha_1$ Branch | | $\alpha_2$ Branch | | |
| | Re | Im | Re | Im | |
| 1. | — | — | — | — | 6.14735 |
| 2.* | −2.746 | 9.44918 | −.6783 | 9.93037 | 18.4421 |
| 3.* | −3.529 | 21.4099 | −.9628 | 21.7742 | 30.7368 |
| 4.* | −3.658 | 39.8147 | −1.490 | 38.6929 | 43.0315 |
| 5. | −3.823 | 53.2242 | −.9151 | 51.9313 | 55.3262 |
| 6. | −4.613 | 64.3926 | −.7899 | 65.1048 | 67.6209 |
| 7. | −4.754 | 77.6284 | −.8924 | 78.0511 | 79.9156 |
| 8. | −4.696 | 90.6127 | −1.224 | 90.7904 | 92.2103 |
| 9. | −4.357 | 103.890 | −1.951 | 102.998 | 104.505 |
| 10. | −4.743 | 114.106 | −1.357 | 114.409 | 116.800 |
| 11. | −4.902 | 127.122 | −.7845 | 127.016 | 129.094 |
| 12. | −4.907 | 139.799 | −.5884 | 139.671 | 141.389 |
| 13. | −4.690 | 152.112 | −.5443 | 152.252 | 153.684 |
| 14. | −4.920 | 164.680 | −.5754 | 164.776 | 165.979 |
| 15. | −4.899 | 177.203 | −.6898 | 177.241 | 178.273 |
| 16. | −4.712 | 189.886 | −.8915 | 189.528 | 190.568 |
| 17. | −4.831 | 201.386 | −.8153 | 201.610 | 202.863 |
| 18. | −4.929 | 214.000 | −.6139 | 213.968 | 215.157 |
| 19. | −4.932 | 226.449 | −.5353 | 226.399 | 227.452 |
| 20. | −4.867 | 238.754 | −.5199 | 238.814 | 239.747 |
| 21. | −4.934 | 251.173 | −.5397 | 251.213 | 252.042 |
| 22. | −4.919 | 263.582 | −.6054 | 263.584 | 264.336 |
| 23. | −4.715 | 276.142 | −.7003 | 275.851 | 276.631 |

| | | | | | |
|---|---|---|---|---|---|
| 24. | −4.903 | 287.988 | −.6383 | 288.060 | 288.926 |
| 25. | −4.937 | 300.421 | −.5490 | 300.399 | 301.220 |
| 26. | −4.933 | 312.795 | −.5165 | 312.766 | 313.515 |
| 27. | −4.926 | 325.094 | −.5131 | 325.126 | 325.810 |
| 28. | −4.937 | 337.458 | −.5304 | 337.478 | 338.104 |
| 29. | −4.923 | 349.829 | −.5784 | 349.809 | 350.399 |
| 30. | −4.439 | 361.815 | −.6217 | 362.065 | 362.694 |
| 31. | −4.929 | 374.314 | −.5671 | 374.329 | 374.989 |
| 32. | −4.939 | 386.676 | −.5232 | 386.661 | 387.283 |
| 33. | −4.929 | 399.021 | −.5095 | 399.000 | 399.578 |
| 34. | −4.937 | 411.318 | −.5110 | 411.335 | 411.873 |
| 35. | −4.937 | 423.655 | −.5277 | 423.665 | 424.167 |
| 36. | −4.916 | 436.013 | −.5653 | 435.973 | 436.462 |
| 37. | −4.870 | 448.131 | −.5756 | 448.231 | 448.757 |
| 38. | −4.937 | 460.525 | −.5344 | 460.523 | 461.052 |
| 39. | −4.940 | 472.857 | −.5122 | 472.846 | 473.347 |
| 40. | −4.914 | 485.173 | −.5068 | 485.171 | 485.641 |
| 50. | −4.791 | 608.231 | −.5448 | 608.213 | 608.588 |
| 60. | −4.929 | 731.223 | −.5045 | 731.223 | 731.535 |
| 70. | −4.869 | 854.207 | −.5240 | 854.214 | 854.482 |
| 80. | −4.935 | 977.194 | −.5038 | 977.195 | 977.429 |
| 90. | −4.940 | 1100.15 | −.5155 | 1100.17 | 1100.38 |
| 100. | −4.938 | 1223.14 | −.5034 | 1223.14 | 1223.32 |
| Asym. Re: | −4.941 | | −.5027 | | |

Table 1A. Numerical eigenfrequencies 1–40, 50, 60, 70, 80, 90 and 100 for the $\alpha_1$ and $\alpha_2$ branches from Example 1. The asymptotic imaginary parts are given in the last column, while the asymptotic real parts appear at the bottom.

| | Numerical | | | | Asymptotic (Im) |
|---|---|---|---|---|---|
| | $\beta_1$ Branch | | $\beta_2$ Branch | | |
| | Re | Im | Re | Im | |
| 1.* | −2.335 | 27.3749 | −1.321 | 26.9029 | 8.80167 |
| 2.* | −3.537 | 35.5415 | −3.172 | 36.5570 | 26.4050 |
| 3.* | −5.139 | 51.3999 | −4.007 | 50.2773 | 44.0084 |
| 4.* | −6.082 | 67.0480 | −3.538 | 67.8770 | 61.6117 |
| 5. | −6.339 | 83.3545 | −3.632 | 83.7187 | 79.2150 |
| 6. | −5.939 | 100.464 | −4.061 | 99.7040 | 96.8184 |
| 7. | −6.716 | 118.544 | −3.738 | 118.957 | 114.422 |
| 8. | −7.386 | 135.540 | −3.642 | 135.511 | 132.025 |
| 9. | −7.623 | 152.566 | −3.874 | 152.736 | 149.628 |
| 10. | −7.641 | 169.729 | −3.682 | 169.794 | 167.232 |
| 11. | −7.510 | 187.131 | −3.879 | 186.797 | 184.835 |
| 12. | −7.625 | 204.784 | −3.775 | 205.036 | 202.438 |
| 13. | −7.843 | 222.194 | −3.695 | 222.184 | 220.042 |
| 14. | −7.919 | 239.557 | −3.762 | 239.624 | 237.645 |
| 15. | −7.900 | 256.965 | −3.713 | 256.973 | 255.248 |
| 16. | −7.830 | 274.497 | −3.920 | 274.214 | 272.852 |
| 17. | −7.903 | 292.103 | −3.738 | 292.187 | 290.455 |

| | | | | | |
|---|---|---|---|---|---|
| 18. | −7.995 | 309.592 | −3.715 | 309.577 | 308.059 |
| 19. | −8.018 | 327.064 | −3.725 | 327.096 | 325.662 |
| 20. | −7.990 | 344.566 | −3.728 | 344.549 | 343.265 |
| 21. | −7.957 | 362.152 | −4.213 | 362.406 | 360.869 |
| 22. | −8.015 | 379.737 | −3.726 | 379.758 | 378.472 |
| 23. | −8.057 | 397.260 | −3.730 | 397.244 | 396.075 |
| 24. | −8.060 | 414.783 | −3.723 | 414.796 | 413.679 |
| 25. | −8.032 | 432.332 | −3.743 | 432.293 | 431.282 |
| 26. | −8.027 | 449.937 | −3.790 | 450.040 | 448.885 |
| 27. | −8.068 | 467.510 | −3.726 | 467.512 | 466.489 |
| 28. | −8.087 | 485.056 | −3.749 | 485.055 | 484.092 |
| 29. | −8.081 | 502.606 | −3.727 | 502.609 | 501.695 |
| 30. | −8.058 | 520.184 | −3.772 | 520.119 | 519.299 |
| 31. | −8.069 | 537.788 | −3.740 | 537.822 | 536.902 |
| 32. | −8.095 | 555.359 | −3.728 | 555.355 | 554.505 |
| 33. | −8.102 | 572.920 | −3.732 | 572.931 | 572.109 |
| 34. | −8.091 | 590.488 | −3.731 | 590.482 | 589.712 |
| 35. | −8.077 | 608.084 | −3.875 | 608.067 | 607.315 |
| 36. | −8.095 | 625.681 | −3.732 | 625.690 | 624.919 |
| 37. | −8.110 | 643.255 | −3.732 | 643.248 | 642.522 |
| 38. | −8.110 | 660.826 | −3.730 | 660.831 | 660.125 |
| 39. | −8.098 | 678.408 | −3.738 | 678.391 | 677.729 |
| 40. | −8.094 | 696.011 | −3.760 | 696.054 | 695.332 |
| 50. | −8.119 | 871.910 | −3.733 | 871.914 | 871.365 |
| 60. | −8.130 | 1047.85 | −3.732 | 1047.85 | 1047.40 |
| 70. | −8.134 | 1223.82 | −3.735 | 1223.82 | 1223.43 |
| 80. | −8.135 | 1399.80 | −3.733 | 1399.80 | 1399.47 |
| 90. | −8.138 | 1575.80 | −3.734 | 1575.80 | 1575.50 |
| 100. | −8.141 | 1751.80 | −3.734 | 1751.80 | 1751.53 |
| Asym. Re: | −8.143 | | −3.734 | | |

Table 1B. Numerical eigenfrequencies 1–40, 50, 60, 70, 80, 90 and 100 for the $\beta_1$ and $\beta_2$ branches from Example 1. The asymptotic imaginary parts are given in the last column, while the asymptotic real parts appear at the bottom.

| | Numerical | | | | Asymptotic (Im) |
|---|---|---|---|---|---|
| | $\alpha_1$ Branch | | $\alpha_2$ Branch | | |
| | Re | Im | Re | Im | |
| 1.* | | | | | 12.2947 |
| 2.* | −2.961 | 9.93687 | −1.610 | 12.4325 | 24.5894 |
| 3. | −3.190 | 29.3701 | −1.397 | 31.1896 | 36.8841 |
| 4. | −4.064 | 46.0071 | −1.634 | 45.0854 | 49.1788 |
| 5. | −4.495 | 57.8442 | −1.657 | 58.5989 | 61.4735 |
| 6. | −4.053 | 71.8202 | −1.621 | 71.7962 | 73.7682 |
| 7. | −4.142 | 85.2854 | −1.523 | 84.8430 | 86.0630 |
| 8. | −4.567 | 98.6461 | −1.323 | 98.0245 | 98.3577 |
| 9. | −5.318 | 105.571 | −1.169 | 106.236 | 110.652 |
| 10. | −4.540 | 119.808 | −1.533 | 120.191 | 122.947 |
| 11. | −4.216 | 133.109 | −1.633 | 133.237 | 135.242 |

| | | | | | |
|---|---|---|---|---|---|
| 12. | −4.218 | 146.000 | −1.647 | 145.953 | 147.536 |
| 13. | −4.179 | 158.522 | −1.639 | 158.542 | 159.831 |
| 14. | −4.174 | 171.218 | −1.612 | 171.110 | 172.126 |
| 15. | −4.406 | 183.978 | −1.517 | 183.787 | 184.421 |
| 16. | −5.446 | 194.130 | −1.130 | 194.144 | 196.715 |
| 17. | −4.348 | 207.405 | −1.571 | 207.533 | 209.010 |
| 18. | −4.164 | 220.088 | −1.624 | 220.138 | 221.305 |
| 19. | −4.165 | 232.619 | −1.631 | 232.603 | 233.599 |
| 20. | −4.143 | 245.025 | −1.628 | 245.027 | 245.894 |
| 21. | −4.161 | 257.514 | −1.613 | 257.457 | 258.189 |
| 22. | −4.348 | 270.059 | −1.540 | 269.988 | 270.484 |
| 23. | −4.753 | 281.519 | −1.352 | 281.352 | 282.778 |
| 24. | −4.211 | 294.071 | −1.603 | 294.130 | 295.073 |
| 25. | −4.138 | 306.548 | −1.623 | 306.566 | 307.368 |
| 26. | −4.151 | 318.951 | −1.625 | 318.946 | 319.662 |
| 27. | −4.130 | 331.320 | −1.623 | 331.313 | 331.957 |
| 28. | −4.159 | 343.734 | −1.611 | 343.694 | 344.252 |
| 29. | −4.363 | 356.177 | −1.518 | 356.192 | 356.547 |
| 30. | −4.346 | 367.985 | −1.536 | 367.969 | 368.841 |
| 31. | −4.155 | 380.426 | −1.614 | 380.456 | 381.136 |
| 32. | −4.129 | 392.820 | −1.622 | 392.824 | 393.431 |
| 33. | −4.138 | 405.165 | −1.622 | 405.168 | 405.725 |
| 34. | −4.128 | 417.522 | −1.620 | 417.511 | 418.020 |
| 35. | −4.164 | 429.902 | −1.607 | 429.873 | 430.315 |
| 36. | −4.414 | 442.219 | −1.394 | 442.404 | 442.609 |
| 37. | −4.204 | 454.269 | −1.595 | 454.287 | 454.904 |
| 38. | −4.133 | 466.654 | −1.618 | 466.669 | 467.199 |
| 39. | −4.127 | 479.008 | −1.621 | 479.006 | 479.494 |
| 40. | −4.127 | 491.332 | −1.621 | 491.334 | 491.788 |
| 50. | −4.211 | 614.282 | −1.579 | 614.263 | 614.735 |
| 60. | −4.122 | 737.376 | −1.619 | 737.377 | 737.682 |
| 70. | −4.162 | 860.323 | −1.600 | 860.315 | 860.630 |
| 80. | −4.120 | 983.347 | −1.619 | 983.347 | 983.577 |
| 90. | −4.138 | 1007.97 | −1.612 | 1007.96 | 1106.52 |
| 100. | −4.120 | 1130.91 | −1.618 | 1130.91 | 1129.47 |
| Asym. Re: | −4.118 | | −1.618 | | |

Table 2A. Numerical eigenfrequencies 1–40, 50, 60, 70, 80, 90 and 100 for the $\alpha_1$ and $\alpha_2$ branches from Example 2. The asymptotic imaginary parts are given in the last column, while the asymptotic real parts appear at the bottom.

| | Numerical | | | | Asymptotic (Im) |
|---|---|---|---|---|---|
| | $\alpha_1$ Branch | | $\alpha_2$ Branch | | |
| | Re | Im | Re | Im | |
| 0.* | −2.767 | 20.7386 | −.3306 | 20.7810 | |
| 1.* | −3.365 | 32.4318 | −.3818 | 31.1907 | 17.6033 |
| 2.* | −4.405 | 43.3261 | −.4055 | 44.5802 | 35.2067 |
| 3. | −4.889 | 59.8085 | −.4281 | 59.3359 | 52.8100 |
| 4. | −5.844 | 74.6045 | −.5117 | 74.9216 | 70.4134 |

| | | | | | |
|---|---|---|---|---|---|
| 5. | −5.926 | 90.1511 | −.7273 | 90.8426 | 88.0167 |
| 6. | −5.396 | 112.518 | −.8981 | 112.008 | 105.620 |
| 7. | −6.281 | 127.780 | −.5491 | 127.584 | 123.223 |
| 8. | −6.594 | 144.098 | −.4656 | 144.153 | 140.827 |
| 9. | −6.736 | 161.038 | −.4744 | 161.084 | 158.430 |
| 10. | −6.648 | 177.856 | −.5675 | 178.041 | 176.033 |
| 11. | −5.663 | 197.373 | −.9579 | 197.391 | 193.637 |
| 12. | −6.785 | 213.821 | −.5208 | 213.750 | 211.240 |
| 13. | −6.945 | 230.896 | −.4746 | 230.912 | 228.843 |
| 14. | −6.990 | 248.188 | −.4806 | 248.215 | 246.447 |
| 15. | −6.852 | 265.397 | −.5525 | 265.465 | 264.050 |
| 16. | −6.465 | 283.725 | −.7417 | 283.906 | 281.653 |
| 17. | −7.012 | 300.977 | −.4935 | 300.948 | 299.257 |
| 18. | −7.061 | 318.329 | −.4777 | 318.336 | 316.860 |
| 19. | −7.077 | 335.745 | −.4853 | 335.769 | 334.464 |
| 20. | −6.892 | 353.121 | −.5783 | 353.106 | 352.067 |
| 21. | −6.916 | 371.147 | −.5602 | 371.174 | 369.670 |
| 22. | −7.105 | 388.547 | −.4836 | 388.536 | 387.274 |
| 23. | −7.118 | 406.017 | −.4793 | 406.019 | 404.877 |
| 24. | −7.109 | 423.484 | −.4901 | 423.506 | 422.480 |
| 25. | −6.866 | 441.016 | −.7033 | 440.833 | 440.084 |
| 26. | −7.079 | 458.824 | −.5031 | 458.814 | 457.687 |
| 27. | −7.144 | 476.297 | −.4809 | 476.297 | 475.290 |
| 28. | −7.150 | 493.819 | −.4809 | 493.822 | 492.894 |
| 29. | −7.114 | 511.317 | −.4975 | 511.331 | 510.497 |
| 30. | −6.958 | 529.040 | −.6926 | 529.212 | 528.100 |
| 31. | −7.142 | 546.621 | −.4875 | 546.611 | 545.704 |
| 32. | −7.161 | 564.142 | −.4806 | 564.143 | 563.307 |
| 33. | −7.164 | 581.683 | −.4828 | 581.690 | 580.910 |
| 34. | −7.098 | 599.211 | −.5157 | 599.200 | 598.514 |
| 35. | −7.092 | 616.952 | −.5196 | 616.974 | 616.117 |
| 36. | −7.169 | 634.487 | −.4829 | 634.483 | 633.720 |
| 37. | −7.173 | 652.039 | −.4810 | 652.039 | 651.324 |
| 38. | −7.167 | 669.587 | −.4853 | 669.595 | 668.927 |
| 39. | −7.070 | 687.169 | −.5948 | 687.091 | 686.530 |
| 40. | −7.153 | 704.848 | −.4913 | 704.845 | 704.134 |
| 50. | −7.188 | 880.714 | −.4823 | 880.711 | 880.167 |
| 60. | −7.195 | 1056.65 | −.4817 | 1056.65 | 1056.20 |
| 70. | −7.199 | 1232.61 | −.4820 | 1232.61 | 1232.23 |
| 80. | −7.197 | 1408.59 | −.4836 | 1408.59 | 1408.27 |
| 90. | −7.206 | 1584.59 | −.4821 | 1584.59 | 1584.30 |
| 100. | −7.176 | 1760.61 | −.4824 | 1760.61 | 1760.33 |
| Asym. Re: | −7.208 | | −.4821 | | |

Table 2B. Numerical eigenfrequencies 1–40, 50, 60, 70, 80, 90 and 100 for the $\beta_1$ and $\beta_2$ branches from Example 2. The asymptotic imaginary parts are given in the last column, while the asymptotic real parts appear at the bottom.

| | $C' = 0$ | $C' = .25$ | $C' = .5$ | $C' = .75$ | $C' = 1$ |
|---|---|---|---|---|---|
| 1. | 2.9458212164 | 1.2833048444 | — | — | — |
| 2. | 2.9458212164 | 2.9458212164 | 2.9458212164 | 2.9458212164 | 2.9458212164 |
| 3. | 10.678276269 | 10.429210089 | 10.173172662 | 9.9095407887 | 9.6376145423 |
| 4. | 10.678276269 | 10.678276269 | 10.678276269 | 10.678276269 | 10.678276269 |
| 5. | 22.186791607 | 22.071993648 | 21.956728997 | 21.840975799 | 21.724712242 |
| 6. | 22.186791607 | 22.186791607 | 22.186791607 | 22.186791607 | 22.186791607 |
| 7. | 26.977657335 | 26.925021146 | 26.872366881 | 26.819700315 | 26.767023640 |
| 8. | 26.977657335 | 26.977657335 | 26.977657335 | 26.977657335 | 26.977657335 |
| 9. | 36.385895938 | 36.330795486 | 36.275730248 | 36.220689467 | 36.165662881 |
| 10. | 36.385895938 | 36.385895938 | 36.385895938 | 36.385895938 | 36.385895938 |
| 11. | 39.346020018 | 39.290455669 | 39.234922527 | 39.179434389 | 39.123994694 |
| 12. | 39.346020018 | 39.346020018 | 39.346020018 | 39.34602002 | 39.346020018 |
| 13. | 50.846627673 | 50.805226033 | 50.763519267 | 50.721507452 | 50.679174702 |
| 14. | 50.846627673 | 50.846627673 | 50.846627673 | 50.846627673 | 50.846627673 |
| 15. | 52.832231307 | 52.792215617 | 52.752514490 | 52.713137027 | 52.674092535 |
| 16. | 52.832231307 | 52.832231307 | 52.832231307 | 52.832231307 | 52.832231307 |
| 17. | 64.810103855 | 64.762969581 | 64.715722238 | 64.668361902 | 64.620894071 |
| 18. | 64.810103855 | 64.810103855 | 64.810103855 | 64.810103855 | 64.810103855 |
| 19. | 67.511260081 | 67.495035827 | 67.478904495 | 67.462874126 | 67.446934409 |
| 20. | 67.511260081 | 67.511260081 | 67.511260081 | 67.511260081 | 67.511260081 |
| 21. | 77.945061900 | 77.902776945 | 77.860470284 | 77.818146754 | 77.775801859 |
| 22. | 77.945061900 | 77.945061900 | 77.945061900 | 77.945061900 | 77.945061900 |
| 23. | 83.358682455 | 83.347923221 | 83.337147524 | 83.326356066 | 83.315549557 |
| 24. | 83.358682455 | 83.358682455 | 83.358682455 | 83.358682455 | 83.358682455 |
| 25. | 90.953565995 | 90.918343817 | 90.883145591 | 90.847963440 | 90.812804258 |
| 26. | 90.953565995 | 90.953565995 | 90.953565995 | 90.953565995 | 90.953565995 |
| 27. | 99.219305787 | 99.204551012 | 99.189697002 | 99.174754327 | 99.159714381 |
| 28. | 99.219305787 | 99.219305787 | 99.219305787 | 99.219305787 | 99.219305787 |
| 29. | 104.36226852 | 104.33801023 | 104.31384569 | 104.28976275 | 104.26577156 |
| 30. | 104.36226852 | 104.36226852 | 104.36226852 | 104.36226852 | 104.36226852 |
| 31. | 113.79274626 | 113.76791681 | 113.74302537 | 113.71805740 | 113.69303441 |
| 32. | 113.79274626 | 113.79274626 | 113.79274626 | 113.79274626 | 113.79274626 |
| 33. | 119.25906418 | 119.24966905 | 119.24033598 | 119.23105702 | 119.22184083 |
| 34. | 119.25906418 | 119.25906418 | 119.25906418 | 119.25906418 | 119.25906418 |
| 35. | 126.98609223 | 126.95906982 | 126.93201897 | 126.90494972 | 126.87787207 |
| 36. | 126.98609223 | 126.98609223 | 126.98609223 | 126.98609223 | 126.98609223 |
| 37. | 135.64225489 | 135.63853685 | 135.63483364 | 135.63113319 | 135.62744782 |
| 38. | 135.64225489 | 135.64225490 | 135.64225489 | 135.64225489 | 135.64225489 |
| 39. | 139.71595073 | 139.69029336 | 139.66462947 | 139.63895904 | 139.61328206 |
| 40. | 139.71595073 | 139.71595073 | 139.71595073 | 139.71595073 | 139.71595073 |

Table 3A. The first 40 imaginary parts of the numerical eigenfrequencies from Example 3, computed for five different values of the Van der Waals constant $C'$. The "real-life" value of the constant is approximately .57.

| | Numerical | Asymptotic |
|---|---|---|
| 49. | 177.211 | 178.283 ($\alpha$-branch) |

| | | |
|---|---|---|
| 50. | 177.293 | 178.283 ″ |
| 99. | 361.308 | 360.869 ($\beta$-branch) |
| 100. | 361.331 | 360.869 ″ |
| 149. | 537.847 | 536.902 ($\beta$-branch) |
| 150. | 537.848 | 536.902 ″ |
| 199. | 718.895 | 719.240 ($\alpha$-branch) |
| 200. | 718.916 | 719.240 ″ |
| 249. | 903.393 | 903.661 ($\alpha$-branch) |
| 250. | 903.410 | 903.661 ″ |
| 299. | 1083.03 | 1082.61 ($\beta$-branch) |
| 300. | 1083.03 | 1082.61 ″ |
| 349. | 1260.06 | 1260.21 ($\alpha$-branch) |
| 350. | 1260.07 | 1260.21 ″ |
| 399. | 1444.46 | 1444.62 ($\alpha$-branch) |
| 400. | 1444.47 | 1444.62 ″ |

Table 3B. Numerical and asymptotic eigenfrequencies (imaginary parts) 49, 50, 99, 100, 149, 150, 199, 200, 249, 250, 299, 300, 349, 350, 399, 400 from Example 3, computed for the Van der Waals constant $C' = 1$.

| | $C' = 0$ | $C' = .25$ | $C' = .5$ | $C' = .75$ | $C' = 1$ |
|---|---|---|---|---|---|
| 1. | 12.98454240 | 12.82401972 | 12.66021095 | 12.49299083 | 12.32222077 |
| 2. | 12.98454240 | 12.98454240 | 12.98454240 | 12.98454240 | 12.98454240 |
| 3. | 20.80444376 | 20.73055727 | 20.65705225 | 20.58390602 | 20.51109394 |
| 4. | 20.80444376 | 20.80444376 | 20.80444376 | 20.80444376 | 20.80444376 |
| 5. | 31.18843099 | 31.12463266 | 31.06111752 | 30.99788409 | 30.93493102 |
| 6. | 31.18843099 | 31.18843099 | 31.18843099 | 31.18843099 | 31.18843099 |
| 7. | 31.24700816 | 31.18715530 | 31.12710108 | 31.06685140 | 31.00640685 |
| 8. | 31.24700816 | 31.24700816 | 31.24700816 | 31.24700816 | 31.24700816 |
| 9. | 44.57678921 | 44.53928646 | 44.50196213 | 44.46481864 | 44.42785126 |
| 10. | 44.57678921 | 44.57678921 | 44.57678921 | 44.57678921 | 44.57678921 |
| 11. | 45.09876440 | 45.04148608 | 44.98405651 | 44.92647256 | 44.86873484 |
| 12. | 45.09876440 | 45.09876440 | 45.09876440 | 45.09876440 | 45.09876440 |
| 13. | 58.59976143 | 58.54891895 | 58.49801214 | 58.44703446 | 58.39599137 |
| 14. | 58.59976143 | 58.59976143 | 58.59976143 | 58.59976143 | 58.59976143 |
| 15. | 59.33988894 | 59.31872578 | 59.29763244 | 59.27660934 | 59.25566107 |
| 16. | 59.33988894 | 59.33988894 | 59.33988894 | 59.33988894 | 59.33988894 |
| 17. | 71.76457338 | 71.72043968 | 71.67628715 | 71.63210155 | 71.58790511 |
| 18. | 71.76457338 | 71.76457338 | 71.76457338 | 71.76457338 | 71.76457338 |
| 19. | 74.95903748 | 74.94532704 | 74.93161035 | 74.91790464 | 74.90420214 |

| | | | | | |
|---|---|---|---|---|---|
| 20. | 74.95903748 | 74.95903748 | 74.95903748 | 74.95903748 | 74.95903748 |
| 21. | 84.79872934 | 84.76134601 | 84.72397393 | 84.68660688 | 84.64925063 |
| 22. | 84.79872934 | 84.79872934 | 84.79872934 | 84.79872934 | 84.79872934 |
| 23. | 90.88275651 | 90.86960300 | 90.85639777 | 90.84314155 | 90.82983509 |
| 24. | 90.88275651 | 90.88275651 | 90.88275651 | 90.88275651 | 90.88275651 |
| 25. | 97.98924692 | 97.96005606 | 97.93091660 | 97.90181859 | 97.87277054 |
| 26. | 97.98924692 | 97.98924692 | 97.98924692 | 97.98924692 | 97.98924692 |
| 27. | 106.2587082 | 106.2388530 | 106.2189144 | 106.1988926 | 106.1787763 |
| 28. | 106.2587082 | 106.2587082 | 106.2587082 | 106.2587082 | 106.2587082 |
| 29. | 111.9946283 | 111.9779469 | 111.9613540 | 111.9448287 | 111.9283915 |
| 30. | 111.9946283 | 111.9946283 | 111.9946283 | 111.9946283 | 111.9946283 |
| 31. | 120.2222931 | 120.1964306 | 120.1705341 | 120.1445952 | 120.1186307 |
| 32. | 120.2222931 | 120.2222931 | 120.2222930 | 120.2222931 | 120.2222931 |
| 33. | 127.5643741 | 127.5578845 | 127.5514182 | 127.5449753 | 127.5385763 |
| 34. | 127.5643741 | 127.5643741 | 127.5643741 | 127.5643741 | 127.5643741 |
| 35. | 133.2540877 | 133.2279431 | 133.2017948 | 133.1756314 | 133.1494527 |
| 36. | 133.2540877 | 133.2540877 | 133.2540878 | 133.2540877 | 133.2540877 |
| 37. | 144.1489244 | 144.1458534 | 144.1427760 | 144.1397215 | 144.1366607 |
| 38. | 144.1489244 | 144.1489244 | 144.1489244 | 144.1489244 | 144.1489244 |
| 39. | 145.9551025 | 145.9304056 | 145.9056787 | 145.8809675 | 145.8562415 |
| 40. | 145.9551025 | 145.9551025 | 145.9551025 | 145.9551025 | 145.9551025 |

Table 4A. The first 40 imaginary parts of the numerical eigenfrequencies from Example 4, computed for five different values of the Van der Waals constant $C'$.

| | Numerical | Asymptotic |
|---|---|---|
| 49. | 183.705 | 184.421 ($\alpha$-branch) |
| 50. | 183.779 | 184.421 ″ |
| 99. | 367.917 | 368.841 ($\alpha$-branch) |
| 100. | 367.955 | 368.841 ″ |
| 149. | 546.609 | 545.704 ($\beta$-branch) |
| 150. | 546.609 | 545.704 ″ |
| 199. | 725.047 | 725.388 ($\alpha$-branch) |
| 200. | 725.068 | 725.388 ″ |
| 249. | 909.548 | 909.808 ($\alpha$-branch) |
| 250. | 909.565 | 909.808 ″ |
| 299. | 1091.80 | 1091.41 ($\beta$-branch) |
| 300. | 1091.80 | 1091.41 ″ |
| 349. | 1267.85 | 1267.44 ($\beta$-branch) |
| 350. | 1267.85 | 1267.44 ″ |
| 399. | 1450.61 | 1450.78 ($\alpha$-branch) |
| 400. | 1450.62 | 1450.78 ″ |

Table 4B. Numerical and asymptotic eigenfrequencies (imaginary parts) 49, 50, 99, 100, 149, 150, 199, 200, 249, 250, 299, 300, 349, 350, 399, 400 from Example 4, computed for the Van der Waals constant $C' = 1$.

| | $C' = 0$ | | $C' = .5$ | | $C' = 1$ | |
|---|---|---|---|---|---|---|
| | Re | Im | Re | Im | Re | Im |
| 1. | — | — | — | — | — | — |
| 2. | — | — | — | — | — | — |
| 3. | −2.746 | 9.449183995 | −2.877 | 9.172044078 | −2.972 | 8.827341448 |
| 4. | −.6783 | 9.930366988 | −.7344 | 9.678929479 | −.8576 | 9.477841762 |
| 5. | −3.529 | 21.40990132 | −3.575 | 21.29275588 | −3.611 | 21.16999982 |
| 6. | −.9628 | 21.77418316 | −.9720 | 21.65493676 | −.9945 | 21.54041229 |
| 7. | −1.321 | 26.90292372 | −1.312 | 26.85129816 | −1.308 | 26.79542808 |
| 8. | −2.335 | 27.37492665 | −2.342 | 27.32751386 | −2.343 | 27.28372611 |
| 9. | −3.537 | 35.54153378 | −3.556 | 35.47898352 | −3.569 | 35.40700136 |
| 10. | −3.172 | 36.55704093 | −3.177 | 36.50955538 | −3.190 | 36.47367372 |
| 11. | −1.490 | 38.69290303 | −1.496 | 38.63128423 | −1.504 | 38.56575342 |
| 12. | −3.658 | 39.81469654 | −3.664 | 39.76185775 | −3.665 | 39.71104768 |
| 13. | −4.007 | 50.27733227 | −4.028 | 50.23454838 | −4.048 | 50.18747340 |
| 14. | −5.139 | 51.39985683 | −5.143 | 51.37684915 | −5.148 | 51.35549893 |
| 15. | −.9151 | 51.93127198 | −.9182 | 51.87409462 | −.9225 | 51.81686142 |
| 16. | −3.823 | 53.22422671 | −3.821 | 53.18663028 | −3.816 | 53.15176710 |
| 17. | −4.613 | 64.39259632 | −4.626 | 64.34564444 | −4.637 | 64.29753890 |
| 18. | −.7899 | 65.10483274 | −.7935 | 65.05533361 | −.7986 | 65.00630011 |
| 19. | −6.082 | 67.04799021 | −6.082 | 67.03545255 | −6.081 | 67.02281832 |
| 20. | −3.538 | 67.87701409 | −3.539 | 67.86205923 | −3.539 | 67.84780896 |
| 21. | −4.754 | 77.62836514 | −4.762 | 77.58540148 | −4.768 | 77.54197597 |
| 22. | −.8924 | 78.05109828 | −.8968 | 78.00762380 | −.9026 | 77.96449011 |
| 23. | −6.339 | 83.35453188 | −6.335 | 83.34816855 | −6.331 | 83.34180146 |
| 24. | −3.632 | 83.71862800 | −3.635 | 83.70997652 | −3.638 | 83.70136960 |
| 25. | −4.696 | 90.61268633 | −4.700 | 90.57580895 | −4.703 | 90.53866701 |
| 26. | −1.224 | 90.79035136 | −1.230 | 90.75096706 | −1.237 | 90.71182474 |
| 27. | −4.061 | 99.70402490 | −4.068 | 99.69160776 | −4.076 | 99.67839224 |
| 28. | −5.939 | 100.4640803 | −5.931 | 100.4692419 | −5.924 | 100.4751321 |
| 29. | −1.951 | 102.9983756 | −1.960 | 102.9543524 | −1.970 | 102.9099689 |
| 30. | −4.352 | 103.8895026 | −4.350 | 103.8631919 | −4.346 | 103.8372742 |
| 31. | −4.743 | 114.1061665 | −4.749 | 114.0804654 | −4.754 | 114.0546217 |
| 32. | −1.357 | 114.4090098 | −1.352 | 114.3757397 | −1.348 | 114.3424705 |
| 33. | −6.716 | 118.5438854 | −6.722 | 118.5429195 | −6.727 | 118.5418301 |
| 34. | −3.738 | 118.9568732 | −3.736 | 118.9487280 | −3.735 | 118.9408092 |
| 35. | −.7845 | 127.0161580 | −.7832 | 126.9880193 | −.7825 | 126.9598517 |
| 36. | −4.902 | 127.1222585 | −4.906 | 127.0951053 | −4.909 | 127.0679252 |
| 37. | −3.642 | 135.5114464 | −3.642 | 135.5080511 | −3.642 | 135.5046704 |
| 38. | −7.386 | 135.5399041 | −7.388 | 135.5375232 | −7.390 | 135.5351432 |
| 39. | −.5884 | 139.6706311 | −.5884 | 139.6447863 | −.5888 | 139.6189351 |
| 40. | −4.907 | 139.7988314 | −4.910 | 139.7733338 | −4.912 | 139.7478086 |

Table 5A. The first 40 numerical eigenfrequencies from Example 5, computed for three different values of the Van der Waals constant $C$.

| | Numerical | | Asymptotic | |
|---|---|---|---|---|
| | Re | Im | Re | Im |
| 49. | −4.902 | 177.162 | −4.941 | 178.283 ($\alpha_1$-branch) |
| 50. | −.6921 | 177.198 | −.5027 | 178.283 ($\alpha_2$-branch) |
| 99. | −7.957 | 362.152 | −8.143 | 360.869 ($\beta_2$-branch) |
| 100. | −4.195 | 362.404 | −3.734 | 360.869 ($\beta_1$-branch) |
| 149. | −8.069 | 537.788 | −8.143 | 536.902 ($\beta_2$-branch) |
| 150. | −3.740 | 537.822 | −3.734 | 536.902 ($\beta_1$-branch) |
| 199. | −.5069 | 718.904 | −.5027 | 719.240 ($\alpha_2$-branch) |
| 200. | −4.941 | 718.909 | −4.941 | 719.240 ($\alpha_1$-branch) |
| 249. | −4.940 | 903.399 | −4.941 | 903.661 ($\alpha_1$-branch) |
| 250. | −.5046 | 903.403 | −.5027 | 903.661 ($\alpha_2$-branch) |
| 299. | −3.734 | 1083.03 | −3.734 | 1082.61 ($\beta_1$-branch) |
| 300. | −8.127 | 1083.03 | −8.143 | 1082.61 ($\beta_2$-branch) |
| 349. | −.5123 | 1260.02 | −.5027 | 1260.21 ($\alpha_2$-branch) |
| 350. | −4.923 | 1260.04 | −4.941 | 1260.21 ($\alpha_1$-branch) |
| 399. | −.5053 | 1444.46 | −.5027 | 1444.62 ($\alpha_2$-branch) |
| 400. | −4.940 | 1444.46 | −4.941 | 1444.62 ($\alpha_1$-branch) |

Table 5B. Numerical and asymptotic eigenfrequencies (imaginary parts) 49, 50, 99, 100, 149, 150, 199, 200, 249, 250, 299, 300, 349, 350, 399, 400 from Example 5, computed for the Van der Waals constant $C' = 1$.

| | $C' = 0$ | | $C' = .5$ | | $C' = 1$ | |
|---|---|---|---|---|---|---|
| | Re | Im | Re | Im | Re | Im |
| 1. | −2.961 | 9.936871385 | −3.079 | 9.795166383 | −3.169 | 9.644727421 |
| 2. | −1.610 | 12.43254005 | −1.618 | 12.28811944 | −1.632 | 12.14607344 |
| 3. | −2.767 | 20.73861326 | −2.794 | 20.64458863 | −2.815 | 20.55053147 |
| 4. | −.3306 | 20.78104522 | −.3395 | 20.70597402 | −0.355 | 20.63293559 |
| 5. | −3.190 | 29.37005728 | −3.187 | 29.32574131 | −3.182 | 29.28091887 |
| 6. | −1.397 | 31.18963235 | −.3840 | 31.12726000 | −.3882 | 31.06474018 |
| 7. | −.3818 | 31.19073440 | −1.392 | 31.12875704 | −1.390 | 31.06625299 |
| 8. | −3.365 | 32.43182622 | −3.369 | 32.36004659 | −3.369 | 32.28961412 |
| 9. | −4.405 | 43.32612319 | −4.402 | 43.28601387 | −4.398 | 43.24625757 |
| 10. | −.4055 | 44.58018737 | −.4065 | 44.54246627 | −.4083 | 44.50537030 |
| 11. | −1.634 | 45.08538379 | −1.632 | 45.02792865 | −1.634 | 44.96962880 |
| 12. | −4.064 | 46.00707085 | −4.068 | 45.95274391 | −4.070 | 45.89830745 |
| 13. | −4.495 | 57.84423832 | −4.479 | 57.81269810 | −4.461 | 57.78159556 |
| 14. | −1.657 | 58.59889870 | −1.656 | 58.54816779 | −1.659 | 58.49709714 |
| 15. | −.4281 | 59.33578065 | −.4286 | 59.31422388 | −.4295 | 59.29277982 |
| 16. | −4.889 | 59.80848155 | −4.905 | 59.76734039 | −4.920 | 59.72601742 |
| 17. | −1.621 | 71.79615374 | −1.620 | 71.75209563 | −1.621 | 71.70774465 |

| | | | | | | |
|---|---|---|---|---|---|---|
| 18. | −4.053 | 71.82019141 | −4.045 | 71.78210831 | −4.037 | 71.74432634 |
| 19. | −5.844 | 74.60448514 | −5.849 | 74.58483733 | −5.854 | 74.56510945 |
| 20. | −.5117 | 74.92157699 | −.5133 | 74.90763370 | −.5152 | 74.89373280 |
| 21. | −1.523 | 84.84297209 | −1.522 | 84.80549984 | −1.522 | 84.76785224 |
| 22. | −4.142 | 85.28536486 | −4.141 | 85.25105512 | −4.140 | 85.21691580 |
| 23. | −5.926 | 90.15112979 | −5.923 | 90.13610573 | −5.920 | 90.12104814 |
| 24. | −.7273 | 90.84264675 | −.7301 | 90.82960542 | −.7331 | 90.81651834 |
| 25. | −1.323 | 98.02449900 | −1.320 | 97.99508254 | −1.319 | 97.96567545 |
| 26. | −4.567 | 98.64607511 | −4.570 | 98.61744170 | −4.573 | 98.58888833 |
| 27. | −5.318 | 105.5709742 | −5.311 | 105.5523351 | −5.304 | 105.5336172 |
| 28. | −1.169 | 106.2355904 | −1.173 | 106.2156779 | −1.177 | 106.1956428 |
| 29. | −.8981 | 112.0084405 | −.8948 | 111.9918049 | −.8917 | 111.9752688 |
| 30. | −5.396 | 112.5178851 | −5.403 | 112.4994593 | −5.409 | 112.4810878 |
| 31. | −4.540 | 119.8083113 | −4.535 | 119.7838877 | −4.530 | 119.7594125 |
| 32. | −1.533 | 120.1912992 | −1.535 | 120.1654629 | −1.536 | 120.1395677 |
| 33. | −.5491 | 127.5842376 | −.5480 | 127.5777146 | −.5471 | 127.5712440 |
| 34. | −6.281 | 127.7796866 | −6.285 | 127.7714673 | −6.288 | 127.7632923 |
| 35. | −4.216 | 133.1088121 | −4.214 | 133.0838601 | −4.211 | 133.0588793 |
| 36. | −1.633 | 133.2370084 | −1.634 | 133.2109568 | −1.635 | 133.1848771 |
| 37. | −6.594 | 144.0977330 | −6.594 | 144.0939736 | −6.594 | 144.0902605 |
| 38. | −.4656 | 144.1532932 | −.4655 | 144.1501251 | −.4654 | 144.1469825 |
| 39. | −1.647 | 145.9530993 | −1.647 | 145.9284497 | −1.648 | 145.9037766 |
| 40. | −4.218 | 145.9995218 | −4.218 | 145.9750728 | −4.217 | 145.9505817 |

Table 6A. The first 40 numerical eigenfrequencies from Example 6, computed for three different values of the Van der Waals constant $C'$.

| | Numerical | | Asymptotic | |
|---|---|---|---|---|
| | Re | Im | Re | Im |
| 49. | −1.516 | 183.750 | −1.618 | 184.421 ($\alpha_2$-branch) |
| 50. | −4.408 | 183.941 | −4.118 | 184.421 ($\alpha_1$-branch) |
| 99. | −1.538 | 367.949 | −1.618 | 368.841 ($\alpha_2$-branch) |
| 100. | −4.344 | 367.965 | −4.118 | 368.841 ($\alpha_1$-branch) |
| 149. | −.4874 | 546.610 | −.4821 | 545.704 ($\beta_2$-branch) |
| 150. | −7.142 | 546.620 | −7.208 | 545.704 ($\beta_1$-branch) |
| 199. | −1.619 | 725.057 | −1.618 | 725.388 ($\alpha_2$-branch) |
| 200. | −4.122 | 725.058 | −4.118 | 725.388 ($\alpha_1$-branch) |
| 249. | −1.618 | 909.557 | −1.618 | 909.808 ($\alpha_2$-branch) |
| 250. | −4.121 | 909.559 | −4.118 | 909.808 ($\alpha_1$-branch) |
| 299. | −.4987 | 1091.80 | −.4821 | 1091.41 ($\beta_2$-branch) |
| 300. | −7.168 | 1091.81 | −7.208 | 1091.41 ($\beta_1$-branch) |
| 349. | −7.173 | 1267.83 | −7.208 | 1267.44 ($\beta_1$-branch) |
| 350. | −.5013 | 1267.84 | −.4821 | 1267.44 ($\beta_2$-branch) |

| | | | | |
|---|---|---|---|---|
| 399. | $-4.120$ | 1450.61 | $-4.118$ | 1450.78 ($\alpha_2$-branch) |
| 400. | $-1.618$ | 1450.61 | $-1.618$ | 1450.78 ($\alpha_1$-branch) |

Table 6B. Numerical and asymptotic eigenfrequencies (imaginary parts) 49, 50, 99, 100, 149, 150, 199, 200, 249, 250, 299, 300, 349, 350, 399, 400 from Example 6, computed for the Van der Waals constant $C' = 1$.

## 6. References


Coleman, M.P. & Schaffer, L. (2010). Asymptotic analysis of the vibration spectrum of coupled Timoshenko beams with a dissipative joint, *Eur. J. Mech. A Solids*, Vol. 29, No. 4, 629-636.

Coleman, M.P. & Schaffer, L. The single Timoshenko beam with general boundary damping, an asymptotic and numerical study, preprint.

Gibson, R.F.; Ayorinde, E.O. & Wen, Y.-F. (2007). Vibrations of carbon nanotubes and their composites: A review, *Comp. Sci. Tech.*, Vol. 67, 1-27.

Gottlieb, D.; Hussaini, M.Y. & Orszag, S.A. (1984). Theory and applications of spectral methods, *Spectral methods for partial differential equations* , pp. 1-54, Hampton, VA, 1982, SIAM, Philadelphia, PA.

Jakobson, B.I.; Brabec, C.J. & Berhold, J. (1996). Nanomechanics of carbon nanotubes: Instabilities beyond linear response, *Phys. Rev. Lett.*, Vol. 76, No. 14, 2511-2514.

Jamieson, V. (2000). Carbon nanotubes roll on, *Phys. World*, Vol. 13, No. 6, 29-30.

Mahan, G.D. (2002). Oscillations of a thin hollow cylinder: Carbon nanotubes, *Phys. Rev. B*, Vol. 65, No. 23, 235402.1-235402.7.

Pantano, A.; Boyce, M.C. & Parks, D.M. (2003). Nonlinear structural mechanics based modeling of carbon nanotube deformation, *Phys. Rev. Lett.*, Vol. 91, No. 14, 145504.1-145504.4.

Pantano, A.; Boyce, M.C. & Parks, D.M. (2004). Mechanics of deformation of single and multi-wall carbon nanotubes, *J. Mech. Phys. Solids*, Vol. 52, No. 4, 789-821.

Qian, D.; Wagner, G.J.; Liu, W.K.; Yu, M.-F. & Ruoff, R.S. (2002). Mechanics of carbon nanotubes, *Appl. Mech. Rev.*, Vol. 55, No. 6, 495-533.

Ru, C.Q. (2000). Effect of Van der Waals forces on axial buckling of a double-walled carbon nanotube, *J. Appl. Phys.*, Vol. 87, 1712-1715.

Ru, C.Q. (2001). Degraded axial buckling strain of multiwalled carbon nanotubes due to interlayer slips, *J. Appl. Phys.*, Vol. 89, No. 6, 3426-3433.

Shubov, M.A. & Rojas-Arenaza, M. (2010a). Vibrational frequency distribution for nonconservative model of double-walled carbon nanotube, *Appl. Math. Comput.*, Vol. 217, No. 3, 1246-1252.

Shubov, M.A. & Rojas-Arenaza, M. (2010b). Mathematical analysis of carbon nanotube model, *J. Comput. Appl. Math.*, Vol. 234, No. 6, 1631-1636.

Shubov, M.A. & Rojas-Arenaza, M. (2010c). Asymptotic distribution of eigenvalues of dynamics generator governing vibrations of double-walled carbon nanotube model, *Asymptotic Anal.*, Vol. 68, No. 1–2, 89-124.

Traill-Nash, R.W. & Collar, A.R. (1953). The effects of shear flexibility and rotatory inertia on the bending vibrations of beams, *Quart. J. Mech. Appl. Math.*, Vol. 6, 186-222.

Wang, C.M.; Tan, V.B.C. & Zhang, Y.Y. (2006). Timoshenko beam model for vibration analysis of multi-walled carbon nanotubes, *J. Sound Vibrations*, Vol. 294, 1060-1072.

Wang, C.Y.; Ru, C.Q. & Mioduchowski, A. (2004). Applicability and limitations of simplified elastic shell equations for carbon nanotubes, *J. Appl. Mech.*, Vol. 71, 622-631.

Wang, C.Y.; Ru, C.Q. & Mioduchowski, A. (2005). Free vibrations of multiwall carbon nanotubes, *J. Appl. Phys.*, Vol. 97, 114323.1-114323.10.

Wang, Q. (2005). Wave propagation in carbon nanotubes via nonlocal continuum mechanics, *J. Appl. Phys.*, Vol. 98, 124301.

Xu, K.Y.; Guo, X.N. & Ru, C.Q. (2006). Vibration of a double-walled carbon nanotube aroused by nonlinear intertube Van der Waals forces, *J. Appl. Phys.*, Vol. 99, 064303.1-064303.7.

Yoon, J.; Ru, C.Q. & Mioduchowski, A. (2003). Vibration of an embedded multiwall carbon nanotube, *Comp. Sci. Tech.*, Vol. 63, 1533-1542.

# 18

# Electronic Band Structure of Carbon Nanotubes in Equilibrium and None-Equilibrium Regimes

Mehdi Pakkhesal and Rahim Ghayour
*Department of Electrical Engineering,*
*School of Engineering,*
*Shiraz University, Shiraz,*
*Iran*

## 1. Introduction

The exploration of CNTs was a great contribution to the world of science and technology. After its exploration in 1991 by Iijima [1], extensive practical and theoretical researches about its nature gradually began to develop [2-6]. Today, we know about CNTs much more about its chemical, mechanical, optical and electrical properties than before. The methods of fabrication have also progressed. Due to their electrical and optical properties, CNTs are the subject of studies about their usage in new electronic and optoelectronic devices. In this chapter we will focus on their electronic band structure, because it is the most important characteristic of a solid that should be studied to be used in determination of its electronic, optical and optoelectronic properties. In order to investigate the electronic band structure of a solid, it is first necessary to have a good understanding of its crystal lattice and atomic structure. Therefore, as the first step of this chapter we will begin with the investigation of the geometry of SWCNTs. Then we will continue with the calculation of allowed wave vectors for the electronic transport. Having finished this step, we will introduce the electronic band structure of SWCNTs.

As is known, single walled carbon nanotube or SWCNT consists of grephene sheet that is rolled into a cylinder over a vector called "chiral vector" (Fig. 1(a)) so that the beginning and the end of this vector join to form the circumstantial circle of the cylinder Fig. 1(b).

As is shown in Fig. 1(a) the chiral vector may be written in terms of unit vectors $\mathbf{a}_1$ and $\mathbf{a}_2$ , therefore **C** may be written as:

$$\mathbf{C} = m\mathbf{a}_1 + n\mathbf{a}_2 \tag{1}$$

Here $|\mathbf{a}_1| = |\mathbf{a}_2| = a_0 = \sqrt{3}a_{C\text{-}C}$ where $a_{C\text{-}C}$ is the bonding distance of the two adjacent carbon atom and is equal to 0.142nm and m > n. Having been familiar with chiral vector, its usage and its relationship with unit vectors $\mathbf{a}_1$ and $\mathbf{a}_2$, one can investigate the geometry of carbon nanotube.

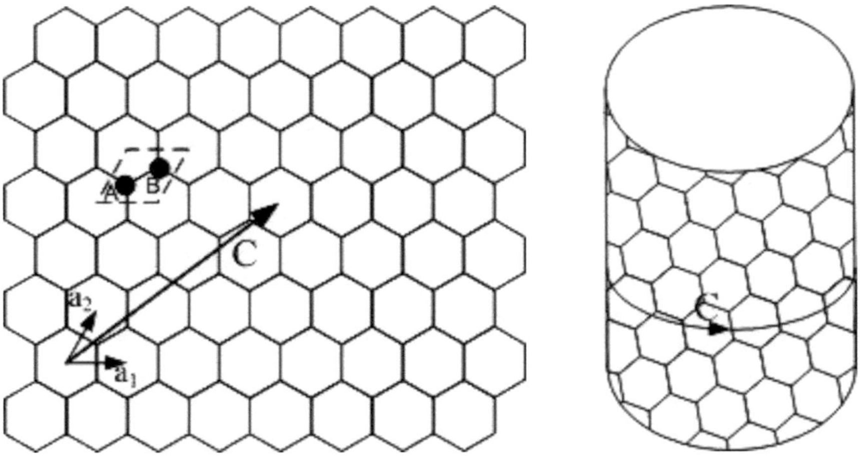


Fig. 1. (a) Illustration of Chiral vector **C** and unit vectors $a_1$ and $a_2$, A and B the two lattice sites of the graphene lattice. (b) The graphene sheet when rolled over Chiral vector **C.**

## 2. Investigation of the geometry of SWCNT

### 2.1 The investigation of radius and the chiral angle

In this section of this chapter we continue with the calculation of some aspects of the geometry of SWCNT, e.g. radius, chiral angle. As is illustrated in Fig. 1(b), the chiral vector **C** coincides the circumference of the cross sectional circle of the cylinder. Now, keeping this reality in the mind, we can easily infer the radius of the cylinder:

$$|\mathbf{C}| = 2\pi r \tag{2a}$$

which yields:

$$r = \frac{a_{C-C}}{2\pi}\sqrt{3\left(m^2 + n^2 + mn\right)} \tag{2b}$$

Next, we are to investigate a quantity called chiral angle. Chiral angle is the angle between chiral vector and the unit vector $\mathbf{a}_1$. The value can simply be calculated as:

$$\theta = \tan^{-1}\left(\frac{\sqrt{3}n}{2m+n}\right) \tag{3}$$

This value is a symbol of the way that the carbon atomic pairs (unit cell of graphene) are arranged.

### 2.2 Translational, helical and rotational symmetries

In this section we explain the three major types of symmetries of SWCNT. As a chiral structure, SWCNT is expected to have a *translational* symmetry. Thus, if we represent this symmetry with the vector **T**, such that $\mathbf{T} = t_1\mathbf{a}_1 + t_2\mathbf{a}_2$ ($t_1$ and $t_2$ are natural numbers) we are faced with shortest symmetry vector that is perpendicular to the vector **C**, so:

$$\mathbf{C}.\mathbf{T} = 0 \tag{4}$$

Therefore:

$$(t_1\mathbf{a}_1 + t_2\mathbf{a}_2).(m\mathbf{a}_1 + n\mathbf{a}_2) = 0 \tag{5}$$

in solving this equation we note that $\mathbf{a}_i.\mathbf{a}_j$ is equal to $0.5a_0^2$ if $i \neq j$ and is equal to $a_0^2$ if $i = j$. Now, solving (5), regarding that $p_1$,$p_2$,m and n are positive natural numbers, m > n and we are seeking for the smallest value of $p_1$ and $p_2$, we will have the following equation:

$$\frac{t_2}{t_1} = \frac{-\dfrac{2n+m}{\gcd(2n+m,2m+n)}}{\dfrac{2m+n}{\gcd(2n+m,2m+n)}} \tag{6}$$

$$\mathbf{T} = -\frac{2n+m}{\gcd(2n+m,2m+n)}\mathbf{a}_1 + \frac{2m+n}{\gcd(2n+m,2m+n)}\mathbf{a}_2 \tag{7}$$

where $gcd$ is standing for Greatest Common Devisor. As described before, **T** is a translational symmetry vector which means that if we move on the surface of the nanotube by **T** vector we catch up similar points.

Now we are to investigate the second and the third types of symmetries on the surface of the SWCNT which are *helical* and *rotational* symmetries [7]. As mentioned before, nanotube's cylinder is formed by rolling graphene on the lattice vector **C**. Thus, we begin our investigation by means of a mapping process. We first, try to map the unit cell of graphene on the surface of the cylinder. We suppose that **d** is a vector such that it begins from the lattice site A and ends to lattice site B. The first atom can be placed on an arbitrary place on the surface of the cylinder. The second atom must be placed at the height of $\frac{|\mathbf{d}\times\mathbf{C}|}{|\mathbf{C}|}$ from the first atom and the azimuthal angle of $2\pi\frac{|\mathbf{d}.\mathbf{C}|}{|\mathbf{C}|^2}$ with respect to the first atom. Until now, we have mapped a unit cell of graphene to the surface the cylinder. Where to place the next atomic pair? Now, we want to find a slice of the cylinder such that it includes the minimum number of graphene unit cells. We know that, the area of this slice is calculated using the formula: $A_M = 2\pi rh$. Where h is the height of the mentioned section. h can be regarded as the magnitude of a vector $\mathbf{H} = p_1\mathbf{a}_1 + p_2\mathbf{a}_2$ ; therefore, $A_M$ can be expressed as:

$$A_M = |\mathbf{H}\times\mathbf{B}| = (p_1 m - p_2 n)|\mathbf{a}_1 \times \mathbf{a}_2| \tag{8}$$

Now we are to minimize the term: $p_1m$-$p_2n$. Mathematically, it can be shown that this term is minimized when:

$$p_1 m - p_2 n = \pm N \tag{9}$$

where $N = gcd(m,n)$. In order to acquire unique values for $p_1$ and $p_2$ we find $p_1$ and $p_2$ such that $p_1 \geq 0$ and $|\mathbf{H}|$ has the minimum value. Knowing that the area of a unit cell of the graphene (which is an atomic pair) is equal to $|\mathbf{a_1} \times \mathbf{a_2}|$, the mentioned slice contains $N$ atomic pairs which are located in the multiples of the azimuthal angle of $\frac{2\pi}{N}$. This implies a symmetry in azimuthal direction which is so called "rotational symmetry". Now, we return to our question which is finding the place of the second atomic pair on the surface of the tubule. After finding the **H** vector with the mentioned conditions, it is clear that it implies a type of symmetry in the helical direction (along the vector **H**) [7]. There for, the second atomic pair should be place at a position which is located by an **H** vector next to the first atomic pair. The third atomic pair is located **2H** from the first one and so on. This "helical motif" should be copied $N$ times in angular space of $\frac{2\pi}{N}$ to construct whole the nanotube's structure. Now that we have known the symmetries of the nanotube, we are ready to investigate the band structure of SWCNT.

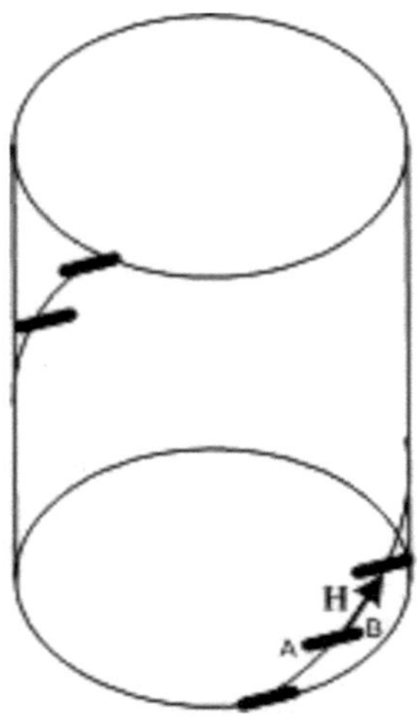


Fig. 2. In this figure the "helical motif" and **H** vector are illustrated.

## 3. The band structure of SWCNT in equilibrium conditions

### 3.1 Bloch function

At this step we are facing the problem of finding the wave function for a crystal lattice. In this situation we are facing periodic boundary conditions. Therefore, it is expected that we acquire a periodic wave function. Using these facts, in 1927 Bloch showed that the electron wave function has the following form for a crystal lattice:

$$\psi_{\mathbf{k}}(\mathbf{r}) = u_{\mathbf{k}}(\mathbf{r})e^{i\mathbf{k}.\mathbf{r}} \tag{10}$$

where $\psi_{\mathbf{k}}(\mathbf{r})$ is the electron wave function, $u_{\mathbf{k}}(\mathbf{r})$ a periodic function with the period of the crystal and **k** is the electron wave vector. After this step, we find the energy of the electron, E, using the Hamiltonian operator, H, as follows:

$$H\psi_{\mathbf{k}}(\mathbf{r}) = E\psi_{\mathbf{k}}(\mathbf{r}) \tag{11}$$

But we don't have $u_k(\mathbf{r})$. Therefore, we don't know the exact form of $\psi_k(\mathbf{r})$. There are a variety of methods to describe the interaction of electron and the crystal lattice. In this chapter we investigate the mentioned interaction according to nearest neighbor π-Tight Binding (π-TB) and the third neighbor π-TB method.

### 3.2 Brillouin zone

Suppose that we have a wave function of the form $e^{i\mathbf{G}.\mathbf{r}}$. We want to find **G** vector such that

$$e^{i\mathbf{G}.(\mathbf{r}+\mathbf{R})} = e^{i\mathbf{G}.\mathbf{r}} \tag{12}$$

or:

$$\mathbf{G}.\mathbf{R} = 2\pi l \tag{13}$$

where $l$ is an arbitrary integer. Now regarding the following equations for **G** and **R**:

$$\mathbf{G} = g_1\hat{\mathbf{k}}_1 + g_2\hat{\mathbf{k}}_2 + g_3\hat{\mathbf{k}}_3 \tag{14a}$$

$$\mathbf{R} = n_1\hat{\mathbf{a}}_1 + n_2\hat{\mathbf{a}}_2 + n_3\hat{\mathbf{a}}_3 \tag{14b}$$

where $\hat{\mathbf{a}}_1, \hat{\mathbf{a}}_2, \hat{\mathbf{a}}_3$ are unit vectors in lattice space and $\hat{\mathbf{k}}_1, \hat{\mathbf{k}}_2, \hat{\mathbf{k}}_3$ are unit vectors in, so called, "reciprocal lattice" space. If we apply (13) we will have:

$$\mathbf{G}.\mathbf{R} = 2\pi(n_1g_1 + n_2g_2 + n_3g_3) \tag{15}$$

which suggests that:

$$\begin{array}{lll} \hat{\mathbf{k}}_1.\hat{\mathbf{a}}_1 = 2\pi & \hat{\mathbf{k}}_1.\hat{\mathbf{a}}_2 = 0 & \hat{\mathbf{k}}_1.\hat{\mathbf{a}}_3 = 0 \\ \hat{\mathbf{k}}_2.\hat{\mathbf{a}}_1 = 0 & \hat{\mathbf{k}}_2.\hat{\mathbf{a}}_2 = 2\pi & \hat{\mathbf{k}}_2.\hat{\mathbf{a}}_1 = 0 \\ \hat{\mathbf{k}}_3.\hat{\mathbf{a}}_1 = 0 & \hat{\mathbf{k}}_3.\hat{\mathbf{a}}_2 = 0 & \hat{\mathbf{k}}_3.\hat{\mathbf{a}}_3 = 2\pi \end{array} \tag{16}$$

Solving above equations [8]:

$$\hat{\mathbf{k}}_1 = 2\pi\frac{\hat{\mathbf{a}}_1 \times \hat{\mathbf{a}}_3}{\hat{\mathbf{a}}_1.(\hat{\mathbf{a}}_2 \times \hat{\mathbf{a}}_3)} \tag{17a}$$

$$\hat{\mathbf{k}}_2 = 2\pi\frac{\hat{\mathbf{a}}_3 \times \hat{\mathbf{a}}_1}{\hat{\mathbf{a}}_1.(\hat{\mathbf{a}}_2 \times \hat{\mathbf{a}}_3)} \tag{17b}$$

$$\hat{\mathbf{k}}_3 = 2\pi\frac{\hat{\mathbf{a}}_1 \times \hat{\mathbf{a}}_2}{\hat{\mathbf{a}}_1.(\hat{\mathbf{a}}_2 \times \hat{\mathbf{a}}_3)} \tag{17c}$$

Now we have unit vectors of the reciprocal lattice. In order to get the Brillouin zone we should we should apply the following condition:

$$(\mathbf{k} - \mathbf{G})^2 = \mathbf{k}^2 \tag{18a}$$

or:

$$\mathbf{k.G} = \frac{1}{2}\mathbf{G}^2 \tag{18b}$$

thus:

$$\mathbf{k} = \frac{1}{2}\mathbf{G} \tag{18c}$$

Using (18-c) we can draw the borders of the Brillouin zone. The inner most area is called the first Brillouin zone and hence, simply it is called "Brillouin zone".
Now we return to our lattice which is graphene sheet, a two dimensional crystal. If we write (16) for this kind of lattice we will have:

$$\begin{array}{ll} \mathbf{k}_1.\mathbf{a}_1 = 2\pi & \mathbf{k}_1.\mathbf{a}_2 = 0 \\ \mathbf{k}_2.\mathbf{a}_1 = 0 & \mathbf{k}_2.\mathbf{a}_2 = 2\pi \end{array} \tag{19}$$

From (19) it is clear that $\mathbf{k}_1$ and $\mathbf{k}_2$ are perpendicular to $\mathbf{a}_2$ and $\mathbf{a}_1$ respectively. Having $\mathbf{a}_1$ and $\mathbf{a}_2$ from Fig. 1(a) we can easily find $\mathbf{k}_1$ and $\mathbf{k}_2$ and draw the Brillouin zone (Fig. 3).

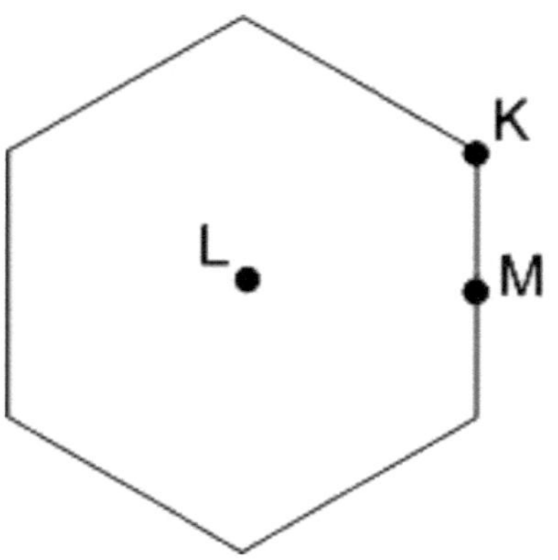


Fig. 3. The Brillouin zone for the graphene lattice is illustrated. L, K and M are high symmetry points.

As mentioned earlier, theoretically, SWCNT can be considered as a graphene lattice that is rolled over into a cylinder. Thus, according to Fig. 1(b) we catch up the following:

$$u_{\mathbf{k}}(\mathbf{r})e^{i\mathbf{k}.(\mathbf{r}+\mathbf{C})} = u_{\mathbf{k}}(\mathbf{r})e^{i\mathbf{k}.\mathbf{r}} \tag{20}$$

Therefore [9]:

$$\mathbf{k.C} = 2\pi l \tag{21}$$

where $l$ is again, an arbitrary integer. This boundary condition which is so called, "Born-von Karman" condition, makes the Brillouin zone to be quantized. Fig. 4 shows this fact. At this point we can begin our investigation about the band structure of SWCNT.

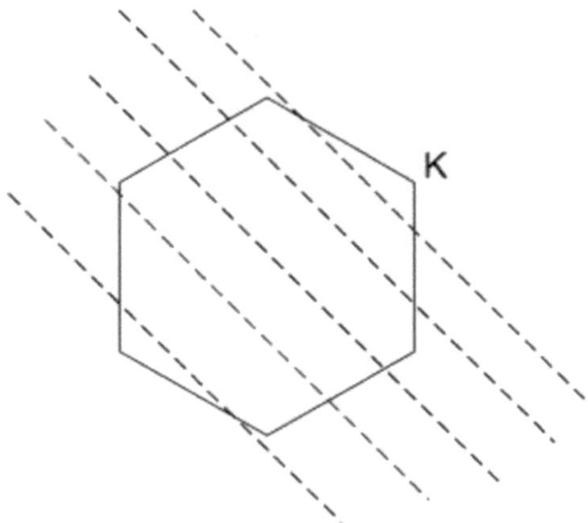


Fig. 4. The Born-von Karman condition makes the SWCNT's Brillouin zone to be quantized.

### 3.3 Tight-binding approximation

As mentioned, there are many methods and approximations that are used to investigate the electronic band structure of a solid. In this section we use the tight-binding approximation. In this approximation we consider the wave function of an electron as the Linear Combination of Atomic Orbitals and hence the method is also called as LCAO.

As is known, the energy of an electron can be estimated using Schrödinger's equation as follows:

$$\left[-\frac{\hbar^2\nabla^2}{2m}+V(\mathbf{r})\right]\psi_{\mathbf{k}}(\mathbf{r})=E\psi_{\mathbf{k}}(\mathbf{r}) \tag{22}$$

where m is the mass of an electron and $\psi_{\mathbf{k}}(\mathbf{r})$ is the wave function of a single electron with the wave vector **k**. Now $\psi_{\mathbf{k}}(\mathbf{r})$ is written as the following:

$$\psi_{\mathbf{k}}(\mathbf{r})=\sum_{\mathbf{r}}c_{\mathbf{kr}}\varphi_{\mathbf{kr}}(\mathbf{r}) \tag{23}$$

where $\varphi_{\mathbf{kr}}(\mathbf{r})$'s are basis functions that are made from atomic orbitals as:

$$\varphi_{\mathbf{kr}}(\mathbf{r})=\frac{1}{\sqrt{N_t}}\sum_{\substack{unit\,cells\,of\\ the\,system}}e^{i\mathbf{k}.\mathbf{R}}\chi_{\mathbf{r}}(\mathbf{R}-\mathbf{r}) \tag{24}$$

where $N_t$ is the total number of unit cells in the system. We regard the single $2p_z$ orbital of the carbon atoms to used in (23); besides, we take into account the interaction of the nearest neighbor atoms (Fig. 5), because they have the most important role in formation of the energy states [10]. We write the wave function $|\psi\rangle$ in terms of basis functions, $|\varphi_1\rangle$ and $|\varphi_2\rangle$ as the following:

$$|\psi\rangle=c_1|\varphi_1\rangle+c_2|\varphi_2\rangle \tag{25}$$

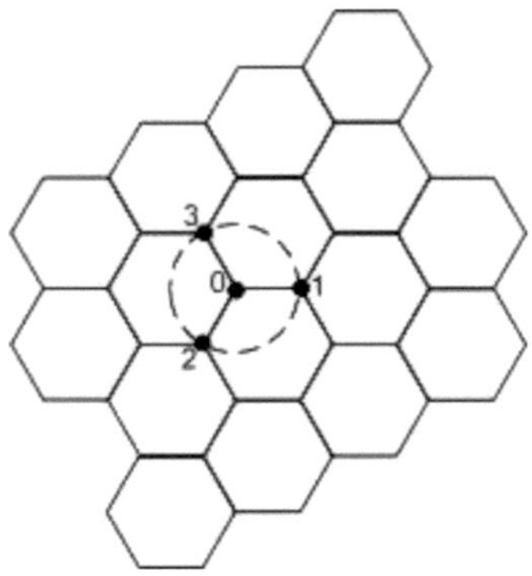


Fig. 5. In this figure the nearest neighbor atoms with respect to atom 0 are illustrated.

$|\varphi_1\rangle$ corresponds to atom 0 and $|\varphi_2\rangle$ corresponds to atoms 1, 2 and 3 in Fig. 5. Now applying (22) to (25) yields:

$$H|\psi\rangle = c_1 H|\varphi_1\rangle + c_2 H|\varphi_2\rangle = c_1 E|\varphi_1\rangle + c_2 E|\varphi_2\rangle \tag{26}$$

and consequently:

$$c_1\langle\varphi_1|H|\varphi_1\rangle + c_2\langle\varphi_1|H|\varphi_2\rangle = c_1 E\langle\varphi_1|\varphi_1\rangle + c_2 E\langle\varphi_1|\varphi_2\rangle \tag{27a}$$

$$c_1\langle\varphi_2|H|\varphi_1\rangle + c_2\langle\varphi_2|H|\varphi_2\rangle = c_1 E\langle\varphi_2|\varphi_1\rangle + c_2 E\langle\varphi_2|\varphi_2\rangle \tag{27b}$$

Now, we define the following values:

$$H_{AA} = \langle\varphi_1|H|\varphi_1\rangle \tag{28a}$$

$$H_{AB} = \langle\varphi_1|H|\varphi_2\rangle \tag{28b}$$

$$S_{AA} = \langle\varphi_1|\varphi_1\rangle \tag{28c}$$

$$S_{AB} = \langle\varphi_1|\varphi_2\rangle \tag{28d}$$

then (27-a) becomes:

$$c_1(H_{AA} - ES_{AA}) + c_2(H_{AB} - ES_{AB}) = 0 \tag{29}$$

knowing that:

$$|\varphi_1\rangle = \frac{1}{\sqrt{N_t}} \sum_{Lattice\,site\,A} e^{i\mathbf{k}.\mathbf{R}_A} \chi_{\mathbf{r}}(\mathbf{r} - \mathbf{R}_A) \tag{30a}$$

$$|\varphi_2\rangle = \frac{1}{\sqrt{N_t}} \sum_{Lattice\,site\,B} e^{i\mathbf{k}.\mathbf{R}_B} \chi_{\mathbf{r}}(\mathbf{r} - \mathbf{R}_B) \tag{30b}$$

Replacing (30-a) and (30-b) in (28-a) to (28-d) yields:

$$H_{AA} = \varepsilon_{2p} \tag{31a}$$

$$H_{AB} = (e^{i\mathbf{k}.\mathbf{R}_{11}} + e^{i\mathbf{k}.\mathbf{R}_{12}} + e^{i\mathbf{k}.\mathbf{R}_{13}})V_{pp\pi} \tag{31b}$$

$$S_{AA} = 1 \tag{31c}$$

$$S_{AB} = (e^{i\mathbf{k}.\mathbf{R}_{11}} + e^{i\mathbf{k}.\mathbf{R}_{12}} + e^{i\mathbf{k}.\mathbf{R}_{13}})s_0 \tag{31d}$$

$$H_{BB} = \langle \varphi_2 | H | \varphi_2 \rangle = H_{AA} \tag{31e}$$

$$H_{BA} = \langle \varphi_2 | H | \varphi_1 \rangle = H^*_{AB} \tag{31f}$$

$$S_{BB} = S_{AA} = 1 \tag{31g}$$

$$S_{BA} = S^*_{AB} \tag{31h}$$

which make (27-b) to become:

$$c_1(H^*_{AB} - ES^*_{AB}) + c_2(H_{AA} - ES_{AA}) = 0 \tag{32}$$

considering (29) and (32) together; to have a non trivial solutions for $c_1$ and $c_2$ we should have:

$$\begin{vmatrix} H_{AA} - ES_{AA} & H_{AB} - ES_{AB} \\ H^*_{AB} - ES^*_{AB} & H_{AA} - ES_{AA} \end{vmatrix} = 0 \tag{33}$$

Solving (33) for $E$ [11]:

$$E(\mathbf{k})^{\pm} = \frac{-(-2E_0 + E_1) \pm \sqrt{(-2E_0 + E_1)^2 - 4E_2E_3}}{2E_3} \tag{34}$$

where:

$$E_0 = H_{AA}S_{AA} \tag{35a}$$

$$E_1 = S_{AB}H^*_{AB} + H_{AB}S^*_{AB} \tag{35b}$$

$$E_2 = H^2_{AA} - H_{AB}H^*_{AB} \tag{35c}$$

$$E_3 = S^2_{AA} - S_{AB}S^*_{AB} \tag{35d}$$

Neglecting the overlap of $2p_z$ orbitals of atomic neighbors, $S_{AB}$, we get:

$$E(\mathbf{k})^{\pm} = \pm V_{pp\pi}\sqrt{3 + 2\cos(\mathbf{k}.\mathbf{a_1}) + 2\cos(\mathbf{k}.\mathbf{a_2}) + 2\cos(\mathbf{k}.(\mathbf{a_1} - \mathbf{a_2}))} \tag{36}$$

Now applying Born von-Karman boundary condition (equation (21)) to (36) one can draw the energy diagram or the electronic band structure of SWCNT. Illustrated in Fig. 6(a) to 6(f) are the electronic band structures for several chiral vectors. At this step of our work, it is necessary to mention a few points. First of all, according to their chiralities, SWCNTs are

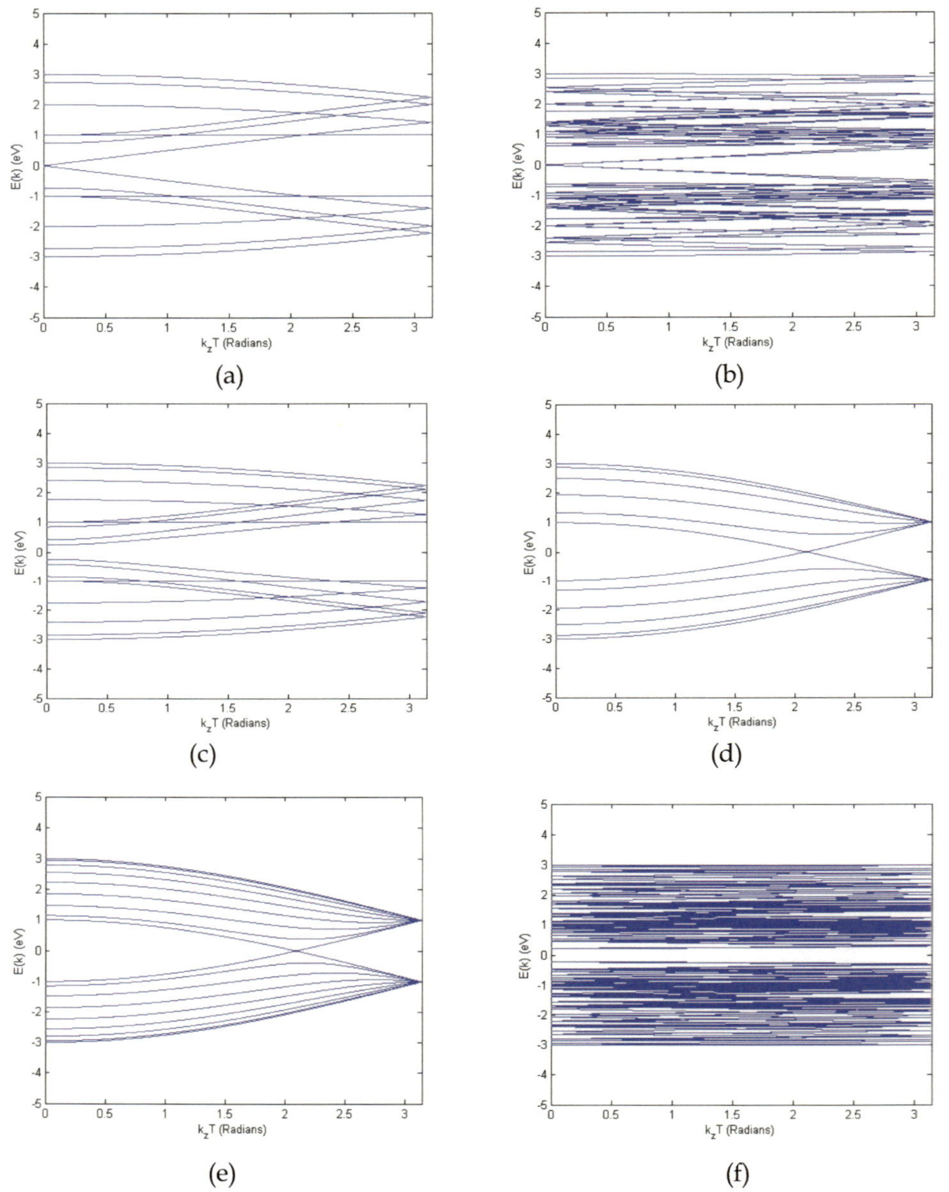


Fig. 6. The electronic band structures of several nanotubes according to (36) are illustrated. (a) is the electronic band structure of chiral vector (6,0), (b) (6,3), (c) (8,0), (d) (5,5), (e) (8,8), (f) (5,4)

roughly divided to three classifications. A nanotube with chirality of (n,0) is called a "zig-zag" nanotube. A nanotube with chirality of (n,n) is called an "armchair" nanotube and a nanotube without the two mentioned chiralities, is called a "chiral" nanotube. As examples, illustrated in Fig. 6(a) and (c) are the band structure of SWCNTs with chiral vectors (6,0) and (8,0) which are zig-zag nanotubes, and Fig. 6(d) and (e) show the band structure of SWCNTs with chiral vectors (5,5) and (8,8) which are armchair nanotubes.
As the Second point, it worth noting that, if we examine (36) with Born-von Karman boundary condition, it is observed that for any chiral vector (n,m) when (n-m) mod 3 is equal to 0, then the band-gap is equal to zero. Two samples of this type are shown in Fig. 6(a) and (b). It is clear that according to this model armchair nanotubes are of this type. At early days it was believed that these nanotubes are metallic, but next, the deeper researches and calculations with other methods and approximations showed that they are "semi-metallic"[12].
Until now, we have performed our analytic calculations with the two assumptions. First, we assumed that the overlap of the two nearest neighbors is zero. Second, we assumed that the $2p_z$ orbitals of the second and the third neighbors have no participation in formation of the band structure. However, in the following lines, we take into account the donation of these neighbors to the formation of the band structure of SWCNT.
Shown in Fig. 7 are the second and the third neighbors of the atom 0 of this figure. According to this figure, one can write:

$$\mathbf{R}_{11} - \mathbf{R}_0 = \frac{2\mathbf{a}_1 - \mathbf{a}_2}{3} \tag{37a}$$

$$\mathbf{R}_{12} - \mathbf{R}_0 = \frac{2\mathbf{a}_2 - \mathbf{a}_1}{3} \tag{37b}$$

$$\mathbf{R}_{13} - \mathbf{R}_0 = -\frac{\mathbf{a}_1 + \mathbf{a}_2}{3} \tag{37c}$$

$$\mathbf{R}_{21} - \mathbf{R}_0 = \mathbf{a}_1 - \mathbf{a}_2 \tag{37d}$$

$$\mathbf{R}_{22} - \mathbf{R}_0 = \mathbf{a}_1 \tag{37e}$$

$$\mathbf{R}_{23} - \mathbf{R}_0 = \mathbf{a}_2 \tag{37f}$$

$$\mathbf{R}_{24} - \mathbf{R}_0 = -(\mathbf{a}_1 - \mathbf{a}_2) \tag{37g}$$

Now, if we apply the formalism of the tight-binding approach, we catch up the following formulae:

$$E_0 = [\varepsilon_{2p} + \gamma_1 u(\mathbf{k})][1 + s_1 u(\mathbf{k})] \tag{38a}$$

$$E_1 = 2s_0 s\gamma_0 f(\mathbf{k}) + (s_0\gamma_2 + s_2\gamma_0)g(\mathbf{k}) + 2s_2\gamma_2 f(2\mathbf{k}) \tag{38b}$$

$$E_2 = [\varepsilon_{2p} + \gamma_1 \mathrm{u}(\mathbf{k})]^2 - \gamma_0^2 \mathrm{f}(\mathbf{k}) - \gamma_0\gamma_2 \mathrm{g}(\mathbf{k}) - \gamma_2^2 \mathrm{f}(2\mathbf{k}) \tag{38c}$$

$$E_3 = [1+s_1u(\mathbf{k})]^2 - s_0^2f(\mathbf{k}) - s_0s_2g(\mathbf{k}) - s_2^2f(2\mathbf{k}) \tag{38d}$$

$$g(\mathbf{k}) = 2u(\mathbf{k}) + u(2k_1\text{-}k_2,k_1\text{-}2k_2) \tag{38e}$$

$$f(\mathbf{k}) = 3+u(\mathbf{k}) \tag{38f}$$

$$u(\mathbf{k}) = 2\cos(\mathbf{k.a_1}) + 2\cos(\mathbf{k.a_2}) + 2\cos(\mathbf{k.(a_1 - a_2)}) \tag{38g}$$

where the hopping parameters $\gamma_0$, $\gamma_1$, $\gamma_2$ and the overlap parameters $s_0$, $s_1$ and $s_2$ are introduced as follows:

$$\gamma_0 = \langle \chi(\mathbf{r}-\mathbf{R_0}) | H | \chi(\mathbf{r}-\mathbf{R_{1i}}) \rangle \tag{39a}$$

$$s_0 = \langle \chi(\mathbf{r}-\mathbf{R_0}) | \chi(\mathbf{r}-\mathbf{R_{1i}}) \rangle \tag{39b}$$

$$\gamma_1 = \langle \chi(\mathbf{r}-\mathbf{R_0}) | H | \chi(\mathbf{r}-\mathbf{R_{2i}}) \rangle \tag{39c}$$

$$s_1 = \langle \chi(\mathbf{r}-\mathbf{R_0}) | \chi(\mathbf{r}-\mathbf{R_{2i}}) \rangle \tag{39d}$$

$$\gamma_2 = \langle \chi(\mathbf{r}-\mathbf{R_0}) | H | \chi(\mathbf{r}-\mathbf{R_{3i}}) \rangle \tag{39e}$$

$$s_2 = \langle \chi(\mathbf{r}-\mathbf{R_0}) | \chi(\mathbf{r}-\mathbf{R_{3i}}) \rangle \tag{39f}$$

Then, (38-a) to (38-g) should be replaced in (34) to get the energy formula. The numerical values for $\gamma_0$, $\gamma_1$, $\gamma_2$ and $s_0$, $s_1$,$s_2$ in addition to a comparison between the results of the mentioned method with the nearest neighbor π-TB can be found in [13].

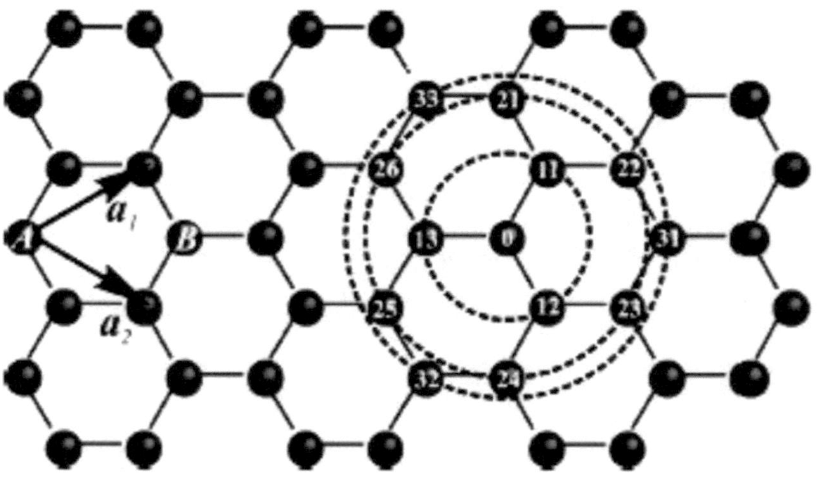


Fig. 7. In this figure the nearest neighboring atoms, the second and the third neighboring atoms are illustrated.

At this point, we continue our work by examining some SWCNTs with different chiral vectors to investigate the effect of radius and chiral angle on the band-gap of these nanotubes. In Table I we have collected chiral vectors that have the same radii but different chiral angles to investigate such an effect. In this table from left, the first column shows the pairs of chiral vectors with the same radii. The second column shows their radii; the third column, their chiral angle; the forth one, the difference between chiral angles; the fifth column indicates the energy gap and finally sixth column shows the difference in band-gap which emanates from the difference between the chiral angle of the nanotubes with the same radii. As can be seen in this table the effect radius on the band-gap is considerable and the band-gap is approximately proportional to $\frac{1}{R}$. On the other hand, as can be concluded from this table, change of the chiral angle has a little effect on the band-gap of SWCNT.

| C:(m,n) | r (nm) | θ (Degrees) | \|Δθ\| (Degrees) | G (eV) | \|ΔG\| (eV) |
|---|---|---|---|---|---|
| (9,1) | 0.373 | 5.20 | 21.78 | 1.091448 | 0.031806 |
| (6,5) | | 26.99 | | 1.059642 | |
| (9,8) | 0.576 | 28.05 | 17.89 | 0.694152 | 0.018414 |
| (13,3) | | 10.15 | | 0.675738 | |
| (14,3) | 0.615 | 9.51 | 13.17 | 0.655092 | 0.010044 |
| (11,7) | | 22.68 | | 0.645048 | |
| (15,2) | 0.630 | 6.17 | 9.43 | 0.617706 | 0.021204 |
| (13,5) | | 15.60 | | 0.63891 | |
| (15,4) | 0.679 | 11.51 | 15.17 | 0.593154 | 0.000558 |
| (11,9) | | 26.69 | | 0.593712 | |
| (18,2) | 0.746 | 5.20 | 21.78 | 0.5219532 | 0.0054126 |
| (12,10) | | 26.99 | | 0.5273658 | |
| (19,2) | 0.785 | 4.94 | 17.89 | 0.5116302 | 0.001953 |
| (14,9) | | 22.84 | | 0.5135832 | |
| (19,3) | 0.808 | 7.22 | 7.34 | 0.483786 | 0.01395 |
| (17,6) | | 14.56 | | 0.497736 | |
| (19,5) | 0.858 | 11.38 | 9.43 | 0.4684968 | 0.0026784 |
| (16,9) | | 20.81 | | 0.4658184 | |
| (23,1) | 0.920 | 2.11 | 21.78 | 0.4248612 | 0.01607 |
| (16,11) | | 23.89 | | 0.4409316 | |
| (23,4) | 0.987 | 7.88 | 16.42 | 0.3977982 | 0.014564 |
| (17,12) | | 24.31 | | 0.412362 | |
| (29,4) | 1.221 | 6.37 | 21.78 | 0.322524 | 0.019139 |
| (19,17) | | 28.16 | | 0.3416634 | |
| (30,4) | 1.260 | 6.17 | 9.43 | 0.3167766 | 0.0071982 |
| (26,10) | | 15.60 | | 0.3095784 | |

Table 1. A comparison between the effects of the radius and the chiral angle on the band-gap of SWCNT. In this table G is the band-gap. ΔG is the difference in band-gap of the two SWCNT with the different chiral angles.

## 4. The electronic band structure of SWCNTs under non-equilibrium conditions

### 4.1 The investigation of the band gap under mechanical strain

In this section of this chapter, we investigate the effect of the two types of mechanical strain, namely uniaxial (tensile) and torsional strains, by means of the two mentioned approximations.

If we denote the amount of uniaxial strain by $\sigma_t$, the angle of shear by $\alpha$ and the bonding lengthes $\mathbf{R}_{11}$-$\mathbf{R}_0$, $\mathbf{R}_{12}$-$\mathbf{R}_0$, $\mathbf{R}_{13}$-$\mathbf{R}_0$ by $\mathbf{r}_1$, $\mathbf{r}_2$ and $\mathbf{r}_3$ respectively, then, under these two type of strain we have the following relations [14]:

$$r_{it} \xrightarrow{\ Tensile\ } r_{it}(1+\sigma_t) \tag{40a}$$

$$r_{ic} \xrightarrow{\ Torsion\ } r_{ic} + r_{it}\tan(\alpha) \tag{40b}$$

where $r_{it}$ is that part of $\mathbf{r}_i$ that is along the axis of the nanotube (with the unit vector $\hat{\mathbf{t}}$) and $r_{ic}$ is that part of $\mathbf{r}_i$ that is in azimuthal direction or along the circumference of the nanotube (with the unit vector $\hat{\mathbf{c}}$). In order to use (40-a) and (40-b) we have to express (37) in terms of $\hat{\mathbf{t}}$ and $\hat{\mathbf{c}}$:

$$\mathbf{r}_1 = \frac{an_1}{2d}\hat{\mathbf{c}} - \frac{1}{\sqrt{3}d}\left(n_2 + \frac{n_1}{2}\right)\hat{\mathbf{t}} \tag{41a}$$

$$\mathbf{r}_2 = \frac{an_2}{2d}\hat{\mathbf{c}} + \frac{1}{\sqrt{3}d}\left(n_1 + \frac{n_2}{2}\right)\hat{\mathbf{t}} \tag{41b}$$

$$\mathbf{r}_3 = -(\mathbf{r}_1 + \mathbf{r}_2) \tag{41c}$$

Using these relations in conjunction with (40-a) and (40-b), we have the following formulae for $\mathbf{r}_1$, $\mathbf{r}_2$ and $\mathbf{r}_3$:

$$\mathbf{r}_1 = \left[\frac{an_1}{2d} - \frac{\tan(\alpha)}{\sqrt{3}d}\left(n_2 + \frac{n_1}{2}\right)\right]\hat{\mathbf{c}} - \frac{(1+\sigma_t)}{\sqrt{3}d}\left(n_2 + \frac{n_1}{2}\right)\hat{\mathbf{t}} \tag{42a}$$

$$\mathbf{r}_2 = \left[\frac{an_2}{2d} + \frac{\tan(\alpha)}{\sqrt{3}d}\left(n_1 + \frac{n_2}{2}\right)\right]\hat{\mathbf{c}} + \frac{(1+\sigma_t)}{\sqrt{3}d}\left(n_1 + \frac{n_2}{2}\right)\hat{\mathbf{t}} \tag{42b}$$

and (41-c) is still valid. At this step, we are to derive the 3rd neighbor π-tight-binding formulation to investigate the effect of uniaxial and torsional strains. We know that, there is the following formula for the interaction energy [14]:

$$\frac{\gamma_{0i}}{\gamma_0} = \frac{\langle \chi(\mathbf{r}-\mathbf{R}_0) | H | \chi(\mathbf{r}-\mathbf{R}_{1i}) \rangle\, with\, strain}{\langle \chi(\mathbf{r}-\mathbf{R}_0) | H | \chi(\mathbf{r}-\mathbf{R}_{1i}) \rangle\, without\, strain} = \left(\frac{a_{C-C}}{r_{1i}}\right)^2 \tag{43}$$

where $a_{C\text{-}C}$ is the bond length in the absence of strain and $r_{1i}$ with $i$ =1,2,3 is $|\mathbf{r}_i|$ in the presence of strain. After performing the formal routine of the deriving of the tight-binding approximation formulae, we find:

$$E_0 = \left[\varepsilon_{2p} + \gamma_1 u(\mathbf{k})\right]\left[1 + s_1 u(\mathbf{k})\right] \tag{44a}$$

$$E_1 = f_{s\gamma}(\mathbf{k}) + \gamma_2 g_s(\mathbf{k}) + s_2 g_\gamma(\mathbf{k}) + 2 s_2 \gamma_2 f(\mathbf{k}) \tag{44b}$$

$$E_2 = \left[\varepsilon_{2p} + \gamma_1 u(\mathbf{k})\right]^2 - f_{\gamma\gamma}(\mathbf{k}) - \gamma_2 g_\gamma(\mathbf{k}) - \gamma_2^2 f(2\mathbf{k}) \tag{44c}$$

$$E_3 = \left[1 + s_1 u(\mathbf{k})\right]^2 - f_{ss}(\mathbf{k}) - s_2 g_s(\mathbf{k}) - s_2^2 f(2\mathbf{k}) \tag{44d}$$

where functions $f(\mathbf{k})$, $f_{s\gamma}(\mathbf{k})$ , $g_s(\mathbf{k})$ ,$g_\gamma(\mathbf{k})$, $f_{\gamma\gamma}(\mathbf{k})$, and $f_{ss}(\mathbf{k})$ in addition to details of calculations are given in [15].
Now, it's time to apply (44-a) to (44-d) and see the results in comparison to other methods. Illustrated in Fig. 8 are the results of application of mentioned method for uniaxial and torsional strains in comparison with the nearest neighbor π-TB and the four orbital tight-binding approximations.

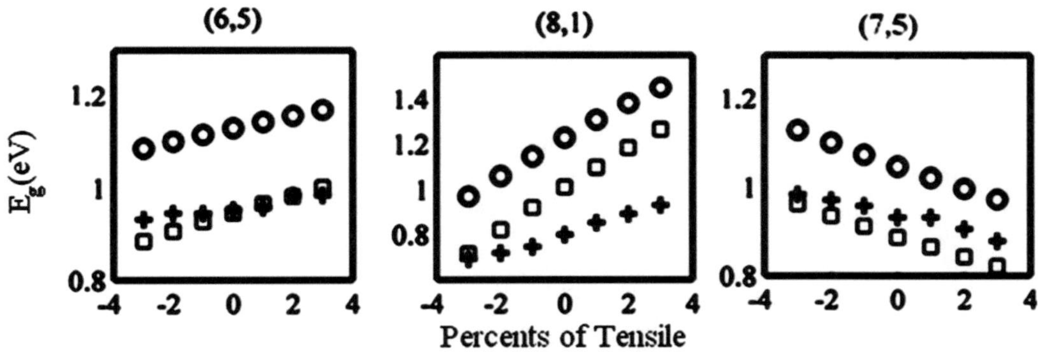


(a)

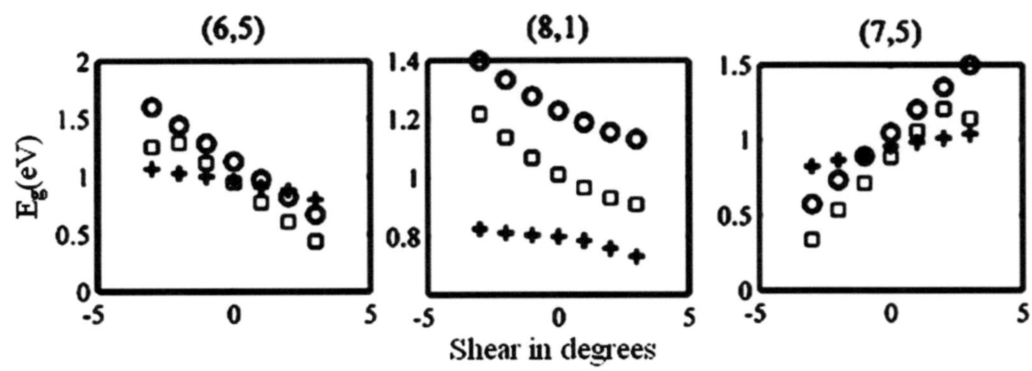


(b)

Fig. 8. A comparison between the results obtained using the nearest neighbor π-TB (circles), the third neighbor π-TB (squares), four orbital TB (plus signs) for (a) -3 to +3 percents of uniaxial strain (b) -3 to +3 degrees of shear [15].

As shown in Fig. 8 the method is examined for three chiral vectors, namely (6,5), (8,1) and (7,5). It can roughly be seen that, the 3rd neighbor π-TB approach yields a better agreement with the four orbital TB than the nearest neighbor π-TB. If we examine the energy formulae for a wide variety of chiral vectors, we find that, there is an approximately, linear relation between the percents of strain (both uniaxial and torsional) and the increase in band-gap [15].

### 4.2 The investigation of the band structure under magnetic field

The effect of magnetic field on the electronic band structure of SWCNT is the second effect that is investigated in this section. The application of $H$ field parallel to the tubule axis is investigated by ***k.p*** method in [16],[17] and an Aharanov-Bohm effect is shown during this investigation. In this section the effect of perpendicular magnetic field is investigated using π-TB model. The investigation is originally performed by R. Saito et al. [18]. The investigation is based on two assumptions: first, the atomic wave function is localized at a carbon site; second, the magnetic field varies sufficiently slowly over a length scale equal to the lattice constant. The vector potential **A** is declared as:

$$\mathbf{A} = (0, \frac{LH_M}{2\pi}\sin\frac{2\pi}{L}x) \tag{45}$$

where L = |**C**|, $H_M$ is the magnetic field and the coordinates $x$ and $y$ are taken along the circumference and the axis of the nanotube, respectively. Under the perpendicular magnetic field the basis functions of (30-a) and (30-b) are changed to:

$$|\varphi_s\rangle = \frac{1}{\sqrt{N_t}} \sum_{Lattice} e^{i(\mathbf{k}.\mathbf{R}_s + \frac{e}{\hbar c}G_\mathbf{R})} \chi_\mathbf{r}(\mathbf{r} - \mathbf{R}_s) \qquad \text{s = A,B} \tag{46}$$

$G_\mathbf{R}$ is the phase factor that is associated with the magnetic field and is expressed as the following:

$$G_\mathbf{R} = \int_\mathbf{R}^\mathbf{r} \mathbf{A}(\xi).d\xi = \int_0^1 (\mathbf{r} - \mathbf{R}).\mathbf{A}[\mathbf{R} + \lambda(\mathbf{r} - \mathbf{R})]d\lambda \tag{47}$$

Under application of magnetic field Hamiltonian operator becomes:

$$H = \left(\frac{1}{2m}\right)\left[\mathbf{p} - \frac{e}{c}\mathbf{A}\right]^2 + V \tag{48}$$

After application of Hamiltonian to (46):

$$H|\varphi_s\rangle = \frac{1}{\sqrt{N_t}} \sum_{Lattice} e^{i(\mathbf{k}.\mathbf{R}_s + \frac{e}{\hbar c}G_\mathbf{R})} \left\{\left(\frac{1}{2m}\right)\left[\mathbf{p} - \frac{e}{c}\mathbf{A}\right]^2 + V\right\} \chi_\mathbf{r}(\mathbf{r} - \mathbf{R}_s) \tag{49}$$

Since $\mathbf{B} = \nabla \times \mathbf{A} = \nabla \times (\mathbf{A} - \nabla G_\mathbf{R})$ and considering (47), then:

$$H|\varphi_s\rangle = \frac{1}{\sqrt{N_t}} \sum_{Lattice} e^{i(\mathbf{k}.\mathbf{R}_s + \frac{e}{\hbar c} G_{\mathbf{R}})} \left\{ \left(\frac{1}{2m}\right) \left[ \mathbf{p} - \frac{e}{c}(\mathbf{A} - \nabla G_{\mathbf{R}}) \right]^2 + V \right\} \chi_{\mathbf{r}}(\mathbf{r} - \mathbf{R}_s) =$$
$$\frac{1}{\sqrt{N_t}} \sum_{Lattice} e^{i(\mathbf{k}.\mathbf{R}_s + \frac{e}{\hbar c} G_{\mathbf{R}})} \left( \frac{\mathbf{p}^2}{2m} + V \right) \chi_{\mathbf{r}}(\mathbf{r} - \mathbf{R}_s) \quad (50)$$

In deriving the equation above the two mentioned assumptions are used, namely, it is assumed that the magnetic field is slowly changing compared with the change of $\chi_r(\mathbf{r} - \mathbf{R_s})$ and $\chi_r(\mathbf{r} - \mathbf{R_s})$ is localized at $\mathbf{r} = \mathbf{R_s}$. Now, we can calculate the matrix elements of Hamiltonian between the two Bloch functions, $|\varphi_1\rangle$ and $|\varphi_2\rangle$ and solve to obtain the eigenvalues. If we examine the π-TB calculated band structure, it is observed that when the magnetic field increases the energy dispersion of each tubule energy band becomes narrower and the total energy bandwidth decreases with increasing magnetic field ,however, when we apply higher magnetic field the total energy bandwidth is found to oscillate as function of $H_M$ [18].

## 5. Conclusion

In this chapter we first described the concept of chiral vector, chiral angle and the radius of SWCNTs and formulated them. Then we explained different symmetries of single walled carbon nanotubes including translational, helical and rotational symmetries. We investigated the Brillouin zone and the electronic band structure of single walled carbon nanotube in the absence of perturbating mechanisms. Our investigation included the nearest neighbor π-TB and the third nearest neighbor π-TB approximations. Next, using these two models we investigated the effect of two types of mechanical strain and perpendicular magnetic field.

## 6. References

[1] S. Iijima, Nature (London) Vol. 354, pp. 56-58, (1991)
[2] R. Saito, M. Fujita, G. Dresselhaus, M. S. Dresselhaus, Appl. Phys. Lett. 60, 2204 (1992)
[3] V. N. Popov, L. Henrard. Phys. Rev. B 70. 115407(2004)
[4] L. Yang , M. P. Anantram, J. Han, J. P. Lu, Phys. Rev. B 60,13874 (1999)
[5] M. Pakkhesal, R. Ghayour, Cent. Eur. Phys. 6, 824 (2008)
[6] M. Pakkhesal, R. Ghayour, Z. Kordrostami, Fullerenes, Nanotubes and Carbon Nanostructures, 17, 99 (2009)
[7] C. T. White, D.H. Robertson, J. W. Mintmire. Phys. Rev. B 47. 5485 (1993)
[8] Kittel, "Solid State Physics", (1999)
[9] R. Saito, M. Fujita, G. Dresselhaus, M. S. Dresselhaus. Phys. Rev. B 46. 1804(1992)
[10] P.R. Wallace Phys. Rev. 71 (1947)
[11] S. Reich, C. Thomsen. Phys. Rev. B 65 .15541(2002)
[12] J. W. Mintmire, B.I. Dunlap, C. T. White. Phys. Rev. B 63. 073408 (1993)
[13] S. Reich, C. Thomsen. Phys. Rev. B 65 .15541(2002)

[14] L. Yang, M. P. Anantram, J. Han, J. P. Lu, Phys. Rev. B, 60, 13874, 1999
[15] M. Pakkhesal, R. Ghayour, Cent. Eur. Phys. 8, 304 (2010)
[16] H. Ajiki, T. Ando, J. Phys. Soc. Jpn. 62, 1255 (1993)
[17] H. Ajiki, T. Ando, J. Phys. Soc. Jpn. 62, 2470 (1993)
[18] R. Saito, G. Dresselhaus, M. S. Dresselhaus. Phys. Rev. B 50. 14698 (1994)

# 19

# An Alternative Approach to the Problem of CNT Electron Energy Band Structure

Ali Bahari
*Department of Physics, University of Mazandaran, Babolsar*
*Iran*

## 1. Introduction

We intended to discuss in this chapter, TBM (tight binding method), APW (augmented-plane-wave), OPW (orthogonalized -plane-wave) methods and corresponding theoretical concepts. In particulars, we pay a great attention to the theory of CNT (Carbon Nano Tube), but discuss in less details some conventional band structure models, unless nearly electron approximation (NFA), TBM, APW and OPW models have been used for determining the electron energy band structure of solids. In fact, this chapter is partly based on the many - electron description of nano transistor - CNTFET (carbon nano tube field effect transistor), which was done with a number of MSC and PhD students for a number of years at university of Mazandaran in Iran (See our published papers [1-7] for more details). We hope this chapter can complete the present book and be of interest for researchers whom work in the nano technology and for beginners. Some part of the material may be used in lection course for students.

There are actually two different approaches for studying the band spectrum of CNT. In the first view, some researchers believe that carbon atoms are as isolated atoms and consider the CNT potential of neighbor's atoms as a perturbation and neglect the intra atomic potential. The second approach is about the density functional theory (DFT), in that the exact exchange energy (EXX) instead of the exchange energy given by the local - density approximation (LDA). The EXX energy, which corresponds to the Fock term in the Hartree-Fock scheme, is treated as a function of electron densities via the eigenfunctions of the Kohn-Sham KS equations [8]. This approach cannot satisfy the electron behavior in CNT due to its self-interaction-free in its construction.

Indeed, this chapter discusses about electronic band energy. It is an energy interval in which electronic states exist in the CNT. This energy structure has been usually obtained by solving the Schrödinger equation for electrons in the CNT. As usual, the electronic wave functions depend on both the wave vector and the spatial coordinates. The eigenvalues and eigenvectors have been determined by Fourier - transforming the differential equation into an algebraic equation. The solution of this equation can be used for some special cases with some reasonable approximation, such as NFE and TB methods. However, these approaches cannot be used for samples with critical dimensions of less than 100 nm due to overlap integrals in nano scale samples.

The reason is that carbon atoms are not in fact stationary, but continually undergo vibrations (like thermal vibrations of ions in a crystal) about their positions, in where, the overlapping between carbon atom functions is of importance, in particularly while the nearest neighbor atoms come close together. In principle a many - electron problem, for the full Hamiltonian of the CNT should be taken into account. It means Hamiltonian should contain not only the one - electron potentials describing the interactions of the electrons with the massive carbon atomic nuclei, but also pair potentials describing the electron - electron interactions in CNTs. But this idea should be included both the exchange and correlation effects into the interaction phenomena due to nearly free electrons. The Schrödinger equation for a many-electron system can be then reduced to the effective one-particle problem for an electron in a self-consistent field.

We therefore need to develop a method of band structure spectrum; because in the conventional method of solution, the unknown functions of Schrödinger equation has usually been expanded in some bases set. The search for the unknown expansion coefficients will be necessarily reduced to the solution of a secular equation which is usually of large dimension and provide high speed of expansion convergence, in order to doing less effort for finding band structure spectrum.

As stated above, in second view, a large majority of the electronic structures and band plots are calculated using DFT [9], which is not a model but rather a theory. It involves the electron-electron many-body problem via the introduction of an exchange-correlation term in the functional of the electronic density. Although, the band shape is typically well reproduced by DFT, there are also systematic errors in DFT bands due to shrinking the CNT size.

In addition, some researchers [10 and references therein] believe OPW can solve this problem, but some critical technological barriers and fundamental limitations to size reduction are threatening the use of OPW method for calculation of band energy. It means that there are some difficulties with current crystalline potentials which reside quite simply in considering, for example, electrons of carbon atoms as independent particles.

Furthermore, the OPW expansion converges poorly for a CNT even when modified by the addition of an atomic like function to the basis set. The APW expansion also converges rapidly, but requires the crystal potential to be approximated by an unphysical spherical muffin-tin potential. However, in many of above methods you need to an ingenious the choice of CNT potential, which is not so easy due to the enormously complicating effects of the interactions between atoms (and electrons). Henceforth, a more accurate calculation of the electronic properties of a CNT should start with modifying of above approaches, in particularly, NFA, TB and OPW methods. We should thus develop a modified APW/OPW expansion and compare its convergence with the other methods.

An alternative approach to the problem of CNT band energy and of constructing exchange - correlation potential uses the calculation of the total energy. However, after describing the conventional methods and/or models, we will see that these models cannot sufficiently describe the electron behaviors in CNT. A new method is presented for finding the band structure of a lattice of potentials which individually are spherically symmetric, but with overlap's functions. The method does not necessitate a division of space into non-overlapping spherical regions. It exploits the properties of the complete set of functions associated with the individual potentials. An expansion of the wave function of the crystal in this set yields a relatively simple determinant secular equation. The present method can be employed to introduce a matrix of CNT band energy.

## 2. Summary of some band structure models

Several efficient methods have been developed in last four decades: Korringa, Kohn and Rostocker ( KKR) model [11], indicates the initials of Korringa (in 1947), Kohn, and Rostoker (in 1954), DFT, Green function methods [12] and *ab intitio* approximation [13] have been used for studying the electronic band structure of CNT, because they lend themselves very well in reproducing the band shape. In this area, we naturally prefer to consider the simplest form of the approximation centers non-overlapping spheres (referred to as muffin tins) on the atomic positions. In one hand, within these regions, the potential experienced by an electron is approximated to be spherically symmetric about the given carbon atoms. In the remaining interstitial region, the potential is approximated as a constant. Continuity of the potential between the atom-centered spheres and interstitial region is enforced. On the other hand, The KKR method is one of the popular methods of electronic structure calculation and is also called Green's function method. Therefore, KKR is actually referred to multiple scattering theory of solving the Schrödinger equation, in where the problem is broken up into two parts: solving the scattering problem of a single potential in free space and then solving the multiple scattering problems by demanding that the incident wave to each scattering centre should be the sum of the outgoing waves from all other scattering centers. The scheme has met great success as a Green function method, within DFT. To calculate the bands including electron-electron interaction many-body effects, one can resort to so-called Green's function methods.

Indeed, knowledge of the Green's function of a system provides both ground (the total energy) and also excited state observables of the system. The poles of the Green's function are the quasiparticle energies, the bands of a solid. Sometimes spurious modes appear. Large problems scaled as $O(n^3)$, with the number of the plane waves (n) used in the problem. This is both time consuming and complex in memory requirements. Its applications range from the full potential *ab initio* treatment of bulk, surfaces, interfaces and layered systems with O(N) scaling to the embedding of impurities and clusters in bulk and on surfaces. In this way, after the single particle Hamiltonian (H) is generated either by empirical pseudo potential method or the charge patching method, it needs to be solved in an order N scaling [14].

As we know, the band plot can obviously show the excitation energies of electrons injected or removed from the system. It can say nothing about energies of a fictive non-interacting system, the Kohn-Sham system, which has no physical interpretation at all. The Kohn-Sham electronic structure must not be confused with the real, quasi particle electronic structure of a system, and there is no Koopman's theorem holding for Kohn-Sham energies, as there is for Hartree-Fock energies, which can be truly considered as an approximation for quasi particle energies. Hence, in principle, DFT is not a band theory, i.e., not a theory suitable for calculating bands and band-plots.

The self-energy can also in principle be introduced variationally [14]. A variational derivation of the self-energies for the electron-electron and electron-phonon interactions are presented in [36].Due to the presence of the strong Coulomb interaction between electrons in the CNT atoms, the differential equations for the single- electron Green functions contain the multi-electron Green functions and all these coupled equations form an infinite system of differential equations for an infinite number of Green functions. In order to find some approximate finite closed system of equations one can either to apply the perturbation theory and retain only some appropriate chain of ladder diagrams or to assume some approximation to decouple the infinite system of equations and obtain a finite closed system. For CNT, the self-energy is a very complex quantity and usually approximations are needed to solve the problem.

In addition to DFT and KKR methods, one of the other popular methods which has been usually used to all band structure calculations and studies, is NFA model. It is a method of approximating the energy levels of electrons in a CNT by considering the potential energy resulting from carbon atomic nuclei and from other electrons in the CNT as a perturbation on free electron states. Although the NFA is able to describe many properties of electron band structures, it can only predict the same number of electrons in each unit cell, which conflict with this result as for materials require inclusion of detailed electron-electron interactions (treated only as an averaged effect on the crystal potential in band theory) known as Mott insulators [15]. The Hubbard model is an approximate theory that can include these interactions and contains a large number of closely spaced molecular orbitals, which appear as a band.

Anyway, we cannot here explain all band structure's model, but they are based on some elementary theory as reflected in Bloch, NFA or NFE and TB idea, which well models useful for illustration of band formation need these idea. The main is that each model describes some types of solids very well and others poorly. The NFE model works well for metals, but poorly for non-metals. The NFE model works particularly well in materials like metals where distances between neighboring atoms are small. In such materials the overlap of atomic orbitals and potentials on neighboring atoms are relatively large. In that case the wave function of the electron can be approximated by a modified plane wave. The TB model is extremely accurate for ionic insulators, such as metal halide salts (e.g. NaCl), but cannot be used for free electrons in a solid, e.g. CNT.

The main difficulty with current graphite potentials resides quite simply in considering electrons of carbon atoms as independent particles. The reason is due to neglecting the wave function's overlapping, i.e., in the independent electron approximation the electron - electron interactions are just represented by an effective one - electron potential.

If we pay somewhat closer attention to the form of potential, recognizing that it will be made up of a sum of atomic potentials centered at carbon atoms, then we can draw some further conclusions that are important in studying the electronic structure of graphite as well as graphene structures. Suppose that the basis consists of identical atoms at positions $d_j$. Then the periodic potential $u_{nk}(r)$ will have the form

$$u_{nk}(\mathrm{r}) = \sum_{\mathrm{R}}\sum_{\mathrm{j}}\phi(r - R - d_j) \tag{1}$$

Where $\Phi(k)$ is the Fourier transform of the atomic potential,

$$\phi(\mathrm{k}) = \int \mathrm{e}^{-\mathrm{ik.R}}\, \phi(r)\, \mathrm{dr} \tag{2}$$

One can see that it has the form of a traveling plane wave, as represented by the factor $e^{ik.r}$, which implies that the electron propagates through the crystal like a free particle. The effect of the function $u_k(\mathrm{r})$ is to modulate this wave so that the amplitude oscillates periodically from one cell to the next. However, it cannot affect the basic character of the state function, which is that of a traveling wave. But the electron in CNT is not completely free. Since electrons in CNT can interact with the other of CNT atom's electrons, the special character of

the periodic function $u_k$ will be varied. Moreover, $\Psi_{nk}$ may be delocalized throughout the CNT atoms and not localized around any particular atom, meaning it may be as NFE wave functions (As an example graphene structure in figure 1. A graphene structure has been considered for determining of CNT band structure. It only includes four nearest neighbors and can be expanded to the other neighbors as well.).
To these notifications, researchers [16 and references therein] have considered some special form of crystalline potentials in calculating of the electronic band energy of the CNT . They have tried to construct Bloch waves from appropriately defined functions (Known as Wannier functions) localized at each lattice site and used **K.P** approximation method. In this view the dispersion around the external points of an energy band can be found, but within these models, the spatial derivatives in the Schrödinger equation of the CNT, are carried out only for the plane wave component of the Bloch function, given by [2];

$$\{\frac{|p|^2}{2m}+\frac{\hbar\, k.p}{m}+\frac{\hbar\,|k|^2}{2m}+V(r)\}\, u_{n,k}(r) = E_n(k)\, u_{n,k}(r) \quad (3)$$

Here, n denotes the band index, V and $E_n$ are lattice potential and eigen state, respectively. For k = 0, it simplifies significantly, and an approximate solution can be found for all band involved. A non - vanishing but small wave vector can then be treated as a perturbation. The term $\propto k^2$ produces an energy shift that depends on k, but does not couple the bands. The term containing **k.p**, however, must be treated with degenerate perturbation theory.

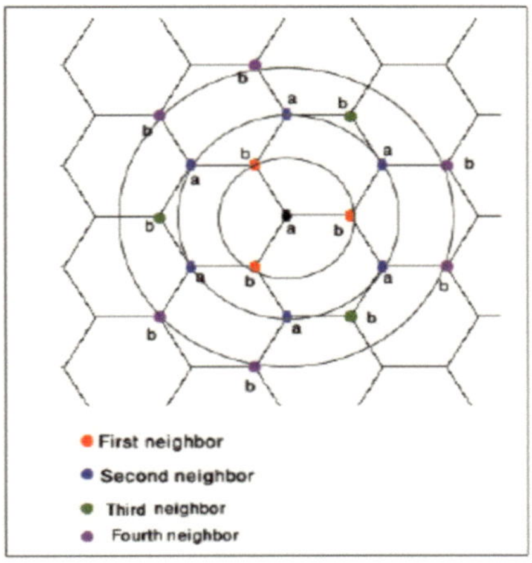


Fig. 1. Graphene structure. There are two different carbon shape atoms in the graphene sheet, where each 'a' atom has 3 'b' atom as the first neighbor's atoms, 6 'a' atom in the second neighborhood and 3 'b' atom as the third neighbor atoms and finally 6 'b' atom in its 4'th neighbors.

However, these wave functions represent a type of plane wave throughout space - a graphene as well as CNT crystal actually has infinite size based on the definition of a lattice (Note: the wave function must be normalized on a finite region of space with volume CNT that usually comes from periodic boundary conditions over the CNT circumference, so that with the definition of;

$$\phi_{nk}(\mathrm{r}) = \frac{e^{ik.R}}{\sqrt{V_{CNT}}} \mathrm{u}_{nk}(r) \tag{4}$$

And explicitly demonstrate the normalization for $u_{nk}$ (r ). As we know, NFE method starts from a free electron gas in a CNT and treats a weak periodic crystal potential within perturbation theory. There is also a different approach TBM, which constructs the electronic eigenstates from those of the individual atoms that form the CNT, which belongs in the independent-electrons framework. Within TB picture, the energy bands and the band gaps are reminders of the discrete atoms. Contrary to the free-electron picture, TB model describes the electronic states starting from the limit of isolated-atom orbitals. It is based on the assumption that the atomic orbitals belonging to an energy eigenvalue are good starting point for constructing Bloch waves. The CNT wave function in this view is usually expanded in the Bloch functions. But there are some assumptions: the energy level is non degenerate and there is no other energy level nearly. In that case, it yields to an approximation of the Bloch waves that emerges from the atomic wave functions.
A more accurate approach using this idea employs Wannier functions, defined by [20]. The Wannier functions are localized near atomic sites, like atomic orbitals, but being defined in terms of Bloch functions they are accurately related to solutions based upon the CNT potential. Wannier functions on different atomic sites are orthogonal. The Wannier functions can be used to form the Schrödinger solution for the *n*-th energy band . The width of the energy bands is determined by the overlap of atomic wave functions at neighbor lattice sites and decreases rapidly for inner shells. As a rule, the bands, which originate from different levels, overlap considerably. This simple model gives good quantitative results for bands derived from strongly localized atomic orbitals, which decay to essentially zero on a radius much smaller than the next neighbor half-distance in the solid.
The size of this matrix eigenvalue problem is clearly as large as the number of eigenstates of the atomic problem, i.e. infinite. It is therefore necessary to do some approximation. In particular, one could hope that all the off-diagonal matrix elements of the matrices could be neglected for some given level. This cannot work for atomic degenerate levels. Due to the exponential decay of the atomic wave functions at large distance, both the overlap integrals and the energy integrals become exponentially small for large distance **R** between the centers of the atoms. It therefore makes sense to ignore all the integrals outside some $\mathbf{R}_{max}$, which would bring in only negligible corrections to the band structure. One may obtain a band structure depending on a minimal number of parameters by making further rather radical approximations [20].

## 3. CNT band structure

According to the definition of SWCNT (single walled carbon nano tube), the energy bands of a SWCNT consist of a set of one-dimensional energy dispersion relations which are cross sections of those of graphene. When graphene sheet is rolled to make a CNT, $K_\perp$ is rolled too. So by using periodic boundary conditions in the circumference direction denoted by the chiral vector $C_h$, the wave vector associated with the $C_h$ direction becomes quantized, while the wave vector associated with the direction of the translational vector T (or along the nanotube axis) remains continuous for a nanotube of infinite length. Since N $K_\perp$ corresponds to a reciprocal lattice vector, two wave vectors which differ by N $K_\perp$ are equivalent. In this view, the wave vector of CNT is a continuum component along tube axis and a discrete value of $K_\perp$, as found before [1,2] and shown in figures 2, 3 (for details see ref. [2]).

$$K_{\nu}^{CNT} = (\mathrm{k}\frac{\mathrm{K_{II}}}{|\mathrm{K_{II}}|} + \nu\,\mathrm{K_{\perp}}),$$
$$\nu = 0, \ldots, \mathrm{N\text{-}1.} \quad , \quad \mathrm{k} = -\frac{\pi}{|\mathrm{T}|}, \ldots, \frac{\pi}{|\mathrm{T}|} \qquad (5)$$

Therefore, the band structure of CNT can be determined via;

$$E = \frac{\pm\beta|\mathrm{f(K)}|}{1 \pm s|\mathrm{f(K)}|} \qquad (6)$$

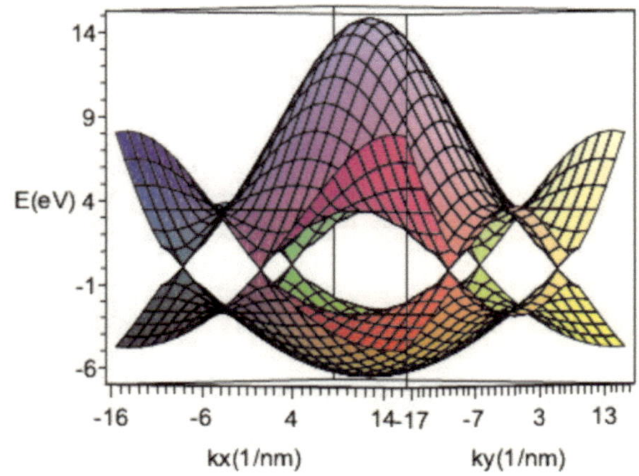


Fig. 2. Graphene band structure.

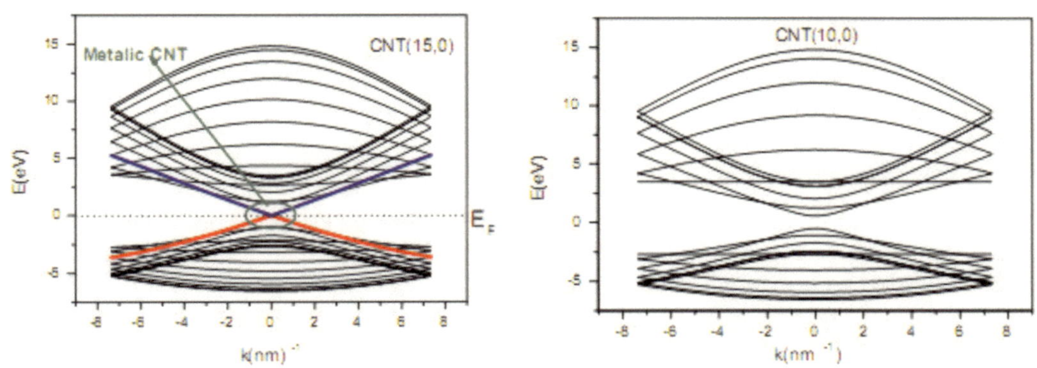


Fig. 3. Electronic band structure of some CNTs based on TB model. The banding energy of π orbital is equal to -3.03 eV and its overlap matrix is equal to 0.129. This figure clearly show that CNT (15,0) is a metallic CNT. We will drive an important relation between the geometry of CNT and its conduction. There is an important note. There are (N/2) + 1 degenerate levels in Zig-Zag CNT (n, 0).

Furthermore, density of State (DOS) of a one dimensional lattice with a lattice vector T and for one level is given by [2]

$$g_i(\varepsilon) = \frac{|T|}{\pi} \cdot \frac{1}{\left|\frac{\partial E_i}{\partial k}\right|_{E_i=\varepsilon}} \tag{7}$$

$$\mathrm{DOS}\ (\varepsilon) = \sum_i g_i(\varepsilon) \tag{8}$$

Figure 4 Shows DOS of some CNT's. As we see in this figure semiconducting Zig-Zag CNT's have not any density of state at the Fermi level but armchair CNT's have a little density of state at Fermi level. If we focused to the armchair CNTs at around Fermi level, we find that the DOS has not treat as a constant value and treat as a parabola curvature.

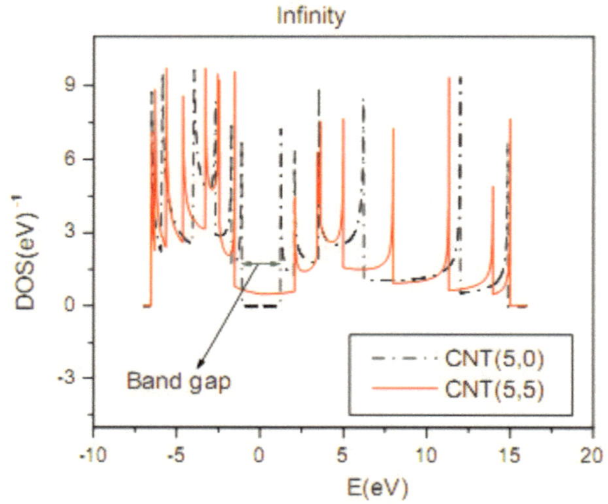


Fig. 4. This figure compares density of state of a Zig-Zag CNT via as an armchair CNT. As you see the Zig-Zag CNT has not any DOS near the Fermi level.

## 4. Augmented Plane Wave (APW) method

Slater introduced the APW method in 1937. Shortly after that researchers have used it for determining the electronic band structure of the rocksalt lattice structure. Although APW method is a sound one for calculating the band structure in metals, it has a great deal in the past few years. In this method the influence of potentials from non - nearest neighbors is taken into account.

As one can see in a schematic view in figure 5, the effective crystal potential is constant in most of the open spaces between the cores. Therefore, we can begin by assuming such a potential, which is referred to as the muffin-tin potential (because the potential is constant there). The potential is that of a free ion at the core, and a plane wave outside the core. Inside the core the function is atom-like, and is found by solving the appropriate free-atom Schrödinger equation. Also, the atomic function is chosen such that it joins continuously to the plane wave at the surface of the sphere forming the core; this is the boundary condition. The wave function does not have the Bloch form, but this can be remedied by forming the linear combination.

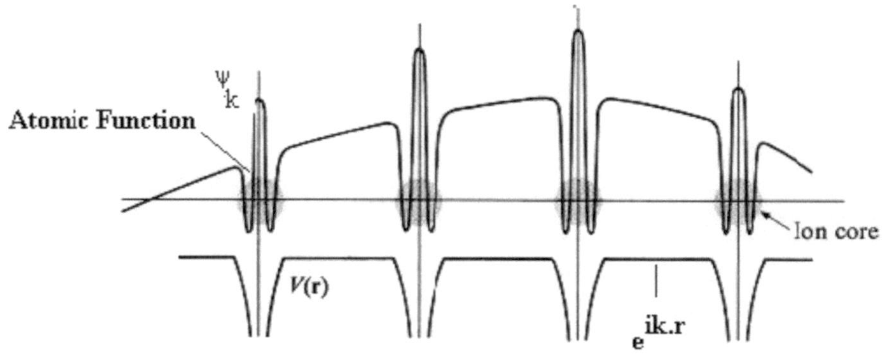


Fig. 5. The potential and wave function in the APW method.

From the point of APW method view, the overlap of the wave functions centered, on the six contact sites cannot be neglected, indicating that the atomic levels should be essentially altered in a CNT. We thus assume that the bound levels of the atomic Hamiltonian are not well localized, meaning the wave function is not small when **r** (like the core radius ) exceeds a distance of the order of the lattice constant.

Therefore, we have to consider a many body system. The Schrödinger equation for a many-electron system can be reduced to the effective one-particle problem for an electron in a self-consistent field. Thus:

$$(-\frac{\nabla^2}{2m} + U(r))\, \psi_k(r) = E_k \psi_k(r) \tag{9}$$

where U(r) is the crystal self-consistent potential, and $\psi_k$ and $E_k$ are the wave function and the eigenvalue of the electron energy in k-state, correspondingly.

One of the main consequences of basis function over completeness is their linear dependence. This means

$$\sum_{G=0}^{G_m} |TPW_{k+G}\rangle\, C_G \rightarrow 0 \tag{10}$$

If $|G_m| \rightarrow 0$, $C_G$ will indicate the numerical coefficients. Moreover, if $C_G \neq 0$, the transformed plane wave (OPW, APW) is denoted by the symbol $|TPW_{k+G}>$, the wave vector k belongs to the first Brillouin zone, and G are reciprocal lattice vectors. In the method of linearized augmented plane waves (LAPW) the linear dependence of the basis set $|APW_{k+G}>$ is manifested for $R_0\, G_{max} \geq 9$, which corresponds to accounting for 70-80 basis functions in APW, where $R_0$ is a muffin tin sphere radius. The OPW linear dependence begins to be manifested if the number of basis functions is more than 100 in (10) .

Although both APW and OPW, as two modern methods of band calculations, use combinations of atomic functions and plane waves, they cannot yet yield to exact results. In fact in the APW method, the wavefunction in general has discontinuous derivatives on the boundary between the interstitial and atomic regions. It means that we have to consider variational method in stead of Schrödinger equation. In this method, the augmenting function corresponds to the exact muffin-tin potential eigenstates of eigenenergy. Because of this energy dependence of the augmenting function the eigenvalue problem will be non-linear in energy and has to be solved iteratively. This is, however, computationally very costly.

In addition, any eigenstate of a different eigenenergy will be poorly described without adapting. Hence, we need to linearized versions of the APW method with modifying the basis functions which gain extra flexibility to cover a larger energy region around their linearization energy. In this view, the linear combinations of energy-independent APW as a trial function and muffin-tin orbitals are inserted in the one-electron Hamiltonian. Then the secular equations are therefore eigenvalue equations, linear in energy in that the energy bands depend on the potential in the spheres through potential parameters which describe the energy dependence of the logarithmic derivatives. Keep in mind that the energy-independent APW inside the sphere is linear combination of an exact solution, which matches continuously and differentiable onto the plane-wave part in the interstitial region.

## 5. Orthogonalized Plane Waves (OPW) method

OPW method, as a simplified version of the pseudo potential method [17], has been used for the calculation of the electronic band structure of almost all types of solids with neglecting nonlocal effects. It has been especially determining the band structure of materials with covalent binding where the potential cannot be approximated by the conventional muffintin construction. Indeed, OPW method is rather practical and time-saving from the computational point of view since it leads to an eigenvalue problem involving matrix elements which do not depend on the eigenvalues, as in other methods of band theory. In contrast to above methods, in OPW method, the eigensolutions can be found easily by conventional methods of linear algebra . However, sometimes it cannot be used for calculating of nano scale materials due to the structure of the secular problem arising in the OPW formalism which can be related to a Born-series expansion, and it is known in scattering theory that resonances cannot be appropriately accounted for in any order of such an expansion.

Two main approaches based on expansion have been used: (i) basis set and (ii) trial wave function. Pseudo potential methods or OPW method use plane waves or modified plane waves as the basis set. The TBM are based on the second concept. There are also approaches which combine both delocalized and localized functions. In this approach the atomic-like functions are squeezed by an additional attractive potential. The extention of the basis functions is tuned by a parameter that can be found self-consistently [49]. The problem of APW and OPW methods for a CNT structure is an abundance of multi- center integrals, which must be performed to arrive at a reasonable accuracy of band structure calculations due to existence of a great number of neighbours within a given distance. To avoid these difficulties, we have tried to introduce an alternative method (see next section).

In our method we consider a lot of plane waves in the basis to decrease the spatial extent of localized valence orbitals in CNT. We could take a method far beyond usual pseudo potentials and improve our plane-wave basis set. The results show a good converged Bloch function for both valence electrons and excited states using a relatively small number of plane waves.

Two separated core orbital contributions and plane wave contributions, which are not OPWs at the outset have been involved in this approach, so that in the basis set three types

of functions are used: true core orbitals, squeezed local valence orbitals and plane waves. If a larger number of plane waves are included in the band structure calculations, there is usually the reason for the over-completeness breakdown of OPW expansions. The local basis function (both core and valence) can be constructed from radial functions, which are solutions of the radial Schrödinger equation. Our approach provides a full interpolation between the APW and OPW approaches adopting pseudo-potential features [18].
It is clear that there is an intense overlapping between electron wave functions of CNT when carbon atoms come close to each other, whilst in OPW method, each electron are imagined as a nearly free electron. Obviously, the above assumptions cannot explain behavior of the electron when carbon atoms come together like d-layer electrons.

## 6. A new method

One approach to overcoming these impending barriers involves finding on evaluating the potential of CNTs as the basis of a future nanoelectroncs technology. Single-walled CNT (SWCNTs) are materials with unique properties. They have several millimeters in length and are strongly bonded covalent materials. Because of their extremely small diameter, the OPW method should be modified and completed with TB method, with considering the overlapping of wave function of electrons. The procedure is to augment the basis set of present method by including wave functions which are OPWs between nuclei of carbon atoms but represent modified Bloch waves near the nuclei. It means that by scaling the CNT dimension, the carbon atoms come close to each other and change the band energy. Thus, by using Ritz variational method, we have modified the band energy.
Nothing said up to now has exploited any properties of the potential U( r) other than its periodicity, and, for convenience, inversion symmetry. If we pay somewhat closer attention to the form of U, recognizing that it will be made up of a sum of atomic potentials centered at the positions of the carbon atoms, then we can draw some further conclusions that are important in studying.
There is the other view, known the electron correlations. In fact, the existence of a unique density function which yields the exact ground state energy may not cause the possibility of reducing the many - electron problem to the one - electron one. This is due to at least the Coulomb interaction among electrons, in where at weaker electron correlations, it can involve a self - consistent potential which depends on electron density. In fact, the correlation effects near to carbon cores in where the strong intrasite Coulomb repulsion may lead to splitting of one - electron bands into many - electron subbands meaning TBM is inapplicable.
On one bands, nearest neighbor carbon atoms may share and/or transport electrons so that electrons become localized at carbon sites. In a such a situation of CNT, we need to modify the eigen functions and introduce a correction term by expressing them in terms of many - electron and/or overlapping functions of the atomic problem. It depends on many electron quantum numbers, s, L occurs in the full Hartree - Fock approximation [8].
Let us have a somewhat closer look at the band structures. The constructions reported so far, imply that there is just an electron as a localized particle (and completely free electron) if there are sufficiently many Bloch waves available. This is not necessarily the case in nano structures due to localization of electron with building a localized wave packet from the Bloch waves. It leads the sharply peaked character of the weight function and the spatial extension of such an electron wave packet which can be larger than the lattice constant.

In the Ritz method, the minimizing element in the n-th approximation is sought in the linear hull of the first n coordinate elements. The Ritz ansatz function is a linear combination of N orbitals. Based on linear combination of atomic orbital (LCAO) approximation, $\Psi(r)$ is

$$\Psi(\mathrm{r}) = \sum_{\mathrm{j}} \Phi_j(r) \tag{11}$$

The eigenvalues and eigenvectors can be found with finding a solution of the Ritz method. It is widely applied when solving eigenvalue problems, boundary value problems and OPW equations in general. The trial wave function will always give an expectation value larger than the ground energy (or at least, equal to it). It is known to be orthogonal to the ground state.

Further development of the OPW method led to the idea of introducing a weak pseudo potential which permits (unlike the real crystal potential) the use of perturbation theory. Because of strong core level potential within it, it may be represented in the form of a new Schrödinger equation where the non-local energy-dependent pseudo potential operator W is defined by

$$\mathrm{W} = \mathrm{V(r)} + \mathrm{V^R} \tag{12}$$

Although OPW method with pseudo potential principle a possibility to eliminate the difficulty pointed before, it is rather complicated and goes far beyond the original concept of the CNT band - structure methods due to requiring exact diagonalization of a matrix of the pseudo potential idea - applicability methods may not provide as a rule sufficiently satisfactory description of CNT.

To overcome of these difficulties, KKR method has been used. The advantage of the KKR method in comparison with the APW one is the decoupling of structural and atomic factors. For the same lattice potentials, the KKR and APW methods yield usually close results. However, the main difficulty of the KKR method is the energy dependence of the structural constants.

In the general APW, KKR and LCMTO (MT: Mofin Tin) methods, the matrix elements are functions of energy. Therefore, at calculating eigenvalues one has to compute the determinants in each point of k-space for large number values of E (of order of 100) which costs much time.

In the present work, according to Andersen theorem [2], we expand the radial wave functions at some energy value to linear terms in E, in which, both Hamiltonian and matrices do not depend on energy. We can get more accuracy by amount of higher - order terms in the expansion. Using this idea, we will be able to improve considerably the accuracy of CNT - band energy and achieve very good results.

In this case we deform new wave function based on OPW and TB methods, in that the carbon atoms cores are placed in the crystal lattice sites R. There are localized electrons and the lattice sites. Henceforth, the core electron wave functions can be assumed to be approximately equal to the corresponding Hartree-Fock functions of a free atom, that is,

$$\Psi_{\mathrm{i}}^{\mathrm{Cr}}(r-R) \approx \Psi_{\mathrm{i}}^{\mathrm{HF}}(r-R) \tag{13}$$

where I = n, I, m is a set of quantum numbers which characterize bound electron states. $\Psi_i^{HF}(r-R)$ are localized to such an extent and the overlap of $\Psi_i^{HF}(r-R)$ centered in different sites should be ignored, which leads:

$$\int \Psi_i^*(r-R)\Psi_{i'}(r-R')dr = \delta(R-R')\delta_{ii'} \tag{14}$$

On the other hand, for N unit cells in CNTs, $u(k_i; r)$ can be written by the following expression:

$$u(k_i,r) = \frac{1}{\sqrt{N}} \sum_j e^{ik_i \cdot R_n} \Psi_{nlm}(r - R_n) \tag{15}$$

where $R_n$ is the distance between two nearest neighbor carbon atoms. It yields new orthogonalized coefficients. We consider a correction term as $L_{ij}$ ;

$$M_{ij} = (k_i \cdot k_j - E)\,\delta_{ij} + L_{ij} \tag{16}$$

By using separable variables method, the atomic wave functions, $\Psi_{nlm}(r)$, split into a set of radial $R_{nl}(r)$, azimuth angle part $\Phi_m(\varphi)$ and associated Legendre equation $P_{lm}(x)$, with x = cos ɵ. After doing some calculations on solving the above equation, the electronic band energy is determined by the following equation;

$$\det |H - EP| = 0 \tag{17}$$

Where

$$H = \begin{pmatrix} H_{ij}^{OPW} & H_{is} \\ H_{is}^{+} & H_{ss} \end{pmatrix} \tag{18}$$

And

$$P = \begin{pmatrix} P_{ij}^{OPW} & P_{is} \\ P_{is}^{+} & P_{ss} \end{pmatrix} \tag{19}$$

Finally, the overlapping wave functions of SWCNTs is demonstrated by H and P matrices,

$$H_{is} = \int_0^\infty \Phi_i(r) H\, u_s(k,r)\, d^3r - \sum_q E_q\, a_q\, b_{sq} \tag{20}$$

$$H_{ss'} = \int_0^\infty \Phi_i(r)\, u_s(k,r)\, H\, u_{s'}(k,r)\, d^3r - \sum_q E_q\, b_{sq}\, b_{s'q} \tag{21}$$

And

$$P_{is} = \int_0^\infty \Phi_i(r)\, u_s(k,r)\, d^3r - \sum_q C_q\, b_{sq} \tag{22}$$

$$P_{ss'} = \int_0^\infty \Phi_i(r)\, u_s(k,r)\; u_{s'}(k,r)\, d^3r - \sum_q b_{sq}\; b_{s'q} \tag{23}$$

Where

$$b_{sq} = \sum_{q} e^{ik.R_n} \int_{0}^{\infty} \Psi_q(r)\ \Phi_s(r - R_n)\, d^3r \tag{24}$$

We found a reliable matrix which can describe the CNT electron behavior with doing the series of calculations based on the Ritz variational, OPW and TB methods. In this method, there is no limitation on crystalline potential of CNT structure, so it can be suggested for evaluating the electronic band energy of SWCNTs.

Therefore, the wave functions, which enter the Slater integrals, are based on self - consistent way from the corresponding integro - differential equations. It means the one - electron Hamiltonian of CNT in the many - electron representations, should take into account the electron transfer owing to matrix elements of electrostatic interaction, which will be more complicated in solving the CNT –atomic problem. Moreover, the general Hartree - Fock approximation may give us the radial one - electron wave functions which depend explicitly on atomic term, whilst these wave functions can not be factorized into one - electron ones, or, the interaction of different carbon electron on the other sites is sometimes required.

Thus we cannot describe unlocalized electron states in CNTs within above methods. In contrast to TBM, the strength of CNT potential can determine the widths of gaps rather than of electron bands ( as addressed in TBM).

## 7. References

[1] A. Bahari, P. Morgen, Z.S. Li and K. Pederson, *J. Vac. Sci. and Tech. B*, 24 (2006) 2119.

[2] A. Bahari and M. Amiri, *Acta physica Polonica A* , 115 (2009) 625.

[3] A. Bahari, U. Robenhagen, P. Morgen and Z. S. Li, *Phys. Rev. B*, 72 (2005) 205323.

[4] A. Bahari, P. Morgen and Z.S. Li, *Surf. Sci.*, 602 (2008) 2315.

[5] F. M. Nakhei and A. Bahari, *Int. J. Phys. Sci* , 4 (2009) 290.

[6] A. Bahari, P. Morgen and Z.S. Li, *Surf. Sc.*, 600 (2006) 2966.

[7] P. Morgen, A. Bahari and. Pederson, *Functional properties of Nano structured Material*, Springer, pp: 229-257, 2006.

[8] R. K. Lake and R. R. Pandey, *Handbook of Semiconductor Nanostructures and Devices*, Los Angles: American Scientific Publishers, 2006.

[9] M. Brandbyge, J.-L. Mozos, P. Ordejón, J. Taylor, and K. Stokbro, *Phys. Rev.B*, 65 (2002) 165401.

[10] L.W. Wang, *Phys. Rev. B*, 49 (1994) 10154.

[11] M. P. Anantram, M. S. Lundstrom, and D. E. Nikonov, *cond-mat/0610247*, 2006.

[12] R. Lake, D. Jovanovic, and C. Rivas, *Nonequilibrium Green's Functions in Semiconductor Device Modeling* (NewJersey), pp. 143-158, World Scientific, 2003.

[13] L.-H. Ye, B.-G. Liu, D.-S. Wang, and R. Han, *Phys.Rev.B*, 69 (2004) 235409.

[14] J. Kohanoff, *Electronic Structure Calculations for Solids and Molecules*, Cambridge University Press, (2006).

[15] Ch. Kittel, *Introduction to Solid State Physics*, Chapter 7. John Wiley & Sons, Inc., (2005).

[16] http://en.wikipedia.org/wiki/Electronic_band_structure.

[17] O. Krogh Andersen, *Phys. Rev. B*, 12 (1975) 3060.

[18] A. Ernst and M. L¨uders , *Methods for Band Structure Calculations in Solids*, Springer-Verlag Berlin Heidelberg 2004.

20

# Electronic Structure and Magnetic Properties of N@$C_{60}$-SWCNT

Atsushi Suzuki and Takeo Oku
*The University of Shiga Prefecture*
*Japan*

## 1. Introduction

Single-walled-carbon nanotube (SWCNTs) encapsulating fullerenes as peapods have been considerably interested to apply organic field effect transistor, magnetic device and quantum device (Bailey et. al. 2007, Khlobystrov et. al. 2005). Electronic structure and physical properties of fullerenes within SWCNT as carbon peapods has been studied. Table 1 lists recent works. For instant, the endohedral metallofullerenes nested in the strained nanotube have the electronic structure with the relative energy levels at the different states and produces a spatial modulation of the energy gap (Cho et. al. 2003). A scalable spin quantum computer that combines aspects of electronic and magnetic properties of endohedral fullerenes that enclose small clusters of metal atoms, such as $Sc_2$@$C_{84}$, La@$C_{82}$ (as shown in Fig. 1) and Gd@$C_{82}$, N@$C_{60}$ and P@$C_{60}$ encapsulated within SWCNT as peapods has been investigated (Tooth et. al. 2008, Cantone et. al. 2008, Warner et. al. 2008, Cho et. al. 2003, Yang et. al. 2010). The key advantages are that magnetic interactions between electronic spins and nuclear spins in carbon peapods are used. As an example using nitrogen endohedral fullerene N@$C_{60}$ encapsulated within SWCNT as peapods (see Fig. 2), the nitrogen atom occupies a high-symmetry site at the center of the cage and retains its atomic configuration, and the cage offers protection of the nitrogen electron paramagnetic moment, which is related to electron spins, S=3/2, coupled to be nuclear spin I=1. This material has a high advantage of the NMR quantum computer to control spin gate with decoupling pulses in relaxation time. Morton reported the NMR quantum-computer controlled un-perturbation Rabi oscillation of spin polarization under influence of decoupling process (Morton, et. al. 2005, 2006, 2011). Additionally, nitrogen endohedral fullerene $^{14}$N@$C_{60}$ encapsulated within SWCNT allow to control quantum qubit-gate under a mixture of biding interaction between electron and nuclear spins in the NMR quantum computing. Recently, Yang reported how to efficiently implement the quantum logical gate operations required for universal quantum computation. Transfer of information between qubits had been considered by direct dipole-dipole couplings or by using a mobile electron spin as the bus qubit (Yang et. al. (2010), Meyer et. al. (2004)).
The band structures for infinite periodic chains of $C_{60}$ and N@$C_{60}$ are shown in Fig. 2. All the energies are given in units of E -$E_F$ for each system. The effects of encapsulated nitrogen on the band structure of $C_{60}$ have been considered. The results reveal that the nitrogen causes the lowering of the $C_{60}$ at the LUMO and HOMO levels together with a splitting of the degenerate levels.

| Contents | References |
|---|---|
| Fullerene-based quantum computer | Morton et. al. (2005), (2006), (2011), Yang et. al. (2010) |
| | Harneit et. al. (2002), Meyer et. al. (2004), Benjamin et. al. (2006) |
| $N@C_{60}$-SWCNT, $P@C_{60}$-SWCNT | Cho et. al. (2003), Bailey et. al. (2007), Jue et. al. (2007), |
| | Simon et. al. (2006) (2007), Tooth et. al,. (2008), Yang et. al. (2010), Suzuki et. al. (2010), Iizumi et. al. (2010) |
| $Sc@C_{82}$-SWCNT, $La@C_{82}$-SWCNT | Cho et. al. (2003), Cantone et. al. (2008), Warner et. al. (2008) |
| Carbon tubes | Rao et. al. (1997), Khlobystrov et. al. (2005) |
| $C_{60}$ | Schiller et. al. (2006) |
| $N@C_{60}$, $P@C_{60}$, $(C_{59}N)_2$, $C_{59}N$ | Buhl et. al. (1997), Fulop et. al. (2001), Kobayashi et. al. (2003), Abronin et. al. (2004), Schulte et. al. (2007) |

Table 1. Electronic structure and physical properties of fullerenes within SWCNT as carbon peapods

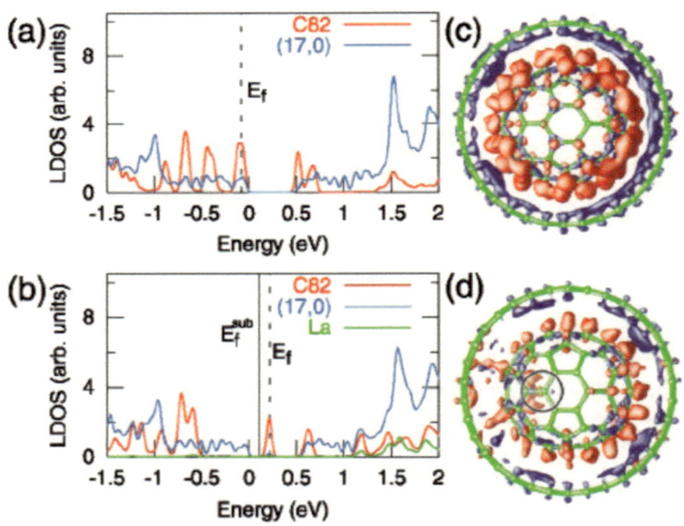


Fig. 1. Local density of states (LDOS) of (a) $C_{82}$@(17, 0) and (b) La@$C_{82}$@(17, 0) decomposed into the constituents. Black vertical lines indicate the Fermi level (*Ef*). $E_{sub}{}^{f}$ represents the Fermi level with the metal substrate in contact. Isodensity surface plots of the electron accumulation (red) and depletion (blue) are shown for $C_{82}$-SWCNT (17, 0) and La@$C_{82}$-SWCNT(17, 0) in (c) and (d), respectively. The values for the red and blue surfaces in (c) are $\pm 0{:}0035 e/$eÅ$^3$. Corresponding values in (d) are $\pm 0{:}0025 e/$eÅ$^3$. The gray circle in (d) indicates the position of the La atom inside $C_{82}$. (Cho et. al. (2003))

The gap between HOMO and LUMO is reduced from 1:8 eV of $C_{60}$ to 1:66 eV of N@$C_{60}$. The effects of majority and minority spins in the nitrogen basis were found not to have any appreciable further contribution. These small changes to the band gap and Fermi level shifts the support to the evidence that the nitrogen atom is well screened within the cage.

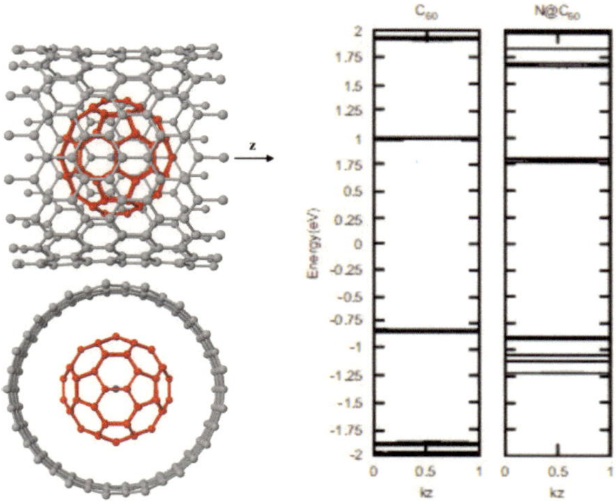


Fig. 2. The relaxed atomic coordinates of N@$C_{60}$-SWCNT (17, 0). For clarity the carbon atoms of the $C_{60}$ are shown in red and the nitrogen in blue. The band structures for infinite periodic chains of $C_{60}$ and N@$C_{60}$ are shown. All the energies are given in units of E - $E_F$ for each system. (Bailey et. al. (2007))

Design of spin labels inside for possible molecular spintronics, which contains of 1D spin chains filling SWCNTs with magnetic endohedral fullerenes of N@$C_{60}$ and P@$C_{60}$ as spin qubit have been proposed (Yang et. al. 2010). Jue reported a scheme to implement the two-qubit gates between the nuclear spins of the encapsulated atoms in endohedral fullerenes $^{15}$N@$C_{60}$ or $^{31}$P@$C_{60}$, within today's magnetic resonance techniques (Jue et. al. 2007). The electronic spin of the N and P ground state is 3/2, while the nuclear spin is either 1/2 ($^{15}$N, $^{31}$P) or 1($^{14}$N). The electronic structure at highest occupied molecular orbital (HOMO), lowest unoccupied molecular orbital (LUMO), the magnetic and optical properties of N@$C_{60}$-SWCNT have been characterized on the basis of magnetic techniques of NMR, ESR and electronic nuclear double resonance (ENDOR) (Benjamin, et. al. 2006) using *ab-nitio* quantum calculation. The relaxation properties of the $^{14}$N nucleus in N@$C_{60}$ were determined using techniques related to Davies ENDOR which have been proposed for measuring $T_1$ and $T_2$. In order to demonstrate the effectiveness of the bang-bang decoupling scheme in N@$C_{60}$, a strong environmental interaction has been simulated by applying a strong RF field to drive Rabi oscillations in the nuclear spin; this strong RF field was then decoupled by applying fast phase 'kicks' using the π phase gate (Morton et. al. (2006)).
Figure 3 shows Bang-bang control of fullerene qubits by ultrafast phase gates. Morton discussed the benefits of a coupled electron spin to the nuclear spin qubit and demonstrated the ideas using the $^{14}$N nuclear spin in the N@$C_{60}$ molecule. In addition to providing a resource for nuclear spin polarization and detection, the electron spin can be exploited to perform ultrafast nuclear spin phase gates, which can in turn be used to dynamically bang-bang decouple the nuclear spin from unwanted interactions. Figure 4 shows arbitrary nuclear phase gates are implemented by driving two electron spin transitions simultaneously (Morton (2005)). The path of the electron magnetization vector was driven

by a microwave field. The corresponding z-magnetizations are displayed as shown in Fig. 4 (a). One transition is driven resonantly, and one strongly detuned. On each cycle, the phase accumulated is proportional to the area enclosed by the path. Both transitions must undergo an integer number of complete cycles to ensure the populations remain unchanged. As shown in Fig. 4 (b), both transitions were detuned by equal and opposite amounts, the population evolution for each is the same, whilst the phase accumulated is of opposite sign. Hence, the relative phase shift can be tuned by controlling the microwave pulse power.
The effect of the non-nearest neighbor couplings has been investigated (Yang et. al. (2010)). Figure 5 show transitions of a single qubit in the presence of couplings to other qubit as a function of the chain length. The uppermost spectrum corresponds to the case of nearest neighbors only [trace (a)]. If non-nearest neighbors are added [see trace (b)], each resonance line splits into seven lines separated by $D_{nn}/8$, where $D_{nn}$ is the coupling between nearest neighbors. The weight of the lines (1:2:3:4:3:2:1) is given by the number of states of the next-nearest neighbors. Adding a third pair of qubits causes an additional splitting of each line into a (1:2:3:4:3:2:1) multiplet, with line separation $D_{nn}/64$. With the resolution of the figure [see trace (c)], this appears as a line broadening. The ESR frequency on transitions of a single qubit was appeared in the presence of couplings to other qubit under nearest neighbor. Especially, electronic spin qubits on five molecules of N@$C_{60}$ in a magnetic field gradient has been simulated. The magnetic parameters of principle g-tensor, hyperfine coupling constant of nitrogen atom and chemical shift of $^{13}$C in $^{14}$N@$C_{60}$ within SWCNT for chiral index (n, m) are important to control a quantum qubit-gate of 1D spin in the NMR quantum-computer. In the present work, geometrical effect of diameter and chiral index on the electronic structure, magnetic and optical properties of N@$C_{60}$-SWCNT have been investigated on the basis of experimental results using *ab-initio* density functional theory.

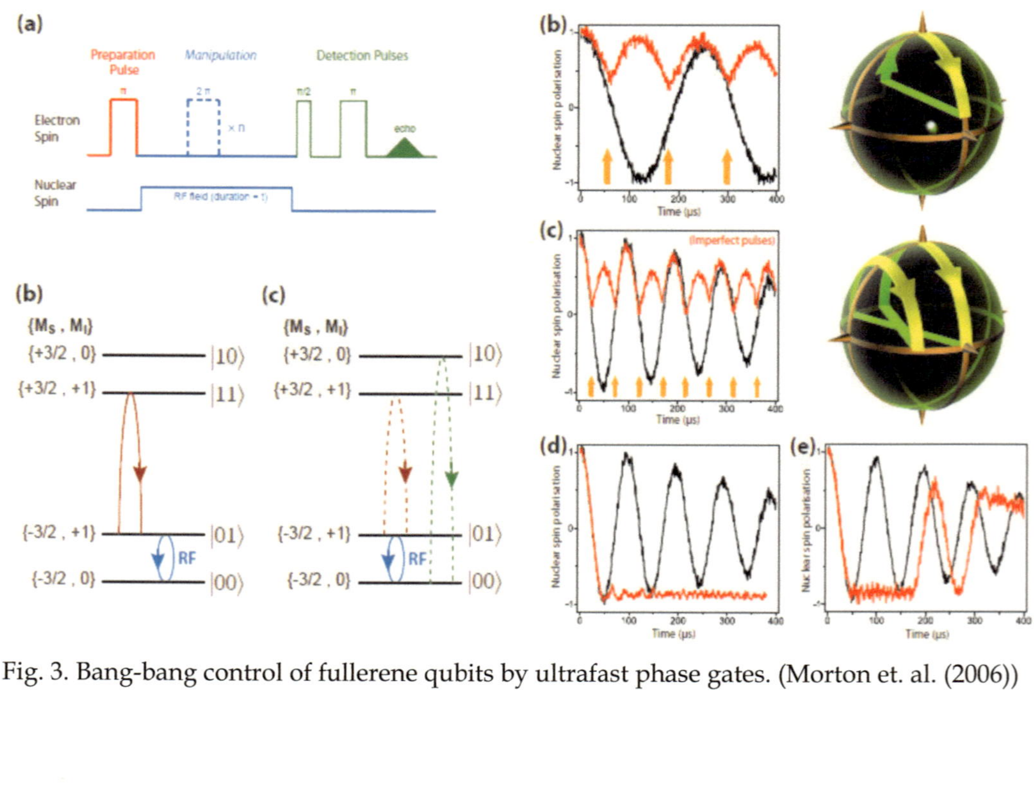


Fig. 3. Bang-bang control of fullerene qubits by ultrafast phase gates. (Morton et. al. (2006))

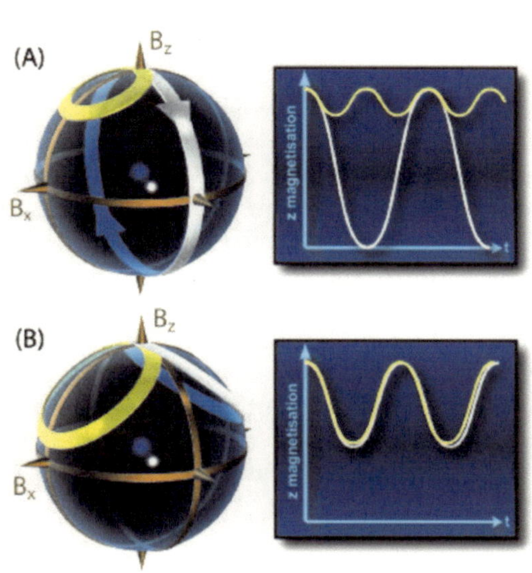


Fig. 4. Arbitrary nuclear phase gates are implemented by driving two electron spin transitions simultaneously (Morton (2005))

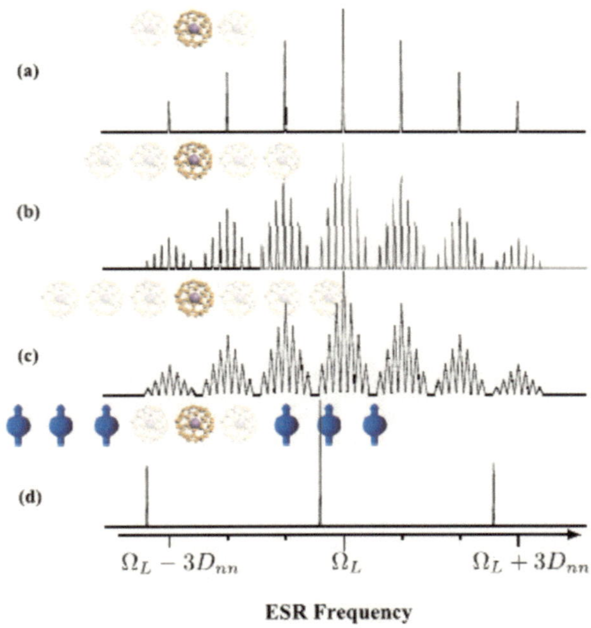


Fig. 5. Transitions of a single qubit in the presence of couplings to other qubits as a function of the chain length. (a) Nearest neighbors only. (b) Nearest neighbors and next-nearest neighbors. (c) Three qubits on each side. (d) Infinite chain, where all but the nearest neighbors are polarized. (Yang et. al. (2010))

## 2. Quantum chemical calculation

The molecular structures were assembled by CS ChemDraw, CS Chem3D (Cambridge Soft) and Nanotube Modeler (JCrystal Soft). Molecular orbital calculations were carried out by molecular mechanics calculations (MM2) and semi-empirical molecular orbital calculations (Hamiltonian: Parameterized Model Revision 3: PM3). In addition, the isolated molecular structures were optimized by *ab-initio* quantum calculation using unrestricted Hartree–Fock (UHF) and DFT using UB3LYP with hybrid function LANL2DZ and STO-3G*, 3-31G* and 6-31G* as basis set (Gaussian 03 Inc.). The electronic structure at HOMO, LUMO, LUMO+1 and the HOMO-LUMO band gap ($E_g$) were calculated. Nitrogen atomic charges of $^{14}N@C_{60}$ within SWCNT were estimated using Mulliken population analysis. Wavelength and the exited transition state were calculated by time-dependence of DFT (TD-DFT) with hybrid function UB3LYP and 3-31G* as basis set. Continuously, chemical shift of $^{13}C$ (δ), principle g-tensor ($g_{xx}$, $g_{yy}$, $g_{zz}$) and principle A-tensor ($A_{xx}$, $A_{yy}$, $A_{zz}$) in hyperfine coupling constant (hfc) of nitrogen atom were calculated by DFT using NMR/GIAO with hybrid function UB3LYP and 3-31G* as basis set.

## 3. Results and discussion

### 3.1 Electronic structure of N@C$_{60}$-SWCNT

Electronic structure of molecular orbital at HOMO, LUMO, next LUMO+1 and energy levels of N@C$_{60}$-SWCNT-armchair have been investigated by DFT with hybrid function UB3LYP using 6-31G* as basis set. Molecular orbitals and energy levels of $^{14}N@C_{60}$-SWCNT armchair (13, 13), N@C$_{60}$ and SWCNT are shown in Fig. 6. The molecular orbital of N@C$_{60}$-SWCNT-zigzag (13, 13) was delocalized on π-electrons at a long axis of SWCNT surface interacted with π-electrons on the N@C$_{60}$ cage surface as hybrid orbital interaction. The energy level between HOMO-LUMO was smaller than that of original SWCNT. This behavior was due to a mixture of binding interaction with spin distribution.

Molecular orbital of $^{14}N@C_{60}$-SWCNT (9, 9) and SWCNT at (a) HOMO, (b) LUMO, (c) LUMO+1 are shown in Fig. 7. The molecular orbital of N@C$_{60}$-SWCNT-zigzag (9, 9) was delocalized on π-electrons at a long axis of SWCNT surface interacted with π-electrons on the N@C$_{60}$ cage surface as hybrid orbital interaction. The energy gap of N@C$_{60}$-SWCNT between HOMO and LUMO was estimated to be 1.16 eV, which was smaller than that of original SWCNT (9, 9) to be 1.24 eV, due to a mixture of binding interaction. Mulliken atomic charge of N in $^{14}N@C_{60}$ within SWCNT was 3.8 x $10^{-5}$e. This result indicates a slight charge transfer from nitrogen atom to C$_{60}$ cage within SWCNT. This supports the assertion that the nitrogen is well screened within the C$_{60}$ cage.

Electronic structures of N@C$_{60}$-SWCNT-zigzag (14, 0) at (a) HOMO, (b) LUMO and (c) charge distribution are shown in Fig. 8. The molecular orbital were well distributed around the SWCNT interacted with C$_{60}$ surface cage. Energy gap between HOMO and LUMO was estimated to be 0.87 eV, which indicates semi-conductive behavior. There existed a wide charge distribution on the inner surface of SWCNT, which had π-electron interaction with hybrid orbital on the C$_{60}$ surface cage.

Figure 9 show molecular orbital of $(^{14}N@C_{60})_2$-SWCNT-armchair (9, 9) at (a) HOMO and (b) LUMO, calculated by DFT/UB3LYP/STO-3G*. In both cases at HOMO and LUMO, molecular orbital of $(^{14}N@C_{60})_2$-SWCNT (9, 9) formed a circle ring to distribute on the SWCNT surface. The band gap between HOMO and LUMO was estimated to be 3.72 eV,

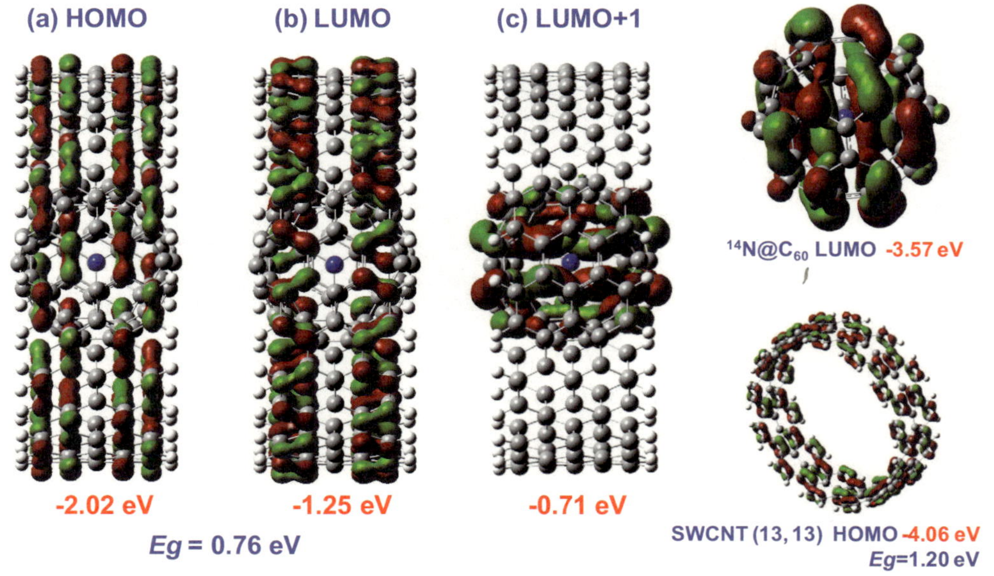


Fig. 6. Molecular orbitals and energy levels of $^{14}N@C_{60}$-SWCNT armchair (13, 13), $N@C_{60}$ and SWCNT (Suzuki, el. al. (2010))

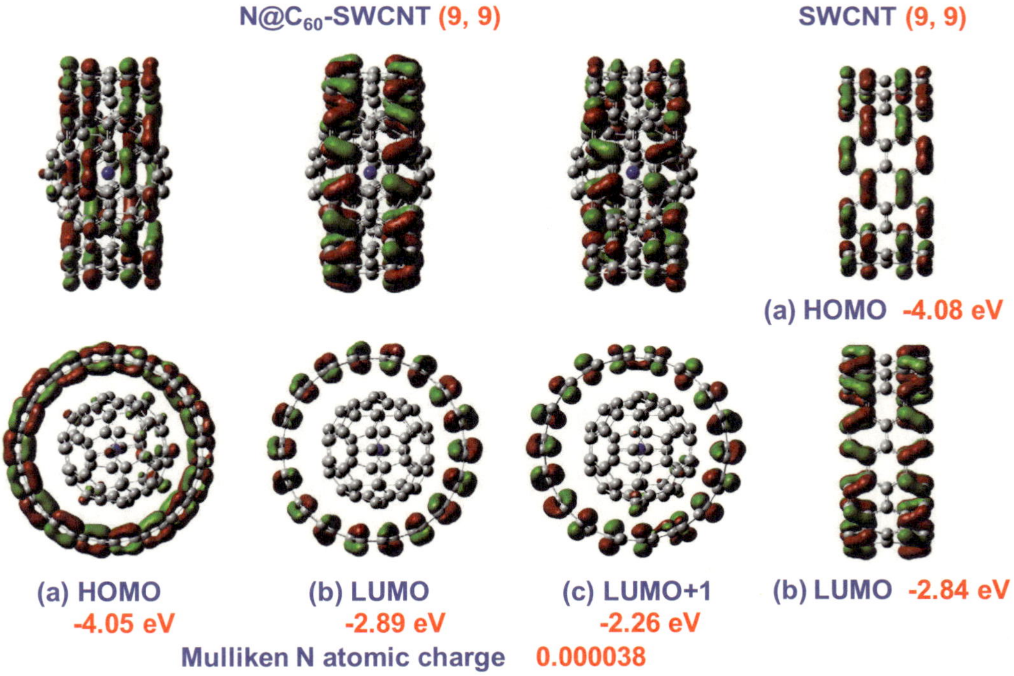


Fig. 7. Molecular orbital of $^{14}N@C_{60}$-SWCNT (9, 9) and SWCNT at (a) HOMO, (b) LUMO, (c) LUMO+1 and energy levels (Suzuki et. al. (2010)).

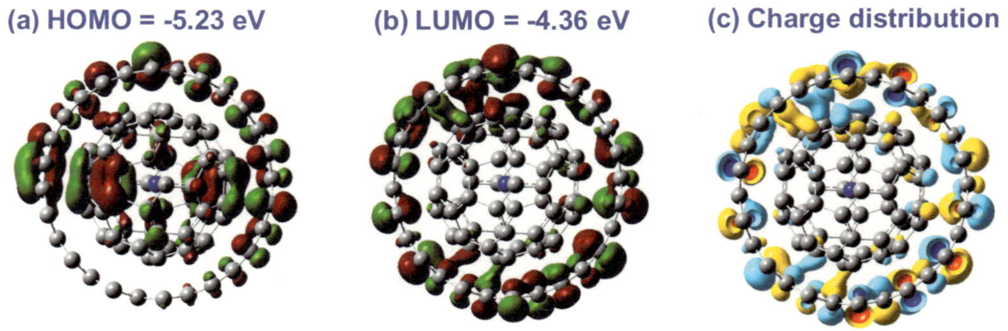


Fig. 8. Molecular orbital of $^{14}N@C_{60}$-SWCNT-zigzag (14, 0) at (a) HOMO, (b) LUMO and (c) charge distribution. (Suzuki et. al. (2010))

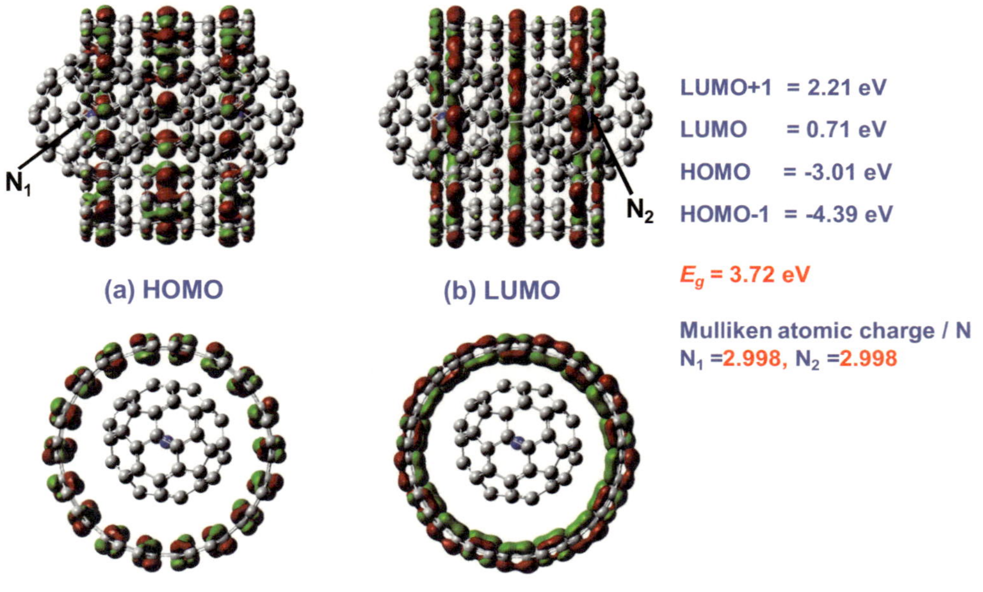


Fig. 9. Electronic structures of $(^{14}N@C_{60})_2$-SWCNT-armchair (9, 9) at (a) HOMO and (b) LUMO. (Suzuki et. al. (2010))

which was larger than 1.72 eV, 3.06 eV and 2.11 eV of SWCNT (9, 9), $N@C_{60}$ and $(N@C_{60})_2$. Mulliken atomic charge of nitrogen atom in $(^{14}N@C_{60})_2$-SWCNT (9, 9) was estimated to be 2.998e. This positive charge indicates a considerable charge transfer from nitrogen atom to $C_{60}$ cage within SWCNT.

### 3.2 Electronic structure of $N@C_{60}$

Electronic structure of $N@C_{60}$ (S=3/2, I=1) at HOMO, LUMO and energy levels are shown in Fig. 10. The molecular orbital was considerably distributed on the $C_{60}$ surface cage. Mulliken N atom charge was 0.21e, which indicates charge transfer from nitrogen atomic to the $C_{60}$

surface cage. The energy gap between HOMO and LUMO was estimated to be 3.0 eV, which made agreement with the experimental energy gap of N@$C_{60}$ (2.7 eV) adsorbed on Cu (887) surface. The molecular structure with the energy levels have been investigated by scanning tunneling microcopy/spectroscopy (STM/STS) and near-edge x-ray absorption fine structure (NEXAFS) spectrum (Schiller, et. al. 2006).

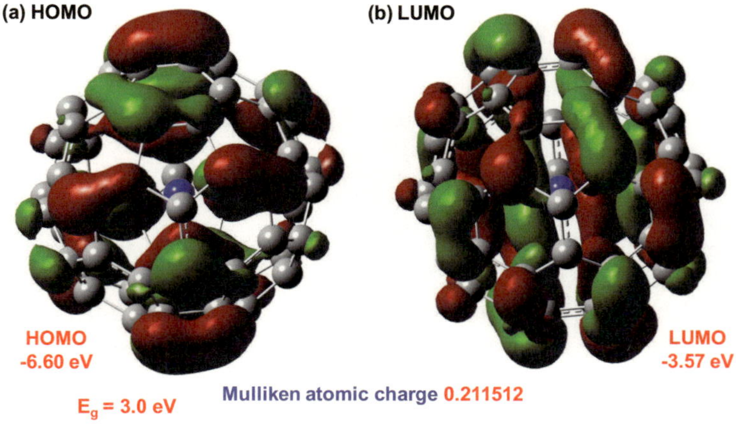


Fig. 10. Electronic structure and molecular orbital of N@$C_{60}$ at (a) HOMO (a) and (b) LUMO (Suzuki et. al. (2010))

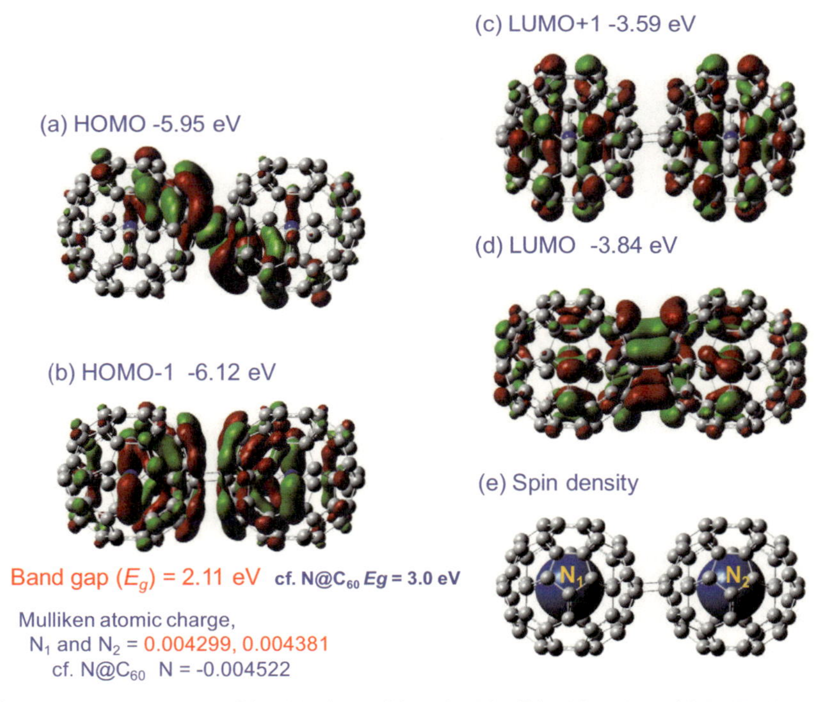


Fig. 11. Electronic structures of (N@$C_{60}$)$_2$ at (a) HOMO, (b) HOMO-1, (c) LUMO+1, (d) LUMO and (e) spin density. (Suzuki et. al. (2010))

Electronic structures of $(^{14}N@C_{60})_2$ as dimmer at (a) HOMO, (b) HOMO-1, (c) LUMO+1, (d) LUMO, (e) spin density and Mulliken atomic charge of $N_1$ and $N_2$ are shown in Fig. 11. The molecular orbital of the $(N@C_{60})_2$ at HOMO and LUMO were partial to distribute between the $C_{60}$ surface cages. The molecular orbital was interacted with each other on π electron of hybrid orbital. The energy gaps between HOMO and LUMO were estimated to be 2.11 eV, which was smaller than that of $N@C_{60}$. The dimerization was stable. The molecular orbital at HOMO-1 and LUMO+1 were well distributed around the $C_{60}$ cage surfaces. The molecular orbital was anti-interacted with each other, due to anti-bonding behavior of isolated molecular orbital of $N@C_{60}$. Mulliken atomic charge of $N_1$ and $N_2$ in $(N@C_{60})_2$ were at 0.004299 e and 0.004381 e. The slight values of positive charge were due to slight charge transfer from N atoms to the surface cage of $C_{60}$.

### 3.3 Electronic structure of $C_{60}N$

Electronic structure of $C_{59}N$ (S=1/2, I=1) at HOMO and LUMO are shown in Fig. 12. The molecular orbital were impartially distributed on the $C_{60}$ cage surface. Mulliken N atomic charge was -0.6e. The negative value indicates considerable charge transfer from $C_{60}$ cage to nitrogen atom. The energy gap between HOMO and LUMO was calculated to be 1.29 eV, which made agreement with the experimental result of $(C_{59}N)_2$ at 1.1 eV by C K NEXAFS spectrum (Schulte, et. al. 2007). The dimerization of $(C_{59}N)_2$ was stable as compared with isolation of $C_{59}N$.

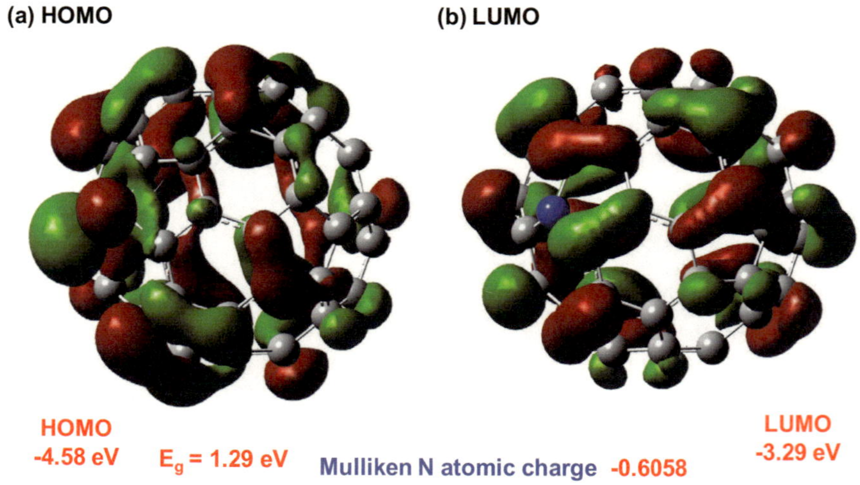


Fig. 12. Electronic structure and molecular orbital of $C_{59}N$ at (a) HOMO and (b) LUMO. (Suzuki et. al. (2010))

The single replacement in $C_{59}N$ keeps the bond lengths close to the $C_{60}$ values (1.40 and 1.45 Å). The dimerization has strong effects as shown in Fig. 13. The charge redistributes itself to weaken the intramolecular bonds of the tetra-coordinated C′ and to strengthen both the N-C bonds on the pentagon and the "double" bonds in the hexagons containing C′-N. This bond turns out to be a relatively weak one, as the electron density distribution emphasizes. In combination with the weak intermolecular C′-C′ bonding, this results in an electronic environment for C′ remarkably different from that of a typical $sp^3$ atom. This fact should account for the missing NMR signal in the $sp_3$ region. Accurate calculations of the carbon chemical shifts would be highly desirable. Their calculations predict a very similar structure for the latter (*anti*-conformation), apart from the much elongated intermolecular distance (1.675Å). The *anti*-conformation has been determined also by less sophisticated calculations.

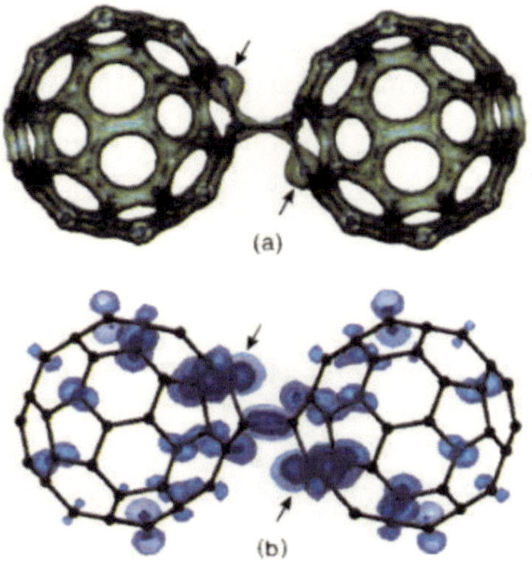


Fig. 13. For $(C_{59}N)_2$: Isodensity hypersurfaces of (a) the electron density and (b) the HOMO. Values (in au) are 0.2 (a); 0.002 (dark) and 0.0005 (light) (b). (Andreoni et. al. (1996))

### 3.4 Electronic structure of $(C_{60})_3$ as trimmer

Electronic structure of the $C_{2h}$ symmetry trimmer, $(C_{60})_3$ based on the P4 peanut has been studied (Beu 2006). The HOMO and LUMO energies obtained with the PBE (HCTH) functional are respectively $E_{HOMO}$=-0.175 (-0.179) a.u. and $E_{LUMO}$=-0.171 (-0.173) a.u. The distribution of the Mulliken charges in the two waist regions are similar to those of the dimmer (between -0.030 e and +0.056 e for the PBE functional). Figure 14 shows geometrical structure and HOMO-orbital (isovalue 0.01) of the $C_{2h}$ symmetry $C_{60}$ trimer, $(C_{60})_3$ based on the P4 peanut. The HOMO-orbital qualitatively showed the same behavior, bridging the cages. A transverse nodal plane through the middle of the central cage seems to limit the delocalization of the electrons. Anyway, for all methods the HOMO-LUMO gap of the trimmer was significantly reduced as compared to the dimmer, $(C_{60})_3$, continuing the monotonous decrease with respect to the $C_{60}$ monomer.

### 3.5 Geometrical effect of N@$C_{60}$-SWCNT, SWCNT, N@$C_{60}$ and $C_{59}N$ on electronic structure

Geometrical effect of N@$C_{60}$-SWCNT, SWCNT, N@$C_{60}$ and $C_{59}N$ on the electronic structure has been investigated by DFT using UB3LYP. Table 2 lists comparison between N@$C_{60}$-SWCNT, $(N@C_{60})_2$-SWCNT, SWCNT, $(N@C_{60})_2$, N@$C_{60}$ and $C_{59}N$ on energy levels at HOMO, LUMO and band gap ($E_g$). The energy levels of N@$C_{60}$-SWCNT were affected by diameter and chiral index. The energy levels of N@$C_{60}$-SWCNT compared with SWCNT had effected with increasing diameters. Especially, the energy gap of N@$C_{60}$-SWCNT was smaller than those of SWCNT, N@$C_{60}$ and $C_{59}N$. This behavior would be originated in narrowing energy gap based on molecular interaction between N@$C_{60}$ and SWCNT. These behaviors of $^{14}N@C_{60}$-SWCNT were considerably closed to those in density of state (DOS), which calculated with a tight binding model for SWCNT armchair. Electronic density of states (DOS) calculated with a tight binding model for SWCNT armchair (8, 8), (9, 9), (10, 10), and (11, 11) are shown in Fig. 16. The HOMO-LUMO gaps of SWCNT (n, n) for n = 9-13 as metal were inverse proportion to the diameters (Rao, et al. 1997). The energy gap of $(N@C_{60})_2$-SWCNT (9, 9) was extend to be 3.72 eV as compared with that of $N@C_{60}$-SWCNT (9. 9), due to instable dimmer within SWCNT. As reference, the energy gap of dimmer, $(^{14}N@C_{60})_2$ was estimated to be 2.11 eV, which was smaller than that of $^{14}N@C_{60}$, due the stable formation.

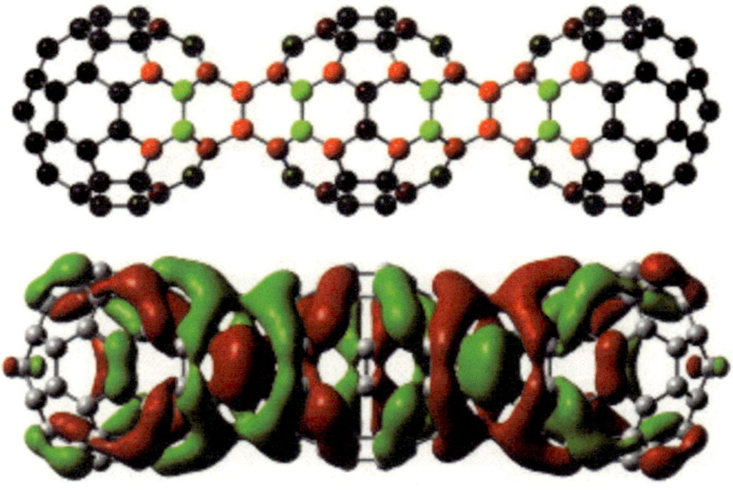

Fig. 14. Geometrical structure and HOMO-orbital (isovalue 0.01) of the $C_{2h}$ symmetry $C_{60}$ trimer, $(C_{60})_3$ based on the P4 peanut. (Rue (2006))

| $^{14}N@C_{60}$-SWCNT | HOMO (eV) | LUMO (eV) | $E_g$ (eV) |
|---|---|---|---|
| (14, 0) | -5.23 | -4.36 | 0.87 |
| (13, 13) | -2.02 | -1.25 | 0.77 |
| (10, 10) | -5.88 | -5.47 | 0.40 |
| (9, 9) | -4.05 | -2.89 | 1.16 |
| SWCNT | | | |
| (13, 13) | -4.06 | -2.86 | 1.20 |
| (10, 10) | -5.90 | -5.49 | 0.41 |
| (9, 9) | -4.08 | -2.84 | 1.24 |
| $(^{14}N@C_{60})_2$-SWCNT | | | |
| (9, 9) | -3.01 | 0.71 | 3.72 |
| $(^{14}N@C_{60})_2$ | -5.95 | -3.84 | 2.11 |
| $^{14}N@C_{60}$ | -6.60 | -3.57 | 3.03 |
| $C_{59}N$ | -4.58 | -3.29 | 1.29 |

Table 2. Comparison of $N@C_{60}$-SWCNT, $(N@C_{60}$-SWCNT$)_2$, SWCT, $(N@C_{60})_2$, $N@C_{60}$, and $C_{59}N$ on energy levels (Suzuki, et. al. (2010))

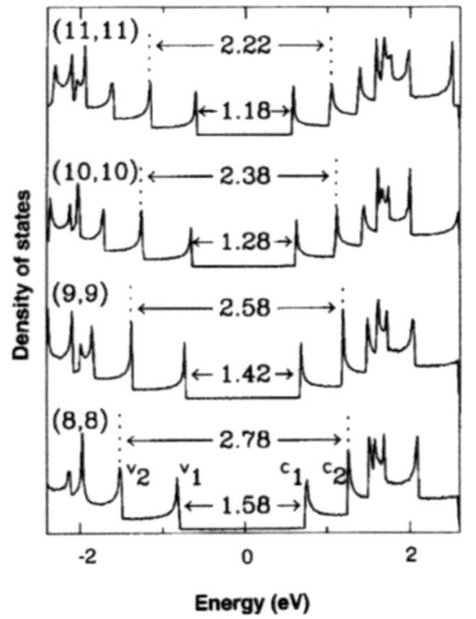


Fig. 15. Electronic density of states (DOS) calculated with a tight binding model for SWCNT armchair (8, 8), (9, 9), (10, 10), and (11, 11). The Fermi energy is located at 0 eV. Wave vector–conserving optical transitions can occur between mirror image spikes, that is, $v_1 \rightarrow c_1$ and $v_2 \rightarrow c_2$. (Rao, et al. (1997))

| Excited state MO | Coefficient | Energy (eV) | Wavelength cal. (nm) | Oscillator Strength |
|---|---|---|---|---|
| αHOMO → LUMO | 0.688 | 0.728 | 1703 | 0.0001 |
| β HOMO → LUMO | 0.687 | | | |
| β HOMO-1→ LUMO | 0.725 | 1.169 | 1061 | 0.0007 |
| β HOMO-1→ LUMO | 0.666 | | | |

Table 3. Excited state transitions of N@$C_{60}$-SWCNT (9, 9), calculated by TD-DFT. (Suzuki, et. al. (2010))

Excited state transitions of N@$C_{60}$-SWCNT were investigated by TD-DFT quantum calculation. Table 3 list excited state transitions of N@$C_{60}$-SWCNT (9, 9), calculated by TD-DFT. The wavelength converted from the exited energy levels were estimated to be 1703 nm and 1061 nm, which were closed to experimental results of $C_{59}N$-SWCNT by UV-*vis* NIR spectra and 2D PL contour maps as shown in Fig. 16 (a) (Iizumi, et. al. 2010). The optical behavior displayed the broad peaks around 2000 nm and 1000 nm, which could be assigned as the first ($E_{11}M$) and second ($E_{22}M$) van Hove transition on Raman spectroscopy with density of state using tight binding model (Rao, et al. 1997). As showed in Fig. 16 (b), the 2D PL contour maps of $(C_{59}N)_2$-SWCNT demonstrated that the emission and excited wavelength were around 1730 nm and 1000 nm. The slight different value between theoretical and experimental results would be originated in narrowing band gap based on molecular interaction between N@$C_{60}$ and SWCNT.

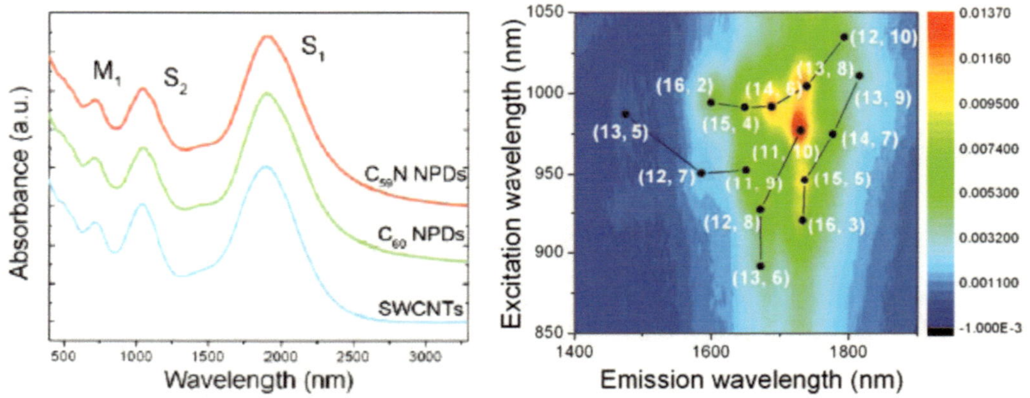


a) b)

Fig. 16. a) UV-*vis* NIR spectra, b) 2D PL contour maps of $NC_{59}$-SWCNT. (Iizumi et. al. (2010))

(a) $^{14}N@C_{60}$–SWCNT (13, 13) (b) SWCNT (13, 13)

| Atom | Chemical shift δ (ppm) / TMS | multi. |
|---|---|---|
| C | 84.86 | 39.0 |
| | 84.58 | 9.0 |
| | 84.39 | 2.0 |
| | 84.26 | 8.0 |
| | 84.14 | 2.0 |
| | 68.00 | 1.0 |
| | 67.77 | 51.0 |
| | 62.58 | 26.0 |
| | 57.71 | 1.0 |
| | 57.51 | 47.0 |
| | 57.30 | 4.0 |

| Atom | Chemical shift δ (ppm) / TMS | multi. |
|---|---|---|
| C | 71.83 | 1.0 |
| | 71.46 | 51.0 |
| | 62.98 | 52.0 |
| | 60.70 | 2.0 |
| | 60.49 | 24.0 |
| (c) $^{14}N@C_{60}$ | | |
| C | 78.14 | 60 |

Table 4. Chemical shifts of $^{13}C$ with multicity in $^{14}N@C_{60}$-SWCNT, SWCNT and $^{14}N@C_{60}$ (Suzuki et. al. (2010))

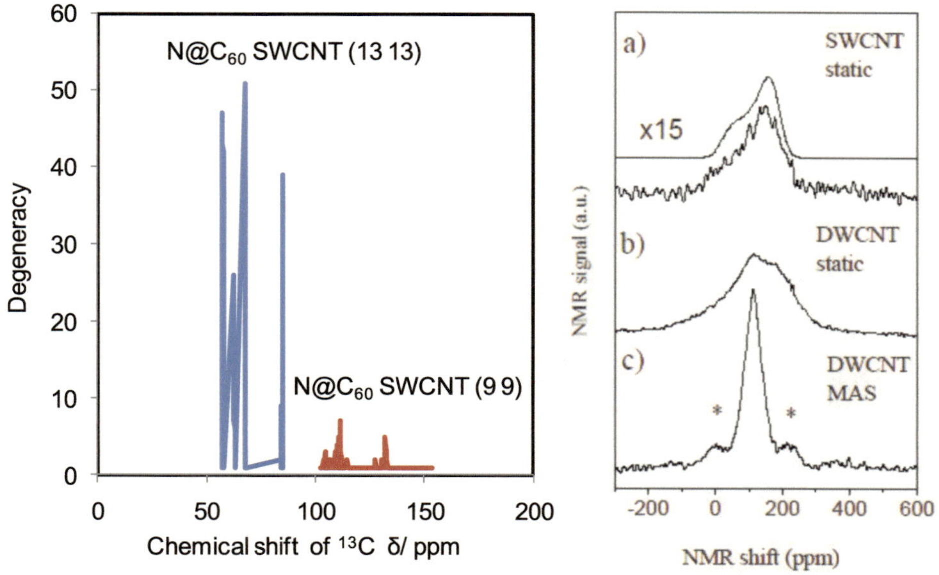


Fig. 17. The measured chemical shift of (a) SWCNT, (b) and (c) DWCNT (Simon et. al. (2007)).

### 3.6 Magnetic properties of $N@C_{60}$-SWCNT, $N@C_{60}$ and $C_{59}N$.

The magnetic parameters of chemical shift, principle g-tensor, A-tensor in hyperfine coupling constant (hfc) of $N@C_{60}$-SWCNT and $N@C_{60}$ have been studied by NMR, electron paramagnetic resonance (EPR) and electron nuclear double resonance (ENDOR) spectroscopy (Simon, et. al. 2006, 2007). Especially, chemical shifts of $^{13}C$ with multicity of $N@C_{60}$-SWCNT (13, 13), $N@C_{60}$ and original SWCNT have been investigated. Table 4 lists chemical shifts of $^{13}C$ with multicity in $^{14}N@C_{60}$-SWCNT, $^{14}N@C_{60}$ and SWCNT. Comparison results of the chemical shift between $N@C_{60}$-SWCNT, SWCNT and $N@C_{60}$ were investigated. As listed in Table 4, line position of $N@C_{60}$-SWCNT shifted out at degenerated position with 39, 51, 26, 47 multicity states in a low magnetic field. Effect of chiral index on the chemical shifts of $^{13}C$ in $^{14}N@C_{60}$-SWCNT is shown in Fig. 17. The chemical shifts positions of $N@C_{60}$-SWCNT shifted out with decreasing the diameters. The calculated results were considerably closed to the experimental results (Simon, et. al. 2007). These results were originated in a lack of 2s spin density distribution, spin-local orbital interaction and π electron interaction between $N@C_{60}$ and inner surface on SWCNT. The electronic spin-nuclear spin interaction based on unsymmetrical distribution of 2p spin orbital influenced the chemical shift position separated with splitting lines under a low magnetic field.

The chemical shifts of $^{13}C$ with multicity in optimized molecular structure will be calculated. The chemical shift will be constructed with a total sum of four terms as below:

$$\sigma = \sigma^{dia} + \sigma^{para} + \sigma^{FC} + \sigma^{SD} \tag{1}$$

where $\sigma^{dia}$ is diamagnetic term, $\sigma^{para}$ is paramagnetic term, $\sigma^{FC}$ is Fermi contact interaction, $\sigma^{SD}$ is spin dipole-interaction with angle dependence on 2p spin density distribution. The magnetic behavior of the chemical shift will be mainly based on the diamagnetic term and the spin-dipole interaction. Especially, the chemical shifts of N@$C_{60}$-SWCNT compared with $C_{59}N$ and $(C_{59}N)_2$ have been quantitatively analyzed on the basis of the experiment using *ab-initio* DFT calculation.

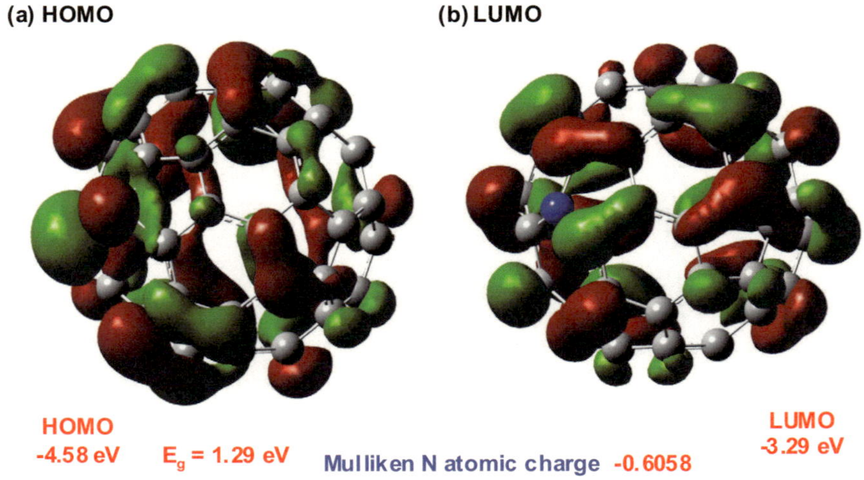


Fig. 18. Electronic structure of $C_{59}N$ at (a) HOMO and (b) LUMO. (Suzuki et. al. (2010))

Electronic structure and calculated chemical shifts of $^{13}C$ and multicity in parentheses for the atoms in the close environment in $C_{59}N$ are shown in Fig. 18 and 19. The calculated chemical shifts of $^{13}C$ in $C_{59}N$ separated in the range of 126 ppm and 152 ppm. The chemical shift of C near N atom suggests at 136.22 ppm with multicity at 5.0. The behavior would be arisen from electron-nuclear spin interaction based on spin density distribution at HOMO. The $^{13}C$ chemical shifts of $(C_{59}N)_2$ compared them with the experimental spectrum. Comparison

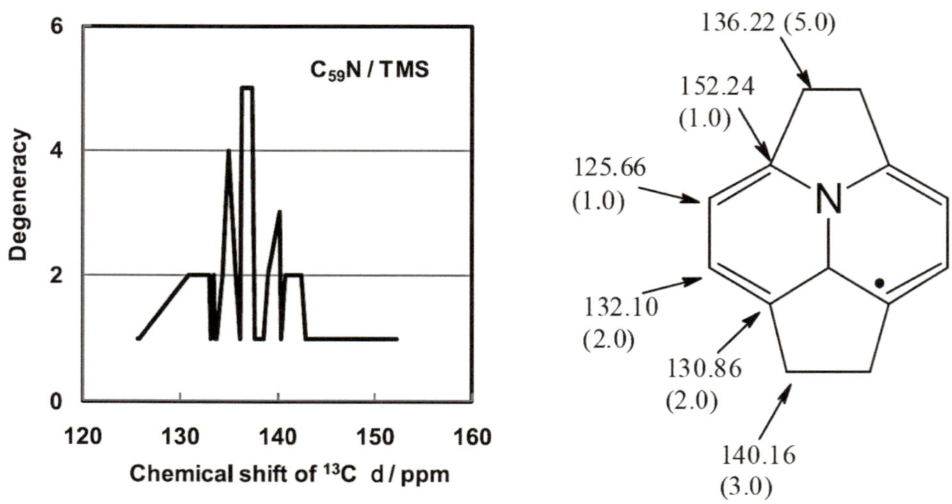


Fig. 19. Calculated chemical shifts of $^{13}C$ and multicity in parentheses for the atoms in the close environment in $C_{59}N$. (Suzuki et. al. (2010))

between the calculated $^{13}C$ chemical shifts of $(C_{59}N)_2$ and the experimental spectrum is showed in Fig. 20. The value for the saturated carbon atom (C′), 83.1 ppm, is refined to 85.3 ppm with the TZP/DZ basis, is in reasonably good agreement with the experiment (90.3 ppm), as is the value for the corresponding carbon atom in $C_{59}NNH$, 68.4 ppm (TZP/DZ) versus 72.1. Figure 21 shows $^{13}C$ chemical shifts for the atoms in the close environment of N: (a) $(C_{59}N)_2$, (b) $C_{59}NH$, and (c) indole. There was a substantial de-shielding of the saturated carbon atom in going from $C_{59}NH$ to $(C_{59}N)_2$, both in the experimental and theoretical data using the electronic density distribution.

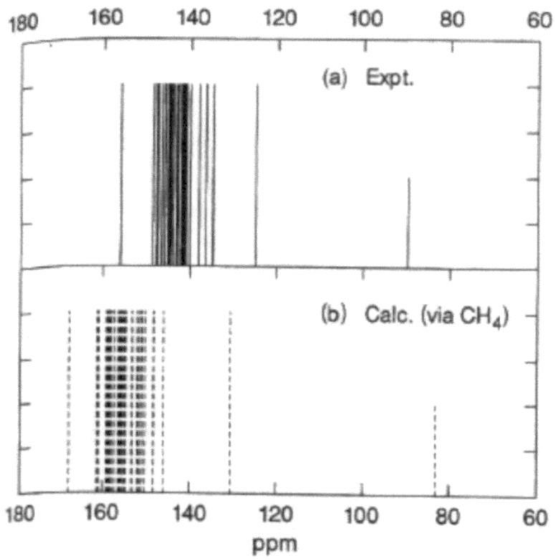


Fig. 20. $^{13}C$ chemical shifts $(C_{59}N)_2$: (a) experimental spectrum and (b) computed values at the DZ level (Buhl et. al. (1997))

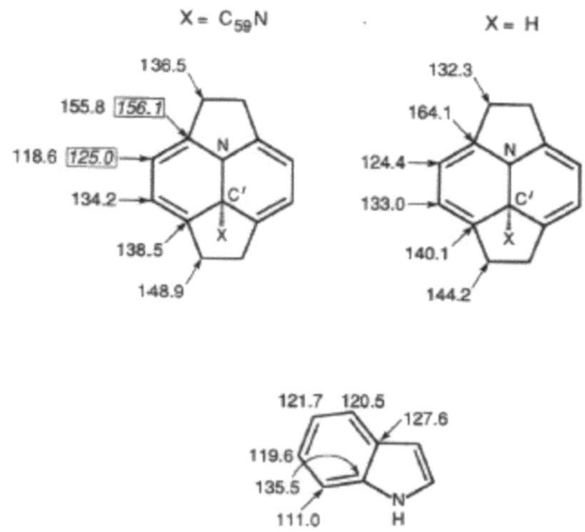


Fig. 21. $^{13}C$ chemical shifts for the atoms in the close environment of N: (a) $(C_{59}N)_2$, (b) $C_{59}NH$, and (c) indole. Experimental data are in boxes. Note that in the three cases the location and magnitude of the "extrema" are similar (Buhl et. al. (1997))

The magnetic interaction with the electron spins will be explained. The Hamiltonian can be formulated by Eq. (2).

$$H = g\,\beta_e\, S\, H - g_N\, \beta_N\, I\, H + S\, A\, I + S\, D\, S + I\, Q\, I \tag{2}$$

The five terms represent the following interactions: electronic and nuclear Zeeman interaction, the hyperfine of the spin nuclear interaction, the fine structure of the spin-spin interaction and the nuclear quadrupole.
Energy level diagram for isotropic atomic $^{14}N@C_{60}$ in a strong external magnetic field is shown in Fig. 22 (A). This Hamiltonian yields the 12-level system illustrated in Fig. 22 (A). The first split shows the four possible values of the electronic spin, and these are each split into the three possible values of I. The splitting of the nuclear energy levels has been investigated. The principle value of g-tensor will be mainly estimated by second perturbation. $\Delta g_{ii}$ related to spin local interaction with energy interaction between paramagnetic and nearest neighbor interaction.

$$g_{ii} = g_e + \Delta g_{ii} \tag{3}$$

$$g_{ii} = g_e \left\{ 1 + \sum_{m} \frac{\left| \varphi_m \middle| L_i \middle| \varphi_p \right|^2}{\epsilon_p - \epsilon_m} \right\} \tag{4}$$

$$\Delta g_{ii} = g_{ii} - 2 = \frac{\zeta a}{\Delta E_i} \tag{5}$$

The hyperfine coupling strength of A-tensor is constructed with a sum of isotropic and anisotropic hyperfine interaction.

$$A\text{-}tensor = A_{iso} + A'_{ii} \tag{6}$$

$$A_{iso} = 8\pi/3\; g_n\, \mu_N\, \rho(0) \tag{7}$$

$$A'_{ii} = g_N\, \mu_N \int \frac{(3\cos^2\theta - 1)}{r^3}\, \rho(r)\, dv \tag{8}$$

The isotropic hyperfine interaction, $A_{iso}$ will be related on the 2s spin density distribution, ρ (0), as described by Eq. (7). The anisotropic hyperfine interaction, $A'_{ii}$–tensor has influence of the 2p spin density distribution with angle dependence on a direction of the electron nuclear interaction, as noted in Eq. (8). A quantitative relation between the π–electron spin density and the isotropic hyperfine coupling constant has been discussed. For planar π–radical (e.g., $CH_3$ radical), this relation has been known as noted by Eq. (9).

$$A_{iso} = Q\, \rho^{\pi} \tag{9}$$

where $\rho^{\pi}$ is the π-electron density on the C atom and Q = 45 G. A future of non-planar radicals including $C_{59}N$ is the "umbrella effect characterized by (i) deviation of the C atom containing the unpaired electron from the plane and (ii) partial change in the hybridization of this C atom, which becomes intermediate between $sp^2$ and $sp^3$ (Abronin et. al. 2004). The spin-spin interaction with zero field constant, S D S is mainly originated in the dipole-dipole interaction, as noted by fifth term in Eq. (2). In the quadruple nuclear interaction, I Q I, the quadruple coupling constant, Q of nitrogen atom with more than I = 1 on nuclear quadruple resonance (NQR) will be dependence on magnetic interaction between electric quadruple moment, e Q and electric field gradient (EFG). The electric quadruple moment, e Q will be related on deviation of nuclear charge distribution. The magnetic parameters of g-tensor, A-tensor in the hyperfine coupling constant and the quadruple coupling constant of nitrogen atom will be under the spin nuclear interaction, e Q, EFG and the spin density distribution.

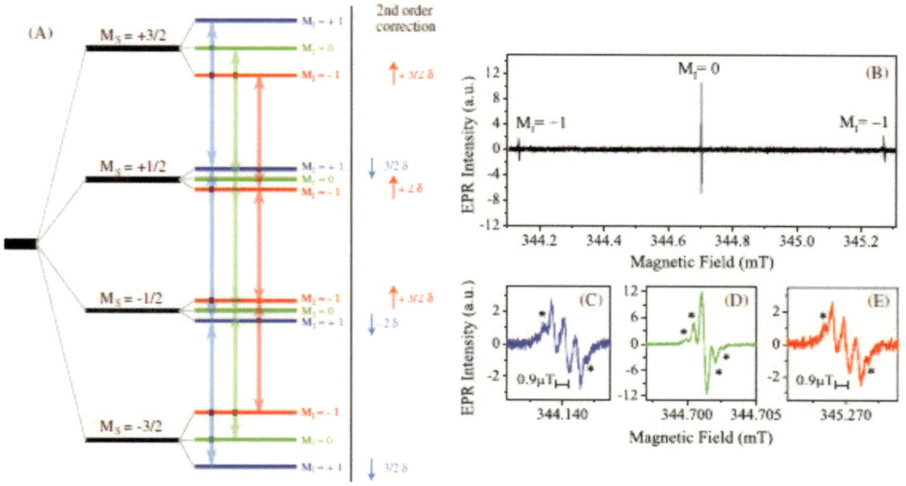


Fig. 22. (A) $^{14}$N@C$_{60}$ has electron spin $S = 3/2$ and nuclear spin $I = 1$ which together provide a rich 12-level structure. Considering only the first-order hyperfine interaction, the three electron transitions associated with a particular nuclear spin projection are degenerate. Adding the second order corrections ($\delta = a2/B$) lifts the degeneracies for the $MI = \pm 1$ lines. (B) Continuous wave EPR spectrum of high purity N@C$_{60}$ in degassed CS$_2$ at room temperature. Each line in the triplet signal is labeled with the corresponding projection $MI$ of the $^{14}$N nuclear spin. (C–E) Zoom-in for each of the three hyperfine lines reveals further structure. Stars (*) mark the line split by $^{13}$C hyperfine interactions with C$_{60}$ cage. Measurement parameters: microwave frequency, 9.67 GHz; microwave power, 0.5 $\mu$W; modulation amplitude, 0.2 $\mu$T; modulation frequency, 1.6 kHz. (Simon et. al. (2006))

### 3.7 Calculated magnetic parameters

The EPR spectrum of $^{14}$N inside N@C$_{60}$ consists of three lines centered at electron g-factor, split by a $^{14}$N isotropic hyperfine interaction. Table 6 lists magnetic parameters of g-tensor, A-tensor in hfc and Mulliken N atomic charge of N@C$_{60}$-SWCNT (13, 13), (9, 9), (N@C$_{60}$)$_2$, N@C$_{60}$ and C$_{59}$N, calculated by *ab-nitio* DFT quantum calculation. The magnetic parameters such as principal g-tensor, hfc and Mulliken N atomic charges in N@C$_{60}$-SWCNT (9, 9) were estimated to be 2.00093-98, 3.8-6.0 and $3.8\times10^{-5}$, respectively. The calculated magnetic parameters of N@C$_{60}$-SWCNT were discussed on the basis of the experimental results as shown in Fig. 23. The slight un-symmetrical values of hfs on N@C$_{60}$-SWCNT would be originated in anisotropy spin-dipole interaction and π electron interaction between N@C$_{60}$ and inner surface on SWCNT. In this case, a few percent of the values of ρ (2 s) and ρ (2 p) on nitrogen in N@C$_{60}$-SWCNT will be expected by the values of $A_N$ (2 s) and (2 p) of $^{14}$N to

be 557 G and 17.8 G at the ρ (2 s) = 1 and ρ (2 p) = 1. The calculated values of hfs in N@$C_{60}$-SWCNT, N@$C_{60}$ and $C_{59}N$ was in proportion to the spin-dipole interaction, a few percent of contribution from the 2p spin density distribution with a lack of 2s spin density distribution, obeyed by Fermi contact parts. The nuclear quadruple interaction of $^{14}N$ (I = 1) in N@$C_{60}$-SWCNT was negligible due to a slight value of EFG based on charge transfer from the CNT to $C_{60}$ with a slight value of the Mulliken atomic charges. The principle g-tensor, hfc and Mulliken nitrogen atomic charges on N@$C_{60}$-SWCNT (9, 9) as compared with N@$C_{60}$ and $C_{59}N$ were decreased. These results would be dependence with geometrical effects of N@$C_{60}$-SWCNT (n, n) for chiral index on π-electron interaction in cooperation with the magnetic interaction between electron spins, S = 3/2 and nuclear spin, I = 1.
Experimental values of A-tensor in hfs on N@$C_{60}$ and $C_{59}N$ are shown in Fig. 23 and 24. The symmetrical structure of N@$C_{60}$ had isotropic value of magnetic parameters. The experimental results of N@$C_{60}$ compared with $C_{59}N$ (Fulop et. at. (2001)) were considerably closed to the calculated results (Harneit et. al. (2002)). The unsymmetrical structure of $C_{59}N$ affected the principle lines at several states, which splitting by anisotropy hyperfine interaction. Spin density distribution of paramagnetic azafullerene, $C_{59}N$ have been calculated (Abronin et. al. 2004, Schrier et. al. 2006). Based on hyperfine coupling constants for $^{14}N$ and $^{13}C$ nuclei in $C_{59}N$, the spin density distribution of the unpaired π-electron in $C_{59}N$ was mainly localized around the nitrogen atom. Mulliken charge transfer of nitrogen atom in $C_{59}N$ was calculated to be -0.6 e. This result indicates a considerable value of charge transfer from $C_{60}$ to N atom. The magnetic parameters of the anisotropic molecular structure would be strongly attributed from the spin-dipole interaction with 2p spin density distribution. Considerably, molecular design of the endohedral fullerenes within SWCNT varied with geometry structure is important to control the quantum spin qubit, splitting the magnetic parameters including g-tensor, A-tensor of hfc and the chemical shifts in the NMR quantum computer.

| Elements | g-tensor ($g_{xx}$, $g_{yy}$, $g_{zz}$) | hfc / G ($A_{xx}$, $A_{yy}$, $A_{zz}$) | Mulliken N atomic charges / e |
|---|---|---|---|
| $^{14}N@C_{60}$ -SWCNT (13, 13) | 2.00119<br>2.00119<br>2.00117 | 2.4<br>2.4<br>1.3 | -0.004521 |
| $^{14}N@C_{60}$ -SWCNT (9, 9) | 2.00098<br>2.00098<br>2.00093 | 6.0<br>6.0<br>3.8 | 0.000038 |
| $(^{14}N@C_{60})_2$ Dimmer | 1.99775<br>1.99776<br>1.99785 | 4.6<br>4.6<br>15.0 | 0.004299, 0.004381 |
| $^{14}N@C_{60}$ | 2.00490<br>2.00490<br>2.00490 | 4.5<br>4.5<br>4.5 | 0.211512 |
| $C_{59}N$ | 2.00545<br>2.00511<br>2.00356 | 2.4<br>2.5<br>7.6 | -0.605804 |

Table 5. g-tensor, hfc and Mulliken N atomic charge of $^{14}N@C_{60}$ –SWCNT (Suzuki et. al. (2010)).

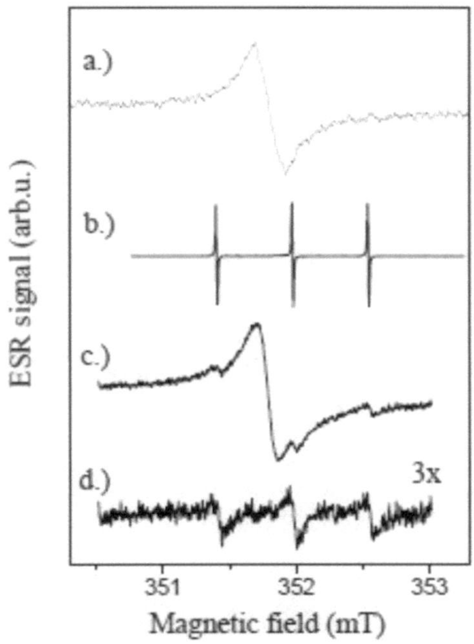


Fig. 23. ESR spectra of (a) SWCNT, (b) N@$C_{60}$ and (c)-(d) N@$C_{60}$-SWCNT at triplet state. (Simon et. al. (2007)).

| Endohedral fullerenes | Nuclear spin I | Hyperfine constant A (MHz) | Electron spin relaxation in dilute powder T=300K $T_1$ (μs) | T=300K $T_2$ (μs) | T=5K $T_1$ (s) | T=5K $T_2$ (μs) |
|---|---|---|---|---|---|---|
| $^{14}N@C_{60}$ | 1 | 15.88 | 120 | 20 | 0.5–9[a)] | 20 |
| $^{15}N@C_{60}$ | 1/2 | 22.26 | 45 | 11 | 1 | 14 |
| $^{31}P@C_{60}$ | 1/2 | 138.4 | 2.7 | 1.3 | 0.5–8[a)] | 14–28[b)] |

a) depending on sample quality
b) depending on endohedral fullerene concentration

Table 6. Spin properties of the endohedral $C_{60}$ fullerenes with electron spin S = 3/2 (Harneit et. al. (2002))

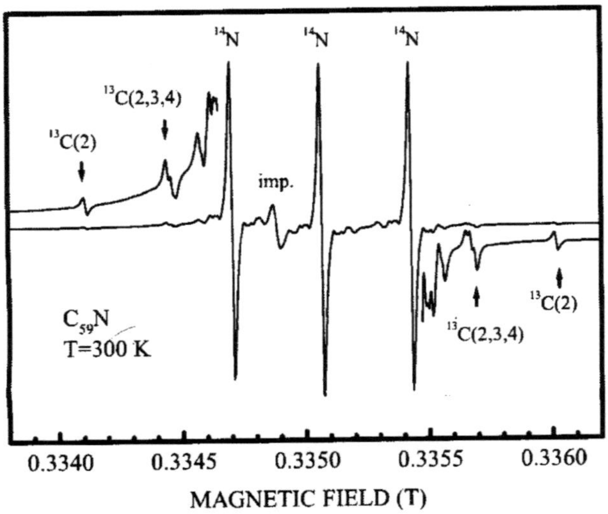


Fig. 24. Motionally narrowed ESR spectrum of tumbling $C_{59}N$ substituted in $C_{60}$ at 290 K and 9 GHz (Fulop et. al. (2001)).

## 4. Conclusion

Design of spin labels for the NMR quantum computer that contained 1D spin chains filling SWCNT with N@$C_{60}$ has been proposed. The electronic structure, chemical shift, g-tensor, A-tensor of hfc would be influenced by geometrical structure, varied with diameter and chiral index. The magnetic properties were originated in spin density distribution with π-electron interaction between N@$C_{60}$ and inner surface on SWCNT and CT. The slight un-symmetrical values of A-tensor in hfs on N@$C_{60}$-SWCNT (9, 9) would be originated in anisotropy spin-dipole interaction and π-electron interaction between N@$C_{60}$ and inner surface on SWCNT. The values of A-tensor in hfc would be in proportion to the 2 p spin distribution based on the spin-dipole interaction with a lack of 2 s spin density distribution, obeyed by Fermi contact parts. The magnetic parameters of the anisotropic structure would be strongly influenced by the spin-dipole interaction with a slight distribution of the 2 p spin density. Molecular design of the endohedral fullerenes within SWCNT as peapods is important to control quantum qubit, splitting the principle lines based on hyperfine interaction as the NMR quantum computer.

## 5. References

### NMR quantum computer using N@$C_{60}$, P@$C_{60}$ and N@$C_{60}$-SWCNT

Benjamin, S. C.; Ardavan, A.; Briggs, G. A. D.; Britz, D. A.; Gunlycke, D.; Jefferson, J.; Jones, M. A. G.; Leigh, D. F.; Lovett, B. W.; Khlobystov, A. N.; Lyon, S. A.; Morton, J. J. L.; Porfyrakis, K.; Sambrook, M. R.; Tyryshkin, A. M.; (2006) Towards a fullerene-based quantum computer, *J. Phys.: Condens. Matter* Vol.18, pp.S867–S883.

Harneit, W.; Meyer, C.; Widinger, A.; Suter, D.; Twamley, J.; (2002) Architectures for a Spin Quantum Computer Based on Endohedral Fullerenes, *phys. stat. sol. (b)* Vol. 233, pp. 453–461.

Ju, C.; Suter, D.; Du, J.; (2007) Two-qubit gates between noninteracting qubits in endohedral fullerene-based quantum computation, *Phys, Rev. A.* Vol. 75, pp. 012318-1-5.

Meyer, C.; Harneit, W.; Naydenov, B.; Lips, K.; Widinger A.; (2004) N@$C_{60}$ and P@$C_{60}$ as quantum bits, *Appl. Magn. Reson.* Vol. 27, pp. 123-132.

Morton, J. J. L.; Tyryshkin, A. M.; Ardavan, A.; Benjamin, S. C.; Porfyrakis, K.; Lyon, S. A.; Briggs, G. A. D.; (2005) Bang-bang control of fullerene qubits using ultra-fast phase gates, Nature Physics Vol. 40, pp. 40-44.

Morton, J. J. L.; Tyryshkin, A. M.; Ardavan, A.; Benjamin, S. C.; Porfyrakis, K.; Lyon, S. A.; Briggs, G. A. D.; (2006) The N@$C_{60}$ nuclear spin qubit: Bang-bang decoupling and ultrafast phase gates, *phys. stat. sol.* (b) Vol.243, pp. 3028–3031.

Brown, R. M.; Tyryshkin, A. M.; Porfyrakis, K.; Gauger E. M.; Lovett, B. W.; Ardavan, A,; Lyon, S. A.; Briggs, G. A. D.; Morton, J. J. L. (2011) Coherent State Transfer between an Electron and Nuclear Spin in $^{15}$N@$C_{60}$, Phys. Rev. Lett. Vol. 106, pp. 110504-1-4.

Yang, W. L.; Xu, Z. Y.; Wei, H.; Feng, M.; Suter D.; (2010) Quantum-information-processing architecture with endohedral fullerenes in a carbon nanotube, *Phys. Rev. A.* Vol. 81, pp.032303-1-8.

**N@$C_{60}$-SWCNT**

Bailey, S. W. D.; Lambert, C. J.; (2007) The electronic transport properties of N@$C_{60}$@(n,m) carbon nanotube peapods, *Phys. E*, Vol.40, pp.99-102.

Cho, Y.; Han, S.; Kim, G.; Lee, H.; Ihm, J.; (2003) Orbital Hybridization and Charge Transfer in Carbon Nanopeapods, *Phys. Rev. Lett.* Vol. 90, 106402-1-4.

Iizumi, Y.; Okazaki, T.; Liu, Z.; Suenaga, K.; Nakanishi, T.; Iijima, S.; Rotas, G.; Tagmatarchis, N.; (2010) Host-guest interactions in azafullerene ($C_{59}$N)-single-wall carbon nanotube (SWCNT) peapod hybrid structures, *Chem. Commun, Vol.* 46, pp.1293-1295.

Simon, F.; Kuzmany, H.; Na´fra´di, B.; Fehe´r, T.; Forro´, L.; Fu¨lo¨p, F.; Ja´nossy, A.; Korecz, L.; Rockenbauer, A.; Hauke, F.; Hirsch, A.; (2006) Magnetic fullerenes inside single-wall carbon nanotubes, *Phys. Rev. Lett.* Vol.97, pp.136801-1-4.

Simon, F.; (2007) Studying Single-Wall Carbon Nanotubes Through Encapsulation: From Optical Methods Till Magnetic Resonance, *J. Nanoscience and Nanotechnology,* Vol.7, pp.1197–1220.

Suzuki, A.; Oku, T.; Kikuchi, K.; (2010) Electronic structure and magnetic properties of $^{14}$N@$C_{60}$ within single-walled carbon nanotube as peapods, *Physica B,* Vol. 405, pp.2418-2422.

Tooth, S.; Quintavalle, D.; Náfrádi, B.; Forró, L.; Korecz, L.; Rockenbauer, A.; Kálai, T.; Hideg, K.; Simon, F.; (2008) Stability and electronic properties of magnetic peapods, *Phys. Stat. Solids. (B)* Vol.245, pp.2034-2037.

**Sc@$C_{82}$-SWCNT**

Cantone, A. L.; Buitelaar, M. R.; Smith, C. G.; Anderson, D.; Jones, G. A. C.; Chorley, S. J.; Casiraghi, C.; Lombardo, A.; Ferrari, A. C.; Shinohara, H.; Ardavan, A.; Warner, J.; Watt, A. A. R.; Porfyrakis, K.; Briggs, G. A. D.; (2008) Electronic transport characterization of Sc@$C_{82}$ single-wall carbon nanotube peapods, *J. App. Phys.* Vol.104, pp.083717-1.

Warner, J. H.; Watt, A. A. R.; Ge, L.; Porfyrakis, K.; Akachi, T.; Okimoto, H.; Ito, Y.; Ardavan, A.; Montanari, B.; Jefferson, J. H.; Harrison, N. M.; Shinohara, H.; Briggs, G. A. D.; (2008) Dynamics of Paramagnetic Metallofullerenes in Carbon Nanotube Peapods, *Nano Lett.* Vol.8, pp.1005-1010.

**Carbon tubes**

Khlobystrov, A. N.; Britz, D. A.; Andrew, G.; Briggs, D.; (2005) Molecules in carbon nanotubes, *Acc. Chem. Res.* Vol.38, pp.901-909.

Rao, A. M.; Richter, E.; Bandow, S.; Chase, B.; Eklund, P. C.; Williams, K. A.; Fang, S.; Subbaswamy, K. R.; Menon, M.; Thess, A.; Smalley, R. E.; Dresselhaus, G.; Dresselhaus, M. S.; (1997) Diameter-Selective Raman Scattering from Vibrational Modes in Carbon Nanotubes Science, Vol. 275, pp.187–191.

**N@$C_{60}$, P@$C_{60}$, $C_{60}$ ,$(C_{59}N)_2$ and $C_{59}N$**

Abronin, I. A.; Breslavskaya, N. N.; Rakitina, V. A.; Buchachenko, A. L.; (2004) Spin density distribution in paramagnetic azafullerene: *Russ. Chem. Bull. Int. Ed.* Vol. 53, pp. 2475-2477.

Andreoni, W.; Curioni, A.; Holczer, K.; Prassides, K.; Keshavarz-K., M.; Hummelen, J.; Wudl, F.; (1996) Unconventional Bonding of Azafullerenes: Theory and Experiment, *J. Am. Chem. Soc.* 118, pp. 11335 - 11336.

Beu, T. A.; (2006) Electronic structure calculations of peanut-shaped $C_{60}$ polymers, *J. Opt. Adv. Mat.* Vol. 8, pp.177 - 180.

Buhl, M.; Curioni A.; Andreoni, W.; (1997) Chemical shifts of diamagnetic azafullerens: $(C_{59}N)_2$ and $C_{59}NH$, *Chem. Phys. Lett.* pp. 231-234.

Fulop, F.; Rockenbauer, A.; Simon, F.;Pekker, S.; Korecz, L.; Garaji, S., Janossy. Andrais; (2001) Azafullerene $C_{59}N$, a stable free radical substituent in crystalline $C_{60}$: *Chem. Phys. Lett.* Vol. 334, pp. 233-237.

Kobayashi K.; Nagase S.; Dinse K.; (2003) A theoretical study of spin density distributions and isotropic hyperfine couplings of N and P atoms in N@$C_{60}$, P@$C_{60}$, N@$C_{70}$, N@$C_{60}(CH_2)_6$, and N@$C_{60}(SiH_2)_6$ : *Chem. Phycs. Lett.* pp 93-98

Schiller, F.; Ruiz-Osés, M.; Ortega, J. E.; Segovia, P.; Martínez-Blanco, J.; Doyle, B. P.; Pérez-Dieste, V.; Lobo, J.; Néel, N.; Berndt, R.; Kröger, J.; (2006) Electronic structure of $C_{60}$ on Au (887), *J. Chem. Phys.* Vol. 125, pp.144719-1-6.

Schrier, J.; Whaley, K. B.; (2006) Hyperfine coupling constants of the azafulleenes $C_{19}N$, $C_{59}N$, $C_{69}N$, and $C_{75}N$: *J. Phys. Chem.* A Vol. 110, pp. 5386-5390.

Schulte, K.; Wang, L.; Moriarty, P. J.; Prassides, K.; Tagmatarchis, N.; (2007) Resonant processes and Coulomb interactions in $(C_{59}N)_2$, *J. Chem. Phys. Vol.* 126, 184707-184713.

# 21

# Carbon Nanotubes Addition Effects on $MgB_2$ Superconducting Properties

Adriana Serquis[1], Gabriela Pasquini[2] and Leonardo Civale[3]
[1]*Instituto Balseiro-UNCuyo, Centro Atómico Bariloche – CNEA, CONICET*
[2]*Dept. de Física – FCEyN - Universidad de Buenos Aires, IFIBA - CONICET*
[3]*MPA-STC, Los Alamos National Laboratory*
[1,2]*Argentina*
[3]*USA*

## 1. Introduction

Since the discovery of superconductivity at 39 K in $MgB_2$ (Nagamatsu et al., 2001), considerable progress has been made in the understanding of the fundamental properties and the development of commercial applications of this material.

The strong potential for technological uses of $MgB_2$ is due to a unique combination of characteristics, such as a high transition temperature $T_c$ ~ 39K, chemical simplicity, lightweight and low cost of the raw materials (Buzea et al., 2001). In addition, the absence of weak-link behavior at grain boundaries in polycrystalline samples (Larbalestier et al., 2001) allows the use of simple Powder in Tube (PIT) methods to fabricate wires and tapes (Flükiger et al., 2003). One of the most important issues for $MgB_2$ magnet applications is the simultaneous enhancement of its critical current density ($J_c$) and the upper critical field ($H_{c2}$). Thus, on one hand, the pinning force may be improved by the incorporation of defects (nano particle doping, chemical substitutions, etc.). On the other hand, the doping level affects the intraband scattering coefficients and the diffusivity of the two bands of this peculiar superconductor, and these changes may cause a significant $H_{c2}$ variation. Carbon or C-compounds additions have been very successful to improve $J_c$ and/or $H_{c2}$, and the effects of carbon doping on superconductivity in $MgB_2$ has been extensively studied. The $J_c$-$H$ performance can be greatly improved by adding different carbon sources, such as carbon doped $MgB_2$ filaments (Wilke et al., 2004, 2005a), nanocarbon (Soltanian et al., 2003; Ma et al., 2006; Yeoh et al., 2006; Häβler et al., 2008), amorphous carbon (Senkowicz et al., 2005), diamond (Cheng et al., 2003), $B_4C$ (Wilke et al., 2005b; Ueda et al., 2005; Yamamoto et al., 2005a, 2005b), carbon nanohorn (Ban et al., 2005) and, particularly, carbon nanotubes (Dou et al., 2003; Yeoh et al., 2004, 2005, 2006; Serquis et al., 2007; Serrano et al., 2008, 2009); Shekhar, 2007; Vajpayee et al., 2010; etc).

In this chapter we present a review of recent developments in the study of the effect of carbon nanotubes (CNT) on the superconducting properties of $MgB_2$ bulk and wire samples, based on the known literature data and our own results.

It is well known that pinning of vortex lines to defects in superconductors plays an extremely important role in determining their properties. CNT inclusions, with diameters close to the $MgB_2$ coherence length ($\xi_{ab}(0)$ ~ 3.7-12nm; $\xi_c(0)$ ~ 1.6-3.6 nm, depending on the doping level, as discussed below in section 2.1.2) (Buzea et al., 2001) may be very good candidates for vortex pinning if they do not completely dissolve in the matrix but remain as tubes acting as columnar defects. Therefore, the presence of CNT may result in a critical current density improvement under applied fields that will depend on the synthesis parameters. Besides, theoretical models predict that the presence of two superconducting gaps could allow tuning of different upper critical fields by controlling diverse defect sublattices relative to orthogonal hybrid bands (Golubov et al., 2002; Gurevich , 2003). These models predict a significant $H_{c2}$ enhancement in the dirty limit and an anomalous $H_{c2}(T)$ upward curvature. Several reports indicate that $H_{c2}$ can be significantly increased by introducing disorder through oxygen alloying, carbon doping or He-ion irradiation (Brinkmann et al., 2002; Gurevich et al., 2004; Putti et al., 2004; Braccini et al., 2005). Thus, it is important to analyze the effect of carbon-doping through CNT or other C-sources to the $H_{c2}(T)$ in bulk $MgB_2$ samples, where extrapolated values of $H_{c2}(0)$ between 29 and 44 T have been reported (Wilke et al., 2004; Senkowicz et al., 2005; Serquis et al., 2007; Serrano et al., 2008).

In section 2 of this work, we describe the influence of C-incorporation into the $MgB_2$ structure by using different CNT type over the microstructural and superconducting properties (critical current density, critical fields and critical temperature). In particular. we present the influence of synthesis parameters in the superconductivity of bulk samples prepared with single-walled (SW), double-walled (DW) and multi-walled (MW). In the first subsection, we compare the evolution of $T_c$, and the lattice parameters with the amount of CNT, analyzing if C is replacing B in the $Mg(B_{1-x}C_x)_2$ structure or remains as CNT in each case. In the following subsections, we present a review of $J_c$ determined by magnetization, and $H_{c2}$(T) by transport measurements, which may require measurements performed using high fields (i.e. 50 T) pulsed magnets. In the last subsection, we also present an analysis of the distinctive effect of C addition in samples with optimum contents of single-walled and double-walled carbon nanotubes SiC, that simultaneously increase $J_c$ and $H_{c2}$.

In section 3, we offer a review of the use of CNT in the production of PIT $MgB_2$ wires and tapes required for applications. The fabrication and processing conditions strongly affect microstructures and current carrying capability of PIT conductors, making the question of grain connectivity more relevant. The standard and low-cost fabrication PIT method involves filling a metallic tube with superconducting powder (ex-situ) or precursors (in-situ) and drawing it into a wire and/or rolling into a tape (Flükiger et al., 2003). We show the results obtained in $MgB_2$ wires and tapes prepared by PIT using different kind of CNT and treated at different temperatures between 600 and 900 °C, using several sheath materials (e.g. iron, stainless steel, Fe/Nb, etc), to establish a relationship among annealing conditions, modifications in the microstructure, and changes in $J_c$.

In Section 4 we go deeper in the role of doping by studying the magnetic relaxation of $MgB_2$ with and without DWCNT bulk samples. In the first subsection, we introduce the main equations contained in a very recent work (Pasquini et al., 2011) necessary to investigate the current decay, that relates parameters such as the magnetic field B, the magnetization $m$, current density $J$ flowing in the sample, and the activation barrier $U(J,T,B)$ that governs the creep of vortices. In the rest of this section we present the experimental relaxation rates (measured using a DC magnetization technique), and the general behaviour is described and analyzed under the Anderson-Kim frame model (Anderson, 1964). The pinning energies $U_c$, true critical current densities $J_{c0}$, and correlation volumes $V_c$ are estimated and compared.

Finally, we summarize the main conclusions in section 6.

## 2. Effect of different kind of CNT addition to superconducting properties of $MgB_2$ bulk samples

There are many reports in the literature that study the effect of CNT on the superconducting properties of $MgB_2$. However, many of them mentioned the effect of CNT-doping without clarifying the meaning of this word. *Doping* is generally the practice of adding impurities to something, but it specifically means the replacement of one element within the same crystalline structure, creating a substitutional point defect. In the case of CNT additions to $MgB_2$ it is possible that part of the carbon replaces boron giving place to the compound $Mg(B_{1-x}C_x)_2$. As a consequence, in many works, the nominal amount of CNTs added during the synthesis is calculated as the stoichiometric amount according to this formula or, in other cases as some extra %at or %wt of carbon added to $MgB_2$. In the later case, if part of carbon replaces boron, some extra magnesium should be added that may also evaporate or form MgO during the synthesis process. In this chapter, unless it is specified, we will use "x" as the nominal amount of carbon added according to $Mg(B_{1-x}C_x)_2$, and the term "doping" when carbon is actually replacing boron.

### 2.1 Carbon doping and critical temperature ($T_c$)

Several systematic carbon doping studies of $Mg(B_{1-x}C_x)_2$ have been performed in single crystals (Lee et al., 2003; Kazakov et al., 2005), polycrystalline wires fabricated by chemical vapor deposition (CVD) (Wilke et al., 2004, 2005a), and $B_4C$-doped (Avdeev et al., 2003; Wilke et al., 2005b). All these studies indicate a monotonic decrease of $T_c$ and the lattice parameter $a$ for increasing $x$, while the lattice parameter $c$ remains constant.

For single crystals, the solubility limit of C in $MgB_2$ was estimated to be about $15 \pm 1\%$, which is substantially larger than that reported for the polycrystalline samples. A dramatic decrease in $T_c$ was also observed with C-substitution, followed by complete suppression of superconductivity for $x>0.125$. This solubility limit is usually not reached in bulk samples. A comparison of the critical temperature as a function of the nominal C-content for different carbon sources is displayed in Figure 1.

The reduction in $T_c$ resulting from C addition using CNT sources is much lower than in the previous cases, suggesting that C substitution is lower than the concentration given by $Mg(B_{1-x}C_x)_2$. The $T_c$ decrease is even lower when using SWCNT as a source of carbon indicating a very low doping level.

As an example, it is possible to calculate the actual C content substituting B into the $MgB_2$ structure, using the lattice parameters obtained from XRD: the shift in the $a$-axis lattice parameter can be used as a calibration for the actual amount of C ($x$) in the $Mg(B_{1-x}C_x)_2$ structure in comparison with the fitting of neutron diffraction data (Avdeev et al., 2003) or single crystal data (Kazakov et al., 2005). The dependence of $a$-axis lattice parameter as a function of the nominal DWCNT %at content is plotted in Figure 2. The corrected $x$ values obtained were used to plot the same data as open squares in Figure 1, indicating that not all C is incorporated into the $MgB_2$ structure. A similar behavior is observed for all CNT samples, indicating that actual C-substitution tends to saturate for nominal CNT contents larger than 10%at. The actual C-doping varies according to the CNT type, synthesis temperature or other synthesis parameters (i.e. pressure or magnetic field). This C-substitution level determines not only the $T_c$ but other superconducting properties such as $H_{c2}$, as will be described in section 2.2. Nevertheless, we will continue using the nominal content as the parameter to describe most effects to make it easier the comparison with other author's results.

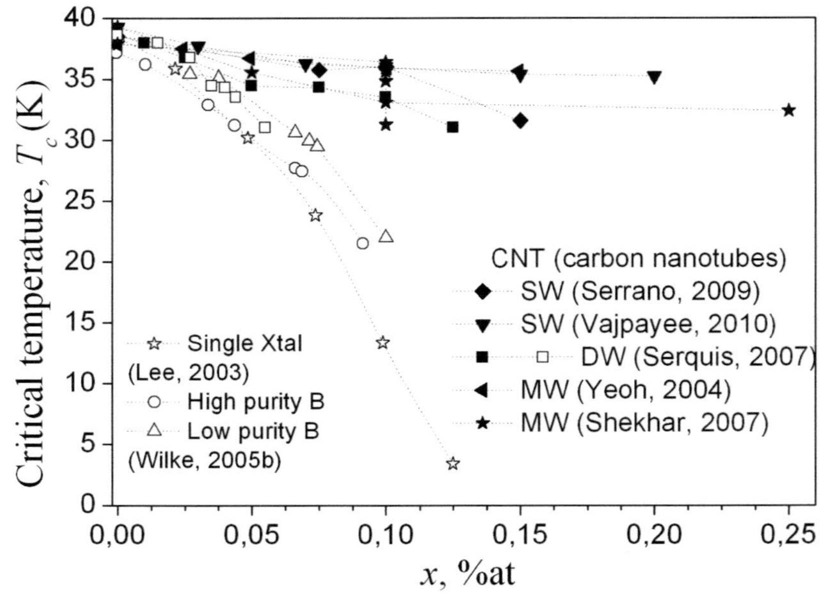


Fig. 1. Critical temperature as a function of the nominal C-content using different carbon sources.

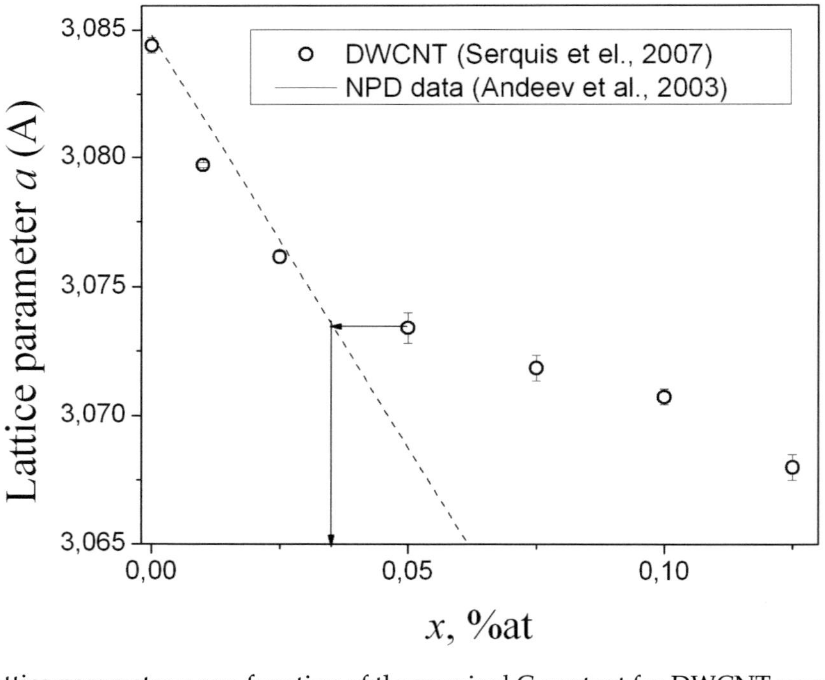


Fig. 2. Lattice parameter $a$ as a function of the nominal C-content for DWCNT samples compared with others with full carbon substitution, measured by neutron diffraction.

It is important to mention that not only the nominal carbon content $x$ but also other synthesis parameters affect the superconducting properties like $T_c$ due to the amount of C substituting B in the structure. All parameters studied up to now are listed at the beginning of next section.

### 2.2 Simultaneous enhancement of critical currents and fields

#### 2.2.1 Critical currents ($J_c$)

Since the first work devoted to the study of the effect of the addition of CNT on the critical current density of $MgB_2$ (Dou et al., 2003), several parameters have been studied:

- the amount of "x" for a particular CNT type : MWCNT (Dou et al., 2003; Shekhar, 2007), DWCNT (Serquis et al., 2007), SWCNT (Serrano et al., 2009 ; Vajpayee et al., 2010)
- the effect of CNT type (Serrano et al., 2008) and size (Yeoh et al., 2005, 2006)
- the effect of sintering conditions such as temperature (Yeoh et al., 2004), pressure (Yuan et al., 2005), ultrasonication of precursors (Yeoh et al., 2006) or the application of magnetic field during sintering (Li et al., 2007)

Most researchers performed magnetization loops measurements to determine $J_c(H)$ in bulk samples using the Bean critical state model (Bean, 1962) assuming full penetrated samples for a long parallelepiped:

$$J_c(H)[A/cm^2] = \frac{20 \times \Delta M(H)[emu/cm^3]}{(a - \frac{a^2}{3b})[cm]} \qquad (1)$$

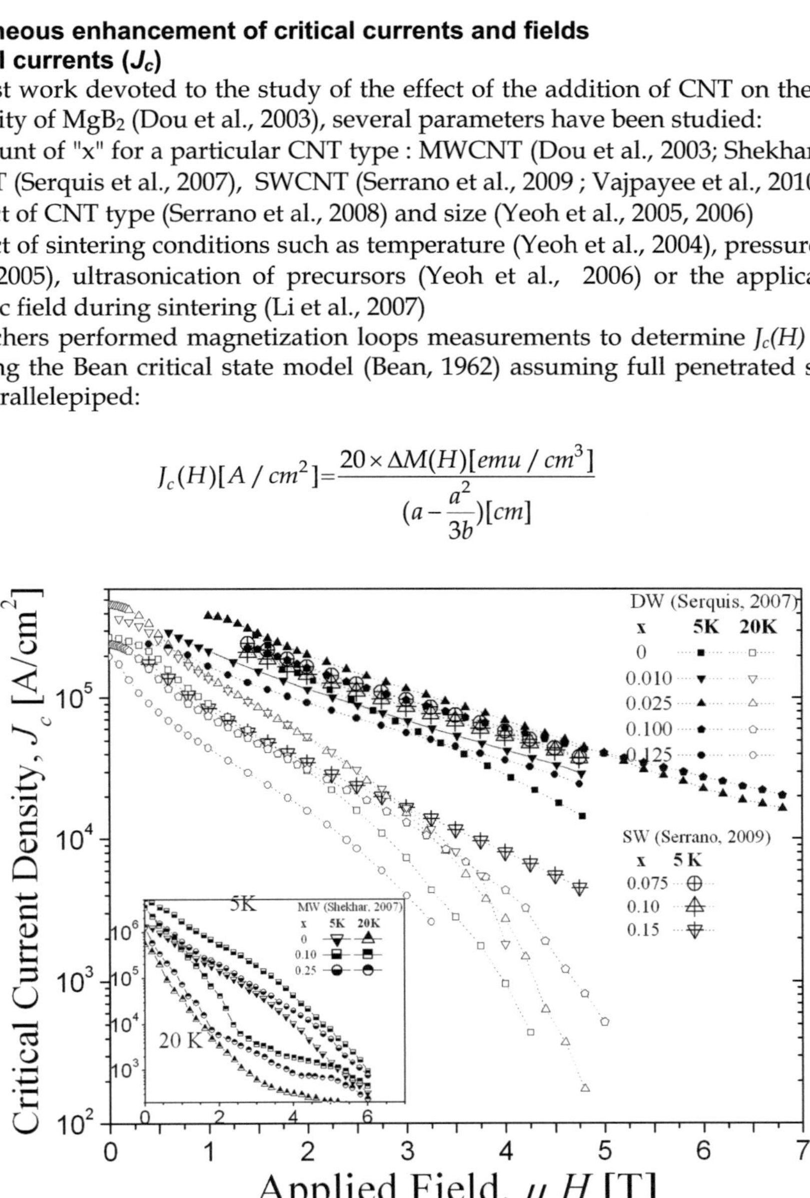


Fig. 3. $J_c$ field dependence determined by magnetization for samples with different amount of SW, DW and MW CNT at 5K and 20 K. Data from MWCNT (Shekhar et al., 2007), DWCNT (Serquis et al., 2007) and SWCNT (Serrano et al., 2009)

where $a$ and $b$ are the lengths of the parallelepiped edges perpendicular to the magnetic field and $\Delta M$ is the width of the magnetization characteristic at the applied magnetic field $H$.
Figure 3 shows the dependence of $J_c$ with the applied magnetic field at 5 and 20 K for samples prepared with different amount of single (SW) (Serrano et al., 2009), double (DW) (Serquis et al., 2007) and multi-walled (MW) (Shekhar et al., 2007) CNT (see inset for the later case). Some flux jumps were frequently reported at 5 K in regions up to 2 T, and these data were not included in the corresponding figures. It can be observed that in all cases the addition of CNT progressively improves the overall $J_c(H)$ performance, reaching an optimum for the x=0.10 sample, where $J_c$ is higher than for any other composition in all the reported temperatures and fields. When the amount of added CNT is over 10%at the performance deteriorates. However, many other authors also reported that the optimum addition is around 10%at, although in some cases the critical current densities are not higher than for other compositions in the whole range of temperatures or applied fields.

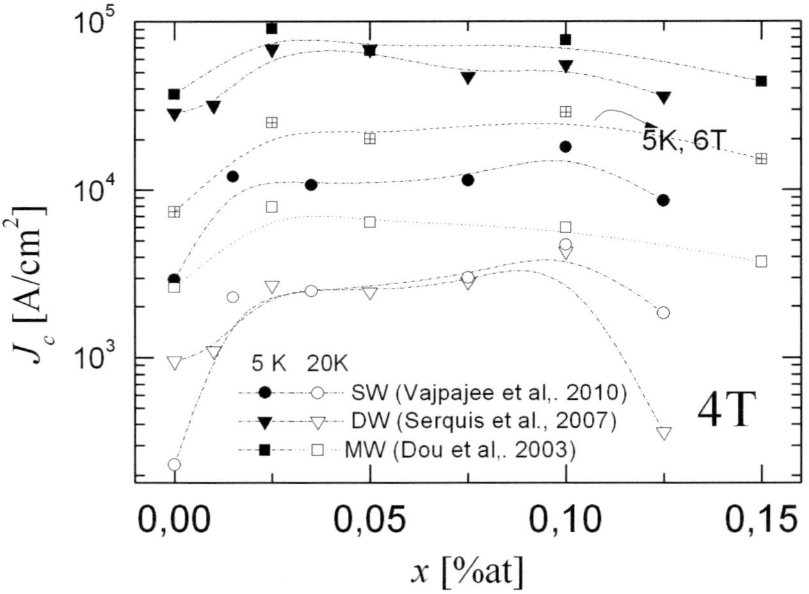


Fig. 4. Critical current densities at 4 T as determined by magnetization as a function of the amount of SW (circles), DW (triangles) and MW (squares) CNT added to $MgB_2$ at 5K (full symbols) and 20 K (open symbols). Data from MWCNT (Dou et al., 2003), DWCNT (Serquis et al., 2007) and SWCNT (Vajpayee et al., 2010)

Figure 4 displays the dependence of $J_c$ with different amount of MWCNT (Dou et al., 2003), DWCNT (Serquis et al., 2007) and SWCNT (Vajpayee et al., 2010) added to $MgB_2$ samples under an applied field of 4 T and temperatures of 5 and 20 K. It is interesting to note that the critical current densities are very similar between x = 0,025 and x = 0,10 for each type of sample, but the $J_c$ improvement tends to be more notorious for x = 0,10 at higher temperatures and fields. The best composition for a certain field and temperature may also change with the sintering temperature. Samples included in Figure 4 were prepared at 800, 900 and 850°C for MW, DW and SW, respectively. A study about the dependence with

sintering temperature (Ts) in $MgB_2$ samples with MW CNT (Yeoh et al., 2004) indicated that when the Ts is 900°C or higher, the nanotubes tend to dissolve and are incorporated into the $MgB_2$ matrix, decreasing $T_c$ but increasing the irreversibility field ($H_{irr}$). This effect will be discussed in the next subsection, since it is correlated with the $H_{c2}$ improvement. Therefore, the larger the amount of C-doping the smaller is the $J_c(H)$ slope improving the performance at higher fields. On the contrary, the $J_c$ decrease for x > 0,10 may be due to both an even larger decrease in $T_c$ and a probable deterioration of interconnectivity between grains. This deterioration of grain connectivity was denoted by a large normal resistivity value (i.e. ρ(40K) ~ 200 μΩcm) for the DW sample with x = 0,125.
Figure 5 shows the dependence of $J_c$ with the applied magnetic field of samples prepared with the same amount of single (SW), double (DW) and multi-walled (MW) CNT.

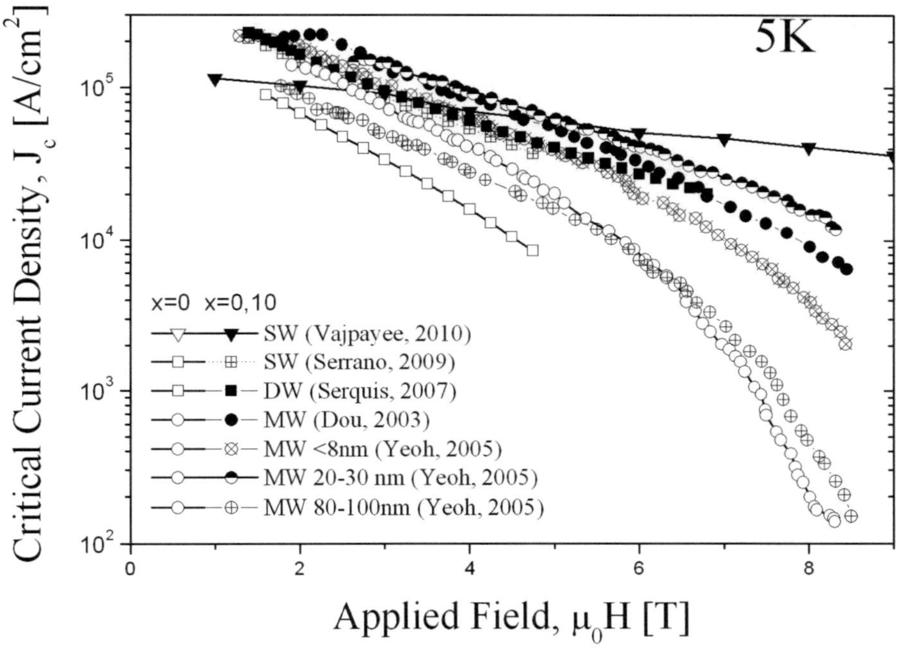


Fig. 5. $J_c$ field dependence determined by magnetization at 5K and 20 K for samples with the same nominal composition (x=0,10) but different CNT types (SW, DW and MW of different sizes). The corresponding pure samples (x = 0) are included as references (open symbols)

For comparing the absolute values it is important to take into account the pure MgB2 sample used as reference since this can be a good parameter to check the influence of variables other than composition, such as connectivity, that may result from the sintering process. Consequently, the best performances (i.e. Jc values and Jc(H) dependences) are obtained for MW (diameter 20-30 nm) (Yeoh et al., 2005) and DW (diameter 1.3–5 nm) (Serquis et al., 2007). These two samples (both heat treated in flowing high purity Ar at 900 for 30 minutes) presented the largest C-doping (smallest *a* lattice parameter) and Yeoh et al. suggested that it is not the diameter but the MWCNT length the responsible for allowing a more homogeneous C-incorporation by avoiding the nanotubes agglomeration. This explanation may not apply to DW CNTs, which have a length ≤ 50 μm. However, it is highly probable that synthesis parameters that promote a more homogeneous CNT distribution and improve connectivity (avoiding agglomerates at grain boundaries) are the key for Jc performance.

### 2.2.2 Upper critical fields ($H_{c2}$)

The $H_{c2}(T)$ dependences are usually determined from four probe transport measurements. Since the involved fields are frequently very high for standard laboratory test equipments, several groups performed measurements at large High Magnetic Field facilities. As an example, measurements of DW and SW CNT samples were performed in the mid-pulse magnet of NHMFL-LANL, capable of generating an asymmetric field pulse up to 50 T. Figure 6 exhibits the temperature dependences of $H_{c2}$ and $H_{irr}$, defined as the onset (extrapolation of the maximum slope up to the normal state resistivity) and the beginning of the dissipation, respectively, of the $R$ versus $H$ data for samples with several DWCNT contents, at temperatures between 1.4 and 34 K. The criteria to define $H_{c2}$ and $H_{irr}$ are exemplified in the inset of figure 6.

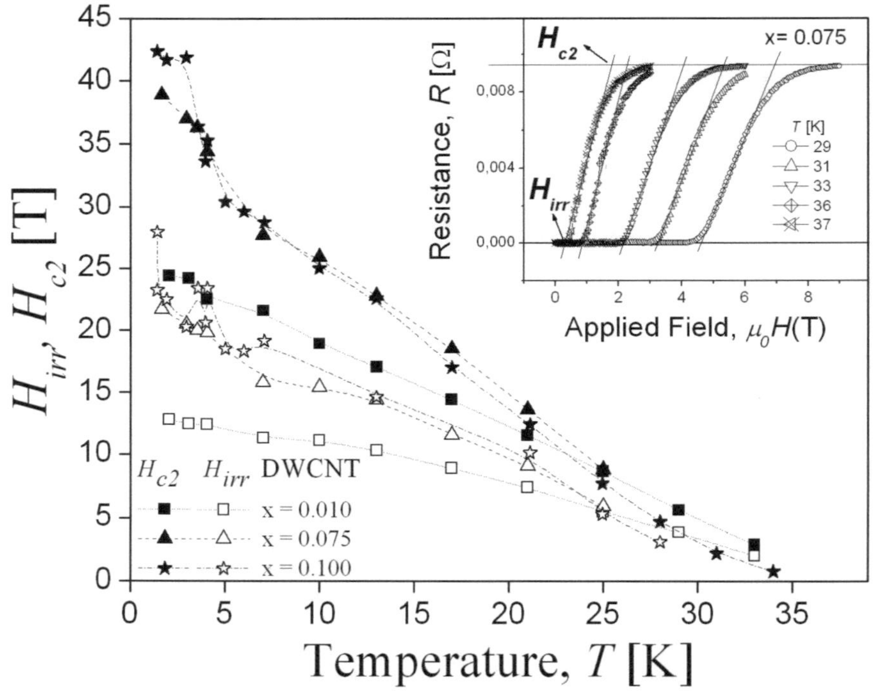


Fig. 6. Transport measurements of the upper critical field ($H_{c2}$) and the beginning of the dissipation ($H_{irr}$) as a function of temperature for samples with DWCNT contents of 0.01, 0.075 and 0.10. The inset shows some of $R(H)$ curves to exemplify the criteria used to determine $H_{c2}$ and $H_{irr}$.

In all superconducting materials, an $H_{c2}$ enhancement is expected when disorder is increased (e.g., by doping). The reason is that the reduction in the electronic mean free path produces a decrease of the effective coherence length, driving the superconductor to the "dirty limit". This effect is especially large in C-doped $MgB_2$, and additionally an anomalous $H_{c2}(T)$ upward curvature is also noted in the present data and reported many times. This positive curvature is related to the existence of two-gaps in $MgB_2$, (π-band and σ-band) and predicted by some theoretical models.

These models for a two gap superconductor in the dirty limit (Golubov et al., 2002; Gurevich, 2003) consider that nonmagnetic impurities affect the *intra*band electron diffusivities $D_\sigma$ and $D_\pi$, and the *inter*band scattering rates $\Gamma_{\pi\sigma}$ and $\Gamma_{\sigma\pi}$. An implicit $H_{c2}$ dependence on temperature can be obtained from the Usadel equations, derived taking into account only the *intra*band effect.

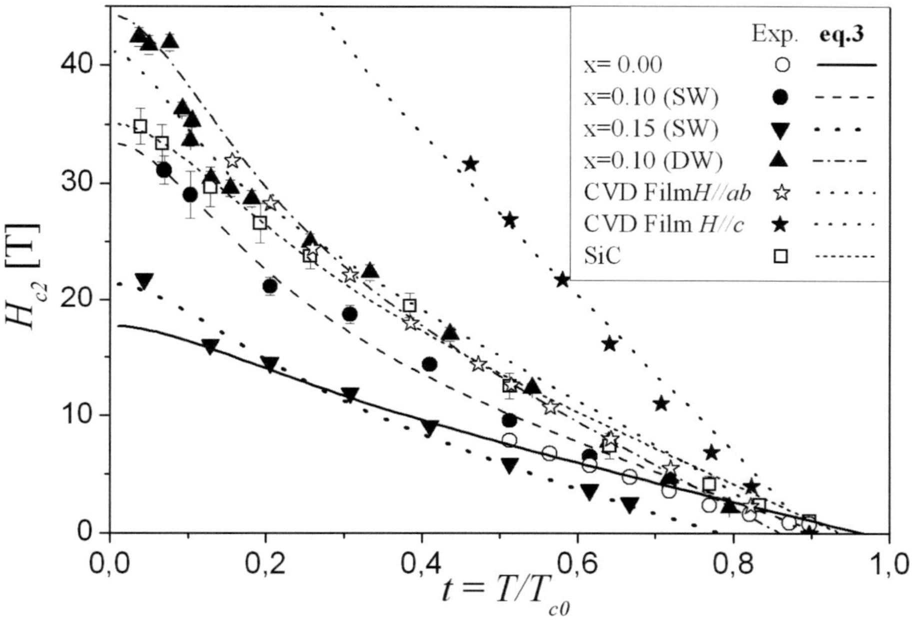


Fig. 7. Transport measurements of the upper critical field ($H_{c2}$) determined from the $R(H)$ curves as a function of the reduced temperature $t = T/T_{c0}$, for selected samples with and without CNT addition and fit to data using equation 2 (lines). Other C-doped samples are included for comparison (Serrano et al., 2008; Braccini et al., 2005)

$$a_0\left[\ln t + U(h)\right]\left[\ln t + U(\eta h)\right] + a_2\left[\ln t + U(\eta h)\right] + a_1\left[\ln t + U(h)\right] = 0 \quad (2)$$

where, $t = T / T_{c0}$ is the reduced temperature[1],

$$\eta = D_\pi / D_\sigma,$$

$$U(x) = \psi(1/2 + x) - \psi(1/2),$$

[1] In Refs. (Gurevich, 2003; Serquis et al., 2007 ; Serrano et al., 2008)) $T_{c0} = T_c(g=0)$ is defined, where $g$ the interband scattering parameter. This value corresponds to the clean limit $MgB_2$ sample.

$$h = H_{c2} D_\sigma / 2\phi_0 T\,,$$

$$a_0 = 0.7806$$
$$a_1 = 1.93$$
$$a_2 = 0.07$$

$a_0$, $a_1$, and $a_2$ were determined in ref. (Gurevich, 2003).
Another equation was derived when taking also into account the *inter*band effects:

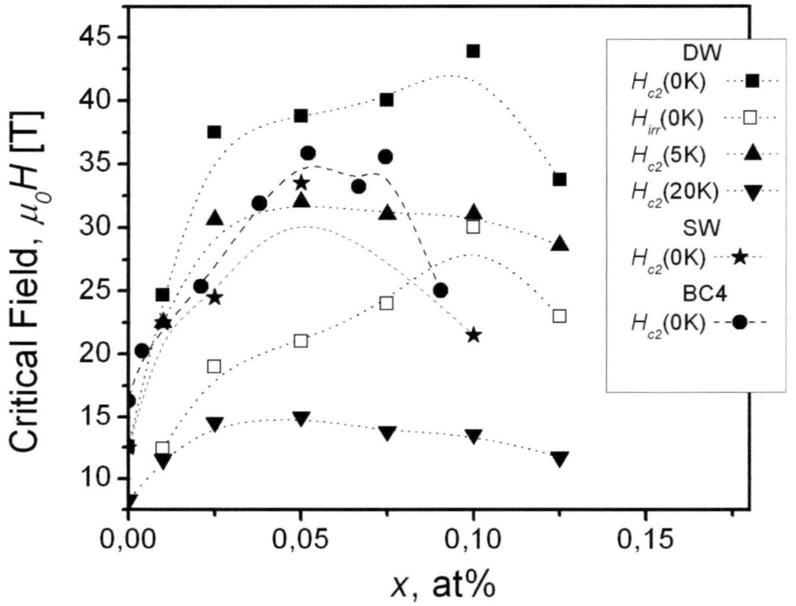


Fig. 8. The extrapolation at 0K of $H_{c2}$ as a function of carbon content $x$ for DW and SW CNT samples (full symbols). $H_{c2}$ data at 5K , 20K and $H_{irr}$(5K) is also included for DW sample (open symbols). Lines are only guides to the eyes.

$$2w(\ln t + U_+)(\ln t + U_-) + (\lambda_0 + \lambda_i)(\ln t + U_+) + (\lambda_0 - \lambda_i)(\ln t + U_-) = 0 \tag{3}$$

where $\lambda_i = (\lambda_- \Gamma_- - 2\lambda_{\pi\sigma}\Gamma_{\sigma\pi} - 2\lambda_{\sigma\pi}\Gamma_{\pi\sigma})/\Gamma_+$, $U_\pm = U_\pm(T, H_{c2}, D_\sigma, D_\pi, \Gamma_{\pi\sigma}, \Gamma_{\sigma\pi})$,
$w$, $\lambda_0$ are constants that depend on $\lambda_{mn}$ ($m,n$ = π,σ) (values obtained from *ab initio* calculations) (Golubov et al., 2002; Braccini et al., 2005).
In this case we optimized the diffusivity ratio $\eta = D_\pi / D_\sigma$ and *inter*band scattering parameter to fit the measurements using the equation (3).
Figure 7 shows the dependence of $H_{c2}$ as a function of the reduced temperature $t = T/T_{c0}$, where $T_{c0}$ = 39 K, of some selected samples with and without CNT addition. Other C-doped samples are included for comparison (Serrano et al., 2008; Braccini et al., 2005). The lines are fits to the data with the model proposed using the fitting parameters described above.

The upward curvature signaled as a characteristic of the presence of two gaps is apparent in these $H_{c2}(T)$ data.
Figure 8 displays the $H_{c2}(0)$ extrapolations as a function of $x$ for DW (Serquis et al. 2007) and SW (Serrano et al. 2007) in comparison with data for other C-doped samples (Wilke et al., 2004). We observe that $H_{c2}(0)$ increases with $x$ and has a maximum for 10 at% for DW and 5 at% for SW. For a larger $x$ values a decrease in $H_{c2}$ was found, probably due to a larger $T_c$ decrease and an increase in the resistivity of the samples. Earlier $MgB_2$ carbon doped data (Wilke et al. 2005) also indicates an initial rapid rise for lower C contents that then slows down, reaches a maximum at intermediate carbon concentrations, and decreases for larger C contents and the same behavior was reported for $MgB_2$ single crystals.
The CNT additions produce a larger C incorporation than SiC, probably because of the higher synthesis temperature, resulting in samples with lower $T_c$. The DWCNT 10 at% sample has the highest C content into the lattice, indicating that using this kind of inclusions is an easier way to incorporate C. This allows to reach a record $H_{c2}$ value for this sample. Earlier $MgB_2$ carbon doped data from Wilke *et al* (Wilke et al., 2005b) also indicates an initial rapid rise for lower C contents that then slows down, reaches a maximum at intermediate carbon concentrations, and decreases for larger C contents and the same behavior was reported for $MgB_2$ single crystals. The $H_{c2}(0)$ of CNTdw10 is close to the maximum $H_{c2}$ value as a function of $x$. A decrease in $H_{c2}(0)$ was also observed for a larger $x$ value (see Fig.8)), in agreement with other reported data (Senkowicz et al., 2007).

## 3. Effect of different kind of CNT addition on the superconducting properties of $MgB_2$ PIT wires and tapes

The standard and low-cost fabrication powder in-tube (PIT) method involves filling a metallic tube with superconducting powder (ex-situ) or precursors (in-situ) and drawing it into a wire and/or rolling into a tape (Flükiger et al., 2004).
$MgB_2$ crystallizes in the hexagonal $AlB_2$ type structure (space group *P6 /mmm*), and the anisotropic structure has given the motivation to investigate formation of texture by different deformation processes. A key issue on which there is no agreement in the literature is the optimization of the heat treatment parameters. Although some post-annealing appears to be necessary to achieve higher $J_c$, some authors reported detrimental effects of heat treatments in the performance of $MgB_2$ wires or tapes (Serquis et al., 2003; Goldacker, 2003; Civale et al., 2003). However, most results obtained in $MgB_2$ wires and tapes prepared by PIT using different kind of CNT and treated at different temperatures using several sheath materials (e.g. iron, stainless steel, Nb/Fe) indicate that the final sintering temperature is very important to improve the superconducting properties allowing carbon to be incorporated in $MgB_2$.
Only one work (Xu et al., 2007) studied the influence of CNT amount, reporting the effect of the "doping level" (x) in the field dependence of critical current density for MWCNT Fe-sheathed $MgB_2$ wires and tapes. Similarly to what was reported for bulk $MgB_2$ samples, they found that there is an optimum composition for all fields and temperatures studied, but the best composition for the nominal $Mg(B_{1-x}C_x)_2$ was x=0.05 and $J_c$ decreased for x = 0.10 (see Figure 9). However, many other researchers focused in which was considered the best composition x=0.10 studying the influence of other parameters:

- the effect of CNT type (Kováč et al., 2007) in comparison with other C-compounds (SiC, graphite)

- the effect of MWCNT size and sintering temperature (Kim et al., 2006a, 2006b, 2006c)
- the effect of sintering temperature when using SW CNT (Kim et al., 2007)
- the possibility of MW CNT alignment by mechanical drawing in the PIT process (Dou et al., 2006).

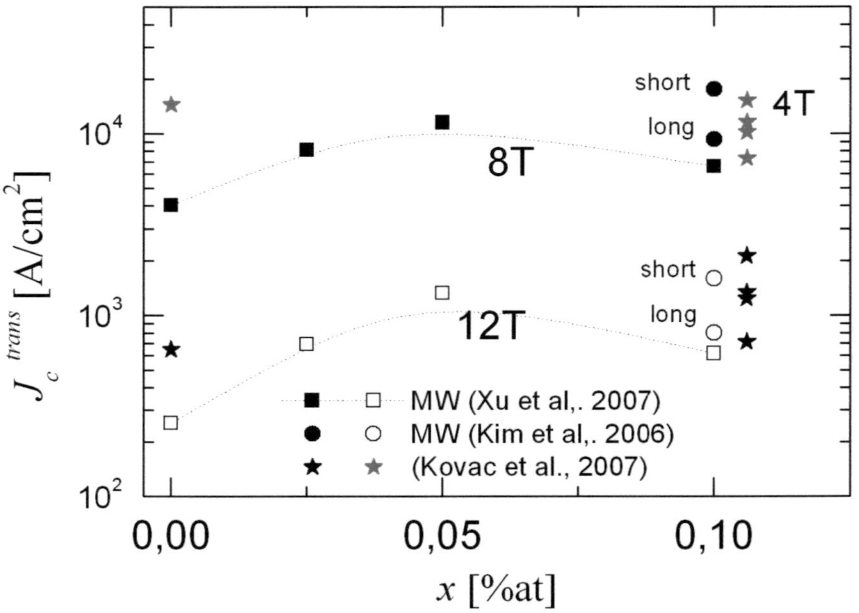


Fig. 9. Critical current density as a function of MW CNT content measured by transport in Fe-sheath $Mg(B_{1-x}C_x)_2$ wires at 8 T (black symbols) and 12 T (open symbols) (Xu et al., 2007). For comparison were included data from Fe/Nb sheath tapes with 5%wt addition of different CNT types (Kováč et al., 2007) and short and long MW CNT $MgB_2$ wires (Kim et al., 2006b)

To illustrate the effect of some of the mentioned parameters we plotted in Figure 10 the critical current density of

- Left side: $MgB_2$ /Fe-Nb tapes with 5%wt addition of different CNT types (Kováč et al., 2007). SW: Arc discharge single wall nanotubes; SW-D: Purified arc discharge single wall nanotubes, Dry-mixed; SW-W: Purified arc discharge single wall nanotubes, Wet-mixed; MW: Multi wall carbon nanotubes (~60 nm × 1 μm). For comparison were included the pure (0) and a $MgB_2$ tapes with 5%wt addition of SiC. This last addition is more effective in the whole field range in agreement with bulk results, while all CNT types only enhance $J_c$ for high fields. The normalized current densities (shown in Figure 11) allow better comparison between the slopes of $J_c(\mu_0 H)$, indicating a positive effect for all additions. This field dependence enhancement is ascribed to the effect of C-incorporation, which increases the upper critical field. The decrease of $T_c$ corresponding to a higher C-doping effect can be observed in the inset of Figure 9.
- Right side: $MgB_2$/Fe wires with aligned MW CNT measured along (a) and perpendicular (c) to the wire (Dou et al., 2006). While the pure sample has little differences between both measurement field directions, there is a clear difference in

samples with MW addition, which increased with field. This is a good indication that CNTs could be aligned along the longitudinal axis of the $MgB_2$ wires by mechanical drawing in the PIT process.

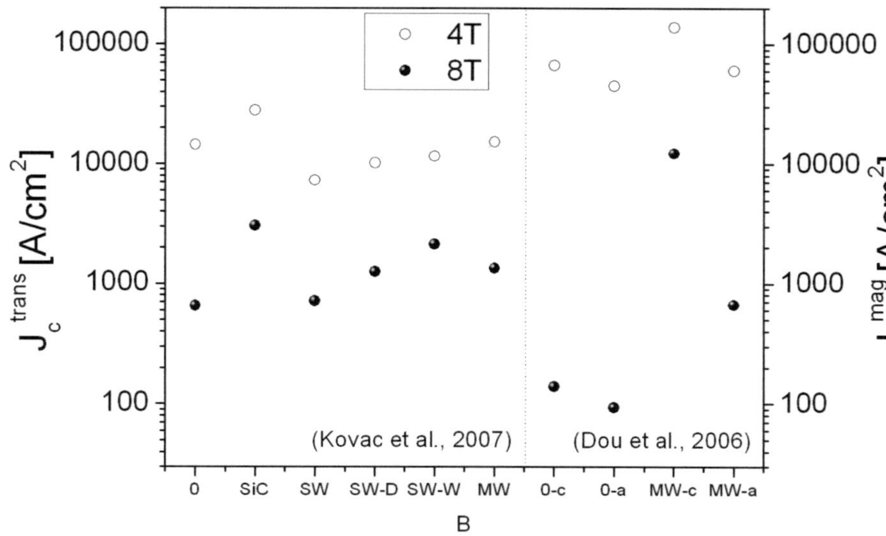


Fig. 10. (Left) Critical current density measured by transport in $MgB_2$ /Fe-Nb sheath tapes with 5%wt addition of different CNT types (Kováč et al., 2007) at 8 T (black symbols) and 4 T (open symbols). (Right) Critical current density measured by magnetization in Fe sheath $MgB_2$ wires with aligned MW CNT measured along (a) and perpendicular (c) to the wire (Dou et al., 2006)

Matsumoto (Matsumoto et al., 2006) reported that $J_c$ in SiC-alloyed $MgB_2$ tapes depends on a complex relation between grain connectivity, $H_{c2}$, and flux pinning induced by grain boundaries and precipitates. However, the distinct effect of C incorporation through different routes in $J_c$ and $H_{c2}$ is still not entirely understood.

Matsumoto *et al* reported a similar $H_{c2}(0)$ to our record value for a SiC-doped $MgB_2$ tape prepared by the PIT method. These results correspond to a PIT tape sample that probably has some texturing and the reported $H_{c2}$ data was measured with the applied field parallel to the tape. Since we measured randomly oriented polycrystalline bulk samples, our results cannot be easily compared with Matsumoto's.

## 4. Effect of CNT addition on the magnetic relaxation of $MgB_2$

### 4.1 Introduction

In Section 2.2 we have shown that the main effect of DW CNTs addition is the simultaneous increase in the critical current density and the upper critical field $H_{c2}$. However, it is not clear if these two effects are connected. In fact, the vortex physics underlying this performance improvement has not been so far understood. In this section, we present a very recent work (Pasquini et al., 2011), where the role of CNT addition is investigated by studying the magnetic relaxation in bulk $MgB_2$ samples.

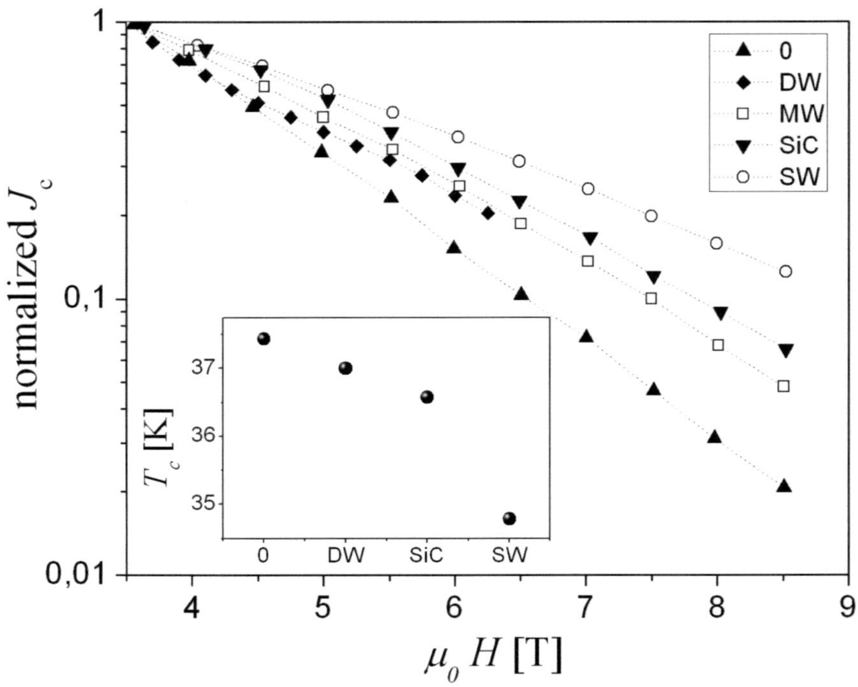


Fig. 11. Normalized critical current densities $J_c(\mu_0 H)/J_c(3.5\ T)$ showing the difference in slope for a pure sample and $MgB_2$ tapes with 5%wt addition of different CNT types (Kováč et al., 2007), including a DW with similar content (Serquis et al., 2011). The inset shows the $T_c$ of the same tapes.

As was mentioned in Section 2.1.1, if a superconducting sample is fully penetrated by the magnetic field B, the magnetization *m* is proportional to the density current *J* flowing in the sample that, in the ideal critical state approximation is the critical current density $J_c$. However, due to the thermal activated motion, this "measurable" critical current is lower than the "true" $J_{c0}$. The main quantity governing the creep of vortices is the activation barrier U(J,T,B). In a magnetic relaxation experiment both the magnetization *m* and $J_c$ decay in time as

$$-\frac{\partial J_c}{\partial t} = C_j e^{\frac{-U(J_c,T,B)}{kT}}$$

and then

$$-\frac{\partial m}{\partial t} = C_m e^{\frac{-U(j_{(m)},T,B)}{kT}} \tag{4}$$

where $C_c \sim J_c/\tau_0$ or $C_m \sim m/\tau_0$ can be approximated as constants.
The activation barrier is expected to be described by the general expression (Blatter et al.,1994; Geshkenbein et al., 1989)

$$U(J;T;B) = \frac{U_c(T,B)}{\mu}[(\frac{J_{c0}(T,B)}{J})^{\mu} - 1] \quad (5)$$

where $U_c(T,B)$ is the pinning energy and $\mu$ is a critical exponent characteristic of the particular creep regime.

In High Temperature Superconductors (HTSs), the creep drastically reduces the measured current density, so $J_c << J_{c0}$. A large number of glassy creep regimes in this limit have been proposed theoretically (Blatter et al., 1994; Yeshurum et al., 1996) and many of them observed experimentally (Thompsom et al., 1994, 1997), where $U(J,T,B) \cong g(T,B)J^{-\mu}$.

In those cases the temperature dependence of the measurable $J_c$ is dominated by creep while the $J_{c0}(T)$ dependence is negligible. On the other hand, in traditional type II superconductors the current decay is very small, so $(J_{c0}-J_c) << J_{c0}$, therefore equation (5) take the traditional Anderson-Kim (A-K) linear dependence (Anderson et al., 1964)

$$U(J,T,B) \cong U_c(T,B)(1 - \frac{J}{J_{c0}(T,B)}) \quad (6)$$

In these very low creep superconductors an experimental confirmation of glassiness, which regardless of the specific approach always involves detecting tiny deviations from the Anderson-Kim model, is extremely challenging.

From this basic point of view, $MgB_2$ is a very particular system, as the intermediate $T_c$ and moderate anisotropy makes the creep effects smaller than in HTS but larger than in conventional superconductors. The influence of thermal fluctuations in the vortex physics is measured by the Ginzburg number $G_i=(1/2)(kT_c/H_c^2\xi^3\gamma^{-1})^2$. For $YBa_2Cu_3O_{7-\delta}$ this is as large as $G_i \sim 10^{-2}$, and even larger for the more anisotropic Bi-based compounds, while for NbTi, the paradigmatic strong-pinning conventional superconductor, $G_i \sim 10^{-8}$. For $MgB_2$, depending on the doping level we have $G_i \sim 10^{-4}$-$10^{-5}$. This is just in the middle between the extreme cases. The first consequence is that in magnetization measurements, the A-K limit is not necessary valid in the whole temperature range, and the measured $J_c$ is expected to be in the intermediate range $(J_{c0}-J_c) \lesssim J_c$.

In the rest of the section we present a creep study in bulk $MgB_2$ samples as-grown and doped with different doses of carbon nanotubes. In subsection 4.2 experimental relaxation rates are presented, and the general behaviour is described. In subsection 4.3 we review some fundamental concepts and formulas concerning creep rates and activation energies to perform in subsection 4.4 a careful analysis of results that allows us to identify the region where the A-K model is valid. In 4.5, the pinning energies $U_c$, true critical current densities $J_{c0}$, and correlation volumes $V_c$ are estimated and compared. Conclusions are summarized in 4.6.

### 4.2 Relaxation rates

Samples used in this study were prepared by solid-state reaction with magnesium (-325 mesh, 99%) and amorphous boron (99%) as starting materials (Serrano et al., 2008). The powders were ground inside a glove box and pressed under 500 MPa into small pellets with dimensions of 6 mm in diameter and 4 mm in thickness, wrapped together with extra 20%at Mg turnings (99:98% Puratronic) in Ta foil and then placed in an alumina crucible inside a tube furnace in flowing Ar=$H_2$ at 900C for 30 min.

The relaxation measurements were carried out in a Quantum Design model MPMS XL 7T SQUID based magnetometer. Time-dependent data were taken with a protocol similar to that described in ref. (Civale et al., 1996). A scan length of 3 cm was used in order to minimize the effects due to the non-uniformity of the applied magnetic field that was applied parallel to the longest axis of the sample. For each relaxation measurement, the samples were first cooled and stabilized at the measurement temperature. Then the field was first raised up to 6T and then lowered to the measuring field to assure that the sample was fully penetrated. Intermediate measurements were performed in the upper and lower magnetization branches to subtract the reversible magnetization. We checked for and ruled out any effects due to the magnet self relaxations ($H$ variations during the measurement time) that could lead to spurious changes in the magnetization of the samples.
The experimental critical current density $J_c(t)$ has been calculated from the measured magnetic moment using equation (1), as in previous sections.
As reported in section 2.2., the response for samples with DWCNT additions between 1% and 10% at temperature far below $T_c$ is very similar and the "measured" critical current density $J_c$ is optimal in this range of doping. This fact is illustrated in Figure 3, where $J_c$ obtained from typical magnetization loops in samples with different doses of DWCNT is plotted as a function of the applied magnetic field H at T = 5K and 20K . The response for samples with 1 at% and 10 at% addition of DWCNT is very similar. At low temperature (5 K), CNT addition enhances $J_c$ more than twice for all the measured range of $H$. On the other hand, at high temperature (20 K) the response of pure and CNT samples is quite similar but CNT addition continues being efficient at fields higher than $H$ = 2T.
The magnetic relaxation has been measured in pure $MgB_2$ samples and samples with DWCNT doses between 1 at% and 10 at% in the range between 5K and 25K. The corresponding relaxation rate $S=-d(\ln(J))/d(\ln(t))$ has been obtained as the slope of the $\ln J$ vs $\ln t$ graph. Results for $H$ =1T (full symbols) and 3T (open symbols) are shown in Figure 12.
Samples with doses of 2.5% and 5% of carbon nanotubes have also been measured and display similar results, not shown in the figure for clarity.
As was described in the introduction, the $S$ values are intermediate between those measured in low and high $T_c$ materials. With this moderate relaxation, the temperature dependence of $J_c(t)$ (Figure 3) cannot be explained assuming a $T$ independent $J_{c0}$ at low temperature. An intrinsic $J_{c0}(T)$ dependence must to be taken into account.
Again, at $H$ = 1 T the decay in time is quite similar in all the doped samples with doses between 1% and 10%. However, same unexpected results appear. The $S(T)$ curve at $H$=3T is similar for pure and CNT samples in all the temperature range. Furthermore, at $H$=1T the pure sample that has a lower $J_c$ (associated with a lower pinning) has also a lower relaxation (associated with a higher pinning energy $U_c$).
We have proceeded to analyze the relaxation data in a pure and a 10 at% CNT samples to get an understanding of the causes of this puzzle. The procedure and results are described in the subsections 4.3 and 4.4.

### 4.3 Creep and activation energies in the Anderson-Kim approximation

In the general case (Blatter et al., 1994), the relaxation rate $S$ is related with the pinning energy by

$$\frac{1}{S} = \frac{U_c(T,B)}{kT} + \mu \ln(1 + t/t_0) \sim \frac{U_c(T,B)}{kT} + \mu \ln(t/t_0) \qquad (7)$$

if $t >> t_0$. This expression should be valid in all the current range, including the A-K limit. Therefore, if $U_c$ is $T$ independent at low temperatures, a linear relationship between *1/S* and *1/T* is expected.

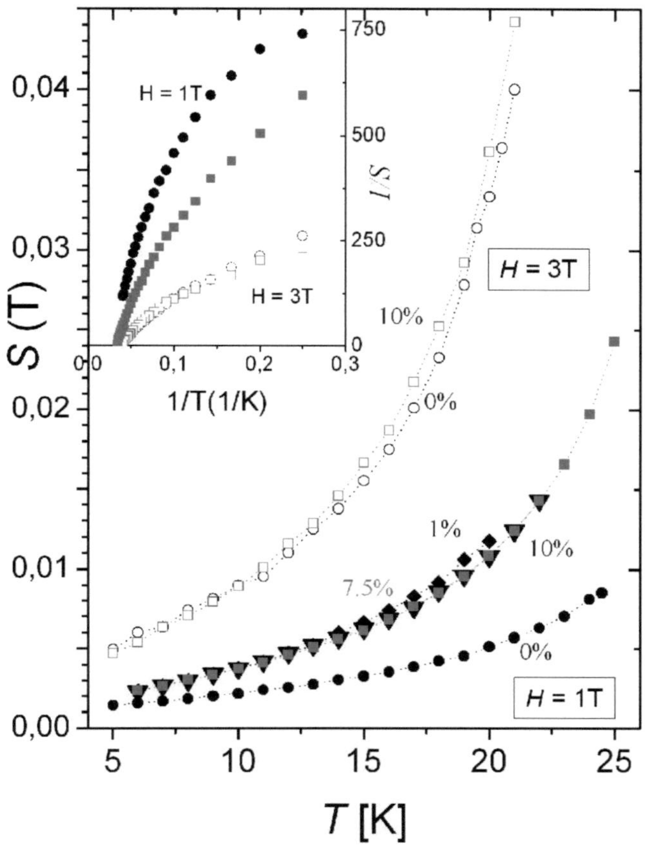


Fig. 12. Relaxation rate $S$ as a function of temperature for a pure $MgB_2$ sample (circles) and for samples with different doses of DWCNT: 1% (diamonds), 7.5% (triangles) and 10%(squares) at $H = 1$ T (full symbols) and $H = 3$T (open symbols). In the inset, the plot of *1/S* vs. *1/T* reveals that it is not possible to assume a temperature independent pinning energy at low temperature in all the cases.

As was mentioned in the introduction, the A-K approximation assumes $(J_{c0} - J) << J_{c0}$ or $(J / J_{c0}) \sim 1$ , that lets to the linear dependence of eq. (6) between the activation barrier $U$ and the density current $J$. The integration of eq. (4) in this limit gives the well known logarithmic decay of the experimental critical current density with time.

$$\frac{J_c(t)}{J_{co}(T,B)} \sim [1 - \frac{kT}{U_c(B,T)} \ln(t/t_o)] \tag{8}$$

and allows to estimate $U_c(T,B) / kT$ under the supposition that $J \sim J_{c0}$.

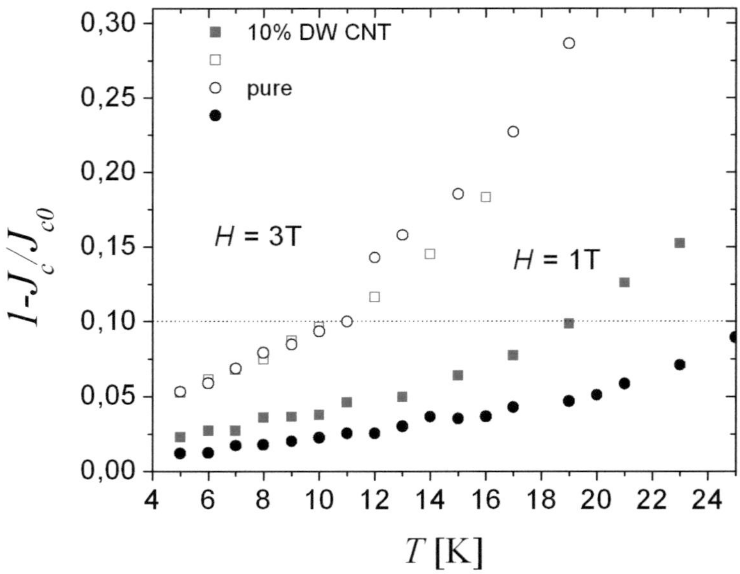


Fig. 13. Ratio $\Delta J/J_{c0}$ as a function of temperature for pure (circles) and 10% CNT (squares) samples at $H$ = 1T (full symbols) and $H$=3T (open symbols). $\Delta J = J_{c0} - J_c$ is the difference between the estimated "true" critical current density (see text) and the measured critical current density. A criterion of 10% (doted line) has been taken to validate the A-K approximation $\Delta J << J_{c0}$.

An alternative method combines eqs. (4) and (6) to obtain a linear dependence between $J$ and $\partial J/\partial t$ that, under the condition $\ln C_J << U_c(T,B)/kT$ results in (Pasquini et al., 2011)

$$J_c(t) = \frac{kT\, J_{co}(T,B)}{U_c(t,B)} \ln\left|\frac{dJ}{dt}\right| + J_{co}(T,B) \tag{9}$$

Due to the numerical differentiation, this method has a greater error in the calculation of the pinning energy, but allows estimating $J_{c0}(B,T)$.
A mayor difficulty to directly decide the validity of the A-K approximation from the relaxation data at a single temperature is the extremely large time needed to reliably determine the linear relationship (8) (i.e. the logarithmic decay with time of the current density). However, the resulting fitting parameter $J_{c0}(B,T)$ must be consistent with the A-K assumption $(J_c / J_{c0}) \sim 1$ and the resulting fitting parameter $U_c(T,B)$ must be consistent with eq. (7). If these two conditions are not fulfilled then some of the assumptions was wrong and we can conclude that the AK description is not valid.
In the following section, we apply the above procedure to $MgB_2$ relaxation data, determining the temperature region where results are consistent.

#### 4.4 Data analysis in $MgB_2$ samples

As can be observed in the inset of Figure 12, it is not possible to identify in all the cases a low T region with a linear relationship between 1/S and 1/T, characteristic of a temperature independent pinning energy. For this reason, experimental data where separately analysed

at each temperature **assuming** a linear dependence between *U* and *J*, i.e. the validity of the A-K approximation.
At each temperature $J_c(t)$ has been plotted has a function of *ln t* and $ln\ (-\partial J/\partial t)$. The fitting parameter $A(T) = (kTJ_{c0}(T))/U_c(T))$ has been extracted from the slope using equations (8) and (9), and $J_{c0}(T)$ from the intercept using eq. (9) (Pasquini et al., 2011).
In Figure 13 the estimated true critical current density $J_{c0}$ is compared with the measured critical current density $J_c$ (the first point measured in the $J_c(t)$ relaxation) by plotting $(J_{c0}-J)<<J_{c0}$ as a function of temperature for a pure and a 10%doped sample, at *H*=1 T and 3T.
Taking a 10% criterion to determine the limit of the A-K condition $(J_{c0}-J)<<J_{c0}$, the resulting $J_{c0}(t)$ is consistent below the temperature where each experimental curve crosses the doted horizontal line.
In figure 14 the resulting pinning energies obtained from the slope of eq. (8) (squares) and eq. (9)(circles) are plotted as a function of temperature and compared with *T/S* (asterisks) for a pure and a 10%doped sample, at *H*=1 T and 3T. Results have been restricted to the consistent T region obtained by the above analysis.

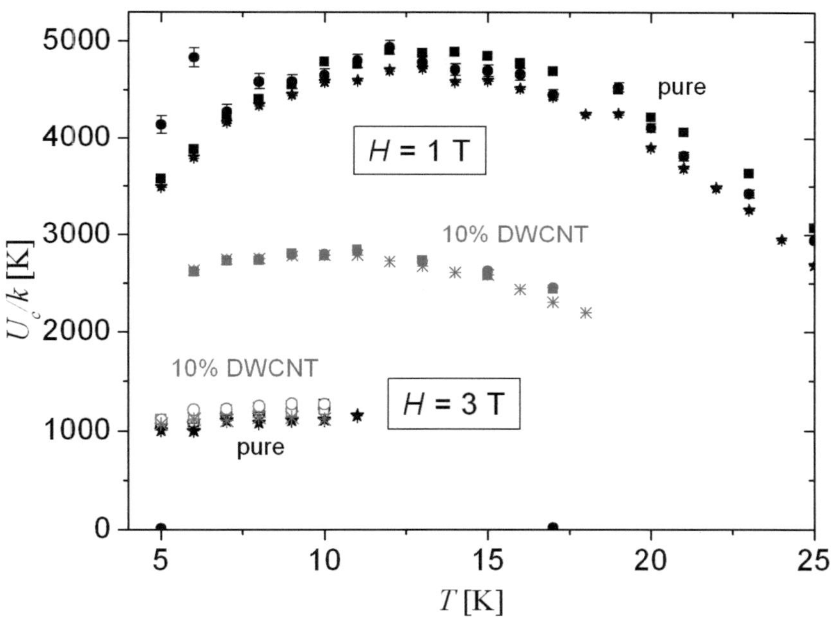


Fig. 14. Estimated pinning energies $U_c/k$ as a function of temperature for pure (black symbols) and 10% doped (gray symbols) samples at *H* = 1T (full symbols) and *H* = 3T (open symbols) from the slope of eq. (8) (circles) and eq. (9) (squares). Asterisks indicate the calculated *T/S* in each case. Results shown are limited to the region where the A-K approximation is consistent.

As can be observed (with exception of the lower temperatures data in the undoped sample at *H*= 1 T) pinning energies obtained from the three methods are very similar, and this validates the fitting procedure, including the estimation of $J_{c0}(T)$ from eq. (9).
Observing the resulting *T* dependence in Figure 14 it seems that, in most of the cases, pinning energies remain nearly constant at low temperatures. The unexpected drop in $U_c$ at

the lower temperatures data in the undoped sample at $H$= 1 T, together with the inconsistence of the analyzed data, could imply the presence of another relaxation mechanism, perhaps associated with macroscopic flux jumps. In the data measured at $H$=1 T a drop at a higher temperature far below the irreversibility line is also observed.
On the other hand, a continuous decrease occurs in the critical current density. This can be observed in figure 15, where the true critical current density is plotted in the studied range for the same samples.
The previous analysis excludes the creep study in the high temperature region using the A-K approximation. Successful procedures have been developed to analyze creep data in high $T_c$ superconductors (Maley et al., 1990; Civale et al., 1996) in the limit $J_c << J_{c0}$, but they are not suitable for intermediate creep regime. However, in the present subsection we showed that, from the present study at low temperatures, a good insight about the role of doping can be obtained.

### 4.5 Discussion

Consistent with the relaxation rates, at $H$=1 T there is a clear decrease of $U_c$ after CNT addition, whereas at $H$=3 T the doping has not an evident effect in the pinning energies.
The parameter $U_c$ is associated with the pinning energy of a pinning volume that, in a single vortex pinning regime, is defined by the disorder parameter $\gamma(T)$ and the superconducting coherence length $\xi(T)$ (Blatter et al., 1994). In section 2.2 we have shown that DWCNT addition enhances $H_{c2}(T)$ by doping the B site, i.e. reduces $\xi(T)$. On the other hand, it is expected that doping increases disorder. Therefore, these two consequences of DWCNT doping will compete and define the effect in the pinning energies and critical currents.

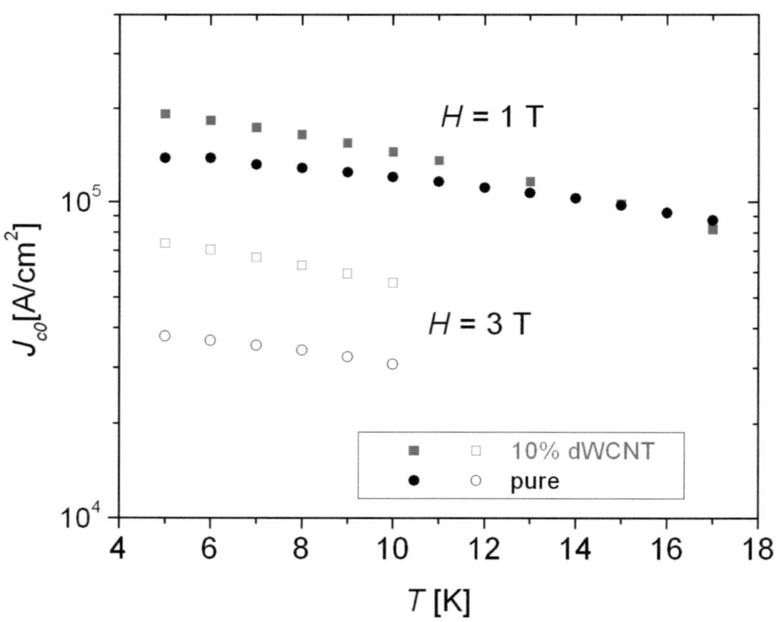


Fig. 15. Estimated critical current density $J_{c0}$ as a function of temperature for pure (circles) and 10% CNT (squares) samples at $H$ = 1T (full symbols) and $H$=3T (open symbols).

The fact that the doping effect in the pinning energies is field dependent is a clear indication of a collective pinning regime.
In a collective regime, the pinning volume is determined by the competition between elastic and pinning energies and each volume $V_c$ of the vortex system is collectively pinned with energy $U_c$. The Lorentz force over a volume $V_c$ is $F_L \sim (1/c)BJ\ V_c$ and, in a rough estimation, the pinning force is $F_p \sim U_c/\xi$. When $J$ reaches the critical current density $J_{c0}$, the pinning and Lorentz forces are balanced and therefore the following relationship is obtained:

$$V_c(T,B) \sim \frac{c\,U_c(B,T)}{B\,\xi(T) J_{co}(B,T)} \tag{10}$$

In section 2.2, we have presented $H_{c2}$ measurements in pure and DWCNT $MgB_2$ samples and we have found (Serquis et al., 2007 ) that $H_{c2}(T)$ fits very well the function proposed in ref (Braccini et al,. 2005). We have estimated $\xi$(T) from $H_{c2}(T)$ for the as grown samples and for samples with 10% DWCNT. Using this $\xi(T)$ dependence we have estimated $V_c(T,B)$ from our data. Results are shown in Figure 16, where $V_c$ is plotted as a function of temperature a $H(T)$ = 1 T (right axis) and 3 T (left axis) for both samples.
The estimated numerical values (~$10^{-15}$ $cm^3$) imply that the correlation radius is larger than the main vortex distance, in agreement with a collective pinning regime of vortex bundles.
Doping additionally reduces the correlation volume, consistent with the increase in the critical current density.

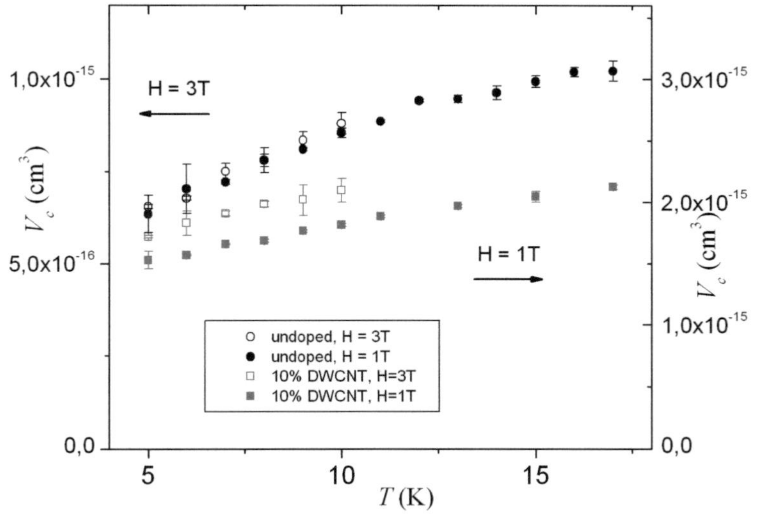


Fig. 16. Estimated collective pinning volume $V_c$ as a function of temperature for pure (circles) and 10% DWCNT (squares) samples at $H$ = 1T (full symbols, left axis) and $H$=3T (open symbols, right axis).

From the comparison of data taken at $H$ = 1 T and 3 T, in both pure and CNT samples, the correlation volume at $H$= 3T is approximately three times smaller than that obtain at $H$ = 1 T and both $J_{c0}$ and $U_c$ decrease with $B$. This field dependence for $U_c$ does not correspond with that predicted by the classical collective pinning theory for the activation energy of vortex bundles when pinning arises from random point defects (Blatter et al., 1994). However, this discrepancy is not surprising, as we know that the strong pinning in these $MgB_2$ samples arises from a variety of larger defects rather than atomic-size disorder.

### 4.6 Conclusions

Magnetic relaxation in bulk $MgB_2$ samples as-grown and with DWCNT doses between 1% and 10% in the range between 5K and 25K has been measured and the corresponding critical current density has been calculated. The current decay in time is quite similar in all the doped samples with doses between 1% and 10%.

A careful creep analysis has been carried out in a pure sample and in one with 10% of carbon nanotubes at $H$= 1 T and 3 T.

The analysis has been performed under the Anderson-Kim approximation, valid in the limit where the measured critical current density $J_c$ is similar to the "true" one $J_{c0}$ that would be present in the absence of creep phenomena. In these samples, even at the loweest temperatures it is necessary to include an explicit field and temperature dependence of both the pinning energies $U_c(T,B)$ and true critical current densities $J_{c0}(T,B)$.

These pinning properties have been obtained as fitting parameters from the experimental data for two different methods. The consistence of the fitting parameters with the A-K limit has been required to delimit the region of validity of the analysis.

In the valid region, the pinning energies and critical current densities have been estimated and compared. The dependence with magnetic field, together with numerical estimations of the pinning volume, indicate the presence of a collective pinning regime of vortex bundles. There is a decrease in the pinning energies that implies an increase in the relaxation rates as a consequence of DWCNT addition. However, the true critical current densities increase, due to a decrease in the collective pinning volumes.

We conclude that the origin of the main changes in the pinning properties with doping are the reduction of the coherence length (that decreases the pinning energies ) and the increase in the disorder parameters (that increase the critical current and decrease the pinning volume).

The strong temperature dependence of the coherence length is probably the main reason to the observed temperature dependence in the pinning properties. At the lower temperatures, thermal instabilities that originate macroscopic flux jumps could also play a role.

A method to analyse the relaxation data in the high temperature region, beyond the A-K approximation, is necessary and will be object of a future work.

## 5. Summary

The effect of carbon substitution is one of the most studied in $MgB_2$ and the results on C solubility and the effects of C-doping on $T_c$, $J_c$ and $H_{c2}$ reported so far vary significantly due to precursor materials, fabrication techniques and processing conditions used. The distinct effect of C incorporation through different routes using various CNT types leads to a simultaneous improvement in $J_c$ and $H_{c2}$, but their effectiveness change with temperature or applied field according to each type of addition. This effect was reported in many works, not only for bulk samples but also for wires and tapes prepared by PIT method.

The reason for this is the dual role of the CNT. They partially dilute into the $MgB_2$ matrix, acting as a source of C that increases $H_{c2}$. In fact, the highest $H_{c2}$ values observed so far in bulk $MgB_2$ correspond to a 10% addition of DWCNT, which present a high level of C doping. At the same time, the fraction of the CNT that retain their structural integrity are ideally suited to act as strong vortex pinning centers due to their tubular geometry and their diameter close to the superconducting coherence length of $MgB_2$, producing a large $J_c$ enhancement. As an additional result, the measured $H_{c2}$ vs $T$ in all samples are successfully described using a theoretical model for a two-gap superconductor in the dirty limit (Gurevich, 2003). This has strong fundamental impact, as it provides clear evidence in support of the basic scenario currently used to describe the superconducting behavior of $MgB_2$, and opens a path for future research in $H_{c2}$ enhancement.
The study of the magnetic relaxation of $MgB_2$ with and without DWCNT bulk samples can give some insight about the possible correlation in the simultaneous increase in the critical current density and the upper critical field. The experimental relaxation rates showed that the pure sample that has a lower $J_c$ (associated with a lower pinning) has also a lower relaxation (associated with a higher pinning energy $U_c$). To understand these results the relaxation data was described and analyzed under the Anderson-Kim frame model (Anderson, 1964). The pinning energies $U_c$, true critical current densities $J_{c0}$, and correlation volumes $V_c$ were estimated and compared. The strong temperature dependence of the coherence length is probably the main reason to the observed temperature dependence in the pinning properties with CNT additions: the reduction of the coherence length (that decreases the pinning energies) and the increase in the disorder parameters (that increase the critical current and decrease the pinning volume).

## 6. Acknowledgment

Research supported by CONICET, UNCuyo, UBACyT, and MinCyT-PICT (AS, GP) and by the U.S. Department of Energy, Basic Energy Sciences, Materials Sciences and Engineering Division (LC).

## 7. References

Anderson, P.W. & Kim, Y.B. (1964). Rev. Mod. Phys. 36, 39

Avdeev, M.; Jorgensen, J. D.; Ribeiro, R. A.; Bud'ko, S. L. & Canfield, P. C. (2003). Physica C 387, 301-306

Ban, E.; Sakaguchi, R.; Matsuoka, Y.; Goto, T.; Watanabe, K. & Nishijima, G. (2005). Physica C, vol. 426–431, pp. 1249–1253

Bean, C. P. (1962). Phys. Rev. Lett 8, 250-253

Blatter, G.; Feigel'man, M. V.; Geshkenbein, V. B.; Larkin, A. I. & Vinokur, V. M. (1994). Rev. Mod. Phys. 66, 1125

Braccini, V.; Gurevich, A.; Giencke, J. E.; Jewell, M. C.; Eom, C. B.; Larbalestier, D. C.; Pogrebnyakov, A.; Cui, Y.; Liu, B. T.; Hu, Y. F.; Redwing, J. M.; Li, Qi; Xi, X. X.; Singh, R. K.; Gandikota, R.; Kim, J.; Wilkens, B.; Newman, N.; Rowell, J.; Moeckly, B.; Ferrando, V.; Tarantini, C.; Marré, D.; Putti, M.; Ferdeghini, C.; Vaglio, R. & Haanappel, E. (2005). Phys. Rev. B 71, pp. 012504-012508

Brinkman A.; Golubov A. A.; Rogalla H.; Dolgov O V.; Kortus J.; Kong Y.; Jepsen O & Andersen O K (2002). Phys. Rev. B 65, 1805171-1805174

Buzea C. & Yamashita T. (2001). Supercond. Sci. Technol. 14, R115-R146

Cheng C. H.; Zhang H.; Zhao Y.; Feng Y.; Rui X. F.; Munroe P.; Zeng H. M.; Koshizuka N. & Murakami M. (2003). Supercond. Sci. Technol. 16, 1–5

Civale L.; Pasquini G.; Levy P.; Nieva G.; Casa D & Lanza H. (1996). Physica C 263, 389

Civale L.; Serquis A.; Hammon D. L.; Liao X. Z.; Coulter J. Y.; Zhu Y. T.; Holesinger T.; Peterson D. E. & Mueller F. M. (2003). IEEE Trans Appl. Supercond .13, 3347-3350.

Dou S. X.; Yeoh W K.; Horvat J & Ionescu M (2003). Appl Phys Lett 83, 4996-4998

Dou S. X.; Yeoh W. K.; Shcherbakova, O.; Wexler, D.; Li Y.; Ren, Z. M.; Munroe P.; Chen, S. K.; Tan, K. S.; Glowack B. A. & MacManus-Driscoll, J. L. (2006). Advanced Materials, 18, 785-788

Flükiger R.; Suo HL.; Musolino N.; Beneduce C.; Toulemonde P & Lezza P (2003). Physica C 385, 286-305, & references therein

Geshkenbein V. B.; Feigelman M.V & Vinokur V. M (1989). Physica C 162-164, 239

Goldacker et al (2004). Supercond. Sci. Technol. 17, S490-S495

Goldacker W et al, in: A. Narlikar (Ed.) (2002). Studies of High Temperature Superconductors Vol. 45, Nova Science Publishers, New York

Golubov A. A.; Kortus J.; Dolgov O. V.; Jepsen O.; Kong Y.; Andersen O. K.; Gibson B. J.; Ahn K. & Kremer R. K. (2002). J. Phys: Condensed Matter 14, 1353-1360

Gurevich A. (2003). Phys. Rev. B 67, 184515-184528

Gurevich A.; Patnaik S.; Braccini V.; Kim K. H.; Mielke C.; Song X.; Cooley L. D.; Bu S. D.; Kim D. M.; Choi J. H.; Belenky L. J.; Giencke J.; Lee M. K.; Tian W.; Pan X. Q.; Siri A.; Hellstrom E. E.; Eom C. B. & Larbalestier D. C. (2004). Supercond. Sci. Technol. 17, 278–286

Häβler W.; Herrmann M.; Rodig C.; Schubert M.; Nenkov K. & Holzapfel B. (2008) Supercond. Sci. Technol. 21, 062001

Kazakov S M.; Puzniak R.; Rogacki K.; Mironov A. V.; Zhigadlo N D.; Jun J.; Soltmann Ch; Batlogg B & Karpinski J (2005). Phys. Rev. B 71, 024533-024543

Kim, J.H.; Yeoh, W.K.; Qin,M.J.; Xu, X.; Dou, S.X.(2006a). J. Appl. Phys. 100, 01390

Kim, J.H.; Yeoh, W.K.; Qin,M.J.; Xu, X.; Dou, S.X.; Munroe,P.; Kumakura,H.; Nakane,T.; Jiang, C.H.(2006b) Applied Physics Letters 89, 122510

Kim, J.H.; Yeoh, W.K.; Qin,M.J.; Xu, X.; Dou, S.X.; Munroe,P.; Rindfleisch, M.; Tomsic, M. (2006c). Physica C, 449, 133–138

Kim, J.H.; Yeoh, W.K.; Xu, X.; Shi M.J. & Dou, S.X. (2007). IEEE Transactions on Applied Superconductivity 17, 2907-2910

Kováč P. ; Hušek I.; Skákalova V.; Meyer J.; Dobročka, E.; Hirscher, M. & Roth, S. (2007). Supercond. Sci. Technol. 20, 105–111

Larbalestier D. C.; Cooley L. D.; Rikel M. O.; Polyanskii A. A.; Jiang J.; Patnaik S.; Cai X. Y.; Feldmann D. M.; Gurevich A.; Squitieri A. A.; Naus M. T.; Eom C. B.; Hellstrom E. E.; Cava R. J.; Regan K. A.; Rogado N.; Hayward M. A.; He T.; Slusky J. S.; Khalifah P.; Inumaru K. & Haas M. (2001). Nature (London) 410, 186-189

Lee S.; Masui T.; Yamamoto A.; Uchiyama H & Tajima S (2003). Physica C 397, 7-13

Ma Y.; Zhang X.; Nishijima G.; Watanabe K.; Awaji S. & Bai, X. (2006). Appl. Phys. Lett, vol. 88, pp. 072502–072504

Maley M. P.; Willis J. O.; Lessure H. & McHenry M. E. (1990). Phys. Rev. B. 42, 2639

Matsumoto A.; Kumakura H.; Kitaguchi H.; Senkowicz B J.; Jewell M C.; Hellstrom E E.; Zhu Y.; Voyles P. M. & Larbalestier D. C. (2006). Appl. Phys. Lett. 89, 2508-2510

Nagamatsu J.; Nakagawa N.; Muranaka T.; Zenitani Y. & Akimitsu J. (2001). Nature (London) 410, 63-64

Pasquini G.; Serquis A.; Moreno A.; Serrano G.; Civale L. (2011). Preprint

Putti M.; Braccini V.; Ferdeghini C.; Pallecchi I.; Siri A. S.; Gatti F.; Manfrinetti P. & Palenzona A. (2004). Phys. Rev. B 70, 052509-052513

Senkowicz B J.; Giencke J E.; Patnaik S.; Eom C B.; Hellstrom E. E. & Larbalestier D. C. (2005). Appl. Phys. Lett .86, 202502-202505

Senkowicz B J.; Polyanskii A.; Mungall R J.; Zhu Y.; Giencke J. E.; Voyles P M.; Eom C B.; Hellstrom E. E. & Larbalestier D. C. (2007). Supercond. Sci. Technol .20, 650

Serquis A.; Civale L.; Hammon D. L.; Coulter J. Y.; Liao X. Z.; Zhu Y. T.; Peterson D. E. & Mueller F M (2003). Appl. Phys. Lett 82, 1754

Serquis A.; Serrano G.; Moreno M S.; Civale L.; Maiorov B.; Balakirev F. & Jaime M (2007). Supercond Sci. Technol. 20, L12–L15

Serrano G.; Serquis A.; Dou S. X.; Soltanian S.; Civale L.; Maiorov B.; Balakirev F. & Jaime M. (2008). Journal of Applied Physics 103, 023907

Serrano G.; Serquis A.; Civale L.; Maiorov B.; Holesinger T.; Balakirev F. & Jaime M. (2009). International Journal of Modern Physics B 23, 3465-3469

Shekhar C.; Giri R.; Malik S. K. & Srivastava O. N. (2007). Journal of Nanoscience and Nanotechnology 7, 1804–1809

Soltanian S.; Horvat J.; L.Wang X.; Munroe P. & Dou S. X. (2003). Physica C, vol. 390, pp. 185–190

Thompson J. R.; Sun Y.; Christen D.; Civale L.; Marwick A. & Holtzberg F (1994). Phys. Rev. B 49, 13287; Thompson J. R.; Krusin Elbaum L.; Civale L.; Blatter G. & Field C. (1997). Phys. Rev. Lett. 78, 3181.

Ueda S.; Shimoyama J. I.; Yamamoto A.; Katsura Y.; Iwayama I.; Horii S. & Kishio K., (2005). Physica C, vol.426–431, pp. 1225–1230

Vajpayee A.; Awana V. P.S.; Yuc S.; Bhalla G.L. & Kishan H. (2010). Physica C 470 S653–S654

Wilke R. H. T.; Bud'ko S. L.; Candfield P. C.; Finnemore D. K.; Suplinskas R. J. & Hannahs S. T. (2004). Phys. Rev. Lett. 92, 217003-217007

Wilke R. H. T.; Bud'ko S. L.; Canfield P. C.; Finnemore D. K.; Suplinskas R. J. & Hannahs S. T. (2005a). Physica C 424, 1-16

Wilke R. H. T.; Bud'ko S. L.; Canfield P. C.; Finnemore D. K.; & Hannahs S T (2005b). Physica C 432, 193-205

Xu A.; Ma Y.; Wang D. L.; Gao Z. S.; Zhang X. P, & Watanabe K. (2007). Physica C, 466, 190-195

Yamamoto A.; Shimoyama J. I.; Ueda S.; Katsura Y.; Horii S. & Kishio K., (2005a). IEEE Trans. Appl. Supercond., vol. 15, pp. 3292–3295

Yamamoto A.; Shimoyama J. I.; Ueda S.; Iwayama I.; Horii S. & Kishio K. (2005b). Supercond. Sci. Technol., vol. 18, pp. 1323–1328

Yeoh W K.; Horvat J.; Dou S. X. & Keast V. (2004). Supercond. Sci. Technol. 17, S572

Yeoh W K.; Horvat J.; Dou S. X. & Munroe P. (2005). IEEE Trans Appl. Supercond. 15, 3284-3287

Yeoh W. K.; Kim J. H.; Horvat J.; Dou S. X. & Munroe P. (2006). Supercond. Sci. Technol. 19, L5-L8

Yeshurun Y.; Malozemoff A.P ; Shaulov A. (1996). Rev. Mod. Phys. 68, 911

# Part 3

## Carbon Nanotube Applications

22

# Carbon Nanotube as VLSI Interconnect

Mayank Kumar Rai[1] and Sankar Sarkar[2]
[1]*Department of Electronics and Communcation Engineering, Thapar University, Patiala, Punjab*
[2]*School of Electronics, Shobhit University, Meerut, U.P.*
*India*

## 1. Introduction

A VLSI interconnect is a thin film of conducting material that provides electrical connection between two or more nodes of the circuit/system formed in the silicon chip. Earlier the most commonly used material was aluminum. The choice was based on its good conductivity and adherence on silicon dioxide. Another useful property of aluminum is that it forms good ohmic contact with silicon. As device density increased with technology scaling, interconnect current density increased. A disadvantage with aluminum is that at high current densities considerable electro migration takes place. Later it was realized that copper a material of higher conductivity is several times more resistant to electro migration than aluminum. In comparison with aluminum, copper can withstand about five times more current density with equal reliability for IC-applications. Due to the advantages that it offers copper became the preferred interconnect material, especially for submicron and deep submicron high density, high performance chips. As the aggressive technology scaling continues a new problem is surfacing. With decrease in cross-section copper interconnect resistivity increases due to surface roughness and grain boundary scattering, causing increase in propagation delay, power dissipation and electromigration [1-2]. To alleviate this problem, for interconnects of future generation chips alternative solutions are under consideration [1-14]. The most promising alternative for copper interconnects turns out to be Carbon Nanotube (CNT).

The CNTs are grown in the form of seamless cylinders with the walls formed by one atomic layer of graphite (graphene). The diameters of these cylinders are of the order of a nanometer. These tubes are either metallic or semiconductor. For interconnect applications the metallic ones are useful. There are two types of CNTs. Single walled CNT (SWCNT) and Multiwall CNT (MWCNT). CNTs constituted by only one thin wall of graphene sheet are SWCNTs. There are some CNTs which consist of a multiple of concentric SWCNT like graphene tubes. These are termed MWCNT. The metallic CNTs are attractive interconnect materials because of their high thermal and mechanical stability, thermal conductivity as high as 5800W/mK, ability to carry current in excess of $10^{14}A/m^2$ current density even at temperatures higher than 200°C and Fermi velocity comparable with that of a metal[15]. It is very difficult to make a good contact with a CNT. The unavoidable contact imperfection increases resistance. CNT resistances in the range 7 KΩ - 100 KΩ have been reported [16, 17]. Such a high resistance is a major disadvantage; if an isolated CNT is used as interconnect.

The problem can be circumvented if for interconnect application CNT bundles are used instead of isolated ones.
A CNT bundle consists of a large number of electrically parallel isolated CNTs. The result of the parallel connection is considerable reduction of resistance between the ends of the bundle. Therefore, a CNT bundle makes a better interconnect than the isolated counterparts. The type of CNTs in a bundle is generally either SWCNT or MWCNT. In a bundle, some of the constituent CNTs are metallic while others are semiconducting. The metallic CNTs contributes to the formation of interconnect. MWCNTs are mostly metallic, whereas a large fraction of CNTs in a SWCNT bundle are semiconducting. One type of MWCNT is a DWCNT (double walled CNT). This form of CNT has been found to be a very useful for application as interconnects.

## 2. CNT growth

The growth of on chip CNT bundle for commercial purposes is very challenging [18]. As interconnect material SWCNT having higher conductivity is more preferable to the relatively lower conductivity MWCNT. The cause of this conductivity difference between the two types of CNTs is the much longer mean free path (about 1μm) of SWCNT as compared to that of MWCNT (a few nm). So far it has been easier to grow MWCNT. The process of CNT growth generally involves some catalyst particles (Fe, Ni, Co or their compound with Mo). The catalysts assist growth process and controls tube diameter. To fill via with CNT the catalyst particles are placed on the metal1 at the bottom of via. CNT is then grown by chemical vapor deposition (CVD) at 450-800°C in presence of a carbon containing gas [19].
Li et al. [20] proposed a bottom-up approach in which MWCNT via is grown on metal1. The carbon fibers grow from Ni catalysts deposited at predefined locations. By means of PECVD and an applied bias voltage the fibers are aligned perpendicular to the wafer surface. Finally $SiO_2$ is deposited and the wafer is planarized. The planarization process also exposes the CNT ends for contact with the metal 2 layer. This method yields high interconnect resistance of the order of a few hundred kilo ohms. This is attributed to imperfections in the structure of the MWCNTs thus grown. By this method high density growth could not be achieved. It is suitable for growing single MWCNT fillings. A different approach is etching via down to metal 1 layer and growing the CNTs in these vias [19, 21]. In [21] dry etching stops at a film of the catalyst (Ni or Co). Arrays of MWCNTs are formed by hot-filament CVD (HF-CVD). The resistance achieved by this method is about 30% of what the method of Li et al. [20] results. Instead of HF-CVD, a pure CVD approach was adopted by Kreupl et al. [19]. This method produced tubes of resistance of nearly 10 Ω. In their approach, care is also taken to ensure that via etching stops exactly on the catalyst layer so that highly reliable MWCNTs are grown with density varying between 100 and 10000/$\mu m^2$.
Earlier, arc discharge and laser ablation methods were used for growth of SWCNT [22, 23]. These processes involve evaporation of carbon from its solid form at temperatures as high as 3000°C or more. The high temperature requirement is not desirable. Another limitation of these approaches is that the nanotubes formed are twisted around each other. This makes it difficult to purify and arrange the CNTs for application purposes. More recently CVD with methane as carbon source and iron oxide nano particles as catalyst has been used to grow high quality SWCNT. The desired result is achieved in temperature range of 850-1000°C.

### 3.1 Problems of existing interconnect systems

With continuous reduction of feature size there has been a parallel increase in die size. The result is more and more increase in length of some of the on chip interconnects as technology scaling continues. Based on their length interconnects are categorized as local, semi-global and global. A global interconnect is very long which is connected to several nodes across the chip for example, clock lines, ground lines etc. The local ones connect nearby nodes and are of shorter lengths. Interconnects of intermediate lengths are the semi-global ones. Interconnect is equivalent to an RLC circuit as shown in Fig. 3.1.

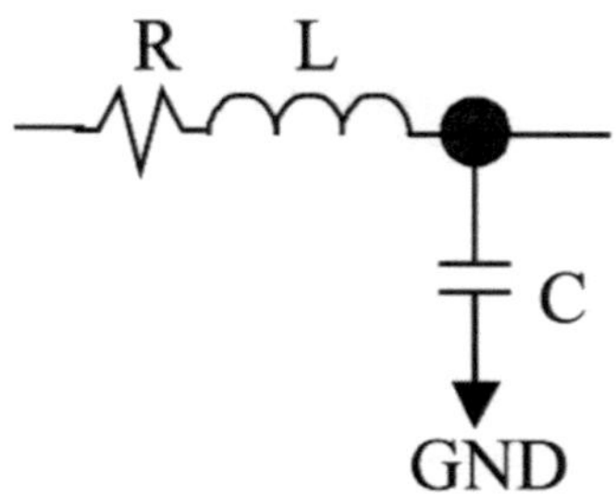


Fig. 3.1. RLC-equivalent of interconnect.

Any increase in interconnect length causes R, L and C to increase. This in turn results in an increase in interconnect signal propagation delay. Thus as technology scaled signal delay caused by interconnect became increasingly significant compared to delay caused by the gate and thus affecting the circuit's reliability. As per ITRS predictions [24], for nanometer size gate lengths interconnect delay is mostly affected by resistive and capacitive parasitics. For decreasing the resistive part of the RC delay, various alternatives to aluminum were considered in early 1990s. As already discussed a metal of much better electrical resistivity and electro migration than those of aluminum is copper. Table 3.1 compares the resistivities of the two metals.

| Metal | Bulk resistivity (μΩ.cm) |
|---|---|
| Al | 2.67 |
| Cu | 1.70 |

Table 3.1. Resistivities of aluminum and copper

Copper has a higher melting point (1,357 K) than aluminum (933 K). This provides more thermal stability to copper. Because of these advantages copper is the most preferred on chip interconnect material for the present day integrated circuits.

With advancement of VLSI technology, the number of on chip interconnects is on the rise. To accommodate more interconnects the cross-sectional dimensions are reduced rapidly resulting in dimensions of the order of mean free path of electrons in copper (~ 40 nm at room temperature). As the dimension approaches electron mean free path grain boundary and surface scattering are enhanced [25, 26]. Consequently, resistivity of interconnect is increased. Another effect of dimension scaling is increase in current density. Thus as technology scales these effects on resistivity together with increase in interconnect resistance with length enhances delay. Besides increase in delay, interconnect power dissipation increases because of increased current density and increase in frequency of operation. The increased heating due to the rise in power dissipation assists electro migration. As these scaling dependent limitations of copper interconnect is going to be more and more severe for the future generation of VLSI chips it is time to look for an alternative material. CNT with all its advantage seems to be the right choice.

### 3.2 CNT interconnect

To analyze and understand the behavior of any interconnect (CNT or else) it is essential to first develop its equivalent circuit (Fig.3.1). The development of an equivalent circuit is complete only when various impedance parameters like resistance, capacitance and inductance are fully defined by means of their analytical expressions. Such an equivalent circuit is then used in analysis and simulation of interconnect performance.

#### 3.2.1 SWCNT interconnect

An isolated SWCNT on ground plane is shown in Fig.3.2. The separation between the nanotube and the ground is y and the diameter of the SWCNT is d. Assuming it to be in cylindrical form on the basis of Luttinger Liquid Theory, Burke [29] developed an electrical equivalent of the structure as shown in Fig. 3.3. If a 1-D system has N conducting channels in parallel then its resistance is $h/Ne^2T$. Where h is Planks constant, e is electron charge and T is electron transmission coefficient. Due to spin and sub lattice degeneracy of electrons there are 4 parallel conducting channel in SWCNT (N=4). Thus assuming perfect contacts (T=1), the resistance of an SWCNT is $h/4e^2$. With the values of the physical constants substituted the resistance assumes the fairly large value of 6.45 KΩ. In the equivalent circuit (Fig.3.3) this resistance ($R_f$) is equally divided between the contacts at the two ends of the nanotube. That is,

$$R_f = h/4e^2 \tag{3.1}$$

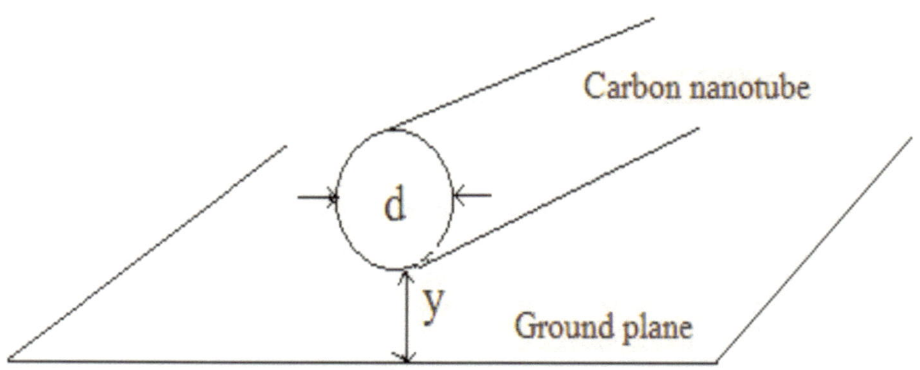


Fig. 3.2. Carbon nanotube, of diameter 'd', distance 'y' below it.

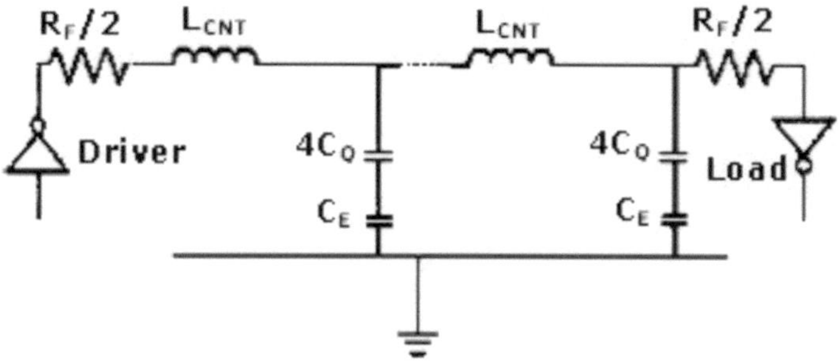


Fig. 3.3. Equivalent RLC circuit for an isolated SWCNT.

of length less than mean free path of electrons and assuming ideal contact. For lengths less than one electron mean free path ($L_0$), the tube resistance equals the fundamental resistance given by (3.1). If the tube length (L) is larger than mean free path then enhancement of scattering gives rise to an addition resistance. This resistance increases with increase in CNT length and is

$$R_{CNT} = (h/4e^2)L/L_0 \tag{3.2}$$

Making metal-CNT contact at the two ends of a tube is very complex. In most cases the contacts are far from being perfect. The contact imperfection leads to very large resistance in series with $R_{CNT}$. CNT resistance of the order of 100 KΩ has been reported [31]. However, resistances around the fundamental resistance are also observed in some cases. Thus the resistance of a CNT has three components: the fundamental resistance $R_f$, the scattering resistance $R_{CNT}$ and the contact resistance at the two ends of the tube. CNT resistance is also bias dependent. At higher bias where electric field is very high current saturates and CNT does not show ohmic behavior. In the low bias regime on the other hand the CNTs show perfect ohmic behavior and are compatible with VLSI interconnect applications.

CNT has two capacitances of different origins. One is electrostatic capacitance and the other quantum capacitance. The electrostatic capacitance ($C_E$) is due to charge stored by the CNT-ground plane system (Fig.3.2) and is given by

$$C_E = 2\pi\varepsilon/ln\ [y/d] \tag{3.3a}$$

This is per unit length of the nanotube.

The quantum capacitance ($C_Q$) accounts for the quantum electrostatic energy stored in the nanotube when it carries current. Considering this energy an effective capacitance (per unit length) may be obtained which is expressed by

$$C_Q = 2e^2/(hv_f) \tag{3.3b}$$

Where $v_f$ is the Fermi velocity. Typically, $C_Q$ is 100aF/μm or so. As CNTs have four conducting channels, total effective quantum capacitance resulting from four parallel channel is $4C_Q$. When current flows both $C_E$ and $4C_Q$ carry same charge. Thus the two capacitances appear in series in the isolated SWCNT equivalent circuit (Fig.3.3).

For a SWCNT, there are two types of inductances termed magnetic inductance and kinetic inductance. Magnetic inductance ($L_M$) is due to the total magnetic energy resulting from the current flowing in the wire. The kinetic inductance ($L_k$) arises from kinetic energy stored in each conducting channel of the CNT. The four parallel conducting channels in a CNT results in an effective kinetic inductance of $L_k/4$. The expressions for $L_M$ and $L_k$ are

$$L_M = \frac{\mu}{2\pi} \ln\left(\frac{y}{d}\right) \tag{3.4}$$

$$L_K = h/\{2e^2 (v_F)\} \tag{3.5}$$

For $d$ = 1nm and $y$ = 1μm, $L_M$ (per unit length) and $L_K$ (per unit length) for a CNT is equal to ≈1.4pH/μm and 16nH/μm, respectively. Since, $L_k >> L_M$, inclusion of $L_M$ does not have significant impact on the delay model for interconnects.
Fig.3.4 and 3.5, shows the equivalent Circuits of an SWCNT-bundle for L<L0 and L>L0 [29-33]. Where L is bundle length. The resistances, inductances and capacitances of a bundle can be obtained from the following expressions [29-33].
The CNT bundle resistance is given by (3.6) and (3.7), where H is thickness and w is the width of CNT bundle interconnect, and $n_{CNT}$ is the total number of CNTs in the bundle.

$$R_{CNT}(Bundle) = \left(\frac{h}{4e^2}\right)\frac{L}{L_0} \bigg/ n_{CNT} \tag{3.6}$$

$$n_{CNT} = \left\lfloor \frac{w-d}{x} \right\rfloor \left( \left\lfloor \frac{H-d}{\left(\frac{\sqrt{3}}{2}\right)x} \right\rfloor + 1 \right) - \left\lfloor \frac{1}{2} \right\rfloor \left( \left\lfloor \frac{H-d}{\left(\frac{\sqrt{3}}{2}\right)x} \right\rfloor + 1 \right) \tag{3.7}$$

If the number of rows in the bundle is even and

$$n_{CNT} = \left\lfloor \frac{n_W - d}{x} \right\rfloor \left( \left\lfloor \frac{H-d}{\left(\frac{\sqrt{3}}{2}\right)x} \right\rfloor + 1 \right) - \left\lfloor \frac{1}{2} \right\rfloor \left( \left\lfloor \frac{H-d}{\left(\frac{\sqrt{3}}{2}\right)x} \right\rfloor \right) \tag{3.8}$$

if number of rows is odd. The number of rows is given by (3.9) and x is the separation between the centres of two neighbouring tubes. For parallel structure x=d.

$$n_H = \left( \left\lfloor \frac{H-d}{\left(\frac{\sqrt{3}}{2}\right)x} \right\rfloor + 1 \right) \tag{3.9}$$

The total effective capacitance of bundle of SWCNT is given by (3.10),Where $C_E^{Bundle}$ and $C_Q^{Bundle}$ are the total electrostatic capacitance and total quantum capacitance of bundle of SWNT and are calculated by (3.11) and (3.12).

$$C(Bundle) = \left( \frac{C_E^{Bundle} \cdot C_Q^{Bundle}}{C_E^{Bundle} + C_Q^{Bundle}} \right) \tag{3.10}$$

$$C_E^{Bundle} = 2\left(\frac{2\pi\varepsilon_{ox}}{\ln\left(\frac{s}{d}\right)}\right) + \left(\frac{\left\lfloor \frac{w-d-2}{x} \right\rfloor}{2}\right)\left(\frac{2\pi\varepsilon_{ox}}{\ln\left(\frac{s+w}{d}\right)}\right) + 3\frac{(n_H - 2)}{5}\left(\frac{2\pi\varepsilon_{ox}}{\ln\left(\frac{s}{d}\right)}\right) \tag{3.11}$$

$$C_Q^{Bundle} = \left(\frac{2e^2}{hv_f}\right) n_{CNT} \tag{3.12}$$

Where S is the separation between adjacent bundle. The inductance of CNT bundle is given by the parallel combination of the inductances corresponding to each CNT forming the bundle, which is

$$L((Bundle)) = \left(\frac{L_M + L_K}{4n_{CNT}}\right) \tag{3.13}$$

Where $L_M$ and $L_K$ are the magnetic and kinetic inductance of an isolated CNT.

The impedance parameters of a SWCNT bundle are calculated from (3.6) – (3.13) [33]. Table-3.2, shows the data used for these calculations [33]. Fig. 3.6, shows how the resistance of long (1mm) interconnects vary with technology scaling. It shows that in case of such long interconnects the resistance of CNT interconnects is several times lower than that of copper based interconnects in advanced technology nodes. The impedance parameters of CNT interconnect are calculated from (3.1)-(3.13). For the sake of comparison the impedance parameters of a copper interconnect are also determined. The Cu-interconnect impedance parameters are determined by using expressions available in literature [34].

As may be seen from (3.6) through (3.8) the CNT resistance is a function of tube diameter d. Fig. 3.7 illustrates the dependence of resistance on tube diameter. This is an example of the dependence of SWCNT-interconnect resistance on tube diameter. It can be seen that a bundle composed of tubes of larger diameters, when used as interconnect will have larger line resistance.

The two CNT capacitances are calculated from (3.11) and (3.12). Fig.3.8 shows the variations of SWCNT-interconnect capacitance with tube diameter. The decrease in capacitance with tube diameter indicates the desirable effect of reducing interconnect delay. Increase in interconnect resistance has the effect of increasing propagation delay. A comparison of Fig.3.7 and Fig. 3.8 therefore shows that two competing effects take place on delay as tube diameter is increased. Much depends on which of the two is the dominating one. When the effect of resistance dominates, delay rises as diameter increases. Otherwise the opposite effect takes place. If the two effects balance each other, delay is independent of change in diameter.

Calculations using (3.4), (3.5) and (3.7) or (3.8) show that inductance in an SWCNT-bundle is negligible in comparison with its resistance and capacitance. It is of the order of a few fH (Fig. 3.9), which is much less compared to the several nano Henry of inductance that is generally found in the copper interconnects. This shows that an SWCNT has negligible inductive effect even in case of high-speed applications. Generally, interconnect delay increases with increase in its equivalent resistance, inductance and capacitance [35]. As inductive effect on CNT impedance is negligible, the dependence of resistance and capacitance on tube diameter can provide insight into the variations of delays of the SWCNT-interconnects with variation in tube diameter.

The impedance parameters of interconnect is known to affect its delay and power dissipation [36]. As the diameter of an isolated tube controls the impedance parameters of a CNT bundle, it is of importance to study the effect of isolated tube diameter on delay and power dissipation. The equivalent circuit shown in Fig.3.5 is used to SPICE-simulate signal propagation down SWCNT-interconnect in 32nm and 22 nm technologies [37]. The clock speed is 1.1MHz. Simulation is also carried out for copper-interconnects of same technologies and clock speed. For simulation purpose copper interconnect is modelled by a $\pi$-equivalent circuit [38].

For both SWCNT and copper interconnects, 90% delay has been extracted from the SPICE simulation results. Copper-interconnect propagation delay is used to normalize corresponding SWCNT-interconnect propagation delays. From now onwards this normalized SWCNT-interconnect delay will be referred to as 'relative delay'. Similarly, relative power dissipation in SWCNT-interconnect will mean normalization by copper power dissipation. Fig.3.10 shows relative delay as function of isolated tube diameter. The variations are simply reflections of the resultant of the effects of diameter variations on resistance and capacitance taken together.

The relatively larger inductance of copper facilitates the lowering of the relative delay, thereby placing SWCNT-bundle as a favourite candidate for future VLSI interconnects. The gradual rise of relative delay in case of 22nm technology indicates dominance of CNT resistance over its capacitance. For 32nm technology as diameter is increased the capacitance dominates till a critical diameter is reached. Beyond this critical diameter resistance takes over. Briefly, for good performance, the optimum tube diameter should be selected if possible (32nm). Otherwise tube diameter should be restricted (22nm).

The diameter dependence of relative power dissipation in SWCNT interconnect is illustrated in Fig. 3.11 at 32nm and 22nm technology nodes. As may be seen, in general SWCNT-interconnect dissipate more power than its copper counterpart due to higher value of tube capacitance. Furthermore, SWCNT interconnect power dissipation improves as technology scales.

Power dissipation in SWCNT-interconnect can be lesser if the tubes are of larger diameter (Fig. 3.12). The diameter dependence of power results from the functional relationship between the bundle impedance parameters and tube diameter. As SWCNT interconnects have negligible inductance the decrease in CNT power dissipation with increase in tube diameter indicates that the effect of capacitance dominates over the effect of resistance. Thus, reduction of interconnect power dissipation by increasing tube diameter should be possible.

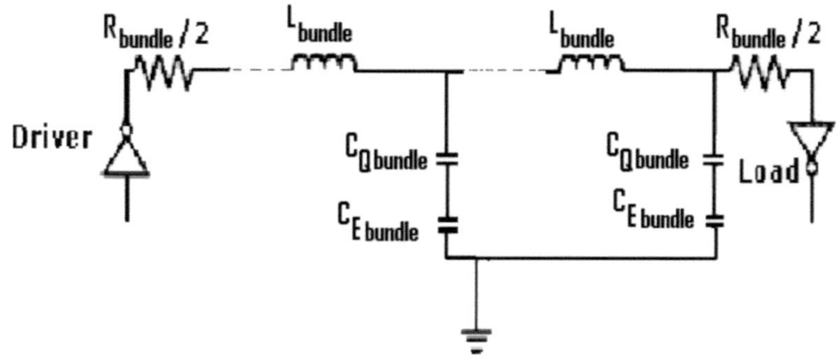


Fig. 3.4. Equivalent circuit diagram for SWCNT bundle interconnects (length $L<L_0$ *where* $L_{bundle}=(L_M+L_k/4)/n_{CNT}$).

| Technology | 32nm | 22nm |
|---|---|---|
| Vdd | 0.9v | 0.7v |
| Width(W) of local and semi global interconnect | 32nm | 22nm |
| A/R* | 2 | 2 |
| Thickness(H) of local and semi global interconnect | 64nm | 44nm |
| Separation(s) between adjacent bundle of local and semi global | 32nm | 22nm |
| Width(W) of Global interconnect | 48nm | 32nm |
| A/R(Global) | 3 | 3 |
| Diameter of SWCNT | 1nm | 1nm |
| Thickness(H) of global interconnect | 144nm | 96nm |
| Separation(s) between adjacent bundle of global interconnect | 48nm | 32nm |
| Oxide Thickness ($t_{OX}$) | 54.4nm | 39.6nm |
| $\varepsilon_{OX}$ (Relative) | 2.25 | 2.05 |

*A/S (Aspect ratio) for local and Intermediate.

Table 3.2. Simulation parameters

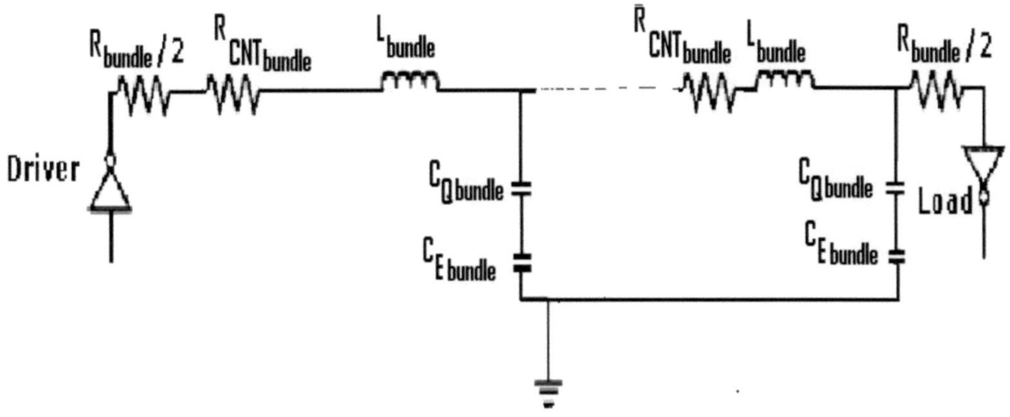


Fig. 3.5. Equivalent circuit diagram for SWCNT bundle interconnects ( length $L>>L_0$ *where* $L_{bundle}=(L_M)/n_{CNT}$).

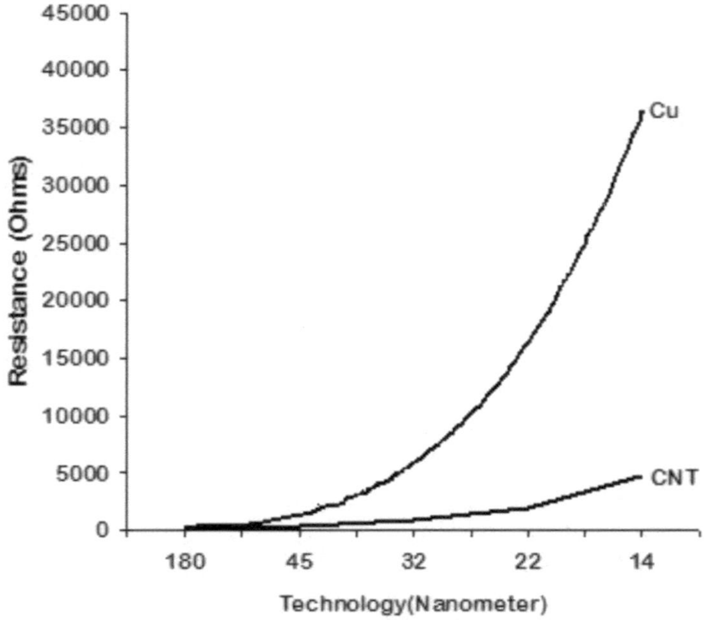


Fig. 3.6. Technology node dependence of resistances of long Cu [33]

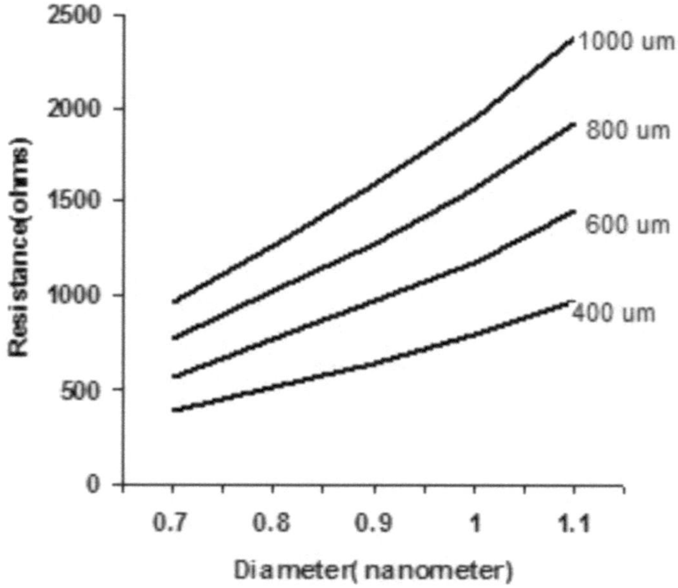


Fig. 3.7. Variation of equivalent resistance with diameters at different interconnect lengths for 22nm technology node [33].

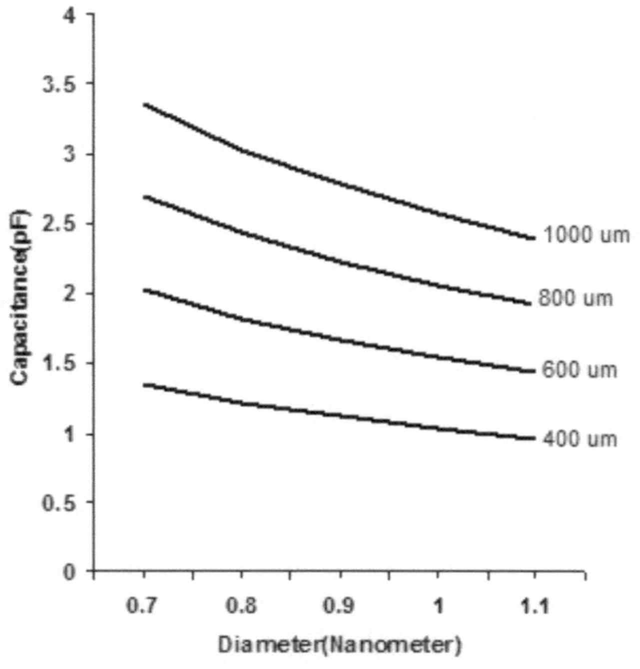


Fig. 3.8. Variation of capacitance with tube diameter at different interconnect lengths for 22nm technology node [33].

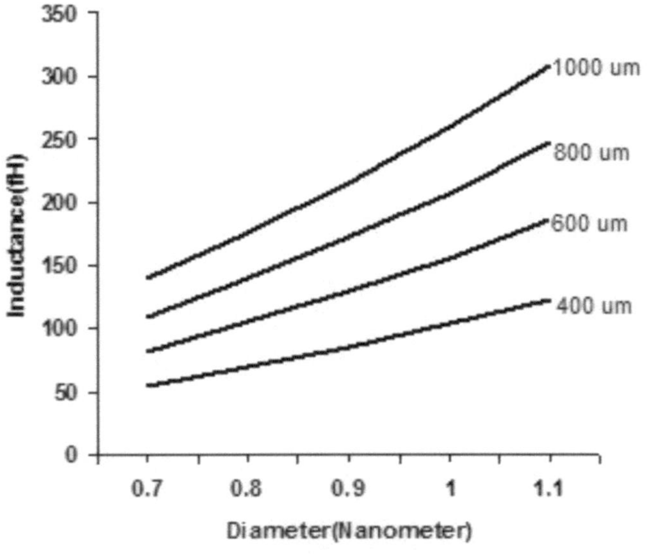


Fig. 3.9. Variation of inductance with tube diameter at different interconnect lengths for 22nm technology node [33].

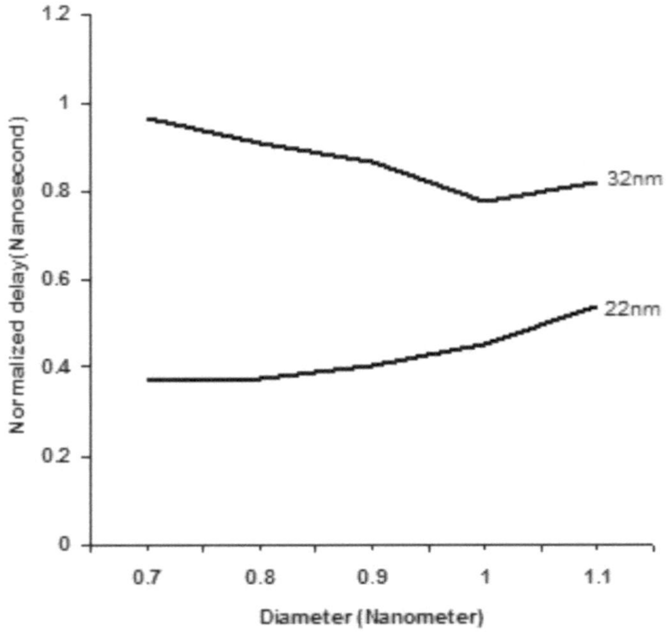


Fig. 3.10. Tube diameter dependence of normalized SWCNT bundle interconnect propagation delay in two different technology nodes with Interconnect length=1mm [33].

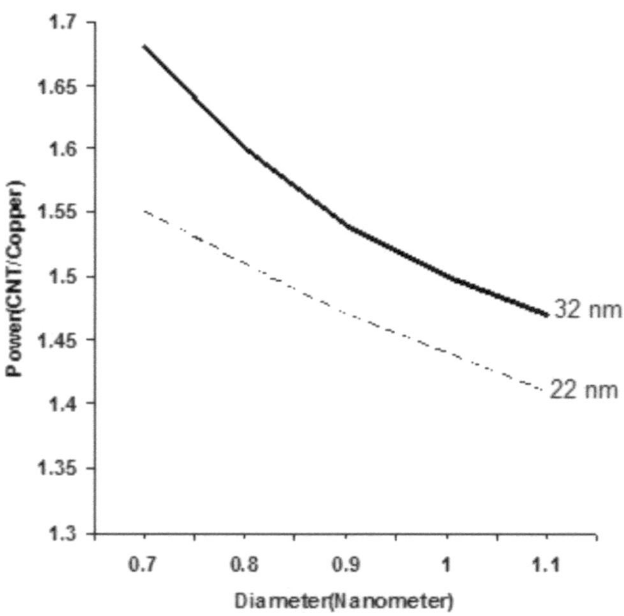


Fig. 3.11. Relative power dissipations as function of tube diameter with length 1mm as parameter for 32nm and 22nm technology node [33].

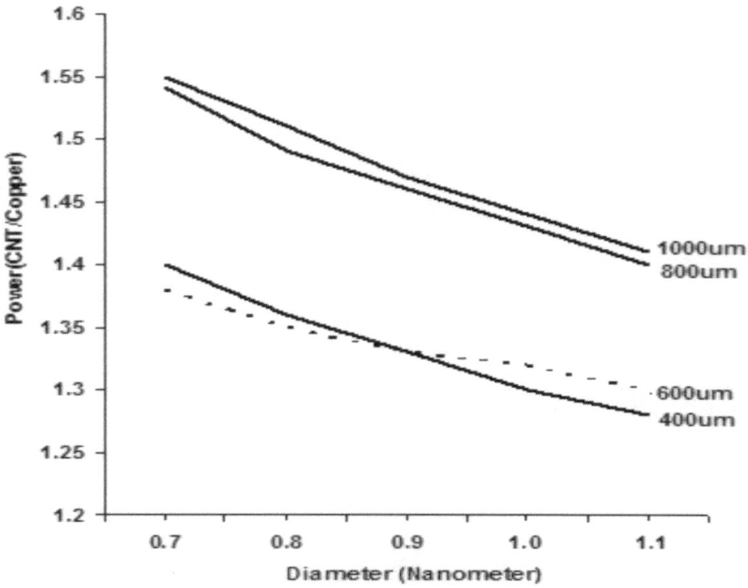


Fig. 3.12. Relative power dissipations as function of tube diameter with length as parameter for 22nm technology node [33].

Srivastava et al.[31], Li et al[39] and Massoud et al[40], analyzed the performance of the CNT interconnects of all three categories of interconnect length.

A comparison of propagation delays of CNT bundle interconnect and Copper interconnect [31, 41] is shown in Fig.3.13. In this figure 38% length corresponds to 0.6μm and 100% corresponds to 1.6μm. The lengths of the local interconnect are of this order and are comparable to mean free path of electrons in CNT. Fig.3.13 shows that the delay of CNT bundle is larger than that of copper interconnect. The difference between the two increases as technology advances. The analyses of simulation results [31,39, 40] also indicate that non-zero metal-nanotube contact resistance degrades CNT-interconnect performance, however, its impact is negligible. Furthermore, in case of sparse bundle the performances of a CNT interconnect is better [39]. The dependence of propagation delay ratio on interconnect length was also studied [31,41].

For the local interconnects, the influence of pitch on delay of CNT bundle is quite significant (Fig. 3.14.). It can be seen that the delay ratio of CNT bundle to copper interconnect is increased when the pitch increases. This, however, does not agree with what is reported in [40]. Fig.3.13 and Fig.3.14 also show that the delay ratio change very slowly as interconnect length is increased. This may be attributed to ballistic transport in CNT because of its low resistance.

Li et al. [39], analyzed propagation delay ratios of CNT and copper semi- global and global level interconnects. Fig.3.15, illustrates a comparison between CNT and Copper interconnect delays at these levels. It can be seen that the propagation delay-ratio of CNT and copper interconnect is very high at the local level, but undergoes a steep decrease as we move to the semi- global level where it first decreases to a minimum value and then increases by a slight amount. Finally it becomes constant at lengths of global level. From Fig.3.15, it is clear that a CNT interconnect gives best performance at the semi-global level because the propagation delay ratio of CNT to Copper interconnect attain low values at lengths in the range of 200μm - 650μm. It can be observed that in the semi global and global regimes the performance of CNT interconnect is better than that of copper interconnects.

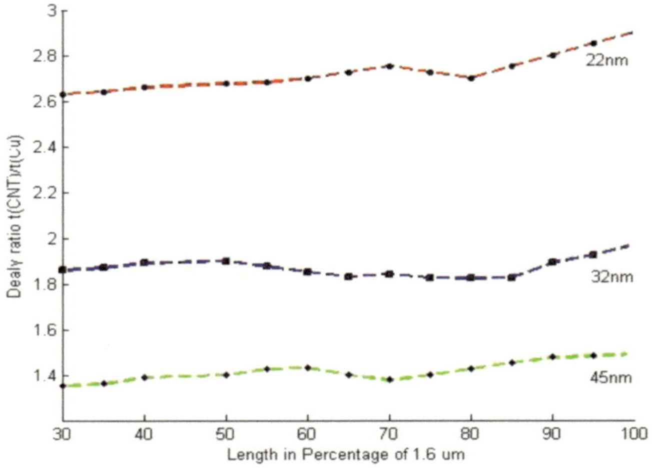


Fig. 3.13. Comparison of propagation delay between CNT bundle and copper interconnects [41].

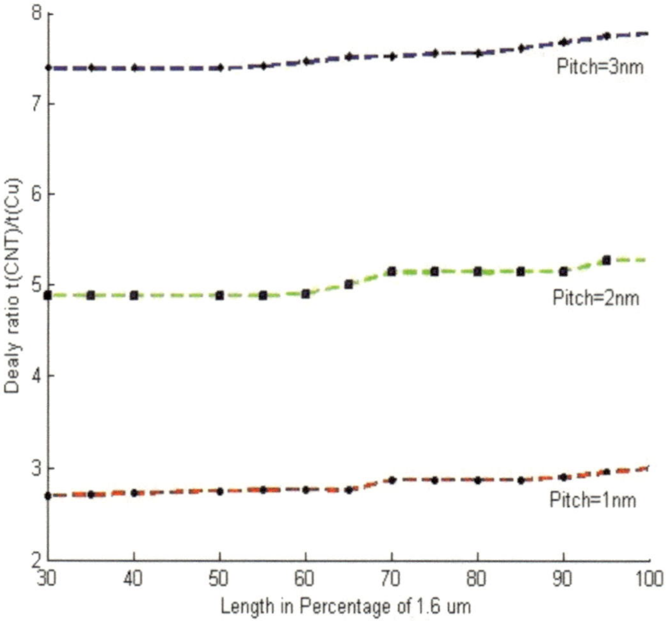


Fig. 3.14. Propagation delay ratio of densely packed CNT bundle to copper interconnects at different pitches [41].

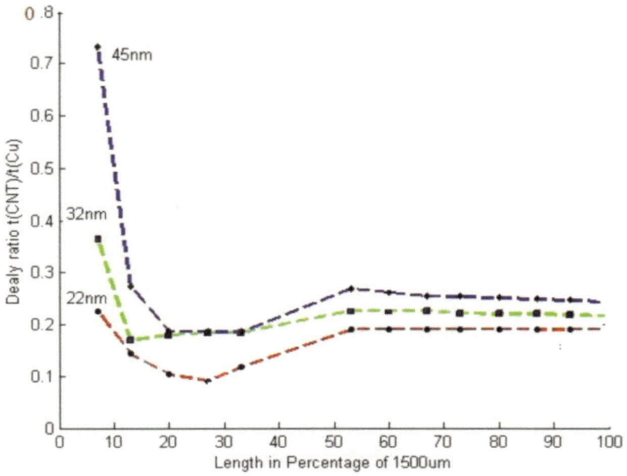


Fig. 3.15. Propagation delay ratio of densely packed CNT bundle to Cu interconnect [41].

### 3.2.2 MWCNT interconnect

The complex structure formed by a large number of concentric tubes makes analysis and design of MWCNT interconnect to be difficult. For analytical purpose Hong et al.[42] proposed an equivalent circuit model as shown in Fig. 3.16.

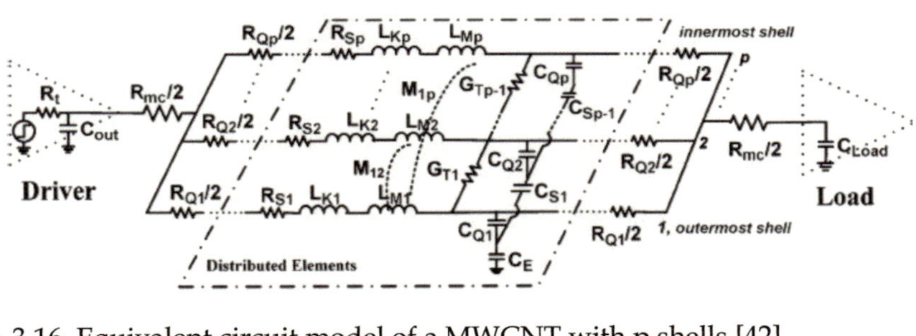


Fig. 3.16. Equivalent circuit model of a MWCNT with p shells [42]

This figure shows the equivalent circuit model of a MWCNT with p shells. $R_{mc}/2$ are contact resistances at the two ends of the MWCNT. Lumped quantum resistance per shell is $R_Q$. $R_S$ is the scattering resistance. $L_K$ is kinetic inductance. Magnetic inductance is $L_M$. The mutual inductance is M. Quantum capacitance is $C_Q$ and $C_E$ is the electrostatic capacitance between the outermost shell and ground. Other shells do not have this capacitance. $C_S$ is shell to shell capacitance. $G_T$ is tunneling conductance that has p - 1 components for the MWCNT. $R_t$ and $C_{out}$ are respectively resistance and output capacitance of the driver. These impedance parameters can be calculated from those available in literature [42].

SPICE simulation using the equivalent circuit (Fig.3.16) shows that as in the case of SWCNT the resistivity of MWCNT bundle is also length dependent. For longer length (> 10 μm), MWCNT resistivity is much lower than that of copper interconnect and is comparable SWCNT interconnect. The simulation results show that for global interconnects delay in MWCNT interconnect is less than delay in copper interconnect. The delay improvement

further enhances with increase in interconnect length. However, increase in interconnect width deteriorates MWCNT delay. A better performance can be achieved if MWCNTs of larger diameters form the MWCNT bundle interconnect.
Hong et al. also studied MWCNT delay at semi global and local levels of length. They observed that in range of semi global length MWCNT interconnects are significantly faster than their copper counterparts. Furthermore, technology scaling improves delay performance. The local interconnects have smallest cross section which results in substantial resistivity. As a result line delay is more in the local interconnects. Over this range of interconnect length the copper interconnect yield better performance.

### 3.2.3 Comparison of SWCNT with MWCNT interconnects

It is amply clear from the foregoing discussions that at highly scaled technologies for global and semi global interconnects both SWCNT and MWCNT can be appropriate replacements for copper. Between the two forms of CNTs SWCNT should have been the choice because of its lower resistivity resulting from longer mean free path. In reality this advantage cannot be taken as random chirality (direction in which graphene sheet is rolled up) leads to a large number of semiconducting tubes in a bundle. These semiconducting tubes do not participate in the conduction process. Growing a totally metallic SWCNT bundle is challenging job. Relatively simpler growth process and the highly metallic nature of the MWCNT interconnect makes it a better option at present. Its disadvantage is the short mean free path.
Simulation results [41-42] show that both SWCNT and MWCNT interconnect performs better than copper interconnect at global and semi-global levels. A highly metallic SWCNT out performs the MWCNT, especially at highly scaled technology. However, in reality existence of a population of semiconducting constituents in SWCNT bundle causes MWCNT to perform better. In local level MWCNT marginally outperform SWCNT.

## 4. Cross talk between CNT interconnects

Because of capacitive and inductive coupling between adjacent interconnects signal propagating down one (aggressor) affects signal propagating down the other (victim). This is commonly referred to as crosstalk. Crosstalk in copper and aluminum interconnects are extensively studied [43-46]. It is now well established that in such interconnects crosstalk induce signal delay, overshoot, undershoot and glitches in the victim. These aberrations travel down interconnect and cause faulty operation of the receiving device. However, very little work has been done in the area of crosstalk in CNT interconnect.
Rossi et al. [47] are the first to study crosstalk in CNT interconnect. They analyzed crosstalk in CNT interconnect implemented bus architectures. Both SWCNT and MWCNT interconnects were considered. It was shown that delay and voltage noise margins in MWCNT busses are much better than SWCNT busses. The crosstalk delay is also lower in the SWCNT busses. In these busses cross delay can be improved by optimizing the spacing between the interacting interconnects. Furthermore, crosstalk induced logic error in the output device can be considerably large in case of MWCNT where as no such problem occurs in SWCNT architecture. Rossi et al. also proposed a crosstalk aware CNT bus architecture. This architecture is formed by double walled carbon nanotubes (DWCNT) in parallel. It is shown to be significantly less susceptible to cross talk produced delay and noise voltage peaks.

Pu et al. [48] developed analyzed crosstalk effects in SWCNT and DWCNT interconnects. Their analysis included coupling inductance along with coupling capacitance. The analytical crosstalk models thus developed capture crosstalk delay, glitches etc. with good accuracy. Crosstalk induced delay in SWCNT and DWCNT bundle interconnects were compared with that of copper interconnect. It is observed that for semi-global and global interconnects CNT especially DWCNT results in much reduced crosstalk induced signal delay. For suppression of crosstalk induced glitch copper interconnect is better.

The crosstalk induced voltage peaks produce stress in oxide layers underlying the victim interconnect. With technology scaling oxide thickness has drastically reduced. As a consequence of oxide thinning a small crosstalk produced overshoot or undershoots causes a prohibitively large electric field to generate across the oxide. With time such electric field weakens the oxide layer and possibility of its damage increases. The possibilities of oxide damage due to crosstalk overshoot and undershoot was studied by Das and Rahaman [49]. They observed that with scaling ratio of overshoot/undershoot voltages to power supply voltage does not vary with scaling in all types of interconnects. However, in case of copper interconnects overshoot and undershoot increases as interconnect length is increased. In case of the CNT based interconnects on the other hand, neither scaling nor increase in length affect crosstalk induced voltage overshoot and undershoot.

## 5. Conclusion

An over view of the exploratory research on CNT as possible VLSI interconnect is presented. The problem of continuing with copper interconnects in highly scaled technologies of future are briefly discussed. The works carried out in finding an alternative solution indicates that the CNT based interconnects have the potential to replace copper in future. The SWCNT bundle is most desirable form of CNT based interconnect provided all constituent CNTs of the bundle are metallic. The SWCNT has been studied extensively and it is found that isolated tube diameter plays an important role in determining delay and power dissipation. Another parameter of importance is the interconnect length. It is observed that both SWCNT and MWCNT perform better than copper in the semi-global and global levels of interconnect length. At local level performance of copper interconnects is better. In brief, the analyses and simulations reported by various authors show that if a CNT technology compatible with present form of IC technology can be developed, then it will be possible to partially or wholly replace copper interconnect by CNT based interconnect.

## 6. References


[1] W.Steinhogl, G.Schindler, G.Steinlesberger, M.Tranving, and M.Engelhardt, "Comprehensive study of the resistivity of copper wires with lateral dimensions of 100nm and smaller," *Journal of Applied Physics*, Vol.97, 023706,2005.

[2] Naeemi et al. "Performance comparison between carbon nanotube and copper interconnects for giga scale integration (GSI)", *Electron Device letters*, vol. 26, No. 2, pp. 84-86, 2005 .

[3] C. Schonenberger et al."Interference and interaction in multiwalled carbon nanotubes", *Applied Physics A*, vol, 69, pp.283-295, 1999.

[4] A. Naeemi and J. D. Meindl, "Monolayer metallic interconnects: promising candidates for short local interconnects", *Electron device letters*,vol. 26, No. 8, pp. 544-546, 2005.

[5] G. Zhang, et.al., "Ultra-high-yield Growth of Vertical Single-Walled Carbon Nanotubes: Hidden Roles of Hydrogen and Oxygen," Proc. Nation Academy of Sciences, Vol.102,No.45, pp.16141-16145,2005.

[6] B.Q.Wei,R. Vajtai and P.M.Ajayan,Applied Physics Letters vol.79,no.8pp.1172-1174,2001.

[7] Dong, C., Haruehanroengra, S. and Wang, W., "Exploring Carbon Nanotubes and NiSi Nanowires as On -Chip Interconnections", *Proceedings of ISCAS*, pp. 3510-3513,*2000*,.

[8] Gayasen, A., Vijaykrishnan, N. and Irwin, M.J., "Exploring technology alternatives for nano- scale FPGA interconnects," *Proc. of DAC'05*, pp. 921-926June, 2005.

[9] Wu, Y., Xiang, J., Yang, C., Lu, W. *et al.*, "Single crystal metallic nanowires and metal/semiconductor nanowire heterostructures", *Nature*, vol. 430, pp. 61-65,July 2004.

[10] Morimoto, T. *et al.*, Self-aligned nickel-mono-silicide technology for high-speed deep submicrometer logic CMOS ULSI. *IEEE Trans. Electron Devices* 42, pp. 915–922, 1995.

[11] Cui, Y., Lauhon, L. J., Gudiksen, M. S., Wang, J. and Lieber, C. M., "Diameter-controlled synthesis of single-crystal silicon nanowires", *Applied Physics Letter* 78, pp. 2214–2216,2001.

[12] Wu, Y. *et al.*, "Controlled growth and structures of molecular-scale silicon nanowires", *Nano Letter 4*, pp. 433– 436,2004.

[13] Toman, K. "The structure of NiSi", *Acta Crystallogr.* 4, pp. 462–4641951.

[14] Meyer, B. *et al.*, "Intrinsic properties of NiSi", J. *Alloys Compounds* 262/263, pp.235–237 1997.

[15] Banerjee, K. and Srivastava, N., "Are carbon nanotubes the future of VLSI interconnections?", *43rd ACM IEEE DAC Conference Proceedings, San Francisco*, CA, pp. 809-14 2006.

[16] Th.Hunger,et al., "Transport in Ropes of Carbon Nanotubes: Contact Barriers and Luttinger Liquid Theory", *PRB*,Vol.69,195406,2004.

[17] W. Liang et al., "Fabri-parrot interference in a nano electron waveguide", *Nature*,vol. 411, pp. 665-669, 2001.

[18] H. Dai, J. Kong, C. Zhou, N. Franklin, T. Tombler A. Cassel, S. Fan and M. Chapline, "Controlled chemical routs to nano architectures, physics and devices", *J.Phys. chem. B*, Vol.103, pp. 11246-11255, 1999.

[19] F. Kreupl, A. P. Graham, M. Liebau, G. S. Duesberg, R. Seidel and E. Unger, "Microelectronic interconnects based on carbon nanotubes", available: http://arxiv.org/ftp/cond-mat/papers/0410/0410630.pdf.

[20] J. Li, Q. Ye, A. Cassel, H. T. Ng, R. Stevens, J. Han and M. Meyyappan, "Bottom-up approach for carbon nanotube interconnects", *Applied Physics Letters*, 82, p. 291, 2003.

[21] M.Nihei,M.Horibe,A Kawabata,Y.Awano, "Simulataneous formation of multiwall carbon nanotubes and their end-bonded ohmic contacts to Ti electrodes for future ULSI interconnects", *japan.J.Appl.Phvs.* Vol.43.No.4B, pp.1856-1859,2004.

[22] D. S. Bethune et al. *Nature*, vol. 363, pp. 605-607, 1993.

[23] A. Thess et al., *Science*, vol. 273, pp. 483-487, 1996.

[24] International Technology Roadmap for Semiconductors, 2007, available: (www.itrs.net/Links/2007ITRS/2007).

[25] Dresselhaus, M.S., Dresselhaus, G. and Avouris, P., *Carbon Nanotubes: Synthesis, Structure, Properties and Applications*. New York: Springer- Verlag, 2001.
[26] Wu, W. and Maex, K., "Studies on size effects of copper interconnect lines," in *Proc. Solid-State and Integrated-Circuit Technology*, Shanghai, China, vol. 1, pp. 416–418 Oct. 2001.
[27] Zhong, G. and Koh, C.-K., "Exact close form formula for partial mutual inductances of on- chip interconnects," In *Proc. Int. Conf. Computer Design*, Freiburg, Germany, pp. 428–433, 2002.
[28] Raychowdhury, A. and Kaushik, R., "Modeling of metallic carbon-nanotube interconnects for circuit simulations and a comparison with Cu interconnects for scaled technologies", *IEEE Transactions on Computer-Aided Design of Integrated Circuits and Systems*, Vol. 25 No.1, pp. 58-65 2006.
[29] P. J. Burke, "Luttinger Liquid Theory as a Model of the Gigahertz Electrical Properties of Carbon Nanotubes", *IEEE Trans. Nanotechnology*, Vol. 1, No. 3, pp. 129-144, 2002.
[30] McEuen, P.L. and Park, J.Y., "Electron Transport in Single-Walled Carbon Nanotubes," *MRS Bulletin*, Vol. 29, no. 4, pp. 272-2752004.
[31] Srivastava, N. and Banerjee, K., "Performance Analysis of Carbon Nanotube Interconnects for VLSI Applications", *ICCAD*, pp. 383-390,2005,.
[32] P. L. McEuen, M. S. Fuhrer and H. Park, "Single-Walled Carbon Nanotube Electronics," *IEEE Trans. Nanotechnology*, Vol. 1, No. 1, pp.78-85, 2002.
[33] M. K. Rai and S. Sarkar, "Influence of tube diameter on C nanotube interconnect delay and power output", *Physica Satus Solidi A* 298, No.3, pp. 735-739, 2011.
[34] M.A. El-Moursy, E.G. Friedman, "Power characteristics of inductive interconnect," *IEEE Trans. Very Large Scale Integration (VLSI) Syst. 12(12)* pp. 1295–1306,2004.
[35] Rajeevan Chandel, S. Sarkar and R.P. Agarwal "Repeater insertion in global interconnects in VLSI circuits", *Microelectronics International journal, Vol.22 · Number 1*,pp. 43–50, 2005.
[36] R. Chandel, S Sarkar and R. P. Agarwal, "Delay and Power Management of Voltage-scaled Repeater Driven Long Interconnects",*Int. J. Modeling and Simulation*, Vol. 27, pp.333-339, 2007.
[37] Predictive Technology Model .[Online] . Available: www.eas.asu.edu/~ptm/.
[38] Brajesh Kumar Kaushik, Sankar Sarkar and R.P. Agarwal, "Waveform analysis and delay prediction for a CMOS gate driving RLC interconnect load", *Integration, the VLSI journal 40*,pp.394–405, 2007.
[39] H. Li, W. Yin, and J. Mao, "Modeling of Carbon Nanotube Interconnects and Comparative Analysis with Cu Interconnects" *Proceedings of Asia-Pacific Microwave Conference,* 2006.
[40] Y. Massoud and A. Nieuwoudt," Performance Analysis of Optimized Carbon Nanotube Interconnect", *IEEE Symposium on Circuits And Systems*, pp.792-795 2008, ISCAS-2008.
[41] Mayank Kumar Rai, Nivedita and Sankar Sarkar, "Carbon Nanotube Based Interconnects for VLSI Application", *IE(I) Journal-ET*, vol. 91, pp. 3-6, 2011.
[42] Hong Li, Wen-Yan Yin, Kaustav Banerjee, and Jun-Fa Mao, "Circuit Modeling and Performance Analysis of Multi-Walled Carbon Nanotube Interconnects", *IEEE Transactions On Electron Devices, Vol.* 55,NO. 6, JUNE 2008.

[43] L.Gal, "On-Chip crosstalk-The new signal integrity challenge," in proc.*Custom-Integrated Circuits Conf.*, pp.12.1.1-12.1.4,May 1995.

[44] A.K.Palit,V.Meyer, W.Anheier, and J.Schloeffel, "Modeling and analysis of Crosstalk coupling effect on the victim interconnect using the ABCD network model," in proc. *IEEE Int. Work. Defect and Fault Tolerance VLSI Syst.*,pp.174-182, 2004.

[45] L.D.Huang,H.M.Cheng,and D.F.Wong, "Global wire bus configuration with minimum delay uncertainty," Proc.*IEEE*,pp.50-55,2003,A.Kanwal,*A Review of Carbon Nanotube Field Effect Transistors.Piscataway, NJ:Rutgers Univ.*,2003.

[46] B.K. Kaushik and S. Sarkar, "Crosstalk analysis for a CMOS-gate driven coupled interconnects", *IEEE Tras. Computer aided design of Integrated circuits and systems*, vol. 27, No. 6, pp.1150-1154, 2008.

[47] Daniele Rossi, José Manuel Cazeaux, Cecilia Metra, and Fabrizio Lombardi, *IEEE Transactions On Nanotechnology* ,vol. 6, no. 2, pp.133-145, MARCH 2007.

[48] Shao-Ning Pu,Wen-Yan Yin,Senior Member,IEEE, Jun-Fa Mao,Senior Member,IEEE,and Qing H.Liu,Fellow,IEEE, "Crosstalk prediction of Single - and Double - walled Carbon-Nanotube (SWCNT/DWCNT) Bundle Interconnects", *IEEE Transactions On Electron Devices, Vol.56.No.4, pp. 560-568,*APRIL 2009.

[49] Debaprassad Das and Hafizur Rahaman, "Crosstalk analysis in carbon Nanotube interconnects And its impact on gate oxide reliability", *IEEE 2nd Asia symposium on quality electronic design*, 2010.

23

# Carbon Nanotube Based Magnetic Tunnel Junctions (MTJs) for Spintronics Application

Elby Titus[1], Manoj Kumar Singh[1], Rahul Krishna[1], Ricardo G. Dias[2], Antonio Ferreira[2] and Jose Gracio[1]

[1]*Nanotechnology Research Division, Center for mechanical technology and automation, Department of Mechanical Engineering, University of Aveiro,*
[2]*Department of Physics and Institute for Nanostructures, Nanomodelling and Nanofabrication (I3N), University of Aveiro*
*Portugal*

## 1. Introduction

Spintronics devices exploiting the spin of the electron [1-10] are prepared to revolutionise the electronics industry. The significance of this new generation device is faster memory and lower power consumption at low electron density. The late 20th century has been considered as an era of microelectronics. However, the avalanche growth of microelectronics is a major threat to Moore's law and spintronics may be a solution for it. From the first transistor to the signally powerful microprocessor in our recent computers, most electronic devices have employed circuits that express data as binary digits, or bits (1 and 0) represented by the existence or absence of electric charge. Unlike microelectronics, spintronics exploits spin (spin up ↑ and spin down ↓) of the electron to carry information between devices.

The discovery of Giant Magnetoresistance (GMR) by Nobel Prize winners Albert Fert and Peter Grünberg had actually led to the birth of novel field spintronics [11]. Currently, most of the existing spintronic devices [12] are based on metallic systems such as magnetic tunnel junctions (MTJs) and single electron transistor [13]. On the other hand, a wealth of intriguing spin phenomena has been observed in nanoscale materials [14]. This triggered an extensive research effort on spin transport in nanoscale MTJs and other interesting phenomena were realised. One of the most important phenomena is tunnel magnetoresistance (TMR) of the MTJs. A MTJ is composed of two ferromagnetic conducting layers separated by an ultra-thin insulating layer [15-20]. The TMR was first demonstrated by M. Jullière [21]. Yakushiji et al. [22] experimentally demonstrated the influence of spin conduction on TMR. The enhancement and oscillation of TMR in ferromagnetic multiple junctions have been predicted by several authors [23-30]. However, there have been only a few experiments on spin-dependent single electron tunnelling (SET) to date [31-36] due to the difficulty in fabricating appropriate sample structures for spin-dependent SET.

The desire to build spintronic devices that show larger spin dependent phenomena has led many researchers to combine single electron tunneling (SET) and spin dependent electron tunneling (SDT). The charge quantization in low capacitance magnetic tunnel junctions

(MTJs) can lead to SET phenomena represented by Coulomb blockade (CB) below a threshold voltage and Coulomb staircase (CS) at higher voltages [37]. Carbon nanotubes (CNTs) can actually be thought of as a spacer in MTJ devices due to its one dimensional nano structure with ballistic conduction. It can control current transport by quantum mechanical spin degree of freedom. The remarkable spin dependent phenomena are expected in these devices due to the interplay between SET and spin dependent electron tunneling (SDT). In this chapter, we explore the state of the art MTJ devices with special emphasis to CNT. Novel phenomena, TMR and SET in spintronics devices are highlighted.

## 2. Spintronic devices

The working principle of a spintronic device follows the steps (i) information is stored into spins as an orientation (i.e. up or down), (ii) spin information is carried by mobile electrons along a path or wire and (iii) the information is then read at a final point. Figure 1 shows the schematic representation of a spintronic device. The spin orientation of conduction electrons will exist for several nanoseconds making them useful in electronic circuit and chip design. The most basic method of creating a spin-polarized current is to transport current through a ferromagnetic material and to transmit the electron spin carrying the information to the receiver point. Spin current is therefore an important tool to detect spin in spintronic devices. The important avenues for the development of spintronics devices are: (i) fabrication of nanoscale nanostructures including novel magnetic materials, thin films, hybrid structures, and functional materials, (ii) research on spin effect (spin injection, and spin transport and detection), (iii) demonstration of spintronic devices including giant magnetoresistance (GMR) and tunnel magnetoresistance (TMR) devices in magnetic tunnel junctions (MTJs) and (iv) study of SET in MTJs.

## 3. Magnetic Tunnel Junctions

A magnetic tunnel junction (MTJ) can be considered as a spintronic device since it is composed of two ferromagnetic materials, such as nickel, cobalt or iron, separated by an ultrathin layer of insulator with a thickness of the order of nanometre ($10^{-9}$m). It exhibits two resistances, low ($R_p$) or high ($R_{ap}$) depending on the relative direction of ferromagnet magnetizations, parallel (P) or antiparallel (AP), respectively. The insulating layer is so thin that electrons can tunnel through the barrier if a bias voltage is applied between the two metal electrodes. The schematic of a magnetic tunnel junction (MTJ) is illustrated in Figure 2. In MTJs the tunneling current depends on the relative orientation of magnetizations of the two ferromagnetic layers, which can be changed by an applied magnetic field. This phenomenon is called tunnel magnetoresistance (TMR). An important factor in TMR is the interaction between the electron spin (S) and angular momentum (L) that is, spin orbit coupling (SOC). An example of SOC is splitting of hydrogen spectrum [38-40]. The SOC deforms the electron shell as the direction of the magnetization rotates. This deformation also changes the amount of scattering undergone by the conduction electrons when traversing the lattice. There will be minimum resistance if the magnetizations are in parallel orientation and it will go to maximum with opposite orientations (Figure 3). Therefore, such kind of junction can be easily switched between two states of electrical resistance, one with low and one with very high resistance.

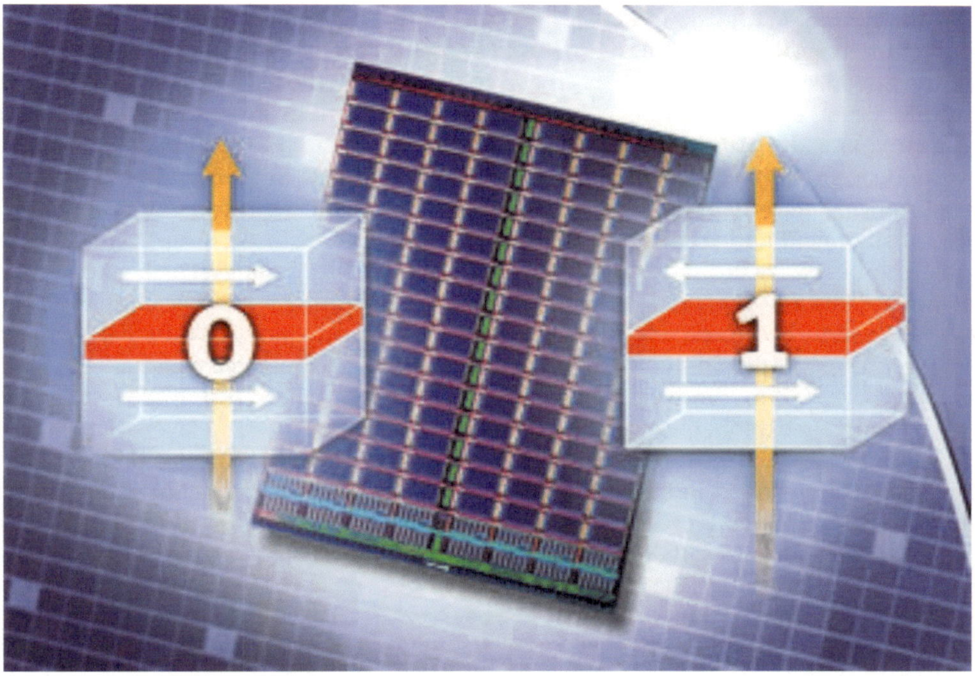


Fig. 1. A Schematic representation of spintronic device.

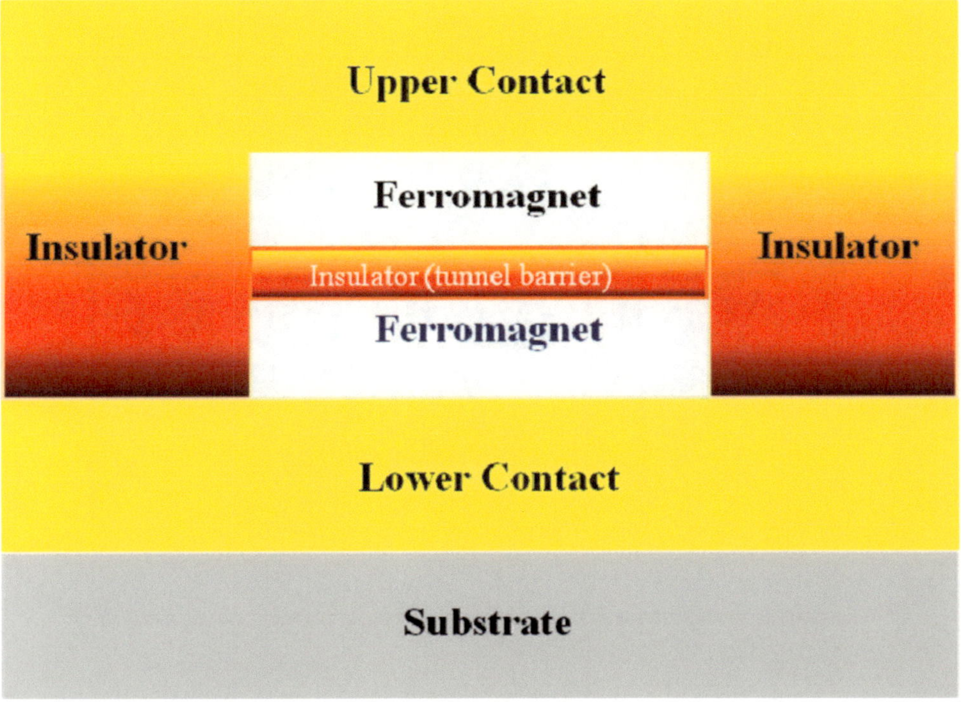


Fig. 2. A Schematic magnetic tunnel junction (MTJ).

### 3.1 Fabrication of MTJs

The fabrication of MTJs with high TMR ratios is crucial in developing spintronic devices. With the advance of nanotechnology, there are various methods to deposit MTJs, such as molecular beam epitaxy (MBE), magnetron sputtering, electron beam evaporation and chemical vapour deposition (CVD), and so on. In detail, the MTJ's main components are ferromagnetic (FM) layer and insulator layer. The FM layers can be fabricated by sputter deposition (magnetron sputtering and ion beam deposition). The fabrication issue is the magnetic alignment and thickness (deposition rates should be in the Angstrom-per-second range). The best way of fabricating insulating layer is still under research. Some of the proven materials are $Al_2O_3$ tunnel barriers made by depositing a metallic aluminium layer in the range of 5-15 Å thickness. In addition, ion beam oxidation, glow discharge, plasma, atomic-oxygen exposure and ultraviolet-stimulated oxygen exposure are also alternate ways of insulator deposition. Since the first report on TMR by Julliere [21], many studies have been performed to explore this property, especially on $Al_2O_3$ insulating layers. The necessity of controlling the magnetic properties of the magnetic layers introduces special requirements on the deposition process. The maintaining of inherent magnetic anisotropy is crucial in the deposition process. This can be set by applying magnetic field during deposition. The thickness & uniformity of the material, the coercivity, magnetorestriction, all are important in controlling the magnetic anisotropy.

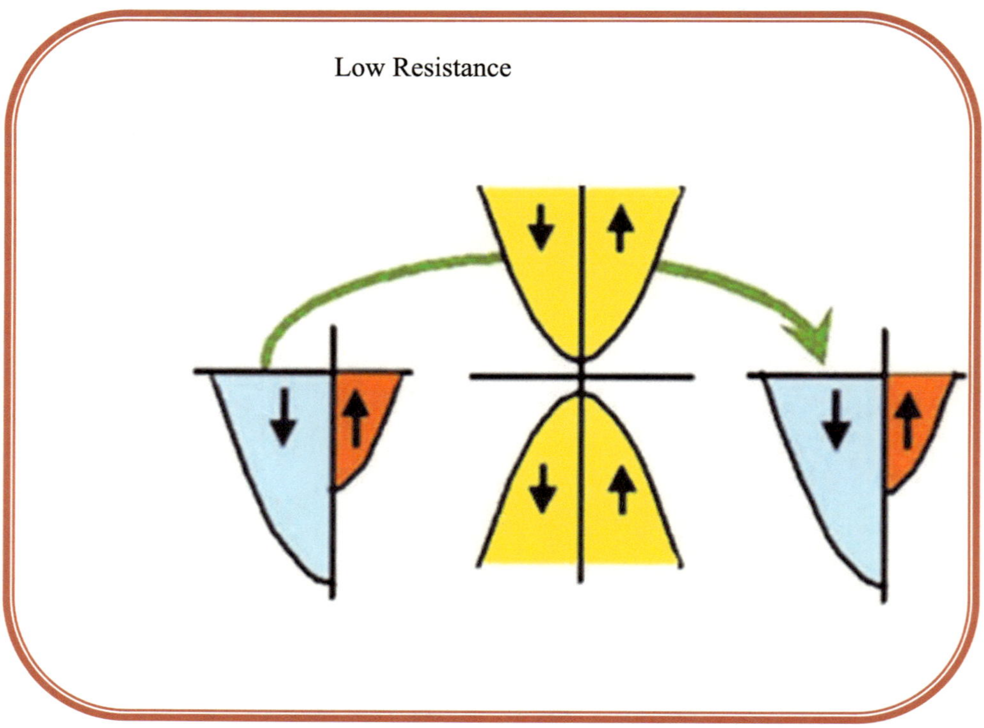


Fig. 3. Resistance goes from minimum to maximum with change in magnetization orientation from parallel to anti-parallel.

There are some theoretical predictions stating the MTJs with maximum TMR ratios could be fabricated by the epitaxial growth of ferromagnetic-non ferromagnetic-ferromagnetic nanoscale sandwich structure, Fe/MgO/Fe and Fe/MgO/FeCo [41, 42]. The high ratios resulted from the effective coupling of the majority spin band of Fe/FeCo into MgO and MgO into Fe/FeCo. TMR ratios up to 180% and 220% were achieved in these structures [43, 44]. The literature also says MTJs can be fabricated using half-metallic ferromagnets [45] and can generate 100% spin polarization at Fermi level (EF) due to the energy gap for one spin direction. The maximum spin polarization up to 100% is highly desirable in spintronic devices for the efficient spin injection from ferromagnetic electrodes into the non-ferromagnetic spacer and for the development of current induced magnetization switching in MTJs. Other MTJs with half-metallic ferromagnets which have shown higher TMR at room temperature are cobalt based alloy thin films including $Co_2Cr_{1-x}Fe_xAl$, $Co_2MnSi$ or Co-Mn-Al. In addition to MgO tunnel barrier, aluminium oxide also been used as tunnel barrier with ferromagnetic electrodes $Co_2Cr_{0.6}$ $Fe_{0.4}$ Al and $Co_2MnSi$. The advantage of these alloys is the high Curie temperature above room temperature. N. Tezuka et al have developed a MTJ of $Co_2FeAl_{0.5}Si_{0.5}$ electrodes and a MgO barrier fabricated by MBE and observed that their device had a TMR ratio of 386% at approximately 300K and 832% at 9K [46]. Nowadays, magnetic data storage technologies prefer perpendicular magnetic anisotropy (PMA) compared to the conventional devices that exploit the magnetization of the layers within the film plane. Figure 4 shows the cell structure with perpendicular magnetic anisotropy. An alloy of cobalt-iron is employed in the magnetic layer, with magnesium oxide in the insulating layer and cobalt-iron-boron in the interface layers. The structure developed by Toshiba is used for recording media. PMA has also been observed in several ferromagnetic materials including multilayers such as Co-Pt, Co-Pd, Co-Ni, CoFe-Pt, and CoFe-Pd, Co-Cr-Pt, alloys such as CoPt, FePt, and CoCr, and rare-earth transition metal alloys [47-50]. However, an ideal system with high thermal stability at nanoscale dimension, low current induced magnetization and high TMR was not successful. In view of this, S.

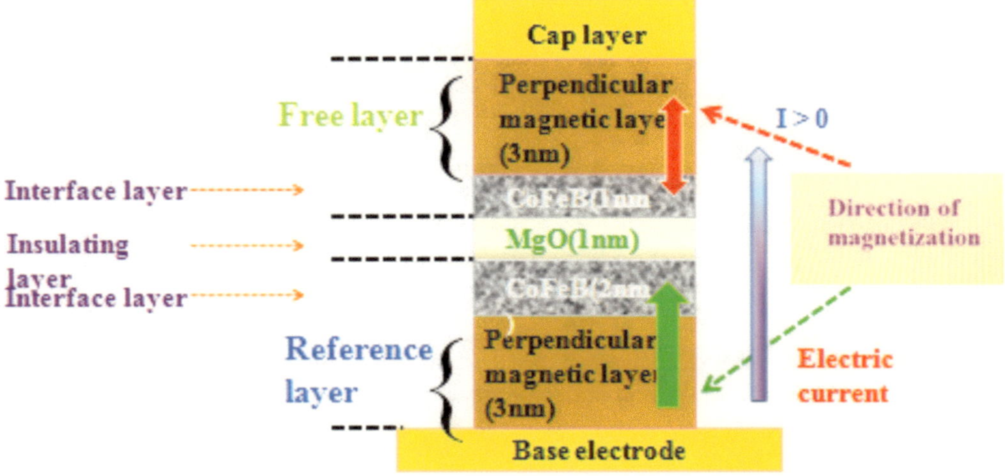


Fig. 4. Magnetoresistive random access memory (MRAM) with giga bits capacity developed by Toshiba.

Ikeda et al [51] employed interfacial PMA between the ferromagnetic electrodes and the tunnel barrier of the MTJ by using the material combination of CoFeB–MgO. The MTJs consisting of Ta/CoFeB/MgO/CoFeB/Ta showed a high TMR ratio, over 120%, high thermal stability at dimension as low as 40 nm diameter and a low switching current of 49 μA.

### 3.2 TMR in MTJs

To continue the discussion, TMR is a magnetoresistive effect that occurs in component consisting of two ferromagnets separated by a thin insulator (MTJ). The interest towards TMR is driven by the fact that MTJs with spin dependent tunneling (SDT) are expected to provide technical promises that will allow the realization of nanoscale devices in more advanced spintronic applications. T. Moodera et al [52] fabricated the first reproducible TMR up to 24% (Figure 5) at room temperature on CoFe/$Al_2O_3$/Co or NiFe junction. Today, reproducible TMR value up to 50% can be obtained with three dimensional ferromagnets making them useful for industrial application [ 53].

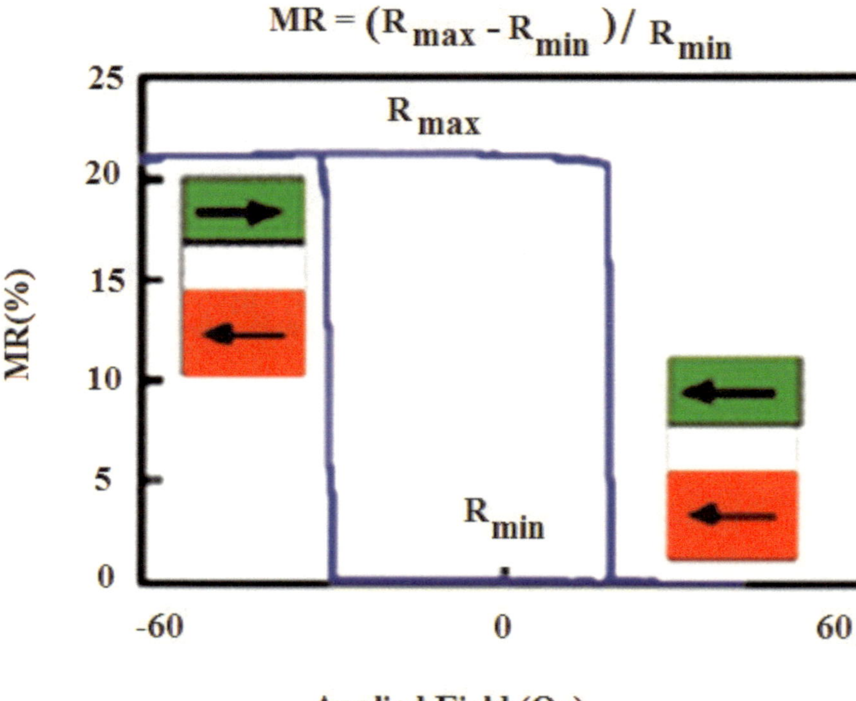


Fig. 5. Ratio of tunnel magnetoresistance (TMR) shown in CoFe/ $Al_2O$/ Co junction., J. S. Moodera, L. R. Kinder, T. M. Wong, and R. Meservey, Phys. Rev. Lett. 74, 3273 **(1995)** © American Physical Society

#### 3.2.1 Theory of TMR

The TMR effect in MTJs can be derived using Julliere's model [21, 51-53]. According to this model the magnetoresistance,

$$\text{TMR} = (R_{AP}\text{-}R_P)/R_P = 2P^2/(1\text{-}P^2) \tag{1}$$

where $R_P$ *and* $R_{AP}$ are the resistance in parallel and antiparallel magnetic configuration, respectively and $P$ is the spin polarization. The Differential TMR can be re-written as

$$\text{TMR} = \frac{(dI/dV_{AP})^{-1} - (dI/dV_P)^{-1}}{(dI/dV_{AP})^{-1}} \tag{2}$$

According to Wiesendanger [54] the derivative of an *I-V* curve (the differential conductivity is expressed as

$$\frac{dI}{dV} = \left(\frac{2\pi e^2}{\hbar}\right) |t|^2 D_1 \ (E_F) D_2 (E_F + eV) \tag{3}$$

where $|t^2|$ is the tunnel probability, $D_1$ and $D_2$ are the DOS of the two electrodes, $E_F$ is the Fermi energy, and $V$ is the bias voltage applied to the electrode at low temperature.
The injection of spin currents from a ferromagnetic manganite, $La_{0.7}Sr_{0.3}MnO_3$, into a single CNT (non-ferromagnetic) and the transformation of spin information into large electrical signal have been demonstrated by L. E. Hueso *et al* [55]. The TMR calculated in their system is 61%. R. S. Liu *et al.* [56] observed TMR (10%) due to the interplay of SDT and CB in a Ni/NiO/Co/NiO/Ni double junction of a submicron scale. Interestingly, both the systems are MTJ which is similar to our system. H.B Peng et al. [57] also studied I-V characteristics on patterned growth of SWCNT arrays from vapour-deposited Fe catalyst. Similar I-V features were observed in Nickel-CNT-Nickel MTJ system fabricated by the current authors [58]. Undoubtedly, Coulomb blockade (CB) effect, the principle of single electron tunnelling (SET), is responsible for the enhanced TMR in nanoscale MTJs. The SET phenomena and the associated MR occur only if the transport of electrons from one electrode to another is inhibited due to the extremely high electrostatic energy $e^2/2C$ (e = charge of electron and C is the capacitance) of a single electron compared to the thermal energy $k_BT$. When the bias voltage increases and exceeds the threshold $V^{th} = e/2C$, the current starts to increase. If the resistance of two junctions are similar (R1≈ R2), the current increases smoothly with bias voltage. T. Niizeki et al [59] could reduce the junction area to 10nm and observed enhanced TMR due to the coulomb blockade (CB) effect.
Motivated by the possibility of spintronics, the current authors have attempted a novel MTJ system using vertically aligned CNTs [58, 60]. The TMR along with Coulomb blockade and Coulomb staircase was observed in our system. Our device is formed of a vertical array of carbon nanotubes bridged between the nickel electrodes. The higher value of CB spacing (~ 2 V) in our system compared to the reported value (0.8V) is expected due to the bulk effect of CNT arrays. It also confirms that there is no short connection in the sample. From the I-V analysis of our system it appears that TMR is originating from the spin polarisation of the electrodes, i.e the difference between the density of states (DOS) of the up-and down-spin sub-bands in the two electrodes. More details of CNT- MTJs are given section 2.2.

### 3.2.2 Characterization

It is well known that the TMR is highly sensitive to the structural and chemical nature of the material. The characterization techniques such as electrical conductivity, magnetoresistance and tunneling microscopy can give detailed information on MTJs properties. The knowledge on magnetic property of the ferromagnetic electrodes is crucial in development of MTJ devices. The superconducting quantum interference device (SQUID) is the most sensitive magnetic field equipment to measure the magnetic property. It has enough sensitivity to measure measure the magnetic fields in nanoscale ferromagnets. The magnetization of the material can also be measured by vibrating-sample magnetometer (VSM) technique. VSM is based on Faraday's law which implies that an emf will be generated in a coil when flux changes in the coil. TMR also can be measured using the four probe method by sweeping the magnetic field. Scanning tunneling spectroscopy (STS)/scanning tunneling microscopy (STM) is another technique which can give precise TMR measurements. The details of all these measurements are given in section 5.

### 3.3 Application of MTJs

With wider knowledge on how to manipulate spins [61], we can build more state of the art spintronic devices with extraordinary properties. Extended research into application possibilities of any spintronic effects is therefore crucial to realise more advanced spintronic devices. These devices made huge impact on computer technology by enabling higher storage of information in hard drives and faster reading of data in random access memories. The first successful application of MTJ was demonstrated in computer read head technology with $Al_2O_3$ barrier and MgO barrier MTJ. The magnetic recording density in hard disk drive increased (300-600 Gbit/Inch$^2$) considerably in these devices [62-65]. Another application of MTJ is to develop magnetic random access memory (MRAM) devices. It has been claimed that MRAM can exceed the speed of SRAM (static RAM), density of DRAM (Dynamic RAM) and non-volatility of flash memory. In addition, the nano-dimension device has low power consumption and less heating. MRAM is an upgrade version of SRAM and DRAM where data is stored using spin instead of electrical charges. It overcomes one of the disadvantages of the conventional RAM, the loss of information by power failure. Leading companies like IBM, Motorola, and Honeywell started the MRAM research in 1995 and they were supported by United States Defense Advanced Research Projects Agency (DARPA). Figure 6 shows the images of MRAM used by leading companies like Thoshiba, IBM and Motorola.

## 4. Spin current in MTJs

In the view of rapid progress in the fabrication of nanoscale MTJs, spin is a subject of great interest. Spin is a purely quantum mechanical quantity which provides an extra degree of freedom for the electron to interact with a magnetic field. In 1922, Stern and Gerlach demonstrated the most direct experimental evidence of the existence and of the quantized nature of the electron spin. The first experimental evidence of spin dependent tunneling was reported by Jullerie [21] in 1975. Later, Berger proposed the idea that spin polarized current act on local magnetization of ferromagnets and leads to giant magnetoresistance [66]. The important property of spin is its weak interaction with the environment and with other spins, resulting in a long coherence or relaxation time, which is a very important parameter in the field of spin-transport and quantum computing. For the successful incorporation of spins into the currently existing electronics, one has to resolve issues such as efficient spin injection, spin transport, control and manipulation of spins and finally detection of spin polarized current.

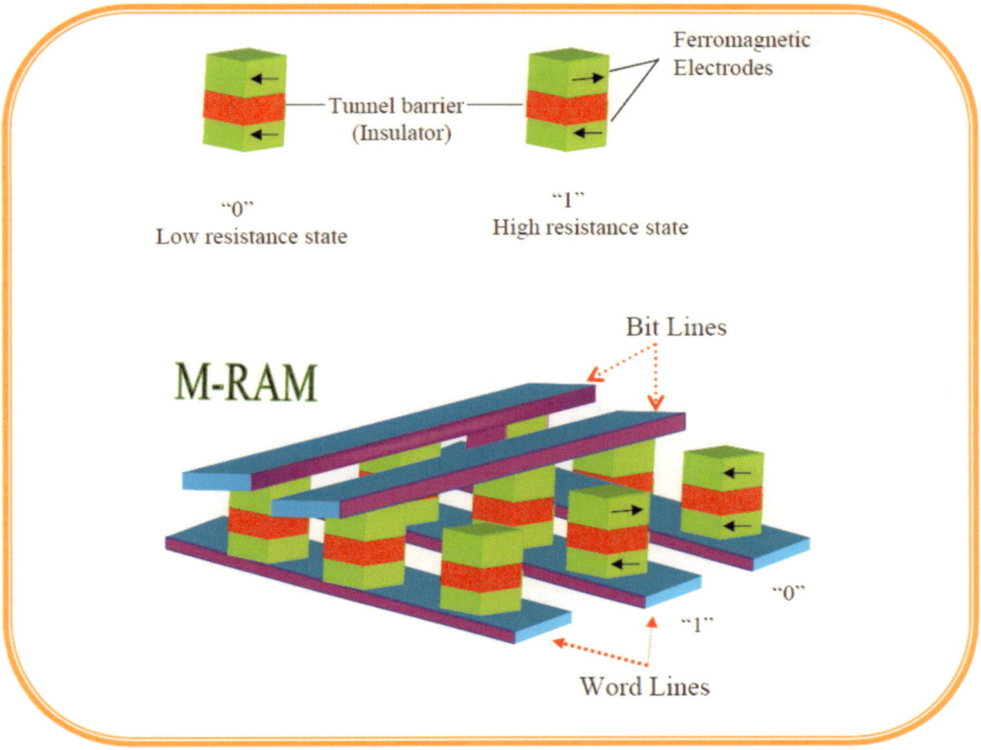


Fig. 6. MRAM memory cells composed of a magnetic tunnel junction (MTJ). States "0" and "1" in the cells correspond to the parallel and antiparallel alignments of the electrode magnetic moments.

Spintronics without magnetism is an attractive pathway for designing semiconductor spintronic devices since spin orbit coupling (SOC) enables that the spin is generated and manipulated merely by electric field. By the application of electric field, the electrons move in the lattice generating a magnetic field which acts up on the spin. The spin orbit interaction on mobile electrons was proved theoretically many decades ago. However, the practical harnessing of this concept is still at an early stage

### 4.1 Spin transport

The influence of spin transfer in MTJs can be observed by measuring resistive loops as a measure of external applied field and applied voltage. By sweeping the magnetic and electrical field, one can observe sharp drop in resistance which is attributed to the switching from parallel to antiparallel and vice versa. The drop of resistance is associated with the TMR. One of the factors that affect drop of resistance and TMR is density of states (DOS) at the interface [67-71].

### 4.2 Spin polarization

In addition to the spin transport, spin injection and spin polarization also an important factors in governing TMR. The spin polarization is a result of a subtle cancellation between two spin channels and is greatly influenced by the atomic, electronic and magnetic structures of the system. While the fundamentals of electron tunnelling are well understood, the quantitative theoretical description is lacking in real systems due to limitations in fabrication. Sophisticated and stable nanofabrication method will solve the problem of interface in MTJs. However, to build up on experimental findings, it is also essential to develop an accurate model of the spin polarization and transport of spin current through the ferromagnetic/nonferromagnetic interface and finally into vacuum which is highly sensitive to the chemical and material details of the device. In this context, Density functional theories [72] of MTJ system that can produce spin polarization effects in the Fermi Energy (FE) are important. Density functional theory (DFT) is a widely used method for modelling charge/spin carrier transport semiconductors. There is plenty of literature on DFT based calculations in studying SDT in MTJs [73-75]. The key components in the modelling are schottky barrier ($\varphi_b$) and the applied voltage $V_A$ against current density. Ab initio simulation of magnetic tunnel junctions has been demonstrated by Derek Waldron *et al* [76]. The effect of schottky barrier profile on spin dependent tunnelling in a ferromagnet-insulator-semiconductor system is reported in N.L.Chung *et al*'s work [77].

### 4.3 Theoretical modelling

Landauer [78] formula is really useful in order to compute spintronic devices. For the simplicity of modelling, the system can be assumed as one-dimensional, composed by a quantum wire with length L and two ferromagnetic reservoirs with electrochemical potential μL and μR which are given by

$\mu L^{\uparrow\downarrow} = E_F +(-)\Delta\mu$,

$\mu R^{\uparrow\downarrow} = E_F +(-)\Delta\mu + V_A$,

$\Delta\mu = (\mu\uparrow - \mu\downarrow)/2$ being half of the spin splitting energy. The effect of the Schottky barrier is important on the SDT and therefore negligible spin relaxation in the tunnel barrier should be considered.

If *f* is a function in the k space, the sum of f over k is given by

$$I = \sum_{\mathbf{k}} f(\mathbf{k}) \tag{4}$$

Let us assume that the transport in the channel is ballistic and the electrons entering the reservoir is in equilibrium. The positive current, carried by the k> 0 state in i subband is then given by

$$I_i^{>} = \frac{q}{L}\sum_{k>0} v_i(k) f(E - \mu_L) \tag{5}$$

where υi is the velocity and f is the Fermi-Dirac distribution function.

Using

$$\Sigma_k f(k) \to 2\,\frac{L}{2\pi}\int f(k)dk \tag{6}$$

eqn 6 can be rewritten as

$$I_i^{>} = \frac{q}{\pi}\int_0^\infty v_i(k) f(E-\mu_L)dk \tag{7}$$

Since $v_i = \frac{2\pi}{h}\frac{\partial E}{\partial k}$, the positive current can be expressed as

$$I_i^{>} = \frac{2q}{h}\int_{\epsilon_i}^\infty f(E-\mu_L)dE \tag{8}$$

where h is the Planck's constant and $\epsilon_i$ is the cutoff energy of the i subband. The negative current is similarly given by

$$I_i^{<} = \frac{2q}{h}\int_{\epsilon_i}^\infty f(E-\mu_R)dE \tag{9}$$

and the total current as

$$I = I^{>} - I^{<} = \frac{2q^2}{h} M \frac{\mu_L - \mu_R}{q} \tag{10}$$

$$M(E) = \sum_i u(E-\epsilon_i), \tag{11}$$

is the number of modes in subband i having energy less than E, and u is a Heaviside function. Low temperature has been assumed so that $f(\epsilon-\mu)\approx u(\mu-\epsilon)$ and M has been assumed to be constant in the energy range

$$\mu_2 < \epsilon < \mu_1 . \tag{12}$$

The total current density will be calculated by adding current of each spin channel,

$$I = I\uparrow + I\downarrow \tag{13}$$

and the spin current is given by the difference between up spin and down spin currents.
The introduction of a 3D system can further increase the consistency of calculation compared to 1D.
In this review, we emphasis on graphene and CNTs based MTJs which are the most recent attractions in MTJ research.

## 5. Graphene based MTJs

Graphene is the building block of graphite and it is made from a hexagonal lattice of carbon atoms with $sp^2$ bonds (Figure 7). Graphene is a promising material for spintronics and MTJs due to its intrinsic spin orbit interaction [79-87]. The electron spin of graphene also has interaction with the carbon nuclei. In addition, its unique properties, especially long spin flip length up to 1μm at room temperature enhances its potential to spintronics. They also show half integer quantum hall effect and high electrical conductivity. Many groups have studied TMR for graphene connected to ferromagnetic leads. They have demonstrated injection of spin polarized current in the system. Graphene is ideal for application of spin valve effect in which the resistance of the devices can be changed by manipulating the relative orientation of the magnetization. Hill et al [88] fabricated a spin valve device consisting of a graphene connected by two FM leads and investigated a 10% change in resistance as the electrodes switch from parallel to antiparallel state.

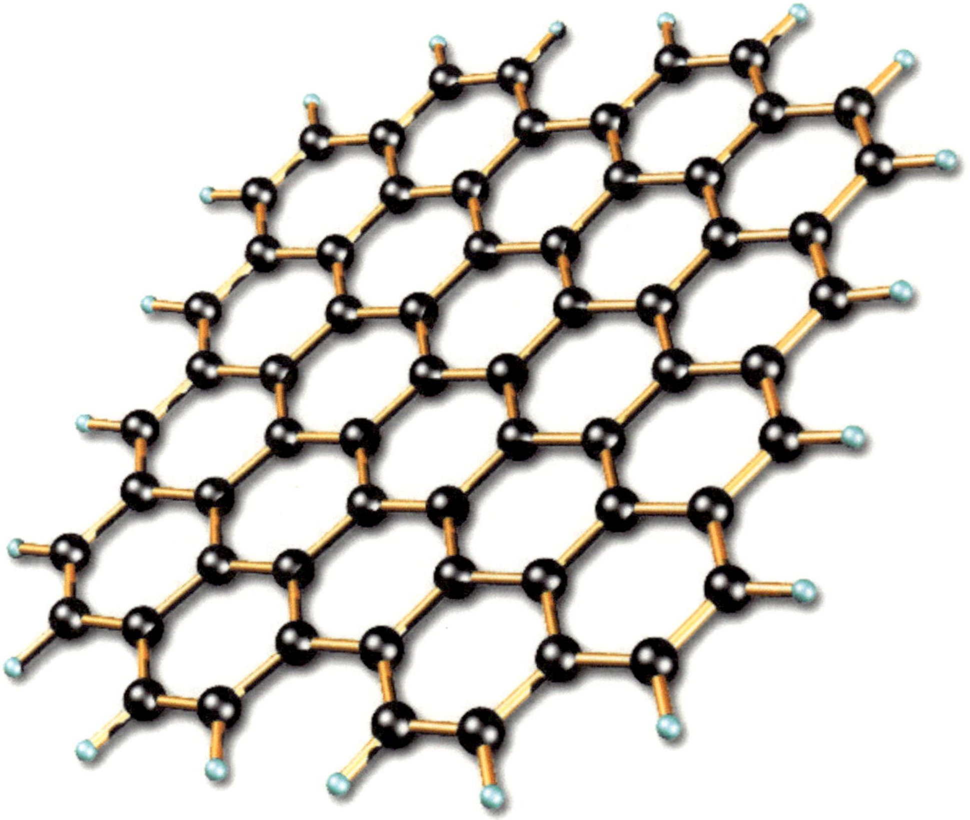

Fig. 7. Graphene showing the 2-D network of $sp^2$ bonded carbon atoms.

Experimentally, there have been few evidences for spin-polarized transport in graphene spin valves [89]. Wang et al investigated the magnetoresistance (MR) properties of quasi-two-dimensional mesoscopic graphite (MG) spin valve devices consisting of flakes with thickness between 1 and 40 nm (3 to 100 layers of graphene) contacted by two ferromagnetic electrodes and observed signatures of spin-polarized transport for MG flakes in the thickness range 10–40 nm. For devices in which an ultrathin magnesium oxide (MgO) tunnel barrier is inserted at the FM/MG interface, the spin valve effect has been observed with MR magnitudes up to 12% at 7 K and signals persisting up to temperatures as high as 60 K. In contrast, the spin valve effect has not been seen in devices without MgO, suggesting the importance of spin-dependent interfacial resistance for spin injection into MG [90-92]. Investigation of the voltage bias dependence of the MR finds a reduction of the MR with increasing voltage and a correlation with the differential conductance. Finally, the spin valve signal exhibits oscillatory MR as a function of gate voltage [93-95].

## 6. Carbon nanotubes based MTJs

Carbon nanotubes (CNTs) [96] are molecular tubes of carbon with outstanding properties (Figure 8). They are among the stiffest and strongest materials known, and have remarkable electronic behaviour and many other unique properties. They are attractive for spintronic devices due to their nanoscale size, extremely large spin flip scattering lengths and because they can behave as one-dimensional ballistic quantum conductors [97-102]. Experimental investigations on coherent spin transport through Co-contacted CNTs showed that spin can be coherently transported over 130 nm through the carbon nanotube [103].

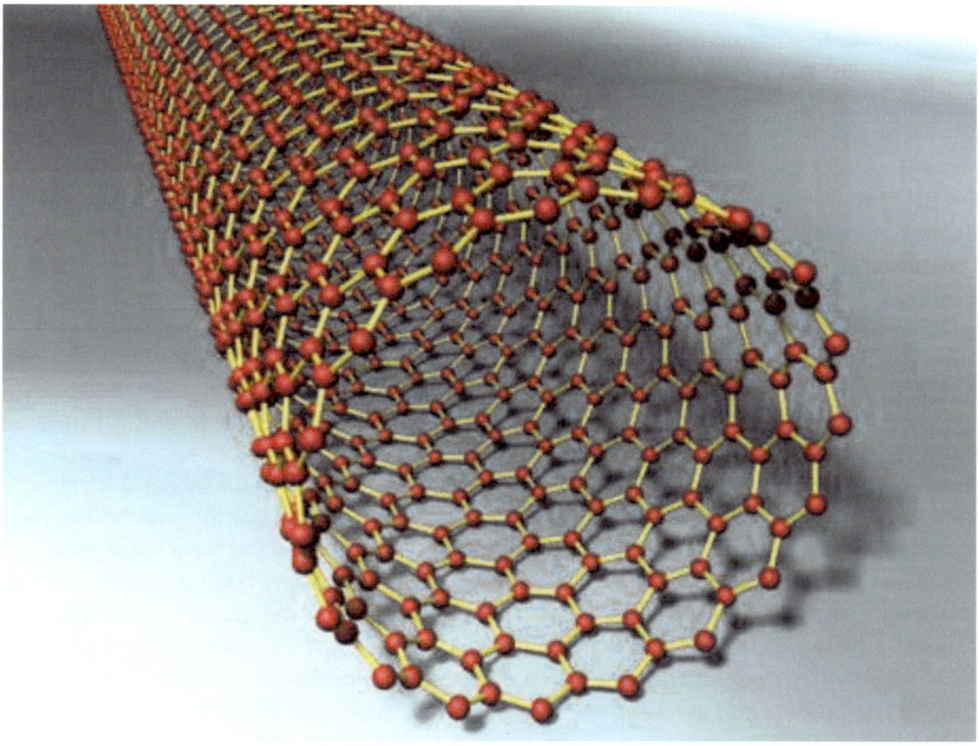

Fig. 8. Image of a carbon nanotube.

### 6.1 Fabrication of vertically aligned carbon nanotubes

Various conventional methods have been used to grow CNTs, including arc-discharge, laser ablation, and chemical vapor deposition (CVD). However, only chemical vapour deposition (CVD) methods allow the controlled growth directly on a substrate, which is important for many applications, especially as the individual manipulation of CNTs is difficult and expensive, due to their reduced size. The orientation of CNTs is highly important in achieving the maximum benefit of its property. The ability to pattern or position carbon nanotubes (CNTs) with a vertical orientation in controlled arrays has been demonstrated by many researchers [104-106]. Conventional plasma enhanced (PE) CVD is an established technique for the synthesis of vertically aligned CNTs[107-110]. Selective, aligned growth of CNTs on silicon and glass substrates has been demonstrated by plasma enhanced chemical vapor deposition. However, despite the high level of control, PECVD growth typically involves processing temperatures over 900 °C, which significantly limits the choice of the possible substrate materials and the integration processes. Bonard et al[6] have demonstrated in their work that the enhanced field emission factors of individual CNTs (30 000-50 000) are very large, when compared to the CNT films (1000-3000) [111-112]. The smaller field emission factors in the CNT film have been considered a consequence of the effect of the planar substrate supporting the CNT film. The field-enhancement factors of dense CNT films were even smaller because the electric field on one tube is screened by the proximity of neighboring tubes.

We reported the selective growth of VACNTs on large area copper substrates using a double plasma hot-filament chemical vapour deposition system (DP-HFCVD) [JNN paper 113]. We employed our modified HF-CVD system (Figure 9) for the deposition of large area (1-1.5 inch$^2$) carbon nanotubes onto nickel electroplated copper substrates. The system employed three independent power supplies, which were used to generate two independent plasma glow discharges within the vacuum chamber. The system consisted of two parallel molybdenum plates sandwiching a resistive tantalum filament. Each of the two molybdenum plates was electrically biased, either positively or negatively, with respect to the filament, in order to generate the two glow discharges. The precursor gases used were methane and hydrogen (gas flow: 46/200 sccm) at 7 mbar. The DC voltage (0-600 V) was applied between the upper substrate electrode and the filament. The filament power was maintained at 600 W and the substrate temperature (650 °C) was measured using a sealed thermocouple placed inside the substrate holder immediately beneath the substrate.

The growth of the CNTs involves three processes: (i) substrate pretreatment, (ii) nucleation and (iii) the actual growth of the CNTs.

(i) A 0.8 mm thick flat copper substrate with 98% purity was treated by sand blasting and then cleaned in ultrasonic acetone bath. The cleaned substrate was subsequently electroplated with nickel in a nickel sulphate bath. The copper substrates were used as the cathode and a nickel rod served as the anode. The 50 nm nickel coated copper substrates were subsequently transferred to the HF-CVD chamber and the chamber was purged by 100 sccm hydrogen gas for a few hours.

(ii) For the nucleation, the substrates were exposed to hydrogen plasma, promoting the etching of the catalyst and enabling the formation of nickel nano-clusters.

(iii) Finally, the CNTs were grown with hot filament power in the presence of precursor gases, methane and hydrogen.

The morphology of the CNTs deposited at different times was examined using SEM microscopy. Figure 10 shows the SEM image of CNTs deposited by DP-HFCVD (a) and conventional CVD (microwave) (b) in 2 hours. Unlike the conventional CVD CNTs, the DP-HFCVD CNTs appeared highly dense and were grown perpendicular to the substrate. It is reported that the growth rate in plasma enhanced (PE)CVD is generally slower than that in thermal CVD, which may be partly due to the atmospheric pressure operation of thermal CVD [114]. The CNTs grown by DP- HFCVD system also showed a slow growth rate and, taking almost 2 hours to form a forest of tubes. Bower *et al* [115] showed the effect of the electric field on the alignment of nanotubes unambiguously in a microwave plasma of acetylene and ammonia. Initially, when the plasma was on, their multiwalled (MW) CNTs were vertical; when growth proceeded with plasma off (in a thermal CVD mode), the nanotubes were found to be curly or randomly oriented. They also found that nanotubes always grew perpendicular to the substrate surface regardless of the substrate position or orientation. Since their work, a number of articles have appeared, considering the general orientation of PECVD-grown nanotubes better than the results from thermal CVD [116]. However, all these works were concentrated only on vertical alignment of the tubes, and the adherent properties of the tubes were ignored.

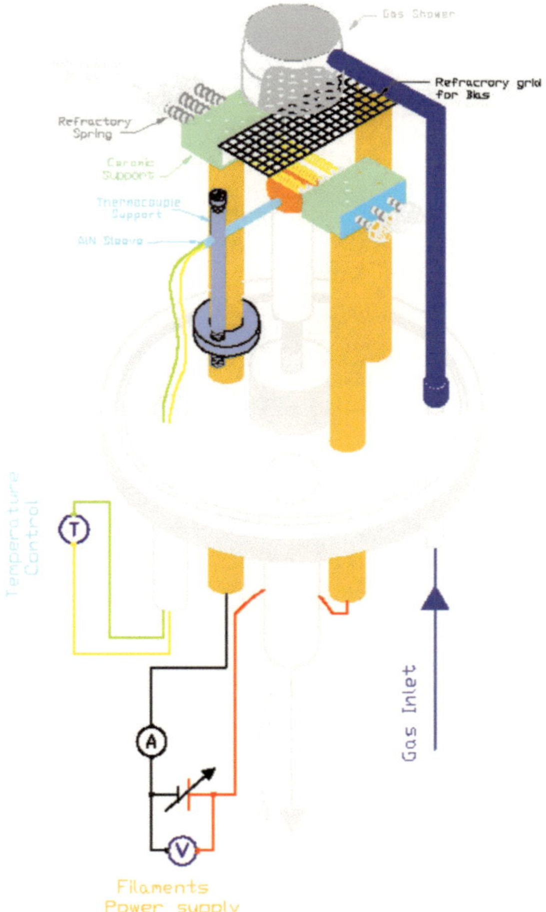


Fig. 9. Computer drawing of DP- HFCVD system.

The quality of the CNT was confirmed by Raman Spectroscopy. Figure 11 shows the micro-Raman spectra for samples (a) VACNT and (b) conventional CVD. The most prominent features in the first-order Raman spectra were observed for both the samples positioned at around 1570cm$^{-1}$ (G band) and 1350cm$^{-1}$ (D band). The peaks positioned at 1350 cm$^{-1}$ represent the disorder-induced phonon mode (D-band) and the strong peak at 1577 cm$^{-1}$ (G-band) can be assigned to the Raman-allowed C-C phonon mode (E2g-band) [117]. A small difference in the D and G peak positioning was noted with these samples and we attribute this to the structural differences in the tubes.

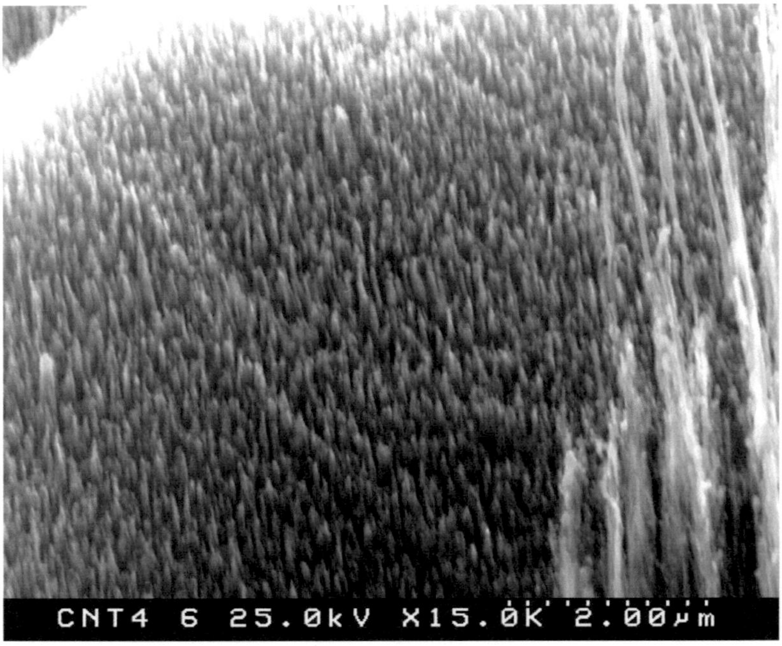


(a)

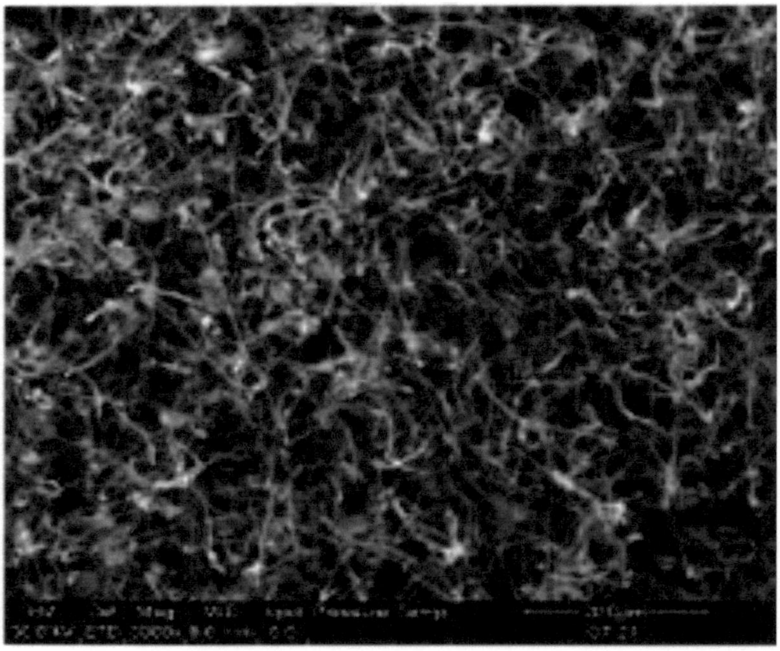

(b)

Fig. 10. SEM images of CNTs deposited by (a) DP-HFCVD and (b) Conventional CVD (micro-wave) in 2 hours.

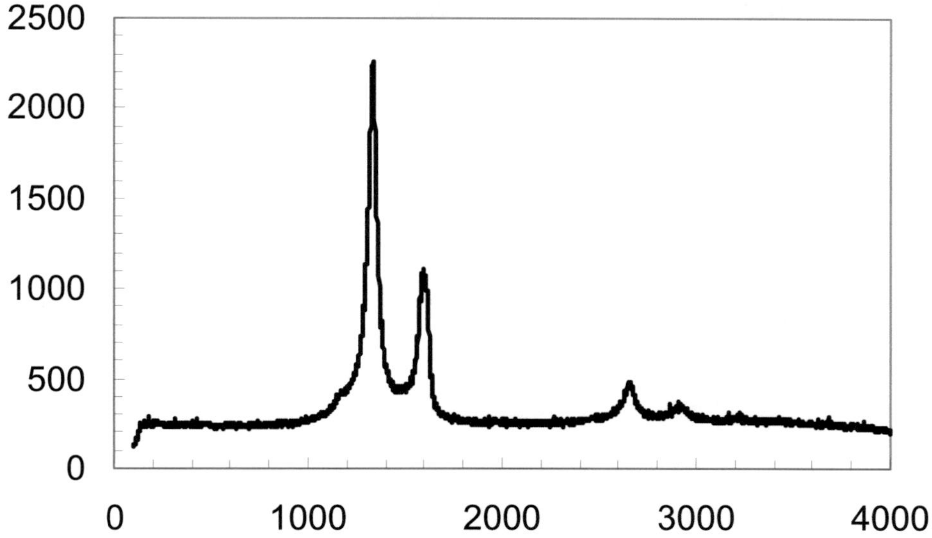


Fig. 11. Micro-Raman spectra of VACNT.

### 6.2 TMR in CNT based spintronic devices

Since its discovery, a large number of nanodevices such as single electron transistor, Light emitting diode, field effect transistor has been demonstrated. However, these devices are based on charge of the electron. TMR characteristics have already been measured in CNTs both experimentally and theoretically [118]. K Tsukagoshi et al [103] demonstrated the MR in a single CNT contacted by ferromagnetic metal electrodes. The spintronic devices exhibiting TMR using ferromagnet-contacted single walled carbon nanotubes (SWCNTs) have been demonstrated by A. Jensen *et al* [118]. Most of the reports on CNT-TMR system are mainly based on single carbon nanotube contacted to bulk ferromagnetic material by an *ex-situ* method [119-121]. The TMR effect is also known to be sensitive to the tunnel barrier/electrode interface. The barrier sensitivity may be more evident in a system with single CNT. De Teresa *et al* [122] studied ferromagnetic MTJ with various barrier materials and found that even the sign of the TMR depends on the barrier materials. S. Yuasa *et al* [123] also investigated the effect of crystal anisotropy of the spin polarisation on MTJ using single crystal iron electrodes of various crystal orientations. They found a clear crystal orientation dependence of the TMR which might reflect the crystal anisotropy of the electronic states in the electrodes.

The current authors followed a different approach for the development of a potential MTJ device. The details are discussed below.

Figure 12 shows the Field-Emission Scanning Electron Microscopy (FE-SEM) image of the MTJ system fabricated by the current authors. The Ni nanoparticles (on top, ~1μm in thickness) deposited on top of the VCNTs are apparent in figure 12. High density well aligned MWCNTs with length larger than 2 μm are apparent in the image. The details of the substrate pre-treatment method, the formation of nanoclusters and the synthesis of vertically aligned MWCNTs on nickel plated copper substrate using a double plasma method is reported in our previous paper [124]. Energy Dispersive X-ray microanalysis (EDS) attached to the FE-SEM reveals Ni nanoparticle deposition on the top surface of VCNTs. The magnified FE-SEM images of the green dashed square areas marked by 1 and 2 in Figure 12 are illustrated in inset. Unlike CNTs grown by conventional techniques, VCNTs grown by DP-HFCVD technique are short (~ 2μm), isolated and vertically aligned to the substrate. The green dotted square area 1 illustrates the VCNTs covered by nano nickel paticle. However, the particles are less in the lower portion of the sample (Green dashed square area 2).

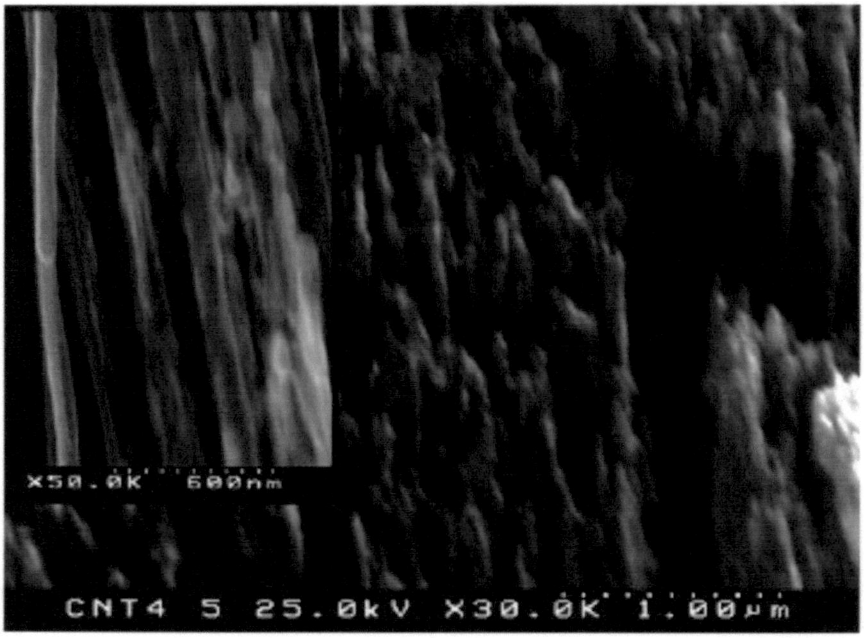


Fig. 12. FE-SEM image of the VCNT sample deposited with nickel nanoparticles. The inset shows a magnified FE-SEM image displaying isolated VCNTs.

TMR/GMR is known to originate from spin interaction between the magnetic and nonmagnetic particle at the interface and are related to the coercivity value [125]. To probe the magnetic properties, the field dependence of the magnetization was measured using a superconducting quantum interface device (SQUID) [126]. Figure 13a shows magnetization *vs.* applied magnetic field for nickel nanoparticle deposited VCNTs at 2 K and 300 K, respectively. It indicates that the sample exhibits ferromagnetic behaviour with a hysteresis loop at 2 K and 300 K, respectively. The saturation magnetization ($M_s$), coercivity ($H_c$), and remnant magnetization ($M_r$) for both temperatures are summarized in Table 1. The saturation magnetization values and coercivity for nickel nanoparticle deposited VCNTs shows a decrease from 5.3 to 4.4 emu $g^{-1}$ and 395 to 115 Oe, respectively, with the increase of temperature, indicating a characteristic ferromagnetic behaviour. To gain a better understanding of the magnetic behaviour of nickel nanoparticle deposited VCNTs, we have performed zero-field-cooled (ZFC) and field cooled (FC) magnetization measurements. For the ZFC measurement, the nickel nanoparticle deposited VCNTs sample is first cooled from 300 to 2 K in zero magnetic field. On the other hand, for FC measurements, the sample is cooled in the magnetic field (25 Gauss) from 300 to 2 K, and later the magnetization is measured in the warming cycle keeping the field on. Figure 13b shows the temperature dependence of ZFC and FC measurements under the applied magnetic field of 25 Gauss for nickel nanoparticle deposited VCNTs which exhibits the main features of ferromagnetic behaviour [127]. The blocking temperature $T_B$ (transition temperature from ferromagnetic to superparamagnetic state) peak can be observed in ZFC curve at about ~44 K. The low value of $T_B$ is directly in agreement with smaller size of nickel nanoparticles randomly deposited on the VCNTs [128,129].

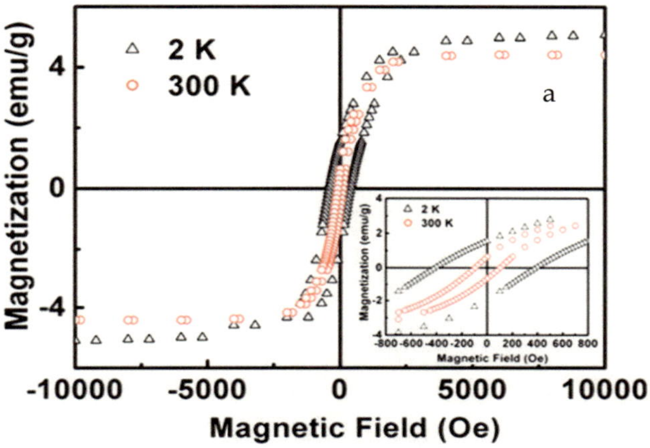


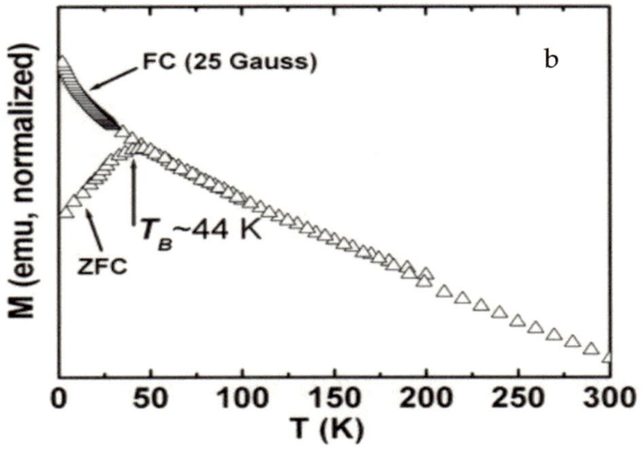


Fig. 13. (a) Magnetization vs. applied magnetic field for nickel nanoparticles deposited VCNTs at 2 K and 300 K. The ferromagnetic behaviour of the sample is apparent from the hysteresis (b) Temperature dependence of ZFC and FC measurements under the applied magnetic field of 25 Gauss for nickel nanoparticle deposited VCNTs.

| T/K | $M_s$/emu g$^{-1}$ | $H_c$/Oe | $M_r$/emu g$^{-1}$ |
|---|---|---|---|
| 2 | 5.3 | 395 | 1.6 |
| 300 | 4.4 | 115 | 0.6 |

Table 1. $M_s$, $H_c$, $M_r$ for Ni nanoparticle deposited VCNTs at 2 K and 300 K.

Figure 14 shows the simplest schematic configuration of the VCNT sample with a STM tip. The nickel substrate act as a source that can generate a current of spin polarised electrons, VCNTs act as a spacer that can transport spin polarised current and the nickel nanoparticle deposit act as a sensor that can detect the spin polarised current. The STM tip acts as contact to the top electrode and the Au to the bottom electrode. The magnetisation of the source defines the spin direction of the charge carriers injected into the CNTs and the nickel coating serves as a spin detector. Upon reversing the magnetisation direction of one of these electrodes, the spin polarisation of the charge carriers arriving at the detector electrode will be changed, giving rise to a change of the device resistivity, i.e., a magnetoresistance.

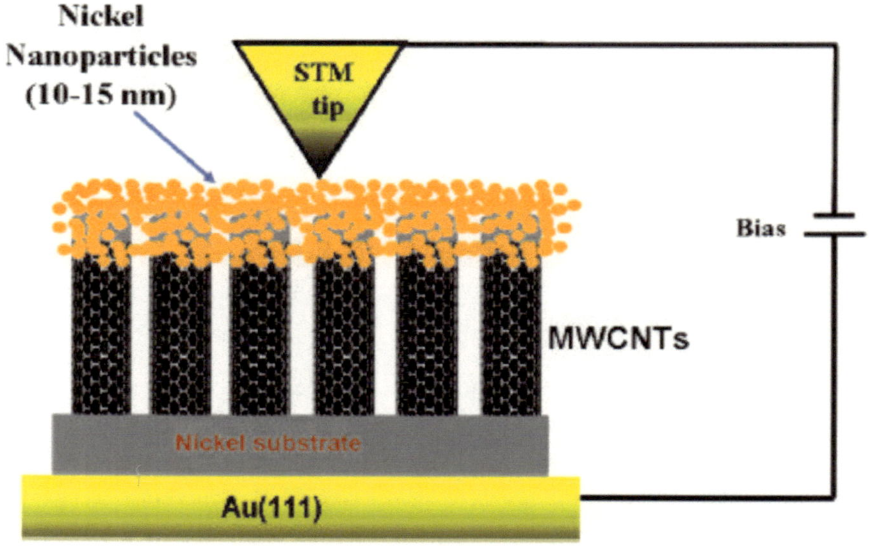


Fig. 14. Schematic diagram of spintronic devices.

Figure 15a shows the current-bias voltage ($I$-$V_b$) curves of the sample at H = 0 T and H = 2 T at 300 K. The Coulomb blockade region is observed between -1 V to 1 V and a stepwise increment of the current (CS) is observed beyond the CB region. The ($I$-$V_b$) curves exhibit similar features at zero and applied magnetic field (parallel and antiparallel configuration). The similarity in ($I$-$V_b$) at zero and applied field is reported in literature. The magnified portion of the green dotted square in the ($I$-$V_b$) curves is shown in Figure 15b. A clear CS is apparent in Figure 15b. The $dI/dV$ *vs.* $V_b$ plot (see Figure 15c) exhibits clear peaks in a periodic pattern beyond CB. The $dI/dV$ *vs.* $V_b$ plot in the range (-1.9 to -1 $V_b$) showing clear periodic pattern is illustrated in the inset of Figure 15c. The bias voltage ($V_b$) dependence of MR derived from the normalised $I$-$V_b$ curves shown in Figure 15d. The oscillation of MR as a function of bias voltage is observed.

The ~ null current in the CB region and the CS beyond CB region are clear indications of the SET phenomena. The SET phenomena and the associated MR occur only if the transport of electrons from one electrode to another is inhibited due to the extremely high electrostatic energy $e^2/2C$ (e = charge of electron and C is the capacitance) of a single electron compared to the thermal energy $k_BT$. When the bias voltage increases and exceeds the threshold $V^{th} = e/2C$, the current starts to increase. If the resistance of two junctions are similar (R1≈ R2), the current increases smoothly with bias voltage. A fluctuation in current is expected if there is difference between the resistances of two junctions. Interestingly, if the difference between the two junction resistances is very large (R1»R2 or R1«R2), the current increases stepwise (CS) with bias voltage depending on the number of electrons accumulated on the spacer. In that case, the tunnelling process in a current-biased junction is no longer random, but becomes more or less coherent (single electron tunnelling), at a frequency $\nu = I/e$. As a result the voltage across the junction will oscillate with the same frequency, and could in principle be used to define a quantum standard for the current.

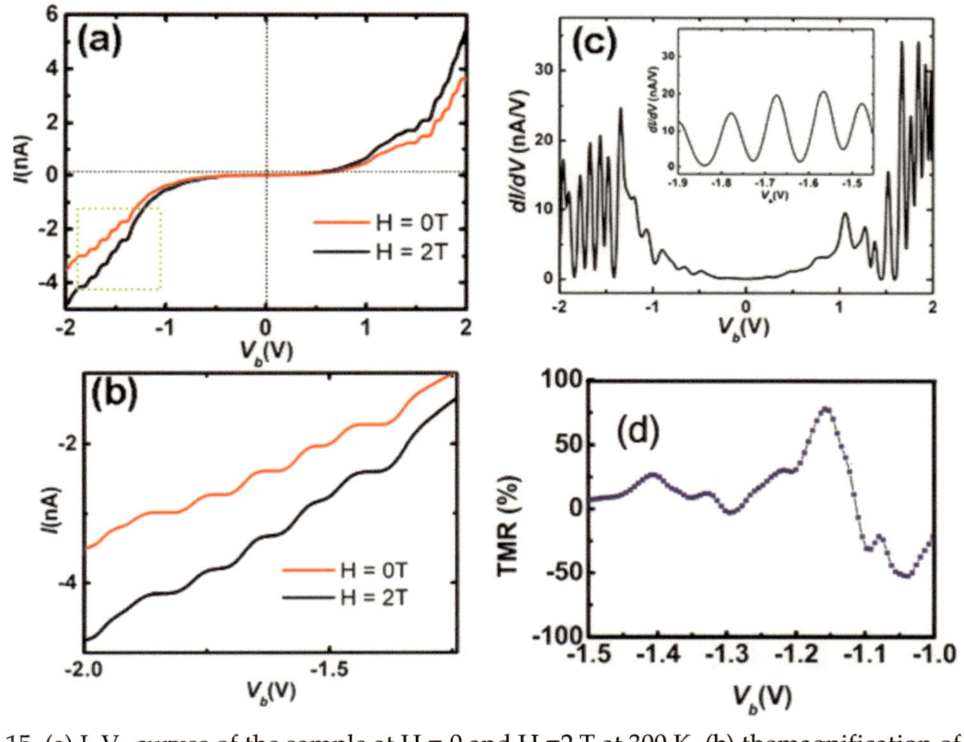


Fig. 15. (a) I–$V_b$ curves of the sample at H = 0 and H =2 T at 300 K, (b) themagnification of the I- $V_b$ (c) the dI/ dV vs. Vb plot of the sample at H= 0 T. (d) The MR% calculated from I-$V_b$ curves. The inset of (c) shows the I–Vb curve in the range -1.9 to -1 Vb.

The calculation of capacitance C is also crucial in determining the TMR effect. We have calculated the capacitance C from the period of $dI/dV$ curve. The peaks show a periodic pattern with a period of $\Delta V$ = 102.0 ±1.8mV. Assuming $\Delta V = e/C$, the capacitance is calculated as 1. 5705 × $10^{-18}$ F. Our result is in agreement with the work of A. B Mantel *et al.*[49] who has studied the spin injection in a single cobalt nanoparticle tunnel junction and they obtained C value as 1.14 $10^{-18}$ F. The C value calculated by us is consistent with the ferromagnetic nanoparticle tunnel junctions which confirm that the MR effect is due to the spin injection from nickel electrode into the non ferromagnetic CNT and spin detection by the nickel nanoparticle. The MR of the sample was also confirmed by low field measurements (Figure 16). The low field measurements were performed from 100 to 100 Oe and back to 100 Oe. The resistance peak appears as the field moves through zero and it shows a change of resistance from the parallel to antiparallel alignment of the magnetizations.

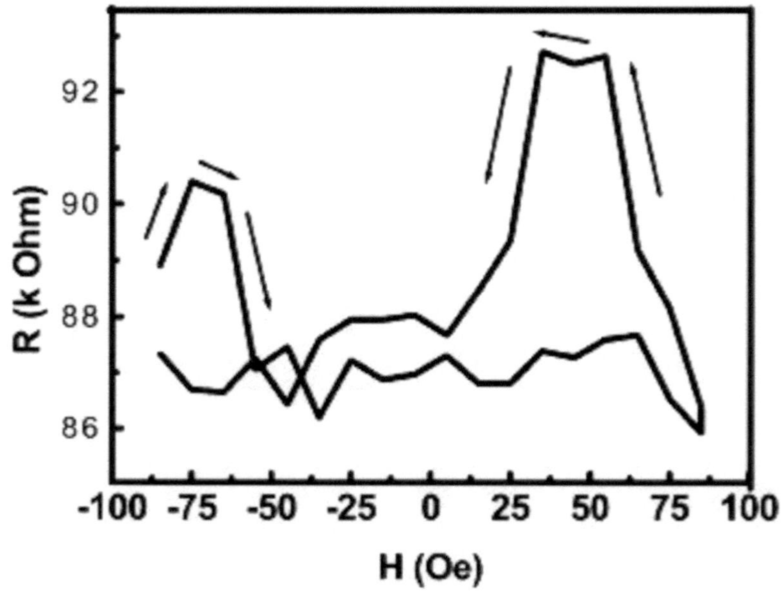


Fig. 16. Schematic of spintronic device showing vertically aligned carbon nanotubes (VCNT) deposited with nickel nanoparticles.

## 7. Conclusion

The development of room temperature spintronic devices is very motivating for future digital world. Spintronic devices can lead to advances in microelectronics creating faster and high capacity data storage-devices with less power consumption. The spin of the electrons can easily be manipulated in MTJ structures achieving novel phenomena of TMR and SET. However, bringing new or improved MTJs to market requires theoretical modeling and rapid assessment of the effect of enhanced material properties. Carbon nanotubes (CNTs) are ideal spin transporters due to their unique electronic properties.

## 8. References

[1] Y. Oshima, T.Takenobu, K. Yanagi, Y. Miyata, H. Kataura, K. Hata, Y. Iwasa, and H. Nojiri, Phys. Rev.Lett. 104, 016803 (2010).
[2] B.W. Alphenaar, K. Tsukagoshi, and M. Wagner, Physica E 10, 499 (2001).

[3] S. L.-Mei, G.Wei, C. S.-Xun, and Z. J.-Cang, Chin. Phys. Lett. 9, 3397 (2008).
[4] S.Roche, and R.Saito, Phys. Rev.Lett. 87, 246803 (2001).
[5] X. Hoffer, C. Klinke, J.-M. Bonard, L. Gravier, and J.-E.Wegrowe, Europhys. Lett. 67, 103 (2004).
[6] G. T. Kim, E. S. Choi, D. C. Kim, D. S. Suh, Y. W. Park, K. Liu, G. Duesberg, and S. Roth, Phys. Rev. B 58, 16064 (1998).
[7] F. A. Zwanenburg, D. W. van der Mast, H. B. Heersche, E. P. A. M. Bakkers, and L. P. Kouwenhoven, Nano Lett. 9, 2704 (2009).
[8] S.Krompiewskia, Acta Physica Polonica A, Proceedings of the European Conference Physics of Magnetism, (2008) June 24-27, Poznan, Poland.
[9] A. Bachtold, C. Strunk, J.-P. Salvetat, J.-M.Bonard, L. Forro, T. Nussbaumer, and C. Schonenberger, Nature, 397, 673 (1999).
[10] A. Bachtold, M. Henny, C. Terrier, C. Strunk, C. Schönenberger, J.-P. Salvetat, J.-M. Bonard, and L. Forró, Appl. Phys. Lett.73, 274 (1998).
[11] P. Baláž, Editor *Mechanochemistry in nanoscience and minerals engineering*, Springer, Berlin Heidelberg (2008).
[12] G. Baumgartner, M. Carrard, L. Zuppiroli, W. Bacsa, Walt A. de Heer, and L. Forró, Phys. Rev. B 55, 6704 (1997).
[13] X. Lou, C. Adelmann, S A.Crooker, E S. Garlid, J. Zhang, K. S.M. Reddy, S. D. Flexner, C. J. Palmstrøm, and P. A. Crowell, Nat. Phys. 3, 197 (2007).
[14] B. Zhao, I. Mönch, H. Vinzelberg, T. Mühl, and C. M. Schneider, Appl. Phys. Lett. 80, 3144 (2002).
[15] R.P. Hunt, IEEE Trans. Mag. 7, 150, (1971).
[16] M. N. Baibich, J. M. Broto, A. Fert, F. N. V. Dau, F. Petroff, P. Etienne, G. Creuzet, A. Friederich, and J. Chazelas, Phys. Rev. Lett. 61, 2472 (1988).
[17] H. Sato, P.A. Schroeder, J. Slaughter, W.P. P. Jr., and W. A.-Razzaq, Superlatt. Microstruct. 4, 45 (1988).
[18] T. Miyazaki, and N. Tezuka, J. Magn. Magn. Mater.139, 231 (1995).
[19] Y. M. Lee, J. Hayakawa, S. Ikeda, F. Matsukura, and H. Ohno, Appl. Phys. Lett. 90, 212507 (2007).
[20] Y. M. Lee, J. Hayakawa, S. Ikeda, F. Matsukura, and H. Ohno, Appl. Phys. Lett. 89, 042506 (2006).
[21] M. Julliere, Phys. Lett. 54 225 (1975).
[22] K. Yakushiji, F. Ernult, H. Imamura, K. Yamane, S. Mitani, K. Takanashi, S. Takahashi, S. Maekawa, and H. Fujimori, Nat. Mater. 4, 57 (2005).
[23] J. Mathon and A. Umerski, Phys. Rev. B 63, 220403 (2001).
[24] G. Baumgartner, M. Carrard, L. Zuppiroli, W. Bacsa, Walt A. de Heer, and L. Forró, Phys. Rev. B 55, 6704 (1997).
[25] R. Fiederling, M. Keim, G. Reuscher, W. Ossau, G. Schmidt, A. Waag, and L. W. Molenkamp, Nature (London) 402, 787 (1999).
[26] B. Zhao, I. Mönch, H. Vinzelberg, T. Mühl, and C. M. Schneider, Appl. Phys. Lett. 80, 3144 (2002).
[27] J. Z. Cai, L. Lu1,W. J. Kong, H. W. Zhu, C. Zhang, B. Q. Wei, D. H. Wu, and F. Liu, Phys. Rev. Lett. 97, 026402 (2006).
[28] S. Krompiewski, R. Gutie´rrez, and G. Cuniberti, Phys. Rev. B 69,155423 (2004).
[29] C. Strunk, Science 306, 63 (2004).
[30] X. Hoffer, Ch. Klinke, J.-M. Bonard, L. Gravier, and J.-E. Wegrowe, Europhys. Lett. 67, 103 (2004).

[31] A. N. Chantis, K. D. Belashchenko, E. Y. Tsymbal, and M. V. Schilfgaarde, Phys. Rev. Lett. 98, 046601 (2007).
[32] J. Velev, K. D. Belashchenko, S. S. Jaswal, and E. Y. Tsymbal, Appl. Phys. Lett. 90, 072502 (2007). "
[33] J. Velev, K. D. Belashchenko, D. Stewart, M. van Schilfgaarde, S. S. Jaswal, and E. Y. Tsymbal, Phys. Rev. Lett. 95, 216601 (2005).
[34] K. D. Belashchenko, J. Velev, and E. Y. Tsymbal, Phys. Rev. B 72, R140404 (2005).
[35] K. D. Belashchenko, E. Y. Tsymbal, I. I. Oleinik, and M. V. Schilfgaarde, Phys. Rev. B 71, 224422 (2005).
[36] K. D. Belashchenko, E. Y. Tsymbal, M. V. Schilfgaarde, D. Stewart, I. I. Oleinik, and S. S. Jaswal, Phys. Rev. B 69, 174408 (2004).
[37] F. Sols, F. Guinea, and A. H. CastroNeto, Phys. Rev. Lett. 99, 166803 (2007).
[38] M. Bratkovsky, Appl. Phys. Lett. 72, 2334 (1998).
[39] J. Zhang and R. M. White, J. Appl. Phys. 83, 6512 (1998).
[40] P.P. Dholabhai, R. Atta-Fynn, and A.K. Ray, Eur. Phys. J. B 61, 261 (2008).
[41] S. Yuasa, J. Phys. Soc. Jpn. 77,031001 (2008).
[42] E. Y. Tsymbal, O. N. Mryasov, and P. R. LeClair, J. Phys.: Condens. Matter 15, R109 (2003).
[43] C. Tiusan, F. Greullet, M. Hehn, F. Montaigne, S. Andrieu and A. Schuhl, J. Phys.: Condens. Matter 19,165201(2007).
[44] S. Ikeda, J. Hayakawa, Y. M. Lee, T. Tanikawa, F. Matsukura, and H. Ohno, J. Appl. Phys. 99, 08A907 (2006).
[45] C. Park, J.-G. Zhu, Y. Peng, D.E. Laughlin, and R. M. White, IEEE Trans. Magn. 40, 182 (2004).
[46] N. Tezuka, N. Ikeda, F. Mitsuhashi, and S. Sugimoto, Appl. Phys. Lett. 94, 162504 (2009).
[47] M. T. Johnson, P. J. H. Bloemen, F. J. A. den Broeder and J. J. de Vries,Rep. Prog. Phys. 59 1409 (1996).
[48] S. Yuasa, Y. Suzuki, T. Katayama, and K. Ando, Appl. Phys. Lett. 87, 242503 (2005).
[49] J. Hayakawa, Y. M. Lee, S. Ikeda, F. Matsukura, and H. Ohno, Appl. Phys. Lett. 89, 232510 (2006).
[50] M. G. Wang, C. Ni, A. Rumaiz, Y. Wang, X. Fan, T. Moriyama, R. Cao, Q. Y. Wen, H. W. Zhang, and J. Q. Xiao, Appl. Phys. Lett. 92, 152501 (2008).
[51] S. Ikeda, K. Miura, H. Yamamoto, K. Mizunuma, H. D. Gan, M. Endo, S. Kanai, J. Hayakawa, F. Matsukur and H. Ohno, Nat. Mat. 9, 721(2010).
[52] J. S. Moodera, L. R. Kinder, T. M. Wong, and R. Meservey, Phys. Rev. Lett. 74, 3273 (1995).
[53] S.S.P. Parkin, K.P. Roche, M.G. Samant, P.M. Rice, R.B. Beyers, R.E. Scheuerlein, E.J. O'Sullivan, S.L. Brown, J. Bucchigano, D.W. Abraham, Y. Lu, M. Rooks, P.L. Trouilloud, R.A. Wanner, and W.J. Gallagher, J. Appl. Phys. 85, 5828 (1999).
[54] R. Wiesendanger, Editor *Scanning Probe Microscopy and Spectroscopy*, New York: Cambridge University Press, 1994.
[55] L. E. Hueso, J. M. Pruneda, V. Ferrari, G. Burnell, J. P. V. -Herrera, B. D. Simons, P. B. Littlewood, E. Artacho, A. Fert, and N. D. Mathur, Nature 445, 410 (2007).
[56] R. S. Liu, H. Pettersson, L. Michalk, C. M. Canali, and L. Samuelson, Nano Lett. 7, 81 (2007).
[57] H. B. Peng and J. A. Golovchenko, Appl. Phys. Lett. 84, 5428 (2004).
[58] E. Titus, M. K. Singh, G. Cabral,V.Paserin, P. R. Babu, W.J. Blau, J. Ventura, J. P. Araujo, and J. Gracio, J. Mater. Chem. 19, 7216 (2009).
[59] T. Niizeki, H. Kubota, Y. Ando, and T. Miyazaki, J. Magn.Magn. Mater. 272, 1947 (2004
[60] N. Tombros, S. J. van der Molen, and B. J. van Wees, Phys. Rev. B 73, 233403 (2006).
[61] C. Chappert, Albert Fert, and F. N. V. Dau, Nat. Mat.6,813 (2007).
[62] S. Sanvito, Nat. Nanotechnol. 2, 204 (2007).

[63] I. Appelbaum, B. Huang, and D. J. Monsma, Nature 447, 295 (2007).
[64] E. Y. Tsymbal, K. D. Belashchenko, J. Velev, S. S. Jaswal, M. V. Schilfgaarde, I. I. Oleynik, and D. A. Stewart, Prog. Mater. Science 52, 401 (2007).
[65] A. N. Chantis, K. D. Belashchenko, D. L. Smith, E. Y. Tsymbal, M. V. Schilfgaarde, and R. C. Albers, Phys. Rev. Lett. 99, 196603 (2007).
[66] L. Berger, J. Appl. Phys. 49, 2156 (1978).
[67] E. Y. Tsymbal, K. D. Belashchenko, J. Velev, S. S. Jaswal, M. V. Schilfgaarde, I. I. Oleynik, and D. A. Stewart, Prog. Mater. Science 52, 401 (2007).
[68] A. N. Chantis, K. D. Belashchenko, D. L. Smith, E. Y. Tsymbal, M. V. Schilfgaarde, and R. C. Albers, Phys. Rev. Lett. 99, 196603 (2007).
[69] J. D. Burton, R. F. Sabirianov, J. P. Velev, O. N. Mryasov, and E. Y. Tsymbal, Phys. Rev. B 76, 144430 (2007).
[70] H. Suzuura, and T.Ando, Physica E 6, 864 (2000).
[71] J.-G. Zhu and C. Park, Mater. Today 9, 36 (2006).
[72] R. Arras, L. Calmels, and B. Warot-Fonrose, IEEE Trans. Magn.46, 1730 (2010).
[73] N. M. Caffrey, T. Archer, I. Rungger, and S. Sanvito, Phys. Rev. B. 83, 125409 (2011).
[74] M. Stilling, K. Stokbro, and K. Flensberg, J. Comp. Aided Mater. Des. 14, 141 (2007).
[75] Y. Ke, K. Xia, and H. Guo, Phys.Rev. Lett. 100, 166805 (2008).
[76] D. Waldron, L. Liu and H. Guo, Nanotechnology 18, 424026 (2007).
[77] N. L. Chung, M. B. A. Jalil, and S. G. Tan, J. Phys. D: Appl. Phys. 42, 195502 (2009).
[78] R. Landauer, IBM J. Res. Dev. 1, 233 (1957).
[79] K.S. Novoselov, A.K. Geim, S.V. Morozov, D. Jiang, Y. Zhang, S.V. Dubonos, I.V. Grigorieva, and A.A. Firsov, Science 306, 666 (2004).
[80] Y. Zhang, Y.-W. Tan, H.L. Stormer, and P. Kim, Nature 438, 201 (2005).
[81] A.K. Geim and K.S. Novoselov, Nature Mater. 6, 183 (2007).
[82] K. I. Bolotin, K. J. Sikes, J. Hone, H. L. Stormer, and P. Kim, Phys. Rev. Lett. 101, 096802 (2008).
[83] M. Maiti and K. Sengupta, Phys. Rev. B 76, 054513 (2007).
[84] S. Yuasa, T. Nagahama and Y. Suzuki, Science 297, 234 (2002).
[85] O.V. Yazyev and L. Helm, Phys. Rev. B 75, 125408 (2007).
[86] C. Bai and X. Zhang, Phys. Lett. A 372, 275 (2008).
[87] K.S. Novoselov, A.K. Geim, S.V. Morozov, D. Jiang, M.I. Katsnelson, I.V. Grigorieva, S.V. Dubonos, and A.A. Firsov, Nature 438, 197 (2005).
[88] E. W. Hill, A. K. Geim, K. Novoselov, F. Schedin, and P. Blake, IEEE Trans. Magn. 42, 2694 (2006).
[89] C. L. Kane and E. J. Mele, Phys. Rev. Lett. 95, 226801 (2005).
[90] F. J. Jedema, A. T. Filip, and B. J. van Wees, Nature 410, 345 (2001).
[91] S. Sahoo, T. Kontos, J. Furer, C. Hoffmann, M. Gräber, and C. Schönenberger, Nat. Phys. 1, 99 (2005).
[92] L. Gammaitoni, P. Hänggi, P. Jung, and F. Marchesoni, Rev. Mod. Phys. 70, 223 (1998).
[93] A. De Martino and R. Egger, J. Phys.: Condens. Matter 17, 5523 (2005).
[94] S. Garzon, I. Zutic, and R. A. Webb, Phys. Rev. Lett. 94, 176601 (2005).
[94] M. Bockrath, D. H. Cobden, P. L. McEuen, N. G. Chopra, A. Zettl, A. Thess, and R. E. Smalley, Science 275, 1922 (1997).
[95] N. Tombros, S. J. van der Molen, and B. J. van Wees, Phys. Rev. B 73, 233403 (2006).
[96] B.-C. Min, K. Motohashi, C. Lodder, and R. Jansen, Nat. Mater. 5, 817 (2006).
[97] L. I. Glazman and K. A. Matveev, JETP Lett. 48, 445 (1988).
[98] D. J. Monsma and S. S. P. Parkin, Appl. Phys. Lett. 77, 720 (2000).
[99] K. P. Kamper, W. Schmitt, G. Guntherodt, R. J. Gambino, and R. Ruf, Phys. Rev. Lett. 59, 2788 (1987).

[100] W. A. Hofer, A. S. Foster, and A. L. Shluger, Rev. Mod. Phys. 75, 1287 (2003).
[101] D. J. Monsma and S. S. P. Parkin, Appl. Phys. Lett. 77, 720 (2000).
[102] W. Liang, M. Bockrath, and H. Park, Phys. Rev. Lett. 88, 126801 (2002).
[103] K. Tsukagoshi, B. W. Alphenaar, and H. Ago, Nature 401, 572 (1999).
[104] J.M. Bonard, F. Maier, T. Stockli, A. Chatelain, W. A. Heer, J.P. Salvetat, and L. Forro, Ultramicroscopy 73 ,7 (1998).
[105] L. Dai, A. Patil, X.Y. Gong, Z.X. Guo, L.Q. Liu, Y. Liu, D.B. Zhu, Chem. Phys. Lett. 4 ,1150 (2003).
[106] W.I. Milne, K.B.K. Teo, E. Minoux, O. Groening, L. Gangloff, L. Hudanski, J.P. Schnell, D. Dieumegard, F. Peauger, I.Y.Y. Bu, M.S.Bell, and P. Legagneux, J.Vacuum Sci. Technol. 24, 345 (2006).
[107] M. Chhowalla, C. Ducati, N.L. Rupesingh, K.B.K. Teo, and G.A.J. Amaratunga, Appl.Phys. Lett. 79 , 2079 (2001).
[108] K.B.K. Teo, M. Chhowalla, G.A.J. Amaratunga, W.I. Milne, G. Pirio, P. Legagneux, F. Wyczisk, D. Pribat, and D.G.Hasko, Appl. Phys. Lett. 80 , 2011 (2002).
[109] J.M. Bonard, F. Maier, T. Stockli, A. Chatelain, W.A. de Heer, J.P. Salvetat, and L.Forro, Ultramicroscopy 73 , 7 (1998).
[110] M.S. Kabir, R.E. Morjan, O.A. Nerushev, P. Lundgren, S. Bengtsson, P. Enoksoni, and E.E.B. Campbell, Nanotechnology 16, 458 (2005).
[111] Z.F. Ren, Z.P. Huang, and J.W.Xu, Science 282 , 1105 (1998).
[112] C. Taschner, F. Pacal, A. Leonhardt, P. Spatenka, K. Bartsch, and A.Graff, Surface Coatings Technol. 81, 174 (2003).
[113] G. Cabral, E.Titus , D. S. Misra, and J.Gracio, J Nanosci Nanotechnol. 8, 4029 (2008).
[114] M. Wienecke, C. Mihaela, K. Deistung, P. Fedtke, and E.Borchartd, Carbon 44 , 718 (2006).
[115] C. Bower, O .Zhou, W. Zhu, D.J. Werder, and S .Jin, Appl. Phys. Lett.77, 2767 (2000).
[116] J.G. Wen, Z.P. Huang, D.Z. Wang, J.H. Chen, S.X. Yang, and Z.F. Ren, J.Mater Res. 16, 3246 (2001).
[117] A.M. Rao, A. Jorio, M.A. Pimenta, M.S.S. Dantas, R. Saito, G. Dresselhaus, M.S. Dresselhaus, Phys. Rev. Lett. 84 , 1820 (2000).
[118] A. Jensen, J. R. Hauptmann, J. Nygard, and P. E. Lindelof, Phys. Rev. B, 72, 035419 (2005).
[119] Z. H. Xiong, D. Wu, Z. V. Vardeny, and J. Shi, Nature 427, 821 (2004).
[120] H. Shimada, K. Ono, and Y. Ootuka, J. Phys. Soc. Jpn. 67, 1359 (1998).
[121] P. L. McEuen, M. Fuhrer, and H. Park, IEEE Trans. Nanotechnol. 1,78 (2002).
[122] J. M. de Teresa, A. Barthélémy, A. Fert, J. P. Contour, F. Montaigne, and P. Seneor, Science, 286,507 (1999).
[123] S. Yuasa, T. Nagahama, A. Fukushima, Y. Suzuki, and K. Ando, Nat.Matr.3, 868 (2004).
[124] E. Titus, M.K. Singh, Gil Cabral, V. Paserin, Ramesh Babu P, W.J. Blau, and J. Gracio, J. Nanosci. Nanotechnol. 10, 2606 (2010).
[125] I. Bergenti, A. Deriu, E. Bosco, M. Baricco, E. Angeli, D. Bisero, A. Da Re, F. Ronconi, F. Spizzo, and P. Vavassori, J. Magn. Magn. Mater. 272–276, 1554 (2004).
[126] S. Kenane, J. Voiron, N. Benbrahim, E. Chainet, and F. Robaut, J. Magn. Magn. Mater. 297, 99 (2006).
[127] Y. Li, T. Kaneko, T. Ogawa, M. Takahashi, and R. Hatakeyamma, Chem. Commun. 254, (2007).
[128] F. C. Fonseca, G. F. Goya, R. F. Jardim, R. Muccillo, N. L. Carre~no, E. Longo, and E. R. Leite, Phys. Rev. B 66, 104406 (2002).
[129] S. Linderoth, L. Balcells, A. Labarta, J. Tejada, P. V. Hendriksen, and S. A. Sethi, J. Magn. Magn. Mater. 124, 269 (1993).

# 24

# Mechanisms of Single-Walled Carbon Nanotube Nucleation, Growth and Chirality-Control: Insights from QM/MD Simulations

Alister J. Page[1], Ying Wang[2], K. R. S. Chandrakumar[1],
Stephan Irle[2] and Keiji Morokuma[1,3]
[1]*Fukui Institute for Fundamental Chemistry, Kyoto University,*
[2]*Institute for Advanced Research and Department of Chemistry, Nagoya University*
[3]*Cherry L. Emerson Centre for Scientific Computation and Department of Chemistry, Emory University,*
[1,2]*Japan*
[3]*U.S.A.*

## 1. Introduction

The experimental characterisations of carbon nanotubes (CNTs) (Iijima, 1991) and in particular single-walled CNTs (SWNTs) (Iijima & Ichihashi, 1993) in the early 1990s were landmark moments in 20th century science. The potential uses of these remarkable nanostructures are now becoming realised, as their synthesis is now routinely performed on the industrial scale. The initial successes in this respect were generally experimental techniques that were previously well established in other fields. This is particularly true of the chemical vapor deposition (CVD) and arc-discharge processes. The original experimental characterisation of SWNTs was in fact accomplished using nanotubes synthesised with the former method (Iijima & Ichihashi, 1993). The understanding of the way in which CNTs nucleate and grow was therefore synergic with the evolution and refinement of these synthetic methods. Indeed, the original mechanisms of CNT nucleation and growth were conceived from experimental observations. The most prevalent of these today is the vapor-liquid-solid (VLS) mechanism (Saito, 1995). According to this mechanism, SWNT nucleation growth is postulated to consist of three distinct stages. The first of these features a mixed carbon/catalyst vapor phase, from which co-condensation yields liquid catalyst-carbide nanoparticles. Typical catalysts in the growth of SWNTs are traditionally transition metals such as Fe, Ni, Co, Mo, and alloys thereof (see (Journet *et al.*, 1997; Moisala *et al.*, 2003; Harris, 2007), and references therein). The precipitation of atomic carbon from this liquid carbide phase takes place once the carbide phase is saturated with carbon. This precipitation yields the formation of solid phase CNTs. Due to the inherent limits in spatial and temporal resolutions that are furnished by experimental techniques and instrumentation, there inevitably remain questions regarding the VLS mechanism and CNT growth that, for now, cannot be answered from an experimental standpoint. There are several infamous examples in this respect. For instance, the mechanism of so-called 'catalyst-free' SWNT nucleation growth remains unknown, following the recent experimental reports demonstrating the growth of CNTs in the absence of a transition metal catalyst. The factors that govern 'chirality-controlled' growth (*i.e.* growth that produces a specific (*n*,*m*) chiral SWNT, as opposed to a broad distribution of (*n*,*m*) SWNTs) have also remained elusive to date.

It is in this respect that theoretical models of CNT growth have recently come to the fore by complementing, and in some cases pre-empting, experimental understanding of SWNT nucleation and growth. The most notable theoretical approach in this respect is molecular dynamics (MD). In this work, we will highlight the recent advances made in our understanding of SWNT nucleation and growth mechanisms gained from quantum mechanical MD (QM/MD) simulations. Following a brief review of experimental SWNT synthesis (§1.1) and previous theoretical investigations of SWNT nucleation and growth (§1.2), we will briefly outline the quantum chemical approach used in our simulations (§2). A discussion of QM/MD simulations of SWNT nucleation on both Fe and Ni catalysts will be presented in §3. SWNT nucleation as the result of gas-phase acetylene, C and $C_2$ adsorption, as well as the decomposition of the Fe- and Ni-carbide phases will be considered here. More recent simulations concerning the mechanism of SWNT nucleation on Si-based catalysts, particularly $SiO_2$, SiC and Si, will then be discussed in §3.4. In §4 we will discuss QM/MD simulations concerning the continued growth of SWNTs on Fe and Ni catalysts. Finally, insights gained from recent QM/MD simulations regarding the issue of chirality-controlled growth will be the subject of §5.

### 1.1 Experimental synthesis of SWNTs

An exhaustive review of experimental techniques of CNT and SWNT synthesis lies beyond the scope of the current work. Instead, we will provide a cursory overview of two relevant experimental techniques pertinent to the simulations presented in this work, *viz.* the CVD and arc-discharge techniques on transition metal catalysts. We will also briefly summarise recent experimental reports concerning the synthesis of CNTs on 'non-traditional' catalysts such as $SiO_2$, SiC and $Al_2O_3$. For more extensive reviews of this area, we direct the reader to the several books (Dresselhaus *et al.*, 1996; Dresselhaus *et al.*, 2001) and reviews (Teo *et al.*, 2004; Yoshinori, 2004) that are concerned with experimental synthesis.

The CVD process is widely used throughout the areas of material science and solid-state physics. Put simply, it involves a substrate material being deposited and subsequently exposed to a chemical reagent. This reagent subsequently decomposes or reacts on the substrate surface, yielding a desired chemical deposit. In the context of SWNT synthesis, the substrate is traditionally a thin (typically $< 100$ nm) layer of transition metal nanoparticles, and the reagent is a carbonaceous gas such as methane, acetylene or ethanol. These gaseous reagents are believed to decompose on the catalyst nanoparticle surface, thus providing a source of atomic carbon, before dissolving and diffusing into the metal catalyst. As will be shown in §3, however, there is currently a lack of consensus over exactly how these initial stages of SWNT nucleation occur. The metal catalyst layer itself may be deposited *via* a number of different techniques. However, SWNT synthesis is only successful with relatively small nanoparticle diameters (*ca.* 5 nm or less) (Teo *et al.*, 2004). The catalytic species that are typically used for the synthesis, or growth, of SWNTs include first row transition metals such as Fe, Ni, Co and Mo (Journet *et al.*, 1997; Moisala *et al.*, 2003; Harris, 2007). These metals, and alloys thereof, are employed today in the industrial scale CVD synthesis of SWNTs. However, the alkaline earth metals Mg and Ca, as well as Ir and W have also shown catalytic capabilities in the context of CVD SWNT synthesis (Esconjauregui *et al.*, 2009). Most recently, so-called 'catalyst-free' synthesis of SWNTs has been reported (see

(Homma *et al.*, 2009), and references therein). In these cases, SWNTs have been synthesised *via* CVD in the absence of the metal catalyst layer. It follows then that the cracking/decomposition of the gaseous reagent takes place on the supporting substrate itself. The precise mechanism underpinning this un-catalysed reagent decomposition is as yet not unknown. While a number of such covalent CVD 'catalysts' have been reported (including Ge (Takagi *et al.*, 2007), $Al_2O_3$ (Liu *et al.*, 2008; Liu *et al.*, 2010a) and even other nanocarbon structures (Homma *et al.*, 2009)) the majority of experimental investigation in this area has so far focused upon $SiO_2$ (Bachmatiuk *et al.*, 2009; Huang *et al.*, 2009; Liu *et al.*, 2009a; Liu *et al.*, 2009b; Liu *et al.*, 2010b).
In the most general terms, the arc-discharge method involves an electrical discharge between two electrodes through a particular gas (in the process of which this gas is broken down). This method was employed in the synthesis of $C_{60}$ fullerene (Krätschmer *et al.*, 1990; Heath, 1992), before becoming a popular synthetic method for SWNTs. Indeed, the first experimental characterisation of SWNTs (Iijima & Ichihashi, 1993) employed arc-discharge synthesis. In both fullerene and CNT synthesis, the arc-discharge method involves the electrical discharge between two carbon electrodes (one of which is vaporised), usually in the presence of some inert buffer/carrier gas. In the case of SWNT synthesis however, it is preferable that a metal-doped anode is used, otherwise multi-walled carbon nanotubes (MWNTs) are formed. Thus, both the carbon and metal vapors may co-condense, forming liquid phase metal carbide particles. In this sense, the arc-discharge technique constitutes a 'pure' example of the VLS mechanism of SWNT growth. The most common catalytic metals employed in this arc-discharge SWNT synthesis are the same as those employed in the traditional CVD synthesis of SWNTs (*i.e.* Fe, Co, Mo *etc.*) (Journet & Bernier, 1998). There is great variability, however, regarding the yield and diameter distribution and of the synthesised nanotubes with different catalyst metals. The same may be said regarding the pressure and composition of the environmental buffer gas (Journet & Bernier, 1998; Farhat *et al.*, 2001).

### 1.2 Theoretical investigations of SWNT growth

The first foray of theoretical investigation aimed at understanding CNT nucleation and growth took place in the 1990s (see (Irle *et al.*, 2009), and references therein). Amongst the first of these undertakings was that of Smalley *et al.*, who demonstrated that a single metal atom (Ni or Co) has the ability to prevent the closure of an extended $sp^2$-hybridised carbon structure, simply by 'scooting' around the open edge (Thess *et al.*, 1996). Despite the simplicity of this 'scooter mechanism', it revealed for the first time this most fundamental property that is now commonly ascribed to the catalytic nanoparticle during SWNT growth. Another early discovery made by Smalley *et al.* concerned the observation of SWNT growth during the Boudouard reaction ($2CO \rightarrow C + CO_2$) on nanoparticle Mo catalysts (Dai *et al.*, 1996). It was in this investigation that the now well-known 'yarmulke' mechanism was proposed. This mechanism postulates that a SWNT 'cap' fragment exists on the catalyst nanoparticle surface prior to the continued growth of the nanotube, a fact that has since been corroborated independently on a number of occasions (see (Irle *et al.*, 2009), and references therein).
It was not until 2002 that MD simulations were employed successfully in simulating the SWNT nucleation and growth processes. In a series of investigations, Shibuta and co-workers simulated Ni-catalysed SWNT nucleation (Shibuta & Maruyama, 2002; 2003). The roles of both Ni vapor and condensed Ni nanoparticles in the earliest stages of the SWNT nucleation process were therefore established. These theoretical investigations proved

remarkably consistent with contemporary experimental work. Moreover, the previous prediction made by Smalley *et al.* regarding the catalyst preventing closure of the nascent SWNT was realized *in situ*. In a somewhat different approach, Bolton and co-workers investigated the relationship between diffusion and precipitation of carbon from Fe nanoparticles, and SWNT nucleation in a series of investigations (Ding *et al.*, 2004a; Ding *et al.*, 2004c; b; Ding *et al.*, 2006a; Ding *et al.*, 2006b). Notably, the nucleation of SWNTs from Fe 'carbide' was demonstrated at temperatures between 800 and 1400 K (Ding *et al.*, 2004a). According to this investigation, the formation of the SWNT cap structure was preceded by three distinct stages: (1) the incorporation of carbon into the Fe nanoparticle bulk; (2) the saturation of the Fe nanoparticle with atomic carbon, and (3) the formation of polyyne chains and small graphitic 'islands' on the Fe nanoparticle surface. It was also noted that at lower temperatures (below 600 K), encapsulation of the nanoparticle, rather than the formation of a well-defined cap structure, took place. These efforts were complemented by investigations in 2007 (Shibuta & Maruyama, 2007a; b) in which an inverse relationship between the melting behaviors of catalyst nanoparticles and the carbon-catalyst interaction was observed. This was in agreement with an earlier independent investigation (Ding *et al.*, 2006b). These authors have also investigated the relative behaviors of Fe-, Ni- and Co-carbide nanoparticles (Shibuta & Maruyama, 2007b), and observed that the Co-C interaction exceeds both the Fe-C and Ni-C interactions. This correlation suggests that a stronger catalyst-carbon interaction may yield more defective synthetic SWNTs.
The pioneering MD investigations in this area relied on the reactive empirical bond order (REBO) force field (Brenner, 1990; 1992; Brenner *et al.*, 2002), which is itself based upon the Tersoff interactive potential (Tersoff, 1988; 1989). While the use of this force-field makes MD simulations on nanosecond timescales possible, it nonetheless has several notable deficiencies with respect to the chemistry of SWNT nucleation. For instance, π-conjugation and aromatic stabilization of carbon (central to the formation and extension of an $sp^2$-hybridised carbon network, such as a CNT), charge transfer effects and the near-degeneracy of transition metal *d*-orbitals (crucial in the case of transition metal catalysts) can not be accurately described by the REBO or Tersoff potentials. One infamous outcome of the former of these deficiencies is the overestimation and underestimation of the $sp^3$- and *sp*-hybridised carbon fractions, respectively, during the self-assembly of fullerenes at high temperature (Zheng *et al.*, 2004; Irle *et al.*, 2006). On a few occasions (Gavillet *et al.*, 2001; Raty *et al.*, 2005), more reliable simulations based on Carr-Parinello MD (CPMD) have been reported. The description of atomic interaction and bonding in these latter investigations relied on density functional theory (DFT), and so therefore significantly exceeded that given by the REBO potential. However, this advantage incurs a substantial increase in the computational cost of the calculation. Thus, these simulations employed generally unphysical assumptions, or model systems, in order to alleviate these computational costs. In addition, the timescales of these simulations were restricted to less than 25 ps, and as such can hardly be considered to be sufficient in the context of SWNT growth. While these simulations nevertheless represent Herculean efforts, relatively few conclusions regarding the mechanisms of SWNT nucleation and growth have been gained as a result. Ideally, the most suitable approach to the simulation of SWNT nucleation and growth would provide a compromise between quantum mechanical accuracy, and the computational efficiency provided by semi-empirical, or classical, force-field based methods. In the subsequent section, we will turn to one such method, the density-functional tight-binding (DFTB) method, and provide a brief picture of its formulation.

## 2. QM/MD method: Theoretical background

In this section we will present a brief account of the main theoretical methods and algorithms employed in the QM/MD simulations that are presented in §3-5. Our approach is based on the DFTB method. In essence this method is a two-centre approximation to the popular DFT method, which has its origins in the 1990s (Porezag *et al.*, 1995; Seifert *et al.*, 1996). Consequently, for systems consisting of hundreds of atoms (such as those considered here), DFTB is *ca.* three orders of magnitude faster than traditional DFT methods. In DFTB theory, the atomic/molecular energy is given as,

$$E_{\text{DFTB}} = \sum_{i}^{\text{occ.}} \varepsilon_{\text{i}} + \frac{1}{2} \sum_{A \neq B}^{\text{atoms}} E_{\text{rep}}^{AB} \tag{1}$$

where $\varepsilon_{\text{i}}$ is the $i^{\text{th}}$ Kohn-Sham eigenvalue (obtained from the diagonalization of the Hamiltonian matrix in the atomic orbital basis), and $E_{\text{rep}}^{AB}$ describes the repulsive force between nuclei $A$ and $B$. It is noted that the Hamiltonian matrix elements from which the $\varepsilon_{\text{i}}$'s are computed *via* diagonalization need only be computed once (*i.e.* during the development of a particular set of DFTB parameters). The Hamiltonian matrix elements and $E_{\text{rep}}^{AB}$ potentials are subsequently stored in memory and recalled for each subsequent DFTB calculation. This results in a significant reduction in the computation time compared to traditional DFT. Since DFTB is based upon DFT, it inherits both the strengths and weaknesses of DFT. Molecular geometries and vibration frequencies calculated using DFTB are therefore generally reliable. On the other hand, DFTB poorly describes systems exhibiting dispersive or multi-reference characters. The description of conduction bands *etc.* is also limited with DFTB, as it is a minimal basis set method. In addition, although the standard DFTB method describes homonuclear and ionic systems very well, it is unable to describe accurately systems exhibiting a degree of charge transfer. To this end, the self-consistent-charge DFTB (SCC-DFTB) method was developed (Elstner *et al.*, 1998). The SCC-DFTB energy is given as,

$$E_{\text{SCC-DFTB}} = \sum_{i}^{\text{occ.}} \varepsilon_{\text{i}} + \frac{1}{2} \sum_{A \neq B}^{\text{atoms}} E_{\text{rep}}^{AB} + \frac{1}{2} \sum_{A \neq B}^{\text{atoms}} \gamma_{AB} \Delta q_A \Delta q_B \tag{2}$$

The SCC-DFTB energy includes a 2nd order contribution to the DFTB energy involving the charge fluctuation, $\Delta q_\alpha = q_\alpha - q_\alpha^0$, where $q_\alpha$ and $q_\alpha^0$ are the molecular and lone-atom nuclear charges, respectively. The SCC-DFTB molecular orbitals (MOs) are iteratively optimized until the corresponding energy of equation (2) becomes self-consistent with respect to $\Delta q_A$ and $\Delta q_B$. Typically this iterative solution incurs an increase in computational time of approximately one order of magnitude with respect to DFTB. In these cases it is common that self-consistency with respect to $\Delta q_A$ and $\Delta q_B$ cannot be attained. However, this convergence issue is improved dramatically by introducing a finite electronic temperature during the convergence of the MOs. In such a case, the variational SCC-DFTB energy becomes,

$$E_{\text{SCC-DFTB}} + T_e 2k_B \sum_{i}^{\text{orbitals}} \left[ f_i \ln f_i + (1 - f_i) \ln(1 - f_i) \right] \tag{3}$$

where $T_e$ is the electronic temperature, and the population of the $i^{th}$ MO is now defined using the Fermi-Dirac distribution,

$$f_i = \frac{1}{\exp\left[\left(\varepsilon_i - \mu\right)/k_B T_e + 1\right]} \tag{4}$$

Note that this occupation is a continuous function of the $i^{th}$ MO energy, $\varepsilon_i$, and $\mu$ is the chemical potential. This function is continuous over [0,1] (and typically varies near the Fermi level).

The MD method essentially involves the discrete integration of Newton's equations of motion as a function of time. Since its conception (Alder & Wainwright, 1957; Rahman, 1964), it has been applied with great success in fields as diverse as molecular physics, materials science and biological sciences. The discretization of time in MD integration may be achieved in a number of different ways. One such method is the Velocity-Verlet algorithm (Swope *et al.*, 1982), which is perhaps the most popular MD integration scheme today. In this algorithm both the nuclear coordinates and velocities are updated at each iteration of the integration, using coordinates/velocities of the previous iteration,

$$\mathbf{x}\left(t+\Delta t\right) \approx \mathbf{x}\left(t\right) + \mathbf{v}\left(t\right)\Delta t - \frac{1}{2\mathbf{m}} \nabla U\left(\mathbf{x}\left(t\right)\right)\Delta t^2 \tag{5a}$$

$$\mathbf{v}\left(t+\Delta t\right) \approx \mathbf{v}\left(t\right) - \frac{1}{2\mathbf{m}}\left[\nabla U\left(\mathbf{x}\left(t\right)\right) + \nabla U\left(\mathbf{x}\left(t+\Delta t\right)\right)\right]\Delta t \tag{5b}$$

where $U$ is the derivative of the electronic potential energy (in this case calculated using DFTB). Discrete integration of the equations of motion in this fashion results in the micro-canonical, or NVE, ensemble (in which the number of atoms, $N$, the volume, $V$, and the total energy, $E$, of the system are held constant). We will limit the present discussion to MD in which $N$, $V$, and the system temperature, $T$, are held constant throughout the simulation. Placing these restrictions on the MD system results in what is otherwise known as the NVT ensemble. There are several popular methods (more commonly known as thermostats) by which the MD temperature is maintained, and each results in the re-scaling of nuclear velocities in some way. Of particular note are the thermostats of Anderson (Andersen, 1980), Berendsen (Berendsen *et al.*, 1984) and the method of velocity scaling (Woodcock, 1971). In the present context, we employ the Nosé-Hoover chain thermostat (Nose, 1984; Hoover, 1985; Martyna *et al.*, 1992; Martyna *et al.*, 1996), in which the Hamiltonian of the system is augmented with a term representing a heat-bath that is coupled to the degrees of freedom of the system. The augmented equations of motion thus sample microcanonical and canonical distributions in the extended and original systems, respectively. However, care must be taken when deciding the strength at which the Nose-Hoover chain thermostat is coupled to the MD system. Coupling that is too weak will result in inadequate temperature control, whereas coupling that is too strong is known to result in high-frequency temperature oscillations, and consequently unreliable dynamics.

## 3. QM/MD simulations of SWNT nucleation

We turn now to a discussion of recent QM/MD simulations of SWNT nucleation. This discussion will focus on the mechanism of SWNT nucleation on a number of different catalyst nanoparticles, including Fe, Ni, $SiO_2$, SiC and Si. SWNT nucleation resulting from both model CVD and arc-discharge processes will also be discussed. In this way we will demonstrate that, at the atomic level, the mechanism of SWNT nucleation is surprisingly invariant to both the experimental method employed, and several pertinent environmental factors. We begin with the case of CVD on Fe catalyst nanoparticles.

### 3.1 Acetylene CVD and SWNT nucleation on Fe catalysts

Today, CVD synthesis of SWNTs is perhaps the most popular method of choice on the commercial scale. The most typically employed gaseous precursors include acetylene, ethanol and methane (almost always in the presence of some inert buffer/carrier gas). Yet little was known regarding the atomistic mechanism of such carbonaceous CVD processes until relatively recently. Such knowledge was furnished entirely by theoretical MD simulations, and in particular QM/MD simulations. Here we will focus on the mechanism of Fe-catalysed acetylene CVD elucidated from such recent QM/MD simulations.

In order to investigate the Fe-catalysed acetylene ($C_2H_2$) CVD process, we employed an $Fe_{38}$ catalyst nanoparticle. The diameter of this nanoparticle is *ca.* 0.7 nm, and so is of comparable diameter to experimental SWNT diameter distributions (Sugai *et al.*, 2003). Gas-phase $C_2H_2$ molecules were initially adsorbed onto the equilibrated catalyst nanoparticle (see Fig. 1a), after which the resultant $30C_2H_2$-$Fe_{38}$ model complex was relaxed at 1500 K for 500 ps. During the $C_2H_2$ adsorption process the occasional abstraction of atomic H by the Fe catalyst surface was observed, thus forming $C_2H$ radicals. Similarly, abstraction of atomic H by adjacent $C_2H_2$ molecules was also observed, resulting in both $C_2H$ and $C_2H_3$ moieties. Both abstraction processes are endothermic, with barriers between *ca.* 20 – 35 kcal $mol^{-1}$. The direct formation of $H_2$ was however not observed, despite the abstraction of atomic H by the catalyst surface. This is not surprising, considering the high endothermicity of the $H_2$ formation process (using SCC-DFTB, this barrier is estimated to be *ca.* 35-50 kcal$mol^{-1}$). Such endothermic processes are inherently difficult to observe in MD simulations on this time scale. The radical products $C_2H$ and $C_2H_3$ are extremely reactive, and therefore rapidly initiated oligomerisation between adjacent $C_2H_x$ species. Such oligomerisation is exothermic by *ca.* 18 kcal $mol^{-1}$ (see Fig. 1b). Following these oligomerisation reactions, extended $sp^2$-hybridised carbon networks ultimately form on the catalyst surface. The cross-linking of neighboring polyyne chains drove this process, and resulted in the formation of pentagonal and hexagonal carbon rings. In all cases, pentagonal rings were formed first – an observation that will frequently recur in §3.2-3.4. Such a cross-linking process is depicted schematically in Fig. 2. Also depicted in Fig. 2 is the polyyne cross-linking mechanism (pertaining to SWNT growth) proposed by Eres (Eres *et al.*, 2009). While both processes are distinctly similar, no hexagonal rings were formed in the cross-linking process in the present work.

Fig. 3 shows the ultimate product of this H-abstraction and polyyne oligomerisation/cross-linking process, *viz.* the formation of an extended $sp^2$-hybridised carbon network. The structure of this network generally fell into one of three categories. The most 'successful' structure regarding SWNT nucleation is structure (i), in which the network extends over the catalyst surface. In essence such a structure constitutes a SWNT cap-fragment, similar to the 'yarmulke' cap proposed by Smalley and co-workers (Dai *et al.*, 1996). However, structure (i) was only observed at relatively low H/C ratios (see §3.2-3.3 for examples of SWNT nucleation in the absence of H). In cases of higher H/C ratios (*i.e.* more H), structure (ii) was typically observed, due to the passivative effect of H at the unsaturated edge of the carbon network. It is assumed that this graphene-like sheet may coalesce to form an open nanotube

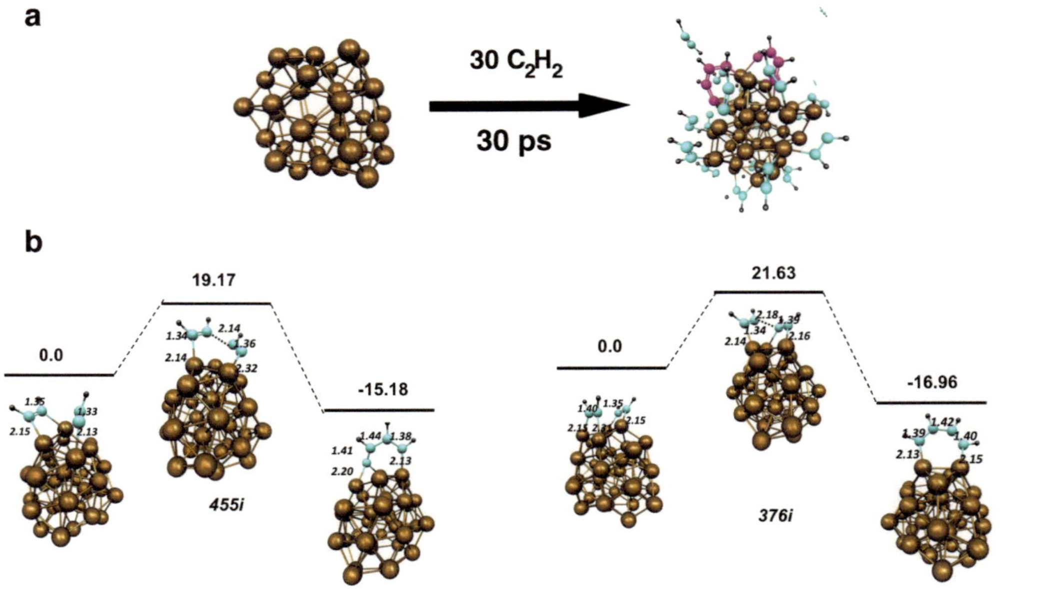


Fig. 1. The initial stages of simulated acetylene CVD on $Fe_{38}$ catalyst nanoparticles at 1500 K. a) Oligomerisation results in the formation of extended polyyne chains on the catalyst surface within 30 ps. Brown, cyan and grey spheres represent Fe, C and H atoms, respectively. *sp*²-hybridised C atoms are represented by magenta spheres. b) Examples of $C_2$ oligomerisation, and their associated energetics. Examples of both the $C_2H_2 + C_2H \rightarrow C_4H_3$ and $C_2H_2 + C_2H_2 \rightarrow C_4H_4$ oligomerisation reactions are depicted. The $C_2H$ precursors here are occasionally produced *via* the abstraction of atomic H onto the catalyst surface. All energies and bond lengths given in kcalmol$^{-1}$ and Å, respectively. Transition state imaginary frequencies are given in cm$^{-1}$.

according to the mechanism proposed by Eres (Eres *et al.*, 2009) (see Fig. 2). Structure (iii) was also observed as a result of the polyyne oligomerisation/cross-linking process. In this case, the process yielded 'islands' of *sp*²-hybridised carbon. Assumedly, QM/MD relaxation of such structures over longer time scales (*i.e.* several nanoseconds) would yield a more consistent network, such as that typical of structure (i). Thus, it is demonstrated here that SWNT nucleation is not *necessarily* preceded by a carbon cap-structure, or a liquid carbide phase. The latter conclusion will be corroborated in §3.2.

### 3.2 SWNT nucleation on Fe & Ni catalysts *via* adsorption of gas-phase $C_2$

It was observed in §3.1 that the removal/sequestration of hydrogen from feedstock acetylene molecules was the most problematic issue in these QM/MD simulations. Presumably the same problem would exist regardless of the type of carbonaceous precursor employed in this respect (be it acetylene, methane, ethanol, *etc.*). This difficulty arises due to a problem inherent to MD methods, since such methods have difficulty overcoming large energy barriers on the global potential energy surface (PES). Although, in the limit of infinite time, an MD simulation will sample all possible geometrical configurations, and consequently will have overcome all such barriers on the global PES, in practice, such sampling is obviously not possible. To this end, we will discuss an alternative QM/MD approach to the problem of SWNT nucleation on Fe catalysts (Ohta *et al.*, 2009).

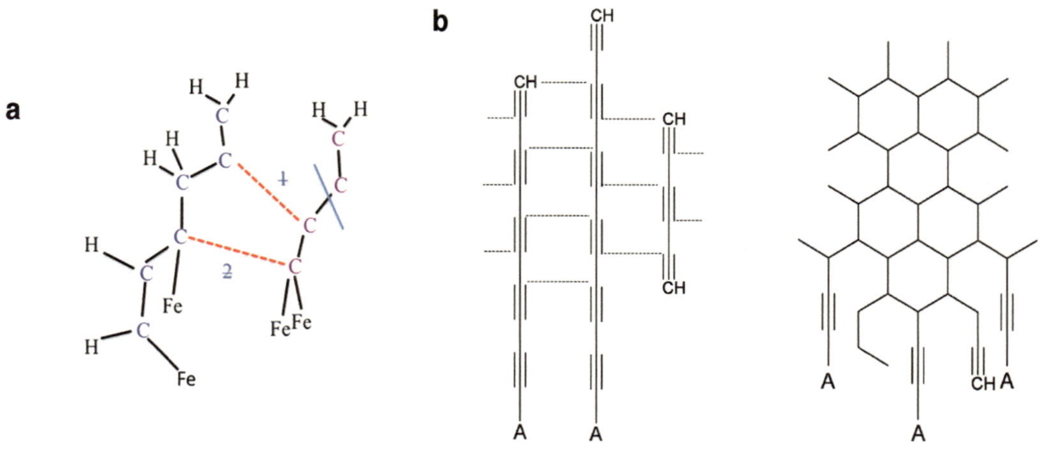


Fig. 2. Oligomerisation, or cross-linking, of extended polyyne chains on the catalyst surface leads to carbon ring formation. a) Explicit example of cross-linking resulting in pentagonal ring formation observed in QM/MD simulations. Red lines indicate newly formed C-C bonds, blue lines indicate broken C-C bonds. Number 1or 2 indicates reaction step. b) Cross-linking reaction proposed by Eres (Eres *et al.*, 2009), resulting in the formation of a graphene-type structure. (Reprinted with permission. © 2009 American Chemical Society)

In the present approach, the hydrogen was simply removed from the gas-phase carbonaceous molecules prior to their interaction with the catalyst nanoparticle. Fig. 4 depicts this adsorption process, and the process of constant temperature annealing that followed. In this case, the MD relaxation of this $Fe_{38}$-carbon system was continued for 410 ps. Here $C_2$ moieties have been employed, which are considered to be essentially equivalent with the gas-phase acetylene feedstock molecules discussed in §3.1. However, the absence of hydrogen here is not unrealistic, considering the known products resulting from the vaporisation of graphite/graphene *via* arc-discharge or laser-ablation processes. In addition, the use of $C_2$ allowed the atomistic mechanism of SWNT nucleation to be probed more easily than before. Fig. 4a shows that this SWNT nucleation process may be partitioned into three distinct stages. To begin with, $C_2$ units adsorbed onto the $Fe_{38}$ catalyst surface from the gas-phase. The relatively weak Fe-C interaction energy facilitated the subsequent diffusion of these $C_2$ units over the catalyst surface. As a natural consequence of this diffusion, $C_2$ units ultimately began to interact with each other, coalescing to form longer polyyne chains (*i.e.* Fe-$C_n$-Fe structures) on/over the catalyst surface. This was apparently the rate-limiting-step of the nucleation process, in that it was *ca.* 100 ps before the second stage of the nucleation mechanism took place (Fig. 4b). This second stage featured the initial ring condensation processes on the catalyst surface, while the subsequent third stage consisted entirely of additional ring condensation events, resulting in the formation of a SWNT cap-fragment. The exact mechanism of this initial ring condensation process is depicted in Fig. 4c. Fig. 4b shows that there was generally a preference for the formation of pentagonal rings, as opposed to hexagonal or heptagonal rings, during the initial stages of SWNT nucleation.

Indeed, the initial ring structure formed was pentagonal. This fact is consistent with knowledge regarding the formation mechanism of fullerenes at high temperatures (Irle *et al.*, 2006). The longevity of these pentagonal rings, however, is attributed to the high positive curvature of the catalyst nanoparticle surface (due to its small diameter) (Fan *et al.*, 2003). In effect, the growing $sp^2$-hybridised carbon network attempts to 'mould' itself to its supporting catalyst substrate from its very beginnings.

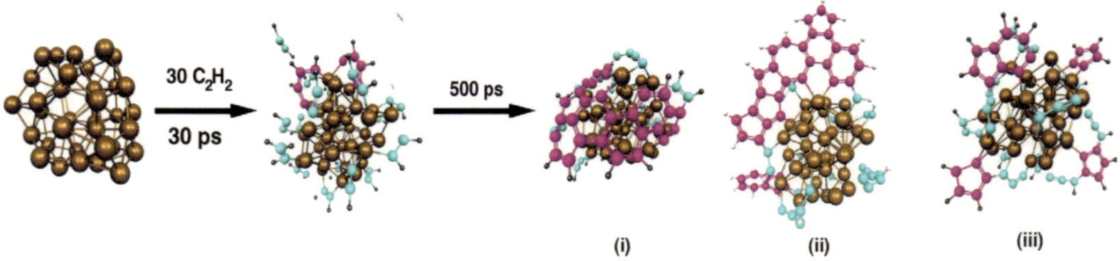


Fig. 3. An extended $sp^2$-hybridised carbon network is formed on the catalyst surface following polyyne oligomerisation. QM/MD simulations indicate that structures (i), (ii) and (iii), are typically formed. Structure (i) is akin to the 'yarmulke' SWNT cap fragment proposed by Smalley *et al.* (Dai *et al.*, 1996). Structure (ii) is typically formed in the presence of higher H concentrations, and points to the possibility that SWNT nucleation may take place in the absence of a SWNT cap fragment. Structure (iii) features a catalyst nanoparticle covered with $sp^2$-hybridised carbon 'islands'. Color conventions as in Fig. 1; pink spheres represent $sp^2$-hybridised carbon atoms.

The initial pentagonal ring observed in Fig. 4 acted as an anchor, or cornerstone, for all subsequent ring condensation events. This period of ring condensation (stage three of the nucleation process) consisted of a periodic process (Fig. 4d) in which adjacent polyyne chains interacted on the catalyst surface due to their diffusion, thus extending the $sp^2$-hybridised carbon structure. This process is best illustrated by the initial ring condensation event, in which two adjacent polyyne chains coalesced, resulting in a 'Y-junction'. This initial $sp^2$-hybridised carbon atom was, in essence, the nucleus of the final SWNT itself, since all subsequent ring condensation was based around it. The first pentagonal ring formed following the sinusoidal-type diffusion of the two arms of the Y-junction. With respect to the original $sp^2$-hybridised carbon atom, the most energetically favorable interaction corresponded to the interaction between the second carbon atoms of each arm. This therefore explains the observed preference for pentagonal ring formation observed during the initial stages of SWNT nucleation. It is noted, however, that this interaction was only the most favorable due to the approximate 120° bond angle provided by the single $sp^2$-hybridised carbon atom. In subsequent ring condensation events, this single carbon atom was often replaced with a C-C moiety, thus this bond angle was modified (or removed entirely). The most energetically favorable interaction therefore corresponded to that between carbon atoms at varying positions along the arms of the Y-junction structures.

As will be discussed in §4, extended polyyne chains play a dominant role not only in SWNT nucleation, but also in the subsequent 'continued' growth of SWNT structures. In this sense, SWNT growth is therefore very similar to the self-assembly of fullerenes (Irle *et al.*, 2006). This leads to the conclusion that such polyyne chains are essential for both the conception and the extension of any $sp^2$-hybridised carbon network. Of course, the crucial difference between SWNT and fullerene nucleation here is the presence of the catalyst nanoparticle. In particular, it is noted here that the QM/MD simulation presented in Fig. 4 once again verifies the original proposal of Smalley *et al.*, *i.e.* that one of the fundamental roles of the catalyst nanoparticle here is preventing the closure of the growing $sp^2$-hybridised structure (Thess *et al.*, 1996).

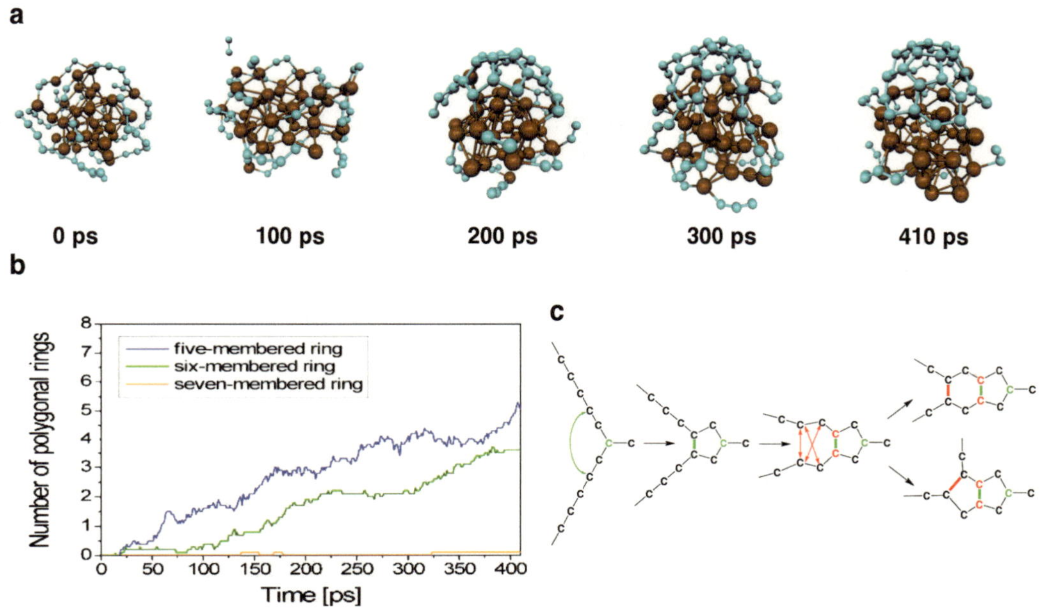


Fig. 4. SWNT nucleation occurs *via* three distinct stages, according to QM/MD simulations. a) QM/MD relaxation of a $30C_2$ – $Fe_{38}$ model system at 1500 K yields a distinct SWNT cap fragment after 410 ps. Color conventions as in Fig. 1. b) SWNT nucleation is driven by successive ring condensation events on the catalyst surface. The preferential formation of pentagonal rings in this structure is attributed to the curvature of the catalyst surface, and the diffusion dynamics of extended polyyne chains. c) The SWNT nucleus. A single $sp^2$-hybridised carbon atom acts as the cornerstone of all subsequent ring formation events in the nascent SWNT structure. (Adapted from (Ohta *et al.*, 2009). Reprinted with permission. © 2009 American Chemical Society)

### 3.3 SWNT nucleation from amorphous Fe & Ni carbide precursors

According to the VLS mechanism, CNT nucleation and growth are preceded by a gaseous carbon/catalyst phase which co-condense forming a catalyst-carbide nanoparticle. Images of transition metal carbide nanoparticles have been obtained using transmission electron microscopy on several occasions (Yoshida *et al.*, 2008; Yoshida *et al.*, 2009) (see Fig. 5). Yet, to date there is no experimental evidence indicating that this carbide phase *necessarily* precedes the nucleation and growth of SWNTs. Indeed, QM/MD simulations discussed in §3.1-3.2 indicate that, for nanoparticle catalysts of *ca.* 1 nm, a carbide phase is not formed at 1500 K. The thermodynamic stability of bulk transition metal carbide nanoparticles have also been drawn into question from a number of independent approaches. Assumedly then, SWNT nucleation may occur in the *absence* of a carbide phase. Such disparities between theoretical and experimental assertions give reason for further study of the role of the carbide phase with respect to SWNT nucleation and growth.

QM/MD simulations of SWNT nucleation from amorphous Fe- and Ni-carbide nanoparticles at 1400 K are depicted in Fig. 6. It is evident from this figure that, upon MD relaxation at constant temperature, the amorphous carbide phase almost immediately decomposes (within *ca* 5 - 10 ps), yielding segregated Fe/Ni-carbon systems. This phenomenon is known to take place regardless of temperature, or the carbon concentration in the amorphous carbide phase (Page *et al.*, 2010d). Fig. 6 also indicates that the SWNT nucleation mechanism in this case (from a Ni-carbide) is the same as that presented in §3.1-3.2. For example, the almost immediate precipitation of carbon from the nanoparticle bulk to the surface leads to the formation of extended polyyne chains over the nanoparticle surface. The oligomerisation/cross-linking *etc.* of these chains then leads to the formation of primarily pentagonal and hexagonal rings (Fig. 6a) as the SWNT cap fragment is formed. The initial ring structure in all cases here is invariably a pentagonal ring. Thus, the SWNT nucleation mechanism on Fe/Ni catalysts is evidently independent of the type of metal catalyst, temperature, and origin/type of the feedstock carbon employed. While the fact that SWNT nucleation originates from a Ni-carbide structure is not unexpected (since it has been predicted in several prior REBO-based MD investigations (Shibuta & Maruyama, 2002; 2003)), what is remarkable is the invariance of the SWNT nucleation mechanism with respect to these pertinent experimental factors.

Despite this invariance, differences in the ultimate product of this nucleation process are evident with respect to both the catalyst type and the simulation temperature. The kinetics of SWNT nucleation was also affected by the type of catalyst employed - explicitly, SWNT nucleation from the decomposition of Ni-carbide proceeded more quickly, compared to that from Fe-carbide (see Fig. 6 and Fig. 7). It was observed that at higher temperatures (2000 K, as opposed to 800 or 1400 K), the populations of pentagonal and hexagonal rings in the SWNT cap fragment were approximately equal. On the other hand, at lower temperatures a distinct preference towards pentagonal ring formation existed. These differing ring populations were ascribed to the effect of temperature on the SWNT nucleation dynamics. At higher temperatures, the growing polyyne chains on the catalyst surface are more thermally excited, and thus exhibited larger amplitude vibrational motion. Considering the pentagonal ring mechanism given in Fig. 4c, this increased motion makes the formation of a C-C bond between tertiary carbon atoms (with respect to the $sp^2$-hybridised 'cornerstone' carbon atom) more likely. Hence, hexagonal ring formation is more probable in this case. Perhaps the most important difference observed between the kinetics of SWNT nucleation from Fe- and Ni-carbide nanoparticles, however, pertains to the relative rates of SWNT nucleation. It was recently established that SWNT nucleation is significantly more labile on Ni catalysts, compared to Fe catalysts. This observation may be directly attributed to the relative strengths of the catalyst-carbon interactions. This point will be a recurring theme throughout the present work, as it dominates many aspects of both SWNT nucleation and growth. For example, the catalyst-carbon interaction strengths, calculated using SCC-DFTB, are 1.78 and 1.06 eV for Fe-C and Ni-C, respectively. For comparison, the C-C interaction strength is 9.14 eV. Therefore, Fe-C bond formation is more favorable than Ni-C bond formation, in a thermodynamic sense. Consequently, C-C bond formation during the decomposition of Fe-carbide nanoparticles is impeded, which in turn impedes the nucleation of the $sp^2$-hybridised carbon network. This argument also explains other phenomena related to

SWNT nucleation, such as the lifetimes of the bulk and subsurface carbide intermediate species (Page *et al.*, 2010d) (see Fig. 7). In particular, the average time required for

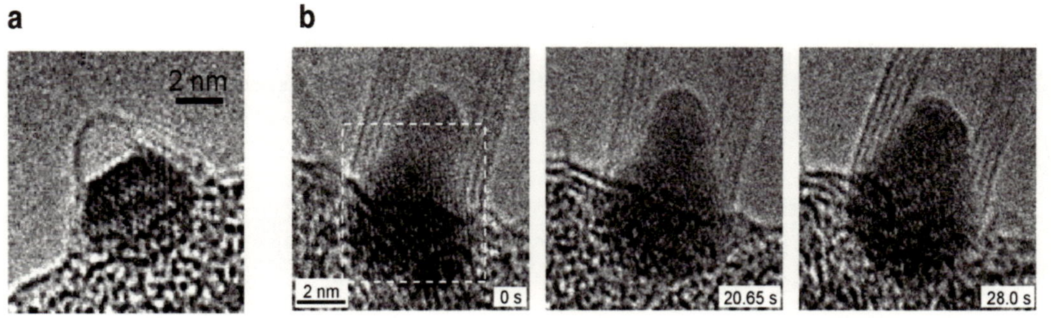


Fig. 5. TEM images of intermediate structures observed during SWNT growth experiments. a) A SWNT cap-fragment bound to a crystalline $Fe_3C$ nanoparticle with an approximate diameter of 2 nm. (Adapted from (Yoshida *et al.*, 2008). Reprinted with permission. © 2008 American Chemical Society) b) A MWNT growing from a $(Fe,Mo)_{23}C_6$ nanoparticle catalyst. In this case the nanoparticle diameter is *ca.* 6 nm. (Adapted from (Yoshida *et al.*, 2009). Reprinted with permission. © 2009 American Chemical Society)

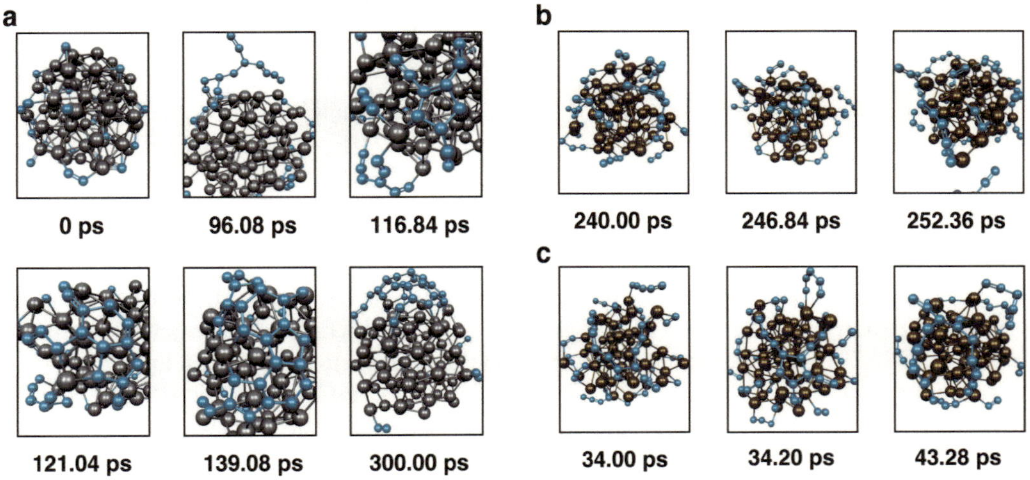


Fig. 6. The kinetics of SWNT nucleation from Ni-carbide are enhanced relative to those from Fe-carbide. Nevertheless, the mechanism of SWNT nucleation is identical in both cases. a) The evolution of a computed SWNT nucleation trajectory from a $Ni_{77}C_{39}$ carbide nanoparticle at 1400 K. The SWNT cap fragment is formed within 300 ps following repeated ring condensation events. b), c) The evolution of two $Fe_{58}C_{58}$ carbide nanoparticles at 1400 K. The stronger Fe-C interaction impedes the formation of C-C bond, and therefore impedes the SWNT nucleation process itself. Both trajectories show examples of the formation, and subsequent destruction, of pentagonal carbon rings. Color conventions as in Fig. 1; grey spheres represent Ni atoms. (Adapted from (Page *et al.*, 2010d). Reprinted with permission. © 2010 American Chemical Society)

precipitation of all carbon from the nanoparticle bulk to the nanoparticle surface/subsurface in the case of Fe-carbide always exceeds that for Ni-carbide, except at low temperatures. At 800 K, the reverse is the case, since the Ni-carbide nanoparticle exists in the solid phase, while the Fe-carbide nanoparticle is liquid. These QM/MD findings therefore support recent claims that a subsurface carbide structure (in which a high density of carbon exists at, or just below, the nanoparticle surface) precedes SWNT nucleation and growth (Amara *et al.*, 2006; 2008b; a; Harutyunyan *et al.*, 2008; Amara *et al.*, 2009).

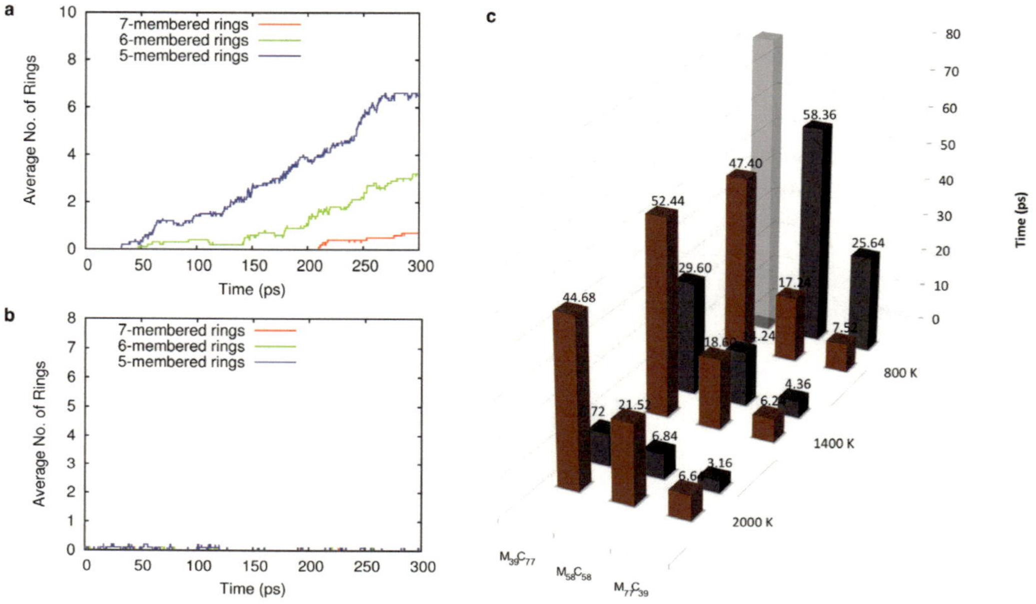


Fig. 7. The kinetics of SWNT nucleation from Ni-carbide are enhanced relative to those from Fe-carbide. Average polygonal carbon rings formed from a) $Ni_{58}C_{58}$ and b) $Fe_{58}C_{58}$ at 1400 K. c) Average carbon precipitation times (in ps) for Fe- and Ni-carbide nanoparticles between 800 and 2000 K. Fe-carbide carbon remains within the nanoparticle bulk for a longer time period, compared to Ni-carbide carbon, due to the stronger Fe-C interaction. At 800 K the trend is reversed since Ni-carbide exists in the solid phase. All data averaged over 10 trajectories. Brown and grey columns refer to Fe- and Ni-carbide data, respectively. Transparent columns indicate precipitation times greater than 300 ps. (Adapted from (Page *et al.*, 2010d). Reprinted with permission. © 2010 American Chemical Society)

### 3.4 A new breed of catalysts: SWNT nucleation on $SiO_2$, SiC and Si

The mechanism of SWNT nucleation on traditional, transition metal catalysts such as Fe, Ni and Co has now been the subject of both experimental and theoretical scrutiny for approximately a decade. Since 2009, however, a number of experimental reports (Takagi *et al.*, 2007; Liu *et al.*, 2008; Bachmatiuk *et al.*, 2009; Homma *et al.*, 2009; Huang *et al.*, 2009; Liu *et al.*, 2009a; Liu *et al.*, 2009b; Liu *et al.*, 2010a; Liu *et al.*, 2010b) have established non-traditional nanomaterials to be catalytically active in the context of SWNT nucleation and growth from methane and ethanol CVD. Si-based materials, and in particular $SiO_2$, have been remarkably successful in this respect. Yet it has only been since 2011 that the atomistic mechanisms of $SiO_2$-, SiC- and Si-catalysed SWNT nucleation have been established. These QM/MD investigations will be the focus of this section.

QM/MD simulations of methane CVD on $SiO_2$ nanoparticles at 1200 K (Page *et al.*, 2011b) is outlined in Fig. 8. Due to the inherently low catalytic activity of $SiO_2$ itself, $CH_x$ radicals ($x$ = 0 - 3 and is chosen randomly) were supplied to the $SiO_2$ instead of $CH_4$. This approach was motivated by the prior conclusion that $CH_4$ decomposes pyrolitically prior to adsorption on the $SiO_2$ surface (Liu *et al.*, 2009b). In contrast to CVD using traditional transition-metal catalysts, a complex chemical process was observed on $SiO_2$. Most notably, CO was produced as the primary chemical product *via* the carbothermal reduction of the $SiO_2$ nanoparticle, a fact that is consistent with recent experimental observations (Bachmatiuk *et al.*, 2009). The production of each CO molecule first required hydrogen-abstraction from neighboring C, Si or O atoms. Ultimately, the insertion of carbon into/removal of oxygen from the $SiO_2$ nanoparticle resulted in the local formation of amorphous SiC. However, this carbothermal reduction was limited to the outer regions of the catalyst, with the core of the particle remaining 'oxygen rich'. The amorphous SiC regions were composed predominantly of extended polyyne chains 'anchored' in place by native Si atoms.

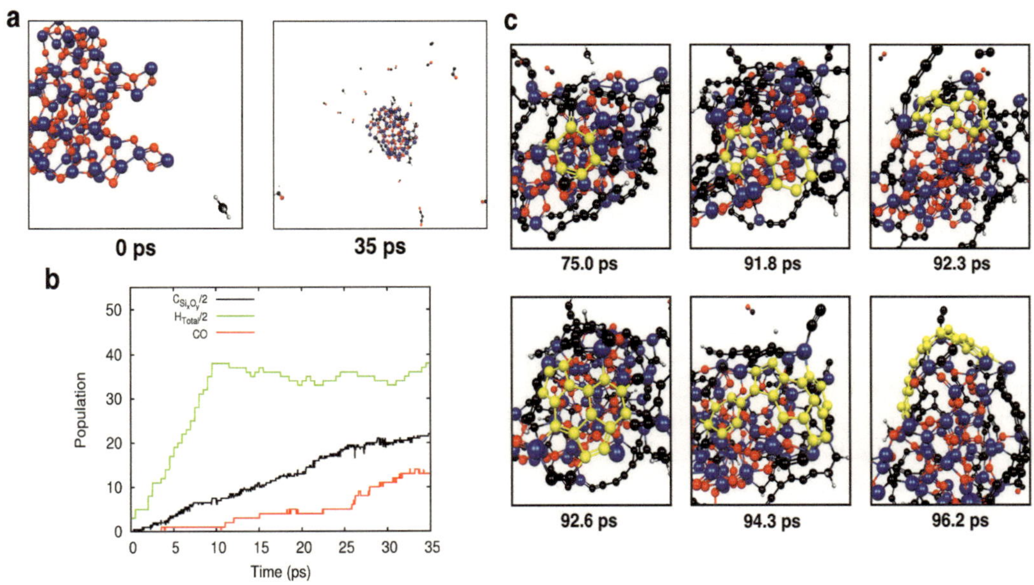


Fig. 8. $CH_4$ CVD on $SiO_2$ nanoparticles at 1200 K leads to SWNT nucleation *via* a VSS mechanism. a) Snapshots at 0 and 35 ps showing the CVD process. b) CO is the major chemical product of the $CH_4$ CVD process. The production of CO first requires the natural removal of H from the CO carbon atom. $C_{SixOy}$ and $H_{SixOy}$ are the concentration of C and H on the $SiO_2$ nanoparticle, respectively. c) Evolution of SWNT nucleation on $SiO_2$ nanoparticles. Contrary to nucleation on transition-metal catalysts, nucleation here requires the saturation of the solid-phase catalyst with carbon. Blue, red and black spheres represent Si, O and C, respectively. Yellow spheres represent C atoms involved in SWNT nucleation. (Adapted from (Page *et al.*, 2011b). Reprinted with permission. © 2011 American Chemical Society)

Consequently, these polyyne chains exhibit restricted vibrational and translational mobility, compared to the equivalent precursor structures observed during transition-metal catalysed SWNT nucleation. A more detailed discussion of the thermodynamic reasons underpinning these phenomena is given below. At high concentrations of surface polyyne chains SWNT nucleation was observed. This observation supports the previous claim by Homma and co-workers that SWNT nucleation on solid, covalent catalysts requires a 'carbon-covered' catalyst nanoparticle in order for nucleation to take place (Homma *et al.*, 2009). The pentagonal-ring-first mechanism, established by QM/MD simulations and discussed in §3.1-3.3, therefore played no role in the current context. Similarly, the liquid carbide phase that is central to the VLS mechanism of SWNT, discussed in §3.3, is absent in the case of $SiO_2$-catalysed SWNT nucleation. This conclusion followed an analysis of the instantaneous Lindemann index (Lindemann, 1910) of the $SiO_2$ nanoparticle during the CVD process. At all times, the Lindemann index revealed that the $SiO_2$ nanoparticle existed as a solid phase structure. Moreover, QM/MD relaxation of this nanoparticle at elevated temperatures (up to 3000 K) indicated that nanoparticle $SiO_2$ decomposes from the solid phase at sufficiently high temperatures (Page *et al.*, 2011a). This sublimative phenomenon here rules out the VLS mechanism as an explanation of $SiO_2$-catalysed SWNT nucleation and growth entirely. Instead, QM/MD simulations point to a vapor-solid-solid (VSS) mechanism explaining SWNT nucleation and growth in this case. The mechanisms of SWNT nucleation and growth on traditional and non-traditional catalysts are therefore of fundamentally different natures. Subsequent experimental results (Liu *et al.*, 2011) have since corroborated this proposed VSS mechanism.

The observation that the catalytically relevant region of the $SiO_2$ nanoparticle is effectively devoid of oxygen motivated the subsequent QM/MD investigation of SWNT nucleation on pure Si nanoparticles. To this end, a $Si_{58}$ nanoparticle of approximate dimension $0.9 \times 0.9 \times 0.9$ nm$^3$ was employed as a CVD catalyst at 1200 and 1800 K. Gas-phase $C_2$ moieties were adsorbed on the surface of this catalyst nanoparticle in the manner described in §3.2. Two different concentrations of carbon, *viz.* 30 and 100, were employed here, following the observation made regarding the dependence of SWNT nucleation on surface carbon concentration using $SiO_2$ catalyst nanoparticles. The structures of these $Si_{58}C_{60}$ and $Si_{58}$ $C_{200}$ model systems, following 100 and 45 ps, are given in Fig. 9. Upon adsorption on the $Si_{58}$ surface, these $C_2$ moieties generally coalesced, forming extended polyyne chains, in an identical fashion to nucleation on Fe, Ni and $SiO_2$ catalysts. However, the mobility of these polyyne chains in the case of $Si_{58}$ was notably restricted, as was observed in the case of $SiO_2$. This was also the case at a higher annealing temperature of 1800 K, leading to the conclusion that the effect of temperature (at least below 2000 K) on this SWNT nucleation process was effectively negligible. It was noted that this was not the case at even higher temperatures, as will be discussed below in the context of SWNT nucleation from SiC. Once formed, these polyyne chains themselves gradually coalesced on the nanoparticle surface, ultimately forming extended branched carbon networks. While this is consistent with the initial steps in SWNT nucleation discussed in §3.1-3.3 in an atomistic sense, it is noted that the kinetics of this coalescence on $Si_{58}$ is significantly slower, compared to traditional, transition metal catalysts. In particular, in the latter case the rate-limiting step of SWNT nucleation may be considered to be the formation of the SWNT 'nucleus' (the initial polygonal carbon ring structure). Following the formation of this structure, the subsequent ring condensation and cap-formation process proceeds relatively quickly. This is not so in the presence of Si nanoparticle catalysts. Fig. 10a shows that, following the formation of the SWNT nucleus on

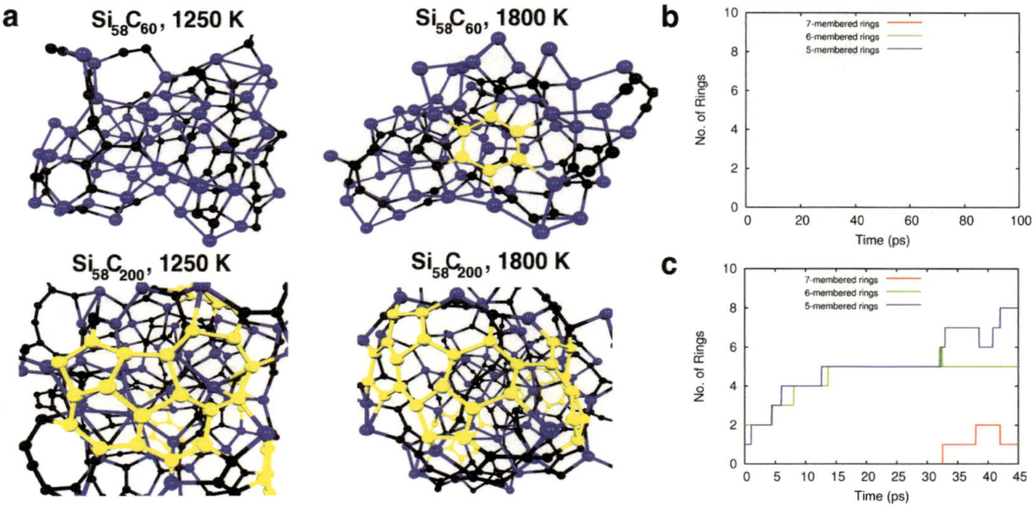


Fig. 9. SWNT nucleation on Si catalyst nanoparticles following the adsorption of gas-phase $C_2$. a) Structures of $Si_{58}C_{60}$ and $Si_{58}C_{200}$ model complexes at 1200 and 1800 K. Color conventions as in Fig. 8. b) Polygonal ring populations observed using low [C] conditions (*i.e.* a $Si_{58}C_{60}$ model system). c) Polygonal ring populations observed using high [C] conditions (*i.e.* $Si_{58}C_{200}$ model system). It is evident that the initial saturation of the Si catalyst surface with carbon is necessary in order for SWNT nucleation to proceed.

$Si_{58}$, the subsequent extension of the $sp^2$-hybridised carbon network proceeded at a significantly slower rate. Fig. 9a also illustrates the effect of surface carbon concentration on SWNT nucleation. For example, the formation of polygonal carbon rings in the $Si_{58}C_{60}$ complex (following the adsorption of 30 $C_2$ species) is limited to a single hexagonal ring structure after 100 ps. Conversely, an extended network of carbon ring structures was formed in the $Si_{58}C_{200}$ model complex after only 50 ps. Thus, as was the case regarding $SiO_2$

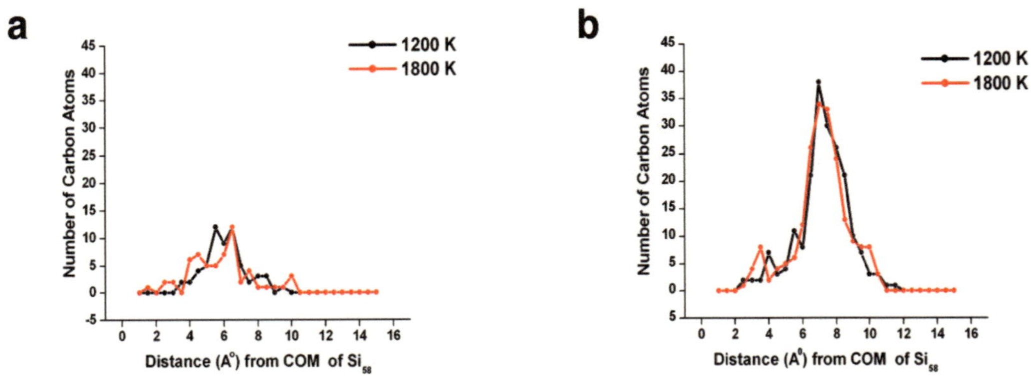


Fig. 10. Radial distributions of carbon in a) $Si_{58}C_{60}$ and b) $Si_{58}C_{200}$ model complexes at 1200 and 1800 K. The inability of carbon to freely diffuse through the bulk region of the Si nanoparticle is evident. Consequently, the majority of the carbon in both cases resides on the nanoparticle surface, the latter of which is solid. SWNT nucleation cannot therefore proceed *via* a VLS mechanism.

catalyst nanoparticles, it is evident that the saturation of the Si nanoparticle surface with carbon is also a prerequisite for SWNT nucleation in this case. In this sense then, SWNT nucleation on $SiO_2$ and $Si_{58}$ seemingly proceeds *via* an identical route – this point will be discussed at greater length below.

Following the adsorption of $C_2$ onto the $Si_{58}$ nanoparticle surface, the resultant surface structure resembled an amorphous SiC phase, while the core of the Si nanoparticle remained pristine. This is evident from Fig. 10a, which shows the radial distribution of carbon within the Si nanoparticle as SWNT nucleation proceeds. This figure also shows that, at higher temperature, the penetration of the Si nanoparticle by adsorbed carbon atoms becomes more probable, and is independent of the surface carbon concentration. Yet, the free diffusion of carbon through the nanoparticle bulk and surface in this case is restricted below 2000 K. The latter observation may be explained with recourse to an analysis of the nanoparticle phase during SWNT nucleation. This is conveniently done in the realm of QM/MD simulations *via* the Lindemann index (Lindemann, 1910), $\delta$,

$$\delta = \frac{1}{N}\sum_{i}\delta_i \tag{6a}$$

where,

$$\delta_i = \frac{1}{N-1}\sum_{j\neq i}\frac{\sqrt{\left\langle r_{ij}^2\right\rangle_T - \left\langle r_{ij}\right\rangle_T^2}}{\left\langle r_{ij}\right\rangle_T} \tag{6b}$$

Here, $N$ is the number of atoms in the relevant system, $r_{ij}$ is the instantaneous distance between atoms $i$ and $j$, and the brackets denote thermal averaging over a finite interval of time at temperature $T$. It is noted here that $\delta$ describes all atoms in the system, and is thus generally referred to as the 'global' Lindemann index. On the other hand, $\delta_i$ pertains only to the motion of atom $i$, and is therefore referred to as the 'atomic' Lindemann index. In the current discussion, we will make reference to both $\delta$ and $\delta_i$. The Lindemann index has been used with particular success in the investigation of transition and main group metal species (both bulk and nanoparticle structures) (Ding *et al.*, 2006b; Puri & Yang, 2007; Neyts & Bogaerts, 2009; Wen *et al.*, 2009). From these investigations, the efficacy of the Lindemann index in the prediction of nanoparticle melting points has been established. For example, it is now generally accepted that the 'threshold' $\delta$ value, which signifies the transition between the solid and liquid phases is between 0.10 – 0.15 (Ding *et al.*, 2006b; Puri & Yang, 2007; Neyts & Bogaerts, 2009; Wen *et al.*, 2009). Thus, any system exhibiting a $\delta$ below this threshold value may be considered to be solid, whereas those with $\delta$ above this threshold value are considered to be liquid. In the case of the pristine $Si_{58}$ catalyst nanoparticle, $\delta$ at 1200 and 1800 K were 0.298 and 0.372, respectively. However, upon the adsorption of $C_2$ on the $Si_{58}$ surface, a dramatic decrease in this Lindemann index was observed. At low carbon concentrations (*i.e.* the $Si_{58}C_{60}$ model complex), these same $\delta$ values were 0.093 and 0.231, while at high concentrations (*i.e.* the $Si_{58}C_{200}$ model complex), they were 0.049 and 0.088, respectively. This decrease indicates that the phase of the catalyst nanoparticle here changes from a liquid (when pristine) to solid (when carbon-doped). This therefore makes SWNT nucleation *via* the VLS mechanism impossible. Considering this impasse, and the atomistic similarity between SWNT nucleation from Si and $SiO_2$ nanoparticles, it is apparent that both proceed *via* the VSS mechanism, as opposed to the VLS mechanism.

The production of an amorphous SiC nanoparticle following the adsorption of $C_2$ on Si nanoparticles warranted the further investigation of the possibility of SWNT nucleation from SiC itself. Moreover, following the conclusion discussed above, *viz.* that SWNT nucleation on both $SiO_2$ and Si occur *via* identical pathways, it is reasonable to anticipate that the same applies in the context of SiC. To this end, we have investigated nucleation of SWNT cap fragments as a result of the constant temperature thermal annealing of SiC nanoparticles alone. In this case, a model $Si_{96}C_{96}$ nanoparticle was annealed between 1000 and 3000 K. An example of SWNT nucleation observed at 2500 K is illustrated in Fig. 11. It is noted that this temperature is approximately that employed in relevant experiments which demonstrate SWNT growth following the decomposition of SiC crystals (Kusunoki *et al.*, 1997). These QM/MD simulations indicated that SWNT nucleation in this case followed the degradation of the SiC crystalline structure. Indeed, upon annealing even at temperatures as low as 1000 K a well-defined crystallinity was not evident in the model SiC nanoparticle even after a relatively short simulation time (*i.e.* 10-20 ps). The result of this degradation was the disruption of C-Si bonds, in favor of C-C bonds, which exhibited free diffusion through/over the SiC nanoparticle. This diffusion immediately lead to the elongation/oligomerisation of these polyyne chains with high frequency. However, the frequency of these interactions was concomitantly slower at lower temperatures, such as 1000 K. As is evident from Fig. 11a,b, polygonal ring formation followed the initial period in which the oligomerisation of polyyne chains took place. In this case, the initial polygonal ring formation was the result of the diffusion and subsequent interaction of neighboring $C_3$ and $C_2$ species. Fig. 11b shows that subsequent ring condensation then proceeded reasonably rapidly, with a definite cap structure being formed within *ca.* 200 ps. However, following the formation of this cap structure, the population of polygonal rings here then decreased - such a phenomenon has not been observed in the case of traditional, transition metal catalyst nanoparticles. In a kinetic sense, therefore, SWNT nucleation resulting from thermal degradation of SiC is anticipated to be less favorable, compared to other traditional catalysts. SWNT nucleation, at the atomic level, is essentially no more than the continual formation of C-C bonds. The origin for these inhibited SWNT nucleation kinetics can therefore be found in thermodynamics, which, at high temperatures, dominate SWNT nucleation. In this sense then, SWNT nucleation is in effect a 'thermodynamic sink'. From §3.3, it is evident that thermal annealing of amorphous Fe- and Ni-carbide nanoparticles yielded well-defined SWNT cap structures, similar to those observed here. However, SWNT nucleation from Fe- and Ni-carbide nanoparticles also resulted in cap structures exceeding the size of those observed using SiC, both on shorter timescales (generally within *ca.* 100 ps) and at lower temperatures (below 2000 K). The strengths of the Fe-C, Ni-C and Si-C interactions are 1.78, 1.06 and 6.29 eV/atom, respectively, at the SCC-DFTB level of theory (Page *et al.*, 2010d). Recall that the strength of the C-C interaction, using SCC-DFTB, is 9.14 eV/atom. The weaker interaction of the Fe/Ni catalyst with carbon therefore correlates directly with an increased rate of SWNT nucleation. Once a C-C bond forms in the latter case, it is rarely broken due to its greater thermodynamic stability (even if it is not the most energetically stable ring structure). On the other hand, the Si-C and C-C interactions are, thermodynamically, comparable to each other. Consequently, C-C bonds are more frequently broken during nucleation on SiC nanoparticles.

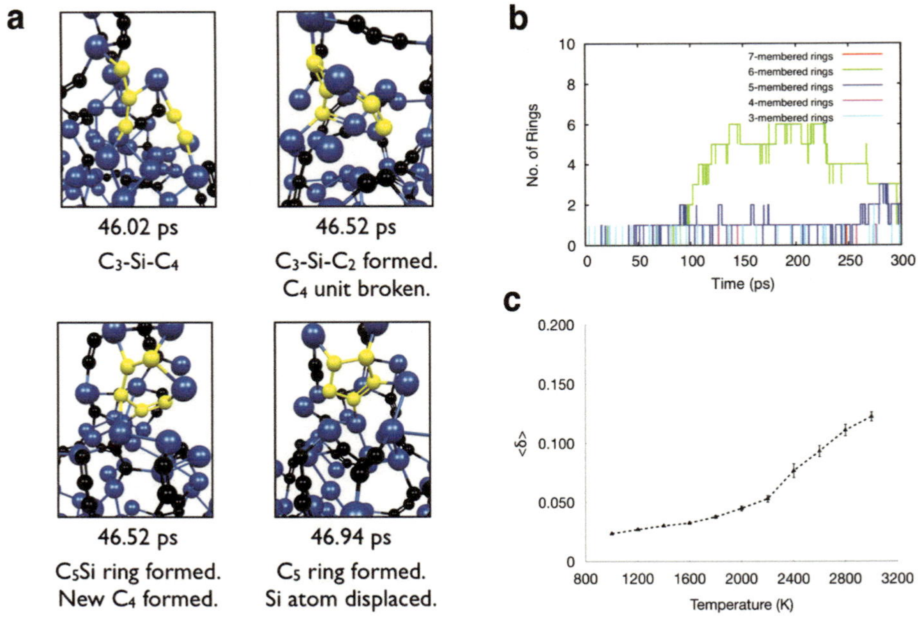


Fig. 11. Thermal annealing at constant temperature (2500 K) leads to the structural deformation of SiC nanoparticles, ultimately producing SWNT nucleation. a) The first polygonal ring formation event due to the free diffusion of $C_n$ units within the SiC nanoparticle. Color conventions as in Fig. 8. b) Polygonal ring formation formed from the structural decomposition of the SiC nanoparticle at 2500 K. c) Time-averaged $\delta$ values of the SiC nanoparticle between 1000 – 3000 K, computed over an interval of 50 ps. SWNT nucleation below 2600 K evidently occurs while the SiC nanoparticle is in the solid phase. Thus, SWNT nucleation can be explained with recourse to a VSS mechanism.

The dependence of $< \delta >$ on simulation temperature for the SiC nanoparticle are depicted in Fig. 11c. From this figure it is evident that the SiC nanoparticle existed in the solid state below 2600 K. However, Fig. 11c suggests that there was undoubtedly some liquid-like character in the SiC nanoparticle at temperatures above 2600 K. In particular, $< \delta_i >$ values (not shown) indicate that, between 1000 and 3000 K, the SiC nanoparticle exhibited three distinct behaviors depending on the temperature. Firstly, at lower temperatures (<1400 K) the SiC nanoparticle were unquestionably solid. At intermediate temperatures (between 1400 - 2600 K) a gradual increase in $< \delta_i >$ for atoms residing close, or near to, the nanoparticle surface was evident. Surface premelting therefore became prevalent at these temperatures, ultimately causing $< \delta >$ to increase slightly. Such surface premelting has been shown to be a prominent phenomenon in the melting dynamics of transition metal nanoparticle species (Neyts & Bogaerts, 2009). In this respect therefore, transition metals and SiC nanoparticles appear to be equivalent. According to established trends regarding transition metal nanoparticle melting, by increasing the temperature further this surface premelting is followed by the complete liquefaction of the nanoparticle. However, rather than undergoing this solid-liquid phase transition, the SiC nanoparticle instead became quasi-solid at temperatures above 2600 K. One probable cause of this unexpected behavior is ascribed to the influence of surface chemistry (*viz.* the formation of C-C bonds, polyyne chains and polygonal carbon rings *etc.*) on the Lindemann index itself. In extreme cases, the formation of an extended $sp^2$-hybridised carbon network on the SiC nanoparticle surface, in part, solidified the SiC nanoparticle surface, therefore retarding the melting process.

A pronounced similarity is therefore observed regarding the SWNT nucleation mechanisms on $SiO_2$, SiC and Si catalysts. The results discussed here constitute the first evidence of a catalyst independent mechanism with respect to Si-based catalysts. In addition, these results indicate the mechanism of SWNT nucleation on these Si-based catalysts is remarkably different to that established for transition metal catalysts, and centres around a solid phase catalyst nanoparticle. Since the independence of the SWNT nucleation mechanism has been established and accepted in the case of transition metal catalysts, this conclusion is seemingly unremarkable. However, we point out here that with respect to the majority of 'non-traditional' catalysts such as $SiO_2$, SiC, Si, $Al_2O_3$, $ZrO_2$, and so on, the precise mechanisms of SWNT nucleation remains are in fact unknown at present. Moreover, at first glance there is no reason to suspect that the SWNT nucleation mechanism on such a diverse range of catalyst species should be in any way related, considering their respective physicochemical properties.

## 4. QM/MD simulations of SWNT growth

We now consider the phenomenon of continued SWNT growth. This is generally defined as the extension of the nanotube sidewall (by the addition of newly created polygonal ring structures) parallel to the axis of growth. Note that this process differs from the process of SWNT nucleation, in which the nascent nanotube cap-fragment is formed. This partitioning of what is actually (in reality) a continuous process is somewhat arbitrary. Nonetheless, it has enabled the precise atomistic mechanism of SWNT growth to be identified and studied.

### 4.1 SWNT growth on Fe catalysts

Continued SWNT growth has been modeled using QM/MD simulations on a number of occasions (see (Page *et al.*, 2010c) and references therein). The approach employed in these investigations typically was similar to that described in §3.2 (see Fig. 12). Fe-catalyst nanoparticles were thus first annealed at 1500 K, after which 'simulated' gas-phase carbon feedstock (in this case, C or $C_2$) was adsorbed at various rates at the base of the growing SWNT, or onto the nanoparticle surface itself. Two such nanoparticles have been employed, *viz.* $Fe_{38}$ and $Fe_{55}$. In both cases, a model SWNT cap fragment (a $C_{40}$ cap of (5,5) chirality), or short SWNT segment (depicted in Fig. 12) were employed to approximate a SWNT cap fragment formed *in situ* (such as that shown in Fig. 4). The effect of the nanoparticle diameter on the mechanism and kinetics of continued SWNT growth has therefore been elucidated. Somewhat unsurprisingly, the increase in nanoparticle diameter from 0.70 nm ($Fe_{38}$) to 0.94 nm ($Fe_{55}$) has no effect on the atomistic mechanism of continued SWNT growth. This mechanism is depicted in Fig. 12. From this figure it is evident that, like SWNT nucleation, the continued SWNT growth process was driven by the extension of the $sp^2$-hybridised carbon network. This extension itself was driven by the formation of polygonal carbon rings at the base of the nanotube structure (at the interface between the nanotube and the catalyst nanoparticle), thereby extending the SWNT cap in a unidirectional manner. From Fig. 12 it can be seen that the SWNT growth process took place almost entirely on the catalyst surface. Only very rarely did carbon penetrate the catalyst surface and diffuse through the subsurface region. Similarly, carbon was never observed to freely diffuse through the bulk of the catalyst nanoparticle. Unsurprisingly, this behavior was no different from the behavior observed during SWNT nucleation on $Fe_{38}$, a fact that is attributed to the nanoparticles relatively small diameter, and consequently relatively high surface energy. It is also noted here that $Fe_{38}$ and $Fe_{55}$ are both 'magic number' metal clusters, and so exhibit unusual stability compared to other nanoparticles of comparable diameter. The SWNT growth depicted in Fig. 12 is an example growth from a 'floating' catalyst (most similar to that observed during pure VLS processes, such as arc-discharge). However, it is likely that the mechanism of SWNT 'root'/'tip' growth on supported catalyst nanoparticles is similar to that depicted in Fig. 12, since the majority of SWNT growth chemistry is mediated by the nanoparticle surface itself.

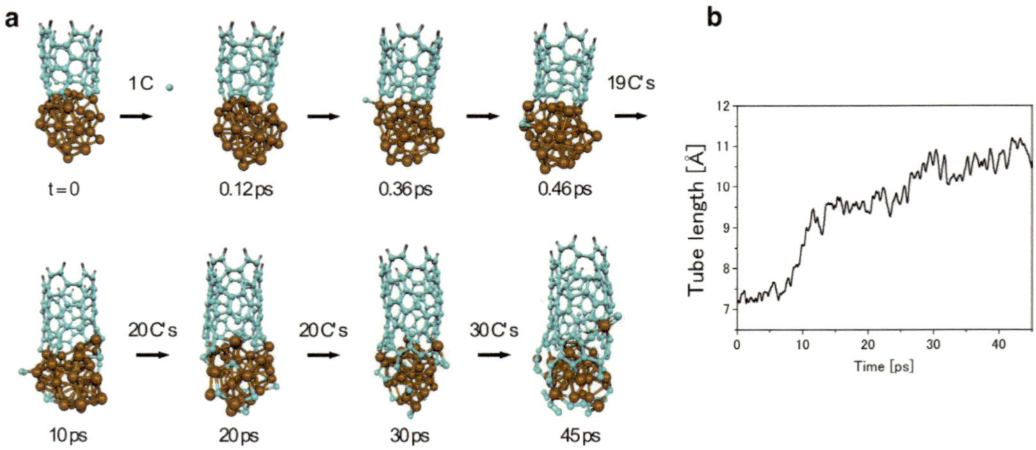


Fig. 12. Continued SWNT growth from a (5,5) SWNT fragment on an $Fe_{38}$ catalyst nanoparticle at 1500 K. a) The adsorption of gas-phase carbon atoms at a rate of 1 C / 0.5 ps at the base of the SWNT structure leads to the extension of the $sp^2$-hybridised carbon network *via* the formation of new polygonal rings at the SWNT base. Growth is mediated entirely by the catalyst surface in this case. Color conventions as in Fig. 1. b) The SWNT length as a function of time at 1500 K. Adsorption of gas-phase carbon atoms results in the addition of *ca.* 4 Å to the base of the SWNT. (Adapted from (Ohta *et al.*, 2008). Reprinted with permission. © 2008 American Chemical Society)

While the SWNT growth mechanisms on $Fe_{38}$ and $Fe_{55}$ were observed to be the same, this is not so with respect to the kinetics of SWNT growth. QM/MD simulations (Page *et al.*, 2010b) indicate that SWNT growth slows with increasing catalyst nanoparticle diameter - a conclusion that parallels others based on experimental evidence (Huang *et al.*, 2002; Cau *et al.*, 2006; Mora & Harutyunyan, 2008). This phenomenon is ascribed primarily to the relative surface areas and volumes of the two catalyst nanoparticles. In particular, although the diameter of $Fe_{55}$ is only slightly larger than that of $Fe_{38}$, the increases in surface area and volume are more substantial. Thus, the domain over/through which adsorbed $C_n$ species may migrate, before being incorporated into the growing SWNT, is concomitantly larger in the case of $Fe_{55}$. SWNT growth employing the former, smaller catalyst nanoparticle is therefore *ca.* 19% faster compared to that on $Fe_{55}$. It is conceded that both of these growth rates exceed those determined experimentally (Puretzky *et al.*, 2002; Futaba *et al.*, 2005; Sharma *et al.*, 2005; Geohegan *et al.*, 2007; Yao *et al.*, 2007; Xiang *et al.*, 2009) by several orders of magnitude. This is a natural consequence of the relatively unnatural carbon adsorption model that has been employed here. Nevertheless, the error thus induced is systematic, and so these relative trends in growth rates remain valid.

### 4.2 The importance of interaction energy: Ni versus Fe catalysts

The fact that different SWNT catalyst materials yield different SWNT growth rates has been established experimentally on numerous occasions (Puretzky *et al.*, 2002; Futaba *et al.*, 2005; Sharma *et al.*, 2005; Geohegan *et al.*, 2007; Yao *et al.*, 2007; Xiang *et al.*, 2009). Nevertheless, no clue was gained as to why this was the case until recently. QM/MD simulations (Page *et al.*, 2010a; Page *et al.*, 2010b) again proved to be of value in this respect, and established the single origin of catalyst-dependent SWNT growth kinetics.

QM/MD simulations of $Ni_{38}$-catalysed growth from a $C_{40}$ SWNT cap fragment are summarised in Fig. 13a. Fig. 13b shows a comparison of $Fe_{55}$- and $Ni_{55}$-catalysed SWNT growth rates. Once again, in all cases growth was induced by the adsorption of gas-phase carbon atoms at the base of the $C_{40}$ SWNT cap structure at a rate of 1 C /0.5 ps. Comparison of Fig. 12a and 13a shows that the mechanism of SWNT growth, at the atomistic scale, exhibits significant differences. Most notably in this respect is the role of the extended polyyne chains which bridge between the SWNT base and the catalyst surface. In the case of $Fe_{38}$ (Fig. 12a), these chains generally consisted of 3-4 carbon atoms, and were formed as individual $C/C_2$ species diffused across the $Fe_{38}$ surface towards the SWNT base. On the other hand, Fig. 13a shows that the polyyne chains bridging between the SWNT base and the catalyst surface in the case of $Ni_{38}$ were far greater in length. Generally, such polyyne chains were observed to be as large as $C_{10}$ for $Ni_{38}$ and $Ni_{55}$ catalyst nanoparticles. In both Fe- and Ni-catalyst cases, continued SWNT growth was driven by the formation of polygonal carbon rings at the base of the SWNT, generally from the interaction of these bridging carbon chains. The length of these carbon chains therefore proved to be a critical factor in the context of the SWNT growth mechanism. For $Ni_{38}$ and $Ni_{55}$ catalysts, the rate of extension of these carbon chains was greater than the rate at which they self-isomerised, or 'collapsed' (Page *et al.*, 2010a). In the case depicted in Fig. 13a, the extension and collapse of a single polyyne chain bound to the base of the growing $C_{40}$ cap structure resulted in the formation of a conjugated 6-5-7-5 carbon ring system. Conversely, the rates of polyyne extension and collapse observed using $Fe_{38}$ and $Fe_{55}$ catalyst nanoparticles were generally more equivalent. SWNT growth was thus limited by the rate of polyyne chain extension. Ultimately these mechanistic differences yield Ni-catalysed SWNT growth rates *ca.* 69 – 106% greater than those found using Fe-catalysed, for equivalent catalyst nanoparticle size.

Somewhat unsurprisingly, the fundamental factor explaining the kinetic differences of Fe- and Ni-catalysed SWNT growth are the same as those which explain the differences in Fe- and Ni-catalysed SWNT nucleation. Fig. 12a and 13a show that, once again, the relative strengths of the Fe-C, Ni-C and C-C interactions correlate exactly with the observed SWNT nucleation kinetics. For example, the rate of SWNT growth is limited by the rate at which the bridging polyyne chains (pictured in Fig. 12a and 13a) can incorporate new carbon. This rate, in turn, is determined by the relative thermodynamics of C-C bond formation in the presence of Fe and Ni atoms. As was discussed in §3.3, the relative weakness of the Ni-C interaction means that, in a thermodynamic sense, the formation of C-C bonds on Ni-catalysts is a more favorable process compared to that on Fe-catalysts. In this sense, therefore, the strength of the catalyst-carbon interaction constitutes a fundamental, guiding principle for understanding the mechanisms and kinetics of SWNT growth on different catalyst materials.

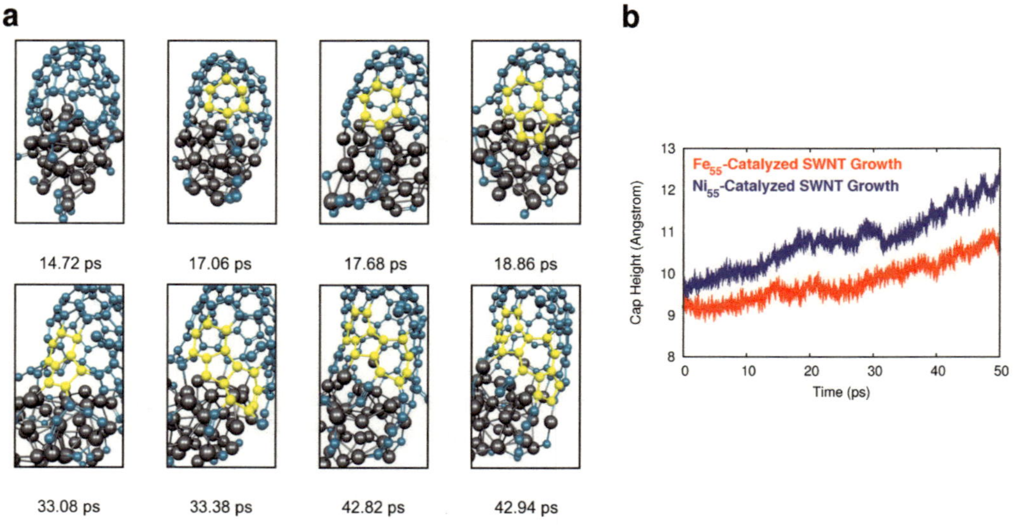


Fig. 13. Continued SWNT growth from a (5,5) $C_{40}$ SWNT cap on a $Ni_{38}$ catalyst nanoparticle at 1500 K. a) In this case, the extension and collapse of a single bridging polyyne chain results in the formation of an extended conjugated system at the base of the SWNT, including a hexagonal, heptagonal and two pentagonal carbon rings. Color conventions as in Fig. 6. b) Depending on the size of the catalyst nanoparticle, Ni-catalysed SWNT growth is found to be *ca.* 69 – 106% faster than Fe-catalysed SWNT growth at 1500 K. (Adapted from (Page *et al.*, 2010a). Reprinted with permission. © 2010 American Chemical Society)

## 5. SWNT defects, healing and chirality-controlled growth

As has been shown in §2 – 4, there have been significant advances in both experimental and theoretical understanding of SWNT nucleation and growth on a number of different catalyst species. Yet there are still outstanding issues regarding phenomena associated with SWNT growth. The most notable phenomenon at present is that of 'chirality-controlled' growth. That is, a method by which a single particular $(n, m)$ chirality SWNT (or, at most a narrow distribution of $(n, m)$ SWNTs) may be synthesised *in situ* remains elusive to date. At the atomistic scale, chirality-controlled growth equates to growth in which only hexagonal rings are incorporated into the growth SWNT structure. The fundamental principles guiding such chirality-specific synthesis are, as yet, largely unknown. Such chirality-controlled growth is extremely desirable, since the physical, electrical and optical properties of a SWNT are determined entirely by its $(n, m)$ chiral indices. Current experimental SWNT synthesis techniques (such as CVD and arc-discharge) are known to produce a broad distribution of $(n, m)$ SWNTs. While it is possible to subsequently isolate a narrow distribution of $(n, m)$ SWNTs, such techniques invariably damage the SWNT structures by either chemical or physical means (Li *et al.*, 2007; Zheng & Semke, 2007). Such damage potentially limits the application of these SWNTs in nanoscale devices. An understanding of how to control a SWNTs chirality *in situ* is therefore critical in this respect.

### 5.1 SWNT growth: An inherently defective process

It was shown in §2 - 4 that SWNT nucleation and growth are far from linear, ordered processes. On the contrary, they proceed *via* extremely complex pathways, resulting in disordered and unpredictable dynamics. Considering the temperature at which SWNTs nucleate and grow in CVD and arc-discharge environments (*i.e.* typically 1000 K or higher), this in itself is not so surprising. Yet it provides the greatest hurdle regarding the control of SWNT chirality, since these nonlinear dynamics result in the formation of a large number of defects in the SWNT structure during nucleation and growth. The high concentration of defect structures must be, to some extent, due to the various assumptions placed on these QM/MD simulations. Nevertheless, the formation of defect structures must also be attributable to the non-equilibrium conditions present during SWNT nucleation and growth. Indeed, by revisiting the discussion of §3.1 - 3.2, it is apparent that the inclusion of defects (such as polyyne chains, non-hexagonal ring structures and vacancies in the $sp^2$-hybridised carbon network) is inherent to the SWNT nucleation process itself. The 'nucleus' of the SWNT itself on a transition metal catalyst is actually a pentagonal ring 'defect'. The subsequent ring condensation process, by which the SWNT cap fragment is formed, also produces a majority of non-hexagonal ring defects. Although this is attributed to the curvature of the catalyst surface imposing itself onto the growing $sp^2$-hybridised carbon network, the further formation of defect structures during growth (see §4) cannot be rationalized in this manner. Nevertheless, the incorporation of defect structures into a growing SWNT effectively alters its chirality, and therefore physical properties. Since it is apparent that SWNT nucleation and growth are inherently defect-inducing processes, it is important to understand the mechanisms by which such defects are removed *in situ*. QM/MD simulation of such defect removal on transition metal catalysts is the subject of §5.2.

### 5.2 SWNT healing: A fundamental aspect of chirality-controlled SWNT growth

The removal of SWNT defect structures during growth has been investigated previously using QM/MD (Page *et al.*, 2009). To induce growth gas-phase carbon atoms were adsorbed at the region between a model $C_{40}$ cap and its supporting $Fe_{38}$ catalyst surface. The hypothesis of this approach took into account the inherent stability of the C-C bond (relative to the Fe-C bond), and therefore the greater stability of the SWNT as a whole. Due to this stability, the removal, or healing, of defects during growth was anticipated to occur over longer time scales than those considered in prior QM/MD simulations (*ca.* 50 ps). Three gas-phase carbon adsorption rates were therefore employed, *viz.* 1 C / 0.5 ps 1 C / 10 ps and 1 C / 20 ps (rates denoted using 'fast', 'slow' and 'very slow'). It is noted here that the former of these adsorption rates is the same as that employed in the simulations discussed in §4.
Comparison of the three carbon adsorption rates is made in Fig. 14. It is immediate from this figure that the ability of the SWNT to heal itself during growth is directly correlated to the rate of carbon adsorption. As this rate decreases, the number of polygonal ring defects in the growing $sp^2$-hybridised carbon network decreases. Moreover, the active removal of defects from the growing SWNT structure, resulting in hexagonal ring formation is observed for slow and very slow carbon adsorption. This suggests that the kinetics of SWNT growth is more favorable compared to those of defect removal. Fig. 15 shows the reason explaining

why this is indeed the case. This figure depicts schematically two examples of defect removal during slow SWNT growth. The first of these defects is a conjugated pentagonal-heptagonal ring defect and the second is an adatom defect. In both cases, the defect is removed solely by the self-isomerization of the SWNT cap structure itself. For example, the adatom defect shown in Fig. 15b is formed following the adsorption of a carbon atom onto an existing hexagonal ring at the SWNT base. This adatom defect quickly converted to a heptagonal ring defect, which is evidently significantly more stable (lasting for *ca.* 15 ps).

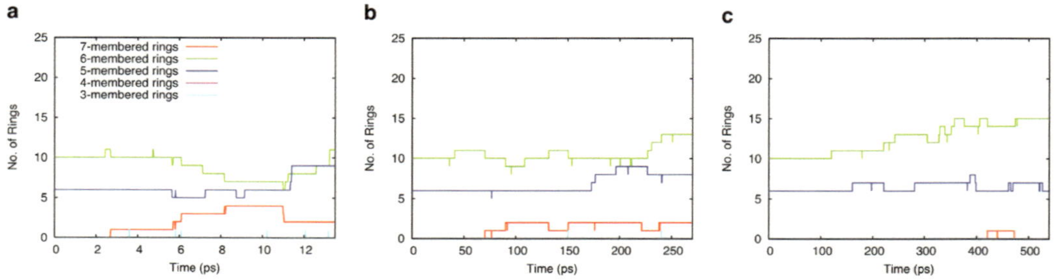


Fig. 14. SWNT healing is directly correlated with the rate of carbon adsorption at the SWNT base. Polygonal rings in formed a single SCC-DFTB/MD trajectory using adsorption rates of 1 C / a) 0.5 ps, b) 10 ps and c) 20 ps. At the fastest adsorption rate considered, SWNT growth incorporates several defect ring structures into the SWNT structure. Slowing the adsorption rate to 1 C / 10 ps, the incorporation of defects is suppressed due to the action of the self-isomerisation of the $sp^2$-hybridised carbon network. At the slowest rate considered, SWNT growth occurs solely due to hexagonal ring addition, thereby illustrating chirality-controlled SWNT growth. (Adapted from (Page *et al.*, 2010c). Reprinted with permission. © 2011 American Chemical Society)

The addition of a second carbon atom results in a heptagonal-hexagonal ring rearrangement, which ultimately forms a $C_2$ defect at the base of the SWNT. Following a further *ca.* 10 ps, this $C_2$ unit detaches from the SWNT cap, and diffuses away over the catalyst surface. Both instances of SWNT defect removal depicted in Fig. 15 occur in the vicinity of the catalyst surface. The assistance of the catalyst surface is therefore implicated in these cases of SWNT healing. The timescales over which these two examples of self-isomerization take place are between 5 and 25 ps, respectively. This observation is indicative of a fundamental principle regarding the *in situ* control of SWNT chirality. That is, the rate at which defect structures are incorporated into the SWNT structure depends on the relative rates of defect addition (due to growth) and defect removal (due to SWNT self-isomerization).

### 5.3 SWNT healing: Dependence on catalyst composition and size

In §3.3 and §4.2 it was established that a number of kinetic and mechanistic phenomena associated with SWNT nucleation and growth can be understood with recourse to the relative carbon-catalyst interaction strengths. For example, a stronger carbon-catalyst interaction leads to slower growth rates, and changes the mechanisms of SWNT nucleation and growth. From the previous section, it was seen that the catalyst nanoparticle is implicated in the SWNT healing process (§5.2). It therefore seems reasonable to hypothesize that the carbon-catalyst interaction may also play some role regarding the relative ability of different catalysts to assist in SWNT healing processes. We will presently discuss such a proposal with respect to Fe and Ni-catalyst nanoparticles.

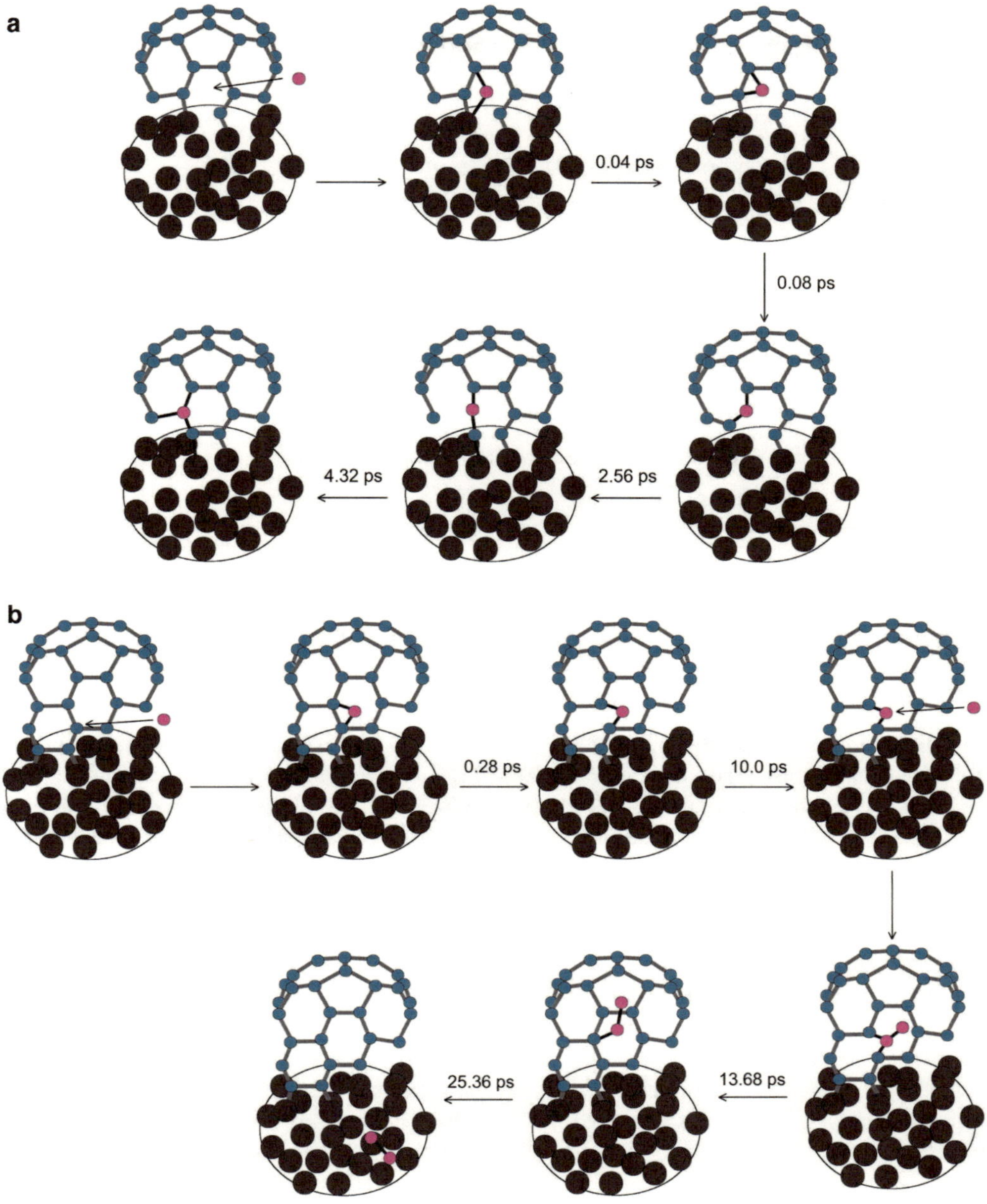


Fig. 15. Examples of SWNT healing observed during SWNT growth on $Fe_{38}$ catalysts at 1500 K. In both cases, defects are removed from the growing SWNT cap solely by the self-isomerisation of the $sp^2$-hybridised carbon network. a) A conjugated pentagonal-heptagonal defect is removed, resulting in the formation of two hexagonal rings at the base of the SWNT cap. b) An adatom defect is removed, forming a hexagonal ring *via* a heptagonal ring defect intermediate structure. (Adapted from (Page *et al.*, 2009). Reprinted with permission. © 2011 American Chemical Society)

| | | $Fe_{38}$ | $Ni_{38}$ | $Fe_{55}$ | $Ni_{55}$ |
|---|---|---|---|---|---|
| Defect Formation | Pentagonal Formation | 3.2 | 6.1 | 2.2 | 4.6 |
| | Heptagonal Formation | 0.2 | 0.4 | 0.3 | 0.2 |
| | Hexagonal→Heptagonal Transformation | 2.7 | 2.8 | 2.3 | 2.9 |
| | Hexagonal→ Deformation | 1.0 | 0.2 | 0.6 | 0.2 |
| | Hexagonal→Pentagonal Transformation | 0.1 | 0.3 | 0.6 | 0.1 |
| | Total Defects Formed ($\Sigma_1$) | 7.2 | 9.8 | 6.0 | 8.0 |
| Defect Removal | Hexagonal Formation | 3.4 | 3.3 | 3.4 | 2.9 |
| | Heptagonal→Hexagonal Transformation | 1.1 | 1.0 | 0.8 | 1.3 |
| | Pentagonal→Hexagonal Transformation | 1.2 | 1.7 | 1.5 | 1.6 |
| | Total Defects Removed ($\Sigma_2$) | 5.7 | 6.0 | 5.7 | 5.8 |
| Net Healing ($\Sigma_2$-$\Sigma_1$) | | -1.5 | -3.8 | -0.3 | -2.2 |

Table 1. SWNT healing statistics on Fe- and Ni-catalyst nanoparticles for a carbon adsorption rate of 1 C / 10 ps. The net rate of SWNT healing may be considered as the difference between the rates of defect formation and defect removal. All data averaged over 10 SCC-DFTB/MD trajectories, following 300 ps of simulation.

QM/MD simulations of SWNT growth were carried out using $Fe_{38}$, $Ni_{38}$, $Fe_{55}$ and $Ni_{55}$ catalyst nanoparticles. Growth was induced at 1500 K using a slow carbon supply rate (*i.e.* 1 C / 10 ps). The average defect formation and defect removal statistics following 300 ps are given in Table 1. For the purpose of this analysis, 'defect formation' is defined here as the formation of a new pentagonal or heptagonal ring, the conversion of a hexagonal ring to a pentagonal/heptagonal ring, or the destruction of a hexagonal ring (*i.e.* ring opening). Conversely, 'defect removal' is defined here as essentially the opposite of defect formation, *i.e.* the formation of new hexagonal rings and the conversion of pentagonal/heptagonal rings to hexagonal rings. From Table 1 it is evident that the rate of defect removal in the case of the four catalysts considered are essentially equivalent after 300 ps. Thus, there is little dependence of the defect removal process on the size, or elemental composition of the catalyst. This is reasonable, since the catalyst nanoparticle was never explicitly involved in the process of healing (as discussed in §5.2). Rather, it plays an implicit role, by saturating dangling bonds at the edge of the SWNT structure, thereby supporting the self-isomerization process. On the other hand, Table 1 shows that the size, and more obviously, the elemental composition of the catalyst nanoparticle directly affects the rate of defect formation during growth. For example, for an equivalent catalyst size, SWNT growth on a Ni catalyst induces *ca.* 30-35% more total defects, compared to SWNT growth on an Fe catalyst. For a particular type of metal, Table 1 also shows that the number of defects formed during SWNT growth decreases with increasing nanoparticle diameter. It is noted here that these two correlations are consistent with the effect of nanoparticle size and composition on the total SWNT growth rate, as discussed in §4.2. That is, faster growth leads to more defects, whereas slower growth leads to a smaller number of defects.

## 6. Conclusion

We have reviewed our own recent investigations into the phenomena of SWNT nucleation and growth using state-of-the-art QM/MD methods. A summary of the primary conclusions discussed herein is provided in Fig. 16. The significance of the QM/MD method in this

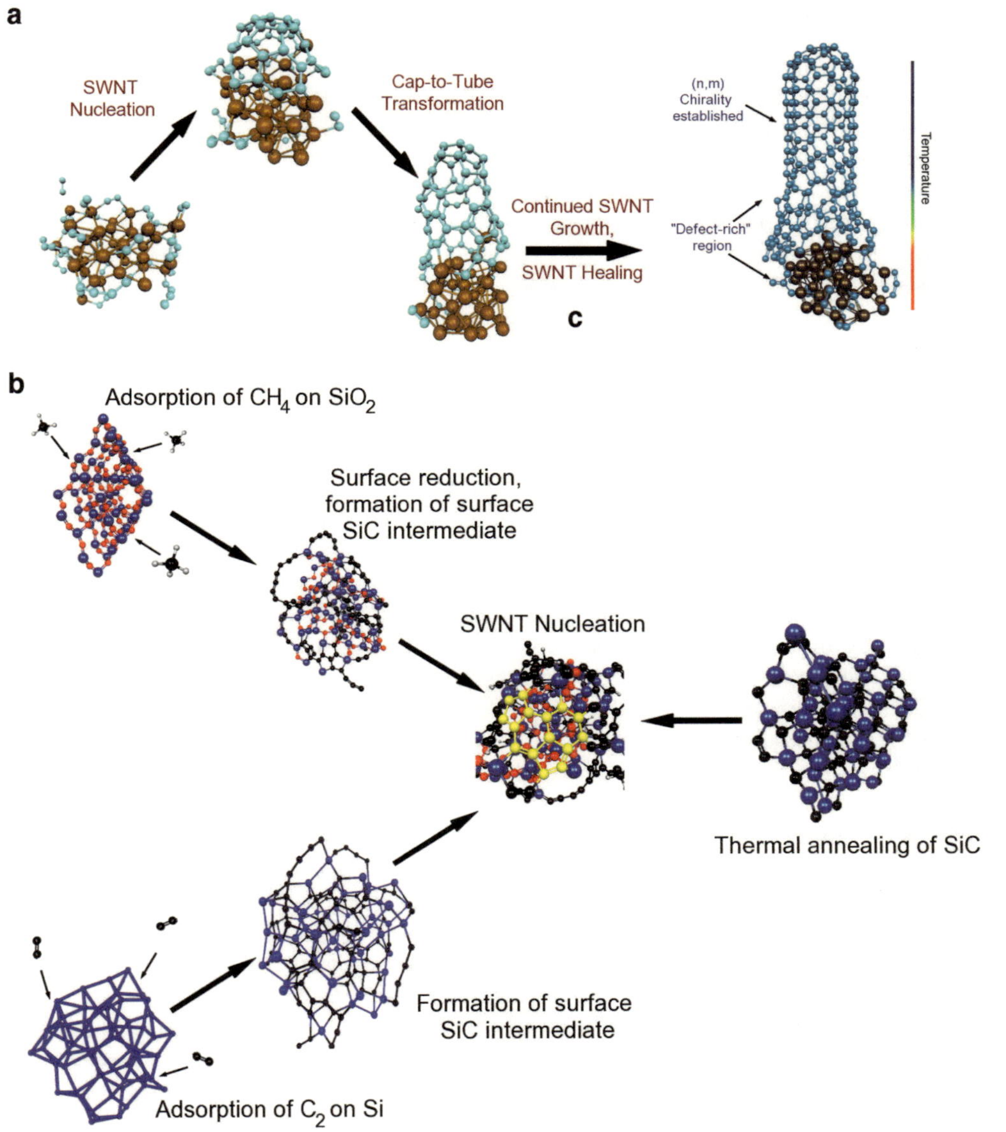


Fig. 16. Insights into the nucleation growth and defect-healing of SWNTs gained from QM/MD simulations. a) SWNT nucleation on transition metal nanoparticles begins with the oligomerisation of small carbon fragments on the catalyst surface. These units subsequently coalesce to form longer, extended polyyne chains which are able to isomerise/interact, thus forming polygonal carbon rings. SWNT nucleation and growth is then the result of

continual polygonal ring condensation on the catalyst surface according to this mechanism. QM/MD simulations also suggest that the removal of defect structures in the growing nanotube occurs *via* a natural process in which the isomerization of the SWNT $sp^2$-hybridised carbon network itself converts defects into hexagonal rings. This defect-healing occurs closest to the catalyst surface where the local temperature is hottest. As root growth continues, the established hexagonal rings in the cooler regions of the SWNT impart a templating effect on the healing process. Brown and cyan spheres represent Fe and C atoms, respectively. (Adapted from (Page *et al.*, 2010c). Reprinted with permission. © 2011 American Chemical Society). b) SWNT nucleation on Si-based catalysts has been elucidated using QM/MD simulations. These simulations have established that a mechanism fundamentally different to that observed for transition metal catalysts is responsible for SWNT nucleation in this case. In particular, these catalyst nanoparticles remain in the solid phase throughout the nucleation process, and the saturation of the catalyst surface with carbon is a necessary prerequisite for SWNT nucleation. Accordingly, it is concluded that these cases of SWNT nucleation are explained with recourse to a VSS mechanism, as opposed to a VLS mechanism. QM/MD simulations have also established that this mechanism is independent of the catalyst employed, at least with respect to $SiO_2$, SiC and Si catalysts. Blue, red and black spheres represent Si, O and C, respectively. Yellow spheres represent C atoms involved in SWNT nucleation.

context has therefore been demonstrated. QM/MD simulations of such non-equilibrium, high-temperature processes can provide fundamental knowledge that complements experimental understanding. Moreover, considering the spatial and temporal resolutions furnished by QM/MD methods (*i.e.* nanometers and picoseconds, respectively), and their physical reliability, such simulations can predate, or correct experimental understanding of these phenomena. This is certainly the case with respect to models of SWNT nucleation and growth. For example, the VLS mechanism of SWNT nucleation and growth on a variety of transition metal catalysts is very widely accepted. Yet it is only since the application of QM/MD in this area that true understanding of various aspects of the VLS mechanism has come to light. One such aspect regards the atomistic processes of SWNT nucleation and growth, which are dominated by the formation and coalescence of extended polyyne chains, and the interaction of these chains with the supporting catalyst surface. Another aspect, which remains under debate at the time of writing, regards the existence and role of the transition metal carbide phase in the context of SWNT nucleation and growth. In particular, recent QM/MD simulations and experiments have challenged the traditional role ascribed to this carbide phase in the SWNT nucleation process. QM/MD methods have also uncovered the phenomenon of defect healing during continued SWNT growth. This phenomenon, by which a SWNT structure consisting entirely of hexagons can be attained, has since been implicated in models of chirality-controlled growth. Most recently, QM/MD simulations have led the way in understanding the manner in which SWNTs nucleate and grow on 'non-traditional' catalysts including $SiO_2$, SiC and Si. They have also revealed the atomistic mechanism underpinning the CVD process on the former catalyst species. In this context, QM/MD methods alone have uncovered the remarkable fact that SWNT nucleation on these solid phase catalysts proceeds according to an entirely different mechanism, compared to the traditional picture of SWNT nucleation/growth on transition metal nanoparticles. Yet we do not acquiesce, and claim that QM/MD can offer nothing more in the understanding of SWNT nucleation and growth. There are still many aspects of these phenomena that remain misunderstood, and others that are currently under debate. The precise atomistic mechanism governing the termination of SWNT growth is once such aspect. Very recently it has been suggested (upon the basis of DFT calculations) that the phenomenon of Ostwald ripening plays an active role in the termination of SWNT growth (Börjesson & Bolton, 2011). Nevertheless, it is clear that, as the computational technology continually advances, the understanding that can be gained from QM/MD simulations of such physical systems can only improve.

## 7. Acknowledgement

This work was in part supported by a CREST (Core Research for Evolutional Science and Technology) grant in the Area of High Performance Computing for Multiscale and Multiphysics Phenomena from the Japanese Science and Technology Agency (JST). Simulations were performed in part using the computer resources at the Research Centre for Computational Science (RCCS), Okazaki Research Facilities, National Institutes for Natural Sciences, and at the Academic Centre for Computing and Media Studies (ACCMS) at Kyoto University. A.J.P. acknowledges the Kyoto University Fukui Fellowship. S.I. acknowledges the Program for Improvement of Research Environment for Young Researchers from Special Coordination Funds for Promoting Science and Technology (SCF) commissioned by the Ministry of Education, Culture, Sports, Science and Technology (MEXT) of Japan for support.

## 8. References

Alder, B. J. & Wainwright, T. E. (1957). Phase Transition for a Hard Sphere System. *The Journal of Chemical Physics*, Vol. 27, No. 5, pp. (1208-1209), 0021-9606

Amara, H., Bichara, C. & Ducastelle, F. (2006). Formation of carbon nanostructures on nickel surfaces: A tight-binding grand canonical Monte Carlo study. *Physical Review B*, Vol. 73, No. 11, pp. (113404/1-113404/4), 1098-0121

Amara, H., Bichara, C. & Ducastelle, F. (2008a). A Tight-Binding Grand Canonical Monte Carlo Study of the Catalytic Growth of Carbon Nanotubes. *Journal of Nanoscience and Nanotechnology*, Vol. 8, No. pp. (6099-6104), 1533-4880

Amara, H., Bichara, C. & Ducastelle, F. (2008b). Understanding the Nucleation Mechanisms of Carbon Nanotubes in Catalytic Chemical Vapor Deposition. *Physical Review Letters*, Vol. 100, No. 5, pp. (056105/1-056105/4), 0031-9007

Amara, H., Roussel, J. M., Bichara, C., Gaspard, J. P. & Ducastelle, F. (2009). Tight-binding potential for atomistic simulations of carbon interacting with transition metals: Application to the Ni-C system. *Physical Review B*, Vol. 79, No. 1, pp. (014109/1-014109/17), 1098-0121

Andersen, H. C. (1980). Molecular dynamics simulations at constant pressure and/or temperature. *The Journal of Chemical Physics*, Vol. 72, No. 4, pp. (2384-2393), 0021-9606

Bachmatiuk, A., Börrnert, F., Grobosch, M., Schäffel, F., Wolff, U., Scott, A., Zaka, M., Warner, J. H., Klingeler, R., Knupfer, M., Büchner, B. & Rümmeli, M. H. (2009). Investigating the Graphitization Mechanism of $SiO_2$ Nanoparticles in Chemical Vapor Deposition. *ACS Nano*, Vol. 3, No. 12, pp. (4098-4104), 1936-0851

Berendsen, H. J. C., Postma, J. P. M., van Gunsteren, W. F., DiNola, A. & Haak, J. R. (1984). Molecular dynamics with coupling to an external bath. *The Journal of Chemical Physics*, Vol. 81, No. 8, pp. (3684-3690), 0021-9606

Börjesson, A. & Bolton, K. (2011). First Principles Studies of the Effect of Ostwald Ripening on Carbon Nanotube Chirality Distributions. *ACS Nano*, Vol. 5, No. 2, pp. (771-779), 1936-0851

Brenner, D. W. (1990). Empirical Potential for Hydrocarbons for Use in Simulating the Chemical Vapor-Deposition of Diamond Films. *Physical Review B*, Vol. 42, No. 15, pp. (9458-9471), 0163-1829

Brenner, D. W. (1992). Empirical potential for hydrocarbons for use in simulating the chemical vapor deposition of diamond films. [Erratum to document cited in CA114(6):53045x]. *Physical Review B*, Vol. 46, No. pp. (1948), 0163-1829

Brenner, D. W., Shenderova, O. A., Harrison, J. A., Stuart, S. J., Ni, B. & Sinnott, S. B. (2002). A second-generation reactive empirical bond order (REBO) potential energy expression for hydrocarbons. *Journal of Physics: Condensed Matter*, Vol. 14, No. pp. (783-802), 0953-8984

Cau, M., Dorval, N., Cao, B., Attal-Tretout, B., Cochon, J. L., Loiseau, A., Farhat, S. & Scott, C. D. (2006). Spatial Evolutions of Co and Ni Atoms During SWNT Formation: Measurements and Modeling. *Journal of Nanoscience and Nanotechnology*, Vol. 6, No. 5, pp. (1298-1308), 1533-4880

Dai, H., Rinzler, A. G., Nikolaev, P., Thess, A., Colbert, D. T. & Smalley, R. E. (1996). Single-wall nanotubes produced by metal-catalysed disproportionation of carbon monoxide. *Chemical Physics Letters* Vol. 260, No. pp. (471-475),

Ding, F., Bolton, K. & Rosen, A. (2004a). Nucleation and growth of single-walled carbon nanotubes: A molecular dynamics study. *Journal of Physical Chemistry B*, Vol. 108, No. 45, pp. (17369-17377), 1520-6106

Ding, F., Bolton, K. & Rosen, A. (2006a). Molecular dynamics study of SWNT growth on catalyst particles without temperature gradients. *Computational Materials Science*, Vol. 35, No. 3, pp. (243-246), 0927-0256

Ding, F., Rosen, A. & Bolton, K. (2004b). Molecular dynamics study of the catalyst particle size dependence on carbon nanotube growth. *The Journal of Chemical Physics*, Vol. 121, No. 6, pp. (2775-2779), 0021-9606

Ding, F., Rosen, A. & Bolton, K. (2004c). Size dependence of the coalescence and melting of iron clusters: A molecular-dynamics study. *Physical Review B*, Vol. 70, No. 7, pp. (075415/1-075415/6), 0163-1829

Ding, F., Rosen, A., Curtarolo, S. & Bolton, K. (2006b). Modeling the melting of supported clusters. *Applied Physics Letters*, Vol. 88, No. pp. (133110/1-133110/3), 0003-6951

Dresselhaus, M. S., Dresselhaus, G. & Avouris, P. (Ed.). (2001). *Carbon Nanotubes: Synthesis, Structure, Properties and Applications*, Springer-Verlag, 3-540-41086-4, Heidelberg

Dresselhaus, M. S., Dresselhaus, G. & Eklund, P. C. (1996). *Science of Fullerenes and Carbon Nanotubes*, Academic Press, 978-0-12-221820-0, San Diego

Elstner, M., Porezag, D., Jungnickel, G., Elsner, J., Haugk, M., Frauenheim, T., Suhai, S. & Seifert, G. (1998). Self-consistent-charge density-functional tight-binding method for simulations of complex materials properties. *Physical Review B*, Vol. 58, No. 11, pp. (7260-7268), 1098-0121

Eres, G., Rouleau, C. M., Yoon, M., Puretzky, A. A., Jackson, J. J. & Geohegan, D. B. (2009). Model for Self-Assembly of Carbon Nanotubes from Acetylene Based on Real-Time Studies of Vertically Aligned Growth Kinetics. *The Journal of Physical Chemistry C*, Vol. 113, No. 35, pp. (15484-15491), 1932-7447

Esconjauregui, S., Whelan, C. M. & Maex, K. (2009). The reasons why metals catalyze the nucleation and growth of carbon nanotubes and other carbon nanomorphologies. *Carbon*, Vol. 47, No. pp. (659-669),

Fan, X., Buczko, R., Puretzky, A. A., Geohegan, D. B., Howe, J. Y., Pantelides, S. T. & Pennycook, S. J. (2003). Nucleation of Single-Walled Carbon Nanotubes. *Physical Review Letters*, Vol. 90, No. 14, pp. (145501/1-145501/4), 0031-9007

Farhat, S., de La Chapelle, M. L., Loiseau, A., Scott, C. D., Lefrant, S., Journet, C. & Bernier, P. (2001). Diameter control of single-walled carbon nanotubes using argon-helium mixture gases. *The Journal of Chemical Physics*, Vol. 115, No. 14, pp. (6752-6759), 0021-9606

Futaba, D. N., Hata, K., Yamada, T., Mizuno, K., Yumura, M. & Iijima, S. (2005). Kinetics of Water-Assisted Single-Walled Carbon Nanotube Synthesis Revealed by a Time-Evolution Analysis. *Physical Review Letters*, Vol. 95, No. 5, pp. (056104/1-4), 0031-9007

Gavillet, J., Loiseau, A., Journet, C., Willaime, F., Ducastelle, F. & Charlier, J.-C. (2001). Root-Growth Mechanism for Single-Wall Carbon Nanotubes. *Physical Review Letters*, Vol. 87, No. 27, pp. (275504/1-275504/4), 0031-9007

Geohegan, D. B., A. A. Puretzky, D. Styers-Barnett, H. Hu, B. Zhao, H. Cui, C. M. Rouleau, G. Eres, J. J. Jackson, R. F. Wood, S. Pannala & J. C. Wells. (2007). In situ time-resolved measurements of carbon nanotube and nanohorn growth. *Physica Status Solidi (b)*, Vol. 244, No. 11, pp. (3944-3949), 1521-3951

Harris, P. J. F. (2007). Solid state growth mechanisms for carbon nanotubes. *Carbon*, Vol. 45, No. 2, pp. (229-239), 0008-6223

Harutyunyan, A. R., Awasthi, N., Jiang, A., Setyawan, W., Mora, E., Tokune, T., Bolton, K. & Curtarolo, S. (2008). Reduced Carbon Solubility in Fe Nanoclusters and Implications for the Growth of Single-Walled Carbon Nanotubes. *Physical Review Letters*, Vol. 100, No. 19, pp. (195502/1-195502/4), 0031-9007

Heath, J., R. (1992). Synthesis of $C_{60}$ from Small Carbon Clusters, In: *Fullerenes: Synthesis, Properties and Chemistry of Large Carbon Clusters*, G. S. Hammond and V. J. Kuck (Ed.), pp. (1-23), American Chemical Society, 0-8412-2182-0,

Homma, Y., Liu, H., Takagi, D. & Kobayashi, Y. (2009). Single-walled carbon nanotube growth with non-iron-group "catalysts" by chemical vapor deposition. *Nano Research*, Vol. 2, No. 10, pp. (793-799), 1998-0124

Hoover, W. G. (1985). Canonical dynamics: Equilibrium phase-space distributions. *Physical Review A*, Vol. 31, No. 3, pp. (1695), 1050-2947

Huang, S., Cai, Q., Chen, J., Qian, Y. & Zhang, L. (2009). Metal-Catalyst-Free Growth of Single-Walled Carbon Nanotubes on Substrates. *Journal of the American Chemical Society*, Vol. 131, No. 6, pp. (2094-2095), 0002-7863

Huang, Z. P., Wang, D. Z., Wen, J. G., Sennett, M., Gibson, H. & Ren, Z. F. (2002). Effect of nickel, iron and cobalt on growth of aligned carbon nanotubes. *Applied Physics A: Materials Science & Processing*, Vol. 74, No. 3, pp. (387-391), 0947-8396

Iijima, S. (1991). Helical Microtubules of Graphitic Carbon. *Nature*, Vol. 354, No. 6348, pp. (56-58), 0028-0836

Iijima, S. & Ichihashi, T. (1993). Single-Shell Carbon Nanotubes of 1-nm Diameter. *Nature*, Vol. 363, No. 6430, pp. (603-605), 0028-0836

Irle, S., Ohta, Y., Okamoto, Y., Page, A. J., Wang, Y. & Morokuma, K. (2009). Milestones in molecular dynamics simulations of single-walled carbon nanotube formation: A brief critical review. *Nano Research*, Vol. 2, No. 10, pp. (755-767), 1998-0124

Irle, S., Zheng, G., Wang, Z. & Morokuma, K. (2006). The $C_{60}$ Formation Puzzle "Solved": QM/MD Simulations Reveal the Shrinking Hot Giant Road of the Dynamic Fullerene Self-Assembly Mechanism. *Journal of Physical Chemistry B*, Vol. 110, No. pp. (14531-14545), 1520-6106

Journet, C. & Bernier, P. (1998). Production of carbon nanotubes. *Applied Physics A: Materials Science & Processing*, Vol. 67, No. 1, pp. (1-9), 0947-8396

Journet, C., Maser, W. K., Bernier, P., Loiseau, A., delaChapelle, M. L., Lefrant, S., Deniard, P., Lee, R. & Fischer, J. E. (1997). Large-scale production of single-walled carbon nanotubes by the electric-arc technique. *Nature*, Vol. 388, No. 6644, pp. (756-758), 0028-0836

Krätschmer, W., Fostiropoulos, K. & Huffman, D. R. (1990). The infrared and ultraviolet absorption spectra of laboratory-produced carbon dust: evidence for the presence of the $C_{60}$ molecule. *Chemical Physics Letters*, Vol. 170, No. 2-3, pp. (167-170), 0009-2614

Kusunoki, M., Rokkaku, M. & Suzuki, T. (1997). Epitaxial carbon nanotube film self-organized by sublimation decomposition of silicon carbide. *Applied Physics Letters*, Vol. 71, No. 18, pp. (2620-2622),

Li, X., Tu, X., Zaric, S., Welsher, K., Seo, W. S., Zhao, W. & Dai, H. (2007). Selective Synthesis Combined with Chemical Separation of Single-Walled Carbon Nanotubes for Chirality Selection. *Journal of the American Chemical Society*, Vol. 129, No. 51, pp. (15770-15771), 0002-7863

Lindemann, F. A. (1910). The calculation of molecular vibration frequencies. *Zeitschrift fur Physik*, Vol. 11, No. pp. (609-612), 0340-2347

Liu, B., Ren, W., Gao, L., Li, S., Pei, S., Liu, C., Jiang, C. & Cheng, H.-M. (2009a). Metal-Catalyst-Free Growth of Single-Walled Carbon Nanotubes. *Journal of the American Chemical Society*, Vol. 131, No. 6, pp. (2082-2083), 0002-7863

Liu, B., Ren, W., Liu, C., Sun, C.-H., Gao, L., Li, S., Jiang, C. & Cheng, H.-M. (2009b). Growth Velocity and Direct Length-Sorted Growth of Short Single-Walled Carbon Nanotubes by a Metal-Catalyst-Free Chemical Vapor Deposition Process. *ACS Nano*, Vol. 3, No. 11, pp. (3421-3430), 1936-0851

Liu, B., Tang, D.-M., Sun, C., Liu, C., Ren, W., Li, F., Yu, W.-J., Yin, L.-C., Zhang, L., Jiang, C. & Cheng, H.-M. (2011). Importance of Oxygen in the Metal-Free Catalytic Growth of Single-Walled Carbon Nanotubes from $SiO_x$ by a Vapor–Solid–Solid

Mechanism. *Journal of the American Chemical Society*, Vol. 133, No. 2, pp. (197-199), 0002-7863

Liu, H., Takagi, D., Chiashi, S., Chokan, T. & Homma, Y. (2010a). Investigation of Catalytic Properties of $Al_2O_3$ Particles in the Growth of Single-Walled Carbon Nanotubes. *Journal of Nanoscience and Nanotechnology*, Vol. 10, No. 6, pp. (4068-4073), 1533-4880

Liu, H., Takagi, D., Chiashi, S. & Homma, Y. (2010b). The growth of single-walled carbon nanotubes on a silica substrate without using a metal catalyst. *Carbon*, Vol. 48, No. 1, pp. (114-122), 0008-6223

Liu, H., Takagi, D., Ohno, H., Chiashi, S., Chokan, T. & Homma, Y. (2008). Growth of Single-Walled Carbon Nanotubes from Ceramic Particles by Alcohol Chemical Vapor Deposition. *Applied Physics Express*, Vol. 1, No. 1, pp. (014001/1-014001/3), 1882-0778

Martyna, G. J., Klein, M. L. & Tuckerman, M. (1992). Nose-Hoover chains: The canonical ensemble via continuous dynamics. *The Journal of Chemical Physics*, Vol. 97, No. pp. (2635-2643), 0021-9606

Martyna, G. J., Tuckerman, M. E., Tobias, D. J. & Klein, M. L. (1996). Explicit reversible integrators for extended systems dynamics. *Molecular Physics: An International Journal at the Interface Between Chemistry and Physics*, Vol. 87, No. 5, pp. (1117 - 1157), 0026-8976

Moisala, A., Nasibulin, A. G. & Kauppinen, E. I. (2003). The role of metal nanoparticles in the catalytic production of single-walled carbon nanotubes - a review. *Journal of Physics: Condensed Matter*, Vol. 15, No. 42, pp. (S3011-S3035), 0953-8984

Mora, E. & Harutyunyan, A. R. (2008). Study of Single-Walled Carbon Nanotubes Growth via the Catalyst Lifetime. *Journal of Physical Chemistry C*, Vol. 112, No. 13, pp. (4805-4812), 1932-7447

Neyts, E. C. & Bogaerts, A. (2009). Numerical Study of the Size-Dependent Melting Mechanisms of Nickel Nanoclusters. *Journal of Physical Chemistry C*, Vol. 113, No. 7, pp. (2771-2776), 1932-7447

Nose, S. (1984). A unified formulation of the constant temperature molecular dynamics methods. *The Journal of Chemical Physics*, Vol. 81, No. 1, pp. (511-519), 0021-9606

Ohta, Y., Okamoto, Y., Irle, S. & Morokuma, K. (2008). Rapid Growth of a Single-Walled Carbon Nanotube on an Iron Cluster: Density-Functional Tight-Binding Molecular Dynamics Simulations. *ACS Nano*, Vol. 2, No. 7, pp. (1437-1444), 1936-0851

Ohta, Y., Okamoto, Y., Page, A. J., Irle, S. & Morokuma, K. (2009). Quantum Chemical Molecular Dynamics Simulation of Single-Walled Carbon Nanotube Cap Nucleation on an Iron Particle. *ACS Nano*, Vol. 3, No. 11, pp. (3413-3420), 1936-0851

Page, A. J., Chandrakumar, K. R. S., Irle, S. & Morokuma, K. (2011a). Do $SiO_2$ and Carbon-Doped $SiO_2$ Nanoparticles Melt? Insights from QM/MD Simulations and Ramifications Regarding Carbon Nanotube Growth. *Chemical Physics Letters*, In press

Page, A. J., Chandrakumar, K. R. S., Irle, S. & Morokuma, K. (2011b). SWNT Nucleation from Carbon-Coated $SiO_2$ Nanoparticles via a Vapor–Solid–Solid Mechanism. *Journal of the American Chemical Society*, Vol. 133, No. 3, pp. (621-628), 0002-7863

Page, A. J., Irle, S. & Morokuma, K. (2010a). Polyyne Chain Growth and Ring Collapse Drives Ni-Catalysed SWNT Growth: A QM/MD Investigation. *Journal of Physical Chemistry C*, Vol. 114, No. 18, pp. (8206-8211), 1932-7447

Page, A. J., Minami, S., Ohta, Y., Irle, S. & Morokuma, K. (2010b). Comparison of single-walled carbon nanotube growth from Fe and Ni nanoparticles using quantum chemical molecular dynamics methods. *Carbon*, Vol. 48, No. 11, pp. (3014-3026), 0008-6223

Page, A. J., Ohta, Y., Irle, S. & Morokuma, K. (2010c). Mechanisms of Single-Walled Carbon Nanotube Nucleation, Growth, and Healing Determined Using QM/MD Methods. *Accounts of Chemical Research*, Vol. 43, No. 10, pp. (1375-1385), 0001-4842

Page, A. J., Ohta, Y., Okamoto, Y., Irle, S. & Morokuma, K. (2009). Defect Healing during Single-Walled Carbon Nanotube Growth: A Density-Functional Tight-Binding Molecular Dynamics Investigation. *Journal of Physical Chemistry C*, Vol. 113, No. 47, pp. (20198-20207), 1932-7447

Page, A. J., Yamane, H., Ohta, Y., Irle, S. & Morokuma, K. (2010d). QM/MD Simulation of SWNT Nucleation on Transition-Metal Carbide Nanoparticles. *Journal of the American Chemical Society*, Vol. 132, No. 44, pp. (15699-15707), 0002-7863

Porezag, D., Frauenheim, T., Kohler, T., Seifert, G. & Kaschner, R. (1995). Construction of tight-binding-like potentials on the basis of density-functional theory: Application to carbon. *Physical Review B*, Vol. 51, No. 19, pp. (12947-12957), 1098-0121

Puretzky, A. A., Geohegan, D. B., Schittenhelm, H., Fan, X. & Guillorn, M. A. (2002). Time-resolved diagnostics of single wall carbon nanotube synthesis by laser vaporization. *Applied Surface Science*, Vol. 197-198, No. 1, pp. (552-562), 0169-4332

Puri, P. & Yang, V. (2007). Effect of Particle Size on Melting of Aluminum at Nano Scales. *Journal of Physical Chemistry C*, Vol. 111, No. 32, pp. (11776-11783), 1932-7447

Rahman, A. (1964). Correlations in the Motion of Atoms in Liquid Argon. *Physical Review*, Vol. 136, No. 2A, pp. (A405), 1050-2947

Raty, J. Y., Gygi, F. & Galli, G. (2005). Growth of carbon nanotubes on metal nanoparticles: A microscopic mechanism from ab initio molecular dynamics simulations. *Physical Review Letters*, Vol. 95, No. 9, pp. (096103/1-096103/4), 0031-9007

Saito, Y. (1995). Nanoparticles and filled nanocapsules. *Carbon*, Vol. 33, No. 7, pp. (979-988), 0008-6223

Seifert, G., Porezag, D. & Frauenheim, T. (1996). Calculations of molecules, clusters, and solids with a simplified LCAO-DFT-LDA scheme. *International Journal of Quantum Chemistry*, Vol. 58, No. 2, pp. (185-192), 1097-461X

Sharma, R., Rez, P., Treacy, M. M. J. & Stuart, S. J. (2005). In situ observation of the growth mechanisms of carbon nanotubes under diverse reaction conditions. *Journal of Electron Microscopy*, Vol. 54, No. 3, pp. (231-237), 0022-0744

Shibuta, Y. & Maruyama, S. (2002). Molecular dynamics simulation of generation process of SWNTs. *Physica B*, Vol. 323, No. pp. (187), 0921-4526

Shibuta, Y. & Maruyama, S. (2003). Molecular dynamics simulation of formation process of single-walled carbon nanotubes by CCVD method. *Chemical Physics Letters*, Vol. 382, No. pp. (381-386), 0009-2614

Shibuta, Y. & Maruyama, S. (2007a). A molecular dynamics study of the effect of a substrate on catalytic metal clusters in nucleation process of single-walled carbon nanotubes. *Chemical Physics Letters*, Vol. 437, No. 4-6, pp. (218-223), 0009-2614

Shibuta, Y. & Maruyama, S. (2007b). Bond-order potential for transition metal carbide cluster for the growth simulation of a single-walled carbon nanotube. *Computational Materials Science*, Vol. 39, No. pp. (842-848), 0927-0256

Sugai, T., Yoshida, H., Shimada, T., Okazaki, T., Shinohara, H. & Bandow, S. (2003). New Synthesis of High-Quality Double-Walled Carbon Nanotubes by High-Temperature Pulsed Arc Discharge. *Nano Letters*, Vol. 3, No. 6, pp. (769-773), 1530-6984

Swope, W. C., Andersen, H. C., Berens, P. H. & Wilson, K. R. (1982). A computer simulation method for the calculation of equilibrium constants for the formation of physical clusters of molecules: Application to small water clusters. *The Journal of Chemical Physics*, Vol. 76, No. 1, pp. (637-649), 0021-9606

Takagi, D., Hibino, H., Suzuki, S., Kobayashi, Y. & Homma, Y. (2007). Carbon Nanotube Growth from Semiconductor Nanoparticles. *Nano Letters*, Vol. 7, No. 8, pp. (2272-2275), 1530-6984

Teo, K. B. K., Singh, C., Chhowalla, M. & Milne, W. I. (2004). Catalytic Synthesis of Carbon Nanotubes and Nanofibers, In: *Encyclopedia of Nanoscience and Nanotechnology*, H. S. Nalwa (Ed.), pp. (665-686), American Scientific Publishers, 1-58883-001-2, Sevenson Ranch, California

Tersoff, J. (1988). New empirical approach for the structure and energy of covalent systems. *Physical Review B*, Vol. 37, No. 12, pp. (6991-7000), 0163-1829

Tersoff, J. (1989). Modeling solid-state chemistry: Interatomic potentials for multicomponent systems. *Physical Review B*, Vol. 39, No. 8, pp. (5566-5568), 0163-1829

Thess, A., Lee, R., Nikolaev, P., Dai, H. J., Petit, P., Robert, J., Xu, C. H., Lee, Y. H., Kim, S. G., Rinzler, A. G., Colbert, D. T., Scuseria, G. E., Tomanek, D., Fischer, J. E. & Smalley, R. E. (1996). Crystalline ropes of metallic carbon nanotubes. *Science*, Vol. 273, No. 5274, pp. (483-487), 0036-8075

Wen, Y.-H., Zhang, Y., Zheng, J.-C., Zhu, Z.-Z. & Sun, S.-G. (2009). Orientation-Dependent Structural Transition and Melting of Au Nanowires. *Journal of Physical Chemistry C*, Vol. 113, No. 48, pp. (20611-20617), 1932-7447

Woodcock, L. V. (1971). Isothermal molecular dynamics calculations for liquid salts. *Chemical Physics Letters*, Vol. 10, No. 3, pp. (257-261), 0009-2614

Xiang, R., Einarsson, E., Okawa, J., Miyauchi, Y. & Maruyama, S. (2009). Acetylene-Accelerated Alcohol Catalytic Chemical Vapor Deposition Growth of Vertically Aligned Single-Walled Carbon Nanotubes. *Journal of Physical Chemistry C*, Vol. 113, No. 18, pp. (7511-7515), 1932-7447

Yao, Y., Liu, R., Zhang, J., Jiao, L. & Liu, Z. (2007). Raman Spectral Measuring of the Growth Rate of Individual Single-Walled Carbon Nanotubes. *Journal of Physical Chemistry C*, Vol. 111, No. 24, pp. (8407-8409), 1932-7447

Yoshida, H., Shimizu, T., Uchiyama, T., Kohno, H., Homma, Y. & Takeda, S. (2009). Atomic-Scale Analysis on the Role of Molybdenum in Iron-Catalysed Carbon Nanotube Growth. *Nano Letters*, Vol. 9, No. 11, pp. (3810-3815), 1530-6984

Yoshida, H., Takeda, S., Uchiyama, T., Kohno, H. & Homma, Y. (2008). Atomic-Scale In-situ Observation of Carbon Nanotube Growth from Solid State Iron Carbide Nanoparticles. *Nano Letters*, Vol. 8, No. 7, pp. (2082-2086), 1530-6984

Yoshinori, A. (2004). Carbon Nanotubes: Synthesis by Arc Discharge Technique, In: *Encyclopedia of Nanoscience and Nanotechnology*, H. S. Nalwa (Ed.), pp. (603-610), American Scientific Publishers, 1-58883-001-2, Sevenson Ranch, California

Zheng, G. S., Irle, S., Elstner, M. & Morokuma, K. (2004). Quantum chemical molecular dynamics model study of fullerene formation from open-ended carbon nanotubes. *Journal of Physical Chemistry A*, Vol. 108, No. 15, pp. (3182-3194), 1089-5639

Zheng, M. & Semke, E. D. (2007). Enrichment of Single Chirality Carbon Nanotubes. *Journal of the American Chemical Society*, Vol. 129, No. 19, pp. (6084-6085), 0002-7863

25

# Multiple Andreev Reflections and Enhanced Quantum Interferences with Reentrant Behavior in NbN/Network-like Carbon Nanotubes/NbN SNS Junctions

Yuan-Liang Zhong[1], Hayato Nakano[2], Tatsushi Akazaki[3] and Hideaki Takayanagi[4]

[1]*Department of Physics, Chung Yuan Christian University*
[2,3]*NTT Basic Research Laboratories, NTT Corporation*
[4]*Department of Applied Physics, Tokyo University of Science*
[1]*Taiwan*
[2,3,4]*Japan*

## 1. Introduction

The superconducting proximity effect has been studied in superconductor (S)/normal conductor (N) junctions or SNS junctions for a few decades. If a clean contact is brought into an SN junction, some cooper pairs will penetrate into the normal conductor from the superconductor to form an incident electron and a retroreflected hole in it. This process is called Andreev refection (AR). If both electrodes are superconducting, two SN junctions with very-low-barrier tunnel junctions, a series of ARs can be observed in the differential resistance at submultiples of gap voltage, $2\Delta/ne$, where n=1,2,.. (Octavio et. al.,1983) and $\Delta$ is superconducting energy gap. These subharmonic gap structures (SGS) occurring at the maximum slopes of differential resistance are due to multiple Andreev reflection (MAR) (Bezuglyi et. al.,2000). If the ARs can be bounded at the two SN boundaries to confine the electron-hole motion spatially in the normal region, this Andreev bound states are localized states carrying finite supercurrent. The AR process has been studied in ballistic and diffusive SNS junctions using metal, the two-dimensional electron gas (2DEG) of semiconductor heterostuctures (Hoss et. al.,2000; Nitta et al.,1994; Octavio et. al.,1983), carbon nanotubes(Bezuglyi et. al.,2000) and graphene (Dirks et al.,2011) as normal conductor. Carbon nanotubes (CNTs) with a diameter of only a few nm as normal conductors are one of the best candidates for studying quasi one-dimensional (1D) proximity effect and AR. Though some experiments and theories have demonstrated 1D Tomonaga-Luttinger liquid behavior in single-wall carbon nantube (SWNT) or even multi-wall CNT (MWNT) (Bockrath et al.,1999), some experiments have shown AR (Morpurgo et al.,1999), MAR (Jarillo-Herrero et al.,2006; Buitelaar et al.,2003; Jorgensen et al.,2006), Andreev bound states (Pillet et al.,2010) and supercurrent in CNTs (Jarillo-Herrero et al.,2006; Kasumov et al.,1999),

and even superconducting quantum interference device made with CNTs (Cleuziou et al.,2006).
The superconducting proximity effect in the network-like structure with quasi-1D CNT will be discussed in the chapter. The enhanced quantum interferences can be given rise to as Aharonov-Bohm(AB)-type oscillations due to the loops as AB ring in the network-like structure by applied magnetic field. We have studied the MAR and enhanced quantum interferences, and the reentrant behavior of conductance in this random network CNT.
The updated superconducting proximity effect theory for reentrant behavior has shown that the proximity correction to the conductance $\triangle G_N(V,T,B)$ disappears at low energies and reaches a maximum value around temperature T or bias voltage V corresponding to the Thouless energy (correlation energy), $E_{th}$ constant (Volkov&Takayanagi,1997; Golubov,Wilhelm&Zaikin,1997; Nakano&Takayanagi,2000) (see Fig. 12). This reentrant behavior in $\triangle G_N$ occurs near $eV \cong E_{th}$, $k_BT \cong E_{th}$, or $B \cong B_c$, where $B_c$ is a correlation magnetic field. The reentrant behavior has also been predicted for magnetoconductance oscillations, but has not been completely observed in experiment yet. In what follows, we also present the MAR in magnetic field and then demonstrate the reentrant behavior of the conductance and magnetoconductance fluctuations.

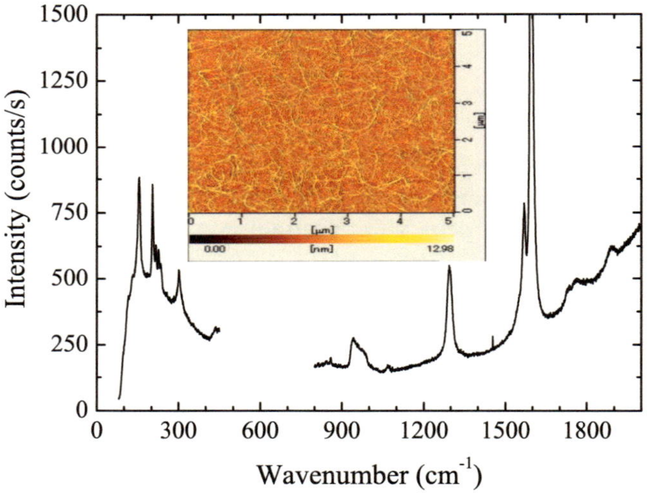


Fig. 1. Raman spectra of carbon nanotubes taken with 785-nm excitation from a sample consisting CNTs on $SiO_2/Si$ substrate. The Raman spectra peaks of carbon nanotubes, RBM modes at $\sim$200 cm$^{-1}$, the high-energy graphitelike mode at $\sim$1600 cm$^{-1}$ and the defect-induced D mode at $\sim$1300 cm$^{-1}$, were observed. Inset: Carbon nanotubes on $SiO_2/Si$ wafer substrate.

## 2. Fabrication of carbon nanotubes bridging superconducting electrodes

### 2.1 Fabrication of carbon nanotubes

CNTs were synthesized on a silicon wafer with 100-nm-thick thermal oxide by the chemical vapor deposition method using Co catalyst. A Co film is about 0.1-0.3 nm thick by

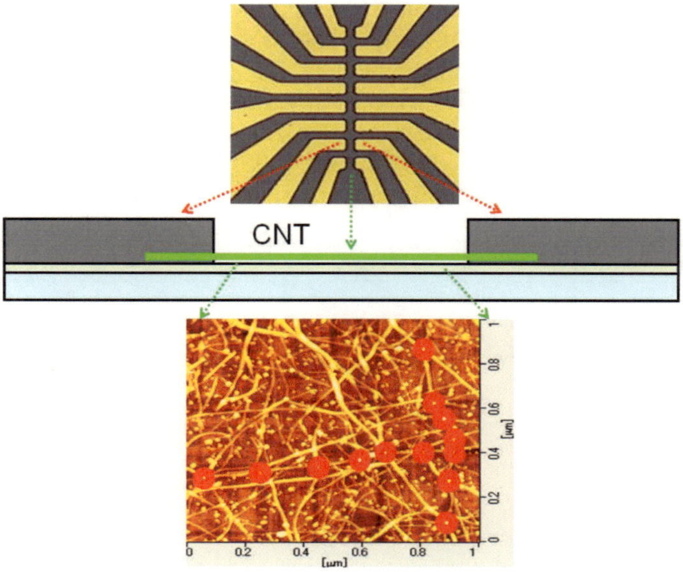


Fig. 2. Schematics of NbN/Network-like Carbon Nanotubes/NbN SNS Junctions (middle). The patterns of NbN electrodes (top) and an AFM image of the CNTs device (bottom) are made on $SiO_2$ substrate. The circle marks indicate the crossed junction of carbon nanotubes (bottom)

electron-beam deposition. The diameter of the Co nanoparticles control the diameter of the CNTs. For the synthesis procedure, argon gas fed into the furnace was heated to reach the growth temperature of 900-1000°C and then replaced by pure methane at a flow rate 300 $cm^3$/min with the pressure of 500 Torr for 5-10 min. This growth condition is optimized for SWNT growth and we observed very few MWNTs in TEM and AFM images. Inset of Fig. 1 shows the carbon nanotubes on the $SiO_2$ substrate. The CNTs belong to the SWNTs of individual nanotubes or bundles, which were measured from the height of each nanotube in an AFM (the CNTs with the diameter of 1-3 nm) and was confirmed by Raman spectra measurement. Raman spectroscopy technique was used to measure the characteristic of carbon nanotubes to show in Fig. 1. The diameter-selective Raman scattering is particularly important for the Raman band at about 200 $cm^{-1}$, which is associated with the radical breathing mode (RBM) of the carbon nanotube (Rao et al.,1997), to indicate the characteristic of SWNTs and sharp high energy graphite-like mode (G band) are also supported this results. The CNTs were freely synthesized to form random network structures, as shown in the inset of Fig. 1.

### 2.2 Fabrication of superconducting electrodes

The CNTs are growth between the electrodes with Co catalyst patterns using lithography technique. The CNTs were directly connected to NbN electrodes as shown in Fig. 2. Between the electrodes were designed to be 1.5 $\mu$m length on the photomask. The number of CNTs between the NbN electrodes was roughly estimated to be about 1000-2000. These network-like CNTs have the characteristic of quasi-diffusive transport, which is due to the network structure including defects, bundles and crossed junctions that marked in the bottom figure of Fig. 2. A excellent superconducting properties for NbN electrodes with 100-nm thickness and 1-$\mu$m wide have typically high superconducting critical temperature $T_c \sim$15 K and high superconducting critical magnetic field $H_{c2} \gtrsim$ 20 T. Optimum ohmic contact in junctions was made by annealing. The samples in vacuum were annealed with an infrared heater at 700°C for ~15 min and contact resistance was reduced nearly 1~ 2 orders of magnitude. After the annealing, the $T_c$ of NbN electrodes is close to 11 K. The samples were measured using the lock-in technique and four-probe method in a helium cryostat.

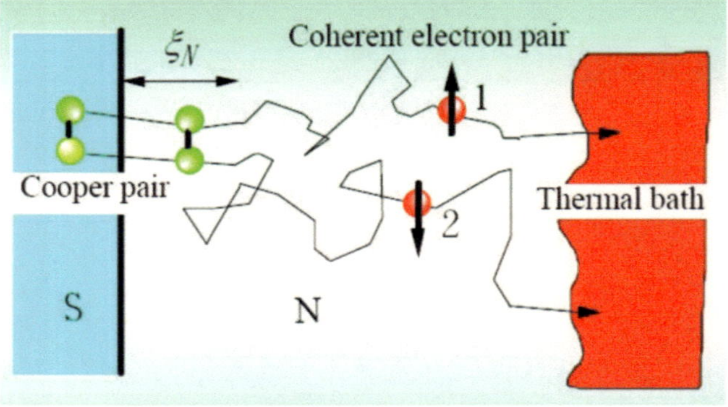


Fig. 3. The schematic process of Cooper pairs into normal conductor given rise by superconducting proximity effect

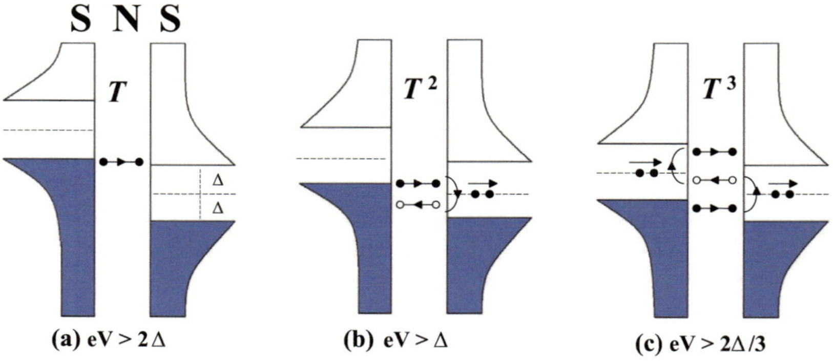


Fig. 4. Schematic process of (a) normal transport, (b) AR, and (c) MAR. There are different electron transparent probability T shown on figure for different transport processes

.

### 3. Superconducting proximity effect and Andreev reflection

The Cooper pairs into normal conductor diffusion process is given rise by superconducting proximity effect shown in Fig 3. In the junction of SN-thermal bath, Cooper pairs into normal diffusive conductor move within a distance called thermal coherent length, $\xi_T = \sqrt{\hbar D/2\pi k_B T}$, before breaking, and then keep coherence in the phase of two electron called coherent electron pairs. Cooper pairs leak into normal-conductor side from superconductor as generating correlated electron-hole pairs at SN interference called Andreev reflection . Figure 4 shows a processes of AR. For electron voltage higher than superconducting energy gap $\triangle$ (see Fig. 4(a)), electron merely transport in normal process. For electron voltage lower than $\triangle$ (see Fig. 4(b)(c)), Cooper pairs leak at SN junction to create Andreev reflection (Fig. 4(b)) and even MAR (Fig. 4(c)).

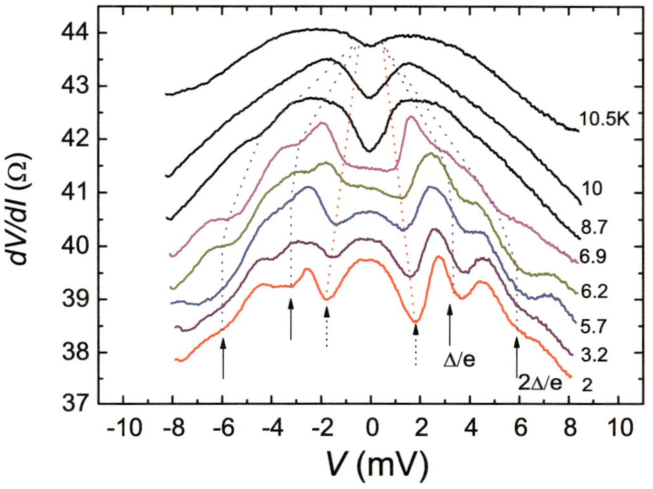


Fig. 5. Differential resistance as a function of applied voltage at different constant temperature between 2 and 11 K. The resistances are offset from each other for clarity. MAR occurs at 3 and 6 meV (solid line arrows) and AR occurs at zero-bias voltage above around 9 K.

The MAR was observed by measuring differential resistance as a function of applied voltage. Figure 5 shows SGS due to the MAR (Bezuglyi et. al.,2000; Hoss et. al.,2000; Nitta et al.,1994) and the submultiples of the gap voltage, $\pm 2\Delta/e$ and $\pm\Delta/e$(solid line arrows). The MAR become smeared gradually at temperature above $\sim$ 7 K and then a dip of the AR is observed at more high temperature but below $T_c$, 11 K. In addition, the MAR and AR were observed in different constant magnetic field between 0 and 4.3 T at temperature 1.75 K (Fig. 6). The SGS and AR dip become smeared above $\sim$450 mT and $\sim$1.5 T, respectively. Pervious experiments on 2D SNS junctions based on clean semiconductor heterostructures have also observed the MAR and AR smeared in the magnetic field (Nitta et al.,1994), but only below a few $\mu$T. The suppression of MAR and AR in a magnetic field can be easily understood by a physical process shown in Fig. 7. The AR takes place at the SN interfaces, and quasi-particles trajectories are deflected and the angle of incidence is increased. Because the parallel momentum becomes larger than the perpendicular momentum, the AR probability is suppressed (Mortensen et al.,1999). Though the smearing mechanism is the same as other dimensions, owing to 1D CNTs with small diameters, the quasiparticles are not easily deflected and their parallel moment not easily enhanced during motion as schematic process. Therefore, the applied magnetic field for seaming AR is larger than other dimensions over two orders of magnitude.

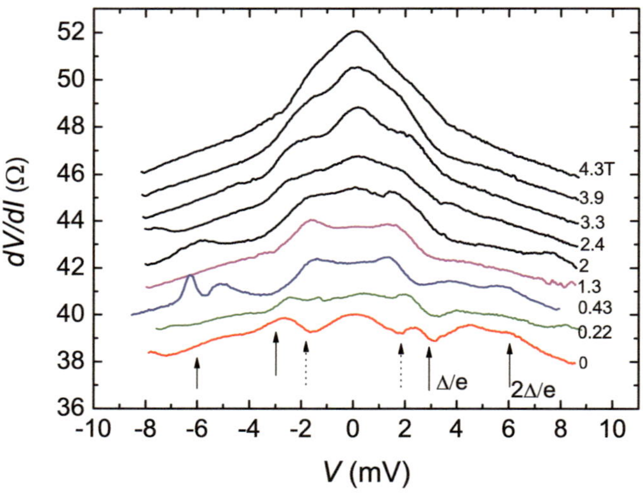


Fig. 6. Differential resistance as a function of applied voltage in constant magnetic field below 4.3 T. MAR processes occur at 3 and 6 meV (solid line arrows) and reentrant behavior occur at close zero zero bias (dash line arrows). The MAR is smeared by applying magnetic field over ~250 mT. Above 0.4 T, the curves of peak are changed to dip at around zero bias voltage. ARs disappear at over 2 T.

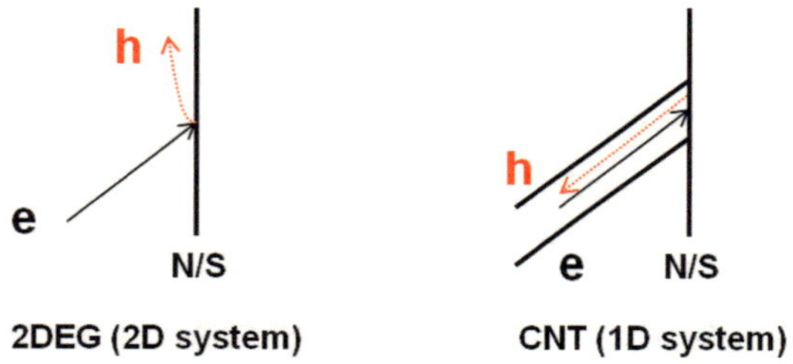


Fig. 7. Schematic process of the AR in 2DEG (2D system) and CNT (1D system) at NS junction with an incident electron (e) and a retroreflected hole (h) in an applied magetic field below mT. The electron and hole can not be easily deflected in CNT compared with 2DEG in applied magnetic field and are still confined in CNT

## 4. Magnetoconductance fluctuations and enhanced quantum interference

A single electronic wave that is split into two propagating waves over different paths as shown in Fig. 8. The quantum interference of electron will be modulated by an applied magnetic field passed through in the ring. The oscillations in conductance due to this AR effect can be observed in the conductor ring of ballistic transport by measuring magnetoconductance. In the metallic wire within mesoscopic scale of diffusive transport, a single electron that scatters around the closed path and interference with itself by impurities scattering to cause AB effect in some loops. Some oscillations by loops in wire will be formed fluctuations in conductance called universal conductance fluctuations (UCF). Some loops in network-like CNT as the interference loops are shown in Fig. 9.

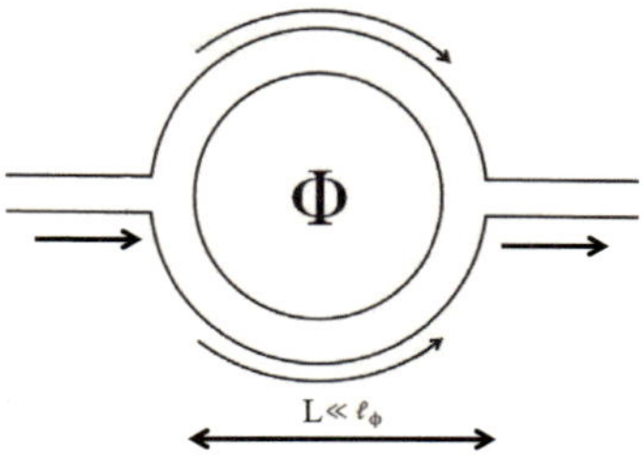


Fig. 8. AB ring with an applied magnetic field passed through caused quantum interferences from electronic wave splitting into two different paths

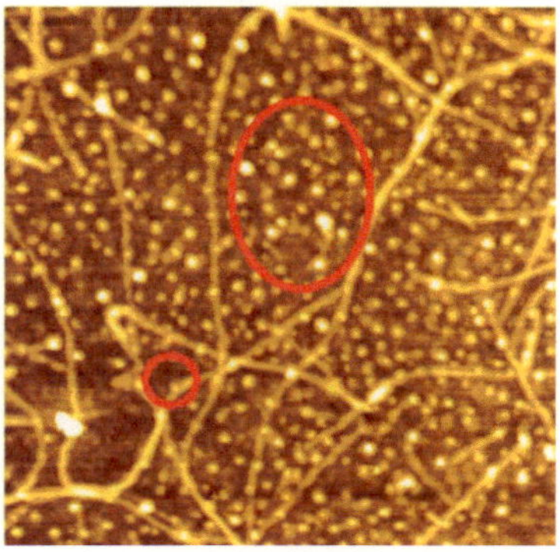

Fig. 9. Some loops in network-like CNTs as the AB rings

We have observed enhanced magnetoconductance fluctuations (EMF) by magnetoconductance measurement. The magnetoconductance fluctuations were measured by applying a magnetic field of up to 3 or 4 T at temperature between 10 and 4 K (Fig. 10). In the inset of Fig. 10, the two curves at 8 K (dash and solid curve) show reproducible fluctuations and those curves at temperatures 8, 9.2 and 9.6 K also show similar fluctuation structures. The fluctuation amplitudes become large from $\sim$10 to 8 K (the inset of Fig. 10). The other similar fluctuation structures become small from 8 to 4 K as shown in Fig. 11. This reentrant behavior of magnetoconductance fluctuations will be discussed in more detail later. These fluctuations are similar to AB-type oscillation caused by the loops in a network-like structure. The AB-type interferences can be formed in the loops by applying magnetic field. Owing to the loops of different size, the magnetoconducance fluctuations look like the UCFs observed in metallic wire (Washburn&Webb,1992).
The fluctuation amplitude ($\triangle G_f$) is larger than $\sim e^2/h$, the normal UCF amplitude. The enhanced magnetoconductance fluctuations can be as a superconducting UCFs. The larger amplitude indicates that the fluctuations are not due to normal quantum interference, but due to Andreev quasiparticle interference. From the evaluated number of loops, about $10^6$ (about 10 crossed junctions between electrodes), we can obtain theoretically the maximum amplitude of $\sim 4e^2/h$ compared with the $\sim 6e^2/h$ in the inset of Fig. 10. (see the details in section 6).

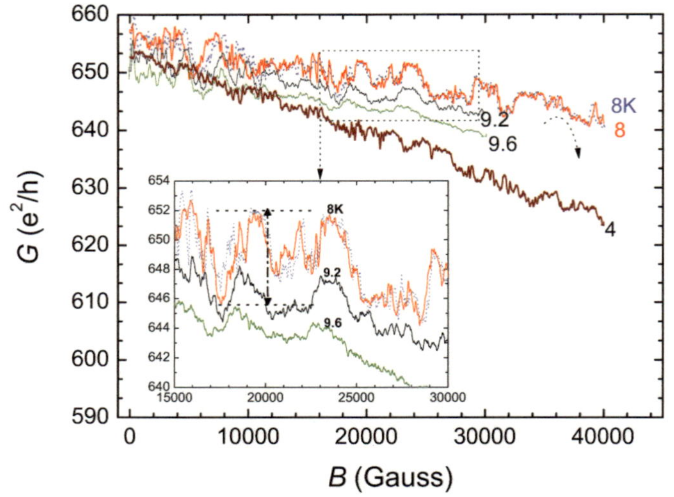


Fig. 10. EMF at different constant temperature between 10 and 4 K. The fluctuations have a maximum amplitude at around 8 K. Inset: reproducible fluctuations at 8 K (the dash and solid curve) and three similar curves of fluctuations at 8, 9.2 and 9.6 K.

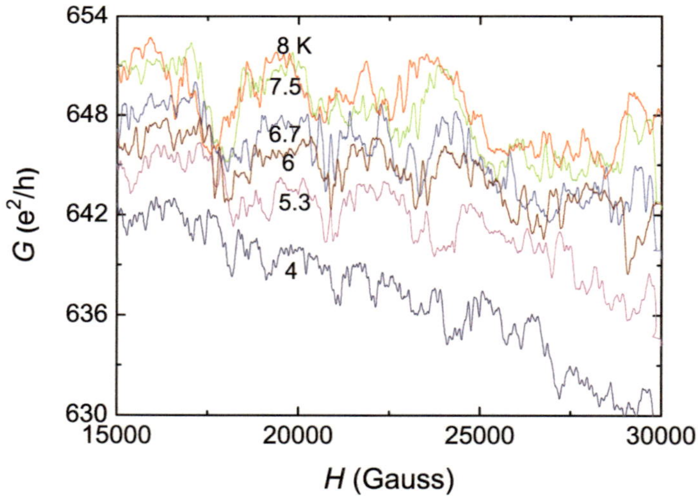


Fig. 11. EMF at the variation of temperatures. The similar curves of fluctuations are shown at 8, 7.5, 6.7, 6, 5.3 and 4 K

## 5. Reentrant behavior caused by superconducting proximity effect

For superconducting proximity effect in the conductor of diffusive transport, a updated superconducting proximity effect theory has shown that the proximity correction to the

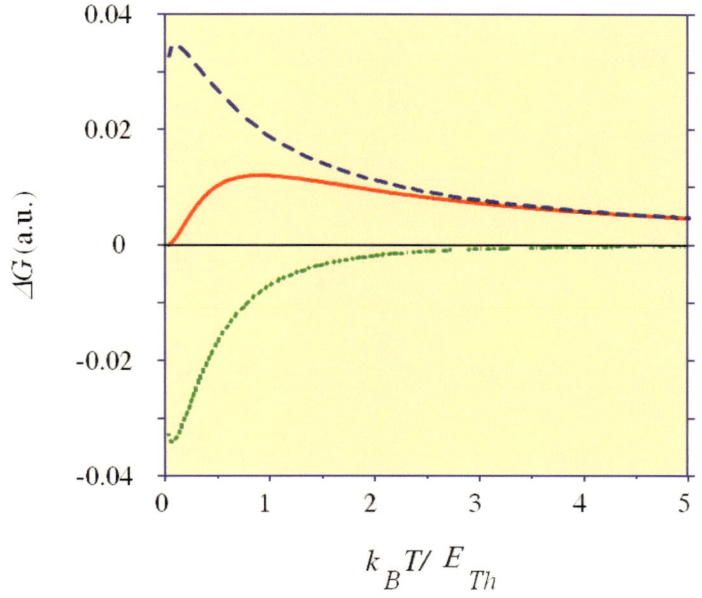


Fig. 12. The reentrant behavior of the red curve predicted by the updated superconducting proximity effect theory (sold curve). The blue curve from MTSF effect (long dash curve) and the green curve from DDOS effect (short dash curve), two effects contribute to reentrant behavior

conductance ($\triangle G_N(V,T,B)$) disappears at low energies and reaches a maximum value around temperature T or bias voltage V corresponding to the Thouless energy (correlation energy), $E_{th} \equiv \hbar D/L^2$, where $D$ and $L$ are the diffusion constant and sample length (Volkov&Takayanagi,1997; Golubov,Wilhelm&Zaikin,1997; Nakano&Takayanagi,2000), as shown in Fig. 12. Two effects, the Maki-Thompson-type of superconducting fluctuation(MTSF) effect and the decreased quasiparticles density of states (DDOS) effect at the Fermi level, contribute to $\triangle G_N(V,T,B)$. These two contributions become equal and cancel each other out exactly at the absolute zero-temperature or at the zero-bias voltage limit as shown in the middle curve (red curve) of Fig. 12. This reentrant behavior has also been predicted for the temperature dependance of magnetoconductance fluctuations $\Delta G_f(T)$ and the magnetic field dependance of magnetoconductance fluctuations $\Delta G_f(B)$.
The differential resistance near zero-bias voltage that has been observed a reentrant behavior. As shown in Fig. 5, the reentrant behavior was observed in the voltage dependance of differential resistance at 2 K. The differential resistance becomes larger as the applied voltage approaches zero voltage and reaches a minimum value (dip), ~1.8 mV (dash line arrows), near zero-bias. This dip is due to the reentrant effect on voltage and is similar to $E_{th}$. The differential resistance of zero-bias voltage from the peak to dip varies with temperature between 2 and 11 K (Fig. 5), which is transformed into the conductance as a function of temperature as shown in the inset of Fig.13. The conductance becomes higher and reaches a maximum at about 8 K related to $E_{th}$, and then becomes lower at low temperature. This reentrant behavior of conductance have been studied in both theory (Golubov,Wilhelm&Zaikin,1997; Volkov&Takayanagi,1997; Nakano&Takayanagi,2000) and

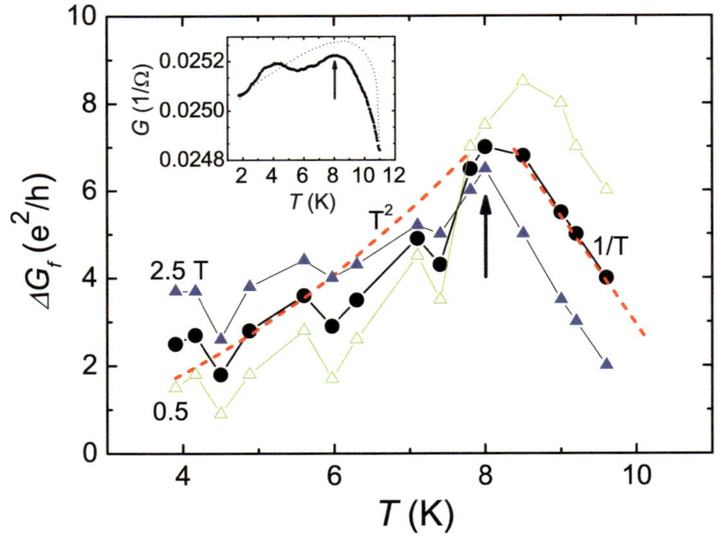


Fig. 13. Fluctuation amplitude as a function of temperature. Two curves for different magnetic field regime are shown in 0.5 (open triangles) and 2.5 T (filled triangles), and the average of the two curves is shown by filled circles. The dashed curve shows the $T^2$ behavior at $T < 8$ K and the $1/T$ behavior at $T > 8$ K in theory. Inset: the temperature dependance of conductance (filled circle curve) and calculated conductance (thin dashed curve)

experiments on metal and 2DEG semiconductor (den Hartog et al.,1996; Akazaki et al.,2004). The higher $E_{th}$ was observed in our experiment because of high $D$. We estimated the diffusion constant $D$ of network-like CNTs by $D = v_F l_e$, where $v_F$ is the Fermi velocity about $8\times 10^5$ m/s (Lee et al.,2004), and $l_e$ is the mean free path. Because pure CNT has the property of ballistic transport, the length between crossed junctions corresponds to $l_e$. The $l_e$ of $\sim$0.15 $\mu$m was obtained by counting the crossed junction number of CNT, about 11$\sim$4, between electrodes as shown in Fig. 2 (bottom figure). Thus, $D$ is calculated to be about 0.07$\sim$0.18 $m^2/s$, which is larger than the previous experiment value by 1$\sim$2 orders of magnitude. Using averaged value $D = 0.12$ $m^2/s$, we calculated the conductance as a function of temperature (dash curve) by the Usadel equation to compare it to our experimental result in the inset of Fig. 13 (see (Akazaki et al.,2004) for detail). The variation in conductance due to the proximity effect is defined as $\triangle G_{N'}=(G_N - G_{N0})/G_{N0}$, where $G_N$ is the conductance of the SN junction measured at zero-bias voltage and $G_{N0}$ is that measured at high temperature, where the proximity effect can be neglected. The details of the function $\triangle G_{N'}$ are given by Golubov et. al (Golubov,Wilhelm&Zaikin,1997). The result of calculation is $G=A\times\triangle G_{N'}\times G_{N0}+G_{N0}$, where free parameter $A$ is 0.165 and $G_{N0}$ is 24.84 mS [the inset of Fig. 13)]. The temperature of maximum conductance in the theoretical prediction, $T_m$ ($\sim E_{th}/k_B$), $\simeq 9$ K, which is very close to the measured value of 8 K.

The fluctuation amplitude is estimated by evaluating the bandwidth in a range of magnetic field such as at 8 K (inset of Fig. 10). Fig. 13 shows three curves of $\triangle G_f(T)$ between 10 to 4 K. The evaluated $\triangle G_f(T)$ at applied magnetic field around 0.5 (open triangles) and 2.5 T (closed triangles) are shown. The averaged $\triangle G_f(T)$ (closed circle)

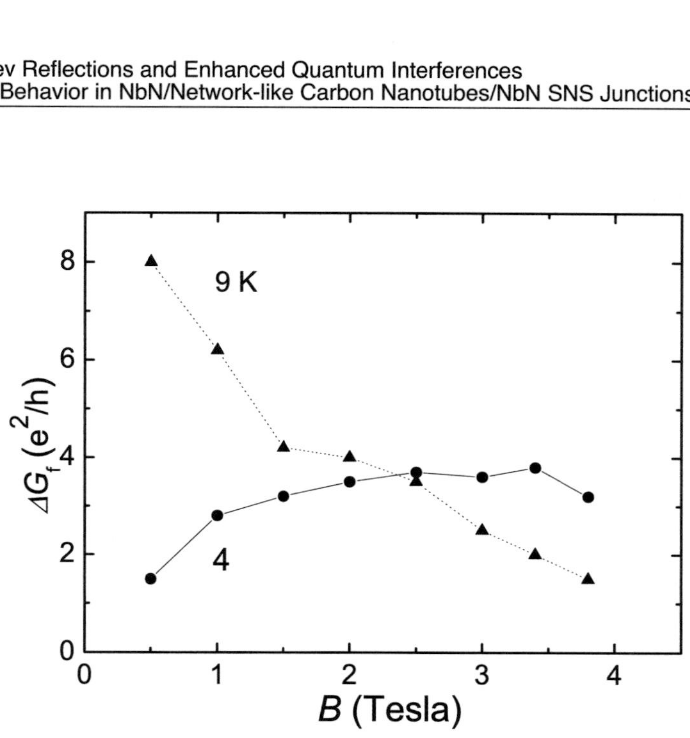


Fig. 14. Fluctuation amplitude as a function of magnetic field. Two curves for different temperature are shown by filled circle (4 K) and filled triangles (9 K)

from the curves of 0.5 and 2.5 T is also shown in between two curves. This averaged $\triangle G_f(T)$ becomes small as low temperature is approached and reaches a maximum amplitude at 8 K, which is related to $E_{th}$, and this $T_m$ of 8 K is the same as the $T_m$ of conductance in the insert of Fig. 13. Theoretically, the model has been constructed by the proximity correction to quasi-1D normal conductor with the loop of AB-type oscillation and the reentrant behavior of $\triangle G_f(T)$ has been predicted (Golubov,Wilhelm&Zaikin,1997; Nakano&Takayanagi,2000). When $T > T_m$, $\triangle G_f(T)$ is proportional to $1/T$. This can be explained by a simple physical interpretation: Superconductivity penetrates a distance with temperature between the two electrodes, whereas the rest keeps normal conductivity and, because the distance of the rest is proportional to $T$, the resistance of the rest is proportional to $T$ (Golubov,Wilhelm&Zaikin,1997). This means that the enhanced $\triangle G_f(T)$ is proportional to $1/T$. The electron coherent length should be shorter than the electrode spacing and the MTSF effect dominates the behavior of $\triangle G_f(T)$. In the limit $T < T_m$, the electron coherent length is longer than the electrode spacing and the DDOS effect is enhanced, and then $\triangle G_f(T)$ exhibits a diminishment behavior of $T^2$ as $T$ approaches 0. Experimentally, for enhanced $\triangle G_f(T)$ behavior had only been observed, but diminishment behavior has never been observed, to the best of our knowledge. As shown in Fig. 13, the reentrant behavior of $\triangle G_f(T)$ are observed in $T < 8$ K regimes, and the dashed curve was predicted in theory.

We also investigated the $\triangle G_f$ with the magnetic field in the case of two limits. The behavior of $\triangle G_f(B)$ is shown at 9 K ($> T_m$) in Fig. 14. The $\triangle G_f(B)$ is reduced with increasing magnetic field and exhibits a monotonic behavior. This behavior due to the MTSF effect is smeared by increasing the magnetic field gradually. At 4 K ($< T_m$), the $\triangle G_f(B)$ with an applied magnetic field exhibits a nonmonotonic behavior (Fig. 14), because this behavior is a reentrant behavior. Both DDOS effect and the MTSF effect have the same contribution in $\triangle G_f(T)$ behavior as $T$ approaches 0, but the DDOS effect is more sensitive to the magnetic field than the MTSF effect. According to the quasiparticles transport process in Nakano's theory (Nakano&Takayanagi,2000), DDOS effect needs time reversal symmetry that is twice as long as that needed by the MTSF effect. Therefore, the DDOS effct should be more sensitive and be smeared more heavily than the MTSF effect. The conductance diminishment (DDOS effect) of the rapid decay with the increase of the energies (T or V) occurs for the same reason and the reentrant behavior on T or V has also been observed in nonmonotonic $\triangle G_f(B)$ behavior. The $\triangle G_f(B)$ becomes small with increasing strength of the field and reaches a maximum amplitude at 2.5 $\sim$ 3.5 T as correlation magnetic field $B_c$, which is related to $E_{th}$. This nonmonotonic behavior can also be seen in the differential resistance of zero-bias voltage with magnetic field in Fig. 6 that shows changes from a peak to a dip and then back to a peak again. The $B_c$ can be estimated by $B_c \sim \Phi_0/L_\phi^2$, where $\Phi_0$ is $h/2e$, because the magnetic field required to destroy the interferences can be determined by one flux quantum through the area of the largest possible phase coherent path (Dikin et al.,2001). For AR in a CNT, an electron is incident into the interface and should be Andreev reflected into the same CNT. Since the AR process is phase coherent, the retroreflected hole can interfere with incident electron. The quasiparticles are induced by localized AR, but are not affected by the network structures of the CNTs. We can obtain $B_c$ = 3.3 T by $B_c \sim h/2eLW$ (Washburn&Webb,1992; Dikin et al.,2001), where the width of the wire $W$ = 1 nm and the length $L$ = 1 $\mu$m are the carbon nanotube diameter and electrode spacing. The $B_c$ is much higher than in previous experiments with only a few microtesla in metallic wires (Dikin et al.,2001). Because this high $B_c$, the reentrant behavior of $\triangle G_f(B)$ can be observed. Here, we emphasize that this result is difficult to be observed in normal metals or 2DEG semiconductor, because the enhanced $\triangle G_f(B)$ could not be easily observed in previous experiments with only low $B_c$. There is a unique characteristic in network-like ultra-thin CNTs. The high $B_c$ is due to the ultra-thin tubes and fluctuation oscillation is due to the network-like structures.

## 6. The theoretical analysis of enhanced quantum interference and reentrant behavior for network structure

Since the total conductance of the junction is the conductance of the series of the interface $G_\mathrm{T}$ and the diffusive $G_\mathrm{N}$ part such that

$$G = \frac{1}{1/G_\mathrm{T} + 1/G_\mathrm{N}},$$

the conductance change is approximately given by

$$\Delta G \simeq \frac{G_\mathrm{N0}^2 \Delta G_\mathrm{T} + G_\mathrm{T0}^2 \Delta G_\mathrm{N}}{(G_\mathrm{N0} + G_\mathrm{T0})^2},$$

where $G_\mathrm{N} = G_\mathrm{N0} + \Delta G_\mathrm{N}, G_\mathrm{T} = G_\mathrm{T0} + \Delta G_\mathrm{T}$.
In this expression, only $\Delta G_\mathrm{N}$ shows so-called"re-entrant behavior" as a function of applied voltage $V$, temperature $T$, or applied magnetic filed $B$. $\Delta G_\mathrm{T}$ is not so sensitive to the applied field, voltage, or temperature, when the measurement temperature is sufficiently lower than the bulk superconducting gap $\Delta$; $V, T \ll \Delta$, and when the applied field is much smaller than

the critical one; $B \ll B_c$. Therefore, we make $\Delta G_T$ a constant. Then,

$$\Delta G(V,T,B) = \frac{G_{N0}^2 \Delta G_T + G_{T0}^2 \Delta G_N(V,T,B)}{(G_{N0}+G_{T0})^2},$$

and, we can thusly discuss the conductance change as the functions of $T, B, V$:

$$\delta G(V,T,B) = K_0 \Delta G_N(V,T,B),$$

with a constant

$$K_0 = \frac{G_{T0}}{(G_{N0}+G_{T0})^2}.$$

The proximity correction to this conductance $\Delta G_N(V,T,B)$ disappears at low energies and reaches a maximum value around temperature $T$ or bias voltage $V$ corresponding to the Thouless energy (correlation energy). In a quasi 1D quantum wire, the correlation magnetic field for breaking the time reversal symmetry (electron-hole symmetry) is given by $B_c \sim h/2eLW$. On the other hand, correlation energy for the Maki-Thompson enhancement and the decreased quasiparticle density of states effect effect is the Thouless energy $E_{Th} = \hbar D/L^2$. Therefore, the reentrant behavior in $\Delta G_N$ happens near $eV = E_{Th}$, $k_B T = E_{Th}$, or $B = B_c$.
Let us consider an AB-type interferometer formed by quasi 1D wires and placed between the S and N electrodes. The AB oscillation amplitude in $\Delta G_N$ is the order of $G_{T0}^2/G_{N0}$, and it should change as a function of $T$, $V$ in the same way it does in a single quasi 1D wire. The explicit form of $\Delta G_N(V,T,B)$ as the function of $V,T,B$ has been given in many literatures (Golubov,Wilhelm&Zaikin,1997; Volkov&Takayanagi,1997; Nakano&Takayanagi,2000). Here, we drop the expression for an AB ring in Nakano and Takayanagi's eq. (51) (Nakano&Takayanagi,2000) because it is convenient for our present purpose. Thus, we have

$$\begin{aligned}\delta G(\Psi=0) - \delta G(\Psi) &\sim \frac{K_0 {G_T}^2}{G_{N0}} \frac{E_{Th}}{k_B T} \exp\left[-\frac{\pi R_0}{\pi R_0 + L_L}\right] \\ &\simeq \frac{{G_{T0}}^4}{G_{N0}(G_{T0}+G_{N0})^2} \frac{E_{Th}}{k_B T} \exp\left[-\frac{\pi R_0}{\pi R_0 + L_L}\right] \qquad (1)\end{aligned}$$

where $R_0$ is the radius of the ring and $L_L = L - \pi R_0$ is the length of the wire except for the loop part (that is, the length of the lead). When the tunnelling conductance is much smaller than the diffusive part, i.e. $G_{T0} \ll G_{N0}$, the original expression is recovered. In contrast, when $G_{T0}$ is comparable to $G_{N0}$, the amplitude of the fluctuation is large.
The oscillation period is $\Psi_0$ with the flux piercing the AB ring, where $\Psi_0$ is the flux quantum. It should be noted that the period is half that of a usual AB oscillation because this is an interference of the proximity-induced quasiparticle.
The experimental results show that the amplitude of the oscillation is for times $e^2/h$ or larger and that the dominant characteristic period is approximately 500 Gauss (See Fig. 10). From the period, we estimate the radius of the dominant loop structure to be about 150 nm. It is postulated that there are many loop structures in the CNT region. Since the distance between the S and N electrode is 1 $\mu$m, can be neglected, the exp factor in eq. (1)
In order to analyze the amplitude of the fluctuation, we need a model of the CNT region
Suppose that the CNT region is a random network formed by many CNTs. For simplicity, let us assume a irregular lattice of CNTs. There are $n$ CNTs bridging the S and N electrodes. We

call them longitudinal CNTs. In the direction perpendicular to the bridge CNTs, $m$ CNTs are distributed, and we call them transverse CNTs. At the first step, we neglect the conductance of the barrier contacting a longitudinal CNT and transverse. The average conductance of the CNT region is the same as that of $n$ times of a single CNT. A single CNT has a conductance of the order of $e^2/h$ or less. Experimentally, the value of the total conductance is about $650e^2/h$; therefore, $n > 650$.

The lattice network has many loops with different areas. We assume that the dominant radius is what we estimated above. Therefore, the period of the oscillation in each loop is almost the same. However, the phase of the oscillation is different in each loop. Then, the oscillation amplitude in the total conductance is enhanced by the factor $\sqrt{s}$, where $s$ is the number of loops with almost the same dominant radius in the lattice. The $\sqrt{s}$ dependence comes as follows. The conductance oscillations of loops are superposed. However, the effect of distributed oscillation causes an averaging effect. So, $s/\sqrt{s}$ gives $\sqrt{s}$.

When we put the conductance of a single CNT as $G_{\mathrm{N0}} = \alpha e^2/h$ and the conductance across the S electrode and a single CNT as $G_{\mathrm{T0}} = \beta G_{\mathrm{N0}}$, the expected amplitude of the oscillation is

$$\sqrt{s}\frac{{G_{\mathrm{T0}}}^4}{G_{\mathrm{N0}}(G_{\mathrm{T0}}+G_{\mathrm{N0}})^2}\frac{E_{\mathrm{Th}}}{k_{\mathrm{B}}T} \simeq \frac{\sqrt{s}\alpha\beta^4}{(1+\beta)^2}\frac{e^2}{h}$$

when $k_{\mathrm{B}}T = E_{\mathrm{th}}$. On the other hand,

$$n \times \frac{1}{\frac{1}{\alpha e^2/h}+\frac{1}{\beta\alpha e^2/h}} \simeq 650e^2/h \tag{2}$$

should be satisfied because this corresponds to the total conductance of the SN junction. When we take into account the barrier conductance across the longitudinal and transverse CNT contact, the amplitude is reduced. We put the factor as $z$. From the observed oscillation amplitude $4e^2/h$, we have

$$z\frac{\sqrt{s}\alpha\beta^4}{(1+\beta)^2} \simeq 4. \tag{3}$$

For example, $n = 2000$, $\alpha = 0.9$, $\beta = 0.56$, $s = 1.8 \times 10^6$ and $z = 0.5$ satisfies both (2) and (3). The number of the loops, $s$, is very large; however, it is not impossible when the number of the transverse CNTs, $m$, is more than 1000.

The conductance of the diffusive region $G_{\mathrm{N0}}$ is much bigger than that of the interface $G_{\mathrm{T0}}$. For example, we put $G_{\mathrm{T0}} = 0.01G_{\mathrm{N0}}$. The measured conductance of the junction, $\sim 650e^2/h$, is mainly determined by $G_{\mathrm{T0}}$ because the interface and the diffusive wire are serially connected. Then, we can put $G_{\mathrm{T0}} \sim 650e^2/h$ and $G_{\mathrm{N0}} \sim 65000e^2/h$. However, the amplitude $\sim 4e^2/h$ is smaller than the expected one:

$$\frac{{G_{\mathrm{T0}}}^4}{G_{\mathrm{N0}}(G_{\mathrm{T0}}+G_{\mathrm{N0}})^2},$$

with $k_{\mathrm{B}}T = E_{\mathrm{Th}}$. When we use $G_{\mathrm{T0}} = 0.01G_{\mathrm{N0}}$, and $G_{\mathrm{N0}} = 65000e^2/h$, the expected fluctuation is about $6\,e^2/h$.

## 7. Conclusion

The multiple Andreev reflection and the enhanced magnetoconductance fluctuations, such as an superconducting UCFs, were observed in S/network-like CNTs/S junctions. The reentrant behavior was observed and consisted with each other in the temperature and voltage dependence of conductance, and in the dependence of fluctuation amplitude on temperature and magnetic field. Especially, the reentrant behavior of $\Delta G_f(T)$ was first observed as diminished $\Delta G_f(T) \sim T^2$ as $T$ approached 0 and the reentrant behavior of $\Delta G_f(B)$ was also observed. We found that the high critical magnetic field destroys Andreev reflection process and the high correlation magnetic field induces the dephasing of interference in CNTs, which are larger than that for 2D SNS junctions by about two orders of magnitude. These results are due to the small diameter of the CNTs with network-like structures.

## 8. References

[Akazaki et al., 2004] Akazaki, T., Yamaguchi, H. and Takayanagi, H. (2004). Nonequilibrium-GaAs(111) a heterostructures coupled with superconducting Nb electrodes. Semicond. Sci. Technol. Vol. 19 (2004)S182-S184.

[Bezuglyi et. al., 2000] Bezuglyi, E. V., Bratus, E. N., Shumeiko, V. S., Wendin G. and Takayanagi, H. (2000). Circuit theory of multiple Andreev reflections in diffusive SNS junctions. Phys. Rev. B Vol. 62, No. 21 (December 2000) 14439-14451.

[Bockrath et al., 1999] Bockrath, M., et al. (1999). Luttinger-liquid behaviour in carbon nanotubes. Nature Vol. 397, (18 February 1999), 598-601.

[Buitelaar et al., 2003] Buitelaar, M. R., Belzig, W., Nussbaumer, T., Babic, B., Bruder, C. and Schonenberger, C. (2003). Multiple Andreev Reflections in a Carbon Nanotube Quantum Dot. Phys. Rev. Lett. Vol. 91, No. 5 (2003) 57005-1 - 57005-4.

[Cleuziou et al., 2006] Cleuziou, J. -P. (2006). Carbon nanotube superconducting quantum interference device. Nature Nanotech. Vol. 1 (October 2006)53-59.

[den Hartog et al., 1996] den Hartog, S. G., Kateyn, C. M. A., van Wees, B. J., Klapwijk, T. M., van der Graaf, W. and Borghs, G. (1996). Sample-Specific Conductance Fluctuations Modulated by the Superconducting Phase. Phys. Rev. Lett. Vol. 76, No. 24, (1996)4592-4595; Charlat, P., Courtois, H., Gandit, Ph., Mailly, D., Volkov, A. F., and Pannertier, B. (1996). Phys. Rev. Lett. Vol. 77, No. 24, (1996)4950-4953; Chien, C. -J. and Chandrasekhar, V. (1999). Reentrance effect in normal-metal/superconducting hybrid loops. Phys. Rev. B Vol. 60, No. 22, (1999)15356-15363.

[Dirks et al., 2011] Dirks, Travis, Hughes, Taylor L., Lal, Siddhartha, Uchoa, Bruno, Chen, Yung-Fu, Chialvo, Cesar, Goldbart, Paul M. and Mason, Nadya. (2011). Transport through Andreev bound states in a graphene quantum dot. Nature Physics (6 February 2011)1-5; Du, Xu, Skachko, Ivan and Andrei, Eva Y. (2008). Josephson current and multiple Andreev reflections in graphene SNS junctions. Phys. Rev. B Vol. 77, (2008)184507-1-184507-5; Heersche, Hubert B., Jarillo-Herrero, Pablo, Oostinga, Jeroen B., Vandersypen, LievenM. K., and Morpurgo, Alberto F. (2007). Bipolar supercurrent in graphene. Nature, Vol. 446, (1 March 2007)56-59.

[Dikin et al., 2001] Dikin, D. A., Black, M. J., and Chandrasekhar, V., (2001). Magnetoresistance of Proximity-Coupled Au Wires. Phys. Rev. Lett. Vol. 87, No. 18, (2001)187003-1-187003-4.

[Golubov, Wilhelm & Zaikin, 1997] Golubov, A. A., Wilhelm, F. K., and Zaikin, A. D. (1997). Coherent charge transport in metallic proximity structures. Phys. Rev. B Vol. 55, No. 2, (1997)1123-1137.

[Hoss et. al., 2000] Hoss, T., Strunk, C., Nussbaumer, T., Huber, R., Staufer, U. and Schonenberger, C. (2000). Multiple Andreev reflection and giant excess noise in diffusive superconductor normal-metal superconductor junctions. Phys. Rev. B Vol. 62, No. 6 (2000) 4079-4085.

[Jarillo-Herrero et al., 2006] Jarillo-Herrero, Pablo, Dam, Jorden A. van and Kouwenhoven, Leo P. (2006). Quantum supercurrent transistors in carbon nanotubes. Nature Vol. 439, (23 February 2006) 953-956.

[Jorgensen et al., 2006] Jorgensen, H. I., et al. (2006). Electron Transport in Single-Wall Carbon Nanotube Weak Links in the Fabry-Perot Regime. Phys. Rev. Lett. Vol. 96, (2006) 2070031-1-2070031-4.

[Kasumov et al., 1999] Kasumov, A. Yu., Deblock, R., Kociak, M., Reulet, B., Bouchiat, H., Khodos, I. I., Gorbatov, Yu. B., Volkov, V. T., Journet, C., and Burghard, M. (1999). Supercurrents Through Single-Walled Carbon Nanotubes. Science Vol. 284, (28 May 1999) 1508-1510.

[Lee et al., 2004] Lee, Jhinhwan, Eggert, S., Kim, H., Kahng, S. -J. Shinohara, H., and Kuk1, Y. (2004). Real Space Imaging of One-Dimensional Standing Waves Direct Evidence for a Luttinger Liquid. Phys. Rev. Lett. Vol. 93, No. 16, (2004)1664031-1-1664031-4.

[Morpurgo et al., 1999] Morpurgo, A. F., Kong, J., Marcus, C. M. and Dai, H. (1999). Gate-Controlled Superconducting Proximity Effect in Carbon. Science Vol. 286 (8 October1999) 263-265.

[Mortensen et al., 1999] Mortensen, N. A., Flensberg, K., and Jauho, A. P. (1999). Angle dependence of Andreev scattering at semiconductorą Vsuperconductor interfaces. Phys. Rev. Vol. 59, No. 15, (1999)10176-10182.

[Nakano & Takayanagi, 2000] Nakano, H., and Takayanagi, H. (2000). Influence of phase quantum fluctuations on superconducting proximity correction in normal-metal wire conductance. Phys. Rev. B Vol. 61, No. 22, (2000)15398.

[Nitta et al., 1994] Nitta, J., Akazaki, T. and Takayanagi, H. (1994). Magnetic-field dependence of Andreev reflection in a clean Nb-InAs-Nb junction. Phys. Rev. B Vol. 49, No. 5 (1994) 3659-3662.

[Octavio et. al., 1983] Octavio, M., Tinkham, M., Blonder, G. E. and Klapwijk, T. M. (1983). Subharmonic energy-gap structure in superconducting constrictions. Phys. Rev. B Vol. 27, No. 11 (1983)6739-6746.

[Pillet et al., 2010] Pillet, J-D., Quay, C. H. L., Morfin, P., Bena, C., Levy Yeyati, A. and Joyez, P. (2010). Andreev bound states in supercurrent-carrying carbon nanotubes revealed. Nature Physics Vol. 6, 965ą V969 (2010).

[Rao et al., 1997] Rao, A. M., Richter, E., Bandow, S., Chase, B., Eklund, P. C., Williams, K. A., Fang, S., Subbaswamy, K. R., Menon, M., Thess, A., Smalley, R. E., Dresselhaus, G., and Dresselhaus, M. S. (1997). Diameter-Selective Raman Scattering from Vibrational Modes in Carbon Nanotubes. Science Vol. 275 (1997)187-191.

[Volkov & Takayanagi, 1997] Volkov, A. F., and Takayanagi, H. (1997). Long-range phase-coherent effects in the transport properties of mesoscopic superconductorą Vnormal-metal structures. Phys. Rev. B Vol. 56, (1997)11184.

[Washburn & Webb, 1992] Washburn S., and Webb, R. A. (1992). Quantum transport in small disordered samples from the diffusive to the ballistic. Rep. Prog. Phys. Vol. 55, (1992)1311.

# 26

# Quantum Calculation in Prediction the Properties of Single-Walled Carbon Nanotubes

Majid Monajjemi[1] and Vannajan Sanghiran Lee[2,3]
[1]*Department of Chemistry, Science and Research Branch, Islamic Azad University, Tehran,*
[2]*Department of Chemistry, Faculty of Science, University of Malaya, 50603, Kuala Lumpur,*
[3]*Computational Simulation and Modeling Laboratory (CSML),*
*Thailand Center of Excellence in Physics, Commission on the Higher Education,*
*Ministry of Education, Bangkok,*
[1]*Iran*
[2]*Malaysia*
[3]*Thailand*

## 1. Introduction

Since the first discovery of single-walled carbon Nanotubes (SWCNTs) by Iijima and Bethune in 1993 (Bethune et al., 1993), many applications as molecular components for nanotechnology including conductivity and high-strength composites; energy storage and energy conversion devices; sensors; field emission displays and radiation sources; hydrogen storage media; and nanometer-sized semiconductor devices, probes, and interconnects are known (Ajayan et al., 1994; Saito et al., 1997; deHeer et al., 1995; Collins et al., 1997; Nardelli et al., 1998; Huang et al., 2006). SWCNTs have been considered as the leading candidate for Nan device applications because of their one-dimensional electronic bond structure, molecular size, biocompatibility, controllable property of conducting electrical current and reversible response to biological reagents. Hence SWCNTs make possible bonding to polymers and biological systems such as DNA and carbohydrates. Most SWCNTs have a diameter of close to 1 nanometer, with a tube length that can be many millions of times longer. The average diameter of a SWNT is 1.2 nm (Spires & Brown, 1996). However, Nanotubes can vary in size, and they aren't always perfectly cylindrical. As in Fig. 1, the average bond length and carbon separation values for the hexagonal lattice were shown. The carbon bond length of 1.42 Å was measured by Spires and Brown in 1996 (Spires & Brown, 1996) and later confirmed by Wilder et al. in 1998 (Wilder et al., 1998).
The structure of a SWNT can be formed by the rolling of a single layer of $sp^2$ carbon, called a graphene layer, into a seamless hollow cylindrical tube with Nan scale dimensions of 1-1.5 nm. The length is usually in the order of microns to centimeters. Besides their unique physical properties (elasticity, tensile strength, stiffness, and deformation), Nano tubes exhibit varying electrical properties (depending on the direction that the graphite structure spirals around the tube (quantified by the "Chiral vector"), and other factors, such as doping), and can be superconductor, conductor (metallic), semiconductor or, insulator. The band structure can even be further manipulated, by introducing defects into a tube. Single-

walled nanotubes exhibit electric properties that are not shared by the multi-walled carbon nanotube (MWNT) variants. In particular, their band gap can vary from zero to about 2 eV and their electrical conductivity can show metallic or semiconducting behavior, whereas MWNTs are zero-gap metals. The C-C tight bonding overlap energy is in the order of 2.5 eV. Wilder *et al.* estimated it to be between 2.6 eV - 2.8 eV (Wilder et al., 1998) while at the same time, Odom *et al.* estimated it to be 2.45 eV (Odom et al., 1998). Multi-walled carbon nanotubes have a layer of carbon shells with differing physics that can all potentially interact. It is shown that only the outer shell of MWCNTs contributes to electrical transport, and so only small diameter MWCNTs could be used to make transistor devices. SWCNTs are the most likely candidate for miniaturizing electronics beyond the micro electromechanical scale currently used in electronics. As this field continues to expand and grow, materials technology will produce products, components and systems that are smaller, smarter, multi-functional, environmentally compatible, more survivable, and customizable. These products will not only contribute to the growing revolutions of information and biology, but will also significantly impact manufacturing, logistics, and our culture as a whole. The development of scanning probe techniques has allowed not only the microscopy of surfaces with atomic resolution, but also the manipulation of atoms and molecules on surfaces, and many analytical techniques have been developed to allow detailed characterization of materials and structures on the atomic level with unprecedented accuracy. The utilization of materials with nanometer-sized structures will lead to innovative products which are smaller, smarter, and more multi-functional. Therefore, understanding of fundamental properties of structures at the nano scale with the aid of computational models is important to design the specific material properties.

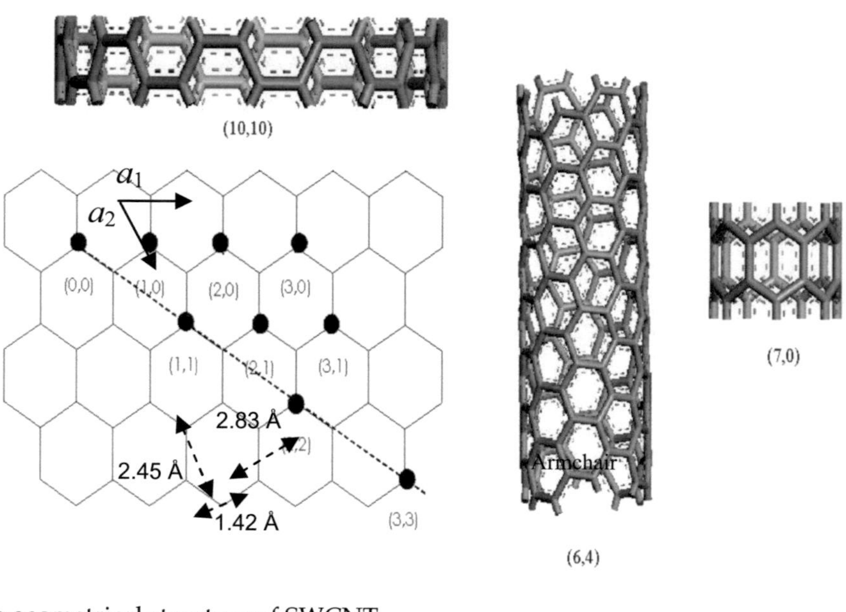


Fig. 1. The geometrical structure of SWCNT

Recently, theoretical and experimental work have predicted that the infinity length SWCNTs are Pi-bonded aromatic molecules that the electrical properties depending upon the tubular diameter and helical angle (Zhou et al., 2004; Baron et al., 2005). SWCNTs can be chiral or nonchiral, again depending on the way of the rolling up vector. As a graphene sheet was rolled in many ways in horizontal, vertical, and diagonal direction represent as arrow vectors, $\vec{a}$, as in Fig. 1, the different types of carbon nanotubes were produced. The three main types are armchair, zig-zag, and chiral nanotube. The geometrical and electronic structure of SWCNT can be described by a chiral vector, the angle between the axis of its hexagonal pattern and the axis of the tube which is presented by a pair of indices ($n_1$,$n_2$) called the chiral vector. The integers $n_1$ and $n_2$ denote the number of unit vectors along two directions in the honeycomb crystal lattice of graphene. When the indices are ($n_1$,0) called zig-zag, ($n_1$, $n_1$) called armchair, and ($n_1$, $n_2$) where $n_1 \neq 0$ and $n_2 \neq 0$ known as chiral SWCNT. For $(2\,n_1 + n_2)/3$ = integer, SWCNTs are metallic and others are semiconductors (Saito et al., 1992a, 1992b). For large diameter SWCNTs defined by $d = \frac{\sqrt{3(n_1^2 + n_2^2 + n_1 n_2)}}{\pi} a_{c-c}$, where $a_{c-c}$ is the distance between neighboring carbon atoms in the flat sheet, armchair SWCNTs are always metallic which is good for nanotechnology application. A zigzag carbon nanotube ($n_1$, 0), is a semiconductor when $n_1/3 \neq$ integer. Such semiconductor zigzag carbon nanotubes have the ability to become base of many nanoelectronic devices and transistors.

Although, scientific efforts focused on the electrostatics properties and commercial applications of these materials (Ouyango et al., 2002; Kane & Mele, 1997; Hartschuh et al. 2005), there have been no experimental structural data sufficiently accurate for the identification of the chirality indices of SWCNTs, especially for the kind of smaller diameter nanotubes. In all experimental methods for the identification commonly utilized so far is Raman spectroscopy and phonon dispersion. Phonon dispersion relations in one dimension of this system have been studied by using zonefolding along one direction of Brillouin zone considering the tube symmetry (Eklund et al., 1995). The tight binding electronic band structure and the reverse of the diameter (1/*d*) dependence of the frequency of the radial breathing mode (RBM) were employed (Jorio et al., 2001; Bachilo et al, 2002; Pfeiffer et al. 2003; Kurti et al., 2004; Maultizsch et al., 2005). The size and chirality of the carbon nanotubes were typical determined from the SWCNT Raman energy spectra of a peak around 150–300 $cm^{-1}$, due to the radial breathing mode (Maultizsch et al., 2005; Jorio et al., 2005). Further Raman studies of SWCNT modified by various reactions e.g. oxidation reactions, ozonolysis, fluorination, residues modification (Srano et al., 2003; Umek et al., 2003; Peng et al., 2003; Bahr et al., 2001; Holzinger et al., 2003; Mickelson et al., 1998; Cai et al., 2002; Banerjee & Wong, 2002; Herrera & Resasco, 2003; Martinez et al., 2003) have revealed that covalent functionalization mainly affects the intensity of the Raman bands. Characterization of nanotube in adsorption gas has been studied by Monte Carlo and Langevin Dynamic Simulation (Monajjemi et al., 2008b). It is also important to investigate the effects of diameter on a SWCNT structure how the diameter depends on geometrical parameters such as the C-C bond lengths and some of the dihedral angles of SWCNTs.

For a better understanding of the physical and electronic properties of SWCNT, a challenging task in theoretical calculation is needed to specify the material properties because of the large size of the SWCNTs and their complicated (and size-dependent) electronic structure. Quantum calculation in prediction the properties of single-walled carbon nanotubes (SWCNTs) will be discussed.

## 2. Vibrational mode of SWCNT

Normal mode analysis has become one of the standard techniques in the study of the dynamics of nanotubes. It is primarily used for identifying and characterizing the slowest motions in a poly system, which are inaccessible by other methods. This text explains what normal mode analysis is and what one can do with it without going beyond its limit of validity. By definition, normal mode analysis is the study of harmonic potential wells by analytic means. The first section of this study will therefore deal with potential wells and harmonic approximations. This study is about normal mode approaches to different physical situations, and it discusses how useful information can be extracted from normal modes. Normalmode coordinates are obtained by a linear combination of Cartesian coordinates. Thus, there are no couplings in the kinetic part; that is, they diagonalize the kinetic energy as well the quadratic part of the potential energy operator. They include simultaneous motion of all atoms during the vibration, which leads to a natural description of molecular vibrations. Therefore, they are good candidates for representation of the molecular Hamiltonian. Since a transformation between different sets of coordinates is possible, the anharmonic terms can be calculated in one representation, and then transformed into another one.

### 2.1 Symmetry of SWCNT

Because a single carbon nanotube may be thought of as a graphene sheet rolled up to form a tube, carbon nanotubes should be expected to have many properties derived from the energy bands and lattice dynamics of graphite. For the very smallest tubule diameters, however, one might anticipate new effects stemming from the curvature of the tube wall and the closing of the graphene sheet into a cylinder. A method for identifying the Raman modes of single-wall carbon nanotubes (SWNT) based on the symmetry of the vibration modes has been widely used. The Raman intensity of each vibration mode varies with polarization direction, and the relationship can be expressed as analytical functions. Each Raman-active mode of SWNT can be distinguished from the group theory principle. The symmetry properties of periodic lattices of carbon nanotubes and the symmetry operations of chiral and achiral nanotubes (Damnjanović et al., 1999; Damnjanović et al., 2001; Alon, 2001, 2003) are usually described in terms of the group of the wavevector (Dresselhaus et al., 2006). However, since nanotubes can be viewed as quasi-1D systems, the line groups approach by Damnjanović et al. is suited to describe nanotube properties (Damnjanović et al., 1999).

As described earlier, the properties of nanotubes are determined by their diameter and chiral angle, both of which depend on $n_1$ and $n_2$. Typically, SWCNT is presented by a pair of integers ($n_1$, $n_2$). Its geometrical structure as shown in diagram can be represented in term of a chiral vector $\vec{C}$ on a two-dimensional sp$^2$-carbon sheet where $\vec{C} = n_1\vec{a}_1 + n_2\vec{a}_2$ with integer $n_1$ and $n_2$. Here, $\vec{a}_1$ and $\vec{a}_2$ represent the unit vectors of the hexagonal graphene lattice. This sheet is then rolled up to a cylinder so that $\vec{C}$ becomes the circumference of the tube. The direction of the nanotube axis is naturally perpendicular to $\vec{C}$. The diameter, $d$, is simply the length of the chiral vector divided by ¼, and $d = \left(\sqrt{3/\pi}\right) a_{c-c}(n_1^2 + n_2^2 + n_1 n_2)^{1/2}$, where $a_{c\text{-}c}$ is the distance between neighbouring carbon atoms in the flat sheet. In turn, the chiral angle (θ) is given by $\tan^{-1}\left(\sqrt{3n/(2n_2 + n_1)}\right)$.

The translational period, *a,* is the shortest possible lattice vector along *z* direction. The translatory unit cell of a nanotube is a cylinder with a length in tube axis direction equal to the magnitude of the translation vector $\vec{T}$ as shown in Fig. 1 which can be calculated as following equation:

$$a = -\frac{2n_2 + n_1}{nR} a_1 + \frac{2n_1 + n_2}{nR} a_2$$

with

$$a = | a | = \frac{\sqrt{3(n_1^2 + n_2^2 + n_1 n_2)}}{nR} a_0$$

where $n$ is the greatest common divisor of $n_1$ and $n_2$,

if $(n1 - n2)/3n \neq$ integer, then R = 1

if $(n1 - n2)/3n =$ integer, then R = 3

and $\vec{a}_1$ and $\vec{a}_2$ form an angle of 60° and their length is $| a_1 | = | a_2 | = a_0 = 2.461$ Å

Since the translational period, *a*, depends inversely on n and R the translation periodicity and thus the number of carbon atoms varies strongly for tubes with similar diameter. The number of graphene cells in the nanotube unit cell ($n_c$) obtained from:

$$n_c = 2q = 4\frac{n_1^2 + n_2^2 + n_1 n_2}{nR}$$

The groups of infinite line $L$ are products $L = ZP$, where $P$ is a point group and $Z$ is the group of translations (screw axis, pure translations, and glide planes). Applying the above symmetry formulation to armchair ($n_1 = n_2$) and zigzag ($n_2 = 0$) nanotubes, such nanotubes with no caps have a isogonal point groups given by $q$ (the number of graphene cells in the unit cell of the nanotubes) namely, $D_{nd}$ when n is odd, $D_{nh}$ when n is even, or $Dqh = D2nh$ for achiral and $Dq$ for chiral tubes. Whether the symmetry groups for armchair and zigzag tubules are taken to be $D_{nd}$ or $D_{nh}$, the calculated vibrational frequencies will be the same; the symmetry assignments for these modes, however, will be different. It is, thus, expected that modes that are Raman or IR-active under $D_{nd}$ (or $D_{nh}$ ) but are optically under $D2nh$ will only show a weak activity resulting from the fact that the existence of caps lowers the symmetry that would exist for a nanotube of infinite length.

### 2.2 Active modes of Raman and IR

The phonon symmetries are found by decomposing the dynamical representation into its irreducible representations using symmetries of carbon and other nanotubes studied for line groups (Damnjanović et al., 1999). One direct set up of the dynamical representation from the atomic and vector representation is to use factor group analysis. A representation can be decomposed into the sum of its irreducible representations by the following formula

$$f_\alpha = \frac{1}{g}\sum_G \chi^{(\alpha)}(G) * \chi^{(\Gamma_{DG})}(G)$$

where $f\alpha$□is the appearance frequency of the irreducible representation α□, $g$ is the order of the symmetry group; the sum is over all symmetry operations G.

The Raman $\Gamma_R$ and infrared active $\Gamma_{IR}$ vibrations transform according to the representation of the second rank tensor and the vector representation, respectively (Damnjanović et al., 1983)

$$\Gamma_R = [\Gamma_{vec} \otimes \Gamma_{vec}] = A_{1g} \oplus E_{1g} \oplus E_{2g} \ (\oplus A_{2g})$$

$$\Gamma_{IR} = \Gamma_{vec} = A_{2u} \oplus E_{1u}$$

According to the symmetries of Raman-active modes (Pelletier, 1999) for the armchair carbon nanotube with the chair vector *(n1, n2)*, the point group for this kind nanotube belongs to D*n*h when *n* is even and its Raman-active modes are denoted by A1g + E1g + E2g. Three flavors of modes are longitudinal, transversal radial (orthogonal to tube surface) and transversal axial (parallel to tube surface). Satio et al. (Satio et al., 1998) pointed out that the low frequency A1g mode is a radial breathing mode and two high frequency is belong to Eg modes, E1g and E2g. E mode has the same displacement pattern with additional standing wave on the circumference.

### 2.3 Projection operators

The zigzag single-walled carbon nanotubes (SWCNTs) with (3,0), (4,0), and (5,0) structure were built using the tool in HyperChem7.0. The symmetries of the nanotube are D3d, D4d, and D5d respectively. Four different systems were studied in this work as follow: (1) gas-phase SWCNT, (2) SWCNT with 23 water molecules in the *a* x *b* x *c* box, (3) SWCNT with 23 methanol molecules in the *a* x *b* x *c* box, and (4) SWCNT with mixed solvent of water and methanol molecules in the *a* x *b* x *c* box. Energy minima of systems (2) – (4) were carried out by Metropolis Monte carlo (MC) calculation which generate random configurations in regions of space that make the important contributions to the calculation of thermodynamic averages. Then the ab initio and semiemperical with AM1 were used to optimize the structure of the nanotubes. All the normal mode frequencies and IR intensity were calculated using the optimized structures.

To find a function or the displacement pattern of eigenvectors transforming as a particular irreducible representation, the projection operators in group theory have been applied. Consider an arbitrary function *F*. This function can, in general, be expanded into several irreducible represent ations $F = \sum_{\alpha}\sum_{n} c_{\alpha}^{n}\zeta_{\alpha}^{n}$ where α labels the irreducible representations, $c_{\alpha}^{n}$ are the coefficients of the expansion, and the $\zeta_{\alpha}^{n}$ are functions transforming according to the representation α. A projection operator defined by $P_{l(n)}^{(\beta)} = \frac{d_{(\beta)}}{g}\sum_{G} D_{\ln}^{(\beta)}(G)^{*}(G)$ applied to *F* picks out the symmetry adapted function $\zeta_{l}^{(\beta)}$. In equation $d_{\beta}$ is the degeneracy of the irreducible representation β, *g* the order of the symmetry group, *G* are the symmetry operations, and $D_{ln}^{(\beta)}$ *ln* is the *ln*th element of the representation matrix $D^{(\beta)}$. From a given function, and its irreducible presentation, functions can be generating if that function has a "component" or a "non-zero projection" along the irreducible presentation of interest. This explains the name of "projector". As an example, if there is an orthonormal set *Li* of the function $\varphi^{i}_{1}, \varphi^{i}_{2}, \ldots, \varphi^{i}_{Li}$ which is used to form the i$^{th}$ irreducible representation of a group by order *h*, for each operator, ***R***, in the group, by definition we can have:

$$\boldsymbol{R}\varphi^{i}_{t} = \Sigma s \varphi^{i}_{s}\, \Gamma(R)^{i}_{st} \tag{1}$$

By producting (1) in $[\Gamma(R)^{i}_{s't'}]^*$ and summing all over the symmetrical functions in the group we will have:

$$\Sigma_R[\Gamma(R)^{i}_{s't'}]^*\, \boldsymbol{R}\varphi^{i}_{t} = \Sigma_R \Sigma_s\, \Gamma\varphi^{i}_{s}\, \Gamma(R)^{i}_{st}\, \Gamma(R)\, ^{i}_{s't'}]^* \tag{2}$$

Considering $\varphi^{i}_{s}$ are functions independent from $R$, the right side of (2) can be written as: $\Sigma_s\varphi^{i}_{s}\ \Sigma_R\Gamma(R)^{i}_{st}\ [\Gamma(R)\ ^{i}_{s't'}]^*$ So we have a series of $Li$ terms and each of them are equal to a production of $\varphi^{i}_{s}$ and a coefficient. These coefficients are following the orthogonality rule:

$$\Sigma_R\Gamma(R)^{i}_{st}\ [\Gamma(R)\ ^{i}_{s't'}]^* = h/(L_iL_j)^{1/2}\, \delta_{ij}\, \delta_{ss'}\, \delta_{tt'} \tag{3}$$

By use of the eq.(3), the eq.(2) is simplified as follows:

$$\Sigma_R\Gamma(R)\ ^{i}_{s't'}]^*\, \boldsymbol{R}\varphi^{i}_{t} = (h/L_j)\ \varphi^{i}_{s'}\, \delta_{ij}\, \delta_{tt'} \tag{4}$$

Now, by introducing

$$\boldsymbol{P}^{j}_{s't'} = L_j/\mathrm{h}\ \Sigma_R\Gamma(R)^{i}_{st}\ [\Gamma(R)\ ^{i}_{s't'}]^*\, \boldsymbol{R} \tag{5}$$

The eq.(4)gives the following form:

$$\boldsymbol{P}^{j}_{s't'}\varphi^{i}_{t} = \varphi^{i}_{s'}\, \delta_{ij}\, \delta_{tt'} \tag{6}$$

The $\boldsymbol{P}^{j}_{s't'}$ is call projection operator. The application of this operator on each $\varphi^{i}_{t}$ is non-zero only when this function or some of its terms is a function of $\varphi^{i}_{s'}$. One of the most important application of this operator is projecting function $\varphi^{i}_{t}$ from any function $\varphi^{i}_{t}$. In other words

$$\boldsymbol{P}^{j}_{t't}\varphi^{i}_{t} = \varphi^{i}_{t'}\, \delta_{ij}\, \delta_{tt'} \tag{7}$$

By use of the projection operator on the base of $L_j$ diagonal elements of a matrix, we can have some $\varphi^{i}_{t}$ functions, which are the bases for the j$^{th}$ irreducible presentation (Wilson, et al., 1955)

### 2.4 The relation between projection and transfer operators

Assume that $\Gamma_k$ (p) $_{ij}$ is the ij$_{th}$ element of the matrix which shows the p$^{th}$ operator ($O$p) in k the k$_{th}$ irreducible presentation. By this assumption the operator $\boldsymbol{O}_{k,\, ij}$ is defined as follows

$$\boldsymbol{O}_{k,ij} = \mathrm{L_k} / \mathrm{h}\ \Sigma_p\Gamma_k\,(\mathrm{p})^*{}_{ij}\, \mathbf{O}\mathrm{p} \tag{8}$$

Where h is group order and $L_k$ is the presentation dimension. If i = j these operators called Projection operators, $\boldsymbol{P}_{k,ii}$, in other words:

$$\boldsymbol{P}_{k,ij} = \boldsymbol{O}_{k,ii} \tag{9}$$

The non- diagonalized operators are called Transfer operators or shift operators,

$$\boldsymbol{T}_{k,ij} = \boldsymbol{O}_{k,ij}\,,\ i \neq j \tag{10}$$

In one-dimensional presentations $\boldsymbol{P}_{k,ij}$ and $\boldsymbol{O}_{k,ii}$ are the same and we have no $\boldsymbol{T}_{k,ij}$. With use of the above definitions, making the irreducible basis becomes possible in the following way:

At first the point group of the molecule is determined.Then the character of the system ($\Gamma_{angles}$ or $\Gamma_{bonding}$) is calculated. By use of the standard reduce formulation these characters can be reduced to give the irreducible presentations:

$$n_{\Gamma} = 1/h\ \Sigma_g\ n_g\ \chi_R\ \chi_\Gamma \qquad (11)$$

Where h is the order of the group, $n_g$ is the number of the symmetry operation in the class of g, $\chi_R$ is the character of reducible presentation and $\chi_\Gamma$ is the character of irreducible presentation for the symmetric operations of class g. In this part there is a note about the reducing the $C_{\infty v}$ and $D_{\infty h}$ point groups. The method of reduce is a different from the normal method of reduce. For more information see from the references (Cotton, 1971; Schafer & Cyvrin, 1971; Strommen & Lippincott, 1972; Alvarino, 1978; Flurry, 1979; Strommen, 1979).
At the next step, the interested function is written by use of the projection operator. A set of the results gives the internal coordinate system for a given point group. There are several examples to illustrate this procedure in Table 1. The geometry and electrical properties of nanotube are very sensitive to dielectric constants. The normal modes also will be changed in the high dielectric constants. With the calculation of the normal modes using the U Matrix it is possible to get the F Matrix from the multiplication of frequency to the U Matrix. Solving the determination of F Matrix versus dielectric can be useful for understanding of the electrical behavior of nanotubes in the quantitative structure activity relationship (QSAR) studies. With use of the resulting coordinate system, the U Matrix (UMAT) can be written easily. These are matrices which perform the linear transformations on the internal coordinates sets (Alvarino & Chammoro, 1980).

### 2.5 Linear combination of primitive's harmonic vibrations and UMAT

The molecules and their internal coordinates of D4d have been given in Fig. 2. By following the above steps a complete set of the linear combinations and their normalization coefficients are achieved. The irreducible representations of the symmetry group are given by A and B. These data are given in Table 1. We use the application of projection operator method in finding the coordinate system, and by using it the U matrix is written and finally the frequencies and distributions of peak position are achieved. The (3, 0), (4, 0), (5, 0) zig-zag nanotubes were investigated. They have 66, 138, and 174 normal modes, respectively.

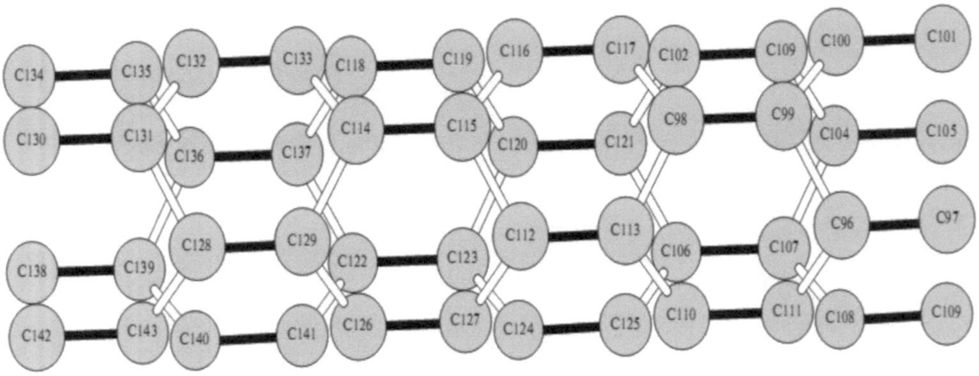


Fig. 2. The structure of (4, 0) nanotube in $D_{4d}$ point group (Lee et al., 2009)

| $C_{2v}$ | $E$ | $C_2$ | $\sigma_v$ | $\sigma'_v$ | Irreduced presentation | Primitives | Linear combination of primitives based on their symmetric $A_1$, $A_2$, $B_1$, $B_2$ |
|---|---|---|---|---|---|---|---|
| $\Gamma_1$ | 3 | 1 | 1 | 3 | $2A_1$<br>$B_2$ | 3 | $S_1(A_1)=\frac{1}{\sqrt{2}}(R_1+R_2)$, $S_2(B_2)=\frac{1}{\sqrt{2}}(R_1-R_2)$, $S_3(A_1)=R_{31}$ |
| $\Gamma_2$ | 3 | 1 | 1 | 3 | $2A_1$<br>$B_2$ | 3 | $S_4(A_1)=\frac{1}{\sqrt{2}}(R_8+R_9)$, $S_5(B_2)=\frac{1}{\sqrt{2}}(R_8-R_9)$, $S_6(A_1)=R_{51}$ |
| $\Gamma_3$ | 3 | 1 | 1 | 3 | $2A_1$<br>$B_2$ | 3 | $S_7(A_1)=\frac{1}{\sqrt{2}}(R_{14}+R_{15})$, $S_8(B_2)=\frac{1}{\sqrt{2}}(R_{14}-R_{15})$, $S_9(A_1)=R_{43}$ |
| $\Gamma_4$ | 3 | 1 | 1 | 3 | $2A_1$<br>$B_2$ | 3 | $S_{10}(A_1)=\frac{1}{\sqrt{2}}(R_{20}+R_{22})$, $S_{11}(B_2)=\frac{1}{\sqrt{2}}(R_{20}-R_{22})$, $S_{12}(A_1)=R_{52}$ |
| $\Gamma_5$ | 3 | 1 | 1 | 3 | $2A_1$<br>$B_2$ | 3 | $S_{13}(A_1)=\frac{1}{\sqrt{2}}(R_{25}+R_{26})$, $S_{14}(B_2)=\frac{1}{\sqrt{2}}(R_{25}-R_{26})$, $S_{15}(A_1)=R_{57}$ |
| $\Gamma_6$ | 3 | 1 | 1 | 3 | $2A_1$<br>$B_2$ | 3 | $S_{16}(A_1)=\frac{1}{\sqrt{2}}(R_{29}+R_{30})$, $S_{17}(B_2)=\frac{1}{\sqrt{2}}(R_{29}-R_{30})$, $S_{18}(A_1)=R_{64}$ |
| $\Gamma_7$ | 14 | 2 | 2 | 2 | $5A_1$<br>$3A_2$<br>$3B_1$<br>$3B_2$ | 14 | $S_{19}(A_1)=\frac{1}{2}(R_5+R_{17}+R_3+R_2)$, $S_{20}(A_2)=\frac{1}{2}(R_5+R_3-R_{17}-R_2)$,<br>$S^*_{20}(A_2)=\frac{1}{2}(R_{17}+R_2-R_5-R_3)$, $S_{21}(A_2)=\frac{1}{2}(R_3-R_5+R_2-R_{17})$,<br>$S_{22}(A_1)=\frac{1}{\sqrt{2}}(R_4+R_{16})$, $S_{23}(A_1)=\frac{1}{\sqrt{2}}(R_4-R_{16})$, $S_{24}(A_1)=\frac{1}{\sqrt{2}}(R_{40}+R_{52})$,<br>$S_{25}(B_2)=\frac{1}{\sqrt{2}}(R_{40}-R_{52})$, $S_{26}(A_1)=\frac{1}{\sqrt{2}}(R_{51}+R_{50})$, $S_{27}(B_2)=\frac{1}{\sqrt{2}}(R_{51}-R_{50})$,<br>$S_{28}(A_1)=\frac{1}{2}(R_{29}+R_{30}+R_{51}+R_{37})$, $S_{29}(B_1)=\frac{1}{2}(R_{29}-R_{50}+R_{53}-R_{37})$,<br>$S^*_{29}(B_1)=\frac{1}{2}(R_{30}+R_{50}-R_{29}-R_{53})$, $S_{30}(A_2)=\frac{1}{2}(R_{29}+R_{37}-R_{53}-R_{30})$ |
| $\Gamma_8$ | 14 | 2 | 2 | 2 | $5A_1$<br>$3A_2$<br>$3B_1$<br>$3B_2$ | 14 | $S_{31}(A_1)=\frac{1}{2}(R_{11}+R_6+R_7+R_9)$, $S_{32}(A_2)=\frac{1}{2}(R_{11}+R_7-R_6-R_9)$,<br>$S^*_{32}(A_2)=\frac{1}{2}(R_6+R_9-R_{11}-R_7)$, $S_{33}(B_1)=\frac{1}{2}(R_7-R_{11}+R_9-R_6)$,<br>$S_{34}(A_1)=\frac{1}{\sqrt{2}}(R_4+R_{10})$, $S_{35}(B_2)=\frac{1}{\sqrt{2}}(R_4-R_{10})$, $S_{36}(A_1)=\frac{1}{\sqrt{2}}(R_{43}+R_{54})$,<br>$S_{37}(B_2)=\frac{1}{\sqrt{2}}(R_{43}-R_{54})$, $S_{38}(A_1)=\frac{1}{\sqrt{2}}(R_{55}+R_{56})$, $S_{39}(B_2)=\frac{1}{\sqrt{2}}(R_{55}-R_{56})$,<br>$S_{40}(A_1)=\frac{1}{2}(R_{33}+R_{34}+R_{57}+R_{31})$, $S_{41}(B_1)=\frac{1}{2}(R_{33}-R_{31}+R_{57}-R_{34})$,<br>$S^*_{41}(B_1)=\frac{1}{2}(R_{31}+R_{34}-R_{33}-R_{57})$, $S_{42}(A_2)=\frac{1}{2}(R_{33}+R_{31}-R_{57}-R_{34})$ |

| | | | | | | | |
|---|---|---|---|---|---|---|---|
| $\Gamma_9$ | 14 | 2 | 2 | 2 | $5A_1$<br>$3A_2$<br>$3B_1$<br>$3B_2$ | 14 | $S_{43}(A_1)=\frac{1}{2}(R_{21}+R_{16}+R_{18}+R_{17}),\ S_{44}(A_2)=\frac{1}{2}(R_{21}-R_{26}+R_{17}-R_{18}),$<br>$S^{*}_{44}(A_2)=\frac{1}{2}(R_{26}+R_{17}-R_{21}-R_{18}),\ S_{45}(B_1)=\frac{1}{2}(R_{18}-R_{21}+R_{17}-R_{26}),$<br>$S_{46}(A_1)=\frac{1}{\sqrt{2}}(R_{19}+R_{25}),\ S_{47}(B_2)=\frac{1}{\sqrt{2}}(R_{19}-R_{25}),\ S_{48}(A_1)=\frac{1}{\sqrt{2}}(R_{48}+R_{51}),$<br>$S_{49}(B_2)=\frac{1}{\sqrt{2}}(R_{48}-R_{51}),\ S_{50}(A_1)=\frac{1}{\sqrt{2}}(R_{58}+R_{59}),\ S_{51}(B_2)=\frac{1}{\sqrt{2}}(R_{58}-R_{59}),$<br>$S_{52}(A_1)=\frac{1}{2}(R_{60}+R_{41}+R_{45}+R_{47}),\ S_{53}(B_1)=\frac{1}{2}(R_{60}-R_{41}+R_{45}-R_{47}),$<br>$S^{*}_{53}(B_1)=\frac{1}{2}(R_{41}+R_{47}-R_{60}-R_{45}),\ S_{54}(A_2)=\frac{1}{2}(R_{60}-R_{41}-R_{45}+R_{47})$ |
| $\Gamma_{10}$ | 14 | 2 | 2 | 2 | $5A_1$<br>$3A_2$<br>$3B_1$<br>$3B_2$ | 14 | $S_{55}(A_1)=\frac{1}{2}(R_{13}+R_{15}+R_{18}+R_{12}),\ S_{56}(A_2)=\frac{1}{2}(R_{13}-R_{15}+R_{18}-R_{12}),$<br>$S^{*}_{56}(A_2)=\frac{1}{2}(R_{15}+R_{12}-R_{13}-R_{18}),\ S_{57}(B_1)=\frac{1}{2}(R_{18}-R_{13}+R_{12}-R_{15}),$<br>$S_{58}(A_1)=\frac{1}{\sqrt{2}}(R_{10}+R_{16}),\ S_{59}(B_2)=\frac{1}{\sqrt{2}}(R_{10}-R_{16}),\ S_{60}(A_1)=\frac{1}{\sqrt{2}}(R_{61}+R_{46}),$<br>$S_{61}(B_2)=\frac{1}{\sqrt{2}}(R_{61}-R_{46}),\ S_{62}(A_1)=\frac{1}{\sqrt{2}}(R_{62}+R_{63}),\ S_{63}(B_2)=\frac{1}{\sqrt{2}}(R_{62}-R_{63}),$<br>$S_{64}(A_1)=\frac{1}{2}(R_{35}+R_{64}+R_{39}+R_{65}),\ S_{65}(B_1)=\frac{1}{2}(R_{35}-R_{65}+R_{39}-R_{64}),$<br>$S^{*}_{65}(B_1)=\frac{1}{2}(R_{64}+R_{65}-R_{35}-R_{39}),\ S_{66}(A_2)=\frac{1}{2}(R_{41}+R_{14}-R_{47}-R_{13})$ |

Table 1. The combination and their normalization coefficients of (3,0) nanotube in D3d point group (Lee et al., 2009)

The character of the system assigned by Γ was calculated from the character tables and the UMAT are written from the application of projection operator. Vibrational Calculation was carried out by the MOLVIB algorithms and by Hyper Chem. Calculation and a few sets of calculation were performed. Molecular motions can be assigned by the potential energy distribution (PED) analysis among internal coordinates by the method of the projection operator. There are good agreements between the most cases.

### 2.6 Normal mode dependence on dielectric

As can be inferred from Table 1 and the Fig. 3, 4, and 5, there are good agreements between the semi and Monte Carlo and even ab initio calculation. In Table 1 the various of intensity and frequency and potential energy from different methods are shown versus the inverse dielectric for some normal modes. From Fig. 3 we have two maximum for both of energy and frequency in the dielectric between 77.40 up to 70.42 and also the third maximum is located in the 61.76. This region range is considered to be the unstable geometry of nanotubes which are very sensitive to dielectric. After these range the frequency, intensity and energy goes toward a stable geometry which are not sensitive to dielectric. The same results are obtained in the insets Fig. B and C of Fig. 3 for D3d of normal mode 61 and 66 respectively. In the Fig 4, similar to normal mode 1 and 131 and 138 are shown with A, B, and C respectively for nano tube (4 0) in D4d point group, a common general behavior is observed in this nanotube as same as (3 0) nanotube, only with a shift in data, this shift is due to the difference between the geometrical structures of two nanotubes.

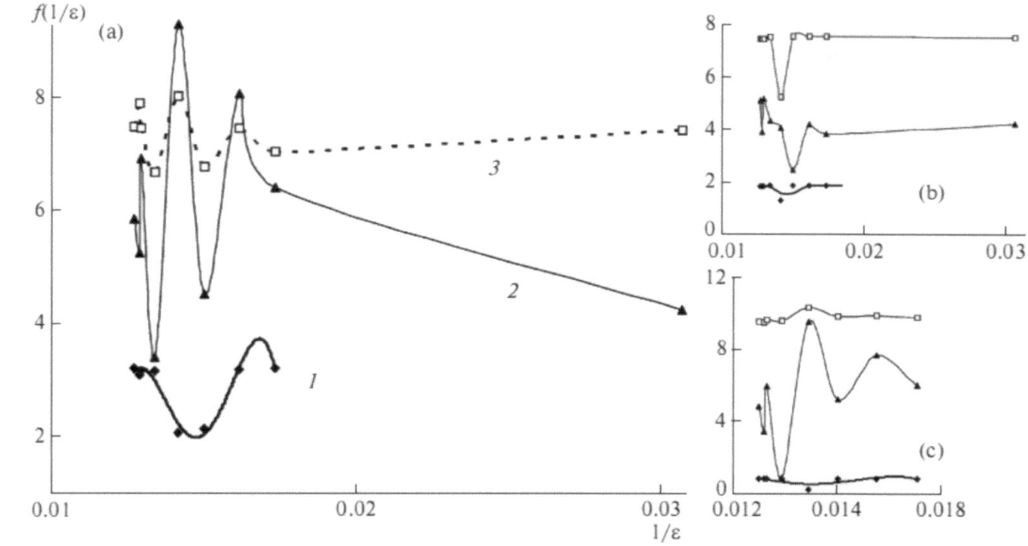


Fig. 3. The natural logarithms of the potential energy (1), intensity (2), and frequency (3) of three normal modes (a) 1, (b) 61, (c) 66 versus inverse of dielectric constant for nanotube (3, 0) with D3d point group by AM1 calculation (Lee et al., 2009)

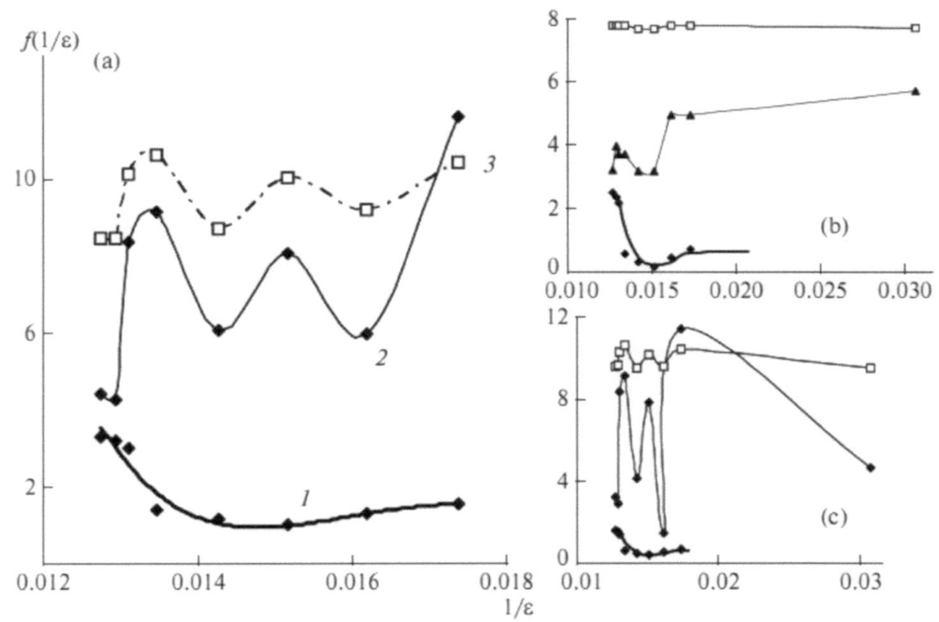


Fig. 4. The logarithms of the potential energy (1), intensity (2), and frequency (3) of three normal modes (a) 1, (b) 131, (c) 138 versus inverse of dielectric constant for nanotube (4, 0) with D4d point group by AM1 calculation (Lee et al., 2009)

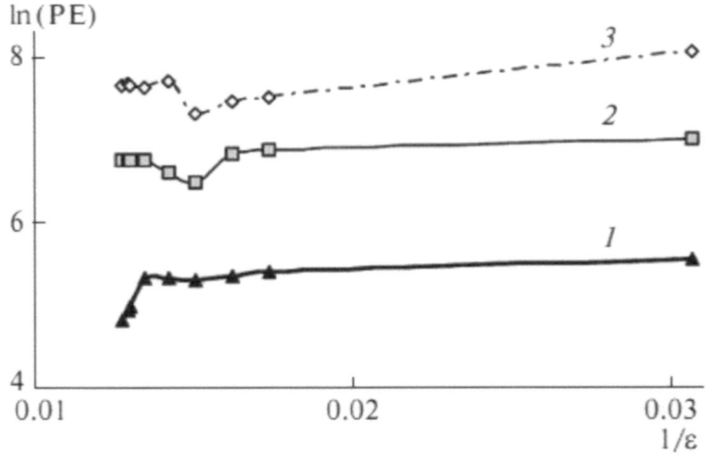


Fig. 5. The logarithms of the potential energy (PE) of three different zigzag nanotubes (1) D3d, (2) D4d, (3) D5d versus inverse of dielectric constant by MC simulations (Lee et al., 2009)

### 2.7 Potential energy dependence on the dielectrics of zigzag nanotubes

In Fig. 3 three line of potential energy for three nanotubes are shown by the Monte Carlo calculation versus dielectric constants. For D3d symmetric nanotube, the logarithm of potential energy increases as the dielectric constant reduces from 78.39 up to 76.10 while the D4d and D5d symmetries show an unchanged potential energy in this region. Beyond this point, the potential energy of D4d and D5d nanotubes drop and rise again to be in a new equilibrium, whereas the potential energy of the D3d mostly constant as the dielectric constant decreases. There are some changing in the energy in variable with the dielectrics above 60, and by decreasing the dielectrics the energy of three nanotubes goes toward constant variables. Similar trends between three figures and there are very good agreement with *ab initio* calculation in the Table 1.

## 3. Stability of SWCNTs: Solvents and temperature effects by molecular dynamics simulation and quantum mechanics calculations

Structural properties of solvents such as water, methanol, and ethanol surrounding single-walled carbon nanotube (SWCNT) and mixtures of them as well have an effects on the relative energies and dipole moment values. Because some of the physicochemical parameters are elated to structural properties of SWCNT, the different force fields can be examined to determine energy and other types of geometrical parameters, on the particular SWCNT. Because of the differences among force fields, the energy of a molecule calculated using two different force fields will not be the same. The structure of SWCNT as well as its dipole moments and relative energies has been studied by molecular dynamics simulation and quantum mechanics calculations (Monajjemi et al., 2010). The term "*Ab Initio*" is given to computations which are derived directly from theoretical principles, with no inclusion of experimental data. The most common type of ab initio calculation is called a Hartree-Fock (HF) calculation, in which the primary approximation is called the central field approximation. A method, which avoids making the HF mistakes in the first place, is called

Quantum Monte Carlo (QMC). There are several favors of QMC variational, diffusion, and Green's functions. These methods work with an explicitly correlated wave function and evaluate integrals numerically using a Monte Carlo integration. These calculations can be very time consuming, but they are probably the most accurate methods known today. In general, *ab initio* calculations give very good qualitative results and can give increasingly accurate quantitative results as the molecules in question become smaller (Monajjemi et al., 2008a). In general, there are three steps in carrying out any quantum mechanical calculation. First, prepare a molecule with an appropriate starting geometry. Second, choose a calculation method and its associated options. Third, choose the type of calculation with the relevant options and finally, analyze the results. We will give a short detail of computational method in the following section.

### 3.1 Molecular mechanics (Monte Carlo simulation)

The Metropolis implementation of the Monte Carlo algorithm has been developed by studying the equilibrium thermodynamics of many-body systems. Choosing small trial moves, the trajectories obtained applying this algorithmagree with those obtained by Langevin's dynamics (Tiana et al., 2007). This is understandable because the Monte Carlo simulations always detect the so-called "important phase space" regions which are of low energy (Liu & Monson, 2005). Because of imperfections of the force field, this lowest energy basin usually does not correspond to the native state in most cases, so the rank of native structure in those decoys produced by the force field itself is poor. In density function theory the exact exchange (HF) for a single determination is replaced by a more general expression of the exchange correlation functional, which can include terms accounting for both exchange energy and the electron correlation, which is omitted from HartreeFock theory:

$$E_{ks}=v+<hp>+1/2<P_j(\rho)>+E_{\chi(\rho)}+E_{C(\rho)}$$

where $E_{\chi(\rho)}$ is the exchange function and $E_{C(\rho)}$ is the correlation functional. The correlation function of Lee et al. includes both local and nonlocal terms (Lee et al., 1988).

### 3.2 Langevin dynamics (LD) simulation

The Langevin equation is a stochastic differential equation in which two force terms have been added to Newton's second law to approximate the effects of neglected degrees of freedom (Wang & Skeel, 2003). These simulations can be much faster than molecular dynamics. The molecular dynamics method is useful for calculating the time-dependent properties of an isolated molecule. However, more often, one is interested in the properties of a molecule that is interacting with other molecules.

### 3.3 Effect of differenct solvents of temperatures of SWCNTusing molecular dynamics simulation and quantum mechanics calculations

Difference in force field is illustrated by comparing the energy calculated by using force fields, MM+, Amber, and Bio+. The quantum mechanics (QM) calculations were carried out with the GAUSSIAN98 program based on HF/3-21G level. In the Gaussian program a simple approximation is used in which the volume of the solute is used to compute the radius of a cavity which forms the hypothetical surface of the molecule (Witanowski et al., 2002; Mora-Diez et al., 2006). The structures in gas phase and different solvent media such as water, methanol, ethanol, and mixtures of them have been compared. The structure of SWCNT as well as its dipole moments and relative energies has been studied by molecular dynamics simulation and quantum mechanics calculations within the Onsager self-consistent reaction field (SCRF) model using a Hartree-Fock method (HF) at the HF/3-21G level and the structural stability of considered nanotube in different solvent media and temperature (between 309K and 327K) have been compared and analyzed.

Since the influence between a molecule in solution and its medium can describe most simply by using Onsager model, in this model we have assumed that the solute is placed in a spherical cavity inside the solvent. The latter is described as a homogeneous, polarizable medium of dielectric constant. We started our studies with HF/3-21G gas phase geometry and water, methanol and ethanol surrounding SWCNT and mixtures of them as well. The results obtained from Onsager model calculations are illustrated using the energy difference between these conformers which are quite sensitive to the polarity of the surrounding solvent. The solvent effect has been calculated using SCRF model. According to this method, the total energy of solute and solvent, which depends on the dielectric constant ε has been listed in Table 2.

| | Dielectric constant | Temperature (K) | | | | | | | | | |
|---|---|---|---|---|---|---|---|---|---|---|---|
| Medium | $\Delta E$ (Kcal/mol) | 309 | 311 | 313 | 315 | 317 | 319 | 321 | 323 | 325 | 327 |
| Gas phase | 1 | −0.1075 | −0.1227 | −0.0916 | −0.1878 | −0.1287 | −0.1575 | — | −0.0511 | 0 | −0.1721 |
| Water | 78.39 | −0.1929 | −0.1965 | −0.1573 | 0 | — | −0.2629 | −0.2316 | −0.2517 | — | — |
| Methanol | 32.63 | −0.0743 | 0 | — | −0.0653 | −0.1936 | −0.0545 | −0.1942 | −0.0056 | −0.0615 | −0.0770 |
| Ethanol | 24.55 | 0 | — | −0.0129 | −0.0120 | −0.0382 | −0.0764 | −0.0901 | — | −0.0918 | — |
| Water−Methanol | 70.763 | −0.0029 | — | −0.0790 | −0.0846 | — | −0.0918 | −0.0483 | −0.0510 | 0 | −0.0485 |
| Water−Ethanol | 69.416 | −0.0840 | −0.0852 | −0.0442 | −0.0215 | — | −0.0276 | −0.0890 | 0 | −0.0144 | −0.0610 |

Table 2. Theoretical relative energies at different temperature and dielectric constant

These energies have been compared with the gas phase total energy CNT at the HF/3-21G level of theory and different solvents, and the graph of energy values versus dielectric constant of different solvents has been displayed at considered temperatures in Fig. 6.

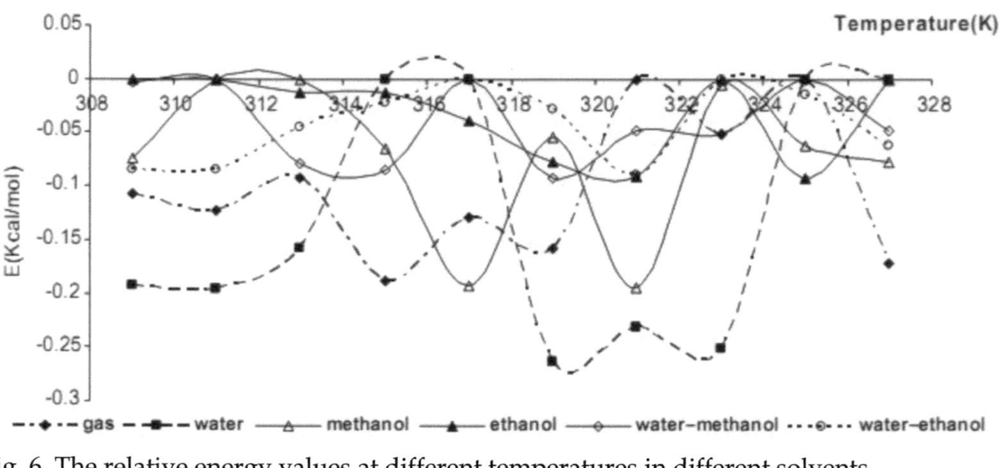


Fig. 6. The relative energy values at different temperatures in different solvents.

Since the solute dipole moment induces a dipole moment in opposite direction in the surrounding medium, polarization of the medium in turn polarizes the charge distribution in the solvent. The dipole moment value of SWCNT in different solvent media and at different temperatures has been reported in Table 3.

| Medium | Temperature (K) | | | | | | | | | |
|---|---|---|---|---|---|---|---|---|---|---|
| Dielectric constant | 309 | 311 | 313 | 315 | 317 | 319 | 321 | 323 | 325 | 327 |
| Gas phase | 14.8595 | 14.9783 | 15.8972 | 1.4569 | 12.7230 | 0.9106 | — | 15.1528 | 1.5045 | 0.7366 |
| Water | 1.3302 | 1.8910 | 1.0015 | 7.0286 | — | 0.7134 | 1.6209 | 0.9655 | — | — |
| Methanol | 1.4854 | 22.4244 | — | 9.5556 | 5.1790 | 9.4344 | 7.7342 | 21.3318 | 7.4751 | 7.2150 |
| Ethanol | 0.7029 | — | 2.4926 | 11.8624 | 14.9195 | 0.9389 | 0.9646 | — | 0.4796 | — |
| Water–Methanol | 12.4020 | 46.8625 | 4.4462 | 4.2048 | — | 4.8710 | 5.7352 | 5.9246 | 3.8414 | 6.1855 |
| Water–Ethanol | 0.8771 | 5.9677 | 6.3670 | 6.6015 | — | 3.8895 | 4.4362 | 6.4769 | 4.9592 | 5.8514 |

Table 3. Theoretical dipole moment values at different temperatures

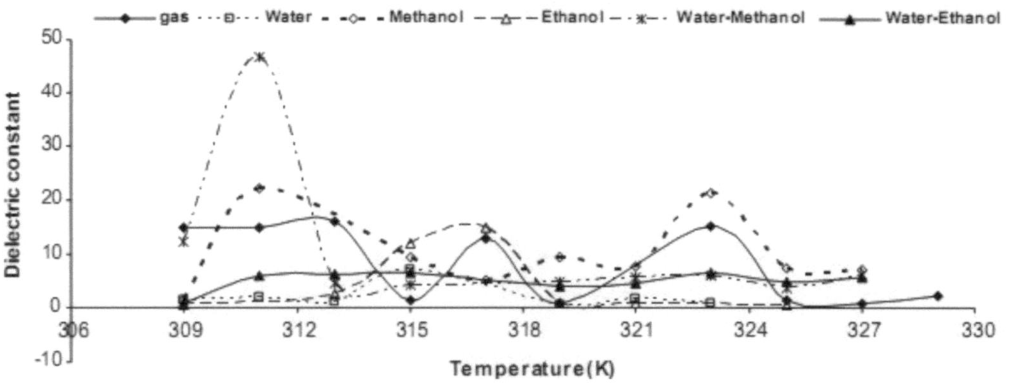


Fig. 7. The dipole moment values at different temperatures

One much more practical approach consists of calculating the molecular volume as defined through the contour of constant electron density, equating this (nonspherical) molecular volume to the radius of an ideally spherical cavity, and adding a constant increment for the closest possible approach of solvent molecules. This latter approach was used in Gaussian when the volume keyword was being used. In this work, we studied the structural properties of water, methanol, and ethanol surrounding SWCNT and mixtures of them as well as using molecular dynamics simulations. We used different force fields for determination of energy and other types of geometrical parameters, on the particular SWCNT. Because of the differences among force fields, the energy of a molecule calculated using two different force fields will not be the same. So, it is not reasonable to compare the energy of one molecule calculated with a particular force field with the energy of another molecule calculated using a different force field. In this study difference in force field illustrated by comparing the energy calculated by using force fields, MM+, AMBER, and BIO+. Theoretical energy values using difference force fields which are the combination of attraction van der Waals forces due to dipole-dipole interactions and empirical repulsive forces due to Pauli repulsion have been demonstrated in Table 4 and Fig. 8.

| | | MM+ | AMBER | BIO+ |
|---|---|---|---|---|
| Medium | Dielectric constant | $E$ (kcal/mol) | | |
| Gas phase | 1 | 487.9812 | 382.4433 | 1628.176 |
| Water | 78.39 | 419.9189 | 316.7664 | 1560.24 |
| Methanol | 32.63 | 535.3614 | 431.2506 | 1533.685 |
| Ethanol | 24.55 | 468.4228 | 359.4595 | 1601.808 |
| Water−Methanol | 70.763 | 413.1738 | 309.1905 | 1484.31 |
| Water−Ethanol | 69.416 | 637.6032 | 539.6583 | 1476.198 |

Table 4. Theoretical energy values using different force fields

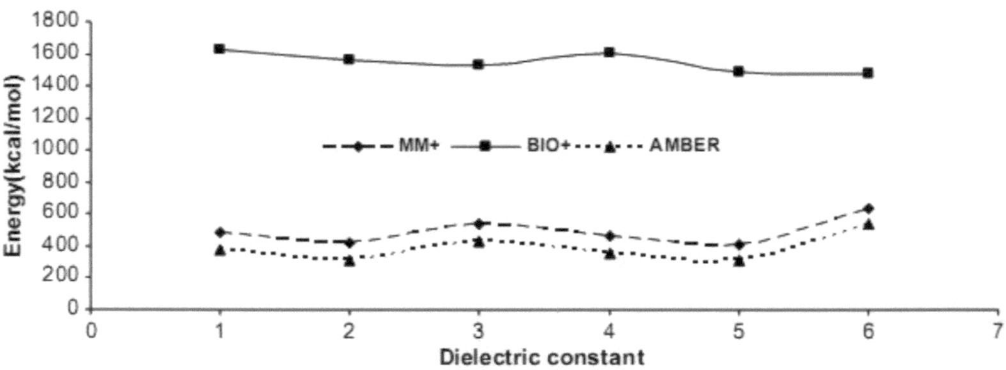


Fig. 8. The energy values using different force fields

The result of the calculated dipole moment, quadrupole moment, octapole moment, and hexadecapole moment values of SWCNT has been reported in Table 5, and optimized structures of nanotube in different media are shown in Fig. 9.

| | | Dipole moment | | | Quadrupole moment | | | Octapole moment | | | | | Hexadecapole moment | | | | | | | | |
|---|---|---|---|---|---|---|---|---|---|---|---|---|---|---|---|---|---|---|---|---|---|
| | | X | Y | Z | XX | YY | ZZ | XXX | YYY | ZZZ | XYY | XXY | XXXX | YYYY | ZZZZ | XXXY | XXXZ | YYYX | YYYZ | ZZZX | |
| Solvent | Temp | | TOT | | XY | XZ | YZ | XXZ | XZZ | YZZ | YYZ | XYZ | ZZZY | XXYY | XXZZ | YYZZ | XXYZ | YYXZ | ZZXY | | ΔE |
| SWCNT[illegible] | 309 | −14.6440 | 0.1793 | −2.5154 | −804.3781 | −656.2325 | −653.9581 | −1365.5072 | 3.5459 | −23.9528 | −69.9815 | 8.0217 | ****** | −15823.3320 | −15239.2153 | −74.7350 | −1492.05 | 31.5585 | 6.0318 | −117.1608 | −0.107814 |
| (4.7) | | | 14.8595 | | −0.5636 | −18.7461 | −1.1821 | −163.5864 | −38.7868 | 3.5363 | −5.9538 | −2.0505 | −15.2039 | −19782.5644 | −19607.8861 | −5110.2134 | −84.6609 | −28.5092 | −1.7157 | | |
| | 311 | 14.7152 | 2.1027 | 1.8413 | −807.3251 | −650.1956 | −657.7436 | 1403.4345 | 8.8112 | 7.6379 | 33.5305 | 182.9625 | ****** | −15394.6418 | −15186.5538 | −1724.0098 | −1252.2866 | −114.3129 | −18.4200 | −123.4675 | −0.122794 |
| | | | 14.9783 | | −21.6388 | −16.3081 | −4.1513 | 155.8165 | 71.8907 | 5.8749 | 8.4026 | 17.0032 | −34.7751 | −19898.2406 | −19811.4622 | −5088.5586 | −296.1877 | −43.5357 | −67.6681 | | |
| | 313 | 15.4846 | −2.4142 | 2.6682 | −813.8514 | −650.2292 | −655.6187 | 1420.1162 | −12.1769 | 16.5637 | 32.9314 | −193.6747 | ****** | −15171.8188 | −15085.1863 | 1683.6953 | −1602.4717 | 154.7971 | 25.2140 | −107.2728 | −0.091625 |
| | | | 15.8972 | | 22.6096 | −20.4078 | 2.4426 | 202.6341 | 68.7774 | −7.0593 | 7.5674 | −12.5116 | −6.1266 | −19383.0292 | −19681.8544 | −5041.8864 | 194.6506 | −17.1304 | 52.2818 | | |
| | 315 | 0.4958 | −0.9083 | 1.0256 | −712.1719 | −659.4462 | −655.5195 | 78.0914 | 1.6508 | 14.0220 | 27.5257 | −66.2198 | ****** | −15677.7207 | −15064.1668 | −70.2858 | −257.0442 | −32.8378 | −14.9118 | −24.9122 | −0.187885 |
| | | | 1.4569 | | −1.4415 | −3.4517 | −0.8489 | 58.2204 | −15.4389 | −4.7577 | 2.5424 | 6.0305 | 5.1805 | −19567.5739 | −19162.4177 | −5122.5846 | −72.8613 | 14.6364 | −6.8946 | | |
| | 317 | 12.7077 | 0.5180 | −0.3458 | −797.7993 | −652.5755 | −657.7653 | 1225.0214 | 1.3741 | −8.1095 | 18.4392 | 40.7788 | ****** | −15462.8319 | −15081.0940 | −801.5745 | 630.3633 | −98.1915 | 10.8582 | 134.9002 | −0.12871 |
| | | | 12.7230 | | −9.5071 | 7.5995 | 1.2156 | −51.5699 | 81.7003 | −0.5721 | 1.3640 | −11.6036 | 0.9670 | −19451.0951 | −19879.3155 | −5102.7808 | 71.1309 | −37.6227 | −25.4519 | | |
| | 319 | −0.3855 | −0.8136 | 0.1365 | −715.8083 | −660.1265 | −656.0085 | −16.8360 | −5.4686 | −1.7159 | −17.5566 | −26.1266 | ****** | −15425.0187 | −15140.8960 | 421.2191 | −16.7640 | 51.4588 | −53.3547 | 9.4804 | −0.15756 |
| | | | 0.9106 | | 6.2135 | 0.7334 | −1.2094 | 4.8922 | 1.8009 | −2.2083 | 1.5087 | −9.8140 | 17.7446 | −19636.7873 | −19185.2741 | −5103.0485 | −32.0774 | −10.7256 | 29.6524 | | |
| | 321 | — | — | — | — | — | — | — | — | — | — | — | — | — | — | — | — | — | — | — | — |
| | | | — | | — | — | — | — | — | — | — | — | — | — | — | — | — | — | — | | |
| | 323 | −15.0621 | −1.6222 | −0.3327 | −807.7053 | −656.2562 | −651.9912 | −1369.0978 | −10.1028 | 2.4310 | −66.1265 | −146.9892 | ****** | −15258.9407 | −14960.8568 | −2210.3063 | −828.5021 | −214.3513 | 1.2022 | −17.4442 | −0.051173 |
| | | | 15.1528 | | −28.7476 | −11.4370 | −3.5739 | −20.9444 | −40.9744 | −4.1223 | 0.1815 | −12.0892 | −40.8385 | −19815.1920 | −19349.2798 | −5033.8180 | −317.6068 | −113.3090 | −68.6385 | | |
| | 325 | −0.1757 | 0.7390 | 1.2987 | −730.6613 | −655.4731 | −659.3065 | −30.5621 | 8.3931 | 22.7699 | −21.7612 | 52.8099 | ****** | −15418.2153 | −15183.3142 | −23.3290 | −483.4387 | 30.5261 | 15.1234 | 52.2549 | 0 |
| | | | 1.5045 | | 2.4355 | −5.6029 | 2.6182 | 104.6224 | 14.7022 | −1.6996 | 1.4374 | −15.3525 | 11.3987 | −19314.5094 | −19508.0924 | −5080.6466 | 219.0057 | −79.8726 | 87.9432 | | |
| | 327 | 0.0947 | −0.6480 | −0.3374 | −713.5693 | −660.0270 | −655.8960 | 44.3072 | −8.3920 | −0.4140 | −21.7778 | −71.8008 | ****** | −15533.6913 | −15042.5782 | 430.0432 | 486.3275 | 79.5713 | −11.3200 | 13.5049 | −0.194347 |
| | | | 0.7366 | | 6.1065 | 6.9554 | 2.2931 | −11.6248 | 29.1555 | −2.4015 | −7.6069 | 30.6341 | 28.2309 | −19600.5284 | −19183.7106 | −5074.8363 | 196.8413 | 28.7602 | 14.5447 | | |
| | 329 | −1.5982 | 1.0704 | −1.1913 | −732.6384 | −657.8583 | −657.8600 | −147.2097 | 12.0277 | −22.5520 | −58.9183 | 54.9599 | ****** | −15395.6207 | −15283.1608 | 868.6320 | 265.3962 | 9.5175 | −44.7107 | 55.0741 | −0.172122 |
| | | | 2.2626 | | 11.4300 | 3.2758 | −6.4204 | −81.1060 | 34.0920 | 5.8612 | 8.1285 | −25.9273 | −51.2774 | −19520.5055 | −19522.6786 | −5114.1251 | −451.6274 | 38.7111 | 106.8356 | | |

| | | Dipole moment | | | Quadrupole moment | | | Octapole moment | | | | | Hexadecapole moment | | | | | | | | |
|---|---|---|---|---|---|---|---|---|---|---|---|---|---|---|---|---|---|---|---|---|---|
| | | X | Y | Z | XX | YY | ZZ | XXX | YYY | ZZZ | XYY | XXY | XXXX | YYYY | ZZZZ | XXXY | XXXZ | YYYX | YYYZ | ZZZX | |
| Solvent | Temp | | TOT | | XY | XZ | YZ | XXZ | XZZ | YZZ | YYZ | XYZ | ZZZY | XXYY | XXZZ | YYZZ | XXYZ | YYXZ | ZZXY | | ΔE |
| SWCNT$_{(10-}$ (4,7)-water | 309 | 0.9078 | 0.2025 | -0.9509 | -721.6414 | -659.6380 | -659.2722 | 93.7872 | -1.7243 | -12.4079 | 20.6372 | 13.9168 | ****** | -15454.1671 | -15158.1740 | 400.6732 | 136.5951 | 48.5804 | 20.3502 | 84.2959 | -0.192953 |
| | | | 1.3302 | | 6.2040 | 3.9204 | 1.2292 | -90.7160 | 0.7236 | 9.2513 | 1.1876 | 14.2028 | 13.6611 | -19780.6831 | -19698.8392 | -5100.6233 | 44.4429 | -81.0372 | 83.6765 | | |
| | 311 | 0.6731 | -1.6782 | 0.5537 | -712.2610 | -660.7967 | -660.8692 | 78.8571 | -14.1268 | -3.2561 | 18.5044 | -114.5149 | ****** | -15816.8973 | -14892.0649 | 737.9869 | -337.9681 | -69.7363 | 26.7012 | -51.9835 | -0.196564 |
| | | | 1.8910 | | 7.4520 | -3.3201 | 0.8200 | 40.5644 | -5.6268 | -3.7160 | -1.7549 | -19.3398 | -7.1261 | -19699.1193 | -19648.7674 | -5120.8106 | 138.7880 | -1.4452 | 80.2318 | | |
| | 313 | -0.9173 | -0.1928 | -0.3526 | -710.3840 | -657.3522 | -660.9073 | -110.0398 | -6.0247 | -3.0210 | 1.3566 | -27.0916 | ****** | -15541.3077 | -14964.6745 | -196.4832 | -416.9805 | -61.2680 | -27.1663 | -64.7470 | -0.157327 |
| | | | 1.0015 | | -2.3197 | -4.0948 | -0.4097 | -27.0978 | -18.2127 | 7.0109 | -0.9826 | 13.9556 | 22.7802 | -19624.5370 | -19709.3499 | -5092.3169 | 3.3751 | 4.6669 | -3.9262 | | |
| | 315 | 2.9644 | -1.9953 | 6.0525 | -919.6398 | -653.4151 | -652.6520 | 233.6732 | -12.6057 | 45.8260 | -27.3419 | -169.0830 | ****** | -15354.1349 | -14558.1040 | -113.2612 | -69.1162 | -26.8119 | 16.1684 | -115.3540 | 0 |
| | | | 7.0286 | | -0.7342 | -2.1383 | 1.7635 | 481.3963 | 65.4141 | -3.6642 | 9.2140 | -2.1691 | -8.7799 | -20571.4992 | -20203.1909 | -4986.6548 | 141.2732 | 85.4590 | 41.1064 | | |
| | 317 | — | — | — | — | — | — | — | — | — | — | — | — | — | — | — | — | — | — | — | — |
| | | | — | | — | — | — | — | — | — | — | — | — | — | — | — | — | — | — | — | |
| | 319 | -0.3132 | 0.6143 | 0.1831 | -717.1320 | -658.6630 | -658.0192 | -47.9089 | 11.7478 | 6.1816 | -19.9581 | 24.6970 | ****** | -15548.7987 | -14819.0157 | 406.7381 | -226.1439 | 66.9447 | 15.9164 | -56.9492 | -0.262943 |
| | | | 0.7134 | | 5.3825 | -3.4315 | 0.9784 | 11.1880 | 13.0846 | 1.0844 | -1.1261 | 15.6369 | -14.5358 | -19602.2749 | -19508.6983 | -5053.6339 | 135.0959 | 2.8647 | 20.2842 | | |
| | 321 | -1.5387 | -0.4985 | -0.1057 | -721.3000 | -662.6148 | -655.6483 | -119.1921 | -6.9063 | 3.8343 | -51.9817 | -23.4512 | ****** | -15812.4436 | -15062.9102 | -251.7614 | 270.4083 | -41.7526 | -7.2658 | 3.2563 | -0.231667 |
| | | | 1.5209 | | -3.7784 | 3.6885 | -2.7907 | 3.6626 | 38.3813 | -4.1944 | -9.7555 | -15.1862 | -15.8953 | -20032.9276 | -19247.4531 | -5149.7668 | -215.4586 | 23.8832 | -22.2197 | | |
| | 323 | -0.0663 | 0.4220 | 0.8658 | -721.9988 | -657.0860 | -663.4265 | 5.9553 | 2.1894 | 1.8781 | 9.1011 | 37.4186 | ****** | -15586.2034 | -14786.5979 | 564.9449 | -516.1939 | -10.3218 | -34.8678 | -29.5529 | -0.251753 |
| | | | 0.9655 | | 2.8352 | -5.8553 | -2.4379 | 64.6385 | -19.6017 | -5.5754 | 9.2236 | 18.8221 | -4.9180 | -19693.1419 | -19890.8235 | -5058.7565 | -86.2293 | -44.8569 | -17.7365 | | |
| | 325 | — | — | — | — | — | — | — | — | — | — | — | — | — | — | — | — | — | — | — | — |
| | | | — | | — | — | — | — | — | — | — | — | — | — | — | — | — | — | — | | |
| | 329 | — | — | — | — | — | — | — | — | — | — | — | — | — | — | — | — | — | — | — | — |
| | | | — | | — | — | — | — | — | — | — | — | — | — | — | — | — | — | — | | |

| | | Dipole moment | | | Quadrupole moment | | | Octapole moment | | | | | Hexadecapole moment | | | | | | | | |
|---|---|---|---|---|---|---|---|---|---|---|---|---|---|---|---|---|---|---|---|---|---|
| | | X | Y | Z | XX | YY | ZZ | XXX | YYY | ZZZ | XYY | XXY | XXXX | YYYY | ZZZZ | XXXY | XXXZ | YYYX | YYYZ | ZZZX | |
| Solvent | Temp | | TOT | | XY | XZ | YZ | XXZ | XZZ | YZZ | YYZ | XYZ | ZZZY | XXYY | XXZZ | YYZZ | XXYZ | YYXZ | ZZXY | | ΔE |
| SWCNT$_{(10-}$ (4,7)-ethanol | 309 | 0.2301 | 0.2337 | -0.6217 | -720.7922 | -657.5045 | -659.7962 | 49.1533 | 3.1361 | 2.3580 | 4.7784 | 3.8344 | ****** | -15479.6634 | -15265.9007 | 859.4473 | 616.7403 | 109.2258 | -16.4335 | 63.7184 | 0 |
| | | | 0.7029 | | 11.1676 | 6.0212 | -1.2966 | -48.9874 | 6.0104 | 7.6096 | -1.3645 | -17.2371 | -4.9074 | -19523.0452 | -19645.2758 | -5184.8998 | -108.3800 | 5.4027 | 7.0261 | | |
| | 311 | — | — | — | — | — | — | — | — | — | — | — | — | — | — | — | — | — | — | — | — |
| | | | — | | — | — | — | — | — | — | — | — | — | — | — | — | — | — | — | | |
| | 313 | 2.1287 | 1.0149 | -0.8073 | -718.0002 | -657.0558 | -661.7489 | 191.5845 | 15.0467 | -2.6289 | 6.4943 | 72.7805 | ****** | -15381.7243 | -15262.0379 | 71.5196 | -373.9308 | 25.1173 | -66.0198 | -53.9676 | -0.0129921 |
| | | | 2.4926 | | 1.0357 | -4.7548 | -7.8347 | -77.6895 | 23.7453 | -2.4513 | 2.4626 | 64.2695 | -52.0330 | -19581.6480 | -19851.1019 | -5115.3538 | -573.4797 | 13.0398 | -4.2161 | | |
| | 315 | -10.8367 | -2.6081 | -4.0597 | -775.4244 | -657.7152 | -658.2211 | -986.0580 | -11.5713 | -33.1305 | -65.8448 | -246.7706 | ****** | -15437.7036 | -14898.7095 | -2298.7436 | -2432.9815 | -225.9123 | -64.3304 | -195.6334 | -0.012007 |
| | | | 11.8624 | | -28.5901 | -30.0023 | -5.9464 | -287.4888 | -53.0196 | -15.2632 | -13.2281 | -63.1000 | -7.0656 | -19909.6624 | -19708.3379 | -5035.6149 | -487.1965 | -121.0565 | -97.7227 | | |
| | 317 | -14.8319 | 0.5370 | -1.5227 | -649.1182 | -654.8977 | -659.1986 | -703.9376 | -3.0525 | -6.3110 | -88.9254 | 58.3950 | ****** | -15463.3263 | -14860.4576 | 123.5850 | 201.0876 | 20.6689 | -30.1734 | 19.2732 | -0.038293 |
| | | | 14.9195 | | 3.0209 | 5.8111 | -4.4153 | -74.9155 | -46.9965 | 5.7715 | -0.2713 | 34.5433 | -30.7131 | -19295.7294 | -19368.4680 | -5039.0416 | -346.1639 | 37.7840 | 0.7480 | | |
| | 319 | -0.0960 | 0.7655 | 0.5350 | -724.1597 | -660.2918 | -661.8909 | 18.2858 | 13.9486 | 8.0818 | -8.1416 | 1.4580 | ****** | -15504.7425 | -15141.2187 | 266.6062 | -284.3215 | 62.2636 | 3.9252 | -28.5704 | -0.076452 |
| | | | 0.9389 | | 5.2301 | -4.1263 | -0.6843 | 39.5798 | 11.4550 | 3.3970 | -2.9303 | -19.4796 | 1.2438 | -19869.2985 | -19912.3756 | -5092.3333 | -155.3072 | -43.6530 | 63.9648 | | |
| | 321 | -0.4662 | 0.8392 | 0.0937 | -712.4423 | -662.3868 | -655.6282 | -25.7609 | 4.5630 | 2.8793 | -24.1112 | 54.9980 | ****** | -15545.0286 | -14984.5662 | -748.6216 | 193.7805 | -17.5819 | 5.4363 | 26.7263 | -0.09019 |
| | | | 0.9646 | | -7.4381 | 3.0980 | 0.7290 | 1.1267 | 15.7079 | 1.5852 | -8.1452 | -17.5562 | 1.1017 | -19968.5852 | -19303.4447 | -5091.9930 | 49.8215 | 24.5039 | -7.9894 | | |
| | 323 | — | — | — | — | — | — | — | — | — | — | — | — | — | — | — | — | — | — | — | — |
| | | | — | | — | — | — | — | — | — | — | — | — | — | — | — | — | — | — | | |
| | 325 | -0.0367 | -0.1718 | 0.4462 | -714.7345 | -661.9806 | -656.1977 | -5.2923 | -1.0334 | -4.9482 | 11.3696 | -32.7107 | ****** | -15390.1851 | -15082.4235 | -398.9528 | 770.0961 | -79.1779 | 12.1014 | 58.2614 | -0.09184 |
| | | | 0.4796 | | -5.6931 | 10.1184 | 0.4237 | 25.9912 | -21.6086 | -1.0686 | 8.0053 | 8.1842 | 3.8388 | -19909.9894 | -19444.8108 | -5078.5129 | 37.7004 | 26.6834 | 29.1106 | | |
| | 327 | — | — | — | — | — | — | — | — | — | — | — | — | — | — | — | — | — | — | — | — |
| | | | — | | — | — | — | — | — | — | — | — | — | — | — | — | — | — | — | | |

| | | Dipole moment | | | Quadrupole moment | | | Octapole moment | | | | | Hexadecapole moment | | | | | | | | |
|---|---|---|---|---|---|---|---|---|---|---|---|---|---|---|---|---|---|---|---|---|---|
| | | X | Y | Z | XX | YY | ZZ | XXX | YYY | ZZZ | XYY | XXY | XXXX | YYYY | ZZZZ | XXXY | XXXZ | YYYX | YYYZ | ZZZX | |
| Solvent | Temp | | TOT | | XY | XZ | YZ | XXZ | XZZ | YZZ | YYZ | XYZ | ZZZY | XXYY | XXZZ | YYZZ | XXYZ | YYXZ | ZZXY | | ΔE |
| SWCNT$_{(10-}$ (4,7)-methanol | 309 | 0.2325 | -0.8428 | 1.2008 | -733.4175 | -661.9906 | -658.1747 | -24.7957 | -7.4988 | 17.2260 | 38.6579 | -46.9186 | ****** | -15591.1687 | -15174.4829 | 56.1877 | 758.0189 | 3.7376 | 21.6306 | 41.0697 | -0.074369 |
| | | | 1.4854 | | -2.7756 | 7.8680 | 4.5457 | 43.8034 | -38.7405 | -9.4967 | -4.1260 | 56.2727 | 58.2795 | -19992.4746 | -19545.9261 | -5122.5088 | 239.7142 | 78.8713 | -58.9477 | | |
| | 311 | -22.1691 | -3.0275 | -1.4895 | -843.0728 | -654.3508 | -659.4412 | -1803.7568 | -27.4327 | -5.1855 | -89.9411 | -237.1913 | ****** | -15713.5048 | -15302.5312 | -1414.9964 | -20.6835 | -71.8516 | 4.7823 | 38.2617 | 0 |
| | | | 22.4244 | | -16.5498 | -1.3655 | 1.4492 | -103.5324 | -107.1319 | -6.7140 | -7.1572 | 23.9161 | 25.1046 | -20006.0502 | -20270.3885 | -5210.0858 | 95.4907 | -11.0198 | -113.2310 | | |
| | 313 | — | — | — | — | — | — | — | — | — | — | — | — | — | — | — | — | — | — | — | — |
| | | | — | | — | — | — | — | — | — | — | — | — | — | — | — | — | — | — | | |
| | 315 | -9.4004 | 0.6453 | 1.5895 | -773.6897 | -660.0144 | -658.6332 | -792.1821 | -6.0824 | 12.0697 | -56.2600 | 65.3777 | ****** | -15549.1408 | -15105.1051 | 564.0678 | 2746.0029 | 101.8697 | 49.6627 | 256.1469 | -0.0653 |
| | | | 9.5556 | | 6.2495 | 32.1931 | 4.2501 | 161.1667 | -59.1913 | -7.1102 | 12.1004 | 37.0965 | 6.7656 | -20050.9552 | -19875.0217 | -5102.7071 | 344.1837 | 121.8207 | -82.2839 | | |
| | 317 | -4.8984 | 1.6719 | 0.1802 | -749.3056 | -659.7450 | -659.8531 | -424.0047 | 29.1717 | 4.3840 | -106.7746 | 119.1897 | ****** | -15497.5748 | -15379.6056 | 1299.1827 | -485.7991 | 218.1838 | -28.1675 | -27.1485 | -0.193631 |
| | | | 5.1790 | | 14.8902 | -6.8753 | -3.2343 | -20.2964 | 43.3589 | -0.3222 | 7.1393 | -1.1054 | -39.6450 | -19966.1746 | -19883.2673 | -5166.4385 | -207.8272 | 48.7896 | 7.8704 | | |
| | 319 | -9.0138 | 2.3724 | -1.4599 | -768.9798 | -660.6028 | -654.0680 | -697.1521 | 32.6762 | -4.4130 | -108.8953 | 167.7784 | ****** | -15546.6875 | -15287.9331 | 1542.6574 | -730.1954 | 249.7533 | -8.6615 | -20.0014 | -0.054531 |
| | | | 9.4344 | | 19.1250 | -9.8784 | 0.9364 | -97.8407 | 6.6126 | 0.5194 | -7.2757 | 15.1337 | 23.8073 | -20156.6136 | -19446.6015 | -5128.8066 | 19.8514 | -20.9376 | 18.3278 | | |
| | 321 | 7.5748 | -0.2093 | -1.5479 | -752.0355 | -657.0198 | -662.1121 | 648.0518 | -7.9626 | -22.1673 | 42.5126 | -11.6903 | ****** | -15516.5820 | -15363.2650 | -135.2120 | 1311.5923 | -76.0997 | -50.6852 | 136.9694 | -0.194206 |
| | | | 7.7342 | | -3.7412 | 14.1483 | -4.4997 | -118.0325 | 54.6421 | -3.7733 | -5.1812 | 39.0464 | -9.9496 | -19654.8568 | -20027.7955 | -5149.3249 | -413.3877 | 39.5904 | 52.1312 | | |
| | 323 | 21.2212 | 1.9895 | 0.8669 | -844.7077 | -660.0454 | -653.9390 | 1778.9615 | -12.7277 | 9.1955 | 103.5071 | 143.1093 | ****** | -15535.2568 | -15158.0960 | -795.3769 | -890.9801 | 14.7380 | -29.8364 | -120.3293 | -0.005656 |
| | | | 21.3318 | | -12.1306 | -13.6211 | -4.4234 | 56.7607 | 98.1184 | 9.8276 | 1.0805 | 1.5679 | -32.3236 | -20394.1903 | -19890.3528 | -5066.4351 | -276.2565 | -47.0784 | -24.0052 | | |
| | 325 | -7.0492 | -1.9355 | 1.5621 | -782.0435 | -662.1841 | -658.5341 | -544.6619 | -18.6346 | 2.9682 | -80.8945 | -198.5488 | ****** | -15597.9551 | -15297.5686 | -1180.3778 | 2290.3445 | -192.0559 | 43.3185 | 94.9529 | -0.061553 |
| | | | 7.4751 | | -15.5308 | 28.4839 | 5.9891 | 167.1361 | -22.6931 | -7.3315 | 5.8047 | 33.2638 | 32.9312 | -20164.9856 | -19771.1611 | -5142.3851 | 445.0417 | 200.5994 | -51.8619 | | |
| | 327 | -7.0087 | 1.5284 | 0.7738 | -752.1815 | -662.1982 | -654.1601 | -715.1820 | 19.4842 | 2.9452 | -86.0838 | 121.9688 | ****** | -15647.8111 | -15124.5612 | 1310.6834 | -511.9812 | 202.0872 | -19.4090 | -99.7560 | -0.077016 |
| | | | 7.2150 | | 16.7621 | -7.6146 | -1.8681 | 97.4825 | -14.4641 | -0.0959 | 4.9120 | -15.8119 | -16.4649 | -20224.8224 | -19477.8204 | -5140.7928 | -182.0415 | 44.3731 | 6.2946 | | |

| | | Dipole moment | | | Quadrupole moment | | | Octapole moment | | | | | Hexadecapole moment | | | | | | | | |
|---|---|---|---|---|---|---|---|---|---|---|---|---|---|---|---|---|---|---|---|---|---|
| | | X | Y | Z | XX | YY | ZZ | XXX | YYY | ZZZ | XYY | XXY | XXXX | YYYY | ZZZZ | XXXY | XXXZ | YYYX | YYYZ | ZZZX | |
| Solvent | Temp | | TOT | | XY | XZ | YZ | XXZ | XZZ | YZZ | YYZ | XYZ | ZZZY | XXYY | XXZZ | YYZZ | XXYZ | YYXZ | ZZXY | | ΔE |
| SWCNT (4,7)-water = ethanol | 309 | 0.6751 | 0.4929 | −0.2656 | −712.5204 | −655.2920 | −663.5865 | 38.9541 | 4.7228 | −3.7045 | 29.8230 | −4.0983 | ****** | −15425.2236 | −14705.8716 | −322.3457 | −335.2780 | −91.5713 | 7.9254 | −2.6228 | −0.084093 |
| | | | 0.8771 | | −5.5049 | −5.4217 | 1.2912 | −2.5657 | −24.0551 | 8.8902 | 4.0472 | −19.9839 | 13.9364 | −19563.9745 | −19988.9281 | −5024.1302 | 129.1434 | −68.4288 | −24.6220 | | |
| | 311 | −5.5325 | −2.0302 | 0.9395 | −732.2581 | −661.7991 | −664.7473 | −456.1862 | −7.5924 | 15.3594 | 34.4629 | −116.0547 | ****** | −16609.5859 | −15464.7467 | −649.8821 | 2036.5391 | −60.5729 | 0.8470 | 321.7151 | −0.085221 |
| | | | 5.9677 | | −10.1534 | 24.7073 | 4.5939 | 122.0901 | −94.0579 | −16.4499 | 3.4873 | 21.2429 | 75.1825 | −19746.7585 | −19888.7274 | −5354.2888 | 308.8526 | 21.0294 | −163.9256 | | |
| | 313 | −5.7577 | −1.9182 | 1.9257 | −729.6999 | −661.8059 | −665.4156 | −470.3752 | −11.8852 | 23.9506 | −2.4557 | −122.0863 | ****** | −16386.2652 | −15508.7953 | −83.8679 | 1097.9675 | −8.2373 | 24.1364 | 154.9812 | −0.044286 |
| | | | 6.3670 | | −2.8586 | 12.3961 | 3.8830 | 182.2235 | −51.1772 | −12.2148 | 0.8413 | 27.6542 | 31.2759 | −19652.9260 | −19877.7251 | −5327.7026 | 295.5603 | 36.9241 | −140.9214 | | |
| | 315 | −6.0967 | −1.5394 | 2.0100 | −739.6855 | −659.4316 | −667.9510 | −456.6056 | −18.0449 | 33.0751 | 18.8187 | −84.3076 | ****** | −16745.0814 | −15503.9051 | 156.6532 | 884.4741 | 34.1365 | 1.8405 | 124.5123 | −0.021558 |
| | | | 6.6015 | | 0.4228 | 10.4290 | 3.6773 | 180.5086 | −74.0004 | −13.6272 | −5.2538 | 47.5439 | 35.8160 | −19587.3032 | −19918.5923 | −5376.6268 | 271.5607 | 19.4347 | −126.5068 | | |
| | 317 | — | — | — | — | — | — | — | — | — | — | — | — | — | — | — | — | — | — | — | |
| | | | — | | — | — | — | — | — | — | — | — | — | — | — | — | — | — | — | | |
| | 319 | 3.6530 | −0.5571 | −1.2138 | −741.7181 | −658.7982 | −669.6092 | 330.4041 | −12.0154 | −31.1413 | 14.1086 | 1.7784 | ****** | −16762.4308 | −15567.5635 | 482.0524 | 939.5850 | 41.7567 | −24.9775 | 177.1722 | −0.027679 |
| | | | 3.8895 | | 7.8311 | 10.2308 | −2.1702 | −120.2621 | 23.0542 | −8.8939 | 4.2285 | 9.7327 | −18.1290 | −19457.3868 | −20016.8812 | −5358.4028 | −136.5317 | −31.1998 | 104.7935 | | |
| | 321 | −4.2737 | −1.1785 | 0.1629 | −734.0236 | −658.9000 | −666.2975 | −353.0907 | −6.9938 | 5.5957 | −11.6927 | −58.6552 | ****** | −16826.9490 | −15529.4392 | −1114.0878 | 2005.0098 | −158.1294 | 18.1706 | 323.9398 | −0.089011 |
| | | | 4.4362 | | −16.3334 | 25.3873 | 3.3399 | 28.4876 | −28.4655 | −17.9712 | 3.9308 | 35.2403 | 11.8985 | −19450.6580 | −19665.0505 | −5406.7469 | 225.7249 | 24.2015 | −159.0748 | | |
| | 323 | −6.4297 | 0.2387 | 0.7430 | −746.4172 | −661.1047 | −663.6051 | −535.8458 | 2.7839 | 16.7989 | −18.4097 | 38.4390 | ****** | −16628.9472 | −15714.0022 | −1193.8838 | 851.6480 | −203.3875 | 16.9704 | 89.6261 | 0 |
| | | | 6.4769 | | −17.1796 | 7.9433 | 4.3720 | 84.8085 | −51.1356 | −15.8860 | 6.1699 | 35.7176 | 52.3093 | −19489.3441 | −19507.3702 | −5415.3007 | 295.9076 | 3.0434 | −139.6450 | | |
| | 325 | −4.4197 | 1.7130 | −1.4580 | −743.2502 | −657.5881 | −668.5018 | −399.0351 | 25.7112 | −15.4201 | −26.5555 | 77.1364 | ****** | −16554.2859 | −15714.6375 | 885.0122 | −1184.6577 | 109.3553 | 32.8134 | −186.7683 | −0.014435 |
| | | | 4.9592 | | 13.3487 | −13.5525 | 4.4219 | −128.3075 | −25.6599 | 13.1474 | −4.6996 | 26.9011 | 47.8881 | −19203.1236 | −19783.0291 | −5399.1027 | 355.0758 | −62.9532 | 178.0209 | | |
| | 327 | 5.7009 | 0.9564 | 0.9078 | −740.9573 | −661.7147 | −664.2185 | 430.0343 | 7.2939 | 12.6935 | 28.5207 | 72.6647 | ****** | −16680.0092 | −15909.1424 | −969.2686 | −1609.9506 | −144.1439 | −42.4670 | −280.6230 | −0.061018 |
| | | | 5.8514 | | −13.8708 | −20.7771 | −6.5811 | 111.0503 | 37.0460 | 11.9591 | 1.3096 | 37.6933 | −72.1748 | −19411.1350 | −19505.5483 | −5452.5271 | −442.1145 | −26.2798 | −122.2933 | | |

| | | Dipole moment | | | Quadrupole moment | | | Octapole moment | | | | | Hexadecapole moment | | | | | | | | |
|---|---|---|---|---|---|---|---|---|---|---|---|---|---|---|---|---|---|---|---|---|---|
| | | X | Y | Z | XX | YY | ZZ | XXX | YYY | ZZZ | XYY | XXY | XXXX | YYYY | ZZZZ | XXXY | XXXZ | YYYX | YYYZ | ZZZX | |
| Solvent | Temp | | TOT | | XY | XZ | YZ | XXZ | XZZ | YZZ | YYZ | XYZ | ZZZY | XXYY | XXZZ | YYZZ | XXYZ | YYXZ | ZZXY | | ΔE |
| SWCNT (4,7)-water = methanol | 309 | −11.3885 | 2.5892 | 4.1724 | −791.4335 | −658.6603 | −656.6492 | −1159.8462 | 14.7805 | 41.5695 | −60.4705 | 217.5594 | ****** | −15635.9319 | −14598.9912 | 1975.8405 | 3524.4047 | 105.1303 | −33.2100 | 347.3424 | −0.002968 |
| | | | 12.4020 | | 22.9786 | 42.5160 | −4.0860 | 330.9982 | −36.8268 | 1.2714 | 3.5941 | −72.4616 | 9.0952 | −20204.2667 | −19726.7142 | −5022.9658 | −330.4378 | 58.5911 | 78.3472 | | |
| | 311 | −8.7794 | −0.5031 | −46.0288 | −629.9688 | −637.5319 | −881.0655 | 621.0819 | −1.8798 | −1430.9166 | 125.4003 | 5.3168 | ****** | −15719.3000 | −23256.2842 | 820.4061 | −8014.6728 | 110.0792 | 16.5485 | −5186.7589 | −36.7489 |
| | | | 46.8625 | | 2.6402 | −176.7598 | −2.7032 | −657.4983 | −879.6485 | −39.4302 | −16.5889 | −37.7133 | −162.6187 | −19483.7004 | −22495.5998 | −5325.4211 | 233.3390 | −209.8276 | −84.6308 | | |
| | 313 | 3.9552 | −0.9676 | 1.7856 | −749.1739 | −662.2205 | −661.3851 | 396.3294 | −13.5840 | −8.3486 | 56.0750 | −32.1804 | ****** | −16025.7085 | −14620.9463 | 753.0033 | −856.6283 | 45.7953 | 65.1172 | −65.0932 | −0.079089 |
| | | | 4.4462 | | 8.6236 | −10.6025 | 6.5365 | 131.0510 | −5.6539 | −8.3570 | 22.5930 | −35.2857 | 56.6873 | −20379.5692 | −19925.2837 | −5098.5564 | 410.0442 | −108.0026 | 162.6425 | | |
| | 315 | 4.0446 | 0.4069 | 1.0750 | −736.7582 | −662.0093 | −660.2534 | 298.1491 | 4.5339 | −14.7818 | 43.8839 | −1.5366 | ****** | −15967.6270 | −14362.3729 | 593.7134 | −365.9110 | 79.3137 | 9.7937 | −46.8671 | −0.084662 |
| | | | 4.2048 | | 6.5857 | −7.9183 | 3.2949 | 113.5393 | 23.2824 | −9.8805 | 16.9016 | −38.3850 | 48.0370 | −20208.1694 | −19745.6223 | −5068.8513 | 337.7629 | −64.1165 | 123.7279 | | |
| | 317 | — | — | — | — | — | — | — | — | — | — | — | — | — | — | — | — | — | — | — | — |
| | | | — | | — | — | — | — | — | — | — | — | — | — | — | — | — | — | — | | |
| | 319 | 4.7164 | −0.0275 | 1.2171 | −745.6118 | −662.4539 | −657.9540 | 471.1509 | −7.2400 | −6.7600 | 38.8919 | −2.0844 | ****** | −15950.7850 | −14382.9191 | 1599.3402 | −1293.7610 | 167.2037 | 24.8207 | −71.3836 | −0.091821 |
| | | | 4.8710 | | 19.1851 | −15.4781 | 0.9449 | 128.4669 | 26.0887 | −5.6816 | 15.2897 | −57.9627 | 3.6201 | −20382.0560 | −19760.1100 | −5081.4305 | 97.2160 | −131.0272 | 171.1821 | | |
| | 321 | −5.0672 | −1.5977 | −2.1595 | −743.1634 | −662.5592 | −660.4124 | −444.2160 | −16.3062 | −15.2540 | −35.7271 | −120.9415 | ****** | −16162.1831 | −14457.6679 | −1034.1554 | −1152.3988 | −195.8397 | −28.9555 | −33.2703 | −0.048331 |
| | | | 5.7352 | | −14.3093 | −15.2335 | −2.0273 | −140.6448 | −35.4025 | −11.6477 | −14.6877 | −66.0855 | −14.7993 | −20416.8735 | −19849.4000 | −5107.8802 | −166.1022 | −183.8515 | −134.0981 | | |
| | 323 | −5.8337 | 0.7194 | 0.7425 | −740.5898 | −662.6042 | −661.2422 | −585.2623 | 5.3709 | −4.4880 | −66.7706 | 57.8137 | ****** | −16088.5196 | −14384.6212 | 1641.1430 | 1672.5287 | 207.0356 | −76.4001 | 95.1824 | −0.051009 |
| | | | 5.9246 | | 21.7821 | 21.3857 | −7.3498 | 105.8218 | −27.4808 | 14.7276v | 11.3174 | −58.9476 | −35.1778 | −20404.3031 | −19718.2706 | −5093.6996 | −606.8721 | 131.5597 | 158.6765 | | |
| | 325 | −0.9894 | −1.8957 | −3.1913 | −758.4672 | −661.0518 | −661.8919 | −61.8386 | −9.0721 | −20.1992 | −28.1379 | −151.3141 | ****** | −15709.9809 | −14643.8703 | −1383.8534 | −1307.0699 | −188.2195 | −17.7332 | −163.1191 | 0 |
| | | | 3.8414 | | −17.4762 | −20.3205 | −0.9768 | −251.5513 | −7.4980 | −20.8865 | −20.9040 | −79.6055 | −9.5396 | −20302.6052 | −19999.3695 | −5111.5054 | −69.5930 | −134.3812 | −116.9775 | | |
| | 327 | 5.5830 | −2.3291 | 1.2907 | −750.2504 | −659.6515 | −659.0377 | 515.0852 | −29.5770 | 2.8401 | 54.7252 | −122.0588 | ****** | −15794.1671 | −14524.9466 | 1899.0896 | −1786.3951 | 255.4472 | 67.2278 | −142.0949 | −0.048575 |
| | | | 6.1855 | | 24.3457 | −25.2385 | 8.4008 | 101.6817 | 16.9234 | −12.7722 | 10.3198 | −24.4762 | 61.8170 | −20121.9397 | −19819.0530 | −5071.2223 | 628.7498 | −184.1491 | 177.3505 | | |

Table 5. The calculated dipole moment, quadrupole moment, octapole moment, and hexadecapole moment values of SWCNT

## 4. NMR and IR theoretical study on the interaction of doping metal with carbon nanotube (CNT)

Numerous electrical measurements on SWCNT ensembles have revealed that chemical doping by donors (Li, K, Cs, or Rb) or acceptors ($Br_2$, $I_2$, or acids) decreases the room temperature electrical resistance by up to two orders of magnitude at saturation doping (Kaaoui et al., 1999; Coluci et al. 2006). The important problem is metals passing through cells membrane. Because, there are barriers for them passing through protein canals in cells membrane. Additionally, upon interaction, changes in activity, stability, and solubility ions compatibility may occur in cells. A lot of studies are for replacing protein canals into cells membrane for passing proteins, drug and ions of metal. Therefore the presence of the SWCNT and its consequences to the biological activity of ions metal are of high impact in the development of biosensors, immunoassays and drug delivery systems (Zhang et al., 2005; Ganjali et al., 2006). This work, we used armchair carbon nanotube (5, 5) and (6, 6). Indeed, vibrational frequencies of finite-length carbon nanotubes were recently examined (Tagmatarchris & Prato, 2004) and another result of 319.9 $cm^{-1}$ is consistent with oscillations along the radial directions (radial modes), although it cannot be assessed accurately due to the sensitivity to the number of rings (Yumura et al., 2005). We suggest that SWCNT intercalate into cells membrane replacing protein canals and are studying passing metal ions (Na, Mg, Al, and Si) in length of SWCNT by Quantum Mechanics (QM).

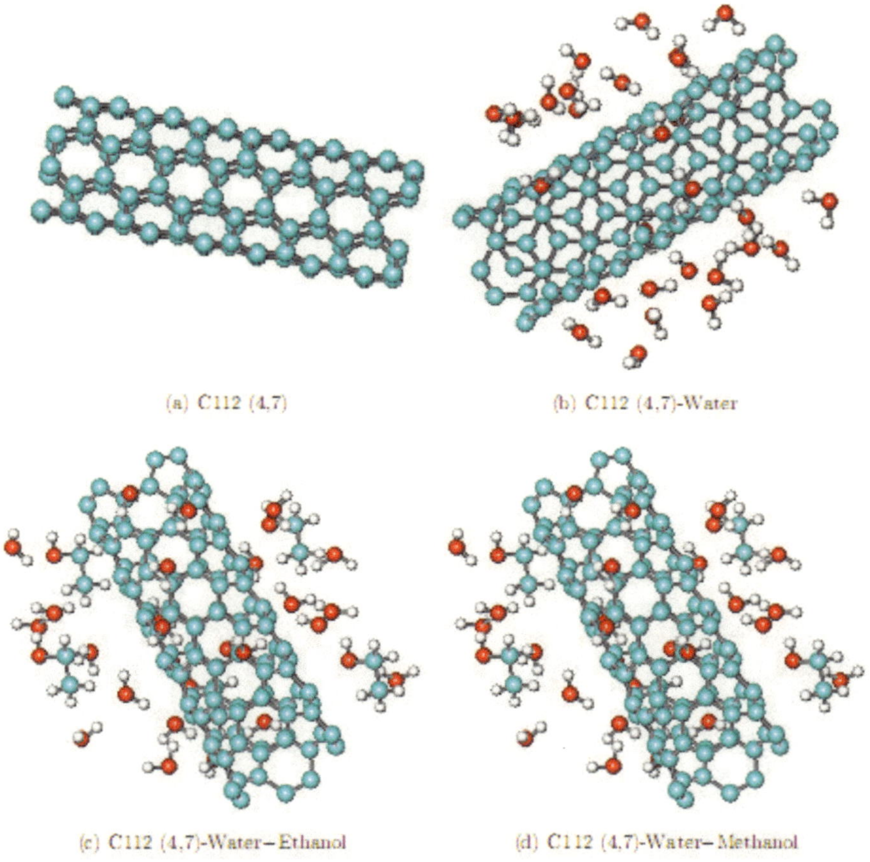


Fig. 9. Optimized structures of nanotube in different media

### 4.1 Computational details

The geometry optimizations were performed using an all-electron linear combination of atomic orbitals Hatree-Fock (HF) and density functional theory (DFT) calculations using the Gaussian A7 package. SWCNTs (100–120) from kind of armchair carbon nanotubes (5, 5) and (6, 6) show in Fig. 10. We are interested in the structural features of single-walled carbon nanotube (SWCNT) in the ground state an atomic and amino acids (His and Ser). In HF theory the energy has from:

$$E_{ks}=v+\langle hp\rangle+1/2\langle P_j(\rho)\rangle-1/2\langle P_k(\rho)\rangle$$

where v is the nuclear repulsion energy, ρ is the density matrix, ⟨*hp*⟩ is the one electron (kinetic plus potential energy). 1/2 $\langle P_j(\rho)\rangle$ is the classical coulomb repulsion of the electrons and -1/2 $\langle P_k(\rho)\rangle$ is the exchange energy resulting from the quantum (fermions) nature of electrons.
In density function theory the exact exchange (HF) for a single determinant is replaced by a more general expression the exchange correlation functional, which can include terms accounting for both exchange energy and the electron correlation, which is omitted from Hartree-Fock theory:

$$E_{ks}=v+\langle hp\rangle+1/2\langle P_j(\rho)\rangle+E_{\chi(\rho)}+E_{C(\rho)}$$

where, $E_{\chi(\rho)}$ is the exchange function and $E_{C(\rho)}$ is the correlation functional. The correlation function of Lee, Yang, and Parr is includes both local and non-local term (Kar et al., 2006). The optimizations of solids are carried out including exchange and correlation contributions using Becks three parameters hybrid and Lee-Yang-Parr (LYP) correlation [B3LYP]; including both local and non-local terms with the program Gaussian A7 package (Lee et al., 1988; Becke, 1993; Becke, 1997).
Compared to Raman spectroscopy, much less information about the vibration properties of carbon nanotubes can be gained from IR spectra. This limitation mainly results from the strong absorption of SWCNTs in the IR range. Accurate predictions of molecular response properties to external fields are of general significance in various areas of chemical physics. This especially refers to the second-order magnetic response properties (NMR), since the magnetic resonance based techniques have gained substantial importance in chemistry and biochemistry that NMR data shown with two parameters isotropic ($\sigma_{iso}$) and an isotropic ($\sigma_{aniso}$) shielding.

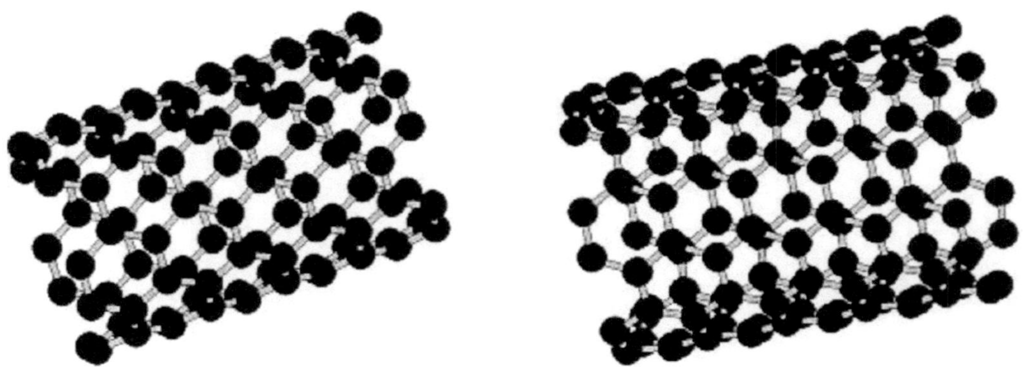

Fig. 10. The optimized configuration Side-view SWCNT: $C_{100}$ (a) and $C_{120}$ (b)

### 4.2 Interaction of Na, Mg, Al, Si with Carbon Nanotube (CNT): NMR and IR Study
The B3LYP and HF by 6-31G and 6-31G* calculaion for the molecular SWCNT models with Na, Mg, Al, and Si considered were validated by the calculated 13C and 1H NMR shifts and thermodynamic properties of an open-ended SWCNT (5, 5) and (6, 6) molecular systems (Monajjemi et al., 2009). The total energy (Etotal) of this interaction is listed in Table 6, which the Etotal increase to converge with an increasing carbon number.

| Energy total (Hartree) | | Na | | Mg | | Al | | Si | |
|---|---|---|---|---|---|---|---|---|---|
| | | HF | B3LYP | HF | B3LYP | HF | B3LYP | HF | B3LYP |
| $C_{100}$ | 6-31G | -2241.98 | -2256.53 | -2401.83 | -2416.63 | -2438.87 | -2453.67 | -2480.80 | -2495.72 | -2527.27 | -2542.30 |
| (5.5) | 6-31G* | -2269.16 | -2284.21 | -2430.75 | -2446.16 | -2468.15 | -2483.58 | -2510.66 | -2526.16 | -2557.67 | -2573.29 |
| $C_{120}$ | 6-31G | -2990.32 | -3009.391 | -3150.15 | -3169.61 | -3187.08 | -3206.55 | -3229.18 | -3248.72 | -3275.58 | -3295.37 |
| (6.6) | 6-31G* | -3026.37 | -3046.26 | -3188.02 | -3208.34 | -3225.34 | -3245.70 | -3267.95 | -3288.37 | -3315.00 | -3335.57 |

Table 6. The total energy calculated in various basis set at HF & B3LYP for SWCNTs (5,5) and (6,6) with ions metal Na, Mg, Al and Si

In this study the metals on the center of a hexagon (HC) and muse are related to competitive interactions between ions metal and SWCNTs. The structural electronic and magnetic properties have been investigated. The most stable configuration for Si adsorbed on SWCNTs is also at the (HC) site at competitive another atoms of SWCNT because the electro negativity is the most great. The calculated amounts of Dipole, Quadrupole, Octapole, and Hexa-decapole moments at the HF and B3LYP levels in various basis set are given in Table 7. Hybridizing Coefficient is different in various methods and basis set.
Calculations of the NMR shifts with the magnetic field perturbation method of GIAO (gauge in dependent atomic orbital) incorporated with the program Gaussian A7 package. The results of the calculations for the carbon nearest neighbors' atoms in SWCNTs are presented in Table 7. The calculated magnetic shielding in Figs. 11, 12 was converted into $\sigma_{iso}$, $\sigma_{aniso}$ chemical shifts by 13C absolute shielding in SWCNT (5, 5). They are worth noting that the last approach leads to a substantial improvement in the calculated magnetic properties. Regarding the method for achievement of gauge invariance for the present case, at the B3LYP and HF levels on the other hand at the hybrid B3LYP level, GIAO is found to be slightly superior. The calculated infrared is for C100 at HF/6–31G with Na, Mg, Al, and Si. They showed in Table 2. The properties thermodynamic are decrease with increase electro negativity atoms.

| | $Cv_{th}$ (cal/mol) | $S_{th}$ (cal/mol) | $H_{th}$ (cal/mol) | $G_{th}$ (cal/mol) | $E_{th}$ (cal/mol) |
|---|---|---|---|---|---|
| $SWCNT_{100}$(5.5) | 104.25 | 146.61 | 245.64 | 201.93 | 245.05 |
| $SWCNT_{100}$-Na | 103.13 | 139.32 | 262.12 | 220.60 | 261.68 |
| $SWCNT_{100}$-Mg | 103.44 | 138.53 | 255.49 | 214.21 | 255.04 |
| $SWCNT_{100}$-Al | 105.85 | 142.81 | 247.06 | 204.50 | 246.61 |
| $SWCNT_{100}$-Si | 105.58 | 143.55 | 243.11 | 200.33 | 242.65 |

Table 7. Calculated thermal energy, thermal enthalpy, total enthalpy, thermal entropy, thermal Gibbs free energy, Gibbs free energy, and heat capacity by IR-HF/6-31G

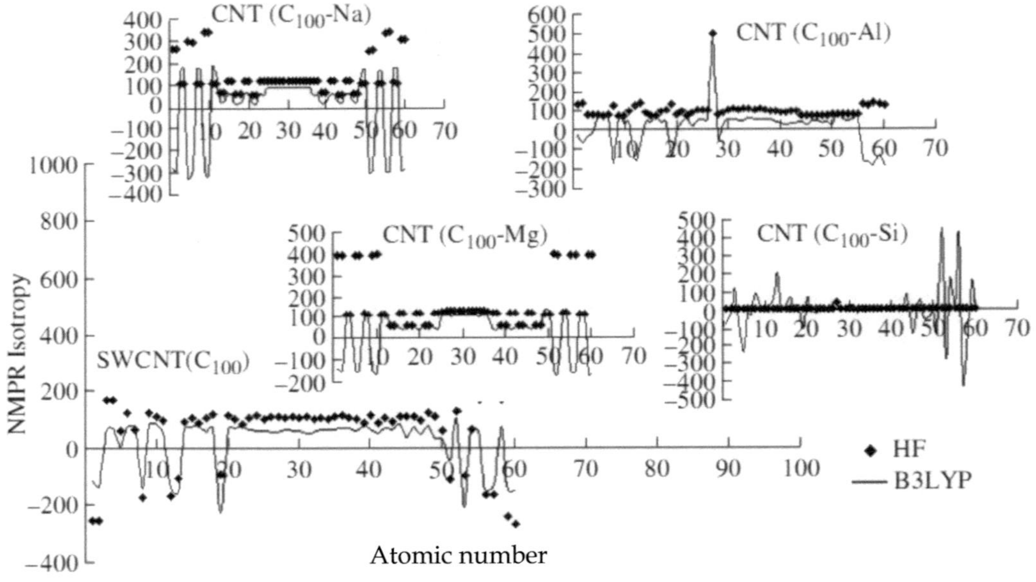


Fig. 11. NMR isotropy diagrams of SWCNT ($C_{100}$) for HF/6–31G (◆) and BLYP/6–31G (–) method

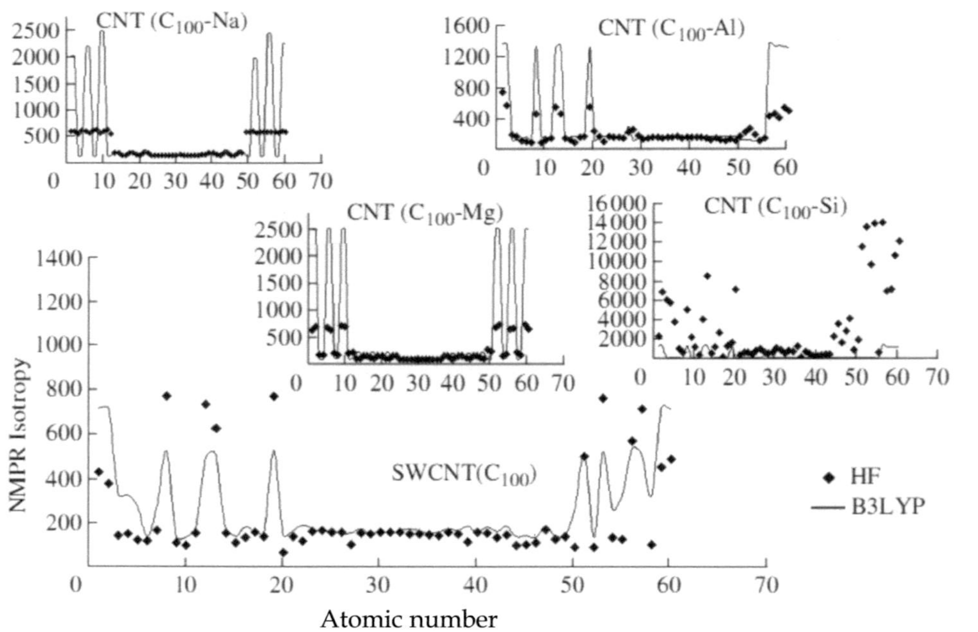


Fig. 12. NMR anisotropy diagrams of SWCNT (C100) for HF/6–31G (◆) and BLYP/6–31G (–) method

## 5. Conclusion

Carbon Nanotubes have been intensively studied due to their importance as building block in nanotechnology. The special geometry and unique properties of carbon Nanotube offer great potential applications, including Nanoelectronic devices, energy storage, gas sensing, chemical probe, electron transport, and biosensors, field emission display, etc. Such devices operate typically on the changes of electrical response characteristics of the Nanowire active component with the application of an externally applied mechanical stress or the adsorption of chemical or bio-molecule. For a better understanding of the physical and electronic properties of single-walled carbon Nanotubes (SWCNTs) at the Nano scale, a challenging task in theoretical calculation is needed in order to design the specific material properties because of the large size of the SWCNTs and their complicated and size dependent electronic structure. Modeling of functionalized Nanotubes and nanostructures for such technologies of SWCNTs can be greatly benefit from the first principles methods based on the density functional theory (DFT). The equilibrium position, adsorption energy, binding energy, charge transfer, and electronic band structures can be computed for different kinds of SWNTs. Effects of surrounding medium and intrinsic structural defects can also be taken into account. In this work we review some recent DFT investigation on the gas-sensing properties and the dielectric properties. Charge transfer and gas-induced charge fluctuation might significantly affect the transport properties of SWNTs. The size and chirality's of the carbon Nanotubes were typical determined from the SWCNT Raman energy spectra of a peak around 150–300 $cm^{-1}$, due to the radial breathing mode. Besides, the geometry and electrical properties of Nanotube are very sensitive to dielectric constants which we can observe from the normal mode analysis. A calculation method for identifying the Raman modes of SWCNTs based on the symmetry of the vibration modes has been discussed. The Raman intensity of each vibration mode varies with polarization direction, and the relationship can be expressed as analytical functions. Each Raman active mode of SWCNT can be distinguished from the group theory principle.

In section 2, with the calculation of the normal modes using the U Matrix it is possible to get the F Matrix from the multiplication of frequency to the U Matrix. Solving the determination of F Matrix versus dielectric can be useful for understanding of the electrical behavior of nanotubes in the quantitative structure activity relationship studies. The geometry and electrical properties of nanotube are very sensitive to dielectric constants. The normal modes also will be changed in the high dielectric constants.

In section 3, *Ab initio* calculations were carried out with GAUSSIAN 98 program at the HF/3-21G level of theory to investigate the effects of polar solvents and different temperatures on the stability of SWCNT in various solvents. The results obtained from Onsager model calculations are illustrated using the energy difference between these conformers which are quite sensitive to the polarity of the surrounding solvent, that the water and methanol solvents can be suggested as the most compatible solvent for studying the structural properties of SWCNT. Also orientation of the water molecules at the CNT-water interface can be affected by the orientation of the water dipole moment. Moreover, among the energy values obtained from different MM+, AMBER, and BIO+ force fields, the AMBER force field is the most proper force field for studying SWCNT.

In section 4, A Quantum Mechanics (QM) is used for investigated the nature of metals transport and interaction with single-walled carbon nanotubes (SWCNTs) inter membranes. Metal species can be transported actively by a combination of SWCNT-membranes conducting channels that have been used for bio-molecular and detection. Ab initio calculations using DFT/B3LYP and HF levels with 6–31G and 6–31G* basis set of theory have allowed the determination of structure electronic, properties thermodynamic, magnetic properties for SWCNTs with Na, Mg, Al, and Si. NMR chemical shielding tensors in the methods framework makes it possible to study the chemical shift of specific group in carbon nanotubes in absence and presence metals. A comprehensive on effects of atoms on SWCNTs were revealed that it is on its electronic structure: 1) transfer of charge from the atom to the SWCNTs; 2) electrostatic interactions between the delocalized e electrons of the SWCNTs and atoms. The basis set used 6–31G and 6–31G* that increasing electronegativity metals increased the total energy. The proportion SWCNTs were changed by them. The results are presented for T = 310 K, the temperature of human's body. In fact, it was determined that SWCNT blocked potassium channels in a dose-dependent manner. Fullerenes were discovered to be less effective channel Blockers than CNT. The mechanism was solely dependent on the size and shape of the nano-particles. They also concluded that electrochemical interactions are between CNT and the ion channels.

## 6. Acknowledgment

The work has been supported by Thailand Research Fund (TRF), Thailand Center of Excellence in Physics (ThEP), Center for Innovation in Chemistry (PERCH-CIC), and the National Research University Project under Thailand's Office of the Higher Education Commission, Thailand for financial support.

## 7. References

Ajayan, P.M.; Stephan, O.; Colliex, C. & Trauth, D. (1994) Aligned carbon nanotube arrays formed by cutting a polymer resin-nanotube composite, *Science,* Vol.265, No.5176 (August 1994), pp. 1212-1214.

Alon, O.E. (2001) Number of raman- and infrared-active vibrations in single-walled carbon nanotubes, *Physical Review B,* Vol.63, pp. 201403-201406.

Alon, O.E. (2003) From spatial symmetry to vibrational spectroscopy of single-walled nanotubes, *Journal of Physics: Condensed Matter,* Vol.15, p. S2489, ISSN 0953-8984.

Alvarino, J.M. (1978) On the complete reduction of representations of infinite point groups, *Journal of Chemical Education,* Vol.55, p. 307.

Alvarino, J.M. & Chamorro, A. (1980) Continuous point groups: A simple derivation of the closed formula for the reduction of representations, *Journal of Chemical Education,* Vol.57, p. 785.

Bachilo, S.M.; Strano, M.S.; Kittrell, C.; Hauge, R.H.; Smalley, R.E. & Weisman, R.B. (2002) Structure-assigned optical spectra of single-walled carbon nanotubes, *Science,* Vol.298, No.5602, (December 2002), pp. 2361-2366.

Bahr, J.L.; Yang, J.; Kosynkin, D.V.; Bronikowski, M.J.; Smalley, R.E. & Tour, J.M. (2001) Functionalization of carbon nanotubes by electrochemical reduction of aryl diazonium salts: a bucky paper electrode, *Journal of the American Chemical Society,* Vol.123, pp. 6536-6542.

Banerjee, S. & Wong, S.S. (2002) Rational sidewall functionalization and purification of single-walled carbon nanotubes by solution-phase ozonolysis, *The Journal of Physical Chemistry B,* Vol.106, pp. 12144-12151.

Barone, V.; Peralta, J.E.; Wert, M.; Heyd, J. & Scuseria, G.E. (2005) Density functional theory study of optical transitions in semiconducting single-walled carbon nanotubes, *Nano Letters,* Vol.5, pp. 1621-1624.

Becke, A.D. (1993) Density-functional thermochemistry. Iii. The role of exact exchange, *The Journal of Chemical Physics,* Vol.98, pp. 5648-5652.

Becke, A.D. (1997) Density-functional thermochemistry. V. Systematic optimization of exchange-correlation functionals, *The Journal of Chemical Physics,* Vol.107, pp. 8554-8560.

Bethune, D.S.; Klang, C.H.; de Vries, M.S.; Gorman, G.; Savoy, R.; Vazquez, J. & Beyers, R. (1993) Cobalt-catalysed growth of carbon nanotubes with single-atomic-layer walls, *Nature,* Vol.363, pp. 605-607.

Cai, L.; Bahr, J.L.; Yao, Y. & Tour, J.M. (2002) Ozonation of single-walled carbon nanotubes and their assemblies on rigid self-assembled monolayers, *Chemistry of Materials,* Vol.14, pp. 4235-4241.

Collins, P.G.; Zettl, A.; Bando, H.; Thess, A. & Smalley, R.E. (1997) Nanotube nanodevice, *Science,* Vol.278, No.5335, (October 1997), pp. 100-102.

Coluci, V.R.; Galvão, D.S. & Jorio, A. (2006) Geometric and electronic structure of carbon nanotube networks: 'super'-carbon nanotubes, *Nanotechnol,* Vol.17, p. 617.

Cotton, F.A. 2nd Ed. (1971) *Chemical application of group theory,* John Wiley & Sons, ISBN 0471175706, New York, USA.

Damnjanovic, M. (1983) Standard components of polar and axial vectors for quasi one-dimensional systems, *Physics Letters A,* Vol.94, pp. 337-339.

Damnjanovic, M.; Miloscaronevic, I.; Vukovic, T. & Sredanovic, R. (1999) Full symmetry, optical activity, and potentials of single-wall and multiwall nanotubes, *Physical Review B,* Vol.60, pp. 2728-2739.

Damnjanovic, M.; Vukovic, T. & Milosevic, I. (2000) Modified group projectors: Tight-binding method, *Journal of Physics A: Mathematical and General,* Vol.33, p. 6561.

Dresselhaus, M.S.; Dresselhaus, G. & Jorio, A. (2006) *Applications of Group Theory to the Physics of Condensed Matter,* Springer, ISBN 978-3-540-32897-1, Heidelberg, Germany.

de Heer, W.A.; Châtelain, A. & Ugarte, D. (1995) A carbon nanotube field-emission electron source, *Science,* Vol.270, No.5239, (November 1995), pp. 1179-1180.

Eklund, P.C.; Holden, J.M. & Jishi, R.A. (1995) Vibrational modes of carbon nanotubes; spectroscopy and theory, *Carbon,* Vol.33, pp. 959-972.

Flurry, R.L. (1979) On the characters and representations of continuous point groups, *Journal of Chemical Education,* Vol.56, p. 638.

Ganjali M.R.; Norouzi, P.; Rezapour, M.; Faridbod, F. & Pourjavid, M. R. (2006). Supramolecular based membrane sensors. *Sensors,* Vol.8, pp. 1018-1086.

Hartschuh, A.; Pedrosa, H.N.; Peterson, J.; Huang, L.; Anger, P.; Qian, H.; Meixner, A.J.; Steiner, M.; Novotny, L. & Krauss, T.D. (2005) Single carbon nanotube optical spectroscopy, *ChemPhysChem,* Vol.6, pp. 577-582.

Herrera, J.E. & Resasco, D.E. (2003) In situ tpo/raman to characterize single-walled carbon nanotubes, *Chemical Physics Letters,* Vol.376, pp. 302-309.

Holzinger, M.; Abraham, J.; Whelan, P.; Graupner, R.; Ley, L.; Hennrich, F.; Kappes, M. & Hirsch, A. (2003) Functionalization of single-walled carbon nanotubes with (r-)oxycarbonyl nitrenes, *Journal of the American Chemical Society,* Vol.125, pp. 8566-8580.

Huang, J.Y.; Chen, S.; Ren, Z.F.; Chen, G. & Dresselhaus, M.S. (2006) Real-time observation of tubule formation from amorphous carbon nanowires under high-bias joule heating, *Nano Letters,* Vol.6, pp. 1699-1705.

HyperChem 7.0 (2001), Hypecube Inc., Florida, USA.

Jorio, A.; Saito, R.; Hafner, J.H.; Lieber, C.M.; Hunter, M.; McClure, T.; Dresselhaus, G. & Dresselhaus, M.S. (2001) Structural ( n, m) determination of isolated single-wall carbon nanotubes by resonant raman scattering, *Physical Review Letters,* Vol.86, pp. 1118-1121.

Jorio, A.; Fantini, C.; Pimenta, M.A.; Capaz, R.B.; Samsonidze, G.G.; Dresselhaus, G.; Dresselhaus, M.S.; Jiang, J.; Kobayashi, N.; Gr; uuml; neis, A. & Saito, R. (2005) Resonance raman spectroscopy (n,m) -dependent effects in small-diameter single-wall carbon nanotubes, *Physical Review B,* Vol.71, pp. 075401-075412.

Kane, C.L. & Mele, E.J. (1997) Size, shape, and low energy electronic structure of carbon nanotubes, *Physical Review Letters,* Vol.78, pp. 1932-35.

Kar, T.; Akdim, B.; Duan, X. & Pachter, R. (2006) Open-ended modified single-wall carbon nanotubes: A theoretical study of the effects of purification, *Chemical Physics Letters,* Vol.423, pp. 126-130.

Kazaoui, S.; Minami, N.; Jacquemin, R.; Kataura, H. & Achiba, Y. (1999) Amphoteric doping of single-wall carbon-nanotube thin films as probed by optical absorption spectroscopy, *Physical Review B,* Vol.60, pp. 13339-13342.

Kürti, J.; Zólyomi, V.; Kertesz, M.; Sun, G.; Baughman, R.H. & Kuzmany, H. (2004) Individualities and average behavior in the physical properties of small diameter single-walled carbon nanotubes, *Carbon,* Vol.42, pp. 971-978.

Lee, C.; Yang, W. & Parr, R.G. (1988) Development of the colle-salvetti correlation-energy formula into a functional of the electron density, *Physical Review B,* Vol.37, pp. 785-789.

Lee, V.; Nimmanpipug, P.; Mollaamin, F.; Kungwan, N.; Thanasanvorakun, S. & Monajjemi, M. (2009) Investigation of single wall carbon nanotubes electrical properties and normal mode analysis: Dielectric effects, *Russian Journal of Physical Chemistry A, Focus on Chemistry,* Vol.83, pp. 2288-2296.

Liu, J.C. & Monson, P.A. (2005) Molecular modeling of adsorption in activated carbon: Comparison of monte carlo simulations with experiment, *Adsorption,* Vol.11, pp. 5-13.

Martinez; M, T.; Callejas; M, A.; Benito; A, M.; Cochet; M; Seeger; T; Anson; A; Schreiber; J; Gordon; C; Marhic; C; Chauvet; O; Maser & W. K. (2003) Modifications of single-wall carbon nanotubes upon oxidative purification treatments, *Nanotechnology,* Vol.14, 691-695.

Maultzsch, J.; Telg, H.; Reich, S. & Thomsen, C. (2005) Radial breathing mode of single-walled carbon nanotubes: Optical transition energies and chiral-index assignment, *Physical Review B,* Vol.72, pp. 205438-205454.

Mickelson, E.T.; Huffman, C.B.; Rinzler, A.G.; Smalley, R.E.; Hauge, R.H. & Margrave, J.L. (1998) Fluorination of single-wall carbon nanotubes, *Chemical Physics Letters,* Vol.296, pp. 188-194.

Monajjemi, M.; Baei, M. & Mollaamin, F. (2008a) Quantum mechanics study of hydrogen chemisorptions on nanocluster vanadium surface, *Russian Journal of Inorganic Chemistry,* Vol.53, pp. 1430-1437.

Monajjemi, M.; Mahdavian, L. & Mollaamin, F. (2008b) Characterization of nanocrystalline silicon germanium film and nanotube in adsorption gas by Monte Carlo and Langevin dynamic simulation, *Bulletin of the Chemical Society of Ethiopia,* Vol.22, No.2, pp. 277-286, ISSN: 1011-3924.

Monajjemi, M.; Mahdavian, L.; Mollaamin, F. & Khaleghian, M. (2009) Interaction of na, mg, al, si with carbon nanotube (cnt): Nmr and ir study, *Russian Journal of Inorganic Chemistry,* Vol.54, pp. 1465-1473.

Monajjemi, M.; Khaleghian, M.; Tadayonpour, N. & Mollaamin, F. (2010) The Effect of Different Solvents and Temperatures on Stability of Single-Walled Carbon Nanotube: a QM/MD Study, *International Journal of Nanoscience,* Vol.9, pp. 517-529.

Mora-Diez, N.; Senent, M.L. & Garcia, B. (2006) Ab initio study of solvent effects on the acetohydroxamic acid deprotonation processes, *Chemical Physics,* Vol.324, pp. 350-358.

Nardelli, M. B.; Yakobson, B.I. & Bernholc, J. (1998) Mechanism of strain release in carbon nanotubes, *Physical Review B,* Vol.57, pp. R4277-4280.

Odom, T.W.; Huang, J.-L.; Kim, P. & Lieber, C.M. (1998) Atomic structure and electronic properties of single-walled carbon nanotubes, *Nature,* Vol.391, pp. 62-64.

Ouyang, M.; Huang, J.-L. & Lieber, C.M. (2002) Fundamental electronic properties and applications of single-walled carbon nanotubes, *Accounts of Chemical Research,* Vol.35, pp. 1018-1025.

Pelletier, M.J. (1999) *Analytical Applications of Raman Spectroscopy,* Kaiser Optical Systems Inc., ISBN 0632053054, Michigan, USA.

Peng, H.; Alemany, L.B.; Margrave, J.L. & Khabashesku, V.N. (2003) Sidewall carboxylic acid functionalization of single-walled carbon nanotubes, *Journal of the American Chemical Society,* Vol.125, pp. 15174-15182.

Pfeiffer, R.; Kuzmany, H.; Kramberger, C.; Schaman, C.; Pichler, T.; Kataura, H.; Achiba, Y.; Kürti, J., and Zólyomi, V. (2003) Unusual high degree of unperturbed environment in the interior of single-wall carbon nanotubes, *Physical Review Letters,* Vol.90, p. 225501.

Saito, R.; Dresselhaus, G. & Dresselhaus, M.S. (1992a) Topological defects in large fullerenes, *Chemical Physics Letters,* Vol.195, pp. 537-542.

Saito, R.; Fujita, M.; Dresselhaus, G. & Dresselhaus, M.S. (1992b) Electronic structure of graphene tubules based on $C_{60}$, *Physical Review B,* Vol.46, pp. 1804-1811.

Saito, Y.; Hamaguchi, K.; Hata, K.; Uchida, K.; Tasaka, Y.; Ikazaki, F.; Yumura, M.; Kasuya, A. & Nishina, Y. (1997) Conical beams from open nanotubes, *Nature,* Vol.389, pp. 554-555.

Saito, R.; Takeya, T.; Kimura, T.; Dresselhaus, G. & Dresselhaus, M.S. (1998) Raman intensity of single-wall carbon nanotubes, *Physical Review B,* Vol.57, pp. 4145-4153.

Spire, T. & Brown R. M. (February 1996). High Resolution TEM Observations of Single Walled Carbon Nanotubes,Availble from http://www.botany.utexas.edu/facstaff/facpages/mbrown/ongres/tspires/nano.htm.

Strano, M.S.; Dyke, C.A.; Usrey, M.L.; Barone, P.W.; Allen, M.J.; Shan, H.; Kittrell, C.; Hauge, R.H.; Tour, J.M. & Smalley, R.E. (2003) Electronic structure control of single-walled carbon nanotube functionalization, *Science,* Vol.301, No.5639, (September 12, 2003), pp. 1519-1522.

Schafer, L. & Cyvin, S.J. (1971) Complete reduction of representations of infinite point groups, *Journal of Chemical Education,* Vol.48, p. 295.

Strommen, D.P. & Lippincott, E.R. (1972) Comments of infinite point groups, *Journal of Chemical Education,* Vol. 49, p. 341.

Strommen, D.P. (1979) Some additional comments on infinite point groups, *Journal of Chemical Education,* Vol.56, p. 640.

Tagmatarchis, N. & Prato, M. J. (2004) Mater. Chem. Vol.14, No.4, p. 437.

Tiana, G.; Sutto, L. & Broglia, R.A. (2007) Use of the metropolis algorithm to simulate the dynamics of protein chains, *Physica A: Statistical Mechanics and its Applications,* Vol.380, pp. 241-249.

Umek, P.; Seo, J.W.; Hernadi, K.; Mrzel, A.; Pechy, P.; Mihailovic, D.D. & Forró, L. (2003) Addition of carbon radicals generated from organic peroxides to single wall carbon nanotubes, *Chemistry of Materials,* Vol.15, pp. 4751-4755.

Wang, W. & Skeel, R. D. (2003) Analysis of a few numerical integration methods for the Langevin equation, *Molecular Phys*ics Vol.101, No.14, pp. 2149-2156.

Wilder, J.W.G.; Venema, L.C.; Rinzler, A.G.; Smalley, R.E. & Dekker, C. (1998) Electronic structure of atomically resolved carbon nanotubes, *Nature,* Vol.391, pp. 59-62.

Wilson, E.B.; Decius, J.C. & Cross, P.C. (1955). *Molecular vibrations : The theory of infrared and raman vibrational spectra,* McGraw-Hill, New York, USA

Witanowski, M.; Biedrzycka, Z.; Sicinska, W. & Grabowski, Z. (2002) A study of solvent polarity and hydrogen bonding effects on the nitrogen nmr shieldings of n-nitramines and ab initio calculations of the nitrogen shieldings of c-nitro, n-nitro and o-nitro systems, *Journal of Molecular Structure,* Vol.602-603, pp. 199-207.

Yumura, T.; Nozaki, D.; Bandow, S.; Yoshizawa, K. & Iijima, S. (2005) End-cap effects on vibrational structures of finite-length carbon nanotubes, *Journal of the American Chemical Society,* Vol.127, Aug 24, pp. 11769- 11776, ISSN 0002-7863.

Zhang, M.; Fang, S.; Zakhidov, A.A.; Lee, S.B.; Aliev, A.E.; Williams, C.D.; Atkinson, K.R. & Baughman, R.H. (2005) Strong, transparent, multifunctional, carbon nanotube sheets, *Science,* Vol.309, August 19, pp. 1215-1219, ISSN 1095-9203.

Zhou, Z.; Steigerwald, M.; Hybertsen, M.; Brus, L. & Friesner, R.A. (2004) Electronic structure of tubular aromatic molecules derived from the metallic (5,5) armchair single wall carbon nanotube, *Journal of the American Chemical Society,* Vol.126, pp. 3597-3607.

# 27

# Single Wall Carbon Nanotubes in the Presence of Vacancies and Related Energy Gaps

Edris Faizabadi
*School of Physics, Iran University of Science and Technology, Tehran*
*Iran*

## 1. Introduction

Carbon nanotube family is one of the most important elements in nanotechnology. High ratio of surface to volume of nano materials cause to appear nanotechnology. This matter is one of the most important properties of produced materials at nano scales. At this scale, materials begin replace their bulky behavior with surface one. Some of physical relations that are used for ordinary materials, also abandoned. In fact, at this scale, laws of quantum physics play a key role and it will be possible to control special properties of material such as melting point, magnetic behavier, charge capacitance and even colour of material with no change in their chemical properties. This text is concentrated on some most important carbon nanotubes.

Carbon atoms can form chemical bonds by hybridizing the atomic orbitals of their valence bonds and assume many structural forms such as graphite, diamond, carbon fibers, fullerenes, and carbon nanotubes. Carbon nanotubes (CNTs) discovered by Sumio Iijima in 1991 [Iijima,1991], are one of the most exciting quasi-1-D solids that exhibit fascinating electrical, optical, and mechanical properties such as high current density, large mechanical stiffness, and field emission characteristics [Choi et al., 1999; Saito & Dresselhaus ,1998]. These properties of CNTs enable a wide range of applications in the various fields such as electron emission [Bonard et al., 1999; Dresselhaus et al., 2001; poole et al. 2003], energy storage [Meyyappan , 2005, Chambers et al., 1998], composites [Dresselhaus et al., 2001; Meyyappan et al., 2005], solar cells [Lee, 2005; Pradhan et al., 2006; Wei et al., 2007], nanoprobes and sensors [Dai et al., 1996], and biomedicine [Sinha et al., 2005].

A single-wall carbon nanotube (SWCNT) is a graphene sheet rolled into a cylindrical shape with a diameter of about 0.7 - 2.0 nm [Saito et al.,1998], but A multiwall carbon nanotube (MWCNT) comprises a number of graphene sheets rolled concentrically with an inner diameter of about 5 nm [Harris, 2005 ]. Since the aspect ratio of the carbon nanotube cylinders (length/diameter) is as large as 104-105 [Saito et al., 1998], these nanotubes can be considered as one-dimensional nanostructures.

CNTs according to their structures are classified to three types of armchair, zig zag, and chiral [Saito et al., 1998]. The terms 'zigzag' and 'armchair' refer to the arrangement of hexagons around the circumference. Armchair and zigzag nanotubes are defined by a carbon nanotube whose mirror image has an identical structure to the original one. On the contrary, Chiral nanotubes in which the hexagons are arranged helically around the tube axis, exhibit a spiral symmetry whose mirror image cannot be superposed on to the original one [Saito et al., 1998 & Harris, 2009].

SWCNTs can be either metallic or semiconducting, depending on their diameter and chirality. All (n, n) armchair nanotubes yield 4n energy subbands with 2n conduction and 2n valence bands. All armchair nanotubes have a degenerated band between the highest valence band and the lowest conduction band, where the bands cross the Fermi level. Thus, all armchair nanotubes are expected to be metallic. For a general (n, 0) zigzag nanotube, if n is a multiple of 3, the nanotube becomes metallic as the energy gap at k=0 becomes zero; however, if n is not a multiple of 3, the nanotube becomes semiconducting because an energy gap which is proportional to the nanotube diameter opens at k=0 [Saito et al., 1998].
One of the important things which plays essential role on electronic properties, is the energy gap which can inform us about metallic and semi-metallic properties of nanotubes. Variety of probes predict that armchair single-wall carbon nanotubes are always metallic [Dresselhaus et al., 1996] and all the other tubes (zigzag and chiral), depend on whether they satisfy n-m=3I or not (where I is an integer), are metallic or semi-metallic [Wildöer et al., 1998]. In the pervious dedicates, some calculations and experiments have been focused on how we can change the electronic properties of SWCNTs for achieving new nanoelectronic [Andriotis & Menon,2007; Lee et al. ,2007; Tans et al. , 1998] and spintronic [Meyyappan, 2005, Chambers , 1998; Lee ,2005, Pradhan and Batabyal, 2006] devices. Some of them have used the external fields such as electric, magnetic, and radiation fields. In these works, the Probes have demonstrated that the presence of a magnetic field perpendicular to the nanotube axis induces a metal-in, 2005, ulator transition for the metallic (9, 0), (12, 0) . . . nanotubes in the absence of disorder and semiconducting nanotubes can become metallic with increasing magnetic strength [Wei et al., 2007]. Some other works showed that for zigzag tubules (n, 0), the gap varies linearly with stress and independently of diameter, so a uniaxial-stress applied parallel to the axis of carbon nanotubes can significantly modify the band gap and induce a semiconductor-metal transition [Dai et al., 1996].
Some other investigations used structural defects such as substitutional disorders, vacancies, and adatoms on SWCNTs [Dresselhaus et al., 1996; Harris , 2005; Sinha et al., 2005]. Among the various kinds of structural defects in SWCNTs, effects of vacancies have been studied more recently [Wildöer, 1998]. Experimental observations showed that carbon atoms in carbon nanotubes can be released under electron or ion irradiation [Ajayan et al., 1998; Zhu et al., 1999] effectively, leaving vacancies in SWCNTs behind. Lately, Yan Li et al. have showed that by symmetry breaking in armchair carbon nanotubes, metal-semiconductor transition can be occurred. They estimate the band gap opening as a function of both the external potential strength and the nanotube radius and suggest an effective mechanism of metal-semiconductor transition by combination of different forms of perturbations [Li et al., 2004].
In addition, the recent probes have illustrated that vacancies can change the electronic properties of SWCNTs, converting some metallic nanotubes to semiconductors and semiconducting ones into metals, also for most of SWCNTs, the electronic properties strongly depend on the configuration of vacancies [Yuchen et al., 2004]. The influence of vacancy defect density on electrical properties of armchair SWCNTs was investigated and the results showed that there is no simple correlation between mono-vacancy defect density and band gap [Tien et al., 2008].
Quite recently, by considering the vacancies as substitutional disorders we investigate effects of vacancy percentage on the energy gap of different zigzag SWCNTs, by using Green's function technique and self-consistent coherent potential approximation (CPA) method [Faizabadi, 2009].

## 2. Electronic density of states

The electronic density of states (DOS) is the number of available electrons for a given energy interval. DOS of a crystalline solid, which dramatically depends on the dimension of the system, is fundamental in describing the electronic transport, electrical, optical, thermal, and mechanical properties of the solid [Lu, Pan, 2004, Zhu et al., 1933]. The density of states enters in the experimental study, the application of the electronic properties, and computation of some useful quantities of a system such as electrical resistance and conductance.The helicity or local symmetry of CNTs, along with the diameter which determines the size of the repeating structural unit introduces significant changes in the electronic density of states, and hence provides a unique electronic character for the nanotubes [Hansson et al., 2000]. The DOS of semiconducting carbon nanotubes near the Fermi level located at E=0 is zero, but the DOS of metallic nanotubes near the E=0 is non-zero. In addition, the DOS of zigzag and armchair carbon nanotubes shows van Hove singularities whose numbers are consistent with the number of hexagons around the circumference of the (n, m) nanotube.

## 3. Synthesis methods of carbon nanotubes

Generally, three techniques are being used to produce carbon nanotubes including the electric arc discharge, the laser ablation, and the chemical vapor deposition (CVD). In each of techniques it is possible we have some vacancies in produced carbon nanotube structures. In the next sections, the method of evaluating the effects of vacancies in the DOS of these structures are discussed, here a brief explanation on Synthesis methods of carbon nanotubes is presented.

### 3.1 Electric arc discharge

The electric arc discharge method as one of the first methods to produce CNTs, employs a chamber filled with an inert gas, two electrodes of pure graphite rods, and a DC power supply. A current of about 50–100A passed though the electrodes causes carbon atoms are vaporized from the graphite anode in the form of crystallites and are deposited on the cathode electrode in the form of small carbon clusters. Next, these carbon clusters rearrange themselves into a tubular shape forming the MWCNTs, which drift toward the cathode and deposit on its surface. MWNTs are obtained when a pure graphite target is used and SWNTs when the target is a mixture of graphite and metallic catalysts such as Ni-Co or Ni-Y mixtures. However a cylindrical and homogenous deposit forms on the cathode with a quite high rate, the presence of "unwanted" graphite crystallites that do not form into nanotubes is the limiting factor for high yield CNTs [Peter, Harris, 2009, Dervish et al., 2009].

### 3.2 Laser ablation

The laser ablation or the evaporation method was introduced by Smalley and his coworkers in 1995. In this technique, a powerful laser is used to ablate a carbon target in an inert atmosphere at low pressure. The laser beam incident results in evaporating carbon from the graphite. The carrier gas sweeps the carbon atoms from the high-temperature zone to a cold copper collector on which they condense into nanotubes. In order to generate SWCNTs using the laser ablation technique, it is necessary to impregnate the graphite target with transition metal catalysts. It is experimentally found that the SWCNT growth time in this technique is only a few milliseconds long. Generally, along with SWCNTs and MWCNTs, fullerenes, amorphous carbon, and other carbon by-products are produced while using the laser ablation technique. MWCNTs produced by this method have a number of layers varying from 4 to 24 and an inner diameter ranging between 1.5 and 3.5nm. Unfortunately, the laser ablation methode is very expensive because it involves high-purity graphite rods and high power lasers [Dervish, 2009; Loiseau et al., 2006].

### 3.3 Chemical vapor deposition (CVD)

Chemical vapor deposition (CVD) is a relatively slow method that produces long CNTs quantities. In this method, CNTs are synthesized by taking hydrocarbons (the commonly used sources are methane, ethylene, and acetylene) and using an energy source, such as electron beam or resistive heating, to impart energy to them. At first, the energy source heats The hydrocarbon at high temperatures, typically between 700 and 1000 ˚C, in the presence of catalytic systems (usually a first-row transition metal such as Ni, Fe, or, Co) and breaks it into hydrogen and carbon. Then, carbon atoms dissolve and diffuse into the metal surface and rearrange themselves into a network containing hexagons of carbon atoms and finally precipitate out in the form of CNTs. Higher yields of MWCNTs and SWCNTs are produced when the catalytic system is composed of two different metals. CNTs produced with the CVD method can be scaled up for large scale and high-quality at a relatively low cost. In addition, the growth of CNTs can be controlled by adjusting the reaction parameters such as the catalyst system, temperature, type of hydrocarbon, and the flow rate of the gases. Another advantage of the CVD method is that it enables the deposition of CNTs on pre-designed lithographic structures, producing ordered arrays of CNTs [Dervish, 2009].

## 4. Coherent potential approximation (CPA)

In 1963, Davis, Langer and Klauder introduced the coherent potential approximation to solve the single-site scattering problem [Yoyozawa,1968]. Then, several authors [Onodera & Yoyozawa, 1968; Soven 1960] used it to calculate DOS in a disordered substitutional alloy in which random elements have short-range potential. In this approximation, random disordered system would be replaced by an effective ordered system which is chosen so that the average of the scattering amplitudes for the actual atoms placed in the effective medium vanishes. Besides, according to this theory, the behavior of a single site or cell specifies the manner of the whole system.

To calculate the averaged Green's function, the single electron Green's function can be expressed as follows,

$$G(\omega) = \frac{1}{\omega - H}, \tag{1}$$

here H is the Hamiltonian of the system. In order to use the multiple scattering theory, H is better to divide into the unperturbed or reference Hamiltonian K and the perturbation term v as,

$$H = v + K. \tag{2}$$

Also, the reference Green's function P is written as,

$$P(\omega) = \frac{1}{\omega - K}, \tag{3}$$

Using P, the t scattering matrix associated with the atom n in the effective medium becomes,

$$t_n = v_n\left[1 - pv_n\right]^{-1}. \tag{4}$$

Next, the average of the total scattering matrix T related to the G given by

$$G = P + PTP, \tag{5}$$

Can be expressed as

$$< T >= \sum_n < T_n > = \sum_n < t_n(1 + P\sum_{m \neq n} T_m) >. \tag{6}$$

On the other hand, the effective Hamiltonian can be defined as,

$$H_{eff} = K + < T > (1 + p < T >)^{-1}, \tag{7}$$

or

$$H_{eff} = K + \sum_n < t_n > (1 + P < t_{n>})^{-1}. \tag{8}$$

Within the single site approximation, the condition of $< t_n >=0$ for any n leads to $< T >= 0$ and as a result $< G >= P$. It means that additional scattering due to a real atom in the CPA environment is averagely zero. Thus, average scattering caused by an impurity is equal to the scattering caused by the environment. In this condition, the effective Hamiltonian is equal to K. Since the K is not exactly specified, it can be computed by the self consistent method.

## 5. Infinite zigzag single wall carbon nanotubes in the presence of vacancies

The single-wall carbon nanotube in tight bonding model is described by [Economou, 2006, Heyd, 1997],

$$\hat{H} = \sum_{i\alpha} \left| i_\alpha \right\rangle \varepsilon_i^\alpha \left\langle i_\alpha \right| + \sum_{ij\alpha\beta} \left| i_\alpha \right\rangle \gamma_{i,j}^{\alpha\beta} \left\langle j_\beta \right| , \tag{9}$$

where $\alpha$ and $\beta$ refer to the $A$ or $B$ sublattices, $\varepsilon_i^\alpha$ refer to on-site energies, and $\gamma_{i,j}^{\alpha\beta}$ is the hopping integral between sites $i$ and $j$ in $A$ or $B$ sublattices. Since we consider the nearest neighbor interaction, the hopping integrals $\gamma_{i,j}^{AA}$ and $\gamma_{i,j}^{BB}$ take zero values while the off-diagonal hopping integrals, $\gamma_{i,j}^{AB}$ and $\gamma_{i,j}^{BA}$, take a constant value $\gamma_0$. The on-site energies associated with the two different sites $A$ and $B$ of a graphene sheet have the same values. We neglect the effect of wall curvature in Hamiltonian. The complete set of $\{\left| i_\alpha \right\rangle\}$ is the atomic wave functions localized on site $i$. in $A$ or $B$ sublattices. The Bloch wave functions of the tube are given by [Heyd, 1997],

$$\left\langle \boldsymbol{X} \middle| \boldsymbol{K} \right\rangle = \frac{1}{\sqrt{N}} \sum_j e^{i\boldsymbol{K}.\boldsymbol{R}_j} \left\langle \boldsymbol{X} \middle| j \right\rangle \tag{10}$$

Here, we just consider the electronic properties of π- bonds since every carbon atom has four valance electrons, one 2$s$ electron and three 2$p$ electrons and the overlap between the $p_z$ wave function with the $s$ or the $p_x$ and $p_y$ electrons is strictly zero, we can treat the $p_z$ electrons, form π-bonds of graphene, independently of the other valance electrons. There are also some high symmetric points in Brillouin zone which are located near the Fermi level, so many of the electronic properties of SWCNTs are related to this bonds [Wallace, 1972]. The associated Green's function operator can be described as [Economou, 2006],

$$\hat{G} = \left[\hat{I}z - \hat{H}\right]^{-1} \tag{11}$$

So the local density of states can be written as

$$DOS(E) = \mp\frac{1}{\pi}\mathrm{Im}\left[Tr\left\{\hat{G}(E \pm is)\right\}\right] = \mp\frac{1}{\pi}\mathrm{Im}\left[\sum_{m}\hat{G}^{(m)}(E \pm is)\right] \tag{12}$$

where $m$, the number of allowed band, is obtained by the half number of at0oms in the first Brillouin zone and $s$ is an infinitesimal number for convergence. We present a schematic plot of our system which is a typical zigzag SWCNT in the presence of vacancies as substitutional disorders in Fig. 1.

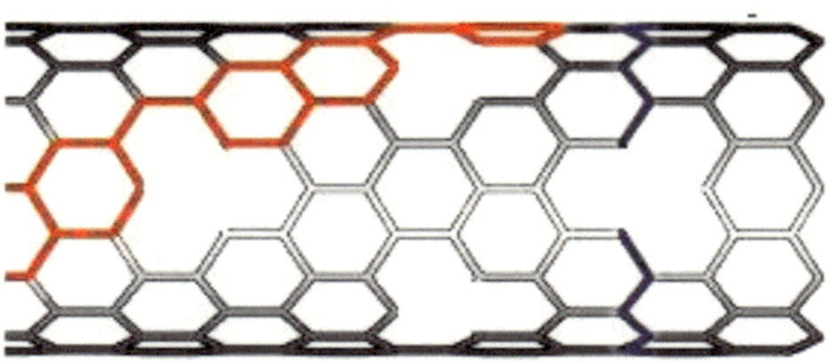

Fig. 1. The schematic plot of zigzag single-wall carbon nanotube with random vacancies

On the other hand, we suppose in the presence of vacancies the structure remain the same as in the absence of them except that some carbon atoms are replaced by holes.
By considering this assumption, in order to investigate effects of vacancy percentage on density of states of SWCNTs, we use coherent potential approximation. As mentioned before, in CPA, the Green's function for a system of scatterers can be written in operator notation as [Györffy, 1972; Korringa & Mills, 1972]:

$$\tilde{G} = G + GTG \tag{13}$$

where $\tilde{G}$ is the Green's function of real nanotube and $G$ is the average effective Green's function and $T$ is the total scattering matrix of the system. The scattering matrix of the system may be written as [Györffy, 1972]:

$$T = \sum_{i,j=1}^{N} t_i\delta_{i,j} + \sum_{i,j=1}^{N}\sum_{k\neq j} T_{ik}Gt_j \;, \tag{14}$$

where $T_{ik}$ is the scattering-path operator which have been introduced by Györffy for the first time and $t_i$ is the $t$ matrix that describes the scattering from the potential on the $i$-th site which can be written as,

$$t_i = V_i(1 - GV_i)^{-1}, \tag{15}$$

Here, $V_i$ is the extra potential that is caused the scattering $i$-th site with respect to an effective medium. The effective medium is introduced by the self energy $\xi$ which can be determined by $\langle T \rangle = 0$. Using the CPA, we impose the condition that for any site $\langle t_i \rangle = 0$ which means the extra effective scattering due to the atom at site $i$ vanishes and the following self-consistent equations can be achieved [Datta & Thakur, 1994],

$$G^{(m)\alpha}(E \pm is) = \frac{\Omega}{2\pi N} \times \lim_{s \to 0^+} \int_{(m)band} \frac{E \pm is}{(E \pm is)^2 - \left| \sum_{ij\alpha} e^{iK.(R_i - R_j)} \gamma_{i,j}^{\alpha} \right|^2 - \xi^{(m)}} \tag{16}$$

and

$$\xi^{(m)}(E \pm is) = \frac{-1}{2G^{(m)}(E \pm is)} + \left[ \frac{1}{4G^{(m)2}(E \pm is)} + \frac{\varepsilon_c}{G^{(m)}(E \pm is)} - \frac{\eta \varepsilon_c}{G^{(m)}(E \pm is)} \right]^{\frac{1}{2}}, \tag{17}$$

where $\varepsilon_c$ is the on-site energy of carbon atoms, $\eta$ is the vacancy percentage, and $\xi^{(m)}$ is the self-energy.

By using the method described above, we investigate the density of states of zigzag (12, 0) SWCNT. All calculations are performed at zero temperature and we use $\varepsilon_c$= 1 eV and $\gamma_0$= 2.75 eV. The density of states of (12, 0) SWCNT as a function of energy in units of $\gamma_0$ are depicted in Fig. 2 for $\eta$ = 0. The energy gap of (12, 0) SWCNT in the absence of vacancy, $\eta$ = 0, is $E_g$= 0.02 eV that is in agreement with local density approximation (LDA) calculations [Miyake & Saito,1998]. Saito et al. showed that by just using the π tight-binding method there is no gap in the band structure, but this is not shown in our calculations. The associated experimental probes showed that the energy gap of (12, 0) SWCNT is 0.04±0.004 eV [Ouyang et al., 2001]. The energy gap of three zigzag SWCNTs is compared in Table 1. In regard to Table 1, by increasing SWCNT's index or increasing the diameter of SWCNTs, our results approach to the experimental results of Min Ouyang et al. The difference is as a result of neglecting the effect of wall curvature in Hamiltonian as mentioned before.

| $(n, 0)$ SWCNT | $E_g$ (eV) | $E_g$ (eV) from Ref. |
|---|---|---|
| (9, 0) | 0.022 | 0.080±0.005 |
| (12, 0) | 0.02 | 0.04±0.004 |
| (15, 0) | 0.02 | 0.029±0.004 |

Table 1. Comparison between our computational energy gaps with experimental energy gaps from Ref. [Ouyang, 2001 ].

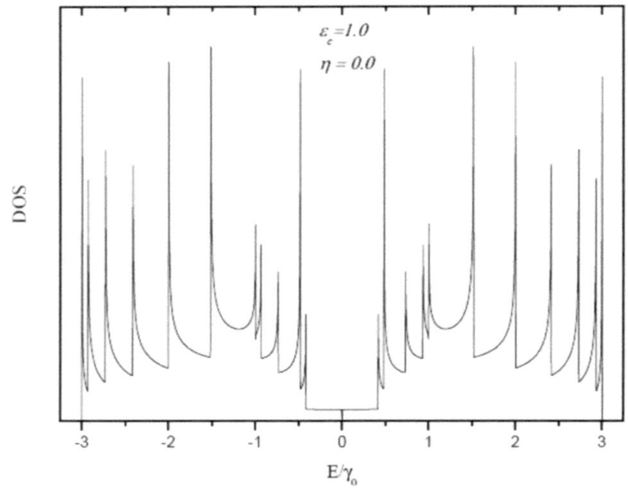


Fig. 2. Density of states for a pure (12, 0) SWCNT

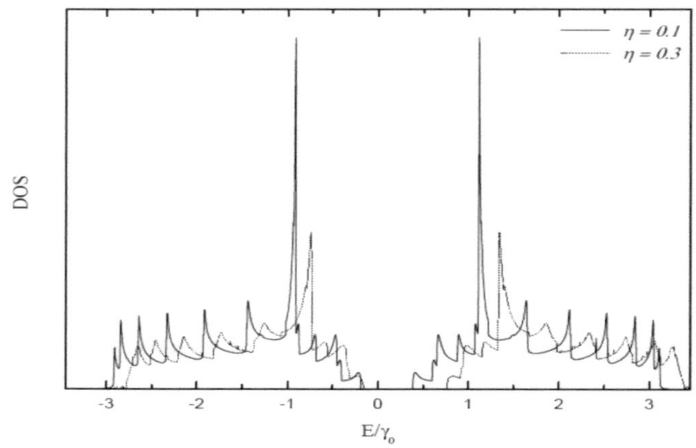


Fig. 3. Density of states for (12, 0) SWCNT in the presence of vacancy for $\eta$ = 0.1(Solid line), $\eta$ = 0.3 (Dot line).

The band gap opening is occurred, as it is depicted in Fig. 3 for $\eta$ = 0.1, 0.3. According to this figure, the maximum value of the peak of one-dimensional (1D) Van Hove singularities in greater values of vacancy concentrations become less, also at greater values of vacancy concentrations, SWCNTs lose their 1D characteristics and become similar to a kind of 2D disordered graphene sheet. In realistic, by increasing the value of vacancy concentrations, the freedom of valance electrons becomes higher, so we expect the feature of 2D structure. Also there is a small shift in DOS because of influence of vacancy percentage on the on-site energy of carbon atoms. By increasing the value of vacancy concentrations, as it is depicted in Fig. 3, the number of peaks becomes less. In addition, by increasing vacancy percentage on (12, 0) SWCNT the energy gap is increased. This fact can be realized by considering the broken lattice symmetries. By breaking symmetry, the valance electrons do not obey Bloch order, so multiple scattering is occurred, then we have gap opening and metallic SWCNTs is changed to semi-metallic SWCNTs.

The variations of the energy gap versus vacancy percentage is shown in Fig. 4(a), 4(b), and 4(c) for (7, 0), (8, 0), (12, 0) SWCNTs, respectively. The results show that, by increasing vacancy percentage for all three types of zigzag SWCNTs, the energy gap is also increased. This treatment is realized by multiple scattering that is occurred in vacancies. Moreover, the slopes of the graphs for the small values of vacancy percentages are not the same and at low vacancy percentages, the growth of gap opening versus vacancy percentage for (12, 0) SWCNT is higher than the other ones. Also by considering Fig 4(a), 4(b), and 4(c), we realize that for $\eta = 0$, the energy gaps of (7, 0), (8, 0), and (12, 0) SWCNTs, are 1.19, 1.12, and 0.02 eV, respectively. Furthermore, in special vacancy percentage, the energy gap of (7, 0) SWCNT is higher than the other ones. We can realize this result by considering the number of carbon atoms in the unit cell of SWCNTs and the diameter of SWCNTs. The number of carbon atoms in the unit cell of $(n, 0)$ SWCNTs, $q$, is derived by $q = 2n$, so the number of carbon atoms in the unit cell of (7, 0), (8, 0), (12, 0) SWCNTs are 14, 16, and 24, respectively. Thus the number of carbon atoms and diameter of (7, 0) SWCNT is less than the other ones, so in special vacancy percentage, the affects of breaking symmetries and backscattering are more dominant respect to the others.

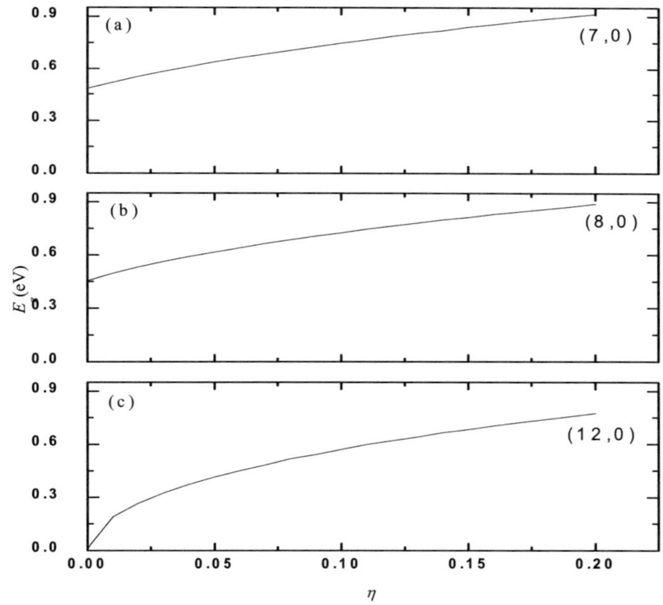


Fig. 4. The energy gap versus vacancy percentage for (a) (7, 0) SWCNT, (b) (8, 0) SWCNT, (c) (12, 0) SWCNT.

The energy gap of different kinds of zigzag SWCNTs as a function of nanotube index $n$ for three different values of vacancy percentage 0, 0.01, and 0.15 is displayed in Fig. 5. This figure shows that for all zigzag SWCNTs the energy gap is also increased by increasing vacancy percentage. Thus metallic to semi-metallic transition is occurred for metallic zigzag SWCNTs that is considered by Yuchen Ma et al. [Yuchen, 2004] by using spin-polarized density functional theory. They investigated the influence of configuration mono-vacancy on electronic properties of zigzag SWCNTs and showed for six kinds of zigzag SWCNTs with $n$ = 5… 10, the vacancies can change the electronic properties of SWCNTs, converting some metallic nanotubes to semiconductors. Also they found that only (5, 0) SWCNT is always semiconductor by changing configuration of mono-vacancy and here according to Fig. 5, we see that by increasing the value of vacancy percentage, (5, 0) SWCNT remain semiconductor. However, this treatment also occurs for all kinds of semiconductor zigzag SWCNTs such as (7, 0), (10, 0), and so on.

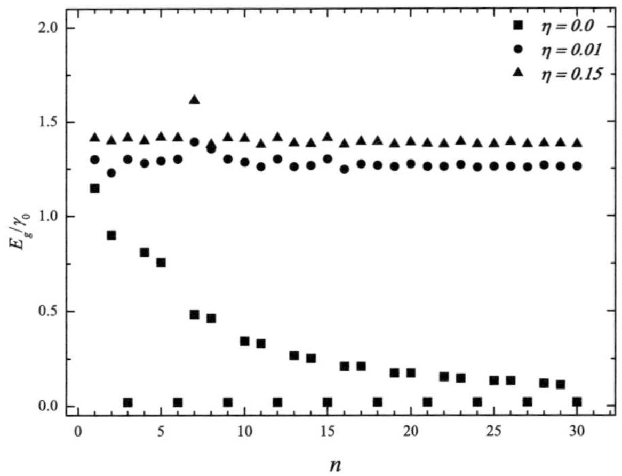


Fig. 5. The energy gap versus the zigzag nanotube index for three values of vacancy percentage $\eta$ = 0.0, 0.01, 0.15.

In summary by using CPA method, effects of vacancy percentage on the energy gap of zigzag SWCNTs were investigated. It is found that the energy gap of pure (12, 0) SWCNT is 0.02 eV that is in agreement with LDA calculations and the experimental work of M. Ouyang et al., Also by increasing the concentration of vacancies, the energy gap is increased and a metallic to semi-metallic transition is occurred for metallic ($n$, 0) zigzag SWCNTs. For semiconductor zigzag SWCNTs any transition does not appear. As mentioned before, at large enough vacancy concentrations, the SWCNTs lose their 1D characteristics and the associated density of states become similar to a kind of 2D disordered graphene sheet. In addition, our calculations show that the maximum energy gap is increased by decreasing zigzag nanotube index ($n$) which is due to decreasing diameter of SWCNT and therefore increasing the effect of vacancies in back scattering. Therefore by creation vacancies in the structure of zigzag SWCNTs the metallic to semi-metallic transition are occurred. In addition, it is possible to achieve the special semiconductor SWCNT with a predetermined energy gap in order to use in semiconductor industry.

## 6. Semiconductor finite zigzag carbon nanotubes in the presence of vacancies

Carbon nanotubes show promise for applications in future electronic systems, and the performance of carbon nanotube devices, in particular, has been rapidly advancing. Semiconducting nanotubes are suitable for transistors. In order to correctly treat carbon nanotube transistors, strong quantum confinement around the tube circumferential direction, quantum tunneling through Schottky barriers at the metal/nanotube contacts, and quantum tunneling and reflection at barriers in nanotube channel need to be considered by the non-equilibrium Green's function (NEGF) formalism. We study the density of states of *(n, 0)* carbon nanotubes in which an atom has been removed to produce a vacancy[Faizabadi et al.,2009](Fig 6). Vacancy is one of the most common defects in carbon nanotubes and affects the physical properties of them. The on-site energy of the vacancy element is equaled high value, essentially repelling the carriers from that particular site.

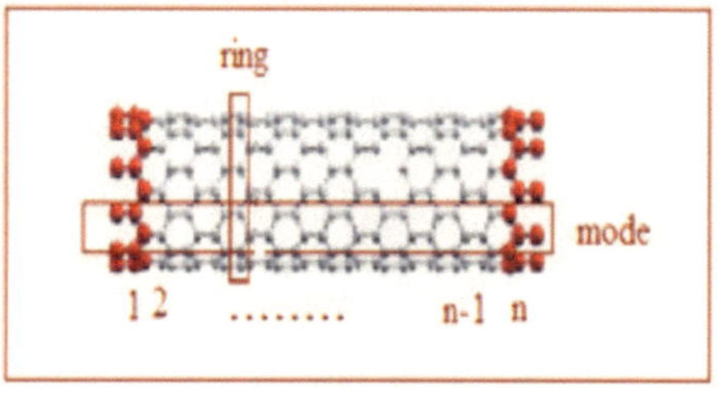


Fig. 6. The schematic of a finite zigzag carbon nanotube in presence a single vacancy. The modes are aligned along the carbon nanotube axis and rings that are around the cylindrical circumference.

Our system consists of a carbon nanotube that coupled to the left L and right R lead described by the following Hamiltonian

$$H = \sum_{i} \varepsilon_i \left| i >< i \right| + \sum_{i \neq j} t_{i,j} \left| i >< j \right| \tag{18}$$

where, $E_i$ is the on-site energy for each carbon atom in the carbon nanotube and is the hopping between the atomic sites which are supposed to be non zero for the nearest neighbors. In this model there are orthogonal basis, corresponding to a spherically symmetric local orbital at each atomic site in the system.
The total Hamiltonian for a system consist of a finite carbon nanotube which is connected into two leads, the left and right ones is

$$\mathrm{H} = \mathrm{H_L} + \mathrm{H_R} + \mathrm{H_C} \tag{19}$$

where, $\mathrm{H_{L(R)}}$ describes the Hamiltonian for left and right leads. $\mathrm{H_C}$ refers to the Hamiltonian for central region. An effective method to describe scattering and transport is the Green's function that is determined by the equation

$$\left[(\mathrm{E} \pm i\eta)\mathrm{I} - \mathrm{H}\right]\mathrm{G_C^{r,a}} = \mathrm{I} \tag{20}$$

here, r and a refer to retarded and advanced Green's function. For an isolated non interacting system the Green function is obtained after the matrix inversion.
The Green function coupled to the leads is determined by the expression

$$\mathrm{G_C^r(E)} = [(\mathrm{E} + i\eta)\mathrm{I} - \mathrm{H_C} - \Sigma_\mathrm{L} - \Sigma_\mathrm{R}]^{-1} \tag{21}$$

where, $\Sigma_{\mathrm{L,R}}$ is called self-energy that describe the interaction between leads and system

$$\Sigma_{\mathrm{L(R)}}(\mathrm{E}) = \tau_{\mathrm{L(R)C}}\, g_{\mathrm{SL(R)}}\, \tau^{\dagger}_{\mathrm{L(R)C}} \tag{22}$$

where, $g_{SL(R)}$ is the surface green function of the decoupled left and right leads that is calculated in the absence of the scattering region and different from the $G_C^r(E)$.
The density of states (DOS) as a function of the incident electron energy for a carbon nanotube is [*Datta,*1995]

$$DOS = -\frac{1}{\pi}\,\mathrm{Im}\left[\mathrm{Trace}\left(G_C^r(E)\right)\right] \tag{23}$$

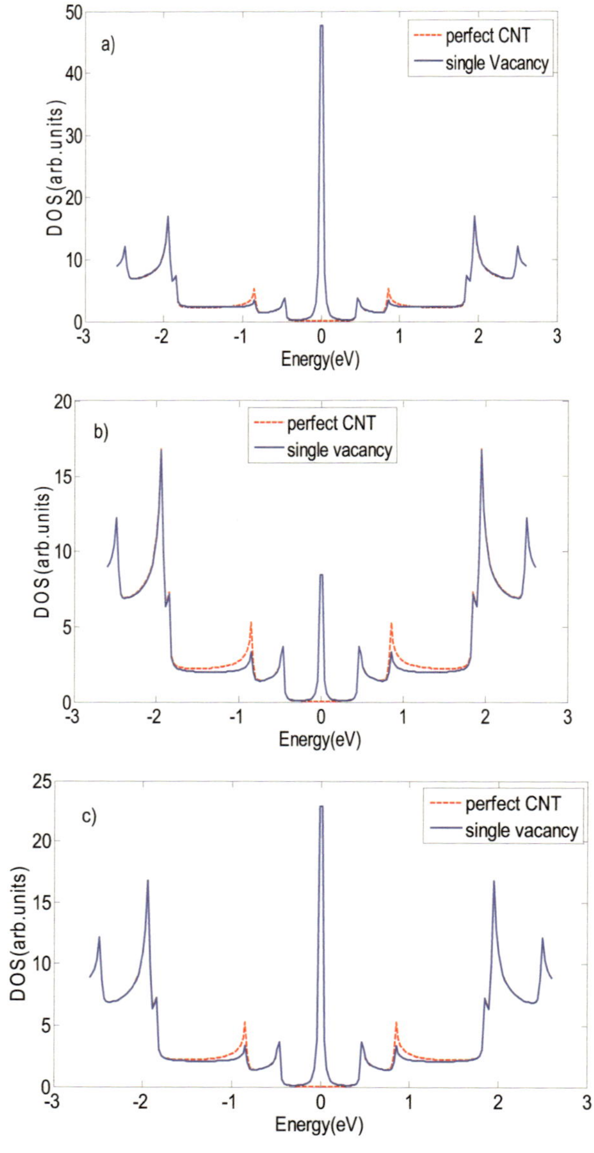


Fig. 7. The density of states of finite zigzag carbon nanotubes (11,0) in presence of the single vacancy in a: 2th ring b: 3th ring c: 4th ring of 4th mode

The hopping parameter for the nearest neighbour is fixed to $V_{pp\pi} = t = -2.66\ [eV]$ and the on-site energies for this system are set to zero.
By using this model the effect of vacancies on the density of states of finite zigzag carbon nanotubes (11, 0), is investigated. This kind of carbon nanotube is semiconductor with an energy gap around the Fermi energy.
In Fig 7 the density of states of finite zigzag carbon nanotubes (11, 0) in presence of the single vacancy is calculated. In these figures we consider the vacancy site in the same mode and different rings. The presence of vacancies breaks the symmetry of the system and creates available states around zero energy, in prohibited band gaps. The density of states around the Fermi level rises due to the broken $\pi$ bands at the vacancy site. These states filled with electrons around the Fermi level that in perfect carbon nanotubes there aren't any states in Fermi energy.
In other work the density of states of finite zigzag carbon nanotubes in the channel of (11, 0) in presence of two vacancies is investigated that they are in different rings and modes (Fig 8 and Fig9). The two vacancies are placed at different sites in the channel. In these figures the states that they are near the Fermi level filled with electrons. The vacancies decrease the density of states of finite semiconductor zigzag carbon nanotubes totally.

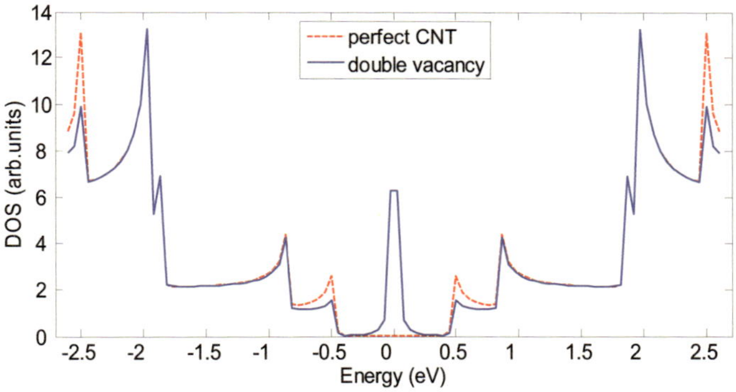


Fig. 8. The density of states of finite zigzag carbon nanotubes (11, 0) in presence of two vacancies in 5th ring of 2th and 8th modes

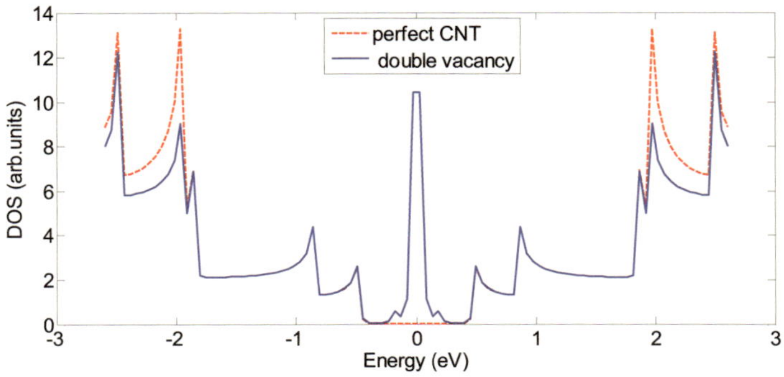


Fig. 9. The density of states of a finite zigzag carbon nanotube (11, 0) in presence of two vacancies in 2th and 7th rings of 6th mode

## 7. Finite armchair carbon nanotubes in the presence of vacancies

Here by using the method which described in the previous section the density of states of finite armchair carbon nanotubes for different finite length are probed [Orlikowski et al.,2007]. Besides the effect of two vacancies on different length of the finite armchair carbon nanotubes (6, 6), is investigated [Faizabadi & Heidaripour, 2010]

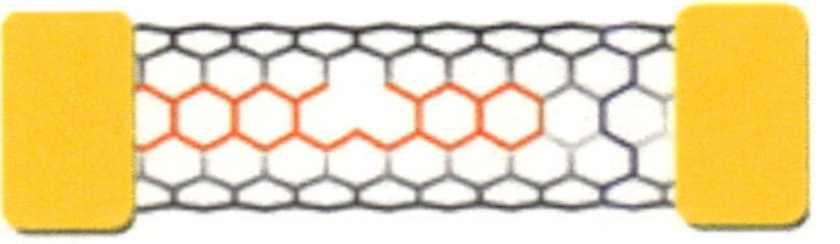


Fig. 10. The finite armchair carbon nanotube with metallic leads in the presence of a vacancy.

Our Probes show that the presence of two vacancies that they are near each other has little affect on the density of states at Fermi energy (Fig 11). When two vacancies are far from each other, by breaking $\pi$ orbital of atoms, create available states around the Fermi energy and increase the density of states (Fig12).
The vacancies by breaking the system symmetry create available states around zero energy and omit the carbon atomic potential in carbon nanotube. The $\pi$ bands break at the vicinity of the vacancy site and density of states around the Fermi level rises and electrons filled the states.

## 8. Carbon nanotube quantum dot in the presence of vacancies

Carbon nanotubes of different chiralities can be joined through a small number of pentagon heptagon defects, forming a nano scale metal/semiconductor or metal/metal carbon nanotube heterojunction (HJ). Carbon nanotube heterojunctions (HJs), which continuously connect nanotubes of different chiral structure using a small number of atomic scale defects, represent the ultimate scaling of electronic interfaces [Chico et al.,1996].
We consider a quantum dot (QD) that is built of zigzag $(2n,0)$ and armchair $(n,n)$ carbon nanotubes connected by n pair of pentagon- heptagon interface defect. The interface itself has n fold rotational symmetry and by connecting the armchair and zigzag nanotubes together a quantum dot is created. We denote this kind of structure by $m(2n,0)/l(n,n)/m(2n,0)$, where $(2n,0)$ on both sides of the quantum dot used as the contacts, one armchair carbon nanotube is central part of the system with specific length. To understand this peculiar behaviour, we plot the structure of a carbon nanotube quantum dot in Fig13.
The junction between two single wall Carbon nanotubes is a pentagon–heptagon (p-h) pair that is the smallest topological defect with minimal local curvature and zero net curvature (Fig 14).

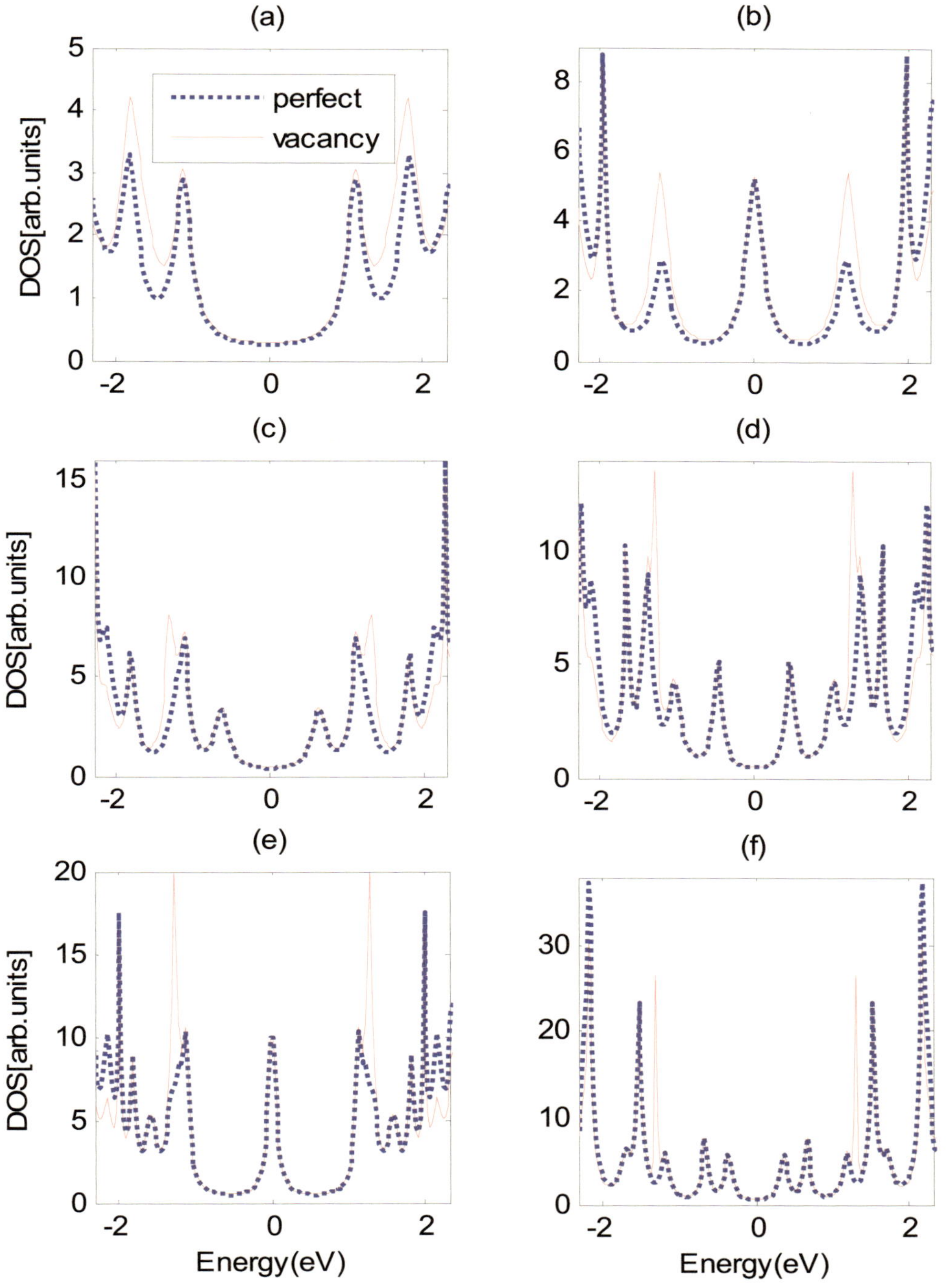


Fig. 11. The density of states of an armchair carbon nanotube (6,6) in the presence of two vacancies which they are near each other for different length of carbon nanotube a: 3rings b: 5 rings c: 7 rings d: 9 rings e: 11 rings f: 13 rings

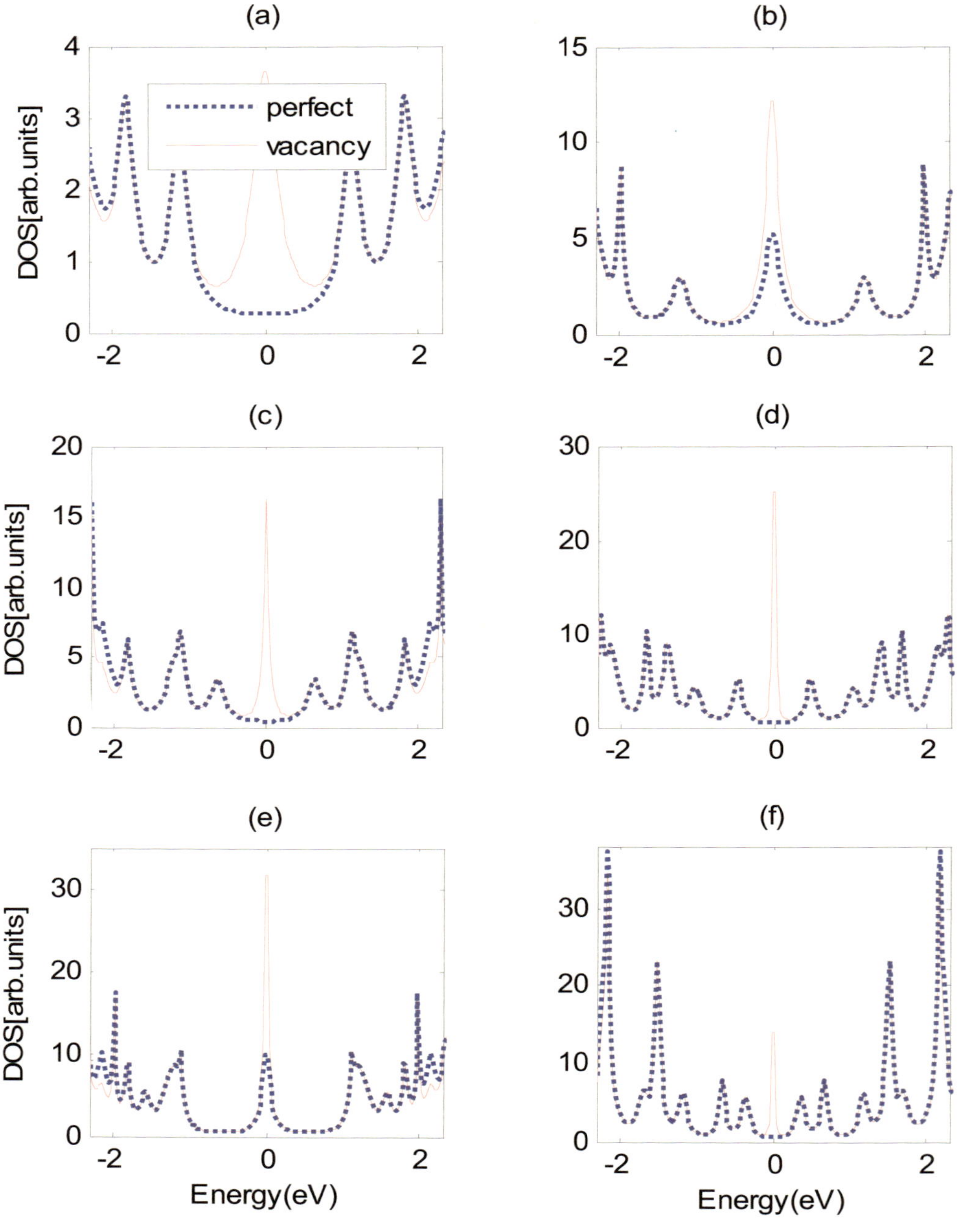


Fig. 12. The density of states of an armchair carbon nanotube (6,6) in the presence of two vacancies which they are far from each other for different length of carbon nanotube a: 3 rings b: 5 rings c: 7 rings d: 9 rings e: 11 rings f: 13 rings

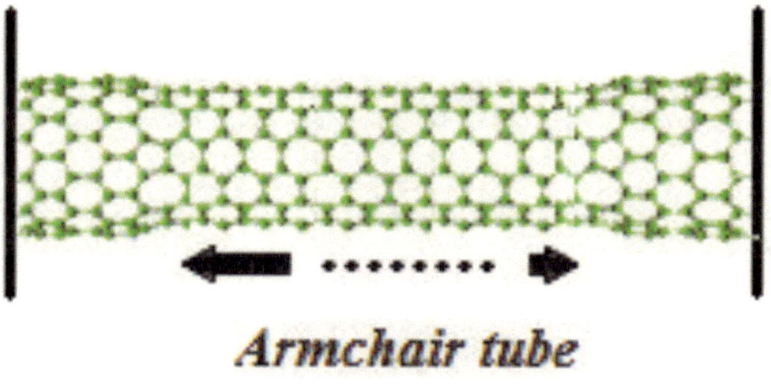


Fig. 13. Schematic view of a carbon nanotube quantum dot.

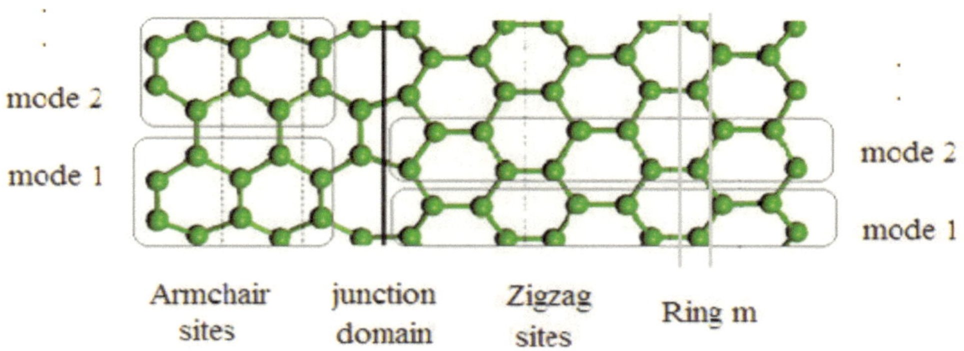


Fig. 14. Schematic representation of Armchair and zigzag modes and rings in real space, a pentagon-heptagon ring as a junction between zigzag and armchair carbon nanotubes quantum dots are shown.

We work in a single π-band tight-binding approximation. The Hamiltonian in the site representation is

$$H = \sum_{r} \varepsilon_r C_r^{\dagger} C_r + \sum_{r \neq s} t_{rs} C_r^{\dagger} C_s \tag{24}$$

where r, s denote carbon sites, C and $C^{\dagger}$ the electron annihilation and creation operators, $\varepsilon_r$ the on-site energies, and $t_{rs}$ the nearest-neighbors hopping integral between the atomic sites . The on-site energies are set equal zero, while t = −2.66 eV as for graphene which yields the Fermi energy at *0 eV*. It is well known that this approximation gives good description of the band structure of pure nanotubes around the Fermi energy [Reich, et al., 2002].

$$H_{QD} = \begin{bmatrix} H_L & \tau_{LC} & 0 \\ \tau_{LC}^{\dagger} & H_C & \tau_{CR}^{\dagger} \\ 0 & \tau_{CR} & H_R \end{bmatrix} \tag{25}$$

where, $H_{QD}$ is the Hamiltonian for the quantum dot where $H_{C,L,R}$ refers to the Hamiltonian for left and right leads and the Hamiltonian for central region between two leads and there are some terms $\tau_{C(R,L)}$ describe the coupling between zigzag and armchair carbon nanotubes. The density of states of the system is

$$DOS = -\frac{1}{\pi} Im\left[Trace\left(G_{QD}^{r}(E)\right)\right] \tag{26}$$

$G_{r,a}$ are the retarded and advanced Green functions inside the system, taking into account the coupling with the electrodes via the self-energies $\Sigma_L$ and $\Sigma_R$

$$G_{QD}^{r}(E) = [(E + i\eta)I - H_{QD} - \Sigma_L - \Sigma_R]^{-1} \tag{27}$$

where, $\Sigma_{L(R)}$ is the self energy due to the left (right) lead and is defined by

$$\Sigma_{L(R)}(E) = \tau_{L(R)}\, g_{s\,L(R)}^{r}\, \tau_{L(R)}^{\dagger} \tag{28}$$

where, $\tau_{L(R)}$ is the coupling matrix between the system and the left (right) lead, and $g_{s\,L(R)} = [\epsilon - H_{L(R)}]^{-1}$, is the Green's function of the semi-infinite leads which we evaluate using an iterative procedure [Datta,1995a, 1995b].

We investigate, the density of states of a single wall Carbon nanotube $m(2n,0)/\ l(n,n), m(2n,0)$ quantum dot with several armchair length with a method combining the Green's Function formalism with tight binding model. By increasing the length of armchair carbon nanotube in metal-metal system $m(12,0)/l(6,6)/m(12,0)$ and metal-semiconductor system $m(8,0)/l(4,4)/m(8,0)$, we observe changing in the density of states and increment of oscillation of the density of states. In the presence of the single vacancy and omission of a carbon atomic potential, we observe small changes in the density of states of the system (Fig15 and Fig16).

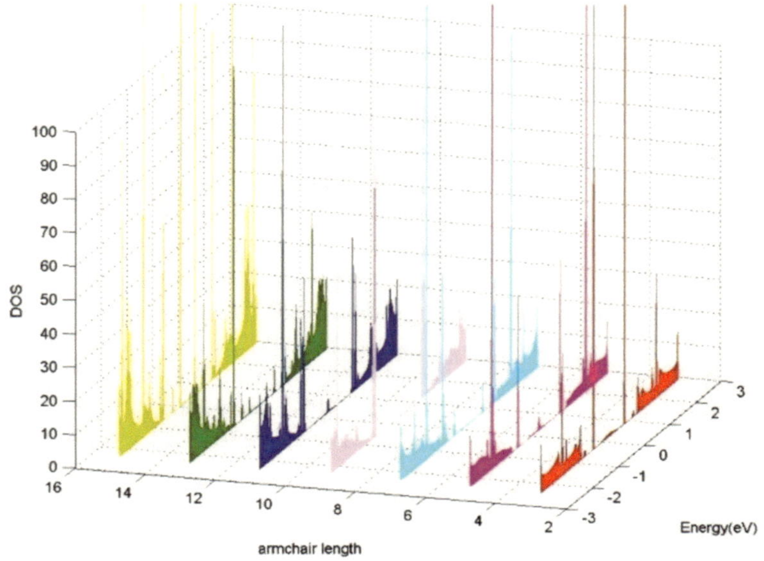


Fig. 15. The density of states of a single wall Carbon nanotube quantum dot with several armchair length

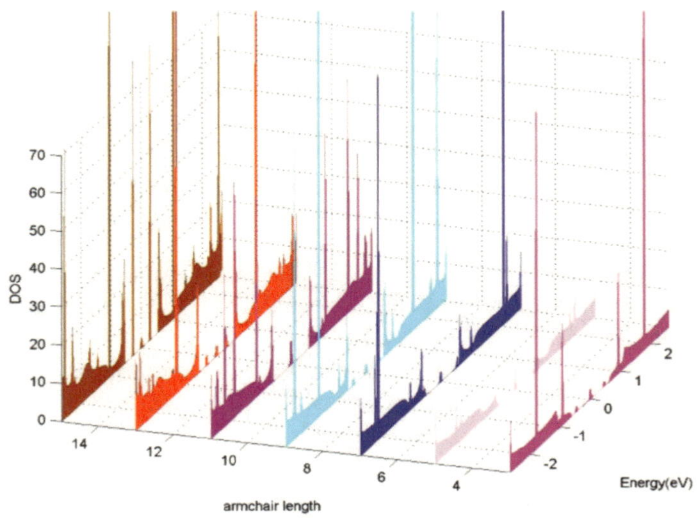


Fig. 16. The density of states of a single wall Carbon nanotube quantum dot with several armchair length.

## 9. References


Antonis N. Andriotis, Madhu Menon, "Structural and conducting properties of metal carbon-nanotube contacts: Extended molecule approximation",Phys. Rev. B 76 (2007) 045412.

Ajayan, P.M., Ravikumar, V., and J. -C. Charlier, "Surface Reconstructions and Dimensional Changes in Single-Walled Carbon Nanotubes", Phys.Rev. Lett. 81 (1998) 1437.

Bonard, J.-M., J.-P. Salvetat, T. Stöckli, L. Forró, and A. Châtelain. . "applications and clues to the emission mechanism" Appl. Phys. A., 69: 245, (1999).

Chambers, A., C. Park, R. T. K. Baker, N. M. Rodriguez, J., Chem. B., "Hydrogen Storage in Graphite Nanofibers", Phys. 102 (22): 4253, (1998).

Choi, W. B., Chung, D. S., Kang, J. H., Kim, H. Y., Jin, Y. W. ,I. T. Han, Y. H. Lee, J. E. Jung, N. S. Lee, G. S. Park, and J. M. Kim, "Fully sealed, high-brightness carbon-nanotube field-emission display," Appl.Phys. Lett., 75, 20, 3129–3131, (1999).

Dai, H. J., Hafner, J. H., Rinzler, A. G., Colbert, D. T., Smalley, R. E., "Nanotubes as nanoprobes in scanning probe microscopy", Nature, 384, 147, (1996).

Datta A. , Thakur, P. K., "The coherent-potential approximation in the tight-binding linearized-muffin-tin-orbital formalism for a single-band model of a solid, "Phys. Condens Matter 6 (1994) 4707.

Dresselhaus, M. S., Dresselhaus, G., Eklund, P. C., "Science of Fullerenes and Carbon Nanotubes", Academic, San Diego, 1996.

Dresselhaus, M. S., G. Dresselhaus, and P. Avouris, "Carbon Nanotubes: Synthesis, Structure, properties, and Applications," Berlin: Springer, (2001).

Economou, E. N. ,Green's Functions in Quantum Physics, Third ed., Springer press, 2006.

Faizabadi , E., Bagheri, A.,"Effects of vacancy percentage on the energy gap of zigzag single-wall carbon nanotubes", Physica E 41 (2009) 1828–1831

Faizabadi, E., Heidaripour, F., Kargar, Z., Saleh Kootahi, M.,"Effect of vacancies on the density of states of finite zigzag carbon nanotubes", Proceeding of Annual Physics Conference of Iran, 2009.

Faizabadi, E., Heidaripour, F., "Effect of two vacancies on the density of states of finite armchair carbon nanotubes", Proceeding of Annual Physics Conference of Iran, 2010

Gun-Do Lee, Cai-Zhuang Wang, Jaejun Yu, Euijoon Yoon, Nong-Moon Hwang, and Kai-Ming Ho, Phys. "Formation of carbon nanotube semiconductor-metal intramolecular junctions by self-assembly of vacancy defects", Rev. B 76 (2007) 165413.

Györffy, B. L. , Phys. Rev. B 5 (1972) 2382. " Coherent-Potential Approximation for a Nonoverlapping-Muffin-Tin-Potential Model of Random Substitutional Alloys ",Phys. Rev. B 5, 2382–2384 (1972)

Hansson, A., Paulsson, M., Stafstrom, S. "Effect of bending and vacancies on the conductance of carbon nanotubes",Phys. Rev. B 62 (2000) 7639.

Harigaya, K., "Electronic states of metallic and semiconducting carbon nanotubes with bond and site disorder", Phys. Rev. B 60 (1999) 1452.

Heyd, R., Charlier, A., McRae, E., "Uniaxial-stress effects on the electronic properties of carbon nanotubes" Phys. Rev. B 55 (1997) 6820.

Korringa J. , Mills, R. L., Phys. Rev. B 5 (1972) 1654. Coherent-Potential Approximation for Random Systems with Short-Range Correlations

Lee, J. U.," Photovoltaic effect in ideal carbon nanotube diodes " , Appl. Phys. Lett., 87: 073101, (2005).

Li-Gan Tien , Chuen-Horng Tsai , Feng-Yin Li , Ming-Hsien Lee , "Influence of vacancy defect density on electrical properties of armchair single wall carbon nanotube", Diamond & Related Materials 17 (2008) 563.

Li, Y., Rotkin, S.V., Ravaioli, U., "Metal-semiconductor transition in armchair carbon nanotubes by symmetry breaking", Appl. Phys. Lett. 85 (2004) 4178.

Iijima, S., "Helical Microtubules of Graphitic Carbon," Nature, 354, 56 58, (1991).

Lu, A. J. and Pan, B. C., "Nature of SingleVacancy in Achiral Carbon Nanotubes", Phys. Rev. Lett. 92 (2004) 105504.

Mariana Weissmann, Griselda García, Miguel Kiwi, Ricardo Ramírez, and Chu-Chun Fu, "Theoretical study of iron-filled carbon nanotubes", Phys. Rev. B 73 (2006) 125435.

Mauro S. Ferreira, and Stefano Sanvito, "Contact-induced spin polarization in carbon nanotubes", Phys. Rev. B 69

Meyyappan, M., "Carbon Nanotubes: Science and Applications," Boca Raton, Fla.: CRC, (2005).

Miyake T.and S. Saito, "Band-gap formation in (n,0) single-walled carbon nanotubes" Phys. Rev. B 72 (2005) 073404. (n=9,12,15,18): A first-principles study

Niraj Sinha and John T.-W. Yeow, "Carbon Nanotubes for Biomedical Applications," IEEE Transactions on nanobioscience, 4, 2, (2005).

Ouyang, M., Huang, J.-L., Cheung, C. L. , Lieber, C. M. , "Energy Gaps in 'Metallic' Single-Walled Carbon Nanotubes," Science 292, 702-705 (2001).

Peter J F Harris, "Carbon Nanotube Science, Synthesis, Properties and Applications", Cambridge university press, (2009).

Poole, C. P., Jr. and F. J. Owens, "Introduction to Nanotechnology," Hoboken, N.J.: John Wiley. 2003

Pradhan, B., S. K. Batabyal, and A. J. Pal, "Functionalized carbon nanotubes in donor/acceptor-type photovoltaic devices" ,Appl. Phys. Lett., 88: 093106–093108, (2006).

Rocha, C. G. , Latgé, A.,Chico, L., "Metallic carbon nanotube quantum dots under magnetic fields", Phys. Rev. B 72 (2005) 085419.

Roche, S., Saito, R., "Effects of magnetic field and disorder on the electronic properties of carbon nanotubes",Phys. Rev. B 59 (1999) 5242.

Saito, R., Dresselhaus, G. , Dresselhaus, M. S.," Physical properties of carbon nanotubes," London, U.K: Imperial College press, (1998).

Solange B. Fagan, R. Mota, A. J. R. daSilva, and A. Fazzio, "Ab initio study of an iron atom interacting with single-wall carbon nanotubes", Phys. Rev. B 67 (2003) 205414.

Tans S et al , "Room-temperaturetransistor based on a single carbon nanotube", Nature 393 (1998) 49.

Tsukagoshi, K. ,Alphenaar, B. W. ,Ago, H., "Coherent transport of electron spin in a ferromagnetically contacted carbon nanotube",Nature 401 (1999) 572.

Víctor M. García-Suáez, Jaime Ferrer, and Colin J. Lambert, "Tuning the electrical conductivity of nanotube-encapsulated metallocene wires", Phys. Rev. Lett. 96 (2006) 106804.

Wei, J., Y. Jia, Q. Shu, Z. Gu, K. Wang, D. Zhuang, G. Zhang, Z. Wang, J. Luo, A. Cao, and D. Wu, Double-walled carbon nanotube solar cells" ,Nano Lett. 7 (2007).

Wildöer, J. W. G., Venema, L. C., Rinzler, A. G. , Smalley, R. E, Dekker, C., "Electronic structure of atomically resolved carbon nanotubes", Nature (London) 391 (1998) 59.

Yuchen Ma, P O Lehtinen, A S Foster and R M Nieminen," Magnetic properties of vacancies in graphene and

Zhu Y., et al.," Reentrant Semiconducting Behavior of Zigzag Carbon Nanotubes at Substitutional Doping by Oxygen Dimers", Appl. Surf. Sci. 137 (1999) 83.

# 28

# STM Observation of Interference Patterns near the End Cap and Its Application to the Chiral Vector Determination of Carbon Nanotubes

Tadahiro Komeda[1] and Masayuki Furuhashi[2]
*[1]Tohoku University*
*[2]Osaka Univversity*
*Japan*

## 1. Introduction

There is a rising demand for the development of alternative materials used for the electronic devices. This is partially due to the difficulties in a further miniaturization of current silicon-based transistors. Carbon nanotube (CNT) is considered as the most prominent candidate for this purpose (Joachim et al., 2000, Tans et al., 1997, Saito et al., 1998). The electronic property of the CNT can be regarded as a combination of those of a molecule and a solid-state bulk material, which appears as a quantum mechanical behavior at a relatively high temperature including the ballistic conductivity (Javey et al., 2003). The realization of these novel phenomena would be the key for the CNT to be widely used in real applications.
For these demands, it requires precise controls of the electronic states whose size reaches to the spatial structure of the electronic wavefunctions. It cannot be realized without a precise measurement of those properties, but there are not many characterization tools with the required resolution.
Scanning tunneling microscopy (STM) has played an important role in atomic scale characterization. It has been known that the CNT becomes metallic or semiconductive depending on the wrapping of the CNT, which is often cataloged by its chirality. Despite its importance, a technique for its precise measurement is not well established. This is partially because conventional diffraction techniques like as low-energy electron diffraction (LEED) or reflection high-energy electron diffraction (RHEED) cannot be applied to the characterization of molecule like CNT tubes.
The atomic scale resolution of STM enables an estimation of the internal structure of the CNT, which has been employed to the determination of the chirality. However, a precise determination of the chirality is not an easy task. STM can also detect the electronic structure in addition to the structural characterization. A clear example can be seen in the observation of the standing wave formed by the interference of an electronic state of CNT, which is an appearance of the wavefunctions. (Lemay et al., 2001, Venema et al., 1999) Using this technique, the band structure of the CNT has been determined. (Ouyang et al., 2002, Lee et al., 2004)
Here in this chapter, we discuss the STM measurements focusing on the superlattice structures detected near the end-cap of the CNT. We combine high-resolution STM imaging

and several types of simulation methods both for metallic and semiconducting CNT (Furuhashi et al., 2007, Furuhashi et al., 2008).
This study is intended partially for the basic understanding of the mechanism how the interference pattern is formed, unveiling the underlying physics of the complex pattern observed at the end of CNT. Other important issue is an application to the precise determination of the chirality of the CNT. For a determination of the atomic ordering, the analysis of the interference pattern often gives more precise information than the real-space image, as has been demonstrated with other microscopes. We like to apply the observation of the interference pattern near the end of the CNT to the precise determination of the chirality utilizing the sensitivity of the reflection patterns. We like to discuss the possibility of the application of this technique to a precise determination of the chirality of CNT.

## 2. STM observation of CNT

In this section we show the STM imaging of the CNT, which is usually executed by transferring the CNT on a metal surface. First we describe the experimental procedure for the STM observations. Second we show the actual observed STM images both on the metallic and semiconductive CNT (Furuhashi et al., 2007, Furuhashi et al., 2008).

### 2.1 Experimental set-up of STM observation of CNT

We performed STM measurement on CNT deposited on an Au(111) surface. The substrate was atomically flat Au(111) thin film which was evaporated on a mica surface at 480°C in a homebuilt evaporation system. The gold surface was cleaned by repeated cycles of $Ar^+$ sputtering and annealing before CNT deposition in ultra high vacuum (UHV). We confirmed characteristic herringbone reconstruction on the cleaned gold surface by STM. Purified single-wall CNTs (SWCNTs) from Carbon Nanotechnologies Incorporated were used as a sample, which was formed by HiPco method (Nikolaev et al., 1999). After the SWCNTs were dispersed in 1, 2-dichloroethane with ultrasonic agitation for 1h, a droplet of the solution was deposited on the Au(111) substrate in air, which was cleaned in the UHV chamber and take out of it just before the deposition. The sample was then dried with pure nitrogen gas and returned to the UHV chamber. In the UHV condition, the sample was annealed at 220 °C for 1 hour in order to remove residual solvent. (Furuhashi et al., 2007) For the STM measurement, we used JEOL JSTM-4500XT in which the observation was done at liquid nitrogen temperature. STM tips were electrochemically etched W wires, which were annealed at 800 °C in the UHV chamber for the removal of tungsten oxides. Topographic images were obtained by a constant current mode. Current-voltage ($I$-$V$) spectra were taken by recording tunneling current as sweeping bias voltage applied to the sample with feedback-loop opened.

### 2.2 STM images of CNT

CNT is a graphene sheet wrapped in a cylindrical shape. Thus, in order to discuss the STM image of CNT, we should check the observed STM images of the materials composed of the graphene sheet.
The atomic structure of the graphene layer is illustrated in Fig. 1(a). The unit cell is defined by the two vectors $\boldsymbol{a}_1$ and $\boldsymbol{a}_2$, which includes two carbon atoms. The lengths of both vectors, depicted $a$ hereafter, are $\sqrt{3}$ times of the nearest neighbor distance (nnd) of the carbon atoms ($a$ = 0.246 nm).

The most studied material is the graphite in which the graphene sheets are stacked.

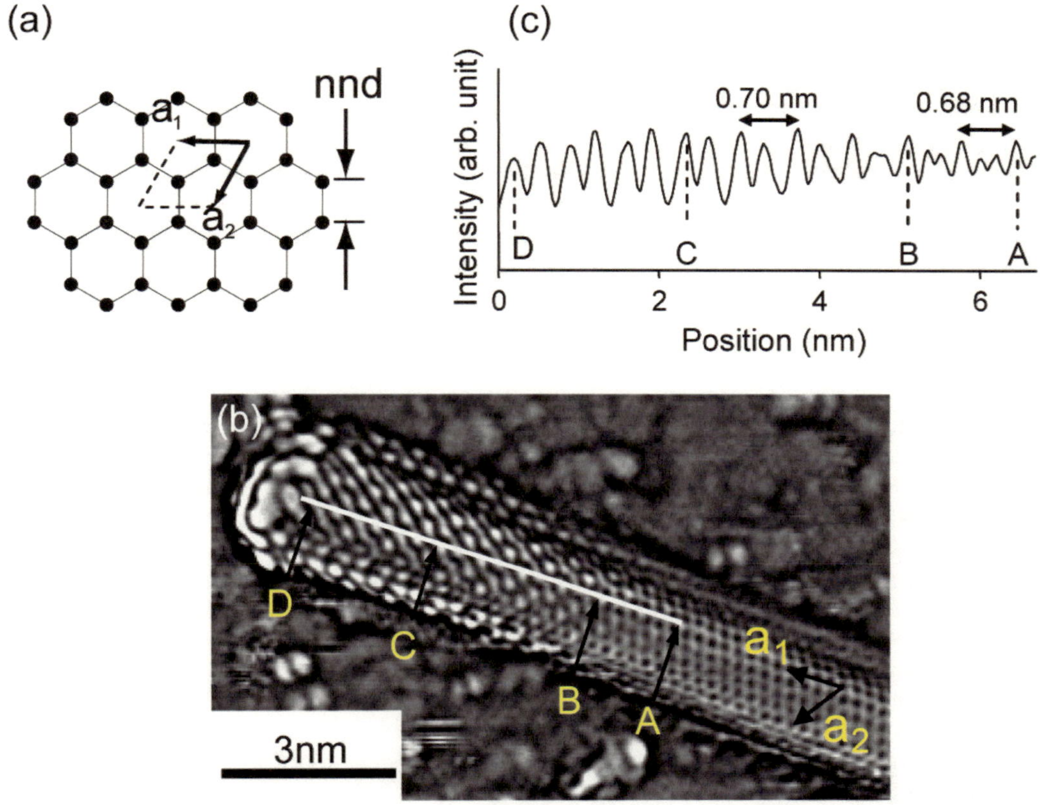


Fig. 1. STM image of armchair-like CNT (a) The atomic structure of the graphene layer. (b) Derivative STM image of a capped (28, -13) nanotube. The letters A, B, C, and D correspond to the positions in (c). (c) The line profile along the highlighted line in (b). A is the starting point of the superlattice structure and two small peaks merge at B. The distance between large peaks in AB is 0.68 nm. The period of paired peaks in BD is 0.70 nm. The intensity of paired two peaks becomes nearly same at C.

Though the carbon atoms are ordered in a honeycomb shape in the graphite layer, the observed STM image shows a close-packed, three-fold symmetry structure. The discrepancy has been explained by the influence of the subsurface layer. The density of states (DOS) at the carbon atom of the top layer changes depending on whether it has a carbon atom just beneath in the second layer. Thus the carbon atoms appear as bright protrusions in an alternative manner in the STM image, forming a three-fold, close-packed structure with a periodicity of $a$ (Batra et al., 1987). In the STM image of the single-layer graphene, such an effect cannot appear and a honeycomb shape structure appears in the STM image.

Figure 1(b) shows an atomically resolved derivative topographic image of a semiconductive CNT, which was taken at 100 pA and 0.6 V. The 'bulk' area shows a periodic grid pattern, where dark spots correspond to the centers of hexagons. The ordering of these spots can be described by using unit vectors (see Fig. 1(a)). Here we define $\boldsymbol{a}_1$ and $\boldsymbol{a}_2$ in the directions shown in Fig. 1(b). The measured length does not really reproduce expected $a$, 0.22 nm along $\boldsymbol{a}_1$ and 0.27 nm along $\boldsymbol{a}_2$. The discrepancy from $a$ of 0.25 nm is due to the geometrical configuration of STM tip and cylindrical CNT surface in which the carbon atoms are positioned on a curved surface (Venema et al., 2000, Meunier et al., 1998). Since a slight geometrical difference is not important in our discussion, no correction of the distortion of the STM images was executed. The end of the CNT has a round shape, and is probably capped by one of the hemispheric fullerene family members.

Characteristic parallel wavy lines appear within 6−7 nm from CNT end. The height-profile is measured along the line A-D which is parallel to $\boldsymbol{a}_1$ (see Fig. 1(c)). In the region between A and B, we see a characteristic features, i.e., two small peaks were sandwiched between larger peaks. These small peaks become merged in the region between B and D. The periodicity of the large peaks is ~0.68-0.70 nm in the whole region, which should correspond to ~$3a$. We consider the periodicity is derived from the Fermi wavelength, which is estimated as three times of the unit vector of the graphene (Saito et al., 1998).

Figure 2(a) shows an enlarged image of Fig. 1(b) at the interface region of the superlattice and the bulk. The positions of carbon atoms in the superlattice can be clearly identified from the carbon hexagons in the bulk region. We see superlattice structures in the CNT image, which is schematically illustrated in Fig. 2(b). The structures can be described by a combination of cranked lines and oval-shape spots, which were sketched by grey lines and white ellipses, respectively, in Fig. 2(b). The trace of the line has the same shape as that can be seen in the edge of the ideal armchair CNT. The direction of the cranked line is parallel to the 'armchair direction' which is marked in Fig. 2(b). The periodicity of the ovals along this direction is $\sqrt{3}a$. A similar superlattice structure with the same periodicity was reported in the STM images obtained near a particle and an edge on graphite surface (Xhie et al., 1991, Kobayashi et al., 2005).

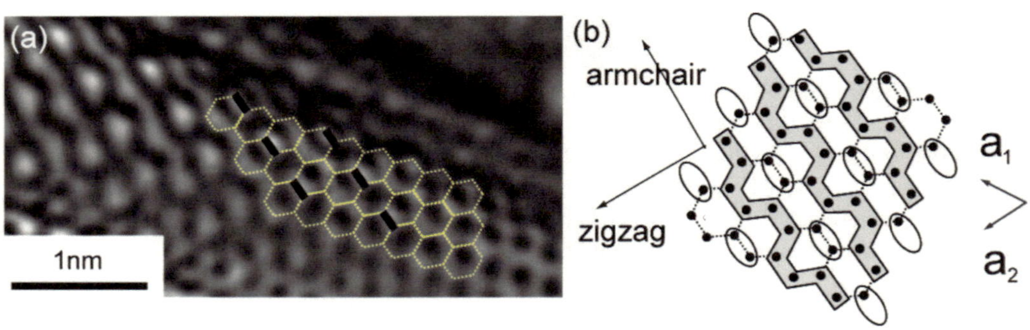


Fig. 2. Magnified image of Fig. 1(b) (a) Enlarged image of the point B in Fig. 1(b). The dotted lines mean the honeycomb structure of primitive graphene. The ellipses are emphasized by thick bars. (b) Diagram of the relation between the positions of carbon atoms and the apparent structure in (a). The black circles represent carbon atoms, and the arrows indicate the $\mathbf{a}_1$ and $\mathbf{a}_2$.

We measured another semiconductive CNT, whose image was obtained with the conditions of 0.6 V and 100 pA (see Fig. 3(a)). In this image, two CNTs are overlapped with each other, and we show the border of two CNTs by the dashed line in Fig. 3(a). Due to the conditions of the STM tip, the two CNTs were convoluted into a single tube in the image. However, low bias dI/dV mapping image can clearly distinguish the two CNTs (inset figure). The mapping image was obtained by plotting a derivative conductance at −0.4 eV which was detected by using a lock-in amplifier, while the tip scanned the surface with a closed feedback-loop (modulation voltage was superimposed on the sample voltage, which was small enough to obtain a decent topographic image). We assign the upper CNT as (11, 4) or (12, 4), whose chiral angle is between armchair and zigzag nanotubes. The ambiguity of the chiral vector is due to the error bar of the measurement of the diameter.

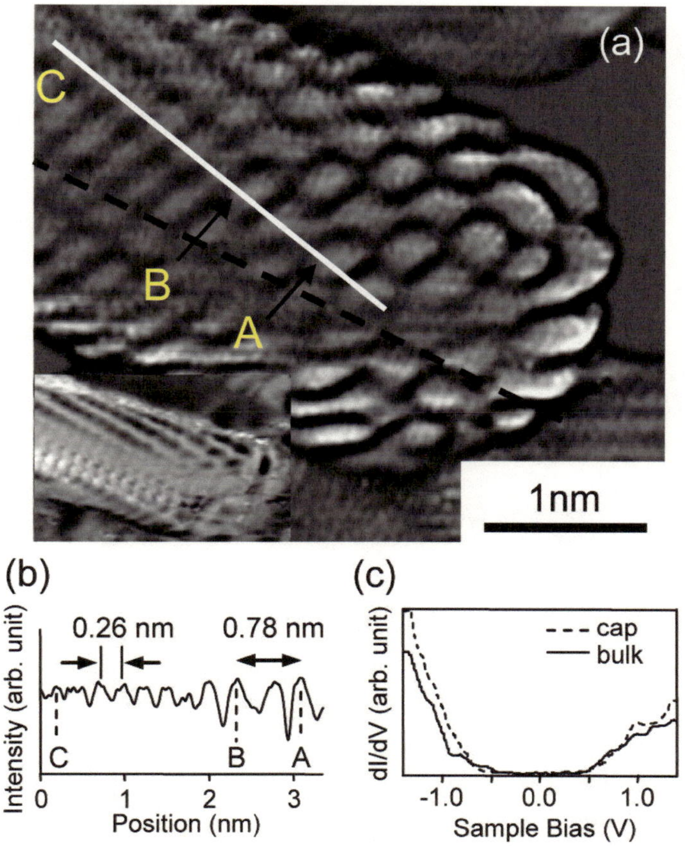


Fig. 3. STM image of semiconductive CNT (a) Derivative STM image of two nanotubes ends. The dashed line represents the border of nanotubes. Ellipse rows and cranked lines are observed. (inset) the dI/dV mapping image of the same nanotubes at -0.4 eV. (b) The line profile along the highlighted line in (a). AB corresponds to the distance between ellipse rows in (a), and the point C is on the normal period. (c) dI/dV-V spectra on the cap (dashed line), and on the bulk (solid line) where is 5 nm far from the end.

The combination of the oval-shape spots and the cranked lines can be observed near the end-cap, which is similar to the ones observed for the CNT of Fig. 2(a).
Figure 3(b) shows the height-profile along the line A-C in Fig. 3(a). The distance between A and B is ~ 0.78 nm, which is close to 3*a*. The superlattice structure appears in a limited area from the end-cap. The dI/dV curves measured on the cap and the bulk are shown in Fig. 3(c). The band gap was observed on both sites, indicating that they are both semiconductive.

The van Hove singularities were weak and can not be assigned. The electronic states detected with the condition of the sample bias of 0.6 eV correspond to the ones at the edge of the conduction band. The band gap of this CNT expected from a theoretical calculation is ~0.7 eV, which was deduced from the equation of $E_{gap}=2|\gamma_0|a_{c\text{-}c}/d$ (Saito et al., 1998), where $\gamma_0$ = -2.7 eV corresponding to the C-C overlap energy, $a_{c\text{-}c}$ = 0.142 nm is the nearest neighbor distance between carbon atoms, and d = 1.1 nm is the diameter of the CNT.

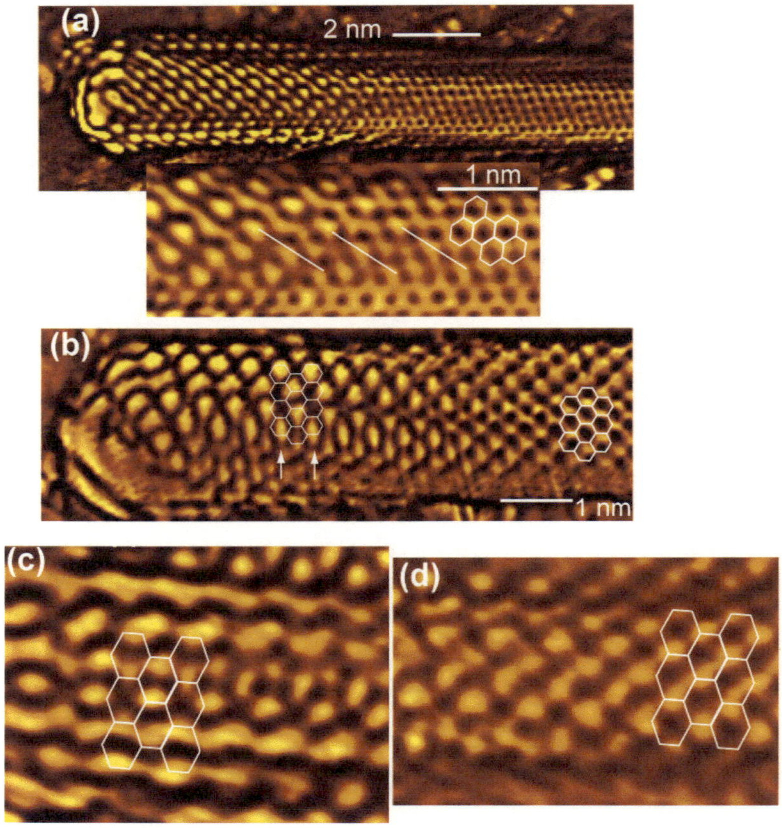


Fig. 4. Comparison of STM images of armchair and zigzag CNT (a) Upper panel: Interference pattern observed near the end-cap of armchair-like semiconducting CNT (chiral vector (28, -13)), ($I_t$ = 100 pA $V_{bias}$ = 0.6 V). Lower panel: Magnified image of the upper panel. The thin lines are included as a guide to the eye for the aligned bright spots. The superlattice structure imposed hexagons are the primitive honeycomb structures. (b) Interference patterns for the zigzag CNT ($I_t$ = 100 pA $V_{bias}$ = 0.7 V). Bright spots separated by 4***a*** along the arrows. (c)(d) Bias dependence of an interference pattern of the zigzag CNT measured with $I_t$ = 100 pA and (c) $V_{bias}$ = 0.7 V , (d) $V_{bias}$ = 1.0 V at an identical position of the same CNT(chiral vector (15,-1)).

The band gap estimated by the onsets of the dI/dV curve is in the range of 0.9-1.0 eV, which is larger than the theoretical prediction. Similar discrepancies are reported previously which should be due to the band broadening.

Near the end-cap of the CNT, we can find a report that detected a gap-state (Kim et al., 2005). However we could not see a similar state. Instead we detect a small energy shift (~ 0.1 eV) to the higher sample bias in the dI/dV spectrum at the end-cap region compared to that observed at the normal region.

Here we compare the structure at the end-cap of the armchair and the zigzag CNT in Fig. 4. The images are shown in a derivative manner to flatten the background. In the upper panel of Fig. 4(a), the end-cap is positioned on the left hand side and the infinite tube extends to the right. We can estimate a chiral angle of ~2.9° and a tube diameter of ~1.9 nm, from which the most plausible chiral vector of (28, -13) can be deduced. Since this CNT is close to an armchair type, the CNT axis is close to the ΓK direction in ***k*** space.

Near the end-cap, we observe a characteristic structure with a complex pattern. This structure can be described as a combination of aligned bright spots and wavy lines sandwiched by the former. The alignment is indicated by thin lines in the figure, which are rotated 60° from the tube axis. The wavy patterns are also aligned along the thin lines. The separation of the bright spots (pointed by upper-ends of the bars) along the CNT axis is ~ 0.68-0.70 nm that is close to 3*a*. This agrees with the interference periodicity discussed in metallic CNTs (Lemay et al., 2001, Ouyang et al., 2002, Rubio et al., 1999 ), wherein the energy band crosses the Fermi level at the K point that corresponds to $\boldsymbol{k} = 2\pi/3a$, which is the origin of the periodicity of 3*a*.

When the chirality of the CNT is close to zigzag, the STM image near the end-cap is like the one shown in Fig. 4(b). Similar to the case of Fig. 4(a), a complex pattern can be observed within ~6 nm from the end-cap. In the right-hand side area, we can see a hexagonal network rotated 30° from the armchair direction. The CNT axis is aligned to the ΓM direction of the Brillouin zone. On the left-hand side, we can notice a pattern of bright spots with a large periodicity perpendicular to the tube-axis. Along the thin-arrows, we note bright spots with a separation of ~4*a* (see superimposed hexagons). Along a line between the two, two bright spots are separated by *a*.

We show STM images of another zigzag CNT in Fig. 4(c) that was imaged with a condition of $V_{bias}$ = 0.7 V, which exhibits more minute features. We assigned the chirality of the CNT as (15, -1) which is close to the zigzag but different from the one of Fig. 4(b).

When the same CNT was observed with $V_{bias}$ = 1.0 V, the image shows a 1×1 pattern even near the end-cap. This is shown in Fig. 4(d), which was obtained in the same area with that of Fig. 4(c). This is caused primarily by a mixture of various electronic states contained in the allowed energy region.

## 3. Simulation of STM images of CNT

The electronic property of the CNT is partially determined by how it is wrapped from the graphene sheet, whose unit vectors of $\boldsymbol{a}_1$ and $\boldsymbol{a}_2$ are shown in Fig. 5(a). The wrapping is characterized with a chiral vector ***C*** and translational vector ***T***. The tube axis is parallel to the latter which is perpendicular to the former.

The chiral vector is expressed with a form of $\boldsymbol{C}_h = n\boldsymbol{a}_1 + m\boldsymbol{a}_2$, which is often denoted as (*n*, *m*) (see Fig. 5(a)). Two symmetry lines are often referred in the chiral vector discussion, and are called armchair and zigzag direction which are shown in Fig.5 (a). The chiral angle (ϕ) is defined as the angle between the chiral vector and the armchair direction. The CNT with ϕ=0° is called an armchair CNT and the one with ϕ=30° is called a zigzag CNT.

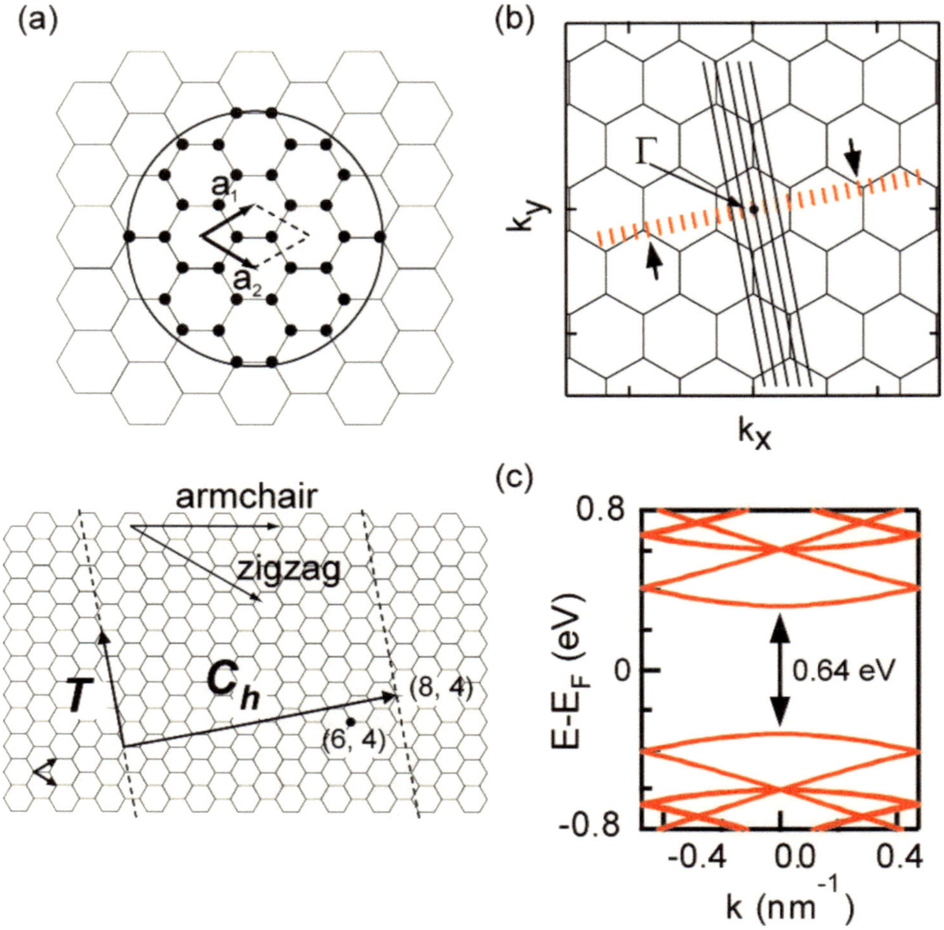


Fig. 5. Chiral vector and band structure of CNT (a) (upper) atomic structure of a graphene layer. The carbon atoms are indicated by solid circles, and a unit cell is indicated by two unit vectors and dashed lines which contains two carbon atoms. 34 atoms in the circle are used for the calculation of electron probability for the STM simulation. (lower) Definition of chiral vector $\boldsymbol{C}_h$ and translational vector $\boldsymbol{T}$ of carbon nanotube. (b) Brillouin zones of graphene and allowed $\boldsymbol{k}$ points for the (4, 2) CNT that are marked by the long black lines. The short red lines are equivalent to them. The k points that give the energy minimum (maximum) in the conduction band (valence band) are indicated by arrows, which are the closest points to K and K'. (c) Calculated band structure for a CNT of (15,-1). The length along the horizontal axis corresponds to the length of the red lines in (b).

Though the electronic structure of a CNT is close to that of graphene, limited $\boldsymbol{k}$ points are allowed due to quantization of an electron in the wrapping direction. The allowed $\boldsymbol{k}$ points can be plotted on straight lines along the tube axis direction, separated by $2\pi/|\boldsymbol{C}_h|$ from each other. In Fig. 5(b), we show an example of such lines for a CNT with a chirality of (4, 2). These are shown by long black lines. An equivalent expression is a series of short bars whose lengths are an inverse of the translational vector, which are also shown in Fig. 5(b) by the short red lines. It has been demonstrated that if the chiral vector (*n*, *m*) satisfies the condition of $n$-$m$=0 (mod 3), the CNT is metallic and the energy band (E(*k*)) curve crosses the Fermi level at *K* or *K*'. For a semiconducting CNT, the lines of the allowed ***k*** vectors never cross the *K* point. The minimum (maximum) energy in the conduction (valence) band is given at $\boldsymbol{k}_{min}$, which is on the red lines and is closest to the *K* or *K*' point (White et al., 1998). In Fig. 5(b), we show the positions of $\boldsymbol{k}_{min}$ with arrows.

The band structure can be calculated using the standard formula, and the result for (15, -1) is shown in Fig. 5(c). The length of the x-axis corresponds to that of the short red lines in Fig. 5(b), and the valence band maximum and conduction band minimum is given at $\boldsymbol{k}_{min}$, which has an energy gap as expected from a semiconducting nature of the CNT.

Kane *et al.* discussed that the symmetry breaking might occur at semiconductive tube end in STM image. This is because that the incoming and back-scattered electron waves might be interfered near the end, and break the rotational symmetry. The expected interference pattern was 1×1 spiral stripes, which ran along the zigzag direction of nanotube (Kane et al., 1999). However, the patterns near the end-caps shown above are not simple stripes. Moreover, they are along the armchair direction instead of the predicted zigzag direction. These superlattice structures could not be explained by a simple electron wave scattering theory. Rubio *et al.* calculated molecular orbitals and STS images of an armchair nanotube with a finite length by density-functional theory method (Rubio et al., 1999 ). One of the orbitals in the Hückel catalog near the Fermi level has a similar pattern to that of Fig 2(b), which was obtained near the interface area. The STM image of the metallic CNT near the Fermi level was calculated by the combination of the four patterns at K point on the Brillouin zone of nanotube. Indeed, the dI/dV mapping image of the nanotube with 30 nm length equaled to the shape of wavefunction because the small length made the energy levels discrete by quantum size effect (Lemay et al., 2001). The observed wavefunctions were the combination of Bloch functions and equaled to the molecular orbitals.

We examine two types of the calculation focusing on the simulation of the symmetry breaking patterns observed neat the cap end.

### 3.1 Simulation by molecular orbital calculation

First we show a simulation with using molecule orbital calculation. We calculated the molecular orbitals of capped nanotube by PM3 method (Stewart, 1989, Stewart, 1989, Stewart, 1991). The calculation was carried out by MOPAC2000 package (Stewart, 1990). We employed a model nanotube (10, -4) instead of (28, -13) latter of which was actually shown in the STM observations. This is because of the limitation of our computational resource. Because the nanotube (10, -4) has an apparent chiral angle ($\phi$) of 6.6° which is close to $\phi$=2.9° observed for the (28, -13) CNT. Thus we believe the calculation can reproduce the experimental results. The calculated HOMO and LUMO orbitals are shown in Fig. 6(a) and (b), respectively. The distribution of LUMO orbital shows clear ellipses and cranked-lines along the armchair direction, which is similar to the image of Fig. 2(b). The distance between the ellipses along the zigzag direction is 3*a*. On the other hand, the distribution of HOMO does not show the cranked-line, whereas circles are visible along the armchair direction. The STM image obtained with a positive sample bias corresponds to the distribution of an unoccupied electronic state; therefore the theoretical simulation above reproduces the observed STM images.

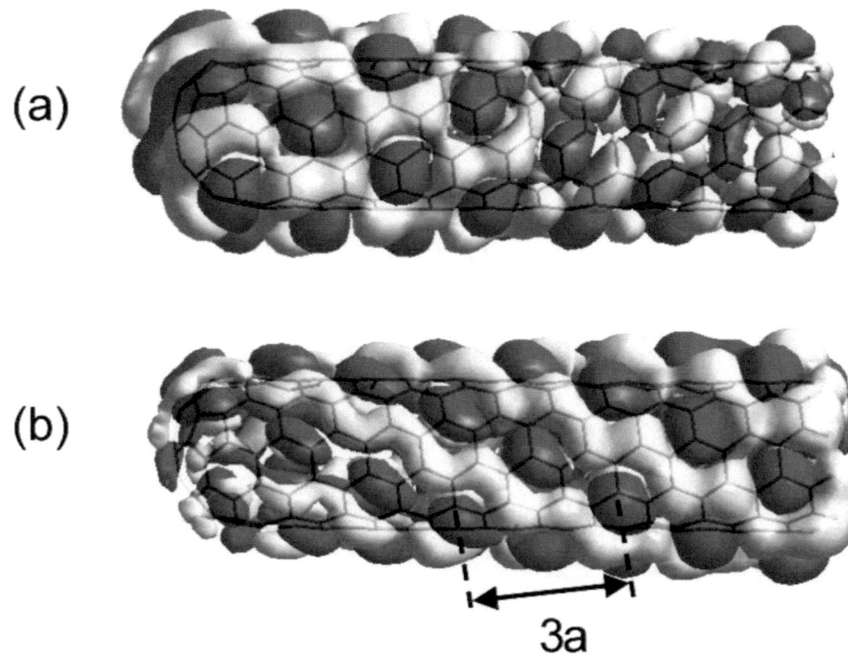


Fig. 6. STM image of CNT simulated by PM3, molecular orbital method (a) HOMO and (b) LUMO calculated by PM3 method. The nanotube consists of 220 carbon and ten hydrogen atoms. The grey-scale tone corresponds to the sign of phase. The left side is capped by a hemisphere of carbon atoms, and the right side is terminated by hydrogen atoms. The wave length along the apparent chiral angle direction is indicated by an arrow.

In addition, further information can be obtained from Fig. 6. The local density of states, $\rho_{LDOS}$, can be expressed by the following equation:

$$\rho_{\mathrm{LDOS}}(\mathbf{r},E)=\sum_{i}\left|\psi_{i}(\mathbf{r})^{2}\right|\delta(E_{i}-E)\,, \qquad (1)$$

where $\psi_i$ are the wavefunction with the eigenvalue $E_i$, and $i$ corresponds to the index of eigenstates. It is obvious that the electron population can be estimated with the LDOS of the atom at the position. It is measured that the HOMO has a larger population at the cap than at the normal part, whereas the LUMO behaves in an opposite manner. This is why the pentagon in the cap works as electron acceptor. It is due to the Hückel rule that the cyclic $\pi$-electron state containing $4n + 2$ electrons is the most stable one ($n$ is an integer). The carbon pentagon contains five $\pi$-electrons in the neutral state, and tends to attract a single electron. Therefore a negative charge is localized near the cap.

The change of the charge distribution causes a band bending, and the occupied states at the end-cap should be shifted downward. We consider that this can account for the energy shift of the dI/dV spectrum shown in Fig. 3(c). We conclude that the modulated pattern was observed when the STM sample bias agreed with the bottom of the conduction band whose energy level was shifted due to the band bending.

### 3.2 Simulation by Bloch theorem

Now we discuss a simulation based on Bloch theorem. We first examine the electronic structure of the graphene sheet (Saito et al., 1998). The unit cell of the graphene includes two carbon atoms (A and B in Fig. 5(a)). In the tight-binding approximation, Bloch function is given by a linear combination of atomic orbitals (LCAO). In a case of the $\pi$-electron system, we only have to consider $p_z$ orbital, and the one-electron wavefunction is obtained by a linear combination of the $p_z$ Bloch functions:

$$\psi(\mathbf{k},\mathbf{r}) = \frac{C_A}{\sqrt{N}} \sum_{\mathbf{R_A}}^{N} e^{i\mathbf{k}\cdot\mathbf{R_A}} \varphi_A(\mathbf{r}-\mathbf{R_A}) + \frac{C_B}{\sqrt{N}} \sum_{\mathbf{R_B}}^{N} e^{i\mathbf{k}\cdot\mathbf{R_B}} \varphi_B(\mathbf{r}-\mathbf{R_B}), \tag{2}$$

where $\boldsymbol{R_A}$ and $\boldsymbol{R_B}$ indicate coordinates of the carbon atom A and B, respectively. $C_A$ and $C_B$ are normalization coefficients, and $\varphi$ depicts atomic orbitals. The details of the calculation about overlap integrals and eigenvalue were discussed by Saito and Dresselhaus (Saito et al., 1998). When we assume the overlap integral to be zero, the eigenvalue becomes $E(\boldsymbol{k}) = \pm t\omega$, where $t$ is the overlap energy integral and $\omega$ is defined below. The normalization coefficient becomes $C_B=\pm(f^*/\omega)C_A$. Here $f^*/\omega$ can be depicted as $Ae^{i\theta}$ $(A \geq 0)$, and $|f^*/\omega|^2 = A^2 = 1$. The wavefunction can be deduced as following:

$$\psi_{\pm}(\mathbf{k},\mathbf{r}) = \frac{C(\mathbf{k})}{\sqrt{N}} \left[ \sum_{\mathbf{R_A}}^{N} e^{i\mathbf{k}\cdot\mathbf{R_A}} \varphi_A(\mathbf{r}-\mathbf{R_A}) \pm \sum_{\mathbf{R_B}}^{N} e^{i(\mathbf{k}\cdot\mathbf{R_B}+\theta)} \varphi_B(\mathbf{r}-\mathbf{R_B}) \right], \tag{3}$$

where

$$\begin{aligned} \cos\theta = \mathrm{Re}\frac{f^*}{\omega} &= \frac{1}{\omega}\left( \cos\frac{k_x a}{\sqrt{3}} + 2\cos\frac{k_x a}{2\sqrt{3}}\cos\frac{k_y a}{2} \right), \\ \sin\theta = \mathrm{Im}\frac{f^*}{\omega} &= \frac{1}{\omega}\left( -\sin\frac{k_x a}{\sqrt{3}} + 2\sin\frac{k_x a}{2\sqrt{3}}\cos\frac{k_y a}{2} \right), \end{aligned} \tag{4}$$

and

$$f = e^{ik_x a/\sqrt{3}} + e^{-ik_x a/2\sqrt{3}} \cos\frac{k_y a}{2}, \tag{5}$$

$$\omega = \sqrt{ff^*} = \sqrt{1 + 4\cos\frac{\sqrt{3}k_x a}{2}\cos\frac{k_y a}{2} + 4\cos^2\frac{k_y a}{2}}, \tag{6}$$

Eq. (3) gives a wavefunction with the wavevector $\boldsymbol{k}$ as a parameter, whose value is determined by the periodic boundary condition of each carbon nanotube. We treat $C$ as a constant value that is independent of $\boldsymbol{k}$ for simplicity.

The tunneling current $I_t$ is proportional to the integrated LDOS between the Fermi level and the applied bias voltage $V$ (here the LDOS of the tip is neglected):

$$I_t(\mathbf{r},eV) \propto \int_{E_F}^{E_F+eV} \rho_{LDOS}(\mathbf{r},E)dE. \tag{7}$$

The thermal broadening of the Fermi level is ignored. Since the electronic states that contribute to the tunneling current are located in the energy region close to the Fermi level and the band edge, we can simplify the integration in eq. (7). For a simulation of the STM images, we estimate a space distribution of the tunneling current in the logarithmic scale assuming the tip scans at a certain constant height. This is due to a limited calculation resource.

In order to estimate the probability that one can find the electron at $\boldsymbol{r}$, we have to consider the contributions from $2N$ atomic orbitals. However, the contribution decreases exponentially with the tip-surface distance. Hence, we consider only the 34 atoms close to the tip position [illustrated in Fig. 5(a)].

First, we show the simulation without the interference for a metallic, (5, 5) armchair CNT. As shown in Fig. 7(a), there are two symmetric sets of crossed energy dispersion curves (E vs $\boldsymbol{k}$) at the Fermi level in the infinite CNT. In the 'normal' region (where the electronic structure is not affected from the end or defect), the topographic image is formed by integrating the tunneling current from the four dispersion curves of Fig. 7(a).

Figure 7(b) shows a topographic image of the CNT at 20 meV above the Fermi level, where the CNT axis is set to the horizontal line. We have confirmed that the image obtained at 20 meV below the Fermi level has a similar distribution. The pattern shows a simple 1×1 hexagonal structure, which seems to reproduce the positions of the carbon atoms. However, this pattern is the superposition of two superlattice structures that have different symmetries. We illustrate two patterns in Fig. 7(c) and (d), which correspond to the square of the wavefunctions on curve I and II, respectively (the wavefunction in III and IV give completely same electron probability with II and I, respectively).

In the images, the real part of the wavefunctions is superimposed as circles whose size and colors correspond to the amplitude and the phase, respectively.

We see a periodic variation in the amplitude of the real part of the wavefunctions along the nanotube axis (horizontal direction in the figures), whose periodicity is ~ $3a$ both in Fig. 7(c) and (d). However, as can be seen in the blue and red distribution, the phase is reversed for half of the carbon atoms in the image. The superposition of the tunneling current makes an isotropic hexagon pattern illustrated in Fig. 7(b).

Similar simulations of the STM image of the metallic CNT have been performed previously. Meunier and Lambin calculated STM image of metallic nanotube and shows a good agreement with our image except for a small anisotropy that is purely originated from geometric distortion around circumferential direction (Meunier et al., 1998, Meunier et al., 2004). As far as we consider defect-free CNT, it is expected that the 1×1 sixfold hexagon structure always appears in the topographic image of metallic CNT. We like to note that this is because STM can only measure the superposition of the multiple contributions, and cannot distinguish the four dispersion curves.

Next, we show the topographic image generated by the interference of the wavefunctions.

One-dimensional interference wavefunction can be expressed by a superposition of the forward and backward waves that has an equivalent electron energy (Ouyang et al., 2002). A two-dimensional interference pattern can be generated by using eq. (3):

$$\rho_{LDOS} \propto \left|\psi(\mathbf{k}_1,\mathbf{r}) + R e^{i\eta}\,\psi(\mathbf{k}_2,\mathbf{r})\right|^2 , \tag{8}$$

where $\eta$ and $R$ are a phase shift and a reflectivity, respectively. These parameters are determined how the waves are reflected at the terminal and the defect.

In Fig. 8(a), we show the simulation result assuming $R = 1$ and $\eta = 0°$. The rows of emphasized dimers can be seen along the vertical direction. Between these rows, there are cranked-lines with a weak intensity along the vertical armchair direction. The distance between dimers along CNT axis is $3a$, which makes a $\sqrt{3}\times\sqrt{3}$ superlattice structure. Figure 8(a) shows a good agreement with previous calculation and measured STM image (Kane et al., 1999, Clauss et al., 1999).

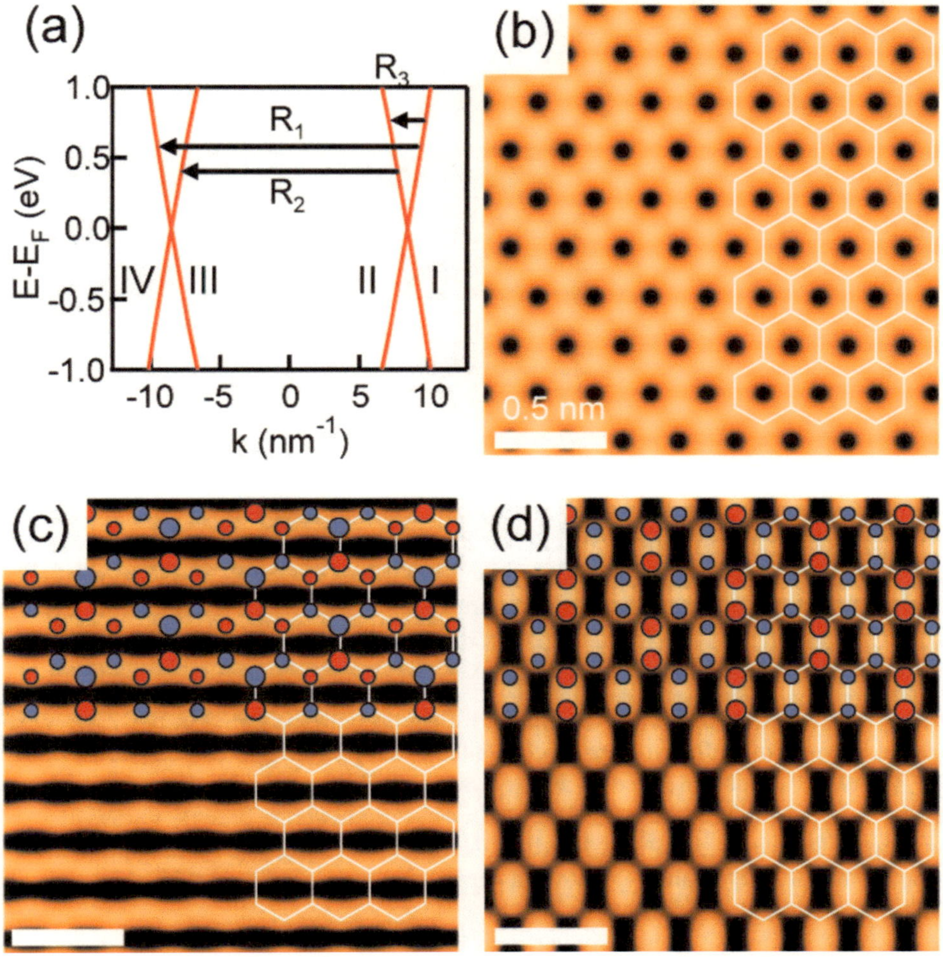


Fig. 7. Simulated STM image of CNT by Bloch theorem. Simulation of (5, 5) armchair carbon nanotube. (a) Energy dispersions near the Fermi level. The $k$ is the 1D wavevector, which is the part of 2D wavevector $\boldsymbol{k}$ along nanotube axis. The R1, R2, and R3 indicate possible reflection paths. (b) Simulated STM image without interference at 20 meV above Fermi level. The highlighted lines represent primitive honeycomb structure. (c) and (d) are electron distributions with schematic images of wavefunctions on curve I and II at 20 meV above Fermi energy, respectively. The uppers are the superposed real parts of wavefunctions, whose circle size and color represent the value of LCAO coefficient and the sign of phase, respectively.

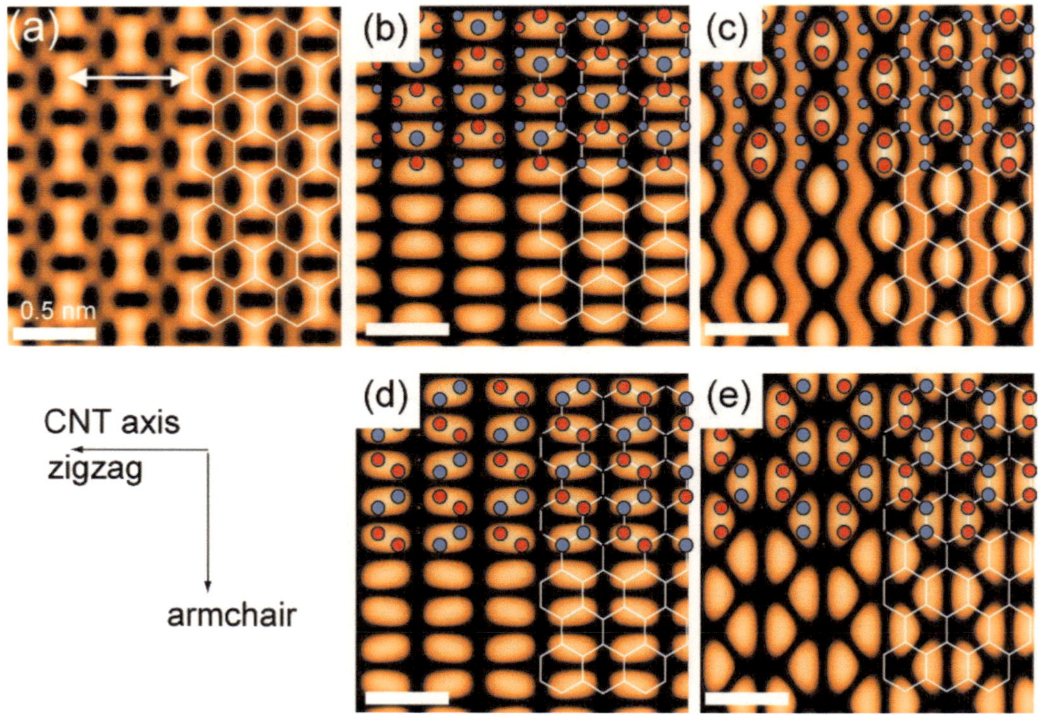


Fig. 8. Interference topographic image of armchair CNT (a) Interference topographic image of a (5, 5) armchair carbon nanotube at $E(k)$ = 20 meV, $R$ = 1 and $\eta = 0$. The white arrow indicates $3a$. The images of (b) and (c) correspond to the reflection $R_1$ and $R_2$ at $\eta = 0$, respectively. The real parts of wavefunction are superposed on upper region. (d) and (e) are the same images with (b) and (c) substituting $\eta = \pi$, respectively. The imaginary parts of wavefunctions are illustrated on upper region.

In order to analyze the contribution from the reflection path indicated in Fig. 7(a), we divide the image of Fig. 8(a) into two interfered waves corresponding to the reflection path of $R_1$ and $R_2$, which are shown in Fig. 8(b) and (c), respectively. ($\eta = 0$) The obtained patterns also have $\sqrt{3}\times\sqrt{3}$ superlattice structures. We also check interference pattern variation with the phase shift $\eta$. We show examples for the case of $\eta = \pi$ in Fig. 8(d) and (e). However, the superposition of Fig. 8(d) and (e) does not reproduce the experimental STM images, whose reason is not clear at this moment.

These interference images give a similarity with molecular orbitals (MOs) in a finite length metallic nanotube near the Fermi level as we have shown above (Lemay et al., 2001, Rubio et al., 1999 ). It suggests that all simulated scattering can occur by the condition of the reflection. In addition, several groups discussed about electron wave scattering in an armchair nanotube (Kane et al., 1999, Yaguchi et al., 2001), and considered two scattering ways, which are inter-valley ($R_1$ and $R_2$) and intra-valley ($R_3$) reflection. Inter-valley reflection occurs with the phase shift of 0 and $\pi$. We consider that the actual interference image becomes complex superposition of the standing wave patterns with various phase shifts.

Moreover, we have simulated the topographic image of a semiconductive CNT. As shown in Fig. 9(a), a semiconductive CNT has a band gap and the energy dispersion curve is symmetric around the top and bottom of the band gap. Figure 9(b) gives a magnified image of the conduction band edge in Fig. 9(a). Figure 9(c) and 9(d) show images at the edges of the conduction band and the valence band of the (6, 4) CNT, respectively. The CNT axis is aligned to the horizontal line. Both images shows a 1×1 striped spiral pattern that runs along a zigzag direction and π/3 rotation with respect to each other. These features are consistent with the report of Kane and Mele (Kane et al., 1999).

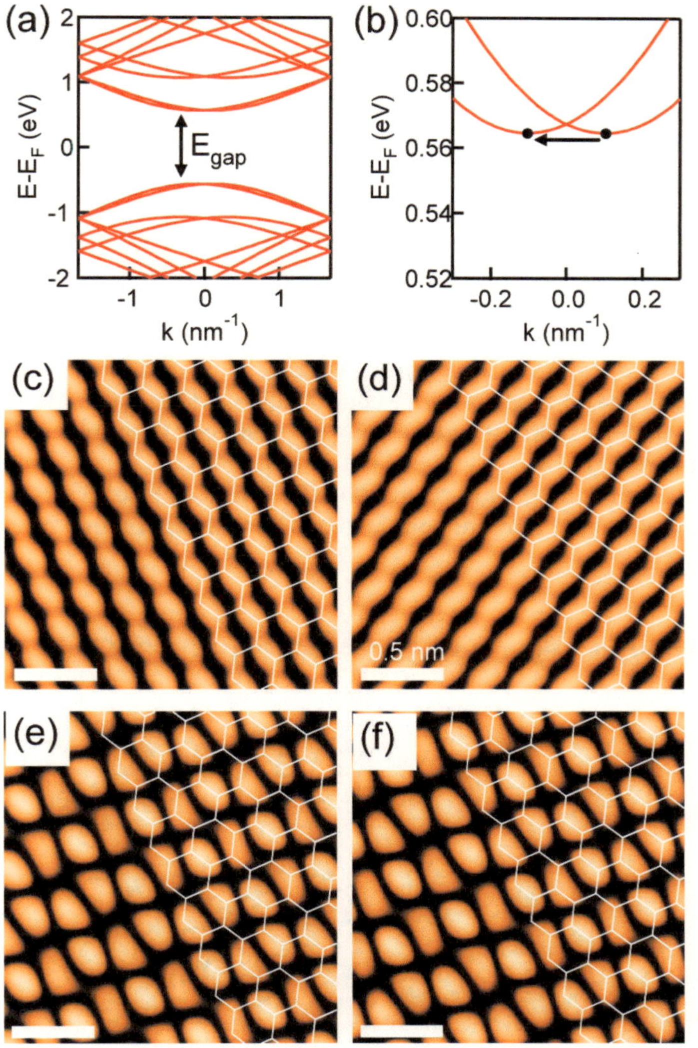


Fig. 9. Dispersion curve of semiconductive CNT and interference pattern (a) Energy dispersion curve of (6, 4) semiconductive nanotube. The band gap is defined as the difference between conduction band bottom and valence band top. (b) Magnified image of (a) near the conduction band bottom. The black dots indicate the energy bottom position. (c) and (d) are simulated topographic images at conduction band edge and valence band edge, respectively. The interference images at (e) $\eta = 0$ and (f) $\eta = \pi$ shows similar patterns.

At the band edge, there are only two wavefunctions that contribute to the tunneling current which have an equivalent electron distribution. This is due to the symmetry in the reciprocal lattice space. Moreover, the LCAO coefficients of carbon B in eq. (2) change by the band (conduction or valence) and 2D wavevector. That is the reason why the spiral pattern appears at the band edge and the running direction change by the sign of bias voltage.
The interference at the band edge occurs for a single condition of the wave vector, which is the reflection from bottom $k$ to $-k$ [Fig. 9(b)]. The images at the conduction band edge are given in Fig. 9(e) and 9(f), which are obtained with $\eta = 0$ and $\pi$, respectively. Similar to the case of the metallic CNT, the real and imaginary parts of the wavefunction appear in these interference images according to the phase shift of the reflection. However, these images show similar distributions and seem to be the mixing of Fig. 8(b)-(e). The reason why the clear pattern is not observed is that the band edge energy is well separated from Fermi level. When the electron energy is close to Fermi level, the wavelength is ~$3a$ (Fermi wavelength). On the other hand, the wavelength at conduction band bottom is shorter than $3a$. Therefore, the periodicity of LCAO coefficient $\exp(i\boldsymbol{k}\cdot\boldsymbol{R})$ does not agree with that of the unit cell of the graphene sheet. Hence, the interference image shows a modulation with a long periodicity.

### 3.3 Comparison of simulated results and STM images

We now compare the observed STM images near the end-cap of a CNT and the simulated images. We concentrate on the images of a conduction band ($V_{sample}>0$ in STM observation).
We consider a standard Bloch state based on the graphene sheet using the eq. (2) shown above and the parameter of $\boldsymbol{k}_{min}$.
The bonding and antibonding band corresponds to + and –, respectively. If we consider the antibonding band at the Γ point, the carbon atoms have an alternative phase that gives – in eq. (2).

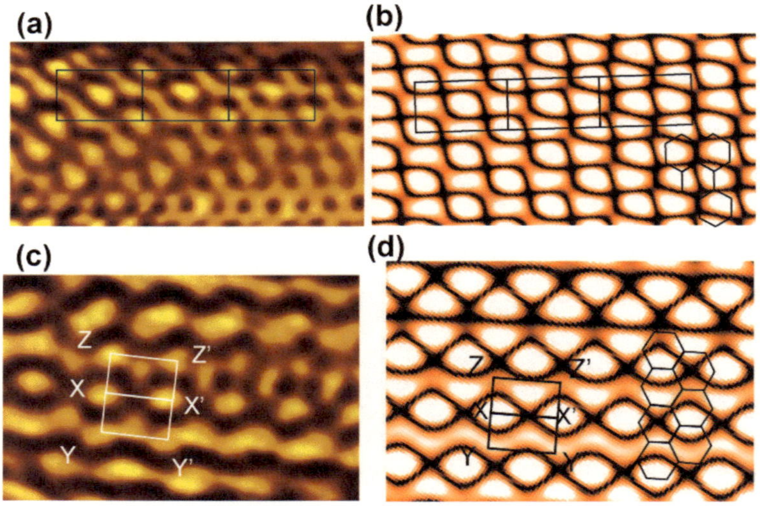


Fig. 10. Comparison of STM image and simulation (a)(b) Comparison of an STM image (same as Fig. 4(a)) and a simulation result for a CNT of (28, -13). Superimposed rectangles are included as a guide to the eye of the equivalent site. (c)(d) Comparison for (15,-1). XX', YY' and ZZ' indicate the comparison.

The interference of the wavefunctions is estimated by using eq. (9) We fixed $\eta$ at $\pi$ for the entire simulation, which is based on the result of a theoretical calculation of an electronic state at end-cap (Yaguchi et al., 2001). However, the calculated interference pattern is not sensitive to the changes of $\eta$. The Bloch states of $\boldsymbol{k_1}$ and $\boldsymbol{k_2}$ should have the same energy for elastic scattering. This condition limits the allowed wavevectors, and the combination of $\boldsymbol{k_{min}}$ and -$\boldsymbol{k_{min}}$ satisfies this condition. Though $\boldsymbol{k_{min}}$ is close to **K**, they do not coincide. A complex interference pattern is produced depending on the relative position of $\boldsymbol{k_{min}}$ around K. The calculated $\rho$ is expressed by mapping the amplitude two dimensionally.

We first compare an STM image and a simulation result for a CNT whose chirality is close to an armchair. The experimental and simulation results are shown in Fig. 10(a) and 10(b), respectively; the former is a part of Fig. 4(a). The scale and alignment of the tube axis are common in two panels. To clarify the similarity between Fig. 10(a) and 10(b), three rectangles ($3\boldsymbol{a}\times\sqrt{3}\boldsymbol{a}$) are superimposed. Bright spots at the four corners and the center of the rectangle are visible both in the image and simulation. In addition, wavy lines are reproduced well.

$\boldsymbol{k_{min}}$ is slightly off **K**, which forms a modulation with a longer periodicity. In the simulation, this can be seen as a gradual change from a wavy line to a dot-like feature in the right-hand side of the panel. This is also seen in the STM image in Fig. 4(a); the wavy lines become dot-like near the end-cap.

The results for a zigzag-type CNT are shown in Figs. 10(c) and 10(d). $\boldsymbol{k_{min}}$ is rotated ~30° from the tube-axis direction. Characteristic wavy lines along the tube-axis direction and aligned bright spots perpendicular to the tube-axis are well reproduced. XX', YY' and ZZ' are shown in both panels as an eye-guide, with a length of $\sqrt{3}\boldsymbol{a}$ and a separation of $\boldsymbol{a}$. Good agreement between the STM image and the simulation results are seen at each point. Since there expected a long-range modulation in the image along YXZ direction, the features along YY' and those along ZZ' in the STM images are not identical.

## 4. Future perspective of interference imaging of CNT

Here we try to observe variations in the interference pattern in a systematic manner; in other words, we attempt to obtain the relationship between $\boldsymbol{k_{min}}$ and the interference pattern. (Furuhashi et al., 2008) $\boldsymbol{k_{min}}$ can be expressed as a relative position from ***K***, *i.e.*, $\Delta\boldsymbol{k}=(\boldsymbol{k_{min}}-\boldsymbol{K})$ which can be specified by the length $|\Delta\boldsymbol{k}|$ and an angle from a symmetric line ($\phi$). In Fig. 11(a), solid circles show five $\boldsymbol{k_{min}}$ points with the same $|\Delta\boldsymbol{k}|$ and different $\phi$ from the direction of ***ΓK***. For these $\boldsymbol{k_{min}}$ points, we calculated the interference pattern by using eq. (3). The results for an unoccupied state are shown in Fig. 11(b).

For $\phi$=0°, $\boldsymbol{k_{min}}$ is located on the ***MK*** line. The Bloch states in the unoccupied state have a *positive* combination in eq (3). This is because they are on the line from ***Γ*** to ***K***, and $\boldsymbol{k_{min}}$ is beyond ***K***. Thus, the unoccupied wave function is positioned on the band that has a bonding character at ***Γ***. The sign of eq. (3) is positive, and consequently, wavy lines are clearly observed. The periodicity along arrow A in Fig. 11(b) (***ΓK*** direction) is deviated from 3***a***, but the periodicity along arrow B (***ΓM*** direction) is fixed at $\sqrt{3}\,\boldsymbol{a}$.

For $\phi$=60°, $\boldsymbol{k_{min}}$ is on the symmetry line of ***ΓK***; this is similar to the $\phi$=0° case but $\boldsymbol{k_{min}}$ does not reach ***K***. As the reverse case of $\phi$=0°, the Bloch state corresponds to a negative combination in eq. (3). With this configuration, the wavy patterns disappear and the periodic lattice is visible. The periodicity in the ***ΓK*** direction is shifted from 3***a***, however, modulation with a long periodicity is not visible.

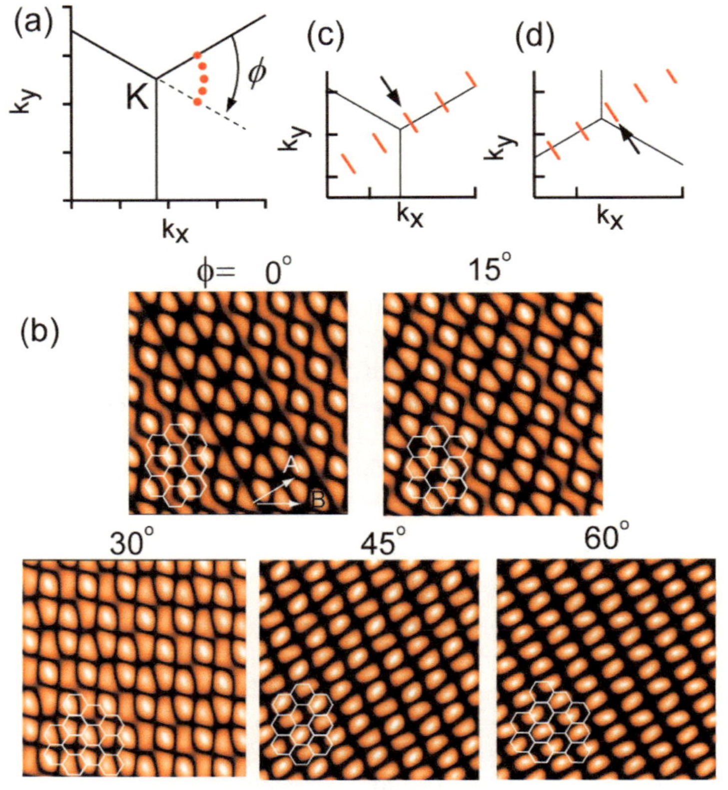


Fig. 11. Interference catalogue with chiral vector (a) Assumed $k_{min}$ positions for a simulation of the interference pattern, specified by the angle from the ΓK direction, $\phi$=0°, 15°, 30°, 45°, and 60° and distance from K. (b) Simulated images for each $\phi$. The $k_{min}$ used for the simulation is set close to the direction shown by the arrow. Though there are six equivalent k points in a Brillouin zone with a same energy, the interference pattern is also aligned in this direction. (c), (d) Brillouin zones and the short lines of allowed $k$ of the CNT of (c) (15, -1) and (d) (16, -1). The points closest to K and K' are depicted by arrows.

The pattern of $\phi$=30° is a mixture of the former two cases. Wavy lines are clearly observed in some parts, but are not symmetric and show long-range modulation. This corresponds to the modulation observed in Fig. 4(a).

If the chirality is close to the armchair, $k_{min}$ expectedly appears at $\phi$~30°. This can be understood by using the Brillouin zone in Fig. 5(b). For armchair CNTs, the translational vector is close to the vertical direction in Fig. 11(a). Since $k_{min}$ is specified by the cross point of the translational vector and its perpendicular line through K, it is located in the horizontal direction from $K$ and corresponds to $\phi$~30°. With similar consideration, $k_{min}$ can be found at $\phi$~0° or 60° for a zigzag CNT.

It is intriguing to see that a large change in the interference pattern is caused by a small change in the chiral vector. A clear example of this can be seen for the zigzag CNT, and we show a comparison for cases of (15, -1) and (16, -1). In Fig. 11(c) and (d), we show the allowed $k$ as short bars for (15, -1) and (16, -1), respectively. The positions of $k_{min}$ for both cases are indicated by arrows, which are on the reverse sides of $K$. In the Brillouin zone, the former and latter are located close to the line of $\phi=0^o$ and $\phi=60^o$, respectively. The expected difference in the interference patterns for $\phi=0^o$ and $\phi=60^o$ is clearly seen in Fig. 11(b); wavy lines are visible in the former case but disappear in the latter case. In chiral vector determination, the observation of interference patterns can play a complementary role to direct observations of the hexagonal network, as electron and photon diffraction techniques have contributed to structure determination of the materials.

We have focused our discussion on semiconducting CNTs. Similar discussions can be applied for metallic CNTs. However, as were pointed out by Kane and Mele (Kane et al., 1999), two branches of E(k) curves exist in the band structure near $E_f$, which makes the interference patterns more complicated than those of semiconducting CNTs.

## 5. Conclusion

We have discussed the internal structure of the CNT by experimental scanning tunneling microscopy observation and theoretical simulations, focusing on the interference pattern observed near the end-cap of the CNT. We analyzed the complex shape of the pattern for the understanding of the underlying physics, which is an appearance of the wavefunctions. It is also intended for the application of the interference observation to the determination of the chirality, which is one of the most important parameters for the CNT, utilizing the sensitivity of the reflection patterns. Simulations were executed both by molecular orbital calculation and the Bloch theorem. In the latter, a two-dimensional interference of the wavefunctions was calculated assuming the superposition of the forward and backward waves that has equivalent electron energy.

The observed STM images showed a complex pattern at the end-cap of the CNT, which was typically observed for the area of ~6 nm from the end. For the armchair CNT, cranked wavy lines and oval shaped spots with the periodicity of $\sqrt{3}\,a$ were commonly observed along the armchair direction ($a$ is the unit vector length of the graphene sheet; a=0.25 nm). The oval spots were also observed in the CNT with the chirality between the armchair and zigzag, whose periodicity was ~$3a$ in the zigzag direction.

These structures were well reproduced both by the molecular orbital and the Bloch theorem calculations. The simulation clearly demonstrated the expected patterns both for metallic and semiconductive CNTs revealing the wavefunctions that constructed the interface patters. In addition, the interference patterns were successfully catalogued as a function of the chirality. An intriguing finding was that a minute change in the chirality modified the interference pattern considerably. The sensitivity of the interference pattern with the chirality differences demonstrated the possibility of the use of this technique for the precise determination of the chirality of the CNTs.

## 6. Acknowledgment

TK acknowledge support from KAKENHI (22241026), and “R&D promotion scheme funding international joint research” promoted by NICT (National Institute of Information and Communication Technology), Japan..

## 7. References

Joachim, C. (2000). Electronics using hybrid-molecular and mono-molecular devices. *Nature*, 408, 541-548

Tans, S. J. (1997). Individual single-wall carbon nanotubes as quantum wires. *Nature*, 386, 474-477

Saito, R. Physical Properties of Carbon Nanotubes (Imperial College Press, London, 1998).

Javey, A. (2003). Ballistic carbon nanotube field-effect transistors. *Nature*, 424, 654-657

Lemay, S. G. (2001). Two-dimensional imaging of electronic wavefunctions in carbon nanotubes. *Nature*, 412, 617-620

Venema, L. C. (1999). Imaging electron wave functions of quantized energy levels in carbon nanotubes. *Science*, 283, 52-55

Ouyang, M. (2002). One-dimensional energy dispersion of single-walled carbon nanotubes by resonant electron scattering. *Phys. Rev. Lett.*, 88, 66804

Lee, J. (2004). Real Space Imaging of One-Dimensional Standing Waves: Direct Evidence for a Luttinger Liquid. *Phys. Rev. Lett.*, 93, 166403-166404

Furuhashi, M. (2007). Direct Observation of Molecular Orbital at Carbon Nanotube End *Japanese Journal of Applied Physics*, 46, L161-164

Furuhashi, M. (2008). Chiral Vector Determination of Carbon Nanotubes by Observation of Interference Patterns Near the End Cap. *Phys. Rev. Lett.*, 101, 185503

Nikolaev, P. (1999). *Chem. Phys. Lett.* , 313, 91

Batra, I. P. (1987). A study of graphite surface with stm and electronic structure calculations. *Surf. Sci.*, 181, 126-138

Venema, L. C. (2000). Spatially resolved scanning tunneling spectroscopy on single-walled carbon nanotubes. *Phys. Rev. B*, 62, 5238-5244

Meunier, V. (1998). Tight-binding computation of the STM image of carbon nanotubes. *Phys. Rev. Lett.*, 81, 5588-5591

Xhie, J. (1991). Superstructures on graphite near platinum particles. *J. Vac. Sci. Technol. B*, 9, 833-836

Kobayashi, Y. (2005). Observation of zigzag and armchair edges of graphite using scanning tunneling microscopy and spectroscopy. *Phys. Rev. B*, 71, 193406

Kim, H. (2005). Local electronic density of states of a semiconducting carbon nanotube interface. *Phys. Rev. B*, 71, 235402

Rubio, A. (1999 ). *Phys. Rev. Lett.*, 82., 3520

White, C. T. (1998). Density of states reflects diameter in nanotubes. *Nature*, 394, 29-30

Kane, C. L. (1999). Broken symmetries in scanning tunneling images of carbon nanotubes. *Phys. Rev. B*, 59, R12759

Stewart, J. J. P. (1989). Optimization of Parameters for Semiempirical Methods .1. Method. *J. Comput. Chem.*, 10, 209-220

Stewart, J. J. P. (1989). Optimization of Parameters for Semiempirical Methods .2. Applications. *J. Comput. Chem.*, 10, 221-264

Stewart, J. J. P. (1991). Optimization of Parameters for Semiempirical Methods .3. Extension of Pm3 to Be, Mg, Zn, Ga, Ge, as, Se, Cd, in, Sn, Sb, Te, Hg, Tl, Pb, and Bi. *J. Comput. Chem.*, 12, 320-341

Stewart, J. J. P. (1990). Special Issue - Mopac - a Semiempirical Molecular-Orbital Program. *Journal of Computer-Aided Molecular Design*, 4, 1-45

Meunier, V. (2004). Scanning tunnelling microscopy of carbon nanotubes. *Philosophical Transactions of the Royal Society of London. Series A: Mathematical, Physical and Engineering Sciences*, 362, 2187-2203

Clauss, W. (1999). Electron backscattering on single-wall carbon nanotubes observed by scanning tunneling microscopy. *EPL (Europhysics Letters)*, 47, 601

Yaguchi, T. (2001). Electronic states in capped carbon nanotubes. *J. Phys. Soc. Jpn.*, 70, 1327-1341

29

# Liquid Crystal - Anisotropic Nanoparticles Mixtures


Vlad Popa-Nita[1], Matej Cvetko[2] and Samo Kralj[3]
[1]*Faculty of Physics, University of Bucharest*
[2]*Regional Development Agency Mura, Murska Sobota*
[3]*Faculty of Natural Science and Mathemetics, University of Maribor*
[1]*Romania*
[2,3]*Slovenia*


## 1. Introduction

Carbon nanotubes (Iijima, 1991) are one of the most interesting new materials which emerges during the last twenty years. They either consist of a single sheet of carbon atoms covalently bonded in hexagonal arrays rolled up into a cylinder with a diameter of about one nanometer (single-walled nanotubes - SWNTs), or are built up of multiple carbon sheets producing rods with diameters ranging from a few to tens or even hundreds of nanometer (multi-walled nanotubes - MWNTs). The aspect ratio of these objects can vary from hundred to many thousands.

Most of CNTs extraordinary properties of potential use in various applications could be realized in relatively well aligned samples. One method of alignment consists in dispersing the CNTs into a nematic phase of a LC (either thermotropic or lyotropic). For recent reviews see e.g., (Lagerwall & Scalia, 2008; Rahman & Lee, 2009; Zakri, 2007; Zhang & Kumar, 2008).

Theoretically, equilibrium orientation of a single elongated particle immersed in a nematic LC phase is rather well explored (Andrienko et al., 2002; 2003; Brochard & de Gennes, 1970; Burylov & Raikher, 1990; 1994; Hung et. al., 2006). Continuum theory predicts alignment of the particle's longer axis along the nematic director $\vec{n}$ for different types of boundary conditions in the strong anchoring limit case (Brochard & de Gennes, 1970; Burylov & Raikher, 1990; 1994). On the contrary, in the weak anchoring limit the particle may orient either along or perpendicular to the nematic director depending on the boundary conditions (Burylov & Raikher, 1990).

The collective behavior of CNTs dispersed in isotropic solvents or in LC is theoretically relatively weakly explored. Due to their structure and behavior, the CNTs can be consider essentially as rigid-rod polymers with a large aspect ratio (Green et. al., 2009). The steric theory for the electrostatic repulsion of long rigid rods has been used to investigate the SWNT phase behavior in their suspensions (Sabba & Thomas, 2004). Calculations have shown that SWNTs in a good solvent is analogous to the classic rigid-rod system if the van der Waals force between CNTs is overcome by strong repulsive interrods potentials. When the solvent is not good, the van der Waals attractive interactions between the rods are still strong and as a result, only extremely dilute solutions of SWNTs are thermodynamically stable and no LC phases form at room temperature. The liquid crystallinity of CNTs with and without van der Waals interactions has been analyzed by using the density functional theory (Somoza & Sagui, 2001). In the presence of van der Waals interaction, the nematic as well as the columnar phases occur in the temperature-packing fraction phase diagram in a wide range of very high temperatures. In the absence of van der Waals interaction the system is dominated only by steric repulsive interactions. With an increase of packing fraction, the system undergoes an isotropic-nematic phase transition via a biphasic region. The isotropic-nematic packing fraction decreases with the increase of the aspect ratio of CNTs. The phase behavior of rodlike particles with polydisperse length and solvent-mediated attraction and repulsion is described by an extension of the Onsager theory for rigid rods (Green et. al., 2009). The main conclusion of these theoretical models is that to obtain liquid crystal phases of CNTs at room temperature the strong van der Waals interaction between them must be screened out. This requires a good solvent with an ability to disperse CNTs down to the level of individual tube.

In two recent papers (van der Schoot et. al., 2008; Popa-Nita & Kralj, 2010), two of us presented a phenomenological theory for predicting the alignment of CNTs dispersions in thermotropic nematic LC in the two limits of the anchoring of LC molecules at the CNT surface. We combined the Landau-de Gennes free energy for thermotropic ordering of the LC solvent and the Doi free energy for the lyotropic nematic ordering of CNTs caused by excluded-volume interactions between them. We have analyzed the phase ordering of the binary mixture as a function of the volume fraction of CNTs, the strength of the coupling and the temperature.

However, coupling between LC molecules and nanoparticles (NPs) of regular geometry could in some circumstances give rise to disordered structures with pronounced memory effects. Namely, LC orientational ordering is extremely sensitive to perturbations due to its soft character (de Gennes & Prost, 1993). For example, if a LC is quenched from an isotropic into a nematic phase a continuous symmetry breaking takes place (Imry & Ma, 1975; Kralj et al., 2008; Zurek, 1996). In the isotropic phase all directions are equivalent while in the nematic phase a preferred orientation is singled out locally. Because of finite speed of information propagation well separated regions are causally disconnected. For this reason a domain-type in orientational ordering is inevitable formed. A domain pattern is well characterized by a single characteristic domain size $\xi_d$. In pure LC the domain size grows with time obeying the scaling law $\xi_d \propto t^\gamma$, where $\gamma = 0.5$ in a bulk sample (Bradac et al., 2002). The sample gradually evolves into a homogeneously aligned sample in order to reduce relative expensive domain wall penalties. In a liquid crystal- NP mixture, the NPs could act as pinning centers and consequently domain pattern could be stabilized (Kralj et al., 2008). Therefore, in certain conditions NPs could introduce disorder into a system.

In the present paper we study both ordering and disordering phenomena in a LC-NP mixture. In the first part of the present paper we focus on LC induced ordering of CNT. We present comparatively the results of our phenomenological model in the two limiting cases: i) **the weak anchoring limit** where the interaction between CNTs and LC molecules is thought to be sufficiently weak not to cause any director field deformations in the nematic host fluid and ii) **the strong (rigid) anchoring limit** where the CNT causes the nematic director field distortions generating topological singularities. In the second part we study conditions where orientational ordering of a NP-LC mixture could be essentially short ranged.

The plan of the paper is as follows. In Sec. 2 we study LC driven orientational ordering of CNTs. In Subsection 2.1 our phenomenological model is introduced. In Sec. 2.2 we analyze ordering of CNT in the isotropic LC phase. The trictitical behavior is analyzed in detail in Sec. 2.3. Possibility of homogeneous alignment of CNTs in investigated in Sec. 2.4. Both weak and strong anchoring LC-CNT interactions are taken into account. In Sec. 3 possible disordering effects in LC-NP mixtures are analyzed using a simple lattice-type semimicroscopic description. The semimicroscopic model is described in Sec. 3.1 and the corresponding simulation method in Sec. 3.2. The NPs induced domain type stabilization of LC ordering is demonstrated in Sec. 3.3. In the last section we summarize our conclusions.

## 2. LC induced CNT ordering

We first discuss conditions under which LC orientational ordering could be used in order to align immersed anisotropic NPs. We confine our interest to CNTs due to their importance in various applications. The ordering of LC-CNT mixtures is treated using a mean field type phenomenological model.

### 2.1 Free energy

The free energy per unit volume of the CNTs-LC binary mixture consists of four contributions

$$f = f_{mix} + f_{CNT} + f_{LC} + f_C. \tag{1}$$

The first term in Eq. (1) is the free energy density of isotropic mixing of CNTs and LC that in the framework of Flory lattice theory is given by (Flory, 1953)

$$f_{mix}/k_B T = v_{CNT}^{-1} \Phi \ln \Phi + v_{LC}^{-1}(1-\Phi)\ln(1-\Phi) + v_0^{-1}\chi\Phi(1-\Phi), \tag{2}$$

where $k_B$ is the Boltzmann constant, $T$ is the absolute temperature, $\Phi$ is the volume fraction of CNT, $1-\Phi$ is the volume fraction of LC, and $\chi \equiv U_0/k_B T$ is the Flory-Huggins interaction parameter related to the isotropic interaction between CNT and LC (Flory, 1953). In the following, we assume $\chi > 0$ (positive free energy of mixing), this being the most usual case, at least when van der Waals interactions are dominant. Here $v_{CNT} \approx \frac{\pi}{4} L D^2$ approximates volume occupied by a carbon nanotube of length $L$ and diameter $D$. The volume of a LC molecule of length $l$ and diameter $d$ is given by $v_{LC} \approx \frac{\pi}{4} l d^2$. In the following we consider that the volume of a LC molecule is equal to the volume of a cell in the Flory lattice (Flory, 1953) ($v_{LC} = v_0$). The first two terms in Eq. (2) represent the entropy of isotropic mixing of LC and CNT components neglecting their orientational degree of ordering.
The free energy density representative of the excluded volume effects responsible for the first order nematic-isotropic transition of CNT is expressed as (Doi & Edwards, 1989)

$$f_{CNT}/k_B T = \frac{\Phi}{v_{CNT}} \left[ \frac{1}{2}\left(1 - \frac{u}{3}\right) S_{CNT}^2 - \frac{u}{9} S_{CNT}^3 + \frac{u}{6} S_{CNT}^4 \right]. \tag{3}$$

The parameter $u$ is related to the volume fraction of CNT by the relation $u = \Phi L/D$. The model neglects the van der Waals attractions between CNTs which are responsible for their tendency to form bundles. The degree of orientational ordering of CNT is given by the order parameter $S_{CNT}$. Perfectly aligned CNT correspond to $S_{CNT} = 1$, and isotropic ordering is signaled by $S_{CNT} = 0$. On increasing $\Phi$ this term enforces the first order orientational phase transition of CNT from the isotropic phase of CNT with $S_{CNT} = 0$, to the nematic phase of CNT phase with $S_{CNT} = (1 + \sqrt{9 - 24/u})/4$. The first order nematic-isotropic phase transition takes place at $u = u_{NI} = 27/10$ and $S_{CNT}(u_{NI}) = 1/3$. The limits of stability of nematic and isotropic phases are given by $u^+ = 8/3$ ($S_{CNT}^+ = 1/4$) and $u^* = 3$ ($S_{CNT}^* = 1/2$).

For the thermotropic uniaxial nematic LC component we use the standard Landau-de Gennes (de Gennes & Prost, 1993) free energy density in terms of the nematic order parameter $S_{LC}$

$$f_{LC} = (1-\Phi)\left[\frac{3}{2}a(T-T^*)S_{LC}^2 - \frac{3}{4}BS_{LC}^3 + \frac{9}{4}CS_{LC}^4\right], \tag{4}$$

where $(1-\Phi)$ accounts for the part of the volume not taken up by LC. The quantities $T^*$ (the undercooling limit temperature of the isotropic phase of the pure liquid crystal), $a$, $B$, and $C$ are material-dependent constants. This free energy density enforces the weakly first order phase transition from the isotropic phase to the nematic phase of LC. At $T = T_{NI} = T^* + B^2/(24aC)$, the two phases of LC, nematic ($S_{nem\,0} = B/6C$) and isotropic ($S_{iso} = 0$) coexist in equilibrium. The detailed derivation of the coupling free energy density term $f_C$ in the two limiting regimes has been presented previously (van der Schoot et. al., 2008; Popa-Nita & Kralj, 2010). Here we give only the final expressions. The two regimes can be defined by the relation between the radius of the particle $R$ and the surface extrapolation length $d_e = K/W$, where $K$ is the Frank elastic constant and $W$ is the anchoring energy.

i) **the weak anchoring limit** ($R << d_e$)- the anchoring is not able to produce large deformations in the surrounding nematic matrix. In this limit the coupling free energy density writes as (van der Schoot et. al., 2008)

$$f_C = -\gamma_w\Phi(1-\Phi)S_{LC}S_{CNT}\left(1-\frac{1}{2}S_{CNT}\right), \tag{5}$$

where the coupling constant is given by $\gamma_w = 4W/6R$. Depending on the value of $R$ and considering a typical value of anchoring energy of $10^{-6}$N/m, the coupling constant of the weak anchoring regime can range between $10^{-3} - 10^{3}$N/m$^2$.

ii) **strong anchoring limit** ($R >> d_e$) - the anchoring is rigid and gives rise to topological singularities of the nematic director which cause strong interaction between the dispersed nanotubes. In this limit the coupling free energy density writes as (Popa-Nita & Kralj, 2010)

$$f_C = -\gamma_s\Phi(1-\Phi)S_{LC}^2S_{CNT}\left(1-\frac{1}{2}S_{CNT}\right), \tag{6}$$

where the coupling constant is given by $\gamma_s = 2K/3R^2$. Depending on the value of $R$ and considering a typical value of elastic constant of $10^{-11}$N the coupling constant of the strong anchoring regime can range between $10^3 - 10^7$N/m$^2$.

Using Eqs. (2), (3), (4), (5), and (6), the total phenomenological free energy density in the weak anchoring limit becomes:

$$\begin{aligned} f = {} & k_BT\left[v_{CNT}^{-1}\Phi\ln\Phi + v_{LC}^{-1}(1-\Phi)\ln(1-\Phi) + v_{LC}^{-1}\chi\Phi(1-\Phi)\right] \\ & + \frac{k_BT\Phi}{v_{CNT}}\left[\frac{1}{2}\left(1-\frac{u}{3}\right)S_{CNT}^2 - \frac{u}{9}S_{CNT}^3 + \frac{u}{6}S_{CNT}^4\right] \\ & + (1-\Phi)\left[\frac{3}{2}a(T-T^*)S_{LC}^2 - \frac{3}{4}BS_{LC}^3 + \frac{9}{4}CS_{LC}^4\right] \\ & - \gamma_w\Phi(1-\Phi)S_{LC}S_{CNT}\left(1-\frac{1}{2}S_{CNT}\right), \end{aligned} \tag{7}$$

while in the strong anchoring limit, only the last term differ:

$$f = k_B T \left[ v_{CNT}^{-1} \Phi \ln \Phi + v_{LC}^{-1} (1-\Phi) \ln(1-\Phi) + v_{LC}^{-1} \chi \Phi (1-\Phi) \right]$$
$$+ \frac{k_B T \Phi}{v_{CNT}} \left[ \frac{1}{2} \left(1 - \frac{u}{3}\right) S_{CNT}^2 - \frac{u}{9} S_{CNT}^3 + \frac{u}{6} S_{CNT}^4 \right]$$
$$+(1-\Phi) \left[ \frac{3}{2} a (T - T^*) S_{LC}^2 - \frac{3}{4} B S_{LC}^3 + \frac{9}{4} C S_{LC}^4 \right]$$
$$-\gamma_s \Phi (1-\Phi) S_{LC}^2 S_{CNT} \left(1 - \frac{1}{2} S_{CNT}\right). \quad (8)$$

In calculations we consider a regime determined by the following geometrical and material parameters. We limit to CNT of characteristic dimensions $L \approx 400$ nm and $R \approx 1$ nm. For LC molecules we set $L \approx 3$ nm and $R \approx 0.25$ nm, corresponding to a typical nematogen. For LC material constants we chose as representative pentylcyanobiphenyl (5CB) LC, for which $T^* = 306$ K, $a \approx 3.5 \cdot 10^4$ J $\cdot$ m$^{-3} \cdot$ K$^{-1}$, $B \approx 7.1 \cdot 10^5$ J $\cdot$ m$^{-3}$, $C \approx 4.3 \cdot 10^5$ J $\cdot$ m$^{-3}$, $K \sim 10^{-11}$ N (Oswald & Pieranski, 2005). This choice yields $S_{nem\,0} \approx 0.28$ and $T_{NI} = T^* + B^2/24aC \approx$ 307.5 K.
Using this phenomenological form of the free energy, in the following sections we present the phase behavior of the binary mixture of CNTs and LC as a function of temperature, volume fraction and the coupling strength.

## 2.2 Dispersion of CNTs in the isotropic phase of LC

In this case, $S_{LC} = 0$, the coupling free energy densities (5) and (6) cancel and the free energy density becomes

$$f/k_B T = v_{CNT}^{-1} \Phi \ln \Phi + v_{LC}^{-1} (1-\Phi) \ln(1-\Phi) + v_{LC}^{-1} \chi \Phi (1-\Phi)$$
$$+ \frac{\Phi}{v_{CNT}} \left[ \frac{1}{2} \left(1 - \frac{u}{3}\right) S_{CNT}^2 - \frac{u}{9} S_{CNT}^3 + \frac{u}{6} S_{CNT}^4 \right]. \quad (9)$$

The equilibrium between the isotropic ($S_{CNT} = 0$) and the nematic phase ($S_{CNT} > 0$) of CNTs is determined by equating the chemical potentials of CNTs ($\mu_{CNT} = v_{CNT}[f + (1-\phi)\partial f/\partial \phi]$) and LC ($\mu_{LC} = v_{LC}(f - \phi \partial f/\partial \phi)$) in the two phases. The value of $S_{CNT}$ is obtained by minimizing the free energy with respect to $S_{CNT}$.
In the absence of the isotropic interaction term ($\chi = 0$), for our set of geometric parameters the nematic-isotropic coexistence is determined by the following values of parameters: ($\Phi_{iso} =$ 0.013487, $S_{CNT} = 0$) and ($\Phi_{nem} = 0.013513$, $S_{CNT} = 0.3365$) and correspondingly the relative variation of volume fraction of CNTs at the transition (the Flory chimney) is about 0.19%. The phase gap in this condition is very narrow indeed less than the 1% predicted by Onsager and Flory for monodisperse lyotropic LC. On the basis of the polarized light microscopy, Song and Windle (Song & Windle, 2005) have estimated a biphasic range between 1% and 4% in an aqueous dispersion of MWNTS. This comparatively large range of the biphasic region is plausibly caused by the polidispersity in terms of length, diameter, and straightness of the nanotubes as well as the possibility of segregation of the more nematogenic tubes.
In Figure 1, we have plotted the relative variation of the volume fraction of the CNTs at the nematic-isotropic phase transition as a function of the isotropic interaction parameter $\chi$.

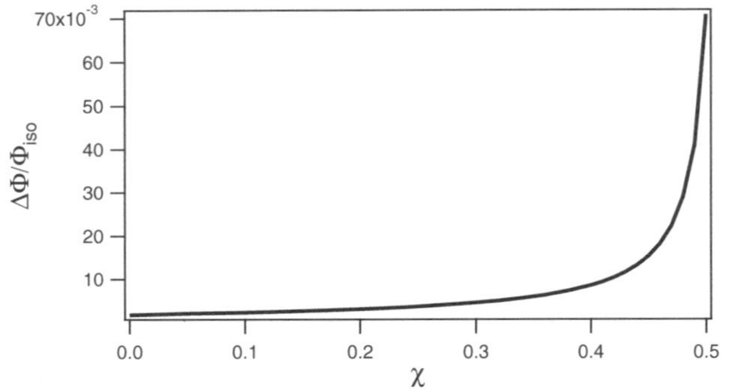


Fig. 1. The relative variation of the volume fraction of CNTs at the nematic-isotropic phase transition in the isotropic phase of LC.

Because the isotropic interaction parameter is proportional with reciprocal temperature (here we did not consider this dependence), the volume fraction gap at the transition increases with lowering the temperature; the poor solvent for CNTs becomes poorer lowering the temperature. For values of $\chi$ larger than $\approx 0.4$ the relative variation of the volume fraction at the transition increases considerably with the value of $\chi$. The predicted value of $\Delta\Phi/\Phi_{iso} = 1\%$ is obtained for $\chi = 0.42$.

### 2.3 Tricritical point

The free energy density is given now by the general forms (7) and (8), respectively.

For very small values of the coupling parameter $\gamma_w$ (or equivalently $\gamma_s$), the nematic-isotropic phase transition of the CNTs is first order and the order parameter jumps at the transition from zero to some non-zero value depending on the value of the coupling parameter. With increasing the interaction parameter, the transition becomes continuous at a tricritical value $\gamma_{w,t}$ (or equivalently $\gamma_{s,t}$). The tricritical point where the discontinuous phase transition becomes continuous is obtained by solving the equations $\partial^2 f/\partial S_{LC}^2 = \partial^3 f/\partial S_{LC}^3 = 0$. They yield at the tricritical point universal values for the order parameter $S_{CNT}^{(t)} = 1/6$ and volume fraction $\Phi_t = 0.01296$. The tricritical values of the coupling parameter $\gamma_t$ as a function of temperature are presented in Figure 2.

With decreasing the temperature, as $S_{LC}$ increases, the external field felt by the CNTs increases and the nematic-isotropic phase transition of CNTs becomes continuous for increasingly lower values of interaction parameter. Our estimates of the coupling constant suggest that for typical value of parameters, $\gamma >> \gamma_t$ for both limiting cases, meaning that in the nematic phase of LC, the nematic-isotropic phase transition of CNTs becomes a continuous transition.

To illustrate how in the limit of low values of coupling parameter the isotropic-nematic phase transition of the CNTs is affected by the nematic host fluid, the phase diagram of the coexisting volume fractions as a function of the coupling parameter is plotted in Figure 3 for both limiting regimes. The binodals are drawn at the undercooling limit temperature of the isotropic phase ($T^* = 307.56K$).

In both limiting regimes, the gaps of the volume fractions decrease with increasing the coupling constant and becomes zero for $\gamma_w = \gamma_{w,t} = 79.6$ N/m$^2$ (for weak anchoring) and for $\gamma_s = \gamma_{s,t} = 191$ N/m$^2$ for strong anchoring case, respectively. The isotropic-nematic phase transition of CNTs takes place at lower volume fraction with increasing the coupling constant.

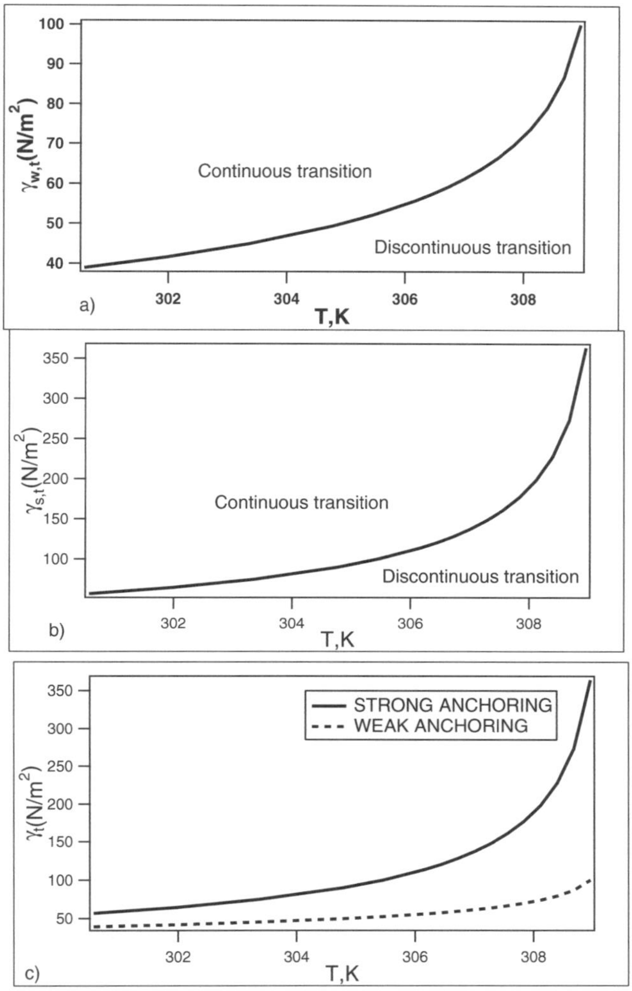


Fig. 2. The tricritical values of the coupling parameter $\gamma_t$ as function of temperature. The weak anchoring limit case is represented in Fig. 2a. In Fig. 2b is shown the strong anchoring limit case, while in Fig. 2c the both regimes are represented.

### 2.4 Homogeneous mixture

One of the most challenging task for realizing the applications in optoelectronics is to homogeneously disperse the CNTs into the nematic host, as CNTs have a tendency to aggregate into networks and fibrils (MWNTs) within the dispersion due to the van der Waals interactions between the nanotubes. The three different standard approaches to obtain the homogeneous dispersion are reviewed in (Lagerwall & Scalia, 2008; Rahman & Lee, 2009).
To calculate the phase diagram of a homogeneous mixture, we use the free energy density given by Eq. (7) (for the weak anchoring case) and the same quantity given by Eq. (8) (corresponding to the strong anchoring case) with a constant volume fraction $\Phi$ of the CNTs. The equilibrium values of the order parameters $S_{CNT}$ and $S_{LC}$ are obtained by minimizing

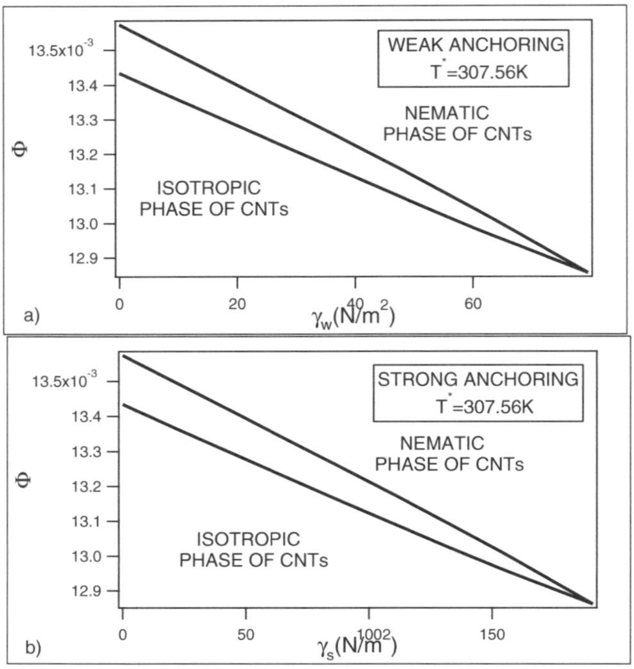


Fig. 3. Phase diagram of CNTs in the nematic host fluid. Indicated are the coexisting volume fractions in the isotropic and nematic phases discussed in the main text. The weak anchoring case is shown in Fig. 3a, while Fig. 3b presents the strong anchoring case.

the free energy density ($\partial f/\partial S_{CNT} = \partial f/\partial S_{LC} = 0$). The corresponding phase diagrams are plotted and discussed in the next two sections.

### 2.4.1 Weak anchoring case

The phase diagram $(\Phi, T)$ in the weak anchoring regime for a typical value of the coupling constant $\gamma = 6.6 \cdot 10^5$ N/m$^2$ is plotted in Fig. 4.

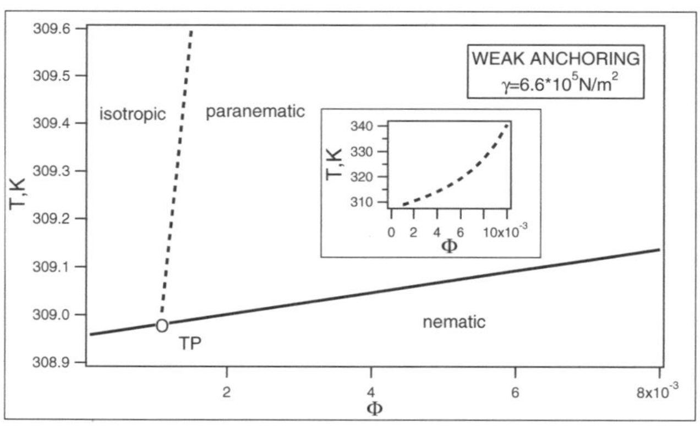


Fig. 4. The $(\Phi, T)$ phase diagram of a homogeneous mixture in the weak anchoring limiting case.

Three different regions are to be distinguished. **The isotropic region** corresponds to a total absence of the orientational order (both components are in the isotropic phase), **the paranematic region** is characterized by a very small degree of the orientational order of both components, while in **the nematic region** both components possess a relatively large degree of orientational order. The dashed line defines a second order (the order parameter is continuous at the transition, while its derivative with respect to temperature has a jump) isotropic-paranematic phase transition (inside the figure we have shown this line for larger values of the temperature). The first segment ($\Phi \leq 0.00108$) of the continuous line corresponds to first order (the order parameter has a jump at the transition) nematic-isotropic phase transition of both components. The second segment ($\Phi \geq 0.00108$) corresponds to the first order nematic-paranematic phase transition of the both components. The circle defines the triple point of the system that has the coordinates $\Phi_t = 0.00108$ and $T = 308.98$ K. The increase of nematic-isotropic phase transition temperature (shown by the continuous line in Fig. 4) is experimentally verifiably (Duran et. al., 2005; Lebovka et. al., 2008) and could be explained by the fact that CNTs act as heterogeneous nucleation agents for LC.
To see in more detail the characteristics of these phase transitions we have plotted in Figure 5 the order parameters profiles as a function of volume fraction of CNTs for two different values of the temperature.

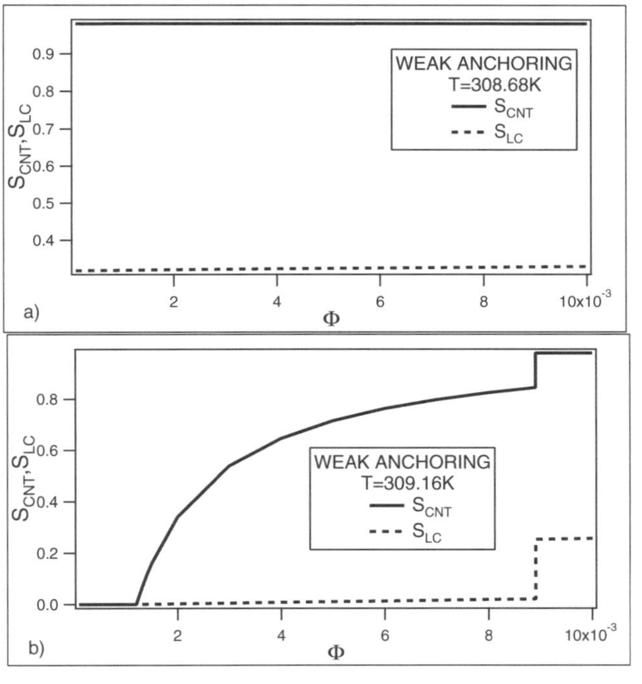


Fig. 5. The order parameters profiles of CNTs (continuous line) and LC (dotted line) for $\gamma = 6.6 \cdot 10^5$ N/m$^2$ and for two different values of the temperature.

In Figure 5a, the order parameters profiles are shown for $T = 308.68$ K, value that belongs to the nematic region in Fig. 4. There is no phase transition for this value of temperature and as consequence the order parameters are continuous and CNTs are almost perfectly aligned ($S_{CNT} \approx 1$) for all values of volume fraction (see the continuous line). On the

contrary, the value of temperature in Figure 5b ($T = 309.16$ K) corresponds to a second order isotropic-paranematic phase transition at a small value of the volume fraction and to a first order paranematic-nematic phase transition at a larger value of the volume fraction. In this case, the CNTs are strongly aligned at relatively large value of the volume fraction.
In Figure 6, we have shown the order parameters profiles as a function of temperature for two different values of the volume fraction of CNTs.

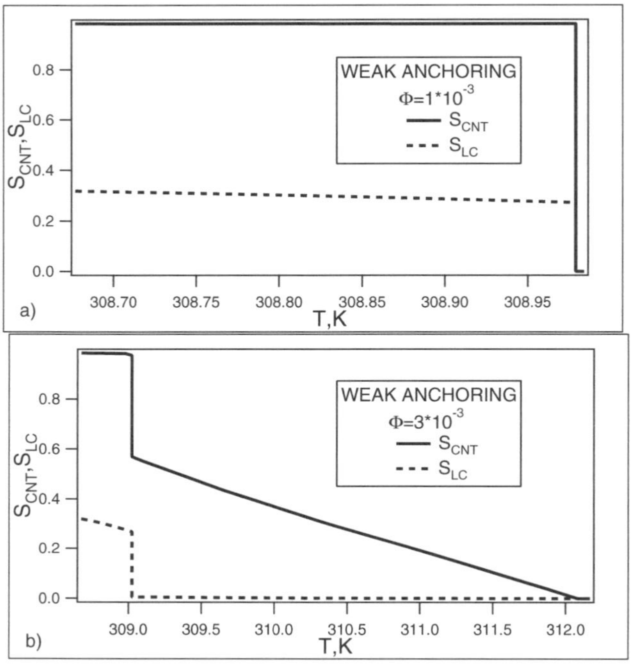


Fig. 6. The order parameters profiles of CNTs (continuous line) and LC (dotted line) for $\gamma = 6.6 \cdot 10^5$ N/m$^2$ and two different values of the volume fraction.

In Figure 6a, the order parameter profiles are shown for a small value of the volume fraction ($\Phi = 1 \cdot 10^{-3}$). For this value of the volume fraction, increasing the temperature the only transition is the first order nematic-isotropic phase transition (see Figure 4) during which both order parameters jump to zero. The larger value of the volume fraction of Figure 6b induces two phase transitions (see Figure 4): a first order nematic-paranematic phase transition at a lower value of the temperature during which both order parameters jump at lower values and a second order paranematic-isotropic phase transition at a larger value of the temperature during which both order parameters decrease continuously to zero.

### 2.4.2 Strong anchoring case

The corresponding phase diagram $(\Phi, T)$ for a typical value of the coupling constant $\gamma = 6.6 \cdot 10^5$ N/m$^2$ is plotted in Fig. 7. Even if the corresponding phase diagram is similar to the weak anchoring limiting case, there are some differences which we discuss in the following.
Also in this strong anchoring case, there are three different regions in the phase diagram. **The region I** corresponds to a total absence of the orientational order (both components are in the isotropic phase), **region II** corresponds to a nematic phase of CNT and an isotropic phase of

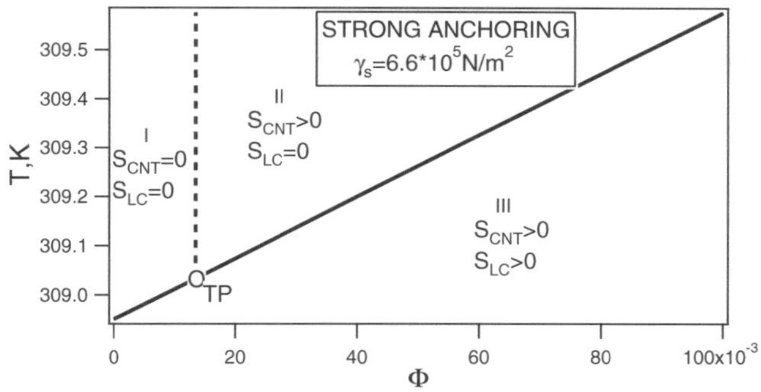


Fig. 7. The $(\Phi, T)$ phase diagram of a homogeneous mixture in the strong anchoring limiting case.

LC, while in **the region III** both components are in the nematic phase. The vertical dashed line defines a second order isotropic (region I)-nematic (region II) phase transition of the CNTs in the isotropic phase of LC. The first segment ($\Phi \leq 0.0135$) of the oblique continuous line corresponds to the first order isotropic-nematic phase transition of CNTs as well as to the first order isotropic-nematic phase transition of LC. The second segment ($\Phi \geq 0.0135$) corresponds to the first order isotropic-nematic phase transition of the LC and to a first order nematic-nematic phase transition of the CNTs (in the region II the degree of orientational order of CNTs is lower that that of region III and in passing from region II to region III there is a jump of the CNTs order parameter). The circle defines the triple point of the system that has the coordinates $\Phi_t = 0.0135, T = 309.03$ K. The experimental fact (Duran et. al., 2005; Lebovka et. al., 2008) that the isotropic-nematic phase transition temperature of LC increases with the volume fraction of CNTs is again theoretically explained by the present model.
In Figure 8 the order parameters profiles as a function of volume fraction of CNTs for two different values of the temperature are shown.
The value of the temperature ($T = 309$ K) in Figure 8a, corresponds to one isotropic-nematic first order phase transition obtained increasing the volume fraction (transition between regions I and III in Figure 7), phase transition during which both order parameters jump from zero to finite values. CNTs becomes almost perfectly aligned at very small values of the volume fractions. The value of temperature in Figure 8b ($T = 309.2$ K) induces a second order isotropic-nematic phase transition of CNTs at a relatively small volume fraction (transition between regions I and II in Figure 7) followed by a first order isotropic-nematic phase transition of both components LC (transition between regions II and III in Figure 7). In this case, the CNTs are strongly aligned at relatively large value of the volume fraction.
The order parameters profiles as a function of temperature for two different values of the volume fraction of CNTs are plotted in Figure 9.
In Figure 9a, the order parameter profiles are shown for a small value of the volume fraction ($\Phi = 1 \cdot 10^{-2}$). For this value of the volume fraction, increasing the temperature, the transition is the first order nematic-isotropic phase transition (transition between regions III and I in Figure 7) during which both order parameters jump to zero. The larger value of the volume fraction of Figure 9b induces a first order phase transition of both components (transition between regions III and II in Figure 7). The order parameter jump of CNTs is very small (in both region III and II, CNTs are in nematic phase but with different degree of orientational order), while the order parameter of LC jumps from a finite value to zero.

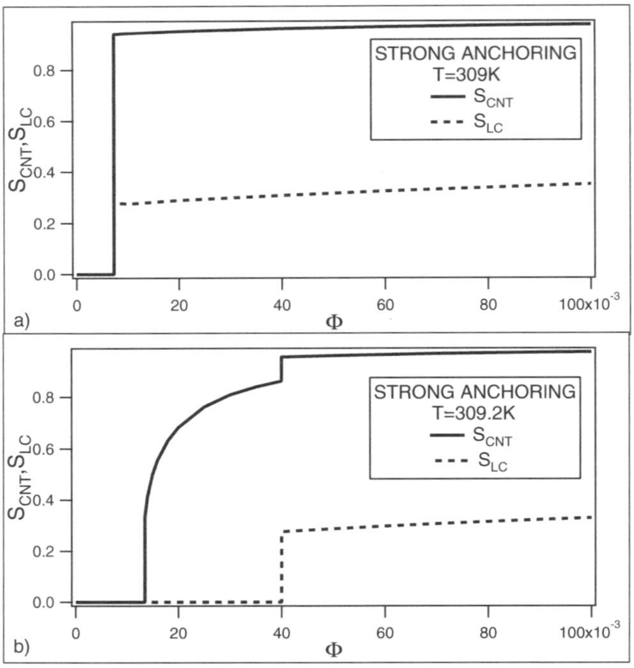


Fig. 8. The order parameters profiles of CNTs (continuous line) and LC (dotted line) for $\gamma = 6.6 \cdot 10^5$ N/m$^2$ and two different values of the temperature.

## 3. Nanoparticle induced disorder

We next consider mixtures of LC molecules and anisotropic NPs. It is of interest to find conditions for which NPs could impose relatively strong disorder to surrounding LC molecules. In order to find simple possible key conditions for such phenomena we constrain our investigations to appropriate type of interactions among NPs and LC molecules. We set LC molecules and NPs to be of comparable size. We do not impose any inherent disorder to systems. Molecules of both types are assumed to have regular shape and interaction potential which are intuitively expected to enforce order on a macroscopic scale. Below we demonstrate that symmetry breaking and NP driven stabilization of LC domains could destroy LC long range order in case of appropriate initial conditions.

### 3.1 Interaction energy

We confine our interest to mixtures of liquid crystals and anisotropic nanoparticles, where both components have comparable characteristic geometrical length. We assume that concentrations of NPs are relatively low. In order to study structural properties of the ensembles we use a lattice-spin type model (Bellini et al., 2000; Lebwohl & Lasher, 1972) in three dimensional space. The lattice points form a three dimensional cubic lattice characterized by the lattice constant $a_0$. The number of sites equals $N^3$, where we typically set $N = 80$. The NPs are randomly distributed within the lattice with probability $p$. For $p = 1$ all the sites are occupied by NPs and a pure LC sample is obtained in the limit $p = 0$. Therefore, value of $p$ is equal to the volume fraction $\Phi$ of NPs.

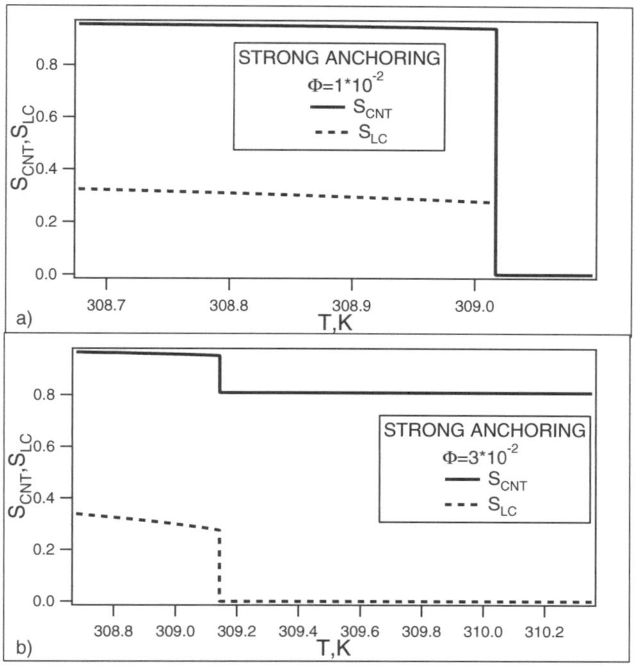


Fig. 9. The order parameters profiles of CNTs (continuous line) and LC (dotted line) for $\gamma = 6.6 \cdot 10^5$ N/m$^2$ and two different values of the temperature.

A key anisotropic property of a LC molecule or a nanoparticle at a site $\vec{r}_i$ is roughly represented by the unit vector $\vec{n}_i$ and $\vec{m}_i$, respectively. We assume the head-to-tail invariance of LC molecules. Therefore, the orientations $\pm\vec{n}_i$ are equivalent due to a rod-like character of molecules. In the continuum limit $\vec{n}_i$ roughly corresponds to the nematic director field. On the other hand sign of $\vec{m}_i$ matters. For example, such property of a NP could be due to its magnetic or electric dipole. We henceforth refer to $\vec{n}_i$ ( $\vec{m}_i$) as the nematic (magnetic) spin. The interaction energy $E$ of the system is given by (Cvetko et al., 2009; Krasna et al., 2010; Lebwohl & Lasher, 1972)

$$E = -\sum_i\sum_j J_{ij}^{(n)}\left(\vec{n}_i\cdot\vec{n}_j\right)^2 - \sum_i\sum_j J_{ij}^{(m)}\left(\vec{m}_i\cdot\vec{m}_j\right) + \sum_i\sum_j J_{ij}^{(nm)}\left(\vec{n}_i\cdot\vec{m}_j\right)^2. \quad (10)$$

The constants $J_{ij}^{(n)}$, $J_{ij}^{(m)}$, and $J_{ij}^{(nm)}$ stand for the coupling strengths LC-LC, NP-NP, and LC-NP, respectively. The sums run over all the sites and only the interactions between first neighbors are different from zero. The first term describes interaction among LC molecules, where $J_{ij}^{(n)} = J > 0$ for neighboring molecules. Therefore, a pair of LC molecules tend to orient either parallel or antiparallel. The coupling between neighboring NPs is determined with $J_{ij}^{(m)} = J_{NP} > 0$, enforcing antiparallel orientation. On the contrary, neighboring LC-NP pairs tend to be aligned perpendicularly due to $J_{ij}^{(nm)} = w > 0$. The latter condition is crucial in order to introduce strong disorder into the system and related memory effects. Note that replacing

the 2nd term by Lebwohl-Lasher type coupling $-\sum\sum J_{ij}^{(m)}\left(\overrightarrow{m}_i \cdot \overrightarrow{m}_j\right)^2$ would not introduce qualitatively different behavior.
In the subsequent work distances are scaled with respect to $a_0$. The interaction energies are measured with respect to $J$, i.e. we set $J = 1$.

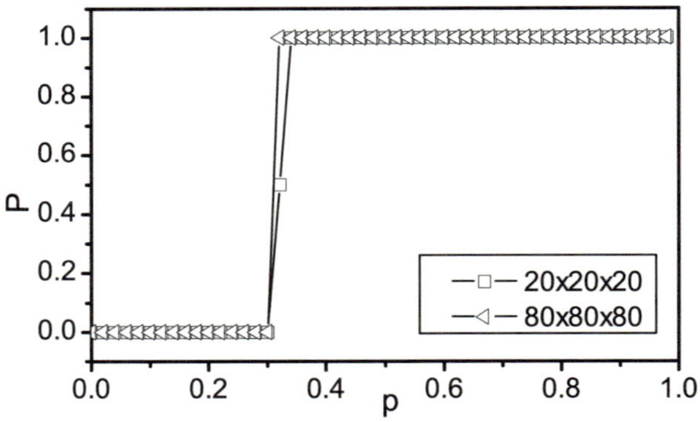


Fig. 10. The percolation probability $P$ as a function of $p$ and system size. For a finite value of $N$ the percolation threshold $p_c$ is defined as $P(p_c) = 0.5$. We obtain $p_c \sim 0.3$.

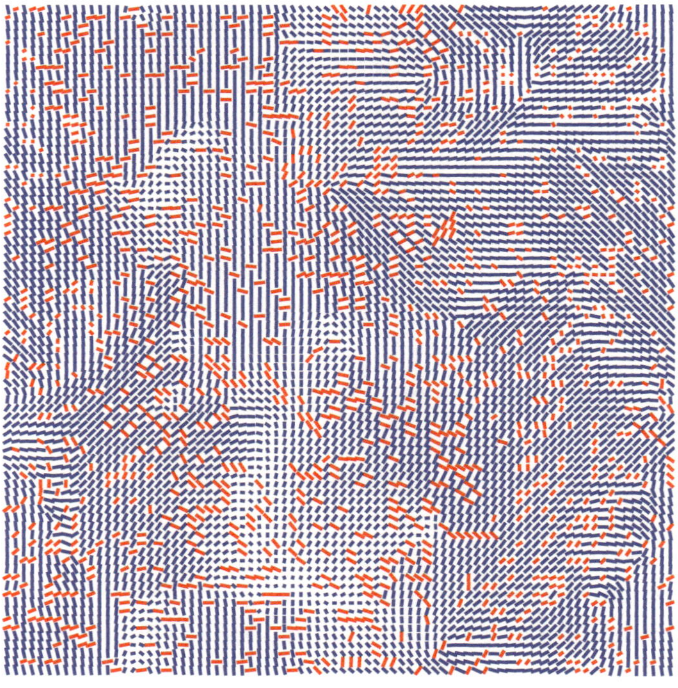

Fig. 11. Representative cross-sections of a LC+NP mixture. NPs and LC molecules are colored with red and blue, respectively. $p = 0.2$.

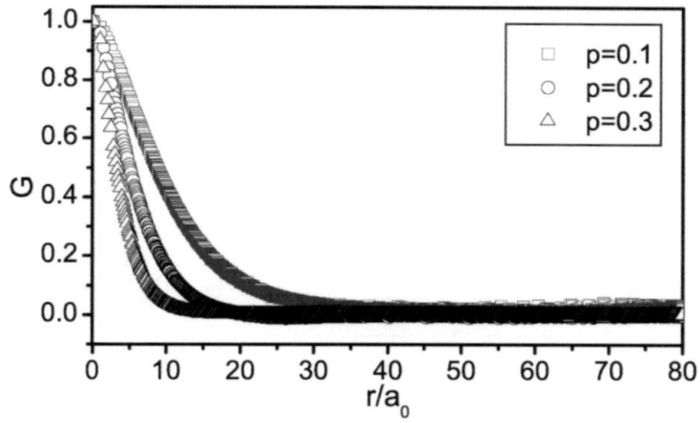


Fig. 12. Typical $G(r)$ plots for different concentrations of NPs.

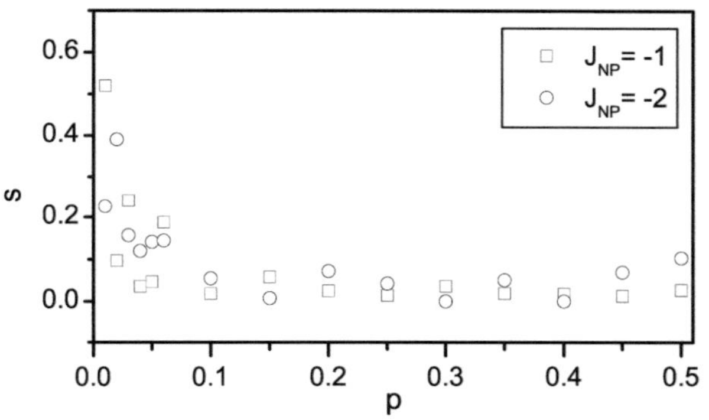


Fig. 13. Characteristic $s(p)$ behavior. The random field regime extends roughly between $p = p_{RF} \sim 0.1$ and $p \sim p_c \sim 0.3$.

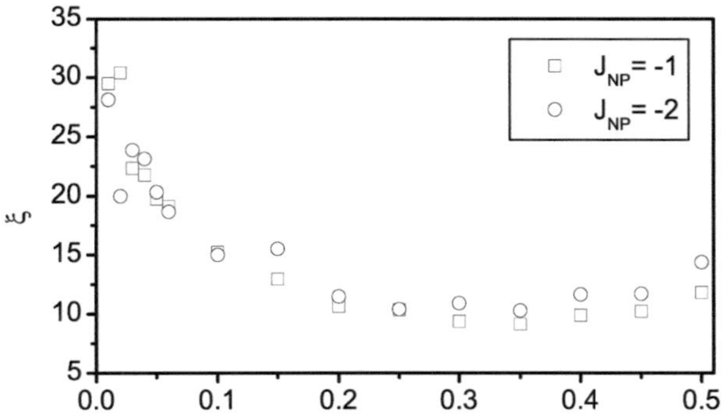


Fig. 14. The correlation length $\xi$ as a function of $p$.

### 3.2 Simulation method and measured quantities

All simulations take place at zero temperature where we minimize the interaction energy given by Eq.(10) with respect to orientation of nematic and magnetic spins. The corresponding

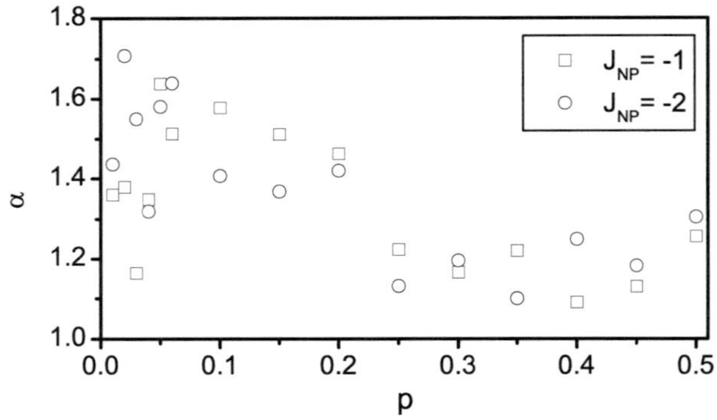


Fig. 15. The coefficient $\alpha$ as a function of $p$.

set of equations have been solved using the Newton's method. Periodic boundary conditions are imposed. Simulations are carried over different values of $p$, $J_{NP}$ and $w$. The concentration regime from $p = 0$ to $p > p_c$ is investigated where $p_c$ stands for the percolation threshold of NPs.

Both nematic and NP spins are initially randomly orientationally distributed. Experimentally such conditions correspond to quenches from high temperature isotropic phase states. Maximum simulation box sizes are set to $N = 80$. In order to diminish the influence of statistical variations ten simulations starting from statistically different initial configurations have been carried out for a given set of parameters.

In simulations we calculate steady state configurations. For obtained steady state configuration we calculate the orientational correlation function $G$ of LC molecules. It measures the orientational correlation of LC spins as a function of their mutual separation $r$ and is defined as (Cvetko et al., 2009)

$$G(r) = \frac{1}{2}\left\langle 3\left(\vec{n}_i \cdot \vec{n}_j\right)^2 - 1\right\rangle. \tag{11}$$

The brackets denote the average over all lattice sites that are separated by a distance $r$. Our test simulations indicate that $G(r)$ does not depend on orientation of $\vec{r}$, therefore we focus on $r$ dependence. The $G(r)$ properties are as follows. For completely correlated nematic spins aligned along a single symmetry breaking orientation it holds $G(r) = 1$. On the contrary $G(r) = 0$ signals completely uncorrelated nematic spins. Because each spin is necessarily parallel with itself it holds $G(0) = 1$. Furthermore, we normally expect the correlation function to be a decreasing function of distance $r$. If a long range LC order exists in the system then $G(r \to \infty) = S^2 > 0$, where $S$ stands for the nematic orientational order parameter. On the contrary, short range order (SRO) or quasi long range order (QLRO) are signaled by $G(r \to \infty) = 0$.

To obtain structural details from the simulation results we fit $G(r)$ to an empirical ansatz Eq.((Cvetko et al., 2009)).

$$G(r) = (1-s)e^{-(r/\xi)^{\alpha}} + s. \tag{12}$$

Here the coherence length $\xi$, the stretched exponential parameter $\alpha$, and $s$ are adjustable parameters. The coherence length $\xi$ measures distance over which nematic spins are relatively strongly correlated. The stretched exponential parameter $\alpha$ is introduced by analogy with the stretched exponential temporal decay which occurs in many glassy systems. The parameter $s$ directly reflects the range of ordering. Namely, $s = 0$ for SRO (Cvetko et al., 2009).

### 3.3 Nanoparticle stabilized domain patterns

We study domain pattern stabilization as a function of $p$ and typical interaction strengths. Qualitative changes in the behavior are expected below and above the percolation threshold of NPs. For this reason we first analyze the percolation characteristics in our systems. For a given concentration $p$ of NPs we calculate the probability $P$ that there exists a connected path of NPs between the oppositely placed boundaries of a simulation cell. The $P(p)$ plot in Figure 10 reveals that on increasing $p$ the percolation threshold is reached at $p_c \sim 0.3$. In the thermodynamic limit the dependence displays a phase transition type of behavior, where $P$ plays the role of order parameter. For a finite simulation cell a pretransitional tail appears below $p_c$, and at the transition steepness of $P(p)$ increases with increasing $N$. In simulations we use large enough values of $N$ so that finite size effects are negligible.

The presence of NPs enforces to LC molecules a certain amount of disorder. If the latter is strong enough LC long range could be destroyed. A typical LC-NP pattern is shown in Fig. 11 where short range order is evident. The range of ordering can be inferred from $G(r)$. In Figure 12 we plot typical correlation functions for different values of $p$. The calculated $G(r)$ dependencies are fitted with the ansatz Eq.(12).

In the regime $p < p_c$ the network of NPs is not percolated. Due to frustrating tendencies (i.e. the LC-LC interaction favors parallel or antiparallel alignment, while the LC-NP interaction enforces perpendicularly oriented configurations) it is anticipated that in this regime disorder strength is apparent. This is clearly shown in the $s(p)$ plot in Fig. 13. We observe $s(p) \sim 0$ in appropriate interval of concentrations, roughly between $p \sim p_{rf} \sim 0.1$ and $p \sim p_c$ (for $w \sim J_{NP}$. Note that on increasing $p$ above the percolation threshold the $s(p)$ dependence begins to increases with $p$. For concentrations, where $s(p) \sim 0$, the system exhibits short range order (at least approximatively). Therefore, the disorder is strong enough to destroy long range nematic LC ordering. We refer to the concentration range $p_{rf} < p < p_c$ as the **random field regime.**

We next study structural behavior across the percolation threshold. Characteristic structural properties are shown in Fig. 14 and 15. As expected a qualitative change in behavior is observed on crossing the percolation thresholds, which is particularly evident from the $\xi(p)$ plots (Fig. 14). We see that above the percolation threshold $\xi(p)$ tends to increase. The reason for this is formation of correlated order in NPs orientational order. Consequently, the LC component also becomes ordered. One sees increase in $s$ in the regime $p > p_c$. The changes in $\alpha$ are shown in Fig. 15. We see that in the **random field regime** $\alpha(p)$ values relatively strongly fluctuate. On approaching the percolation threshold values of $\alpha$ tend to assemble closer to $\alpha = 1$, which would be reached in absence of disorder.

## 4. Conclusions

In the first part of the present paper, using a mean field type phenomenological model, we have examined the phase and structural behavior of a binary mixture composed of CNTs dispersed in thermotropic LC in two anchoring regimes. In the weak anchoring limit of the nematic LC molecules at the nanotube's surface, the CNTs alignment is caused by the anisotropic interfacial tension of the nanotubes in the nematic host fluid. In this

case, the interaction between CNTs and LC molecules is sufficiently weak to not cause any director deformations in the nematic and the coupling parameter is given by $\gamma_w = 4W/6R$ (independent of $L$). In the strong anchoring limit, the nematic ordering around nanotube is distorted (this case includes two different situations, either the nematic director is nonsingular or topological defects are present). Now, the coupling parameter is given by $\gamma_s = 2K/3R^2$.
First, we have considered the dispersion of CNTs in the isotropic phase of LC (in this case $S_{LC} = 0$ and the interaction free energy densities given by Eqs. (5) and (6) cancel. By fitting the relative variation of the volume fraction of CNTs $(\Phi_{nem} - \Phi_{iso})/\Phi_{iso}$ at the transition with the experimental value, we have obtained the value of Flory-Huggins interaction parameter $\chi = 0.42$.
Second, in both anchoring cases, the first-order nematic - isotropic phase transition of CNTs dispersed in the nematic phase transforms into a continuous transition for a strong enough coupling to the nematic host fluid. The corresponding tricritical value of the coupling parameter increases with increasing temperature being larger in the strong anchoring limit case (see Figure 2). The numerical estimate of the coupling constants in the two anchoring regimes indicates that the coupling is so strong that CNTs are far above the tricritical point, meaning that the nematic-isotropic phase transition is a continuous one.
Third, in both anchoring cases, we have plotted the phase diagram of the homogeneous mixture for the same value of the coupling parameter. In both cases, three regions of the phase diagram could be distinguished and correspondingly the existence of triple points are shown. We mention that in both anchoring cases, the nematic-isotropic phase transition temperature of LC increases with the volume fraction of CNTs, a well-known experimental result.
It is to be stressed that the model presented does not consider two important features of the system: the van der Waals interaction between CNTs and the polydispersity of CNTs. In the future work we intend to include this features into the model.
In the second part of the paper we have shown that in a LC-NP mixture frozen domain type structure of orientational order might appear, yielding (at least approximate) short range orientational ordering. In order to find out key ingredients giving rise to such phenomena we use a semimicroscopic lattice model. The essential ingredients of the model are simple and regular. Pure systems of LC molecules (i.e. $\Phi = p = 0$) or NPs ( $p = 1$) tend to form regular structures. We assume that phase separation is absent and that NPs are essentially homogeneously distributed within a liquid crystal phase. If the LC-NP coupling tends to orient molecules perpendicularly (or close to such configuration), the resulting configurations might be trapped in domain type structures exhibiting short range order. The main reason for this is softness of the LC component and consequently relatively strong susceptibility to various perturbations in orientational ordering. We considered concentration ranges between $p = 0$ and slightly above the percolation threshold $p_c$. In case of comparable sizes of LC molecules and NPs we obtained $p_c \sim 0.3$. If a mixture was quenched from an isotropic phase, than structures exhibiting essentially short range order were obtained in the concentration range between $p \sim 0.1$ and $p \sim p_c$. Below $p \sim 0.1$ NPs are too diluted to destroy LC long range order. On the other for $p > p_c$ the NPs become percolated and consequently at least partially orientationally ordered. This order is transferred also to the LC component. Therefore, within the interval $p \sim 0.1$ and $p \sim p_c$ mixtures could display glass like states with pronounced memory effects. In our future study we plan to investigate in detail glassy characteristics and memory effects within this interval.

## 5. Acknowledgments

V.P.-N. is very grateful to P. Poulin and P. van der Schoot for very helpful discussions. Matej Cvetko acknowledges support of the EU European Social Fund. Operation is performed within the Operative program for development of human resources for the period 2007-2013.

## 6. References

Andrienko D., Allen M. P., Skacej G., and Žumer S. (2002) Defect structures and torque on an elongated colloidal particle immersed an a liquid crystal host. *Phys. Rev. E* Vol. 65, 041702, 7 pp.

Andrienko D., Tasinkevych M., Patricio P., Allen M. P., and Telo da Gamma M. M. (2003) Forces between eleongated particles in a nematic colloid. *Phys. Rev. E* Vol. 68, 051702, 5pp.

Bellini T., Buscagli M., Chiccoli C., Mantegazza F., Pasini P., and Zannoni C. (2000) Nematics with quenched disorder: What is left when long range order is disrupted?, Phys. Rev. Lett. *Small* Vol 85, No 5, pp.1008-11.

Bradac Z., Kralj S., and Zumer S., (2002) Molecular dynamics study of isotropic-nematic quench, Phys.Rev.E *Small* 65, 021705-1-10.

Brochard F. and de Gennes P. G. (1970) Theory of magnetic suspensions in liquid crystals. *J. Phys. (Paris)* Tome 31, pp. 691-699 .

Burylov S. V. and Raikher Yu. L. (1990) On the orientation of an anisometric particle suspended in a bulk uniform nematic. *Phys. Lett. A* Vol. 149, number 5,6, pp. 279-283.

Burylov S. V. and Raikher Yu. L. (1994) Orientation of a solid particle embedded in a monodomain nematic liquid crystal. *Phys. Rev. E* Vol. 50, Number 1, pp 358-367.

Cvetko M., Ambrozci M. and Kralj S. (2009) Memory effects in randomly perturbed systems exhibiting continuous symmetry breaking, Liq. Cryst. *Small* Vol 36, No. 1, pp. 33-41.

Doi M. and Edwards S. F. (1989), *Theory of Polymer Dynamics* Clarendon, Oxford.

Duran H., Gazdecki B., Yamashita A., Kyu T.(2005) Effect of carbon nanotubes on phase transitions of nematic liquid crystals. *Liq. Cryst.*, Vol. 32, No. 7, pp. 815-821.

Flory P. J. (1953) *Principles of Polymer Chemistry* Cornell University Press, Ithaca.

de Gennes P. G. Prost J. (1993) *The Physics of Liquid Crystals* Oxford University Press, Oxford.

Green M. J., Behabtu N., Pasquali M., Adams W. W. (2009) Nanotubes as polymers. *Polymer*, Vol. 50, pp. 4979-4997.

Green M. J., Parra-Vasquez A. N. G., Behabtu N., Pasquali M. (2009) Modeling the phase behavior of polydisperse rigid rods with attractive interactions with applications to single-walled carbon nanotubes in superacids. *J. Chem. Phys.*, Vol. 131, 084901, 10 pp.

Hung F. R., Guzman O., Gettelfinger T., Abbott N. L., and de Pablo J. J. (2006) Anisotropic nanoparticles immersed in a nematic liquid crystal: Defect structures and potential of mean force. *Phys. Rev. E* Vol. 74, 011711, 12 pp.

Iijima S. (1991), Helical microtubules of graphitic carbon. *Nature*, Vol 354, pp. 56-58.

Imry Y. and Ma S. (1975) Random-filed instability of ordered state of continuous symmetry, Phys. Rev. Lett. *Small* Vol 35, No. 21, pp. 1399-1401.

Kralj S., Bradac Z., and Popa-Nita V., (2008) The influence of nanoparticles on the phase and structural ordering for nematic liquid crystals, J. Phys.: Condens. Matter *Small* Vol 20, pp. 244112-244122 .

Krasna M., Cvetko M., and Ambrozic M., (2010) Symmetry breaking and structure of a mixture of nematic liquid crystals and anisotropic nanoparticles, Bel. J. Org. Chem. *Small* Vol 6, Article No. 74.

Lagerwall J. P. F. and Scalia G (2008) Carbon nanotubes in liquid crystals. *J. Mater. Chem.* Vol. 18, pp. 2890-2898.

Lebovka N., Dadakove T., Lysetskiy L., Melezhyk O., Puchkovska G., Gavrilko T., Baran J., Drodz M. (2008) Phase transitions, intermolecular interactions and electrical conductivity in carbon multiwalled nanotubes/nematic liquid crystal composites. *J. Molecular Structure* Vol. 887, pp. 135-143.

Lebwohl P. A., Lasher G. (1972) Nematic-Liqid-Crystal Order - Monte Carlo calculation, Phys. Rev. A *Small* Vol 6, No. 1, pp. 426

Oswald P. and Pieranski P. (2005), *Nematic and Cholesteric Liquid crystals; concepts and physical properties illustrated by experiments* Taylor and Francis Group, CRC Press, in Liquid Crystals Book Series, Boca Raton.

van der Schoot P., Popa-Nita V., and Kralj S. (2008) Alignment of Carbon Nanotubes in Nematic Liquid Crystals. *J. Phys. Chem. B* Vol. 112, pp. 4512-4518.

Rahman M. and Lee W. (2009) Scientific duo of carbon nanotubes and nematic liquid crystals. *J. Phys. D: Appl. Phys.* Vol. 42, 063001, 12 pp.

Sabba Y. and Thomas E. L. (2004) High-Concentration Dispersion of Single-Wall Carbon Nanotubes. *Macromolecules* Vol. 37, pp. 4815-4820.

Somoza A. M., Sagui C., and Roland C. (2001) Liquid-crystal of capped carbon nanotubes. *Phys. Rev. B* Vol. 63, 081403 4 pp.

Song W. and Windle A. H. (2005) Isotropic-Nematic Phase Transition of Dispersion of Multiwall Carbon Nanotubes. *Macromolecules* Vol. 38, pp. 6181-6188.

Popa-Nita V., and Kralj S. (2010) Liquid crystal-carbon nanotubes mixtures. *J. Chem. Phys.* Vol. 132, 024902, 8 pp.

Zakri C. (2007) Carbon nanotubes and liquid crystalline phases. *Liquid Crystals Today*, Vol. 16, No. 1, pp 1-11.

Zhang S. and Kumar S. (2008) Carbon Nanotubes as Liquid Crystals. *Small* Vol. 4, No. 9, pp. 1270-1283.

Zurek W. (1996) Cosmological experiments in condensed matter systems, Phys. Rep. *Small* Vol 276, No. 4, pp. 177-221.

# 30

# Strategies to Successfully Cross-Link Carbon Nanotubes

Steve F. A. Acquah, Darryl N. Ventura and Harold W. Kroto
*Florida State University*
*United States*

## 1. Introduction

Since the inception of the research field on carbon nanotubes (CNTs), there has been an enormous effort to understand how the tubes form and how to best garner their unique electronic and mechanical properties. It soon became apparent that in order to develop the next generation of functional materials, a way to modify the surface of the tubes and connect them was required. The development of the oxidation process with acids was the first revolution in the field of CNTs, potentially opening the door to an extensive library of modifications. Research progressed by integrating the nanotubes into composites at low concentrations with some success, but the goal of producing high nanotube component covalently cross-linked materials was still problematic. Two decades after the report by Sumio Ijima on their discovery, cross-linked CNT materials are still difficult to produce, and this has shifted the field towards a back-to-basics approach to try and solve the problem. One key problem identified was the presence of lattice fragments immobilized on the surface of the CNTs (Fig. 1.). The current methods of characterization such as X-ray photoelectron, Infrared and Raman spectroscopy are indirect and generally fail to distinguish between the surface attached functional groups and oxidized lattice fragments. A CNT washing technique has been developed to remove these fragments and any electrostatically attached products to allow pure covalent interactions with the surface of the nanotube (Wang *et al.*, 2010). With an industry now thriving on the production of cheap functionalized carbon vapor deposition (CVD) CNTs, priced according to the percentage surface functionalization, and the decline in published materials on arc-produced CNTs, the need for effective characterization and quality control increases.

It is the intention of this chapter to review some of the successful approaches used to cross-link CNTs with a focus on the importance of the chemistry and techniques involved, and highlight two areas of research we are currently investigating at Florida State University.

### 1.1 The characterization dilemma

As the basic unit for nanotubes, single-walled carbon nanotubes (SWCNT) have been envisaged as a solution in areas such as molecular wires to biological transport vectors, however in order to reach this potential, we need to be able to modify the surface structure. The solubility of SWCNTs is on average 0.1 mg/ml, however this can be increased to 1 mg/ml with surface modification. The problem with this form of modification is that the inherent properties of the nanotubes, both the mechanical and electronic properties, can be significantly altered, questioning the reasons for modifying the tubes.

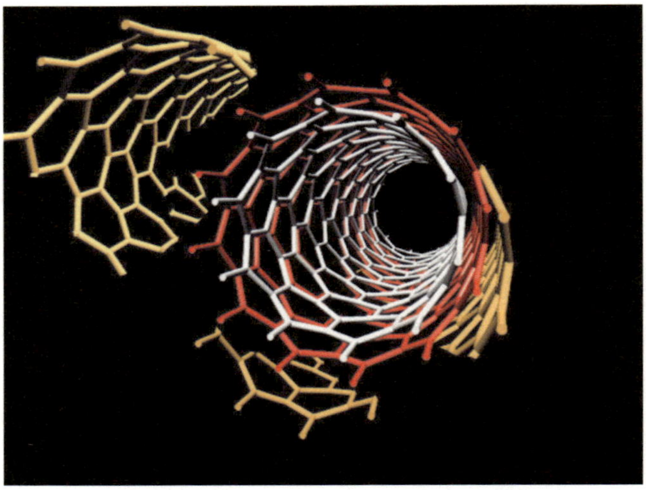

Fig. 1. A carbon nanotube with lattice fragments on the surface. These fragments can be easily oxidized and result in an incorrect assessment of the degree of CNT surface functionalization.

One aspect of CNT research that can prove discouraging at times is the difficulty in the direct characterization of functional moieties after a chemical process. Sidewall functionalization is important for the use of the tubes in cross-linked composites. One of the most popular techniques used in journals was Fourier Transform Infrared Spectroscopy (FTIR). Whilst it is a powerful technique in organic chemistry, for CNTs it is difficult to prove the presence of covalently attached groups although there are good indications of hydroxyls, carbonyls and amine derivatives. The low concentration of functional groups on the surface is also a problem for acquiring a sufficient signal and the generation of a good baseline requires careful preparation. Raman spectroscopy is extremely popular due to the characteristic D-band that measures the degree of disorder and the G-band that provides a measure of the $sp^2$ character. While the ratio of the bands can provide an insight into the quality of the tubes, it can be difficult to form a correlation between the distribution of the functional groups and the ratio of the bands. UV spectroscopy (UV) requires the nanotubes to be dispersed at a low concentration which is difficult, as the CNTs tend to sediment during acquisition. Visual techniques such as transmission electron microscopy (TEM) and scanning electron microscopy (SEM) can be useful for identifying regions and trends, but the sample area is small for TEM and there are charging issues with SEM. Table 1 gives an overview of the types of issues encountered with characterization.

## 2. Cross-linking methods

In this section, we will discuss some of the successful methods employed to facilitate the cross-linking of CNTs. There are a variety of methods available and theoretical studies have shown the possibility of the assembly of higher ordered structures, however the examples listed here differ in the approach to cross-linking from aspects of defunctionalization to nanocrystal interactions.

| Methods | Type | Information | Limitations of Technique |
|---|---|---|---|
| TGA | Solid | Functionalization ratio | No information on covalent modification. |
| XPS | Solid | Elements present and the functionalization ratio | Inference of covalent attachment but quantification is restricted |
| Raman | Solid | Degree of disorder and $sp^2$ character | No direct chemical information and interpretation is problematic |
| Infra red (IR) | Solid & Liquid | Groups | Analytical quantification not advisable |
| UV/visible | Liquid | Sidewall functionalization | Solubility of sample is difficult |
| AFM | Solid | Topography | Sample size is small. No information on the covalent functionalization and no chemical identity |
| TEM | Solid | Transmission Image, Lattice | Sample size is small. Inference of covalent functionalization only. |
| SEM | Solid | Secondary electrons | No chemical identity |

Table 1. A list of the limitations of the characterization techniques typically applied to CNTs. Table adapted from a module by Liling Zeng and Andrew R. Barron.

### 2.1 Cyclo-addition reactions

The side-wall functionalization of CNTs is of great importance, primarily for increasing the solubility of tubes, but equally important is the ability to process the CNTs to form composites. One successful method is that of nitrene chemistry. Nitrenes (R-N:) are structurally similar to carbenes (RR'C:) and are electron-deficient uncharged molecules, which depending on the side groups, can facilitate addition and rearrangement reactions. The singlet nitrenes can react with the sidewall of CNT's by electrophilic [2+1] cyclo-additions or by inter-system crossing. The triplet state reacts with the π system of the CNT with both the singlet and triplet states resulting in the formation of aziridine rings (Fig. 2A).

Cross-linking was achieved by using a di-azidocarbonate, based on poly-ethylene glycol (PEG) (Holzinger *et al.*, 2004). The preparation of the cross-linked nanotubes is a simple process using SWCNTs dispersed in 1,1,2,2-tetrachloromethane (TCE) by sonication. The suspension is then heated to 160 °C and a 20-fold excess of diluted di-azidocarbonate in TCE is added over a period of 30 mins. After cooling, the mixture is filtered and washed with TCE and ethanol. This process is highly effective in cross-linking, but because of the length and flexibility of the cross-linker it is possible for the linker to attach to the same tube forming a loop (Fig. 2B).

The characterization of the composite was performed by using a range of techniques including transmission electron microscopy, atomic force microscopy, Raman spectroscopy and X-ray photoelectron spectroscopy. This cyclo-addition technique holds huge promise for the development of further films using long chain nitrene based cross-linkers.

### 2.2 Ion beam and irradiation techniques

One other method of cross-linking carbon nanotubes is through electron or ion beam irradiation. It has been theorized that cross-linking nanotubes could improve the overall characteristics of nanotubes on the bulk scale. While this method can be achieved on both SWCNTs and MWCNTs, this technique of cross-linking has both its advantages and disadvantages. One advantage is that the setup is simple and there are no chemical reactions that need to be performed. Another advantage is that the bonds formed between the tubes are much stronger than the van der Waals interactions that are sometimes used to link nanotubes. In addition to this, not only can individual tubes be cross-bonded, but it has been demonstrated that it should be possible to link macroscopic carbon structures such as CNT mats and fibers. According to simulations, ion irradiation will affect SWCNTs and MWCNTs differently. The incident energy from irradiation will scatter carbon fragments from a SWCNT, and a percentage of these fragments will be redistributed along the nanotube surface. In the end, these fragments will form the cross-links between the nanotubes. It was predicted that a much higher percentage (~50%) of the fragments will be redistributed between the inner walls of MWCNTs. Therefore, cross-linking via irradiation is more suitable for SWCNTs but it can still be used to reinforce the inner walls of a MWCNTs. These theoretical predictions for cross-linking nanotubes have been confirmed experimentally by researchers. An improvement in electron transport properties in bundles of SWCNTs due to increased intertube coupling, after exposure to an $Ar^+$ beam, has also been demonstrated (Stahl *et al.*, 2000). It has been shown that electron irradiation of MWCNTs can reinforce the inner walls (Fig. 3.) and stiffen the tubes by up to five times (Duchamp *et al.*, 2010). Studies have also looked into the possible mechanisms involved in the radiation induced modification of CNTs (Kis *et al.*, 2004). Similar results have been reported (Peng *et al.*, 2008) and have demonstrated improvements in fracture strength.

Despite all these advantages and promising results for irradiation cross-linking, there are a few drawbacks. One disadvantage is that it destroys the $sp^2$ bonding of the nanotube which could be detrimental to the tubes' intrinsic properties. Another disadvantage of this technique is that the cross-linking capabilities are dependent on where the nanotube can be exposed to the electron or ion beam. If you wanted to produce a cross-linked nanotube mat (similar to buckypaper) only the surface layers of the CNT mat would be cross-linked as the interior tubes would not be exposed to the incident beam.

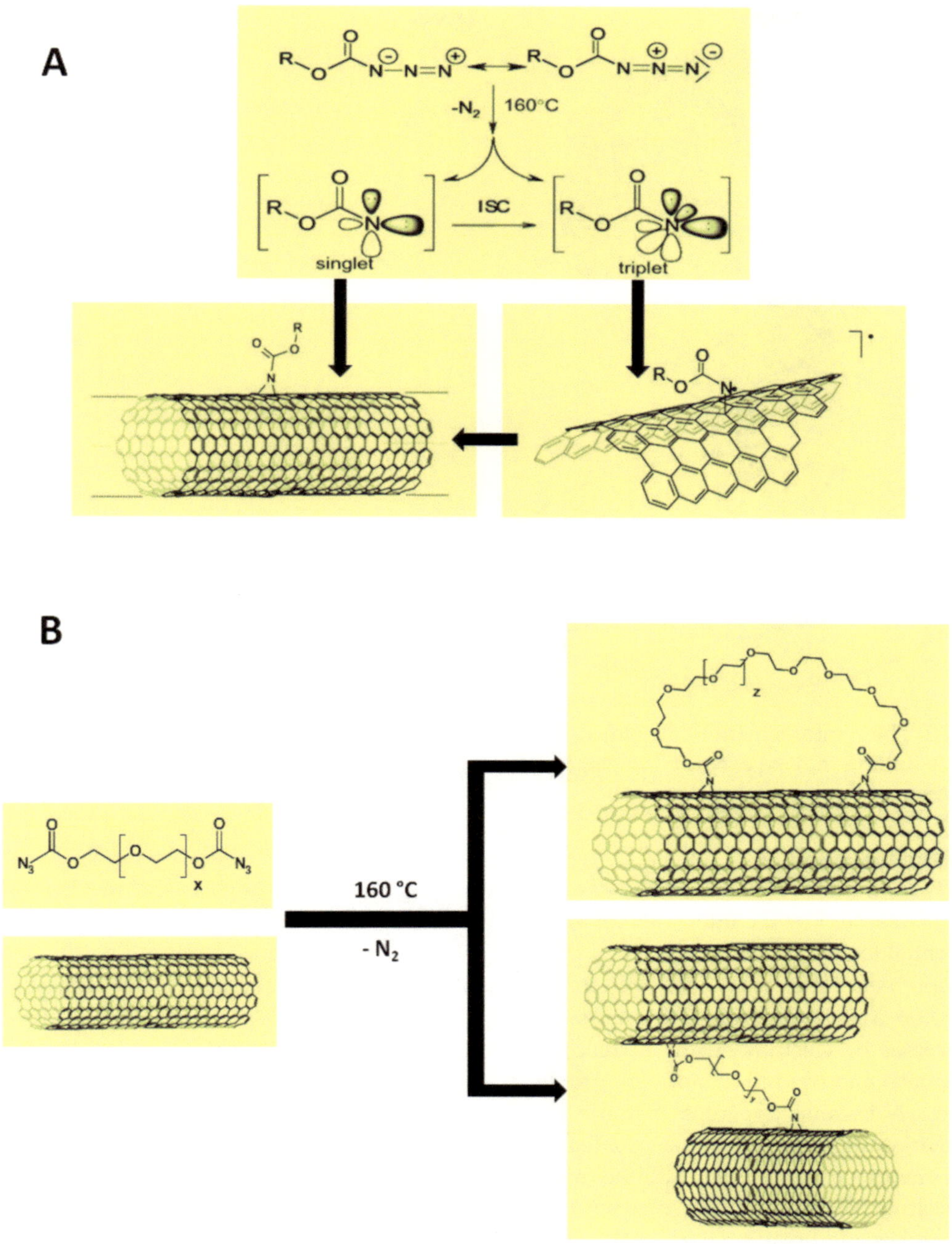


Fig. 2. A) Schematic presentation of the reaction of nitrenes with the nanotube sidewall. B) Reaction of di-nitrenes with the nanotubes using diazidocarbonate polyglycolesters as precursors. The addition of the di-functional molecule can also happen on the sidewall of the same CNT resulting in the formation of a loop. Adapted from (Holzinger *et al.*, 2004)

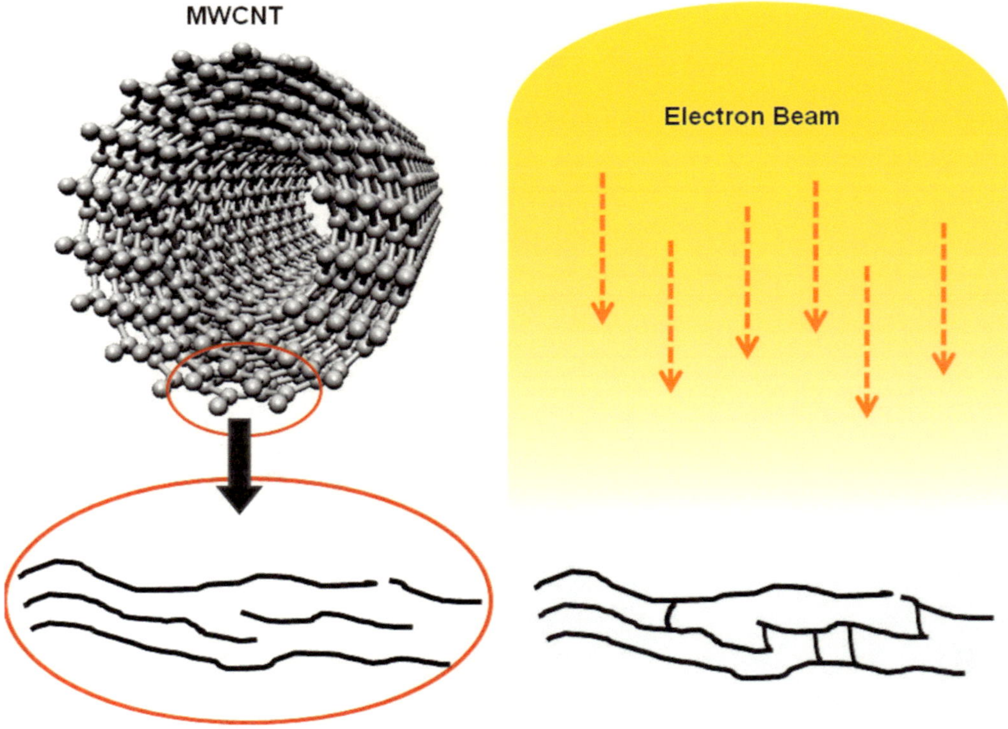


Fig. 3. The reinforcement of the sidewalls of a MWCNT by an electron beam. This form of cross-linking may be extended to neighboring tubes forming a covalently linked intertube junction.

**2.3 Michael addition**

At Florida State University we have been exploring the nature of cross-linking between CNTs. Inspired by the interaction of maleimides with cysteine for biological labeling and the potential for the covalent interaction of CNTs, we used benzoquinone to cross-link thiolated carbon nanotubes (MWCNT-SH) to form mats similar to buckypaper. We wanted to develop a way of producing a cross-linked mat without the need to use high pressure processes or electron beams to fuse the tubes together, so we applied a back-to-basics approach and tried to identify the problems associated with poor cross-linking between the tubes. It became apparent that the inability to control reaction conditions during the formation of the mats was a problem so we attempted to maintain the temperature and dispersion of the tubes until the mat was ready to be cast. The procedure involved sonicating MWCNT-SH (1g, Nanocyl) with an excess of dithiothreitol (DTT) to separate the nanotubes and break up the disulfide bonds. The MWCNT-SH were then washed with DMF and dried for 12 h. From this batch, MWCNT-SH (20 mg) were dispersed in DMF (15 ml) and sonicated. In a separate vial, benzoquinone (100 mg) was dissolved in DMF (10 ml). The benzoquinone solution was slowly added to the nanotube suspension and the mixture was stirred at 75 °C for 12 hours before being vacuum filtered and washed with excess DMF. It was noted that the use of a heat gun to maintain the temperature of the mixture at 75 °C aided cross-linking during filtration (Ventura *et al.*, 2010).

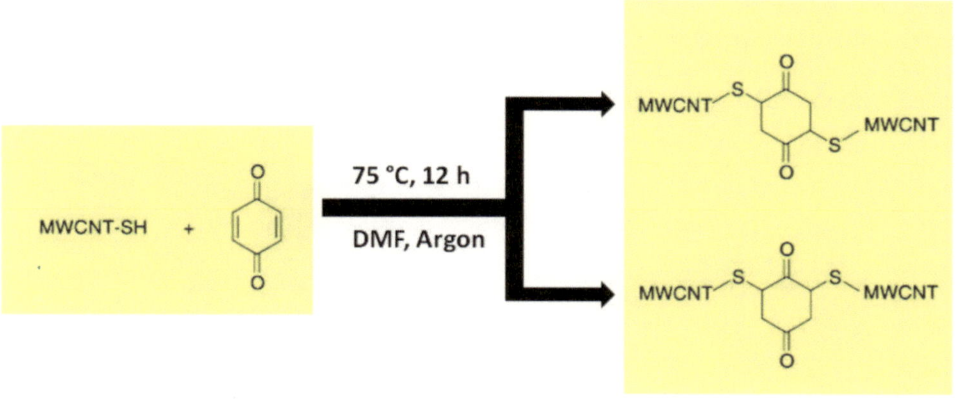


Fig. 4. Reaction scheme for cross-linking MWCNT-SH with benzoquinone.

The products formed during the reaction are of two forms, either the 2,5-dithioMWCNT 1,4-cyclohexanedione adduct or the 2,6-dithioMWCNT 1,4-cyclohexanedione adduct depending on the steric interactions (Fig. 4.). Several concentration ratios of benzoquinone : MWCNT-SH were tested to obtain the optimum flexibility of the mat produced. Tensile strength measurements indicated that the optimum concentration of benzoquinone : MWCNT-SH was 5:1 and the SEM images inferred the that there was sufficient cross-linking (Fig. 5A) The 10:1 composite produced a brittle mat confirming the link between the increasing strength of the mat and the degree of cross-linking.

The surface of the 5:1 nanotube film also contained unreacted thiol groups that were used to attach nanocrystals to enhance the functionality. As a demonstration of this principle, we attached 5.7 ± 0.3 nm gold particles to the surface (Fig 5B).

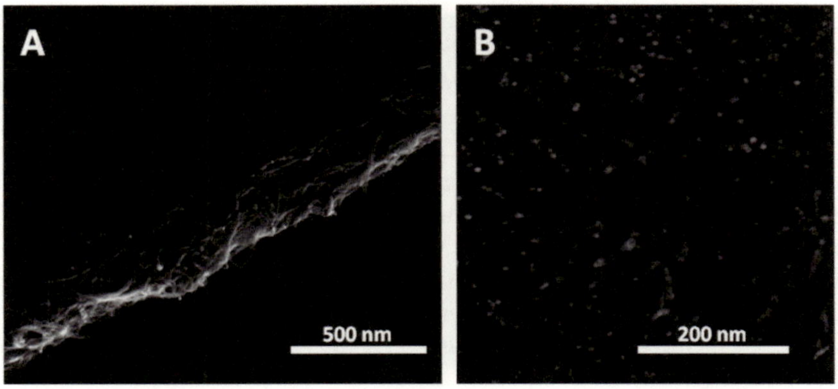


Fig. 5. A) The SEM images of a cross-linked bundle of nanotubes and B) a nanotube mat decorated with 5.7 ± 0.3 nm gold particles.

Once the film was produced we wanted to see if we could create a die-cast composite using a 10:1 ratio of benzoquinone : MWCNT-SH. During this procedure, we injected a predrilled cast, in the letters "FSU", with the cross-linking mixture of benzoquinone and MWCNT-SH and placed the cast on a hotplate heated to 75 °C. We periodically injected more of the mixture as the liquid level dropped until the cast was full, and left the cast at 75 °C for 9 hours before cooling to room temperature (Fig. 6A). The cross-linked FSU blocks were removed with the aid of a scalpel in smaller blocks (Fig. 6B).

Fig. 6. A photograph of A) the letters FSU cast into a stainless steel container with a 10:1 mixture of benzoquinone : MWCNT-SH injected into the cast. B) The cross-linked composite removed from the cast after heating.

Although the removal of the composite from the cast was difficult, it was primarily due to the design of the cast. We plan to explore the use of bismaleimide groups to generate similar composites. We are also exploring alternative methods to modify the carbon nanotubes. One method that we find intriguing is the use of a mechano-chemical approach. This is a novel method in the field of CNTs and provides an alternative route to attach functional groups to the surface of CNTs by ball-milling the tubes in the presence of a reactive gas. The milling creates defects on the tubes which react with the gas forming covalent attachments (Konya *et al.*, 2002).

For cross-linking, it is preferential to have a fairly uniform CNT size and distribution of functional groups on the surface. Both of these requirements were addressed in research into the mechano-chemical functionalization of CNTs with a ball mill, for the generation of surface thiols, chlorides, acyl chlorides, amines and amides.

The process involved placing purified CNTs into a ball-mill and degassing in a heated $N_2$ environment or in a vacuum. The reactant gas was then pumped through the chamber untill the milling process was complete. The excess reactant is removed by the evacuation of the chamber, and the tubes are subsequently washed with ethanol. The research noted that extended periods of milling resulted in the generation of amorphous carbon and lattice fragments with a 30-35% amorphous content formed after a period of 2 weeks. The milling process shortened the tubes and this was also carefully controlled by the cumulative milling time.

The results inferred that the CNTs obtained had a high degree functional groups around the surface as indicated by IR and XPS, and the process can be scaled up depending on the size of the mill.

## 2.4 MWCNT-Titania films

There are many theoretical ways to improve the photocatalytic efficiency of titania ($TiO_2$) including increasing the surface area of $TiO_2$ or creating methods to promote charge separation, but there needs to be a better way of promoting electron transport. Therefore, the development of new materials that are capable of increasing the efficiency of electron mobility are required. CNTs are known to be a good candidate for use with $TiO_2$ because of the semiconducting and metallic behavior depending on the diameter of the tubes and helicity. Enhancing the photocatalytic properties of titania is extremely important, especially when it is considered as a potential for the photocatalytic reduction of $CO_2$ with $H_2O$ (Xia *et al.*, 2007). Titania itself exhibits good photostability properties; however the photocatalytic reaction with $CO_2$ is insufficient for applications. This problem was reduced significantly by the application of carbon nanotubes as a mediator of electron transfer and MWCNTs have been investigated for their charge transfer properties with titania, but a suitable composite needs to be constructed.

We have constructed a nanotube film which uses aminated titania particles as a cross-linker for CNT films (Fig. 7.). In a departure from the standard cross-linking theory, we wanted to examine the potential for these beads to act as cross-linkers in CNT films. In addition to the formation of amide bonds, the nature of interaction allows for the potential of electron transfer from $TiO_2$ to the carbon nanotubes.

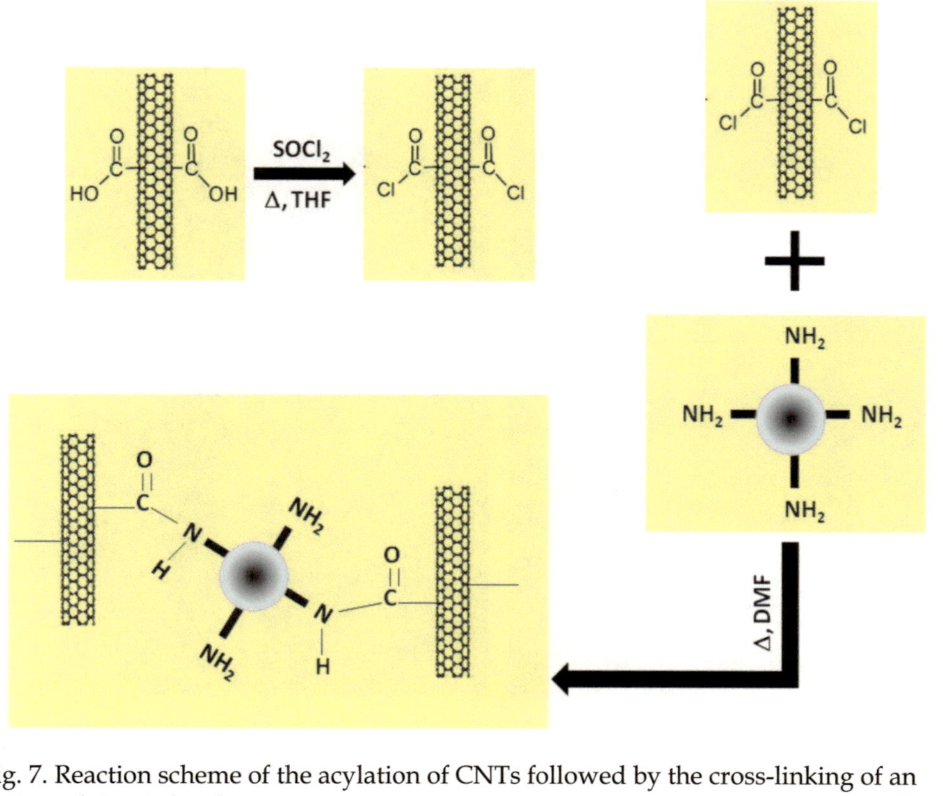


Fig. 7. Reaction scheme of the acylation of CNTs followed by the cross-linking of an aminated titania bead.

MWCNT-COOH (1 g, 50-80 nm OD, Nanostructured & Amorphous Materials Inc.) were converted to MWCNT-COCl by heating the nanotubes under reflux for 3 days in THF with thionyl chloride. The MWCNT-COCl were washed with excess THF under vacuum and the tubes were dried over a period of 24 h. The MWCNT-COCl (100 mg) were sonicated for 10 minutes whilst $TiO_2$-$NH_2$ (25 mg/ml, Corpuscular Inc.) 100 nm particles (Fig. 8A) were added. The mixture was further heated under reflux and slowly vacuum filtered whilst hot to produce a thick film. The films were brittle but the SEM images indicated a sufficient dispersion throughout the composite. The particles are amorphous but can be easily converted to anatase or rutile by heating, making the film a good candidate for photocatalytic activity.

Using particles and nanocrystals as cross-linking agents is of great interest for increasing the surface area for reactivity. Examples can found in literature of the use of such constructs in the formation of photoreactive composites of CNTs and titania (Yao *et al.*, 2008) and Pt nanoclusters/titanium dioxide nanotube composites (Dong *et al.*, 2010). In order to increase the utility of these composites, they would need to be processed into films to fully take advantage of their unique properties.

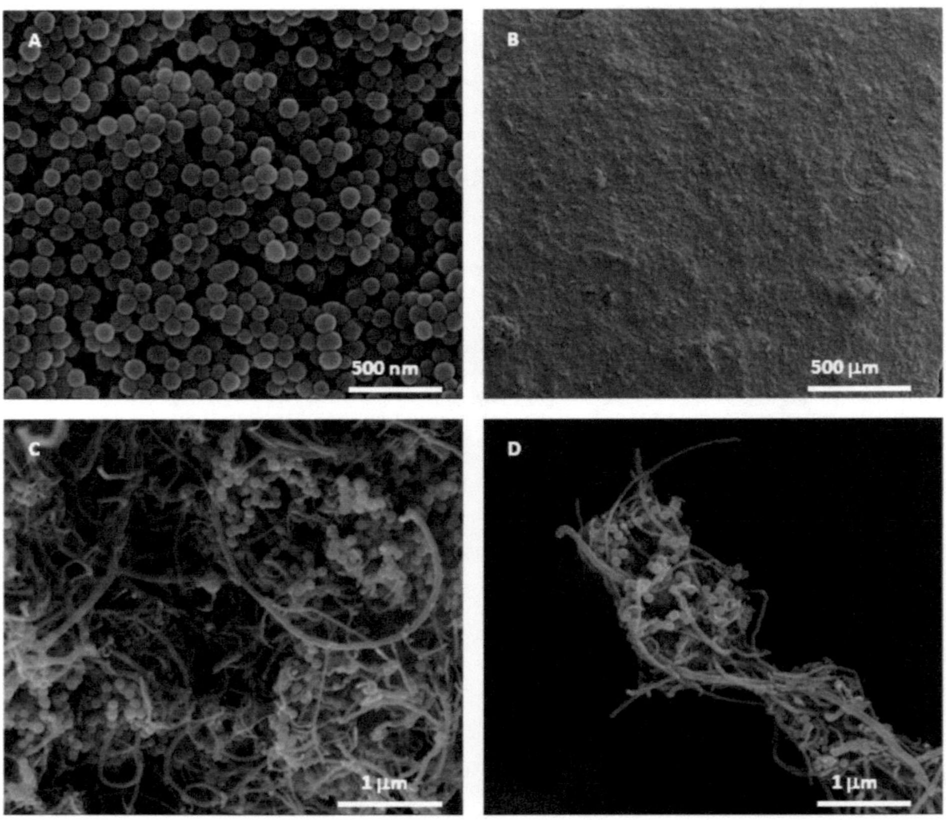


Fig. 8. SEM images of A) aminated titania particles B) low magnification image of the film C) a high magnification section of the film surface D) A highly cross-linked fiber at the edge of the film.

### 2.5 De-fluorination

The ultimate goal of cross-linking carbon nanotubes is to find a way of incorporating the effectiveness of the electron beam techniques with a chemical approach. This concept was highlighted by the work demonstrating the production of binder free cross-linked carbon nanotube composites (Sato *et al.*, 2008). The problem addressed in their research focused on an additional issue with CNT chemistry. CNTs with smaller diameters are typically more reactive than larger MWCNTs because the strain energy is inversely proportional to the diameter of a tube. The surface of larger MWCNTs act like a graphite sheet making surface modifications extremely difficult. One solution to this was the fluorination of CNTs, and this was achieved by reacting purified MWCNTs with a mixture of 20% $F_2$ and 80% $N_2$ at 523 K for a period of 2 hours followed by thermal annealing at 523 K for 6 hours under $N_2$ (Sato *et al.*, 2008). The cross-linking of the CNTs was achieved by using a spark plasma sintering system that produces a significant amount of $sp^3$ carbons, which connect the tubes when operated for 10 minutes below 80 MPa. Fig. 9A shows a picture of the defunctionalized block. The size and dimensions can be tailored depending on the vessel used to cast the block.

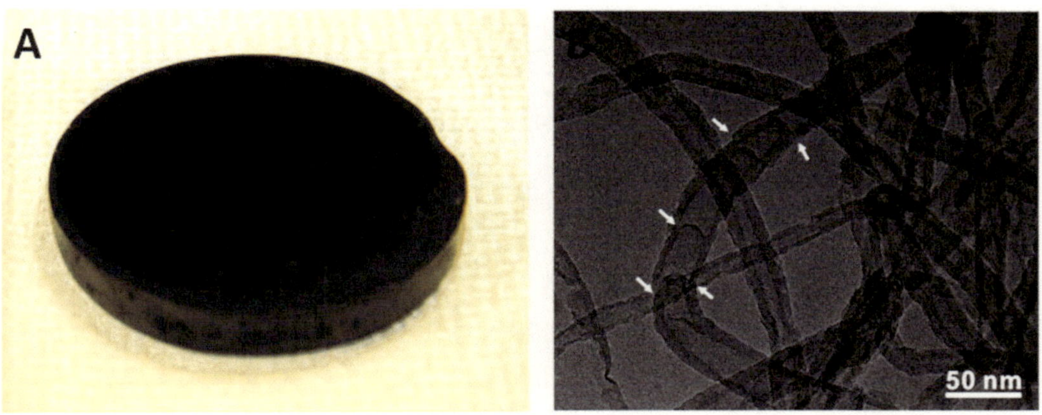


Fig. 9. A) The photograph of a defunctionalized MWCNT block and B) a TEM image of a cross-linked composite showing bamboo-like structures, at the white arrow positions. Adapted from (Sato *et al.*, 2008)

Fig. 9B shows a characteristic high magnification image of the block. The bamboo-like structures have been seen in nitrogen doped carbon nanotubes, however in this case they are the result of the CVD process and beneficial for exploiting defect sites.

The spark plasma sintering solidification process also more accurately relies on the liberation of carbon fluorine gases which generate defects and vacancies on the MWCNT surface, effectively increasing the reactivity. The comparison of the block with commercially available graphite (Table 2.) shows a greater percentage porosity which may indicate a future avenue to tailoring the material for applications such as hydrogen storage. The Vickers Hardness has a value of 46.9 MPa which is much higher than that of graphite at 16.8 MPa, but lower than the minimum value for carbon steel calculated to be approximately 539.4 MPa. It is surprising that the conductivity is slightly more than that of graphite given the increase in $sp^3$ character, but this may be a result of the bulk properties of the composite rather than the dominance of individual tubes. The process is still highly destructive for the outer layer of MWCNTs and detrimental to that of SWCNTs, which means the fluorination process would have to be carefully controlled to prevent extensive damage. Future work on this type of cross-linking may require the relationship between the fluorination time and de-fluorinated cross-linking ability to be established.

| Characteristic | De-F-MWCNT Blocks | Commercial Graphite |
|---|---|---|
| Bulk Density, $\rho (g/cm^3)$ | 1.44 | 1.74 |
| Porosity (%) | 36.2 | 23.0 |
| Young's Modulus, $E_b$ *(GPa)* | 14.2-16.3 | 6.5-8.6 |
| Fracture Bending Strength, $\sigma_b$ (MPa) | 92.4-123.0 | 42.1-43.7 |
| Vickers Hardness, $H_v$*(MPa)* | 46.9 | 16.8 |
| Conductivity, $\sigma$ (S/cm) (four-probe) | $2.1 \times 10^2$ | $6.0 \times 10^2$ |

Table 2. The properties of the de-fluorinated MWCNT blocks in comparison to commercially available graphite. Adapted from (Sato *et al.*, 2008).

In general, the cross-linking by de-fluorination is possibly one of the best methods for producing strong-robust carbon materials. If the process can be further refined, it may also be possible to create thin films for filtration or flexible films for catalytic substrates.

## 3. Applications of cross-linked CNTs

CNTs have been linked to diverse fields from biomedical drug delivery vectors to nanoscale computing. This section highlights some of the applications envisaged for the tubes, focusing on the development of non-covalent and covalently intertwined architectures, and the importance of the techniques used for their assembly.

### 3.1 Buckypaper

One of the more promising avenues for carbon nanotubes is the development of buckypaper. This material is made from aggregates of carbon nanotubes and may be held together by a variety of methods. In the simplest case, buckypaper can be made by the acid functionalization of the tubes and washing with deionised water. After drying, the tubes are suspended in a mixture of solvent and surfactant. A stable suspension is usually achieved after sonication and is then filtered and compressed. The advent of double-walled carbon nanotubes (DWCNTs) provided a means to create even stronger non cross-linked buckypaper due to the coaxial nature of the tubes, that is reflected in a higher stability when compared to SWCNTs. DWCNTs are in fact perfectly situated to provide the benefits of both SWCNTs and MWCNTs and this has been demonstrated by the production of strong flexible DWCNT buckypaper (Endo *et al.*, 2005). In this case, buckypaper was produced by sonicating 15 mg of DWCNT in 100 ml of ethanol for 30 mins. The suspension was then carefully poured onto a polytetrafluroetylene (PTFE) filter and dried for a period of 24 hours before being peeled off. It is important to note that the paper was produced without the use of a surfactant in order to increase the purity of the paper, and a quick but significant mention of the careful pouring with filtration to form the paper was made. This research highlights an important issue for the processability of CNTs. Is the application of technique equally or more important than the chemistry of cross-linking?

### 3.2 Photovoltaic devices

The addition of CNTs to organic photovoltaic (OPV) cells is of great interest due to the inherent properties of the tubes (Li *et al.*, 2010). Organic solar cells have garnered a lot of interest since the discovery of dye sensitized solar cells, and have been envisioned as a cheaper alternative to silicon based cells. The main drive behind the current research is to create a robust, flexible cell capable of sustaining many cycles as well as an increase in light absorption. One of the hurdles that need to be overcome is the relatively low carrier mobility that is found even in thin film organic polymers.

CNTs present and an attractive addition to thin film composites due to high charge mobility and extended conjugation and mechanical strength. SWCNTs provide an advantage to MWCNTs due to the diverse band gaps, which may be used to fine tune optical absorption, and reduce the effect of carrier scattering. Even with the advantages of SWCNTs in polymers, both SWCNTs and MWCNTs have been integrated into photovoltaic composites. These cells did improve the carrier mobility even when the doping level was low but the issue of solubility and processability still limits its potential efficiency. Another limitation is the combination of both the polymer donor and molecular acceptor in the same layers (bulk heterojunction design) which results in a highly disordered system with no clear phase boundaries (Yun *et al.*, 2008). One way to circumvent these issues is to follow an in situ polymerization method for a series of SWCNTs–poly[(2-methoxy,5-octoxy)1,4-phenylenevinylene] (MO-PPV) nanocomposites using different weight ratios. These are then cross-linked to acylated SWCNTs resulting in MO-PPV/SWCNTs nanocomposites with a defined interface (Fig. 10.). The result can be seen in the increase in electron mobility from the SWCNT to the polymer (Yun *et al.*, 2008).

Photoluminescence studies showed that there was significant quenching with the addition of SWCNTs, and further solid-state photoluminescence spectra of the thin films indicated that there was charge transfer from MO-PPV to SWCNTs. The MO-PPV/SWCNTs bulk molecular heterojunction solar cells produced exhibited an improvement in the efficiency which can be attributed to the nanophase separation which helps to not only enable carrier transport and exciton dissociation, but also reduce the recombination of photogenerated charge carriers in the thin films.

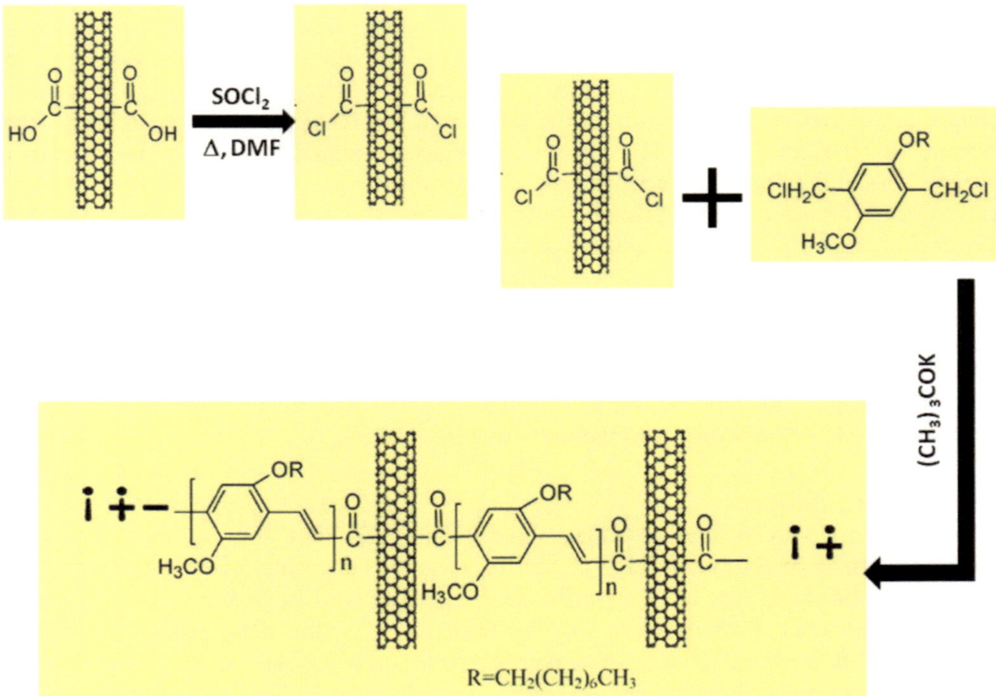


Fig. 10. Reaction scheme of the acylation of CNTs and the cross-linking of the tubes with MO-PPV. Figure adapted from (Yun *et al.*, 2008).

### 3.3 Artificial muscles

In the list of potential applications for carbon nanotube films, the nature of biological interactions is far more curious, following trends more commonly attributed to science fiction. One such concept is that of artificial muscles. The research into using CNTs as muscles has been ongoing with various noteworthy achievements of the last few years (Vohrer *et al.*, 2004). The work by Ray Baughman at the University of Texas has shown that cross-linking vertically aligned CNTs can produce a material with intriguing properties. The films act as CNT actuators and are produced from aerogel sheets which are drawn from forests of aligned carbon nanotubes (Aliev *et al.*, 2009). In this design, actuation, which is shown as a rapid spreading out of the width of the film after an insertion of charge, is followed by the contraction of the length of the film. The actuation across the length creates an isometric specific stress that can be up to 4.0 MPa $cm^3/g$. Mammalian skeletal muscle can typically withstand a stress of 0.1 - 0.35 MPa (Madden *et al.*, 2004), which in comparison to the isometric stress-generation capability of the sheet which was measured at 3.2 MPa, many times higher than the stress-generation capability of mammalian skeletal muscle.

The film is also being explored in relation to the behavior of the strain in extreme temperatures, and the advantages of controlling the structural changes. The research group highlight the possibility of tuning the density of the film for use in electrodes and light emitting displays.

## 4. Conclusion

In this chapter, we have discussed some of the successful approaches to cross-linking CNTs. For the formation of highly cross-linked CNT composites, the de-fluorination process is clearly an advantage. The fluorination of CNTs produces $C_2F$ stoichiometries which translate across the surface of the tubes, and can be beneficial for further processing. The tubes can then be fully defunctionalized by thermal treatment producing volatile fluorinated carbonaceous molecules ($CF_x$). The resulting CNTs have active sites all over the surface.

For the formation of flexible thin films, the production of buckypaper or nanotubes mats from thiolated CNTs are possible and may serve to be good electrodes or sensors for future applications.

Although the characterization of the composites formed needs to be extensive, it is possible to prove the nature of interactions between the tubes and the cross-linker. It is clear that the recent Nobel Prize for the work on graphene is helping to stimulate interest in the field, with many aspects envision for carbon nanotubes being translated to novel graphene architectures. It is expected that for the immediate future, the research into graphene and CNTs will take a synergistic approach especially in the area of cross-linking. There is already some promising research available that explores the nature of cross-linking between the related graphene oxide sheets (Park *et al.*, 2009).

The application of artificial muscles really emphasizes the potential for the future of carbon nanotube research, demonstrating our ability to unlock the skills required to control nanoscale assembly.

## 5. Acknowledgments

We acknowledge support from the Florida State University Research Foundation and Linda Hirst and Lam Nguyen for providing support on the titania-CNT project. We also acknowledge Kimberly Riddle and Thomas Fellers for their continued support in the Biological Sciences Imaging Resource at Florida State University.

## 6. References

Wang, Z., Korobeinyk, A., Whitby, R. L. D., Meikle, S. T., Mikhalovsky, S. V., Acquah, S. F. A. , Kroto, H. W. (2010). Direct confirmation that carbon nanotubes still react covalently after removal of acid-oxidative lattice fragments. *Carbon*, 48: 916-918.

Holzinger, M., Steinmetz, J., Samaille, D., Glerup, M., Paillet, M., Bernier, P., Ley, L. , Graupner, R. (2004). [2+1] cycloaddition for cross-linking swcnts. *Carbon*, 42: 941-947.

Stahl, H., Appenzeller, J., Martel, R., Avouris, P. , Lengeler, B. (2000). Intertube coupling in ropes of single-wall carbon nanotubes. *Phys. Rev. Lett.*, 85: 5186-5189.

Duchamp, M., Meunier, R., Smajda, R., Mionic, M., Magrez, A., Seo, J. W., Forro, L., Song, B. , Tomanek, D. (2010). Reinforcing multiwall carbon nanotubes by electron beam irradiation. *J. Appl. Phys.*, 108: 6.

Kis, A., Csanyi, G., Salvetat, J. P., Lee, T. N., Couteau, E., Kulik, A. J., Benoit, W., Brugger, J. , Forro, L. (2004). Reinforcement of single-walled carbon nanotube bundles by intertube bridging. *Nat. Mater.*, 3: 153-157.

Peng, B., Locascio, M., Zapol, P., Li, S. Y., Mielke, S. L., Schatz, G. C. , Espinosa, H. D. (2008). Measurements of near-ultimate strength for multiwalled carbon nanotubes and irradiation-induced crosslinking improvements. *Nat. Nanotechnol.,* 3: 626-631.

Ventura, D. N., Stone, R. A., Chen, K. S., Hariri, H. H., Riddle, K. A., Fellers, T. J., Yun, C. S., Strouse, G. F., Kroto, H. W. , Acquah, S. F. A. (2010). Assembly of cross-linked multi-walled carbon nanotube mats. *Carbon,* 48: 987-994.

Konya, Z., Vesselenyi, I., Niesz, K., Kukovecz, A., Demortier, A., Fonseca, A., Delhalle, J., Mekhalif, Z., Nagy, J. B., Koos, A. A., Osvath, Z., Kocsonya, A., Biro, L. P. , Kiricsi, I. (2002). Large scale production of short functionalized carbon nanotubes. *Chem. Phys. Lett.,* 360: 429-435.

Xia, X. H., Jia, Z. H., Yu, Y., Liang, Y., Wang, Z. , Ma, L. L. (2007). Preparation of multi-walled carbon nanotube supported tio2 and its photocatalytic activity in the reduction of co2 with h2o. *Carbon,* 45: 717-721.

Yao, Y., Li, G., Ciston, S., Lueptow, R. M. , Gray, K. A. (2008). Photoreactive tio2/carbon nanotube composites: Synthesis and reactivity. *Environmental Science & Technology,* 42: 4952-4957.

Dong, B., He, B. L., Chai, Y. M. , Liu, C. G. (2010). Novel pt nanoclusters/titanium dioxide nanotubes composites for hydrazine oxidation. *Mater. Chem. Phys.,* 120: 404-408.

Sato, Y., Ootsubo, M., Yamamoto, G., Van Lier, G., Terrones, M., Hashiguchi, S., Kimura, H., Okubo, A., Motomiya, K., Jeyadevan, B., Hashida, T. , Tohji, K. (2008). Super-robust, lightweight, conducting carbon nanotube blocks cross-linked by de-fluorination. *ACS Nano,* 2: 348-356.

Endo, M., Muramatsu, H., Hayashi, T., Kim, Y. A., Terrones, M. , Dresselhaus, N. S. (2005). 'buckypaper' from coaxial nanotubes. *Nature,* 433: 476-476.

Li, C., Chen, Y. L., Ntim, S. A. , Mitra, S. (2010). Fullerene-multiwalled carbon nanotube complexes for bulk heterojunction photovoltaic cells. *Appl. Phys. Lett.,* 96: 3.

Yun, D. Q., Feng, W., Wu, H. C., Li, B. M., Liu, X. Z., Yi, W. H., Qiang, J. F., Gao, S. , Yan, S. L. (2008). Controllable functionalization of single-wall carbon nanotubes by in situ polymerization method for organic photovoltaic devices. *Synth. Met.,* 158: 977-983.

Vohrer, U., Kolaric, I., Haque, M. H., Roth, S. , Detlaff-Weglikowska, U. (2004). Carbon nanotube sheets for the use as artificial muscles. *Carbon,* 42: 1159-1164.

Aliev, A. E., Oh, J. Y., Kozlov, M. E., Kuznetsov, A. A., Fang, S. L., Fonseca, A. F., Ovalle, R., Lima, M. D., Haque, M. H., Gartstein, Y. N., Zhang, M., Zakhidov, A. A. , Baughman, R. H. (2009). Giant-stroke, superelastic carbon nanotube aerogel muscles. *Science,* 323: 1575-1578.

Madden, J. D. W., Vandesteeg, N. A., Anquetil, P. A., Madden, P. G. A., Takshi, A., Pytel, R. Z., Lafontaine, S. R., Wieringa, P. A. , Hunter, I. W. (2004). Artificial muscle technology: Physical principles and naval prospects. *IEEE J. Ocean. Eng.,* 29: 706-728.

Park, S., Dikin, D. A., Nguyen, S. T. , Ruoff, R. S. (2009). Graphene oxide sheets chemically cross-linked by polyallylamine. *Journal of Physical Chemistry C,* 113: 15801-15804.